Lecture Notes in Computer Science 5164

Commenced Publication in 1973
Founding and Former Series Editors:
Gerhard Goos, Juris Hartmanis, and Jan van Leeuwen

Věra Kůrková
Roman Neruda
Jan Koutník (Eds.)

Artificial Neural Networks – ICANN 2008

18th International Conference
Prague, Czech Republic, September 3-6, 2008
Proceedings, Part II

Springer

Volume Editors

Věra Kůrková
Roman Neruda
Institute of Computer Science
Academy of Sciences of the Czech
Republic, Pod Vodarenskou vezi 2
182 07 Prague 8, Czech Republic
E-mail: {vera, roman}@cs.cas.cz

Jan Koutník
Department of Computer Science
Czech Technical University in Prague
Karlovo nam. 13
121 35 Prague 2, Czech Republic
E-mail: koutnij@fel.cvut.cz

Library of Congress Control Number: 2008934470

CR Subject Classification (1998): F.1, I.2, I.5, I.4, G.3, J.3, C.2.1, C.1.3

LNCS Sublibrary: SL 1 – Theoretical Computer Science and General Issues

ISSN 0302-9743
ISBN-10 3-540-87558-1 Springer Berlin Heidelberg New York
ISBN-13 978-3-540-87558-1 Springer Berlin Heidelberg New York

Springer is a part of Springer Science+Business Media

springer.com

Printed in Germany

Typesetting: Camera-ready by author, data conversion by Scientific Publishing Services, Chennai, India
Printed on acid-free paper SPIN: 12529186 06/3180 5 4 3 2 1 0

Preface

This volume is the second part of the two-volume proceedings of the 18th International Conference on Artificial Neural Networks (ICANN 2008) held September 3–6, 2008 in Prague, Czech Republic. The ICANN conferences are annual meetings supervised by the European Neural Network Society, in cooperation with the International Neural Network Society and the Japanese Neural Network Society. This series of conferences has been held since 1991 in various European countries and covers the field of neurocomputing and related areas. In 2008, the ICANN conference was organized by the Institute of Computer Science, Academy of Sciences of the Czech Republic together with the Department of Computer Science and Engineering from the Faculty of Electrical Engineering of the Czech Technical University in Prague. Over 300 papers were submitted to the regular sessions, two special sessions and two workshops. The Program Committee selected about 200 papers after a thorough peer-review process; they are published in the two volumes of these proceedings. The large number, variety of topics and high quality of submitted papers reflect the vitality of the field of artificial neural networks.

The first volume contains papers on the mathematical theory of neurocomputing, learning algorithms, kernel methods, statistical learning and ensemble techniques, support vector machines, reinforcement learning, evolutionary computing, hybrid systems, self-organization, control and robotics, signal and time series processing and image processing.

The second volume is devoted to pattern recognition and data analysis, hardware and embedded systems, computational neuroscience, connectionistic cognitive science, neuroinformatics and neural dynamics. It also contains papers from two special sessions, "Coupling, Synchronies, and Firing Patterns: From Cognition to Disease," and "Constructive Neural Networks," and two workshops, New Trends in Self-Organization and Optimization of Artificial Neural Networks, and Adaptive Mechanisms of the Perception-Action Cycle.

It is our pleasure to express our gratitude to everyone who contributed in any way to the success of the event and the completion of these proceedings. In particular, we thank the members of the Board of the ENNS who uphold the tradition of the series and helped with the organization. With deep gratitude we thank all the members of the Program Committee and the reviewers for their great effort in the reviewing process. We are very grateful to the members of the Organizing Committee whose hard work made the vision of the 18th ICANN reality. Zdeněk Buk and Eva Pospíšilová and the entire Computational Intelligence Group at Czech Technical University in Prague deserve special thanks for preparing the conference proceedings. We thank to Miroslav Čepek for the conference website administration. We thank Milena Zeithamlová and Action M Agency for perfect local arrangements. We also thank Alfred Hofmann, Ursula

Barth, Anna Kramer and Peter Strasser from Springer for their help with this demanding publication project. Last but not least, we thank all authors who contributed to this volume for sharing their new ideas and results with the community of researchers in this rapidly developing field of biologically motivated computer science. We hope that you enjoy reading and find inspiration for your future work in the papers contained in these two volumes.

June 2008

Věra Kůrková
Roman Neruda
Jan Koutník

Organization

Conference Chairs

General Chair	Věra Kůrková, Academy of Sciences of the Czech Republic, Czech Republic
Co-Chairs	Roman Neruda, Academy of Sciences of the Czech Republic, Czech Republic
	Jan Koutník, Czech Technical University in Prague, Czech Republic
	Milena Zeithamlová, Action M Agency, Czech Republic
Honorary Chair	John Taylor, King's College London, UK

Program Committee

Włodzisław Duch	Nicolaus Copernicus University in Torun, Poland
Luis Alexandre	University of Beira Interior, Portugal
Bruno Apolloni	Università Degli Studi di Milano, Italy
Timo Honkela	Helsinki University of Technology, Finland
Stefanos Kollias	National Technical University in Athens, Greece
Thomas Martinetz	University of Lubeck, Germany
Guenter Palm	University of Ulm, Germany
Alessandro Sperduti	Università Degli Studi di Padova, Italy
Michel Verleysen	Université catholique de Louvain, Belgium
Alessandro E.P. Villa	Universite jouseph Fourier, Grenoble, France
Stefan Wermter	University of Sunderland, UK
Rudolf Albrecht	University of Innsbruck, Austria
Peter Andras	Newcastle University, UK
Gabriela Andrejková	P.J. Šafárik University in Košice, Slovakia
Bartlomiej Beliczynski	Warsaw University of Technology, Poland
Monica Bianchini	Università degli Studi di Siena, Italy
Andrej Dobnikar	University of Ljubljana, Slovenia
José R. Dorronsoro	Universidad Autónoma de Madrid, Spain
Péter Érdi	Hungarian Academy of Sciences, Hungary
Marco Gori	Università degli Studi di Siena, Italy
Barbora Hammer	University of Osnabrück, Germany

Tom Heskes	Radboud University Nijmegen, The Netherlands
Yoshifusa Ito	Aichi-Gakuin University, Japan
Janusz Kacprzyk	Polish Academy of Sciences, Poland
Paul C. Kainen	Georgetown University, USA
Mikko Kolehmainen	University of Kuopio, Finland
Pavel Kordík	Czech Technical University in Prague, Czech Republic
Vladimír Kvasnička	Slovak University of Technology in Bratislava, Slovakia
Danilo P. Mandic	Imperial College, UK
Erkki Oja	Helsinki University of Technology, Finland
David Pearson	Université Jean Monnet, Saint-Etienne, France
Lionel Prevost	Université Pierre et Marie Curie, Paris, France
Bernadete Ribeiro	University of Coimbra, Portugal
Leszek Rutkowski	Czestochowa University of Technology, Poland
Marcello Sanguineti	University of Genova, Italy
Kateřina Schindler	Austrian Academy of Sciences, Austria
Juergen Schmidhuber	TU Munich (Germany) and IDSIA (Switzerland)
Jiří Šíma	Academy of Sciences of the Czech Republic, Czech Republic
Peter Sinčák	Technical University in Košice, Slovakia
Miroslav Skrbek	Czech Technical University in Prague, Czech Republic
Johan Suykens	Katholieke Universiteit Leuven, Belgium
Miroslav Šnorek	Czech Technical University in Prague, Czech Republic
Ryszard Tadeusiewicz	AGH University of Science and Technology, Poland

Local Organizing Committee

Zdeněk Buk	Czech Technical University in Prague
Miroslav Čepek	Czech Technical University in Prague
Jan Drchal	Czech Technical University in Prague
Paul C. Kainen	Georgetown University
Oleg Kovářík	Czech Technical University in Prague
Rudolf Marek	Czech Technical University in Prague
Aleš Pilný	Czech Technical University in Prague
Eva Pospíšilová	Academy of Sciences of the Czech Republic
Tomáš Siegl	Czech Technical University in Prague

Referees

S. Abe
R. Adamczak
R. Albrecht
E. Alhoniemi
R. Andonie
G. Angelini
D. Anguita
C. Angulo-Bahón
C. Archambeau
M. Atencia
P. Aubrecht
Y. Avrithis
Ľ. Beňušková
T. Beran
Z. Buk
G. Cawley
M. Čepek
E. Corchado
V. Cutsuridis
E. Dominguez
G. Dounias
J. Drchal
D. A. Elizondo
H. Erwin
Z. Fabián
A. Flanagan
L. Franco
D. François
C. Fyfe
N. García-Pedrajas
G. Gnecco
B. Gosselin
J. Grim
R. Haschke
M. Holena
J. Hollmen
T. David Huang
D. Húsek
A. Hussain
M. Chetouani
C. Igel
G. Indiveri
S. Ishii
H. Izumi
J.M. Jerez
M. Jiřina
M. Jiřina, jr.
K.T. Kalveram
K. Karpouzis
S. Kasderidis
M. Koskela
J. Kubalík
M. Kulich
F.J. Kurfess
M. Kurzynski
J. Laaksonen
E. Lang
K. Leiviskä
L. Lhotská
A. Likas
C. Loizou
R. Marek
E. Marchiori
M. A. Martín-Merino
V. di Massa
F. Masulli
J. Mańdziuk
S. Melacci
A. Micheli
F. Moutarde
R. Cristian Muresan
M. Nakayama
M. Navara
D. Novák
M. Olteanu
D. Ortiz Boyer
H. Paugam-Moisy
K. Pelckmans
G. Peters
P. Pošík
D. Polani
M. Porrmann
A. Pucci
A. Raouzaiou
K. Rapantzikos
M. Rocha
A. Romariz
F. Rossi
L. Sarti
B. Schrauwen
F. Schwenker
O. Simula
A. Skodras
S. Slušný
A. Stafylopatis
J. Šťastný
D. Štefka
G. Stoilos
A. Suárez
E. Trentin
N. Tsapatsoulis
P. Vidnerová
T. Villmann
Z. Vomlel
T. Wennekers
P. Wira
B. Wyns
Z. Yang
F. Železný

Table of Contents – Part II

Pattern Recognition and Data Analysis

Hardware, Embedded Systems

Computational Neuroscience

Connectionistic Cognitive Science

Neuroinformatics

Neural Dynamics

Special Session: Coupling, Synchronies and Firing Patterns: from Cognition to Disease

Special Session: Constructive Neural Networks

Workshop: New Trends in Self-organization and Optimization of Artificial Neural Networks

Workshop: Adaptive Mechanisms of the Perception-Action Cycle

Table of Contents – Part I

Mathematical Theory of Neurocomputing

Learning Algorithms

Kernel Methods, Statistical Learning, and Ensemble Techniques

Support Vector Machines

Reinforcement Learning

Evolutionary Computing

Hybrid Systems

Self-organization

Control and Robotics

Signal and Time Series Processing

Image Processing

Image Processing – Recognition Systems

Investigating Similarity of Ontology Instances and Its Causes

Anton Andrejko and Mária Bieliková

Institute of Informatics and Software Engineering
Faculty of Informatics and Information technologies
Slovak University of Technology, Ilkovičova 3, 842 16 Bratislava
{andrejko,bielik}@fiit.stuba.sk

Abstract. In this paper we present a novel method of comparing instances of ontological concepts in regard to personalized presentation and/or navigation in large information spaces. It is based on the assumption that comparing attributes of documents which were found interesting for a user can be a source for discovering information about user's interests. We consider applications for the Semantic Web where documents or their parts are represented by ontological concepts. We employ ontology structure and different similarity metrics for data type and object type attributes. From personalization point of view we impute reasons that might have caused user's interest in the content. Moreover, we propose a way to enumerate similarity for the particular user while taking into account individual user's interests and preferences.

1 Introduction

Applications providing information from large information spaces can provide a user more relevant content if personalization is used. Personalization of visible aspects is usually based on user characteristics represented in the user model. To provide proper personalization the user model needs to be reasonably populated with user characteristics that are up to date and relevant to the information space being accessed. Several approaches are used to obtain user characteristics. Some information can be acquired when the user is asked explicitly or from observing one's behavior while working with the application. Mining user characteristics from activity logs can be helpful to establishing patterns of needs or interests.

Analyzing content that is presented to a user is a good source of information about the user [1]. If we know user's rating given to displayed content (e.g. user's interest) we can acquire some characteristics by analyzing the content. Since the rating varies we need to understand possible reasons for why it is low or high. For instance, one can stumble upon hundreds of job offers on the Web that advertise a position for Java programmers requiring high school education, at least three years of previous experience, knowing basics of Web technologies, offering motivating salary, etc. Let us have two such offers that have most features similar and differ only in the job location. Assume we get different ratings for

V. Kůrková et al. (Eds.): ICANN 2008, Part II, LNCS 5164, pp. 1–10, 2008.

these two offers. The range or variety of the evaluation rating that was derived could have been caused by the job location attribute.

In this paper we present a novel method for comparing instances of ontological concepts aimed at identification of common and different aspects to be used for personalization purposes. Examples used in the paper are from job offers domain that is the subject of a research project NAZOU[1] [2].

2 Related Work

Semantic Web applications typically use ontology methodologies as a base for metadata representation and reasoning. Several approaches to comparison of ontology concepts, or their instances, were mainly developed for the purpose of ontology management. Similarly this problem is also known in ontology mapping, matching or alignment. Their aim is to increase reusability and interoperability between different ontologies covering the same domain. In [3] an approach is described that is aimed at identification of changes in ontology versions on the level of ontology schema and ontology instances using various heuristics.

The approach described in [4] uses three independent similarity assessments. It deals with synonyms to ensure that synonyms refer to the same objects. Semantics are then incorporated and lastly semantic relations (e.g. *is-a*) are used to determine whether connected entities are related to the same set of entity classes. Finally, distance between two concepts is measured by the shortest path.

In [5] an approach is described that conceptualizes ontology mapping in four stages that include similarity of labels, instances, structures and previous mapping results verified by the application. While comparing instances the Edit-Distance method is used in conjunction with a Glue approach based on machine learning techniques [6]. It uses predefined similarity function to compute a similarity value for each pair of concepts and generates the similarity matrix.

A method that accomplishes comparing instances of tourism ontology concepts in two phases is described in [7]. The first phase is focused on preprocessing the concepts. Two graphs are built – the *inheritance graph* organizes ontological concepts according to a generalization hierarchy and the *similarity graph* in which nodes relate to concepts and edges have assigned similarity degree. Similarity is enumerated in the second phase using a three step process. First, structural attributes are used, then hierarchical structure is exploited, and finally a similarity measure is computed as a result of combination of two previous steps.

Comparison with ideal instance related to the particular domain (here job offers) is used in searching based on user's criteria [8]. The method allows searching also offers that do not fulfill criteria fully. The user is allowed to specify for each criterion, whether it has to be fulfilled, its importance, and precision.

A common characteristic for all the mentioned approaches is that they do not investigate causes of similarity. Automated similarity enumeration mimics to human similarity measure if different strategies are used according to clusters

[1] NAZOU – Tools for acquisition, organization and maintenance of knowledge in an environment of heterogeneous information resources, http://nazou.fiit.stuba.sk

of users [9]. Users gave reasons for their assessments which become the basis for machine learning algorithm that assigns users to a cluster. We use an automated approach to quantify and define reasons of similarity, what also contributes to scrutiny of the user model.

3 Similarity Enumeration

Similarity of two objects is expressed as a number from interval $\langle 0, 1\rangle$ where similarity of entirely different objects equals zero and similarity of identical objects equals one. Similarity characteristics are also characterized in *reflexivity* (where an object is identical to itself) and *symmetry* (where if object X equals Y, then Y reciprocally equals X).

For similarity enumeration any aggregation function can be used. We use mean value to enumerate similarity between instances of concepts. The similarity of instances *InstA* and *InstB* is evaluated as follows:

$$sim\left(InstA, InstB\right) = \frac{\sum_{i=0}^{|A\cap B|} GeneralSM_i\left(SetA, SetB\right)}{|A \cup B|} \tag{1}$$

where $GeneralSM_i$ encapsulates all similarity measures that are available (e.g. according to attribute type), A and B are sets of attributes instances consist of, respectively. Since an attribute can appear as a multiple, *SetA* and *SetB* are used as a possible set of objects that can be connected to the particular attribute.

When using aggregation of partial similarities the computed result is the same at all the times no matter what is the context. Since each user has different preferences related to similarity, we consider this in the similarity enumeration. It is useful, especially in cases when a user model that holds user's preferences is available. Therefore, we introduce weights to personalize enumeration what allows computing similarity taking into account user's individuality. Now, similarity is evaluated as follows:

$$sim\left(InstA, InstB\right) = \frac{\sum_{i=0}^{|A\cap B|} weight_i \times GeneralSM_i\left(SetA, SetB\right)}{\sum weight} \tag{2}$$

where the assigned meaning of variables is the same as in Eq. 1. The variable *weight* is computed for each attribute that two instances have in common. It gets a value from range $\langle 1, w\rangle$ according to the match with corresponding characteristic in the user model. We assume that user's likes should result in more influence on total similarity in our similarity assessment model. If there is a corresponding characteristic in the user model to an attribute of the instance and also the value of the characteristic equals the value of the attribute, the *weight* is set to w. In cases where no match between values is detected, *weight* is selected from the range (1,w) according to the computed closeness to preferred value in the user model, e.g. a city belongs to the same region as the city preferred by the user in the user model but it is not that specific city. Our experiments showed that $weight = 2.0$ is a worthy selection value (see Sect. 5).

4 Method for Ontology Instances Similarity Evaluation

In the Semantic Web applications documents or their parts are represented by ontological concepts. A concept describes a set of real objects [10]. Concepts can be ordered in a hierarchy. Instances of concepts reflect objects from real world. An example of an instance representing a job offer is depicted in Fig. 1.

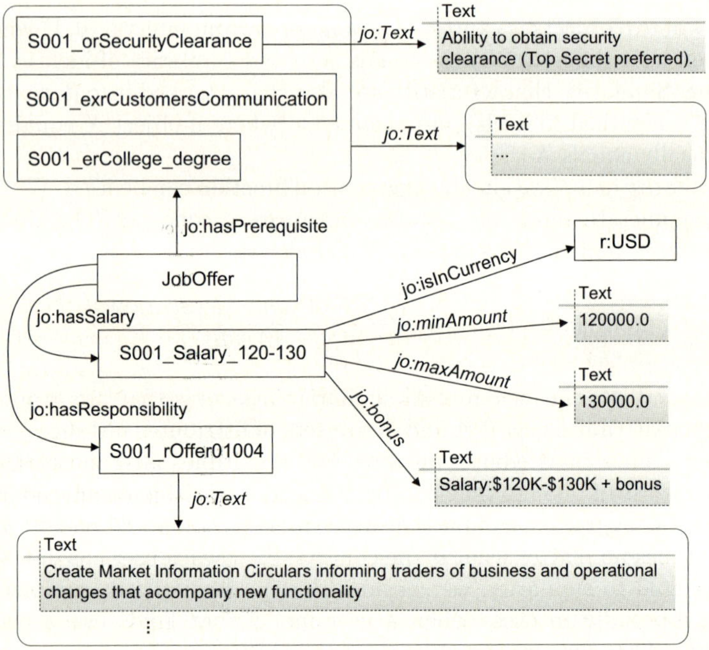

Fig. 1. Example of an instance representing a part of job offer. Each object has its unified identifier, here we present only object's label. *JobOffer* is an identifier of the instance. We use italic font for data type attributes to distinguish them from object type attributes. For simplicity, multiple attributes are surrounded by a rounded box.

If we think about an ontology statement as a triple in form *subject – predicate – object*, an attribute represents predicate. In general, there are *data type* and *object type* attributes. A data type attribute is connected to a literal value that can be of several types defined according XML Schema. An object type attribute expresses the relationship of a concept to another concept, or to an instance.

4.1 Recursive Evaluation of Ontology Instances Similarity

To evaluate similarity we have proposed a method based on recursive evaluation of the attributes and component objects an instance consists of. The main idea is based on looking for common pairs in both attributes and their sequential processing. Basic steps of the method are depicted in Fig. 2.

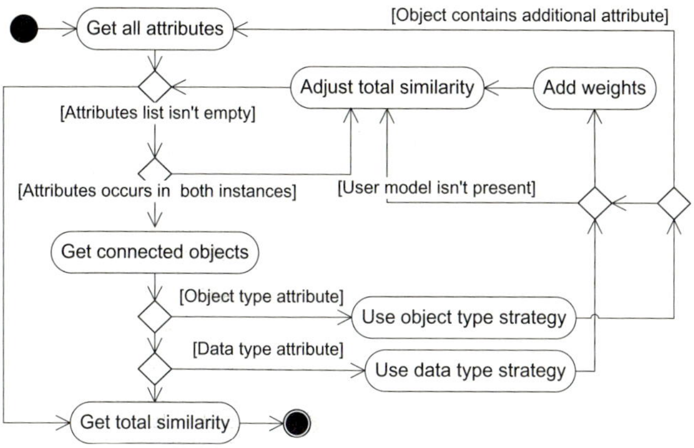

Fig. 2. Basic steps of the method for recursive traversing of instance

The process of comparison begins with acquiring all the attributes from both instances. An attribute can have single or multiple occurrences in both instances or single/multiple occurrence in one instance only. When the attribute has a single occurrence in both instances, objects (literals) refered to, are evaluated for their similarity. Variety of similarity metrics can be used. If the attribute is data type, the comparison for the attribute terminates after a metrics is used to evaluate similarity between connected literals. Resulting computed similarity measure(s) is aggregated to a total similarity measure. In the case of an object type attribute, a metrics for connected object is used. Furthermore, the comparison is being launched recursively on that object until literals are achieved.

A multiple occurrence is the most specific case we have to cope with. We move solution of this problem to the lower level. Anytime a multiple attribute is acquired only its one occurrence in the instance is considered. Afterwards, all objects (literals) connected to that attribute are acquired from both instances. Instead of dealing with attributes we now have to deal with two sets of objects (or literals) possibly with different cardinalities. Here, a problem of how to figure out which object from the first set should be compared with an object from another set with the contribution to the total similarity emerges (see Sect. 4.2).

In the situation, when single or multiple occurrence of an attribute is present in only one instance we use an assumption that instances are entirely different in the attribute if there is no presence of that attribute in both instances. In regard to similarity definition, the similarity equals zero if two objects have nothing in common. In this case we estimate similarity for such an occurrence of the attribute as equal zero.

4.2 Comparison Metrics and Similarity Measure

We proposed two groups of metrics according to an attribute's type: data type and object type. To evaluate similarity between literals connected to a data type attribute, any string based metrics can be used[2]. To achieve better results, the semantic type of the literal content is taken into account (e.g. string, date, number are each treated differently). When evaluating the similarity of objects connected to the object type attribute their other characteristics can be considered (e.g. number of attributes and their types, position in taxonomy tree) [11].

Taxonomy distance is a heuristic similarity measure for evaluating similarity between objects connected to the object type attribute. The edge-counting method computes the shortest path between nodes. Distance is defined as the shortest path linked through a common ancestor or as the general shortest path [9]. Since we do not need a result what is closer or further, but a float number between 0 and 1, we proposed our taxonomy distance metrics. It assumes that the more nodes have two objects in common in the taxonomy tree the more they are similar. Similarity is computed as the number of common nodes in the taxonomy divided by number of nodes in the longer path leading to the object (see Fig. 3).

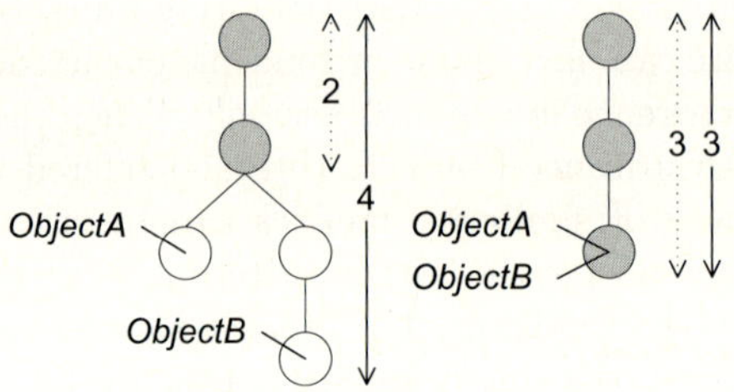

Fig. 3. Taxonomy distance for objects *ObjectA* and *ObjectB* is computed. Common part (nodes) in the taxonomy is emphasized by dotted arrow; solid arrow is used to show longer distance from the root node. For left example $sim(ObjectA, ObjectB) = 2/4 = 0.5$, for right example $sim(ObjectA, ObjectB) = 3/3 = 1.0$.

Identification of relevant pairs using only the object's label is not satisfactory. Each object in the ontology can have a label that could be compared using selected data type metrics. Since the label is optional and does not have to necessarily express any semantics we avoid using it. It should be noted that for automatically acquired instances it is obvious that meaningful labels are not present. We proposed the similarity measure to identify pairs of objects, therefore, a relevance matrix is constructed which size is specified by cardinalities of sets of objects.

The matrix holds similarities for each pair of objects from the sets. In the case of literals, data metrics are used. For objects the recursive algorithm is employed as for the entire instance. Afterwards, an identification of pairs can start. Number of pairs is given by the set with the lower cardinality. Finding pairs with very

[2] A collection of methods suitable for string comparing is implemented in the open source library SimMetrics, http://www.dcs.shef.ac.uk/~sam/simmetrics.html

low similarity measure can be restricted by using a critical threshold as a filter. The algorithm for finding relevant pairs follows these steps:

```
WHILE count(pairs) < count(getSmallerSet(setA, setB)) DO
  SET maxValue to getMaxValue(matrix)
  STORE maxValue in List
  SET coordinates of maxValue to X and Y
  FOR each item in matrix
    IF item.row = X OR item.column = Y
      SET item to -1
    END IF
  END FOR
END WHILE
```

Leftover objects are handled in the same way as described above for attributes that have occurrence in one instance only. An example is shown in Fig. 4.

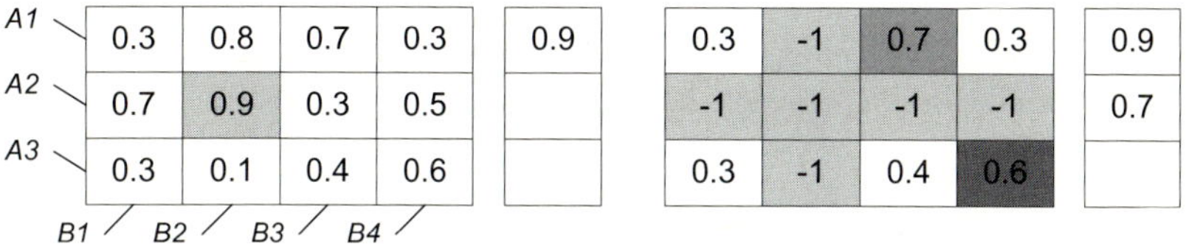

Fig. 4. Identifying relevant pairs from the sets. Similarities in the matrix are random numbers. In first iteration (left) at [A2,B2] is maximal value 0.9 and it is stored. Second row and second column are set to -1. In the next iteration at [A1,B3] is maximal value 0.7. The last coordinate is [A3,B4]. Object B1 is evaluated as a leftover.

Our experimental results show that the way we find related pairs (in case of object type attributes in combination with taxonomy distance) leads to meaningful results. First, identities were found (maximal possible value 1.0). Other found pairs were interpreted as semantically similar by a human. The number of multiple attributes in job offers is usually small (less than 10). Therefore, threshold 0.3 for deciding which pairs are still meaningful is reasonable.

4.3 Investigating Similarity Causes

Our goal is not only to compute the similarity between instances but also to investigate reasons that caused the similarity or difference to be used later for personalization purposes. From the user's evaluation given to content we can deduce user's likes or dislikes. We assume that if the instance includes an attribute that the user likes, it will likely influence his/her rating towards higher (or positive) values. On the other hand, attributes of the content that the user dislikes will influence rating towards lower (or negative) values.

Therefore, we introduced two threshold values that divide attributes into three sets according to their similarity values. Since we are interested in attributes that significantly influence the user's evaluation, we give up splitting outcome interval in the equal parts. An attribute exhibiting similarity greater than the *positive threshold* would be assigned to the positive interval set and the similarity exhibiting lower than the *negative threshold* to negative interval set.

Thresholds were specified experimentally for this job offer domain. We evaluated 55 000 attributes. Attributes with similarity equal 0.0 or 1.0 were not considered to eliminate identities and attributes with no occurrence in both instances. The rest of the attributes were ordered according to similarity measure and the Pareto principle (also known as 80/20 rule) was used. We split the 20% segment in half to select 10 % of highest and 10 % of lowest values. This way, the *positive threshold* was set to 0.65 and *negative threshold* to 0.25. Domain independence of thresholds is subject of further experiments.

Attributes classified by this method can be transformed into user characteristics and then used for filling or updating existing characteristics in the user model. A transformation of attributes to user characteristics as well as their updating in the user model is not included in the scope of this paper. The presented method only prepares inputs for further processing. Using positive and negative set of attributes in combination with user's feedback for characteristics update in the user model would improve user characteristics estimation.

5 Method Evaluation and Conclusions

We described a method for comparing instances of ontological concepts based on recursive traversing of instance's structure. Final similarity is a result of mean aggregation of similarities computed for particular attributes while their type is considered. Introducing similarity computed for individual attributes allows employing semantics from ontology representation. It allowed us to extend similarity enumeration with weights to compute similarity for particular user to be used for personalization purposes. Moreover, we investigate reasons (attributes) that influenced user's evaluation (e.g. interest) of the content. We introduced two thresholds dividing attributes in three sets. From personalization point of view we are interested in only two outer sets (positive and negative). These can be used by other tools for actualization of characteristics in the user model.

We have evaluated proposed method using developed software tool called *ConCom* (Concept Comparer) implemented in Java. Sesame framework was used to access the ontological models represented in OWL DL. Evaluation was processed on an experimental job offer ontology developed in the course of the research project NAZOU. In the experiment, similarity for 10 000 pairs was computed. The experiments showed that computed results fulfill all criterions requested for similarity. In Fig. 5 there is a depiction of a sample of 80 pairs where similarity was computed by *ConCom* for (1) all attributes and (2) for common attributes only. Computed values were sorted out according to the computed similarities in the first way.

In the following experiment a user was involved. A sample of 300 job offer pairs was used where 30 randomly selected pairs appeared twice as a check sample. We

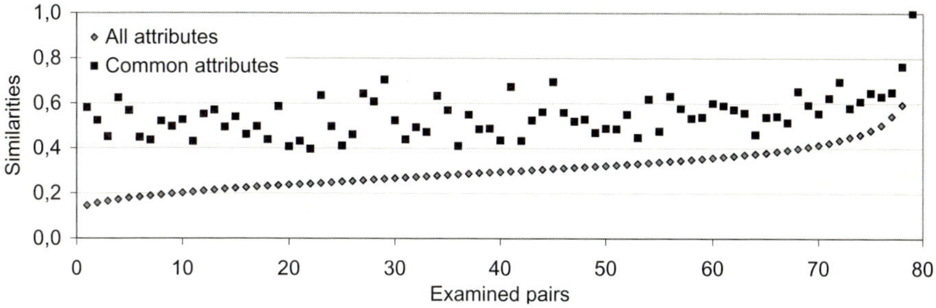

Fig. 5. Similarity computed by *ConCom* in regard to considered attributes

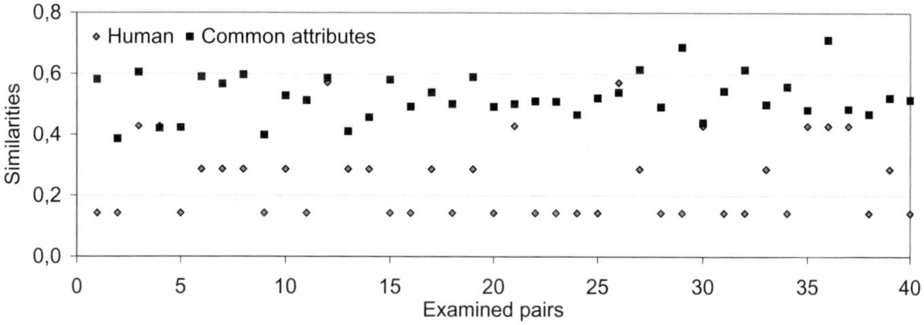

Fig. 6. Similarity estimated by a human and by *ConCom* for common attributes

asked the user to assess similarity on a scale from 0 to 7. Afterwards, acquired values were recounted to similarity interval. The result for a randomly selected set of 40 pairs is depicted in the Fig. 6.

We used similarity computed for common attributes to compare with our test subject human evaluation since its values mimic values from evaluation given by a human better. This result could have been caused by the fact, that a human user can easier evaluate lower amount of attributes and especially common attributes. Therefore, for further experiments with the user model we use similarity computed for common attributes. On the other hand, using only common attributes in our experiments resulted in narrow range of similarity values – in 94.1 % computed similarities were from range 0.35 to 0.7, what makes it not very useful for discovering user's characteristics.

To figure weights for personalized similarity a user model was involved consisting of one characteristic only (*hasDutyLocation*). Job offers used in the experiment consisted of an average of sixteen attributes in averaged and contained that attribute with the same value as in the user model. Already doubled weights cause noticeable change in the similarity – from 0.06 up to 0.10 depending on the number of attributes job offers consist of.

We have started exploring the interest of scientific publications to further investigate domain independence of the method. The achieved results can be useful in user model creation in combination with other methods [12,13], as a support for clustering algorithms, semantic annotation or repository maintenance tools as well as for recommending similar content in recommending systems. The aim here is to improve semantic search using the method for personalized navigation within ontology instances that represent metadata of large information space.

Acknowledgement. This work was partially supported by the State programme of research and development "Establishing of Information Society" under the contract No. 1025/04 and by the Scientific Grant Agency of Slovak Republic, grant No. VG1/3102/06.

References

1. Brusilovsky, P., Tasso, C.: Preface to special issue on user modeling for web information retrieval. UMUAI 14(2-3), 147–157 (2004)
2. Návrat, P., Bieliková, M., Rozinajová, V.: Acquiring, organising and presenting information and knowledge from the web. In: Comp. Sys. Tech. 2006 (2006)
3. Tury, M., Bieliková, M.: An approach to detection ontology changes. In: ICWE 2006: Workshop Proc. of 6th Int. Conf. on Web Eng. ACM Press, Palo Alto (2006)
4. Rodríguez, M.A., Egenhofer, M.J.: Determining semantic similarity among entity classes from different ontologies. IEEE Transactions on Knowledge and Data Engineering 15(2), 442–456 (2003)
5. Liu, X., Wang, Y., Wang, J.: Towards a semi-automatic ontology mapping. In: Proc. of 5th Mexican Int. Conf. on Artificial Intelligence (MICAI 2006). IEEE, Los Alamitos (2006)
6. Doan, A., Madhavan, J., Domingos, P., Halevy, A.: Learning to map between ontologies on the semantic web. In: WWW 2002: Proceedings of the 11th international conference on World Wide Web, pp. 662–673. ACM, New York (2002)
7. Formica, A., Missikoff, M.: Concept similarity in symontos: An enterprise management tool. The Computer Journal 45(6), 583–595 (2002)
8. Pázman, R.: Ontology search with user preferences. In: Tools for Acquisition, Organisation and Presenting of Information and Knowledge, pp. 139–147 (2006)
9. Bernstein, A., Kaufmann, E., Burki, C., Klein, M.: How similar is it? towards personalized similarity measures in ontologies. In: 7th International Conference Wirtschaftsinformatik (WI-2005), Bamberg, Germany, pp. 1347–1366 (2005)
10. Ding, L., Kolari, P., Ding, Z., Avancha, S.: Using ontologies in the semantic web: A survey. In: Sharman, R., et al. (ed.) Ontologies: A Handbook of Principles, Concepts and Applications in Information Systems, pp. 79–113. Springer, Heidelberg (2007)
11. Resnik, P.: Semantic similarity in a taxonomy: An information-based measure and its application to problems of ambiguity in natural language. Journal of Artificial Intelligence Research 11, 95–130 (1999)
12. Andrejko, A., Barla, M., Bieliková, M.: Ontology-based user modeling for web-based inf. systems. In: Advances in Information Systems Development New Methods and Practice for the Networked Society, vol. 2, pp. 457–468. Springer, Heidelberg (2007)
13. Barla, M., Bieliková, M.: Estimation of user characteristics using rule-based analysis of user logs. In: Data Mining for User Modeling Proceedings of Workshop held at the International Conference on User Modeling UM 2007, pp. 5–14 (2007)

A Neural Model for Delay Correction in a Distributed Control System

Ana Antunes[1], Fernando Morgado Dias[2], and Alexandre Mota[3]

[1] Escola Superior de Tecnologia de Setúbal do Instituto Politécnico de Setúbal
Departamento de Engenharia Electrotécnica, Campus do IPS, Estefanilha,
2914-508 Setúbal, Portugal
Tel.: +351 265 790000
aantunes@est.ips.pt

[2] Centro de Ciências Matemáticas - CCM and Departamento de Matemática e Engenharias, Un. da Madeira Campus da Penteada, 9000-390 Funchal, Madeira, Portugal
Tel.: +351 291-705150/1
morgado@uma.pt

[3] DETI/Universidade de Aveiro, 3810 Aveiro, Portugal
Tel.: +351 34 370381
alex@det.ua.pt

Abstract. Distributed control systems frequently suffer from variable sampling to actuation delay. This delay can degrade the control quality of such systems. In some distributed control systems it is possible to know, at control time, the value of the delay. The work reported in this paper proposes to build a model of the behavior of the system in the presence of the variable delay and to use this model to compensate the control signal in order to avoid the delay effect. This model can be used as a compensator that can easily be added to an existing control system that does not account for the sampling to actuation delay effect. The compensator can be applied to distributed systems using online or off-line scheduling policies provided that the sampling to actuation delay can be evaluated. The effectiveness of the neural network delay compensator is evaluated using a networked control system with a pole-placement controller.

1 Introduction

Distributed or networked control systems are widely used in embedded applications. The distribution of the controller and the use of a communication network to connect the nodes of the control loop induce variable sampling to actuation delay.

The delays are introduced due to the medium access control of the network, the processing time and the processor scheduling in the nodes and the scheduling mechanism used to schedule the bus time. They are usually variable from iteration to iteration of the control loop. These delays degrade the performance of the control system and can even destabilize the system [1], [2], [3], [4], [5] and [6].

There are several ways to deal with the variable sampling to actuation delay in distributed control systems. One possibility is to use a delay compensator to improve the control performance of the loop. Different kinds of compensators have been used to

V. Kůrková et al. (Eds.): ICANN 2008, Part II, LNCS 5164, pp. 11–20, 2008.

improve the control performance of networked control systems like the ones presented in [7], [8] and [9].

This work proposes a neural network based implementation to the delay compensator approach presented in [9].

2 The Principle of the Delay Compensator

The delay compensator approach proposes to add a delay compensator to an existing controller that does not take into account the effects of the variable sampling to actuation delay.

The principle is similar to the Additive feedforward Control that is represented in figure 1.

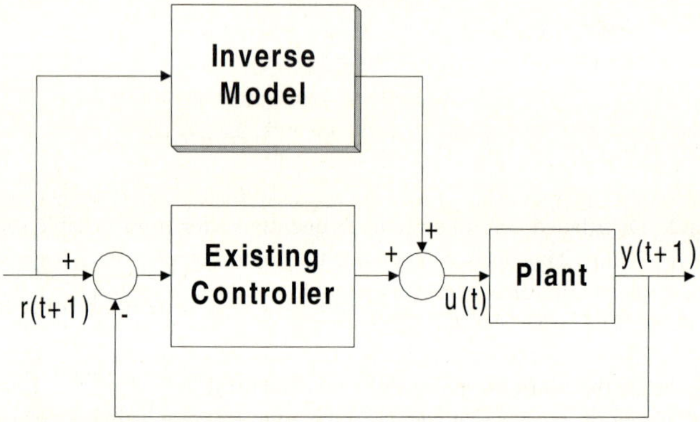

Fig. 1. Block diagram of the Additive Feedforward Controller

The principle of additive feedforward control is quite simple: add to an existing (but not satisfactory functioning) feedback controller an additional inverse process controller.

The additive feedforward control strategy offers the following important advantages [10]:

- Data collecting can be done using the existing closed loop, avoiding plant stopping for data collection and facilitating the access to good quality data.
- There is no need for opening the existing control loop nor during training neither during the introduction of the additive controller.

The delay compensator is slightly different from the Additive Feedforward control since the introduced block is not a full controller but intends only to compensate for the effect of the delay. This way, the compensator proposes a correction to the control action, based in the delay information, in order to improve the overall control performance.

Figure 2 shows the block diagram of the delay compensator approach.

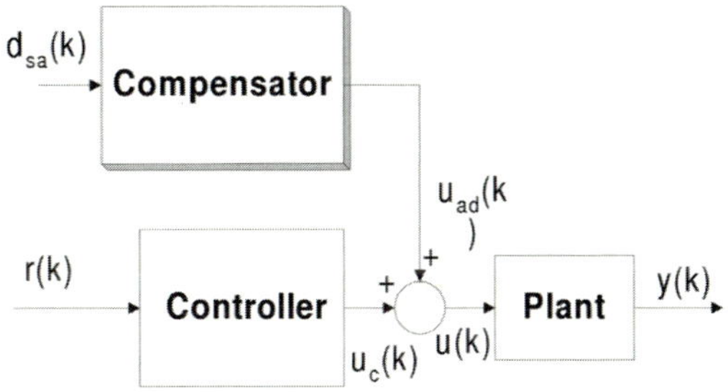

Fig. 2. Block diagram of the delay compensator approach

The output of the compensator is based on the sampling to actuation delay that affects the system at each control cycle and can have any other input.

This principle can be applied to any distributed control system provided that the sampling to actuation delay is known for each control cycle. The determination of the sampling to actuation delay can be done either by an online measurement or by off-line computations depending on the a priori knowledge of the overall system operation conditions. The online measurement of the sampling to actuation delay provides a more generic and flexible solution that does not require the knowledge of the details of the operation conditions of the global system in which the distributed controller is inserted.

In order for the compensator to operate with the correct value of the delay affecting the network the controller and the actuator must be implemented in the same network node.

The delay compensator principle is generic and can be implemented using different techniques. In [9] a fuzzy implementation is proposed based in the empirical knowledge of the system, while here the delay compensator is a neural model, trained in supervised mode.

The compensator can easily be turned on or off and the operation of the existing controller is not disturbed.

2.1 Neural Implementation of the Delay Compensator

The model for the delay compensator is not a regular model. The information available is the output of a certain system with ($y_d(k)$) and without (y(k)) the effect of sampling to actuation delay. The objective is to produce a model that can compensate the effect of this delay (knowing the delay for the iteration) in order to correct the control signal to avoid the degradation of the control system.

The information available is depicted in figure 3. It is now necessary to find a way of producing a model that can perform the requested task.

The authors considered two possibilities: calculating an error between the two outputs: $e_y(k)$=y(k)-$y_d(k)$ and reporting this error to the input through an inverse model or calculating the equivalent input of a system without sampling to actuation delay that would have resulted in the output $y_d(k)$. This is obtained through an inverse model. The second approach is illustrated in figure 4.

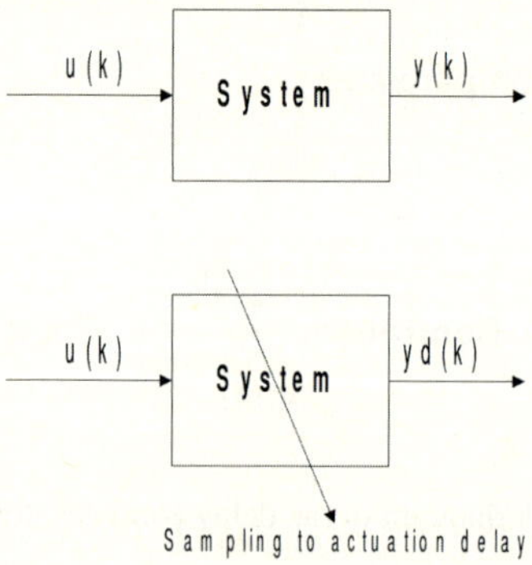

Fig. 3. Representation of the information used to create the delay compensator model

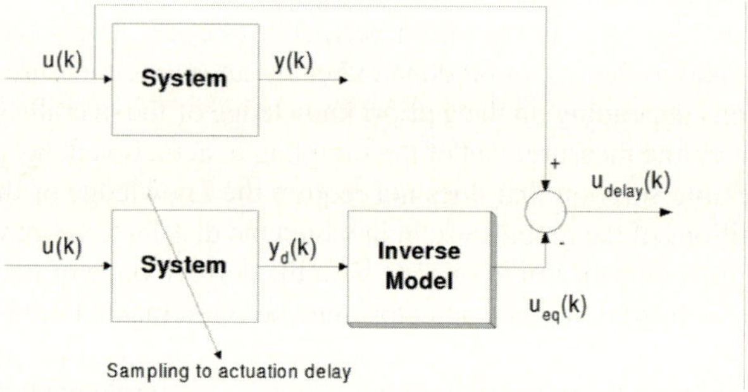

Fig. 4. Determining the delay effect in terms of control signal

The first alternative would only be valid in the case of a linear system, since for non-linear systems there is no guarantee that the output error could be reported to the corresponding input, since the error range $y - y_d$ is very different from the output signal range. Using the difference between y and y_d and applying it to the inverse model could result in a distortion due to the non-linearity.

Producing the required model is somehow similar to the specialized training for inverse models that is shown in figure 5.

The specialized training solution was developed to avoid the problem of training generic inverse models. These models were considered not to be goal directed since they were trained in a situation different from the one where they would be used in a control loop.

Although solving the goal directed problem this solution creates another one: the error obtained here is $e_y(k)$ which reports to the output of the loop, but the error needed to train the inverse model should be $e_u(k)$. To propagate the output error to the input it is necessary to know the Hessian or Jacobian matrices of the system (depending on the training algorithm used). Such information is rarely available which led to an approximation: using the direct model as a replacement of the system and propagate through it the error $e_y(k)$.

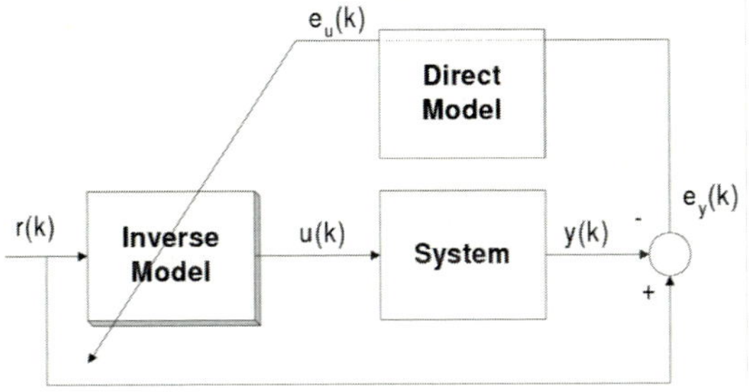

Fig. 5. Specialized training of inverse models

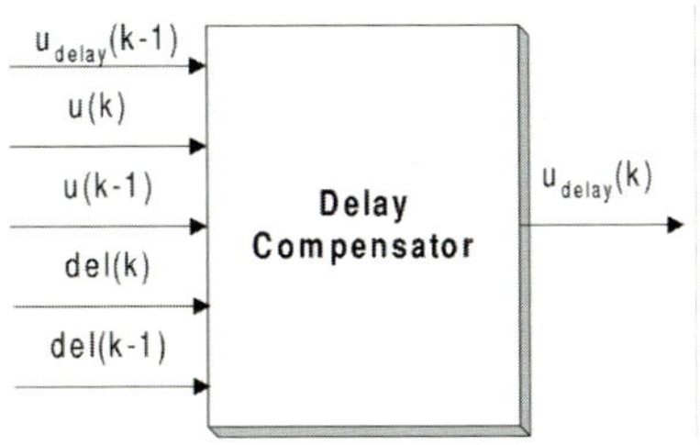

Fig. 6. Delay compensator model

Using the proposal from figure 4 and studying the lag space results in the model represented in figure 6.

This model has, as inputs, a past sample of the output, two past samples of the control signal and two past samples of the delay information.

The model is composed of ten neurons with hyperbolic tangents in the hidden layer and one neuron in the output layer with linear activation function and was trained for 15000 iterations with the Levenberg-Marquardt algorithm.

3 The Test System

The architecture of the test system and the existing controller will be presented in the following subsections.

3.1 Distributed System Architecture

The test system is composed of 2 nodes: the sensor node and the controller and actuator node connected through the Controller Area Network (CAN) bus. The controller and the actuator have to share the same node in order to be possible to measure accurately the value of the sampling to actuation delay that affects the control loop at each control cycle. The block diagram of the distributed system is presented in figure 7.

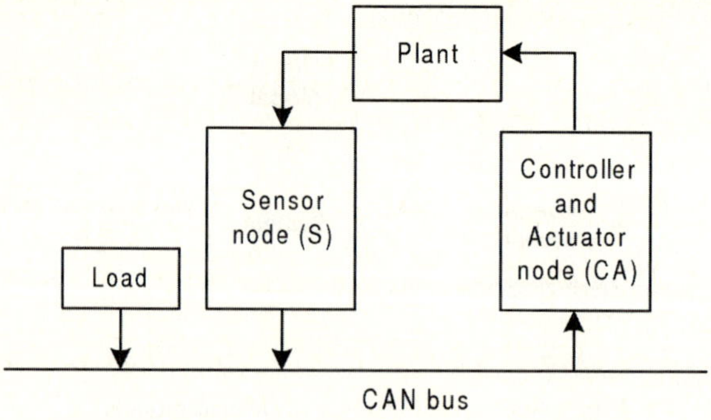

Fig. 7. Block diagram of the test system

The transfer function of the plant is presented in 1.

$$\frac{Y(s)}{U(s)} = \frac{0.5}{s + 0.5} \tag{1}$$

The system was simulated using TrueTime, a MATLAB/Simulink based simulator for real-time distributed control systems [1].

3.2 Existing Controller

The existing controller is of the pole-placement type. It does not take into account the sampling to actuation delay. The controller parameters are constant and computed based on the discrete-time transfer function given by equation 2.

$$G(q^{-1}) = \frac{bq^{-1}}{1 - aq^{-1}} \tag{2}$$

The pole-placement technique allows the complete specification of the closed-loop response of the system by the appropriate choice of the poles of the closed-loop transfer function. In this case the closed-loop pole is placed at α_m=2Hz. An observer pole was also used with α_0=4Hz.

The sampling period is equal to 280ms and was chosen according to the rule of thumb proposed by [11].

The identification of the system was based in the discrete model in 2 and the parameters were computed off-line.

The parameters of the control function were obtained by directly solving the Diophantine equation for the system. The control function is given by 3.

$$u_c(k) = t_0(r(k) - a_0 r(k-1)) - s_0 y(k) - s_1 y(k-1) + u_c(k-1) \tag{3}$$

where t0=3.2832, ao= 0.3263, s0=7.4419 and s1= -5.2299.

4 Tests and Simulation Results

The present work includes three different tests. Test 1 is the reference test, where the sampling to actuation delay is constant and equal to 8ms. It corresponds to the minimum value of the MAC and processing delays.

In tests 2 and 3 additional delay was introduced to simulate a loaded network. The sampling to actuation delay introduced follows a random distribution over the interval [0,h] for test 2 and a sequence based in the gamma distribution that concentrates the values in the interval [h/2, h] for test 3.

The sampling to actuation delay obtained for tests 2 and 3 is depicted in figure 8 and 9.

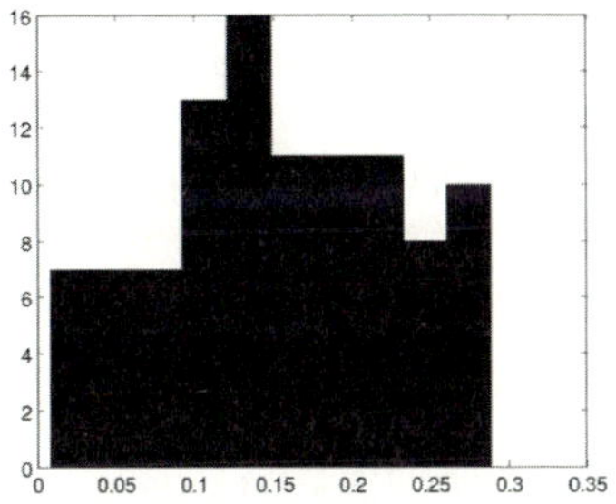

Fig. 8. Sampling to actuation delay for test 2

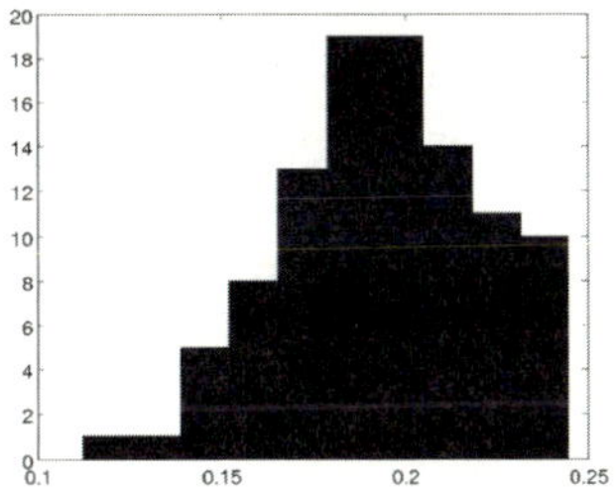

Fig. 9. Sampling to actuation delay for test 3

The results obtained for test 1 for the system without the compensator (PP) are presented in 10.

The tests were performed for the system without the compensator (PP) and for the system with the neural network delay compensator (NNDC). The control performance was assessed by the computation of the Integral of the Squared Error (ISE) between t= 5s and 30s. The results obtained for ISE are presented in Table 1.

The percentage of improvement obtained compared to the reference test for ISE is presented in Table 2.

The improvement is calculated as the amount of error induced by the sampling to actuation delay that the NNDC controller was able to reduce.

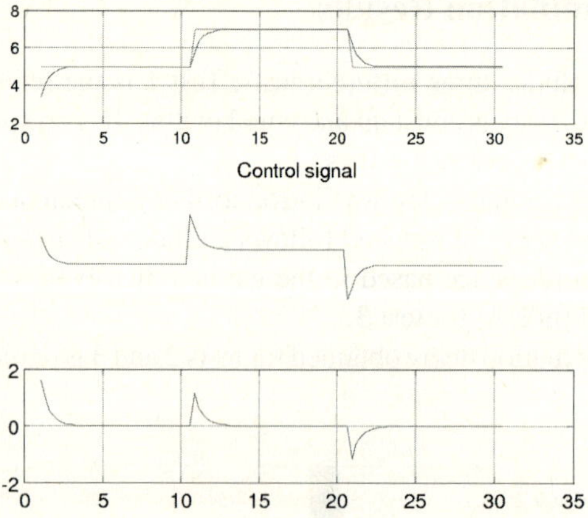

Fig. 10. Control results for test 1

Table 1. ISE report

Test	PP	NNDC
1	3.3	3.3
2	4.1	3.7
3	4.6	3.7

Table 2. ISE report

Test	NNDC
2	50%
3	69%

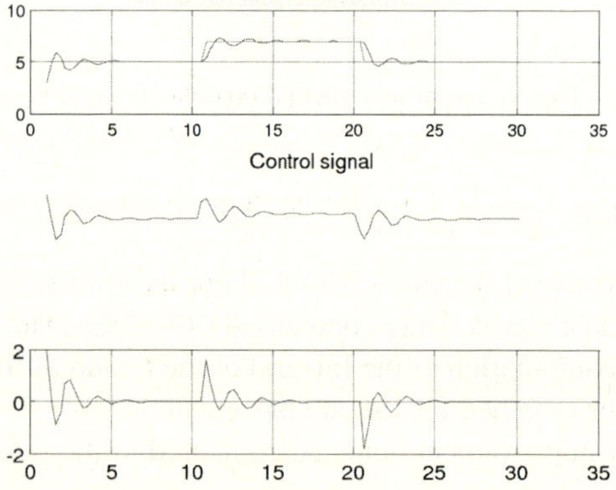

Fig. 11. Control results for test 3 without the NNDC

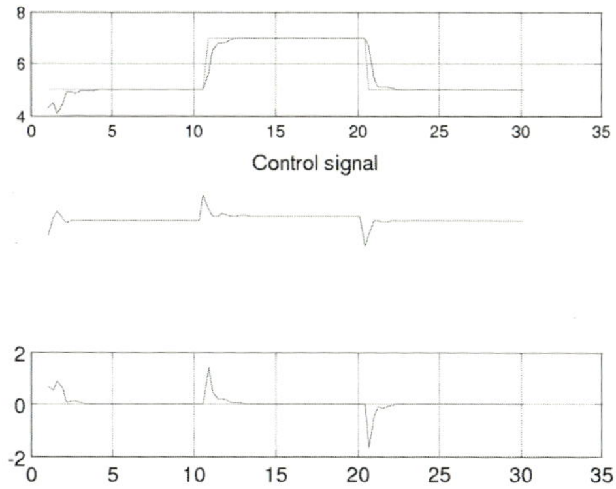

Fig. 12. Control results for test 3 with the NNDC

The control results for test 3 with and without the compensator are presented in figures 11 and 12.

The results show the effectiveness of the NNDC. For tests 2 and 3 the control performance obtained using the NNDC is better than the ones obtained without compensation. For test 2 the compensator reduced by 50% the effects of the variable sampling to actuation delay and for test 3 the reduction is equal to 69%.

The reference test (test 1) was also performed with the compensator. The results show that the modeling of the effect of the sampling to actuation delay is good since the results obtained is the same as the one obtained without the compensator.

5 Conclusion

This work presents a neural network implementation for the delay compensator.

The NNDC can be added to any existing distributed control system provided that the sampling to actuation delay can be determined for each control cycle.

The delay compensator is implemented through a model that describes the effect of the sampling to actuation delay in terms of control signal. This neural model is not a regular model and its training situation is similar to the specialized training.

The neural model is used to compensate the control signal according to the delay at the actuation moment.

The compensator was able to model the effect of the sampling to actuation delay on the distributed control system as the results show and it allowed the improvement of the control performance of the system.

References

1. Cervin, A.: Integrated Control and Real-Time Scheduling. PhD thesis, Lund Institute of Technologie (2003)
2. Tipsuwan, Y., Chow, M.Y.: Control methodologies in networked control systems. IFAC Control engineering practice 11, 1099–1111 (2003)

3. Colom, P.: Analysis and Design of Real-Time Control Systems with Varying Control Timing Constraints. PhD thesis, Universitat Politècnica de Catalunya (2002)
4. Sanfridson, M.: Timing problems in distributed real-time computer control systems. Technical Report ISSN 1400-1179, Royal Institute of Techology (2000)
5. Antunes, A., Dias, F.: Influence of the sampling period in the performance of a real-time distributed system under jitter conditions. WSEAS Transactions on communications 3, 248–253 (2004)
6. Antunes, A., Mota, A.: Control performance of a real-time adaptive distributed control system under jitter conditions. In: Proc. of the Control 2004 Conference (2004)
7. Almutairi, N., Chow, M.Y., Tipsuwan, Y.: Networked-based controlled DC motor with fuzzy compensation. In: Proc. of the 27th annual conference of the IEEE Industrial Electronics Society, vol. 3, pp. 1844–1849 (2001)
8. Lin, C.L., Chen, C.H., Huang, H.C.: Stabilizing control of networks with uncertain time varying communication delays. Control Engineering Practice 16, 56–66 (2008)
9. Antunes, A., Dias, F., Vieira, J., Mota, A.: Delay compensator: An approach to reduce the variable sampling to actuation delay effect in distributed real-time control systems. In: 11th IEEE International Conference on Emerging Technologies and Factory Automation, Prague, Czech Republic (2006)
10. Sørensen, O.: Neural networks in control applications. PhD Thesis, Department of Control Engineering, Institute of Electronic Systems, Aalborg University, Denmark (1994)
11. Astrom, K.J., Wittenmark, B.: Computer Controlled Systems: Theory and Design, 3rd edn. Prentice Hall, Englewood Cliffs (1997)

A Model-Based Relevance Estimation Approach for Feature Selection in Microarray Datasets

Gianluca Bontempi and Patrick E. Meyer

Machine Learning Group, Computer Science Department,
ULB, Université Libre de Bruxelles,
Brussels, Belgium

Abstract. The paper presents an original model-based approach for feature selection and its application to classification of microarray datasets. Model-based approaches to feature selection are generally denoted as wrappers. Wrapper methods assess subsets of variables according to their usefulness to a given prediction model which will be eventually used for classification. This strategy assumes that the accuracy of the model used for the wrapper selection is a good estimator of the relevance of the feature subset. We first discuss the limits of this assumption by showing that the assessment of a subset by means of a generic learner (e.g. by cross-validation) returns a biased estimate of the relevance of the subset itself. Secondly, we propose a low-bias estimator of the relevance based on the cross-validation assessment of an unbiased learner. Third, we assess a feature selection approach which combines the low-bias relevance estimator with state-of-the-art relevance estimators in order to enhance their accuracy. The experimental validation on 20 publicly available cancer expression datasets shows the robustness of a selection approach which is not biased by a specific learner.

1 Introduction

The demand for classification in microarray datasets with a large number of features and a small number of samples gave recently a new impetus to research in feature selection [1,2]. This led to the development of many dimensionality reduction techniques which can be grouped into three main approaches: (i) filter methods which assess the merits of features directly from data, ignoring the effects of the selected feature subset on the performance of a learning algorithm [3], (ii) wrapper methods which assess subsets of variables according to their usefulness to a given prediction model [4] and (iii) embedded methods that perform variable selection as part of the learning procedure and are usually specific to given learning machines [5]. All these strategies aim at finding the subset of features which has the highest predictive power or relevance to the output class. Quantitative measures of relevance of a subset of features are the information gain [6,7], the variance of the target probability distribution conditioned on the given subset [4] or the related Gini index [8]. Several filter approaches [6,9,10] rely on learner independent relevance estimators to perform feature selection.

V. Kůrková et al. (Eds.): ICANN 2008, Part II, LNCS 5164, pp. 21–31, 2008.

In this paper we will focus instead on model-based strategies to select highly relevant features and will show that a cross-validated assessment of the relevance can both outperform wrappers and improve existing filters approaches. A wrapper algorithm demands the definition of a specific learning algorithm and is typically made of three steps: (i) a search procedure in a high dimensional space which selects the candidate feature subsets to be assessed (ii) the assessment of the candidate feature subsets by means of the learning algorithm (e.g. by cross-validation) (iii) the choice of the best feature subset according to the assessment. Note that the second step is particularly demanding in microarray classification tasks because of the high ratio between the dimensionality of the problem and the number of measured samples. This paper argues the biasedness of a wrapper procedure which relies on a given learning algorithm to assess the relevance of a set of variables. Given a binary classification task and a learning algorithm h, Wolpert and Kohavi [11] showed that the expected misclassification error M^h can be decomposed into three components: the variance of the noise, the squared bias of the learner and the variance of the learner. Note that while the first term refers to the difficulty of the classification task or better to the amount of information or relevance that the set of input variables have on the output, the two remaining terms refer to the learner accuracy. A common way to estimate the quantity M^h in a wrapper procedure is to adopt a cross-validation procedure (like leave-one-out). It follows that the assessment step of a wrapper strategy does not estimate the relevance of the feature subset but rather a combination of the relevance with learner related quantities. In other words, although cross-validation is an almost unbiased estimator of the generalization accuracy of a learner h [12], it is a biased estimator of the relevance of a set of variables. We deem that this bias may have a strong negative impact on the selection procedure mainly if it is not the accuracy of a specific classifier h which is at stake but the selection of a biologically meaningful and robust signature of genes. The negative impact of this problem can be mitigated by properly choosing the learning algorithm. This means that we should choose a learner with two properties (i) a low bias term and (ii) a variance term which can return some insight on the conditional variance. A well known algorithm which satisfies both properties is the k-Nearest Neighbour (k-NN) algorithm [13]. The use of k-NN to estimate posterior probabilities has been largely discussed in the pattern recognition litterature [14,15]. In particular for $k = 1$ it is known that the misclassification error of this algorithm asymptotically converges to a term which depends on the conditional variance [12]. Note that this consideration leads to a perspective change in the common practice of wrapper selection. So far, it is common to use the same learner for the selection and the classification task according to the principle that the selection procedure will return the set of features for which the considered learner will perform the best. What we advocate here is that this procedure does not aim to return the best set of variables as a whole but the best set of variables for that specific learner. In fact, what is demanded by biologists is not a learner dependent signature but a robust signature able to shed light on the prognostic value of gene expressions. The use of a k-Nearest Neighbour estimator with small k returns a less biased information on the relevance of the set of variables for the classification task and consequently enables a more general and more robust assessment step. However, a potential shortcoming related to the use of a low biased estimator for feature assessment lies in its presumably high variance. The second contribution of the paper aims to address such a problem by combining the cross-validated estimation returned by a low biased estimator with an independent estimation of the relevance of the feature subset. Two combination schemes will be taken into consideration: i) a combination of the low-biased cross-validated estimator with a direct estimator of the conditional probability inspired to [15] and ii) a combination of the leave-one-out estimator with the filter relevance estimator implemented by the Maximum Relevance Minimum Redundancy (MRMR) algorithm [9]. This algorithm proved to be very effective in dealing with gene selection and we will show how its performance can be further enhanced by combining it with a low bias cross-validated assessment. The proposed model-based approach has been tested on 20 cancer classification tasks where a systematic investigation of the dependency between expression patterns of thousands of genes and specific phenotypic variations is expected to provide an improved taxonomy of cancer. In this context, the number of features corresponds to the number of expressed gene probes (up to several thousands) and the number of observations to the number of tumor samples (typically in the order of hundreds). The experimental session supports two main conclusions about the benefit of a low bias cross-validation estimator: 1) conventional wrappers are biased and are outperformed by a selection strategy which exploits the assessment returned by a low-bias cross-validation estimator, 2) state-of-the-art filter techniques can be improved by integrating their relevance assessment with a low-bias cross-validation estimator.

2 Relevance and Feature Selection

Consider a multi-class classification problem [16] where $\mathbf{x} \in \mathcal{X} \subset \mathbb{R}^n$ is the n-variate input[1] with distribution function $F_{\mathbf{x}}$, $\mathbf{y} \in \mathcal{Y}$ is the target variable which belongs to a set $\mathcal{Y}$ of K distinct classes $\{y_1, \dots, y_K\}$. Let $\mathbf{s} \subset \mathbf{x}$, $\mathbf{s} \in \mathcal{S}$ be a subset of the input vector having the marginal distribution $F_{\mathbf{s}}(\cdot)$. In a feature selection task it is important to define a measure of relevance of a subset $\mathbf{s}$ of variables to the output $\mathbf{y}$. This measure accounts for the predictive power that the input $\mathbf{s}$ has on the target $\mathbf{y}$. Here we define the relevance of the subset $\mathbf{s}$ of input variables to the output $\mathbf{y}$ as

$$R_{\mathbf{s}} = \int_{\mathcal{S}} \left(\sum_{j=1}^{K} \text{Prob}\left\{\mathbf{y} = y_j | s\right\}^2 \right) dF_{\mathbf{s}}(s) = \int_{\mathcal{S}} r(s) dF_{\mathbf{s}}(s) \tag{1}$$

Note that $r(s) = 1 - g(s)$ where $g(s)$ is Gini index of diversity [8] of the conditional probability $\text{Prob}\{\mathbf{y}|s\}$ and that in the zero-one case ($K = 2$) the following

[1] Boldface denotes random variables.

relation holds $\text{Var}\{\mathbf{y}|s\} = \frac{1}{2}[1 - r(s)]$ Another well-known measure of relevance of a subset $\mathbf{s}$ is the mutual information $I(\mathbf{s};\mathbf{y}) = H(\mathbf{y}) - H(\mathbf{y}|\mathbf{s})$ [17] where $H(\mathbf{y})$ is the entropy of the random variable $\mathbf{y}$. Note that the quantities $H(\mathbf{y}|s)$ and $r(s)$ are related: both attain their maximum value when one of the conditional probabilities $\text{Prob}\{\mathbf{y} = y_j|s\}$ is equal to one and attain their minimum when $\text{Prob}\{\mathbf{y} = y_j|s\}$ is uniform for all $j = 1, \ldots, K$. It is possible then to introduce a monotone function $G_H(\cdot) : [0,1] \to [0, 0.5]$ which maps the entropy $H(\mathbf{y})$ of a binary variable $\mathbf{y}$ to the related Gini index g.

Now, given an upper bound d on the number $|\mathbf{s}|$ of features, we may formulate a feature selection problem as the maximisation problem

$$\mathbf{s}^* = \arg \max_{\mathbf{s} \subseteq \mathbf{x}, |\mathbf{s}| \leq d} R_{\mathbf{s}} \tag{2}$$

where the goal is to select the subset $\mathbf{s}^*$ which has the highest relevance in the space of all subsets of $\mathbf{x}$ having size less or equal to d. Unfortunately the quantity $R_{\mathbf{s}}$ to be maximized is unknown if the joint distribution is not available. The wrapper approach [4] consists in attacking the feature selection problem by (i) learning a classifier h on the basis of a training set for each subset $\mathbf{s}$, (ii) assessing it (e.g. by cross-validation) and (iii) using the assessment to estimate the relevance of the subset $\mathbf{s}$. Given a learner h trained on dataset of size N, the wrapper approach translates the (learner independent) maximization problem (2) into the (learner dependent) minimization problem

$$\arg \min_{\mathbf{s} \subseteq \mathbf{x}, |\mathbf{s}| \leq d} M_{\mathbf{s}}^h = \arg \min_{\mathbf{s} \subseteq \mathbf{x}, |\mathbf{s}| \leq d} \left\{ \int_S \text{MME}^h(s) dF_{\mathbf{s}}(s) \right\} \tag{3}$$

where MME^h is the Mean Misclassification Error (MME) of h in the point s. For a zero-one loss function L, since $\mathbf{y}$ and $\hat{\mathbf{y}} = \mathbf{h}(s)$ are independent, $\text{MME}^h(s) = \text{Prob}\{\mathbf{y} \neq \hat{\mathbf{y}}|s\}$ and its bias/variance decomposition [11] is:

$$\begin{aligned}\text{MME}^h(s) &= \frac{1}{2}\left(1 - \left(\sum_{j=1}^{K} \text{Prob}\{\mathbf{y} = y_j|s\}^2\right)\right) + \\ &+ \frac{1}{2}\sum_{j=1}^{K} \left(\text{Prob}\{\mathbf{y} = y_j|s\} - \text{Prob}\{\hat{\mathbf{y}} = y_j|s\}\right)^2 + \\ &+ \frac{1}{2}\left(1 - \left(\sum_{j=1}^{K} \text{Prob}\{\hat{\mathbf{y}} = y_j|s\}^2\right)\right) = \frac{1}{2}\left(n(s) + b(s) + v(s)\right)\end{aligned} \tag{4}$$

where $n(s) = 1 - r(s)$ is the noise variance term, $b(s)$ is the learner squared bias and $v(s)$ is the learner variance. Note that (i) the lower the integral $\int n(s) dF_{\mathbf{s}}(s)$, the higher is the relevance of the subset $\mathbf{s}$ and that (ii) this quantity does not depend on the learning algorithm nor on the training set but only on the feature subset $\mathbf{s}$. In the following, for the sake of simplicity, we will focus our discussion on the zero-one classification task. Note however that this is not restrictive since several techniques exist in literature to transform a multi-class problem into a set of binary classification problems [18].

3 Towards an Unbiased Estimator of Relevance

In real classification tasks, the one-zero misclassification error $M_{\mathbf{s}}^h$ of a learner h for a subset $\mathbf{s}$ cannot be derived analytically but only estimated. Let us denote by $\widehat{\mathrm{M}}_{\mathbf{s}}^h$ the estimate of the misclassification error of the learner h computed on the basis of the observed dataset D_N^S (e.g. by cross-validation) and

$$\mathbf{s}^h = \arg \min_{\mathbf{s} \subset \mathbf{x}, |\mathbf{s}| \leq d} \widehat{\mathrm{M}}_{\mathbf{s}}^h \tag{5}$$

the optimization procedure implemented by a wrapper strategy which relies on the learner h to assess a feature subset. If we use a generic learner h, that is a learner where the bias term $b(s)$ is significantly different from zero, the $\widehat{\mathrm{M}}_{\mathbf{s}}^h$ quantity will be a biased estimate of the term $\int n(s) dF_{\mathbf{s}}(s)$ and consequently of the term $R_{\mathbf{s}} = \int (1 - n(s)) dF_{\mathbf{s}}(s)$ which measures the relevance of the input set $\mathbf{s}$ to the output. Intuitively, the bias would be reduced if we adopted a learner having a small bias term. Examples of low bias, yet high variance, learners are k-nearest neighbour classifiers (kNN) for small values of k where k is the number of neighbours taken into consideration [13]. In particular, it has been shown in [12] that for a 1NN learner and a binary classification problem ($K = 2$) $\lim_{N \to \infty} M_{\mathbf{s}}^{1NN} = 1 - R_{\mathbf{s}}$ where $M_{\mathbf{s}}^{1NN}$ is the misclassification error of a 1NN learner. Consider now a cross-validated estimation $\widehat{\mathrm{M}}_{\mathbf{s}}^{1NN}$ of the quantity $M_{\mathbf{s}}^{1NN}$. Since cross-validation returns a consistent estimation of M^h [14] and since the quantity M^h asymptotically converges to one minus the relevance $R_{\mathbf{s}}$, we have that $1 - \widehat{\mathrm{M}}_{\mathbf{s}}^{1NN}$ is a consistent estimator of the relevance $R_{\mathbf{s}}$ [19]. This property supports the intuition that the cross-validated misclassification

$$\hat{R}_{\mathbf{s}}^{CV} = 1 - \widehat{\mathrm{M}}_{\mathbf{s}}^{kNN} \tag{6}$$

of a kNN learner with low k is a low biased estimator of $R_{\mathbf{s}}$. Note that though $\hat{R}_{\mathbf{s}}^{CV}$ does not tell us much about the accuracy of the learner h which will be used for the classification, it returns an unbiased, yet high variance, estimate of the relevance of the subset $\mathbf{s}$.

The low-bias high-variance nature of the the estimator (6) suggests that the best way to employ this estimator is by combining it with other relevance estimators. We will take into consideration two possible estimators to combine with (6):

1. a direct model-based estimator $\hat{p}_1$ of the conditional probability $\mathrm{Prob}\{\mathbf{y} = y_1 | s\}$ and consequently of the quantity $r(s)$. Similarly to [15] we propose a non-parametric estimator of $R_{\mathbf{s}} = \int_S r(s) dF_{\mathbf{s}}(s)$ which first samples a set of N'' unclassified input vectors s_i'' according to the empirical distribution $\hat{F}_{\mathbf{s}}$ and then computes the quantity

$$\hat{R}_{\mathbf{s}}^D = 1 - \frac{2}{N''} \sum_{i=1}^{N''} \hat{p}_1(s_i'')(1 - \hat{p}_1(s_i'')) \tag{7}$$

where the size N'' is set by the user. Estimations of $\hat{p}_1(s)$ may be returned by classifiers like kNN [15], Naive Bayes, SVM or, in the case of a 0-1 classification task, by any kind of regression algorithm. Note that if we employ a low bias estimator for $\hat{p}_1$ the resulting estimation $\hat{R}^D_{\mathbf{s}}$ will have low bias and presumably high variance.

2. a filter estimator based on the notion of mutual information: several filter algorithms exploit this notion in order to estimate the relevance. An example is the MRMR algorithm [9] where the relevance of a feature subset $\mathbf{s}$, expressed in terms of the mutual information $I(\mathbf{s}, \mathbf{y}) = H(\mathbf{y}) - H(\mathbf{y}|\mathbf{s})$, is approximated by the incremental formulation

$$I_{MRMR}(\mathbf{s}; \mathbf{y}) = I_{MRMR}(\mathbf{s}_i; \mathbf{y}) + I(\mathbf{x}_i, \mathbf{y}) - \frac{1}{m-1} \sum_{\mathbf{x}_j \in \mathbf{s}_i} I(\mathbf{x}_j; \mathbf{x}_i) \quad (8)$$

where $\mathbf{x}_i$ is a feature belonging to the subset $\mathbf{s}$, $\mathbf{s}_i$ is the set $\mathbf{s}$ with the $\mathbf{x}_i$ feature set aside and m is the number of components of $\mathbf{s}$. Now since $H(\mathbf{y}|\mathbf{s}) = H(\mathbf{y}) - I(\mathbf{s}, \mathbf{y})$ and $G_{\mathbf{s}} = 1 - R_{\mathbf{s}} = G_H(H(\mathbf{y}|\mathbf{s}))$ we obtain that

$$\hat{R}^{MRMR}_{\mathbf{s}} = 1 - G_H\left(H(\mathbf{y}) - I_{MRMR}(\mathbf{s}, \mathbf{y})\right)$$

is a MRMR estimator of the relevance $R_{\mathbf{s}}$ where $G_H(\cdot)$ is the monotone mapping between H and Gini index. Note that the computation of $\hat{R}^{MRMR}$ requires the estimation of I_{MRMR} (e.g. by assuming that $\mathbf{x}$ and $\mathbf{y}$ are normally distributed) and the knowledge of the $G_H(\cdot)$ function (e.g. by numerical interpolation).

According to the ensemble of estimators principle, an improved estimator can be derived by simply averaging $\hat{R}^{CV}(\mathbf{s})$ with the two estimators discussed above. This is motivated by the fact that the combination of two unbiased and independent estimators is still an unbiased but lower variance estimator [13]. We obtain then two aggregated estimators

$$\hat{R}'_{\mathbf{s}} = \frac{\hat{R}^{CV}_{\mathbf{s}} + \hat{R}^D_{\mathbf{s}}}{2}, \qquad \hat{R}''_{\mathbf{s}} = \frac{\hat{R}^{CV}_{\mathbf{s}} + \hat{R}^{MRMR}_{\mathbf{s}}}{2} \quad (9)$$

Note that less trivial combination mechanisms [20] could be adopted, if we had access to the variance of the two estimates in (9).

On the basis of the quantitites defined in (9) we can define two novel feature selection algorithms: the algorithm R' where the selected subset is

$$\mathbf{s}^{R'} = \arg \max_{\mathbf{s} \subset \mathbf{x}, |\mathbf{s}| \le d} \hat{R}'_{\mathbf{s}} \quad (10)$$

and the algorithm R" where the selected subset is

$$\mathbf{s}^{R''} = \arg \max_{\mathbf{s} \subset \mathbf{x}, |\mathbf{s}| \le d} \hat{R}''_{\mathbf{s}} \quad (11)$$

4 Experiments

The experimental session aims to show: 1) the improvement in classification accuracy with respect to a conventional wrapper approach when the selection is done according to the strategy R' (Equation (10)) 2) the improvement in classification accuracy with respect to a filter approach like MRMR when the selection is done according to the strategy R"(Equation 11).

Since, as discussed in the introduction, each feature selection algorithm relies on a specific search strategy and given that the search strategy is not the focus of this paper we restrict here to consider a conventional forward search exploration strategy [4].

The experiment uses 20 public domain microarray expression datasets detailed in the left side of Table 1[2]. A first dimensionality reduction step is carried out by hierarchically clustering the variables into 1000 compressed features obtained by averaging the probes of the same cluster. The estimators $\hat{R}^{CV}_{\mathbf{s}}$ and $\hat{R}^{D}_{\mathbf{s}}$ are implemented by a locally constant model with the automatic adaptation of the number of neighbours (in the range $[3,5]$) [21]. Two wrapper strategies are considered: the strategy denoted by WSVM which employs a linear Support Vector Machine (SVMLIN) and the strategy denoted by WNB which employs a Naive Bayes (NB) learner. A three-fold cross-validation strategy is used to measure the generalization accuracy of the feature selection strategies. In order to avoid any dependency between the learning algorithm employed by the wrapper and the classifier used for prediction, the experimental session is composed of two parts:

- Part 1 (summarized in Table 1): here we perform the comparison with the wrapper WSVM and we use the set of classifiers $\mathcal{C}_1$ ={TREE, NB, SVM-SIGM, LDA, LOG} which does not include the SVMLIN learner,
- Part 2 (summarized in Table 2): here we perform the comparison with the wrapper WNB and we use the set of classifiers $\mathcal{C}_2$ ={TREE, SVMSIGM, SVMLIN, LDA, LOG} which does not include the NB learner.

Note that TREE stands for a classification tree, SVMSIGM for sigmoidal SVM, LDA for linear discriminant analysis and LOG for logistic regresssion. All the classifiers are implemented by public domain packages of the R statistical software.

An external cross-validation scheme [22] is used to prevent feature selection bias in our assessment. For each fold of the cross-validation, for each selection approach and for each classifier, once selected features are returned, the generalization accuracy is assessed by (i) training the classifier on the same dataset used for feature selection and (ii) testing the trained classifier on the remaining third. Note that because of the scarcity of the data and to avoid the bias related to the selection of the feature set size, we average the performance over all the classifiers and over all the feature sets whose size is ranging from $d = 3$ to $d = 20$. For the family $\mathcal{C}_1$ of classifiers we compare in Table 1 the accuracy of: the strategy R', the

[2] For reasons of limited space the complete reference list of the datasets is available in `http://www.ulb.ac.be/di/map/gbonte/icann_suppl.pdf`

Table 1. Left: name of the dataset, number N of samples, number n of features and number K of classes. Right: misclassification errors (three-fold cross-validation) averaged over the classifiers of the family $\mathcal{C}_1$. The left side of the table compares the strategy R' with a wrapper selection based on linear SVM. The right side of the table compares the strategy R" with MRMR and the RANK strategy. On the left (right) side the bold notation stands for significantly different from the accuracy of the R' (R") strategy in terms of an adjusted p-value (Holm criterion). The AVG line returns the average of the misclassification percentages. The W/B line returns the number of times that the technique is worse/better than R' (R").

Name	N	n	K
Golub	72	7129	2
Alon	62	2000	2
Notterman	36	7457	2
Nutt	50	12625	2
Shipp	77	7129	2
Singh	102	12600	2
Sorlie	76	7937	2
Wang	286	22283	2
Van't Veer	65	24481	2
VandeVijver	295	24496	2
Sotiriou	99	7650	2
Pomeroy	60	7129	2
Khan	63	2308	4
Hedenfalk	22	3226	3
West	49	7129	4
Staunton	60	7129	9
Su	174	12533	11
Bhattacharjee	203	12600	5
Armstrong	72	12582	3
Ma	60	22575	3

Name	R'	WSVM	R"	MRMR	RANK
Golub	0.0917	**0.1177**	0.1	**0.1079**	**0.1225**
Alon	0.2704	0.2658	0.2267	**0.1996**	0.2281
Notterman	0.1966	**0.0985**	0.1494	0.1472	0.1432
Nutt	0.3798	**0.4171**	0.3873	0.3847	**0.4189**
Shipp	0.1429	**0.1319**	0.1322	0.1362	**0.1873**
Singh	0.1619	**0.1517**	0.1266	**0.1374**	0.1328
Sorlie	0.3835	**0.4314**	0.3963	0.4004	0.3987
Wang	0.4282	**0.4111**	0.4218	0.4232	0.4181
Van't Veer	0.2786	**0.2638**	0.2492	**0.2217**	**0.2277**
VandeVijver	0.454	**0.4724**	0.4365	**0.4636**	**0.4482**
Sotiriou	0.5279	**0.5796**	0.5351	**0.5708**	0.5339
Pomeroy	0.428	0.4191	0.4141	**0.3876**	0.4181
Khan	0.0878	**0.1143**	0.0582	**0.0686**	**0.131**
Hedenfalk	0.5475	0.5263	0.452	**0.5273**	**0.5389**
West	0.6463	**0.6109**	0.6186	**0.5746**	0.6109
Staunton	0.6822	**0.71**	0.6511	**0.6865**	**0.7407**
Su	0.2568	**0.307**	0.2549	**0.3772**	**0.3352**
Bhattacharjee	0.1232	**0.1347**	0.1105	**0.1057**	**0.1515**
Armstrong	0.1082	**0.1199**	0.1306	**0.115**	**0.1122**
Ma	0.2456	**0.2041**	0.2257	**0.2413**	0.2317
AVG	0.323	**0.331**	0.310	**0.326**	**0.331**
W/B than R' (R")		10/7		9/6	9/2

conventional wrapper WSVM, the strategy R", the conventional MRMR with Gaussian estimation of the entropy and a ranking based on the cross-validated misclassification of the univariate learner (RANK). For the family $\mathcal{C}_2$ of classifiers we compare in Table 2 the accuracy of: the strategy R', the conventional wrapper WNB, the strategy R", the conventional MRMR with Gaussian estimation of the entropy and a ranking based on the cross-validated misclassification of the univariate learner (RANK). Each Table is partitioned in two sides: the left side refers to the comparison of the R' algorithm with the wrapper strategy while the right side refers to the comparison of the R" algorithm with filter strategies. The AVG line returns the average of the misclassification percentages. On the left (right) side of each Table, for a given dataset and selection strategy the figure is reported in bold notation if the accuracy of the selection method is significantly better than the one of R' (R"). We assume a difference as significant if the Holm adjusted p-value returned by the permutation paired test is lower than 0.05. On the left (right) side of each Table, the last line contains the number of times that the selection criterion is significantly worse/better than the algorithm R' (R").

Two major considerations can be done

- the strategy R' (i.e. the combination of the cross-validated assessment criterion with the direct relevance estimation) outperforms the wrapper algorithm

Table 2. Misclassification errors (three-fold cross-validation) averaged over the classifiers of the family $\mathcal{C}_2$.The left side of the table compares the strategy R' with a wrapper selection based on Naive Bayes. The right side of the table compares the strategy R" with MRMR and the RANK strategy. On the left (right) side the bold notation stands for significantly different from the accuracy of the R' (R") strategy in terms of an adjusted p-value (Holm criterion). The AVG line returns the average of the misclassification percentages. The W/B line returns the number of times that the technique is worse/better than R' (R").

Name	R'	WNB	R"	MRMR	RANK
Golub	0.0886	**0.1114**	0.0971	0.1019	0.0904
Alon	0.2376	**0.2568**	0.2181	0.2109	0.221
Notterman	0.1852	**0.2059**	0.1491	0.1512	**0.1645**
Nutt	0.3929	**0.3402**	0.36	**0.3898**	**0.4258**
Shipp	0.1261	0.127	0.1198	**0.1338**	**0.1734**
Singh	0.1495	0.1454	0.1297	**0.1377**	0.1245
Sorlie	0.3848	**0.4254**	0.3808	**0.3953**	0.3838
Wang	0.4363	0.4345	0.4298	0.4281	0.4255
Van't Veer	0.2747	0.2715	0.2421	**0.2253**	0.2325
VandeVijver	0.4626	**0.44**	0.4763	0.4721	**0.4358**
Sotiriou	0.5126	**0.5578**	0.5505	**0.5732**	0.5611
Pomeroy	0.4367	0.4389	0.4007	0.3902	**0.4224**
Khan	0.0804	0.0896	0.0628	0.0631	**0.0901**
Hedenfalk	0.5379	0.5187	0.4369	**0.4904**	**0.4949**
West	0.6413	**0.6696**	0.5542	**0.5882**	**0.6728**
Staunton	0.6689	**0.8298**	0.6981	**0.6661**	**0.83**
Su	0.2544	**0.3096**	0.2646	**0.3739**	**0.3529**
Bhattacharjee	0.1235	0.1209	0.101	**0.1061**	**0.1186**
Armstrong	0.1079	**0.1668**	0.125	**0.1148**	**0.1034**
Ma	0.2565	0.2635	0.2335	**0.2443**	**0.2681**
AVG	0.322	**0.3335**	0.315	**0.327**	**0.331**
W/B than R' (R")		9/2		10/3	11/2

(as shown by the 10 wins against 7 losses in Table 1 and by the 9 wins against 2 losses in Table 2);

– the strategy R" (i.e. the combination of the cross-validated assessment criterion with the MRMR criterion) outperforms the MRMR algorithm (as shown by the 9 wins against 6 losses in Table 1 and by the 10 wins against 3 losses in Table 2) as well as the conventional ranking strategy.

In summary, the cross-validated estimate of the relevance brings useful additional information about the subset to be selected in a microarray context characterized by high dimensionality, large noise and low number of samples.

5 Conclusion and Future Work

The paper proposes a low bias cross-validated estimator which can be effectively integrated with other mesures of relevance. The estimated quantity can be used by a generic search algorithm to find the subset of variables expected to provide the highest classification accuracy. The paper assessed the effectiveness of the approach by comparing it to a conventional wrapper and a state-of-the-art filter approach. Future work will extend the approach in two directions: the study of additional relationships between cross-validation and relevance and the integration of the cross-validated relevance measure with other filter algorithms.

References

1. Guyon, I., Elisseeff, A.: An introduction to variable and feature selection. Journal of Machine Learning Research 3, 1157–1182 (2003)
2. Saeys, Y., Inza, I., Larranaga, P.: A review of feature selection techniques in bioinformatics. Bioinformatics 23(19), 2507–2517 (2007)
3. Alter, O., Brown, P., Botstein, D.: Singular value decomposition for genome-wide expression data processing and modeling. Proceedings of the National Academy of Sciences 97, 10101–10106 (2000)
4. Kohavi, R., John, G.H.: Wrappers for feature subset selection. Artificial Intelligence 97(1-2), 273–324 (1997)
5. Ghosh, D., Chinnaiyan, A.M.: Classification and selection of biomarkers in genomic data using lasso. J. Biomed. Biotechnol. 2, 147–154 (2005)
6. Bell, D.A., Wang, H.: A formalism for relevance and its application in feature subset selection. Machine Learning 41(2), 175–195 (2000)
7. Yu, L., Liu, H.: Efficient feature selection via analysis of relevance and redundancy. Journal of Machine Learning Research 5, 1205–1224 (2004)
8. Breiman, L., Friedman, J.H., Olshen, R.A., Stone, C.J.: Classification and Regression Trees. Wadsworth International Group, Belmont (1984)
9. Peng, H., Long, F., Ding, C.: Feature selection based on mutual information: criteria of max-dependency, max-relevance, and min-redundancy. IEEE Transactions on Pattern Analysis and Machine Intelligence (2005)
10. Meyer, P.E., Bontempi, G.: On the use of variable complementarity for feature selection in cancer classification. In: Rothlauf, F., Branke, J., Cagnoni, S., Costa, E., Cotta, C., Drechsler, R., Lutton, E., Machado, P., Moore, J.H., Romero, J., Smith, G.D., Squillero, G., Takagi, H. (eds.) EvoWorkshops 2006. LNCS, vol. 3907, pp. 91–102. Springer, Heidelberg (2006)
11. Wolpert, D.H., Kohavi, R.: Bias plus variance decomposition for zero-one loss functions. In: Prooceedings of the 13th International Conference on Machine Learning, pp. 275–283 (1996)
12. Devroye, L., Györfi, L., Lugosi, G.: A Probabilistic Theory of Pattern Recognition. Springer, Heidelberg (1996)
13. Hastie, T., Tibshirani, R., Friedman, J.: The elements of statistical learning. Springer, Heidelberg (2001)
14. Cover., T.M.: Learning in Pattern Recognition. In: Methodologies of Pattern Recognition. Academic Press, London (1969)
15. Fukunaga, K., Kessel, D.L.: Nonparametric bayes error estimation using unclassified samples. IEEE Transactions on Information Theory 19(4), 434–440 (1973)
16. Duda, R.O., Hart, P.E.: Pattern Classification and Scene Analysis. Wiley, Chichester (1976)
17. Cover, T.M., Thomas, J.A.: Elements of Information Theory. John Wiley, New York (1990)
18. Li, T., Zhang, C., Ogihara, M.: A comparative study of feature selection and multiclass classification methods for tissue classification based on gene expression. Bioinformatics 20(15), 2429–2437 (2004)

19. Dudoit, S., van der Laan, M.J.: Asymptotics of cross-validated risk estimation in estimator selection and performance assessment. Statistical Methodology 2(2), 131–154 (2005)
20. Tresp, V., Taniguchi, M.: Combining estimators using non-constant weighting functions. In: NIPS. MIT Press, Cambridge (1995)
21. Birattari, M., Bontempi, G., Bersini, H.: Lazy learning meets the recursive least-squares algorithm. In: Kearns, M.S., Solla, S.A., Cohn, D.A. (eds.) NIPS 11, pp. 375–381. MIT Press, Cambridge (1999)
22. Ambroise, C., McLachlan, G.: Selection bias in gene extraction on the basis of microarray gene-expression data. PNAS 99, 6562–6566 (2002)

Non-stationary Data Mining: The Network Security Issue

Sergio Decherchi, Paolo Gastaldo, Judith Redi, and Rodolfo Zunino

Dept. of Biophysical and Electronic Engineering (DIBE), Genoa University
Via Opera Pia 11a, 16145 Genoa, Italy
{sergio.decherchi,paolo.gastaldo,judith.redi,rodolfo.zunino}@unige.it

Abstract. Data mining applications explore large amounts of heterogeneous data in search of consistent information. In such a challenging context, empirical learning methods aim to optimize prediction on unseen data, and an accurate estimate of the generalization error is of paramount importance. The paper shows that the theoretical formulation based on the Vapnik-Chervonenkis dimension (d_{vc}) can be of practical interest when applied to clustering methods for data-mining applications. The presented research adopts the K-Winner Machine (KWM) as a clustering-based, semi-supervised classifier; in addition to fruitful theoretical properties, the model provides a general criterion for evaluating the applicability of Vapnik's generalization predictions in data mining. The general approach is verified experimentally in the practical problem of detecting intrusions in computer networks. Empirical results prove that the KWM model can effectively support such a difficult classification task and combine unsupervised and supervised.

Keywords: Clustering, Data Mining, K-Winner Machine, Intrusion detection systems, Network security.

1 Introduction

Data mining exploits clustering methods to arrange huge amounts of data into a structured representation, which eventually support the search for relevant information. The vast datasets and the heterogeneous descriptions of patterns set stringent requirements on the algorithms adopted; in such data-intensive applications, empirical learning methods aim to optimise prediction on unseen data [1]. In this regard, the formulation based on the Vapnik-Chervonenkis dimension (d_{vc}) [2] exhibits a general, theoretical foundation endowed with the widest validity for the accurate estimate of the run-time generalization error.

This paper shows that, in the specific case of clustering methods for data-mining applications, those theoretical results can yet have practical significance. Several aspects seem to favor clustering-based approaches [1]: first, data mining applications are typically rich with patterns and can offer the sample size required to tighten theoretical bounds. Secondly, the associate classifiers prove much simpler than other approaches [1]; finally, a clustering-based classifier, the

V. Kůrková et al. (Eds.): ICANN 2008, Part II, LNCS 5164, pp. 32–41, 2008.

K-Winner Machine (KWM) model [3], has been fully characterized in compliance with Vapnik's theory. Theoretical results [3] proved that the KWM model can set very tight bounds to generalization performance. Moreover, the model is independent of the data dimensionality and inherently supports multi-class classification tasks [3]; these features boost the model's application to large masses of high-dimensional data.

The method presented in this paper applies the hybrid paradigm of the KWM model to the complex Anomaly Intrusion Detection (AID) problem, in which 'abnormal' events in computer networks traffic are identified by dynamically modelling 'normal' traffic. Clustering-based implementations of AID [4,5,6,7,8,9] typically map the network activity into a feature space, and the cluster-based model identifies those space portions that support the distribution of normal traffic, whereas outliers will mark abnormal traffic activities. This research shows that applying the hybrid paradigm of the KWM model to AID can lead to some intriguing results from both a theoretical and an applicative viewpoint.

From a theoretical perspective, the research addresses some issues related to the possible non-stationary distribution of observed data [10], which ultimately gives rise to a discrepancy between the pattern distribution in the training and the test phases. The paper formulates a general criterion to evaluate the consistency and consequent applicability of Vapnik's approach, by measuring the discrepancy between the empirical training set and the test distribution used at run time to control generalization performance.

From an applicative viewpoint, the major result consisted in showing that a KWM could effectively support such a complex classification task, mainly thanks to the model's ability to handle multi-class data distributions. The "KDD Cup 1999" dataset [10] provided the experimental domain for testing the proposed framework. This reliable benchmark is a common testbed for comparing the performances of anomaly-detection algorithms; indeed, experimental results proved that the KWM-based classifier outperformed other clustering-based AID's. The following conventions will be adopted throughout the paper:

1. $C = \{c^{(h)}, h = 1, ..., N_c\}$ is the set of N_c possible pattern classes;
2. $W' = \{(\mathbf{w}_n, c_n), \mathbf{w}_n \in R^D, c_N \in C, n = 1, ..., N_h\}$ is a set of N_h labeled prototypes;
3. $\mathbf{w}^*(\mathbf{x}) = \arg\min_{w \in W'}\{||\mathbf{x} - \mathbf{w}||^2\}$ is the prototype that represents a pattern, $\mathbf{x}$.

2 The K-Winner Machine Model

The training strategy of the KWM model develops a representation of the data distribution by means of an unsupervised process, then builds a classifier on top of that via some calibration process. The basic design criterion is to model the data distribution by Vector Quantization. In the research presented here, the Plastic Neural Gas (PGAS) [11] algorithm for Vector Quantization models the data distribution. The PGAS approach extends the 'Neural Gas' model [12] by some crucial advantages: first, both the number and the positions of prototypes are adjusted simultaneously [11]; secondly, the training process prevents the occurrence of 'dead vectors' (void prototypes covering empty partitions). After VQ training positions the codebook prototypes a calibration process [3] labels the resulting Voronoi tessellation of the data space induced by the positions of the prototypes. Such a process categorizes each partition/prototype according to the predominant class; tie cases can be solved by choosing any class from among the best candidates. A detailed outline of the KWM training algorithm is given in [3]. Here we present in algorithm 1 the runtime operation of KWM.

Algorithm 1. The K-Winner Machine run-time operation

procedure KWM-FORWARD(test pattern $\mathbf{x}$; a calibrated set of N_h prototypes W'; error bounds, $\pi(k)$, for agreement levels, $k = 1, ..., N_h$)

`(Pattern Vector Quantization)` Build a sorted set of prototypes, $W''(\mathbf{x})$, arranged in increasing order of distance from $\mathbf{x}$

`(Count concurrences)` Determine the largest value of $K, 1 \leq K \leq N_h$, such that all elements in the sequence $\{(\mathbf{w}_k, c_k) \in W^*, k = 1, 2, ..., K\}$ share the same calibration, c^*

`(Determine generalization error bound)` Assigne risk bound to the classification outcome of pattern $\mathbf{x}$ $\pi^* = \pi(K)$

`(Output)`

– Categorize $\mathbf{x}$ with class c^*

– Associte the prompted output with an error bound, π^*

end procedure

Statistical Learning Theory can apply to this framework because the KWM model allows one to compute Vapnik's bound to the predicted generalization performance at the local level. In this regard, it has been proved [3] that:

1. The VC-dimension of a Nearest-Prototype Classifier that includes a codebook of N_h prototypes is $d_{vc}^{(1)} = N_h$.
2. The Growth function of a KWM using a codebook of N_h prototypes, N_c classes and a sequence of K concurring elements is: $GF^{(K)}(N_p) = N_c^{\lfloor N_h/K \rfloor}$.

Generalization theory proves that a classifier's performance is upper-bounded by the empirical training error, ν, increased by a penalty. In the latter term, the Growth Function [2], $GF^{(K)}(N_p)$, measures the complexity of the fact that the classifier has been trained with a set of N_p patterns. This theory derives a worst-case bound, π, to the generalization error of the considered classifier:

$$\pi \leq \nu + \frac{\epsilon}{2}\left(1 + \sqrt{1 + \frac{4\nu}{\epsilon}}\right) \tag{1}$$

where $\epsilon = \frac{4}{N_p}[\ln GF(N_p) - ln\frac{\eta}{4}]$, and η is a confidence level. KWM theory [3] proves that one can compute a error bound, $\pi(k)$, for each agreement level, k, and more importantly, that such a bound is a non-increasing function when k increases. This confirms the intuitive notion that the risk in a classification decision about a given point should be reduced by the concurrence of several neighboring prototypes. As a consequence of 1) and 2), unsupervised prototype positioning sharply reduces the bounding term in (1). By contrast, the KWM training algorithm does not provide any a-priori control over the first term (the empirical training error). This brings about the problem of model selection, which is usually tackled by a tradeoff between accuracy (classification error in training) and complexity (number of prototypes).

3 Statistical Learning Theory in Data-Mining Applications

Vapnik's theory adopts a worst-case analysis, hence the predicted bound (1) often falls in a very wide range that eventually lessens the practical impact of the overall approach. However, data-mining environments typically involve very large datasets, whose cardinality ($N_p >> 10^5$) can actually shrink the complexity penalty term down to reasonable values. Moreover the KWM model intrinsically prevents an uncontrolled increase in the classifier's d_{vc}. Thus data-mining domains seem to comply with basic Statistical Learning Theory quite well. On the other hand, data-intensive applications raise the crucial issue of the stationary nature of the pattern distribution, which is basic assumption for the applicability itself of Statistical Learning Theory. In fact, non-stationary environments are quite frequent in data mining, as is the case for the present research dealing with network traffic monitoring: new attack patterns are continuously generated and, as a result, it is impossible to maintain a knowledge base of empirical samples up to date. Theoretical predictions, indeed, are often verified on a test set that should not enter the training process. This is done for a variety of reasons: either because one uses cross-validation for model selection, or because the test set is partially labeled, or because the test set was not available at the time of training. The stationary-distribution assumption may be rephrased by asserting that the training set instance, $T = \{(\mathbf{x}_l^T, c_l), \mathbf{x}_l^T \in R^D, c_l \in C, l = 1, ..., N_p\}$, and the test set instance, $S = \{(\mathbf{x}_j^S, c_j), \mathbf{x}_j^S \in R^D, c_j \in C, j = 1, ..., N_u\}$, are identically and independently drawn from a common probability distribution, $P(\mathbf{x}, c)$; otherwise, the training set is not representative of the entire population, hence expression (1) may not provide the correct estimate of classification accuracy.

The present work proposes a general yet practical criterion to verify the consistency of the generalization bounds; the method uses the VQ-based paradigm of the KWM model to check on the stationary-distribution assumption, and completes in three steps. First, one uses the prototypes in the trained codebook, W, to classify training and test data. Secondly, one estimates the discrete probability distributions, $\hat{T}$ and $\hat{S}$, of the training set and of the test set, respectively; this is easily attained by counting the number of training/test patterns that lie within the data-space partition spanned by each prototype. Finally, one computes the Kullback-Leibler (KL) divergence to measure the mutual information between $\hat{T}$ and $\hat{S}$, and therefore decides whether the two associate samples have been drawn from the same distribution. In the discrete case, the KL divergence of probability distribution, $\hat{S}$, from the reference distribution, $\hat{T}$, is defined as:

$$D_{KL}(\hat{S},\hat{T}) = \sum_{n=1}^{N_h} s_n \log \frac{s_n}{t_n} \tag{2}$$

where s_n and t_n denote the normalized frequencies associated with $\hat{S}$ and $\hat{T}$, respectively. Using the result, $\hat{T}$, of the training process as a reference distribution offers some advantages: it is consistent from a cognitive perspective, since it seems reasonable to adopt an empirical model of the data distribution; in addition, the PGAS algorithm prevents the occurrence of dead vectors during training, hence one has: $t_n > 0, \forall n$; finally, the partitioning schema sets a common ground for comparing the distributions of training and test patterns. The minimum (zero) value of $D_{KL}(\hat{S},\hat{T})$ marks the ideal situation and indicates perfect coincidence between the training and test distributions. Non-null values, however, typically occur in common practice, and it may be difficult to interpret from such results the significance of the numerical discrepancies measured between the two distributions.

The present research adopts an empirical approach to overcome this issue by building up a 'reference' experiment setting. First, one creates an artificial, stationary distribution, J, that joins training and test data: $J := T \cup S$. Secondly, one uses the discrete distribution J to draw at random a new training set, T_J, and a new test set, S_J, such that $T_J \cap S_J = \emptyset$. Both these sets have the same relative proportions of the original samples. Third, using these sets for a session of training and test yields a pair of discrete distributions, $\hat{S_J}, \hat{T_J}$; finally, one measures the divergence (2) between the new pair of data sets, by computing $D_{KL}(\hat{S_J},\hat{T_J})$. The latter value provides the numerical reference for assessing the significance of the actual discrepancy value $D_{KL}(\hat{S},\hat{T})$ by comparison. If the original sample had been drawn from a stationary distribution, then the associate discrepancy value, $D_{KL}(\hat{S},\hat{T})$, should roughly coincide with the value, $D_{KL}(\hat{S_J},\hat{T_J})$, computed on the artificial distribution J. In this case, the theoretical assumptions underlying Statistical Learning Theory hold, and the bound formulation (1) can apply. Otherwise, if one verifies that: $D_{KL}(\hat{S_J},\hat{T_J}) << D_{KL}(\hat{S},\hat{T})$, then one might infer that the original sampling process was not stationary, hence a direct application of theoretical results (1) is questionable. The overall algorithm for the validation criterion is outlined in Algorithm 2.

4 Semi-supervised Anomaly Detection in Network Security

Commercial implementations of Intrusion Detection Systems (IDS's) typically rely on a knowledge base of rules to identify malicious traffic. The set of rules, however, is susceptible to inconsistencies, and continuous updating is required to cover previously unseen attack patterns. An alternative approach envisions adaptive systems that maintain a model of 'normal' traffic and generate alerts in

Algorithm 2. Criterion for validating the applicability of theoretical bounds

1: **procedure** VALIDATE(a training set including N_T labeled data, $\mathbf{x}_i, c(\mathbf{x}_i)$; a test set including N_S labeled data, $\mathbf{x}_j, c(\mathbf{x}_j)$)

2: `(Training)` Apply a VQ algorithm on the training set and position the set of prototypes: $W' = \{(\mathbf{w}_n, c_n), \mathbf{w}_n \in R^D, c_N \in C, n = 1, ..., N_h\}$

3: `(Probability distribution)`

- Estimate the training discrete probability distribution, $\hat{T}$ as follows: $\hat{T} := \{P_n^{(T)}; n = 1, ..., N_h\}$; where: $P_n^{(T)} = \{\mathbf{x}_i^{(T)} \in R^D : \mathbf{w}^*(\mathbf{x}_i^{(T)}) = \mathbf{w}_n\}$;
- Estimate the test discrete probability distribution, $\hat{S}$ as follows: $\hat{S} := \{P_n^{(S)}; n = 1, ..., N_h\}$; where: $P_n^{(S)} = \{\mathbf{x}_i^{(S)} \in R^D : \mathbf{w}^*(\mathbf{x}_i^{(S)}) = \mathbf{w}_n\}$;

4: `(Measuring mutual information)`

- Compute normalized frequencies: $t_n = \frac{|P_n^{(T)}|}{N_T}$; $s_n = \frac{|P_n^{(S)}|}{N_S}$; $n = 1, ..., N_h$
- Compute the KL divergence between $\hat{T}$ and $\hat{S}$: $D_{KL}(\hat{S}, \hat{T}) = \sum_{n=1}^{N_h} s_n \log \frac{s_n}{t_n}$

5: `(Applicability of generalization theory)`

- Form an artificial discrete distribution by joining training and test data $J := T \cup S$;
- Draw from J at random a training set, T_J, and a test set, S_J, having the same relative proportions of the original data sets;
- Repeat steps (2,3,4) by using the new pair of sets;
- If $D_{KL}(\hat{S}_J, \hat{T}_J) \approx D_{KL}(\hat{S}, \hat{T})$ (ideally$\approx$ 0): than Stationary nature is verified and generalization bounds are validated; else Stationary nature is not verified and generalization bounds are not supported empirically

6: **end procedure**

the occurrence of 'abnormal' events. Thus, Anomaly Intrusion Detection (AID) systems do not use sets of rules and are capable of time-zero detection of novel attack strategies; to do that, they require a continuous modeling of normal traffic in a dynamic, data-intensive environment. Several approaches based on data mining techniques have been adopted for that purpose [4,5,6,7,8,9], which typically map network traffic into vector patterns spanning a D-dimensional 'feature' space. The research presented in this paper tackles the anomaly-detection problem by means of the KWM paradigm, mainly because the hybrid KWM model, combining unsupervised clustering with supervised calibration, seems to fit the problem representation that characterizes the anomaly-detection task. Crucial properties contribute to these benefits in the data-mining scenario: 1) the Growth Function of a KWM does not depend on the number of patterns (Theorem 2); 2) the performance properties of the classifiers do not depend on the dimension of the data space: this mitigates the 'curse of dimensionality' and boosts applications in high-dimensional spaces; 3) the KWM paradigm is inherently a multi-class model. In practice, the KWM model supports the anomaly-detection framework as follows. The off-line KWM training algorithm processes an empirical set, P, of N_p traffic data; each datum includes a D-dimensional feature vector and a multiclass indicator: $P = \{(\mathbf{x}_l, c_l), \mathbf{x}_l \in R^D, l = 1, ..., N_p\}$. Class labels, c_l, indicate whether pattern $\mathbf{x}_l$ derives from normal or abnormal traffic; suspect patterns may be further sub-classified according to the various typologies of attacks. This results in both a codebook W' of N_h labeled prototypes and a set of error bounds, $\pi(k)$, associated with increasing values of the prototype-agreement parameter, k.

5 Performance Measurements in Network Security

The well-known "KDD Cup 1999" dataset [10, KDD] provided the experimental domain for the proposed framework. The data spanned a 41-dimensional feature space; each pattern encompassed cumulative information about a connection session. In addition to "normal" traffic, attacks belonged to four principle macro-classes, namely, "DoS" (denial-of-service), "R2L" (unauthorized access from a remote machine), "U2R" (unauthorized access to local "super user" privileges), "probing" (surveillance and other probing such as port scanning). For simplicity, the experimental sessions in this research involved the "10% training set," provided by the KDDCup'99 benchmark, which had been obtained by subsampling original training data at a 10% rate. The resulting training set included 494,021 patterns and preserved the original proportions among the five principal macro-categories cited above. The test set provided by the KDD challenge contained 311,029 patterns, and featured 17 'novel' attack schemes that were not covered by the training set. The pattern descriptors that took on categorical values, most notably "Protocol" and "Service", were remapped into a numerical representation. "Protocol" could assume three different values (TCP, UDP, ICMP) and was therefore encoded by a triplet of bits; each element of the triplet was associated to a protocol, and only one of those could be non-null. The "Service" descriptor took on eleven possible values, and was remapped accordingly into eleven mutually exclusive coordinates. In summary, the patterns forming the eventual dataset used in the experiments included 53-dimensional feature vectors.

5.1 Experimental Validation of Generalization Bounds

The procedure described in Section 3 allows one to verify the stationary nature of the observed data distribution. Thus, the coverage spanned by the original distribution (T, S) was compared with the representation supported by the exhaustive distribution, $J = T \cup S$, that approximated a stationary situation. The artificial, reference training and test sets, T_J and S_J, were obtained by random resampling J. The KL divergence between the training and test coverages for both distributions (T, S) and (T_J, S_J) completed the validation process. Figure 1 reports on the empirical results obtained for increasing codebook sizes.

The results highlights two main aspects: first, the KL divergence for the original distribution (T, S) always resulted to be much larger than the divergence measured when training and test data were drawn from the stationary distribution (T_J, S_J).

Secondly, the sizes of the codebooks differed significantly in the two situations: when training and test data were drawn from a common distribution, J, the

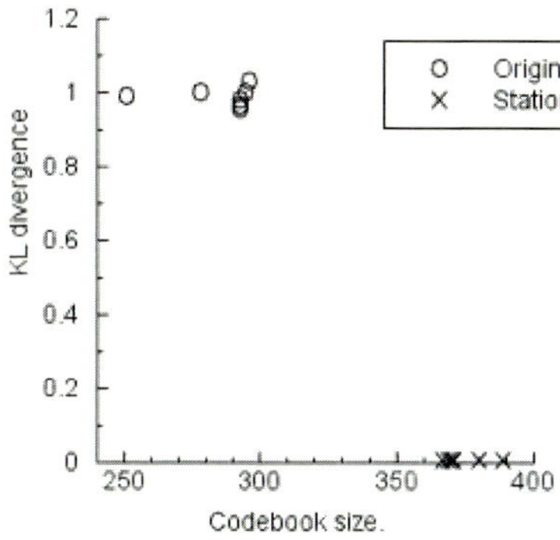

Fig. 1. Stationary Vs Original dataset

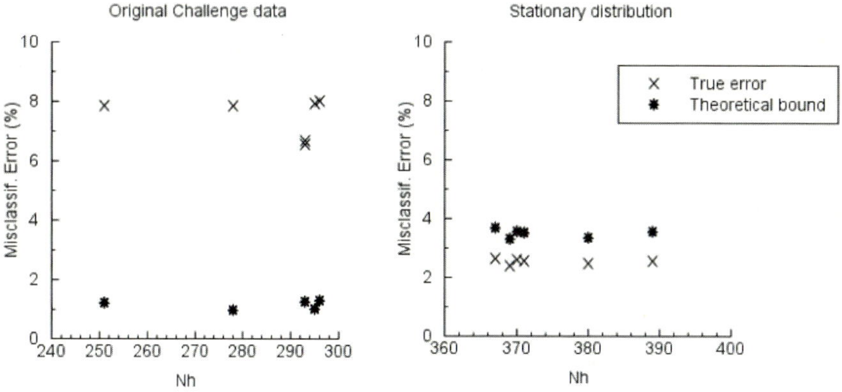

Fig. 2. Error bounds assessment: original data, stationary distribution

probability support was wider, hence the VQ algorithm required a larger number of prototypes to cover the data space. Conversely, original training data, T, were drawn from a limited sector of the actual support region, thus a smaller codebook was sufficient to represent the sample distribution.

5.2 Prediction Accuracy in Intrusion Detection

A complementary experimental perspective aimed to assess the effectiveness of the KWM method in terms of the classification accuracy on the actual dataset from the KDDCup'99 competition. The Vector Quantization set-up phase by using the PGAS clustering algorithm on training data indicated best performance with a codebook of 293 prototypes and an associated digital cost of 0.54%.

Such empirical evidence, mainly due to the marked discrepancies between training and test data sets, clearly seemed to invalidate the applicability of the

Table 1. KWM results

ACTUAL	PREDICTED					
	Normal	Probing	DoS	U2R	R2L	% correct
Normal	59118	152	1261	1	61	97.57%
Probing	720	3179	215	0	52	76.31%
DoS	1274	154	228425	0	0	99.38%
U2R	85	136	0	4	3	1.75%
R2L	14838	25	1317	0	9	0.06%
%correct	77.75%	87.19%	98.79%	80%	7.20%	

Table 2. KDD99 winner results

ACTUAL	PREDICTED					
	Normal	Probing	DoS	U2R	R2L	% correct
Normal	60262	243	78	4	6	99.5%
Probing	511	3471	184	0	0	83.3%
DoS	5299	1328	223226	0	0	97.1%
U2R	168	20	0	30	10	13.2%
R2L	14527	294	0	8	1360	8.4%
%correct	74.6%	64.8%	99.9%	71.4%	98.8%	

theoretical bounds from Statistical Learning Theory for the KDD'99 dataset. As a result, the outcome of the validation criterion was that Vapnik's bound would not hold for the original challenge data. For the sake of completeness, Figure 2 compares the actual classification error with the theoretical bound for the original and the stationary distribution. The obtained results show that theoretical predictions fail in bounding the generalization performance for the original data sets, whereas provide good approximations when the data distribution is artificially reduced to the stationary case.

Such a conclusion gave both an empirical support and a numerical justification to a fact that has often been reported in the literature, namely, the considerable discrepancy between training and test patterns in the KDD dataset. Such a critical issue had been hinted at by the proponents themselves of the competition dataset [13,14], and possibly explains the intrinsic difficulty of the challenge classification problem. In the subsequent validation phase, test data were classified, and the resulting performance was measured by using the scoring rules adopted for the KDD99 competition. In spite of the very low training error, the test error rate was 6.52%. In view of the discussion presented in the previous Section, such a phenomenon seems depend on the non-stationary nature of the data distribution underlying the original challenge datasets. Table 1 compares the confusion matrix for the obtained results with the corresponding matrix for the winning method [13]. When applying the error-weighting scheme of the KDD99 competition [14], the KWM-based approach achieved a score of 0.2229, which slightly improved on the result attained by the winner method, which scored 0.2331.

Such a result seems to be interesting especially in view of the semi-supervised nature of the proposed approach, as compared with the winning method that adopted a fully supervised strategy (decision trees) and therefore was subject to less stringent bounds to the predicted generalization performance.

References

1. Mirkin, B.: Clustering for Data Mining: a Data-recovery Approach (2006)
2. Vapnik, V.: Estimation of Dependences Based on Empirical Data. Springer, Heidelberg (1982)
3. Ridella, S., Rovetta, S., Zunino, R.: K-winner machines for pattern classification. IEEE Trans. on Neural Networks 12, 371–385 (2001)
4. Kemmerer, R., Vigna, G.: Intrusion detection: a brief history and overview. Computer 35, 27–30 (2002)
5. Portnoy, L., Eskin, E., Stolfo, S.J.: Intrusion detection with unlabeled data using clustering. In: Proc. ACM CSS Workshop on Data Mining Applied to Security, pp. 123–130 (2001)
6. Eskin, E., Arnold, A., Prerau, M.: A geometric framework for unsupervised anomaly detection: Detecting intrusions in unlabeled data. Applications of Data Mining in Computer Security (2002)
7. Oh, S.H., Lee, W.S.: An anomaly intrusion detection method by clustering normal user behavior. Computers and Security 22, 596–612 (2003)
8. Lee, W., Stolfo, S., Mok, K.: Adaptive intrusion detection: a data mining approach. Artificial Intelligence Review 14, 533–567 (2000)
9. Zheng, J., Hu, M.: An anomaly intrusion detection system based on vector quantization. IEICE Trans. Inf. and Syst. E89-D, 201–210 (2006)
10. KDD Cup 1999 Intrusion detection dataset, `http://kdd.ics.uci.edu/databases/kddcup99/kddcup99.html`
11. Ridella, S., Rovetta, S., Zunino, R.: Plastic algorithm for adaptive vector quantization. Neural Computing and Applications 7, 37–51 (1998)
12. Tm, M., Sg, B., Kj, S.: Neural gas network for vector quantization and its application to time-series prediction. IEEE Trans. Neural Networks 4, 558–569 (1993)
13. Pfahringer, B.: Winning the kdd99 classification cup: bagged boosting. SIGKDD Explorations 1, 65–66 (2000)
14. Results of the KDD 1999 Classifier Learning Contest, `http://www-cse.ucsd.edu/users/elkan/clresults.html`

Efficient Feature Selection for PTR-MS Fingerprinting of Agroindustrial Products

Pablo M. Granitto[1], Franco Biasioli[2], Cesare Furlanello[3], and Flavia Gasperi[2]

[1] CIFASIS, CONICET/UNR/UPC,
Bv 27 de Febrero 210 Bis – 2000 Rosario – Argentina
granitto@cifasis-conicet.gov.ar
[2] FEM-IASMA Research Center – Agrifood Quality Department,
Via E. Mach 1 – 38010 San Michele allAdige (TN) – Italy
{franco.biasioli,flavia.gasperi}@iasma.it
[3] FBK-irst,
Via Sommarive 18 – 38100 Povo (TN) – Italy
furlan@itc.it

Abstract. We recently introduced the Random Forest - Recursive Feature Elimination (RF-RFE) algorithm for feature selection. In this paper we apply it to the identification of relevant features in the spectra (fingerprints) produced by Proton Transfer Reaction - Mass Spectrometry (PTR-MS) analysis of four agro-industrial products (two datasets with cultivars of Berries and other two with typical cheeses, all from North Italy). The method is compared with the more traditional Support Vector Machine - Recursive Feature Elimination (SVM-RFE), extended to allow multiclass problems. Using replicated experiments we estimate unbiased generalization errors for both methods. We analyze the stability of the two methods and find that RF-RFE is more stable than SVM-RFE in selecting small subsets of features. Our results also show that RF-RFE outperforms SVM-RFE on the task of finding small subsets of features with high discrimination levels on PTR-MS datasets.

1 Introduction

Proton Transfer Reaction - Mass Spectrometry (PTR-MS) [1] is a spectrometric technique with a growing number of applications ranging from medical diagnosis to environmental monitoring [2]. It allows fast, non-invasive, time-continuous measurements of volatile organic compounds (VOCs). These compounds play a relevant role in food and agro-industrial applications. They are related to the real or perceived quality of food and to its sensory characterisation, and they are emitted during most transformation/preservation processes. Among the applications of PTR-MS based classification in food science and technology, we can cite the detection of the effect of different pasteurisation processes of fruit juices [3], the classification of strawberry cultivars [4] or the characterisation of Italian 'Grana' cheeses [5].

Here PTR-MS is used to produce a fingerprint of each sample in the form of a spectrum vector whose components are the intensities of the spectrometric peaks

V. Kůrková et al. (Eds.): ICANN 2008, Part II, LNCS 5164, pp. 42–51, 2008.

at different m/z ratios. Each PTR-MS spectrum can contain up to 500 m/z values. Although this is a relatively low number compared with other spectrometric or spectroscopic approaches, the number of analysed samples per class is usually low in the experimental practice, introducing issues similar to the ones faced in the classification of high-throughput microarray and proteomic data. Moreover, due to the absence of separation, each peak in the spectrum can be related to one or more compounds. The identification of small sets of relevant features for the food product under analysis is of interest for several operative reasons: in particular, we focus on the identification of few relevant 'quality' markers that can be measured in a simple, fast and cheap way or to concentrate to a few relevant masses the identification efforts needed to compensate for the lack of separation. There are indications that PTR-MS features can be related to genetic aspects [4] or to sensory characteristics of food [6] and thus classification based on PTR-MS data could provide a tool to better investigate these fields, possibly providing a link between sensory and genetics.

In this application domain, we introduce instruments from the recent feature selection literature [7,8,9]. As a general taxonomy, the feature selection mechanism may be implemented into the learning algorithm in embedded methods, while wrapper methods directly consider the classifier outputs as in a black box approach. In both cases, care is required to avoid overfitting during the selection process (the bias selection problem [10]), particularly in real applications on small datasets. The use of resampling methods within a complete validation setup is a typical strategy to avoid these problems [11].

The SVM-RFE algorithm [9] introduced a ranking of the features within Support Vector Machine classifiers by Recursive Feature Elimination (RFE). This strategy found several applications in Bioinformatics[12] and also in Quantitative Structure Activity Relationship (QSAR)[13]. The SVM-RFE is often used in practice with linear SVMs, and it can easily be extended from binary to multiclass classification problems. We developed the alternative RF-RFE method[14], which basically replaces SVM with Breiman's Random Forest (RF)[15] into the core of the RFE method. RF is a natural multiclass algorithm with an internal unbiased measure of feature importance, and we may use such internal measure for ranking masses for relevance in discrimination. In this paper, we apply the two feature selection and classification methods and compare their performances in indicating highly discriminative masses for PTR-MS multiclass data.

A usually neglected problem in feature selection methods is the instability of the selection process [16]. Quite different ranked lists of features may be obtained for classifiers developed on slightly different data replicates, as typically observed in functional profiling from microarray data. In the last part of this paper we compare the stability of the two proposed versions of the RFE algorithm.

The article is organized as follows: in Section 2, we describe the full feature selection schemes for RF-RFE and SVM-RFE. In Section 3 we compare both methods on the four real PTR-MS datasets and in Section 4 we discuss the stability of the solutions. Finally, we draw some conclusions in Section 5.

2 The Feature Selection Setup

A feature selection method that uses (in any way) information about the targets may lead to overfitting, in particular with the very low samples-to-features ratios typical of spectrometric experiments. Thus, in order to obtain unbiased estimates of the prediction error with small PTR-MS datasets, feature ranking and selection should be included in the modelling, and not treated as a pre-processing step; moreover, we need to appropriately decouple selection from error estimation [10].

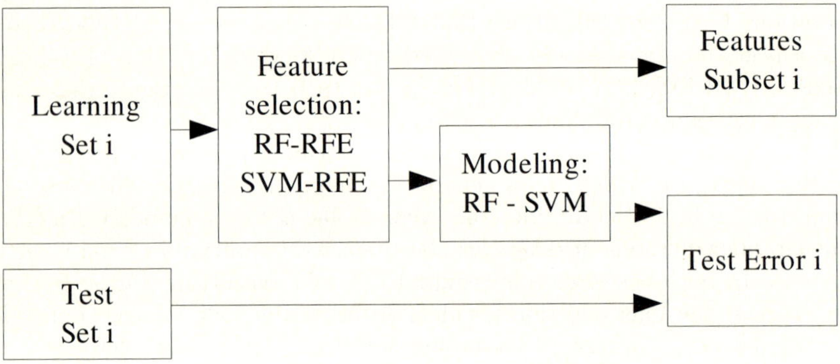

Fig. 1. The computational setup used for the feature selection process

We use a computational setup consisting of two nested processes. The outer loop performs n times a random split of the dataset in a training set (used to develop the models – including the feature selection step), and in a test set, used to estimate the accuracy of the models. The inner process (Figure 1) supports the selection of nested subsets of features and the development of classifiers over these subsets (using only the learning subset provided by the outer loop). The results of the n replicated experiments are then aggregated to obtaine a comprehensive feature ranking and accuracy estimation.

The RFE selection method [9] is basically a recursive process that ranks features according to some measure of their importance. At each iteration feature importances are measured and the less relevant one is removed. The (inverse) order in which features are eliminated is used to construct a final ranking. The feature selection process itself consists only in taking the first n features from this ranking.

The original SVM-RFE method was developed to select features in a binary classification problem. Between the various strategies for solving multiclass problems with binary classifiers [17,18], we choose the One-vs-One method to extend SVM-RFE to handle multiclass datasets. In this case, a problem with c classes

Table 1. Details of the four dataset. The columns 'min #' and 'max #' show the min or max number of samples per class in the corresponding dataset. The last column shows the number of production years included in the dataset.

Dataset	min m/z	max m/z	Samples	Classes	min #	max #	Years
Strawberry	20	250	233	9	21	30	3
Raspberry	20	250	92	5	17	19	2
Nostrani	20	259	48	6	8	8	1
Grana	20	259	60	4	15	15	1

is decomposed into $p = c(c-1)/2$ binary problems. To solve each problem we train a linear SVM [19], obtaining p decision functions

$$D_i(\mathbf{x}) = \mathbf{x}\mathbf{w_i} \qquad i = 1...p. \tag{1}$$

The weight vectors $\mathbf{w}_i$ corresponding to all binary problems are then averaged

$$\mathbf{W} = \frac{1}{p}\sum_{i=1}^{p} \mathbf{w_i} \tag{2}$$

and the components of $\mathbf{W}$ are used for ranking the features. In all our experiments we use a fixed value of $C = 100$, following [9]. We performed a series of experiments on PTR-MS datasets using different C values, finding that on our particular data the results are almost independent of the value of C. It must be noted that other datasets could require a full tunning of this parameter [20].

In a previous work [14] we introduced Random Forest - Recursive Feature Elimination (RF-RFE). We showed how RF's internal measure of features importance can replace SVM weights for features ranking. Also, as RF makes use of Out-of-Bag subsets to estimate the importances, computational efforts are not increased. Moreover, RF was developed as a multiclass algorithm, which suggests it could provide a better measure of importance for this kind of problems.

3 Results

We considered four datasets. The first two refer to cultivar characterization of berry fruits (Strawberries[4] and Raspberries) and the last two to typicality assesment of cheeses (Nostrani[21] and Grana[6]). All products come from Trento Province, North Italy, or other places in the same area. Table 1 shows details of each dataset. In all cases the headspace composition of the samples has been measured by direct injection in a PTRMS apparatus (experimental details can be found in previous papers[3,4]). Each sample was then associated to its PTR-MS spectrum normalised to unit total area.

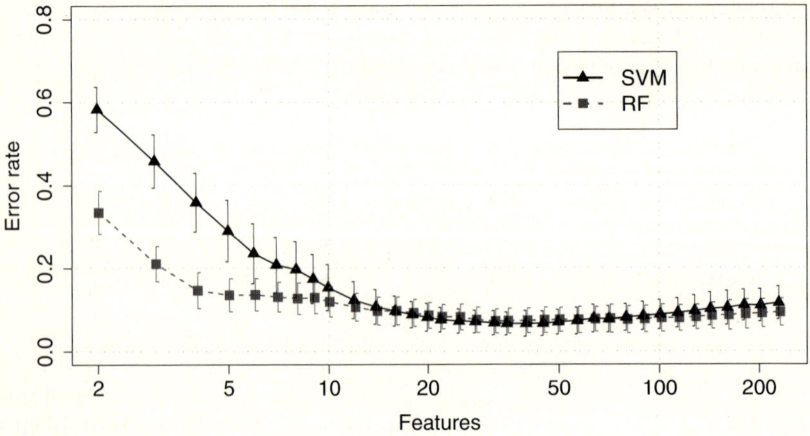

Fig. 2. Mean classification errors for SVM-RFE and RF-RFE on the Strawberry dataset. Bars show one standard deviation evaluated over 100 replications.

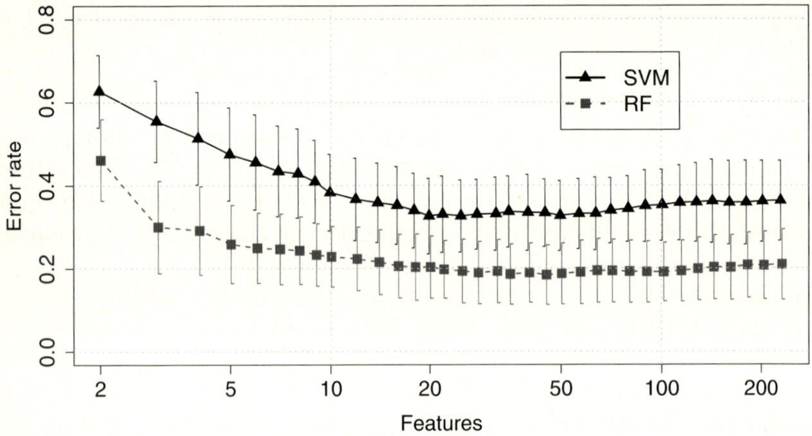

Fig. 3. Mean classification errors for SVM-RFE and RF-RFE on the Raspberry dataset. Bars show one standard deviation evaluated over 100 replications.

For all cases we replicated the feature selection process on $n = 100$ runs. For each run, we split the dataset at random into train/test sets with a 75%/25% proportion, stratifying on class frequencies. The train set is used for RF-RFE and SVM-RFE to select features and to develop models, which are then evaluated on the test set. It is important to note that the 100 runs are not completely independent of each other, because there is a considerable overlap amongh any pair of train sets or any pair of test sets. The results obtained from this kind of replicated

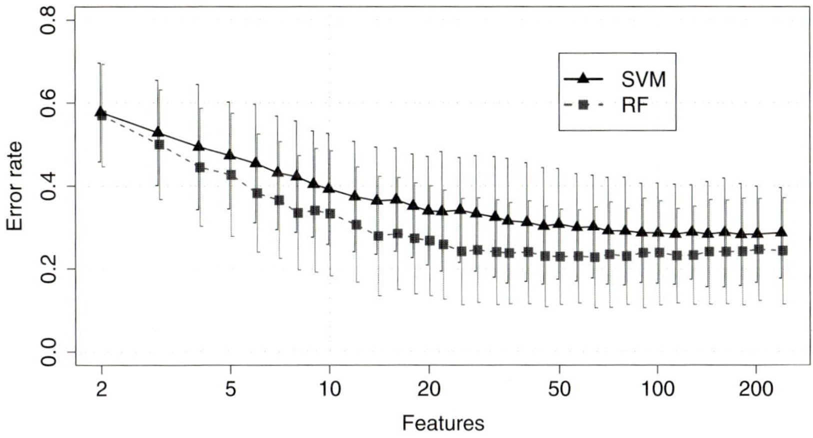

Fig. 4. Mean classification errors for SVM-RFE and RF-RFE on the Nostrani dataset. Bars show one standard deviation evaluated over 100 replications.

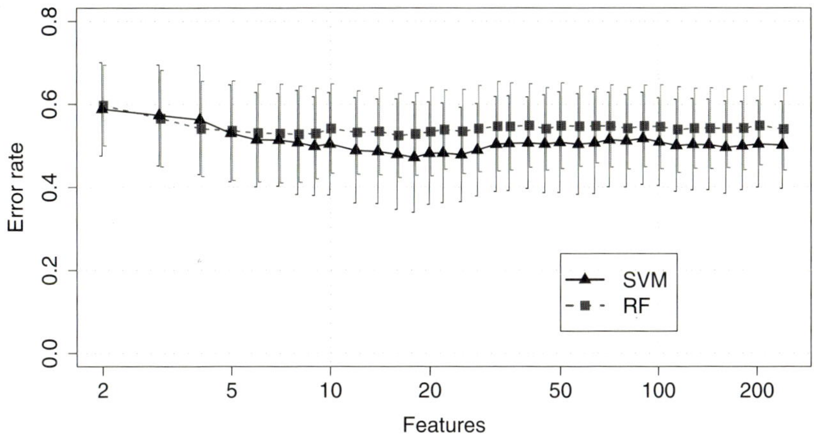

Fig. 5. Mean classification errors for SVM-RFE and RF-RFE on the Grana dataset. Bars show one standard deviation evaluated over 100 replications.

experiments on wide datasets usually do not have statistical significance, they are only strong indications of the espected behaviour of the different methods.

3.1 Modeling Error

In Figure 2 we compare both selection methods on the Strawberry dataset. We show mean classification errors ($\pm$ one standard deviation) for RF or SVM

Table 2. Mean inter-quartile distances (MIQ) ($\times 10^2$) for different models and subsets size. The first 2 columns show RF and SVM MIQ for the 10 features with the highest median ranking. The following columns show the same results for 50 m/Z values and for the full sets.

m/Z values	10		50		All	
	RF	SVM	RF	SVM	RF	SVM
Strawberry	1.4	1.9	3.6	5.7	10.9	9.7
Raspberry	2.5	5.5	12.1	10.0	30.3	17.0
Nostrani	3.9	4.9	10.4	13.2	21.6	19.0
Grana	4.6	6.4	18.3	19.1	30.8	26.3

models adjusted on subsets of different sizes selected with the corresponding RFE methods. In this case RF-RFE clearly outperforms SVM-RFE when using only a few masses. Both methods have a similar behaviour for more than 11 features, reaching their minimum modeling error with around 35 features. This minimum error level is very similar for both methods, with a small edge for SVM. In Figure 3 we show the corresponding results for the Raspberry dataset. RF-RFE shows lower mean errors than SVM-RFE for all subset sizes in this case. The differences between both methods are the biggest of the four datasets under evaluation.

The same analysis was repeated for the cheeses datasets, Figures 4 and 5. The results for the Nostrani dataset are similar to the Raspberry ones. RF-RFE again shows lower mean errors than SVM-RFE for all subset sizes. It reaches the minimum mean error with ~ 50 features. For the Grana dataset (Figure 5), both methods show the same behaviour for small size subsets. But in this case SVM-RFE outperforms RF-RFE with bigger subsets, reaching the minimum mean error for 17 features subsets.

3.2 Complexity

Both algorithms showed comparable execution times. In our experiments on PTR-MS datasets, RF-RFE was in all cases slightly heavier than SVM-RFE, with running times ranging from 1.2 to 1.9 of the corresponding times for SVM-RFE, but these ratios are clearly problem-dependent. Also, any tunning of the C parameter of SVMs should increase the running time of SVM-RFE considerably.

3.3 Stability

As feature selection methods are unstable[16], each replicate of the selection process gives a different ranking. This means that the error levels showed in

Figures 2 to 5 are only indications of the expected behavior of both methods, which cannot be associated with a particular subset. Of course, a higher stability of the selection method helps in the identification of the most relevant features, because rankings are more similar in that case. In order to measure the stability of both methods we assign relative ranking positions to each feature with a linear scale between 1 (first) and 0 (last). An ideal (totally stable) selection method should returns the same value for each feature in all replicates. On the opposite, a completely unstable method should return a random value in [0 : 1]. Thus, the dispersion of the distribution of this relative ranking (measured over the 100 replications) is correlated with the instability of the selection method.

In Table 2 we show the mean inter-quartile distances (MIQ) of these distributions evaluated on different subsets of features. For the 10 and 50 most relevant features for classification RF-RFE values are clearly smaller than SVM-RFE ones. Only in the Raspberry dataset with 50 features SVM-RFE shows a smaller MIQ than RF-RFE. For the full sets, SVM selections are always more stable, but this fact has low influence in the selection of the most relevant m/Z values.

4 Conclusion

In this paper we used RF-RFE (coupled with replicated experiments) for feature selection on PTR-MS datasets, and compared it with SVM-RFE. Feature selection methods can be evaluated at least on two aspects, their capacity to find the smallest subset with a given error level, or to find the minimum possible error without caring about the number of selected features. For the first task we showed that RF-RFE has similar or better performance than SVM-RFE in all four datasets. For the second one, RF-RFE showed similar or better performance than SVM-RFE in 3 out of the 4 datasets under analysis. Furthermore, we compared the stability of the selected features, an usually neglected aspect of the feature selection process. We showed that RF-RFE is more stable then SVM-RFE in selecting the most relevant features for discrimination. Overall, RF-RFE seems to be more appropriate than SVM-RFE for fingerprinting agroindustrial products with PTR-MS.

Work in progress includes the use of other multiclass strategies or non-linear extensions in the SVM-RFE method, the analysis of more agroindustrial products and the identification of the compounds associated with the selected masses.

Acknowledgements

We acknowledge partial support for this project from ANPCyT, Argentina (PICT 643) and from PAT projects MIROP, RASO, INTERBERRY, QUALIFRAPE and SAMPPA (Trento, Italia).

References

1. Hansel, A., Jordan, A., Holzinger, R., Prazeller, P., Vogel, W., Lindinger, W.: Proton transfer reaction mass spectrometry: on-line trace gas analysis at the ppb level. Int. J. Mass. Spectrom. Ion Procs. 149/150, 609–619 (1995)
2. Lindinger, W., Hansel, A., Jordan, A.: On-line monitoring of volatile organic compounds at ppt level by means of Proton-Transfer-Reaction Mass Spectrometry (PTR-MS): Medical application, food control and environmental research. Int. J. Mass. Spectrom. Ion Procs. 173, 191–241 (1998)
3. Biasioli, F., Gasperi, F., Aprea, E., Colato, L., Boscaini, E., Märk, T.D.: Fingerprinting mass spectrometry by PTR-MS: heat treatment vs. pressure treatments of red orange juice - a case study. Int. J. Mass. Spectrom, 223-224, 343-353 (2003)
4. Biasioli, F., Gasperi, F., Aprea, E., Mott, D., Boscaini, E., Mayr, D., Märk, T.D.: Coupling Proton Transfer Reaction-Mass Spectrometry with Linear Discriminant Analysis: a Case Study. J. Agr. Food Chem. 51, 7227–7233 (2003)
5. Boscaini, E., Van Ruth, S., Biasioli, F., Gasperi, F., Märk, T.D.: Gas Chromatography-Olfactometry (GC-O) and Proton Transfer Reaction-Mass Spectrometry (PTR-MS). Analysis of the Flavor Profile of Grana Padano, Parmigiano Reggiano, and Grana Trentino Cheeses. J. Agr. Food Chem. 51, 1782–1790 (2003)
6. Biasioli, F., Gasperi, F., Aprea, E., Endrizzi, I., Framondino, V., Marini, F., Mott, D., Märk, T.D.: Correlation of PTR-MS spectral fingerprints with sensory characterisation of flavour and odour profile of Trentingrana cheese. Food Qual. Prefer. 17, 63–75 (2006)
7. Guyon, I., Elisseeff, A.: An Introduction to Variable and Feature Selection. J. Mach. Learn. Res. 3, 1157–1182 (2003)
8. Kohavi, R., John, G.H.: Wrappers for feature subset selection. Artif. Intell. 97, 273–324 (1996)
9. Guyon, I., Weston, J., Barnhill, S., Vapnik, V.: Gene Selection for Cancer Classification using Support Vector Machines. Mach. Learn. 46, 389–422 (2002)
10. Ambroise, C., McLachlan, G.: Selection bias in gene extraction on the basis of microarray gene-expression data. P. Natl. Acad. Sci. USA 99, 6562–6566 (2002)
11. Furlanello, C., Serafini, M., Merler, S., Jurman, G.: Entropy-Based Gene Ranking without Selection Bias for the Predictive Classification of Microarray Data. BMC Bioinformatics 4, 54 (2003)
12. Ramaswamy, S., et al.: Multiclass cancer diagnosis using tumor gene expression signatures. P. Natl. Acad. Sci. USA 98, 15149–15154 (2001)
13. Li, H., Ung, C.Y., Yap, C.W., Xue, Y., Li, Z.R., Cao, Z.W., Chen, Y.Z.: Prediction of Genotoxicity of Chemical Compounds by Statistical Learning Methods. Chem. Res. Toxicol. 18, 1071–1080 (2005)
14. Granitto, P.M., Furlanello, C., Biasioli, F., Gasperi, F.: Recursive feature elimination with random forest for PTR-MS analysis of agroindustrial products. Chemometr. Intell. Lab. 83, 83–90 (2006)
15. Breiman, L.: Random Forests. Mach. Learn. 45, 5–32 (2001)
16. Breiman, L.: Heuristics of instability and stabilization in model selection. Ann. Stat. 24, 2350–2383 (1996)
17. Hsu, C.-W., Lin, C.-J.: A comparison of methods for multi-class support vector machines. IEEE T. Neural Networ. 13, 415–425 (2002)
18. Allwein, E., Schapire, R., Singer, Y.: Reducing Multiclass to Binary: A unified Approach for Margin Classifiers. J. Mach. Learn. Res. 1, 113–141 (2000)
19. Vapnik, V.: The Nature of Statistical Learning Theory. Springer, New York (1995)

20. Huang, T.-M., Kecman, V.: Gene extraction for cancer diagnosis by support vector machines. Artif. Intell. Med. 35, 185–194 (2005)
21. Gasperi, F., Biasioli, F., Framondino, V., Endrizzi, I.: Ruolo dell´analisi sensoriale nella definizione delle caratteristiche dei prodotti tipici: l´esempio dei formaggi trentini / The role of sensory analysis in the characterization of traditional products: the case study of the cheese from Trentino. Sci. Tecn. Latt.-Cas. 55, 345–364 (2004)

Extraction of Binary Features by Probabilistic Neural Networks

Jiří Grim

Institute of Information Theory and Automation of the ASCR,
P.O. BOX 18, 18208 PRAGUE 8, Czech Republic
grim@utia.cas.cz
http://www.utia.cas.cz/RO

Abstract. In order to design probabilistic neural networks in the framework of pattern recognition we estimate class-conditional probability distributions in the form of finite mixtures of product components. As the mixture components correspond to neurons we specify the properties of neurons in terms of component parameters. The probabilistic features defined by neuron outputs can be used to transform the classification problem without information loss and, simultaneously, the Shannon entropy of the feature space is minimized. We show that, instead of dimensionality reduction, the decision problem can be simplified by using binary approximation of the probabilistic features. In experiments the resulting binary features improve recognition accuracy but also they are nearly independent - in accordance with the minimum entropy property.

Keywords: Probabilistic neural networks, Feature extraction, Multivariate Bernoulli mixtures, Subspace approach, Recognition of numerals.

1 Introduction

It is a common experience that statistical classification methods usually do not perform well in multidimensional spaces. For this reason feature selection or feature extraction is generally accepted to reduce dimensionality of multivariate problems of pattern recognition. However, the description reduced to a small number of highly informative variables is not the only way to simplify decision making. In biological neural networks, the higher level neurons of ascending neural pathways tend to respond to increasingly complex and specific input patterns. In this way the meaning of strongly interrelated low-level input signals may be coded by labels (indices) of output channels which respond nearly exclusively. In such a case any further decision making becomes superfluous since the classification problem is actually solved by the extracted features alone.

Formally, in multi-layer neural networks, the output of each neuron can be viewed as a feature extracted from the respective input layer. Despite the obvious importance of the extracted features, the decision-making aspects are rarely subject of neural network research. If necessary the choice of features for neural networks is usually optimized independently by means of standard approaches.

V. Kůrková et al. (Eds.): ICANN 2008, Part II, LNCS 5164, pp. 52–61, 2008.

It appears that, in the framework of neural networks, feature extraction is rather considered as a problem of mapping data from a high-dimensional space onto a lower-dimensional one. Let us mention in this connection e.g. the self-organizing map of Kohonen [7,14] or the distance preserving mapping of Sammon [17]. On the other hand the most widely used back-propagation training algorithm extracts the optimal features implicitly without any special criteria.

In this paper we discuss the properties of features automatically produced by probabilistic neural networks (PNN). The standard reference to PNN is the paper of Specht [19] which is closely related to Parzen estimates of probability density functions. The non-parametric Parzen estimates are asymptotically unbiased and consistent but, in multidimensional applications, their properties strongly depend on the choice of smoothing parameters. The PNN concept of Specht may considerably save training time but, on the other hand, the resulting PNN has to include one neuron for each training pattern. For this reason the computation may become awkward in case of large training data sets but also the principle is hardly interpretable from biological point of view.

In the following we refer mainly to our papers on PNN (cf. [3] - [11]) based on finite mixtures of product components. The main idea of mixture-based PNN is to view the components of mixtures as formal neurons. Estimating the mixture parameters by means of EM algorithm [1,15,18] we define the outputs of neurons and the properties of the corresponding features. We have shown that the features extracted by PNN can be used to transform the classification problem without information loss [3,4,21]. Simultaneously the Shannon entropy of the feature space is minimized.

In multidimensional spaces the features defined in terms of *a posteriori* probabilities tend to behave as binary variables. Typically, the maximum *a posteriori* probability is near to one and the others are close to zero. We have used this property for a natural binary approximation of the probabilistic features by means of Bayes decision function. It has been verified in [9] that the information loss caused by the binary approximation of *a posteriori* probabilities is bounded by the approximation error [9]. In the paper [8] we have shown that independently computed binary feature vectors can be combined in parallel in order to increase the recognition accuracy. In Section 3 we show that a more liberal binary approximation based on a threshold value is even more successful in terms of recognition accuracy. Moreover, the resulting binary features appear to be of low statistical complexity. As opposed to standard feature selection methods our approach tends to emphasize simplicity of features rather than to reduce their number. The properties of binary features have been verified in several numerical experiments with differently complex mixtures. The recognition accuracy in the feature space was better in all experiments. Moreover, the resulting binary features appear to be nearly independent.

The paper is organized as follows. In Section 2 we first summarize basic properties of the mixture-based PNN. In Section 3 we recall the concept of information preserving transform and introduce the proposed binary probabilistic features.

The properties of the features are illustrated by a numerical example in Section 4. Finally, we summarize the results in the Conclusion.

2 Mixture Based Probabilistic Neural Networks

The concept of mixture-based PNN has been proposed in the framework of statistical pattern recognition (cf. [3] - [11]). Considering a finite set of classes Ω we assume that some vectors $\boldsymbol{x}$ from a multidimensional space $\mathcal{X}$ occur randomly with a priori probabilities $p(\omega)$ according to the class-conditional probability distributions $P(\boldsymbol{x}|\omega)$. Thus, given the probabilistic description

$$\{P(\boldsymbol{x}|\omega)p(\omega), \omega \in \Omega\}, \quad \boldsymbol{x} = (x_1, \dots, x_N) \in \mathcal{X}, \quad \Omega = \{\omega_1, \omega_2, \dots, \omega_K\} \tag{1}$$

we can compute the *a posteriori* probabilities of classes

$$p(\omega|\boldsymbol{x}) = \frac{P(\boldsymbol{x}|\omega)p(\omega)}{P(\boldsymbol{x})}, \quad P(\boldsymbol{x}) = \sum_{\omega \in \Omega} P(\boldsymbol{x}|\omega)p(\omega) \tag{2}$$

for the sake of Bayesian decision making. In this way the problem reduces to estimating the class-conditional distributions $P(\boldsymbol{x}|\omega), \omega \in \Omega$. Unlike similar papers on PNN (cf. [16,20,22]) we approximate the unknown distributions by finite mixtures of product components

$$P(\boldsymbol{x}|\omega) = \sum_{m \in \mathcal{M}_\omega} F(\boldsymbol{x}|m)f(m) = \sum_{m \in \mathcal{M}_\omega} f(m) \prod_{n \in \mathcal{N}} f_n(x_n|m), \quad \mathcal{N} = \{1, \dots, N\}. \tag{3}$$

Here $F(\boldsymbol{x}|m)$ denote the component specific product distributions, $f(m) \geq 0$ are probabilistic weights, $\mathcal{M}_\omega$ are the component index sets of different classes and $\mathcal{N}$ is the index set of variables. For the sake of simplicity we assume consecutive indexing of components. Hence, for each component index $m \in \mathcal{M}_\omega$ the related class $\omega \in \Omega$ is uniquely determined and therefore the parameter ω can be omitted in the components $F(\boldsymbol{x}|m)f(m)$.

A typical feature of the probabilistic approach to neural networks is the complete interconnection of neurons with all input variables. In order to avoid the biologically unnatural complete interconnection property we have proposed the concept of structural mixtures [3,6,8,10]. In particular we define the product mixture in the form

$$P(\boldsymbol{x}|\omega) = \sum_{m \in \mathcal{M}_\omega} F(\boldsymbol{x}|0)G(\boldsymbol{x}|m, \phi_m)f(m), \quad F(\boldsymbol{x}|0) = \prod_{n \in \mathcal{N}} f_n(x_n|0) \tag{4}$$

where $F(\boldsymbol{x}|0)$ is a fixed "background" probability distribution usually defined as a product of unconditional marginals and the component functions $G(\boldsymbol{x}|m, \phi_m)$ include binary structural parameters ϕ_{mn} :

$$G(\boldsymbol{x}|m, \phi_m) = \prod_{n \in \mathcal{N}} \left[\frac{f_n(x_n|m)}{f_n(x_n|0)}\right]^{\phi_{mn}}, \quad \phi_{mn} \in \{0, 1\}. \tag{5}$$

The main motivation for the structural mixture model is the possibility to cancel the background probability distribution $F(\boldsymbol{x}|0)$ in the Bayes formula since then the decision making may be confined only to "relevant" variables. Introducing notation

$$w_m = p(\omega)f(m), \quad m \in \mathcal{M}_\omega, \quad \omega \in \Omega \tag{6}$$

and using substitution (3), we can write

$$P(\boldsymbol{x}) = \sum_{m \in \mathcal{M}} F(\boldsymbol{x}|0)G(\boldsymbol{x}|m, \phi_m)w_m, \quad \mathcal{M} = \bigcup_{\omega \in \Omega} \mathcal{M}_\omega = \{1, \ldots, M\}. \tag{7}$$

Further, considering the conditional component weights

$$q(m|\boldsymbol{x}) = \frac{F(\boldsymbol{x}|m)w_m}{P(\boldsymbol{x})} = \frac{G(\boldsymbol{x}|m, \phi_m)w_m}{\sum_{j \in \mathcal{M}} G(\boldsymbol{x}|j, \phi_j)w_j}, \quad m \in \mathcal{M} \tag{8}$$

we can write

$$p(\omega|\boldsymbol{x}) = \frac{P(\boldsymbol{x}|\omega)p(\omega)}{P(\boldsymbol{x})} = \frac{\sum_{m \in \mathcal{M}_\omega} G(\boldsymbol{x}|m, \phi_m)w_m}{\sum_{j \in \mathcal{M}} G(\boldsymbol{x}|j, \phi_j)w_j} = \sum_{m \in \mathcal{M}_\omega} q(m|\boldsymbol{x}). \tag{9}$$

Thus the posterior probabilities $p(\omega|\boldsymbol{x})$ become proportional to the respective weighted sums of the component functions $G(\boldsymbol{x}|m, \phi_m)$, each of which can be defined on a different subspace. Consequently, the input connections of a neuron can be confined to an arbitrary subset of input neurons or, in biological terms, the "receptive fields" of neurons can be specified without any constraint.

The structural optimization of PNN can be included into the EM algorithm. Confining ourselves to binary variables $x_n \in \{0, 1\}$, we can write:

$$f_n(x_n|m) = \theta_{mn}^{x_n}(1 - \theta_{mn})^{1-x_n}, \quad n \in \mathcal{N}, \quad m = 0, 1, \ldots, M, \tag{10}$$

$$F(\boldsymbol{x}|0) = \prod_{n \in \mathcal{N}} \theta_{0n}^{x_n}(1 - \theta_{0n})^{1-x_n}, \quad (\theta_{0n} = \mathcal{P}\{x_n = 1\}), \tag{11}$$

$$G(\boldsymbol{x}|m, \phi_m) = \prod_{n \in \mathcal{N}} \left[\left(\frac{\theta_{mn}}{\theta_{0n}}\right)^{x_n} \left(\frac{1 - \theta_{mn}}{1 - \theta_{0n}}\right)^{1-x_n} \right]^{\phi_{mn}}. \tag{12}$$

The mixture parameters $f(m), \theta_{mn}$, and the structural parameters ϕ_{mn} can be optimized simultaneously by means of the EM algorithm (cf. [2,6,8,10]). Given a training set of independent observations from the class $\omega \in \Omega$

$$\mathcal{S}_\omega = \{\boldsymbol{x}^{(1)}, \ldots, \boldsymbol{x}^{(K_\omega)}\}, \quad \boldsymbol{x}^{(k)} \in \mathcal{X},$$

we obtain the maximum-likelihood estimates of the mixture (2) by maximizing the log-likelihood function

$$L = \frac{1}{|\mathcal{S}_\omega|} \sum_{x \in \mathcal{S}_\omega} \log P(\boldsymbol{x}|\omega) = \frac{1}{|\mathcal{S}_\omega|} \sum_{x \in \mathcal{S}_\omega} \log \left[\sum_{m \in \mathcal{M}_\omega} F(\boldsymbol{x}|0)G(\boldsymbol{x}|m, \phi_m)f(m) \right]. \tag{13}$$

For this purpose the EM algorithm can be modified as follows:

$$f(m|\boldsymbol{x}) = \frac{G(\boldsymbol{x}|m,\phi_m)f(m)}{\sum_{j\in\mathcal{M}_\omega} G(\boldsymbol{x}|j,\phi_j)f(j)}, \quad \boldsymbol{x}\in\mathcal{S}_\omega, \quad m\in\mathcal{M}_\omega, \quad n\in\mathcal{N}, \tag{14}$$

$$f^{'}(m) = \frac{1}{|\mathcal{S}_\omega|}\sum_{x\in\mathcal{S}_\omega} f(m|\boldsymbol{x}), \quad \theta^{'}_{mn} = \frac{1}{|\mathcal{S}_\omega| f^{'}(m)}\sum_{x\in\mathcal{S}_\omega} x_n f(m|\boldsymbol{x}), \tag{15}$$

$$\gamma^{'}_{mn} = f^{'}(m)\left[\theta^{'}_{mn}\log\frac{\theta^{'}_{mn}}{\theta_{0n}} + (1-\theta^{'}_{mn})\log\frac{(1-\theta^{'}_{mn})}{(1-\theta_{0n})}\right], \tag{16}$$

$$\phi^{'}_{mn} = \begin{cases} 1, \gamma^{'}_{mn}\in\Gamma^{'}, \\ 0, \gamma^{'}_{mn}\notin\Gamma^{'}, \end{cases} \quad \Gamma^{'}\subset\{\gamma^{'}_{mn}\}_{m\in\mathcal{M}_\omega\; n\in\mathcal{N}}, \quad |\Gamma^{'}| = r. \tag{17}$$

Here $f^{'}(m), \theta^{'}_{mn}$, and $\phi^{'}_{mn}$ are the new iteration values of mixture parameters and $\Gamma^{'}$ is the set of a given number of highest quantities $\gamma^{'}_{mn}$. The iterative equations (14)-(17) generate nondecreasing sequence of values of the log-likelihood function (13) converging to a possibly local maximum (cf. [2,6]).

3 Information Preserving Features

Let us recall that the component distributions $F(\boldsymbol{x}|m)$ naturally introduce an additional "descriptive" decision problem [3,4] with a priori probabilities w_m (cf. (6), (7)). In this sense each component of the mixture (7) may correspond to an elementary situation or property. Given a vector $\boldsymbol{x}\in\mathcal{X}$, the presence of elementary properties can be characterized by the conditional probabilities $q(m|\boldsymbol{x})$ (cf. (8)) and also there is a simple relation (9) between the *a posteriori* probabilities of the primary- and descriptive decision problem.

In the framework of statistical pattern recognition the most essential aspect of any transform is to preserve original decision information contained in the input space $\mathcal{X}$. It has been shown (cf. [3]) that the Shannon information $I(\mathcal{X},\mathcal{M})$ about the descriptive decision problem on $\mathcal{X}$ is automatically preserved by any transform

$$\boldsymbol{T}:\mathcal{X}\to\mathcal{Y}, \; \mathcal{Y}\subset R^M, \;\; \boldsymbol{T}(\boldsymbol{x}) = (T_1(\boldsymbol{x}), T_2(\boldsymbol{x}),\ldots, T_M(\boldsymbol{x}))\in\mathcal{Y} \tag{18}$$

defined by Eqs.

$$y_m = T_m(\boldsymbol{x}) = \varphi_m(q^*(m|\boldsymbol{x})), \;\; \boldsymbol{x}\in\mathcal{X}, \;\; m\in\mathcal{M} \tag{19}$$

where $q^*(m|\boldsymbol{x})$, (cf. (8)) are the true conditional probabilities given $\boldsymbol{x}\in\mathcal{X}$ and φ_m are one-to-one mappings of the closed interval $[0,1]$ into the real line R. The transformation (19) preserves Shannon information in the sense that

$$I(\mathcal{X},\mathcal{M}) = I(\mathcal{Y},\mathcal{M}) \tag{20}$$

and simultaneously the entropy of the transformed distribution

$$H(\mathcal{Y}) = \sum_{\boldsymbol{y} \in \mathcal{Y}} -Q(\boldsymbol{y}) \log Q(\boldsymbol{y}), \quad Q(\boldsymbol{y}) = P(\boldsymbol{T}^{-1}(\boldsymbol{y})) \tag{21}$$

is minimized over the class of all transforms $\boldsymbol{T}$ satisfying (20).

It is easily verified that, in view of (9), the transform (19) also preserves the decision information $I(\mathcal{X}, \Omega) = I(\mathcal{Y}, \Omega)$. Roughly speaking, the transform $\boldsymbol{T}$ "unifies" the points $\boldsymbol{x} \in \mathcal{X}$ with the same posterior distributions $q^*(.|\boldsymbol{x})$. In the paper [4] the proof of the above assertions relates to discrete variables, a generalization to continuous case can be found in [21]. In the paper [5] we have shown that, choosing φ_m as logarithm, we obtain information preserving transform $\boldsymbol{T}$ which is fault-tolerant in the sense that bounded approximation inaccuracy may cause only bounded information loss. In particular, let the coordinate functions $T_m(\boldsymbol{x})$ satisfy for some $\delta > 0$ and $\epsilon > 0$ the inequality

$$|T_m(\boldsymbol{x}) - \ln[q^*(m|\boldsymbol{x}) + f^*(m)\delta]| < \epsilon, \quad m \in \mathcal{M}, \quad \boldsymbol{x} \in \mathcal{X} \tag{22}$$

where asterisk denotes true values. Then the information loss accompanying the transform $\boldsymbol{T}$ is bounded by the inequality

$$I(\mathcal{X}, \mathcal{M}) - I(\mathcal{Y}, \mathcal{M}) = H(\mathcal{M}|\mathcal{Y}) - H(\mathcal{M}|\mathcal{X}) < \delta + 2\epsilon \tag{23}$$

where $H(\mathcal{M}|\mathcal{X}), H(\mathcal{M}|\mathcal{Y})$ are the corresponding conditional entropies (for proof cf. [5]). It can be seen that for ϵ and δ approaching zero inequality (22) implies the information preserving property (20).

The condition (22) suggests the fault-tolerant transform

$$y_m = T_m(\boldsymbol{x}) = \log\left[q(m|\boldsymbol{x}) + f(m)\delta\right], \quad \boldsymbol{x} \in \mathcal{X}, \quad m \in \mathcal{M} \tag{24}$$

with the estimated values $q(m|\boldsymbol{x}), f(m)$. Let us note that the positive constant $f(m)\delta$ avoids possible singularities occurring near the zero point $q(m|\boldsymbol{x}) = 0$. However, even if we use the more practical form (24) of the transform $\boldsymbol{T}$, the computational properties of the resulting features appear to be poor. As the probabilistic features y_m are real, we have to use a probability density model. However, the otherwise efficient normal mixture does not seem to be suitable to describe the properties of the transformed space $\mathcal{Y}$. In our numerical experiments the recognition accuracy based on the probabilistic feature vectors (24) was always essentially worse than in the input space $\mathcal{X}$.

In multidimensional spaces the conditional probabilities $q(m|\boldsymbol{x})$ behave nearly binary, i.e. one of them is near one and the remaining values are near zero. This property suggests a natural possibility of binary approximation. In the previously published experiments (cf. [8,9]) the binary approximation was defined by means of the Bayes decision function with only one unit per output vector. Consequently, in this case, the recognition accuracy based on the binary features is approximately the same as in the sample space. In the paper [8] we succeeded to improve the recognition accuracy by combining independent solutions in parallel.

In this paper we apply a less rigorous principle to define the binary output variables. Instead of the Bayes decision function we use a very low threshold with the aim to include additional information about possible ties. In particular, we define

$$y_m = T_m(\boldsymbol{x}) = \begin{cases} 1, \ q(m|\boldsymbol{x}) \geq \theta, \\ 0, \ q(m|\boldsymbol{x}) < \theta, \end{cases} \quad 0 < \theta \ll 1. \tag{25}$$

The positive threshold θ has been chosen very low ($\theta \approx 10^{-6} - 10^{-8}$) in order to detect possible latent alternatives of the given maximum. In this way we succeeded to improve the recognition accuracy in all numerical experiments with handwritten numerals.

4 Numerical Example

In computational experiments we have applied the proposed PNN to the NIST Special Database 19 (SD19) containing about 400000 handwritten digits (see also [11]). The SD19 numerals have been widely used for benchmarking of classification algorithms. As it appears, in literature there is no generally accepted partition of SD19 into the training and testing data set. Unfortunately, the original report [12] recommends for test purposes the data set which differs from the training data in origin and size. For this reason, in order to guarantee the same statistical properties of both the training and testing data sets, we have used the odd samples of each class for training and the even samples for testing.

In our experiments we have normalized all digit patterns to a 32x32 binary raster. The conditional distributions $P(\boldsymbol{x}|\omega)$ have been estimated in the original sample space (dimension N=1024) without any preliminary feature extraction or feature selection. In order to increase the natural variability of data we have extended the training data set four times by making three differently rotated variants of each digit pattern (by -10,-5,+5 degrees). The same principle has been applied to independent test sets. For the sake of the final test of accuracy each digit pattern has been classified by summing first the four *a posteriori* distributions computed for the respective pattern variants.

By using the training data we have estimated in several experiments the class-conditional mixtures of different complexity by means of the EM algorithm of Sec. 2. The EM algorithm has been stopped by the relative increment threshold 10^{-3} implying several tens ($10 \div 40$) of EM iterations. The initial number of mixture components was always the same in all classes but some of the components have been suppressed in the course of iterations. For each experiment (first row) the final number of components is given in the second row of Tab.1. Unlike Eq. (17) the structural parameters ϕ_{mn} have been chosen in each iteration by using the computationally more simple thresholding:

$$\phi'_{mn} = \begin{cases} 1, \ \gamma'_{mn} \geq 0.1\gamma'_0, \\ 0, \ \gamma'_{mn} < 0.1\gamma'_0, \end{cases}, \qquad \gamma'_0 = \frac{1}{|\mathcal{M}_\omega| N} \sum_{m \in \mathcal{M}_\omega} \sum_{n \in \mathcal{N}} \gamma'_{mn}. \tag{26}$$

The threshold value $0.1\gamma'_0$ is relatively low and therefore in each component only the really superfluous variables have been omitted. The resulting total number

Table 1. Classification accuracy. Recognition of numerals from the NIST SD19 database. Classification error in % obtained by different methods and for differently complex mixtures, in comparison with feature space.

Experiment No.:	**1**	**2**	**3**	**4**	**5**	**6**	**7**
Number of Components	100	200	357	494	695	1119	1382
Number of Parameters	96046	174547	243293	272302	533628	574159	1027691
Number of Parameters [in %]	93.8	85.2	66.5	53.8	75.0	50.1	72.6
Mean Number of Units in $\boldsymbol{y}$	1.22	1.27	1.32	1.26	1.44	1.39	1.50
Log-Likelihood (Input Space)	-295.8	-277.1	-265.8	-259.4	-242.0	-239.8	-235.3
Log-Likelihood (Features)	-6.21	-7.15	-7.95	-8.14	-9.19	-9.48	-10.09
Recognition Accuracy							
Error [in %] (Input Space)	5.46	3.94	3.24	2.95	2.52	2.21	2.12
Error [in %] (Features)	5.21	3.82	3.17	2.74	2.46	2.10	2.08

of component specific variables is given in the third row of Table 1. In the fourth row the same number is expressed relatively in % of the "full" mixture model.

The next row of Table 1 illustrates the functioning of the threshold θ in Eq. (25). It can be seen that the value $\theta \approx 10^{-6} - 10^{-8}$ generates the second unit in about 20% - 50% of binary feature vectors. The influence of the chosen threshold value on the resulting classification accuracy is rather limited.

Another important aspect of the information preserving features is the minimum entropy property (cf. (20), (21)). Let us recall in this connection that the log-likelihood function (13) can be viewed as a sample estimate of the corresponding entropy. In particular, we can write

$$\lim_{|\mathcal{S}_\omega|\to\infty} \frac{1}{|\mathcal{S}_\omega|} \sum_{\boldsymbol{x}\in\mathcal{S}_\omega} \log P(\boldsymbol{x}) = \lim_{|\mathcal{S}_\omega|\to\infty} \sum_{\boldsymbol{x}\in\mathcal{X}} \tilde{P}(\boldsymbol{x}) \log P(\boldsymbol{x}) \approx -H_P(\mathcal{X}) \tag{27}$$

where $\tilde{P}(\boldsymbol{x})$ denotes the relative frequency of $\boldsymbol{x}$ in $\mathcal{S}_\omega$. Thus, with the aim to illustrate the decrease of entropy, we have computed the log-likelihood criterion for the independent test set both in the sample space $\mathcal{X}$ (row 6) and in the space of transformed features $\mathcal{Y}$ (row 7). It can be seen that in the binary feature space the log-likelihood values are much greater, i.e. the corresponding entropy is much less than in the sample space $\mathcal{X}$. Moreover, it appears that for each class the considered binary features are nearly independent. In all experiments we have used only one component to estimate the class conditional binary distributions $Q(\boldsymbol{y}|\omega)$ in the feature space $\mathcal{Y}$. In case of two or more components the classification accuracy was the same or even worse. This observation corresponds well with the assumption of class conditional independence of variables.

The last two rows of Tab.1 compare the recognition performance in the sample space $\mathcal{X}$ and in the feature space $\mathcal{Y}$. As it can be seen, the recognition accuracy based on the proposed binary features is slightly better in all experiments and improves with increasing model complexity.

5 Conclusion

Considering PNN in the framework of statistical pattern recognition we obtain formal neurons defined by means of component parameters of the estimated class-conditional mixtures. The output variables of neurons can be viewed as features having some useful theoretical properties. The probabilistic features extracted by PNN can be used to transform the classification problem without information loss while keeping the entropy of the output space to be minimum. We have shown that even a simple binary approximation of probabilistic features is capable of improving the recognition accuracy - when estimating the class-conditional distributions of binary features again. The method simplifies decision making by reducing the feature complexity rather than dimensionality of the problem. In numerical experiments the resulting binary features appear to be almost conditionally independent with respect to classes since the conditional mixtures with one component yield the best recognition accuracy.

The purpose of the numerical experiments has been to illustrate the properties of the proposed approach, no validation data sets have been considered for optimization of the model complexity and of the underlying feature thresholds. In general, the recognition accuracy increases with the model complexity and is slightly better in the space of the extracted features.

Acknowledgement. This research was supported by the project GAČR No. 102/07/1594 of Czech Grant Agency and partially by the projects 2C06019 ZIMOLEZ and MŠMT 1M0572 DAR.

References

1. Grim, J.: On numerical evaluation of maximum-likelihood estimates for finite mixtures of distributions. Kybernetika 18(3), 173–190 (1982)
2. Grim, J.: Multivariate statistical pattern recognition with non-reduced dimensionality. Kybernetika 22(6), 142–157 (1986)
3. Grim, J.: Maximum-likelihood design of layered neural networks. In: International Conference on Pattern Recognition (Proceedings), pp. 85–89. IEEE Computer Society Press, Los Alamitos (1996)
4. Grim, J.: Design of multilayer neural networks by information preserving transforms. In: Pessa, E., Penna, M.P., Montesanto, A. (eds.) Third European Congress on Systems Science, pp. 977–982. Edizioni Kappa, Roma (1996)
5. Grim, J.: Discretization of probabilistic neural networks with bounded information loss. In: Rojicek, J., et al. (eds.) Computer-Intensive Methods in Control and Data Processing (Preprints of the 3rd European IEEE Workshop CMP 1998), September 7 – 9, 1998, pp. 205–210. UTIA AV CR, Prague (1998)
6. Grim, J.: Information approach to structural optimization of probabilistic neural networks. In: Ferrer, L., Caselles, A. (eds.) Fourth European Congress on Systems Science, SESGE, Valencia, pp. 527–539 (1999)
7. Grim, J.: Self-organizing maps and probabilistic neural networks. Neural Network World 10(3), 407–415 (2000)
8. Grim, J., Kittler, J., Pudil, P., Somol, P.: Multiple classifier fusion in probabilistic neural networks. Pattern Analysis & Applications 5(7), 221–233 (2002)

9. Grim, J., Pudil, P.: On virtually binary nature of probabilistic neural networks. In: Amin, A., Dori, D., Pudil, P., Freeman, H. (eds.) Advances in Pattern Recognition. LNCS, vol. 1451, pp. 765–774. Springer, Berlin (1998)
10. Grim, J., Pudil, P., Somol, P.: Recognition of handwritten numerals by structural probabilistic neural networks. In: Bothe, H., Rojas, R. (eds.) Proceedings of the Second ICSC Symposium on Neural Computation, pp. 528–534. ICSC Wetaskiwin (2000)
11. Grim, J., Hora, J.: Recurrent Bayesian Reasoning in Probabilistic Neural Networks. In: de Sá, J.M., Alexandre, L.A., Duch, W., Mandic, D.P. (eds.) ICANN 2007. LNCS, vol. 4669, pp. 129–138. Springer, Heidelberg (2007)
12. Grother, P.J.: NIST special database 19: handprinted forms and characters database, Technical Report and CD ROM (1995)
13. Haykin, S.: Neural Networks: a comprehensive foundation. Morgan Kaufman, San Mateo (1993)
14. Kohonen, T.: The Self-Organizing Maps. Springer, New York (1997)
15. McLachlan, G.J., Peel, D.: Finite Mixture Models. John Wiley and Sons, New York (2000)
16. Palm, H.Ch.: A new method for generating statistical classifiers assuming linear mixtures of Gaussian densities. In: Proc. of the 12th IAPR Int. Conf. on Pattern Recognition, Jerusalem, 1994, vol. II, pp. 483–486. IEEE Computer Soc. Press, Los Alamitos (1994)
17. Sammon, J.W.: A Nonlinear Mapping for Data Structure Analysis. IEEE Transactions on Computers 18(5), 401–409 (1969)
18. Schlesinger, M.I.: Relation between learning and self-learning in pattern recognition. Kibernetika (Kiev) 6, 81–88 (1968) (in Russian)
19. Specht, D.F.: Probabilistic neural networks for classification, mapping or associative memory. In: Proc. of the IEEE International Conference on Neural Networks, I, pp. 525–532 (1988)
20. Streit, L.R., Luginbuhl, T.E.: Maximum-likelihood training of probabilistic neural networks. IEEE Trans. on Neural Networks 5, 764–783 (1994)
21. Vajda, I., Grim, J.: About the maximum information and maximum likelihood principles in neural networks. Kybernetika 34, 485–494 (1998)
22. Watanabe, S., Fukumizu, K.: Probabilistic design of layered neural networks based on their unified framework. IEEE Trans. on Neural Networks 6(3), 691–702 (1995)

Correlation Integral Decomposition for Classification

Marcel Jiřina[1] and Marcel Jiřina Jr.[2]

[1] Institute of Computer Science AS CR, Pod vodarenskou vezi 2,
182 07 Prague 8 – Liben, Czech Republic
marcel@cs.cas.cz
http://www.cs.cas.cz/~jirina
[2] Faculty of Biomedical Engineering, Czech Technical University in Prague,
Nam. Sitna 3105, 272 01, Kladno, Czech Republic

Abstract. In this paper we show that the correlation integral can be decomposed into functions each related to a particular point of data space. For these functions, one can use similar polynomial approximations as used in the correlation integral. The essential difference is that the value of the exponent, which would correspond to the correlation dimension, differs in accordance to the position of the point in question. Moreover, we show that the multiplicative constant represents the probability density estimation at that point. This finding is used for the construction of a classifier. Tests with some data sets from the Machine Learning Repository show that this classifier can be very effective.

1 Introduction

The correlation dimension [7], [11] as well as other effective dimensions - Hausdorff, box-counting, information dimension [9], [14] - are used to study features of different fractals and data generating processes. For estimation of the value of the correlation dimension in a particular case, linear regression is often used for logarithms of variables [7]. We write it in the form

$$\ln(s) = \ln(C) + v\ln(r_s), s = 1, 2, \ldots \quad (1)$$

Here, v is a correlation dimension, s is index of binate distances such that $s = 1$ for the shortest, $s = 2$ for the second shortest binate distance and so on, and C is a multiplicative constant in original polynomial relation

$$s = Cr_s^v, \quad s = 1, 2, ... \quad (2)$$

Constant C has no particular meaning.

In this paper we show that the correlation integral can be decomposed into functions related to particular points of data space. For these functions, one can use similar polynomial approximations as given by (2). The value of exponent q, which corresponds to the correlation dimension v, differs in accordance with the position of point x in question. Moreover, we show that the multiplicative

V. Kůrková et al. (Eds.): ICANN 2008, Part II, LNCS 5164, pp. 62–71, 2008.

constant C in these cases represents the probability density estimation at point x. This finding is used to construct a classifier. Tests with some data sets from the Machine Learning Repository [10] show that this classifier can have a very low classification error.

2 Decomposition of Correlation Integral

We work in the n-dimensional metric space with L_2 (Euclidean) or L_1 (taxicab or Manhattan) metrics.

2.1 Correlation Integral

The correlation integral, in fact, a distribution function of all binate distances in a set of points in a space with a distance was introduced in [7]. The correlation integral $C_I(r)$ is defined by

$$C_I(r) = \lim_{N\to\infty} \frac{1}{N^2} \times \{number\ of\ pairs\ (i,j) : \|X_i - X_j\| \ < r\}.$$

In a more comprehensive form, one can write

$$C_I(r) = \Pr(\|X_i - X_j\| < r).$$

Grassberger and Procaccia [7] have shown that for a small r, the $C_I(r)$ grows like the power $C_I(r) \sim r^\nu$ and that the "correlation exponent" ν can be taken as a most useful measure of the local structure of the strange attractor. This measure allows one to distinguish between deterministic chaos and random noise [2].

The correlation integral can be rewritten in the form [2]

$$C_I(r) = \lim_{N\to\infty} \frac{1}{N(N-1)} \sum_{1\le i<j\le N} h(r - \|X_j - X_i\|),$$

where $h(.)$ is the Heaviside step function. From it

$$\nu = \lim_{r\to\infty} \frac{\ln C_I(r)}{\ln r}.$$

There are methods for estimating the correlation dimension ν. One of the most cited is Taken's estimator [13], [1], [8].

2.2 Probability Distribution Mapping Function

Let a query point x be placed without loss of generality in the origin. Let us build balls with their centers at point x and with volumes V_i, i =1, 2, ... Individual balls are in one another, the (i-1)-st inside the i-th like peels of an onion. Then the mean density of the points in the i-th ball is $\rho_i \;=\; m_i/V_i$. The volume of a ball of radius r in n-dimensional space is $V(r) = const.r^n$. Thus we have constructed a mapping between the mean density ρ_i in the i-th ball and its radius r_i. Then $\rho_i = f(r_i)$. Using the analogy between the density $\rho(z)$ and the probability density $p(z)$ one can write $p(r_i) = f(r_i)$, and $p(r_i)$ is the mean probability density in the i-th ball with radius r_i. This way, a complex picture of the probability distribution of the points in the neighborhood of a query point x is simplified to a function of a scalar variable. We call this function the probability distribution mapping function $D(x, r)$, where x is a query point, and r is the distance from it. More exact definitions follow.

Definition 1. *The probability distribution mapping function $D(x, r)$ in the neighborhood of the query point x is the function$D(x,r) = \int_{B(x,r)} p(z)dz$, where r is the distance from the query point and $B(x, r)$ is a ball with center x and radius r.*

Definition 2. *The distribution density mapping function $d(x, r)$ in the neighborhood of the query point x is the function$d(x,r) = \frac{\partial}{\partial r} D(x,r)$, where $D(x, r)$ is the probability distribution mapping function of the query point x and radius r.*

One can write the probability distribution mapping function in the form

$$D(x,r) = \lim_{N\to\infty} \frac{1}{N-1} \sum_{j=1}^{N-1} h(r - r_j), \tag{3}$$

where $h(.)$ is the Heaviside step function.

2.3 Power Approximation of the Probability Distribution Mapping Function

Let us introduce a simple polynomial function in the form $D(x, r) = C.r^q$. We shall call it a power approximation of the probability distribution mapping function $D(x, r)$.

Definition 3. *The power approximation of the probability distribution mapping function $D(x, r^q)$ is the function r^q such that $\frac{D(x,r^q)}{r^q} \to C$ forr $\to 0+$. The exponent q is a distribution-mapping exponent.*

Using this approximation of the probability distribution mapping function $D(x, r)$ we, in fact, linearize this function as a function of the variable $z = r^q$ in the neighborhood of the origin, i.e. in the neighborhood of the query point. The distribution density mapping function $d(x, r)$ as a function of variable $z = r^q$ is approximately constant in the vicinity of the query point. This constant includes the true distribution of the probability density of the points as well as the influence of the boundary effects.

2.4 Decomposition of the Correlation Integral to Local Functions

We show in this section that the correlation integral is the mean of the distribution mapping functions and that the correlation dimension can be approximated by the mean of the distribution mapping exponents as shown in the theorem below.

Theorem 1. *Let there be a learning set of N points (samples). Let the correlation integral, i.e. the probability distribution of the binate distances of points from the data set, be* $C_I(r)$ *and let* $D(x_i, r)$ *be the distribution mapping function corresponding to point* x_i*. Then* $C_I(r)$ *is the mean value of* $D(x_i, r)$*:*

$$C_I(r) = \lim_{N\to\infty} \tfrac{1}{N} \sum_{i=1}^{N} D(x_i, r). \tag{4}$$

Proof. Let $h(x)$ be the Heaviside step function and l_{ij} be the binate distances of points from the data set. Then the correlation integral is

$$C_I(r) = \lim_{N\to\infty} \tfrac{1}{N(N-1)} \sum_{i=1}^{N} \sum_{j=1}^{N-1} h(r - l_{ij})$$

and also

$$C_I(r) = \lim_{N\to\infty} \tfrac{1}{N} \sum_{i=1}^{N} \left(\tfrac{1}{N-1} \sum_{j=1}^{N-1} h(r - l_{ij}) \right). \tag{5}$$

Comparing (5) with (3) we get (4) directly.

2.5 Distribution Mapping Exponent Estimation

Let U be a learning set composed of points (patterns, samples) x_{cs}, where $c = \{0, 1\}$ is the class mark and $s = 1, 2, \ldots, N_c$ is the index of the point within class c. N_c is the number of points in class $c$and let $N = N_0 + N_1$ be the learning set size.

Let point $x \notin U$ be given and let points x_{cs} of one class be sorted so that index $s = 1$ corresponds to the nearest neighbor, index $s = 2$ to the second nearest neighbor, etc. In Euclidean metrics, $r_s = ||x - x_{cs}||$ is the distance of the s-th nearest neighbor of class c from point x.

We look for the exponent q so, that $r_s^q$is proportional to index s, i.e.

$$s = Cr_s^q, \quad s = 1, 2, ..., N_c, c = 0 \text{ or } 1, \tag{6}$$

where C is a suitable constant. Using a logarithm we get

$$\ln(s) = \ln(C) + q\ln(r_s), \quad s = 1, 2, ..., N_c. \tag{7}$$

On one hand, we exaggerate the distances nonlinearly to make small differences in the distance appear much larger for the purposes of density estimation. On the other hand, there is a logarithm of distance in (7), which decreases large influences of small noise perturbations on the final value of q. Note that it is the same problem as in the correlation dimension estimation where equations of the same form as (6) and (7) arise. In [7] a solution by linear regression was proposed. In [5], [8], [11] and others, different modifications and heuristics were later proposed. Many of these approaches and heuristics can be used for distribution mapping exponent estimation, e.g. use a half or a square root of N_c nearest neighbors instead of N_c to eliminate the influence of the limited number of points of the learning set.

The system of N_c (or $N_c/2$ or $\surd N_c$ as mentioned above) equations (7) with respect to an unknown q can be solved using standard linear regression for both classes. Thus, for two classes we get two values of q, q_0 and q_1 and two values of C', C'_0 and C'_1.

At this point we can say that q_c is something like a local effective dimensionality of the data space including the true distribution of points in each class. At the same time, we get constant C'_c. The values of q_c and C'_c are related to each particular point $x$and, thus, vary from one point x to another.

2.6 Probability Density Estimation

Let $n'_c(r)$ be a number of points of class c until distance r from the query point x. Let q_c be the distribution mapping exponent for points of class c and let

$$z_c = r^{qc}. \tag{8}$$

Also, let $n_c(z_c) = n'_c(r) = n'_c(z_c^{1/qc})$. Then$P_c(z_c) = n_c(z_c)/N$ is a percentage of all points of class c until distance $r = z_c^{1/qc}$ from the query point x, i.e. until a "distance" measured by z_c from point x.

Due to polynomial approximation (8), $n_c(z_c) = C'_c.z_c$. This is a number of points up to distance r, which is related to z_c according to (8). The derivative according to z_c is $\mathrm{d}n_c(z_c)/\mathrm{d}z_c = C'_c$ and it represents a number of points of class c on a unit[1] of the z_c, i.e., in fact, a density of points with respect to z_c.

By dividing by the total number of points N there is a percentage of points of class c on a unit of z_c. This percentage is equal to $p(c|x, z_c) = C'_c/N$. In a limit case for $r \to 0$ (and z_c as well), $\mathrm{p}(c|x, 0) = p(c|x) = C'_c/N = C_c$.

Finally, if there are two classes, there must be $p(0|x) + p(1|x) = 1$ and then $C'_0 + C'_1 = N$.This result includes a priori probabilities N_c/N for both classes. When we need to exclude a priori probabilities we use the formula

$$p(c|x) = \frac{C'_c/N_c}{C'_0/N_0 + C'_1/N_1}. \tag{9}$$

A generalization to a many class case is straightforward. For k classes

$$p(c|x) = \frac{C'_c/N_c}{\sum_{i=1}^{k} C'_i/N_i}, c = 1, 2, ..k. \tag{10}$$

[1] We cannot say "unit length" here as the dimensionality of z_c is $(length)^{qc}$.

A more exact development follows.

Definition 4. *Let* $\boldsymbol{N}$ *be n-dimensional space with metrics* ρ. *Let there be a subset* $\boldsymbol{Q} \subseteq \boldsymbol{N}$ *and a number* $q \in$ g$+, 1 \leq q \leq n$ *be associated with subset* $\boldsymbol{Q}$. *A q-dimensional ball with center at point* $x \in$ g *and radius* r *is* $B_q = B(q, x, r, \rho) = \{y \in \text{g} : \rho(x, y) < r\}$. *The volume of* B_q *is* $V(q, x, r, \rho) = S(q, \rho).r^q$, *where* $S(q, \rho)$ *is a function independent of* r.

The metrics ρ can be omitted when it is clear which metrics we are dealing with.

Lemma 1. *Let* $B(q,\ x,\ R)$ *be a q-dimensional ball with center at point* $x \in$ g *and radius* R*, and let* $V(q,\ x,\ r)$ *be its volume. Let points in* $\boldsymbol{Q}$ *in the neighborhood of point* x *up to distance* R *be distributed with a constant probability density* $p = C$. *Then for* $r < R$, *where* r *is the distance from point* x, *the distribution function is given by*

$$P(x, r) = \int\limits_{B(q,x,R)} pdr = \int pdV(q, x, r) = C \cdot V(q, x, r).$$

The Proof is obvious.

Conversely, let in **Q** holds $P(x, r) = C \cdot V(q, x, r)$, where C is a constant as long as $r < R$. It is obvious that this can be fulfilled even when the distribution density is not constant. On the other hand it is probably a rare case. Then we can formulate an assumption.

Assumption 1. If in **Q** holds $P(x, r) = C \cdot V(q, x, r)$, the corresponding probability density p can be approximated by constant C in $B(q,\ x,\ R)$ and then

Illustration. A sheet of white paper represents 2 dimensional subspace embedded in 3 dimensional space. Let point x be in the center of the sheet. White points of paper are uniformly distributed over the sheet with the some constant (probability) density and a distribution function (the number of white points) is proportional to the circular area around point x. Thus, the distribution function grows quadratically with distance r from point x, and only linearly with the size of the circular area. Furthermore, the size of the circular area is nothing more than the volume of the 2 dimensional ball embedded in 3 dimensional space.

Theorem 2. *Let, in a metric space, each point belong to one of two classes* $c = \{0,\ 1\}$. *Let, for each point* x *and each class* c, *a distribution mapping function* $D(x,\ c,\ r)$ *exist where* r *is the distance from point* x. *Let Assumption 1 hold and the power approximation of the distribution mapping function be* $C_c r^{q_c}$, *where* q_c *is the distribution mapping exponent for point* x *and class* c. *Then* $p(c|x) = C_c$ *holds.*

Proof. Let $z_c = r^{q_c}$ be a new variable. Then $D(x,\ c,\ r) = D(x,\ c,\ z_c)$ and when using z_c $D(x,\ c,\ z_c) = C_c\ z_c$. The $D(x,\ c,\ z_c)$ is, in fact, a distribution function of points of class c. This distribution function corresponds to the uniform distribution in a subspace of dimension q_c and, according to Lemma 1 and Assumption 1, the proportionality constant C_c is equal to the probability density in the neighborhood of point x including this point.

2.7 Classifier Construction

Using formulas (9) or (10) we have a relatively simple method for estimating the probabilities $p(c|x)$. First, we sort points of class c according to their distances from the query point x. Then, we solve linear regression equations

$$q_c \ln(r_s) = \ln(C'_c) + \ln(s), \quad s = 1, 2, ..., K \tag{11}$$

for the first K points, especially with respect to an unknown C'_c. The number K may be a half or a square root of total number N_c of points of class c. We found that one half of the number of samples from the learning set can be an optimal value of K from a point of view of simplicity, reliability and classification quality. This is made for all k classes, c = 1, 2, .. k. Finally we use formula (9) for $k = 2$ or formula (10) for more than two classes. Formula (9) or (10) gives a real number. For a two class classification, a discriminant threshold (cut) θ must be chosen, and then if $p(1|x) > \theta$, then x belongs to class 1 or else to class 0. A default value of θ is 0.5.

In testing a classifier, we found the results a bit better when L_1 metrics were used rather than with L_2 metrics.

3 Experiments

The applicability of the method "QCregre1" demonstrates some tests using real-life tasks from the UCI Machine Learning Repository [10]. Tasks of classification into two classes for which data about previous tests are known were selected: "Heart", "Ionosphere", and "Iris".

The task "Heart" indicates an absence or presence of heart disease for patient.

For the task "Ionosphere", the targets were free electrons in the ionosphere. "Good" radar returns are those showing evidence of some type of structure in the ionosphere. "Bad" returns are those that do not; their signals pass through the ionosphere.

The task "Iris" is to determine whether an iris flower is of class Versicolor or Virginica. The third class, Setoza is deleted, as it is linearly separable from the other two. 100 samples, four parameters and ten-fold cross validation were used as in [15].

We do not describe these tasks in detail here as all of them can be found in descriptions of individual tasks of the Repository and also the same approach to testing and evaluation was used. Especially splitting the data set into two disjoint subsets, the learning set and the testing set and the use of cross validation were the same as in [10] or – for Iris database – [15].

We also checked some standard methods for comparison as follows:

- 1-NN – standard nearest neighbor method [3]
- Sqrt-NN – the k-NN method with k equal to the square root of the number of samples of the learning set [4]
- Bay1 – the naïve Bayes method using ten bins histograms [6]
- LWM1 – the learning weighted metrics by [12].

For k-NN, Bayes, LWM and our method the discriminant thresholds θg were tuned accordingly. All procedures are deterministic (even Bayes algorithm) and then no repeated runs were needed.

In Table 1, the results are shown together with the results for other methods.

Table 1. Classification errors for three different tasks shown for different methods presented in the Machine Learning Repository. The note [fri] refers to the results according to paper [15]. The results shown in bold were computed by authors.

Heart		**Ionosphere**		**Iris**	
Algorithm	Test	Algorithm	Error	Algorithm	Test
QCregre1	**0.1779**	**QCregre1**	**0.02013**	scythe[fri]	0.03
LWM1	0.1882	**Bay1**	0.02013	**QCregre1**	**0.04878**
Bayes	0.374	**LWM1**	0.02649	**sqrt-NN**	0.04879
Discrim	0.393	IB3 (Aha & Kibler, IJCAI-1989)	0.033	mach:ln [fri]	0.05
LogDisc	0.396	backprop an average of over	0.04	mach-bth [fri]	0.05
Alloc80	0.407	**sqrt-NN**	0.05369	CART	0.06
QuaDisc	0.422	Ross Quinlan's C4 algorithm	0.06	mach [fri]	0.06
Castle	0.441	nearest neighbor	0.079	mach:ds [fri]	0.06
Cal5	0.444	"non-linear" perceptron	0.08	**1-NN**	0.06098
Cart	0.452	"linear" perceptron	0.093	**LWM1**	0.06863
Cascade	0.467			**Bay1**	0.08537
KNN	0.478			k-NN	0.8
Dipol92	0.507				
BayTree	0.526				
BackProp	0.574				
LVQ	0.6				
IndCart	0.63				
Kohonen	0.693				
Cn2	0.767				
Radial	0.781				
C4.5	0.781				

4 Discussion

The main goal of this paper is to show that the correlation integral can be decomposed to local functions – the probability distribution mapping functions (PDMF). Each PDMF corresponds to a particular point of data space and characterizes the probability distribution in some neighborhood of a given point. We have also shown that – similarly to the correlation integral – the PDMF can be approximated by a polynomial function. This polynomial approximation is governed by two constants, the distribution mapping exponent, which can be considered as a local analog to the correlation dimension, and a multiplicative constant. It is proven here, that this multiplicative constant is just an estimate of probability density at the given point. This estimate is used to construct a classifier.

This classifier is slightly related to various nearest neighbor methods. It uses information about the distances of the neighbors of different classes from the query point and neglects information about the direction where the particular neighbor lies. On the other hand, nearest neighbor methods do not differentiate individual distances of nearest points. E.g. in the k-NN method the number of points of one and the other class among k nearest neighbors is essential, but not the individual distances of the points. The method proposed here takes individual distances into account even if these distances are a little bit hidden in regression equations. The method outperforms 1-NN, k-NN as well as LWM (learning weighted metrics) [12] in practically all the cases and can be found as the best for some tasks shown in Table 1.

There is an interesting relationship between the correlation dimension and the distribution mapping exponent q_c. The former is a global feature of the fractal or the data generating process; the latter is a local feature of the data set and is closely related to a particular query point. On the other hand, if linear regression were used, the computational procedure is almost the same in both cases. Moreover, it can be found that the values of the distribution mapping exponent usually lie in a narrow interval <-10, +10> percentage around the mean value.

Acknowledgements

This work was supported by the Ministry of Education of the Czech Republic under the project Center of Applied Cybernetics No. 1M0567, and No. MSM6840770012 Transdisciplinary Research in the Field of Biomedical Engineering II.

References

1. Camastra, F.: Data dimensionality estimation methods: a survey. Pattern Recognition 6, 2945–2954 (2003)
2. Camastra, F., Vinciarelli, A.: Intrinsic Dimension Estimation of Data: An Approach based on Grassberger-Procaccia's Algorithm. Neural Processing Letters 14(1), 27–34 (2001)

3. Cover, T.M., Hart, P.E.: Nearest Neighbor Pattern Classification. IEEE Transactions on Information Theory IT-13(1), 21–27 (1967)
4. Duda, R.O., Hart, P.E., Stork, D.G.: Pattern classification, 2nd edn. John Wiley and Sons, Inc., New York (2000)
5. Dvorak, I., Klaschka, J.: Modification of the Grassberger-Procaccia algorithm for estimating the correlation exponent of chaotic systems with high embedding dimension. Physics Letters A 145(5), 225–231 (1990)
6. Gama, J.: Iterative Bayes. Theoretical Computer Science 292, 417–430 (2003)
7. Grassberger, P., Procaccia, I.: Measuring the strangeness of strange attractors. Physica 9D, 189–208 (1983)
8. Guerrero, A., Smith, L.A.: Towards coherent estimation of correlation dimension. Physics letters A 318, 373–379 (2003)
9. Lev, N.: Hausdorff dimension. Student Seminar, Tel-Aviv University (2006), www.math.tau.ac.il/~levnir/files/hausdorff.pdf
10. Merz, C. J., Murphy, P. M., Aha, D. W.: UCI Repository of Machine Learning Databases. Dept. of Information and Computer Science, Univ. of California, Irvine (1997), `http://www.ics.uci.edu/~mlearn/MLSummary.html`
11. Osborne, A.R., Provenzale, A.: Finite correlation dimension for stochastic systems with power-law spectra. Physica D 35, 357–381 (1989)
12. Paredes, R., Vidal, E.: Learning Weighted Metrics to Minimize NearestNeighbor Classification Error. IEEE Transactions on Pattern Analysis and Machine Intelligence 20(7), 1100–1110 (2006)
13. Takens, F.: On the Numerical Determination of the Dimension of the Attractor. In: Dynamical Systems and Bifurcations. Lecture Notes in Mathematics, vol. 1125, pp. 99–106. Springer, Berlin (1985)
14. Weisstein, E. W.: Information Dimension. From MathWorld–A Wolfram Web Resource (2007), `http://mathworld.wolfram.com/InformationDimension.html`
15. Friedmann, J.H.: Flexible Metric Nearest Neighbor Classification. Technical Report, Dept. of Statistics, Stanford University, p. 32 (1994)

Modified q-State Potts Model with Binarized Synaptic Coefficients

Vladimir Kryzhanovsky

Center of Optical Neural Technologies of
Scientific Research Institute for System Analysis of
Russian Academy of Sciences
44/2 Vavilov Street, 119333 Moscow, Russian Federation
Vladimir.Krizhanovsky@gmail.com, iont.niisi@gmail.com

Abstract. Practical applications of q-state Potts models are complicated, as they require very large RAM ($32N^2q^2$ bits, where N is the number of neurons and q is the number of the states of a neuron). In this work we examine a modified Potts model with binarized synaptic coefficients. The procedure of binarization allows one to make the required RAM 32 times smaller (N^2q^2 bits), and the algorithm more than q times faster. One would expect that the binarization worsens the recognizing properties. However our analysis shows an unexpected result: the binarization procedure leads to the increase of the storage capacity by a factor of 2. The obtained results are in a good agreement with the results of computer simulations.

Keywords: Recognition, Potts model, storage capacity.

1 Introduction

At present the problems of identification (classification) of a very large array of vectors of high dimensionality frequently occur. For instance, to identify an attack on a large computer network one has to deal with vectors of the dimensionality $N \sim 4000$ and with $q \sim 32$ states for each coordinate.

One type of neural networks, which can solve such a problem, is a q-state neural network. A well-known network of this type is the Potts-glass neural network [1]. The properties of this model have been examined by the methods of statistical physics [2]-[6], and its recognizing characteristics have been investigated mainly by means of computer modeling.

The Potts model is characterized by a large storage capacity $M \sim Nq(q-1)/4\ln q$. However a practical application of this model is complicated, because it requires very large RAM due to the fact that $\sim 32N^2q^2$ bits are needed for the storage of the connection matrix. In particular, the aforementioned problem of the attack identification requires more than 16 Gb. One can reduce the required RAM by the factor of 32 making it $\sim N^2q^2$ bits by a binarization of the synaptic connections of the neural network (i.e. replacing all positive matrix elements by 1, and leaving zero matrix elements as they are). Then the problem under discussion would require 500 Mb instead of 16 Gb.

V. Kůrková et al. (Eds.): ICANN 2008, Part II, LNCS 5164, pp. 72–80, 2008.

Our attempt to apply the binarization procedure to the standard q-state Potts model [1] led to a negative result: the storage capacity decreased strongly. That is why we applied the binarization procedure to a modified q-state Potts model. Similar models of neural networks (the so-called parametrical neural networks) were examined in a series of works [7]-[16]. They were designed for a realization as an optical device. For these models relatively simple analytical expressions, which describe their storage capacity and noise immunity, are obtained.

An analysis of the modified Potts model with binarized synaptic coefficients shows an unexpected result: instead of the expected worsening of the recognizing properties, the storage capacity of the neural network increases by a factor of 2 as a result of the binarization. It can be accounted for by the fact that the dispersion of the input signal on a neuron decreases faster than it's average value. Besides that, the binarized model works more than q times faster than the standard Potts model [1] with the same N and q. Such an improvement can be important for real-time systems. The results obtained correlate with those of the computer simulations. Further it will be shown that in the range of $q > 150$ and $q < N < q^2$ binarization of synaptic coefficients results in decreasing the storage capacity. However, this range is of no practical value.

The paper is organized as follows. In Section 2 we describe our modified model. In Section 3 we derive the expressions for the storage capacity. In Section 4 we analyze the obtained results and compare them with the results of computer simulations.

2 The Description of the Model

Let us start with the description of the modified Potts model. This is a fully connected network of N spin-neurons, each of which can be in one of q different discrete states ($q \geq 2$). A unit vector $\mathbf{e}_k$ in the space R^q is associated with the state with the number $k = 1, 2, ..., q$. The state of the network as a whole is described by an N-dimensional set $X = (\mathbf{x}_1, \mathbf{x}_2, ..., \mathbf{x}_N)$ of q-dimensional vectors $\mathbf{x}_i$. Each $\mathbf{x}_i$ corresponds to the state of i-th neuron in the pattern:

$$\mathbf{x}_i \in \{\mathbf{e}_k\}^q . \tag{1}$$

The difference between the modified model and the standard Potts model [1] is as follows. In the modified model one of the components of the vector $\mathbf{x}_i$ is equal to 1, and all the others are equal to 0. In the standard Potts model one component of the vector representing the state of a neuron is equal to $1 - 1/q$, and all the other components are equal to $-1/q$.

The synaptic connection between i-th and j-th neurons is given in this model by a $q \times q$-matrix $\hat{\mathbf{T}}_{ij}$, whose elements are obtained according to the Hebb rule with M stored patterns $\mathbf{X}_\mu = (\mathbf{x}_{\mu 1}, \mathbf{x}_{\mu 2}, ..., \mathbf{x}_{\mu N})$, $\mu = \overline{1, M}$:

$$\hat{\mathbf{T}}_{ij} = \sum_{\mu=1}^{M} \mathbf{x}_{\mu i} \mathbf{x}_{\mu j}^{+}, \tag{2}$$

and the diagonal elements are equal to zero: $\hat{\mathbf{T}}_{ii} = 0$. The local field acting on i-th neuron is given by the standard expression

$$\mathbf{H}_i = \sum_{j \neq i}^{N} \hat{\mathbf{T}}_{ij} \mathbf{x}_j. \tag{3}$$

The dynamics of the network are defined in a natural way: we suppose that under the action of the local field $\mathbf{H}_i$ i-th spin-neuron orients along a direction, which is most close to the direction of the field (as the neuron states are discrete, it cannot be oriented exactly along the field $\mathbf{H}_i$). In other words, a neuron orients along that unit vector, which has a maximal projection on the vector $\mathbf{H}_i$. This rule can be formalized by the following algorithm: for each t one calculates the projections of the vector $\mathbf{H}_i$ on all the unit vectors of the space R^q and chooses the largest one. Let it be, for instance, the projection on the unit vector $\mathbf{e}_r$. Then the state of i-th spin-neuron in the next moment $t + 1$ is given by the following rule:

$$\mathbf{x}_i(\mathrm{t} + 1) = \mathbf{e}_r. \tag{4}$$

This procedure should be successively applied to all the neurons (asynchronous dynamics) until the stable state of the system is reached.

Now let us describe the binarized model. Its synaptic matrix $\hat{\tau}_{ij}$ is obtained by the binarization of the initial Hebb matrix $\hat{\mathbf{T}}_{ij}$:

$$\tau_{ij}^{kl} = \mathrm{sgn} T_{ij}^{kl}, \tag{5}$$

where $k, l = \overline{1, q}$, $\quad i, j = \overline{1, N}$, and the local field takes the form:

$$\mathbf{h}_i = \sum_{j \neq i}^{N} \hat{\tau}_{ij} \mathbf{x}_j. \tag{6}$$

The dynamics of the binarized model are the same as of the original (non-binarized) model. Below we examine the properties of these two models (the binarized model and the original non-binarized model) and compare them.

3 Effectiveness of the Pattern Recognition

Let us estimate the storage capacity of the neural network for the binarized model. To do this let us find the probability for a memorized pattern to be a fixed point. Let the initial state of the network correspond to a pattern X_1. The pattern X_1 is a fixed point if for each neuron the following condition is satisfied: The projection of the local field on the unit vector, which corresponds to the state of the neuron in the stored pattern, is maximal. Let us consider 1-st neuron and suppose for simplicity that the first component of X_1 equals $\mathbf{e}_1$. Then the condition of the correct orientation of 1-st neuron is that the following $(q - 1)$ inequalities are simultaneously true:

$$\eta_k = h_1^1 - h_1^k > 0, k = \overline{2, q}. \tag{7}$$

These inequalities follow from the rule (4) and they mean that the projection h_1^1 of the local field on the unit vector $\mathbf{e}_1$ is larger than its projections on any of the other $(q-1)$ unit vectors. From (6) we obtain the projections on 1-st and on an arbitrary k-th unit vector $(k \neq 1)$:

$$h_1^1 = N, \tag{8}$$

$$h_1^k = \sum_i^N \operatorname{sgn}\left(\sum_{\mu\neq 1}^M x_{\mu 1}^k x_{\mu i}^{l_i}\right), \tag{9}$$

where l_j is the number of the non-zero coordinate of the vector $\mathbf{x}_{1j}$.

The probability of the event that all $(q-1)$ inequalities (7) are imultaneously true,

$$P = \Pr\left[\bigcap_k^{q-1} \eta_k > 0\right], \tag{10}$$

can be calculated assuming that the quantities η_k are random Gaussian variables with the same mean value $< \eta_k >$, the dispersion σ^2 and the covariation $\text{cov}(\eta_k, \eta_r)$:

$$< \eta_k >= NP_0, \tag{11}$$

$$\sigma^2 = NP_0\left[1 - NP_0 + (N-1)P_0^{\frac{q}{q+1}}\right], \tag{12}$$

$$\text{cov}(\eta_k, \eta_r) = N^2 P_0^2 \left(1 - P_0^{-\frac{1}{q^2-1}}\right). \tag{13}$$

Here P_0 is the probability that a matrix element of the matrix $\hat{\tau}_{ij}$ is equal to zero:

$$P_0 = \left(1 - \frac{1}{q^2}\right)^M. \tag{14}$$

In what follows we analyze the case $q^2 >> 1$ (the case $q \sim 1$ is not interesting, because then the recognizing properties of the modified network become comparable with the recognizing properties of the Hopfield network).

Using the parameters (11)-(13) one can write down the probability P as a standard error function for multivariate Gaussian distribution. We do not show here the general expression because it is too cumbersome. Omitting intermediate calculations, we write down the expression for the probability P (10) for the most interesting case $P \to 1$ in the limit $\gamma >> 1$:

$$P = 1 - \frac{Nq}{\sqrt{2\pi}\gamma}\, e^{-\frac{1}{2}\gamma^2}, \tag{15}$$

where $\gamma =< \eta_k > /\sigma$ is the so-called signal-noise ratio. The parameter γ is the main indicator of the recognizing reliability of the network: the larger γ, the larger the probability P of the correct recognizing and the larger the storage capacity of the network.

The storage capacity $\overline{M}$ can be determined from the condition $P \to 1$. In the asymptotic limit $N \to \infty$ this condition can be transformed to the following form:

$$\gamma^2 = 2 \ln Nq. \tag{16}$$

When $N << q$ or $N >> q$ we obtain from (16)

$$\overline{M} = \frac{q^2 N}{2(1 + N/q) \ln Nq}. \tag{17}$$

For an arbitrary ratio between N and q an expression for the storage capacity cannot be obtained, because the equation (17) is transcendental. However, numerical simulations show that (18) estimates the storage capacity reasonably well (with the accuracy up to 20 – 40%) for any ratio between N and q.

For a comparison we write down the estimate for the storage capacity of the original non-binarized network (we omit here analogous calculations):

$$\overline{M}_0 = \frac{q^2 N}{4(1 + N/q) \ln Nq}. \tag{18}$$

Before we start further analysis we note that the quantity $\overline{M}$ is 2 times larger than $\overline{M}_0$. However this ratio is not true for some values of N and q (see Fig.2.).

4 Comparative Analysis

Let us compare the storage capacities of two networks, the one is the binarized network (5)-(6), the other is the original model (2)-(3). In Fig.1 we show the dependence of the value of $\overline{M}$ on the dimensionality of the problem N for a fixed q. This dependence was obtained as a numerical solution of equation (17). As is seen from the figure, the storage capacity of the binarized network increases rapidly with the growth of N, reaches its maximum value and then decreases slowly (as $\sim 1/\ln Nq$). The maximal value $\overline{M} = M_{\max}$ is reached at $N = N_{\max}$:

$$M_{\max} = \frac{q^{5/2} \ln q}{8\sqrt{2}}, N_{\max} = \frac{(2 \ln q)^3}{q} \exp\left(\sqrt{q/2}\right). \tag{19}$$

For practical applications such a behavior means the following: if $N > N_{\max}$, then one can use the stored patterns with a reduced size $N = N_{\max}$. Then the storage capacity of the network and the operating speed of the algorithm increase, while the requirements imposed on RAM are lowered.

The dependence $\overline{M}_0 = \overline{M}_0(N)$ obtained from (19) is shown in Fig.1 for a comparison. As one can see, $\overline{M}_0$ also increases at first and then decreases as $\sim 1/\ln Nq$. The maximal value $\overline{M}_0 = M^0_{\max}$ is reached at $N = N^0_{\max}$:

$$M^0_{\max} = \frac{q^3}{8 \ln q}, N^0_{\max} = 2q \ln q. \tag{20}$$

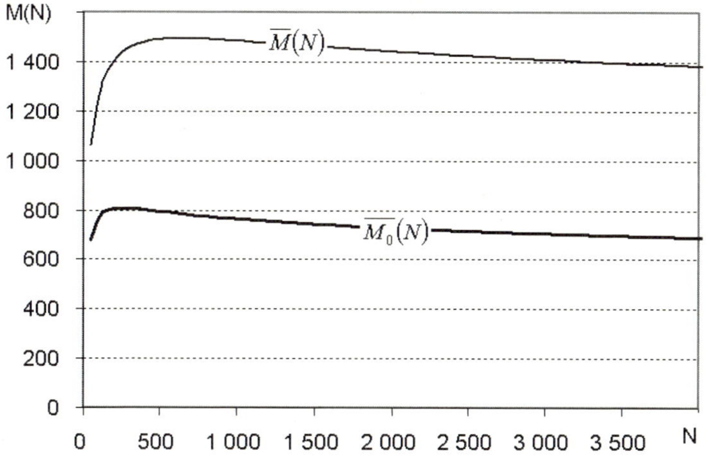

Fig. 1. The dependence of the storage capacity on the network size for binarized (upper curve) and original (lower curve) models for $q = 32$

Equations (19) and (20) determine the maximal storage capacity, which can be reached at a given value of q. The curves in Fig.1 are plotted for $q = 32$. As one can see, for this value of q the storage capacity of the binarized network is about 2 times larger than the storage capacity of the non-binaraized network. Such a ratio is true for moderate values of q.

The dependence of the ratio $\overline{M}/\overline{M}_0$ on the value of N for different values of q is shown in Fig.2. As we see, in the case $q \leq 150$ we have $\overline{M}/\overline{M}_0 > 1$ for almost all parameters. Moreover, $\overline{M}/\overline{M}_0 \sim 2$ for $N < q$ or $N >> q$. However for large enough q ($q > 150$) there is an interval of the values of N ($q < N < q^2$), in which $\overline{M}/\overline{M}_0 < 1$.

Let us compare the characteristics of those networks, which can be used for practical applications, i.e. of the networks which can be modeled with relatively small RAM (for instance, less than 2 Gb). Such networks belong to the part of Fig.2 situated above the dotted line. As one can see, in this region the binarization procedure results in the increase of the neural network storage capacity.

We carried out a series of computer experiments in order to check the obtained estimates. The results of one experiment with $N = 30$ and $q = 10$ are shown in Fig.3. The obtained dependencies show that a network with a binarized connection matrix breaks down at a noticeably larger value of the loading parameter. Furthermore, one can see that the storage capacities of the binarized model and of the standard Potts model are of the same order. However the operating speed of the binarazied model is more than q times larger than the operating speed of the Potts model (10 times in this case), and the binarazied model requires 32 times less RAM. In our experiments the parameters were varied within the ranges $4 \leq q \leq 128$ and $10 \leq N \leq 500$. The results of the experiment show that the binarization always leads to the increase of the storage capacity by a

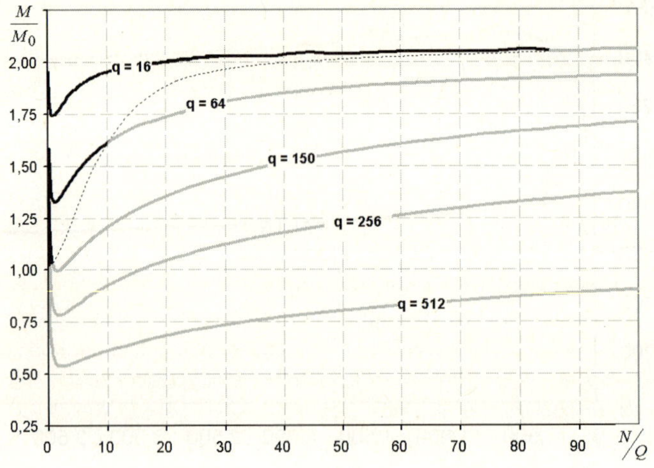

Fig. 2. The figure shows the ratio as a function of N/q for different values of the parameter q. For the points below the dotted line (the curves marked with grey color) the size of the required RAM exceeds 2 Gb.

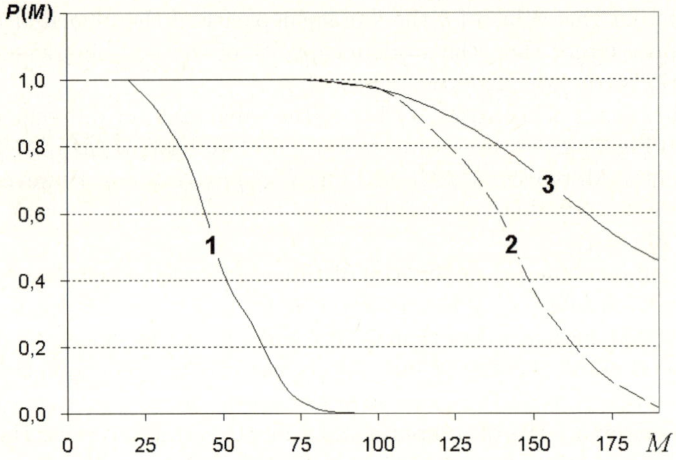

Fig. 3. The fraction of the fixed points P as a function of the number of memorized patterns M for different models. 1 - the non-binarized model; 2 - the binarized model; 3 - the standard Potts model. The experiment is done for $N = 30$ and $q = 10$.

factor of 2. Due to a restricted RAM of our computer we were unable to realize a network with the values of the parameters, at which the binarization would worsen the characteristics of the network ($q > 150$, $N > 32$).

5 Conclusions

In this work a modified Potts model with binarized synaptic coefficients is considered. It is shown that the storage capacity of the network is increased by a factor of 2 after the binarization of the matrix elements. It can be accounted for by the fact that the dispersion of the input signal on a neuron decreases faster than it's average value. For the networks with reasonable values of the external parameters ($4 \leq q \leq 128$ and $10 \leq N \leq 500$) this result is verified in many experiments. Moreover, the binarized model has the storage capacity of the same order of magnitude as the standard q-state Potts model. However, the binarized model requires 32 times less RAM and has a q times larger operating speed. Due to all this the binarized model can be very attractive for applications in identification systems, which work with very large arrays of vectors of high dimensionality in the real-time operation mode.

As the application the one-layer q-state perceptron has been constructed on the base of the described binarized model (2700 input neurons, number of their states $q = 256$; 2 output neurons, number of their states is 325). Such a peceprton can reliably identify any of 10^5 stored patterns under 75% of distortion. To store the matrix of synaptic coefficients, 56 Mb of RAM is required. The time needed for identification is about 200 ms. An analogous perceptron based on the standard Potts-model would require 1.8 Gb of RAM and would work 325 times slower.

Acknowledgments. I would like to express my special gratitude to Marina Litinskaia for help in preparation of the present paper. This work is supported by RFBR grant 06-01-00109.

References

1. Kanter, I.: Potts-glass models of neural networks. Physical Review A 37(7), 2739–2742 (1988)
2. Cook, J.: The mean-field theory of a Q-state neural network model. Journal of Physics A 22, 2000–2012 (1989)
3. Vogt, H., Zippelius, A.: Invariant recognizing in Potts glass neural networks. Journal of Physics A 25, 2209–2226 (1992)
4. Bolle, D., Dupont, P., van Mourik, J.: Stability properties of Potts neural networks with biased patterns and low loading. Journal of Physics A 24, 1065–1081 (1991)
5. Bolle, D., Dupont, P., Huyghebaert, J.: Thermodynamics properties of the q-state Potts-glass neural network. Phys. Rew. A 45, 4194–4197 (1992)
6. Wu, F.Y.: The Potts model. Review of Modern Physics 54, 235–268 (1982)
7. Kryzhanovsky, B.V., Mikaelyan, A.L.: On the Recognizing Ability of a Neural Network on Neurons with Parametric Transformation of Frequencies. Doklady Mathematics 65(2), 286–288 (2002)
8. Kryzhanovsky, B.V., Litinskii, L.B., Fonarev, A.: Parametrical neural network based on the four-wave mixing process. Nuclear Instruments and Methods in Physics Research A 502(2-3), 517–519 (2003)
9. Kryzhanovsky, B.V., Mikaelyan, A.L.: An associative memory capable of recognizing strongly correlated patterns. Doklady Mathematics 67(3), 455–459 (2003)

10. Kryzhanovsky, B.V., Litinskii, L.B., Fonarev, A.: An effective associative memory for pattern recognizing. In: 5-th Int. sympos. on Idvances in Inteligent Data Analisis 2003, Germany, pp. 179–187. Springer, Berlin (2003)
11. Mikaelyan, A.L., Kryzhanovsky, B.V., Litinskii, L.B.: Parametrical Neural Network. Optical Memory & Neural Network 12(3), 227–236 (2003)
12. Kryzhanovsky, B.V., Litinskii, L.B., Mikaelyan, A.L.: Vector-neuron models of associative memory. In: Proc. of Int. Joint Conference on Neural Networks IJCNN 2004, Budapest 2004, pp. 909–1004 (2004)
13. Kryzhanovsky, B.V., Mikaelyan, A.L., Fonarev, A.B.: Vector Neural Net Identifing Many Strongly Distorted and Correlated Patterns. In: Int. conf on Information Optics and Photonics Technology, Photonics Asia-2004, Beijing-2004. Proc. of SPIE, vol. 5642, pp. 124–133 (2004)
14. Alieva, D.I., Kryzhanovsky, B.V., Kryzhanovsky, V.M., Fonarev, A.B.: Q-valued neural network as a system of fast identification and pattern recognizing. Pattern Recognizing and Image Analysis 15(1), 30–33 (2005)
15. Kryzhanovsky, B.V., Kryzhanovsky, V.M., Mikaelian, A.L., Fonarev, A.B.: Parametrical Neural Network for Binary Patterns Identification. Optical Memory & Neural Network 14(2), 81–90 (2005)
16. Kryzhanovsky, B.V., Kryzhanovsky, V.M., Fonarev, A.B.: Decorrelating Parametrical Neural Network. In: Proc. of IJCNN Montreal, pp. 1023–1026 (2005)

Learning Similarity Measures from Pairwise Constraints with Neural Networks

Marco Maggini, Stefano Melacci, and Lorenzo Sarti

DII, Università degli Studi di Siena
Via Roma, 56 — 53100 Siena (Italy)
{maggini,mela,sarti}@dii.unisi.it

Abstract. This paper presents a novel neural network model, called Similarity Neural Network (SNN), designed to learn similarity measures for pairs of patterns exploiting binary supervision. The model guarantees to compute a non negative and symmetric measure, and shows good generalization capabilities even if a small set of supervised examples is used for training. The approximation capabilities of the proposed model are also investigated. Moreover, the experiments carried out on some benchmark datasets show that SNNs almost always outperform other similarity learning methods proposed in the literature.

1 Introduction

In many pattern recognition tasks, an appropriate definition of the distance function on the input feature space plays a crucial role. Generally, in order to compare patterns, the input space is assumed to be a metric space, and Euclidean or Mahalanobis distances are used. In some situations, this assumption is too restrictive, and a similarity measure could be learnt from examples.

In the last few decades, the perception of similarity received a growing attention from psychological researchers [1], and, more recently, the problem of learning a similarity measure has attracted also the machine learning community. In particular, in a wide set of research fields, ranging from bioinformatics to computer vision, the supervision on the relationships between two entities is expressed in the form of similarity/dissimilarity constraints. In those contexts, the main requirement is just to find a way to compare entity pairs coherently with the given supervision, whereas the estimation of a metric is not always needed. A similarity measure can be considered as a distance function where not all the metric properties necessarily hold, as discussed in [2].

In the literature many algorithms that perform distance function or similarity measure learning from pairwise constraints are proposed. Some of them are strictly related to semi–supervised clustering, and the presence of a few class labels or pairwise constraints on a subset of data is used to improve the clustering process. Some approaches exploit an iterative computation of the similarity measure, solving a convex optimization problem, based on a small set of pairs that represent the problem constraints [3,4]. Other techniques use EM–like algorithms, as in [5] where MPCK–Means is proposed, or Hidden Markov Random

V. Kůrková et al. (Eds.): ICANN 2008, Part II, LNCS 5164, pp. 81–90, 2008.

Fields (HMRF–KMeans) [6]. In [7] Relevant Component Analysis is described, a method that aims at identifying and down–scaling global unwanted variability within the data, exploiting only similarity constraints. A technique for the estimation of Gaussian mixture models considering pairwise constraints is used by a boosting algorithm (DistBoost) for similarity measure learning [8,9]. Other approaches try to estimate kernel matrices with the given supervision in order to solve classification tasks [10,11]. Both learning similarity measures and learning kernels are problems for which there exists a growing literature, and they are commonly faced quite differently.

In this paper Similarity Neural Networks (SNN), a novel neural network model for the learning of similarity measures, is presented. The SNN architecture guarantees to learn a non negative and symmetric function, showing extremely promising results compared with four state of the art techniques on some datasets from the UCI repository. Moreover, this is the first neural model that is applied to the general similarity measure learning task exploiting pairwise constraints.

The paper is organized as follows. In the next section, the network architecture and its properties are presented. In Section 3 experimental results are reported, and finally, some conclusions are drawn in Section 4.

2 Similarity Neural Network

An SNN consists in a feed–forward multilayer perceptron trained to learn a similarity measure for pairs of patterns $\boldsymbol{x}, \boldsymbol{y} \in I\!\!R^n$.

Humans are generally able to provide supervision about the similarity of object pairs in binary form (similar/dissimilar), instead of specifying continuous degrees of similarity that can not be easily defined in a coherent way. Hence, SNNs are trained using binary supervisions. In details, given a set of objects described by feature vectors $V = \{\boldsymbol{x}_1, ...\boldsymbol{x}_N\}$, $\boldsymbol{x}_i \in I\!\!R^n$, a training set $T = \{(\boldsymbol{x}_{i_k}, \boldsymbol{y}_{j_k}, t_{i_k,j_k}) | \, k = 1, ..., m; \; \boldsymbol{x}_{i_k}, \boldsymbol{y}_{j_k} \in V\}$, collects a set of m triples $(\boldsymbol{x}_{i_k}, \boldsymbol{y}_{j_k}, t_{i_k,j_k})$, being $t_{i_k,j_k} \in \{0, 1\}$ the similarity/dissimilarity label of the pair $(\boldsymbol{x}_{i_k}, \boldsymbol{y}_{j_k})$, and $[\boldsymbol{x}_{i_k}{}', \boldsymbol{y}_{j_k}{}']' \in I\!\!R^{2n}$ is the input vector of the SNN. Learning a similarity measure is a regression problem, but due to the binary supervisions the learning task is actually specified as a two–class classification problem. The training of the SNN is performed using backpropagation for optimizing the squared error criterion function, leading to a network that is able to approximate the similarity measure for patterns within the distribution described by the training set.

The SNN is an MLP with a single hidden layer containing an even number of units and an output neuron with sigmoidal activation function, that constraints the output range in the interval $[0, 1]$. If $f_{SNN}(\boldsymbol{x}, \boldsymbol{y}, \theta)$ is the function computed by a trained SNN for a pair $(\boldsymbol{x}, \boldsymbol{y})$, and a set of network parameters θ, then the following properties hold:

- $f_{SNN}(\boldsymbol{x}, \boldsymbol{y}, \theta) \geq 0$ (non negativity)
- $f_{SNN}(\boldsymbol{x}, \boldsymbol{y}, \theta) = f_{SNN}(\boldsymbol{y}, \boldsymbol{x}, \theta)$ (symmetry).

Non negativity is guaranteed by using the sigmoidal activation function in the output unit. Symmetry is obtained by exploiting a peculiar network architecture that is based on two independent networks that share the corresponding weights, as shown in Fig.1. The output of an SNN can be written in matrix form as follows:

$$f_{SNN}(\boldsymbol{x}, \boldsymbol{y}, \theta) = \sigma \left(\begin{bmatrix} \boldsymbol{W_{ho}}_1 \\ \boldsymbol{W_{ho}}_2 \end{bmatrix}' \cdot \boldsymbol{a} \left(\begin{bmatrix} \boldsymbol{W_{ih}}_{1,1} & \boldsymbol{W_{ih}}_{1,2} \\ \boldsymbol{W_{ih}}_{2,1} & \boldsymbol{W_{ih}}_{2,2} \end{bmatrix} \cdot \begin{bmatrix} \boldsymbol{x} \\ \boldsymbol{y} \end{bmatrix} + \begin{bmatrix} \boldsymbol{b_h}_1 \\ \boldsymbol{b_h}_2 \end{bmatrix} \right) + b_o \right)$$

where $[\boldsymbol{x}', \boldsymbol{y}']' \in \mathbb{R}^{2n}$ represents the network input, σ, and $\boldsymbol{a}$ are the activation functions of the output and hidden units respectively, while the parameter set θ collects $\boldsymbol{W_{ih}}_{1,1}, \boldsymbol{W_{ih}}_{1,2}, \boldsymbol{W_{ih}}_{2,1}, \boldsymbol{W_{ih}}_{2,2} \in \mathbb{R}^{q,n}$, and $\boldsymbol{W_{ho}}_1, \boldsymbol{W_{ho}}_2 \in \mathbb{R}^q$ that are input to hidden and hidden to output weight matrices, $\boldsymbol{b_h}_1, \boldsymbol{b_h}_2 \in \mathbb{R}^q$ and b_0 that represent the hidden and output biases. If we set, as shown in Fig. 1, $\boldsymbol{W_{ih}}_{1,1} = \boldsymbol{W_{ih}}_{2,2}$, $\boldsymbol{W_{ih}}_{1,2} = \boldsymbol{W_{ih}}_{2,1}$, $\boldsymbol{W_{ho}}_1 = \boldsymbol{W_{ho}}_2$, and $\boldsymbol{b_h}_1 = \boldsymbol{b_h}_2$, it is easy to verify that:

$$f_{SNN}(\boldsymbol{x}, \boldsymbol{y}, \theta) = f_{SNN}(\boldsymbol{y}, \boldsymbol{x}, \theta).$$

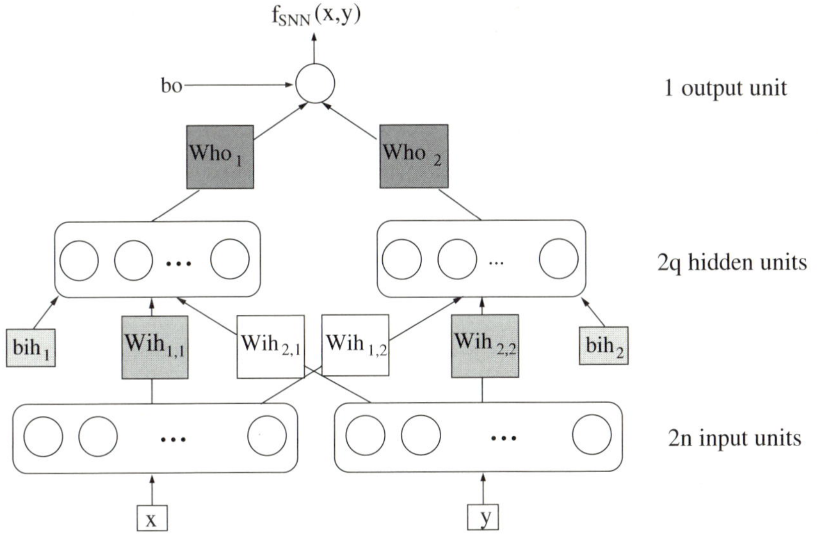

Fig. 1. SNN architecture. The left and right half of the input layer collect the features of the first and second pattern of the compared pair, respectively. Notice that the grey boxes represent all the connections between the groups of linked neurons, and that boxes of the same gray level denote shared weights.

SNNs are universal approximators for symmetric functions. For sake of simplicity, we first consider the case of SNNs with a linear output neuron, that compute a parametric function $f_{LSNN}(\boldsymbol{x}, \boldsymbol{y}, \theta)$, then we extend the result to the case of a sigmoidal output unit.

Theorem 1. *Let $CF(K)$ be the set of all the symmetric functions $l : \mathbb{R}^{2n} \to \mathbb{R}$, that are defined on a compact set $K \subset \mathbb{R}^{2n}$. Suppose that $CF(K)$ is equipped with the $L_\infty(K)$ norm. Then for any compact set K, any function $l \in CF(K)$, and any precision $\epsilon > 0$, there is an SNN, whose input-output function $f_{LSNN}(\boldsymbol{x}, \boldsymbol{y}, \theta)$ is such that*

$$||l(\boldsymbol{x}, \boldsymbol{y}) - f_{LSNN}(\boldsymbol{x}, \boldsymbol{y}, \theta)||_\infty \leq \epsilon$$

Proof. Considering that feed–forward neural networks are universal approximators [12], given the function $\frac{l(\boldsymbol{x},\boldsymbol{y})}{2}$, we can approximate it with an MLP with a linear output unit, that compute the function $g(\boldsymbol{x}, \boldsymbol{y}, \theta)$. Due to the symmetry of $l(\boldsymbol{x}, \boldsymbol{y})$, $g(\boldsymbol{x}, \boldsymbol{y}, \theta)$ can be chosen such that the following statements hold:

$$\begin{aligned} \left|\left|\frac{l(\boldsymbol{x},\boldsymbol{y})}{2} - g(\boldsymbol{x}, \boldsymbol{y}, \theta)\right|\right|_\infty &\leq \frac{\epsilon}{2} \\ \left|\left|\frac{l(\boldsymbol{y},\boldsymbol{x})}{2} - g(\boldsymbol{y}, \boldsymbol{x}, \theta)\right|\right|_\infty &\leq \frac{\epsilon}{2} \end{aligned} \tag{1}$$

Since the SNN architecture is composed by two distinct parts that share the correspondent weights, and due to the linear output assumption, we can consider the function computed by the SNN as the sum of two contributions, $f_{LSNN}(\boldsymbol{x}, \boldsymbol{y}, \theta) = g(\boldsymbol{x}, \boldsymbol{y}, \theta) + g(\boldsymbol{y}, \boldsymbol{x}, \theta)$. The resulting function is symmetric and guarantees that:

$$\begin{aligned} &||l(\boldsymbol{x}, \boldsymbol{y}) - f_{LSNN}(\boldsymbol{x}, \boldsymbol{y}, \theta)||_\infty = \\ &\left|\left|\frac{l(\boldsymbol{x},\boldsymbol{y})}{2} + \frac{l(\boldsymbol{y},\boldsymbol{x})}{2} - (g(\boldsymbol{x}, \boldsymbol{y}, \theta) + g(\boldsymbol{y}, \boldsymbol{x}, \theta))\right|\right|_\infty \leq \\ &\left|\left|\frac{l(\boldsymbol{x},\boldsymbol{y})}{2} - g(\boldsymbol{x}, \boldsymbol{y}, \theta)\right|\right|_\infty + \left|\left|\frac{l(\boldsymbol{y},\boldsymbol{x})}{2} - g(\boldsymbol{y}, \boldsymbol{x}, \theta)\right|\right|_\infty \leq \epsilon \end{aligned} \tag{2}$$

■

Theorem 2. *Let $CF(K)$ be the set of all the symmetric functions $h : \mathbb{R}^{2n} \to [0, 1]$, that are defined on a compact set $K \subset \mathbb{R}^{2n}$. Suppose that $CF(K)$ is equipped with the $L_\infty(K)$ norm. Then for any compact set K, any function $h \in CF(K)$, and any precision $\epsilon > 0$, there is an SNN with sigmoidal output unit, whose input-output function $f_{SNN}(\boldsymbol{x}, \boldsymbol{y}, \theta)$ is such that*

$$||h(\boldsymbol{x}, \boldsymbol{y}) - f_{SNN}(\boldsymbol{x}, \boldsymbol{y}, \theta)||_\infty \leq \epsilon$$

Proof. The function h returns values in $[0, 1]$, while the sigmoid function has values in $(0, 1)$. However, we can assume to approximate the function h with a function $\bar{h} : \mathbb{R}^{2n} \to (0, 1)$ to any certain degree of precision $\frac{\epsilon}{2}$:

$$||\bar{h}(\boldsymbol{x}, \boldsymbol{y}) - h(\boldsymbol{x}, \boldsymbol{y})||_\infty \leq \frac{\epsilon}{2} \tag{3}$$

Since we can write $f_{SNN}(\boldsymbol{x}, \boldsymbol{y}, \theta) = \sigma(f_{LSNN}(\boldsymbol{x}, \boldsymbol{y}, \theta))$, we can consider the condition of Theorem 1 written as:

$$||\sigma^{-1}(\bar{h}(\boldsymbol{x}, \boldsymbol{y})) - \sigma^{-1}(f_{SNN}(\boldsymbol{x}, \boldsymbol{y}, \theta))||_\infty \leq \frac{\epsilon}{2} \tag{4}$$

because $\sigma(x) = \frac{1}{1+\exp(-x)}$ is continuous and invertible in $(0, 1)$. The derivative of the sigmoid is always less than 1, and reaches its maximum value for $x = 0$, $\sigma(0)' = \frac{1}{4}$. Due to this fact, we have guarantee that $||\sigma(x + \epsilon) - \sigma(x)||_\infty < ||x + \epsilon - x||_\infty = \epsilon$, and the application of the sigmoidal function to the left side of the inequality reported in Eq. 4 leads to:

$$||\bar{h}(\boldsymbol{x}, \boldsymbol{y}) - f_{SNN}(\boldsymbol{x}, \boldsymbol{y}, \theta)||_\infty < \frac{\epsilon}{2}. \tag{5}$$

Considering Eq. 3 and Eq. 5, and exploiting triangular inequality:

$$||h(\boldsymbol{x}, \boldsymbol{y}) - f_{SNN}(\boldsymbol{x}, \boldsymbol{y}, \theta)||_\infty \leq$$

$$||h(\boldsymbol{x}, \boldsymbol{y}) - \bar{h}(\boldsymbol{x}, \boldsymbol{y})||_\infty + ||\bar{h}(\boldsymbol{x}, \boldsymbol{y}) - f_{SNN}(\boldsymbol{x}, \boldsymbol{y}, \theta)||_\infty < \tfrac{\epsilon}{2} + \tfrac{\epsilon}{2} = \epsilon \quad \blacksquare$$

Remark 1. The learned function is a similarity measure but not a metric, since it is not guaranteed to satisfy the triangular inequality and the self–similarity $f_{SNN}(\boldsymbol{x}, \boldsymbol{x}, \theta) = 0$. These properties can be approximated by learning from data, but they are not forced in any way by the structure of the network.

3 Experimental Results

In order to evaluate the performances of SNNs we selected 5 datasets from the UCI data repository [13], whose patterns are organized in K distinct classes (see Table 1).

With the aim of approaching the similarity learning task we need a binary supervision associated to pairs of patterns. Hence, we generated such supervision by labeling pairs of patterns that belong to the same class as similar, while the dissimilar examples were created choosing patterns from different classes.

Given a dataset $V = \{\boldsymbol{x}_1, ..., \boldsymbol{x}_N\}$, the set $C = \{(\boldsymbol{x}_i, \boldsymbol{x}_j, t_{i,j}) | \, \boldsymbol{x}_i, \boldsymbol{x}_j \in V;\ t_{i,j} \in \{0, 1\};\ i = 1, ..., N;\ j = i, ..., N\}$ is generated, where $t_{i,j} = 0$ if $\boldsymbol{x}_i$ and $\boldsymbol{x}_j$ have the same class label in the original dataset, $t_{i,j} = 1$ otherwise. Then C is

Table 1. Details of the datasets from the UCI repository used in the experiments

Name	*Size*	*Dimension*	*Classes*
Balance	625	4	3
Boston[a]	506	13	3
Ionosphere	351	34	2
Iris	150	4	3
Wine	178	13	3

[a] The 14th attribute of the original dataset, the median value (MV) of owner–occupied homes in thousands of dollars, was divided in three intervals leading to 3 highly unbalanced classes, in order to generate difficult conditions to approach the similarity learning task with the described randomly generated supervision: $MV \leq 13.1$ (53 patterns), $13.1 < MV \leq 25.1$ (329 patterns) and $MV > 25.1$ (124 patterns).

randomly sampled and the sets $S \subset C$ and $D \subset C$, of similar and dissimilar pairs are built. The number of examples selected from C depends on the connected components belonging to the similarity and dissimilarity graphs, $G_S = (V, S)$ and $G_D = (V, D)$[1]. In particular, the sets S and D are iteratively created adding to each set a new pair from C until a target number of connected components is reached. For instance, in the reported experiments $|S|$ and $|D|$ are chosen such that the number of connected components of G_S and G_D are roughly equal to the 90% or 70% of N, in the case of *small* or *large* side–information. This is the same approach proposed in [3], with the difference that in our experimental setup a limit to the number of supervised dissimilarity constraints is posed.

The sets S and D are enriched exploiting the transitive property of the similarity relationship[2]. The set S can be augmented by considering the transitive closure over G_S. Differently, the set D can be enriched by considering both G_S and G_D. Given two nodes $\boldsymbol{a}, \boldsymbol{b}$ involved in a dissimilarity relationship, such that $\boldsymbol{a} \in con(G_S)_i$ and $\boldsymbol{b} \in con(G_S)_j$, where $con(G_S)_i$ is the i–th connected component of G_S, then each element in $con(G_S)_i$ can be linked to any element of $con(G_S)_j$ in G_D. In other words, if there is a dissimilarity relationship between two nodes, this relationship can be extended to all the nodes belonging to the corresponding connected components of G_S. Finally, the set S can be enriched by exploiting the self–similarity of each data point.

The described augmenting technique can produce unbalanced training sets for the SNN model, due to the different sizes of S and D. Even if perfectly balanced training sets are not required, S and D must be augmented while maintaining an overall balancing of the supervision size. This result has been obtained by considering not all the newly generated pairs but just a randomly sampled subset of them.

The constraint sets S and D were realized 20 times in both the *small* and *large* cases for each dataset. The performances were evaluated using the cumulative neighbor purity index [9,14], that measures the percentage of correct neighbors up to the K-th neighbor, averaged over all the data points. The maximum number of neighbors has been chosen such that $K_{max} \approx \frac{N}{3}$. For this evaluation, the value of similarity produced by the SNN is used to order the neighbors, i.e. for any given point $\boldsymbol{x}_i \in V$ the other points $\boldsymbol{x}_j \in V \setminus \{\boldsymbol{x}_i\}$ are ordered for increasing values of $f_{SNN}(\boldsymbol{x}, \boldsymbol{y}, \theta)$.

The quality of the measure produced by the SNN model has been compared against classical distance functions (Euclidean and Mahalanobis distances), and against four state of the art techniques designed to learn a similarity measure or a distance function exploiting pairwise constraints:

- Xing's metric [3], using a full distance matrix estimated with gradient ascent and multiple iterative projections.
- Relevant Component Analysis (RCA), proposed in [7], that exploits positive (similarity) constraints only.

[1] Single nodes are considered as connected components of size one.

[2] The similarity relationship is not generally transitive, but if supervision is synthetically generated from data belonging to K distinct classes, this property holds.

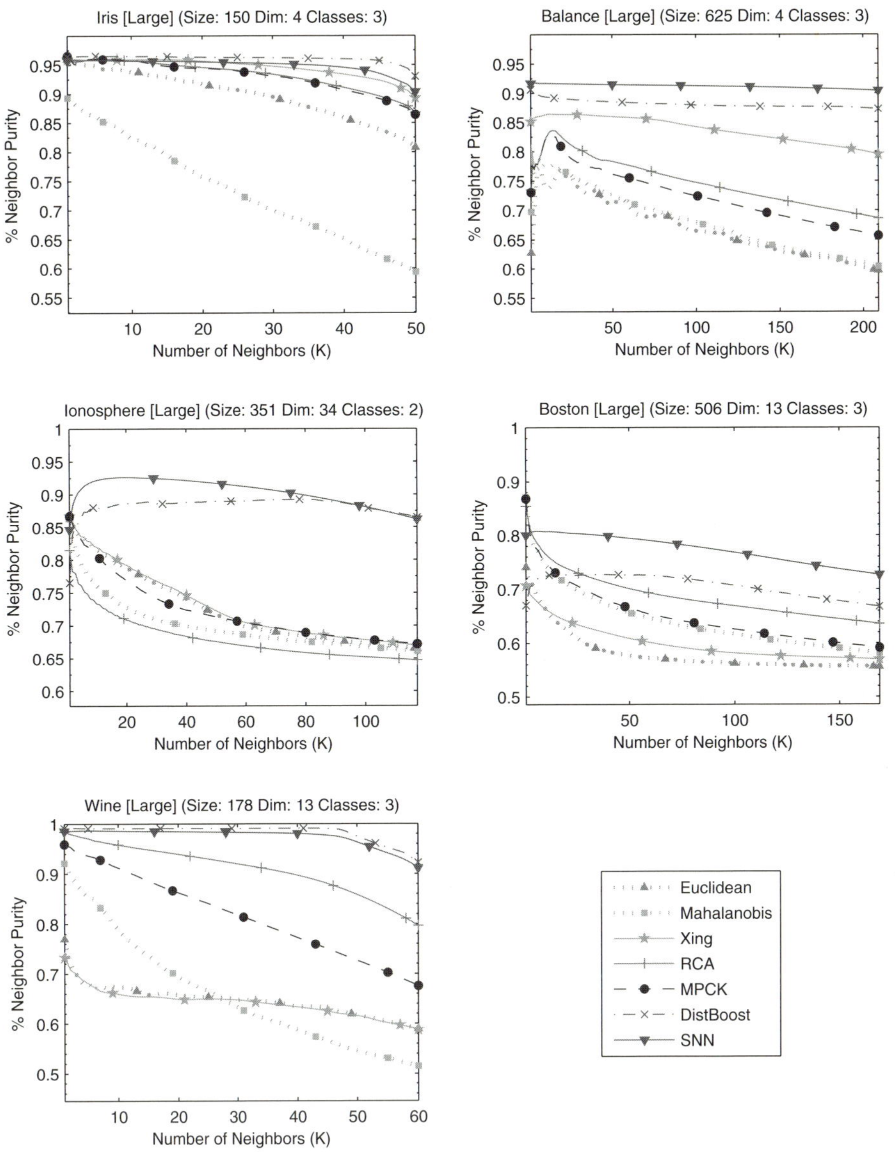

Fig. 2. Cumulative neighbor purity computed on 5 datasets from the UCI repository, in case of *large* side–information. Constraint size, expressed by the number of connected components of the similarity and dissimilarity graphs, is roughly equal to the 70% of the dataset size. Each result is averaged on 20 random realizations of the constraints.

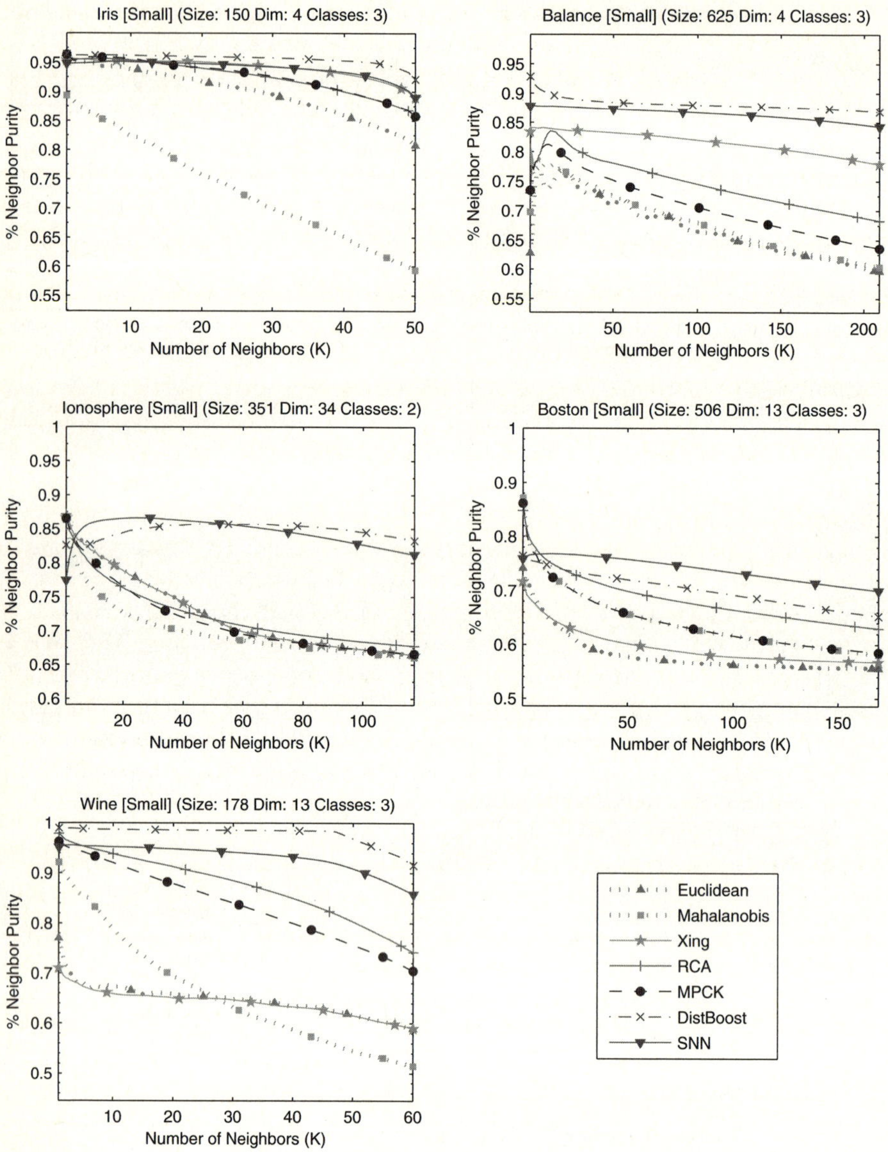

Fig. 3. Cumulative neighbor purity computed on 5 datasets from the UCI repository, in case of *small* side–information. Constraint size, expressed by the number of connected components of the similarity and dissimilarity graphs, is roughly equal to the 90% of the dataset size. Each result is averaged on 20 random realizations of the constraints.

- Metric learned by the MPCK-Means algorithm [5], in the case of a single metric parameterized by a full matrix. The metric learning process is exploited in a K-means like clustering algorithm, which uses supervised constraints both in the cluster assignation and in the algorithm initialization phases (seeding).
- Non metric distance learning with DistBoost [9,14], which combines boosting hypotheses on the product space with a weak learner based on constrained EM [8]. The number of boosting iterations was set to 50, as in [14].

All experiments were carried out in the same conditions among the different techniques, using the software provided by the respective authors on their websites.

Results are reported in Fig. 2 and Fig. 3 and are averaged on 20 random generations of the constraints. Each of the described distance learning algorithms was given exactly the same supervision (with the exception of RCA, as already stated).

As reported in Fig. 2, in the *large* case SNNs outperform all the other distance functions for 3 datasets (Balance, Ionosphere, Boston), while they show comparable results to DistBoost for the Wine and Iris benchmarks, outperforming all the remaining functions. In the *small* scenario of Fig. 3, SNNs demonstrate a better behavior with respect to the other techniques for the Ionosphere and Boston datasets, while for the remaining ones they show comparable or slightly worse results with respect to DistBoost, outperforming all the other distance functions. It can be noticed that SNNs tend to make a better use of the supervision while its size increases, whereas not all the other techniques show this capability.

4 Conclusions and Future Work

In this paper a neural network approach to similarity learning has been presented, showing promising overall results compared to classical distance functions and some state of the art distance learning techniques, even with a small number of supervisions. The proposed architecture assures to learn symmetric and non negative similarity measures, and can also be trained to incorporate other properties of the data. Thus, this is the first neural model that is applied to the similarity measure learning task from pairwise constraints. Moreover, the approximation capabilities of the proposed model are assessed, showing that SNNs are universal approximators for symmetric functions. Future work includes the application of the learned function to clustering tasks.

References

1. Tversky, A.: Features of Similarity. Psychological Review 84(4), 327–352 (1977)
2. Santini, S., Jain, R.: Similarity measures. Pattern Analysis and Machine Intelligence, IEEE Transactions on 21(9), 871–883 (1999)

3. Xing, E., Ng, A., Jordan, M., Russell, S.: Distance metric learning, with application to clustering with side-information. Advances in Neural Information Processing Systems 15, 505–512 (2003)
4. De Bie, T., Momma, M., Cristianini, N.: Efficiently learning the metric using side-information. In: Proc. of the International Conference on Algorithmic Learning Theory, pp. 175–189 (2003)
5. Bilenko, M., Basu, S., Mooney, R.: Integrating constraints and metric learning in semi-supervised clustering. In: Proceedings of the International Conference on Machine Learning, pp. 81–88 (2004)
6. Basu, S., Bilenko, M., Mooney, R.: A probabilistic framework for semi-supervised clustering. In: Proceedings of the International Conference on Knowledge Discovery and Data Mining, pp. 59–68 (2004)
7. Bar-Hillel, A., Hertz, T., Shental, N., Weinshall, D.: Learning a Mahalanobis Metric from Equivalence Constraints. The Journal of Machine Learning Research 6, 937–965 (2005)
8. Shental, N., Hertz, T., Jerusalem, I., Bar-Hillel, A., Weinshall, D.: Computing Gaussian Mixture Models with EM using Equivalence Constraints. Advances in Neural Information Processing Systems 16, 465–472 (2003)
9. Hertz, T., Bar-Hillel, A., Weinshall, D.: Boosting margin based distance functions for clustering. In: Proceedings of the International Conference on Machine Learning, pp. 393–400 (2004)
10. Hertz, T., Bar-Hillel, A., Weinshall, D.: Learning a kernel function for classification with small training samples. In: Proceedings of the International Conference on Machine learning, pp. 401–408 (2006)
11. Tsang, I., Kwok, J.: Distance metric learning with kernels. In: Proceedings of the International Conference on Artificial Neural Networks, pp. 126–129 (2003)
12. Leshno, M., Lin, V.Y., Pinkus, A., Schocken, S.: Multilayer feedforward networks with a nonpolynomial activation function can approximate any function. Neural Networks 6(6), 861–867 (1993)
13. Asuncion, A., Newman, D.: UCI Machine Learning Repository (2007)
14. Hertz, T., Bar-Hillel, A., Weinshall, D.: Learning distance functions for image retrieval. In: Proceedings of the IEEE Computer Society Conference on Computer Vision and Pattern Recognition, vol. 2, pp. 570–577 (2004)

Prediction of Binding Sites in the Mouse Genome Using Support Vector Machines

Yi Sun[1], Mark Robinson[2], Rod Adams[1], Alistair Rust[3], and Neil Davey[1]

[1] Science and technology research school, University of Hertfordshire,
United Kingdom, AL10 9AB
{comrys,r.g.adams,n.davey}@herts.ac.uk

[2] Department of Biochemistry and Molecular Biology, Michigan State University,
East Lansing MI 48824, USA
blobby@msu.edu

[3] Institute for Systems Biology,
1441 North 34th Street,
Seattle, WA 98103, USA
arust@systemsbiology.org

Abstract. Computational prediction of *cis*-regulatory binding sites is widely acknowledged as a difficult task. There are many different algorithms for searching for binding sites in current use. However, most of them produce a high rate of false positive predictions. Moreover, many algorithmic approaches are inherently constrained with respect to the range of binding sites that they can be expected to reliably predict. We propose to use SVMs to predict binding sites from multiple sources of evidence. We combine random selection under-sampling and the synthetic minority over-sampling technique to deal with the imbalanced nature of the data. In addition, we remove some of the final predicted binding sites on the basis of their biological plausibility. The results show that we can generate a new prediction that significantly improves on the performance of any one of the individual prediction algorithms.

1 Introduction

In this paper, we address the problem of predicting transcription factor (TF) binding sites (binding motifs) within sequences of regulatory DNA. Currently, experimental methods for characterising the binding sites found in regulatory sequences are both costly and time consuming. Computational predictions are therefore often used to guide experimental techniques. Computational prediction of *cis*-regulatory binding sites is widely acknowledged as a difficult task [12]. Binding sites are notoriously variable from instance to instance and in higher eukaryotes they can be located considerable distances, both upstream and downstream, from the gene being regulated.

There are many different algorithms for searching for binding sites in current use, such as those proposed in [1] and [2]. However, most of them produce a high rate of false positive predictions. The use of algorithmic predictions prone to high rates of false positives is particularly costly to experimental biologists using the predictions to guide experiments. Moreover, many algorithmic approaches are inherently constrained with respect to the range of binding sites that they can be expected to reliably predict.

V. Kůrková et al. (Eds.): ICANN 2008, Part II, LNCS 5164, pp. 91–100, 2008.

Given the differing aims of these algorithms it is reasonable to suppose that an efficient method for integrating predictions from these diverse strategies should increase the range of detectable binding sites. Furthermore, an efficient integration strategy may be able to use multiple sources of information to remove many false positive predictions, while also strengthening our confidence about many true positive predictions. In [6], five popular motif discovery algorithms are run multiple times with different parameters, then multiple results are collected and grouped by a score rank. The final predictions are obtained based on voting, smoothing and extracting methods. In [7], a software tool, *MultiFinder*, was developed. It performs automated motif searching using four different profile-based motif finders (algorithms), and results from each motif finder are ranked according to the user specified scoring function. The user can select any combination of motif prediction tools.

The nature of the problem allows domain specific heuristics to be applied to the classification problem. Instead of applying voting as discussed in [6], and merging multiple predictions according to the user specified scoring function mentioned in [7], we attempt to reduce these false positive predictions using classification techniques taken from the field of machine learning. In [10] and [11], we found that the integrated classifier, or *meta classifier*, when using a support vector machine (SVM) [9] outperformed each of the original prediction algorithms. In particular the integrated classifier has a better tradeoff between recall and precision.

In this work, we extend our work in [11] by making a major change to the way the training sets are constructed. Previously we have only used proximal annotated DNA sequences close to a gene as both positive and negative examples of binding sites. However a potential problem with this approach is that the nucleotides labelled as not being part of a binding site may be incorrectly labelled, due to unreliable biological evidence. Here we introduce a new *background* dataset which draws negative examples from sequences that are 5000-4500 base pair (bp) away from any gene. In this way we hope to ensure that our negative examples are much less likely to be regulatory.

We use a 6-ary real valued vector, each element of which is a prediction result from one of the algorithms, for a particular nucleotide position, as the input of the system. The data consists of a merger of promoters from the mouse genome (*M.musculus*), annotated with transcription factor (TF) binding sites taken from the *ABS*[1] and *ORegAnno*[2] databases. In total there are 47 promoter sequences (regulatory region containing transcriptional start site), including 142 TF binding sites. The data also includes 250 upstream, non-coding sequences from which negative examples may be taken (*background*). The background sequences were extracted using the UCSC genome website[3].

In this work, one challenging aspect is the imbalanced nature of the data and that has led us to explore some powerful techniques to address this issue. The data has two classes: either binding sites or non-binding sites, with about 97% being non-binding sites. We combine random selection under-sampling and SMOTE [3] over-sampling techniques. In addition, we remove some of the final predicted binding sites on the basis of their biological plausibility. The proposed method can be seen in Figure 1.

[1] http://genome.imim.es/datasets/meta2005/index.html

[2] http://www.oreganno.org/oregano/Index.jsp

[3] http://genome.ucsc.edu/

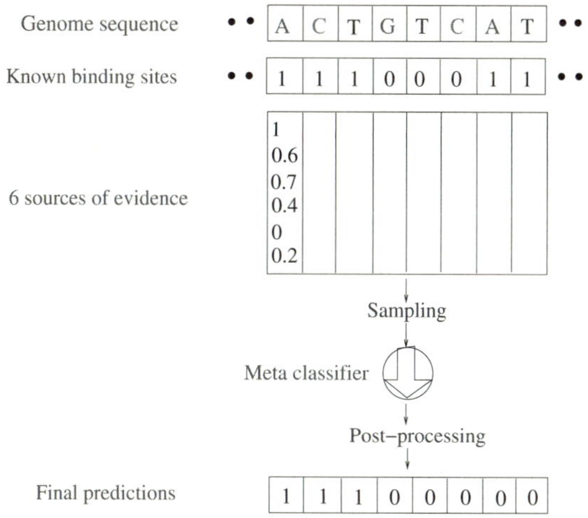

Fig. 1. The integration, sampling and classification of the data. The 6 algorithms give their own value for each sequence position and one such column is shown. The 6 results are combined into a 6-ary real valued vector. The data was under and over sampled, and then classified using a meta-classifier.

2 The Description of the Dataset

As mentioned in Section 1, the data consists of a merger of promoters annotated with transcription factor binding sites for mouse from the *ABS* and *ORegAnno* databases. This data is denoted as *ABS-And-OReg* data. The data also includes 250 upstream, non-coding sequences, denoted as *background* data.

- *ABS-And-OReg*
  There are 47 annotated promoter sequences in total. Sequences extracted from *ABS* are typically around 500 base pairs *(bp)* in length and those taken from *ORegAnno* are typically around 2000bp in length. Most of the promoters are upstream of their associated gene although a small number extend over the first exon and include intronic regions: where promoters were found to overlap they were merged. The total dataset is comprised of 60851 nucleotides, each of which may be part of a binding site.
- *Background*
  250 regions were randomly picked from across the mouse genome (forward strand genes only). The first 500bp from each sequence were selected i.e. the nucleotides that are 5000-4500 away from the gene with which they are associated. The idea is to extract non-coding sequences that are also probably non-regulatory.

  A check is also made that the selected region is indeed at least 4500 base pairs away from any neighbouring gene. It is common that a neighbouring gene can be close by and/or overlapping. The data is a sequence of 124467 nucleotides, and is believed to contain no TF binding sites.

For each nucleotide there is a real valued result from each of the six sources of evidence. Each nucleotide also has a label denoting whether it is part of a known binding site.

Six sources of evidence were generated from UCSC genome website, and were used as input in this study. Computational predictions of binding sites were generated using *MotifLocator* and *EvoSelex*. *MotifLocator* uses the PHYLOFACTS matrices from the JASPAR database[4] to scan for stringent matches in the sequences. *EvoSelex* uses motifs from [4] and the *Fuzznuc* algorithm to search for consensus sequences. A number of sources of genomic annotation evidence were extracted from the UCSC genome browser[5]: *Regulatory Potential* (RP) is used to compare frequencies of short alignment patterns between known regulatory elements and neutral DNA. The RP scores were calculated using alignments from the genomes of human, chimpanzee, macaque, rat, mouse, cow and dog. *PhastCons* is an algorithm that computes sequence conservation from multiple alignments using a phylo-HMM strategy. The algorithm was used with two levels of stringency. The *CpGIsland* algorithm finds 'CG' nucleotide sub-sequences in the regulatory region which are typically found near transcription start sites and are rare in vertebrate DNA.

3 Methods

3.1 Sampling

In our dataset, there are less than 2.93% binding positions amongst all the vectors, so this is an extremely *imbalanced* dataset [8]. Since the dataset is imbalanced, the supervised classification algorithms will be expected to over predict the majority class, namely the non-binding site category. There are various methods of dealing with imbalanced data [13]. In this work, we concentrate on the data-based method [3]: using under-sampling of the majority class (non-binding sites) and over-sampling of the minority class (binding site examples). We combine both over-sampling and under-sampling methods in our experiments.

For under-sampling, we randomly selected a subset of data points from the majority class. In [8], the author addresses an important issue that the class imbalance problem is only a problem when the minority class contains very small subclusters. This indicates that simply over sampling with replacements may not significantly improve minority class recognition. To overcome this problem, we apply a synthetic minority over-sampling technique (SMOTE) as proposed in [3]. For each member of the minority class its nearest neighbours in the same class are identified and new instances are created, placed randomly between the instance and its neighbours.

3.2 Biologically Constrained Post-processing

We propose a two-step post-processing over the SVM predictions. First, since TF binding sites are almost never found within an *exon*, an exon prediction can be considered

[4] http://jaspar.genereg.net/

[5] http://genome.ucsc.edu/

to be negative evidence for a TF binding site at a given position. Although exon predictions are still not perfect, they are much more robust than TF binding site predictions by several orders of magnitude. There is much less noise in the signals that delimit them in the sequence. Therefore, predicted components of a TF binding site will be removed if they are within a predicted exon position.

One important concern when applying classifier algorithms to the output of many binding site prediction algorithms is that the classifier decisions could result in biologically unfeasible results. The original algorithms only predict reasonable, contiguous sets of base pairs as constituting complete binding sites. However when combined in our meta-classifier each base pair is predicted independently of the neighbouring base pairs, and it is therefore possible to get lots of short predicted binding sites of length one or two base pairs. In this and a previous study, it was observed that many of the predictions made by the classifiers were highly fragmented and too small to correspond to biological binding sites. It was not clear whether these fragmented predictions were merely artifacts or whether they were accurate but overly conservative.

Since the limits of biologically observed binding site lengths are typically in the range 5-30 bp, we simply remove any predicted TF binding site with a length smaller than 5bp. It was found that removal of the fragmented predictions had a considerable positive effect on the performance measures, most notably for *Precision*.

3.3 Classifier Performance

In cases such as the imbalanced data simple error rates are inappropriate - an error rate of 2.93% can be obtained by simply predicting the majority class. Therefore it is necessary to use other metrics. Several common performance metrics, such as Recall (also known as Sensitivity), Precision, False Positive rate (FP-Rate) and F-Score, can be defined using the confusion matrix (see Table 1) computed from the test results:

Table 1. A confusion matrix

	Predicted Negatives	**Predicted Positives**
Actual Negatives	True Negatives (TN)	False Positives (FP)
Actual Positives	False Negatives (FN)	True Positives (TP)

$$\text{Recall} = \frac{TP}{(TP+FN)}, \qquad \text{Precision} = \frac{TP}{(TP+FP)},$$

$$\text{F-Score} = \frac{2 \cdot \text{Recall} \cdot \text{Precision}}{\text{Recall}+\text{Precision}}, \qquad \text{FP-Rate} = \frac{FP}{FP+TN}.$$

Furthermore the Correlation Coefficient (CC) [12], is given below:

$$CC = \frac{TP \cdot TN - FN \cdot FP}{\sqrt{(TP+FN)(TN+FP)(TP+FP)(TN+FN)}}.$$

Note that for all the measures except FP-Rate a high value is desirable. Precision is the proportion of the positively categorised samples that are actually part of a binding site.

Increasing the Precision of the prediction is one of the main goals of our meta-classifier. However increasing Precision is normally accompanied by a decrease in the Recall, so the F-Score, which takes into account both Recall and Precision, is a useful measure of overall performance. The Correlation Coefficient (at nucleotide level) measures the correlation of the prediction with the target. The FP-Rate is the proportion of all the negative samples that are incorrectly predicted. The original algorithms generally have a high FP-Rate and reducing this is another major goal of our classifier.

4 Experiments: Binding Sites Prediction

4.1 Simulation Setup

First the *ABS-And-OReg* data was divided into a training set that consisted of $2/3$ of the data, the remaining $1/3$ including 20 promoter sequences was used as the test set. We consider the following cases: 1) all non-binding site examples are selected from the *ABS-And-OReg* data; 2) all non-binding site examples are selected from the *background* data; 3) we repeat case 2) using only 4 features, that is without the two prediction algorithms *MotifLocator* and *EvoSelex* as inputs. In the last two cases, the training sets are actually a combination of *ABS-And-OReg* and *background* data, since all training examples of components of TF binding sites are from *ABS-And-OReg* and all non-binding site examples are from *background*.

Amongst the data, there are repeated vectors, some with the same label (repeated items), and some with contradictory labels (inconsistent items). These items are unhelpful in the training set and were therefore removed. The training datasets are then consistent. However, in the case of the test set, the full set of data is considered.

In the *ABS-And-OReg* data, there are fewer than 2.93% binding positions amongst all the vectors, so this is imbalanced data. To cope with this problem we used sampling. For under sampling, a subset of data points from the majority class is randomly selected. In this work, we apply SMOTE for over sampling, where we take 9 nearest neighbours, and increase the number of items in the minority class by a factor of 7. The final ratio of majority to minority class is set to 1 in all the following experiments. Note that we normalise the consistent training set before sampling so that each feature has zero mean and unit standard deviation.

After sampling, there are 3 different training sets based on each case mentioned above.

Case 1: original data from *ABS-And-OReg* denoted *orig*.

Case 2: postive examples from *ABS-And-OReg* and negative examples from *background* denoted *orig+bg*.

Case 3: As case 2 but using only four features, denoted by *orig+bg_4f*.

Table 2 gives the size of these datasets.

In the following experiments, we apply an SVM for classification. The SVM software is publicly available at http://www.csie.ntu.edu.tw/~cjlin/libsvm/. The *radial basis kernel* is employed. Therefore the SVM has two free parameters: the *cost* C and γ related to the radial basis kernel function. The range for C is set to $[20,\ 250,\ 500,\ 1000,\ 2000,\ 5000]$ and for γ is $[0.001,\ 0.01,\ 0.1,\ 1,\ 10]$. In all the following experiments,

Table 2. Description of datasets used in this work (bp denotes base pair)

Type	Dataset	Negative (bp)	Positive (bp)	Size (bp)
Original	*ABS-And-OReg*	59070	1782	60851
Original	*Background*	124467	0	124467
Training	*orig*	5446	5446	10892
	orig+bg	5757	5757	11514
	orig+bg_4f	5264	5264	10528
Test	test (from *ABS-And-OReg*)	18124	784	18908

the best values of C and γ are $C = 5000$ and $\gamma = 10$, selected by standard 5-fold cross validation.

4.2 Experimental Results

Before presenting the main results we should point out that predicting binding sites accurately is extremely difficult. The best individual original algorithm (*EvoSelex*) produces over 11 times as many false positives as true positives on the test set. This makes the predictions almost useless to a biologist as most of the suggested binding sites will need expensive experimental validation and most will not be useful. Therefore the key aim of our combined classifier is to reduce the number of false positives while increasing the number of true positives given by the original algorithms.

Table 3 shows experimental results without post-processing. For comparison, we also give results of the two original prediction algorithms: *MotifLocator* and *EvoSelex*, over the test set.

Table 3. Classification results without post-processing (in percentage %)

	Recall	*Precision*	*F-Score*	*FP-Rate*	*CC*
MotifLocator	42.5	7.1	12.1	24.2	8.4
EvoSelex	34.8	8.0	13.0	17.2	9.1
orig	43.1	12.5	19.4	13.0	17.2
orig+bg	66.1	13.3	22.1	18.7	23.4
orig+bg_4f	60.5	16.3	25.7	13.4	26.0

The first notable feature of these results is that the meta classifiers have produced stronger Recalls and Precisions than those of the two original algorithms. Therefore, the F-Score, which can be viewed as an average of the Recall and Precision, has also been increased. The nucleotide level correlation coefficient has been significantly improved. As for the FP-Rate, the meta classifiers trained on *orig+bg_4f* and *orig*, have reduced the FP-Rate by 22.1% and 24.4%, respectively, compared with *EvoSelex*, while the meta classifier trained on *orig+bg* has increased the FP-Rate by 8.7%.

The second notable feature of these results is that the meta classifier trained on *orig+bg_4f*, which used only 4 features, produced a better performance than the one that used all 6 features when looking at the F-Score and CC values, which assess the overall performance of a classifier.

One more notable feature of these results is that one can obtain a better overall performance when using non-binding site examples from the *background* set rather than from the *ABS-and-OReg* dataset.

Finally we investigate how the results can be further improved by removing those predictions of base-pairs being part of a binding site that are not biologically plausible. As described earlier we find that removing predictions that are either within exons or not part of a contiguous predicted binding site of at least five nucleotides gives a better result. So here we take the predictions of our experimental results and remove all those that do not meet the criteria. The results can be seen in Table 4.

Table 4. Classification results with and without post-processing (in percentage %)

	Recall	*Precision*	*F-Score*	*FP-Rate*	*CC*
EvoSelex	34.8	8.0	13.0	17.2	9.1
orig+bg_4f	60.5	16.3	25.7	13.4	26.0
orig+post_processing	40.6	13.7	20.4	11.1	17.9
orig+bg+post_processing	61.0	14.8	23.8	15.2	24.2
orig+bg_4f+post_processing	58.0	17.5	26.9	11.8	26.8

It shows that all FP-Rates are reduced when compared with the best original algorithm *EvoSelex*. In addition, comparing *orig+bg_4f+post_processing* with *orig+bg_4f*, one can see that the FP-Rate has been further reduced to 11.8%. Looking at two overall performance values, F-Score and CC, it shows that the accuracy of predictions is further improved after post-processing. Interestingly, *orig+bg+post_processing* has a larger number of true positives (Recall) than *orig+bg_4f+post_processing*. However, *orig+bg_4f+post_processing* has a lower FP-Rate and better overall performance on F-Score and CC. Specifically, *orig+bg_4f+post_processing* has increased the Recall by 66.7%, the Precision by 118.8%, the F-Score by 106.9% and CC by 194.5%, while reduced the FP-Rate by 31.4% when compared with the original prediction algorithm *EvoSelex*.

To further analyse our method, we investigate in more detail the predictions on each test promoter. Figure 2 shows the nucleotide level correlation coefficient within each promoter between the known nucleotide positions and the predicted nucleotide positions for each prediction algorithm.

It can be seen that there are more higher correlation (bright patterns) between the known nucleotide positions and the predicted nucleotide positions based on each promoter given by the 3 meta classifiers. It indicates that the two original prediction algorithms can only successfully find few parts of binding sites, while the meta classifiers can detect more parts of binding sites by integrating several diverse sources. In addition, although both meta classifiers *orig* and *orig+bg* include *MotifLocator* and *EvoSelex* as part of the input, these two prediction algorithms do not contribute significantly in the final decision to the meta classifiers. For example, there is a relatively high CC value in both *MotifLocator* and *EvoSelex* predictions within test promoter 6, but all 3 SVM

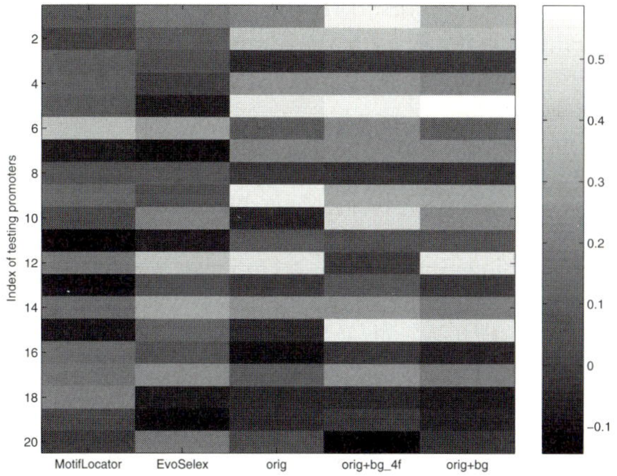

Fig. 2. Correlation coefficients between predicted positions and the known positions in each test promoter. Note that predictions from 3 meta classifiers are post-processed. High correlations are associated with brighter cells. Poor correlations are associated with darker cells.

meta classifiers produce a lower CC value. One more example is test promoter 5. Both *MotifLocator* and *EvoSelex* predictions have low correlation with the known nucleotide positions, but the 3 meta classifiers give a very high CC value. It suggests that those 4 suggestive evidences rather than the two original prediction algorithms are much more important for the classification.

5 Discussion

The identification of regions in a sequence of DNA that are regulatory binding sites is a very difficult problem. Here we have confirmed our earlier results showing that a meta classifier using multiple sources of evidence can do better than any of the original algorithms individually. In particular it was possible to reduce the number of false positive predictions.

Importantly we have also shown that using negative data that is very probably correctly labelled leads to a better prediction results. This is perhaps unsurprising, but it does suggest that some of the original data in the promoter sequences may be incorrectly labelled. This suggests that more binding sites exist on the promoter sequences than have been found by the expensive experimental techniques currently needed to produce such predictions.

Finally results that the meta classifier trained on only 4 features can produce a better performance than the one used all 6 features demonstrate the importance of feature selection. One needs to choose sources of complementary evidences which are in fact the most useful to consider. In the future, we intend to cope with this by applying the SVM classification based on *Recursive Feature Elimination* [5].

Much further work is needed to extend our current methods. The technique needs to be evaluated on other species and the biological significance of the predictions needs close examination. However it seems likely that the use of background data, as demonstrated here, will facilitate generally improved predictions.

References

1. Bailey, T.L., Elkan, C.: Fitting a Mixture Model by Expectation Maximization to Discover Motifs in Biopolymers. In: Proceedings of the Second International Conference on Intelligent Systems for Molecular Biology, pp. 28–36. AAAI Press, Menlo Park (1994)
2. Blanchette, M., Tompa, M.: FootPrinter: A Program Designed for Phylogenetic Footprinting. Nucleic Acids Research 31(13), 3840–3842 (2003)
3. Chawla, N.V., Bowyer, K.W., Hall, L.O., Kegelmeyer, W.P.: Smote: Synthetic Minority Oversampling Technique. Journal of Artificial Intelligence Research 16, 321–357 (2002)
4. Ettwiller, L., Paten, B., Souren, M., Loosli, F., Wittbrodt, J., Birney, E.: The Discovery, Positioning and Verification of a Set of Transcription-associated Motifs in Vertebrate. Genome Biol. 6(12) (2005)
5. Guyon, I., Weston, J., Barnhill, S., Vapnik, V.: Gene Selection for Cancer Classification using Support Vector Machines. Machine Learning 46, 389–422 (2002)
6. Hu, J.J., Yang, Y.F.D., Kihara, D.: EMD: an Ensemble Algorithm for Discovering Regulatory Motifs in DNA Sequsences. BMC Bioinformatics (2006)
7. Huber, B.R., Bulyk, M.L.: Meta-analysis Discovery of Tissue-specific DNA Sequence Motifs from Mammalian Gene Expressin Data. BMC Bioinformatics (2006)
8. Japkowicz, N.: Class Imbalances: Are We Focusing on the Right Issure? In: Workshop on learning from imbalanced datasets, II, ICML (2003)
9. Scholköpf, B., Smola, A.J.: Learning with Kernels: Support Vector Machines, Regularization, Optimization, and Beyond. MIT Press, Cambridge (2002)
10. Sun, Y., Robinson, M., Adams, R., Kaye, P., Rust, A.G., Davey, N.: Using Real-valued Meta Classifiers to Integrate Binding Site Predictions. In: Proceedings of International Joint Conference on Neural Network (2005)
11. Sun, Y., Robinson, M., Adams, R., Davey, N., Rust, A.: Predicting Binding Sites in the Mouse Genome. In: Proceedings The Sixth International Conference on Machine Learning and Applications (ICMLA 2007) (2007)
12. Tompa, M., et al.: Assessing Computational Tools for the Discovery of Transcription Factor Binding Sites. Nature Biotechnology 23(1) (2005)
13. Wu, G., Chang, E.: Class-boundary Alignment for Imbalanced Dataset Learning. In: Workshop on learning from imbalanced datasets, II, ICML (2003)

Mimicking Go Experts with Convolutional Neural Networks

Ilya Sutskever and Vinod Nair

Department of Computer Science
University of Toronto, Toronto, Ontario, Canada
{ilya,vnair}@cs.utoronto.ca

Abstract. Building a strong computer Go player is a longstanding open problem. In this paper we consider the related problem of predicting the moves made by Go experts in professional games. The ability to predict experts' moves is useful, because it can, in principle, be used to narrow the search done by a computer Go player. We applied an ensemble of convolutional neural networks to this problem. Our main result is that the ensemble learns to predict 36.9% of the moves made in test expert Go games, improving upon the state of the art, and that the best single convolutional neural network of the ensemble achieves 34% accuracy. This network has less than 10^4 parameters.

Keywords: Go, Move prediction, Convolutional Neural Networks.

1 Introduction

Go is an ancient and popular two-player board game in which players take turns to place pieces on the board, aiming to capture as many of the opponent's pieces and as much board territory as possible. The pieces are unmovable, but can be captured and removed from the board if surrounded by the opponent's pieces. Go is played on a 19 × 19 board; its rules are described by van der Werf [1, ch. 2]

Existing Go-playing computer programs are still not competitive with Go professionals on 19×19 boards (e.g., [2], [1], [3]). In contrast, some Chess-playing programs can play on par with world champions, even though Chess is not obviously easier than Go, except with respect to board size. Some checkers-playing programs are even more extreme and can play the *optimal* checkers strategy [4].

The techniques used for successful Chess play do not work for Go for several reasons. First, the nature of the rules of Chess makes it easy to estimate the player with the advantage, simply by counting the pieces. This and similar heuristics allow minimax search to compute a good move reasonably rapidly. In contrast, there is no easy way to determine the player with the advantage in Go. In particular, counting captured pieces (which works for Chess) does not work for Go because the value of a piece depends more heavily on its surroundings. The inability to rapidly evaluate Go positions prevents minimax searches from finding good moves. Second, typical Chess positions have 40 legal moves, while

V. Kůrková et al. (Eds.): ICANN 2008, Part II, LNCS 5164, pp. 101–110, 2008.

typical Go positions have 200 legal moves, so the search space is substantially larger. This partly explains why there are still no strong computer Go players.

The difficulty of Go suggests that it may be useful to start by solving a different problem whose solution would help create a strong Go player. In this paper we consider such a problem, which is to predict the moves made in Go experts' games. Predicting moves of Go experts is related to creating a good Go player, because an extremely accurate move predictor (e.g., a Go expert) ought to play Go well. However, constructing a fairly accurate move predictor is potentially easier than playing Go well, because an accurate move predictor can occasionally make devastating mistakes which would make it a poor Go player. For an illustration of the importance of the problem, consider an observer of a professional Go tournament. If the observer is able to predict, on average, every third move the players make, then it is plausible that the observer is a Go expert. Our convolutional neural networks can do precisely this. In addition, the move predictor can be used for educational purposes, since by outputting a probability distribution over all possible moves, it allows the student of expert Go games to see alternative good moves that could have been made (but obviously, were not). Finally, the move predictor can be used to narrow the search done by a full Go player.

We report the prediction accuracy of several convolutional neural networks [5], [6] on the problem of predicting Go experts' moves and also the accuracy when the networks are combined. Our main result shows that an ensemble of convolutional neural networks is able to predict 36.9% of the moves in the test Go games, while the most accurate previous method by Stern et al. [7] predicts 34% of the moves correctly. The best convolutional neural network of our ensemble has less than 10^4 parameters and achieves over 34% accuracy on the test set, while a very small convolutional neural network, with less than 2,000 parameters, achieves 30% prediction accuracy. Furthermore, the convolutional networks learn rapidly and reach 30% accuracy after learning on a small fraction of the training set.

We now outline the difference between Stern et al.'s approach [7] and ours. While our move predictor uses an ensemble of convolutional neural networks the best of which have a small number of parameters, Stern et al.'s predictor uses a 10^7-dimensional feature vector to represent the board, where each feature represents an exact pattern match (up to a translation and a symmetry) and is computed very efficiently. It copes with such high-dimensional feature vectors by using Bayesian methods, and is trained on 180,000 expert games. The nature of its exact pattern matching may make it difficult to generalize to non-expert games.

In addition to using fewer parameters, we used a smaller training set consisting of approximately 45,000 expert games [8]. Since the convolutional neural networks, especially the small ones, have relatively few parameters, and do not perform exact pattern matching, they have the potential to generalize to non-expert games.

The method of Stern et al. has an advantage over ours: it does not use the previously made moves as additional inputs. This is significant, because our results show that the moves made in previous timesteps greatly improve the performance of our convolutional neural networks (see table 1 for the extent to which it helps).

2 Convolutional Neural Networks

Convolutional neural networks [5] form a subclass of feedforward neural networks that have special weight constraints. They enable the network to learn much more efficiently from fewer examples, provided that the learning problem exhibits two-dimensional translation invariance.

Convolutional neural networks have been successfully applied to various problems, and have obtained the best classification accuracy on the MNIST digit dataset [9], [10], the NORB image dataset [11], and on the problem of handwritten zipcode understanding [12].

Convolutional neural networks are well suited for problems with a natural translation invariance, such as object recognition ([9], [11], [12]). Go has some translation invariance, because if all the pieces on a hypothetical Go board are shifted to the left, then the best move will also shift (with the exception of pieces that are on the boundary of the board). Consequently, many applications of neural networks to Go have used convolutional neural networks ([6], [13], [14], among others). A convolutional neural network is depicted in figure 1. Its construction is motivated by the observation that images (and Go boards) are expected to be processed in the same way in every small image patch; therefore, the weights of the convolutional neural network have a replicated structure which applies the same weights to every subrectangle of the image (the size of the subrectangles is always the same; fig. 1), producing the total inputs to the next layer. The weights of a convolutional neural network are also called *the convolutional kernel*, K, and its size, n, is the size of the subrectangles it considers. In particular, a fully connected feedforward neural network has many more parameters than a convolutional network of the same size. The relatively small number of weights makes parameter estimation require many fewer examples, so a good solution (if exists) can be found rapidly.

As in feedforward networks, applying the sigmoid function $(1 + e^{-t})^{-1}$ to the total inputs produces the values of the units in the hidden layer, which form several 2-dimensional laid out "images" that are processed in the same convolutional manner.

More concretely, a layer of a convolutional neural network is governed by the following equation

$$y_{x,y} = \left(1 + \exp\left(-\sum_{u=-(n-1)/2}^{(n-1)/2} \sum_{v=-(n-1)/2}^{(n-1)/2} x_{x+u,y+v} K_{u,v}\right)\right)^{-1} \quad (1)$$

where y is the activity of a hidden unit at position (x, y), and x are the input units.

A convolutional neural network can have several convolutional kernels; for instance, the network of figure 1 has three convolutional kernels in the first hidden layer. The convolutional kernel works in such a way that the sizes of the maps in the input, hidden, and output layers are the same. To accomplish this, some rectangles need to be partially outside of the board; in this case, the weights of the kernels corresponding to the outside-the-board region are unused.

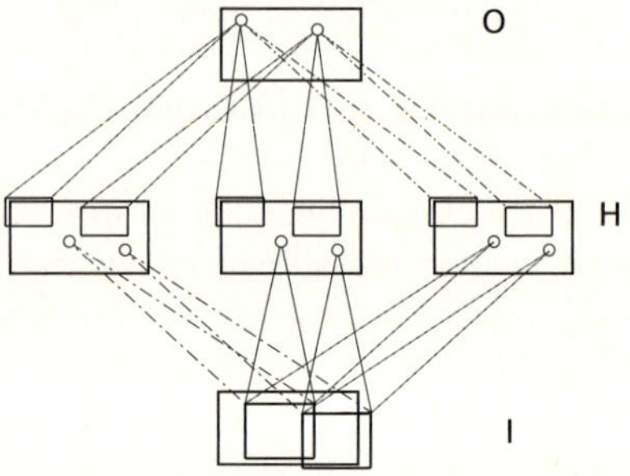

Fig. 1. A convolutional neural network with one hidden layer, and three convolutional kernels; the applications of the kernels to two rectangular regions of their inputs are shown. I, H, and O are the input, hidden, and output layers, respectively.

3 Experiments

In this section we describe the details of the convolutional neural networks that we used for our experiments and report their performance.

For our experiments, we used the Gogod collection of games [8] which contains about 45,000 expert games. D. Stern [7] generously provided us with a quarter of their test set for accurate comparisons. We processed the games in a simple manner so that it is the black's player move on each board, which involved reversing the colors of every second board (in a way that depends on first player's color).

The structure of the convolutional neural network used are as follows: The size of the convolutional kernels of the first layers are 9×9 or 7×7, and the size of the hidden-to-output convolutional kernels are 5×5. The hidden layer consists of 15 convolutional kernels. As usual, the output layer is the softmax of its input from the hidden layer, and the objective function is the log probability of the model given the labels.

Another variable in the experiments is the encoding of the input. We tried a raw input encoding, where each intersection on a board takes three possible values: empty, white and black, which is given to the convolutional layer in the form of two 19×19 bitmaps corresponding to the black and the white pieces.

We also tried an encoding that represents the number of liberties of each group (the liberty representation). A group is a connected set of pieces of the same color, and the number of liberties of a group is equal to the number of pieces the opponent needs to place in order to capture this group, without intervention.

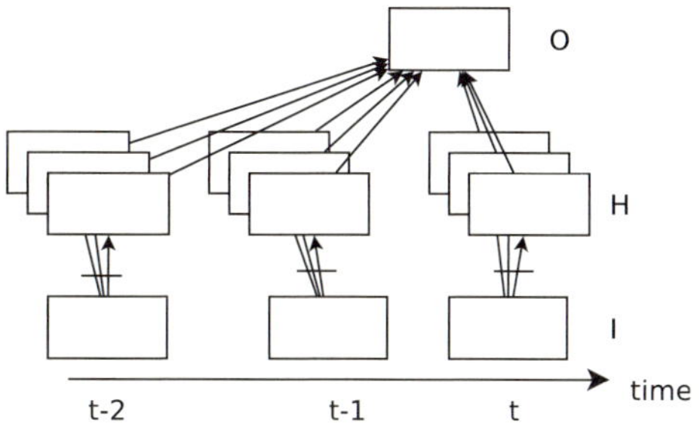

Fig. 2. A convolutional neural network that uses the moves made in previous timesteps. It receives, as inputs, several of the previous board positions. The marked convolutional maps are the same; thus every timestep is proceed in the same way, at first. The output layer consists of 361 softmax units, whose activities are the network's belief in the correctness of the given move.

This feature was used earlier ([7], [13], [14]) and was shown to be useful. The number of liberties of a non-empty intersection is defined to be the number of liberties of the intersection's group, which can take the values $1, 2, \geq 3$ (it cannot be 0 if the intersection is nonempty). Combining the number of liberties with the color of the piece, each nonempty board intersection can take values in the set $Vels = \{1, 2, \geq 3\} \times \{black, white\}$ of size 6. Thus, the input representation is given in the form of 6 bitmaps of size 19×19 (cf. figure 3). An empty intersection causes every bitmap be 0 in this intersection; if the intersection is nonempty, then the bitmap corresponding to the feature in $Vels$ is set to 1 in the corresponding input intersection.

The final variable of the experiment is the number of previous boards used as inputs: instead of using only the current board as an input, it is also possible to use the board configurations of several previous timesteps. We used 0 and 4 previous timesteps. See figure 2.

We also tried using a much smaller a convolutional neural networks that have only 3 convolutional kernels of size 7×7, which achieved 30% accuracy, as well as one that has uses only one previous timestep.

The learning details are as follows: A learning rate of size 0.1 is used for the first 3000 weight updates, which is then reduced to 0.01. The momentum of size 0.9 is used. Learning proceeds for 10^5 weight updates, where each weight update corresponds to processing a single game in the Gogod dataset; thus, learning makes less than three passes over the training set. Finally, the gradient is always divided by 200 (before being multiplied by the learning rate) since the mean length of a Go game in our dataset is 206. No weight decay was used.

3.1 Averaging the Predictions of Different Neural Networks

In this subsection we describe the results that are obtained when all the networks are combined to make a large predictor. We train a neural network that takes

Table 1. Results of the small convolutional neural networks. Each simulation is described by there parameters. The first, $k \times k$, is the size of the input-to-hidden convolutional kernel; t is the number of previous timesteps used; l is written if the liberty representation is used, and n if the raw representation is used. Every network in this table has less than 10^4 parameters and 15 convolutional kernels, except for the last row, which has only 3 convolutional kernels and less than 2,000 parameters. Notice that the best network that does not make use of the previous timesteps achieves 22% accuracy, showing that the previous timesteps are extremely helpful for convolutional neural networks.

Network Structure	Accuracy	Mean log loss
$7 \times 7, t = 0, l$	22.0%	4.9
$7 \times 7, t = 0, n$	17.5%	5.2
$7 \times 7, t = 4, l$	**34.1%**	4.0
$7 \times 7, t = 4, n$	31.7%	4.2
$9 \times 9, t = 0, l$	21.8%	4.9
$9 \times 9, t = 0, n$	18.2%	5.2
$9 \times 9, t = 4, l$	**34.6%**	4.0
$9 \times 9, t = 4, n$	32.3%	4.1
$7 \times 7, t = 4, l, h = 3$	30.0%	4.4

the predictions of the individual nets listed in table 1 as inputs, and computes a single move prediction as output.

In addition to the networks in table 1, we also include a few simple move predictors in the ensemble that improve the accuracy of the ensemble, despite having low individual accuracy. The simple move predictor is a neural network that estimates the probability of an expert making a move at a particular location, given a local "patch" of the board centred at that location as input. For example, the input to the network would be a 9×9 patch of the board, and the output would be the probability of an expert making a move at the center of the patch. Such a model can be used for full-board move prediction by independently applying it to all possible candidate move locations on a board, and then selecting the one with the highest probability as the predicted expert move location. Such a purely local predictor will have limited accuracy because many expert moves in Go are decided based on knowledge about the entire board. Nevertheless, it will be good at predicting moves that are fully determined by local configurations of stones. We train networks on patches of size 9×9 (100 hidden units), 13×13 (100 hidden units), and 17×17 (200 hidden units), with test set move prediction accuracies of 18.4%, 18.9%, and 19.8%, respectively. Although these networks perform poorly, they still improve the accuracy of the ensemble.

We selected a subset of the training set of size 1000 games to learn the weights of the ensemble network. We tried two averaging methods. The first method computes the weighted arithmetic means the predictions of all the networks, with the objective to maximize the log probability. This yields 35.5% accuracy. The second method computes the weighted geometric means of the predictions

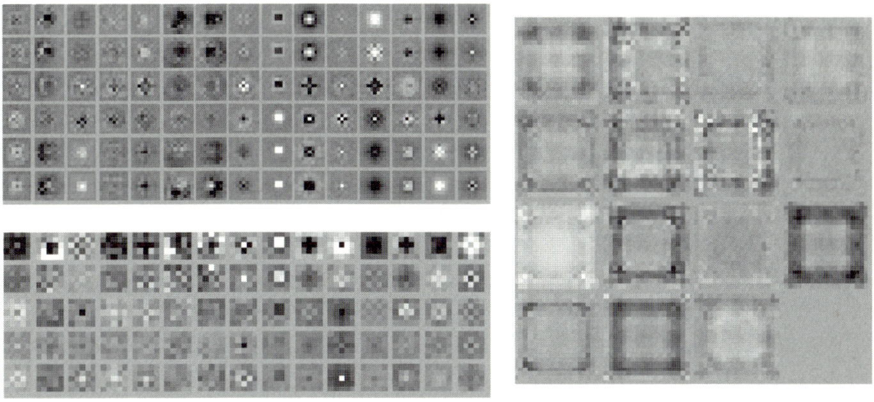

Fig. 3. The weights of a neural network achieving 34.6% accuracy that uses the liberties as an additional input. Recall that the input vector is represented by 6 bitmaps. Each column of the top-left image shows the convolutional kernels corresponding to each such bitmap, for each of the 15 convolutional maps. The image on the right shows the biases of the hidden units; it is displayed as 15 19×19 images. The image at the bottom shows the convolutional kernels mapping the 15 from the current and the 4 previous timesteps. The images are clipped at -1.5, 1.5, to enhance their contrast.

of all the networks and renormalizes the predictions, with the same objective. This approach, although yielding a lower average log probability, obtains 36.9% accuracy.

4 Related Work

A popular research direction for Go is to apply the idea that produced the Backgammon program [15]. The Backgammon playing program uses Reinforcement Learning [16] with self-play to learn a neural network that computes the Q-values of the different moves. This approach was attempted in [6] (but using $TD(0)$ [16]). It is not a straightforward application of $TD(0)$-learning to Go, because the neural network makes 361 predictions, one per intersection, on the identity of its owner in the end of the game. Doing this increases the amount of learning signal obtained from a single game. More elaborate variants of Q-learning with neural networks are also used [14], where the neural network is highly specialized and uses Q-learning to learn to predict many aspects of the final position (e.g., determine whether a given position will be an eye, or whether two positions will belong to the same group in the end of the game). This is motivated by the belief that a good Go player should be able to predict all these aspects of the final position, so the resulting hidden units will be more helpful for accurate board evaluation.

The most successful recent approaches at computer Go used Monte Carlo simulations to estimate the position's value [17]. While this approach yields

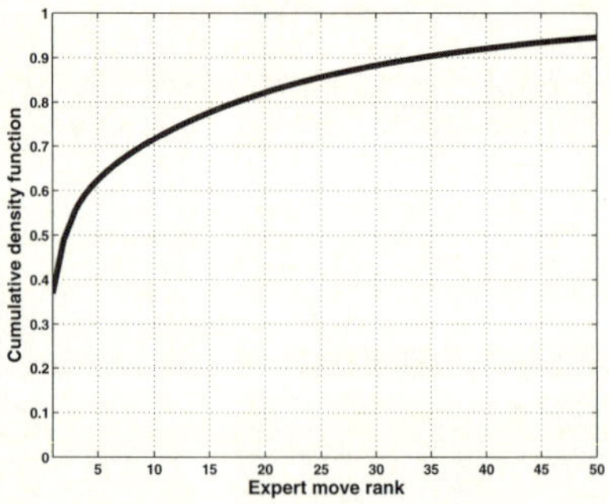

Fig. 4. The curve shows the fraction of the boards (vertical axis) whose correct move is among the top k predictions of the ensemble, where k ranges the horizontal axis

the strongest existing computer Go player [18], we do not directly compare our move predictor to it, since a strong move predictor is not necessarily a strong Go player, and vice versa.

There has also been some work on move prediction for computer Go [1], [13] using a small neural network resembling ours. It used some features, such as the number of liberties, as well as a feature that determines the distance to the previously made move, and were trained with a specially-chosen loss function. Its accuracy was 25%. A different specialized convolutional neural network trained with moderately strong play on 9×9 boards was able to generalize to 19×19 boards, obtaining 10% accuracy in expert's move prediction [19].

There has been more work [20] on move prediction that used high-dimensional feature vectors similar to those that Stern et al. used. In addition to using high-dimensional feature representation, Araki et al. [20] also used the previously made moves as additional features. They were only able to obtain prediction accuracy of 33%, having trained on half of the Gogod dataset.

5 Conclusions

We reported experimental results using neural networks for move prediction for Go. Our results show that small convolutional neural networks are a viable method for predicting Go expert moves. Our main discovery that knowledge of the previous timesteps combined with the convolutional neural networks produce a particularly accurate move predictor. Araki et al.'s [20] experience suggests that using the previous timesteps will be less helpful for feature-vector based approaches.

In particular, the small convolutional neural networks could be used to direct the search very efficiently, and their small number of parameters makes them easy to learn.

Acknowledgements

This research was partially supported by the Ontario Graduate Scholarship and by the Natural Sciences and Engineering Council of Canada.

References

1. van der Werf, E.: AI Techniques for the Game of Go. UPM, Universitaire Pers Maastricht (2004)
2. Müller, M.: Review: Computer Go 1984-2000. Lecture Notes In Computer Science, 405–413 (2000)
3. Bouzy, B., Cazenave, T.: Computer Go: An AI oriented survey. Artificial Intelligence 132(1), 39–103 (2001)
4. Schaeffer, J., Burch, N., Bjornsson, Y., Kishimoto, A., Muller, M., Lake, R., Lu, P., Sutphen, S.: Checkers Is Solved. Science 317(5844), 1518 (2007)
5. LeCun, Y., Boser, B., Denker, J., Howard, R., Habbard, W., Jackel, L., Henderson, D.: Handwritten digit recognition with a back-propagation network. Advances in neural information processing systems 2 table of contents, 396–404 (1990)
6. Schraudolph, N., Dayan, P., Sejnowski, T.: Temporal Difference Learning of Position Evaluation in the Game of Go. Advances in Neural Information Processing Systems 6, 817–824 (1994)
7. Stern, D., Herbrich, R., Graepel, T.: Bayesian pattern ranking for move prediction in the game of Go. In: Proc. of the 23rd international conference on Machine learning, pp. 873–880 (2006)
8. Hall, M.T., Fairbairn, J.: The Gogod Database and Encyclopaedia (2006), www.gogod.co.uk
9. Simard, P., Steinkraus, D., Platt, J.: Best practices for convolutional neural networks applied to visual document analysis. Document Analysis and Recognition, 958–963 (2003)
10. Ranzato, M., LeCun, Y.: A sparse and locally shift invariant feature extractor applied to document images. In: Proc. International Conference on Document Analysis and Recognition (ICDAR) (2007)
11. LeCun, Y., Huang, F., Bottou, L.: Learning methods for generic object recognition with invariance to pose and lighting. Computer Vision and Pattern Recognition 2 (2004)
12. LeCun, Y., Bottou, L., Bengio, Y., Haffner, P.: Gradient-Based Learning Applied to Document Recognition. Proceedings of the IEEE 86(11) (1998)
13. van der Werf, E., Uiterwijk, J., Postma, E., van den Herik, J.: Local Move Prediction in Go. Computers and Games (2003)
14. Enzenberger, M.: Evaluation in Go by a Neural Network using Soft Segmentation. Advances in Computer Games 10 (2003)
15. Tesauro, G.: Temporal difference learning and TD-Gammon. Communications of the ACM 38(3), 58–68 (1995)
16. Sutton, R., Barto, A.: Reinforcement Learning: An Introduction. MIT Press, Cambridge (1998)
17. Brugmann, B.: Monte Carlo Go (1993)
18. Gelly, S., Wang, Y.: Exploration exploitation in Go: UCT for Monte-Carlo Go. In: NIPS-2006: On-line trading of Exploration and Exploitation Workshop, Whistler, Canada (2006)

19. Wu, L., Baldi, P.: A Scalable Machine Learning Approach to Go. Neural Information Processing Systems, 1521–1528 (2007)
20. Araki, N., Yoshida, K., Tsuruoka, Y., Tsujii, J.: Move Prediction in Go with the Maximum Entropy Method. In: Proceedings of the 2007 IEEE Symposium on Computational Intelligence and Games (2007)

Associative Memories Applied to Pattern Recognition

Roberto A. Vazquez and Humberto Sossa

Centro de Investigación en Computación-IPN
Av. Juan de Dios Batz, esq. Miguel Othón de Mendizábal.
Mexico City, 07738. Mexico
ravem@ipn.mx, hsossa@cic.ipn.mx

Abstract. An associative memory (AM) is a special kind of neural network that allows associating an output pattern with an input pattern. In the last years, several associative models have been proposed by different authors. However, they have several constraints which limit their applicability in complex pattern recognition problems. In this paper we gather different results provided by a dynamic associative model and present new results in order to describe how this model can be applied to solve different complex problems in pattern recognition such as object recognition, image restoration, occluded object recognition and voice recognition.

1 Introduction

An associative memory AM can be seen as a particular kind of neural network specially designed to recall output patterns in terms of input patterns that might appear altered by some kind of noise, refer for example to [1], [2], [3], [4], [7] and [9]. Most of these associative models have several constraints that limit their applicability in real life problems. For example input patterns can only be distorted by additive or subtractive noise, but not both or only by mixed noise. Other associative models can only store a limit number of patterns, or can only store binary and bipolar patterns. There are a limited number of associative models that can store real patterns.

A particular problem with images obtained from real life situations is that they could suffer pronounced alterations. Examples of these alterations are image transformations such as illumination changes, scale chances, rotations and different orientations. Working with these kinds of alterations under models like those described in [1], [2], [3], [4] and [9] could produce unexpected results due to their limitations. The same occurs with sound patterns, they suffer different transformations that make almost impossible applying an AM for voice recognition for example.

Recently in [7] a new associative model was proposed. This model can be used to recall a set of images even if these images suffer affine transformations. This model also has been applied to different pattern recognition problems, refer for example to [7], [14] and [15]. In this paper we propose to apply them to object recognition, occluded object recognition and voice recognition.

V. Kůrková et al. (Eds.): ICANN 2008, Part II, LNCS 5164, pp. 111–120, 2008.

2 Description of the Dynamic Associative Model

This proposed model is not an iterative model as Hopfield's model [1]. The model emerges as an improvement of the model proposed in [4] which is not an iterative model. Let $\mathbf{x} \in \mathbb{R}^n$ and $\mathbf{y} \in \mathbb{R}^m$ an input and output pattern, respectively. An association between input pattern $\mathbf{x}$ and output pattern $\mathbf{y}$ is denoted as $(\mathbf{x}^k, \mathbf{y}^k)$, where k is the corresponding association. Associative memory $\mathbf{W}$ is represented by a matrix whose components w_{ij} can be seen as the synapses of the neural network. If $\mathbf{x}^k = \mathbf{y}^k \forall k = 1, \ldots, p$ then $\mathbf{W}$ is auto-associative, otherwise it is hetero-associative. A distorted version of a pattern $\mathbf{x}$ to be recalled will be denoted as $\widetilde{\mathbf{x}}$. If an associative memory $\mathbf{W}$ is fed with a distorted version of $\mathbf{x}^k$ and the output obtained is exactly $\mathbf{y}^k$, we say that recalling is robust.

2.1 Building the Dynamic Associative Memory (DAM)

This model is inspired in some biological ideas of the human brain. Humans, in general, do not have problems recognizing patterns even if these are altered by noise. Several parts of the brain interact together in the process of learning and recalling. This model defines several interacting areas, one per association we would like the memory to learn. Also integrate the capability to adjust synapses in response to an input stimulus. Before an input pattern is learned or processed by the brain, it is hypothesized that it is transformed and codified by the brain. This process is simulated using the procedure introduced in [5].

This procedure allows computing *codified patterns* from input and output patterns denoted by $\bar{\mathbf{x}}$ and $\bar{\mathbf{y}}$ respectively; $\hat{\mathbf{x}}$ and $\hat{\mathbf{y}}$ are *de-codifying patterns*. Codified and de-codifying patterns are allocated in different interacting areas. On the other hand, d defines of much these areas are separated and determines the noise supported by the model. In addition a simplified version of $\mathbf{x}^k$ denoted by s_k is obtained as:

$$s_k = s\left(\mathbf{x}^k\right) = \text{mid } \mathbf{x}^k \tag{1}$$

where mid operator is defined as mid $\mathbf{x} = x_{\frac{n+1}{2}}$.

When the brain is stimulated by an input pattern, some regions of the brain (interacting areas) are stimulated and synapses belonging to those regions are modified. In this model, the most excited interacting area is call *active region* (AR) and could be estimated as follows:

$$ar = r(\mathbf{x}) = \arg\left(\min_{i=1}^{p} |s(\mathbf{x}) - s_i|\right) \tag{2}$$

Once computed the *codified patterns*, the *de-codifying* patterns and s_k we can compute the synapses of the DAM as follows:

Let $\{(\bar{\mathbf{x}}^k, \bar{\mathbf{y}}^k)\,\forall k = 1, \ldots, p\}$, $\bar{\mathbf{x}}^k \in \mathbb{R}^n$ and $\bar{\mathbf{y}}^k \in \mathbb{R}^m$ a fundamental set of associations (codified patterns). Synapses of $\mathbf{W}$ are defined as:

$$w_{ij} = \bar{y}_i - \bar{x}_j \tag{3}$$

In short, building of the DAM can be performed in three stages as:

1. Transform the fundamental set of association into codified and decodifying patterns by means of Procedure 1 described in [5].
2. Compute simplified versions of input patterns by using equation 1.
3. Build $\mathbf{W}$ in terms of codified patterns by using equation 3.

2.2 Modifying Synapses of the Associative Model

In this model, synapses could change in response to an input stimulus; but which synapses should be modified? There are synapses that can be drastically modified and they do not alter the behavior of the DAM. In the contrary, there are synapses that only can be slightly modified to do not alter the behavior of the DAM; we call this set of synapses *the kernel* of the DAM and it is denoted by $\mathbf{K_W}$. In this model are defined two types of synapses: synapses that can be modified and do not alter the behavior of the DAM and synapses belonging to the kernel of the DAM. These last synapses play an important role in recalling patterns altered by some kind of noise.

Let $\mathbf{K_W} \in \mathbb{R}^n$ the kernel of a DAM $\mathbf{W}$. A component of vector $\mathbf{K_W}$ is defined as:

$$kw_i = \min\left(w_{ij}\right), j = 1, \ldots, m \tag{4}$$

Synapses that belong to $\mathbf{K_W}$ are modified as a response to an input stimulus. Input patterns stimulate some ARs, interact with these regions and then, according to those interactions, the corresponding synapses are modified. Synapses belonging to $\mathbf{K_W}$ are modified according to the stimulus generated by the input pattern. This adjusting factor is denoted by Δw and can be computed as:

$$\Delta w = \Delta\left(\mathbf{x}\right) = s\left(\bar{\mathbf{x}}^{ar}\right) - s\left(\mathbf{x}\right) \tag{5}$$

where ar is the index of the AR.

Finally, synapses belonging to $\mathbf{K_W}$ are modified as:

$$\mathbf{K_W} = \mathbf{K_W} \oplus \left(\Delta w - \Delta w_{old}\right) \tag{6}$$

where operator $\oplus$ is defined as $\mathbf{x} \oplus e = x_i + e \forall i = 1, \ldots, m$. As you can appreciate, modification of $\mathbf{K_W}$ in equation 6 depends of the previous value of Δw denoted by Δw_{old} obtained with the previous input pattern. Once trained the DAM, when it is used by first time, the value of Δw_{old} is set to zero.

2.3 Recalling a Pattern Using the Proposed Model

Once synapses of the DAM have been modified in response to an input pattern, every component of vector $\bar{\mathbf{y}}$ can be recalled by using its corresponding input vector $\bar{\mathbf{x}}$ as:

$$\bar{y}_i = \text{mid}\left(w_{ij} + \bar{x}_j\right), j = 1, \ldots, n \tag{7}$$

In short, pattern $\bar{\mathbf{y}}$ can be recalled by using its corresponding key vector $\bar{\mathbf{x}}$ or $\tilde{\mathbf{x}}$ in six stages as follows:

1. Obtain index of the active region ar by using equation 2.
2. Transform $\mathbf{x}^k$ using de-codifying pattern $\hat{\mathbf{x}}^{ar}$ by applying the following transformation: $\breve{\mathbf{x}}^k = \mathbf{x}^k + \hat{\mathbf{x}}^{ar}$.
3. Compute adjust factor $\Delta w = \Delta(\breve{\mathbf{x}})$ by using equation 5.
4. Modify synapses of associative memory $\mathbf{W}$ that belong to $\mathbf{K_W}$ by using equation 6.
5. Recall pattern $\breve{\mathbf{y}}^k$ by using equation 7.
6. Obtain $\mathbf{y}^k$ by transforming $\breve{\mathbf{y}}^k$ using de-codifying pattern $\hat{\mathbf{y}}^{ar}$ by applying transformation: $\mathbf{y}^k = \breve{\mathbf{y}}^k - \hat{\mathbf{y}}^{ar}$.

The formal set of prepositions that support the correct functioning of this dynamic model and the main advantages again other classical models can be found in [18].

3 Associative Memories Applied to Pattern Recognition

In this section we describe different pattern recognition problems and we provide a solution to this problem using the dynamic associative model already described.

In previous papers we have described how a DAM could be use to solve some of the most important problems in pattern recognition: 3D object recognition [14] and face recognition [15]. Instead of use several statistical computationally expensive techniques such as principal component analysis [10] and [11], or view based object recognition [12] and [13], we used the DAM already described combined with two biological hypotheses: the role of the response to low frequencies at early stages, and some conjectures concerning how an infant detects subtle features (stimulating points) from a face or a toy.

Now we will address how a DAM could be used to solve some interesting problems in pattern recognition such as object recognition, image restoration, occluded object recognition and voice recognition.

3.1 Image Restoration (Filtering)

The most common application of an AM is as a filter; refer for example to [3], [4] and [6]. In this case, the AM is fed with a pattern, probably altered by noise; at the output the original image (without noise) should be obtained.

In this experiment we firstly trained the DAM with the set of images A, see Fig. 1, as in section 2.1. Each image was firstly transformed into a raw vector reading the image from left -right and up-down and transforming each RGB pixel (hexadecimal value) into a decimal value and finally, we stored the information into an array. In this case, each image was associated with itself (40 associations). Once trained the DAM, we proceed to test the accuracy of the proposal. First we verified if the DAM is able to recall the fundamental set of association using set of image A, and then we verified if the DAM recall noisy

(a) (b)

Fig. 1. (a) Set of images A is composed by 40 different images of flowers and animals. (b) Set of image B is composed by 40 classes of flowers and animals, 15 images per class. For each class the pixels of each image are altered by Gaussian noisy (altering the 5% of the pixel until 75% of the pixels in step of 5).

version of the images using the set of images B. In both cases the DAM recalled the corresponding image, even when the 75% of the pixels where modified. In summary, the accuracy of the proposal was, in both cases, of 100%. This result supports the applicability of the proposal for the cleaning up of images from noisy versions of them.

3.2 Object Recognition

Now suppose that you want to develop a nurse robot capable of recognizing the different surgeon tools used in a surgery. One possible solution could be to use the methodology described in [17] combined with the dynamic associative model.

First to all, we have to train the DAM using a set of images of the objects that we would like the nurse robot learn. Once trained the DAM we expected to recognize the objects learned by the nurse robot (DAM). For training, we used a set of images composed by 4 objects, 20 images per object at different rotations and positions, see Fig. 2(a). In order to test the accuracy of the proposal we used a set of 100 images containing the objects already trained, see Fig. 2(b).

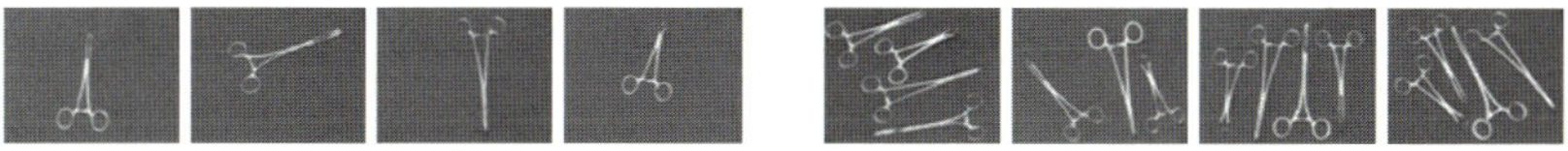

(a) (b)

Fig. 2. (a) Some images used in training phase. (b) Some images used in recognition phase.

Training phase is given as follows: For each image, apply a threshold, apply a labeling-component algorithm, compute Hu descriptors for each labeled component and use them to train the DAM as in section 2.1. In this case, we trained the DAM using the average descriptor vector of each object associated with a vector filled with 0 and with an 1 in the position that indicates the index class of the object.

Recognition phase could be performed as follows: For a given image, apply a threshold, apply a labeling-component algorithm, compute Hu descriptors for

Table 1. Accuracy of the proposal compared against other results (adapted from [17])

Object	Angled-based	Canberra	Manhathan	Euclidian	Bayesian	**DAM**
1	85%	98%	88%	88%	100%	99%
2	76%	76%	76%	76%	99%	98%
3	98%	99%	98%	97%	99%	99%
4	75%	95%	79%	74%	89%	90%
% recognition	83.5%	92%	85.25%	83.75%	96.75%	**96.75%**

each labeled component and operate the DAM using each descriptor already obtained as described in section 2.3.

The accuracy of this proposal was compared with the results obtained using other distance classifiers, for the details, refer to Table 1. As you can appreciate, the accuracy of the proposal using DAM is comparable to that obtained using the Bayesian classifier and better than those obtained using other distance classifiers.

3.3 Object Recognition under Occlusions

If an object is being partially occluded, the corresponding invariant description (as those used in section 3.2) will be distorted or, in the worst case, unable to reflect the identity of the object. One possible solution to this problem could be use the methodology described in [16] combined with the DAM. In this methodology, essential parts (EP), which are parts that characterize an object, are first detected. These parts are then described in terms of invariant features for further processing; this allows object recognition also under image transformations (translations, rotations, and so on). Finally object part descriptions are next used to train the DAM.

For training we used a set of images composed by 5 objects, 20 images per object at different rotations and positions, see Fig. 3(a). In order to test the accuracy of the proposal we used a set of 100 images containing the objects already trained, see Fig. 3(b). Training phase is given as follows (for details refers to [16]): For each image, select and EP part, apply a threshold, apply a labeling-component algorithm, compute Hu descriptors [8] and train the DAM as was described in section 2.1. In this case, the DAM was trained by using the average descriptor vector of each EP part associated with a vector filled with 0 with an 1 in the position that indicates the index class of the object.

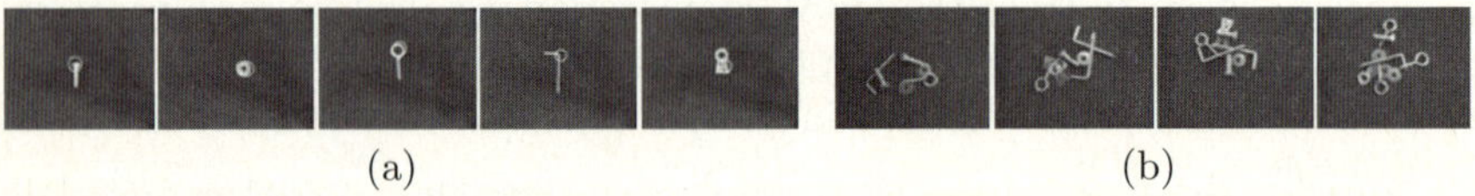

Fig. 3. (a) Some images of the five objects used in training phase. (b) Some of the images used in recognition phase.

Recognition phase is performed as follows (for details refers to [16]): For a given image, extract an EP part, apply a threshold, apply a labeling-component algorithm, compute Hu descriptors and operate the DAM as was described in section 2.3. During object detection, EP parts are searched by means of swapping window connected to the DAM. Each time an EP part of an object is detected a vote is given to that object telling that this object appears in the image.

The accuracy of this proposal was compared with the results obtained using other distance classifiers and AMs, see Table 2. As you can appreciate, the accuracy of the proposal using the DAM is slightly better compared against that obtained using the Bayesian classifier and other associative models.

Table 2. Accuracy of the proposal compared against other results (adapted from [16])

Object	PAM	MAM	Euclidian	Bayesian	SAM	**DAM**
Bolt	76%	70%	74%	90%	86%	88%
Washer	67%	60%	64%	89%	70%	75%
Eyebolt	82%	49%	80%	60%	82%	80%
Hook	52%	52%	52%	51%	52%	53%
Dovetail	81%	78%	83%	92%	95%	93%
% of recognition	71.6%	61.8%	71.0%	76.4%	76.8%	**77.8%**

3.4 Voice Recognition

Until here, the most common characteristic of the previous problems was that they use images to train the DAM. Now, we will show how this model can be applied to other kind of problems which do not involve the use of images.

In the previous problems we associated an image with other image, now we will associate a voice signal with another voice signal, a voice signal with text and a voice signal with an image. Each voice signal was recorded in a wav file (PCM format, 44.1 KHz, 16 bits and mono).

Before training the DAM, each voice signal has to be transformed into a voice signal pattern. In order to build a voice signal pattern from the wav file, we only read the wav information chuck of the file and then we store it in an array. Note that we did not use any sophisticated techniques for extracting voice features. In order to associate a voice signal with an image belonging to set of image A used in section 3.1, each image was first transformed into an array. The same occurs when we associate a voice signal with a word, each word in ASCII was first transformed into their hexadecimal value and then stored in an array. In order to test the accuracy of the proposal we perfomed three experiments using several sets of sounds.

Set of sounds A is composed by 40 different sounds that when reproduced you can hear the name of different flowers and animals. Set of sounds B1 is composed by 40 classes of flowers and animals, 15 sounds per class, see Fig 4. For each class the wav information of each sound was altered by additive noise (we begun with 5% of the wav information until 75% of the total of the wav

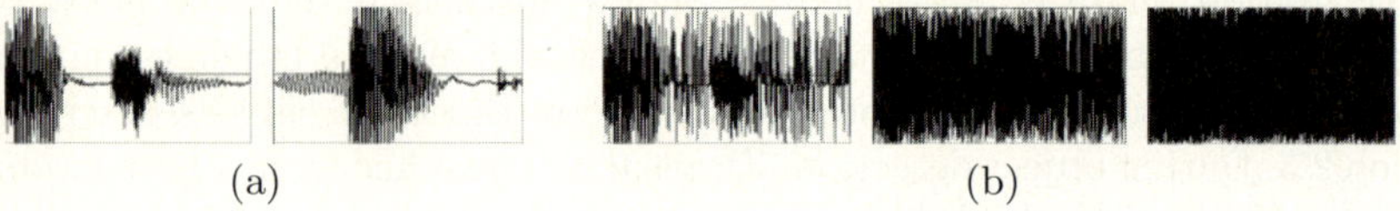

Fig. 4. (a) Some voice signals used in training phase. (b) Some noisy versions of voice signals altered with Gaussian noise used in recognition phase.

information in steps of 5), this set is composed by 600 sounds. Set of sounds B2, B3 and B4 are composed by 600 sounds like B1 but altered by subtractive, mixed and Gaussian noise respectively.

In [15], the authors suggested to compute a simplified version of the DAM model by using a random selection of stimulating points. In order to increase the accuracy of the proposal we adopted this technique. Stimulating points SP are used by the DAM to determine an active region and are given by $\mathbf{sp} \in \{\mathbb{Z}^+\}^c$ where c is the number of SP used. $sp_i = random(n), i = 1, \ldots, c$ where n is the size of the pattern. To determine the active region, the DAM stores during training phase an alternative simplified version of each pattern $\mathbf{x}^k$ given by:

$$\mathbf{ss}^k = ss\left(\mathbf{x}^k\right) = \mathbf{x}^k|_{\mathbf{sp}} = \{x^k_{sp_1}, \ldots, x^k_{sp_c}\} \tag{8}$$

During recalling phase, each element of an input simplified pattern $\tilde{\mathbf{x}}^k|_{\mathbf{sp}}$ excites some of these regions and the most excited region will be the active region. To determine which region is excited by an input pattern we use:

$$b = \arg\left(\min_{k=1}^{p} \left|[ss\left(\mathbf{x}\right)]_i - \mathbf{ss}^k_i\right|\right) \tag{9}$$

For each element of $\tilde{\mathbf{x}}^k|_{\mathbf{sp}}$ we applyed eq. 9 and the most excited region (the region that more times was obtained) will be the active region.

Finally, building of the DAM is done as follows: Let $\mathbf{I}^k_x$ and $\mathbf{I}^k_y$ an association of sounds, images or words, and c the number of stimulating points.

1. Take at random c stimulating points sp_i.
2. For each association transform the input patterns into an array.
3. Train the DAM.

Pattern $\mathbf{I}^k_y$ can be recalled by using its corresponding key input pattern $\mathbf{I}^k_x$ or distorted version $\tilde{\mathbf{I}}^k_x$ as follows:

1. Use the same c stimulating points sp_i as in building phase.
2. Transform the input pattern into an array.
3. Operate the DAM.

Experiment 1. Voice-to-voice. In this experiment we trained the DAM with the set of sounds A. Each sound was associated with itself (40 associations). Once trained the DAM, we proceeded to test the accuracy of the proposal. Firstly we verified if the DAM was able to recall the fundamental set of associations using set of sounds A, and then we verified if the DAM could recall the sounds from noisy versions of them by using the sets of sounds B1, B2, B3 and B4. For the first case, the fundamental set of associations was correctly recalled. For the second case, using 100 stimulating points, the DAM recalled the corresponding sound, even when the 75 % of the wave information was modified. In summary, the accuracy of the proposal for this experiment was of 100%.

Experiment 2. Voice-to-word. In this experiment each sound (from set of sounds A) was associated with the word that reproduce the voice signal. For example, the voice signal that reproduces the word "tiger" was associated with the word "tiger". Once trained the DAM with the 40 associations, we used the sets of sounds B1, B2, B3 and B4 expecting recall the corresponding describing word associated to the sound. Instead of recalling a sound, the trained DAM was used to recall the describing word. Using 100 stimulating points, as equal as the previous experiment, the DAM recalled the corresponding word, even when the 75 % of the wave information was modified. In summary, the accuracy of the proposal, for this experiment was of 100%.

Experiment 3. Voice-to-image. In this experiment each sound was associated with an image. Once trained the DAM with 40 associations, we proceeded to test the accuracy of the proposal. Firstly we verified if the DAM was able to recall the fundamental set of associations using set sounds A, and then we verified if the DAM could recall the images associated to the sounds by using the set of sounds B1, B2, B3 and B4. For the first case, the fundamental set of associations was correctly recalled. For the second case, using 100 stimulating points, the DAM recalled the corresponding image (using a sound), even when the 75 % of the wave information was modified. In summary, the accuracy of the proposal, for this experiment, was of 100%.

4 Conclusions and Directions for Further Research

In this paper we have described how a DAM can be applied to different problems in pattern recognition. We have demonstrated the robustness of this model compared with other associative models and other techniques. We have shown the capabilities of this dynamic model not only in problems with images but in problems that uses other information resources such as sound.

The results obtained in the different pattern recognition problems presented in this research are very encouraged and suggest that DAM could be considerate as a robust classification tool useful in several kind of pattern recognition problems.

These results can be considered as preliminary results for more complex problems. Nowadays we are applying this result to more complex problems such as robot control with voice commands using DAM, images retrieval using voice commands and DAM, voice encryption using DAM.

Acknowledgment. This work was economically supported by SIP-IPN under grant 20071438, 20082948 and CONACYT under grant 46805.

References

1. Hopfield, J.J.: Neural networks and physical systems with emergent collective computational abilities. Proc. Natl. Acad. Sci. 79, 2554–2558 (1982)
2. Kohonen, T.: Correlation matrix memories. IEEE Trans. on Comp. 21, 353–359 (1972)
3. Ritter, G.X., Sussner, P., Diaz de Leon, J.L.: Morphological associative memories. IEEE Trans. Neural Netw. 9, 281–293 (1998)
4. Sossa, H., Barron, R., Vazquez, R.A.: New associative memories to recall real-valued patterns. In: Sanfeliu, A., Martínez Trinidad, J.F., Carrasco Ochoa, J.A. (eds.) CIARP 2004. LNCS, vol. 3287, pp. 195–202. Springer, Heidelberg (2004)
5. Sossa, H., Barron, R., Vazquez, R.A.: Transforming Fundamental set of Patterns to a Canonical Form to Improve Pattern Recall. In: Lemaître, C., Reyes, C.A., González, J.A. (eds.) IBERAMIA 2004. LNCS (LNAI), vol. 3315, pp. 687–696. Springer, Heidelberg (2004)
6. Sussner, P.: Generalizing operations of binary auto-associative morphological memories using fuzzy set theory. J. Math. Imaging Vis. 19, 81–93 (2003)
7. Vazquez, R.A., Sossa, H.: Associative Memories Applied to Image Categorization. In: Martínez-Trinidad, J.F., Carrasco Ochoa, J.A., Kittler, J. (eds.) CIARP 2006. LNCS, vol. 4225, pp. 549–558. Springer, Heidelberg (2006)
8. Hu, M.K.: Visual pattern recognition by moment invariants. IRE Transactions on Information Theory 8, 179–187 (1962)
9. Sussner, P., Valle, M.: Gray-Scale Morphological Associative Memories. IEEE Trans. Neural Netw. 17, 559–570 (2006)
10. Jolliffe, I.: Principal Component Analysis. Springer, Heidelberg (1986)
11. Turk, M., Pentland, A.: Eigenfaces for recognition. J. Cognitive Neurosci. 3, 71–86 (1991)
12. Poggio, T., Edelman, S.: A network that learns to recognize 3d objects. Nature 343, 263–266 (1990)
13. Pontil, M., Verri, A.: Support vector machines for 3d object recognition. IEEE Trans. Pattern Anal. Mach. Intell. 20, 637–646 (1998)
14. Vazquez, R.A., Sossa, H., Garro, B.A.: 3D Object recognition based on low frequencies response and random feature selections. In A. Gelbukh, A. F. Kuri (Eds.) MICAI 2007. LNAI, vol. 4827, pp. 694–704. Springer, Heidelberg (2007)
15. Vazquez, R.A., Sossa, H., Garro, B.A.: Low frequency responses and random feature selection applied to face recognition. In: Kamel, M., Campilho, A. (eds.) ICIAR 2007. LNCS, vol. 4633, pp. 818–830. Springer, Heidelberg (2007)
16. Vazquez, R.A., Sossa, H., Barron, R.: Invariant descriptions and associative processing applied to object recognition under occlusions. In: Gelbukh, A., de Albornoz, Á., Terashima-Marín, H. (eds.) MICAI 2005. LNCS (LNAI), vol. 3789, pp. 318–327. Springer, Heidelberg (2005)
17. Sossa, H., Vazquez, R.A., Barron, R.: Reconocimiento y localización de instrumental medico usando análisis automatizado de imágenes. Revista Mexicana de Ingeniería Biomédica 26, 75–85 (2005)
18. Vazquez, R.A., Sossa, H.: A new associative memory with dynamical synapses (submitted to Neural Processing Letters, 2007) (2007)

MLP-Based Detection of Targets in Clutter: Robustness with Respect to the Shape Parameter of Weibull-Disitributed Clutter

Raul Vicen-Bueno*, Eduardo Galán-Fernández, Manuel Rosa-Zurera, and Maria P. Jarabo-Amores

Signal Theory and Communications Department
Superior Politechnic School, University of Alcalá
Ctra. Madrid-Barcelona, km. 33.600, 28805, Alcalá de Henares - Madrid (Spain)
raul.vicen@uah.es

Abstract. Obtaining analytical expressions for coherent detection of known signals in Weibul-distributed clutter and white Gaussian noise has been a hard task since the last decades. In fact, nowadays, these expressions have not been found yet. This problem lead us to use suboptimum solutions to solve this problem. Optimum approximations can be done by using Multilayer Perceptrons (MLPs) trained in a supervised way to minimize the mean square error. So, MLP-based detectors are constructed and compared with one of the suboptimum detectors commonly used to solve the detection problem under study. First, a study of the dimensionality of the MLP is done for typical values of the target and clutter conditions. And finally, a deep study is done according to the variations of the most important parameters of the target and clutter signals. The last study let us to be conscious about the importance of the selection of the parameters to design both detectors. Moreover, the difference of performances between each other and the superiority of the MLP-based detector against the suboptimum solution is emphasized.

1 Introduction

Coherent detection of known targets [1] in clutter and noise is a difficult task, which can be solved by Artificial Neural Networks (ANNs). Many clutter models have been proposed in the literature [2], although one of the commonly accepted models is the Weibull one [3,4]. So, if the ANN training is supervised, they can approximate the Neyman-Pearson detector [5,6,7], which is usually used in radar systems design. This detector maximizes the probability of detection (Pd) maintaining the probability of false alarm (Pfa) lower than or equal to a given value [8].

The research shown in [9] set the optimum detector for target and clutter with Gaussian Probability Density Functions (PDFs). Due to the impossibility to

* This work has been supported by the *Ministerio de Educación y Ciencia* under project TEC2007-64458.

V. Kůrková et al. (Eds.): ICANN 2008, Part II, LNCS 5164, pp. 121–130, 2008.

obtain analytical expressions for the optimum detector under Weibull-distributed clutter, only suboptimum solutions are proposed. The Target Sequence Known A Priori (TSKAP) detector is one of them and is taken as reference for the experiments, which conveys some implementation problems.

In this work, MLPs are trained to approximate the Neyman-Pearson detector for known targets in coherent Weibull clutter and white Gaussian noise. Due to this approximation, it is expected that this MLP-based detector outperforms the suboptimum solutions (TSKAP detector) commonly used in the literature [4,9], which need to have a priori some knowledge of the environment. A study of the MLP size is carried out for typical values of target and clutter parameters. This paper shows how the MLP-based detector outperforms the suboptimum solutions, but, what does it happen if the target and clutter conditions vary across the time? This question is answered in the paper, where a deep study of both detectors (MLP-based and TSKAP) is done according to their robustness with respect to variations of target and clutter parameters.

2 Models of Signal: Target, Clutter and Noise

A radar system explores a zone at a certain pulse repetition frequency (PRF) or sampling rate of the process, obtaining an observation map of this zone for each radar scan. This map is analyzed in blocks of N cells or pulses (it is considered that for each cell only the information of one pulse transmitted and received exists). So observation vectors ($\mathbf{z}$) composed of N complex samples are presented to the detector. Under hypothesis H0 (target absent), $\mathbf{z}$ is composed of N samples of clutter and noise. Whereas under hypothesis H1 (target present), a known target characterized by a fixed amplitude (A) and phase (θ) (Swerling V model [1]) for each of the N pulses is present in clutter and noise samples. Also, a doppler frequency in the target model (f_s) is assumed.

In the radar system under study, the noise is modeled as a coherent white Gaussian complex process of unity power, i.e., a power of $\frac{1}{2}$ for the quadrature and phase components. The clutter is modeled as a coherent correlated sequence with Gaussian AutoCorrelation Function (ACF), whose complex samples have a modulus distributed with a Weibull PDF, which is an statistical distribution commonly used in the literature:

$$p(|\mathbf{w}|) = ab^{-a}|\mathbf{w}|^{a-1}e^{-\left(\frac{|\mathbf{w}|}{b}\right)^a} \quad (1)$$

where $|\mathbf{w}|$ is the modulus of the coherent Weibull sequence and a and b are the skewness (shape) and scale parameters of the Weibull distribution, respectively.

The NxN autocorrelation matrix of the clutter is given by

$$(\mathbf{M_c})_{h,k} = P_c \rho_c^{|h-k|^2} e^{j(2\pi(h-k)\frac{f_c}{PRF})} \quad (2)$$

where the indexes h and k varies from 1 to N, P_c is the clutter power, ρ_c is the one-lag correlation coefficient and f_c is the doppler frequency of the clutter.

The relationship between the Weibull distribution parameters and P_c is

$$P_c = \frac{2b^2}{a} \Gamma \left(\frac{2}{a} \right) \tag{3}$$

where $\Gamma()$ is the *Gamma function.*

The model used to generate coherent correlated Weibull sequences consists of two blocks in cascade: a correlator filter and a NonLinear MemoryLess Transformation (NLMLT) [3,10]. To obtain the desired sequence, a coherent white Gaussian sequence is correlated with the filter designed according to (2) and (3). The NLMLT block, according to (1), gives the desired Weibull distribution to the sequence.

Taking into consideration that the complex noise samples are of unity variance (power), the following power relationships are considered for the study: Signal to Noise Ratio (SNR=$10log_{10}\left(A^2\right)$) and Clutter to Noise Ratio (CNR=$10log_{10}\left(P_c\right)$).

3 Optimum and Suboptimum Neyman-Pearson Detectors

The problem of optimum radar detection of targets in clutter is explored in [3] when both signals are time correlated and have arbitrary PDFs. The optimum detector scheme is built around two non-linear estimators of the disturbances in both hypothesis, which minimize the mean square error (MSE). The study of Gaussian correlated targets detection in Gaussian correlated clutter plus noise is carried out, but for the cases where the hypothesis are non-gaussian distributed, only suboptimum solutions are obtained.

The proposed detectors basically consist of two channels. The upper channel is matched to the conditions that the sequence to be detected is the sum of the target plus clutter in presence of noise (hypothesis H1). While the lower one is matched to the detection of clutter in presence of noise (hypothesis H0).

For the detection problem considered in this paper, the suboptimum detection scheme (TSKAP) shown in the fig. 1 is taken. Considering that the CNR is very high (CNR >> 0 dB), the inverse of the NLMLT is assumed to transform the Weibull clutter in a Gaussian one, so the Linear Prediction Filter (LPF) is a N-1 order linear filter. Then, the NLMLT transforms the filter output in a

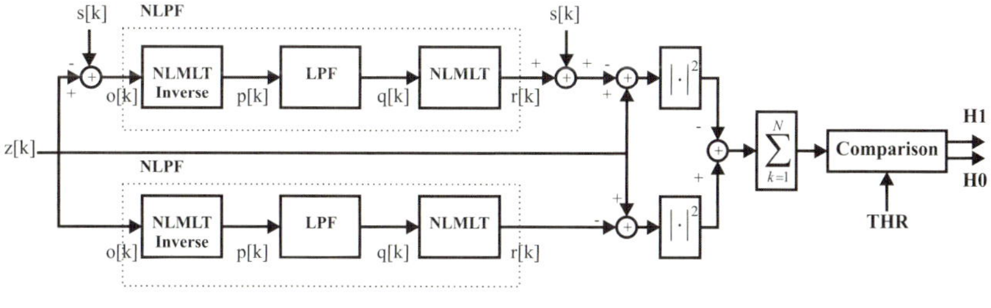

Fig. 1. Target Sequence Known A Priori Detector

Weibull sequence. Besides being suboptimum, this scheme presents two important drawbacks. First, the prediction filters have N-1 memory cells that must contain the suitable information to predict correct values for the N samples of each observation vector. So N+(N-1) pulses are necessary to decide if the target is present or not. And second, the target sequence must be subtracted from the input of the H1 channel. As a conclusion, there is no sense in subtracting the target component before deciding if this component is present or not. So, in practical cases, it makes this scheme non-realizable.

4 MLP-Based Detector

The alternative detector proposed for the experiments is presented in fig. 2. In this case, a detector based on an MLP with log-sigmoid activation function in its hidden and output neurons with hard limit threshold (THR') after its output is proposed. The final thresholding is varied in order to get the performance (Pd Vs Pfa) of the detector. This MLP-based detector tries to overcome the drawbacks of the suboptimum detector taken as reference (see Section 3). Also, as MLPs have been probed to approximate the Neyman-Pearson detector when minimizing the MSE [5], it can be expected that the MLP-based detector outperforms the suboptimum detector proposed in [3,4].

In our case of study, MLPs are trained to minimize the MSE using the LM backpropagation algorithm with adaptive parameter [11]. As this algorithm is based on the Newton method, MLPs with few hundred of weights (W) are able to achieve good performances converging in few epochs.

Three sets of patterns are generated for the design and simulation of these detectors: *train*, *validation* and *test*. The *train* and *validation* sets are used to train the MLPs with an external validation of the process in order to avoid overfitting. To improve the generalization of the trained MLPs, the training is stopped if the estimated MSE with the validation set increases during the last ten epochs of the training. Finally, the *test* set is used to obtain the performance of the MLPs trained working as radar detectors, i.e., to obtain the Pfa and Pd estimation by Moltecarlo simulations.

Before the training process starts, MLPs are initialized using the Nguyen-Widrow method [12] and, in all cases, the training process is repeated ten times. Once all the MLPs are trained, the best MLP in terms of the estimated MSE with the validation set is selected. With this selection, the problem of keeping in local minima at the end of the training is practically eliminated.

The architecture of the MLP considered for the experiments is $I/H/O$, where I is the number of MLP inputs, H is the number of hidden neurons in its hidden layer and O is the number of MLP outputs. As the MLPs work with real arithmetic, if the observation vector ($\mathbf{z}$) is composed of N complex samples, the MLP will have 2N inputs (N in phase and N in quadrature components of the N complex samples), as it is shown in fig. 2. The number of MLP independent elements (weights) to solve the problem is $W = (I+1) \cdot H + (H+1) \cdot O$, considering the bias of each neuron.

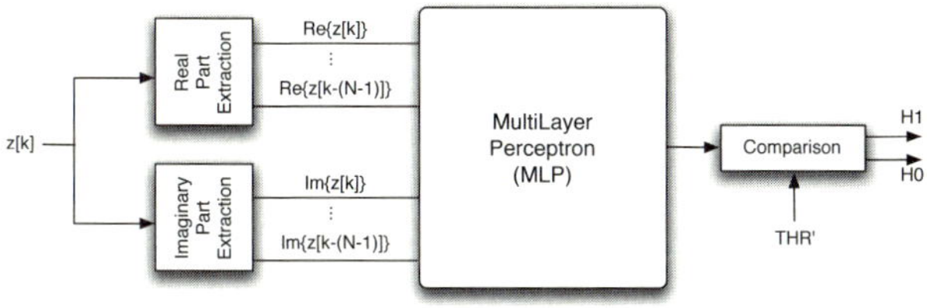

Fig. 2. MLP-based Detector

5 Results

The performance of the detectors exposed in the previous sections is shown in terms of the Receiver Operating Characteristics (ROC) curves. This curves are plotted in two ways. The first one (Pd Vs Pfa) relates the Pd achieved by the detector for a certain threshold, which is obtained according to the desired Pfa. The second one (Pd Vs SNR) relates the Pd obtained for a certain threshold obtained for a desired Pfa taking into account the SNR of the environment. The experiments are developed for an integration of two pulses ($N = 2$). So, in order to test correctly the TSKAP detector, observation vectors of length 3 complex samples ($N+(N-1)$) are generated, due to memory requirements of the TSKAP detector ($N-1$ pulses).

The MLP architecture used in the MLP-based detector is $6/H/1$. The number of MLP outputs (1) is established by the problem (binary detection). The number of hidden neurons (H) is a parameter under study in this work. And the number of MLP inputs (6) is established because of a comparison with the reference radar detector under the same conditions, as it is exposed above.

The a priori probabilities of observation vectors under $H0$ and $H1$ hypothesis is supposed to be the same. Three sets of patterns (*train*, *validation* and *test*) are generated. The first two sets are generated under the same conditions, which are called the design conditions of the experiments. Whereas the third one is usually generated under different conditions than in the design stage. It involves that the designed detectors are tested with a set of conditions different than the design ones in order to analyze the robustness of the detectors. The *train* and *validation* sets are composed of $5 \cdot 10^3$ observation vectors each. The *test* set is composed of $5 \cdot 10^6$ observation vectors, so the error in the estimations of the Pfa and the Pd is lower than 10% of the estimated values in the worst case (Pfa=10^{-4}). Attending to previous studies of detection of targets in clutter [3,4,9], typical values of the target and Weibull-distributed clutter are taken to study the dimensionality of the MLP-based detector and to study the robustness of the TSKAP and MLP-based detectors.

For the case of study of the MLP size, the following conditions are considered for the design and simulation conditions (being the same in both cases):

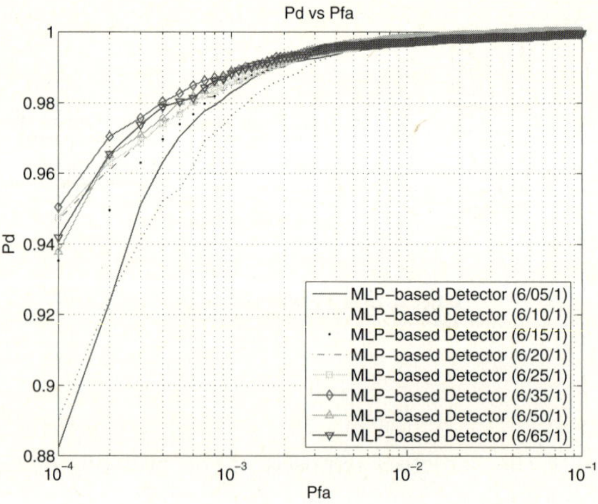

Fig. 3. MLP-based detector performances for different MLP sizes (6/H/1)

- Radar parameters: PRF=1000Hz
- Target conditions: Swerling V model, f_s=500Hz, ρ_s=1.0 and SNR=20dB.
- Clutter conditions: f_c=0Hz, ρ_c=0.9, CNR=30dB and a=1.2.

The results obtained are shown in fig. 3. As can be observed, an MLP size greater than 6/20/1 improves lowly its performance and increases its computational cost. So, an MLP size of 6/20/1 ($W > 121$ weights) is selected for the next experiments because of the tradeoff between performance improvement and computational cost. Moreover, greater MLP sizes than 6/65/1 where probed but very low improvements were achieved. Experiments under different conditions were carried out and the same conclusions were obtained about this MLP size study.

For the cases of study of the TSKAP and MLP-based detectors performances and robustness, the following conditions are fixed:

- Radar parameters: PRF=1000Hz
- Target conditions: Swerling V model, f_s=500Hz and ρ_s=1.0.
- Clutter conditions: f_c=0Hz, ρ_c=0.9 and CNR=30dB.

whereas other conditions are varied in order to study the robustness of both kind of detectors against them:

- Target conditions (design): DSNR=$[5, 15, 25]$dB.
- Clutter conditions (design): a_d=$[0.8, 1.2]$.
- Target conditions (simulation): SSNR=$[5 - 40]$dB.
- Clutter conditions (simulation): a_s=$[0.4, 0.6, 0.8, 1.0, 1.2, 1.6, 2.0]$.

Figures 4-6 and 7-9 show the performance characterization (Pd Vs SSNR) of the TSKAP and MLP-based (6/20/1) detectors created with the different

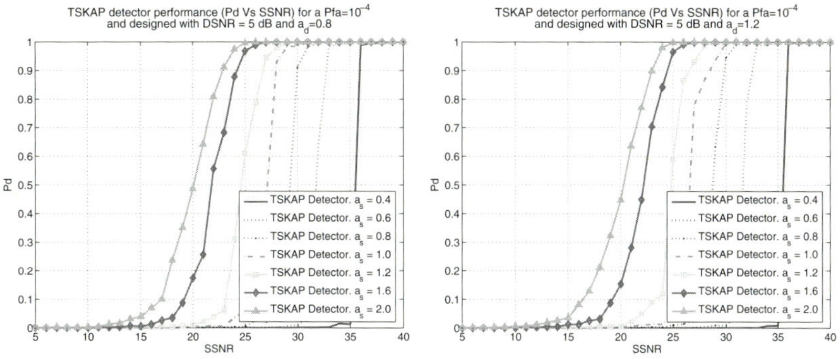

(a) (b)

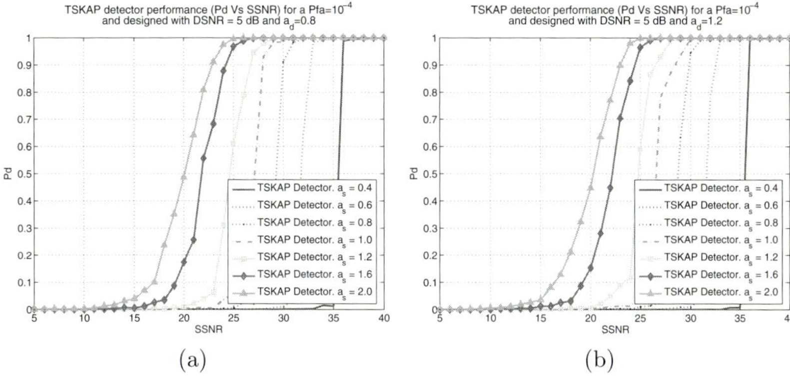

Fig. 4. Performance under different clutter conditions driven by a (Pd Vs SNR) of TKAP detectors designed with DSNR=5dB and (a) a_d=0.8 or (b) a_d=1.2

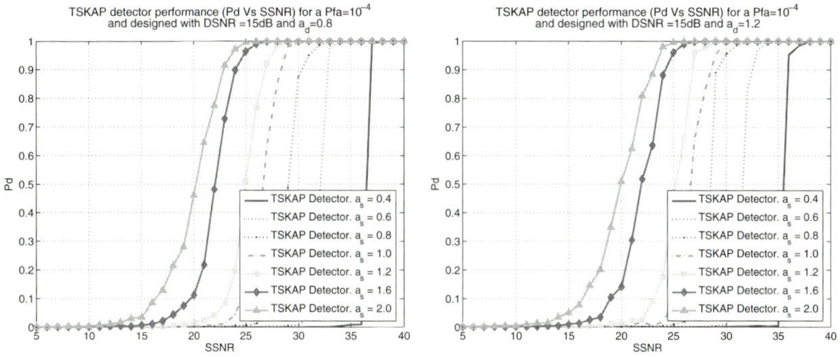

(a) (b)

Fig. 5. Performance under different clutter conditions driven by a (Pd Vs SNR) of TKAP detectors designed with DSNR=15dB and (a) a_d=0.8 or (b) a_d=1.2

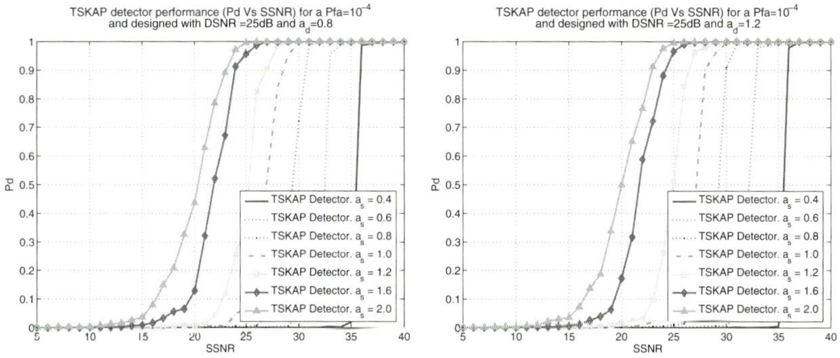

(a) (b)

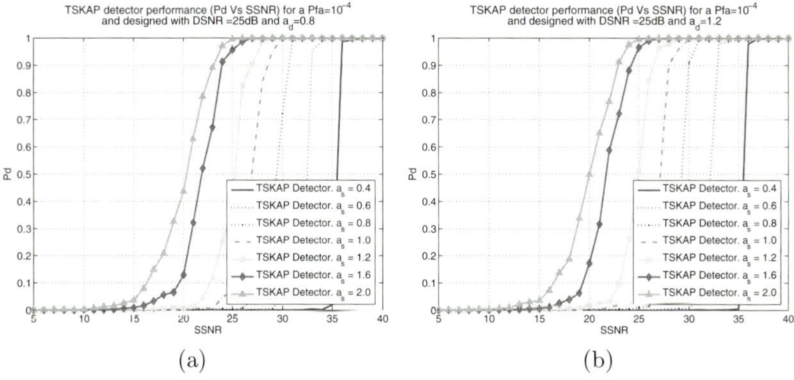

Fig. 6. Performance under different clutter conditions driven by a (Pd Vs SNR) of TKAP detectors designed with DSNR=25dB and (a) a_d=0.8 or (b) a_d=1.2

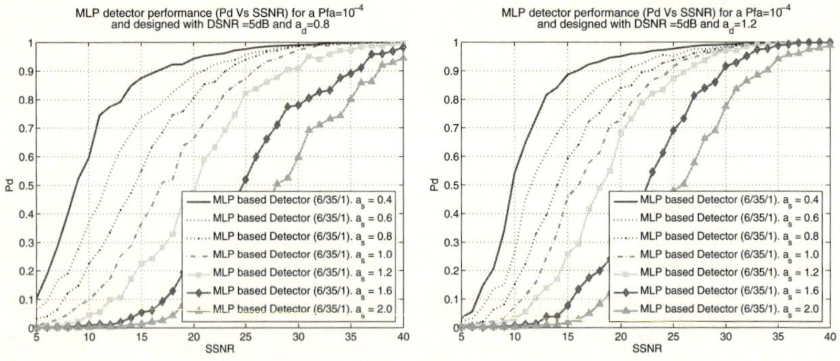

(a) (b)

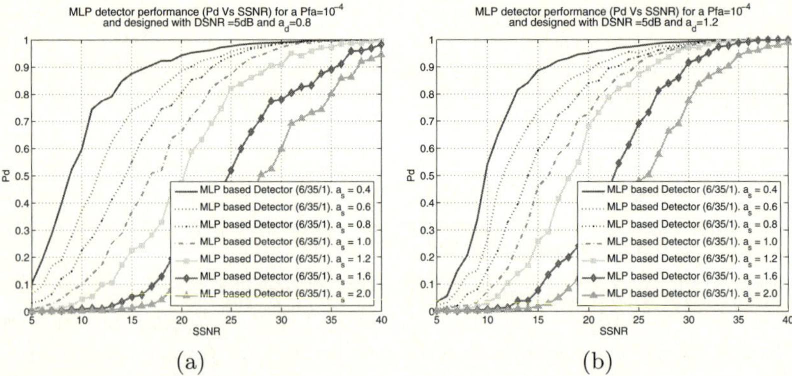

Fig. 7. Performance under different clutter conditions driven by a (Pd Vs SNR) of MLP-based detectors designed with DSNR=5dB and (a)a_d=0.8 or (b)a_d=1.2

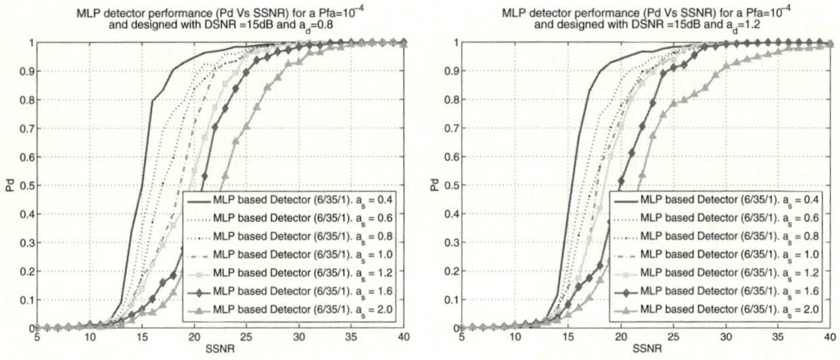

(a) (b)

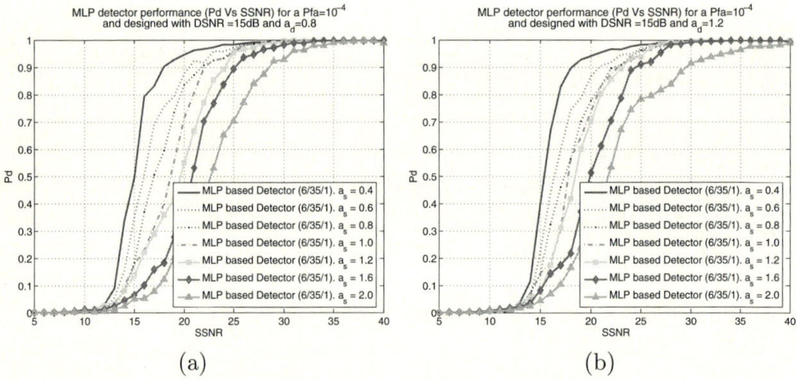

Fig. 8. Performance under different clutter conditions driven by a (Pd Vs SNR) of MLP-based detectors designed with DSNR=15dB and (a)a_d=0.8 or (b)a_d=1.2

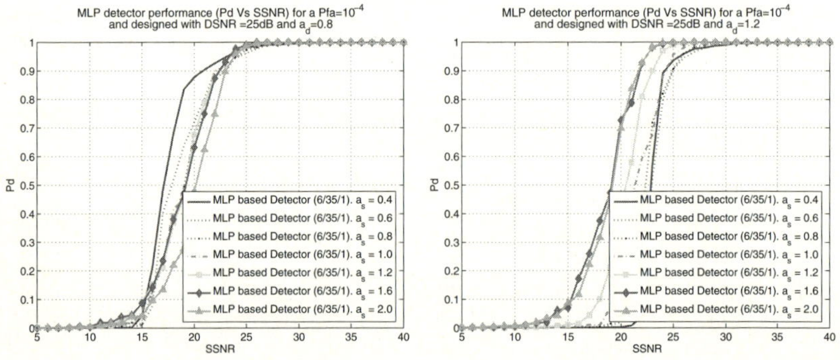

(a) (b)

Fig. 9. Performance under different clutter conditions driven by a (Pd Vs SNR) of MLP-based detectors designed with DSNR=25dB and (a)a_d=0.8 or (b)a_d=1.2

design conditions and tested with the different simulation parameters exposed above for a Pfa=10^{-4}, respectively.The establishment of the these variations of the conditions is done according to the study of how the performances of the detectors under study varies with respect to clutter conditions (Weibull shape parameter variations) and target conditions (power received in the radar due to the target signal). The variation of the clutter conditions is very usual in real radar systems due to the weather conditions and the variation of the received target power is also important due to the observed target size (radar cross section, RCS) and the distance from the target to radar.

Analyzing the results obtained for the TSKAP detector simulations, several aspects can be emphasized. First, the design SNR (DSNR) is practically irrelevant for the TSKAP detector design due to the signal is known a priori. Second, the shape parameter of the Weibull clutter (variation on its PDF) is not relevant too. And third, the performance of the TSKAP detector is better for simulation shape parameter values (a_s) greater than the design shape parameter values (a_d). The opposite effect occurs when a_s is lower than a_d.

Analyzing the results obtained for the MLP-based detector simulations, several aspects can be emphasized. First, the DSNR is a relevant parameter in the detector design. In this way, the greater is the DSNR, the lower is the dispersion of the ROC curves (Pd Vs SSNR) for different simulation shape parameters and the higher is the robustness of the detector against changes in the simulation conditions with respect to clutter conditions changes. Second, the shape parameter of the Weibull distribution in the design stage is not very relevant because of the low variation of the results obtained for different values of a_d with respect to the SSNR. And third, the performance of the MLP-based detector is better when the values of a_s are lower than in the design stage (a_d). The opposite effect occurs when a_s is greater than a_d.

6 Conclusions

In real radar environments, it is very usual that their conditions fluctuate across the time. This fluctuations or variations in the environment concern parameters related to the target and the clutter signals. In our studies, a variation of the power received related to the target and the statistical distribution of the clutter are considered.

The influence of the MLP size in MLP-based radar detectors is studied. This study avoid us to select an MLP-based detector with a structure of 6/20/1, although a detector with lower MLP size (lower computational cost) than this is able to outperform the detector took as reference, the TSKAP one.

After analyzing the performances of the detectors under study, it is appreciated that the MLP-based detector achieves better performance than the TSKAP one, specially for low SSNRs. But not only the MLP-based detector is better than the TSKAP one in terms of performance, it is also better in terms of robustness against variations of clutter and target conditions, specially for high values of DSNR (25dB). Moreover, different behavior in their performances are observed.

On one hand, when the simulation shape parameter of the Weibull clutter is greater than the design one, the TSKAP performance increases, whereas in the MLP-based detector decrease. On the other hand, this behavior is inverted when the simulation shape parameter of the Weibull clutter distribution is lower than the design shape parameter.

As a final conclusion, the MLP-based detector is proposed to detect known targets in a Weibull-distributed clutter and white Gaussian noise radar environments, but considering several aspects as relevant. If the clutter conditions varies across the time and its variation is huge in terms of the shape parameter, an MLP-based detector design with high DSNR values is recommended because of its robustness against the variation of the shape parameter. On the other hand, if this variation is low, an MLP-based detector designed with medium DSNR values is recommended. In all the cases of study, it is necesary to have a priori knowledge of the ranges of variation of the shape parameter in order to select its best value in the design stage.

References

1. Swerling, P.: Radar probability of detection for some additional fluctuating target cases. IEE Trans. on Aerospace and Electronic Systems 33(2), 698–709 (1997)
2. Cheikh, K., Faozi, S.: Application of Neural Networks to Radar Signal Detection in K-distributed Clutter. In: First Int. Symp. on Control, Communications and Signal Processing Workshop Proc., pp. 633–637 (2004)
3. Farina, A., et al.: Theory of Radar Detection in Coherent Weibull Clutter. In: Farina, A. (ed.) Optimised Radar Processors. IEE Radar, Sonar, Navigation and Avionics, vol. 1, pp. 100–116. Peter Peregrinus Ltd., London (1987)
4. Sekine, M., Mao, Y.: Weibull Radar Clutter, Radar, Sonar, Navigation and Avionics. Institution of Engineering and Technology (1990)
5. Ruck, D.W., et al.: The Multilayer Perceptron as an Approximation to a Bayes Optimal Discriminant Function. IEEE Trans. on Neural Networks 1(11), 296–298 (1990)
6. Gandhi, P.P., Ramamurti, V.: Neural Networks for Signal Detection in Non-Gaussian Noise. IEEE Trans. on Signal Processing 45(11), 2846–2851 (1997)
7. Vicen-Bueno, R., et al.: NN-Based Detector for Known Targets in Coherent Weibull Clutter. In: Corchado, E., Yin, H., Botti, V., Fyfe, C. (eds.) IDEAL 2006. LNCS, vol. 4224, pp. 522–529. Springer, Heidelberg (2006)
8. Van Trees, H.L.: Detection, Estimation and Modulation Theory. Part I. John Wiley and Sons, New York (1997)
9. Farina, A., et al.: Radar Detection in Coherent Weibull Clutter. IEEE Trans. on Acoustics, Speech and Signal Processing ASSP-35(6), 893–895 (1987)
10. Li, G., Yu, K.B.: Modelling and Simulation of Coherent Weibull Clutter. IEE Proc. Radar and Signal Processing 136(1), 2–12 (1989)
11. Hagan, M.T., Menhaj, M.B.: Training Feedforward Networks with Marquardt Algorithm. IEEE Trans. on Neural Networks 5(6), 989–993 (1994)
12. Nguyen, D., Widrow, B.: Improving the Learning Speed of 2-layer Neural Networks by Choosing Initial Values of the Adaptive Weights. In: Proc. of the Int. Joint Conf. on Neural Networks, pp. 21–26 (1999)

Modeling and Synthesis of Computational Efficient Adaptive Neuro-Fuzzy Systems Based on Matlab

Guillermo Bosque[1], Javier Echanobe[2], Inés del Campo[2], and José M. Tarela[2,⋆]

[1] Department of Electronics and Telecommunications, University of the Basque Country, Bilbao, Vizcaya, 48012, Spain
[2] Department of Electricity and Electronics, University of the Basque Country, Leioa, Vizcaya, 48940, Spain
guillermo.bosque@ehu.es, {javi,ines,tarela}@we.lc.ehu.es

Abstract. New potential applications for neural networks and fuzzy systems are emerging in the context of ubiquitous computing and ambient intelligence. This new paradigm demands sensitive and adaptive embedded systems able to deal with a large number of stimulus in an efficient way. This paper presents a design methodology, based on a new Matlab tool, to develop computational-efficient neuro-fuzzy systems. To fulfil this objective, we have introduced a particular class of adaptive neuro-fuzzy inference systems (ANFIS) with piecewise multilinear (PWM) behaviour. Results obtained show that the PWM-ANFIS model generates computational-efficient implementations without loss of approximation capabilities or learning performance. The tool has been used to develop both software and hardware approaches as well as special architectures for hybrid hardware/software embedded systems.

Keywords: Neuro-fuzzy systems, ANFIS model, neuro-fuzzy CAD tool, function approximation, Matlab environment, embedded systems.

1 Introduction

Neuro-fuzzy systems (NFS) combine artificial neural networks and fuzzy logic in a synergetic way. Fuzzy systems provide a framework to represent imprecise information and to reason with this kind of information, while neural networks enhance fuzzy systems with the capability of learning from input-output samples; learning is used to adapt parameters of the fuzzy system as membership functions or rules. Some example representative application areas of NFSs are: automatic control, robotics, adaptive signal processing, pattern recognition, and system identification (see, for example, [1]). In addition, new potential applications can be found in the field of ubiquitous computing and ambient intelligence [2]. These kinds of applications involve reasoning and learning algorithms that have to deal with signals from a large number of distributed sensor nodes.

⋆ The authors would like to thank the Basque Country Government for support of this work under Grants SA-2006/00015 and IT-353-07.

V. Kůrková et al. (Eds.): ICANN 2008, Part II, LNCS 5164, pp. 131–140, 2008.

NFSs are rather complex because they integrate many different tasks working in a cooperative way. Hence, neuro-fuzzy implementations must be developed carefully in order to fulfil all the requirements that a real-life application can demand, such as cost, size, power, etc. These implementations can be carried out by means of either hardware or software platforms. Software (SW) implementations are more flexible and economical than hardware implementations, but they provide lower speed processing. Hardware (HW) implementations are generally addressed to real-time applications where high performance is always an essential requirement. In addition hybrid HW/SW approaches provide an optimal solution for many applications where a trade-off between versatility and performance is required. Owing to the above requirements, efficient algorithms and tools are required in order to support the whole development cycle of NFSs, from the design specification to the final prototype.

This paper presents a design methodology based on a new Matlab tool to develop computational efficient NFSs. The Matlab tool, constructed by the authors, is the mainstay of the proposed methodology. It is a user friendly environment for modelling, analyzing and simulating computational efficient NFS applications. The proposed methodology is specially suited for real time applications that involve a large number of inputs. It deals with efficient implementations of a class of NFSs, the adaptive neuro-fuzzy inference system (ANFIS) [3], that has been widely used to develop NFSs in the above application areas. ANFIS is a network representation of different types of fuzzy inference models, endowed with the learning capabilities of neural networks. In the last decade, ANFIS has become very popular mainly due to the powerful capabilities as universal function approximator that it exhibits, even when simple membership functions like trapezes or triangles are used [4,5,6]. Our work focuses specifically on an ANFIS-like model that is functionally equivalent to the Takagi-Sugeno inference system [7]. With regard to the computational cost, some different restrictions are applied to the system in order to reduce drastically the complexity of its inference mechanism which becomes a Piecewise Multilinear (PWM) function. In what follows we will refer to this model as PWM-ANFIS.

The paper is structured as follows: Section 2 overviews the generic ANFIS model and introduces the PWM-ANFIS used in this paper; the computational efficiency is analyzed in order to highlight the advantages of the PWM model. In Section 3 the proposed methodology is described along with the principal features of the tool. Section 4 presents some test examples and the results obtained. Finally, Section 5 outlines the main conclusion of this work.

2 Generic ANFIS Model and PWM-ANFIS

In this section, first we are going to describe the proposed PWM-ANFIS. This model is a NFS of the ANFIS type with some restrictions on its membership functions that lead to a very simple and rapid inference mechanism. These restrictions, as will be seen below, do not affect the learning performance or the approximation capability of the system.

2.1 Anfis Model

First let us introduce the basics of the generic ANFIS model [3], for the case of a zero-order Takagi-Sugeno inference system [4]. Consider a rule-based n-input, one-output fuzzy system:

$$R_j : IF\, x_1\, is\, A_{1j_1}\, and\, x_2\, is\, A_{2j_2} \ldots and\, x_n\, is\, A_{nj_n}\, THEN\, y\, is\, r_j \tag{1}$$

where R_j is the jth rule $(1 \leq j \leq p)$, x_i $(1 \leq i \leq n)$ are input variables, y is the output, r_j is a constant consequent, and A_{ij_i} are linguistic labels (antecedents) with each one being associated with a membership function $\mu_{A_{ij_i}}(x_i)$, where $(1 \leq j_i \leq m_i)$ being m_i the number of antecedents of the variable x_i . In a zero-order Takagi-Sugeno fuzzy model the inference procedure used to derive the conclusion for a specific input $\mathbf{x} = (x_1, x_2, .., x_i, .., x_n)$, $\mathbf{x} \in \mathbb{R}^n$, consists of two main steps. First the firing strength or weight ω_j of each rule is calculated as

$$\omega_j = \prod_{i=1}^{n} A_{ij_i}(x_i) \tag{2}$$

After that, the overall inference result, y, is obtained by means of the weighted average of the consequents

$$y = (\sum_{j=1}^{p} \omega_j \cdot r_j)/\sum_{j=1}^{p} \omega_j) \tag{3}$$

(2) and (3) provide a compact representation of the inference model.ANFIS consists of a representation of different types of fuzzy inference models as adaptive networks. To be precise, the above fuzzy model can be viewed as an adaptive network with the following layers (see Fig. 1):

Layer 1 computes the membership functions

$$A_{ij_i}(x_i) \tag{4}$$

Layer 2 contains m neurons. Neuron j in this layer generates the firing strength of the j-th rule by computing the algebraic product

$$\omega_j = \prod_{i=1}^{n} A_{ij_i}(x_i) \tag{5}$$

Layer 3 is an m-neuron normalization layer. This layer performs the normalization of the activation of the rules; the output of the j-th neuron is the ratio of the j-th rules weight to the sum of the weights of all the rules:

$$\bar{\omega}_j = \omega_j / \sum_{k=1}^{p} \omega_k \tag{6}$$

Layer 4 contains m neurons. The neuron outputs produce the weight of the corresponding consequents:

$$y_j = \bar{\omega}_j \cdot r_j \tag{7}$$

Layer 5 contains only one neuron. The neuron output is the weighted sum of the consequents:

$$y = \sum_{j=1}^{p} \bar{\omega}_j \cdot r_j \tag{8}$$

To train the above network, a hybrid algorithm has been proposed in [3]. The algorithm is composed of a forward pass which is carried out by a Least Squares Estimator (LSE) process, followed by a backward pass which is carried out by a Back Propagation (BP) algorithm. Each epoch of the hybrid procedure is composed of a forward pass and a backward pass. In the forward pass the consequent parameters are identified by the LSE method and in the backward pass the antecedent parameters are updated by the Gradient Descent Method (GDM) [1]. From (2) and (3), it can be deduced that the evaluation of an inference requires one division, $m^n(n-1)$ products, $2(m^n-1)$ sums and mn membership evaluations, where it has been assumed that $m = m_1 = m_2 = \quad = m_n$.

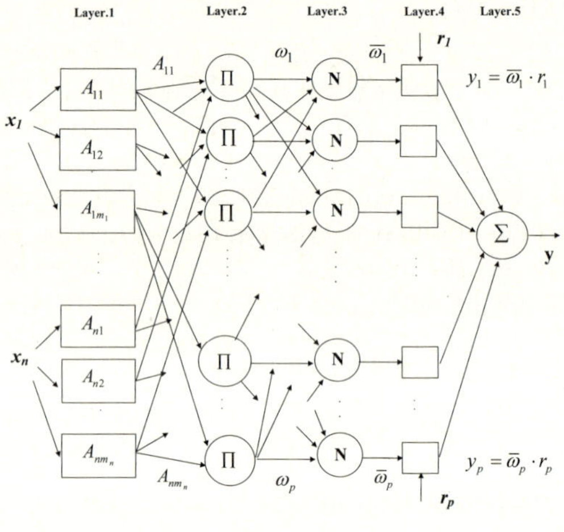

Fig. 1. Architecture of a generic n-input ANFIS

2.2 The Proposed PWM-ANFIS Model

In order to reduce the complexity of the above ANFIS model, let us introduce the following restrictions on the antecedents: (i) the membership functions are overlapped by pairs, (ii) they are triangular shaped, and (iii) they are normalized in each input dimension (see Fig. 2(a)). Similar constraints have been successfully used by many designers because of the useful properties of triangular membership functions.

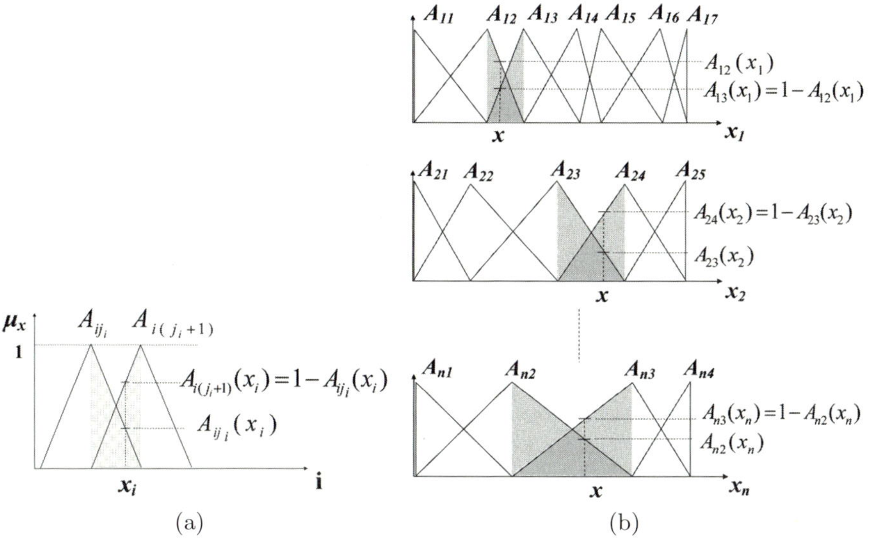

Fig. 2. (a) Triangular membership functions verifying constraints (i) to (iii)., (b) Triangular membership functions for an n-input PWM-ANFIS case example.

In the following we will analyze the advantages of constraints (i) to (iii) on the simplicity of the layered representation of the fuzzy system. First, let us consider some immediate consequences of these restrictions. The first restriction forces the overlapping degree of the antecedents to be two. Therefore, given an input vector $\mathbf{x} = (x_1, x_2, .., x_i, .., x_n)$ only two antecedents per input dimension provide membership values different from zero (i.e. active antecedents). To be exact, due to (ii) and (iii), only one half of the triangles concerned become active (see shadow region in Fig. 2(a)). Fig. 2(b) depicts typical membership functions for a particular case of an n-input system with mi triangular antecedents per input verifying the above constraints ($m_1 = 7, m_2 = 5, .., m_n = 4$). It can also be seen in this figure that the vertex of the mi triangles delimits $(m_i - 1)$ intervals per axis. Note that these intervals induce a partition of the input space into $\prod_{i=1}^{n}(m_i - 1)$ polyhedral cells or regions.

Only one of these cells is involved in the calculus of the system output at each time. The whole system can, therefore, be implemented as a single inference kernel. The parameters of the kernel depend on the concrete cell where the input vector falls (i.e. active cell).

Let us illustrate with the example of Fig. 2(b) in what way the restrictions imposed on the antecedent membership functions simplify the network architecture.

Layer 1. Every neuron in this layer computes one active membership function. In virtue of (i), each input is concerned with two membership functions per dimension. Therefore, we have only $2n$ active neurons in this layer,

$$A_{12}(x_1),\ A_{13}(x_1),\ A_{23}(x_2),\ A_{24}(x_2), .., \ A_{n2}(x_n),\ A_{n3}(x_n)$$

Layer 2. Each neuron in this layer generates a multilinear output which represents the firing strength of a rule; each multilinear term consists of the product of n linear terms as in (5). The rules with non-zero firing strength are only those associated with the active neurons in layer 2 (only one pair of complementary neurons per input), that is, 2^n active rules. We will use ω_k^* $(1 \leq k \leq 2^n)$ to denote the firing strength of active rules.

$$\begin{array}{c} \omega_1^* = A_{12}(x_1) \cdot A_{23}(x_2) \cdot \ldots \cdot A_{n2}(x_n) \\ \omega_2^* = A_{12}(x_1) \cdot A_{23}(x_2) \cdot \ldots \cdot A_{n3}(x_n) \\ \ldots\ldots\ldots\ldots\ldots \\ \omega_{2^n-1}^* = A_{13}(x_1) \cdot A_{24}(x_2) \cdot \ldots \cdot A_{n2}(x_n) \\ \omega_{2^n}^* = A_{13}(x_1) \cdot A_{24}(x_2) \cdot \ldots \cdot A_{n3}(x_n) \end{array} \tag{9}$$

Layer 3. Taking into account constraint (iii), it can easily be proved that the normalization layer disappears because the division is unnecessary [6].

$$\sum_{k=1}^{2^n} \omega_k^* = 1 \tag{10}$$

Layer 4. The neuron outputs produce the weight of the consequents. We will use $(1 \leq k \leq 2n)$ to denote the consequents associated with the active rules.

$$y_k = \omega_k^* \cdot r_k^* \tag{11}$$

Layer 5. Finally, generalizing the expression, the output layer is reduced to the sum of *2n* product terms,

$$y = \omega_1^* r_1^* + \omega_2^* r_2^* + \ldots + \omega_{2^n-1}^* r_{2^n-1}^* + \omega_{2^n}^* r_{2^n}^* \tag{12}$$

In sum, the main benefits of constraints (i) to (iii) on the general ANFIS architecture are a reduction of the number of neurons per layer (layers 1 and 2) due to the activation of a reduced number of antecedents, the elimination of the normalization layer, and a simplification of the network arithmetic.

Concerning the learning procedure, the advantages of restrictions (i) to (iii) are twofold. Firstly, as has been seen, the PWM-ANFIS limits the activation of the system each time to a single cell or region of the input space. This cellular nature of the feed-forward network (fuzzy inferences) also reduces the computational complexity of the learning algorithm because both LSE and GDM equally require the evaluation of the feed-forward network. Secondly, the constraints imposed on the input partition reduce not only the active set of parameters for a concrete input (parameters of the active cell), but also the total set of antecedent parameters.

With the above restrictions, a total of $(2^n + 1)n$ products and $2^n + 2n - 1$ sums is required to perform an inference. This cost is considerably less than that required by the generic ANFIS and it does not depend on the number of membership functions m. Moreover, as n and m increase, the difference between both costs (generic ANFIS and PWM-ANFIS) becomes extremely large.

3 The PWM-ANFIS Tool

The PWM-ANFIS Tool, developed by the authors, provides a compact design flow that encompasses all the steps involved in the development of NFSs. The tool allows the designer to define multivariable systems without limitations either in the number of membership functions or in the number of rules. Triangular and Gaussian membership functions are available to model both PWM-ANFIS systems and generic ANFIS systems, respectively. For purposes of analysis, the tool obtains the Sum Squared Error (SSE) between the target function and their corresponding approximations during the learning process. After training it can also provide the Generalized Sum Square Error (GSSE), which is the SSE but evaluated with a collection of non-training data. These error parameters give measures about the approximation capability of the developed system. The simulation process takes advantage of the Matlab resources, fundamentally in matrix treatments of complex numerical calculations, to give a high speed response. The most popular debugging facilities of the Matlab environment are also available to refine the system designs. The tool also provides a user-friendly graphic interface to monitor the evolution of learning processes. In addition, it generates object code for developing off-line training applications. The tool has been tested extensively by means of several nonlinear functions [8] and it has been used to develop efficient SW solutions, high performance HW solutions, and hybrid HW/SW approaches [9,10].

Fig. 4 shows a block diagram of the PWM-ANFIS Tool. The main block is a Matlab program that gives support off-line learning applications. In the off-line learning mode the program accepts the following files: a file that contains the parameters of the process, a file that contains the Input/Output training data pairs, and a file that contains test data pairs (optional). The parameters of the process are the number of antecedents for each variable, the learning rate, the target error, and the maximum number of iterations.

The first step in the development cycle consists in loading the parameters of the process. Then the tool extracts the system size and generates an initial (non-trained) PWM-ANFIS and the process commences. The training data pairs are read from the text file. Meanwhile, the user is able to monitor the evolution of the SSE. The training process finishes when either the specified error is achieved or the maximum number of iterations is reached. If the GSSE option is enabled, the test data file is loaded and the GSSE is calculated. In addition, the graphical interface provides a representation of both the trained function and the target function. Moreover, the membership functions (initial and final triangle partitions) are represented.

As a result of the trained process the PWM-ANFIS Tool generates a text file with the trained parameters (antecedents and consequents) which can be exported to develop different kinds of implementations. In addition, Matlab is able to compile the m-file program (script) to provide C code. Hence, an executable file can be implemented, for instance, on a microprocessor, a digital signal processor (DSP), or a System-on-a-Programmable Chip (SoPC).

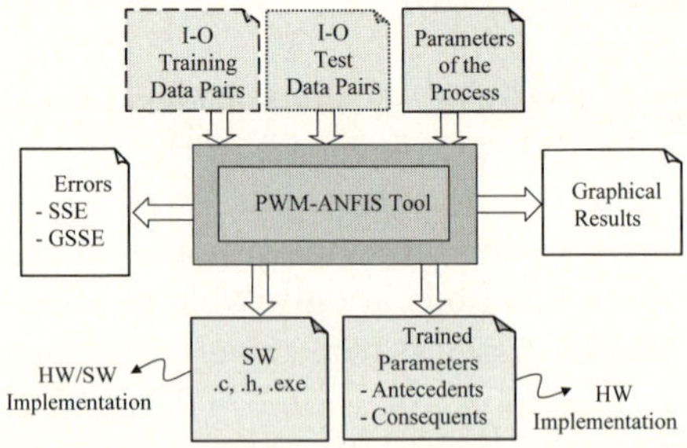

Fig. 3. Block diagram of the PWM-ANFIS environment

4 Application of the Development Methodology to Some NFS Case Examples

In this section, the proposed PWM-ANFIS tool will be used to develop some NFS examples. The examples deal with the approximation of non-linear functions. In particular, we have selected some functions that are commonly used as test examples in related works [1,11]. All these examples clearly illustrate the proposed methodology and they also show how the PWM-ANFIS has good approximation capability despite the imposed restrictions.

4.1 Approximation of Non-linear Functions

Two-Variables: The first example deals with the non-linear function $y = \sin(x_1 \cdot \pi) \cdot \sin(x_2 \cdot \pi)$, with $x_1 \in [-1, 1]$ and $x_2 \in [0, 1]$. This function is graphically shown in Fig. 4(a). A collection of training points is obtained by sampling x_1 and x_2 with 21 and 11 points respectively; hence, we have a total of 231 training points. Another 200 intermediate points are also used as test points. The parameters for the system are the following: 1) number of antecedents for each variable $m_1 = 7$ and $m_2 = 5$; 2) learning rate for each variable $\eta_1 = \eta_2 = 2.5$; 3) maximum number of iterations $i_{max} = 15$; and 4) the target error $SSE_t = 0.0001$.

The tool starts by importing the input files from which the size and dimensionality of the NFS are calculated. Then, it builds up the NFS and begins the training process. After the 15-th iteration, the process stops and displays the *SSE* curve and the *GSSE* value (see Fig. 5(a)). We can see that the system reaches an error as low as *SSE=0.000171.* The tool also provides a graphical representation of the obtained surface (i.e., the output for the training points) (see Fig. 4(b)). Note the close similarity between the target function and the learned function.

Three-variables and Four-variables: In the example for three variable, the approximation of a polynomical 3-input function is presented. The target funcion is $y = (1 + x_1^{0.5} + x_2^{-1} + x_3^{-1.5})^2$, with x_1, x_2 and x_3 in [1,6]. In this experiment

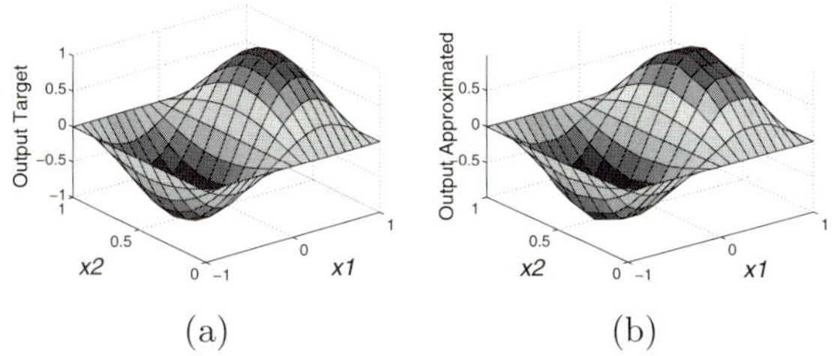

Fig. 4. (a) Two-variable target function, (b) Two-variable learned function

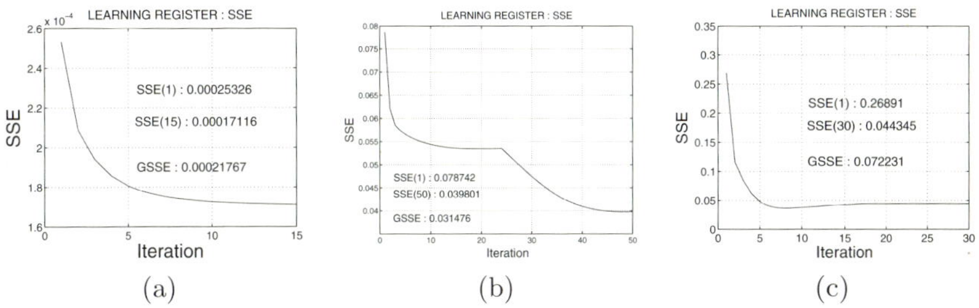

Fig. 5. SSE and GSSE errors (a) 2-variables, (b) 3-variables, (c) 4-variables

a total of 1331 training points (11 samples per dimension) and another 1000 intermediate test points have been collected. The input parameters are now the following: $m_1 = m_2 = m_3 = 5$; $\eta_1 = \eta_2 = \eta_3 = 0.99$; $i_{max} = 50$; and $SSE_t = 0.0001$. Fig. 5(b) shows the error values once the 50 iterations have been completed. We see again that the system reaches very low error values; i.e., *SSE*=*GSSE*=0.0398.

To analyze the response of the system for a large number of inputs, a 4-input function has also been tested. The function is given by $y = 4(x_1 - 0.5)(x_4 - 0.5)\sin(2\pi(x_2^2 + x_3^2)^{1/2})$, where x_1, x_2, x_3 and x_4 in [-1,1]. The number of training points and test points are respectively 4096 (84) and 2041 (74). The parameters for the system are the following: $m_1 = m_2 = m_3 = m_4 = 5$; $\eta_1 = \eta_2 = \eta_3 = \eta_4 = 0.5$; $i_{max} = 50$; and $SSE_t = 0.0001$. Results obtained are depicted in Fig. 5(c). Once again we can see that the values of the errors are very small; i.e., *SSE=0.044* and *GSSE=0.0722.*

5 Concluding Remarks

A design methodology, based on a new Matlab tool, to develop computational-efficient neuro-fuzzy systems has been presented. The proposed methodology is based on a particular class of adaptive neuro-fuzzy inference systems (ANFIS) with piecewise multilinear (PWM) behaviour. The proposed methodology has been tested extensively with well-known nonlinear functions. Results obtained show that the PWM-ANFIS model generates computational-efficient implementations without loss of approximation capabilities or learning performance. The tool has been used to develop both software and hardware approaches as well as special architectures for hybrid hardware/software embedded systems.

References

1. Jang, J.-S.R., Sun, C.-T., Mizutani, E.: Neuro-Fuzzy and Soft Computing, Part VII. Prentice Hall, Upper Saddle River (1997)
2. Acampora, G., Loia., V.: A Proposal of Ubiquitous Fuzzy Computing for Ambient Intelligence. Inf. Science 178, 631–646 (2008); 5, 3005–3009 (2003)
3. Jang, J.S.R.: ANFIS: Adaptive-network-based fuzzy inference system. IEEE Trans. Systems, Man, and Cybernetics 23(3), 665–685 (1993)
4. Buckley, J.J.: Sugeno type controllers are universal controllers. Fuzzy Sets Systems 53(3), 299–303 (1993)
5. Kosko, B.: Fuzzy systems as universal approximators. IEEE Transactions on Computers 43(11), 1329–1333 (1994)
6. Rovatti, R.: Fuzzy piecewise multilinear and piecewise linear systems as universal approximators in Sobolev norms. IEEE Transactions on Fuzzy Systems 6(2), 235–249 (1998)
7. Zeng, X.-J., Singh, M.G.: Approximation accuracy analysis of fuzzy systems as function approximators. IEEE Transactions on Fuzzy Systems 4(1), 44–63 (1996)
8. Basterretxea, K., del Campo, I., Tarela, J.M., Bosque, G.: An Experimental Study on Nonlinear Function Computation for Neural/Fuzzy Hardware Design. IEEE Transaction on Neural Networks 18(1), 266–283 (2007)
9. Echanobe, J., del Campo, I., Bosque, G.: An Adaptive Neuro-Fuzzy System for Efficient Implementations. Inf. Science 178(9), 2150–2162 (2008)
10. del Campo, I., Echanobe, J., Bosque, G., Tarela, J.M.: Efficient Hardware/Software Implementation of an Adaptive Neuro-Fuzzy System. IEEE Transactions on Fuzzy Systems 16(3), 761–778 (2008)
11. Lee, S.J., Ouyang, Ch.S.: A Neuro-Fuzzy System Modeling With Self-Constructing Rule Generation and Hybrid SVD-Based Learning. IEEE Transactions on Fuzzy Systems 11(3), 341–353 (2003)

Embedded Neural Network for Swarm Learning of Physical Robots

Pitoyo Hartono and Sachiko Kakita

Department of Media Architecture, Future University-Hakodate
Kamedanakanocho 116-2, Hakodate, Japan
hartono@fun.ac.jp
http://www.fun.ac.jp/~hartono/hartono.html/

Abstract. In this study we ran real time learning of multiple physical autonomous robots situated in a real dynamic environment. Each robot has an onboard micro controller where a simple neural network is embedded. The neural network was built with the consideration of the power and calculation resources limitation which is a general characteristic of simple robots. In the experiments, several autonomous robots were placed in one environment, where each of them was given a specific task which was expressed as the evaluation function for the robot's neural network. The learning processes of the robots were started simultaneously from their randomized initial conditions. The presence of several robots consequently formed a dynamic environment, in which an action of one robot affected the learning process of others. We demonstrated the efficiency of the embedded learning mechanism with respect to different environmental factors.

1 Introduction

In the past decade, several multi-robot systems have been proposed. This relatively young field encompasses a number of interesting research topics [1]. The fact that so far we have not succeeded in building highly autonomous singular robots capable of reliably and continuously operating in dynamic environment, is one of the factors that motivates the study of multi-robot systems. The emergence of swarm intelligences in nature such as in ants' or bees' colonies is one of the main inspirations for these studies[2]. In [3] a decentralized control theory for multiple cooperative robotic system was developed. A group of robots, each with a simple predefined behavior, capable of cooperating in self-organized manner was proposed in [4]. A study in [5] implemented a social reinforcement learning to encourage the emergence of social behaviors in multi-robot system. Most of the previous studies agreed that bottom-up behavior-based mechanism is a more realistic approach compared to the centralized top down control in running multi-robot systems for achieving sophisticated group behaviors applicable to real world task. It is also becoming obvious that development of learning mechanism of a single robot within a multi-robot system to obtain intelligent strategies in supporting the group behavior is one of the core topics.

V. Kůrková et al. (Eds.): ICANN 2008, Part II, LNCS 5164, pp. 141–149, 2008.

In this research we developed a simple neural network that can be embedded on the onboard processor of a physical simple autonomous robot which is limited in power, computational resources and sensors' precision. The efficiency of the learning mechanism is demonstrated with the execution of a collective learning scheme where several robots are placed in a same environment. Initially the robots did not have any moving strategy, hence they moved in a random manner. A learning target is then given to each robot in the form of an evaluation function for the embedded neural network. There is no particular limitation on the learning task, hence it is possible that all of the robots have different individualistic goals. However, in our research we set a common primitive goal for each of the robots, which is the acquirement of obstacle avoidance strategy while random walking in the presence of each other. Although simple, we consider that an effective obstacle avoidance strategy is strongly correlated with self-preservation capability of autonomous robots situated in dynamic environments. In this research, because each robot must execute its learning process in the presences of other robots, the learning environment becomes highly dynamics, where the learning process of one robot is affected by the behaviors of others. So far, this kind of collective learning is rarely executed in physical multi-robot systems, at least without any partial help from simulators. One of our near future interests is to learn about the collective behavior of the multi-robot system at various stages of the learning process of the respective individual robots. This will lead to a clearer understanding in the relation between the individual behavior and the collective behavior of the multi-robot system. We believe this research contributes in providing experimental platform not only the further multi-robot systems studies but also in understanding the emergence of the swarm intelligence in nature.

2 Robot's Structure and Behavior

For this study we built several robots with two different structures shown in Fig.1. The robot in the right side of the figure is built based on a commer-

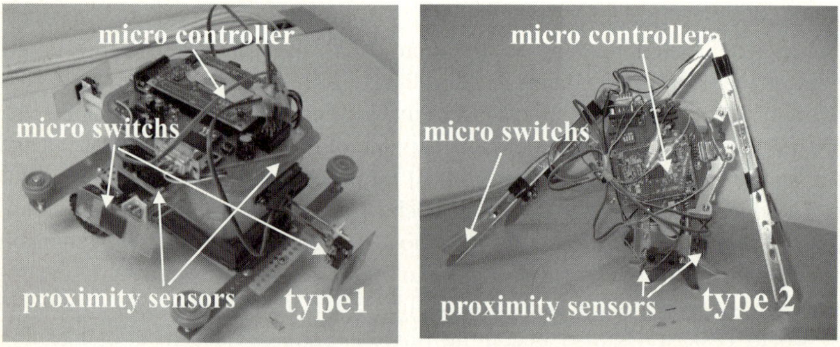

Fig. 1. Simple Robots

cially available toy. The size and weigh of each robot are approximately 230 x 180 x 130 [mm], and 0.5 [kg] (without batteries), respectively. For each robot, we attached four proximity sensors which measure distance from obstacles and four micro-switch to detect collisions. Each robot has two DC motors, one micro controller(H8/3052) where a simple neural network is implemented and two EEPROMs to store the internal state of the neural network. Using ten 1.2[V] nickel hydrogen batteries, a robot can operate for up to 3.5 hours. The internal structure, common for all the robots, is shown in Fig.2.

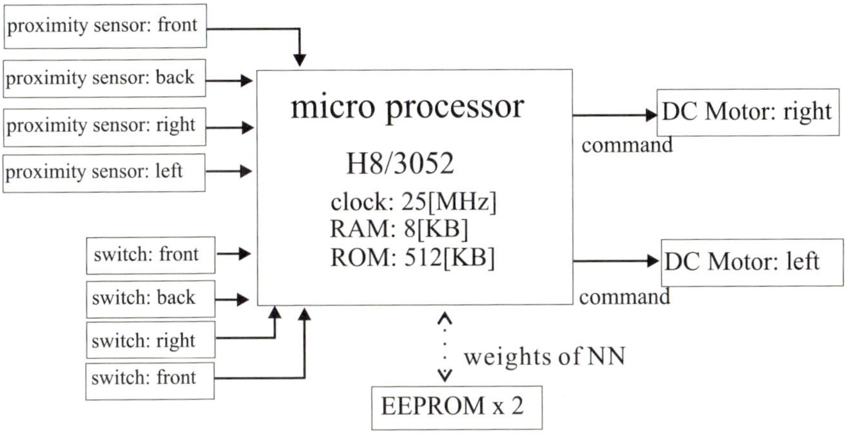

Fig. 2. Internal Structure

3 Learning Algorithm

For the controller of each robot, a simple two-layered neural network is fully embedded onboard. In the input layer, the current sensory values from the sensors, the delayed values of the past sensors' measurements, the past actions and the associated evaluation values are given as the current inputs. Each neuron in the output layer is associated with a primitive behavior of the robot. In this study we designate four primitive behaviors, which are $forward$, *backward*, $turn-left-foward$, and $turn-right-forward$. The execution time of each behavior is the same. The last two behaviors are designated to prevent the robot from rotating in the same position. It should be noted that *stop* is not one of the primitive behaviors, hence the robot is forced to continuously move.

In this research, there is no distinction between a learning phase and a running phase. In its life time, a robot continuously utilizes the embedded neural network to generate movements while also updating its neural network based on the feedback from the environment. Consequently, the robot continuously adapts the gradual change in the environment.

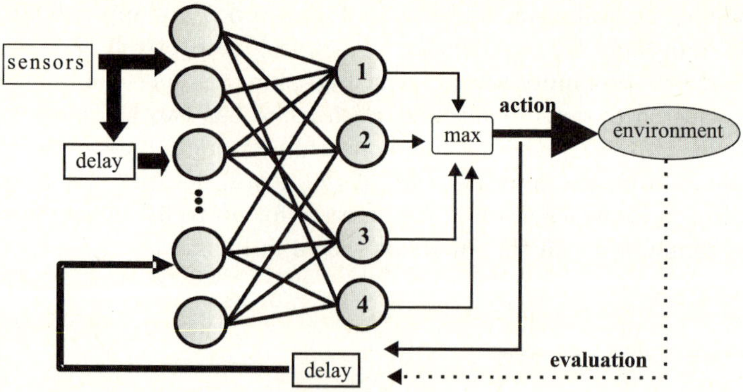

Fig. 3. Neural Network

For the neural network of a robot, the value of the j-th output neuron at time t, $O_j(t)$, can be shown as follows.

$$O_j(t) = f(\sum_{i=1}^{N} w_{ij}(t)x_i(t)) \tag{1}$$

$x_i(t)$, and $w_{ij}(t)$ are the value of the i-th input neuron and the connection weight between the i-th input neuron and the j-th output neuron at time t, respectively, while N is the number of the input neurons. All input x is normalized so that, $x \in [-1, 1]$. The activation function is shown as follows.

$$\begin{aligned} f(y) &= y \quad if \quad \max_j \|y_j(t)\| < 1 \\ &= \frac{y}{\max_j \|y_j(t)\|} \quad otherwise \\ j &\in \{1, 2, 3, 4\} \end{aligned} \tag{2}$$

The normalization in Eq.2 is needed so that the past behaviors that are given as feedbacks to the input layer have the same scale as other inputs.

After processing a given input vector, the values of all of the output neurons are calculated, and then the output neuron that generates the largest value is designated as a winner neuron, as follows.

$$win(t) = arg \max_j \{O_j(t)\} \tag{3}$$

The robot then executes a primitive behavior associated the winner neuron as follows.

$$a(t) = A(win(t)) \tag{4}$$

$a(t)$ denotes the primitive behavior executed at time t, while A is a function to map an output neuron into its associated primitive behavior.

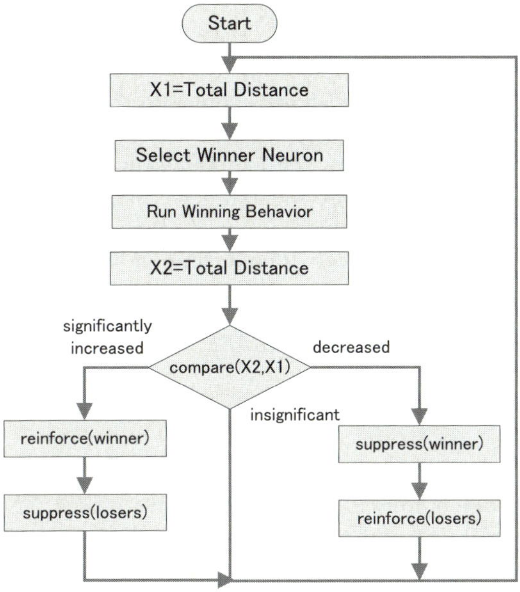

Fig. 4. Flowchart of Behavior of Individual Robot

After the robot executed a winning primitive behavior, it calculates the cumulative value of the distance measurements of all the proximity sensors, as follows.

$$D(t+1) = \sum_{i=1}^{K} p_i(t+1) \tag{5}$$

In Eq.5, $D(t+1)$ shows the cumulative distance value at time t. $p_i(t)$ is the value of the i-th proximity sensor at time t, while K is the number of proximity sensors.

The evaluation for the robot's movement at time t, $E(t)$ is defined as follows.

$$E(t) = D(t+1) - D(t) \tag{6}$$

The evaluation function is simply the difference between the cumulative distances before and after the robot executed its primitive behavior. A positive $E(t)$ is an indication that the robot moved away from obstacles in the environment as the implication of its recently executed behavior. In this case the winning neuron will be rewarded by correcting the connection weights so that its output increases with respect to the same input. At the same time the losing neurons will be punished by correcting the connection weights so that their values decrease with regards to the same input. Hence, in this case the learning process works in a positive feedback manner for the winning neuron, and a negative feedback manner for other neurons. A negative $E(t)$ indicates that the robot executed a wrong behavior because it moved towards obstacles. In this case, the winning neuron is punished by correcting the connection weights so that its value decreases. At the same time, the robot has to be encouraged to execute another behavior, hence the connection weights should be corrected so that the values of the non-winning neurons increase. In this case, the learning process works in a negative feedback manner for the winning neuron and a positive feedback manner of others. In the case that the primitive behavior of the robot did not bring a significant improvement or deterioration, no correction is made to the internal state of the neural network. The outline of the robot's learning process is shown in Fig.4 and the weight correction algorithm is formulated as follows.

$$w_{ij}(t) = w_{ij}(t-1) + \alpha(E(t))\ \phi(E(t), x_i(t), j) \tag{7}$$

$$\begin{aligned} \alpha(E(t)) &= \eta \ \ if \ \ |E(t)| > T \\ &= 0 \ \ otherwise \end{aligned} \tag{8}$$

T and η in Eq.8 are empirically decided positive values. The direction of the correction is decided by the function $\phi()$ which is defined in the table below.

	$E(t) \geq T$		$E(t) < T$	
	$x_i > 0$	$x_i \leq 0$	$x_i > 0$	$x_i \leq 0$
$j = win(t)$	1	-1	-1	1
$j \neq win(t)$	-1	1	1	-1

The correction rule in Eq.7 guarantees that when the executed behavior at time t improved the evaluation of the robot, for the same input vector X,

$$\begin{aligned} O_j^{n+1}(X) &> O_j^n(X) \ \ j = win \\ O_j^{n+1}(X) &< O_j^n(X) \ \ j \neq win \end{aligned} \tag{9}$$

In the case that the executed behavior deteriorated the evaluation of the robot,

$$\begin{aligned} O_j^{n+1}(X) &< O_j^n(X) \ \ j = win \\ O_j^{n+1}(X) &> O_j^n(X) \ \ j \neq win \end{aligned} \tag{10}$$

$O_j^n(X)$ denotes the output of the j-th output neuron in reaction to input vector X given for the n-th time.

Equation 9 implies that the neural network reinforces a good behavior while simultaneously suppresses other alternatives, while Eq.10 shows that the neural network suppresses a proven harmful behavior and explores alternative ones. While the implemented learning mechanism has similar properties with the conventional Reinforcement Learning, it operates in continuous value domain and requires significantly less calculation resources (especially memory). The characteristics of the learning method allow it to be fully implemented on an onboard processor that supports the real time learning process of a robot situated in a dynamic real world environment.

4 Experiments

In this research we run experiments in two types of environments. The first environment is shown in the left side of Fig.5. In this experiment, three robots were arbitrary placed in a straight line surrounded by four walls, and the primitive behaviors are limited to *forward* and *backward*. The connection weights of these robots' neural networks were randomly initialized.

The result of the first experiment is shown in Fig.6. The left graph shows the average of cumulative distance measured by the proximity sensors. From this graph it is obvious that the cumulative distance increases with the progress of the learning process. The unit for the ordinate of this graph is the unit of the distance measurement of the proximity sensor, while the unit of the abscissa is the learning step. The total learning time for 500 learning steps are equivalent to approximately 5 minutes. The right graph shows the average number of collisions of the three robots. This graph indicates that the number of collisions decreases with the progress of the learning process. From these two graphs, it is clear that, through the proposed learning process, the embedded neural

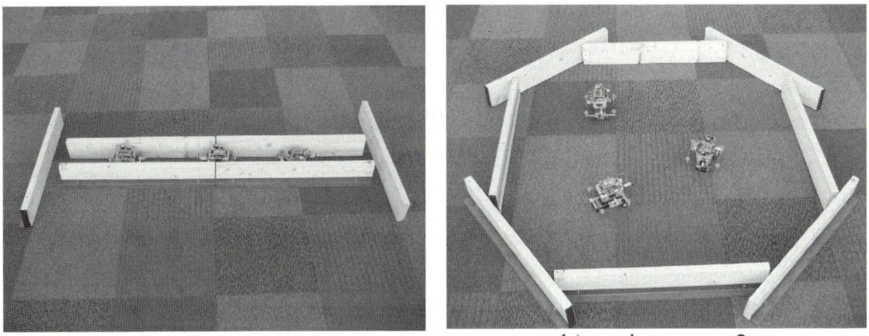

a)environment1 b)environment2

Fig. 5. Environments

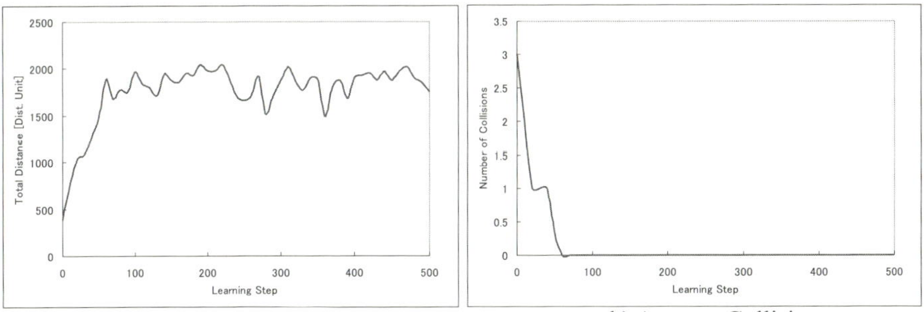

a) Average Distance b) Average Collisions

Fig. 6. Experimental Result in Environment 1

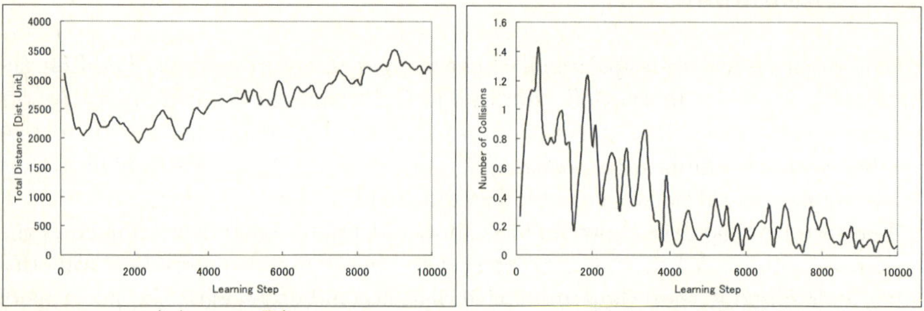

a) Average Distance b) Average Collisions

Fig. 7. Experimental Result in Environment 2

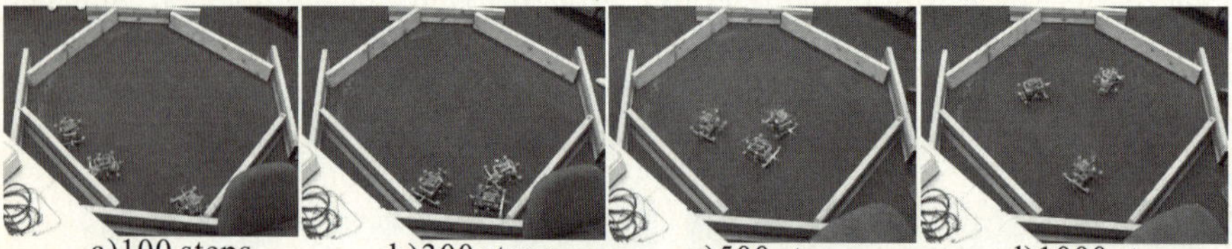

a)100 steps b)300 steps c)500 steps d)1000 steps

Fig. 8. Multi-Robot Learning

networks gradually obtained a strategy to control the robots in achieving the given task (obstacle avoidance) in a dynamic environment. As the consequence of the learning process, we observed that the three robots were able to evenly intervaled themselves in this environment while random walking, thus forming a simple formation.

In the second experiment, three robots are randomly placed inside an environment shown in the right side of Fig.5. In this experiment, all of the primitive behaviors are activated. Figure 7 shows the result of this environment. The two graphs also show the efficiency of the proposed learning method in real time learning. We ran 10000 learning steps which are equivalent to approximately 75 minutes. Although from this experiment we cannot observe any formation generation, the three robots were able to achieve the task of obstacle avoidance while random walking in a two dimensional environment. The shown graphs are the average over 5 iterations of this experiment. Some snapshots of multi-robot learning are shown in Fig.8. The learning parameter (in Eq.8) η is set to 0.1 for the first experiment, and 0.001 for the second experiment, T is commonly set to 10.

5 Conclusions

In this research we proposed a simple learning algorithm that can be fully implemented on onboard processor for low cost autonomous robots with very limited calculation and power resources. Although the proposed learning mechanism is very basic and shares similar properties with Hebbian Learning and Reinforcement Learning, it is sufficiently powerful to be implemented for strategy-acquisition real time learning in dynamic environment.

We run several experiments with two types of simple but dynamic environments, in which three robots must learn the obstacle-avoidance strategy while random walking in the presence of each other, based only on local information and local evaluation. To show the structure-independent property of the learning mechanism, it is also implemented on two types of robots with different structures, which gave similar results. Altough it is not reported in this paper, scalability of the learning algorithm is also tested on the swarm learning containing 10-100 robots on a simulator which also gave good results.

Although very simple, we believe that the proposed learning mechanism including the internal hardware configuration can contribute in setting a low cost standard experimental platform not only in the field of swarm robotics but also for helping to understand the emergence of swarm intelligences in nature.

References

1. Parker, L.E.: Current state of the art in distributed autonomous mobile robotics. Distributed Autonomous Robotic System, vol. 4, pp. 3–12. Springer, Heidelberg (2000)
2. Bonabeau, E., Theraulaz, G.: Swarm Smarts. Scientific American, 82–90 (March 2000)
3. Feddema, J., Lewis, C., Schoenwald, D.: Decentralized control of cooperative robotic vehicles: theory and application. IEEE Trans. on Robotics and Automation 18(5), 852–864 (2002)
4. Parker, C., Zhang, H., Kube, R.: Blind bulldozing: Multiple Robot Nest Construction. In: 2003 Int. Conf. on Robots and Systems, pp. 2010–2015 (2003)
5. Mataric, M.: Learning social behavior. Robotics and Autonomous System 20, 191–204 (1997)

Distribution Stream of Tasks in Dual-Processor System

Michael Kryzhanovsky and Magomed Malsagov

Center of Optical-Neural Technologies of
Scientific Research Institute for System Analysis of
Russian Academy of Science
44/2 Vavilov Str., 119333, Moscow, Russian Federation
iont.niisi@gmail.ru, magomed.malsagov@gmail.com

Abstract. The paper offers an associative-neural-net method to optimize resource allocation between independent tasks in a multiprocessor system. In the case of a dual-core CPU the method allows the task to be fully solved in $O(M)$ operations.

Keywords: Multiprocessor systems, Distributed computing, Neural Networks.

1 Introduction

An important characteristic of today's multi-processor systems is their multitasking ability. Servers, real-time computers and multi-core CPU systems imply a multitasking mode of operation. One of the problems that arise when a multiple-CPU machine tries to compute many tasks simultaneously is efficient allocation of workload between the CPU cores. Inefficient use of multi-CPU systems was the reason for making special resource-controlling systems. The typical design of a system like that is given in Fig. 1. Computation process planning plays an important role in resource management [1]-[6].

However, computational complexity of conventional mathematical methods makes them inefficient in computation process planning. Then the problem is to develop efficient algorithms of computational resource allocation.

The paper offers a load balancing algorithm using methods of associative Hopfield neural nets.

2 Setting the Problem

Let a system of N processors of equal productivity V handle M independent tasks of computational complexity H_i $(i = 1 \ldots M)$. Then the processing of the i-th task takes time

$$\tau_i = H_i/V. \tag{1}$$

Optimization of load allocation among multiple processors suggests minimization of execution time T for all tasks [1], [7]. Time T is defined as

$$T = \max\{T_1, T_2, \ldots, T_N\}, \tag{2}$$

V. Kůrková et al. (Eds.): ICANN 2008, Part II, LNCS 5164, pp. 150–158, 2008.

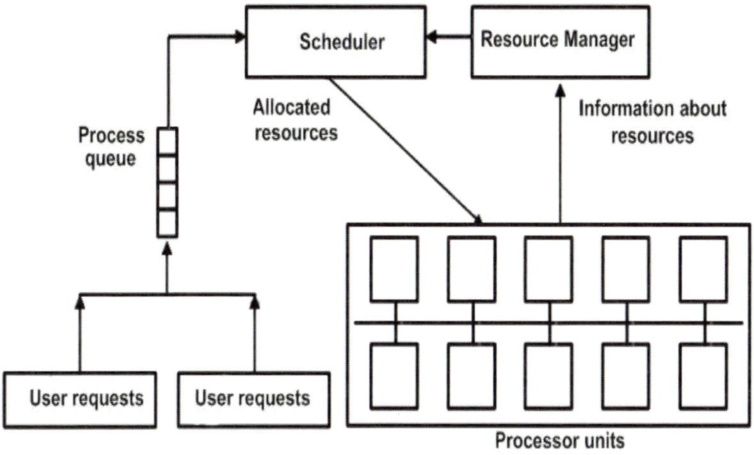

Fig. 1. The typical resource management system

where T_j is the time the j-th processor unit takes to complete its own subset of tasks. For example, if $a, b, \ldots c$ are the numbers of the tasks to be executed by the j-th processor, this time is found to be

$$T_j = \tau_a + \tau_b + \ldots + \tau_c. \tag{3}$$

It is necessary to allocate all the tasks among processor units so that time T is smallest.

3 Setting the Problem in a Different Way

Let us denote an instance of task allocation among processor cores as G. This allocation is characterized by time parameter

$$T = T(G) = \max\{T_1, T_2, \ldots, T_N\}. \tag{4}$$

For each G it is possible to find the variance of execution times

$$\sigma^2 = \frac{1}{N}\sum_{j=1}^{N}\left(T_j - \overline{T}\right)^2, \tag{5}$$

where

$$\overline{T} = \frac{1}{N}\sum_{j=1}^{N} T_j = \frac{1}{N}\sum_{i=1}^{M} \tau_i. \tag{6}$$

Then there is a relationship between σ^2 and T, which is proved by experiment results. In particular, Figures 2 give the results of a numerical experiment, in which a set of 20 tasks of different computational complexity $H = [1, \ldots, 10]$ were randomly allocated among 4 identical processors 10000 times. Execution time T and variance σ^2 were computed for each instance of task allocation and the corresponding graphs were drawn. It is seen (Fig.2) that the same T has several different values of variance. However, the mean value of variance $\overline{\sigma^2} = \frac{1}{K}\sum_{l=1}^{K}\sigma_l^2$ grows steadily with T. Left picture on Figure 2 shows the mean variance $\overline{\sigma^2}$ as a function of execution time. As seen from the figure, the dependence is monotonous.

Over 1000 computer simulation experiments were carried out to make sure this fact. In the experiments the number of processors varied from 2 to 10, and the complexity factor $M \in [6; 20]$. That the experiments gave similar results allows us to turn to minimization of variance σ^2 rather than minimization of T.

Setting the problem this way (i.e., optimization of σ^2 rather than minimization of T) leads us to load balancing algorithms whose idea is to allocate computational load among processing units most evenly. In turn this allows us to resort to optimization algorithms used in vector neural networks. Task allocation in the case of a dual-CPU machine is particularly interesting because:

1. in this event σ^2 is strictly dependent on execution time (Fig. 3) repeating the behavior of T. Below in the paper we deal with a system of two equal-productivity CPUs;

2. the algorithm can be used for resource management in the case of multiple-CPU systems;

3. using methods of associative neural nets, the algorithm allows us to cope with the problem in $O(M)$ operations.

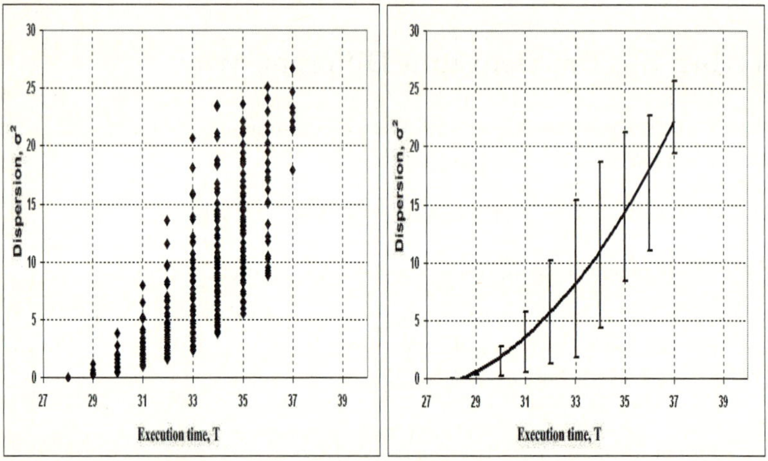

Fig. 2. Variance versus execution time: four processors of equal power; 20 processes; the number of runs 10000 and on right figure mean variance of σ^2 and its mean square deviation versus execution time

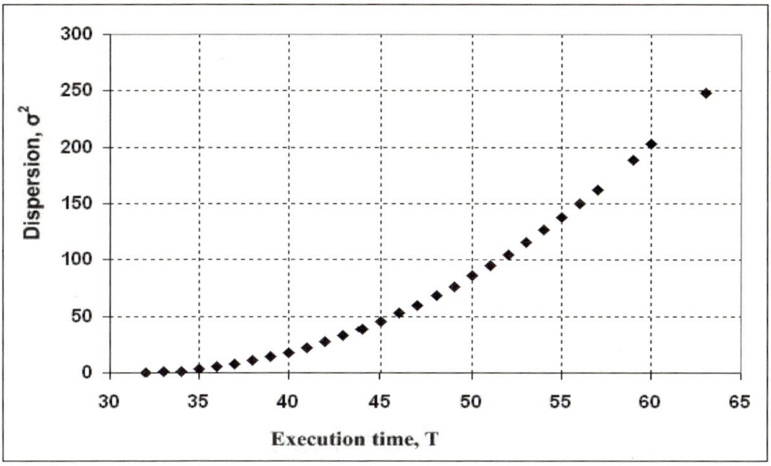

Fig. 3. Dependence of σ^2 on execution time: two equal-productivity processors; ten processes whose computational complexity H=7;4;7;6;7;10;8;8;3;3;

4 The Neural Net Model

Let T_1 and T_2 be the times that the first and second processors take to complete its own task. Let binary variable (coordinate) y_i denote the i-th task belonging to either processor; namely, if $y_i = 1$, the task is run on the first processor, if $y_i = 0$, the second executes the task. In these terms variance σ^2 takes the form:

$$\sigma^2 = [T_1 - T_2]^2 = \left(\sum_{j=1}^{M} (2y_j - 1) H_j \right)^2 , \tag{7}$$

After some transformations, we get

$$\sigma^2 = const + \sum_{i,j=1}^{M} A_{ij} x_i x_j = \left(\sum_{i=1}^{M} H_i x_i \right)^2 - \sum_{i=1}^{M} (H_i x_i)^2 , \tag{8}$$

where

$$A_{ij} = \begin{cases} H_i H_j, & i \neq j \\ 0, & i = j \end{cases} , \tag{9}$$

$$x_i = 2y_i - 1, x_i = \{-1, +1\}, y_i = \{0, 1\} .$$

Notice that expression (8) for variance σ^2 is similar to definition of energy in the Hopfield model of neural net, where matrix A_{ij} stands for $(M \times M)$-dimensional interconnection matrix. For this reason a standard neural-net algorithm is used for finding the minimum of quadratic functional (8) with respect to binary variables x_i. For this kind of interconnection matrix A_{ij} the behavior of the Hopfield network is as follows: the state of the k-th neuron at time $(t+1)$ is determined in accordance with the Hopfield rule:

$$x_k\,(t+1) = -sign\,(S(t) - H_k x_k(t))\,, \quad k = 1, \ldots, M, \tag{10}$$

followed by modification $S = \mathrm{H} \cdot \mathrm{x}$

$$S(t+1) = S(t) - H_k x_k(t) + H_k x_k(t+1). \tag{11}$$

The algorithm like that allows an M-times decrease of the number of operations, which is equal to $O\,(M)$ now.

5 Computer Simulation

That the neural-net algorithm optimization is fast brings much advantage in tackling the process management for multiple-PU systems. However, as the algorithm falls into the category of iteration methods, its effectiveness depends on what initial state we choose. That is why it is necessary to define the method of choosing the initial state. In our computer simulation experiments we used three methods.

1. The leveling method allows us to choose the initial state of the neural network. It works as follows: a) the rough value T_0 of minimal execution time is first found by halving the sum of all run times; b) the first processor is fed with tasks until their total execution time exceeds T_0; c) the other tasks go to the second processor.
2. The random search method involves multiple search for the minimum of functional (8) with the help of Hopfield neural net. The state with the smallest σ^2 is chosen in the process. The initial task allocation is chosen by chance.
3. The modified method of load balancing suggests the initial task allocation being determined by the leveling method.

To get statistically reliable results, runs of the simulation experiments involved 1000 sets of M tasks (M = 15, 17, 20, 60). The computational complexity of the tasks varied in the interval from 1 to D (D was either 10 or 100).

Initially the exhaustive search method was used to find the global minimum of functional (8) for each task set. The result was compared to that obtained with the aid of the leveling method, random start method and modified method. Fig.4-7 give the results of the experiments.

Left graphs in Fig. 4 shows the probability of finding solutions as a function of departure from the global minimum for $D = 10$, $M = 20$. By way of comparison, right graphs in Fig. 4 gives similar graphs for $D = 10$, $M = 15$.

Variable A which is laid off as abscissa of all the graphs is the distance to the global minimum in units of D, that is, $A = (dT - dT_0)\,/D$, where dT_0 is the lag of the execution times for the optimal allocation, dT is the time lag for the given task allocation. Clear that A characterizes asymmetry in processor loading.

The experiments showed (Figure 4) that use of the Hopfield network in the random start method gives better results than when used in the leveling method.

At the same time, combining the leveling method and the Hopfield net algorithm gives still better results. The fact is very interesting because the Hopfield algorithm has 1000 starts with one task set, and direct application of "leveling" is not effective.

Indeed, the probability of finding the optimum at the distance $A \leq 0.4$ is ≈ 0.35 for the leveling method, ≈ 0.9 for the random start method, ≈ 1 for the modified method. Further, these probabilities for $x \leq 0.2$ are 0.22, 0.8, 1. All the runs in these experiments had the number of tasks larger than D ($M > D$).

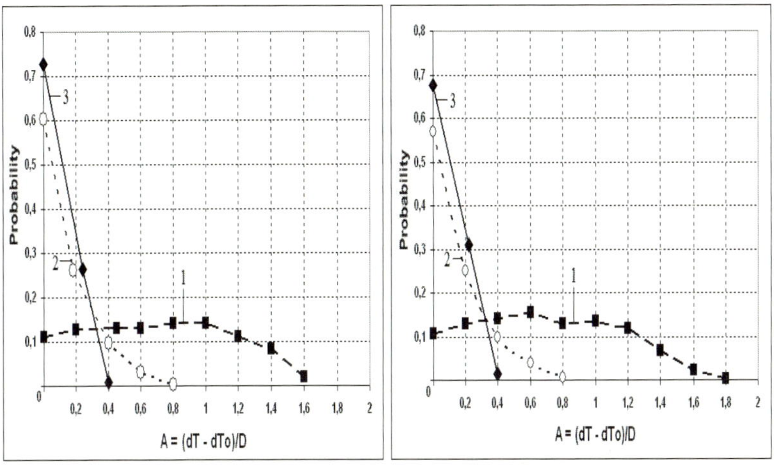

Fig. 4. Probability of finding the minimum by the leveling method 1, by the random search method 2, the modified load-balancing method 3 for $M = 20$, $D = 10$ (left fig.) and $M = 15$, $D = 10$ (right fig.)

Figure 5 gives the results of experiments when the number of tasks ($M = 20$) is much fewer than D ($D = 100$), - another limiting case. As in previous plots, the X-axis has units of dimensionless quantity $A = (dT - dT_0)/D$ (see above). The probability that the distance to the global minimum is less than X ($P\{A < X\}$) is unit of the Y-axis. Plots 1, 2, 3 correspond to different methods: leveling method, random-start method and modified method. Again in this case the modified method (curve 3) shows better results than the other two. In particular, the energy departure from the global minimum is no greater than $\approx 1.6D$ for the leveling method.

To see what happens between the limiting cases, we ran some experiments for M/D ($M = 60$,$D = 100$) and obtained similar results. To illustrate this, Figure 6 reports the data received in the experiments using the modified method. The experimental results allow us to determine expectations of quantity A:

$dT - dT_0 = 0.700D$ for the leveling method;
$dT - dT_0 = 0.126D$ for the random start method;
$dT - dT_0 = 0.065D$ for the modified method.

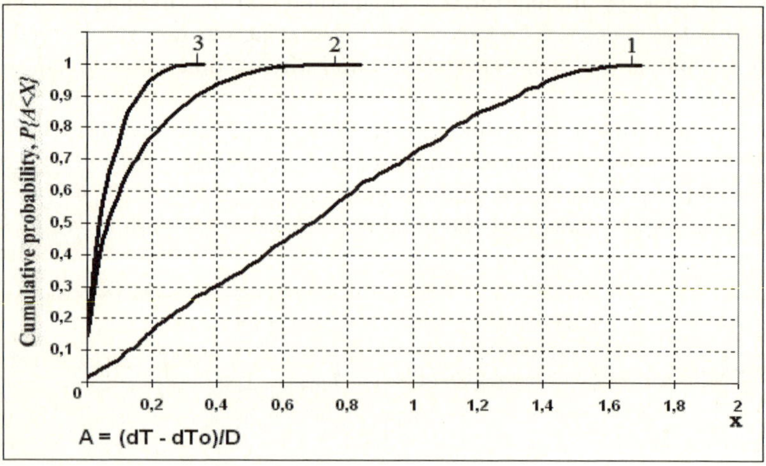

Fig. 5. Cumulative probability $P\{A < X\}$, that is, probability that loading asymmetry is less than X (curve 1 - the leveling method, curve 2 - the random start method, curve 3 - the modified method)

Modified load balancing method				
	M=20 D=10	M=15 D=10	M=20 D=100	M=60 D=100
A	Cumulative probability	Cumulative probability	Cumulative probability	Cumulative probability
0,00	0,728	0,676	0,172	0,355
0,04	-	-	0,507	0,73
0,08	-	-	0,701	0,885
0,12	-	-	0,834	0,98
0,16	-	-	0,904	0,995
0,20	0,992	0,985	0,951	1
0,24	-	-	0,973	-
0,28	-	-	0,994	-
0,32	-	-	0,998	-
0,36	-	-	0,999	-
0,40	1	1	1	-

Fig. 6. The results of using the modified load balancing method for different number of tasks M and complexity range boundary D

Let us look at Figure 7 to understand the reason why determination of the starting point yields better results.

The figure gives the experimentally found distribution of probable solutions of the neural net (a set of 60 tasks whose complexity varies from 1 to 100). The X-axis of the both graphs is the probable solutions of the neural net (energy levels). In the first graph, the degree of degeneration (multiplicity) of the level is laid out along the Y-axis. In the second, it is the probability of level being populated. One million of chance starts were made. The degree of degeneration and the number of hits were computed for each level. The solid vertical line on the graphs denotes the level which we come to when using the leveling method (for this particular experiment).

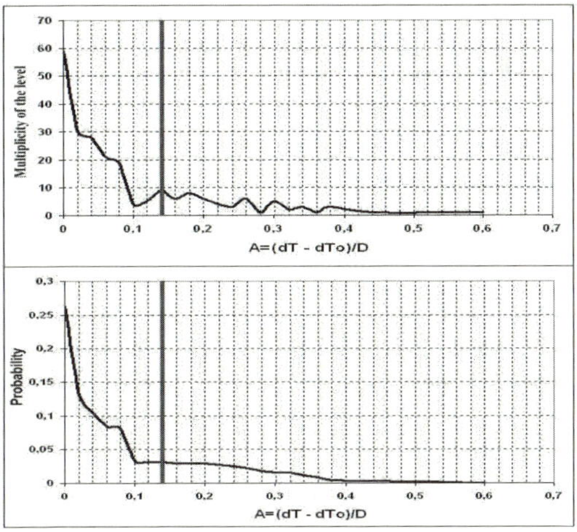

Fig. 7. Experimentally found energy spectrum for $M = 60$, $D = 100$

From the first graph we see that the degeneration degree grows with the depth of the energy level, i.e. the smaller quantity A, the higher the degeneration of the level. It allows a conclusion that the probability of the neural net coming to a particular level also increases with the level depth. The second graph proves the conclusion. It means that the closer to the optimal solution the initial state leads us, the higher the probability of getting to the optimal distribution and the lower the chances of sticking at less deeper minima. The data agree well with the results of earlier researches [8].

6 Conclusions

When the process starts from a random point in the configuration space of vector $\mathbf{x}$, application of neural-net optimization does not guarantee success.

Modification of the load balancing method (when neural-net optimization is accompanied with purposeful choice of the starting point) allows noticeably better results.

A small amount of computations $O(M)$ (M is the dimensionality of the task) is needed to cope with the task. In the most Hopfield network-based methods the amount of computations is $O\left(M^2\right)$.

Acknowledgments. Part of research program "Fundamentals of Information Systems and Techniques" (project 1.8), the work is financially supported by grant 06-01-00109 of the Russian Foundation for Basic Research.

References

1. Casavant, T.L., Kuhl, J.G.: A taxonomy of scheduling in general-purpose distributed computing system. IEEE Trans. on Software Engineering 14(2) (1988)
2. Bruno, J., Coffman, E.G., Sethi, R.: Scheduling independent tasks to reduce mean finishing time. Communications of the ACM 17(7) (1974)
3. Coffman, E.G., Graham, R.L.: Optimal scheduling for two processor systems. Acta Informatica 1(3) (1972)
4. Coffman, E.G., Sethi, R.: Algorithms minimizing mean flow-time. Acta Informatica 6(1) (1976)
5. Horn, W.A.: Minimizing average flow time with parallel machines. Opera- Operations Research 21(3) (1973)
6. Kleinrock, L., Nikon, A.: On optimal scheduling algorithms for time-shared systems. Journal of the ACM 28(3) (1981)
7. Haykin, S.: Neural Networks. A Comprehensive Foundation. Prentice Hall Inc., Englewood Cliffs (1999)
8. Kryzhanovsky, B.V., Magomedov, B.M., Mikaelian, A.L.: Relationship between Depth of Local Minimum and Probability of His Detection in Generalized Hopfield Model. Lectures of Academy of Sciences 40(3) (2005) (in Russian)

Efficient Implementation of the THSOM Neural Network

Rudolf Marek and Miroslav Skrbek

Department of Computer Science and Engineering, Faculty of Electrical Engineering,
Czech Technical University in Prague,
Karlovo namesti 13, 121 35, Prague, Czech Republic
marekr2@fel.cvut.cz,
skrbek@fel.cvut.cz
http://cig.felk.cvut.cz

Abstract. Recent trends in microprocessor design clearly show that the multicore processors are the answer to the question how to scale up the processing power of today's computers. In this article we present our C implementation of the Temporal Hebbian Self-organizing Map (THSOM) neural network. This kind of neural networks have growing computational complexity for larger networks, therefore we present different approaches to the parallel processing – instruction based parallelism and data-based parallelism or their combination. Our C implementation of THSOM is modular and multi-platform, allowing us to move critical parts of the algorithm to other cores, platforms or use different levels of the instruction parallelism yet still run exactly the same computational flows – maintaining good comparability between different setups. For our experiments, we have chosen a multicore x86 system.

1 Introduction

This paper presents a C implementation of THSOM [3] neural network. The THSOM neural network is enhanced SOM [1] neural network which is capable of spatiotemporal clusterization.

In the past, parallel SOM implementations either hardware or software were developed. Multi-threaded parallel implementation of SOM appeared already in 2000 [7], utilizing SGI Power Challenge and SGI Origin 2000 and dual Celeron systems. There were numerous hardware implementations of the SOM, either as part of re-programmable platform like RAPTOR2000 [6] or as IP core [4] or [5].

Nowadays, yet still mainstream x86 architecture offers enhanced SIMD floating point instruction set (SSE) and even multiple processors (cores) in the single chip delivering great amount of parallel processing power to common desktop/server systems. We will present a way how to utilize such parallel power, and create efficient implementation of THSOM neural network.

V. Kůrková et al. (Eds.): ICANN 2008, Part II, LNCS 5164, pp. 159–168, 2008.

2 THSOM Neural Network

The Temporal Hebbian Self-organizing Map (THSOM) were introduced in [2]. Unlike classic SOM [1] based networks, they contain spatial and temporal map, extending the clustering to space and time. The THSOM architecture consists of one layer (usually connected as the grid) of hybrid neurons as depicted Figure 1. They are fully connected to input vector of dimension d, connections make up the spatial map. The neurons are connected to each other in the grid using recurrent temporal synapses (temporal map). Hybrid neurons contain two types of similarity measures, Euclidean metric for measuring similarities in input spatial space and scalar product for measuring similarities in temporal space. The activation (output) of a neuron is defined as follows:

$$y_i^{t+1} = \sqrt{D} - \sqrt{\sum_{j=1}^{d}(x_j^t - w_{ij}^t)^2 + \sum_{k=1}^{n}(y_k^t m_{ik}^t)} \tag{1}$$

where y_i^{t+1} is activation (output) of i-th neuron in time $t+1$ (further time step), $\sqrt{D}$ is a constant square root of input vector dimension D, x_j^t is input of j-th vector in time t, w_{ij}^t is the spatial weight for j-th input in time t, y_k^t is output of k-th neuron in time t, m_{ik}^t is temporal weights for k-th neuron in time t (from neuron k to neuron i), n is number of of neurons in network, and d is a dimension of input space (vector).

The spatial weights are updated using classical SOM (Kohonen's Self-organizing Map) rules [1]. The weights are updated in order to move the neuron in the space closer to the input vector. Temporal weights use modified Hebbian learning rule according to following formula [3] (after new Best Matching Unit (BMU) b is computed):

$$m_{bk}^{t+1} = \begin{cases} \min(\max(m_{bk}^t + \alpha(1 - m_{bk}^t + \beta), K_l), K_h) & \text{for } k = \text{argmax}_k(y_k^t) \\ \min(\max(m_{bk}^t - \alpha(m_{bk}^t + \beta), K_l), K_h) & \text{otherwise} \end{cases} \tag{2}$$

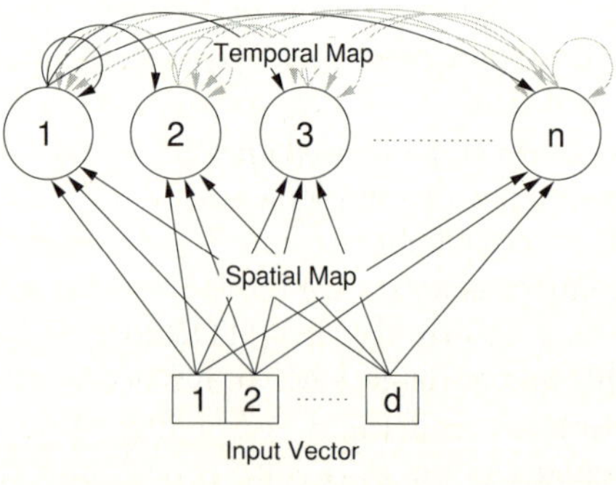

Fig. 1. THSOM network architecture

where y_k^t is an output of k-th neuron of previous time step, β and α control temporal synapses learning rate, α can start on 1.0 and is decreased in time to slow down oscillation around desired temporal weight value, K_l is a low boundary of temporal weights, usually ($K_l = 0$), K_h is a high boundary of temporal weights, usually ($K_h = 1$).

3 THSOM Implementation Architecture

Our C based THSOM implementation consists of two modules. Main module controls the THSOM operations and data handling and the other implements the executive functions.

The main module implements data loading (function loaddata), network initialization (function randomW) and the learning logic (function learn). The executive function module implements the functions which find actual best matching unit (function findBMU_normal) and Gaussian neighbourhood of winning neuron (cutgauss2D). There are also functions which update the neural weights in the neighborhood of winning neuron and the winning neuron itself (functions updateSOM and updateTSOM).

4 Optimizing Reference C Implementation

As a benchmark, we have chosen artificial GPS data containing a simulated car ride across the city. The data were sampled with constant rate. There are 2216 data samples (normalized 2D coordinates). The learning phase of the THSOM neural network was fixed at 1000 Epochs.

The reference C implementation was profiled using the gprof tool. We used the GCC 4.1 compiler (32-bit) with standard optimizing options (`-O2 -msse2`). The Table 1 summarizes the gprof profiler results obtained from the THSOM network with 64 neurons. Each sample has been counted as 0.01 seconds. The results of the benchmark were used for further optimizations as described later.

The profiling results reveal, that our program spends most of the time inside a findBMU_normal routine, which computes the BMU of SOM part of the network. During the weight modification phase, the program computes the Gaussian neighborhood using function cutgauss2D. Although the function is rated as third, it is called many times, therefore it is worth to optimize it too. The functions with _accel postfix are just wrappers for future accelerated versions of findBMU_normal and updateTHSOM.

5 Parallel Processing and Optimization

Our effort will focus on both data parallel processing as well as multi-threaded processing. The most computational intensive function of our reference implementation is the findBMU_normal function. It happens to be good candidate for

Table 1. The summary of profiled functions, findBMU_normal as most complex and cutgauss2D as most called

% time	cumulative seconds	self seconds	calls	name
92.50	47.35	47.35	2216001	findBMU_normal
3.03	48.90	1.55	2216000	updateSOM
2.72	50.29	1.39	141824000	cutgauss2D
1.15	50.88	0.59	15750045	gamma2D
0.18	50.97	0.09	231860	updateTSOM
0.12	51.03	0.06	1	learn
0.10	51.08	0.05	2216000	updateTHSOM_accel
0.08	51.12	0.04	14839040	max
0.08	51.16	0.04	2216001	findBMU_accel
0.06	51.19	0.03	14839040	min
0.00	51.19	0.00	4224	randf
0.00	51.19	0.00	64	dump_neuron_s
0.00	51.19	0.00	64	dump_neuron_t
0.00	51.19	0.00	1	findBMU_accel_init
0.00	51.19	0.00	1	loaddata
0.00	51.19	0.00	1	randomW

data and even for instruction based parallelism. The search for the best matching unit on all neurons is independent and can run parallel in different threads. Moreover, each thread might compute the BMU of group of neurons utilizing the SIMD instruction set of the processor (SSE in our case).

5.1 Data Parallelism on Data Structures

The findBMU_normal function operates on the temporal and spatial maps of neurons, and delivers the proportional “similarity” of input vector to best matching neuron, traditionally called best matching unit (BMU). In order to use SIMD instructions on different neurons we grouped the spatial and temporal maps accordingly as depicted on code fragment bellow:

```
struct neuron4 {
        float m[NEURONS][4]; /* temporal map */
        float w[VDIM][4];    /* spatial map */
}
struct neuron4 neurons4[NEURONS / 4];
```

The grouping by four is architecture specific, because the SIMD SSE instruction can operate on four floats concurrently. The spatial layout of arrays `m` and `w` group each neuron weight elements to the consecutive memory space, allowing to load the vector register without a stride.

5.2 Optimization on x86 Using SSE Intrinsic

We wanted to avoid the direct assembly code for its obvious disadvatages, therefore we used compiler vector extensions. The GCC vector extensions operate on special data type which holds the elements of the vector. The SSE instruction set allows to have 4 floats per vector, the data type is defined as follows:

```
#include <xmmintrin.h>
typedef float v4sf __attribute__ ((vector_size(16)));
union f4vector {
        v4sf v;
        float f[4];
};
```

We declared the v4sf type, which tells the compiler to pack up to four floats together. We created the f4vector union data type which allows to operate on vector or access the elements of the vector individually. The findBMU_normal routine was re-coded using the vector data types. From the implementation point of view it lost the inner-most iteration cycle for all neurons packed in `neuron4` type.

Some vector operations cannot be coded as standard C operators. The GCC solves that with intrinsic function to load resp. store the vector from resp. to memory with functions _mm_load_ps resp. _mm_store_ps. The load of same value to all vector members is accomplished with function _mm_load1_ps. Zeroing whole vector with _mm_setzero_ps. Even the square root function is available – _mm_sqrt_ps. This optimization is labeled as `-DSSE`in further text.

5.3 Optimization on x86 Using Threads

Even when using SSE optimized version, the 76% of the time is still spent in the findBMU_SSE. Up to now, our implementation was just single threaded. In this chapter we propose multi-threaded version of findBMU function.

Most natural way is to use as many threads as CPU cores available. The OS scheduler will plan threads on distinct cores. Each core will compute proportional part of neurons divisible by 4 to ease the additional SSE acceleration. After each core finishes the findBMU_SSE on the subsets, final BMU needs to be found by simple comparison of the results returned from all participating threads. A synchronization barrier needs to be inserted before and after the BMU computation, so that all threads start to compute for the same input vector and it is sure that all threads finished computation, before the final BMU is searched.

Traditionally, the barrier implementation uses the mutexes and sleeping-waking primitives for the thread execution handling. Unfortunately this is not feasible for the very intensive thread synchronization like in ours implementation. Due to context switching and waking/sleeping thread overhead the computations will take much longer than in single thread version.

Therefore we propose our own barrier implementation, which uses the busy waiting and atomic operations for data manipulation and arithmetic. We use the libatomic-ops library which provides platform independent atomic operations. See the following code:

```
int barrier_wait(struct barrier *barrier, int tid) {
    AO_fetch_and_add1(&barrier->b); /* atomicaly incr barrier */
    if (!tid) {
    /* we are thread 0, wait for all other threads to join */
    while ((*((volatile AO_t *) (&barrier->b))) < NUM_THREADS) ;
        AO_store(&barrier->b, 0); /* unblock waiting threads */
    return 0;
    } else {
        while ((*((volatile AO_t *) (&barrier->b))));
        return 0;
    }
    return 0;
}
```

Each thread calls the barrier_wait function with the ID of calling thread (tid) and with the pointer to actual barrier shared between the threads (*barrier). The AO_fetch_and_add1 function atomically increases the barrier. All other threads than thread zero are waiting until the barrier is zero again. Thread zero spins in a loop until all other threads has entered the barrier. Hence the barrier value is equal to number of threads which are trying to synchronize.

For performance reasons, it is feasible to align whole barrier structure to one cache line size, so there are no other data affected with the exclusive state of the cached data.

5.4 Optimization Using cutGauss2D Lookup Table

The profiling revealed the cutGauss2D function as good candidate for improvement. It computes the neighbourhood for the weight adoption in the SOM layer of the network. We implemented a simple on-demand lookup table. The result of the cutGauss2D is stored in the 2D array for each two neurons. If the size of neighbourhood changes, whole lookup table is invalidated. This optimization is labeled as `-DnyCache`in further text.

6 Measurements of Optimized THSOM Implementation

This chapter summarizes all measurements. The data and number of epochs is the same as during the profiling phase to maintain comparability of all results.

6.1 The Influence of Cutgauss2D Lookup Table

The Table 2 summarizes the speedup achieved with additional cutgauss2D optimization. The benefit of lookup table will disappear if the space occupied by lookup table is too large and does not fit well into the cache. This happened for the network with 1024 neurons.

Table 2. The benefit of cutGauss2D lookup table for different network sizes (NS)

NS	Speedup
4	1.45
16	1.30
64	1.39
256	1.08
1024	0.99

6.2 Measurements of Optimized Single Thread THSOM Implementation

The measurements were performed for various levels of compiler based optimizations and include above proposed optimizations. The platform was equipped with Intel based microprocessor Intel(R) Core(TM)2 CPU 6420 running on 2.13GHz with 4MB of L2 cache for each core. The installed compiler was GCC version 4.2.3 (Debian 4.2.3-3) and generated 64-bit code. The run-time results are given in seconds. The table 3 and 4 summarizes the measurements for THSOM network with 4 up to 1024 neurons.

The `-O0` denotes optimization disabled (GCC default). The `-O1` turns some optimizations that are safe and does not prolong the compilation time. The `-O2` turns nearly the rest of available optimization options which do not involve speed versus size trade-off. The `-O3` adds some remaining optimization – this option is

Table 3. The duration of learning phase with various optimizations and network sizes (NS) in seconds

NS	-O0	-O1	-O2	-O2 -DSSE	-O3 -DSSE
4	2.29	1.57	1.48	1.32	1.31
16	9.67	6.14	4.85	3.51	3.20
64	78.57	49.35	25.48	14.52	14.43
256	963.94	662.46	258.00	108.51	107.98
1024	14459.65	10048.41	3832.65	1511.96	1537.86

Table 4. The duration of learning phase with various optimizations and network sizes (NS) in seconds

NS	-O2 -DSSE -DnyCACHE	-O3 -DSSE -DnyCACHE
4	0.25	0.24
16	1.24	0.95
64	7.74	7.74
256	88.76	88.36
1024	1462.03	1442.48

sometimes considered unsafe. As indicated earlier, our optimizations are denoted `-SSE` and `-DnyCache`.

The speedup between fully optimized version and unoptimized is in average 10.2 times. The speedup of compiler generated optimization from `-O0` to `-O2` is in average 2.8 times. Our optimizations `-DSSE` and `-DnyCache` contributes to speedup in average by factor 3.7.

Our `-DnyCache` optimization is only efficient when the lookup table fits the cache size. Figure 2 gives in logarithmic coordinates complete overview of performed optimizations.

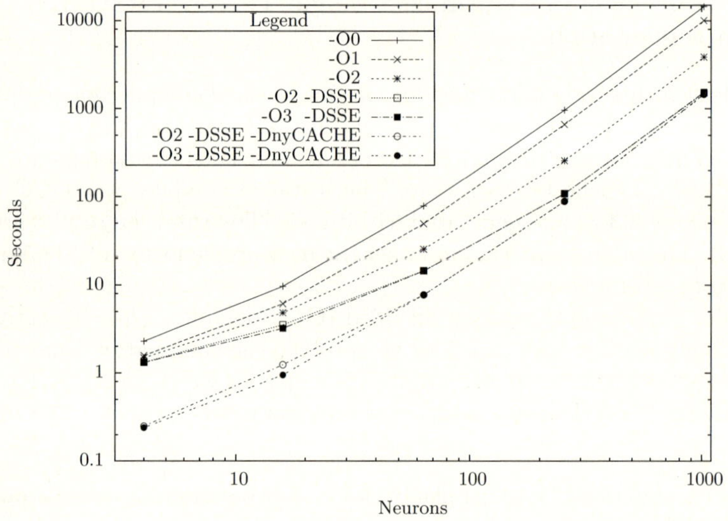

Fig. 2. The x86 optimized versions on single core with clearly visible effect of - DnyCache and other compiler optimizations

6.3 Measurements of Optimized Multi-threaded Version

The multicore version was measured on four core Intel Core2 Quad CPU running on 2.4GHz, with 4MB of L2 cache on each two cores. Table 5 summarizes the achieved results.

The fastest compilation flags were used (`-DSSE -DnyCache -O3 -msse2`). The system uses GCC 4.2.3 (32bit). The Figure 3 depict the achieved parallel speedup for different network sizes. The multicore version becomes faster for networks whose size is bigger then 64 neurons. For smaller networks the BMU computation is faster then the costs for mutual synchronization of the threads. However the parallel speedup does not scale till infinity. The networks with 4096 neurons have speedup just 1.3x. The main impact plays the memory subsystem, which is unable to handle so much bandwidth between caches and main memory. The total memory size of 4096 neurons is 256MB, which is far more then all caches can handle.

Table 5. Computation times in seconds for different network sizes

NS	1 Core	2 Cores	4 Cores
4	0.47	0.01	0.06
16	1.62	3.32	5.01
64	8.83	10.21	11.12
256	89.96	60.37	44.42
1024	1654.37	680.23	364.96
4096	29359.41	21711.38	22141.99

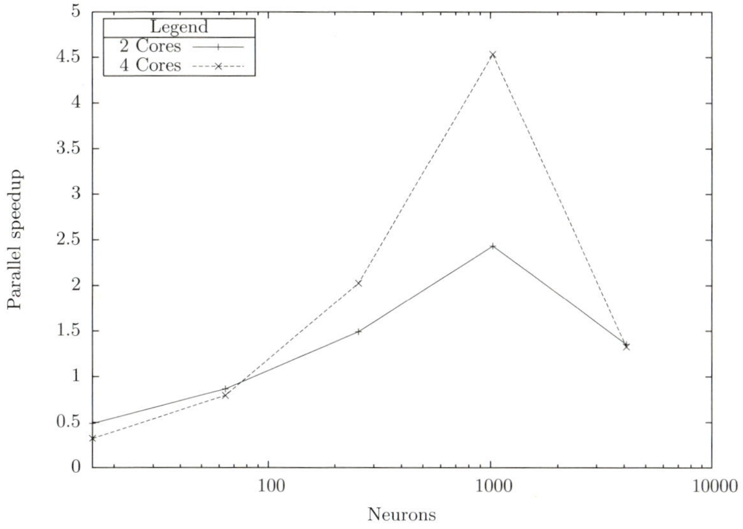

Fig. 3. Parallel speedup with two and four cores

The network with 1024 neurons occupies just 16KB per neuron, that is total 16MB of memory. The total CPU cache size is 8MB and this scales up very well. The parallel speedup of four thread version is even more then 4x. This is caused by slower computation for single thread version, which jumps between cores of CPUs and hits cold caches.

7 Conclusion

We have shown that use of compiler based optimization gains speedup of 2.7 times and additional vector optimization gains additional speedup of 3.7 times. The total speedup for the single threaded program with no optimization to fully optimized is approximately 10 times, which is significant especially when using large networks.

The multicore version using busy waiting synchronization shows significant performance increase if all data memory is well cachable. The best speedup achieved on four cores compared to optimized single threaded version was four times.

Due to single running thread migration between the cores, the single threaded version shall be 1.13x faster then noted. If we compare the slowest unoptimized version with 1024 neurons and the fastest four threads version and project the little different CPU speed factor, 1.012 we get total speedup 39x.

If we just compare single threaded, with compiler based optimization without any "hand-made" optimization with our fastest four threads version, fully optimize, we get total speedup of 10x.

Acknowledgments

This work has been supported by the internal grant of Czech Technical University in Prague, IG: CTU0812213 - "Hardware Acceleration for Computational Intelligence" and the research program "Transdisciplinary Research in the Area of Biomedical Engineering II" (MSM6840770012) sponsored by the Ministry of Education, Youth and Sports of the Czech Republic.

References

1. Kohonen, T.: Self-Organizing Maps3, 3rd edn. Springer, Heidelberg (2001)
2. Koutnik, J.: Inductive Modelling of Temporal Sequences by Means of Self-organization. In: Proceeding of Internation Workshop on Inductive Modelling (IWIM 2007), CTU in Prague, pp. 269–277 (2007) ISBN 978-80-01-03881-9
3. Koutnik, J.: Self-organizing Maps for Modeling and Recognition of Temporal Sequences, A thesis submitted to Faculty of Electrical Engineering, Czech Technical University in Prague (2008)
4. Porrmann, M., Witkowski, U., Rueckert, U.: A massively parallel architecture for self-organizing feature maps. IEEE Transactions on Neural Networks 14(5), 1110-1121 (2003)
5. Hendry, D., Duncan, A., Lightowler, A.: Ip core implementation of a self-organizing neural network. IEEE Transactions on Neural Networks 14(5), 1085–1096 (2003)
6. Franzmeier, M., Pohl, C., Porrmann, M., Ruekert, U.: Hardware Accelerated Data Analysis Parallel Computing in Electrical Engineering. In: International Conference on PARELEC 2004, September 7-10, 2004, pp. 309–314 (2004)
7. Rauber, A., Tomsich, P., Merkl, D.: A Parallel Implementation of the Self-Organizing Map Exploiting Cache Effects: Making the SOM Fit for Interactive High-Performance Data Analysis Neural Networks. In: IJCNN 2000, Proceedings of the IEEE-INNS-ENNS International Joint Conference (2000)

Reconfigurable MAC-Based Architecture for Parallel Hardware Implementation on FPGAs of Artificial Neural Networks

Nadia Nedjah[1], Rodrigo Martins da Silva[1], Luiza de Macedo Mourelle[2], and Marcus Vinicius Carvalho da Silva[1]

[1] Department of Electronics Engineering and Telecommunication
[2] Department of System Engineering and Computation,
Engineering Faculty, State University of Rio de Janeiro
{marcos,nadia,ldmm}@eng.uerj.br

Abstract. Artificial Neural Networks (ANNs) is a well known bio-inspired model that simulates human brain capabilities such as learning and generalization. ANNs consist of a number of interconnected processing units, wherein each unit performs a weighted sum followed by the evaluation of a given activation function. The involved computation has a tremendous impact on the implementation efficiency. Existing hardware implementations of ANNs attempt to speed up the computational process. However these implementations require a huge silicon area that makes it almost impossible to fit within the resources available on a state-of-the-art FPGAs. In this paper, we devise a hardware architecture for ANNs that takes advantage of the dedicated adder blocks, commonly called MACs to compute both the weighted sum and the activation function. The proposed architecture requires a reduced silicon area considering the fact that the MACs come for free as these are FPGA's built-in cores. The hardware is as fast as existing ones as it is massively parallel. Besides, the proposed hardware can adjust itself on-the-fly to the user-defined topology of the neural network, with no extra configuration, which is a very nice characteristic in robot-like systems considering the possibility of the same hardware may be exploited in different tasks.

1 Introduction

Artificial Neural Networks (ANNs) are useful for learning, generalization, classification and forecasting problems [3]. They consists of a pool of relatively simple processing units, usually called artificial neurons, which communicates with one another through a large set of weighted connections. There are two main network topologies, which are feed-forward topology [3], [4] where the data flows from input to output units is strictly forward and recurrent topology, where feedback connections are allowed. Artificial neural networks offer an attractive model that allows one to solve hard problems from examples or patterns. However, the computational process behind this model is complex. It consists of massively parallel non-linear calculations. Software implementations of artificial

V. Kůrková et al. (Eds.): ICANN 2008, Part II, LNCS 5164, pp. 169–178, 2008.

neural networks are useful but hardware implementations takes advantage of the inherent parallelism of ANNs and so should answer faster.

Field Programmable Gate Arrays (FPGAs) [7] provide a re-programmable hardware that allows one to implement ANNs very rapidly and at very low-cost. However, FPGAs lack the necessary circuit density as each artificial neuron of the network needs to perform a large number of multiplications and additions, which consume a lot of silicon area if implemented using standard digital techniques.

The proposed hardware architecture described throughout this paper is designed to process any fully connected feed-forward multi-layer perceptron neural network (MLP). It is now a common knowledge that the computation performed by the net is complex and consequently has a huge impact on the implementation efficiency and practicality. Existing hardware implementations of ANNs have attempted to speed up the computational process. However these designs require a considerable silicon area that makes tem almost impossible to fit within the resources available on a state-of-the-art FPGAs [1], [2], [6]. In this paper, we devise an original hardware architecture for ANNs that takes advantage of the dedicated adder blocks, commonly called MACs (short for Multiply, Add and Accumulate blocks) to compute both the weighted sum and the activation function. The latter is approximated by a quadratic polynomial using the least-square method. The proposed architecture requires a reduced silicon area considering the fact that the MACs come for free as these are FPGA's built-in cores. The hardware is as fast as existing ones as it is massively parallel. Besides, the proposed hardware can adjust itself on-the-fly to the user-defined topology of the neural network, with no extra configuration, which is a very nice characteristic in robot-like systems considering the possibility of the same piece of hardware may be exploited in different tasks.

The remaining of this paper is organized as follows: In Section 2, we give a brief introduction to the computational model behind artificial neural networks; In Section 3, we show how we approximate the sigmoid output function so we can implement the inherent computation using digital hardware; In Section 4, we provide some hardware implementation issues about the proposed design, that makes it original, efficient and compact; In Section 5, we present the detailed design of the proposed ANN Hardware; Last but no least, In Section 6, we draw some useful conclusions and announce some orientations for future work.

2 ANNs Computational Model

We now give a brief introduction to the computational model used in neural networks. Generally, is constituted of few layers, each of which includes several neurons. The number of neurons in distinct layers may be different and consequently the number of inputs and that of outputs may be different [3].

The model of an artificial neuron requires n inputs, say $I_1, I_2, \ldots, I_n$ and the synaptic weights associated with these inputs, say $w_1, w_2, \ldots, w_n$. The weighted sum a, which, also called activation of the neuron, is defined in (1). The model usually includes an output function *nout*(.) that is applied to the neuron activation before it is fed forwardly as input to the next layer neurons.

$$a = \sum_{j=1}^{n} w_j \times f_j \tag{1}$$

The non-linearity of the neuron is often achieved by the output function, which may be the hyperbolic tangent or sigmoid [3]. In some cases, $nout(a)$ may be linear.

A typical ANN operates in two necessary stages: *learning* and *feed-forward computing*. The learning stage consists of supplying known patterns to the neural network so that the network can adjust the involved weights. Once the network has learned to recognize the provided patterns, the network is ready to operate, performing the feed-forward computing. In this stage, the network is supplied with an input data or pattern, which may or not be one of those given in learning stage and verify how the network responds with output results. This allows one to know whether the neural network could recognize the input data. The precision of the net in recognizing the new input patterns depends on the quality of its learning stage and on its generalization. As we have previously mentioned, here we are only concerned with the implementation of feed-forward computing stage.

3 Approximation of the Output Function

Unlike the activation function, which includes operations that can easily and efficiently implemented in hardware, the out function requires a special care before the computation involved can be modeled in hardware. Without loss of generality, we chose to use the sigmoid output function. Note that the same treatment applies to the hyperbolic function too. To allow an efficient implementation of the sigmoid function defined in (2), in hardware, we proceeded with a parabolic approximation of this function using the least-square estimation method.

$$sigmoid(a) = \frac{1}{1 + e^{-a}} \tag{2}$$

The approximation proceeds by defining $nout(a) = C \times a^2 + B \times a + A$ as a parabolic approximation of the sigmoid of (2), just for a short range of the variable a. We used the least-square parabola to make this approximation feasible. Many attempts were performed to try to find out the best range of a for this approximation, so that the parabola curves fits best that of $sigmoid(a)$. We obtained the range $[-3.3586, 2.0106]$ for variable a, taking into account the calculated coefficients $C = 0.0217$, $B = 0.2155$ and $A = 0.4790$ for the parabolic approximation. Thus, the approximation of the sigmoid function is:

$$nout(a) = \begin{cases} 0 & a < -3.3586 \\ 0.0217 \times a^2 + 0.2155 \times a + 0.4790 & a \in [-3.3586, 2.0106]. \\ 1 & a > 2.0106 \end{cases} \tag{3}$$

4 Implementation Issues

An Artificial Neural Network is a set of several interconnected neurons arranged in layers. Let L be the number of layers. Each layer has its own number of neurons. Let m_i be the number of neurons in layer i. The neurons are connected by the synaptic connections. Some neurons get the input data of the network, so they are called *input* neurons and thus compose the input layer. Other neurons export their outputs to the outside world, so these are called output neurons and thus compose the output layer. Neurons placed on the layer 2 up to layer $L-1$ are called the hidden neurons because they belong to the hidden layers. In Fig. 1, we show a simple example of an ANN. The output of each neuron, save output neurons, represents an input of all neurons placed in the next layer.

The computation corresponding to a given layer starts only when that of the corresponding previous layer has finished. Our ANN hardware has just one *real* layer of neurons, constitutes of k neurons, where k is maximum number of neurons per layer, considering all layers of the net. For instance, for the net of Fig. 1, this parameter is 3. This single real layer or physical layer is used to implement all layers of the network. As only one layer operates at a time, this allows us to minimize drastically the silicon area required without altering the response time of the net. For instance, considering the net of Fig. 2, the first stage of the computation would use only 2 neurons, then in the second stage all three physical neurons would be exploited and in the last stage, only one neuron would be useful. So instead of having 6 physically implemented neurons, our hardware requires only half that number to operate. ANN hardware treats the nets layers as virtual.

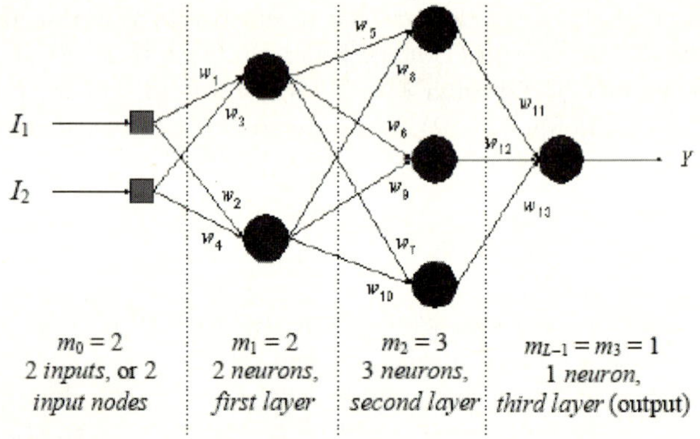

Fig. 1. Neural network with two intpus, three layers and one output Y

Besides reducing the number of neurons that are actually implemented in hardware, our design takes advantage of some built-in cores that come for free in nowadays FPGAs. This blocks are called MACs (Multiply, add and Accumulate), which are usually used in DSPs (Digital Signal Processing) and their architecture is shown in Fig. 2. The MACs blocks are perfect to perform the weighted sum.

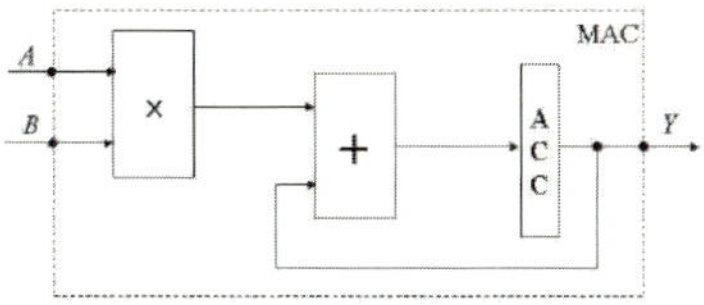

Fig. 2. Built-in MACs blocks in FPGAs

Recall that $nout(a)$ of (3) is the actual neuron output function our ANN hardware will perform. Observe that the computation involved in this function is sum of products (quadratic polynomial) and so the MACs can be used in this purpose to. Actually we use the same block of the neuron to compute the output function.

5 ANN Hardware Architecture

The ANN hardware interface is illustrated in Fig. 3, wherein two other components are included: LOAD CONTROLLER and CLOCK SYSTEM. The former may be any outside system able to setup the neural network topology and to load all necessary data in the ANN hardware. This include the number of inputs, the number of layers, the number of neurons per layer and that of outputs, besides, the net inputs and the definitive weights. Our ANN hardware is organized in a neural control unit (UC) and Neural arithmetic and logic unit (ALU).

Neural UC encompasses all control components for computing all neural network feed-forward computation. It also contains the memories for storing the net's inputs in the INPUT MEMORY, the weights in the WEIGHT MEMORY, the number of inputs and neurons per layer in the LAYER MEMORY and the coefficients of the output function in the OUTPUT FUNCTION MEMORY as described in (3). Fig. 4 and Fig. 5 depict, respectively, two parts of the neural UC.

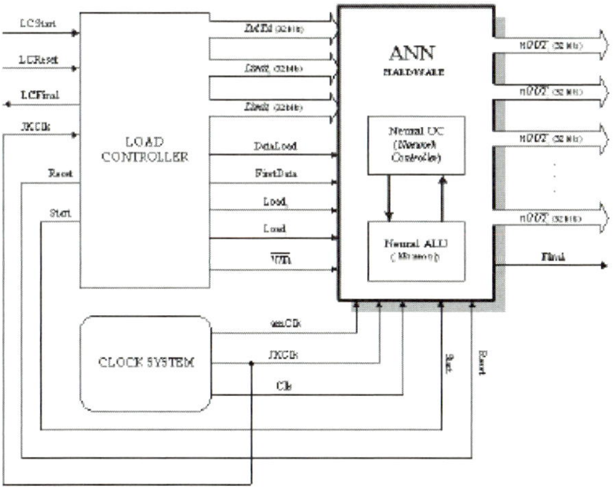

Fig. 3. Interface of the ANN hardware

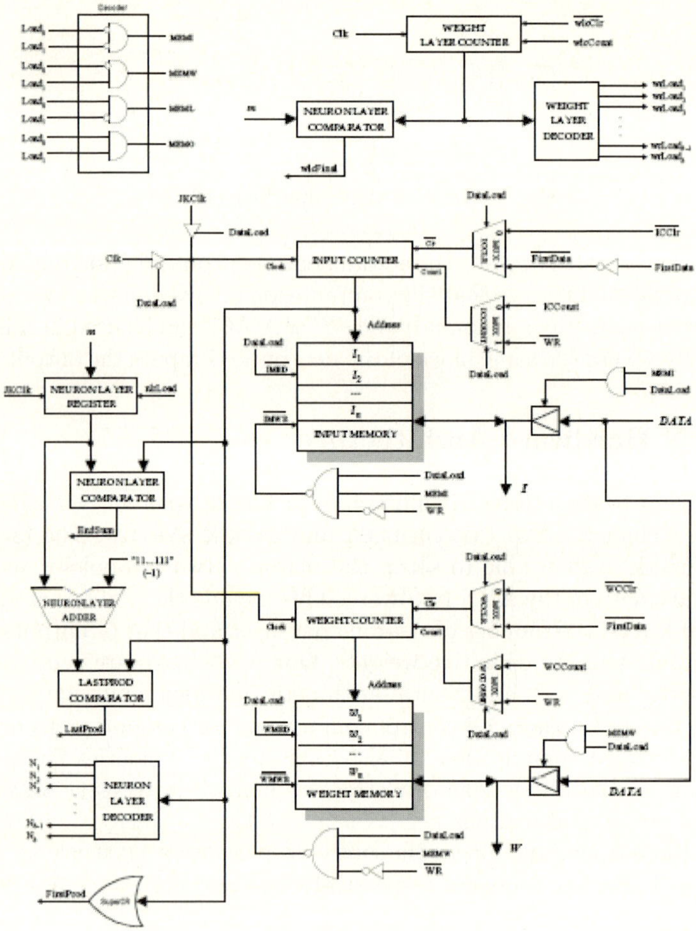

Fig. 4. First part of the neural UC

During the loading process, which commences when $LCStart = 1$, the LOAD CONTROLLER sets signal *DataLoad* and selects the desired memory of the neural UC by signals $Load_0$ and $Load_1$ (see Fig. 3, Fig. 4 and Fig. 5).

The counters that provide addresses for memories are entirely controlled by the LOAD CONTROLLER. Signal *JKClk* is the clock signal (from CLOCK SYSTEM in Fig. 4) that synchronizes the actions of those Counters and of the LOAD CONTROLLER. This one fills each memory through the 32-bit *DATA* in loading process.

When the loading process is finished ($LCFinal = 1$), in Fig. 3, signal *DataLoad* can be turned off and the LOAD CONTROLLER can set signal *Start* for the commencing of the feed-forward Neural Network computing. When $Start = 1$ (and *DataLoad* $= 0$), the ANN hardware gets the whole control of its components; so the LOAD CONTROLLER can no longer interfere in the neural UC. This one has a main controller called *Network Controller* (Fig. 3) that controls all components of the neural UC (Fig. 4 and Fig. 5) and also the neural ALU, which is depicted in Fig. 9.

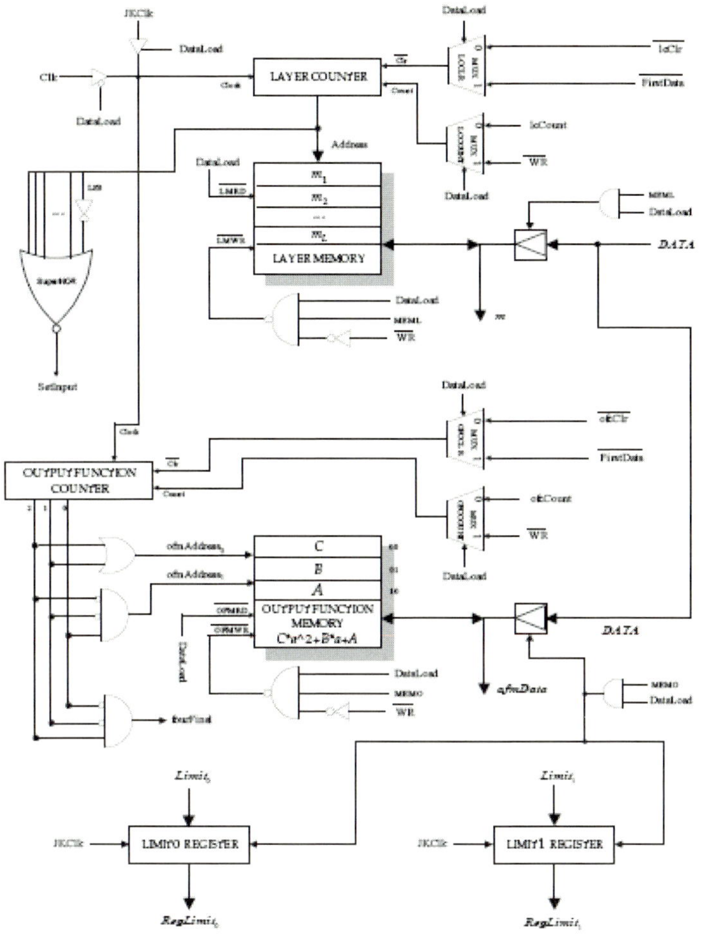

Fig. 5. Second part of the neural UC

During the ANN hardware operation, neural UC by the mean of the network controller, controls the computation of each layer per stage. For each layer of the neural network, all k hardware neurons of the neural ALU of Fig. 9 work in parallel even though not necessarily all physical neurons are needed in the layer. Recall that some layers in the ANN hardware may have fewer neurons than k. At this stage, signal Clk is now the active clock of the ANN hardware, not signal $JKClk$ anymore.

In Fig. 9, ADDER MUX decides the actual input for all hardware neurons and it is exploited to multiplex a network input, from the INPUT MEMORY in Fig. 4 or the output of a hardware neuron $nOUT_i$, which is an output of a neuron placed in a layer i of the net. While all physical neurons are in operation, the WEIGHT REGISTERS of Fig. 9 are already being loaded using signal W (see Fig. 4). These are the

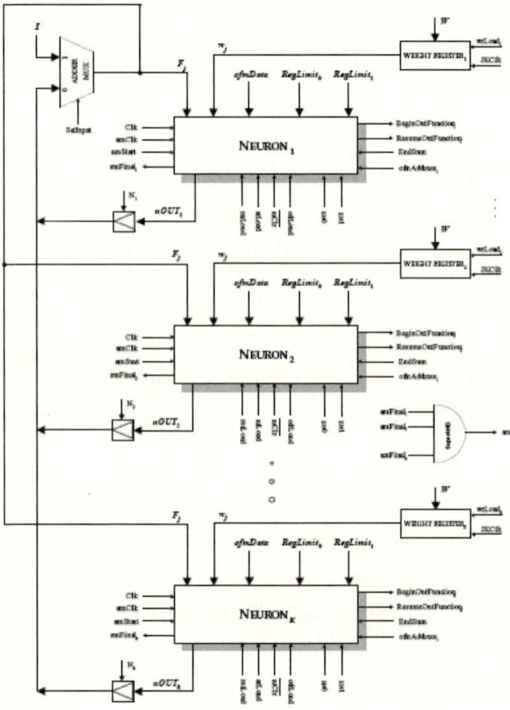

Fig. 6. Overall hardware architecture of the Neural ALU

new weights, which are the weights of the next layer in the network. Furthermore, in Fig. 9, we see a set of tri-state buffers, each of which is controlled by signal N_i, issued by the NEURON LAYER DECODER, in the neural UC of Fig. 4. Fig. 9 shows the neuron architecture. Each hardware neuron performs the weighted sum followed by the application of the output function $nout(a)$.

Observing Fig. 4 (neural UC), the INPUT COUNTER, together with NEURON LAYER REGISTER, NEURON LAYER COMPARATOR, NEURON LAYER ADDER and LASTPROD COMPARATOR control the computation of the weighted sum: signal $FirstProd$ indicates the first product of the weighted sum and $LastProd$, the last one. SuperOR component is an OR of all input bits. Signal *EndSum* (Fig. 4, Fig. 9 and Fig. 9) flags that the weighted sum has been completed. It also triggers the start of the output function computation. During this stage, the OUTPUT FUNCTION COUNTER (see Fig. 5) provides the address to the OUTPUT FUNCTION MEMORY in order to release the coefficients ($C = 0.0217$, $B = 0.2155$ or $A = 0.4790$), through $ofmData$, to the hardware neurons. Signal $fourFinal$, in Fig. 5, indicates that the computation of $nout(a)$ has finished.

Each hardware neuron encloses a MAC block, which consists of a FLOAT MULTIPLIER and a FLOAT ADDER to perform products and sums, respectively. The MULTIPLIER REGISTER allows the LOAD ADDER to works in parallel with FLOAT

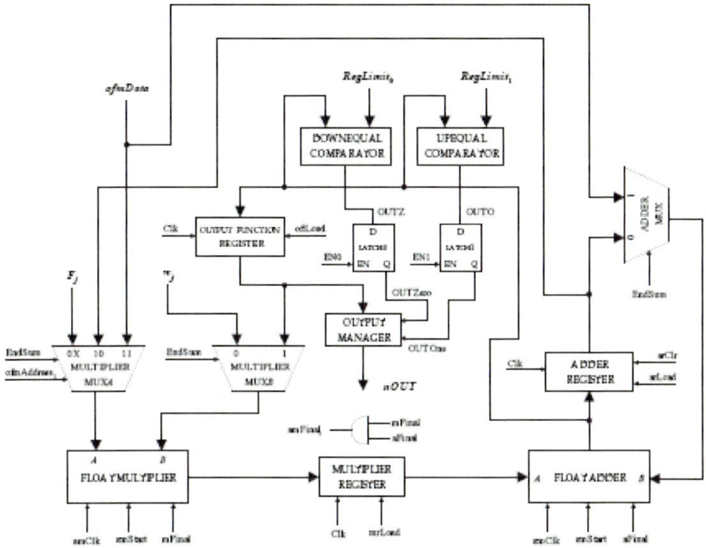

Fig. 7. Hardware architecture of the Neuron

MULTIPLIER. The ADDER REGISTER accumulates the weighted sum. Recall that all hardware neurons work in parallel (see Fig. 9).

At an earlier stage, the LOAD CONTROLLER has loaded $Limit_0 = -3.3586$ and $Limit_1 = 2.0106$ in neural UC so that $RegLimit_0 = -3.3586$ and $RegLimit_1 = 2.0106$ have been obtained. Those float numbers refer to (3), wherein $nout(a)$ is 0 if $a < -3.3586$ and 1 if $a > 2.0106$.

In Fig. 9, which shows the hardware neuron, DOWNEQUAL COMPARATOR sets $OUTZ = 1$, if $a < -3.3586$ and UPEQUAL COMPARATOR sets $OUTO = 1$, if $a > 2.0106$. These components, intermediated by two latches, control the OUTPUT MANAGER, which decides as to the output of the hardware neuron ($nOUT$): *(i)* If $a \in [-3.3586, 2.0106]$, then $nOUT$ is the result of the second degree polynomial as described in (3), which is the content of the OUTPUT FUNCTION REGISTER; *(ii)* If $a < -3.3586$, then the OUTPUT MANAGER provides $nOUT = 0$; *(iii)* If $a > 2.0106$, then $nOUT$ is 0. Components $LATCH_0$ and $LATCH_1$ are used to maintain $nOUT$ stable. Signal $nOUT$ have to be kept during the computation of the weighted sum of a next layer neuron. Furthermore, in Fig. 9, signal $amFinal_i$ indicates the end of both a product and sum performed by the neuron. The multiplier and the adder operate in parallel, i.e. when the adder is accumulating the freshly computed product to the partial weighted sum obtained so far, the multiplier is computing the next product. In Fig. 5, signal $amFinal$ indicates the end of all the computation in all neurons.

In Fig. 3, signal $Final$ indicates that all computation required in all the layers of the network are completed and the outputs of the network have been obtained.

These outputs are available signals $nOUT_1$, $nOUT_2$, …, $nOUT_h$ (see Fig. 3 and 8), where h is the number of neurons placed in the output layer of the Network, with $h \leq k$.

6 Conclusions

In this paper, we presented novel hardware architecture for processing an artificial neural network, whose topology can be changed on-the-fly without any extra reconfiguration effort. The design takes advantage of the built-in MACs block that come for free in modern FPGAs. The model was specified in VHDL [5], simulated to validate its functionality. We are now working on the synthesis process to evaluate time and area requirements. The comparison of the performance result of our design will be then compared to both the binary-radix straight forward design and the stochastic computing based design.

Acknowledgments

We are grateful to FAPERJ (Fundação de Amparo à Pesquisa do Estado do Rio de Janeiro, `http://www.faperj.br`) and CNPq (Conselho Nacional de Desenvolvimento Científico e Tecnológico, `http://www.cnpq.br`) for their continuous financial support.

References

1. Bade, S.L., Hutchings, B.L.: FPGA-Based Stochastic Neural Networks – Implementation. In: IEEE Workshop on FPGAs for Custom Computing Machines, pp. 189–198. IEEE Press, Napa (1994)
2. Brown, B.D., Card, H.C.: Stochastic Neural Computation II: Soft Competitive Learning. IEEE Transactions On Computers 50(9), 906–920 (2001)
3. Hassoun, M.H.: Fundamentals of Artificial Neural Networks. MIT Press, Cambridge (1995)
4. Moerland, P., Fiesler, E.: Neural Network Adaptation to Hardware Implementations. In: Fiesler, E., Beale, R. (eds.) Handbook of Neural Computation. Oxford, New York (1996)
5. Navabi, Z.: VHDL: Analysis and Modeling of Digital Systems, 2nd edn. McGraw Hill, New York (1998)
6. Nedjah, N., Mourelle, L.M.: Reconfigurable Hardware for Neural Networks: Binary radix vs. Stochastic. Journal of Neural Computing and Applications 16(3), 155–249 (2007)
7. Xilinx, Inc. Foundation Series Software, `http://www.xilinx.com`

Implementation of Central Pattern Generator in an FPGA-Based Embedded System

Cesar Torres-Huitzil[1] and Bernard Girau[2]

[1] Information Technology Department, Polytechnic University of Victoria
Ciudad Victoria, Tamaulipas, Mexico
Cesar.Torres@ieee.org
[2] CORTEX team, LORIA-INRIA Grand Est
Campus Scientifique, B.P. 239, Vandoeuvre-les-Nancy Cedex, France
Bernard.Girau@loria.fr

Abstract. This paper proposes an embedded system on a chip to generate locomotion patterns of periodic rhythmic movements inspired by biological neural networks called Central Pattern Generators (CPGs) found in animal nervous system. The proposed system contains a custom digital module, attached to an embedded processor, that mimics the functionality and organization of the fundamental Amari-Hopfield CPG. In order to reduce the CPG hardware integration complexity as well as to provide flexibility, an embedded linux operating system running on a processor is used to deal with the CPG interfacing in a high level transparent way for application development. The system is implemented on a Field Programmable Gate Array (FPGA) device providing a compact, flexible and expandable solution for generating periodic rhythmic patterns in robot control applications. Results show that the obtained waveforms from the FPGA implementation agree with software simulations and preserve the easiness of CPG parameter setting for adaptive behavior.

1 Introduction

It has been shown, by means of neurophysiological studies, that many activities vital to animal locomotion such as walking, swimming and flying are produced by repetitive and rhythmic activations of the animal muscles. These rhythmic movements are controlled by specialized biological neural networks called central pattern generators situated in ganglion or spinal cord [1][2]. Researchers have studied CPGs for decades and some principles have been derived to model their functionality and structure. CPGs consist of sets of neural oscillators that receive inputs from command neurons. These oscillators produce rhythmic patterns of neural activity that activate motor neurons. Though sustained rhythmic outputs can be produced by a CPG without need of sensory feedback, the feedback mechanism plays an important role to regulate the frequency and phase of oscillators to change the locomotion patterns depending on the situation and the environment. Furthermore, CPGs induce some coordination among the several physical parts of animals to produce smooth locomotion patterns.

V. Kůrková et al. (Eds.): ICANN 2008, Part II, LNCS 5164, pp. 179–187, 2008.

In recent years, the computational modeling of CPGs has demonstrated the utility of this theory in the development of autonomous locomotion control systems in robotics such as legged walking robots where control is distributed throughout the body [2][3]. CPGs are an important example of distributed processing where the amount of computations for movement control is reduced as a result of the coordination of the distributed physical parts induced by the rhythmic movements. Engineering CPG-based control systems has been difficult since the simulation of even rather simple neural network models requires a significant computational power that exceeds the capacity of general embedded microprocessors. As a result, CPG dedicated hardware implementations, both analog and digital, have received more attention [1][3]. On one hand, CPGs have been implemented using digital processors providing high accuracy and flexibility but those systems consume high power and occupy a large area restricting their utility in embedded applications. On the other hand, analog circuits have been already proposed, being computation- and power- efficient but they usually lack flexibility and dynamics and they involve large design cycles [3].

With the advent of technological improvements and the high level design methodologies to build embedded systems, it is now plausible to design control systems that counterbalance the analog and digital drawbacks by providing custom efficient hardware attached to embedded processors in a single chip such as FPGA devices [4]. FPGAs have shown to be suitable for neural processing, i.e., processing where massively distributed elementary units interact through dense interconnection schemes [5][6]. In this paper, an FPGA-based embedded implementation of an artificial CPG is presented. Our approach may take advantage of both the spatial computing model of custom logic for efficient neural processing and the flexibility of time-multiplexed processing offered by embedded processors available in FPGAs. Having a dedicated coprocessor to generate the rhythmic patterns makes possible to perform general real-time control on the main processor. In this work, a custom implementation of the Amari-Hopfield CPG is attached to a Xilinx microblaze processor with an embedded linux operating system allowing application development that use CPG in a transparent way. The microprocessor could handle higher functions such as CPG parameter adaptation or robot decision making, leaving the neural processing and coordination to a custom CPG hardware module.

2 Central Pattern Generator Model

Artificial CPGs emerge from the study and modeling of biological oscillatory neurons that generate rhythmic patterns of locomotive motion. Different artificial neural networks have been proposed to model CPGs, such as relaxation oscillators and continuous-time recurrent neural networks [1]. The most fundamental CPG model consists of two neurons with reciprocal inhibition. Although its organization is simple, it is essential since it is used as the elementary component of

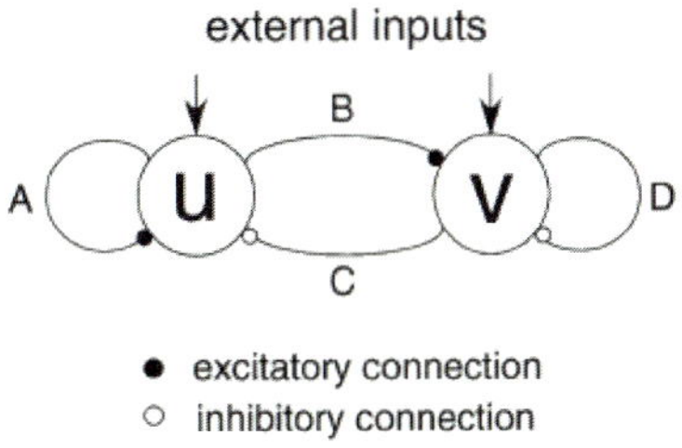

Fig. 1. Amari-Hopfield neuron model [1]

more complex CPG configurations. For instance in [1], a CPG controller for interlimb coordination in quadruped locomotion is presented so as to generate typical gaits of mammals such walking, trotting and galloping. The model consists of four coupled neural oscillators with twelve excitatory and inhibitory interneurons for quadruped locomotion but the generator can be extended to more than four articulations by adding more oscillators.

In this paper, the Amari-Hopfield neuron model [1] is used as the neural oscillator. It consists of an excitatory neuron and inhibitory neuron with excitatory inhibitory interactions as shown in figure 1. The dynamics of the Amari-Hopfield model is given by equation 1:

$$\frac{du}{dt} = -u + Af_{\mu}(u) - Cf_{\mu}(v) + S_u(t) \tag{1}$$
$$\frac{dv}{dt} = -v + Bf_{\mu}(u) - Df_{\mu}(v) + S_v(t)$$

where u and v stand for the internal state of the excitatory and inhibitory neurons, respectively. The model has four tunable parameters A-D that determine the dynamics of the system. S_u and S_v are the external inputs and $f_{\mu}(x)$ is the transfer function given by equation 2:

$$f_{\mu}(x) = \frac{1 + \tanh(\mu x)}{2} \tag{2}$$

where μ is the slope control parameter. Several other possibilities exist for the transfer function used in the model such as the tan-sigmoid and the log-sigmoid functions among others.

3 FPGA-Based Embedded Implementation

The Amari-Hopfield oscillator is suitable for CPG implementation as a digital circuit, however some factors for an efficient and flexible FPGA-based implementation should be taken into account:

- *Area-greedy nonlinear operators*: neural operators require a significant amount of digital logic. Usually, piecewise linear approximations to the nonlinear operators are preferable at the cost of precision [7]. As a first approximation, in this

work, $f(x) = \frac{|x+1|-|x-1|}{2}$ is used as the activation function without reducing significantly the quality and generality of the rhythmic patterns.
- *Arithmetic representation*: neural computations when implemented in general microprocessor-based systems use floating point arithmetic. A preferred approach for embedded implementations is the use of 2's complement fixed point representation with a dedicated wordlength that better matches the FPGA computational resources and that saves further silicon area at the cost of precision.
- *Efficiency and flexibility*: embedded hard processor cores or configurable soft processors developed by FPGA vendors add the software programmability of optimized processors to the fine grain parallelism of custom logic on a single chip [8][9]. In the field of neural processing, several applications mix real-time or low-power constraints with a need for flexibility, so that FPGAs appear as a well-fitted implementation solution for the target application [6].

This paper proposes an FPGA-based embedded system where an efficient digital implementation for the Amari-Hopfield model is attached to the Xilinx microblaze microprocessor running the uclinux embedded operating system. We have chosen to encapsulate neural processing using uclinux wrapping libraries and drivers to avoid the time-consuming integration work and the partial lack of flexibility of a custom architecture for applications development. This work aims at creating a CPG-based neural control embedded in FPGAs, so that this platform might be used more likely and easily than microprocessors for developing robot applications. This work is a preliminary step towards this goal, since it implements the fundamental CPG building block as an embedded and scalable coprocessor.

3.1 Central Pattern Generator Digital Architecture

Figure 2 shows a simplified block diagram of the proposed digital architecture for the discretized Amari-Hopfield equation. All the internal components use 2's complement fixed-point arithmetic representation. Four multipliers are used to weight the output of the activation functions with the excitatory/inhibitory values. The components labeled as ACC stand for accumulators that hold the internal state of the neural oscillators. The activation function was implemented under a piecewise linear approach using two comparators and a 4-input multiplexer. The input value to the activation function module is compared to +1 and -1, and the comparator outputs are used for the control word of the 4-input multiplexer which selects the appropriate value (+1, -1, or the input value) according to the control word. Two separate register banks are used to provide communication channels to an embedded processor, one for the A-D parameters and the external inputs, and other one for the outputs. These banks are accessible in the microblaze-uclinux system and their values are set or read by the running application. From a given uclinux application, the CPG is seen as a memory-mapped device.

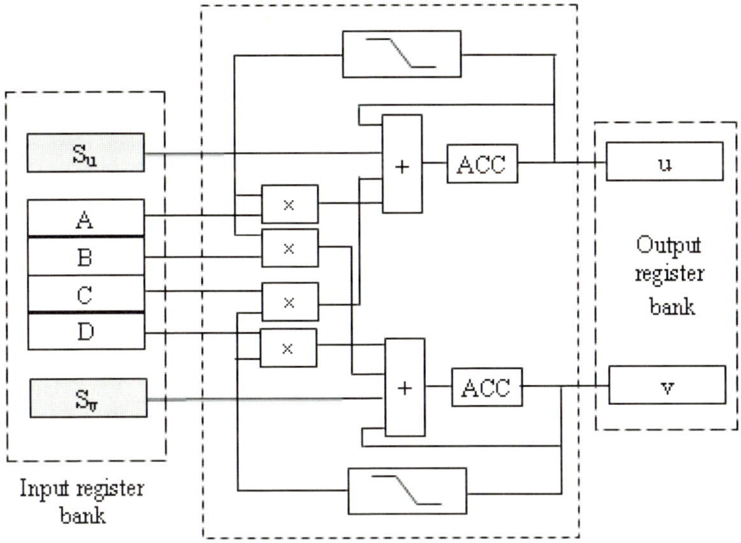

Fig. 2. Digital hardware architecture for the Amari-Hopfield model

3.2 Integration in an Embedded Linux System

In order to overcome the partial lack of flexibility of the CPG digital architecture, it has been attached as a specialized coprocessor to a microblaze processor following an embedded system design approach so as to provide a high level interface layer for application development. Microblaze is a parameterized 32-bit soft-processor engineered by Xilinx for low silicon area utilization without a memory management unit (MMU) with several bus interfaces for adding co-processors and peripherals [10]. The whole system is composed of a microblaze processor and the specialized digital architecture that generates the rhythmic patterns according to the tunable CPG parameters specified by the current

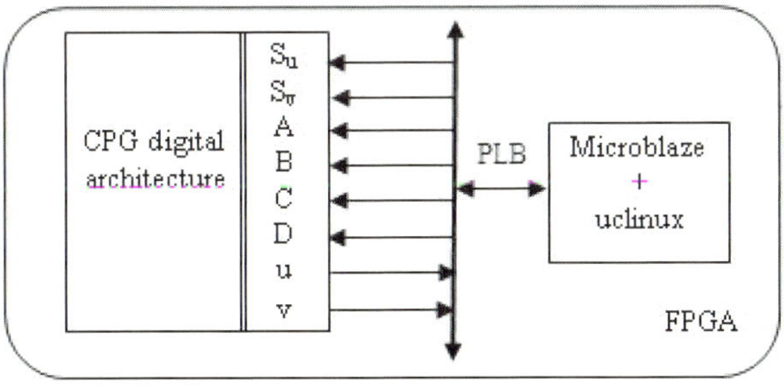

Fig. 3. Interfacing of the CPG digital architecture as a coprocessor to the microblaze-uclinux system

application. The uclinux operating system running on the processor allows to manage the system and to integrate it easily in the perception-action loop of mobile robotic applications. uclinux is a Linux fork for embedded systems without a MMU, which has been ported to microblaze by the open-source petalinux distribution [11]. Figure 3 shows a simplified block diagram of the CPG coprocessor interfacing scheme to the microblaze-uclinux system. The CPG receives six input data and it sends two output data through the microblaze Processor Local Bus (PLB) to the application.

4 Implementation Results

The CPG digital architecture has been modeled using the Very High Speed Integrated Circuits Hardware Description Language (VHDL). The FPGA-based embedded system has been targeted to Xilinx FPGA devices. A 16-bit fixed point precision, 6-bit for the integer part and 10-bit for the fractional part, has been used for testing and validation purposes, but the model is parameterized and the precision may be changed easily. The CPG module has been attached as a slave coprocessor to the microblaze soft-processor using the PLB bus and a set of wrapping libraries according to the Xilinx design flow for embedded systems. The system has been synthesized using ISE Foundation and EDK tools from Xilinx targeted to a Spartan-3E device.

The uclinux embedded operating system was ported to microblaze using the petalinux distribution. Software drivers, wrappers and a simple application example were developed to test the functionality of the CPG architecture using the microblaze cross tool chain from petalinux. A Matlab implementation of the fundamental Amari-Hopfield CPG using Simulink was written to compare the hardware and software results.

4.1 Experimental Results

In order to test the FPGA-based embedded system, a C-based application was written running on the microblaze-uclinux to set the values of the parameters in the Amari-Hopfield digital hardware model and to capture the results for a given period of time, and to send the results to a host computer through a serial connection to visualize the waveforms. Figure 4 shows the obtained sinusoidal-like waveforms for the FPGA implementation in the Spartan-3E starter kit prototyping board. The following parameter values for the Amari-Hopfield model were used: $S_u = -0.3, S_v = 0.3, A = 1.5, B = 1.5, C = 1, D = 0$. These values were obtained from [12], where a genetic algorithm was used to tune the CPG to produce the rhythmic patterns. Figure 5 shows the attractor of the Amari-Hopfield model in the u - v phase plane. For this implementation, a 2's complement fixed-point representation of 16 bits was used with 6 and 10 bits for the integer and fractional part, respectively.

4.2 FPGA Implementation

The whole system was synthesized to a Spartan-3E device using Xilinx ISE and EDK tools and tested in the Spartan-3E starter kit development board. A summary of the FPGA resource utilization of the Amari-Hopfield digital architecture is shown in table 1. For the 20-bit implementation the FPGA hardware utilization is less than seven per cent of the total available resources in the FPGA.

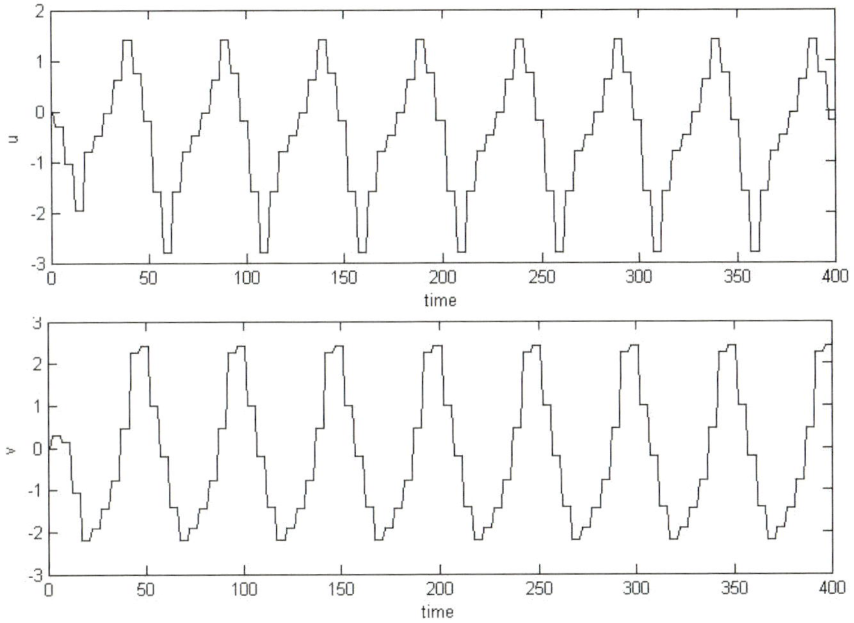

Fig. 4. Rythmic patterns for the excitatory and inhibitory neurons, top and bottom respectively, generated by the CPG FPGA-based embedded implementation

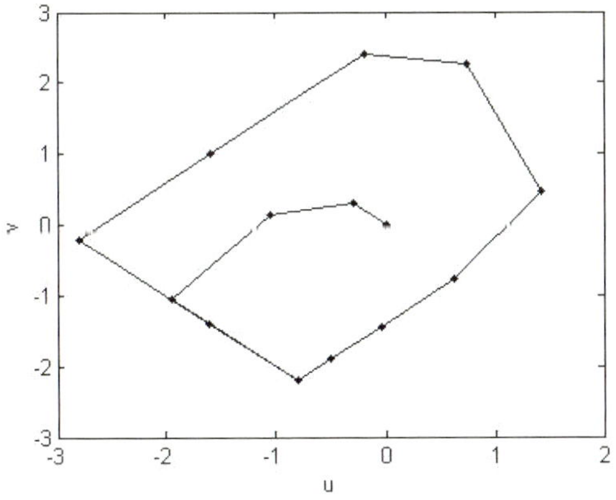

Fig. 5. Attractor of the Amari-Hopfield model in the u - v phase plane

Table 1. Hardware resource utilization for the Amari-Hopfield digital architecture for different precisions targeted to a Xilinx XC3S500e-5fg320 device

Resources	16-bit fixed point precision	20-bit fixed point precision
4-input LUTs	298	514
Flip-flops	188	272
Slices	232	347
Embedded multipliers	6	16
Maximum clock frequency	74 MHz	60 MHz

According to the reported results the CPG architecture could operate at a maximum clock frequency of 74 MHz for a 16-bit fixed point representation and 60 MHz for a 20-bit representation. The architecture was attached to the microblaze and the whole system was tested using a 50 MHz clock frequency. The hardware resource utilization for the complete system, including the microblaze processor and the cores for memory controllers and peripherals, is around 70 percent (3400 slices out of 4656) of the available resources in the FPGA. Better results both in performance and resource utilization might be obtained if most up to date FPGAs are used, such as the Virtex4-FX family devices where a hard-embedded PowerPC processor is available and more complex CPG architectures [1] could be mapped to the dedicated coprocessor.

5 Concluding Remarks and Further Considerations

This paper has presented an implementation of the fundamental Amari-Hopfield CPG in an FPGA-embedded system providing performance and flexibility to generate rhythmic patterns suitable for mobile robotic applications. A specialized digital architecture for the Amari-Hopfielfd model was developed and attached as a co-processor to a microblaze embedded processor so as to provide an integrated platform for developing robot control applications. The system can be improved and extended in different ways. More complex topologies for CPG must be analyzed and generated in order to apply the system in applications involving a large number of freedom degrees, i.e., a large number of oscillators such as those described in [1]. In order to improve further the hardware resource utilization, a serial or bit-stream pulse-mode implementation of CPG oscillators might be addressed since these approaches have been proved to be resource saving in neural network hardware implementations [6][13]. The microblaze-uclinux system opens the possibility to explore dynamic reconfiguration in the FPGA so as to adapt the parameters and topology of the CPG coprocessor according to different needs at runtime.

Acknowledgment. Authors want to thank the financial support received from INRIA, France, through the CorTexMex Research Associate Team and the support from Xilinx Inc. through the Xilinx University Program.

References

1. Nakada, K., Asai, T., Amemiya, Y.: An analog cmos central pattern generator for interlimb coordination in quadruped locomotion. IEEE Transaction on Neural Networks 14(5), 1356–1365 (2003)
2. Kier, R., James, J.C., Beer, R.D., Harrison, R.R.: Design and implementation of multipattern generators in analog vlsi. IEEE Transactions on Neural Networks 17(4), 1025–1038 (2006)
3. Still, S., Hepp, K., Douglas, R.J.: Neuromorphic walking gait control. IEEE Transactions on Neural Networks 17(2), 496–508 (2006)
4. DeHon, A.: The density advantage of configurable computing. IEEE Computer 33(2), 41–49 (2000)
5. Dan, H.: Digital VLSI Neural Networks. In: The Handbook of Brain Theory and Neural Networks, 2nd edn. (2002)
6. Girau, B., Torres-Huitzil, C.: Massively distributed digital implementation of an integrate-and.fire legion network for visual scene segmentation. Neurocomputing 50, 1186–1197 (2007)
7. Vassiliadis, S., Zhang, M., Delgado-Frias, J.: Elementary function generators for neural network emulators. IEEE Transactions on Neural Networks 11(6), 1438–1449 (2000)
8. Torres-Huitzil, C., Girau, B., Gauffriau, A.: Hardware/software co-design for embedded implementation of neural networks. In: Diniz, P.C., Marques, E., Bertels, K., Fernandes, M.M., Cardoso, J.M.P. (eds.) ARCS 2007. LNCS, vol. 4419, pp. 167–178. Springer, Heidelberg (2007)
9. Lee, G., Milne, G.: Programming paradigms for reconfigurable computing. Microprocessors and Microsystems 29, 435–450 (2005)
10. Inc., X.: Microblaze Processor Reference Guide. Version 8.0 (2007)
11. Rahavan, P., Lad, A., Neelakandan, S.: Embedded Linux System Design and Development. Averback Ed. (2005)
12. Muthuswany, B.: Implementing central pattern generators for bipedal walkers usisng cellular neural networks. Master's thesis, Electrical Engineering, and University of California, Berkeley (2005)
13. Torres-Huitzil, C.: A bit-stream pulse-based digital neuron model for neural networks. In: King, I., Wang, J., Chan, L.-W., Wang, D. (eds.) ICONIP 2006. LNCS, vol. 4234, pp. 1150–1159. Springer, Heidelberg (2006)

Biologically-Inspired Digital Architecture for a Cortical Model of Orientation Selectivity

Cesar Torres-Huitzil[1], Bernard Girau[2], and Miguel Arias-Estrada[3]

[1] Information Technology Department, Polytechnic University of Victoria
Ciudad Victoria, Tamaulipas, Mexico
Cesar.Torres@ieee.org
[2] CORTEX team, LORIA-INRIA Grand Est
Campus Scientifique, B.P. 239, Vandoeuvre-les-Nancy Cedex, France
Bernard.Girau@loria.fr
[3] Computer Science Department, INAOE
Apdo. Postal 51 and 216, Puebla, Mexico
ariasmo@inaoep.mx

Abstract. This paper presents a biologically inspired modular hardware implementation of a cortical model of orientation selectivity of the visual stimuli in the primary visual cortex targeted to a Field Programmable Gate Array (FPGA) device. The architecture mimics the functionality and organization of neurons through spatial Gabor-like filtering and the so-called cortical hypercolumnar organization. A systolic array and a suitable image addressing scheme are used to partially overcome the von Neumann bottleneck of monolithic memory organization in conventional microprocessor-based system by processing small and local amounts of sensory information (image tiles) in an incremental way. A real-time FPGA implementation is presented for 8 different orientations and aspects such as flexibility, scalability, performance and precision are discussed to show the plausibility of implementing biologically-inspired processing for early visual perception in digital devices.

1 Introduction

Biological studies indicate that neurons in the mammalian primary visual cortex are selective along several visual stimulus features such as retinal position, spatial frequency, orientation and speed. Neurons tuned to the same retinal position but different stimulus features are grouped into retinotopical arrays of hypercolumns that provide relevant information for further neural mechanisms in the visual pathways [1][2]. The orientation selectivity property is believed to have an important role in the bottom-up mechanisms for visual perception, such as motion estimation and selective visual attention [3]. Thus, its computational relevance has lead to several software implementations to support visual perceptual tasks. Designing and engineering effective systems inspired by such biological mechanisms to process in real-time in the perception-action loop through software techniques on conventional computing systems is a difficult task. It is

V. Kůrková et al. (Eds.): ICANN 2008, Part II, LNCS 5164, pp. 188–197, 2008.

mostly due to the vast amounts of data and processing needed on the limited time-multiplexed computational resources of microprocessor-based systems.

Analog Very Large Scale Integration (VLSI) implementations have been proposed to alleviate the computational load by using specific datapaths. In this context, neuromorphic engineering [4] has emerged as a research area for designing systems that replicate the capabilities of biological systems by mimicking both functional and structural characteristics of neural processing for perceptual tasks through analog and asynchronous computations of electronic devices [5][6], whereas digital hardware implementations appear as quite unable to handle such processing for lack of density. However, devices such as FPGAs, and their associated methodologies and tools have achieved a level of maturity that allow designing and implementing complex systems on a single chip with high computational density and performance by tailoring the underlying architecture to the algorithmic nature [7]. Instead of temporally distributing computations across time shared compute units using a monolithic and linear memory organization, computations in FPGAs are distributed spatially across dedicated compute units exploiting fine grain parallelism of the computation and distributed memory blocks (spatial computing model). In spite of the ability of FPGAs to deal with strong implementation and application constraints, a major current research domain is to find ways to map neural connectivity and hardware-demanding nonlinear functionality onto digital devices through architectures inspired by such massively parallel processing for visual perception [8][9].

Unlike analog neuromorphic implementations of orientation selectivity [5][6], a real-time FPGA-based architecture is presented in this paper. The architecture relies on a set of processing elements organized on a 2-D systolic array providing a high modularity in the sense that the coverage of orientations can be extended by replicating the array for each additional orientation. The architecture constitutes one of our first attempts to make real-time implementations of biologically inspired vision models in fully flexible and functional systems. The rest of the paper is organized as follows. Section 2 provides the biological and mathematical foundations of the Hubel and Wiesel orientation selectivity model in the primary visual cortex. Section 3 describes the proposed architecture and gives details of its hardware organization at different levels in analogy to the biological model. Section 4 presents and discusses the experimental results of synthesis and performance of the architecture targeted to an FPGA device.

2 Orientation Selectivity Model

Hubel and Wiesel suggested that neurons tuned to the same orientation and retinal location are grouped together into what has been called a column [5][6]. In their model, neighboring columns at the same retinal position but with different preferred orientations are grouped into hypercolumns. Conceptually, the primary visual cortex can be though as a two dimensional sheet of neurons where at any point, all neurons are tuned to the same orientation and location [1][2]. Figure 1 (a) shows the hypercolumnar organization in the model where each

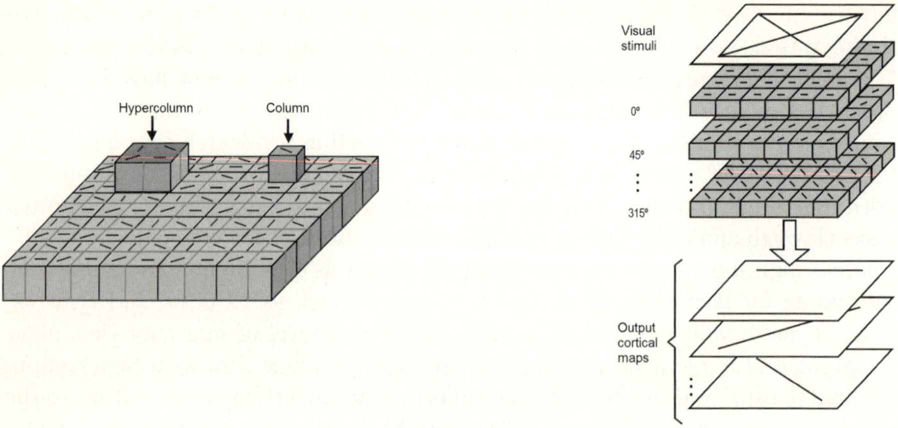

Fig. 1. (a) Conceptual diagram of the hypercolumnar organization of neurons, and (b) a computational model used to implement the functionality of orientation selectivity. Each layer represents neurons tuned to the same orientation, adapted from [6].

block represents a neuron tuned to a specific orientation. The neuron response is strongest when the visual stimulus is applied at a preferred retinal location and with a preferred orientation. Figure 1 (b) shows a computational analogy used to implement this model by means of multiple independent arrays of neurons tuned at the same orientation as proposed in [6].

A functional approximation for a large number of neuron responses in the primary visual cortex consists of a linear spatio-temporal filtering stage followed by several nonlinear mechanisms. Particularly, the spatial filtering determines the orientation selectivity and a set of spatial Gabor filters is commonly used to mimic the response profile of neurons. A 2-D Gabor filter is an oriented complex sinusoidal grating usually modulated by a 2-D Gaussian function [10]:

$$G_{\sigma,\phi,\theta}(x,y) = \frac{1}{2\pi\sigma} e^{\frac{-(x^2+y^2)}{2\sigma^2}} e^{2\pi i\phi(x\cos\theta + y\sin\theta)} \tag{1}$$

The parameters ϕ, θ are real values that specify the frequency and orientation of the span-limited sinusoidal grating, x and y are the spatial coordinates and σ is the scale parameter of the Gaussian function. By changing these values, the filters can be tuned to different orientations, bandwidth and spatial scales of the visual stimuli. In some models, a phase parameter is used in the complex exponential to control the type of stimulus, a bar or edge, that best excites the filter [11]. Neurons display a full range of phases, although there is some evidence that phases cluster around 0°, 90°, 180°and 270°[1][2]. The Gabor-filtered outputs of a gray-level image $f(x,y)$ are obtained by the convolution of the image with the Gabor filter bank. The Gabor filter is a complex valued function that can be decomposed in its real and imaginary parts. Given a neighborhood window with size WxW and $W = 2k+1$, the discrete convolutions of $f(x,y)$ with the real component, $R_{\sigma,\phi,\theta}$, of $G_{\sigma,\phi,\theta}(x, y)$ (the convolution with the imaginary component $I_{\sigma,\phi,\theta}$ is computed in a similar way) is given by:

$$G_R(x, y, \sigma, \phi, \theta) = \sum_{l=-k}^{k} \sum_{m=-k}^{k} f(x + l, y + m) R_{\sigma,\phi,\theta}(l, m) \qquad (2)$$

An image is filtered with a set of Gabor filters tuned to different orientation and spatial frequencies that cover appropriately the spatial frequency domain. The obtained results are feature fields or maps used for further processing in different applications. In this work, only the real part of the discrete Gabor filters is considered for the orientation selectivity [11][12], but the architecture can be applied to compute the imaginary part or both. Particularly, a biologically-inspired digital architecture for discrete Gabor-like filtering of gray-level images in eight different orientations using 7x7 window masks is presented.

3 Biologically Inspired Digital Architecture

Since a fully parallel implementation over a complete image is not viable, despite improvements in FPGA densities [13][14], the implementation aims at reducing hardware resource utilization without compromising processing speed by using systolic arrays of limited size to filter incrementally over input image tiles [15]. Figure 2 shows a block diagram of the proposed architecture that filters the input image in different orientations. The input image is stored in an external memory which is read in a column-based scheme and pixels are sent to all the Gabor Window Processors (GWPs). The elemental processing unit is the GWP, which is organized at different levels of interconnection and modularity. The memory management unit uses efficiently the memory bandwidth by reusing data inside the arrays, exploiting data parallelism, and reducing the input/output buses. A control module is used to coordinate computations and the data exchange among GWPs and, a temporal storage module to pack several results in one location in the output memory. The architecture follows a very regular and modular structure inspired by the organization of neurons in the orientation selectivity model as described in the following subsections.

3.1 Gabor Window Processor

Each GWP computes the convolution of an image tile with the Gabor window mask centered at a specific position in the input image. The block diagram of the GWP and its analogy to orientation selectivity model are shown in figure 2 (right). The GWP has two internal registers to support computational and data pipelines among neighboring GWPs. The computational pipeline computes a partial product between a pixel and a Gabor coefficient on each clock cycle and accumulates previous results. The data pipeline distributes the Gabor coefficients, the received coefficient is delayed one clock cycle and then transmitted to the neighboring GWP in the array.

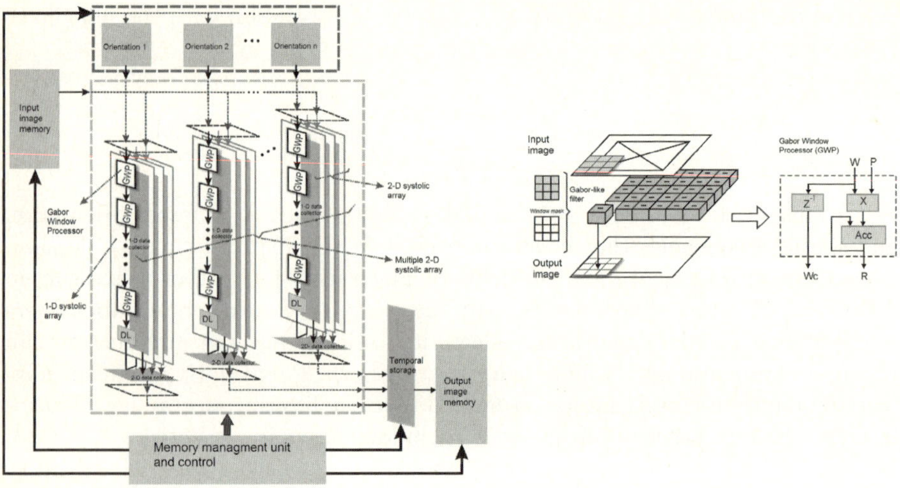

Fig. 2. General organization of the multiple 2-D systolic arrays in the biological-inspired hardware architecture and the Gabor Window Processor

3.2 1-D Systolic Array

In figure 3 (a), each block denoted by GWP represents an orientation-tuned neuron located in the same column on the input image, the block labeled by DL stands for a delay line. The 1-D data collector module collects results of GWPs and sends them to the 2-D data collector. Data flows from top to bottom in the array and the GWPs work progressively in a systolic pipeline. The current pixel is broadcast to all GWPs and the coefficients are transmitted between adjacent GWPs. A delay line, a serial-input serial-output shift register, is required in the boundary of the GWPs array for window coefficient synchronization among columns of GWPs when they are connected in a 2-D systolic array. The window coefficients are temporally stored in the delay line, and then transmitted to the next column of GWPs. The stage number in the delay line depends on the number of rows processed in parallel by the architecture.

3.3 2-D Systolic Array

Figure 3 (b) shows a block diagram of the 2-D systolic organization of GWPs. The array exploits the 2-D parallelism through the incremental computation of neuron responses tuned to the same orientation as image tiles are extracted from the input memory. The interconnection between two contiguous columns of GWPs is straightforward and it can be considered a cascaded connection of 1-D systolic arrays. In the 2-D array, data flows from top to bottom of a column, and then left to right through columns. The 2-D data collector collects progressively results from the 1-D systolic arrays and then sends the results to the temporal storage module.

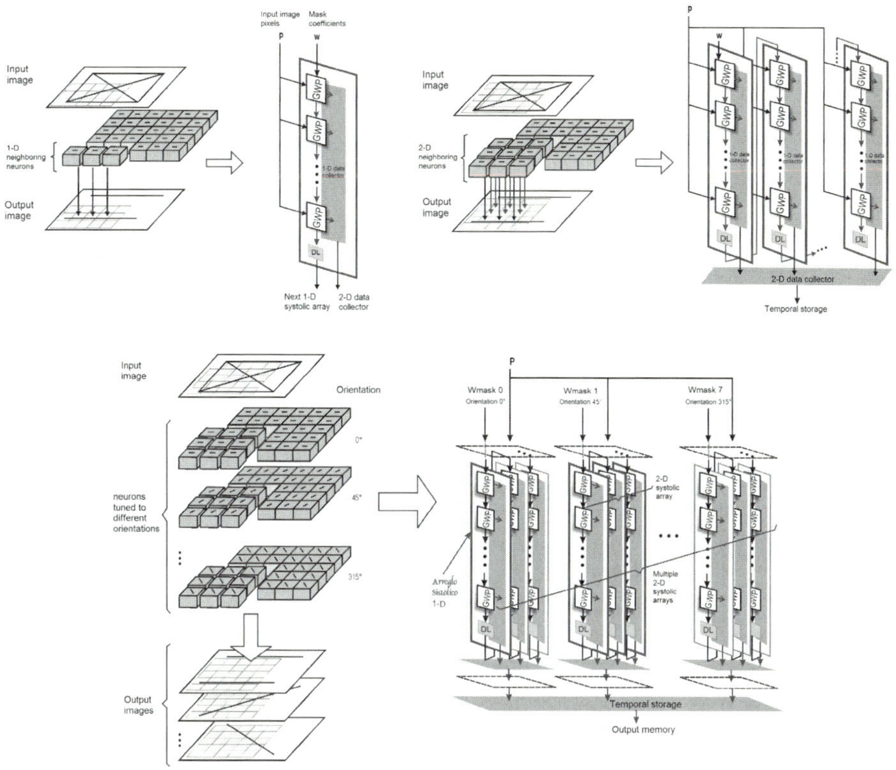

Fig. 3. Biological analogy and organization of GWPs in the architecture, (a) 1-D systolic array of GWPs, (b) 2-D systolic array of GWPs, and (c) Multiple 2-D systolic array for different orientations

3.4 Multiple 2-D Systolic Arrays

In order to filter the input image in different orientations, multiple 2-D systolic arrays are replicated in the architecture as shown in figure 3 (c). Each 2-D array receives the same pixels from the input image but works with different convolution coefficients tuned to a specific orientation. In this sense, the architecture is flexible to include a wide rage of orientations to filter the input image by using as many arrays as orientations. Also the number of GWPs in the 2-D systolic array might be adjusted to different application constraints but a good compromise between area and performance is to use the same number of GWPs as the size of the convolution mask [14].

4 Experimental Results and Discussion

An architecture configuration to compute response in eight different orientations, by 7x7 convolution masks, was used for simulation and validation purposes on

255	255	14	3	0	247	254	255	253	255	255	255	255
255	255	0	5	10	2	255	252	248	254	255	255	255
255	255	1	8	1	0	14	255	245	255	255	255	255
255	255	5	4	0	25	0	0	255	252	255	255	255
255	255	9	0	15	0	5	0	14	242	255	245	255
255	255	0	0	16	0	0	4	0	14	242	255	242
255	255	7	1	0	0	0	0	12	0	7	251	254
255	255	4	0	0	20	4	0	10	6	0	0	255
255	255	0	3	3	0	2	4	0	0	14	3	0
255	255	0	4	5	0	0	13	0	15	0	5	10
255	255	0	0	6	9	5	0	0	2	1	8	1
255	255	11	0	0	4	0	0	12	0	5	4	0
255	255	0	0	0	0	0	0	0	0	9	0	15

0.028056	0.011407	−0.18524	0.33725	−0.13002	−0.31745	0.44917
0.011407	−0.20247	0.40289	−0.16977	−0.45306	0.70067	−0.31745
−0.18524	0.40289	−0.18556	−0.54125	0.91491	−0.45306	−0.13002
0.33725	−0.16977	−0.54125	1	−0.54125	−0.16977	0.33725
−0.13002	−0.45306	0.91491	−0.54125	−0.18556	0.40289	−0.18524
−0.31745	0.70067	−0.45306	−0.16977	0.40289	−0.20247	0.011407
0.44917	−0.31745	−0.13002	0.33725	−0.18524	0.011407	0.028056

7	3	−47	86	−33	−81	115
3	−52	103	−43	−116	179	−81
−47	103	−48	−139	234	−116	−33
86	−43	−139	255	−139	−43	86
−33	−116	234	−139	−48	103	−47
−81	179	−116	−43	103	−52	3
115	−81	−33	86	−47	3	7

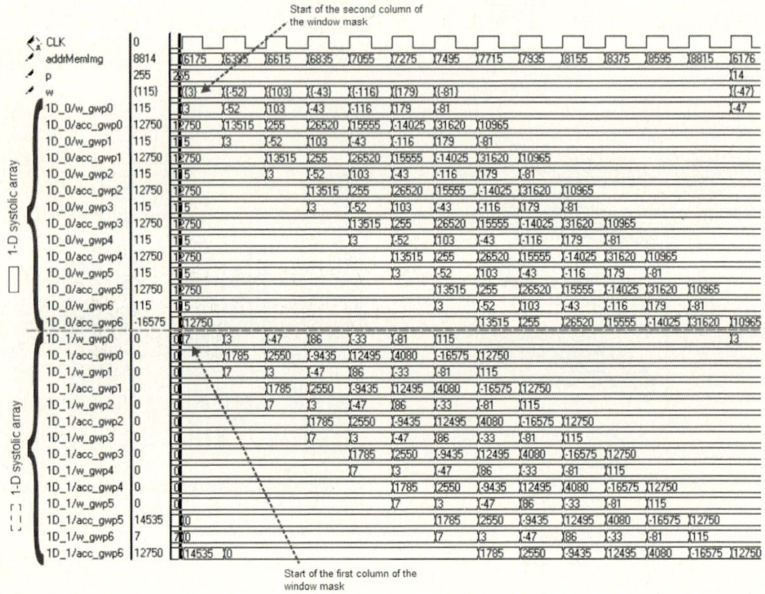

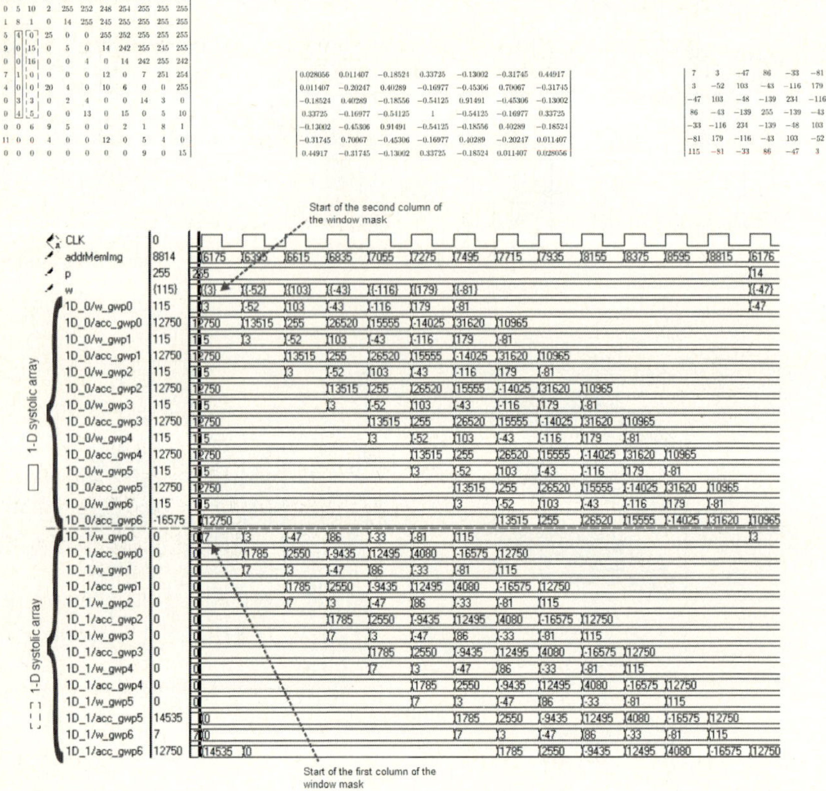

Fig. 4. (a) An input image tile used in the simulation (left), the floating point representation of the window coefficients (middle), and the corresponding fixed point representation interpreted as signed integers (right). (b) Data and computational pipelines for two 1-D systolic arrays associated to the positions in the input image enclosed by solid and broken line rectangles.

8-bit gray level images. A 10-bit fixed point precision, with 9 bits for the fractional part, was selected after some experimental tests on the coefficients numeric range, but generics were used to cope with several precision choices.

To show the main data and computational pipelines inside the systolic arrays, a timing diagram is presented for a limited number of clock cycles. For simplicity, a small tile of an input image and a 7x7 window mask are considered in the simulation. The gray-level pixels, the window coefficients, both in their floating point and fixed point representation (values interpreted as signed integers) are shown in figure 4 (a). Figure 4(b) shows internal computations of GWPs in the 1-D systolic arrays enclosed by solid and broken line rectangles. At the start of the timing diagram, the array enclosed by the solid line rectangle is processing the second column of its associated window mask, and the array in broken line rectangle is processing the first column of the input image. Note that the

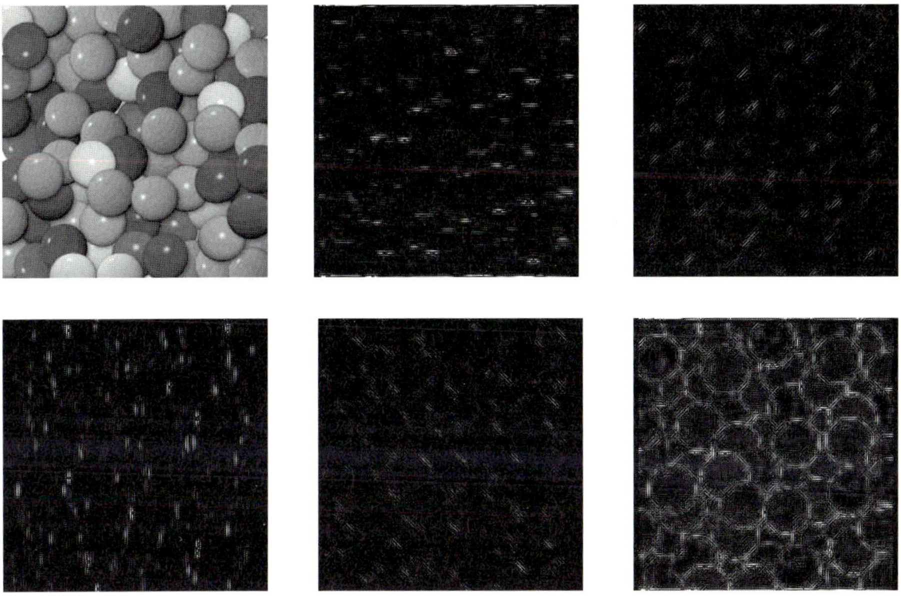

Fig. 5. Results of the orientation selectivity response of the architecture for a gray-level input image. (a) the input image, (b) 0°, (c) 45°, (d) 90°, (e) 135°, (f) the addition of 0°to 135°responses.

addressing scheme reads pixels in a column-based order with 13 pixels per column. Since a GWP processes only seven pixels of this column, an idle period of 6 clock cycles for each GWP is required. The window coefficients are transmitted from the first array to the second one by the delay line with a delay of 6 clock cycles as shown in figure 4(b). Note that the window coefficients are transmitted from one GWP to the next GWP array on each clock cycle. A VHDL testbench was written to test and validate the model with input images. Figure 5(a) is a 8-bit gray level test image with a 512x512 resolution. Figures 5(b)-(e) show the obtained responses for different orientations, 0°, 45°, 90°, and 135°, respectively. Figure 5 (f) shows an image obtained by adding the responses from 0°to 135°. Though the tested architecture configuration can compute the responses in eight different orientations from 0°to 315°, for simplicity only four orientation responses are shown. The scaling and frequency parameters to generate the Gabor coefficients are 0.239 and 0.5, respectively.

The VHDL model was synthesized and implemented into a Virtex device using the Xilinx ISE 8.2i tool suite. Results are summarized in table I. Eight 2-D systolic arrays of 7x7 to compute the response in eight orientations fitted in an XCV3200E-6 device. The architecture compares favorably to other digital architectures reported in the literature for image convolution [16]. Our implementation is able to handle medium and high resolution images with a limited number of GWPs compared to the number of pixels in the image without sacrificing performance. According to the performance analysis presented in [13]

Table 1. Post-place and route synthesis results for the proposed architecture targeted to a XCV3200E-6 device

Resources	number
4-input LUTs	49883
Flip-flops	16811
Slices	31563
Maximum clock frequency	40 MHz

and the reported maximum clock frequency, the processing time for a 512x512 gray-level image is around 15 milliseconds, i.e., around 66 images per second, which fulfill the standard metric for real-time performance of 30 frames per second. Note that the frequency is moderately low due to the high requirements of interconnection resources for the design. A good comprise between performance and hardware resource utilization is setting the size of the 2-D systolic array to the same size as the window mask. Increasing the number of GWPs far beyond does not provide substantially larger performance improvements [16].

5 Conclusion and Further Work

This paper has presented a biologically inspired digital architecture for an efficient implementation of a cortical model of orientation selectivity. The experimental evaluations have shown that a compact architecture allows the real-time computation of different orientation responses over an input image suitable for embedded processing with the additional benefits of flexibility and scalability compared to analog implementations. These benefits allow to adapt the architecture to different application constraints: variable data precision, performance and resource utilization tradeoff, and orientation coverage. The architecture can be extended to other applications related to Gabor processing, such as image analysis, feature extraction, texture segmentation and motion detection. Particularly, the architecture might be useful in a class of motion detection models that arise from a spatio-temporal conceptualization of motion estimation [17] where spatiotemporal energy filters are used to selectively respond to motion in a particular orientation. To extract spatiotemporal energy, filters are chosen as quadrature pairs and their outputs squared and summed. In the case of Gabor filters, this can be done by using the real and imaginary parts of the same filter. As the presented architecture can compute concurrently the quadrature outputs of Gabor filters, it can be used as the main computational core of an embedded digital system for motion perception.

Acknowledgment. Authors want to thank the financial support received from INRIA, France, through the CorTexMex Research Associate Team, the support from CONACyT, Project No. 59474, Mexico, and the contribution made by Sayde Alcantara-Santiago in the experiments.

References

1. Ferster, D., Miller, K.D.: Neural mechanisms of orientation selectivity in the visual cortex. Annu. Rev. Neurosci. 23, 441–471 (2000)
2. Rust, N.C., Shwartz, O., Movshon, J.A., Simoncelli, E.: Spatiotemporal elements of macaque v1 receptive field. Neuron 46, 945–956 (2005)
3. Itti, L., Koch, C.: Computational modeling of visual attention. Nature Reviews NeuroscienceNature Reviews Neuroscience 2(3), 194–203 (2003)
4. Mead, C.A.: Neuromorphic electronic systems. Proc. IEEE 78, 1629–1636 (1990)
5. Choi, T.Y.W., Shi, B.E., Bohanen, K.A.: A multi-chip implementation of cortical orientation hypercolumns. In: Proceedings ISCAS, pp. 13–16 (2004)
6. Choi, T.Y.W., Merolla, P.A., Arthur, J.V., Bohanen, K.A., Shi, B.E.: Neuromorphic implementation of orientation hypercolumn. IEEE Transactions on Circuits and Systems -I 52(6), 1049–1060 (2005)
7. Herbordt, M.C., VanCourt, T., Gu, Y., Sukhwani, B., Conti, A., Model, J., DiSabello, D.: Achieving high performance with fpga-based computing. IEEE Computer Magazine, 50–57 (March 2007)
8. Girau, B., Torres-Huitzil, C.: Massively distributed digital implementation of an integrate-and.fire legion network for visual scene segmentation. Neurocomputing 50, 1186–1197 (2007)
9. Torres-Huitzil, C., Girau, B., Gauffriau, A.: Hardware/software co-design for embedded implementation of neural networks. In: Diniz, P.C., Marques, E., Bertels, K., Fernandes, M.M., Cardoso, J.M.P. (eds.) ARCS 2007. LNCS, vol. 4419, pp. 167–178. Springer, Heidelberg (2007)
10. Tsi, D.M., Lin, C.P., Huang, K.T.: Defect detection in coloured texture surfaces using gabor filters. The Imaging Science Journal 53, 27–37 (2005)
11. Petkov, N., Kruzinga, P.: Computational models of visual neurons specialised in the detection of periodic and aperiodic oriented visual stimuli: Bar and gratings cells. Biological Cybernetics 75, 83–96 (1997)
12. Grigorescu, S.E., Petkov, N., Kruizinga, P.: Comparison of texture features based on gabor filter. IEEE Transactions on Neural Networks 11(10), 1160–1167 (2002)
13. Torres-Huitzil, C., Arias-Estrada, M.: Fpga-based configurable hardware architecture for real-time window-based image processing. EURASIP Journal on Applied Signal Processing 7, 1024–1034 (2005)
14. Himavathi, S., Anitha, D., Muthuramalingam, A.: Feedforward neural network implementation in fpga using layer multiplexing for effective resource utilization. IEEE Transactions on Neural Networks 18(3), 880–888 (2007)
15. Kung, H.T.: Why systolic architectures? IEEE Computer 15(1), 37–46 (1982)
16. Savich, A.W., Moussa, M., Areibi, S.: The impact of arithmetic representation on implementing mlp-bp on fpgas: A study. IEEE Transactions on Neural Networks 18(1), 240–252 (2007)
17. Adelson, E.H., Berge, J.R.: Spatiotemporal energy models for the perception of motion. J. Opt. Soc. Am. A 2(2) (February 1985)

Neural Network Training with Extended Kalman Filter Using Graphics Processing Unit

Peter Trebatický and Jiří Pospíchal*

Slovak University of Technology in Bratislava, Faculty of Informatics and Information Technologies, Ilkovičova 3, 842 16 Bratislava, Slovakia
{trebaticky,pospichal}@fiit.stuba.sk

Abstract. The graphics processing unit has evolved through the years into the powerful resource for general purpose computing. We present in this article the implementation of Extended Kalman filter used for recurrent neural networks training, which most computational intensive tasks are performed on the GPU. This approach achieves significant speedup of neural network training process for larger networks.

1 Introduction

The graphics processing unit (GPU) was and still is used mainly for speedup of graphical operations. It has recently evolved into the powerful resource for general purpose computing.

Recurrent neural network learning is a very difficult numerical problem, which approaches very poorly and slowly to satisfactory results when being solved with the classic gradient optimization methods on longer input sequences. In this paper we present the already studied [9,10] better alternative which is called Extended Kalman filter (EKF) and how to make it faster using modern but generally available graphics processing unit.

In the first section of this paper we explain the concept of Kalman filtering and how to apply it to the task of neural network training. The remaining sections describe in detail how to map the equations involved onto the graphics processing unit. On the chosen problem we demonstrate two things. Firstly, the already known fact that the EKF achieves better results than classical gradient descent methods and secondly, more importantly, the speedup of our implementation on the graphics processing unit as opposed to implementation on CPU.

2 Extended Kalman Filter

Kalman filter (which is the set of mathematical equations) is considered one of the important discoveries in the control theory principles. E. Kalman's paper [5] was published in the year 1960. Its most immediate applications were in control

* This work was supported by Scientific Grant Agency of Slovak Republic under grants #1/0804/08 and #1/4053/07.

V. Kůrková et al. (Eds.): ICANN 2008, Part II, LNCS 5164, pp. 198–207, 2008.

of complex dynamic systems, such as manufacturing processes, aircrafts, ships or spaceships (it was part of the Apollo onboard guidance system). However, the Extended Kalman filter started to appear in the neural network training applications only relatively recently, which was caused by the progress of computer systems development.

Original Kalman filter is targeted at the *linear* dynamic systems. However, when the model is *nonlinear*, which is the case of neural networks, we have to extend Kalman filter using linearization procedure. Resulting filter is then called extended Kalman filter (EKF) [3]. Neural network is a nonlinear dynamic system, that can by described by equations:

$$\mathbf{x}_k = \mathbf{x}_{k-1} + \mathbf{q}_{k-1} \tag{1}$$

$$\mathbf{y}_k = g(\mathbf{x}_k, \mathbf{u}_k, \mathbf{v}_{k-1}) + \mathbf{r}_k \tag{2}$$

The process equation expresses the state of neural network as a stationary process corrupted by the process noise $\mathbf{q}_k$, where the state of the network $\mathbf{x}$ consists of network weights. Measurement equation expresses the desired output of the network as a nonlinear function g of the input vector $\mathbf{u}_k$, of the weight vector $\mathbf{x}_k$ and for recurrent networks also of the activations of recurrent neurons from the previous step $\mathbf{v}_{k-1}$. This equation is augmented by a random measurement noise $\mathbf{r}_k$. The covariance matrix of the noise $\mathbf{r}_k$ is $\mathbf{R}_k = E\left[\mathbf{r}_k\mathbf{r}_k^T\right]$ and the covariance of the noise $\mathbf{q}_k$ is $\mathbf{Q}_k = E\left[\mathbf{q}_k\mathbf{q}_k^T\right]$.

The basic idea of the Extended Kalman filter lies in the linearization of the measurement equation at each time step around the newest state estimate $\hat{\mathbf{x}}_k$. We use for this purpose just the first-order Taylor approximation of non-linear equation, because of computational complexity.

We can express the neural network training as a problem of finding the state estimate $\mathbf{x}_k$ that minimizes the least-squares error, using all the previous measurements. We can express the solution of this problem as:

$$\mathbf{K}_k = \mathbf{P}_k\mathbf{H}_k^T[\mathbf{H}_k\mathbf{P}_k\mathbf{H}_k^T + \mathbf{R}_k]^{-1} \tag{3}$$

$$\hat{\mathbf{x}}_{k+1} = \hat{\mathbf{x}}_k + \mathbf{K}_k[\mathbf{y}_k - g(\hat{\mathbf{x}}_k, \mathbf{u}_k, \mathbf{v}_{k-1})] \tag{4}$$

$$\mathbf{P}_{k+1} = \mathbf{P}_k - \mathbf{K}_k\mathbf{H}_k\mathbf{P}_k + \mathbf{Q}_k \tag{5}$$

where $\hat{\mathbf{x}}$ is a vector of all the weights, $g(\cdot)$ is a function returning a vector of actual outputs, $\mathbf{y}$ is a vector of desired outputs, $\mathbf{K}$ is the so called Kalman gain matrix, $\mathbf{P}$ is the error covariance matrix of the state and $\mathbf{H}$ is the measurement matrix (Jacobian). Matrix $\mathbf{H}$ contains partial derivatives of ith output with respect to jth weight. One can use for this purpose one of two main methods – *Real-Time Recurrent Learning* (RTRL) or *Backpropagation Through Time* (BPTT), or its truncated version BPTT(h) [8].

The RTRL is computationally intensive, therefore we will use the BPTT(h) method according to the recommendation in [8]. With this method for appropriately chosen depth h we obtain derivatives that are close to those obtained by the RTRL, and we significantly reduce the computational complexity.

At the beginning of the recursion (3–5) it is necessary to assign the initial values of $\hat{\mathbf{x}}_0$, $\mathbf{P}_0$, $\mathbf{Q}_0$ and $\mathbf{R}_0$. The initial values of weights should be set randomly e.g. from interval $[-0.5, 0.5]$. According to recommendations in [3,8] $\mathbf{P}_0$ should be set in orders of magnitude $100\,\mathbf{I}$–$1000\,\mathbf{I}$, where $\mathbf{I}$ is identity matrix. $\mathbf{Q}_0$ should have values from $10^{-6}\,\mathbf{I}$ to $0.1\,\mathbf{I}$. Its nonzero value helps to override divergency [3]. For the parameter $\mathbf{R}_0$, values in orders of magnitude $10\,\mathbf{I}$–$100\,\mathbf{I}$ should be chosen.

3 Computational Complexity

In order to express the computational complexity of EKF and BPTT(h), let us summarize the notation and dimensions of every matrix and vector:

n_i	Number of input neurons, i.e. number of inputs to the network
n_o	Number of output neurons, i.e. number of outputs of the network
n_h	Number of hidden neurons
n_x	Number of weights in the network
$\mathbf{x}_{[n_x \times 1]}$	Vector of weights in the network
$g(\cdot)_{[n_o \times 1]}$	Function returning vector of actual outputs
$\mathbf{y}_{[n_o \times 1]}$	Vector of desired outputs
$\mathbf{K}_{[n_x \times n_o]}$	Kalman gain matrix
$\mathbf{P}_{[n_x \times n_x]}$	Error covariance matrix
$\mathbf{Q}_{[n_x \times n_x]}$	Process noise covariance matrix
$\mathbf{R}_{[n_o \times n_o]}$	Measurement noise covariance matrix
$\mathbf{H}_{[n_o \times n_x]}$	Measurement matrix (Jacobian)

Number of weights n_x in the Elman's architecture of recurrent neural network can be expressed as $(n_i+1)n_h+n_h^2+(n_h+1)n_o$, where ones represent bias weights.

Using given notation, we can express the computational complexity of the EKF (3–5) as $O\left(\left[n_o n_x^2 + n_o^2 n_x + n_o^2 + n_o n_x + n_x^2\right] + n_o^3 + h n_o n_x\right)$. The first term $\left[n_o n_x^2 + n_o^2 n_x + n_o^2 + n_o n_x + n_x^2\right]$ comes from matrix multiplications and additions, the second term n_o^3 is the matrix inversion in (3) and the third term $h n_o n_x$ represents the complexity of BPTT(h) used to compute measurement matrix $\mathbf{H}$, where h is the truncation depth. Typically $h, n_o \ll n_x$, so the complexity of the EKF is then $O(n_o n_x^2)$. As BPTT(h) alone has complexity $O(h n_o n_x)$, the speed difference between EKF and BPTT(h) is significant for large n_x.

4 Reducing Complexity Using GPU

In this paper we aim to reduce the impact of the most computationally intensive tasks of the EKF by exploiting the parallel nature of the graphics processing unit. As the number of parallel processors is fixed – concretely 128 in G8800 GTX – this does not have the impact on asymptotical complexity of EKF. But any constant factor speedup makes the EKF more useable and for smaller numbers of n_x can actually make it faster than BPTT(h).

For implementation we chose CUBLAS library readily available for nVidia GPUs (as part of CUDA Toolkit[12]), because it provides basic linear algebra operations, especially matrix multiplication. This library is a significant step in general purpose computing on the GPU as one does not have to overcome steep learning curve of detailed GPU functionality. As it turns out, the provided functionality of CUBLAS library is all we need to speed up matrix equations (3–5).

The only thing that is not readily available is matrix inversion needed to compute $n_o \times n_o$ matrix in (3). This is a symmetric positive definite matrix, that is why we chose to use Cholesky factorization to compute its inverse. Cholesky factorization on GPU has already been studied [4], but because it is usually small matrix ($n_o \ll n_x$) we can simply compute it on the CPU. This requires an additional transfer of data between GPU and CPU which is a costly operation, but we already have to transfer a much bigger measurement matrix and weight vector, so this should not have a big impact.

Cholesky factorization of a symmetric positive definite matrix $\mathbf{M}$ is a lower triangular matrix $\mathbf{N}$ for which holds $\mathbf{NN}^T = \mathbf{M}$. Inverse of matrix $\mathbf{M}$ can then be efficiently and numerically stable computed:

$$\mathbf{MM}^{-1} = \mathbf{I} \tag{6}$$

$$\mathbf{NN}^T\mathbf{M}^{-1} = \mathbf{I} \tag{7}$$

$$\mathbf{N}^T\mathbf{M}^{-1} = \mathbf{X} \tag{8}$$

$$\mathbf{NX} = \mathbf{I} \tag{9}$$

where $\mathbf{I}$ is an identity matrix and $\mathbf{X}$ is used for substitution. The inverse of $\mathbf{M}$ is computed by first solving (9) for unknown $\mathbf{X}$ and then by solving (8) for unknown $\mathbf{M}^{-1}$.

5 Implementation Details

This section describes in detail the implementation of EKF on GPU as well as on CPU. In order to explain the meaning of auxiliary matrices $\mathbf{A}$, $\mathbf{B}$, $\mathbf{C}$ and $\mathbf{Z}$ used in pseudocode we rewrite equation (3):

$$\mathbf{K} = \mathbf{PH}^T\left[\mathbf{HPH}^T + \mathbf{R}\right]^{-1} = \mathbf{C}\left[\mathbf{AH}^T + \mathbf{R}\right]^{-1} = \mathbf{CB}^{-1} = \mathbf{CZ}$$

The pseudocode of initialization of the EKF on GPU follows:

On CPU Initialize matrices $\mathbf{P}$, $\mathbf{R}$, $\mathbf{Q}$, $\mathbf{I}$
Note: we store only diagonal elements of $\mathbf{R}$, $\mathbf{Q}$ and $\mathbf{I}$
On CPU Fill vector $\mathbf{x}$ with uniformly random weights from $[-0.5, 0.5]$
Transfer $\mathbf{P}$, $\mathbf{R}$, $\mathbf{Q}$, $\mathbf{I}$ and $\mathbf{x}$ from system memory to GPU

The pseudocode of one time step of the EKF on GPU follows. The mentioned function names are those that were used from CUBLAS library.

On CPU Set weights to $\mathbf{x}$ and propagate network with actual input
On CPU Compute measurement matrix $\mathbf{H}$ using BPTT(h)
On CPU Compute $\mathbf{yg} = \mathbf{y} - g(\cdot)$
Transfer $\mathbf{H}$ from system memory to GPU

On GPU $\mathbf{A} = \mathbf{HP}$, $\mathbf{P}$ is symmetric – function `cublasSsymm()`
On GPU $\mathbf{C} = \mathbf{PH}^T$ – function `cublasSgemm()`
On GPU $\mathbf{B} = \mathbf{AH}^T$ – function `cublasSgemm()`
On GPU $\mathbf{B} = \mathbf{B} + \mathbf{R}$ – function `cublasSaxpy()`
Transfer $\mathbf{B}$ from GPU to system memory
On CPU Compute Cholesky factor of $\mathbf{B}$
Note: it is stored in the lower triangular part of $\mathbf{B}$
Transfer $\mathbf{B}$ and $\mathbf{yg}$ from system memory to GPU
On GPU $\mathbf{Z} = \mathbf{I}$ – function `cublasScopy()`
On GPU $\mathbf{BZ} = \mathbf{Z}$ – function `cublasStrsm()`
Note: solves $\mathbf{Z}$ in equation $\mathbf{BZ} = \mathbf{I}$
On GPU $\mathbf{B}^T\mathbf{Z} = \mathbf{Z}$ – function `cublasStrsm()`
Note: $\mathbf{Z}$ now contains inverse of $\mathbf{HPH}^T + \mathbf{R}$
On GPU $\mathbf{K} = \mathbf{CZ}$, $\mathbf{Z}$ is symmetric – function `cublasSsymm()`
On GPU $\mathbf{x} = \mathbf{Kyg} + \mathbf{x}$ – function `cublasSgemv()`
On GPU $\mathbf{P} = -\mathbf{KA} + \mathbf{P}$ – function `cublasSgemm()`
On GPU $\mathbf{P} = \mathbf{P} + \mathbf{Q}$ – function `cublasSaxpy()`
Transfer $\mathbf{x}$ from GPU to system memory

We used our own implementation of BPTT(h) and the pseudocode from [4] to compute Cholesky factorization on CPU. Each variable that was transferred to GPU used single precision floating point, which is the limitation of present GPUs. We used double precision floating point for everything else.

The implementation on CPU was essentially the same. The only difference was removal of transfers between CPU and GPU and the substitution of CUBLAS library functions by corresponding ATLAS library functions [11], which is straightforward as they both comply with Basic Linear Algebra Subprograms (BLAS) standard [2]. When configured to support threading, ATLAS library automatically takes advantage of multiple CPU cores to speed up its functions.

The hardware and software specifications used for all the tests:

- Intel Core2 Quad CPU Q6600 2.4 GHz – 4 cores
- nVidia GeForce 8800 GTX GPU – 128 parallel processors
- Ubuntu 7.10
- gcc 4.2.1
  compile flags: `-fomit-frame-pointer -mfpmath=sse -mmmx -msse -msse2 -msse3 -O3`
- ATLAS 3.8.1
  configured with `-b 32 -t 4 -D c -DPentiumCPS=2400`
- nVidia Toolkit 1.1
- nVidia Graphics Driver 169.12

ATLAS library was configured to support threading on CPU. The number of maximum threads was chosen to be 4 which is the number of cores on our test machine.

6 Experiment Description

The main goal of the experiment was to compare the performance of various implementations of the EKF and for comparison also of the BPTT(h) in a task typically used for training of recurrent neural networks.

We trained the recurrent neural network on the next symbol prediction. The predicted sequence[1] is based on real data obtained by quantization of activity changes of laser in chaotic regime. The bounds for quantization were chosen for positive small and big activity change and for negative small and big activity change. One symbol is assigned for each of these categories. The sequence therefore consists of four distinct symbols. This sequence contains relatively predictable subsequences followed by much less predictable events. The sequence length is 10000 symbols, we can therefore predict 9999 symbols.

The next symbol prediction procedure in general is the following: we present in every time step the first symbol in order, and the desired network's output is the next symbol in sequence order. The predictive performance was evaluated by means of a normalized negative log-likelihood (NNL), calculated over symbolic sequence from time step $t = 1$ to T [1]:

$$\mathrm{NNL} = -\frac{1}{\mathrm{T}} \sum_{\mathrm{t}=1}^{\mathrm{T}} \log_{|\mathrm{A}|} \mathrm{p}^{(\mathrm{t})}(\mathrm{s}^{(\mathrm{t})}) \tag{10}$$

where the base of the logarithm $|A|$ is the number of symbols in the alphabet A and $p^{(t)}(s^{(t)})$ is the probability of predicting symbol $s^{(t)}$ in the time step t. If NNL $= 0$, then the network predicts next symbol with 100% accuracy, while NNL ≥ 1 corresponds to a very inaccurate prediction (random guessing).

We chose the following initial values of parameters for the EKF: covariance matrix $\mathbf{P}_0 = 1000\ \mathbf{I}$, measurement noise covariance matrix $\mathbf{R}_0 = 100\ \mathbf{I}$ and process noise covariance matrix $\mathbf{Q}_0 = 0.0001\ \mathbf{I}$. We have not altered the covariance matrices $\mathbf{R}$ and $\mathbf{Q}$ during the training process. These values were inspired by [1] and were chosen also because they correspond with recommendations for EKF.

For the BPTT(h) method, we chose the learning parameter $\alpha = 0.2$ and parameter $h = 10$. The same value of parameter h was used also for the EKF. This choice is in accordance with the recommendations in [7].

We have used the training and testing procedure for the next symbol from this sequence prediction from [1]. We do not update the weights for the first 50 steps, in order to lessen the impact of initial recurrent neurons output values. The training then takes place during next 7949 symbols. The remaining 2000 symbols form the test data set, through which we compute the NNL. That terminates one cycle. A few cycles are usually sufficient for the EKF, much more for the BPTT(h) to converge to its best result (for details see [9,10]). However, we chose 20 training cycles in order to compare the precision of various EKF implementations in longer run.

[1] This sequence is available at `http://www2.fiit.stuba.sk/~cernans/main/download.html`

We have used one-hot encoding for inputs as well as for outputs, so we have 4 inputs as well as outputs – one input/output for each of 4 symbols. With this setup the prediction probability of the desired symbol is determined as the value of that element of the output vector which is reserved for given symbol, after normalizing the output vector (i.e. the sum of its elements equals 1). We chose the Elman's network architecture – i.e. the network with one hidden layer, which is recurrent.

7 Results

We conducted the described experiment with different implementations of EKF and with BPTT(h). We will use following abbreviations of corresponding implementations in this section:

EKF GPU EKF implemented using CUBLAS library
EKF ATLASf EKF implemented using ATLAS library with single precision functions and utilizing single thread
EKF ATLASft EKF implemented using ATLAS library with single precision functions and utilizing four threads
EKF ATLASd EKF implemented using ATLAS library with double precision functions and utilizing single thread
EKF ATLASdt EKF implemented using ATLAS library with double precision functions and utilizing four threads
BPTT(h) Truncated BPTT with double precision – h is truncation depth

Table 1. Elapsed time in seconds for 10 cycles of training and testing of recurrent neural network with various numbers of hidden neurons and thus weights

hidden neurons	4	8	12	16	30	60
number of weights	56	140	256	404	1174	4144
EKF GPU	67s	108s	179s	279s	951s	4752s
EKF ATLASf	35s	96s	187s	325s	2218s	19858s
EKF ATLASft	35s	103s	199s	332s	1483s	13614s
EKF ATLASd	35s	96s	192s	340s	2005s	28677s
EKF ATLASdt	40s	107s	197s	354s	2026s	23315s
BPTT(h)	31s	76s	136s	214s	621s	2184s

Firstly, we present the results for the elapsed time during training of neural networks with various numbers of hidden neurons by each method in absolute numbers in Table 1, as well as relative to EKF GPU in Fig. 1. From these results we can see that the implementation of EKF on GPU provides significant speedup for larger networks, but is not beneficial for smaller networks. This stems from

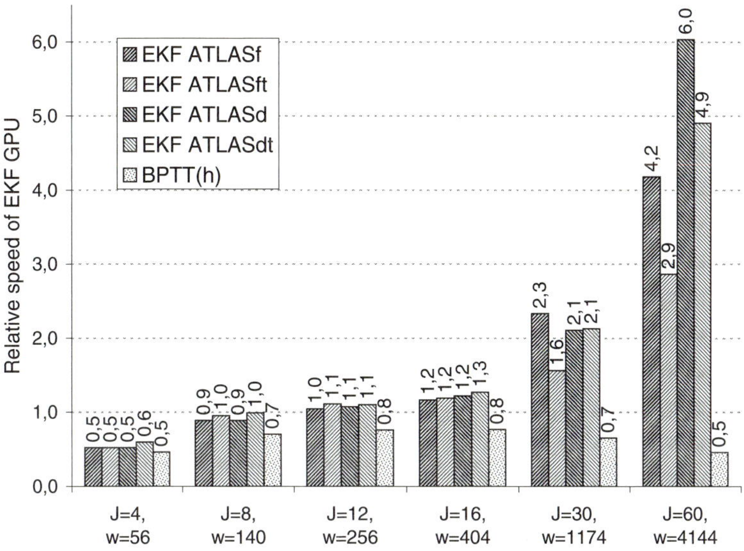

Fig. 1. Graphical representation of Tab. 1 – the speed comparison of EKF GPU relative to other implementations. J is number of hidden neurons and w the number of weights in recurrent neural network. The speedup is significant for networks with many weights even when compared to the threaded CPU version with single floating point precision. On the other hand it is not beneficial for small networks. Gradient descent method BPTT(h) is still faster but converges slowly and achieves worse results [9,10].

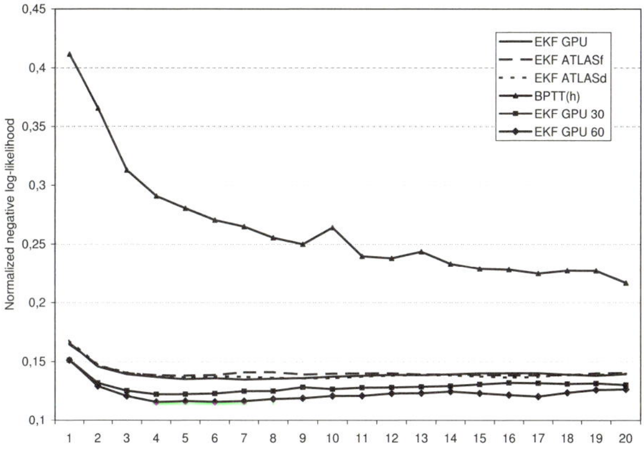

Fig. 2. The NNL dependence on the training cycles for various implementations and methods used for recurrent network with 16 hidden neurons. The exception is *EKF GPU 30* and *EKF GPU 60* which is a result for 30 and 60 hidden neurons respectively. The results show that (a) EKF is superior to gradient descent method BPTT(h) (for details see [9,10]); (b) the various EKF implementations achieve comparable results (see Fig. 3); (c) the increasing of hidden neurons is beneficial for this problem and was made more feasible by achieved speedup.

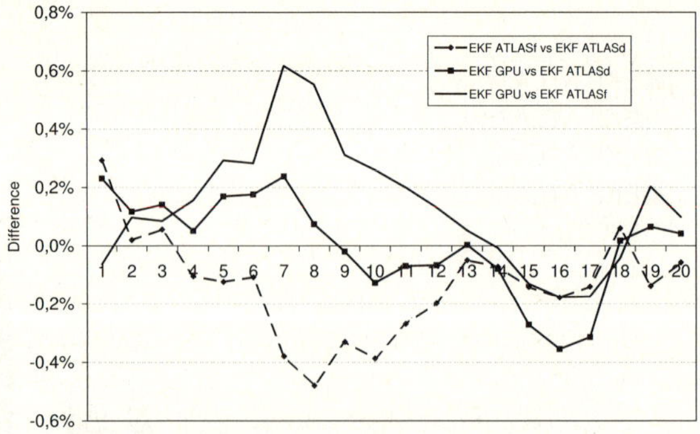

Fig. 3. Detailed view on results depicted in Fig. 2. Difference between achieved results of used implementations of EKF in every training cycle as a percentage of valid NNL interval length (i.e. 1). Positive values mean better achievement of first implementation than second implementation. The differences are negligible, all within 0.6%.

the fact, that the overhead of copying data from system memory to GPU and of parallelization alone is not worth when there is not much to compute.

The most "fair" comparison of EKF GPU is with EKF ATLASft, because both utilize parallelization and work with single precision floating point arithmetic. EKF GPU achieves nearly 3 times speedup for largest network when compared to EKF ATLASft. The most significant acceleration by using computation with EKF GPU – 6 times on largest network – is achieved when compared to EKF ATLASd, which is probably most similar to existing implementations of EKF.

In Fig. 1 we can further see that method BPTT(h) is consistently the fastest, whereas EKF ATLASd is generally the slowest one. The obvious question is if the achieved results are on the one hand worth the speed degradation when compared with BPTT(h) and on the other hand worth the significant speedup when compared with potentially more precise EKF ATLASd. The answer is in Fig. 2 which for each method shows the average results when used for training 10 randomly initialized networks. In this graph we can see the superior convergence and achieved result of EKF when compared with BPTT(h) (see also [9,10]). We can also see the comparable results of various implementations of EKF, which is more obvious from Fig. 3. It means the used floating point arithmetic does not play a significant role in EKF performance in this experiment.

The achieved speedup of EKF GPU made it also more feasible to conduct thorough experiments with larger networks, as seen in Fig. 2. This also justifies the increasing of number of hidden neurons for this problem, as the best achieved result is by networks with 60 hidden neurons (NNL=0.1156 in training cycle 4).

8 Conclusion

In this paper we have shown that the implementation of the most computational intensive tasks in the Extended Kalman filter using the CUBLAS library for nVidia graphics processing units can significantly speed up the recurrent neural network training process. What is important, the speedup was achieved only by means of available library functions, one does not have to overcome steep learning curve of GPU architecture and its efficient parallel programming.

The drawback in current GPUs is the lack of double precision floating point arithmetic. This affects the numerical stability as well as overall achieved result. In our experiments we experienced the numerical stability problem when computing Cholesky factorization of matrix which became non positive definite. This was remedied by restarting the training process with different initial weights. Since the problem usually arose in the early training cycles, and the training is fast, the restarting did not cause significant delays.

The reduced precision was not significant problem in our experiments. However, the recommended practice for the time being would be to tune the training parameters using EKF on GPU and to use CPU only version with double precision for final experiments.

The presented method can be further enhanced by implementing more parts of algorithm on the GPU and reduce thus the need to copy data between CPU and GPU which is a costly operation. This will be the topic of our further research, namely implementing Cholesky factorization [4], neural network propagation [6] and BPTT(h) on graphic processing units.

References

1. Čerňanský, M., Makula, M., Beňušková, L.: Processing Symbolic Sequences by Recurrent Neural Networks Trained by Kalman Filter-Based Algorithms. In: SOFSEM 2004 (2004) ISBN 80-86732-19-3 58–65
2. Dongarra, J.: Basic Linear Algebra Subprograms Technical Forum Standard. Int. J. of High Performance Applications and Supercomputing 16(1), 1–111 (2002)
3. Haykin, S.: Kalman Filtering and Neural Networks. John Wiley & Sons, Inc., New York (2002) ISBN: 0-471-36998-5
4. Jung, J.: Cholesky Decomposition and Linear Programming on a GPU. Sholarly Paper. University of Maryland
5. Kalman, R.E.: A New Approach to Linear Filtering and Prediction Problems. Trans. of the ASME, Series D, Journal of Basic Engineering 82, 35–45 (1960)
6. Kyoung-Su, O., Keechul, J.: GPU Implementation of Neural Networks. Pattern Recognition 37, 1311–1314 (2004)
7. Patel, G.S.: Modeling Nonlinear Dynamics with Extended Kalman Filter Trained Recurrent Multilayer Perceptrons. McMaster University (2000)
8. Prokhorov, D.V.: Kalman Filter Training of Neural Networks: Methodology and Applications. Ford Research Laboratory (2002)
9. Trebatický, P.: Recurrent Neural Network Training with the Kalman Filter-based Techniques. Neural network world 15(5), 471–488 (2005)
10. Trebatický, P.: Recurrent Neural Network Training with the Extended Kalman Filter. IIT SRC: Proc. In: Informatics and Info. Technologies, 57–67 (2005)
11. Automatically Tuned Linear Algebra Software (ATLAS), `http://math-atlas.sourceforge.net`
12. Compute Unified Device Architecture (CUDA), `http://www.nvidia.com/object/cuda_home.html`

Blind Source-Separation in Mixed-Signal VLSI Using the InfoMax Algorithm

Waldo Valenzuela, Gonzalo Carvajal, and Miguel Figueroa

Department of Electrical Engineering
Universidad de Concepción
Concepción, Chile
{wvalenzu,gcarvaja,miguel.figueroa}@udec.cl

Abstract. This paper describes a VLSI implementation of the InfoMax algorithm for Independent Component Analysis in mixed-signal CMOS. Our design uses on-chip calibration techniques and local adaptation to compensate for the effect of device mismatch in the arithmetic modules and analog memory cells. We use our design to perform two-input blind source-separation on mixtures of audio signals, and on mixtures of EEG signals. Our hardware version of the algorithm successfully separates the signals with a resolution within less than 10% of a software implementation of the algorithm. The die area of the circuit is 0.016mm^2 and its power consumption is 15μW in a 0.35μm CMOS process.

1 Introduction

Independent Component Analysis (ICA) [1] is a signal processing technique used to recover the original sources from unknown linear combinations of independent signals captured by spatially-distributed sensors. ICA makes only weak assumptions about the nature of the sources, namely that the signals are statistically independent and that at most one of them exhibits a Gaussian distribution. This allows ICA algorithms to adapt without an external reference, performing blind source-separation in applications such as speech recognition, face classification, and data communications [2, 3].

Despite the advantages described above, most algorithms for ICA pose computational requirements that make them unsuitable for portable, low-power applications. Low-power embedded processors are unable to provide the throughput required for computation and adaptation, while high-performance digital signal-processors (DSPs) are too large and power-hungry. Even custom-VLSI digital implementations can be unsuited to ultra low-power applications, mainly due to the size and power requirements of digital multipliers. For problems that require only moderate arithmetic resolution, analog and mixed-signal circuits offer an attractive tradeoff between performance, die area and power dissipation. Unfortunately, large-scale analog implementations of signal processing algorithms are extremely difficult to design due to limitations such as device mismatch, charge leakage, charge injection, signal offsets, and noise. Often the design techniques used to overcome these problems significantly increase circuit size and power consumption, reducing the advantages that make analog circuits attractive in the first place.

V. Kůrková et al. (Eds.): ICANN 2008, Part II, LNCS 5164, pp. 208–217, 2008.

This paper describes a novel approach to the implementation of adaptive signal processing algorithms in analog and mixed-signal VLSI. We take advantage of the adaptive nature of the algorithm to intrinsically compensate for signal offsets and gain mismatch with negligible impact on die area and power consumption. We also selectively use on-chip calibration techniques to compensate for learning-rate mismatch in the analog memory cells used to store the adaptive coefficients. We illustrate our techniques on the implementation of the InfoMax [4, 5] algorithm for ICA. We describe the computational primitives used by the algorithm and their hardware implementation and test them using a circuit emulator that uses experimental circuit data to evaluate the performance of the implementation on audio and biomedical Electroencephalography (EEG) signals. Our results show that our design is able to blindly retrieve the original sources from the linear mixtures with a resolution within less than 10% of a software implementation of InfoMax. A 2-input circuit dissipates less than 15μW with a settling time of 8μs, a die area of 0.016mm^2 in a 0.35μm CMOS process. We obtain similar performance from 4- and 6-input versions of the network.

2 Independent Component Analysis and InfoMax

Independent Component Analysis (ICA) is a signal processing technique used to separate linear mixtures of signals received from spatially-distributed sensors in statistically-independent components, without using an external reference. ICA makes only weak assumptions about the statistical properties of the original components and applies a linear transformation on the available mixtures, adaptively updating its coefficients to minimize the statistical dependence between the outputs.

The convenient properties of ICA make this technique suitable for performing blind-source separation in a relatively simple neural network. There are several algorithms for ICA, but one of the most widely used is the InfoMax [4, 5], which separates the sources by minimizing the mutual information between the network outputs. Fig. 1 shows the typical structure of a neural network used to implement InfoMax. Like all

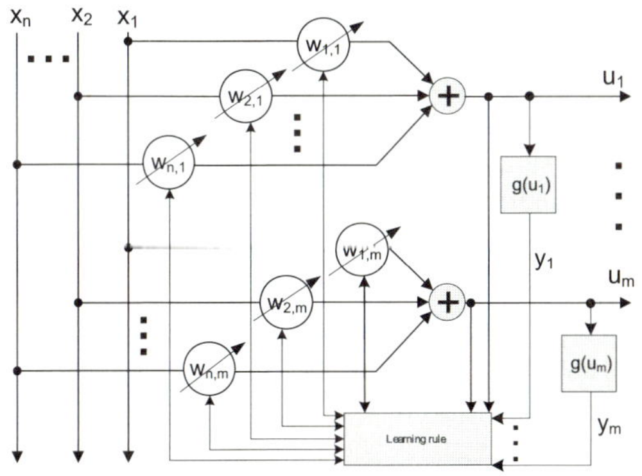

Fig. 1. ICA network with Infomax

ICA algorithms, InfoMax computes the output vector $\mathbf{u}(k)$ as the product of the n-input vector $\mathbf{x}(k) = [x_1(k) \dots x_n(k)]^{\mathrm{T}}$ and a weight matrix $\mathbf{W}(k)$:

$$\mathbf{u}(k) = \mathbf{x}(k)^{\mathrm{T}}\mathbf{W}(k) \tag{1}$$

where k is the time step. The algorithm then applies a nonlinear learning rule to update the coefficients of $\mathbf{W}$. InfoMax maximizes the entropy of the output [4], using the learning rule $\mathbf{W}(k+1) = \mathbf{W}(k) + \mu\mathbf{\Delta W}(k)$, where μ is the learning rate and

$$\mathbf{\Delta W}(k) = (\mathbf{I} - \varphi(\mathbf{u}(k))\mathbf{u}^{\mathrm{T}})\mathbf{W}(k) \tag{2}$$

is the weight increment at time k, where $\mathbf{I}$ is the identity matrix, $\varphi(u) = \frac{\partial g(u)}{\partial u}$ and $\mathrm{g}(\cdot)$ is an invertible nonlinear function [6].

Both the learning rate and the function $\mathrm{g}(\cdot)$ are parameters chosen by the designer, and affect the dynamic and stationary behavior of the algorithm. Some typical nonlinear functions are shown in [6], and one of the most widely used is $\mathrm{g}(u) = \tanh(u) \Rightarrow \varphi(u) = 2\tanh(u)$, which has the advantage of being easily implemented in analog VLSI.

The original version of InfoMax published by Bell and Sejnowski [6] operates on super-Gaussian sources. Later, Lee, Girolami, and Sejnowski [5] generalized the algorithm to also work on sub-Gaussian signals:

$$\mathbf{\Delta W}(k) = (\mathbf{I} - K\tanh(\mathbf{u})\mathbf{u}^{\mathrm{T}} - \mathbf{u}\mathbf{u}^{T})\mathbf{W}(k) \tag{3}$$

where $K = 1$ and $K = -1$ for super-Gaussian and sub-Gaussian sources, respectively. In this paper, we implement this extended version of the algorithm, because of its wider applicability.

3 Mixed-Signal VLSI Implementation of InfoMax

In this section, we describe the circuits used for the forward computation and adaptation phases, as required by the algorithm.

3.1 Multipliers

The main limitations of digital hardware implementations of signal processing algorithms are the large die area and power dissipation required by multipliers. In order to achieve low power and area, we use analog multipliers to perform the linear transformation on the inputs shown in Eqn. 1. The current outputs of the multipliers are simply summed across common wires to form each neuron's output.

A systems analysis of similar adaptive filters in analog VLSI [7] shows that the reconstruction error is sensitive to the multiplier linearity with respect to the synaptic input x, but relatively robust to their linearity of the weight w. Consequently, we use Gilbert-style multipliers with differential current inputs for x and a differential voltage representation for w. We use long transistors and above-threshold operation (1μA-input range for each multiplier) to maximize the linearity of x, but favor a larger range in w to allow space for intrinsic compensation of weight linearity and offsets. Fig. 2 shows the transfer function of eight different multipliers on a single chip.

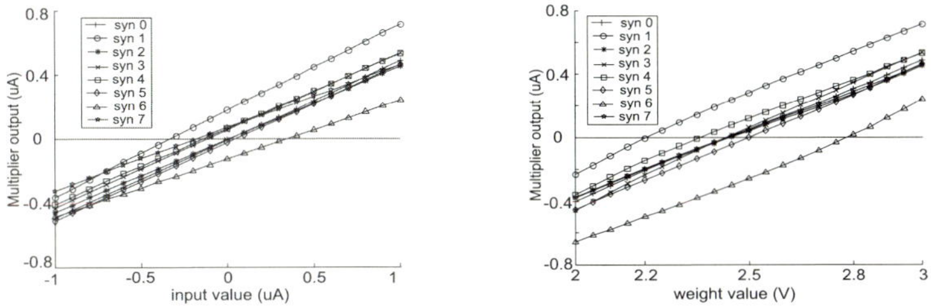

(a) Output versus input value.

(b) Output versus weight value.

Fig. 2. Multiplier transfer functions

3.2 Weight Storage and Updates

Using analog multipliers in the forward path requires that we provide on-chip analog weight storage. Because the performance of the learning algorithm depends directly on the accuracy of the stored weights and update rules, conventional VLSI capacitors are inadequate: Charge leakage and charge injection limit weight resolution, particularly when used with digital pulse-based updates which provide otherwise accurate and compact learning rules [8].

We use *synapse transistors* [9] to store and update our analog weights. These devices use charge on a floating gate to provide compact and accurate nonvolatile analog storage. Fowler-Nordheim tunneling adds charge to the floating gate and hot-electron injection removes charge. The dynamics of tunneling and injection are exponential on the control voltages. Therefore, we use a pulse-based approach [9] to provide linear weight updates as the basis to implement the learning rule depicted in Eqn. 3. Fig. 3(a) shows our memory cell. Negative feedback around a small operational amplifier pins the floating-gate FG to a fixed voltage. We apply digital pulses of fixed width to the control terminals Pinc and Pdec, which add or remove fixed amounts of charge to the floating gate, updating the output voltage Vout in constant increments. As a result, the weight updates are linearly dependent on the density of the update pulses. Previous publications [9] describe the operation of the memory cell in detail.

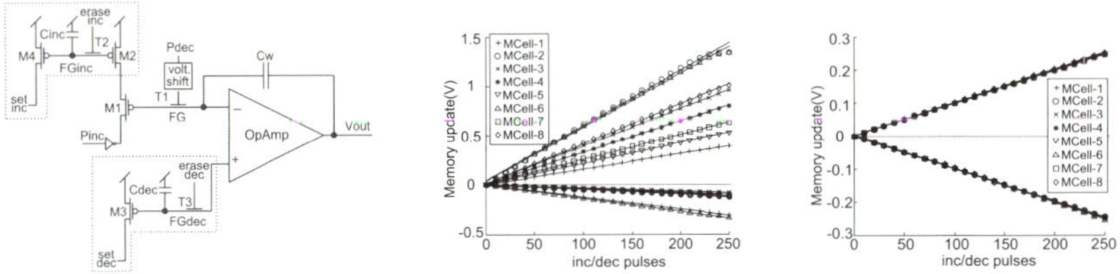

(a) Memory cell with on-chip calibration.

(b) Memory updates before calibration.

(c) Memory updates after calibration.

Fig. 3. Pulse-density modulated memory cell with on-chip calibration

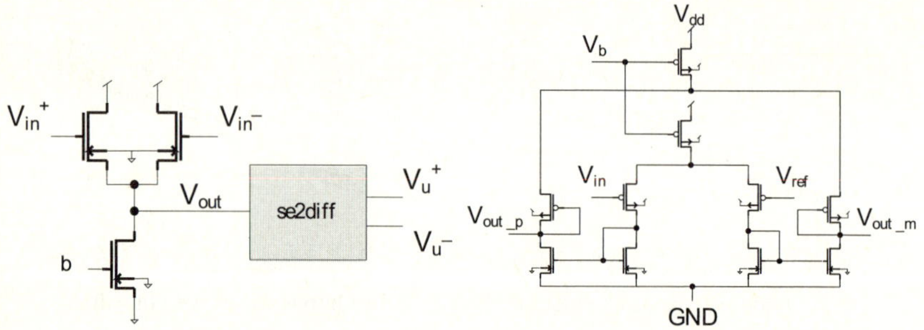

(a) Tanh circuit.

(b) Conversion to differential current.

Fig. 4. Hyperbolic tangent implementation

Fig. 3(b) shows experimental data depicting the magnitude of the voltage updates for 8 different cells on a single chip (in a 0.35μm process) as a function of the density of positive and negative update pulses. The cell achieves linear updates, but device mismatch results in asymmetrical and nonuniform updates across multiple cell. This translates into variations in the learning rates of different synapses, which affect the convergence of the InfoMax algorithm. We solve this problem by electronically calibrating each memory cell, using the boxed circuits in Fig. 3(a). Fig. 3(c) shows the transfer functions of the same 8 memory cells after calibration. The learning rates are now linear, symmetric within a single cell and uniform across cells to within a resolution of 12 bits.

3.3 Learning Rule

The learning rule in Eqn. 3 requires the implementation of a hyperbolic tangent function at the output of each neuron. Fig. 4(a) shows a circuit that computes the $\tanh$ of a differential current. The circuit outputs a single-ended voltage, which is converted into differential current using the operational transconductance amplifier shown in Fig. 4(b).

To implement the InfoMax learning rule, we convert the differential analog-current signals into asynchronous digital pulse signals, using pulse-density modulators (off-chip in our design) [9]. We compute the products and sums between pulse trains using simple logical operations (AND and OR gates), effectively reducing the weight update to a correlational learning rule. As long as the pulses are asynchronous and sparse, this technique can efficiently trade off time for die area, and implement the learning rule with an equivalent resolution up to 8 bits [9] in each synapse.

3.4 Effects of Device Mismatch

The local calibration technique described in Section 3.3 successfully compensates for the effects of device mismatch on the learning rules and weight resolution. However, the analog multipliers used in the forward path are also sensitive to mismatch, resulting in gain variations, offsets, and nonlinear behavior.

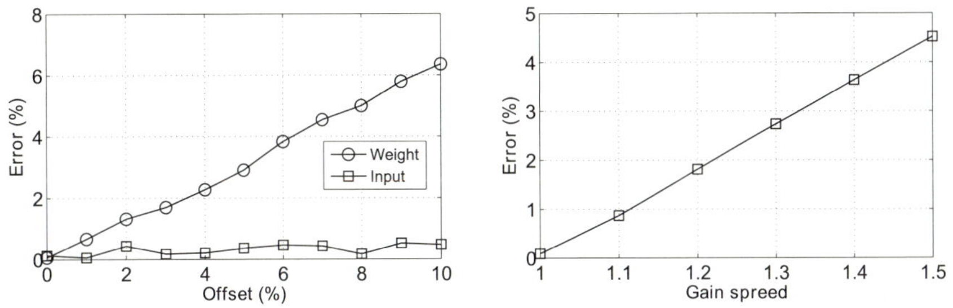

(a) Reconstruction error versus multiplier offset.

(b) Reconstruction error versus multiplier gain spread.

Fig. 5. Impact of device mismatch on the performance of InfoMax

In previous work [7] we have shown a detailed analysis of the effects of mismatch on the performance of linear filters with the LMS learning rule. We have demonstrated that combining local adaptation with simple design techniques and on-chip calibration, it is possible to implement signal-processing systems that adapt with resolutions of up to 10 bits in the presence of device mismatch.

Unfortunately, the more complicated learning rules of Infomax and its intrinsic nonlinear dynamics make it very difficult to formally analyze the effects of device mismatch. Instead, we performed a set of simulations to assess the expected impact of mismatch on the performance of the algorithm.

First, we analyzed the sensitivity of the algorithm to multiplier offsets in the external inputs and weights. We simulated the network with offsets up to 10% of the dynamic range of each multiplier. After each weight update, we stopped adaptation, run the entire signal through the network, and computed the *reconstruction error* as ratio between the Root-Mean Square (RMS) value of the difference between the original source and the network output, and the RMS value of the output. Fig. 5(a) shows that the performance of InfoMax is robust against input offsets. Indeed, fixed input offsets introduced by device mismatch add a nonzero mean to the input, which InfoMax can compensate through adaptation. Fig. 5(a) also shows that the reconstruction error increases linearly with weight offsets. This is mainly because the limited linearity of the multipliers with respect to the weights (Fig. 2(b)) means that small weight offsets can push the target values out of the linear range, thus degrading the quality of the reconstructed signals.

Next, we randomly varied between 1.0 and 1.5 the gains of the multipliers within the network. We define the *gain spread* as the ratio between the maximum and minimum multiplier gain. Fig. 5 shows that the reconstruction error increases linearly with the multiplier gain. It is possible to reduce the error by reducing the learning rate as the gain spread increases, at the cost of increased convergence time. This solution has a weak impact in the overall performance of the algorithm, because the multipliers do not exhibit a large gain spread (Fig. 2), so the reconstruction error is dominated by weight offsets. Our experiments also show that the algorithm is sensitive to the linearity of the multipliers, therefore their dynamic range must be carefully chosen to operate in the linear range.

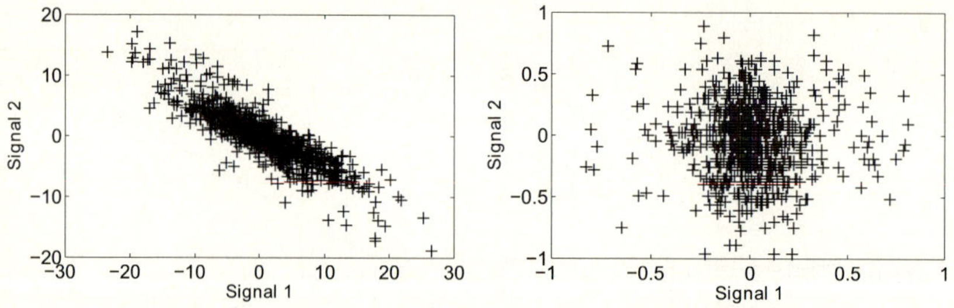

(a) Mixed signals (b) Separated signals.

Fig. 6. Joint PDF of mixed audio signals before and after separation

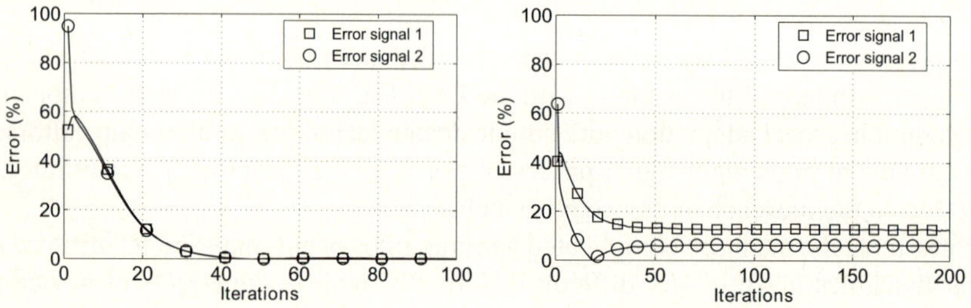

(a) Software implementation. (b) Hardware implementation.

Fig. 7. Evolution of reconstruction error for audio signals

4 Experimental Results

We designed and fabricated arithmetic circuits for the building blocks described in the previous section using a 0.35μm CMOS process, including analog memory cells, multipliers, and weight-update rules for InfoMax. Spice simulations show that a 2-input circuit dissipates less than 15μW with a settling time of 8μs. The die area of the circuit is 0.016mm^2 in a 0.35μm CMOS process.

We characterized these circuits in the lab and built a software emulator that allows us to test the static performance of the network with less than 0.5% error [10], simulating the pulse-density modulators in software. Using the emulator, we performed blind-source separation with the network for up to 6 inputs, both for real audio and electroencephalogram (EEG) signals. In this section, we present the results obtained for a 2-input version of the network. We obtain similar performance results for more inputs, though it becomes necessary to reduce the learning rate to compensate for the larger spread in multiplier gains.

We mixed two independent audio signals and fed them to the 2-input InfoMax network. We used the techniques applied in the EEGLAB toolbox [11] to determine the sub- or super-gaussianity of the signals and choose the appropriate value of K in Eqn . 3. Fig. 6 depicts the joint probability density function (PDF) of the signals before and

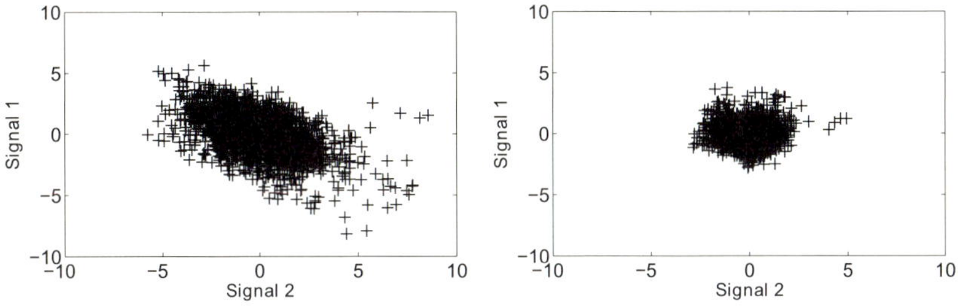

(a) Mixed signals (b) Separated signals.

Fig. 8. Joint PDF of measured EEG signals before and after separation

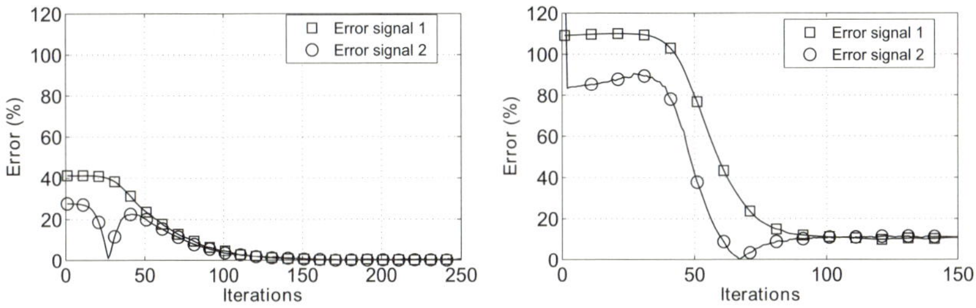

(a) Software implementation. (b) Hardware implementation.

Fig. 9. Evolution of reconstruction error for EEG signals

after separation, showing that the hardware network successfully removes the statistical dependencies found in the mixtures.

Fig. 7 compares the reconstruction performance of our hardware network to that of a software implementation of InfoMax. Fig. 7(a) shows the reconstruction error of the software implementation of the algorithm, and Fig. 7(b) shows the performance of our hardware implementation, as a function of the original signal range. The hardware error settles at 9%–12% above the residual error of the software version. This difference is due mainly to weight offsets and the nonlinearity of the multipliers in the forward path.

We repeated the above experiment, now using experimental EEG data. Because we only had access to the measured mixtures and not the original sources, we used the results of the Matlab Toolbox EEGLAB as a reference in our experiments, limiting the mixtures to 2 EEG signals. Fig. 8 depicts the joint PDF of the signals before and after separation, demonstrating that the circuit outputs independent components. Using the EEGLAB software to evaluate the quality of the reconstruction of the original signals, Fig. 9 shows that the hardware network settles at a reconstruction error of 10% above the results of software separation, which is consistent with the results achieved with audio signals. Again, the reconstruction performance is limited by the linearity of the forward-path multipliers and the weight offsets. This situation imposes a tradeoff between the dynamic range of inputs and weights: If we restrict the input dynamic range of the

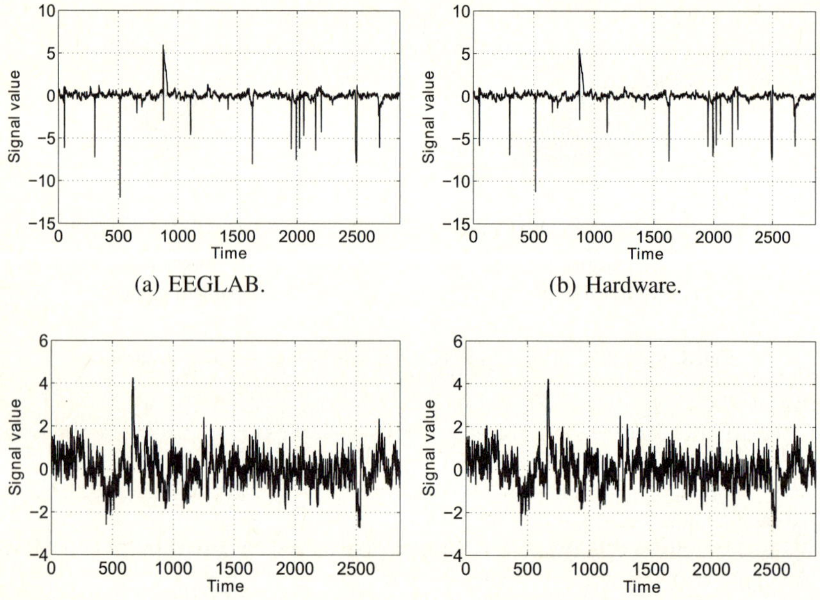

(a) EEGLAB. (b) Hardware.

(c) EEGLAB. (d) Hardware.

Fig. 10. EEG signal separated by EEGLAB and hardware InfoMax

multipliers to improve their linearity, the weights converge to larger values, degrading the results.

Fig. 10 compares the same EEG signal reconstructed by EEGLAB (10(a) and 10(c)) and our circuit (10(b) and 10(d)). Despite the 10% reconstruction difference, both signals are visually similar, suggesting that the hardware results could be suitable for medical analysis and diagnostics.

5 Conclusions

We presented a mixed-signal VLSI network for ICA using the InfoMax algorithm. We designed and tested hardware primitives to implement the arithmetic operations, weight storage and learning rules required by the algorithm. We analyzed the impact of device mismatch on the performance of the algorithm, and applied a combination of design criteria, closed-loop operation and on-chip calibration techniques to compensate for these effects. Using a software emulator based on experimental data with device mismatch, we tested the performance of our circuit performing blind source separation. On both audio and EEG signals, the circuit reconstructs the original sources with an RMS error 10% above that of a software implementation of the algorithm. After calibration and compensation, the main limitations are the weight offsets and the linearity of the multipliers used in the forward path, which restrict the dynamic range of the inputs and weights. We are currently working on a more robust implementation based on adjustable-gain multipliers to extend their linear range, as well as to test other learning rules used to implement ICA. We are also developing an FPGA-based hardware accelerator of the circuit emulator, which will allow us to perform more extensive tests before circuit fabrication.

Acknowledgments

This work was partially funded by grants Fondecyt 1070485 and Milenio ICM P06-67F.

References

1. Hyvärinen, A., Karhunen, J., Oja, E.: Independent Components Analysis, vol. 1. John Wiley and Sons, INC., New York (2001)
2. Comon, P.: ndependent Component Analysis - A New Concept? Signal Processing 3(36), 214–287 (1994)
3. Bell, A.J., Sejnowski, T.J.: The Independent Components of Natural Scenes are Edge Filters. Vision Research 37(23), 3327–3338 (1997)
4. Cardoso, J.: Infomax and Maximum Likelihood for Blind Source Separation. IEEE Signal Processing Letters 4, 112–114 (1997)
5. Lee, T.W., Girolami, M., Sejnowski, T.J.: Independent Component Analysis Using an Extended Infomax Algorithm for Mixed Subgaussian and Supergaussian Sources. Neural Computation 11(2), 417–441 (1999)
6. Bell, A.J., Sejnowski, T.J.: An Information-Maximization Approach to Blind Separation and Blind Deconvolution. Neural Computation 7(6), 1129–1159 (1995)
7. Carvajal, G., Figueroa, M., Bridges, S.: Effects of Analog-VLSI Hardware on the Performance of the LMS Algorithm. In: International Conference on Artificial Neural Networks (ICANN), Athens, Greece, 10-14 September 2006, pp. 963–973 (2006)
8. Hairai, Y., Nishizawa, K.: Hardware Implementation of a PCA Learning Network by an Asynchronous PDM Digital Circuit. In: Proceedings of the IEEE-INNS-ENNS International Joint Conference on Neural Networks (IJCNN), vol. 2(2), pp. 65–70 (2000)
9. Figueroa, M., Bridges, S., Diorio, C.: On-Chip Compensation of Device-Mismatch Effects in Analog VLSI Neural Networks. In: Advances in Neural Information Processing Systems 17. MIT Press, Cambridge (2005)
10. Carvajal, G., Valenzuela, W., Figueroa, M.: Subspace-Based Face Recognition in Analog VLSI. In: Platt, J., Koller, D., Singer, Y., Roweis, S. (eds.) Advances in Neural Information Processing Systems 20. MIT Press, Cambridge (2008)
11. Delorme, A., Makeig, S.: EEGLAB: An Open Source Toolbox for Analysis of Single-Trial EEG Dynamics Including Independent Component Analysis. Journal of Neuroscience Methods 134, 9–21 (2004)

Synaptic Rewiring for Topographic Map Formation

Simeon A. Bamford[1], Alan F. Murray[2], and David J. Willshaw[3]

[1] Doctoral Training Centre in Neuroinformatics
sim.bamford@ed.ac.uk
[2] Institute of Integrated Micro and Nano Systems
[3] Institute of Adaptive and Neural Computation,
University of Edinburgh

Abstract. A model of topographic map development is presented which combines both weight plasticity and the formation and elimination of synapses as well as both activity-dependent and -independent processes. We statistically address the question of whether an activity-dependent process can refine a mapping created by an activity-independent process. A new method of evaluating the quality of topographic projections is presented which allows independent consideration of the development of a projection's preferred locations and variance. Synapse formation and elimination embed in the network topology changes in the weight distributions of synapses due to the activity-dependent learning rule used (spike-timing-dependent plasticity). In this model, variance of a projection can be reduced by an activity dependent mechanism with or without spatially correlated inputs, but the accuracy of preferred locations will not necessarily improve when synapses are formed based on distributions with on-average perfect topography.

1 Introduction

The development of topographic mappings in the connections between brain areas is a subject that continues to occupy neuroscientists. There have been a number of investigations of the development of maps through networks with fixed connectivity and changes to synaptic weights [1,2,3,4,5]. Other models have considered the formation and elimination of synapses with fixed weight [6]. Indeed a mathematical equivalence between such models has been demonstrated for certain conditions [7]. There have been few attempts to include both forms of plasticity in a model (though see [8,9]) however since both forms of plasticity are known to exist, we have created a model of topographic map development which combines both forms of plasticity and we explore some of the consequences of this model. This work is part of a project to implement synaptic rewiring in neuromorphic VLSI [10], however the results presented here are purely computational.

Theories of topographic map formation can be divided by the extent to which activity-dependent processes, based on Hebbian reinforcement of the correlated activity of neighbouring cells, are deemed responsible for the formation of topography. Some assume that activity-independent processes, based on chemoaffinity

V. Kůrková et al. (Eds.): ICANN 2008, Part II, LNCS 5164, pp. 218–227, 2008.

[11] provide an approximate mapping, which is then refined [12]. Others [5] show how activity-independent processes may fully determine the basic topography, thus relegating the role of activity-dependent processes to the formation of "functional architecture" e.g. oculardominance stripes etc. [13]. Our model is in the latter of these categories, assuming that synapses are placed with on-average perfect topography by an activity-independent process. Miller [7] gives evidence that the decision whether newly sprouted synapses are stabilised or retracted may be guided by changes in their strengths; this is a basis for our model.

2 Model

This generalised model of map formation could equally apply to retino-tectal, retino-geniculate or geniculo-cortical projections. There are 2 layers (i.e. 2D spaces on which neurons are located), the input layer and the network layer. Each location in one layer has a corresponding "ideal" location in the other, such that one layer maps smoothly and completely to the other. For simplicity neural areas are square grids of neurons and the 2 layers are the same size (16 x 16 in the simulations presented here). We have worked with small maps due to computational constraints; this has necessitated a rigorous statistical approach. Periodic boundaries are imposed to avoid edge artefacts.

Each cell in the network layer can receive a maximum number of afferent synapses (32 in our simulations). Whilst we acknowledge arguments for the utility of inhibitory lateral connections in building functional architecture [8] we simplified our model using the finding [4] that a topographic projection could form in the absence of long-range lateral inhibition. Thus, two excitatory projections are used, a feed-forward and a lateral projection; these projections compete for the synaptic capacity of the network neurons. We assume that an unspecified activity-independent process is capable of guiding the formation of new synapses so that they are distributed around their ideal locations. We assume a Gaussian distribution, since a process which is initially directed towards a target site and then randomly branches on its way would yield a Gaussian distribution of terminations around the target site. Our model does not specify the underlying mechanisms that cause an axon to be guided towards an ideal location. Thus it is not fundamentally incompatible with lesion studies which show shifts or compression of maps, but rather, to achieve such reorganisation some mechanism for specifying and changing ideal locations would need to be added.

To implement the Gaussian distributions, where a network cell has less than its maximum number of synapses, the remaining slots are considered "potential synapses". At a fixed "rewiring" rate a synapse from the neurons of the network layer is randomly chosen. If it is a potential synapse a possible pre-synaptic cell is randomly selected and synapse formation occurs when:

$$r < p_{form}.e^{-\frac{\delta^2}{2\sigma_{form}{}^2}} \tag{1}$$

where r is a random number uniformly distributed in the range $(0, 1)$, p_{form} is the peak formation probability, δ is the distance of the possible pre-synaptic cell from the ideal location of the post-synaptic cell and ${\sigma_{form}}^2$ is the variance of the connection field. In other words, a synapse is formed when a uniform random number falls within the area defined by a Gaussian function of distance, scaled according to the peak probability of synapse formation, (which occurs at $\delta = 0$). This is essentially a rejection sampling process.

Lateral connections are formed by the same means as feed-forward connections though σ_{form} is different for each projection and p_{form} is set correspondingly to allow the same overall probability of formation for each projection. In the absence of a general rule for the relative numbers of feed-forward vs lateral connections formed, starting with equal numbers of each is a good basis for observing the relative development of these projections; $\sigma_{form-feedforward}$ is given a larger value than $\sigma_{form-lateral}$, in line with generic parameters given in [8].

If the selected synapse already exists it is considered for elimination. In general we propose that the probability of elimination should be some monotonically decreasing function of weight. Due to the nature of the learning rule we have chosen (STDP; see below in this section), which tends to deliver a bimodal weight distribution, we have simplified probability of elimination to one of 2 values with a higher value for synapses with weights below a certain threshold ($p_{elim-dep}$) and vice versa ($p_{elim-pot}$). Data is scarce on appropriate values for these probabilities, however dendritic spines have been imaged extending and retracting over periods of hours compared with others stable over a month or more [14]. We have used much higher rates so that synapses have several chances to rewire during the short periods for which it was tractable to run simulations, while maintaining a large difference between these probabilities (in fact we used a factor of 180 representing the difference between 4 hours and 1 month).

The rest of our model is strongly based on [4]. We use integrate and fire neurons, where the membrane potential V_{mem} is described by:

$$\tau_{mem}\frac{\delta V_{mem}}{\delta t} = V_{rest} - V_{mem} + g_{ex}(t)(E_{ex} - V_{mem}) \tag{2}$$

where E_{ex} is the excitatory reversal potential, V_{rest} is the resting potential and τ_{mem} is the membrane time constant. Upon reaching a threshold V_{thr}, a spike occurs and V_{mem} is reset to V_{rest}. A presynaptic spike at time 0 causes a synaptic conductance $g_{ex}(t) = ge^{\frac{-t}{\tau_{ex}}}$ (where τ_{ex} is the synaptic time constant); this is cumulative for all presynaptic spikes. Spike-Timing-Dependent Plasticity (STDP) is known to occur in biology at least in vitro, and has been used recently to explain map reorganisation in vivo [15]. STDP is implemented such that a presynaptic spike at time t_{pre} and a post-synaptic spike at time t_{post} modify the corresponding synaptic conductance by $g \rightarrow g + g_{max}F(\Delta t)$, where $\Delta t = t_{pre} - t_{post}$ and:

$$F(\Delta t) = \left\{ \begin{array}{ll} A_{+}.e^{(\frac{\Delta t}{\tau_{+}})}, & if \Delta t < 0 \\ -A_{-}.e^{(\frac{-\Delta t}{\tau_{-}})}, & if \Delta t \geq 0 \end{array} \right\} \tag{3}$$

where $A_{+/-}$ are magnitudes and $\tau_{+/-}$ are time constants for potentiation and depression respectively. This is cumulative for all pre- and post-synaptic spike pairs. g is bounded in the range $0 \leq g \leq g_{max}$.

Parameters were set starting from parameters given in [4]. A+ was increased 20-fold as a concession to limited computational resources for simulations (this should not qualitatively change the model since many plasticity events are still needed to potentiate a depressed synapse). Then key parameters were changed; namely g_{max} (the peak synaptic conductivity), $\tau - / \tau +$ (the ratio of time constants for depression and potentiation) and B (the ratio of potentiation to depression, i.e. A_+/A_-) were changed to maintain key conditions, being: the total weight should be approximately 50% of the maximum possible; the average network neuron firing rate should approximately match the average input firing rate; and the total weight of lateral synapses should roughly match the weight of feed-forward ones. In the interests of simplicity we did not allow for different values of B for different projections feedforward vs recurrent). An unjustified simplification is that new synapses start strong and then get weakened; the opposite case seems more likely. We have used this for simplicity because it avoids the need for any homeostatic mechanisms to kick-start the network.

Each input cell was an independent Poisson process. A stimulus location was chosen and mean firing rates were given a Gaussian distribution around that location based on a peak rate f_{peak} and variance ${\sigma_{stim}}^2$ which was added to a base rate f_{base}. The stimulus location changed regularly every 0.02s. This regularity is a move away from biologically realistic inputs (c.f. [4]); this was a necessary concession to provide stronger correlation cues given the smaller number of synapses per neuron. A further concession was the more extreme values of f_{base} and f_{peak}. σ_{stim} was chosen to be between the values of $\sigma_{form-feedforward}$ and $\sigma_{form-lateral}$ and f_{peak} was set so as to keep the overall mean firing rate at a mean value f_{mean} which gave sufficient difference between f_{base} and f_{peak}.

3 Results

Simulations were run with a C++ function, with initial conditions created and data analysis carried out with Matlab. Simulations used a time step of 0.1ms. Parameters are given in table 1. The mean frequency of rewiring opportunities per potential synapse was 1.22Hz (depressed synapses were therefore eliminated after an average of 33s). Initial placement of synapses was performed by iteratively generating a random pre-synaptic partner and carrying out the test for formation described in section 2. Initially feed-forward and lateral connections were placed separately, each up to their initial number of 16 synapses. Weights were initially maximised. Runs were for 5 minutes of simulated time.

For calculating the preferred location for each target cell, the use of the "centre of mass" measure as in [6] would be erroneous because the space is toroidal and therefore the calculation of preferred location would be skewed by the choice of reference point from which synapses' coordinates are measured. In [6] the reference point for calculating centre of mass of the dendritic synapses of a target cell

Table 1. Simulation parameters

for STDP	for rewiring	for inputs
$g_{max} = 0.2$	$\sigma_{form-feedforward} = 2.5$	$f_{mean} = 20Hz$
$\tau_m = 0.02s$	$\sigma_{form-lateral} = 1$	$f_{base} = 5Hz$
$\tau_+ = 0.02s$	$p_{form-lateral} = 1$	$f_{peak} = 152.8Hz$
$\tau_- = 0.064s$	$p_{form-feedforward} = 0.16$	$\sigma_{stim} = 2$
$A_+ = 0.1s$	$p_{elim-dep} = 0.0245(= 0.5*\text{mean formation rate})$	
$B = 1.2$	$p_{elim-pot} = p_{elim-dep}/180 = 1.36 * 10 - 4$	

was chosen as the predefined ideal location, therefore the measures of distance of preferred location were skewed towards the ideal locations dictated by the model. We avoided this by the novel method of searching for the location around which the afferent synapses have the lowest "weighted variance" (${\sigma_{aff}}^2$), i.e.:

$$\sigma_{aff}^2 = \underset{\boldsymbol{x}}{\operatorname{argmin}} \frac{\sum_i w_i.|\boldsymbol{p}_{xi}|^2}{\sum_i w_i} \tag{4}$$

where i is a sum over synapses, $\boldsymbol{x}$ is a candidate preferred location, $|\boldsymbol{p}_{xi}|$ is the minimum distance from that location of the afferent for synapse i and wi is the weight of the synapse (if connectivity is evaluated without reference to weights, synapses have unitary weight). We implemented this with an iterative search over each whole number location in each dimension and then a further iteration to locate the preferred location to 1/10th of a unit of distance (the unit is the distance between two adjacent neurons). Note that in the non-toroidal case this measure is equivalent to the centre of mass, as used in [3].

Having calculated the preferred location for all the neurons in the network layer we took the mean of the distance of this preferred location from the ideal location to give an Average Absolute Deviation (AAD) for the projection. By reporting both AAD and mean σ_{aff} for a projection we have a basis for separating its variance from the deviation of its preferred location from its ideal location. However AAD and mean σ_{aff} are both dependent on the numbers and strengths of synapses and these can change during development. Therefore to observe the effect of the activity-dependent development mechanism irrespective of changes in synapse number and strength we made comparison in two ways. Firstly, for evaluating change in mapping quality based only on changes in connectivity without considering the weights of synapses we created a new map taking the final number of synapses for each network neuron and randomly placing them in the same way as the initial synapses were placed. We then calculated σ_{aff} and AD for each neuron in each of the maps and compared the averages of these (i.e. mean σ_{aff} and AAD), applying significance tests between the values of two populations of neurons, i.e. all the neurons on the final map vs all those on the reconstructed map. Having established what effect there was on connectivity we considered the additional contribution of weight changes by creating a new map with the same

Table 2. Summary of simulation results: Case 1: Rewiring and input correlations; Case 2: Input correlations and no rewiring; Case 3: Rewiring and no input correlations

Case	**1**	**2**	**3**
Network neuron mean spike rate	24.7	17.4	10.5
Final mean no. feed-forward incoming synapses per network neuron	14.1	NA	12.5
Weight as proportion of max for the initial no. of synapses	0.60	0.36	0.33
Mean $\sigma_{aff-init}$	2.36	2.36	2.36
Mean $\sigma_{aff-final-con-shuffled}$	2.32	NA	2.32
Mean $\sigma_{aff-final-con}$	1.95	2.36	2.17
Mean $\sigma_{aff-final-weight-shuffled}$	1.88	2.10	1.99
Mean $\sigma_{aff-final-weight}$	1.70	1.98	1.95
AAD_{init}	0.78	0.78	0.78
$AAD_{final-con-shuffled}$	0.89	NA	0.90
$AAD_{final-con}$	0.83	0.78	0.93
$AAD_{final-weight-shuffled}$	0.92	1.36	1.21
$AAD_{final-weight}$	0.95	1.58	1.34

topology, taking the final weights of synapses for each network neuron and randomly reassigning these weights amongst the existing synapses for that neuron. We then compared the two maps as described above.

Three main experiments were carried out: Case 1 had both rewiring and input correlations, as described in section 2; case 2 had input correlations but no rewiring; case 3 had rewiring but no input correlations (i.e. all input neurons fired at fmean). The results are given in table 2. For comparisons, mean σ_{aff} and AAD were each calculated for the feed-forward connections of the following networks: (a) The initial state with weights not considered (recall that all weights were initially maximised) these results are suffixed "*ini*", i.e. AAD_{ini}; (b) the final ("*fin*") network with weights not considered but only connectivity ("*con*") with all synapses weighted equally, i.e. $AAD_{fin-con}$; (c) for comparison with $AAD_{fin-con}$, the final number of synapses for each network neuron, randomly placed ("*shuf*") in the same way as the initial synapses (not applicable for simulations with no rewiring), i.e. $AAD_{fin-con-shuf}$; (d) the final network including weights, i.e. $AAD_{fin-weight}$; (e) for comparison with $AAD_{fin-weight}$, the final connectivity for each network neuron with the actual weights of the final synapses for each network neuron randomly reassigned amongst the existing synapses, i.e. $AAD_{fin-weight-shuf}$. Results were compared using Wilcoxon Signed-Rank (WSR) tests on AD and σ_{aff} for incoming connections for each network neuron over the whole network layer for a single simulation of each of the two conditions under consideration.

4 Discussion

We observe the effect of rewiring by comparing case 1 (with rewiring) and case 2 (without rewiring). Considering topology change, in case 1 mean $\sigma_{aff-fin-con}$

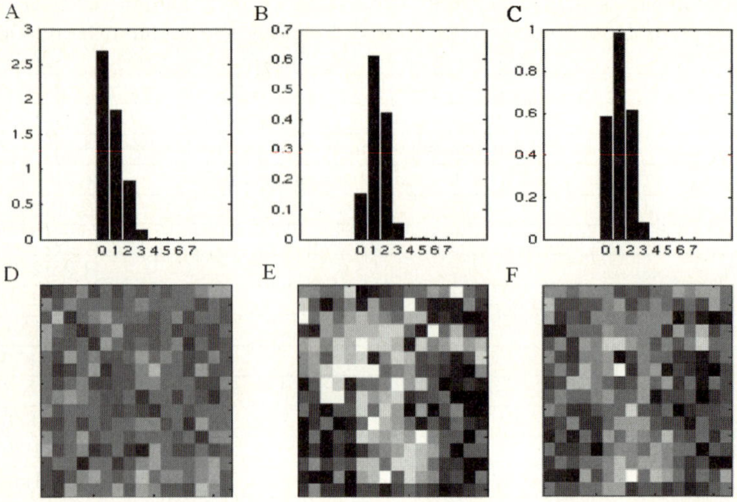

Fig. 1. A-C: Normalised weight density of incoming lateral synapses (weight/unit area; y-axis) radially sampled and interpolated at given distances of pre-synaptic neuron from post-synaptic neuron (x-axis), averaged across population. D-F: ocular preference, i.e. preference for cells from the two intra-correlated input spaces interspersed in the input space, for each network cell on a scale from white to black. A,D: initial. B,E: final, considering synaptic weights. C,F: final, all synapses with unitary weight.

drops to 1.95, c.f. 2.32 for mean $\sigma_{aff-fin-con-shuf}$; this drop is significant (WSR, p=2.4e-25). In case 2 mean $\sigma_{aff-fin-con}$ is constrained to remain at mean $\sigma_{aff-ini} = 2.36$. Considering weight change, in case 1 mean $\sigma_{aff-fin-weight}$ drops to 1.70, c.f. 1.88 for mean $\sigma_{aff-fin-weight-shuf}$. In case 2, mean $\sigma_{aff-fin-weight}$ drops to 1.98, c.f. 2.10 for mean $\sigma_{aff-fin-weight-shuf}$. Both drops are significant (WSR, p=2.7e-27 and 8.7e-6 respectively).

Mean $\sigma_{aff-fin-weight}$ appears to be lower in case 1 than case 2. We cannot say for sure that this superior reduction of variance is due to the effect of the rewiring mechanism because the different numbers of final synapses in each case make a comparison impossible, however there is a good reason to believe that this is so: the drop in mean $\sigma_{aff-fin-con}$. This drop on its own indicates that the rewiring mechanism has helped to reduce variance and would also lay the groundwork for different final measures of σ_{aff} when weights are considered.

We can also see qualitatively that the effect of rewiring is to embed in the connectivity of the network input preferences which arise through the weight changes mediated by the learning rule. STDP favours causal inputs with the lowest latency and local excitatory lateral connections tend to lose the competition with excitatory feed-forward connections as they have a higher latency [4]. The extreme of this effect can be seen in synapses from a network neuron back to itself (recurrent synapses). The placement rule allows these synapses to form, however these synapses only ever receive a pre-synaptic spike immediately following a post-synaptic spike and therefore they are always depressed by the learning rule. Figure 1A shows the initial density of incoming lateral synapses from pre-synaptic partners at given distances out from the post-synaptic neuron. It can be seen that the average neuron receives more synapses from itself (those at x-position 0) than from any of its closest neighbours. Figure 1B shows the final distribution where synapses are weighted. The recurrent synapses have been depressed much more than their neighbours. Figure 1C shows the final distribution only considering numbers of synapses and not their weights. The proportion of recurrent synapses to lateral synapses with neighbours has reduced from the initial state, due to the preferential elimination of the weak recurrent synapses.

As a further demonstration of the effect of rewiring a simulation was carried out with the input neurons divided into two groups, mimicking the effect of binocular inputs. The groups were interspersed in a regular diagonal pattern, i.e. each input neuron is in the opposite group to its 4 adjacent neurons; the stimulus location switched between the two groups every time it changed. To keep the overall input rate the same the peak firing rate was doubled. Figure 1D shows the initial preference of each network neuron for input neurons in the two groups. Figure 1E shows the final ocular dominance map where synapses are weighted. Although the space used was too small and the result of the learning rule with a small number of synapses too random for familiar striped ocular dominance patterns to emerge (c.f. [3]) ocular dominance zones can be seen. This pattern is reflected in the final map of connectivity in Figure 1F, where synaptic weights are not considered; another example of weight patterns caused by input activity becoming embedded in connectivity patterns.

Considering the effect of the algorithm on AAD, in case 2 $AAD_{fin-weight}$ is significantly increased c.f. $AAD_{fin-weight-shuf}$ (WSR, p=0.0012). In case 1 the corresponding change is not significant (WSR, p=0.48). In case 1 the drop in $AAD_{fin-con}$ c.f. $AAD_{fin-con-shuf}$ is not significant (WSR, p=0.31).

The basic action of weight-independent STDP on a set of incoming synapses for a single neuron is to deliver a bimodal weight distribution [4]. Where there are input correlations these cause the more correlated inputs to be maximised and the less- or un-correlated inputs to be minimised. The effect of both the input correlations and the local excitatory lateral synapses on each individual incoming connection field then should be to cause a patch of neighbouring synapses to become potentiated and for outliers from this patch to be depressed. The location of the patch will be random; it is likely to form near the ideal location because there should be a denser concentration of synapses there, however the centre of the patch is unlikely to fall exactly on the ideal location but rather a certain mean distance from it. This introduces a shift of preferred location from the ideal location. Rewiring cannot be expected to eliminate this error but it might be expected to allow the patch to move towards the centre as σ_{aff} reduces due to the preferential placement of synapses towards the centre. However in our simulations AAD did not improve. The slight drop in $AAD_{fin-con}$ c.f. $AAD_{fin-con-shuf}$ is not significant but in any case a drop in AAD could only be a result of the reduction in mean σ_{aff} because $AAD_{fin-weight}$ does not decrease, rather it stays the same (as in case 1) or increases (as in case 2). That is to say, the result of the weight changes is not to drive the preferred location towards the ideal. Rather, the improvement of topography is driven by the continued placement of synapses towards the ideal location; the activity-dependent mechanism simply facilitates by allowing the incoming connection field to be narrowed by the preferential elimination of outliers.

Considering the role of input correlations, in case 3 (rewiring but no input correlations) mean $\sigma_{aff-fin-con} = 2.17$, vs 2.31 for mean $\sigma_{aff-fin-con-shuf}$; this is significant (WSR, p=5.0e-6). Mean $\sigma_{aff-fin-weight} = 1.95$ vs 1.99 for mean $\sigma_{aff-fin-weight-shuf}$; this is significant (WSR, p=0.028).

The slight drop in mean $\sigma_{aff-fin-weight}$ is a sufficient cue to drive the narrowing of the incoming connection fields, as evidenced by the drop in mean $\sigma_{aff-fin-con}$. It was shown [8] that functional architecture could form in the absence of any input except uncorrelated random noise. We show that this applies to topographic map refinement as well, although our explanation differs: A spike from a single input neuron will excite a given network neuron and any other of its neighbours which have a synapse from that input. Thus the neuron will also tend to receive some excitation from lateral connections because of that spike. Network neurons sample afferent neurons more densely around their ideal locations so they are more likely to share an afferent with a neighbour if that afferent is close to their ideal location. Thus synapses from afferents closer to the ideal location are more likely to be potentiated. Therefore the gradient of connection density set up by activity-independent placement acts as a cue which allows the preferential elimination of outliers, giving a reduction in variance.

5 Conclusions

We have presented a model of topographic development including both weight and wiring plasticity, which follows the reasonable assumptions that synapses preferentially form in locations to which their axons are guided and that weaker synapses are more likely to be eliminated. We have shown that spatially correlated inputs help to create patterns of synaptic weights which favour narrower projections, but the spatial correlations are not necessary for some reduction of variance to occur (extending a result from [8]). A weight-change mechanism and a rewiring mechanism can work together to achieve a greater effect than the weight changes alone, with the rewiring mechanism acting to embed patterns of synaptic strengths in the network topology; this is as one would expect, though it has not been demonstrated quantitatively before, to our knowledge. The accuracy of preferred locations for network neurons however may not necessarily improve when synapses are formed based on distributions with on-average perfect topography to start with. The novel division of mapping quality into the quantities of mean σ_{aff} and AAD is therefore a useful means for investigating these effects, and we have demonstrated a method of applying statistical significance tests to extract highly significant effects from small-scale simulations. Future work will include the introduction of weight-dependent STDP and of noise in the measures of proximity and weight used for formation and elimination.

Acknowledgements

We are grateful to Guy Billings for providing the basis of the simulator code. This work was funded by EPSRC.

References

1. Willshaw, D., von der Malsburg, C.: How patterned neural connections can be set up by self-organisation. Proc. R. Soc. Lond. B. 194, 431–445 (1976)
2. Miller, K., Keller, J., Stryker, M.: Ocular dominance column development: analysis and simulation. Science 245, 605–615 (1989)
3. Goodhill, G.: Topography and ocular dominance: a model exploring positive correlations. Biological Cybernetics 69, 109–118 (1993)
4. Song, S., Abbott, L.: Cortical development and remapping through spike timing-dependent plasticity. Neuron 32, 339–350 (2001)
5. Willshaw, D.: Analysis of mouse Epha knockins and knockouts suggests that retinal axons reprogramme target cells. Development 133, 2705–2717 (2006)
6. Elliott, T., Shadbolt, N.: A neurotrophic model of the development of the retinogeniculocortical pathway induced by spontaneous retinal waves. Journal of Neuroscience 19, 7951–7970 (1999)
7. Miller, K.: Equivalence of a sprouting-and-retraction model and correlation-based plasticity models of neural development. Neural Computation 10, 529–547 (1998)
8. Miikkulainen, R., Bednar, J., Choe, Y., Sirosh, J.: Computational Maps in the Visual Cortex. Springer, New York (2005)
9. Willshaw, D., von der Malsburg, C.: A marker induction mechanism for the establishment of ordered neural mappings. Philosophical Transactions of the Royal Society of London. Series B, Biological Sciences 287, 203–243 (1979)
10. Bamford, S., Murray, A., Willshaw, D.: Large developing axonal arbors using a distributed and locally-reprogrammable address-event receiver. In: International Joint Conference on Neural Networks (IJCNN) (2008)
11. Sperry, R.: Chemoaffinity in the orderly growth of nerve fiber patterns and connections. Proc. Natl. Acad. Sci. USA 50, 703–709 (1963)
12. Ruthazer, E., Cline, H.: Insights into activity-dependent map formation from the retinotectal system. Journal of Neurobiology 59, 134–146 (2004)
13. Swindale, N.: The development of topography in the visual cortex: a review of models. Network: Computation in Neural Systems 7, 161–247 (1996)
14. Grutzendler, J., Kasthuri, N., Gan, W.: Long-term dendritic spine stability in the adult cortex. Nature 420, 812–816 (2002)
15. Young, J., Waleszczyk, W., Wang, C., Calford, M., Dreher, B., Obermayer, K.: Cortical reorganisation consistent with spike timing- but not correlation-dependent plasticity. Nature Neuroscience 10, 887–895 (2007)

Implementing Bayes' Rule with Neural Fields

Raymond H. Cuijpers[1,*] and Wolfram Erlhagen[2]

[1] Nijmegen Institute for Cognition and Information, Radboud University, 6500 HE Nijmegen, P.O. Box 9104, The Netherlands
Tel.: +31 24 3612608; Fax: +31 24 3616066
r.cuijpers@nici.ru.nl

[2] Department of Mathematics for Science and Technology, University of Minho, 4800-058 Guimares, Portugal
wolfram.erlhagen@mct.uminho.pt

Abstract. Bayesian statistics is has been very successful in describing behavioural data on decision making and cue integration under noisy circumstances. However, it is still an open question how the human brain actually incorporates this functionality. Here we compare three ways in which Bayes rule can be implemented using neural fields. The result is a truly dynamic framework that can easily be extended by non-Bayesian mechanisms such as learning and memory.

Keywords: Bayesian statistics, Neural fields, Decision making, Population coding.

1 Introduction

Bayesian statistics has become a popular framework for describing various kinds of human behaviour under circumstances of uncertainty [1,2,3]. Generally, it is assumed that populations of neurons could encode probability distributions and indeed this can be used to predict overt behaviour in monkey [4]. One way to implement Bayes' rule is to let each neuron represent the likelihood of an entire probability distribution [5,6]. However, in these studies Bayes' rule is implicit in the decoding mechanism. Another way is to represent probability distributions in the log domain [7], so that Bayes' rule can be implemented by simply adding neuronal activities.

The neural field architecture has been quite successful in explaining behavioural data [8,9,10] and as a control architecture in cognitive robots [11]. Neural fields can support bi-stable activity patterns [12,13], which makes them suitable for memory and decision mechanisms [14,9]. However, these neural field properties are non-Bayesian. For one, the reported neural fields cannot support sustained multi-modal activity patterns in a way that is necessary for representing arbitrary probability distributions. On the other hand, Bayesian statistics does not incorporate any temporal dynamics. We wondered whether it is possible to combine the best of both worlds. In particular, we wanted to know whether it is possible to implement Bayes' rule with neural fields.

* Corresponding author.

V. Kůrková et al. (Eds.): ICANN 2008, Part II, LNCS 5164, pp. 228–237, 2008.

2 Neural Fields of Leaky Integrate-and-Fire Neurons

A common model of a neuron in computational neuroscience is the so-called sigma node representing the average behaviour of leaky integrate-and-fire neurons [15]. Each sigma node has an internal state $u_i(t)$, which is analogous to a biological neuron's membrane potential, and an output activity $r_i(t)$, which is analogous to a biological neuron's firing rate [15]. For ease of reference we will use the names of the biological analogies. The sigma node reflects the average behaviour of one or more biological neurons. The firing rate is related to the membrane potential by:

$$r_i(t) = f(u_i(t))$$

where f is a thresholding function such as the Heaviside step function or the sigmoid function. The function f is often called the activation function.

The membrane potential changes dynamically depending on the input from other neurons and the leaking current.

$$\tau \frac{du_i(t)}{dt} = h - u_i(t) + \sum w_{ij} f(u_j(t)) \,, \tag{1}$$

where τ is a time constant, h is the equivalent of the resting potential, and w_{ij} are the connection weights between neuron i and neuron j. The first term on the right hand side results in an exponential decay of the membrane potential with time constant τ and the second term on the right hand side reflects the weighted sum of inputs from other sigma node neurons (Fig. 1A).

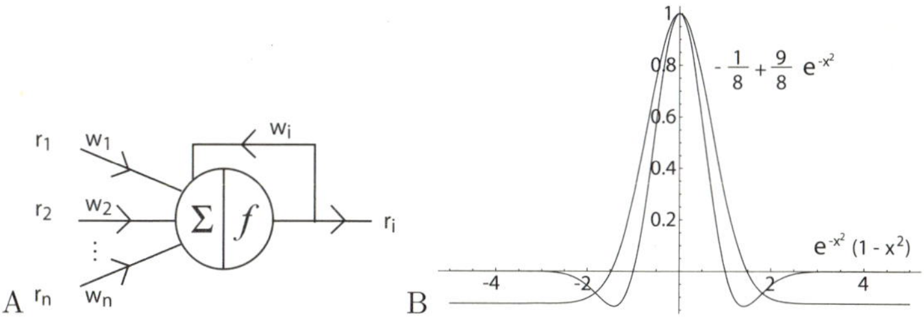

Fig. 1. a) Schematic diagram of a sigma node neuron i. b) Graphs of the lowered Gaussian and Mexican hat function with long and short inhibitory tails, respectively.

The structure of a neural network is determined by the connection weights w_{ij} in (1). It has been observed that in biological neural tissues neurons excite and inhibit nearby neurons in a centre-surround fashion (see e.g. [4]). In the continuous limit we obtain a neural field, whose dynamics are governed by Amari's equation for a neural field [16]:

$$\tau \frac{\partial u(x,t)}{\partial t} = h - u(x,t) + (w * f(u))(x,t) + S(x,t) \,. \tag{2}$$

Here $u(x,t)$ denotes the membrane potential of a neuron at location x, and w is the kernel imposing the centre-surround structure within the neural field and $*$ denotes spatial convolution, which is defined as $f * g = \int f(x-y)g(y)dy$. The term $S(x,t)$ denotes the external input of the neural field. The constant h in Amari's equation can easily be absorbed in the definitions of $u(x,t)$ and f, so we will omit it hereafter. The centre-surround excitation-inhibition is commonly modelled by a lowered Gaussian (Fig. 1B):

$$w(x-y) = A \exp\left(-\frac{(x-y)^2}{2\sigma^2}\right) - w_{\text{inhib}}\,, \tag{3}$$

where A and σ denote the amplitude and width of the lowered Gaussian, respectively. The constant w_{inhib} determines the amount by which the Gaussian is lowered. The Gaussian shaped kernel effectively smoothes the neural field activity locally because it averages the activity of nearby neurons. Other commonly used kernels, such as the Mexican hat function (Fig. 1B), have similar properties. With a suitable choice of parameters A and w_{inhib} neural fields governed by (2)(3) can sustain a self-stabilised local bump of activity [12,14,13]. This is clearly useful for memory-like functions especially because the self-stabilising capability makes the neural field very robust against noise. The global inhibition w_{inhib} prevents the co-existence of two or more self-stabilising bumps. The Mexican hat function (Fig. 1B) does admit multiple bumps [14], but their amplitudes cannot vary independently. This rules out the possibility to represent multi-modal probability distributions in this way.

3 Amari's Equation and Robustness against Noise

3.1 Amari's Equation with Linearised Recurrent Connection Kernel

Robustness against noise is a key concept that leads to Amari's equation (2). In order to obtain robustness against noise, the field activity must be smoothed spatially as time progresses. A simple way to achieve this is by replacing at every time step Δt a fraction of the field activity by the smoothed field activity. In the presence of sustained external input, another fraction is replaced by the external input. The latter is necessary to prevent exponential growth of the total field activity. If $u(x,t)$ denotes the activity of neuron x at time t and $S(x,t)$ represents external input, we can formalise this in the following way:

$$u(x,t+\Delta t) = (1-\varepsilon)u(x,t) + \alpha\varepsilon\,(k * u)(x,t) + (1-\alpha)\varepsilon S(x,t)\,, \tag{4}$$

where $u(x,t+\Delta t)$ is the updated neural field activity, $k(x)$is the smoothing kernel and α, ε are constants. The constant α controls the balance between smoothing and external input, and ε controls the updating rate. We can rewrite this as a partial differential equation by having $\varepsilon = \frac{\Delta t}{\tau}$ and taking the limit $\Delta t \to 0$ to obtain:

$$\tau \frac{\partial u(x,t)}{\partial t} = -u(x,t) + \alpha\,(k * u)(x,t) + (1-\alpha)S(x,t)\,. \tag{5}$$

where τ is some time constant.

When the external input $S(x,t) = s(x)$ is constant over time, the neural field activity $u(x,t)$ decays to a stationary solution. Let $u_\infty(x)$ denote the stationary solution, then it is easy to show that the external input must have the following form:

$$s(x) = (k_{ext} * u_\infty)(x) \equiv \frac{1}{1-\alpha}((\delta - \alpha k) * u_\infty)(x)\,, \tag{6}$$

where $\delta(x)$ is Dirac's delta function. The kernel defined in (6) has an interesting property. Suppose that a second neural field receives external input from the first using connection weights given by $k_{ext}(x) = (\delta(x) - \alpha k(x))/(1-\alpha)$, then the stationary solution of the second field equals the firing rate of the first field. In other words, the neural field activity $u(x,t)$ decays exponentially to $s(x)$. To make this statement more precise, we can show that the total activity decays exponentially over time. If we define the total activity of a neural field $z(x,t)$ as:

$$\bar{z}(t) = \int z(x,t)dx\,, \tag{7}$$

then we can rewrite 5 using 7 to:

$$\tau\frac{d\bar{u}(t)}{dt} = -\bar{u}(t) + \alpha K\bar{u}(t) + (1-\alpha)\bar{S}(t)\,. \tag{8}$$

Here we have used the shift invariance of the kernel and defined $K = \int k(x-y)dx$. If the total external input $\bar{S}(t) = \bar{s}$ is constant, the solution of (8) is simply:

$$\bar{u}(t) = A\exp(-(1-\alpha K)\frac{t}{\tau}) + \frac{1-\alpha}{1-\alpha K}\bar{s}\,. \tag{9}$$

This shows that the total activity $\bar{u}(t)$ decays exponentially with time constant $(1-\alpha K)/\tau$ to a constant times $\bar{s}$. This constant equals 1 independently of the value of α when $K = 1$.

3.2 Non-linear Dynamics of Amari's Equation

Equation (5) resembles Amari's original equation (2): the constants α and $1-\alpha$ can easily be absorbed in the definitions of the kernel and the external input, but the activation function f introduces a non-linearity that cannot be removed so easily. However, when the field activity is in the linear range of f, the dynamics of (5) will approximate that of (2). Amari's original equation can thus be obtained by replacing the convolution term $k * u \rightarrow k * f(u)$. This gives:

$$\tau\frac{\partial u(x,t)}{\partial t} = -u(x,t) + \alpha(k * f(u))(x,t) + (1-\alpha)S(x,t)\,. \tag{10}$$

We obtain the stationary solution of the external input for the non-linear Amari's equation in a similar way as in (6):

$$s(x) = \frac{1}{1-\alpha}\left(u_\infty(x) - \alpha(k * r_\infty)(x)\right) \tag{11}$$

$$\approx (k_{\text{ext}} * r_\infty)(x)\,, \tag{12}$$

where $r_\infty(x) = f(u_\infty(x))$. Clearly, the latter approximation is valid only when the activation function f is approximately linear.

4 A Probabilistic Interpretation for Neural Fields

In Bayesian decision making the likelihood of an event y for a given hypothesis x is transformed into a belief in that hypothesis. This inference is governed by Bayes' rule, which is defined as:

$$p(x|y) = \frac{p(y|x)p(x)}{p(y)}, \tag{13}$$

where $p(x)$ is the prior probability, $p(x|y)$ the posterior probability, and $p(y|x)$ the likelihood of x. Typically, x is variable and y is constant so that the probability $p(y)$ is an arbitrary normalisation constant. In many decision problems, making a decision amounts to finding the value of x that maximises the posterior probability - the so-called *maximum-a-posteriori.*

In order to implement Bayes' rule with neural fields we need to represent the probabilities in such a way that Bayes' rule can be computed using simple addition of neural field activities. This can be achieved by representing probability distributions in the log domain [7]. However, if the dynamics of the neural fields are governed by Amari's equation (2) the activation function introduces a non-linearity in the system. In order to obtain similar dynamics as in the linearised version of Amari's equation (5) we need to scale the log-probabilities to the linear range of the activation function. In particular, if we use the following sigmoid function:

$$f(x) = \frac{1}{1 + \exp\left(-4(x - \frac{1}{2})\right)}, \tag{14}$$

then $f(x)$ is approximately linear within the range (0,1) with slope $f'(\frac{1}{2}) = 1$ at $x = \frac{1}{2}$. Thus, we wish the function $g(x)$ to linearly remap the log-probabilities to the range $(0, 1)$. However, the log-probabilities lie in the range $(-\infty, 0)$, so that such a linear transformation does not exist. Therefore, we define a minimal probability value, say $p_{min} = 10^{-16}$, below which we ignore approximation errors. Then, we can define the function g as:

$$g(x) = 1 - \frac{x}{\ln p_{min}}, \tag{15}$$

which maps the range $(\ln p_{min}, 0)$ to $(0, 1)$. We can now make the following identification:

$$u(x, t) = g(\ln(p_t(x))), \tag{16}$$

where $u(x, t)$ is the neural field activity of neuron representing x at time t, and $p_t(x)$ is some probability distribution of x at time t. With this identification we can incorporate Bayes' rule in the following way. Suppose neural field A is encoding the likelihood $p_t(y|x)$ of x at time t and neural field B is encoding the prior probability $p_t(x)$. We need to construct a third neural field C receiving inputs from neural fields A and B as external input, so that it encodes the posterior probability $p_t(x|y)$. Let

$$\begin{aligned} u_A(x,t) &= g(\ln p_t(y|x))\,, \\ u_B(x,t) &= g(\ln p_t(x))\,, \\ h_C &= -g(\ln p(y))\,, \end{aligned} \tag{17}$$

where $u_A(x,t)$ and $u_B(x,t)$ are the activities of neural fields A and B and h_C is a constant reflecting homogeneous global inhibition of neural field C. In case of the linearised Amari's equation , the stationary activity of neural field C can be shown to encode the posterior probability $p_t(x|y)$ by setting the external input to:

$$S_{C,\text{linear}}(x,t) = (k_{\text{ext}} * u_A)(x,t) + (k_{\text{ext}} * u_B)(x,t) + \frac{1-\alpha K}{1-\alpha} h_C\,, \tag{18}$$

where k_{ext} is the kernel defined in (6). Here we have used the shift-invariance of the kernel k (see (8)) to simplify the constant term involving h_C. By construction the external input of neural field C equals $k_{ext} * g(\ln p_t(x|y))$ and it follows from (6) that the steady state activity encodes the posterior probability as desired.

In a similar way we obtained the expression for the non-linear Amari's equation without approximating the kernel (11) and with approximated kernel (12):

$$\begin{aligned} S_{C,\text{non-linear}}(x,t) = \frac{1}{1-\alpha}&(u_A(x,t) - \alpha(k * f(u_A))(x,t) \\ &+ u_B(x,t) - \alpha(k * f(u_B))(x,t) \\ &+ h_C - \alpha K f(h_C))\,, \end{aligned} \tag{19}$$

$$S_{C,\text{approximate}}(x,t) = (k_{\text{ext}} * r_A)(x,t) + (k_{\text{ext}} * r_B)(x,t) + \frac{1-\alpha K}{1-\alpha} f(h_C) \tag{20}$$

In the remaining part of this paper we will refer to neural fields with external inputs given by (18), (19) and (20) by linear, non-linear and approximate, respectively.

5 Neural Field Simulations

For the simulations we used a scenario where the field position x represents the angular position subdivided in 100 intervals. The simulations were done using discrete versions of (5),(2) using Euler discretisation of both space and time. We used $n = 100$ neurons for encoding field position ($x_i = i\Delta x$ with $i = 0, \ldots, n-1$ and $\Delta x = 1$). Time was discretised as shown in (4) with $\Delta t = 1$. The parameters τ and α were set to $\tau = 10$ and $\alpha = \frac{1}{2}$. To prevent edge effects we used circular neural fields so that neuron $i = 99$ neighbours neuron $i = 0$. For the recurrent connections kernel we used a Von Mises distribution - the circular equivalent of the Gaussian normal distribution - so that $K = 1$. The Von Mises distribution is given by:

$$\text{vonmises}(x,\kappa) = \frac{1}{2\pi I_0(\kappa)} \exp \kappa \cos(\frac{2\pi x}{n\Delta x})\,, \tag{21}$$

where κ is a width parameter and $I_m(\kappa)$ is the modified Bessel function of the first kind with $m = 0$. The width parameter was set to $\kappa = 3^{-2}$ which approximately corresponds to a Gaussian width of $\sigma = 3$. For the likelihood $p(y|x)$ and the prior $p(x)$ we used Von Mises distributions centred on $x = 60$ and $x = 30$, respectively. The corresponding widths were set to $\kappa = 2^{-2}$ and $\kappa = 3^{-2}$. In order to simulate Bayes' rule the external inputs of the neural fields representing the posterior distribution are given by (18)-(20). White noise was added to the external input directly with an amplitude of ± 0.05.

How the neural field activities change over time is shown in the top row of Fig. 2 for linear (left), non-linear (middle), and approximate (right) neural fields. In the bottom row the corresponding firing rates are shown. Both the linear and non-linear neural fields converge to the posterior field activity, but the approximate neural field shows noticeable differences: the neural field activities are restricted to the range (0,1) due to the sigmoid activation function in (14). Consequently, the neural field activity due to the prior and the associated firing rates are not fully suppressed as is evident by the bumps at $x = 30$.

Fig. 3 shows the decoded probability distributions of the linear (left), non-linear (middle) and approximate neural fields (right). All neural fields build up activities at approximately the same location on similar time scales, but their amplitudes differ considerably. The amplitude of the approximate neural field (right) is much less than the other two. This is also evident if we plot the cross-section at different times in comparison to the true posterior probability distribution (bottom). It can

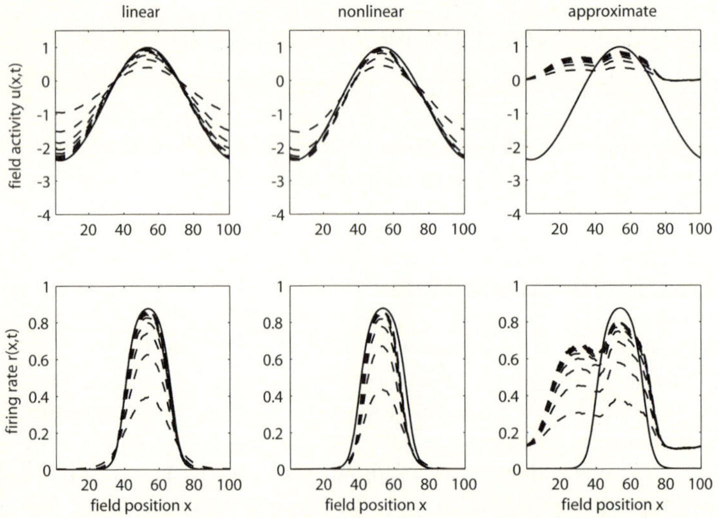

Fig. 2. Neural field activities $u(x, t)$ (top) and firing rates $r(x, t)$ (bottom) for linear (left), non-linear (middle) and approximate kernel. Neural field activities and firing rates are shown for every 10th of 100 iterations (dashed curves). The solid curves indicate the stationary state corresponding to the true posterior.

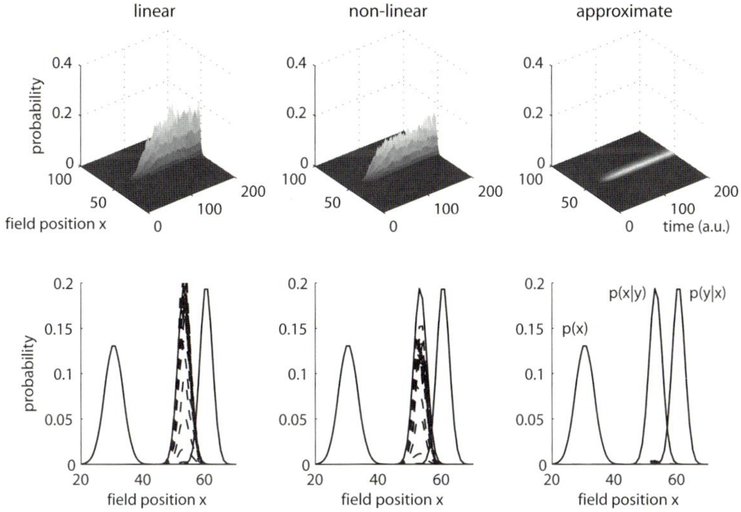

Fig. 3. Decoded probability distributions as a function of time (top) and for every 10th time step (bottom; dashed curves)

be seen that the approximations of the non-linear and approximate neural fields mainly affect the amplitudes of the decoded probability distributions.

The error in amplitude is easily amended by normalising the decoded probability distributions. To address this we compared the location and width of the normalised decoded probability distribution with those of the true posterior. In Fig. 4 the average error in peak location (left) and peak width (right) is shown as a function of time. For the peak location we used the expected value of the posterior distribution. For the peak width we used the standard deviation. The averages were taken from 200 randomly generated Von Mises priors and likelihoods. The location of the Von Mises distributions of both prior and likelihood were randomly drawn from a uniform distribution covering the entire neural field ($x \in [0, 99]$). The width of the distributions was uniformly drawn from $\sigma \in [1, 25]$ with $\kappa = 1/\sigma^2$. From Fig. 4 it is clear that the average error in peak location lies within 1 neuronal unit ($\frac{2\pi}{100} = 3.6°$) irrespective of which type of neural field was used. The average error in peak width is initially largely overestimated after which it decays to a constant value. Both the linear and non linear neural fields approximately converge to the true posterior width within about 20 iterations. The rate of convergence is somewhat faster for the non-linear field. The approximate neural field does not converge to zero but stabilises on an overestimate of 3 neuronal units (10.8°). The angular error can be reduced by increasing the number of neurons. This is shown by the thin dashed line in Fig. 4. The error in decoded peak location for n=1000 is the same as for n=100 showing that the angular error is decimated (left). The error in decoded peak width is about ten

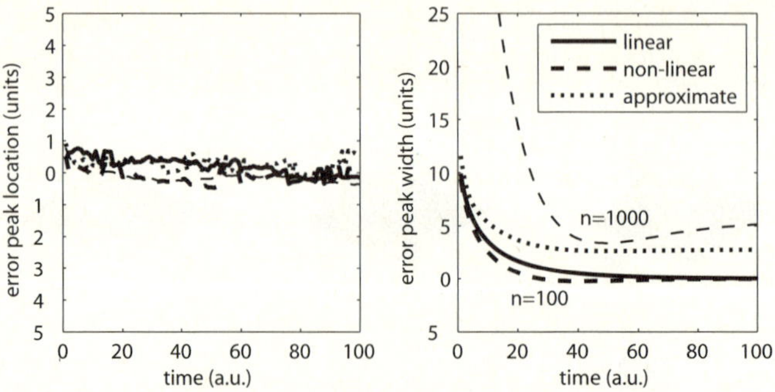

Fig. 4. The mean error in peak location (left) and width (right) of the decoded peak. The lines indicate the errors obtained for linear (solid line), non-linear (dashed line) and approximate (dotted lines) neural field. The time constant was set to $\tau = 10$.

times larger (initial value 100.2) reflecting the fact that the probability distributions are ten times wider when expressed in neuronal units.

6 Discussion and Conclusions

We have developed a way in which Bayes' rule can be represented using neural fields. To do so we superimposed the neural field activities representing the log-likelihood and the log-prior distribution. Due to the non-linear activation function in Amari's equation the superposition of field activities leads to an approximation error that mainly affects the decoded amplitude of the posterior distribution. The approximation error can be fixed by normalisation, but the approximate neural field overestimates the width of the decoded posterior distribution. The approximation error depends on how well the neural field activities fall in the linear range of the activation function. If necessary, the linear approximation could be reduced by adjusting this mapping, but the signal-to-noise ratio will deteriorate.

The proposed implementation of Bayes' rule can be used to build a dynamic version of Bayesian decision making as was used in [17]. Whether the dynamics is in agreement with experimental findings is still an open issue. An additional benefit is that Bayesian inference can now be smoothly coupled with non-Bayesian mechanisms such as Hebbian learning, memory and decision mechanisms.

Acknowledgements

The present study was supported by the EU-Project "Joint Action Science and Technology" (IST-FP6-003747).

References

1. Gold, J.I., Shadlen, M.N.: The neural basis of decision making. Annu. Rev. Neurosci. 30, 535–574 (2007)
2. Hillis, J.M., Watt, S.J., Landy, M.S., Banks, M.S.: Slant from texture and disparity cues: optimal cue combination. J Vis 4, 967–992 (2004)
3. Kuss, M., Jäkel, F., Wichmann, F.A.: Bayesian inference for psychometric functions. J. Vis. 5, 478–492 (2005)
4. Chen, Y., Geisler, W.S., Seidemann, E.: Optimal decoding of correlated neural population responses in the primate visual cortex. Nat. Neurosci. 9, 1412–1420 (2006)
5. Ma, W.J., Beck, J.M., Latham, P.E., Pouget, A.: Bayesian inference with probabilistic population codes. Nat. Neurosci. 9, 1432–1438 (2006)
6. Zemel, R.S., Dayan, P., Pouget, A.: Probabilistic interpretation of population codes. Neural Comput. 10, 403–430 (1998)
7. Rao, R.P.N.: Bayesian computation in recurrent neural circuits. Neural Comput. 16, 1–38 (2004)
8. Cisek, P.: Integrated neural processes for defining potential actions and deciding between them: a computational model. J. Neurosci. 26, 9761–9770 (2006)
9. Wilimzig, C., Schneider, S., Schner, G.: The time course of saccadic decision making: dynamic field theory. Neural Netw. 19, 1059–1074 (2006)
10. Erlhagen, W., Schöner, G.: Dynamic field theory of movement preparation. Psychol. Rev. 109, 545–572 (2002)
11. Erlhagen, W., Mukovskiy, A., Bicho, E.: A dynamic model for action understanding and goal-directed imitation. Brain Res. 1083, 174–188 (2006)
12. Guo, Y., Chow, C.C.: Existence and stability of standing pulses in neural networks: I. existence. SIAM J. Appl. Dyn. Sys. 4, 217–248 (2005)
13. Taylor, J.G.: Neural bubble dynamics in two dimensions: foundations. Bioligal Cybernetics 80, 393–409 (1999)
14. Erlhagen, W., Bicho, E.: The dynamic neural field approach to cognitive robotics. J. Neural. Eng. 3, R36–R54 (2006)
15. Trappenberg, T.P.: Fundamentals of computational neuroscience. Oxford University Press, New York (2002)
16. Amari, S.: Dynamics of pattern formation in lateral-inhibition type neural fields. Biol. Cybern. 27, 77–87 (1977)
17. Cuijpers, R.H., van Schie, H.T., Koppen, M., Erlhagen, W., Bekkering, H.: Goals and means in action observation: a computational approach. Neural Netw. 19, 311–322 (2006)

Encoding and Retrieval in a CA1 Microcircuit Model of the Hippocampus

Vassilis Cutsuridis[1,*], Stuart Cobb[2], and Bruce P. Graham[1]

[1] Department of Computing Science and Mathematics, University of Stirling, Stirling, FK9 4LA, U.K.
{vcu,b.graham}@cs.stir.ac.uk

[2] Division of Neuroscience and Biomedical Systems, University of Glasgow, Glasgow, G12 8QQ, U.K.
s.cobb@bio.gla.ac.uk

Abstract. Recent years have witnessed a dramatic accumulation of knowledge about the morphological, physiological and molecular characteristics, as well as connectivity and synaptic properties of neurons in the mammalian hippocampus. Despite these advances, very little insight has been gained into the computational function of the different neuronal classes; in particular, the role of the various inhibitory interneurons in encoding and retrieval of information remains elusive. Mathematical and computational models of microcircuits play an instrumental role in exploring microcircuit functions and facilitate the dissection of operations performed by diverse inhibitory interneurons. A model of the CA1 microcircuitry is presented using biophysical representations of its major cell types: pyramidal, basket, axo-axonic, bistratified and oriens lacunosum-moleculare cells. Computer simulations explore the biophysical mechanisms by which encoding and retrieval of spatio-temporal input patterns are achieved by the CA1 microcircuitry. The model proposes functional roles for the different classes of inhibitory interneurons in the encoding and retrieval cycles.

Keywords: Hippocampus, CA1, microcircuit, computer model, pyramidal cell, basket cell, bistratified cell, axo-axonic cell, OLM cell, STDP.

1 Introduction

The hippocampus has been studied extensively, yielding a wealth of data on network architecture, cell types, the anatomy and membrane properties of pyramidal cells and interneurons, and synaptic plasticity [1]. It contains principal excitatory neurons (pyramidal (P) cells in CA3 and CA1 and granule cells (GC) in the DG) and a large variety of inhibitory interneurons [2]. Its basic functional role is hypothesized to be the intermediate-term storage of declarative memories [3].

Many computational models have been advanced over the years trying to understand how memories are stored and retrieved in the hippocampus [12], [14], [23], [21], [6], [5], [9].

* Corresponding author.

V. Kůrková et al. (Eds.): ICANN 2008, Part II, LNCS 5164, pp. 238–247, 2008.

A conceptual model of how GABAergic interneurons might provide the control structures necessary for phasing storage and recall in the hippocampus has been proposed recently [15]. Building on this idea, we construct a functional model of the CA1 microcircuit, including a number of different neuronal types (pyramidal (P) cells, basket (B) cells, bistratified (BS) cells, axo-axonic (AA) cells and oriens-laconosum-moleculare (OLM) cells) and their specific roles in storage and recall. The recall performance of the model is tested against different levels of pattern loading and input presentation period.

2 Materials and Methods

2.1 Model Architecture and Properties

Figure 1 illustrates the simulated microcircuit model of the CA1 network. The model consists of 100 pyramidal (P) cells, 2 basket (B) cells, 1 bistratified (BS) cell, 1 axo-axonic (AA; chandelier) cell and 18 oriens lacunosum moleculare (OLM) cells. The cell percentages matched those found in rat hippocampus [4]. All simulations were performed using NEURON [8] running on a PC under Windows XP. The morphology of each cell in the model was adapted from experimental studies [24], [25]. The biophysical properties of each cell were adapted from cell types reported in the literature [16], [17], [19], [18].

Pyramidal Cells. Each P cell was modeled as 15 anatomical compartments. Membrane properties included a calcium pump and buffering mechanism, a calcium activated mAHP potassium current, an LVA L-type Ca^{2+} current, an HVA L-type Ca^{2+} current, an MVA R-type Ca^{2+} current, an HVA T-type Ca^{2+} current, an h current, an HH current that includes both a sodium and a delayed rectifier current, a slow Ca^{2+} - dependent potassium current, a slow non-inactivating potassium channel with HH style kinetics and a K^+ A current [16],[17]. Less than 1% recurrent connections between pyramidal cells in the network was assumed [26].

Each pyramidal cell received nine somatic synaptic inhibition contacts from the population of basket cells [27], mid-dendritic excitation from CA3, distal apical excitation from the entorhinal cortex (EC), proximal excitation from other pyramidal cells in the network (recurrent collaterals) [1], eight axonic synaptic inhibition contacts from the population of chandelier cells [27], [28], six mid-dendritic synaptic inhibition contacts from the bistratified cells population [27] and two distal synaptic inhibition contacts from each OLM cell.

Axo-Axonic Cells. Each AA cell was modeled with 17 compartments. Membrane properties included a leak conductance, a sodium current, a fast delayed rectifier K^+ current, an A-type K^+ current, L- and N-type Ca^{2+} currents, a Ca^{2+} -dependent K^+ current and a Ca^{2+}- and voltage-dependent K^+ current [18]. No recurrent connections between AA cells were assumed [2].

Axo-axonic cells received excitatory inputs from the EC perforant path to their SLM dendrites and excitatory inputs from the CA3 Schaffer collateral to their SR dendrites. In addition, the axo-axonic cells received inputs from active

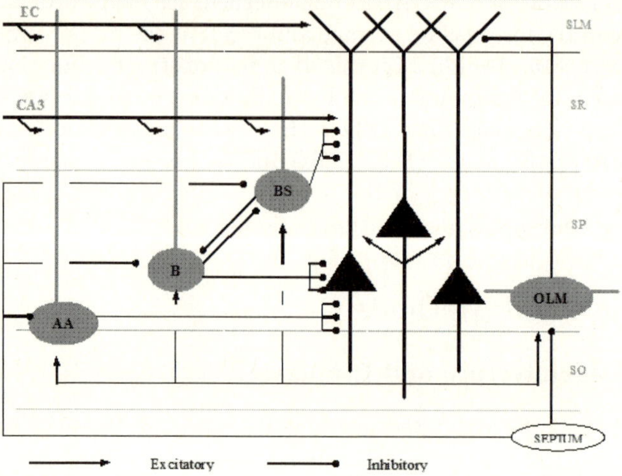

Fig. 1. CA1 microcircuit model of the hippocampus. EC: entorhinal cortex input; AA: axo-axonic cell; B: basket cell; BS: bistratified cell; OLM: oriens-lacunosum-moleculare cell; SO: stratum oriens; SP: stratum pyramidale; SR: stratum radiatum; SLM: stratum lacunosum-moleculare.

pyramidal cells in their SR medium and thick dendritic compartments as well as inhibitory input from the septum in their SO thick dendritic compartments [1].

Basket Cells. Each B cell was modeled with 17 compartments. Membrane properties included a leak conductance, a sodium current, a fast delayed rectifier K^+ current, an A-type K^+ current, L- and N-type Ca^{2+} currents, a Ca^{2+}-dependent K^+ current and a Ca^{2+}- and voltage-dependent K^+ current [18]. Recurrent connections between all B cells and between all B and BS cells in the network were assumed [2].

All B cells received excitatory connections from the EC to their distal SLM dendrites, from the CA3 Schaffer collaterals to their medium SR dendrites and from active pyramidal cells to their medium and thick SR dendritic compartments and inhibitory connections from neighboring B and BS cells in their soma and from the medial septum in their SO thick dendritic compartments.

Bistratified Cells. Each BS cell was modeled with 13-compartments. Membrane properties included the same ionic currents as the B and AA cells. Recurrent connectivity between all BS and between BS and B cells in the network was assumed [2]. All BS cells received excitatory connections from the CA3 Schaffer collaterals in their medium SR dendritic compartments and from the active pyramidal cells in their thick SO dendritic compartments and inhibitory connections from the medial septum in their thick SO dendritic compartments and from neighboring B and BS cells in their somas.

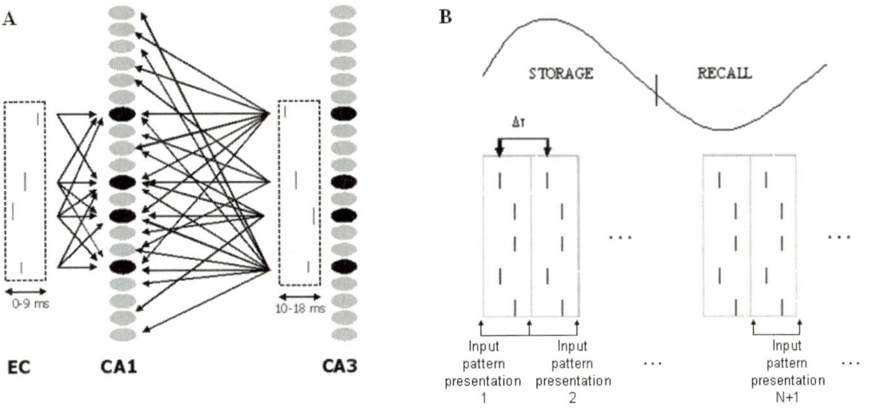

Fig. 2. (A) Model inputs to CA1 microcircuit. EC: entorhinal cortex input. The input arrives asynchronously in CA1 LM P cell dendrites (not shown) between 0-9 ms. A randomly selected subset of P cells receive the EC input. The CA3 input arrives asynchronously in CA1 SR P cell dendrites between 10-18ms (that is, 10ms after the EC input). All P cells non-selectively receive the CA3 input. (B) Input pattern presentation in the model. An input pattern was defined as the spatio-temporal sequence of asynchronously arriving spikes to P cells. The presentation of an input pattern is repeated every $\Delta\tau$ (5ms, 7ms, 8ms, 10ms, 11ms) continuously throughout the encoding and retrieval sub-cycles of the theta rhythm.

OLM Cells. Each OLM cell was modeled as a reduced 4-compartment model [19], which included a sodium (Na+) current, a delayed rectifier K^+ current, an A-type K^+ current and an h-current. No recurrent connections were assumed between OLM cells.

Each OLM cell received excitatory connections from the active pyramidal cells in their basal dendrites as well as inhibitory connections from the medial septum in their soma.

Model Inputs. Inputs to CA1 came from the medial septum (MS), entorhinal cortex (EC) and CA3 Schaffer collaterals. All P cells received the CA3 input, whereas a randomly selected subset of P cells received the EC layer III input (see figure 2A). All B, AA and BS in the network received the CA3 input, whereas only the AA and B cells received the EC input. The conduction latency of the EC-layer III input to CA1 LM dendrites is less than 9 ms (ranging between 5-8 ms), whereas the conduction latency of EC-layer II input to CA1 radiatum dendrites via the di/tri-synaptic path is greater than 9 ms (ranging between 12-18 ms) [13].

In the model, an *input pattern* was defined as the spatio-temporal sequence of asynchronously arriving spikes to corresponding P cells. The size of the input pattern was equal to the percentage of P cells that received the EC input. Both EC and CA3 input patterns were presented to P cell apical LM and medial SR dendrites respectively at various repeated time window shifts (i.e. each input was repeatedly presented with a period of 5 ms, 7 ms, 8 ms, 10 ms, or 11 ms) (see figure 2B). MS input provided GABA-A inhibition to all INs in the model (strongest to B and AA). MS input was phasic at theta rhythm and was on for 70 ms during the retrieval phase.

Synaptic Properties. In the model, AMPA, NMDA, GABA-A and GABA-B synapses were included. GABA-A were present in all strata, whereas GABA-B were present in medium and distal SR and SLM dendrites. AMPA synapses were present in strata LM (EC connections) and radiatum (CA3 connections), whereas NMDA were present only in stratum radiatum (CA3 connections).

Synaptic Plasticity. A local spike-timing dependent plasticity (STDP) learning rule was applied at medial SR AMPA synapses on P cells [22]. Pre-synaptic spike times were compared with the maximal postsynaptic voltage response at a synapse. If the interval is positive (a pre-synaptic spike arrives before the post-synaptic neuron response), then the synapse is potentiated by increasing the maximum AMPA conductance. If the interval is negative, the synapse is depressed by reducing the AMPA conductance.

Network Training and Testing. During encoding the maximum synaptic conductances of the SR AMPA synapses were allowed to change according to the learning rule explained above. During retrieval, the conductances from the last time window of input presentation were kept fixed throughout the entire retrieval cycle.

3 Results

In the model, we make two important assumptions, which are supported by experimental evidence: (1) Encoding and retrieval are two functionally independent sub-cycles of theta rhythm [7], and (2) During the storage sub-cycle the pyramidal cells that receive the EC input (i.e. the input pattern) do not fire action potentials [10], [11], [29] and hence the stored pattern does not "leak" out from the CA1.

3.1 Encoding Cycle

Previous modelling studies emphasized the role of a *feedforward association* of the incoming EC and CA3 inputs as the means of storing patterns in the CA1 P cells [31]. However, recent experimental evidence [13] has shown that in P cells of CA1 the conduction latency of the EC-layer III input to their LM dendrites is less than 9 ms (ranging between 5-8 ms), whereas the conduction latency of EC-layer II input to their radiatum dendrites via the di/tri-synaptic path is greater than 9 ms (ranging between 12-18 ms). That means that forward association of the EC- and CA3-inputs is *not* feasible, given that the information to be stored is contained in the coincident activity of cells in layers II and III of EC. Therefore, a different mechanism is required to associate the two inputs.

During the simulated storage cycle of the theta, we propose the following: an EC input pattern arrives to the apical SLM dendrites of the B, AA and P cells at time t_i, whereas an indirect CA3 input pattern via the di/trisynaptic loop arrives to the medium SR dendrites of the B, AA, BS and P cells at time $t_i+\delta t$ ($\delta t > 9$ms) [13]). In the B and AA cells, the EC input is strong enough to induce an action potential in their soma. Furthermore, the GABAergic cell population of the medial septum is minimally active and therefore transmits the least amount of inhibition to the CA1 inhibitory interneurons. Having the least amount of inhibition impinging on them, the CA1 inhibitory cells are free to do several things: First, the axo-axonic and basket cells exert tight inhibitory control on the axons and somas of the pyramidal cells, thus preventing them from firing during the storage cycle [10]. Second, the basket cells exert powerful inhibitory control to neighbouring basket cells and to bistratified cells, which prevents the later from firing during the storage cycle. As mentioned earlier, the bistratified cells are 180 degrees out-of-phase with the basket and axo-axonic cells and hence not active during the storage cycle [11].

The CA3 input to P cells provides the contextual information, whereas the EC input to P cells provides the sensory information, because place cells in CA1 were found to be maintained solely by direct input from EC and CA1 [20]. Since there is no topography in CA1 [26], during the storage cycle, 20% of the P cells in the network are randomly selected to receive the EC input pattern in their apical SLM dendrites (see figure 2). The summed postsynaptic potentials (PSP) generated in the SLM dendrites are attenuated on their way to the soma and axon [32], where they are finally "stopped" by the B and AA cell inhibition. Due to the strong B and AA cell inhibition on their soma and axon, non-specific hyperpolarizing h-activated cation channels are activated, which send a rebound back-propagating post-synaptic response (BPPR) towards the SR and SLM dendrites. In the model, to induce the BPPR, the conductance of the I_h current is increased 10-fold in the proximal SR dendrites ($g_{hprox} = 0.0005$ mS/cm^2) and 20-fold in the medium and distal SR dendrites ($g_{hmed,dist} = 0.001$ mS/cm^2) compared to the soma ($g_{hsoma} = 0.00005$ mS/cm^2) [30]. In contrast to the EC input, all P cells in the network are activated by the CA3 input in their medial SR dendrites. Careful timing between the incoming CA3 Schaffer collateral spike, the EC spike and the BPPR will induce potentiation (LTP) or depression (LTD) via a local STDP rule [22] applied in the medium SR dendrites of the pyramidal cells.

3.2 Retrieval Cycle

The retrieval cycle begins as the GABAergic cells of the septum approach maximum activity. Because of this septal input, the basket and axo-axonic cells are inhibited, releasing pyramidal cells, bistratified cells and OLM cells from inhibition. Pyramidal cells may now fire more easily, thus allowing previously learned patterns to be recalled. During the retrieval cycle, the CA3 Schaffer collateral input plays the role of a cueing mechanism: If the CA3 input excited a pyramidal cell during this time, any synapses that were strengthened during the storage cycle will be activated, recalling the memory. Because the CA3 input is directed to all P cells, which potentially activates unwanted P cells and hence spurious memories are recalled, the role of the bistratified cells is to ensure that these spurious cells will be silenced by broadcasting a non-specific inhibitory signal to all P cells in the network.

In our model during recall the entorhinal cortical input provides a weak background excitation to the CA1 region that aids the recall process, causing depolarized cells to fire. However, this excitation can potentially give rise to unwanted or similar memories. In our model, P cells after being released by the basket and axo-axonic cell inhibition excite the OLM cells. This excitation was assumed strong enough to overcome the OLM septal inhibition. In return, the OLM cells strongly inhibit the distal SLM dendrites of the P cells [2], where the direct entorhinal input arrives, thus preventing unwanted or similar memories from being recalled.

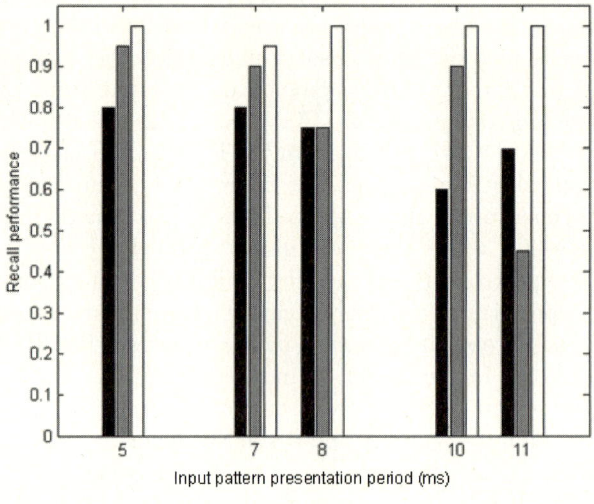

Fig. 3. Normalized recall performance during the 'many-trials' learning paradigm as a function of input pattern loading (10%, 50% and 75%) and input pattern presentation period (a pattern is repeatedly presented every 5ms, 7ms, 8ms, 10ms, 11ms). White bars: 75% pattern loading; Grey bars: 50% pattern loading; Black bars: 10% pattern loading.

3.3 Recall Performance

Twenty percent of all P cells in the network received the EC input pattern (*EC input pattern:* a group of twenty spikes arriving asynchronously within a time window of 0-9ms (see figure 2B)) in their SLM dendrites, whereas all P cells received the CA3 input pattern (*CA3 input pattern:* the same group of twenty asynchronous spikes delayed by at least 9 ms) arrive in their SR dendrites. As we said previously, both the EC and CA3 input patterns were continuously presented with a period $\Delta\tau$ ($\Delta\tau$ = 5-11ms) throughout the encoding and retrieval cycles of the theta rhythm. To estimate the recall performance of our network, we counted the fraction of cells belonging to the stored pattern that were active during the retrieval cycle.

Figure 3 depicts the model's recall performance for a particular input pattern. Different time shifts between pattern presentations during encoding and retrieval were tried ($\Delta\tau$ = every 5 ms, 7 ms, 8 ms, 10 ms or 11 ms), and different levels of cue pattern loading (10%, 50% and 75% of EC input pattern vector is presented to P cells) were used during recall. When 75% of the EC input pattern was presented during recall (i.e. 75% pattern loading), the recall performance was nearly perfect (100%) regardless of input presentation period with the exception at 7 ms (95%). At 50% and 10% pattern loading, the recall performance dropped by 5% and 20% respectively when the input presentation period was 5 ms. At larger input presentation periods, the recall performance degraded progressively for both 50% and 10% pattern loadings reaching a minimum of 45% and 70% respectively at 11 ms.

4 Conclusion

A detailed model of the CA1 microcircuit has been presented. The model proposes functional roles for a variety of CA1 cells in the encoding and retrieval of memories in the hippocampus. The performance of the model is tested against different levels of pattern cueing during recall and different input pattern presentation time shifts during learning. These initial tests indicate that this circuitry can successfully store and recall patterns of information within a theta cycle. The quality of storage does depend on the temporal pattern presentation sequence, as recall performance drops when the time between presentations is increased and only a small cue is used during recall. Much work remains to be done to further explore the temporal constraints on this process and to assess more widely the capacity of this network to operate as an associative memory.

Acknowledgement. This work was funded by an EPSRC project grant to B. Graham and S. Cobb.

References

[1] Andersen, P., Morris, R., Amaral, D., Bliss, T., O'Keefe, J.: The hippocampus book. Oxford University press, Oxford (2007)
[2] Freund, T.F., Buzsaki, G.: Interneurons of the hippocampus. Hippocampus 6, 347–470 (1996)
[3] Eichenbaum, H., Dunchenko, P., Wood, E., Shapiro, M., Tanila, H.: The hippocampus, memory and place cells: is it spatial memory or a memory of space? Neuron 23(2), 209–226 (1999)

[4] Baude, A., Bleasdale, C., Dalezios, Y., Somogyi, P., Klausberger, T.: Immunoreactivity for the GABAA receptor $\alpha 1$ subunit, somatostatin and connexion36 distinguishes axoaxonic, basket and bistratified interneurons of the rat hippocampus. Cerebral Cortex 17(9), 2094–2107 (2007)
[5] Cutsuridis, V., Hunter, R., Cobb, S., Graham, B.P.: Storage and Recall in the CA1 Microcircuit of the Hippocampus: A Biophysical Model. In: Sixteenth Annual Computational Neuroscience Meeting CNS 2007, Toronto, Canada, July 8th - 12th, 2007, vol. 8 (Suppl 2), p. 33. BMC Neuroscience (2007)
[6] Graham, B.P., Cutsuridis, V.: Dynamical Information Processing in the CA1 Microcircuit of the Hippocampus. In: Computational Modeling in behavioral neuroscience: Closing the gap between neurophysiology and behavior. Psychology Press, Taylor and Francis Group, London (December 2008) (to be published)
[7] Hasselmo, M., Bodelon, C., Wyble, B.: A proposed function of the hippocampal theta rhythm: separate phases of encoding and retrieval of prior learning. Neural. Comput. 14, 793–817 (2002)
[8] Hines, M.L., Carnevale, T.: The NEURON simulation environment. Neural. Comput. 9, 1179–1209 (1997)
[9] Hunter, R., Cutsuridis, V., Cobb, S., Graham, B.: Improving Associative Memory in a Model Network of Two-Compartment Spiking Neurons. In: Fourth Annual Scottish Neuroscience Group Meeting, University of Edinburgh (August 31, 2007)
[10] Klausberger, T., Magill, P.J., Marton, L.F., David, J., Roberts, B., Cobden, P.M., Buzsaki, G., Somogyi, P.: Brain-state- and cell-type-specific firing of hippocampal interneurons in vivo. Nature 421, 844–848 (2003)
[11] Klausberger, T., Marton, L.F., Baude, A., Roberts, J.D., Magill, P.J., Somogyi, P.: Spike timing of dendrite-targeting bistratified cells during hippocampal network oscillations in vivo. Nat. Neurosci. 7(1), 41–47 (2004)
[12] Kunec, S., Hasselmo, M.E., Kopell, N.: Encoding and retrieval in the CA3 region of the hippocampus: a model of theta-phase separation. J. Neurophysiol. 94(1), 70–82 (2005)
[13] Leung, L.S., Roth, L., Canning, K.J.: Entorhinal inputs to hippocampal CA1 and dentate gyrus in the rat: a current-source-density study. J. Neurophys. 73(6), 2392–2403 (1995)
[14] Menschik, E.D., Finkel, L.H.: Neuromodulatory control of hippocampal function: towards a model of Alzheimer's disease. Artificial Intelligence in Medicine 13, 99–121 (1998)
[15] Paulsen, O., Moser, E.: A model of hippocampal memory encoding and retrieval: GABAergic control of synaptic plasticity. TINS 21, 273–279 (1998)
[16] Poirazzi, P., Brannon, T., Mel, B.W.: Arithmetic of subthreshold synaptic summation in a model CA1 pyramidal cell. Neuron 37, 977–987 (2003a)
[17] Poirazzi, P., Brannon, T., Mel, B.W.: Pyramidal neuron as a 2-layer neural network. Neuron 37, 989–999 (2003)
[18] Santhakumar, V., Aradi, I., Soltetz, I.: Role of mossy fiber sprouting and mossy cell loss in hyperexcitability: a network model of the dentate gyrus incorporating cells types and axonal topography. J. Neurophysiol. 93, 437–453 (2005)
[19] Saraga, F., Wu, C.P., Zhang, L., Skinner, F.K.: Active dendrites and spike propagation in multicompartmental models of oriens-lacunosum/moleculare hippocampal interneurons. J. Physiol. 552(3), 673–689 (2003)
[20] Brun, V.H., Otnass, M.K., Molden, S., Steffenach, H.A., Witter, M.P., Moser, M.B., Moser, E.I.: Place cells and place recognition maintained by direct entorhinal-hippocampal circuitry. Science 296, 2243–2246 (2002)

[21] Sommer, F.T., Wennekers, T.: Associative memory in networks of spiking neurons. Neural Networks 14, 825–834 (2001)
[22] Song, S., Miller, K., Abbott, L.: Competitive "hebbian" learning through spike-timing-dependent synaptic plasticity. Nat. Neurosci. 3, 919–926 (2000)
[23] Wallestein, G.V., Hasselmo, M.E.: GABArgic modulation of hippocampal population activity: sequence learning, place field development and the phase precession effect. J. Neurophys. 78, 393–408 (1997)
[24] Megias, M., Emri, Z.S., Freund, T.F., Gulyas, A.I.: Total number and distribution of inhibitory and excitatory synapses on hippocampal CA1 pyramidal cells. Neuroscience 102(3), 527–540 (2001)
[25] Gulyas, A.I., Megias, M., Emri, Z., Freund, T.F.: Total number and ration of excitatory and inhibitory synapses converging onto single interneurons of different types in the CA1 area of the rat hippocampus. J. Neurosci. 19(22), 10082–10097 (1999)
[26] Amaral, D., Lavenex, P.: Hippocampal neuroanatomy. In: Andersen, P., Morris, R., Amaral, D., Bliss, T., O'Keefe, J. (eds.) The Hippocampus Book, pp. 37–114. Oxford University press, Oxford (2007)
[27] Buhl, E.H., Halasy, K., Somogyi, P.: Diverse sources of hippocampal unitary inhibitory postsynaptic potentials and the number of synaptic release sites. Nature 368, 823–828 (1994a)
[28] Buhl, E.H., Han, Z.S., Lorinczi, Z., Stezhka, V.V., Kapnup, S.V., Somogyi, P.: Physiological properties of anatomically identified axo-axonic cells in the rat hippocampus. J. Neurophys. 71(4), 1289–1307 (1994b)
[29] Somogyi, P., Klausberger, T.: Defined types of cortical interneurons structure space and spike timing in the hippocampus. J. Physiol. 562(1), 9–26 (2005)
[30] Migliore, M., Hoffman, D.A., Magee, J.C., Johnston, D.: Role of an A-type K^+ conductance in the back-propagation of action potentials in the dendrites of hippocampal pyramidal neurons. J. Comp. Neurosci. 7, 5–15 (1999)
[31] O'Reilly, R.C., McClelland, J.L.: Hippocampal conjunctive encoding, storage, and recall: avoiding a trade-off. Hippocampus 4(6), 661–682 (1994)
[32] Stuart, G., Spruston, N.: Determinants of voltage attenuation in neocortical pyramidal neuron dendrites. J. Neurosci. 18(10), 3501–3510 (1998)
[33] Bi, G.Q., Poo, M.M.: Synaptic modifications in cultured hippocampal neurons: dependence on spike timing, synaptic strength and postsynaptic cell type. J. Neurosci. 18, 10464–10472 (1998)

A Bio-inspired Architecture of an Active Visual Search Model

Vassilis Cutsuridis

Department of Computing Science and Mathematics, University of Stirling, Stirling, FK9 4LA, U.K.
vcu@cs.stir.ac.uk

Abstract. A novel brain inspired cognitive system architecture of an active visual search model is presented. The model is multi-modular consisting of spatial and object visual processing, attention, reinforcement learning, motor plan and motor execution modules. The novelty of the model lies on its decision making mechanisms. In contrast to previous models, decisions are made from the interplay of a winner-take-all mechanism in the spatial, object and motor salient maps between the resonated by top-down attention and bottom-up visual feature extraction and salient map formation selectively tuned by a reinforcement signal spatial, object and motor representations, and a reset mechanism due to inhibitory feedback input from the motor execution module to all other modules. The reset mechanism due to feedback inhibitory signals from the motor execution module to all other modules suppresses the last attended location from the saliency map and allows for the next gaze to be executed.

Keywords: Visual search, cognitive system, dopamine, saliency, ART, decision making, attention, perception, action, reinforcement learning.

1 Introduction

Visual search is a type of perceptual task requiring attention. Visual search involves an active scan of the visual environment for a particular object or feature (the target) among other objects or features (the distracters). Visual search can take place either with (active visual search) or without (passive visual search) eye movements.

Attention has been described as the control system in prefrontal cortex (PFC) whose role is to generate a top-down signal which will amplify specific target (spatial and/or object) representations in the posterior dorsal and ventral cortex, while at the same time will inhibit those of distracters [26]. Many computational theories of visual attention have been proposed over the years [13,14,15,16,17,18,19,20,21,22]. Some of these models [14,21] emphasized the formation of a bottom-up saliency-map that biases attention. According to these models scanpaths (sequences of saccadic eye movements) are generated according to the value of saliency in the map. That means that saccades are generated first

V. Kůrková et al. (Eds.): ICANN 2008, Part II, LNCS 5164, pp. 248–257, 2008.

towards the most salient object in the map, next towards the second most salient object and so forth. On the other hand, other models emphasized the need of the interplay between a bottom-up visual feature extraction module (the "what" and "where" pathways) and a top-down attention selective module (the PFC) to drive attention to specific regions-of-interest (ROIs) [13,17,18,20,22]. However, most of these studies failed to show the mechanisms by which the PFC attentive processes are recruited in order to guide attention in posterior and lower-level cortical areas.

The goal of the present study is to present a biologically plausible cognitive system architecture of active visual search model. The model comprises of many modules with specific as well as distributed functions and it is heavily supported by neuroscientific experimental evidence. Its novelty lies on its proposed distributed decision making mechanisms, whose functions rely on the coordinated actions of its visual, attention, reinforcement teaching, motor plan and motor execution modules.

2 Proposed Architecture

The proposed architecture of the active visual search model (see Figure 1) is multi-modular, consisting of a visual processing module, an attention module, a

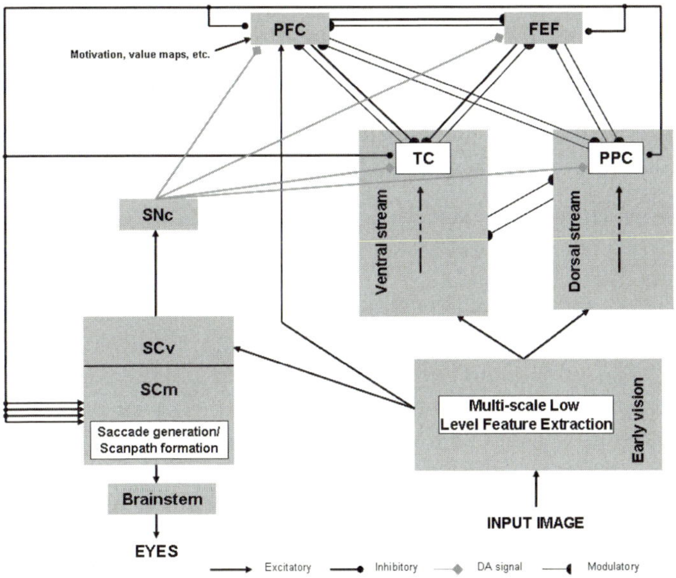

Fig. 1. Proposed multi-modular and their corresponding brain areas active visual search model. PFC: prefrontal cortex; FEF: frontal eye fields; PPC: posterior parietal cortex; TC: temporal cortex; SCv: visual superior colliculus; SCm: motor superior colliculus; SNc: substantia nigra pars compacta; DA: dopamine. See text for the corresponding to the model's modular functionality of each brain area.

decision-making module, a reinforcement learning module, a motor plan module and a motor execution module. In the following section, I will describe each module, its functionality and will provide experimental evidence for its validity. Finally, I will describe the stages of information processing from the presentation of the input image till the execution of an eye movement (see Figure 2).

2.1 Modules

Spatial and object visual processing module. The visual processing module up to the formation of global saliency maps in both the dorsal (space) and ventral (object) streams is the same as in [13,21]. Its functionality is to decompose an input image through several pre-attentive multi-scale feature detection mechanisms (sensitive to color, intensity and orientation) found in retina, lateral geniculate nucleus (LGN) of the thalamus and primary visual cortical area (V1) and which operate in parallel across the entire visual scene, into two streams of processing, that is the dorsal for space and the ventral for object. Neurons in the feature maps in both streams then encode the spatial and object contrast in each of those feature channels. Neurons in each feature map spatially compete for salience, through long-range connections that extend far beyond the spatial range of the classical receptive field of each neuron. After competition, the feature maps in each stream are combined into a global saliency map, which topographically encodes for saliency irrespective of the feature channel in which stimuli appeared salient [13,14]. In the model, the global spatial saliency map is assumed to reside in the posterior parietal cortex (PPC), whereas the global object saliency map resides in the ventral temporal cortex (TC). The speed of visual information processing from the early multi-scale feature extraction in the retina till the formation of global saliency maps in the dorsal PPC and ventral TC is 80-100ms [[27] and references therein].

Attention module. The attention module is represented by prefrontal cortex (PFC) cells. It receive a direct input visual signal from the early stages of visual processing (retina, LGN, V1) as well as from the FEF (motor plans), PPC (spatial representations), TC (object representations) and other brain areas (motivation (medial PFC), value representations (orbito-frontal cortex (OFC); medial PFC and OFC neuronal responses are not modeled in this study). Its role is to send feedback signals to every stage of the visual processing module, which will amplify specific neuronal responses throughout the visual hierarchy [26] as well as to the selectively tuned via the reinforcement learning DA signals target (spatial and object) and motor plan representations in the PPC, TC, and frontal cortices (FC), while at the same time will inhibit those of distracters.

Reinforcement learning module. At the same time and in a parallel manner, the retinal multi-scale low level features propagate to the upper layers of the superior colliculus (SC), which in turn provide the sensory input to the substantia nigra pars compacta (SNc) and ventral tegmental area (VTA). Recent neuroanatomical evidence has reported a direct tectonigral projection connecting the deep layers of the superior colliculus to the SNc across several species

[7,10,11]. This evidence is confirmed by neurophysiological recordings in alive animals [8,23].

The SNc and VTA comprise the reinforcement learning module of the model. Both SNc and VTA contain the brain's dopaminergic (DA) neurons, which have been implicated in signaling reward prediction errors used to select actions that will maximize the future acquisition of reward [9] as well as the progressive movement deterioration of patients suffering from Parkinson's disease [1,2,5]]. The conduction latency of the signal from the retina to SC and from there to SNc is 70-100ms, whereas the duration of the DA phasic response is approximately 100ms [12].

The SC activated SNc DA neurons broadcast reinforcement traching signals to neurons in prefrontal cortex (PFC), frontal eye fields (FEF), posterior parietal (PPC) and temporal cortices (TC), but not to visual cortices [[1] and references therein]. An extensive review of the dopaminergic innervation of the cerebral cortex has been recently published by [1]). Briefly, the source of the dopaminergic (DA) fibers in cerebral cortex were found to be the neurons of the substantia nigra pars compacta (SNc) and the ventral tegmental area (VTA). DA afferents are densest in the anterior cingulate (area 24) and the motor areas (areas 4, 6, and SMA), where they display a tri-laminar pattern of distribution, predominating in layers I, IIIa, and V-VI. In the granular prefrontal (areas 46, 9, 10, 11, 12), parietal (areas 1, 2, 3, 5, 7), temporal (areas 21, 22), and posterior cingulate (area 23) cortices, DA afferents are less dense and show a bilaminar pattern of distribution in the depth of layers I, and V-VI. The lowest density is in area 17, where the DA afferents are mostly restricted to layer I.

The role of the DA broadcasting signals is to act as the vigilant parameter of an ART network, which reinforce via selective tuning [1] the relevant according to previously learned experiences to the visual scene responses of cells in the areas they target. All other cells that don't receive or receive reduced DA signals "perish" as their signal-to-noise ratio responses are extremely low (see Figure 7 in [1]).

Motor plan module. In this module, the global spatial and object saliency maps formed in the PPC and TC respectively are transformed in their corresponding global saliency motor plan maps. The motor saliency plan module is assumed to reside in the frontal eye fields (FEF) of the frontal lobes [28]. Reciprocal connections between the PPC, TC and FEF ensure the sensorimotor groupings of the spatial and object representations with their corresponding motor plans [[29] and references therein].

Decision module. The decision to where to gaze next is determined by the coordinated actions of the Attention, Reinforcement Learning, Visual Processing, Motor Plan and Motor Execution modules in the model. More specifically, bottom-up, top-down and reset mechanisms represented by the complex and intricate feedforward, feedback and horizontal circuits of PFC, PPC, TC, FEF, motor SC and the brainstem are making decisions. Adaptive reciprocal connections between (1) PFC and PPC, (2) PFC and TC, (3) PFC and FEF, (4) FEF and PPC, (5) FEF and TC, and (6) PPC and TC operate exactly as the comparison and recognition fields of an ART (Adaptive Resonance Theory) system [24,25].

Briefly, an ART system consists of a comparison field and a recognition field composed of neuronal populations, a vigilance parameter, and a reset module. The vigilance parameter has considerable influence on the system: higher vigilance produces highly detailed memories (many, fine-grained categories), while lower vigilance results in more general memories (fewer, more-general categories). The comparison field takes an input vector (a one-dimensional array of values) and transfers it to its best match in the recognition field. Its best match is the single neuronal population whose set of weights (weight vector) most closely matches the input vector. Each recognition field neuronal population outputs a negative signal (proportional to that neuron's quality of match to the input vector) to each of the other recognition field neuronal populations and inhibits their output accordingly. In this way the recognition field exhibits lateral inhibition, allowing each neuronal population in it to represent a category to which input vectors are classified. After the input vector is classified, the reset module compares the strength of the recognition match to the vigilance parameter. If the vigilance threshold is met, training commences. Otherwise, if the match level does not meet the vigilance parameter, the firing recognition neuronal population is inhibited until a new input vector is applied; training commences only upon completion of a search procedure. In the search procedure, recognition neuronal populations are disabled one by one by the reset function until the vigilance parameter is satisfied by a recognition match. If no committed recognition neuronal population's match meets the vigilance threshold, then an uncommitted neuronal population is committed and adjusted towards matching the input vector.

In this model, as I mentioned before, the ART's vigilance parameter is represented by the broadcasted DA reinforcement teaching signals. High and intermediate levels of DA ensure the formation of fine and coarse categories respectively, whereas low values of DA ensure that non-relevant representations and plans perish.

The reciprocal connections between (1) PFC, PPC and TC, and (2) PFC and FEF allow for the amplification of the spatial, object and motor representations pertinent to the given context and the suppression of the irrelevant ones, whereas the reciprocal connections between the FEF, PPC and TC ensure for their groupings.

Decisions in the model are made from the interplay of a winner-take-all mechanism in the spatial, object and motor salient maps between the selectively tuned by DA and resonated spatial, object and motor representations [3,4,6] and a reset mechanism due to a feedback signal from the SC to FEF [30], PFC, PPC, TC and SNc [12] analogous to the IOR in [13], which suppresses the last attended location and executed motor plan from their saliency maps and allows for the next salient motor plan to be executed.

Motor execution module. The motor plan that has won the winner-take-all competition in the FEF field propagates to the intermediate and deep layers of superior colliculus (SC) and the brainstem (motor execution module), where the final motor command is formed. This final motor command instructs the eyes about the direction, amplitude and velocity of movement. Once, the motor plan arrives in the SC, inhibitory feedback signals propagate from the SC to PFC, FEF, PPC and TC in order to reset these fields and set the stage for the salient point to gaze to. The speed of processing from the input image presentation till the generation of an eye movement is approximately 200-220ms [3].

2.2 Information Processing

Once an input image is presented three parallel and equally fast processing modes of actions are initiated. In the first mode of action (visual processing), pre-attentive multi-scale feature detection and extraction mechanisms sensitive to color, intensity and orientation operating in parallel at the level of the retina, LGN and V1 start to work. From the level of V1 and on the features are separated into two streams: the dorsal for space processing and the ventral for object processing. At the end levels of the visual hierarchy, the PPC and TC lie, where global saliency maps for space and object are formed. In the second mode of action (reinforcement learning), the retinal signal activates the phasic reinforcement teaching (dopamine) signals via the visual layers of the SC. In turn, the phasic DA teaching signals will be broadcasted to the whole cortex (PFC, FEF, PPC and TC) and will selective tune the responses of different neuronal populations in these areas according to previous similar acquired experiences. In the third mode of action (attention), the retinal signal will travel a long distance to PFC, where will activate the recognition neuronal populations. The recognition neuronal populations will send/receive top-down/bottom-up feedback/feedforward signals to/from the spatial, object and motor saliency maps of the PPC, TC and FEF. All three modes of action take the same amount of time (approximately 100ms) [12,27].

In the next step, the spatial and object salient maps will go through a sensory-motor transformation to generate their corresponding motor salient maps at the FEF level. Reciprocal connections between the PPC, TC and FEF will bind the perceptual and motor salient maps together. While this transformation and grouping is taking place, attentional and reinforcing teaching signals from the PFC and SNc respectively will amplify/selectively tune the neuronal responses at the PFC, PPC, TC and FEF levels. A winner-take-all mechanism in these fields will select the most salient and resonated spatial, object and motor plan representation. The selected motor plan will then be forwarded to the motor execution areas (SC and brainstem) where the final motor command will be formed and the eye movement will be generated. The speed of processing from the start of the attentive resonance, selective tuning and motor plan formation, selection and execution takes another approximately 100-120ms (a total of approximately 200-220ms from input image presentation to eye movement execution) [3].

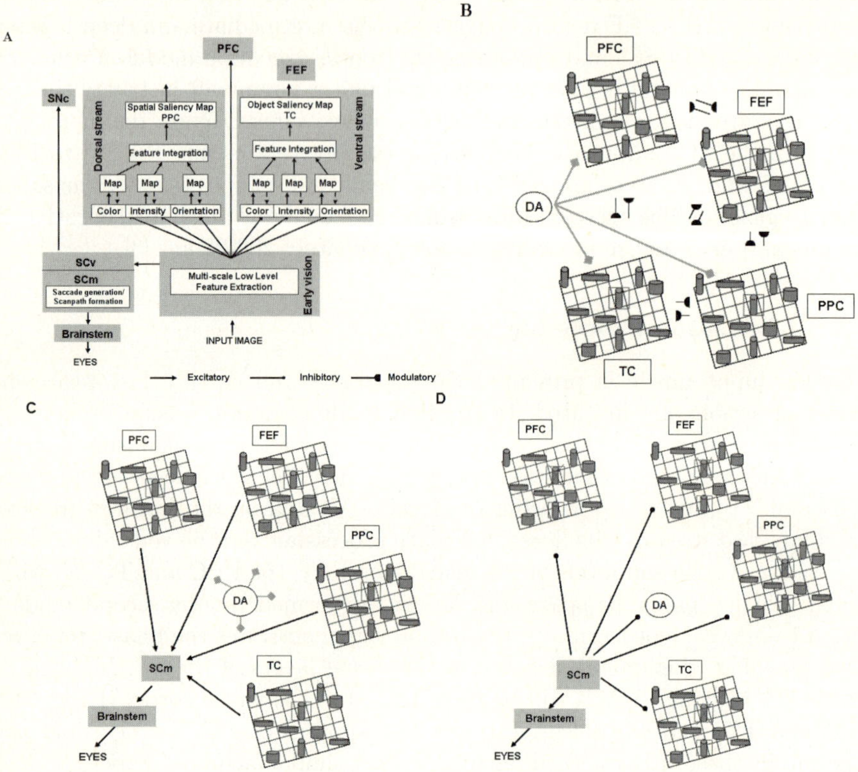

Fig. 2. Information processing stages of the active visual search model. (A) Visual processing stage. Once an input image is presented three parallel and equally fast processing pathways get activated: (1) Visual hierarchy pathways till the level of PPC (space) and TC (object), (2) sensory activated by the SCv SNc (dopamine) system, and (3) direct visual input to PFC. (B) DA broadcasting teaching signals to PFC, PPC, TC and FEF. Different neuronal populations receive different levels of DA. High and intermediate DA values result in "sharp tuned" neuronal responses, whereas low DA values result in "broadly tuned" neuronal responses. Neuronal responses are depicted by gray-colored towers in each brain area. The height of each tower represents the neuronal amplitude activation, whereas the width of each tower represents the degree of tuning. (C) Feedforward activation of the SCm by FEF, PFC, PPC and TC. Red square surrounding the response of a neuronal population represents the winner salient and resonated according to some value of vigilance (DA signal) representation in each brain area. (D) Reset mechanism by feedback inhibitory projections from the SCm to SNc, FEF, PFC, PPC and TC. Reset mechanism prevents previously selected representation (red crossed square) and allows all other resonated neuronal population responses to compete each other for selection. Bottom tower surrounded by red square represents the winner salient and resonated representation. PFC: prefrontal cortex; PPC: posterior parietal cortex; TC: temporal cortex; FEF: frontal eye fields; DA: dopamine; SC: superior colliculus; SCv: visual superior colliculus; SCm: motor superior colliculus; SNc: substantia nigra pars compacta.

Coincidently, Redgrave and Gurney [12] recently reported that the duration of the phasic DA signal (reinforcement teaching signal in this model) is 100ms and it precedes the first eye movement response. That means that the model's assumption about a co-active reinforcing teaching signal with the resonated attention and motor plan selection is valid. All these mechanisms are reset by a feedback excitatory signal from the SC (motor execution module) to the inhibitory neurons of the FEF, PFC, PPC, TC and SNc (all other model modules), which in turn inhibit and hence prevent the previously selected targets, objects and plans from being selected again (see Fig. 2D).

3 Comparison with Other Models and Discussion

A very influential model of visual attention has been put forward by Itti and Koch [13,21]. The model postulated that the decision to where to gaze next is determined by the interplay between a bottom-up winner-take-all network in a saliency map, which detected the point of highest saliency at any given time, and the inhibition-of-return (IOR), which suppressed the last attended location from the saliency map, so that attention could focus onto the next most salient location. However, top-down attentional mechanisms were not explicitly modeled in their model.

A biologically plausible computational model for solving the visual binding problem was developed over the years by John Tsotsos and his colleagues [20,31,32,33,34]. Their "Selective Tuning" model relied on the reentrant connections in the primate brain to recover spatial information and thus to allow features represented in a unitary conscious percept. However, the model fails to show the neural mechanisms by which attention, recognition and grouping work together. An engineering control approach to attention was applied in the Corollary Discharge of Attention Movement (CODAM) model of Taylor and colleagues [18,35,36,37,38]. Their model is multi-modular and has successfully simulated a wide range of visual, working memory, attention and higher level cognitive phenomena.

My model presented herein utilizes similar features with above mentioned models and extends them. Its visual processing module operates the same way as in the Itti and Koch model [13,21]. In contrast to their model, it consists also of an attentional module, whose function is through its re-entrant connections between the PFC, PPC, TC and FEF to amplify the relevant targets, objects and plans to the given context. Re-entrant connections also exist, as in the Selective Tuning model [20], between FEF, PPC and TC whose function is to bind together to a single percept the visuo-motor salient, attentionally resonated and selectively tuned by reinforcement teaching signals representations. The novelty of my model lies on the newly discovered sensory driven DA phasic signal [12], which operates as a reinforcement teaching signal to the neural responses of the PFC, PPC, TC and FEF. The scalar value of this reinforcing signal (vigilant parameter) "determines" by selective tuning [1,2,3] the degree of participation of the various resonated spatial and object salient representations to the formation, selection and execution of motor plans.

References

1. Cutsuridis, V., Perantonis, S.: A Neural Network Model of Parkinson's Disease Bradykinesia. Neural Networks 19(4), 354–374 (2006)
2. Cutsuridis, V.: Does Reduced Spinal Reciprocal Inhibition Lead to Co-contraction of Antagonist Motor Units? A Modeling Study. Int. J. Neur. Syst. 17(4), 319–327 (2007)
3. Cutsuridis, V., Smyrnis, N., Evdokimidis, I., Perantonis, S.: A Neural Network Model of Decision Making in an Antisaccade Task by the Superior Colliculus. Neural Networks 20(6), 690–704 (2007b)
4. Cutsuridis, V., Kahramanoglou, I., Smyrnis, N., Evdokimidis, I., Perantonis, S.: A Neural Variable Integrator Model of Decision Making in an Antisaccade Task. Neurocomputing 70(7-9), 1390–1402 (2007a)
5. Cutsuridis, V.: Neural Model of Dopaminergic Control of Arm Movements in Parkinson's Disease Bradykinesia. In: Kollias, S.D., Stafylopatis, A., Duch, W., Oja, E. (eds.) ICANN 2006. LNCS, vol. 4131, pp. 583–591. Springer, Heidelberg (2006)
6. Cutsuridis, V., Kahramanoglou, I., Perantonis, S., Evdokimidis, I., Smyrnis, N.: A Biophysical Model of Decision Making in an Antisaccade Task Through Variable Climbing Activity. In: Duch, W., et al. (eds.) ICANN 2005. LNCS, vol. 3695, pp. 205–210. Springer, Berlin (2005)
7. Comoli, E., Coizet, V., Boyes, J., Bolam, J.P., Canteras, N.S., Quirk, R.H., Overton, P.G., Redgrave, P.: A Direct Projection from the Superior Colliculus to Substantia Nigra for Detecting Salient Visual Events. Nat. Neurosci. 6(9), 974–980 (2003)
8. Coizet, V., Comoli, E., Westby, G.W., Redgrave, P.: Phasic Activation of Substantia Nigra and the Ventral Tegmental Area by Chemical Stimulation of the Superior Colliculus: An Electrophysiological Investigation in the Rat. Eur. J. Neurosci. 17(1), 28–40 (2003)
9. Schultz, W.: Predictive Reward Signal of Dopamine Neurons. J. Neurophysiol. 80, 1–27 (1998)
10. McHaffie, J.G., Jiang, H., May, P.J., Coizet, V., Overton, P.G., Stein, B.E., Redgrave, P.: A Direct Projection from Superior Colliculus to Substantia Nigra Pars Compacta in the Cat. Neurosci. 138(1), 221–234 (2006)
11. May, P.J., et al.: Projections from the Superior Colliculus to Substantia Nigra Pars Compacta in a Primate. Soc. Neurosci. Abstr. 450(2) (2005)
12. Redgrave, P., Gurney, K.: The Short Latency Dopamine Signal: A Role in Discovering Novel Actions. Nat. Neurosci. 7, 967–975 (2006)
13. Itti, L., Koch, C.: Computational Modelling of Visual Attention. Nat. Neurosci. 2, 194–203 (2001)
14. Koch, C., Ullman, S.: Shifts in Selective Visual Attention: Towards the Underlying Neural Circuitry. Hum. Neurobiol. 4, 219–227 (1995)
15. Didday, R.L., Arbib, M.A.: Eye Movements and Visual Perception: A Two Visual System Model. Int. J. Man-Machine Studies 7, 547–569 (1975)
16. Lee, D.K., Itti, L., Koch, C., Braun, J.: Attention Activates Winner-Take-All Competition Among Visual Filters. Nat. Neurosci. 2, 375–381 (1999)

17. Deco, G., Rolls, E.T.: Attention, Short-Memory and Action Selection: A Unifying Theory. Prog. Neurobiol. 76, 236–256 (2005)
18. Korsten, N.J.H., Fragopanagos, N., Hartley, M., Taylor, N., Taylor, J.G.: Attention as a Controller. Neural Networks 19, 1408–1421 (2006)
19. Desimone, R., Duncan, J.: Neural Mechanisms of Selective Visual Attention. Ann. Rev. Neurosci. 18, 193–222 (1995)
20. Tsotsos, J.K., Culhane, S., Wai, W., Lai, Y., Davis, N., Nuflo, F.: Modeling Visual Attention via Selective Tuning. Art. Intel. 78(1-2), 507–547 (1995)
21. Itti, L., Koch, C.: A Saliency Based Search Mechanism for Overt and Covert Shifts of Visual Attention. Vision Res. 40, 1489–1506 (2000)
22. Grossberg, S., Raisada, R.D.: Contrast-Sensitive Perceptual Grouping and Object-Based Attention in the Laminar Circuits of Primary Visual Cortex. Vision Res. 40, 1413–1432 (2000)
23. Dommett, E., Coizet, V., Blaha, C.D., Martindale, J., Lefebre, V., Walton, N., Mayhew, J.E., Overton, P.G., Redgrave, P.: How Visual Stimuli Activate Dopaminergic Neurons at Short Latency. Science 307(5714), 1476–9 (2005)
24. Grossberg, S.: Competitive Learning: From Interactive Activation to Adaptive Resonance. Cogn. Sci. 11, 23–63 (1987)
25. Carpenter, G.A., Grossberg, S.: Adaptive Resonance Theory. In: Arbib, M.A. (ed.) The Handbook of Brain Theory and Neural Networks, 2nd edn., pp. 87–90. MIT Press, Cambridge (2003)
26. Reynolds, J.H., Desimone, R.: The Role of Neural Mechanisms of Attention in Solving the Binding Problem. Neuron 24(1), 19–29 (1999)
27. Thrope, S., Fize, D., Marlot, C.: Speed of Processing in the Human Visual System. Nature 381(6582), 520–522 (1996)
28. Thompson, K.G., Bichot, N.P.: A Visual Saliency Map in the Primate Frontal Eye Field. Prog. Brain Res. 147, 251–262 (2005)
29. Hamker, F.H.: The Re-entry hypothesis: The Putative Interaction of the Frontal Eye Field, Ventrolateral Prefrontal Cortex and Areas V4, IT of Attention and Eye Movement. Cereb. Cortex 15, 431–447 (2005)
30. Brown, J.W., Bullock, D., Grossberg, S.: How Laminar Frontal Cortex and Basal Ganglia Circuits Interact to Control Planned and Reactive Saccades. Neural Networks 17, 471–510 (2004)
31. Tsotsos, J.K., Culhane, S.M., Wai, W.Y.K., Lai, Y.H., Davis, N., Nublo, F.: Modelling Visual Attention via Selective Tuning. Artif. Intell. 78(1-2), 506–545 (1995)
32. Tsotsos, J.K., Liu, Y., Martinez-Trujillo, J.C., Pomplun, M., Simine, E., Zhou, K.: Attending to Visual Motion. Comput. Vis. Image Und. 100(1-2), 3–40 (2005)
33. Cutzu, F., Tsotsos, J.K.: The Selective Tuning Model of Attention: Psychophysical Evidence for a Suppressive Annulus Around an Attended Item. Vis. Res. 43(2), 205–219 (2003)
34. Rothenstein, A.L., Tsotsos, J.K.: Attention Links Sensing to Recognition. Image and Vision Computing 26(1), 114–126 (2008)
35. Taylor, J.G.: Race for Consciousness. MIT Press, Cambridge
36. Taylor, J.G., Fragopanagos, N.: Simulations of Attention Control Models in Sensory and Motor Paradigms. In: Proc. Int. J. Conf. Neur. Netw., pp. 298–303 (2003)
37. Taylor, J.G., Rogers, M.: A Control Model of the Movement of Attention. Neural Networks 15, 309–326 (2002)
38. Taylor, J.G.: Paying Attention to Consciousness. TICS 6, 206–210 (2003)

Implementing Fuzzy Reasoning on a Spiking Neural Network

Cornelius Glackin, Liam McDaid,
Liam Maguire, and Heather Sayers

University of Ulster,
Faculty of Engineering,
School of Computing & Intelligent Systems,
Magee Campus, Londonderry, BT48 7JL, Northern Ireland
{glackin-c1,lj.mcdaid,lp.maguire,hm.sayers}@ulster.ac.uk

Abstract. This paper presents a supervised training algorithm that implements fuzzy reasoning on a spiking neural network. Neuron selectivity is facilitated using receptive fields that enable individual neurons to be responsive to certain spike train frequencies. The receptive fields behave in a similar manner as fuzzy membership functions. The network is supervised but learning only occurs locally as in the biological case. The connectivity of the hidden and output layers is representative of a fuzzy rule base. The advantages and disadvantages of the network topology for the IRIS classification task are demonstrated and directions of current and future work are discussed.

Keywords: Spiking Neuron Model, Dynamic Synapse, Supervised Learning, Receptive Field, Fuzzy Reasoning.

1 Introduction

The history of neural network research is characterised by a progressively greater emphasis paid to biological plausibility. The evolution of neuron modelling with regard to the complexity of computational units can be classified into three distinct generations [1]. The third generation of neuron modelling (spiking neurons) is based on the realisation that the precise mechanism by which biological neurons encode and process information is poorly understood. The spatio-temporal distribution of spikes in biological neurons holds the key to understanding the brains neural code.

There exists a multitude of spiking neuron models that can be employed in spiking neural networks (SNNs). The extensive amount and variety of neuron models exist in acknowledgement of the fact that there is a trade-off between the individual complexity of spiking neurons and the number of neurons that can be modelled in a neural network. In addition to the variety of neuron models, biological neurons can have two different roles to play in the flow of information within neural circuits. These two roles are excitatory and inhibitory respectively. Excitatory neurons are responsible for relaying information whereas inhibitory

V. Kůrková et al. (Eds.): ICANN 2008, Part II, LNCS 5164, pp. 258–267, 2008.

neurons locally regulate the activity of excitatory neurons. Ongoing physiological experiments continue to illuminate the underlying processes responsible for the complex dynamics of biological neurons. Therefore it is imperative to determine which biological features improve computational capability whilst enabling an efficient description of neuron dynamics. Ultimately neuro-computing seeks to implement learning in a human fashion. In any kind of algorithm where human expertise is implicit, fuzzy IF-THEN rules provide a language for describing this expertise [2]. In this paper, the rationale for the distribution of biologically-inspired computational elements is prescribed by the implementation of fuzzy IF-THEN rules.

In Section 2, unsupervised and supervised learning methods, dynamic synapses and receptive fields are reviewed. Section 3 includes a brief discussion of how fuzzy reasoning can provide a basis for structuring the network topology and introduces a generic network topology outlining the specific models and algorithms used to implement fuzzy reasoning. Experimental results and remarks for the complex nonlinear Iris classification problem are given in Section 4, and conclusions and future research directions are presented in Section 5.

2 Review

The modelling of the neurons synapse is an essential aspect for an accurate representation of real neurons, and one of the key mechanisms to reproducing the plethora of neuro-computational artefacts in SNNs. From a biologically plausible point-of-view synaptic modification in spiking neurons should be based on the temporal relationship between pre and post-synaptic neurons, in accordance with Hebbian principles . In fact, Hebbian learning and its ability to induce long-term potentiation (LTP) or depression (LTD) provide the basis for most forms of learning in SNNs. Hebbian learning gains great computational power from the fact that it is a local mechanism for synaptic modification but also suffers from global stability problems as a consequence [3].

2.1 Unsupervised and Supervised Learning

There are several learning algorithms that can be used to evoke LTP or LTD of synaptic weights. Spike-timing dependent plasticity (STDP) is arguably the most biologically plausible means of inducing LTP and LTD learning since it is a temporal interpretation of Hebbs well-known first generation learning rule. In terms of the temporal coding of spikes it is the order of individual pre and post-synaptic spikes that determines whether the weight is increased or decreased using STDP. BCM and STDP are of course unsupervised learning algorithms, and as such they do not obviously lend themselves to applications requiring a specific goal definition, since this requires supervised learning.

Table 1 presents a review of supervised learning algorithms and their encoding schemes. Time-to-first-spike encoding schemes [4] [5] [6] [7] have the disadvantage that they cannot be used to learn sequences involving multiple spikes. Additionally, the evolutionary strategy [5] and the linear algebra approach are only suitable for offline learning [8]. With the statistical approach [9] the reported findings have been limited to networks consisting of two neurons so it is difficult to know how robust the technique will be for larger networks. Supervised Hebbian Learning [7] is arguably the most biologically plausible of the learning algorithms. However, it suffers from the limitation that even after the goal firing pattern has been achieved the algorithm will continue to change the weights. Of all the supervised learning algorithms presented in Table 1, the ReSuMe [10] approach is perhaps the most efficient. With this approach synaptic weights can be altered as training progresses in an accurate and stable manner.

Table 1. Comparison of Supervised Learning Approaches [10]

Approach	Coding Scheme	References
SpikeProp	Time-to-first-spike	[4]
Statistical approach	Precise spike timing	[9]
Linear algebra formalisms	Precise spike timing	[8]
Evolutionary Strategy	Time-to-first-spike	[5]
Synfire Chains	Relative spike time	[6]
Supervised Hebbian Learning	Time-to-first-spike	[7]
Remote Supervision	Precise spike timing	[10]

2.2 Dynamic Synapses and Receptive Fields

Synaptic efficacy changes on very short-time scales as well as over the longer timescale of training. The rate at which synaptic efficacy changes, is determined by the supply of synaptic resources such as neuro-transmitter and the number of receptor sites. Dynamic synapse models are typically either deterministic or probabilistic. However, it is important that the modelled magnitude of the post-synaptic response (PSR) changes in response to pre-synaptic activity [11]. Further, biological neurons have synapses that can either facilitate or depress the synaptic transmission of spikes [11].

Modelling the dynamics of limited synaptic resources makes neurons selective to particular spike frequencies. The filtering effects of dynamic synapses occur because there is a frequency of pre-synaptic spike trains that optimise the post-synaptic output [12]. A likely explanation for this specificity of frequency is that for certain presynaptic spike train frequencies the synapse will not run out of resources whereas for another it probably will. Between these two pre-synaptic spike frequencies there will be an optimum state where the post-synaptic spike frequency is maximised. This means that certain neurons and synapses can potentially be targeted by specific frequencies of pre-synaptic spike trains. This phenomenon has been described as 'preferential addressing' [12]. Constructing a network of neurons using synapses that operate at different frequency bands is desirable from the perspective of promoting neuron selectivity and richness of information flow. However, it is particularly difficult to tune dynamic synapse models to operate at specific frequency bands by changing the various model parameters. One way to guarantee that synapses are responsive to certain frequencies is with the use of receptive fields (RF).

As far back as 1953, experiments with retinal ganglion cells in the frog showed that the cells response to a spot of light grew as the spot grew until some threshold had been reached [13]. The part of the visual world that can influence the firing of a neuron is referred to as the RF of the neuron [13]. In Barlow's work [13], it was demonstrated that a spot of light within the centre of the RF produces excitation of the neuron whereas when the spot of light is larger than the RF or outside the RF inhibition occurs. The implications for SNNs are that RFs can be used in conjunction with neuron models to promote feature selectivity and hence enhance the 'richness' of information flow.

3 Fuzzy SNN Topology

Biological neuron dynamics are determined by the relationships between spike trains, synaptic resources, post-synaptic currents and membrane potentials. Neuron selectivity can be further strengthened using RFs. The dilemma is in the way in which all these various elements can be combined in a logical way resulting in SNNs that provide insight into the biological neurons code, and are useful from an engineering perspective. Biological neurons obviously implement a form of human reasoning. Human reasoning is fuzzy in nature and involves a much higher level of knowledge representation [2]. Fuzzy rules are typically defined in terms of linguistic hedges, e.g. low, high, excessive, reduced, etc. Taking a cue from fuzzy reasoning, the aim of this paper is to demonstrate how the components necessary to define a fuzzy rule in turn dictate the distribution of the various

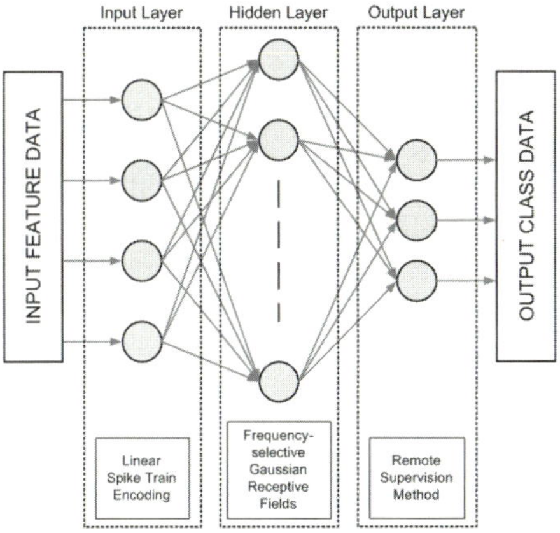

Fig. 1. Generic FSNN Topology

biologically plausible computational elements in an SNN. Fuzzy IF-THEN rules are of the form:

$$IF(x_1 is A_1) AND ... AND(x_N is A_N) THEN(y is Z) \quad (1)$$

where x_1 to x_N represent the network inputs, A_1 to A_N represent hidden layer RFs and y is the network output. Figure 1 presents the generic Fuzzy SNN (FSNN) topology for a typical classification problem. Each layer uses various computational elements to manage the information flow and implement fuzzy reasoning.

3.1 Input Layer

The function of the input neurons is to simply encode feature data into an appropriate frequency range. Spike trains are then generated from the data using a linear encoding scheme. The encoding scheme takes the frequency data points and converts them into an inter-spike interval (ISI) which was then used to create linear input spike trains.

3.2 Hidden Layer

All of the synapses in the FSNN are dynamic. The dynamic synapse model used in this research has kinetic equations that describe inactive, active and recovered states, see [11] for further information. Each pre-synaptic spike arriving at a particular time activates a fraction of synaptic resources, which then quickly inactivate with a time constant of a few milliseconds and then recover slowly. The post-synaptic current is taken to be proportional to the fraction of resources in the active state. The post-synaptic membrane potential is calculated using a leaky-integrate and fire (LIF) passive membrane function. The hidden layer neurons in the FSNN are connected to the input layer using facilitating synapses. Careful choice of parameters in the synapse model determine the type of synapse, again for further details refer to [11]. Figure 2. A shows an example of the Post-synaptic Response (PSR) of a facilitating synapse in response to a spike train of 20 Hertz, whereas Figure 2. B shows the PSR of a depressing synapse to the same spike train. Gaussian RFs are placed at every synapse between the input and the hidden neurons. The frequency dependent RFs determine where an input frequency f_i is in relation to the central operating frequency of the RF F_O. The weight is then scaled by an amount k_i when calculating the PSR.

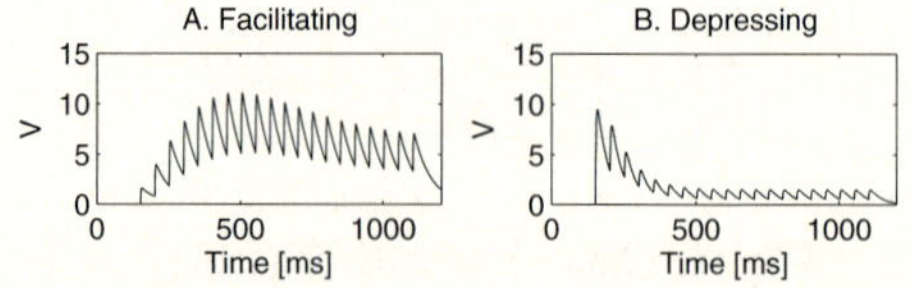

Fig. 2. A. Facilitating and B. Depressing Membrane Potentials

The response of the synapse to an input frequency is determined by the RF. This whole process relates to the $IF(x_i \text{ is } A_i)$ part of the fuzzy rule, where x_i is the input and A_i represents the RF. In this way the RF makes the synapse frequency selective in such a way that the numerous parameters in the dynamic synapse model do not have to be carefully tuned.

The function of each hidden layer neuron is to impose the remaining part of the antecedent fuzzy IF-THEN rule, namely the conjunctive AND. Simply summing the post-synaptic potentials is tantamount to performing a disjunctive OR. The collective aim of the RFs connecting to each hidden layer neuron is to only allow spikes to filter through to the output layer when all of the input frequencies presented at each synapse are within the Gaussian RFs. This can be ensured by making the RF excitatory within the RF and inhibitory outside. The excitatory (positive) part of the Gaussian RF scales the weight in the range [0,1]. The inhibitory (negative) part of the RF scales the weight in the range $[0, (1 - m)]$, where m is the number of input neurons. If even one input frequency lies outside the RF, the resultant inhibitory post-synaptic potential will have sufficient magnitude to negate the post-synaptic potentials from all the other synapses connecting to the same hidden layer neuron. Figure 3 illustrates this process.

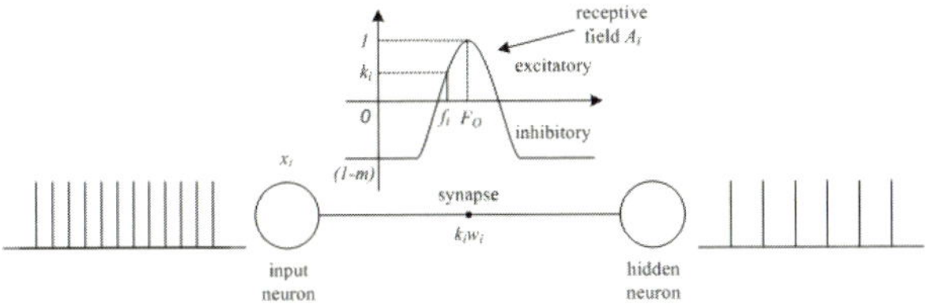

Fig. 3. The conjunctive **AND** part of the fuzzy rule using excitatory/inhibitory RFs

3.3 Output Layer

Depressing synapses model the fact that biological neurons often only respond to the first few spikes in a spike train before running out of synaptic resources. This type of synapse is illustrated by Figure 2. B. The action potential with these types of synapses is only significantly high in magnitude for a very short interval. In this way, this type of synapse can be described as being a coincidence detector. The aim of the depressing synapses connecting the hidden layer to the output layer is to produce spikes in the output in a regimented stable manner. It is then the task of the ReSuMe supervised learning algorithm [10] to associate the hidden layer neurons to the output layer neurons. Thus performing the fuzzy inferencing between the hidden layer (antecedents), and the output layer (consequents).

With ReSuMe weights are changed whenever there is a correlation (or lack of correlation) between the spike frequency coming from the hidden layer neurons (analogous to the 'rule firing strength') and the target teacher signal at a particular class output neuron. Refer to [10] for a detailed explanation of the ReSuMe algorithm. In summary, the presented FSNN topology provides a rationale for the use of RFs, excitatory and inhibitory neurons, as well as facilitating and depressing synapses. The next section of the paper describes how such a network may be used to solve the complex non-linear Iris benchmark classification problem [14].

4 Iris Classification Results

The Iris classification problem [14] is well-known in the field of pattern recognition. The data-set contains 3 classes of 50 types of Iris plant. The 3 species of plant are Iris Setosa, Iris Versicolour, and Iris Virginica. In the interests of clarity the three classes shall be referred to as class 1, class 2, and class 3 respectively. Class 1 is linearly separable from classes 2 and 3. Classes 2 and 3 are not linear separable and make Iris a complex non-linear classification problem.

The Iris data was sorted into two sets, one of 90 samples and the second of 60, to be used for training and testing respectively. Each set contained an equal number of instances of each class, 30 of each in the training set and 20 of each in the testing. The Iris training data was normalized and then linearly scaled into the frequency range [10, 40] (Hertz). The scaled data was then converted into ISIs and the linear input spikes were generated. A sample length of 1 second was used to encode each data point, which adequately satisfies Nyquist sampling criteria. For best results, introducing padding (samples with no spikes) ensures that the order of the data does not affect the filtering by the hidden layer.

4.1 Positioning Receptive Fields

The task of determining the number, position and spread of RFs is an important step in tuning a fuzzy system. This is because the maximum number of possible rules governing the fuzzy inferencing process is determined by $R = IM$ where R is the number of rules, M is the number of membership functions and I is the number of inputs. For the Iris classification task there are 4 inputs (4 features), therefore the number of rules (number of hidden layer neurons) is given by $R = 4M$. The number of possible rules grows exponentially as the number of membership functions increases, this phenomenon is known as 'rule explosion'. There are many methodologies for optimising RF placement [15]. For this dataset Fuzzy C-Means (FCM) clustering was used to calculate the optimum positions for RF placement.

Fuzzy clustering is distinct from hard clustering algorithms such as K-means in that a data point may belong to more than one cluster at the same time. FCM clustering was first developed by Dunn in 1973 [16]. FCM is better at avoiding local minima than K-means but it is still susceptible to the problem in some cases.

The Iris training set was clustered with FCM in a five-dimensional sense (4 features and the class data). Careful selection of appropriate cluster widths can ensure, where possible, that hidden layer clusters are associated with a single

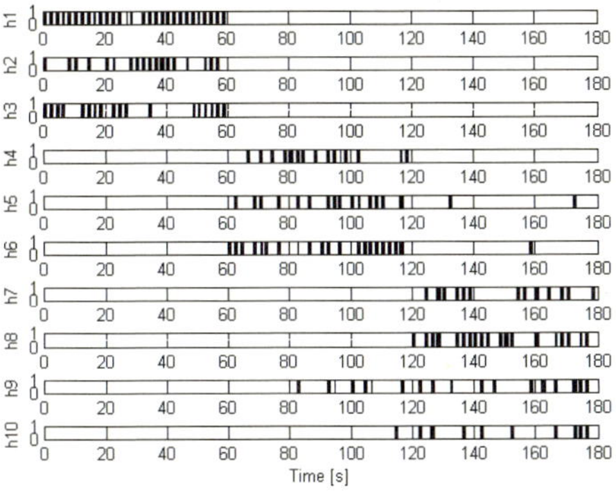

Fig. 4. Hidden Layer Output

class. Matlab's FCM algorithm was used to perform the clustering. The FCM program returns the cluster positionss and a fuzzy membership matrix for all the data samples in the training set. By setting appropriate thresholds on the fuzzy memberships to each cluster, it is possible to determine which data samples should be within each cluster and which should be excluded. Once the thresholding has determined which data samples should belong to each cluster, the variances for each feature can be calculated from the features of each included data sample. These feature variances are then used to specify the widths of each Gaussian RF. Figure 4 shows the output of the hidden layer in response to the training data. As can be seen from Figure 4 there are 3, 3 and 4 clusters (and hence hidden neurons) associated with classes 1, 2 and 3 respectively. The original ordering of the training data is used for clarity, with each successive 60 second interval corresponding to classes 1, 2 and 3. The thresholding technique could not exclusively associate hidden neurons 5 and 6 to class 2. Similarly, hidden neurons 9 and 10 could not be exclusively associated with class 3 without the hidden layer excluding some points all together. In all there were 9 data points (10% of the training set) that were not uniquely associated with one particular class. This means that 10% of the data in the training set is unresolved in terms of the FSNN associating the data correctly with the class data. It is the task of the ReSuMe supervised training regime to resolve this error.

4.2 ReSuMe Training

The first step in implementing the ReSuMe training was to specify the supervisory spike trains to be used. The ReSuMe algorithm was 'forgiving' in this respect producing good convergence for a wide range of supervisory frequencies. Therefore supervisory spike trains of a 50 Hertz frequency were delivered to the appropriate class output neuron whenever the data sample belonged to that class. The supervisory signal for the other class neurons for the given data

Table 2. Comparison of FSNN to other Supervised Learning Algorithms

Algorithm	Training	Testing
SpikeProp	97.4%	96.1%
Matlab BP	98.2%	95.5%
Matlab LM	99.0%	95.7%
Weight Limit Learning	100%	96.6%
FSNN	97.8%	95.0%

sample was of zero frequency. The learning windows were equal and opposite in magnitude producing a maximum possible weight update of 0.05.

Each training data sample is deemed correctly classified when it produces the maximum number of output spikes at the correct output class neuron. The training error smoothly converges to the minimum training error in as few as 6 epochs. The training resulted in 2/90 misclassified data samples (97.78% accuracy), improving the FCM clustering accuracy by 7.33% . Once training was completed, all weights were then fixed and the unseen testing data (60 samples) were presented to the network. During the testing phase 3/60 data points were misclassified (95% accuracy), showing good generalisation. Table 2 compares the training results for FSNN against some other well-known supervised learning algorithms for the Iris classification task. As can be seen from the table, FSNN compares well with the other algorithms. It may be possible to improve on the accuracy of the FSNN algorithm in any number of ways. More hidden layer neurons and different techniques are among the multitude of modifications to the FSNN that could be attempted to improve the accuracy. However, for a preliminary attempt the results are encouraging.

5 Conclusions

This paper presents an FSNN topology which was demonstrated to produce comparable results to existing ANN and SNN techniques for the Iris classification problem. The FSNN topology presented provides a rationale for the assembly of biological components such as excitatory and inhibitory neurons, facilitating and depressing synapses, and receptive fields. In particular, the paper demonstrates how receptive fields may be configured in terms of excitation and inhibition to implement the conjunctive AND of the antecedent part of a fuzzy rule. Currently, the work relies on the use of FCM clustering as a strategy for combating the rule-explosion phenomenon of fuzzy logic systems. Ideally, it would be preferable for the FSNN to determine the number, placement and spread of fuzzy clusters without relying on an external statistical or clustering technique. For this reason, future work in recognition of the growing interest in biological receptive fields, will involve the development of dynamic receptive fields. The ultimate aim is to develop a biologically plausible FSNN that tunes itself using strictly biological principles, and in that regard this work represents a significant first step.

References

1. Maass, W.: Networks of Spiking Neurons: The Third Generation of Neural Network Models. Electronic Colloquium on Computational Complexity (ECCC) 3(31) (1996)
2. Zadeh, L.A.: Fuzzy sets. Information and Control 8(3), 338–353 (1965)
3. Abbott, L.F., Nelson, S.B.: Synaptic plasticity: taming the beast. Nature Neuroscience 2, 1178–1183 (2000)
4. Bohte, S.M., Kok, J.N., La Poutre, H.: Error-backpropagation in Temporally Encoded Networks of Spiking Neurons. Neurocomputing 48, 17–37 (2002)
5. Belatreche, A., Maguire, L.P., McGinnity, T.M., Wu, Q.X.: A Method for Supervised Training of Spiking Neural Networks. In: IEEE Cybernetics Intelligence. Challenges and Advances (CICA), pp. 39–44 (2003)
6. Sougne, J.P.: A learning algorithm for synfire chains. Connectionist Models of Learning, Development and Evolution, pp. 23–32 (2001)
7. Ruf, B., Schmitt, M.: Learning temporally encoded patterns in networks of spiking neurons. Neural Processing Letters 5(1), 9–18 (1997)
8. Carnell, A., Richardson, D.: Linear algebra for time series of spikes. In: 13th European Symposium on Artificial Neural Networks (ESANN) (2005)
9. Pfister, J.P., Barber, D., Gerstner, W.: Optimal Hebbian Learning: A Probabilistic Point of View. In: ICANN/ICONIP Lecture Notes in Computer Science, vol. 2714, pp. 92–98 (2003)
10. Kasinski, A., Ponulak, F.: Comparison of Supervised Learning Methods for Spike Time Coding in Spiking Neural Networks (2005), `http://matwbn.icm.edu.pl/ksiazki/amc/amc16/amc1617.pdf`
11. Tsodyks, M., Pawelzik, K., Markram, H.: Neural Networks with Dynamic Synapses. Neural Computation 10(4), 821–835 (1998)
12. Natschlager, T., Maass, W., Zador, A.: Efficient temporal processing with biologically realistic dynamic synapses. Network: Computation in Neural Systems 12, 75–87 (2001)
13. Barlow, H.B.: Summation and inhibition in the frogfs retina. J. Physiol. 119, 69–88 (1953)
14. Fischer, R.A.: The Use of Multiple Measurements in Taxonomic Problems. Annals of Eugenics 7, 179–188 (1936)
15. Abdelbar, A.M., Hassan, D.O., Tagliarini, G.A., Narayan, S.: Receptive Field Optimisation for Ensemble Encoding. Neural. Comput. & Applic. 15(1), 1–8 (2006)
16. Dunn, J.: A fuzzy relative of the ISODATA process and its use in detecting compact wellseparated clusters. Journal of Cybernetics 3, 32–57 (1973)

Short Term Plasticity Provides Temporal Filtering at Chemical Synapses

Bruce P. Graham[1] and Christian Stricker[2]

[1] Department of Computing Science and Mathematics, University of Stirling, Stirling, FK9 4LA, U.K.
b.graham@cs.stir.ac.uk
[2] Division of Neuroscience, JCSMR, Australian National University, GPO Box 334, A.C.T., Canberra, Australia
Christian.Stricker@anu.edu.au

Abstract. Chemical synapses exhibit a wide range of mechanisms that modulate the postsynaptic response due to presynaptic action potentials (APs) as a function of the arrival pattern of the presynaptic APs. Here we demonstrate that combinations of depression, facilitation and frequency-dependent recovery will give synapses that are either sensitive to the steady-state input signal, or to the onsets and offsets of signal transients. In addition, a synapse may be tuned to preferentially respond to low frequency modulation of an input signal.

Keywords: Synaptic transmission, short term plasticity, depression, facilitation.

1 Introduction

Arrival of an action potential (AP) at a presynaptic terminal causes probabilistic release of neurotransmitter, which in turn results in a conductance change in the postsynaptic membrane due to binding of neurotransmitter to postsynaptic receptor molecules. Both the probability that neurotransmitter is released and the amplitude of the postsynaptic conductance change actually depend on the precise arrival times of presynaptic APs. Thus the postsynaptic response to a particular AP depends on the history of previous arrivals, ranging back in time from milliseconds to minutes. Many molecular mechanisms contribute to the depression or facilitation of the postsynaptic response (for a review of such mechanisms see [7]). So, in complication of the notion of the long-term strength of a synapse as determined by mechanisms of long-term depression (LTD) and long-term potentiation (LTP), a synapse also acts as a temporal filter of a dynamical input signal on short time scales, up to minutes [5].

Synapses between excitatory neurons in the neocortex tend to display a rapid depression to repeated stimuli arriving at frequencies from a few Hertz upwards. The magnitude of depression becomes proportional to the inverse of the stimulation frequency at higher frequencies [6]. As a consequence, the temporally and spatially summed postsynaptic current from such synapses becomes independent of stimulus frequency, and the resultant spiking output of the postsynaptic

V. Kůrková et al. (Eds.): ICANN 2008, Part II, LNCS 5164, pp. 268–276, 2008.

neuron carries no information about the input stimulus rate [1,6]. However, the time course of depression is dependent on the stimulus rate and the receiving neuron remains sensitive to changes in the input stimulus frequency. In fact the postsynaptic neuron may effectively signal the percentage rate change [1].

Recent work has identified both release-independent depression (RID) and frequency-dependent recovery (FDR) at such neocortical synapses [2]. A major impact of the FDR is to restore the ability of the synaptic pathway to transmit information about the steady state input stimulus rate through to the spiking output of the receiving neuron. Thus different combinations of short term plasticity (STP) mechanisms affect how a receiving neuron ultimately responds to steady state and changing input signals.

Here we use a mathematical model of a synapse containing mechanisms that contribute to depression and facilitation to explore the synapse's ability to transmit steady state and transient information about an input signal. We show that it is not just the combination of STP mechanisms, but also the configuration of synaptic release sites that determines the postsynaptic response to changing input signals.

1.1 Short Term Plasticity

Short term plasticity at a chemical synapse is the result of molecular mechanisms acting in the presynaptic terminal or in the postsynaptic membrane. Presynaptic mechanisms affect the occupancy of release sites with readily-releasable vesicles, and the probability that these vesicles will release their neurotransmitter on arrival of a presynaptic AP. Postsynaptic mechanisms include receptor desensitization that limits the number of receptor channels that will open on binding neurotransmitter. Most presynaptic mechanisms ultimately can be traced back to calcium [7]. Presynaptic STP results from alterations in calcium influx due to an AP, or due to alterations in downstream molecular targets. In turn these alterations affect the rates of vesicle endocytosis and exocytosis. Postsynaptic receptor desensitization is a property of the particular receptor molecules and how their conformation changes on binding transmitter molecules.

To summarise, STP may increase (facilitate) or decrease (depress) the amplitude of the postsynaptic response to a presynaptic AP. The effect of a mechanism usually is increased by the arrival of an AP, and then recovers back to a baseline level with a certain time course, which typically ranges from milliseconds to minutes. STP mechanisms act essentially through three different variables:

n - the number of vesicles occupying release sites and available for release
p - the probability that an individual vesicle will release
q - the amplitude of the postsynaptic current (PSC) on release of a single vesicle.

2 Model of Short Term Plasticity

We consider a synaptic model containing three presynaptic mechanisms for STP:

1. depletion of the readily releasable vesicle pool (RRVP) by exocytosis (release-dependent depression: RDD),

2. frequency-dependent replenishment of the RRVP from an effectively infinite reserve pool (FDR), and
3. facilitation of the probability that a vesicle will release (fac).

The release pool size, n, and release probability, p, are treated as continuous, deterministic quantities. Vesicle release sites are assumed to be limited in number, so that a release pool size of $n = 1$ corresponds to all release sites being occupied. The fractional release pool size, n, is given by:

$$\frac{dn}{dt} = \frac{1-n}{\tau_n} - \sum_s p_v.n.\delta(t - t_s) \tag{1}$$

Presynaptic spikes arrive at times t_s. The RRVP is assumed to recover between spikes with a particular time course. The recovery time may be frequency-dependent, decreasing with each presynaptic spike [2,4]:

$$\frac{d\tau_n}{dt} = \frac{\tau_n^0 - \tau_n}{\tau_{nr}} - \sum_s \Delta\tau_n.\tau_n.\delta(t - t_s) \tag{2}$$

Vesicle release probability facilitates on the arrival of each presynaptic spike and recovers back to baseline between spikes:

$$\frac{dp}{dt} = \frac{p^0 - p}{\tau_f} - \sum_s \Delta p.(1-p).\delta(t - t_s) \tag{3}$$

The amplitude of the postsynaptic response (PSR) is simply $p.n$ in arbitrary units. How this should be interpreted depends on the particular synaptic configuration. For large numbers of synchronously activated release sites, as found at giant synapses such as the calxy of Held [3], the PSR represents the average response to a presynaptic AP. For small synapses, as found in neocortex, the PSR represents the average found over many trials at one synapse, or a spatial average over many synapses from different neurons onto a single neuron, provided those synapses are synchronously active.

For a single trial at a small synapse, the PSR should be treated as the probability that a vesicle undergoes exocytosis. The above model can be made stochastic by determining vesicle release as a random process with probability $p.n$ (which includes the probability that a vesicle is available for release, n, and the probability that an individual vesicle will release, p) on the arrival of a presynaptic AP. Instead of decreasing n by $p.n$, n is set to 0 if a stochastic release occurs, otherwise it remains unchanged [4].

This model is in essence that proposed by [2] and extended by [4] and is strongly related to previous models [1,6] (but with the addition of frequency-dependent recovery). Each equation can be solved iteratively to give values for n, τ_n and p at spike time t_s in terms of their values at the previous spike time t_{s-1} [4]. The model has been implemented in Matlab and NEURON.

3 Results

3.1 Step Changes in Stimulus Frequency

In the initial experiments we considered the average PSR during a step change from a regular stimulation frequency of 5Hz to a higher frequency (Fig. 1). Consistent with previous models, when the only STP mechanism is vesicle depletion (RDD), the PSR depresses as a function of the step frequency, with higher frequencies leading to greater depression (Fig. 1a). Facilitation (fac) amplifies the magnitude of depression and also speeds the recovery of the PSR back to the 5Hz baseline (Fig. 1b). Frequency-dependent depression (FDR) strongly reduces the magnitude of depression (Fig. 1c). There is now a transient depression at the step onset, and a transient over-recovery on stimulus offset. These transients are strongly amplified (particularly the offset) by facilitation (Fig. 1d).

To test the effect of STP on the spiking output of a receiving neuron, a single compartment neuronal model, containing Hodgkin-Huxley style sodium and potassium channels to produce action potentials, was constructed in NEURON.

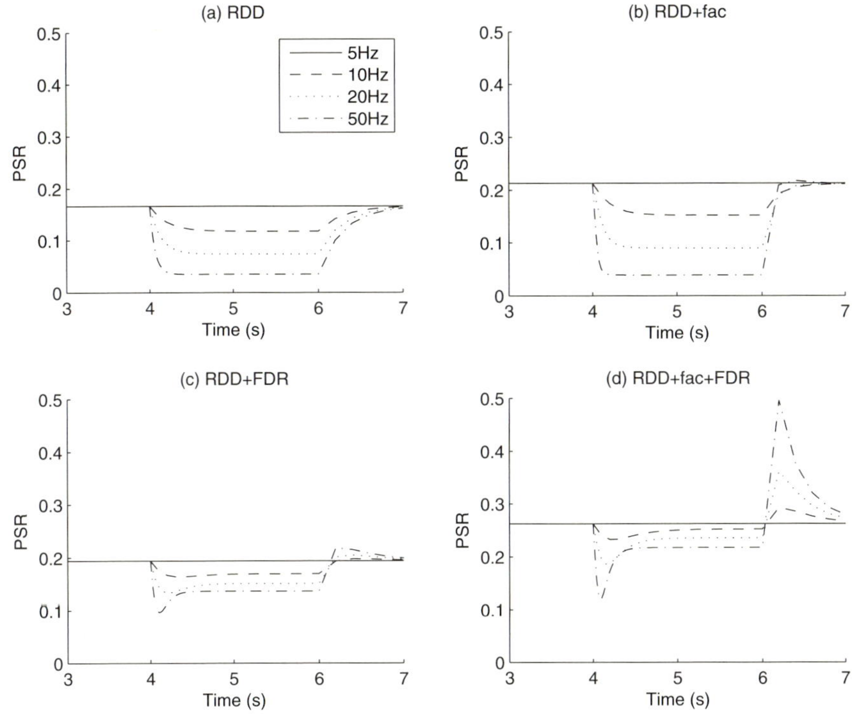

Fig. 1. Step changes after 4s from a base frequency of 5Hz to increasing frequencies, followed by a step back to base 2s later. Synapse contains (a) pure RDD ($\tau_n = 0.5s$), (b) RDD plus facilitation ($p^0 = 0.25$, $\Delta p = 0.1$), (c) RDD plus FDR ($\Delta t_n = 0.2$, $\tau_{nt} = 0.5s$) and (d) RDD plus FDR and facilitation. The model output is PSR$= p.n$.

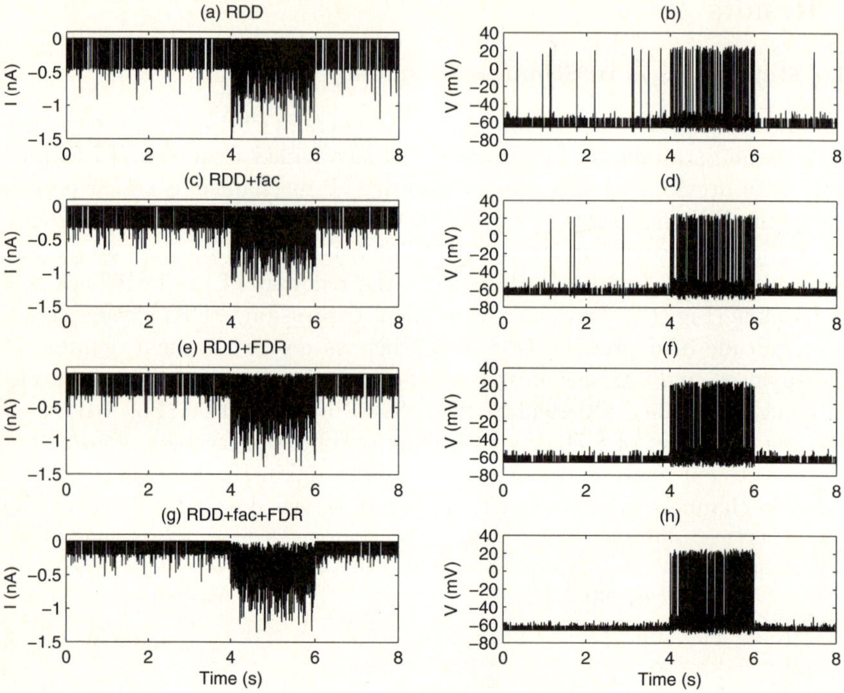

Fig. 2. Neuron driven by 100 stochastic synapses, each driven asynchronously by Poisson-distributed spike trains of the same mean frequency. Stimulus trains undergo a step change after 4s from a base frequency of 5Hz to 50Hz, followed by a step back to base 2s later. Each synapse contains (a,b) pure RDD ($\tau_n = 0.5s$; synaptic conductance $g_{sy} = 7nS$), (c,d) RDD plus facilitation ($p^0 = 0.25$, $\Delta p = 0.1$; $g_{sy} = 5nS$), (e,f) RDD plus FDR ($\Delta t_n = 0.2$, $\tau_{nr} = 0.5s$; $g_{sy} = 5nS$) and (g,h) RDD plus FDR and facilitation ($g_{sy} = 3nS$). Left-hand column shows postsynaptic current and right-hand column shows membrane voltage (spiking output).

This neuron was driven by 100 stochastic excitatory synapses, each containing a single release site. Release of a vesicle resulted in a dual time course EPSC ($\tau_1 = 0.1$ms and $\tau_2 = 1$mS; reversal potential 0mV). Each synapse was driven by a 5Hz stimulus train that stepped to 50Hz after 4 seconds. Results for different synaptic configurations are shown in Figures 2 and 3.

In Fig. 2 the synaptic configuration is equivalent to excitatory input onto a neocortical pyramidal cell. Each synapse is driven by an independent, Poisson-distributed spike train, with all trains having the same mean frequency. Hence releases from all synapses are independent and asynchronous. Spatial and temporal integration of the postsynaptic currents (EPSCs) results in a larger mean current at 50Hz, and faster postsynaptic spiking as a consequence. Though these synapses can transmit steady state information about the input stimulus rate, synaptic depression reduces the possible difference between the mean postsynaptic current at 5Hz and 50Hz (Fig. 2a), and this difference is further reduced

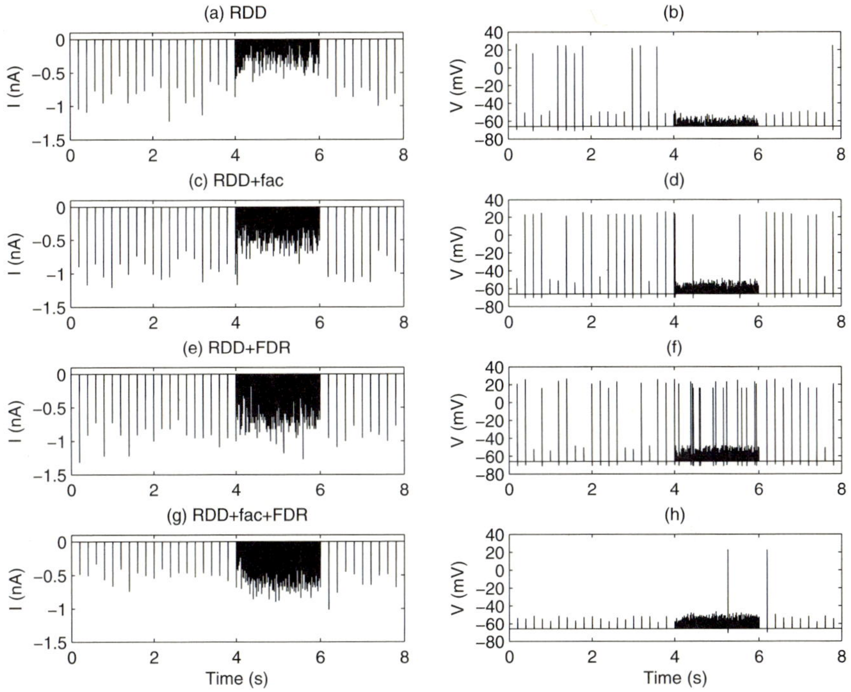

Fig. 3. Neuron driven by 100 stochastic synapses, each driven synchronously by regular spike trains of the same mean frequency. Stimulus trains undergo a step change after 4s from a base frequency of 5Hz to 50Hz, followed by a step back to base 2s later. Each synapse contains (a,b) pure RDD ($\tau_n = 0.5s$; synaptic conductance $g_{sy} = 0.7nS$), (c,d) RDD plus facilitation ($p^0 = 0.25$, $\Delta p = 0.1$; $g_{sy} = 0.6nS$), (e,f) RDD plus FDR ($\Delta t_n = 0.2$, $\tau_{nr} = 0.5s$; $g_{sy} = 0.7nS$) and (g,h) RDD plus FDR and facilitation ($g_{sy} = 0.3nS$). Left-hand column shows postsynaptic current and right-hand column shows membrane voltage (spiking output).

by facilitation (Fig. 2c). However, frequency-dependent recovery acts to amplify this difference, enhancing the ability of the spiking output of the neuron to carry information about the mean stimulus rate. In this example the synaptic conductance has been set such that the neuron filters out (does not respond to) the 5Hz input. These results are in line with previous models [1,2,6].

The situation changes, however, if the synaptic configuration is such that all 100 synapses are driven synchronously by the same stimulus train (Fig. 3). This is equivalent to giant synaptic connections, such as calyces and neuromuscular junctions. Now the postsynaptic current matches the changes in release probability, so that depression results in a lower mean current for a stimulus rate of 50Hz, compared to 5Hz (Fig. 3a). In this case the postsynaptic neuron can act to filter out the higher frequency (Fig. 3b), rather than the lower frequency as with asynchronous synapses. Facilitation and frequency-dependent recovery both act

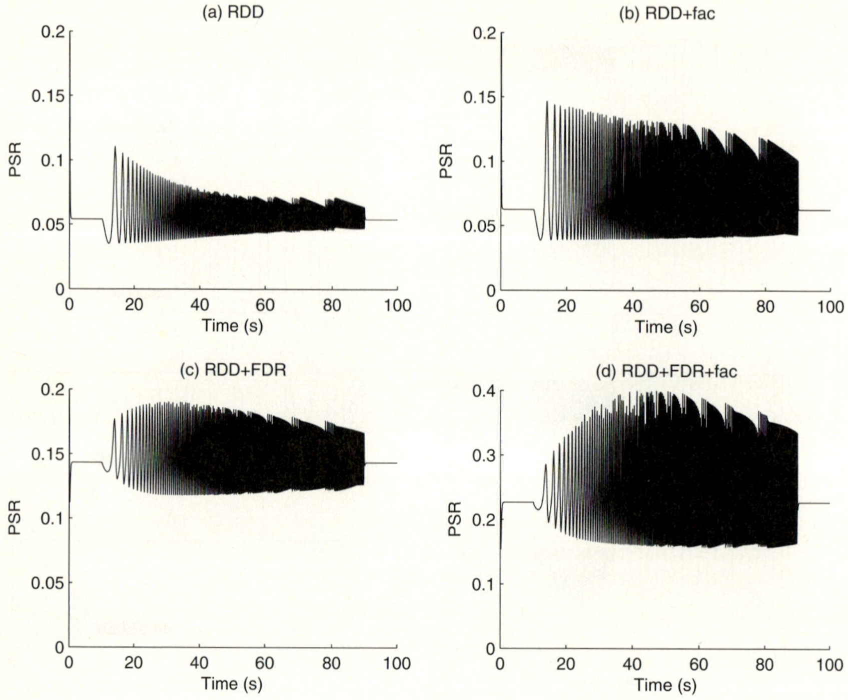

Fig. 4. Sinusoidal modulation of a base frequency of 30Hz between 10Hz and 50Hz. Modulation frequency increases linearly from 0.05Hz to 3Hz over 100s. Synapse contains (a) pure RDD ($\tau_n = 0.5s$), (b) RDD plus facilitation ($p^0 = 0.25$, $\Delta p = 0.1$), (c) RDD plus FDR ($\Delta t_n = 0.2$, $\tau_{nr} = 0.5s$) and (d) RDD plus FDR and facilitation. The model output is PSR= $p.n$.

to minimise the difference between the mean currents at different stimulus frequencies (Fig. 3c,e). Consequently the spiking output can begin to follow the stimulus frequency (Fig. 3d,f). With facilitation and FDR in combination, the mean current at 50Hz slightly exceeds that at 5Hz, and the transient enhanced release probability on the step back from 50 to 5Hz (see Fig. 1d) becomes evident (Fig. 3g). If the synaptic conductance is reduced sufficiently the postsynaptic cell becomes an offset detector, spiking reliably only on the step from a high to low frequency (Fig. 3h).

3.2 Continuous Modulation of Stimulus Frequency

In a test of synaptic filtering of a continuously changing signal we examined the PSR when the input stimulus varied sinusoidally from 10Hz to 50Hz (mean 30Hz). The rate of sinusoidal modulation of the input frequency was varied from 0.05Hz to 3Hz (frequency increasing linearly with time over 100 seconds) (Fig. 4).

With pure depression, the PSR decreases as the stimulation frequency increases and vice versa, roughly tracking the sinusoidal modulation of the carrier signal frequency (Fig. 4a). At high modulation frequencies, the time courses of depression and recovery are too slow for the PSR to reach its steady state value at the limiting frequencies of 50Hz and 10Hz, respectively. Consequently the amplitude of the PSR modulation decreases with increasing modulation frequency. The addition of facilitation amplifies the PSR modulation and reduces the dropoff with modulation frequency as it speeds both depression and recovery (Fig. 4b).

Frequency-dependent recovery, on the other hand, results in an optimal modulation frequency at which the amplitude modulation of the PSR is maximal (Fig. 4c). At low modulation frequencies FDR compensates for the changing stimulus frequency, resulting in only a small change in PSR amplitude. At higher modulation frequencies FDR is too slow to compensate and the PSR is limited by the time courses of depression and recovery. The suppression of PSR modulation amplitude at low and high frequencies results in an optimum modulation frequency on the order of 1Hz at which the time courses of depression and recovery are able to track the stimulus frequency without undue contribution from FDR. This effect is amplified by the addition of facilitation (Fig. 4d).

4 Conclusions

We have used a mathematical model to explore the temporal filtering properties generated by mechanisms of short-term plasticity at a chemical synapse. Extending previous work we highlight that whether or not a synapse can transmit information about the stimulus rate it is receiving depends both on the particular STP mechanisms and the synaptic configuration. For asynchronous synapses depression leads to a reduction in frequency information [1,6], which is restored by a frequency-dependent recovery mechanism [2]. However, for synchronous synapses depression may still be evident in the postsynaptic current, leading to a reduction in spiking output for higher stimulus rates. Facilitation and FDR make the receiving neuron particularly sensitive to the onset and offset of step changes in stimulus rate.

Further, we show that a synapse will also filter continuously modulated stimulus rates, with the combination of depression and FDR leading to an optimal modulation frequency of around 1Hz. This may enhance detection of slowly changing environmental stimuli and is in the frequency range of saccadic eye movements.

Acknowledgement

This work was funded by a Royal Society International Short Visit grant to B. Graham and C. Stricker and a BBSRC project grant to B. Graham.

References

1. Abbott, L., Varela, J., Sen, K., Nelson, S.: Synaptic depression and cortical gain control. Science 275, 220–224 (1997)
2. Fuhrmann, G., Cowan, A., Segev, I., Tsodyks, M., Stricker, C.: Multiple mechanisms govern synaptic dynamics at neocortical synapses. J. Physiol. 557, 415–438 (2004)
3. Schneggenburger, R., Forsythe, I.: The calyx of Held. Cell Tissue Res. 326, 311–337 (2006)
4. Scott, P.C.: Information transfer at dynamic synapses: effects of short-term plasticity. Honours thesis, Australian National University (2005)
5. Thomson, A.: Molecular frequency filters at central synapses. Prog. Neurobiol. 62, 159–196 (2000)
6. Tsodyks, M., Markram, H.: The neural code between neocortical pyramidal neurons depends on neurotransmitter release probability. Proc. Nat. Acad. Sci. 94, 719–723 (1997)
7. Zucker, R., Regehr, W.: Short-term synaptic plasticity. Ann. Rev. Physiol. 64, 355–405 (2002)

Observational Versus Trial and Error Effects in a Model of an Infant Learning Paradigm

Matthew Hartley[1], Jacqueline Fagard[2], Rana Esseily[2], and John Taylor[1]

[1] Department of Mathematics, King's College London
[2] Laboratoire Cognition et Dévelopement Institut de Psychologie
Université Paris 5, France

Abstract. We present results from a paradigm of infant motor skill learning, involving selecting the correct approach to grasp a ball placed on a low stability stand. We also construct a theoretical model, based on the concept of observational learning as the discovery of non-perceived affordances and the transfer of actions on affordances, which can explain some of the experimental data presented. We discuss the underlying concepts of the model (including its basis in neuroanatomy) and how they relate to the paradigm, and suggest possible model extensions.

1 Introduction

Imitation and observation are key components of learning in humans and other species. Recent experimental results suggest that one of the components of observational learning may be the transfer of actions on affordances (meaning that known parts of a motor repetoire become associated with "new" affordances, we will discuss this further later), and the discovery of non-perceived affordances. A very broad definition of affordances is that they provide information about how to interact with objects, so learning actions on affordances is a critical part of development.

1.1 Observational Learning

The capacity to learn by observation has important survival value. Trial and error learning is very inefficient and potentially dangerous in situations where the cost of failure is high or it is important to learn rapidly. Learning by observation, however, provides the opportunity to rapidly transfer existing knowledge. Imitation and observation also seem to be closely connected with how we learn languages.

Recent experiments on both human and monkey subjects have shown that during observation of certain actions there is extensive brain activation in areas usually thought to be involved only with the production of those actions [1]. These results suggest the hypothesis that understanding and recognition of observed actions recruits brain areas involved in producing those actions, possibly for mental simulation.

V. Kůrková et al. (Eds.): ICANN 2008, Part II, LNCS 5164, pp. 277–289, 2008.

1.2 Affordances

The concept of affordances has considerable value in the study of interactions with objects. Gibson's original definintion of affordances was "What [the environment] offers the animal, what it provides or furnishes, either for good or ill." [2] For the purposes of this study, we will define an affordance to be "the set of features of an object that are required to be known before a particular goal based action can be taken on it". The action used by the actor to fulfill that affordance is the affordant action or alternatively, the action on that affordance. The affordances are properties of the objects, which may or may not be known to the observer (see the next section), while the affordant actions are part of the actor's motor repetoire.

One interesting question is that of how initial affordances are "learned". In the Gibsonian view, affordances are part of the properties of the interaction between actor and object, and therefore are not learned. However, under this point of view we can distinguish between "perceived" and "non-perceived" affordances. Non-perceieved affordances are those that are possible given the state of the object, but not known to the actor because the actor does not recognise the affordance. We define a perceived affordance to be an affordance known to the observer, and non-perceived affordance to be a potentially learnable affordance for the observer, given either increased motor skill repetoire or discovery of environmental circumstances.

Associating actions with affordances. As well as the discovery of non-perceived affordances, there is the question of how we learn to associate actions with affordances. In situations where the action in question is already part of the observer's motor repetoire, this involves learning to use an existing action on an affordance not already associated with the action. By observational learning, this requires the observer to recognise the action used, involving both the mechanical characteristics (such as contact points of a grip), and the affordance it is used on, then to associate the two.

Various imaging studies have shown that observation of the actions of others activates considerable areas of the brain previously thought to be used only in the production of actions [1]. It is possible that this occurs due to attempts to understand the observed actions by a process of mental simulation, which uses many of the same pathways as production of actions.

1.3 Model Hypotheses and Construction

As we will show later, one of the things infants appear to be learning by observation is the objects' affordances and the way to act upon them. Transfer of affordances to novel objects/situations by observation would provide a powerful learning mechanism. We therefore present a hypothesis on which we base our model:

That one component of observation learning is the discovery of affordances and of the means to act upon them. Another way of looking at this is that actions on affordances (or action/affordance pairs) are learned.

Effectively what is being learned is the observer's knowledge of the affordances available to it (the discovery of unperceived affordances), and the actions to use on those affordances (and the linking of action-affordance pairs).

To model the process of transfer of the means to act upon newly discovered affordance, our model needs to incorporate the following components:

- Some initial set of affordances and actions.
- A system for recognising the affordances acted upon by the observed actor and associating the actions used with that affordance and the task.
- Recall of the action used and integration of this information into a motor plan.

In addition, to be able to consider the parallel mechanism of trial and error learning, the model also needs to include some form of motor error monitor and system to correct motor errors.

2 Paradigm and Model

We now turn to look at developing the model and applying it to experimental paradigms.

2.1 The Simulation Model

Our model comprises several components. These are:

Vision. This represents the basic visual input available to the model. We model this in an extremely simple manner - the region has dedicated nodes each of which holds a possible view available to the model, and they are activated depending on the simulation setup. The possible views represent the initial setup (the ball on the stand), the ball grasped, and the ball knocked off the stand.

Goals. Here the simulation's current goal is stored and used to influence behaviour. In the simple paradigm we present here, the only possible goals are to grasp the object or push it over. Goals are represented by dedicated nodes that code for an action and an object (i.e grasp-ball/push-ball). In the brain, these are coded in separate areas, but we simplify here.

Object representation. This module picks out object features to identify objects. In our simple simulation, the only possible object to be recognised is a ball. The weights of the projections from the vision module are chosen so that feature nodes present here can be activated, which then project to the object codes module where individual objects are represented.

Object codes. Here objects already known to the system are stored, such that they can be activated by the object representation module.

Stored affordance/action pairs. This contains the set of affordances and related actions known to the system. These are primed by the object codes module based on existing learning. The module contains nodes representing

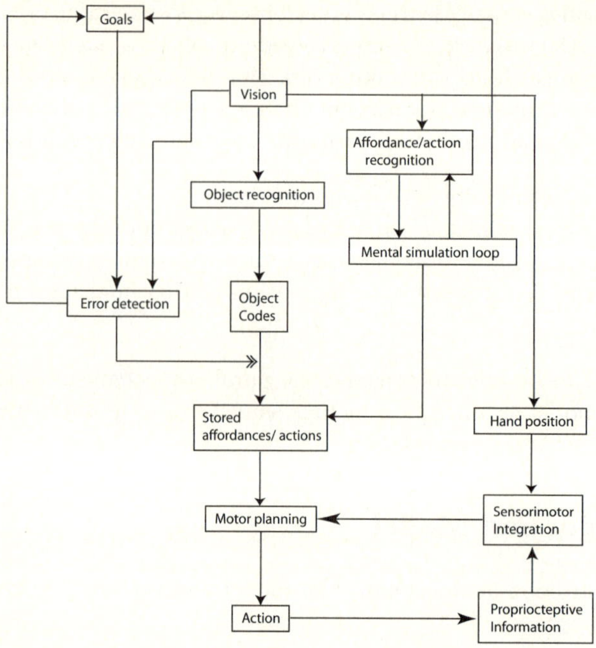

Fig. 1. Modular architecture for execution of grasping actions

actions, and nodes representing affordances, and the connections between these are plastic, such that they can be learned by a Hebbian learning mechanism (concurrent activation of an affordance and an action causes an increase in the connection between them).

Sensorimotor integration. Here proprioceptive information about the current state of the system's hand is fused with visual information to update the motor plan.

Motor planning. Here the affordance information is combined with sensorimotor information about current physical state, to form a motor plan. This involves calculating movements necessary to bring current finger positions to contact points, give the angle of approach.

Action. This module represents the current action the system is attempting to perform, consisting of dedicated nodes. We code the possible approaches to the ball as separate actions.

Proprioceptive feedback. Here the action taken by the system is converted into feedback about the current state of the system's arm/hand. Proprioceptive feedback is coded as a vector update indicating the change in hand position.

Affordance/action recognition. The module processes visual input to attempt to extract information representing the features of an observed affordance and the action used on that affordance, coding here is the same as in the stored action/affordance pairs module.

Affordance/action comparison. Here a candidate action/affordance pair can be compared to the system's stored list of affordances and actions to see whether a match occurs. It works with the recognition and storage systems for affordances and actions to compare and find closest matches.

Mental simulation loop. This module extracts the goals of observed affordance/action pairs by mental simulation - effectively planning out the observed action to determine which goals are associated with it. The module contains a simple internal fprward model which simulates use of the action, to determine variables associated with the action that are not directly observable, in particular the action goal (which is part of the affordance to be recognised).

2.2 Experimental Paradigm – The Ball and Stand

We now apply our model to a specific paradigm to test observational learning. In this paradigm, infants are presented with a ball (40 mm in diameter) placed on a support with a narrow base, such that the ball is easily knocked off the stand. The stand diameter varies with age as follows:

Table 1. Variation in size of base with age

Age (months)	Base diameter (mm)	Notes
8	16	
10	14	
12	10	
15	6	
18	6	Narrow cylinder placed on stand

The setup is different from that usually encountered by infants at the relevant developmental stage, where grasping motions for similar sizes of object can be executed by taking a direct approach path. Here, the top approach is more successful since it confers less danger of knocking the ball off the support.

2.3 Experimental Results – Trial and Error and Observational Learning for Ball and Stand Paradigm

We can firstly examine whether the infants significantly improve their perfomance at the task by repeated trial and error. The following table shows the percentage of success on three sequential trials.

Only the infants at the age of 8 months improve their performance over repeated trials, other ages show no significant increase. For the 8 months olds, the critical increase comes between the first and second trials. Infants at age 15 months actually decrease in performance between the first and second trials, the reasons for this are not clear.

Table 2. Change in success at paradigm with repeated attempts, showing percentage of success on each trial

Age (months)	Trial 1	Trial 2	Trial 3
8	12.5	62.5	50
10	44.4	44.4	44.4
12	44.4	44.4	44.4
15	57.1	14.3	57.1
18	50	57.1	50

Table 3. Effects of observational learning on ball/stand paradigm

Age (months)	C Success (percentage)	E success	Significant
8	12.5	75	Y
10	44.4	55.56	N
12	44.4	50	N
15	57.1	70	N
18	50	66.7	N

Next, we can look at how the infants improve in their performance at the task after a single demonstration by an experimenter. The control group (C) is given the ball and stand with no demonstration, the experimental group (E) sees the demonstrator grasp the ball once, then attempts to grasp it themselves.

Considering the performance of the control group compared to the experimental group, we see that the 8 month old infants demonstrate considerable increase in success rate at the task after a single demonstration. Infants at all other ages show a performance increase after the demonstration, however the results are statistically significant only for the 8 month olds.

Given that the 8 month old infants demonstrate observational learning, we can consider what the infants are learning. Our hypothesis suggests that the infants are discovering non-perceived affordances and the ways in which to act upon them. This might manifest in two ways, one general the other more specific. The general discovery is that the ball is unstable, and so affords particular types of grip not immediately obvious (or rather, the obvious types of grip are likely to fail because of the ball's instability. The specific mechnism is the way in which the experimenter demonstrates grasping the ball (which involves a grasp from above).

2.4 Encoding of Affordance/Action Pairs

In the model, we encode the actions to be used on the affordance of grasping the ball as a direction of approach to the ball, moving to meet the contact points defined by the affordance. The contact points are assumed to be already learned as part of recognising the affordance. The angle of approach can either by vertical or horizontal. The simulation has an initial bias towards the horizontal angle of approach. For both grasps, we assume that the infant already has the action as part of its motor repetoire, and can recognise it when used by the demonstrator (such that no new action is being learned). In the model, this translates to existing representations for the actions in the affordance/action pairs module, but low weights to the initially less preferred action.

We need to add a component of affordances (the contact points) into the action activity going to the motor planning module), although the contact point information would come from a separate affordance map in a more complex model, since this is not really a component of the action in its pure state.

2.5 Simulation Steps

The simulation of trial and error learning occurs with the following steps:

1) Present object. Object module activated.
2) The projection from the objects module to the affordances/actions module is used to stochastically select an affordant action from those known.
3) The affordance/action pair (contact points and a direction of approach) is combinded with information about the current state of the system's arm/hand to form a motor plan
4) The motor plan is executed. During execution, sensorimotor feedback updates the current hand position
5) The plan is recognised to have succeeded (object grasped) or failed (object knocked over)
6) If the plan failed, reduce the weighting of the affordance used, if it succeeded, increase the weighting
7) Repeat

We measure the success rate for each trial. For the observational learning model, the steps are as follows:

1) Present object. Object module activated.
2) Teacher performs action on object. Affordance and action used are matched to existing affordance/action pairs
3) The weight of the object to the matched affordant action is increased
4) Object presented to simulated infant
5) Simulation proceeds as per trial and error task (results compared to that task)

2.6 Simulation Results – Trial and Error Learning

We firstly look at our results for trial and error learning, simulating the 8 month old infants improvement in performance by trial and error over a single trial.

We see in Figure 2 that the simulation shows a rapid gain in success rate after the first trial, from the trial and error learning.

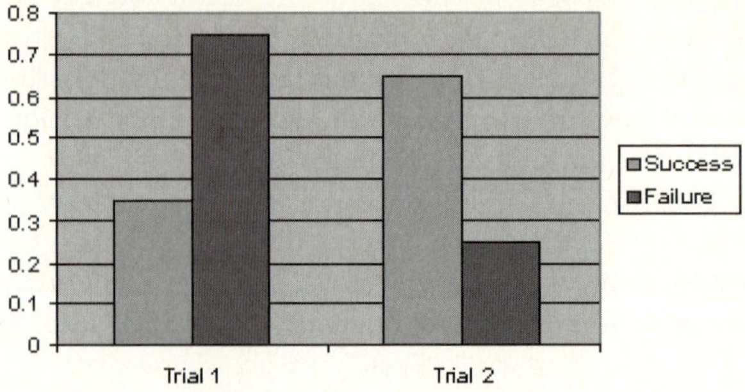

Fig. 2. Simulation results for the ball and stand paradigm showing improvement by trial and error between trials

2.7 Simulation Results – Observational Effects

Next we can look at the simulated observation learning, again for the 8 month old infants. Here we simulate a single demonstration of the task by a teacher using the grasp from above.

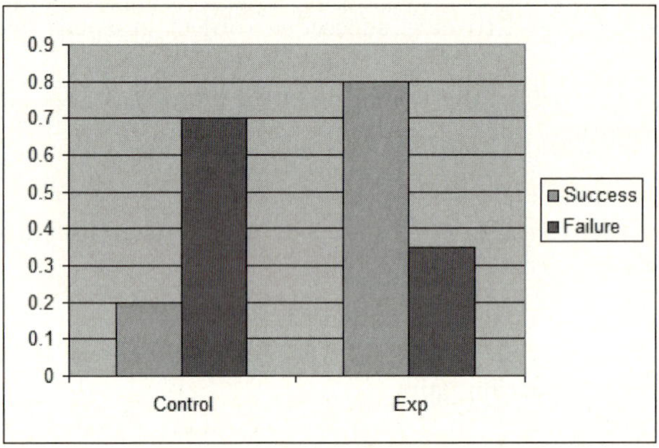

Fig. 3. Simulation results for the ball and stand paradigm showing improvement by observational effects

In this figure we can see that a single observation allows rapid increase in success percentage.

3 Methods

For the ball and stand paradigm we define an affordant action (for the affordance of grasping the ball) to consist of a set of three contact points, each defined in physical space coordinates, and a variable specifying the angle of approach to the ball, which takes two possible values V (for a vertical approach) or H (for a horizontal approach). The object module is automatically activated with the ball object during the simulation trial. The projection to the affordance/action pairs module then activates one of the two possible affordant actions. Initially the probability of selecting the vertical approach angle action is p_V and that of selecting horizontal p_H.

Once an affordance is selected, a motor plan is produced to move the arm to the contact points involved via the approach specified in the affordance. We introduce fine motor control error, such that the fingers arrive at the location of the contact point plus a normally distributed (in three dimensions) error of magnitude E_{motor} and standard deviation σ_{motor}. For the horizontal angle of approach, any errors large enough to move the center of gravity of the ball over the edge of the stand causes it to fall off, failing the task. For the vertical angle of approach, the center of gravity of the ball must be moved more than twice the radius of the base away from its center for it to fall off (a simple way to model the lower chance of knocking the ball off the stand when the approach is from above).

If the ball is knocked off the stand, the probability of using that action on the affordance is adjusted downwards by T_{reduc} and the probability of using the other affordant action is increased correspondingly (such that the probability of using either remains 1). If the grasp is successful, the probability of using the affordant action selected is increased by T_{gain} and the other affordant action reduced correspondingly. During the observation section of the simulation, there is probability p_{recog} of the system correctly recognising the affordant action used. The probability of using this affordant action is then increased by T_{obs}, and the probability of the other affordant action decreased correspondingly.

3.1 Neuronal Basis

We describe here the mechanisms by which the model is implemented.

Dedicated representations - graded neurons. Areas where dedicated representations are needed use graded neurons, the activation of which represents whether that representation is active. These graded neurons have a membrane potential, V, following the equation:

$$C\frac{dV}{dT} = g_{leak}(V - V_leak) + I_{input} \tag{1}$$

and produce output:

$$O = \frac{1}{1 + exp(-V/V_{scale})} \tag{2}$$

Goals are coded with two graded neurons - one for the goal "Get ball" and one for the goal "Push ball over". These nodes are mutually interconnected with inhibitory connections, such that only one can be active concurrently.

Affordances are also coded with dedicated nodes, for "Grasp" and "Push". Affordant actions for the grasp affordance are coded with nodes for H and V, representing the horizonal and vertical grasps. Again, nodes have mutual inhibition such that only one can be active at a time.

The connections between affordances and actions are subject to plasticity, according to the equation:

$$w_{ij} = LV_iV_j \tag{3}$$

where w_{ij} is the weight between nodes i and j, L is a learning rate constant and V_i and V_j are the membrane potentials of neurons i and j. This connection can also be modified by the error monitor when a failed attempt to grasp the ball is detected.

Sensorimotor integration and motor planning. Both the sensorimotor integration and motor planning modules rely on vector subtraction. The motor planning module takes as input an action represented as a direction of approach to contact points of final finger positions and an approach vector, and the current position of the hand, and so performs the calculation:

$$Act = C - H \tag{4}$$

Where Act is the vector representing the action to be performed (as a trajectory of finger movements), H is the vector representing hand position, and C is the position of the contact points to be reached.

Mental simulation - goal inference. The mental simulation loop is used to extract the goal of an observed action on an affordance from the observed movements. It takes as input an encoding of the movements of the observed actor. This movement consists of a set of trajectory points, which are nine dimensional vectors. These are then used to, activate one of two goal nodes, the output of which is one of two possible goals, corresponding to those described above (get ball or push ball).

Actions. Actions are simply output as the positions of the end effectors of the system, resulting in triplet of three dimensional vectors.

3.2 Tube and Stick Paradigm

For the tube and stick paradigm, we use a simpler form of simulation. The system is presented with the setup, and then chooses a strategy to achieve the goal of retrieving the object from the tube. Initially the tube presents the affordance of trying to extract the object manually, A_{manual}. The stick provides the affordance of reaching into the tube A_{insert}, but there is only a probability $p_{A_{initial}}$ of this being recognised.

The system then chooses an affordance on which to act, either A_{manual} or A_{insert}. If the former is selected, the probability of success is p_{manual}, while if the latter is chosen, p_{insert}. The probability of success with the insertion affordance is much higher.

During observation, the demonstrator shows usage of the affordance A_{insert} which may then be perceived by the system.

4 Brain Basis

Development of the infant brain occurs in a topographic manner, from lower to higher areas: a hierarchy of visual areas is brought on stream or enabled to become active by axonal myelination in a successive manner. Such a process has been modelled, for example, in detailed neural network models of the hierarchical visual system V1→V2→V3→V4→TEO→TE. Lower level synaptic connections were trained first in the model and the resulting outputs of the low-level feature detectors used to train the further higher-level feature detectors. In this way it proved possible to explain the sensitivity of V2 neurons to angles, of V4 neurons to partial edges of shapes, up to spatially invariant object representations in TE [3].

Such successive training of a hierarchy of modules, on top of the initial skeleton framework, arising from a genetic inheritance, also applies to prefrontal cortex starting with motor cortex and pre-motor cortices. Thus we expect that affordances, guiding the set of actions that can be taken on a given object, would also be developed in such a two-stage manner.

There are two sorts of affordances objects provide: those based on present experience and those learnt from past experience. The memorised affordances can be regarded as guiding possible actions that can be made, having been made in the past. The perceived affordances would then be those which arise through observation of trial-and-error processing or observational learning, also guided by the genetic skeleton. The perceived affordances would be stored in some short-term memory initially, but would, if salient enough, become part of the set of (long-term) memorised affordances.

We thus arrive at a picture of the learning of affordances as part of the overall learning of sensori-motor actions. The stored affordances module in figure 1 is to be regarded as the set of memorise affordances activated by the input of a given object code. It is this one-to-many mapping which is memorised by an infant in the process of babbling or making more goal oriented actions. The influence of goals coded in the goal module of figure 1 will begin to play their influence as goal creation occurs in the infant brain. Thus the stored affordance module would be part of long-term memory, very likely associated with developing object codes in the infant brain.

Results of brain imaging during observational learning and response indicate that considerable areas of the cortex are activated by observing an action being performed. In particular parietal and temporal areas are activated in this process, as well as those in pre-motor cortex; also STS is activated when an object is observed [4], corresponding to the possible significant points on the object by which action can be taken on it (such as the contact points).

Besides such features of an object there are also the relevant limb to be used and its orientation in making an action on the object (such as the horizontal versus vertical grip on the ball in the paradigm we have discussed). We expect these to be encoded in pre-motor cortex (PMC) as part of the action plan, and in the parietal region as to the limb (associated with a body map)..

The model of figures 1 & 2 is thus related to possible brain sites as follows:

Vision: The hierarchy of visual cortical modules
Object recognition/object codes: The temporal lobe
Goals: Prefrontal cortices (especially DLPFC and VLPFC)
Error detection: Cingulate cortex (and possibly DLPFC)
Contact points of stored affordances: STS
Motor plans for acting on the contact points: PMC
Proprioceptive information: S1, S2
Part of body to be used: Parietal (such as PRR)
Sensorimotor integration: SPL, IPS (parietal reach area, etc)
Hand position: Posterior parietal, such as V6A

5 Discussion

We have shown that the simulation can reproduce experimental results showing improvement at the ball and stand task from both trial and error experimentation and observational learning. The observational learning improvement comes from allowing the transfer of the action on the affordance used.

5.1 Why Eight Months?

One of the most interesting questions posed by the experimental results is why the infants of 8 months of age demonstrate observational learning so clearly, while older infants do not (additionally the 8 month olds seem to learn by trial and error in a manner not shared by older infants). Part of this may be due to the unusually low success rate of 8 month infants attempting the task for the first time, which could be low due to the sample size. However the increase is certainly statistically significant. Another possibility is that the differences in task difficulty are such that strategy becomes more important than motor ability only for the 8 month olds, and that strategy is the key component of observational learning.

5.2 Learning Affordant Actions and Discovering Affordances

Another question we have already raised is that of how initial affordant actions are learned (and to some extent how affordances are discovered). Our model covers only the transfer of existing affordance/action pairs to new situations, not the learning of those affordant actions. This transfer assumes that the actions are already known to the infant, either in a different context or in the abstract form, however it is interesting to address how these actions are developed, and also how they might be modified during transfer.

Whether these actions can be learned by observation or require trial and error learning is unclear and yet to be determined. Since many actions are learned very early, this may be difficult to test. We can assume that there are some action primitives learned during early infant motor development. These are used to form basic actions such as finger movements. Experience and trial and error then allow these to form more complex actions, such as grasps. It may then be possible to develop these actions by observation.

For developing/associating the affordant actions we see in the experiment and simulation, we can consider a two stage process. In the first stage, the action is observed and roughly classified (type of grip, angle of approach, hand shaping). At this point, fine details such as the exact contact points of fingers must be inferred if they cannot be observed (possibly by mental simulation). In the second stage, trial and error allows the representation to be refined, such that contact points become more established, hand shaping is better, speed of approach becomes better tuned and so on. Further experiment is needed to determine whether this two stage mechanism is necessary, particularly looking at the second stage.

6 Conclusions

Based on the model results and other factors, we can draw some conclusions.

- That transfer/discovery of affordances and actions on those affordances can explain how observational learning can operate in this paradigm.
- That several other factors involving the infant's state and state of the object are important, and a more complex understanding of affordances is likely needed.
- That the demonstration of non-perceived affordances may be a more effective method of observational learning than the transfer of actions on affordances.

References

1. Raos, V., Evangeliou, M.N., Savaki, H.E.: Observation of action: grasping with the mind's hand. Neuroimage 23, 193–201 (2004)
2. Gibson, J.J.: The theory of affordances. In: Shaw, R., Bransford, J. (eds.) Perceiving, acting and knowing: towards an ecological psychology. Lawrence Erlbaum, Mahwah (1977)
3. Taylor, N.R., Panchev, C., Kasderidies, S., Hartley, M., Taylor, J.G.: Occlusion, attention and object representations. In: Lecture notes in computer science, pp. 592–601. Springer, Berlin (2006)
4. Miall, R.C.: Connecting mirror neurons and forward models. Neuroreport 14, 2135–7 (2003)

Modeling the Effects of Dopamine on the Antisaccade Reaction Times (aSRT) of Schizophrenia Patients

Ioannis Kahramanoglou[1,⋆], Stavros Perantonis[1], Nikolaos Smyrnis[2,3], Ioannis Evdokimidis[3], and Vassilis Cutsuridis[4]

[1] Computational Intelligence Laboratory, Institute of Informatics and Telecommunications, National Center for Scientific Research "Demokritos", Agia Paraskevi, Athens GR-15310
sper@iit.demokritos.gr
[2] Cognition and Action Group, Neurology Department, National University of Athens, Aeginition Hospital, 72 Vas Sofias Ave, Athens GR-11528
smyrnis@med.uoa.gr
[3] Cognition and Action Group, Neurology Department, National University of Athens, Aeginition Hospital, 72 Vas Sofias Ave, Athens GR-11528
[4] Department of Computing Science and Mathematics, University of Stirling, Stirling, FK9 4LA, U.K.
vcu@cs.stir.ac.uk

Abstract. In the antisaccade task, subjects are instructed to look in the opposite direction of a visually presented stimulus. Controls can perform this task successfully with very few errors, whereas schizophrenia patients make more errors and their responses are slower and more variable. It has been proposed the fundamental cognitive dysfunction in schizophrenia involves prefrontal dopaminergic hypoactivity. We examine via computer simulations the effects of dopamine on the variability of aSRTs in a neural cortico-collicular accumulator model with stochastic climbing activity. We report the simulated aSRTs for the hypo-DA level have higher standard deviation and mean values than in the normal and hyper DA level. The simulated higher mean and standard deviation for the hypo-DA group resemble the performance differences in the antisaccade task observed in patients with schizophrenia and are in accordance with the theory of a hypo-DA state in the frontal cortical areas of patients with schizophrenia.

Keywords: Accumulator model, schizophrenia, antisaccade task, reaction times, dopamine, cortex, superior colliculus, pyramidal cells, inhibitory interneurons

1 Introduction

It has been hypothesized that patients with schizophrenia are hypo-dopaminergic in their prefrontal cortex and this state has been related to their poor performance in decision making and working memory tasks [10]. In this study, we

⋆ Corresponding author.

V. Kůrková et al. (Eds.): ICANN 2008, Part II, LNCS 5164, pp. 290–299, 2008.

examine the effects of dopaminergic state of a cortical input to a collicular neural model by simulating its performance in the antisaccade task. In particular, we extend a biophysical cortico-collicular neural model [3], [4], [5], [6] and model the DA-D_1 effects on the generation of slowly varying climbing activity of the reactive and planned input decision signals of a SC model in the antisaccade task [1], [2]. Psychophysical parameters such as the decrease of mean and standard deviation, as well as the independence of the mean of error prosaccades observed among two groups of individuals are explained qualitatively through the variation of D_1 component of dopamine (DA) in the modeled prefrontal cortex (PFC). The groups that are compared to the simulations are: (1) a group 2006 normal subjects and (2) a group of 30 patients suffering from DSM-VI schizophrenia. This work combines and extends previous biophysical models [3], [4], [5], [6]. A much larger set of modeled empirical findings have been presented in [30].

2 Materials and Methods

2.1 Basis of the Model

In a modeling attempt of the antisaccade task [1],[2], Cutsuridis and colleagues [3], [8], [9] hypothesized that the preparation of an antisaccadic eye movement consisted of two cortically independent and spatially separated decision signals representing the reactive and planned saccade signals, whose linearly rising phases were derived from two normal distributions with different means and standard deviations. These two cortical decision signals were then integrated at opposite colliculi locations, where they competed against each other via lateral excitation and remote inhibition. A saccade was initiated when these decision processes, represented by the neuronal activity of SC buildup neurons with nonlinear growth rates varying randomly from a normal distribution, gradually build up their activity until reaching a preset criterion level. The crossing of the preset criterion level in turn released the "brake" from the SC burst neurons and allowed them to discharge resulting in the initiation of an eye movement. One of the model's predictions was that there is no need of a top-down inhibitory signal that prevents the error prosaccade from being expressed, thus allowing the correct antisaccade to be released. Moreover, the model offered a functional rationale at the SC neuronal population level of why the antisaccadic reaction times are so long and variable and simulated accurately the correct and error antisaccade latencies, the shape distributions and the error probabilities.

In a follow up of this study Cutsuridis and colleagues [4], [5], [6] modeled the decision signals with adjustable slopes with the population activities of two cortical networks of pyramidal neurons and inhibitory interneurons and predicted that only the currents NaP, AMPA and NMDA can produce the range of slope values of these decision signals in the collicular model.

In this study, we extend the latter model to incorporate and study the effects of dopamine on the conductances of the predicted NaP, AMPA and NMDA currents and simulate qualitatively a number of psychophysical parameters in the antisaccade task. Preliminary results of this study have been published in [7], [31].

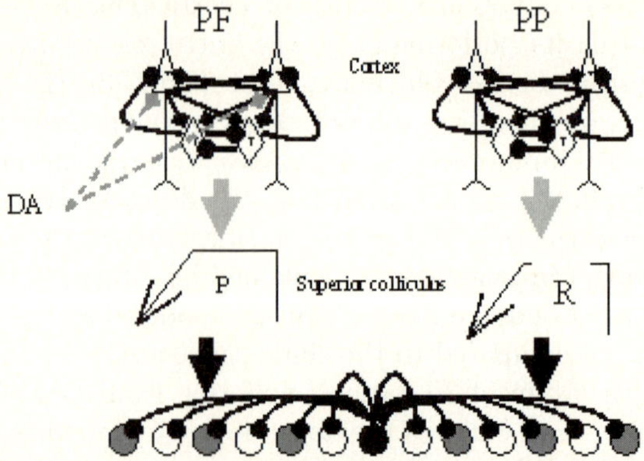

Fig. 1. Composite model architecture of cortical modules and superior colliculus module with reactive and planned inputs. Cortex: triangular neurons symbolize pyramidal cells and diamond shaped neurons symbolize GABAergic inhibitory inteneurons. Superior colliculus: black nodes are fixation cells, gray nodes are buildup cells, and white nodes are burst cells. The inputs to this layer are classified as reactive (R) and planned (P). DA: mesocortical dopamine innervation of prefrontal cortex (PFC). Their respective time course is shown schematically corresponding to an onset and offset.

2.2 Cortical Network Architecture

Standard Hodgkin-Huxley modeling techniques were used to simulate networks of single compartmental models of cortical pyramidal neurons and cortical inhibitory interneurons (IN). Pyramidal neuron membrane potential obeyed

$$CdV/dt = -(I_{leak}+I_{Na}+I_k+I_{NaP}+I_{k(Ca^{2+})}+I_{AMPA}+I_{NMDA}+I_{GABAA}+I_{inj})$$

with Cm = 1.2 μF/cm^2. GABAergic inhibitory interneuron membrane potential obeyed

$$CdV/dt = -(I_{leak} + I_{Na} + I_k + I_{AMPA} + I_{NMDA} + I_{GABAA} + I_{inj})$$

with Cm = 1.2 μF/cm^2. The synaptic currents I_{AMPA}, I_{NMDA}, I_{GABA-A} are represented by the equation $I_X = g_X$ (V - E_X), where the conductances are measured in mS/cm^2 and the reversal potentials in mV (see fig. 2 for numerical values). In addition I_{NMDA} contained a voltage dependent Mg^+ gate s(V)=1.08(1+0.19exp(-0.064V))$^{-1}$. The persistent sodium current (NaP) was modeled as in [14]. The calcium activated potassium current is given by

$$I_{k(Ca^{2+})} = g_{k(Ca^{2+})}Cai(V - E_{k(Ca^{2+})})$$

where the conductance $g_k(Ca^{2+})$=0.07 mS/cm^2 and the reversal potential $E_k(Ca^{2+})$=-95 mV. The variable Cai was the intracellular calcium concentration measured in μM [24]. Because very little is known about the detailed connectivity of neurons and the associated synaptic strengths in the frontal and posterior parietal cortices, we intentionally kept the network model as general as possible. Two networks of 20 pyramidal cells and 4 GABAergic interneuron each were simulated. In each network, we assumed that all pyramidal cells and GABAergic interneurons were fully connected [5]. The output of each network was the average population activity of a homogenous population of neurons with identical connections. These outputs were then used as the input drives of the superior colliculus (SC) model [3].

2.3 Superior Colliculus Model

The superior colliculus model is a one-dimensional on-center off-surround leaky competitive integrator of the intermediate layer of the superior colliculus developed by [13] and extended by our group [3]. The neural architecture of the model is described in figure 1. Self-excitation and lateral inhibition is assumed between all neurons in both superior colliculi.

2.4 Dopamine Modification

The effects of DA-D_1 on the ion conductances were the following:

1. Enhancement of I_{NMDA} (replacing g_{NMDA} with $r_{NMDA}g_{NMDA}$) [15],[16]
2. Enhancement of I_{NMDA} for the inhibitory inteneurons (replacing g_{NMDA} with $r_{NMDA}g_{NMDA}$) [17]
3. Suppression of I_{AMPA} (replacing g_{AMPA} with $r_{AMPA}g_{AMPA}$) [15]
4. Enhancement of I_{NaP} (replacing V_{NaP} and E_{NaP} with $V_{NaP+\delta v}$ and $E_{NaP+\delta v}$) [18]
5. Reduction of $I_K(Ca^{2+})$ (replacing $g_K(Ca^{2+})$ with $r_K(Ca^{2+})g_K(Ca^{2+})$) [28]
6. No effect on the GABA current [16]

Symbol	Value
C_p (μF/cm^2)	1.2
C_{inh} (μF/cm^2)	1.2
g_L (mS/cm^2)	0.02
g_{nmda} (mS/cm^2)	0.17-0.25
g_{ampa} (mS/cm^2)	4
g_{gaba} (mS/cm^2)	1.45
g_{NaP} (mS/cm^2)	4
$g_{K(Ca)}$ (mS/cm^2)	0.07
E_L (mV)	-65
E_{nmda} (mV)	0
E_{ampa} (mV)	0
E_{gaba} (mV)	-70
E_{Nap} (mV)	-55
$E_{K(Ca)}$ (mV)	-95

Fig. 2. Synaptic and ionic model parameters. All the other values of the parameters used in the model were obtained from [5], [6].

	low	optimum	high
r_{NMDA}(pyramidal cell)	0.300-0.400	1.000-1.010	1.180-1.200
r_{NMDA}(inteneuron)	0.001-0.080	1.000-1.300	2.500-3.000
r_{AMPA}	1.110-1.100	1.000-0.950	0.930-0.900
$r_{K(Ca)}$	1.110-1.100	1.000-0.950	0.930-0.900
δv(mV)	2.0-1.5	0-(-0.5)	-2.0-(-2.5)

Fig. 3. The effects of D_1 receptor activation on the ion channels

The detailed values of the above parameters for the three levels of dopamine (D_1) for which the network was simulated are given in fig. 3. The factor r represents the amount of available receptors for a given level of dopamine and it is incorporated in the simplified kinetic model theory [19]. All of the r-values given in fig. 3 are experimentally measured values of the given currents, in patch clamp experiments, under the action of D_1 agonists-antagonists [20],[14]. In the case of the persistent sodium current no effect on the amplitude of the current was observed but rather a shift of the potential at which the maximum amplitude occurred. The GABA current remained unchanged under the action of D_1 agonists-antagonists [20].

3 Experimental Setup

The control data used in this study were collected in an antisaccade task [1],[2]. Details of the experimental procedure used for the collection of these data are described therein [1],[2]. The same experimental setup as in [1],[2] was used for collecting the reaction times of the group with schizophrenia. The mean and standard deviation of the aSRT's for each group was calculated using Ratcliff

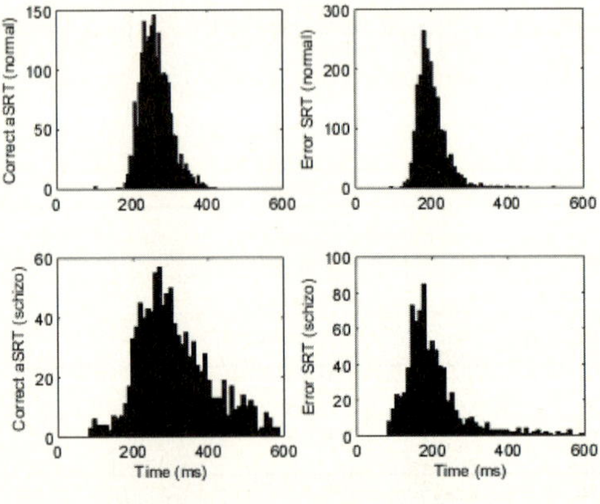

Fig. 4. (Top-left)Histogram of correct antisaccade reaction times (aSRT) of normal subjects. (Top-right) Histogram of error prosaccade reaction times (SRT) of normal subjects. (Bottom-left) Histogram of correct antisaccade reaction times (aSRT) of schizoophrenia subjects. (Bottom-right) Histogram of error prosaccade reaction times (SRT) of schizophrenia subjects.

analysis [12]. The aSRT's for each subject were sorted and then divided into four quantiles, the mean for each quantile was calculated among each group. This procedure produced a distribution of aSRT's, which was considered to be the average distribution of the group under question. From these distributions all the features of the group reaction times could be calculated, mean and standard deviation of the whole set of answers as well as mean and standard deviation of the correct antisaccades and error prosaccades separately. Figure 4 depicts the correct and error SRT histograms of the normal and schizophrenia populations.

4 Results

In a network of 20 pyramidal neurons and 4 inhibitory interneuron and for each of the three dopamine levels, for each simulation, 480 reaction times were obtained from the collicular model. The values of the rising phase of the signals were obtained from the cortical model described above. A variation for the conductance of the NMDA current in the pyramidal neurons gave rise to a bell-shaped rising phase (slope) in the spiking of the cortical network. The spiking of the cortical network was linearly fit with a line from the beginning of the spiking to its maximum value. Then this bell-shaped distribution of gradients was fed into the superior colliculus model to produce the antisaccade reaction times for each dopamine level. In these simulations we assumed that the signal of the error prosaccades was not affected by dopamine, since it is not being produced by the prefrontal cortex, whereas the signal of the correct antisaccades is modified by dopamine (D_1).

4.1 Effects on the Mean of the Antisaccade Reaction Times

The effects of (D_1) dopamine modification of the cortical signal was to decrease the mean of the aSRT's as indicated by Fig. 5 (Left) of the simulated results for

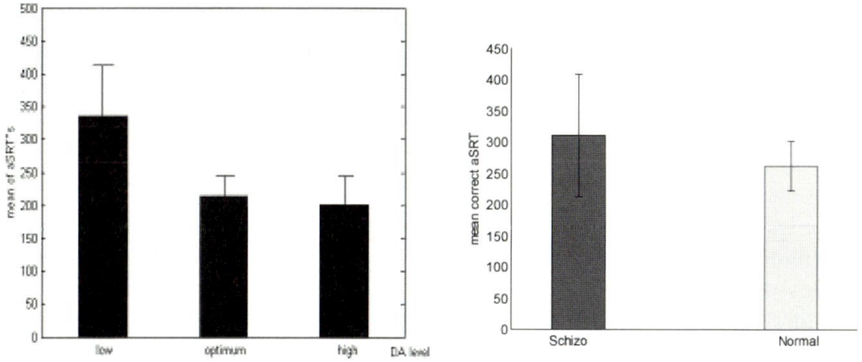

Fig. 5. (Left) Qualitative simulation results indicating a decrease of mean of the aSRT's as D_1 level increased. (Right) The experimental data of the mean of the aSRT's for the two groups (normal and patients) show a similar decrease.

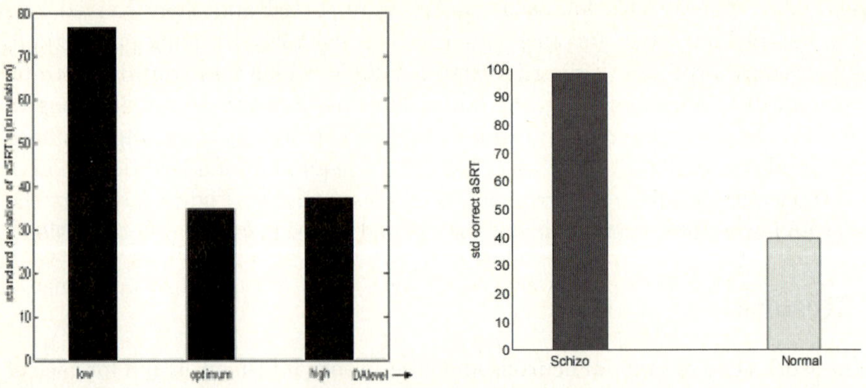

Fig. 6. (Left) Qualitative simulation results showing a decrease of the standard deviation of the aSRT distribution as dopamine level is increased. (Right) The experimental data for the groups under study show a similar decrease.

the three levels of dopamine. In Fig. 5 (Right) the experimental data of the two groups show a similar decrease in mean aSRT [21],[22].

4.2 Effects on the Standard Deviation of the Antisaccade Reaction Times

The effect of (D_1) modification of the cotical signal was an increase in the standard deviation of the aSRT's as the dopamine level was decreased (Fig. 6 Left). This observation was resembled qualitatively the observed increase of the standard deviation in the patient group (Fig 6 Right) [21].

5 Conclusion

The psychophysical data and results of the simulations presented in this study provide evidence that there is a modification of the cortical signal (prefrontal cortex) due to dopamine component D_1, and that this modification can explain the increase in the mean and variance in the antisaccade reaction times produced by patients.

Other models concerning the modification of cortical signals by a D_1 action of dopamine can be found in the literature concerning working memory tasks, the inverted U-shape of firing of the PFC neurons as well as the operation of the PFC under the action of psychostimulants [24],[14],[25],[26]. In all of the above studies the action of D_1 on the cortical currents was modeled and in all of the studies the qualitative action of dopamine on ionic currents was the same.

The effect of dopamine on D_2 receptors was not investigated in this study. This effect could be significant at high dopamine levels [28] and may interpret other sets of data regarding the function of the prefrontal cortex when high levels of dopamine are observed. Hence we could conclude that the lack of a principal neurotransmitter in the prefrontal cortex can have very significant effects in decision making, generating poorer performance in such tasks.

References

[1] Evdokimidis, I., Smyrnis, N., Constantinidis, T.S., Stefanis, N.C., Avramopoulos, D., Paximadis, C., Theleritis, C., Efstratiadis, C., Kastrinakis, G., Stefanis, C.N.: The Antisaccade Task in a Sample of 2006 Young Men I. Normal Population Characteristics. Exp Brain Res. 147, 45–52 (2002)

[2] Smyrnis, N., Evdokimidis, I., Stefanis, N.C., Constantinidis, T.S., Avramopoulos, D., Theleritis, C., Paximadis, C., Efstratiadis, C., Kastrinakis, G., Stefanis, C.N.: The Antisaccade Task in a Sample of 2006 Young Males II. Effects of Task Parameters. Exp. Brain Res. 147, 53–63 (2002)

[3] Cutsuridis, V., Smyrnis, N., Evdokimidis, I., Perantonis, S.: A Neural Model of Decision Making by the Superior Colliculus in an Antisaccade Task. Neural Networks 20(6), 690–704 (2006)

[4] Cutsuridis, V., Kahramanoglou, I., Smyrnis, N., Evdokimidis, I., Perantonis, S.: A Neural Variable Integrator Model of decision making in an Antisaccade Task. Neurocomputing 70(7-9), 1390–1402 (2007)

[5] Cutsuridis, V., Kahramanoglou, I., Perantonis, S., Evdokimidis, I., Smyrnis, N.: A Biophysical Model of decision making in an Antisaccade Task through variable climbing activity. In: Duch, W., Kacprzyk, J., Oja, E., Zadrożny, S. (eds.) ICANN 2005. LNCS, vol. 3696, pp. 205–210. Springer, Heidelberg (2005)

[6] Cutsuridis, V., Kahramanoglou, I., Smyrnis, N., Evdokimidis, I., Perantonis, S.: Parametric Analysis of Ionic and Synaptic Current Conductances in an Acuumulator model with variable Climbing Activity. In: Book of abstracts of 19th Conference of Hellenic Society for Neuroscience, Patra, Greece (2005)

[7] Kahramanoglou, I., Cutsuridis, V., Smyrnis, N., Evdokimidis, I., Perantonis, S.: Dopamine Modification in a Neural Accumulator Model of the Antisaccade Task. In: Book of abstracts of the 1st Computational Cognitive Neuroscience Conference, New Orleans, USA, November 11-13 (2005)

[8] Cutsuridis, V., Smyrnis, N., Evdokimidis, I., Kahramanoglou, I., Perantonis, S.: Neural network modeling of eye movement behavior in the antisaccade task: validation by comparison with data from 2006 normal individuals, Program No. 72.13. 2003 Abstract Viewer/Itinerary Planner. Society for Neuroscience, Washington (2003)

[9] Cutsuridis, V., Evdokimidis, I., Kahramanoglou, I., Perantonis, S., Smyrnis, N.: Neural Network model of eye movement behavior in an antisaccade task. In: Book of abstracts of the 18th Conference of Hellenic Society for Neuroscience, Athens, Greece (2003)

[10] Revheim, N., Schechter, I., Kim, D., Silipo, G., Allingham, B., Butler, P., Javitt, D.C.: Neurocognitive and symptom correlates of daily problem-solving skills in schizophrenia. Schizophr Res. (2006)

[11] Shampine, L.F., Reichelt, M.W.: The MATLAB ODE Suite. SIAM Journal on Scientific Computing 18, 1–22 (1997)

[12] Ratcliff, R.: Group reaction time distributions and an analysis of distribution statistics. Psychol. Bulletin 86(3), 446–461 (1979)

[13] Trappenberg, T.P., Dorris, M.C., Munoz, D.P., Klein, R.M.: A model of saccade initiation based on the competitive integration of exogenous and endogenous signals in the superior colliculus. J. Cogn. Neurosci. 13(2), 256–271 (2001)

[14] Durstewitz, D., Seamans, J.K., Sejnowski, T.J.: Dopamine-Mediated Stabilization of Delay-Period Activity in a Network Model of Prefrontal Cortex. J. Neurophys. 83, 1733–1750 (2000)

[15] Chen, L., Yang, C.R.: Interaction of dopamine D1 and NMDA receptors mediates acute clozapine potentiation of glutamate EPSPs in rat prefrontal cortex. J.Neuphysiology 87, 2324–2336 (2002)

[16] Seamans, J.K., Durstewitz, D., Christie, B.R., Stevens, C.F., Sejnowski, T.J.: Dopamine D1/D5 receptor modulation of excitatory synaptic inputs to layer V prefrontal cortex neurons. PNAS 98(1), 301–306 (2001)

[17] Gorelova, N., Seamans, J., Yang, C.R.: Mechanisms of dopamine activation of fast spiking interneurons that exert inhibition in rat prefrontal cortex. J.Neuroph. 88, 3150–3166 (2002)

[18] Gorelova, N.A., Yang, C.R.: Dopamine D1/D5 receptor activation modulates a persistent sodium current in rat prefrontal cortical neurons in vitro. J.Neurophysiol. 84, 75–87 (2000)

[19] Destexhe, A., Mainen, Z.F., Sejnowski, T.J.: An efficient method for computing synaptic conductances based on a kinetic model of receptor binding. Neural Comp. 6, 14–18 (1994)

[20] Seamans, J.K., Yang, C.R.: The principal features and mechanisms of dopamine modulation in the prefrontal cortex. Progress in Neurobiology 74, 1–57 (2004)

[21] Brownstein, J., Krastishevsky, O., McCollum, C., Kundamal, S., Matthysse, S., Holzman, P.S., Mendell, N.R., Levy, D.L.: Antisaccade performance is abnormal in schizophrenic patients but not their biological relatives. Schizophrenia Research 63, 13–25 (2003)

[22] Manoach, D.S., Lindgren, K.A., Cherkasova, M.V.: Schizophrenic subjects show deficit inhibition but intact task switching of saccadic tasks. Biol. Psychiatry 51, 816–826 (2002)

[23] Reuter, B., Rakusan, L., Kathmanna, N.: Poor antisaccade performance in schizophrenia: An inhibition effect? Psychiatry Research 135, 1–10 (2005)

[24] Tanaka, S.: Dopamine controls fundamental cognitive operations of multi-target spatial working memory. Neural Networks 15, 573–582 (2002)

[25] Yamashita, K., Tanaka, S.: Parametric study of dopaminergic neuromodulatory effects in a reduced model of the prefrontal cortex. Neurocomputing 65-66, 579–586 (2005)

[26] Tanaka, S.: State-dependent alteration of dopamine and glutamate transmission in the prefrontal cortex by psychostimulants. Neurocomputing 65-66, 587–594 (2005)

[27] Yamashita, K., Tanaka, S.: Circuit properties of the cortico-mesocortical system. Neurocomputing 52-54, 969–975 (2003)

[28] Davidson, H.T., Neely, L.C., Lavin, A., Seamans, J.K.: Mechanisms underlying differential D1 versus D2 dopamine receptor regulation of inhibition in prefrontal cortex. J. Neurosci. 24(47), 10652–10659 (2004)

[29] Yamashita, K., Tanaka, S.: Circuit simulation of memory field modulation by dopamine D1 receptor activation. Neurocomputing 44-46, 1035–1042 (2002)
[30] Kahramanoglou, I., Cutsuridis, V., Smyrnis, N., Evdokimidis, I., Perantonis, S.: Dopamine modification of the climbing activity in a neural accumulator model of the antisaccade task. Neural Networks, under review
[31] Kahramanoglou, I., Cutsuridis, V., Smyrnis, N., Evdokimidis, I., Perantonis, S.: Dopamine effect on climbing activity of a cortico-tectal model: Simulating the performance of patients with DSM-IV schizophrenia in the antisaccade task. In: 2nd Annual Conference on Computational Cognitive Neuroscience, Houston, USA, November 16-19 (2006)

Fast Multi-command SSVEP Brain Machine Interface without Training

Pablo Martinez Vasquez[1], Hovagim Bakardjian[1], Montserrat Vallverdu[2], and Andrezj Cichocki[1]

[1] Laboratory for Advanced Brain Signal Processing, RIKEN Brain Science Institute 2-1, Hirosawa, Wako-shi Saitama, Japan
[2] Universidad Politécnica de Cataluña, Barcelona, Spain

Abstract. We propose a new multi-stage procedure for a real time brain machine/computer interface (BMI) based on the Steady State Visual Evoked Potentials (SSVEP) paradigm elicited by means of flickering checkerboards. The developed system work in asynchronous mode and it does not require training phase and its able to detect fast multiple independent visual commands. Up to 8 independent commands, were tested at the presented work and the proposed BMI system could be extended to more independent commands easily. The system has been extensively experimented with 4 young healthy subjects, confirming the high performance of the proposed procedure and its robustness in respect to artifacts.

1 Introduction

A brain-machine interface (BMI) is a system that acquires and analyzes brain signals to create a new communication channel in real-time between the human brain and a computer or a machine [1].

We propose a new multi-stage procedure for a real time brain machine/computer interface (BMI) based on the Steady State Visual Evoked Potentials (SSVEP) paradigm. Applications of the SSVEP to BMI were proposed first by Middendorf et al. [6] and investigated by many researches [3, 7].

A visual stimulation event elicites a visual evoked potential activity (VEP), originated on the occipital cortex, that is embedded into the spontaneous brain activity, measured with the Electro-encephalogram (EEG). When visual stimulation events are repeated and the stimulation frequencies are higher than approximately 4 Hz the consecutive individual elicited potentials (VEP) overlap leading to periodic response, named as Steady State Evoked Potential (SSVEP) [2].

The proposed BMI system allows a fast estimation of the energies of the periodic visual evoked responses elicited when the subject gazes a checkerboard flickering at a specific frequency. The users can select between different independent commands, associated with checkerboards flickering at different frequencies (in our case 8 commands) that could belong simultaneously to different frequency visual stimulation ranges, from low stimulation range [4-9 Hz] to middle range [11-25 Hz]. There is no restriction that the stimulation frequencies are multiple

V. Kůrková et al. (Eds.): ICANN 2008, Part II, LNCS 5164, pp. 300–307, 2008.

to each other because the algorithm does not depend on the estimation of the higher harmonics energies but rather relays on the fundamental stimulation frequency [6,3,7]. The developed BMI system allows on asynchronous mode i.e. no event cue process is required like in synchronous mode.

Although some aspects of the SSVEP paradigm for BCI have been already exploited in a number of studies [3,6,7,8], our design is innovative, suitable for real-time applications and robust to artifacts due to the use of a Blind Source Separation (BSS) method in the preprocessing step, as also applied by us [4,5]. Moreover, the presented approach does not require a classifier, thus making possible to be used without initial training phase (plug and play).

To evaluate the performance of the algorithm we have designed a layout on a computer screen with 8 small checkerboards images flickering and moving along with a controlled object. Extensive experiments with 4 young healthy subjects with 5 electrodes placed on occipital area CPz, Pz, POz, P1, P2 and one in the forehead Fz were performed.

2 Methods

Our BMI system consists of a visual stimulation unit designed as a multiple choice table in the form of an array of eight small checkerboard images flickering with different frequencies and moving continuously along with the controlled object (see Fig. 1(a)). The eight small checkerboards were flickering at frequencies {8.8, 9.4, 11.55, 12.5, 13.65, 15 16.7, 18.8 Hz} using a 21-inch CRT monitor with refresh rate of 170 Hz observed at a distance of 80-100 cm from the center. Each

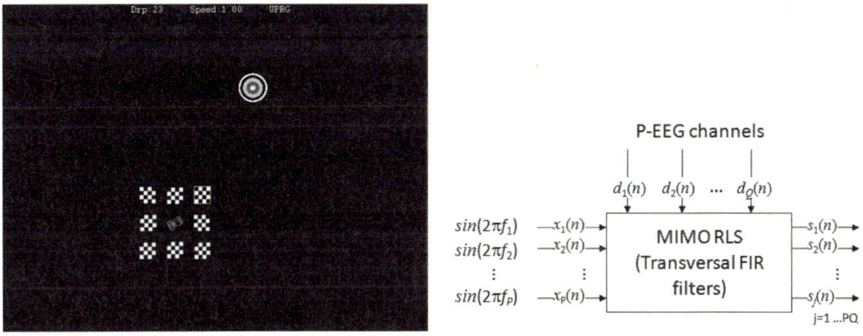

(a) (b)

Fig. 1. (a) Eight small checkerboards flickering at different but fixed frequencies moving along with object. Object can move in eighth different directions. The subject should move the object to reach a target that is located randomly on the screen. (b) Multi-Input Multi-output model for feature extraction stage. Q-EEG channels are adaptively filtered using P-references, sinus waves with the same flickering frequencies, to estimate the SSVEP amplitudes.

checkerboard/frequency is assigned to a specific direction of movement of the controlled object separated 45°.

The BCI multistage analysis BMI unit consist in:

Acquisition. A data acquisition amplifier Biosemi© recording 6 EEG channels sampled at 512 Hz and prefiltered by means of a high pass filter with a cutoff frequency of 1 Hz.

Artifact rejection. A second Order Blind Source Separation (BSS) algorithm AMUSE [10] was applied for artifact rejection. A sliding window of 3 seconds with 0.120 ms of displacement for all channels was selected to apply AMUSE. The implemented BSS-AMUSE algorithm can be considered as consisting of two consecutive PCA (principal component analysis) blocks. First, PCA is applied to the input data of N points; and then a second PCA (SVD) is applied to the time-delayed covariance matrix (in our case, the delay is set to one sample or four milliseconds) of the output from the previous stage. In the rst step standard or robust prewhitening (sphering) is applied as a linear transformation $\mathbf{z}(n) = \mathbf{Q}\mathbf{x}(n)$ where $\mathbf{Q} = \mathbf{R}_{xx}^{1/2}$ of the standard covariance matrix of the observed data vector $\mathbf{x}[n]$. Next, Singular value decomposition(SVD) is applied to a time-delayed covariance matrix of prewhitened data, $\mathbf{R}_z = E\{z^T(t)z^T(t-1)\} = \mathbf{USV}^T$ where $\mathbf{S}$ is a diagonal matrix with decreasing singular values and $\mathbf{U}$, $\mathbf{V}$ are matrices of eigenvectors. Then, an unmixing (separating) matrix is estimated as $\mathbf{W} = \mathbf{U}^T\mathbf{Q}$ The estimated independent components were obtained as $\mathbf{Y} = \mathbf{WX}$ where $\mathbf{X} = [\mathbf{x}(1), \mathbf{x}(2), ..., \mathbf{x}(N)]^T$. The AMUSE BSS algorithm ranks the components (EEG components), thus undesired components corresponding to artifacts (*blinking, high frequency noise*) were removed at each time automatically. Once removed undesired components, the significant useful components were projected back to scalp level using the inverse of $\mathbf{W}$ to get the cleaned signals $\hat{\mathbf{X}} = \mathbf{W}^{-1}\mathbf{Y}$.

The first and the last components were rejected: the first component is related with to ocular and other low frequency artifacts while the last component is related mainly to muscular activity. This procedure enhances the overall performance of the system preventing rejection of many EEG epochs, see our previous publications [4, 5] for more details.

Feature extraction. To estimate the elicited SSVEP amplitude a MIMO mixing model was used. We have applied a Recursive Least Square (RLS) algorithm [9] to estimate it. The cleaned signals from the BSS stage where filtered by means of a MIMO RLS,(see Fig. 1(b)), where the reference signals were sinusoidal signals with the same frequencies as the flickering checkerboards. The output of each FIR transversal filter, of order L, correspond to the estimated sinusoid/SSVEP per channel and frequency.

Therefore, we assume that we have a MIMO system with Q inputs $d_q[n]$(EEG channels) with P reference inputs $x_p[n]$(sinusoidal signals) and PQ outputs $s_{pq}[n]$ representing the estimated SSVEP per channel and frequency, (see Fig.2(a)).

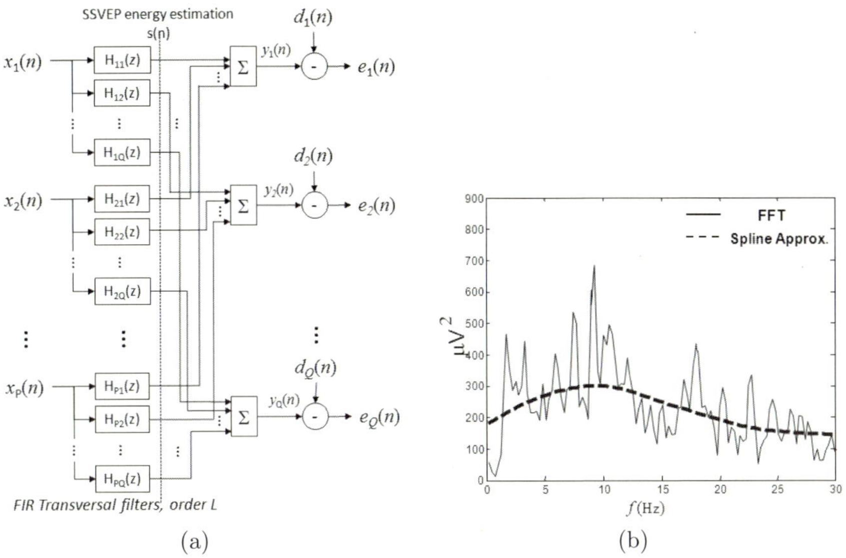

Fig. 2. (a) A MIMO system consisting P references $x_P(n)$(sinusoids or SSVEP approximations) and Q input signals (SSVEP embedded in EEG noise). The outputs $s_{pq}(n)$ of each transversal filter $\mathbf{H}_{pq}(z)$ are the estimations of the SSVEP amplitudes per frequency p and channel q (b)One channel EEG spectra estimation by FFT and the smoothing cubic spline approximation function. $p = 10^{-4}$.

The multivariate model is defined as equation:

$$\mathbf{e}(n) = \mathbf{d}(n) - \mathbf{H}^T\mathbf{x}(n) \tag{1}$$

where

$$\begin{aligned} \mathbf{d}(n) &= [d_1^T(n) d_2^T(n) \cdots d_Q^T(n)]^T \\ \mathbf{e}(n) &= [e_1^T(n) e_2^T(n) \cdots e_Q^T(n)]^T \\ \mathbf{x}(n) &= [\mathbf{x}_1^T(n) \mathbf{x}_2^T(n) \cdots \mathbf{x}_P^T(n)]^T \end{aligned} \tag{2}$$

with buffered reference input $\mathbf{x}_p(n)$

$$\mathbf{x}_p(n) = [x_p(n) x_p(n\text{-}1) \cdots x_p(n\text{-}L\text{+}1)]^T \tag{3}$$

and with the transfer function $\mathbf{h}_{pq}$ between channel q and reference p defined as a transversal L-order FIR, equation 4, included in the multivariate transfer function $H \in \mathbb{R}^{PL\times Q}$, equation 5.

$$\mathbf{h}_{pq} = [h_{pq,0} h_{pq,1} \cdots h_{pq,L-1}]^T \tag{4}$$

$$\mathbf{H} = \begin{vmatrix} \mathbf{h}_{11} & \mathbf{h}_{12} & \cdots & \mathbf{h}_{1Q} \\ \mathbf{h}_{21} & \mathbf{h}_{22} & \cdots & \mathbf{h}_{2Q} \\ \vdots & \vdots & \ddots & \vdots \\ \mathbf{h}_{P1} & \mathbf{h}_{P1} & \cdots & \mathbf{h}_{PQ} \end{vmatrix}. \tag{5}$$

The estimation of the multivariate transfer function $\mathbf{H}$ via RLS algorithm can be performed as follows, (see Eqs.6, 6).

$$e_q(n) = d_q - \mathbf{h}_q^T \mathbf{x}(n)$$
$$\mathbf{h}_q^T(n) = \mathbf{h}_q^T(n-1) + \mathbf{R}_{xx}^{-1}(n)\mathbf{x}(n)e_q(n) \qquad q = 1 \ldots Q$$

$$\mathbf{R}_{xx}^{-1}(n) = \lambda^{-1}\mathbf{R}_{xx}^{-1}(n-1) - \frac{\lambda^{-2}\mathbf{R}_{xx}^{-1}(n-1)\mathbf{x}(n)\mathbf{x}^T(n)\mathbf{R}_{xx}^{-1}(n-1)}{1+\lambda^{-1}\mathbf{x}(n)^T\mathbf{R}_{xx}^{-1}(n-1)\mathbf{x}(n)} \tag{6}$$

The estimation of the SSVEP per channel p and frequency q is given by

$$s_{pq}(n) = \mathbf{h_{pq}}^T \mathbf{x}(n) \tag{7}$$

With the multichannel RLS the Kalman gain is defined as $\mathbf{K}(n) = \mathbf{R}_{xx}^{-1}(n)\mathbf{x}(n)$ is the same for all the Q EEG channels. It should be noted that we need to computed it only one time per new sample.

The variance of the last hundred of milliseconds of the estimated sinusoidal per each frequency and channels were evaluated as the estimation of the energies of the SSVEP, $E_{pq} = Var\{s_{pq}(n)\}$.

The spectral distribution of the spontaneous of the energies varies depending on each band and the presence of other activities, i.e. the high frequencies have lower activities than lower bands and frequencies close to the α band. At the same time, the response level of the SSVEP could be affected depending on the frequency range. In the presented BMI we combine together different stimulation frequencies, some below α band. To diminish the intra and inter subject variability we propose to compensate each estimated SSVEP energy values by the estimation of the spontaneous EEG energies estimated by approximating the FFT of the analysis window by a smoothing cubic spline. It consist on approximating a series of data points $(x_i, y_i))$ by piecewise cubic polynomials $g(x)$, minimizing the the residual sum of squares, (Eq.8), with a restriction on smoothness of the resulting cubic spline [13, 12].

$$S(p) = p\sum_i \{y_i - g(x_i)\}^2 + (1-p)\int g''(x)^2 dx \tag{8}$$

The parameters p of smoothing splines $g(x)$ which controls the level of smoothness trade off between minimizing the residual error and minimizing local variation. In the limits of $p \in [0, 1]$, for a p value of 1 the function obtained is a interpolating cubic spline while when p is equal to zero the spline function is equal to a linear regression. The value of the p is 10^{-4} was

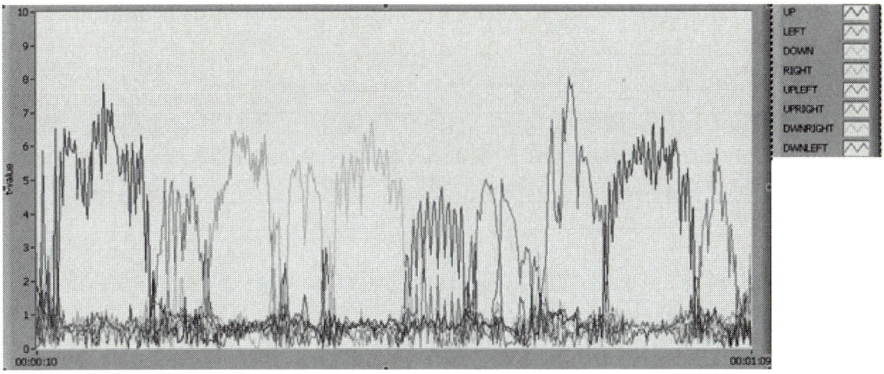

Fig. 3. t-statistic evolution when subject is gazing at one of the 8 different checkerboards/commands. One of each 8 curves represent the t-statistic associated with each simulation frequency. One t-value increases for a specific frequency/command when the subject is gazing at the corresponding checkerboard. This picture corresponds to 1 minute of experiment with subject is looking at different checkerboards for few seconds.

found by experiments to be good enough to give a continuous-wise spectral distribution of the spontaneous EEG, not robust against FFT variations, (see Fig. 2(b)).

The vales of the energies are compensated by a value the relative values of the spline function g for the selected P checkerboard flickering frequencies.

Statistical test. At each 120 ms, the evaluation of which checkerboard the subject is gazing or which frequency elicits the stronger respect to the other frequencies is evaluated by multiple two-samples *t-test* statistical comparisons. The t-statistics values were used as a decision parameters to determine which is the desired output, (see Fig. 3), establishing a fix threshold.

3 Results

We tested our SSVEP-based BCI system with 4 healthy young subjects and 8 flickering frequencies. Without any training an experimental evaluation mode using comparison of voice command requests and the brain responses, in which the success rate and the transfer rates were determined. The computer generated randomly requests for movement in each direction using voice messages. The subject was requested to move the object in one of the 8 directions at intervals of 7 s in 36 trials (4 request per direction). The experimental results are shown in table 1.

The presented results exhibit that this multi-stage procedure allow to have a good performance, with mean delay times lower than 2.5 s and with high success rates.

Table 1. Performance mean values for 4 subject during evaluation mode

	Subjects				Mean results
	1	2	3	4	
Success rate (%)	97	89	96.5	100	95.6
Execution Time [s]	2.1±0.8	3.1±1.5	2.3±0.9	2.4±1.2	2.4±1.0
Bit rate (bit/min)	54	30	48	50	45.5

The used bit rate measure [11] integrates the accuracy of the signal-classification and the number of classes and the time delays:

$$B = V\left(log(N) + Plog(P) + (1-P)log\left(\frac{1-P}{N-1}\right)\right) \quad (9)$$

where B= bit rate [bits/s] P=proportion correct, N=number of classes/targets and V: speed [classification/s].

Once the evaluation experimental mode was performed; in order to observe the subjects ability to control and navigate cursor for demo, it was requested to the subjects for several minutes to move the object toward a small target located randomly on the screen.

4 Conclusions

Although the SSVEP paradigm is well known in the BCI community since the studies performed by several research groups [3, 6, 7, 4, 5], we believe that this proposed approach offers several advantages for improved usability and efficiency. This algorithm offers a faster detection of the desired command without losing classification accuracy.

Also the estimation of the spontaneous EEG activity and noise by mean of cubic spline approximation allowed to mix different flickering frequencies and at the same time. Furthermore, a multiple statistical comparison analysis method indicates that the proposed method does not need any initial training phase (plug and play concept), because we do not need a classifier.

The system was tested with stimulation using small and close flickering checkerboards which elicited a sinusoidal-wise response. Future research will concentrate on other types of visual stimulations were the response are weaker and maybe not close to sinusoidal waves, e.g. Light Emitting Diode (LED). This type of BMI system could have special implications to help paralyzed subjects to control objects in different environments. Especially, in the design of intelligent house (domotic house), where distributed flickering visual stimulators can be placed in different positions to control different house devices or instruments. Without the lack of multiple cameras combined for eye tracking, where these BMI systems can allow to detect the subject desired command with a proper detection related to the visual responses.

References

1. Wolpaw, J., Birmbaumer, N., McFarland, D., Pfurtscheller, G., Vaughan, T.: Brain-computer interfaces for communication and control. Clinical Neurophysiology 113, 767–791 (2002)
2. Niedermeyer, E., Lopes da Silva, F.L.: Electroencephalography: Basic Principles, Clinical Applications and Related Fields, Ch. 54, Lippincott Williams and Wilkins, Philadelphia, Pa, USA (1982)
3. Trejo, L.J., Rosipal, R., Matthews, B.: Brain-Computer Interfaces for 1-D and 2-D Cursor Control: Designs Using Volitional Control of the EEG Spectrum or Steady-State Visual Evoked Potentials. IEEE Transactions of Neural Systems and Rehabilitation Engineering 14(2) (2006)
4. Martinez, P., Bakardjian, H., Cichocki, A.: Fully Online Multicommand Brain-Computer Interface with Visual Neurofeedback Using SSVEP Paradigm. Computational Intelligence and Neuroscience, Article ID 94561, 9 (2007), doi:10.1155/2007/94561
5. Martinez, P., Bakardjian, H., Cichocki, A.: Multicommand Real-Time Brain Machine Interface Fully Online Multicommand MultiBrain-Computer Interface with Visual Neurofeedback Using SSVEP Paradigm. Computational Intelligence and Neuroscience, Article ID 94561, 9 (2007), doi:10.1155/2007/94561
6. Middendorf, M., McMillan, G.R., Calhoun, G.L., Jones, K.S.: Brain-computer interfaces based on the steady-state visual-evoked response. Neural Systems and Rehabilitation Engineering, IEEE Transactions on 8(2), 211–214 (2000)
7. Wang, Y., Wang, R., Gao, X., Hong, B., Gao, S.: A practical VEP-based brain-computer interface. Neural Systems and Rehabilitation Engineering 14(2), 234–240 (2006)
8. Wang, R., Gao, X., Gao, S.: Frequency Selection for SSVEP-based Binocular Rivalry. In: Conference Proceedings. 2nd International IEEE EMBS Conference on Neural Engineering, March 16-19, 2005, pp. 600–603 (2005)
9. Bellanger, M.G.: Adaptive Digital Filters and Signal Analysis. Marcel Dekker, New York (1987)
10. Cichocki, A., Amari, S.: Adaptive Blind Signal and Image Processing: Learning Algorithms and Applications. John Wiley & Sons, West Sussex (2003)
11. Kronegg, J., Voloshynovskiy, S., Pun, T.: Analysis of bit-rate definitions for brain-computer interfaces. In: 2005 Int. Conf. on Human-computer Interaction (HCI 2005), Las Vegas, Nevada, USA (2005)
12. Reinsch, C.: Smoothing by spline functions. Numerische Mathematic 11, 141–156 (1983)
13. Rice, J., Rosenbblatt, M.: Smoothing splines: regression, derivatives and deconvolution. Annals of Statistics 11, 141–156 (1983)

Separating Global Motion Components in Transparent Visual Stimuli – A Phenomenological Analysis

Andrew Meso and Johannes M. Zanker

Computational vision Laboratory, Psychology, Royal Holloway University of London, TW20 0EX, England
a.i.meso@rhul.ac.uk

Abstract. When two distinct movements overlap in the same region of the visual field, human observers may perceive motion transparency. This perception requires the visual system to separate informative and non informative motion signals into transparent components. In this study, we explored the computational constraints in solving this signal separation task - particularly for the stimulus configuration where two grating components move in the same direction at different speeds. We developed a phenomenological model which demonstrates that separation can be achieved only for stimuli with a broadband Fourier spectrum. The model identifies the informative component signals from the non informative signals by considering edges. This approach is shown to be limited by an edge sensitive spatial filtering of the image sequence, the threshold tolerance for local signals considered and the number of iterative computational steps.

Keywords: Motion transparency, Fourier spectra, Global motion, Signal separation.

1 Introduction

Human observers perceive motion transparency when they simultaneously "see" several moving components corresponding to separate objects within the same region of visual field. The computational task of separating local signals into these components is not trivial [1]. To simultaneously represent multiple moving components in the same region of an image requires a mechanism that successfully groups the local motion signals from each component and excludes non informative signals from spatial regions where local components directly overlap to annihilate or make motion responses ambiguous. Separation therefore requires that local signals are sufficiently sparse for a large number of the component signals with minimal spatial overlap so that the relative number of non informative locations are minimal [2,3].

Perceived motion transparency is computed by networks of neurons with a topography of the input image, which exploit the hierarchical structure of cortex thought to perform local motion detection in primary Visual Cortex (V1)

V. Kůrková et al. (Eds.): ICANN 2008, Part II, LNCS 5164, pp. 308–317, 2008.

followed by global pooling of responses for component separation in the motion sensitive Medial Temporal area (MT)[4,5]. A number of computational models generally detect local motion, identify signals most likely to come from the components (informative signals) and separate the resulting distribution into the components. The approach has been tried with transparently moving oriented gratings or plaids [6,7], random dot kinematograms [2,8] and other less regularly structured stimuli [9]. The critical model step is the "gating" of local motion signals to select the informative from the non informative signals. Machine learning algorithms [6] and Bayesian inference [7] based on responses of nearby motion detectors both show some success for this with plaid patterns but require expensive computation.

In this study a phenomenological model is developed to explore space-time (x-t) representations of transparently moving gratings. The model exploits the initial observation in x-t that edges are important for separation in its gating mechanism. Distributions of these local signals contain peaks corresponding to transparent components and therefore partially solve the information processing task of motion transparency perception.

2 Transparently Moving Components in Space-Time

The transparent motion of two linearly superimposed vertical gratings is fully described by a space time (x-t) function $I(x,t)$. In x-t, horizontal motion determines the orientation of the grating, illustrated for a sine wave moving at a constant speed in fig. 2b. The motion dx/dt is related to the orientation angle θ by eq. 1.

$$\frac{dx}{dt} = \tan(\theta) \tag{1}$$

When more than one vertical grating is linearly superimposed to produce a transparently moving stimulus, the x-t representation illustrates the signal separation problem of transparent motion separation as shown in fig. 2. The visibility of the individual component oriented contours seen in the first two columns of fig. 2 within the third column determines whether separation can be achieved for the given stimulus. It can be seen that edges enable the contours to remain during superposition. For single sine wave gratings, the superposition in x-t does not retain the component contour orientation and so the stimulus components cannot be separated. In terms of Fourier spectra, this suggests that broadband stimuli with edges (superposition of several harmonics in odd phase) would appear most transparent.

3 Phenomenological Model

The task for the model is to find orientated contours in x-t and return a global representation of these in which components can be identified. To achieve this, the input $I(x,t)$ is transformed into a global histogram $\mathrm{D}(\theta)$ in eq. 2 to 7. For

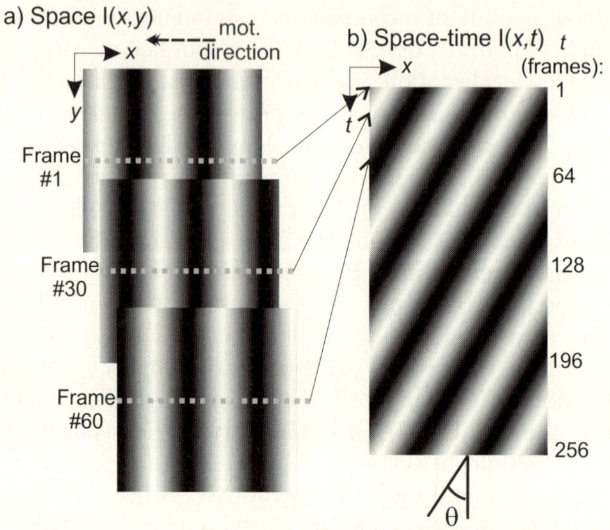

Fig. 1. (a) A representation in space $I(x, y)$ of three frames of a sine wave moving leftwards. The dotted lines show sections through x for the 3 frames in the x-t representation of the motion. (b) The x-t representation of the motion of a single leftward moving sine wave. The orientation θ seen in x-t is constant indicating a constant speed.

sensitivity to edges, a numerical differentiation of $I(x, t)$ is implemented using neighbouring pixels in a three point Langrangian interpolation approximated by eq. 2. The method yields poorer estimates for lines tending to the horizontal orientation and therefore gives an inherently higher sensitivity to slower moving component edges.

$$I'(x, t) \simeq \frac{I(x-1, t) - I(x+1, t)}{2} \tag{2}$$

The estimates of the differential given by $I'(x, t)$ vary with the contrast of the input gratings and this is removed with operator K_T to return a binary image $G(x, t)$ in eq. 3.

$$G(x, t) = K_T \left[I'(x, t) \right]_{I'(x,t)>T; G(x,t)=1}^{I'(x,t)<T; G(x,t)=0} \tag{3}$$

K_T sets $I'(x, t)$ values below T to 0. T is the first model parameter and determines the sensitivity of the edge detection. It is set as a fraction of the maximum of $I'(x, t)$ with $0 < T < 1$. From $G(x, t)$ oriented lines are identified in an iterative template matching process for orientations θ between 0° and 180°. The number of discrete θ is the second model parameter N. θ originates from time $t = 0$ so that the faster speeds result in angles close to the horizontal, likely to wrap around beyond the edges of the $x - t$ representation hence less likely to

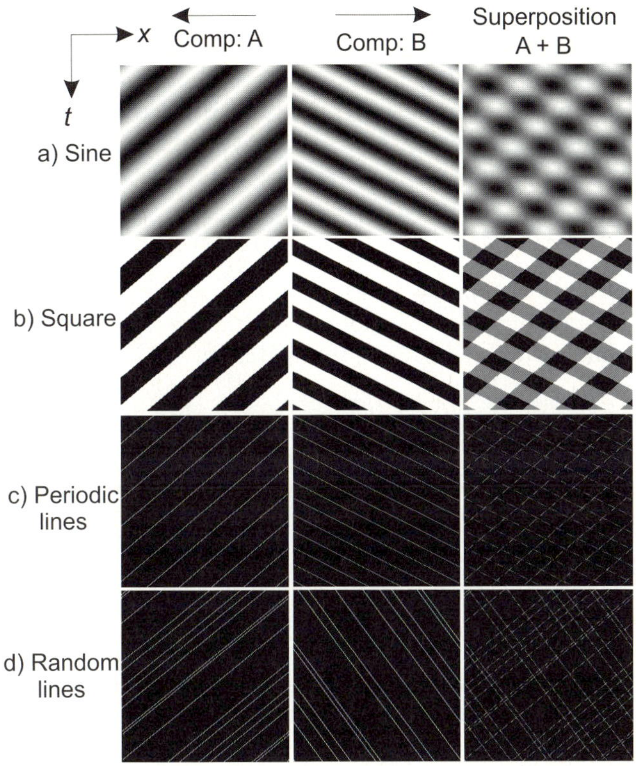

Fig. 2. Single moving gratings shown in the left 2 columns and their linear superposition in the third. (a) Individual sine waves show orientation in x-t, superposition shows no continuous contours in x-t. (b) Square waves, (c) Periodic lines and (d) Random lines all show oriented lines in x-t for both the individual components and the superposition.

be identified by the algorithm. This could be tackled by starting at later t for faster speeds. This is not considered here.

$$H(x,\theta) = \sum_{n=1}^{L_{x,\theta}} G(x + r_n \cos(\theta), r_n \sin(\theta)) \tag{4}$$

Eq. 4 performs a summation of pixel values along an oriented line r which runs from x at $t = 0$ at an angle of θ from the horizontal. $L_{x,\theta}$ is the total number of pixels in the x-t image crossed along the path of r_n. The cosine and sine terms calculate the intermediate orthogonal spatial (x) and temporal (t) coordinates of the line.

$$T(x,\theta) = \sum_{n=1}^{L_{x,\theta}} 1 \tag{5}$$

Eq. 5 is the template with the expected path length of continuous lines with orientation θ. A template matching operator (K_E) compares the image output $H(x,\theta)$ with the template $T(x,\theta)$ within a tolerance given by the term M, the third model parameter. Alternatively, the ratio H/T could be used to quantify the energy of a given orientation. M is used for this explorative model to simplify the output into a binary image. The result is the operation given by eq. 6.

$$D(x,\theta) = K_E \left[H(x,\theta)\right]^{H(x,\theta)\neq T(x,\theta)\pm M; D(x,\theta)=0}_{H(x,\theta)=T(x,\theta)\pm M; D(x,\theta)=1} \tag{6}$$

Eq. 6 maintains local spatial information at $t = 0$ by its x dependence. For a global consideration of the orientations, a position independent histogram function $D(\theta)$ is generated by eq. 7. X is the spatial extent in pixels of the input x-t image.

$$D(\theta) = \sum_{i=1}^{X} I(x_i, \theta) \tag{7}$$

4 Simulations

4.1 Simulation Methods

The model was tested with input 8-bit greyscale functions $I(x,t)$ consisting of 256 spatial pixels (x) and 256 temporal frames (t). Different grating types were compared (sinusoidal, square, periodic line & random line) with variable speeds, relative amplitudes of the superimposed components (superposition is normalised to full range of the greyscale) and directions of the components (limited in the x dimension to either right or left in the same or opposite directions). The routine is written in IDL (by ITT) and run on a single core windows XP system. The output is eq. 7.

4.2 Simulation Results

The outputs for the four grating types were compared for components moving in the same and opposite directions. The model's responses $D(\theta)$ for the transparently moving sine wave gratings given on the left columns of fig. 3 are uninformative of the motion of the components for both low and a high values of T (low value of $T = 0.002$ and a high $T = 0.1$. For all stimuli, $N = 180$, $M = 0.4$). As sine waves have no edges, their components cannot be separated using this edge sensitive method. Square waves responses show two prominent peaks corresponding to the two moving components.

The peaks are seen in the histogram outputs for same direction (left panels fig. 3a & b) and opposite direction stumuli (fig. 3c). Periodic and random line stimuli only allow the identification of the slower moving component when both move in the same direction (right columns fig. 3a & b). The faster grating is not detected and T does not change separation performance for the tested conditions. Simulations show greater susceptibility to noise for random line gratings

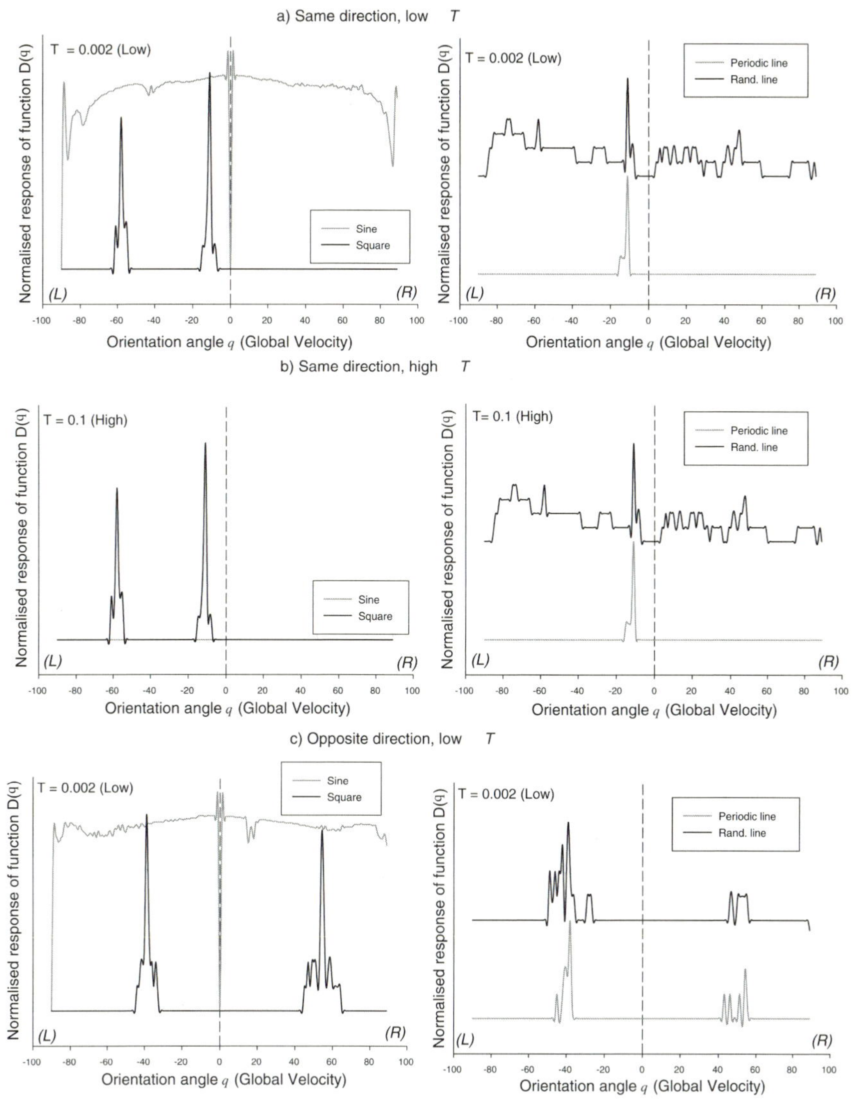

Fig. 3. Model outputs for gratings transparently moving in the same and opposite directions. (a) Same direction; low T. The model output does not detect separate component peaks of the sine wave and detects only the slower component of the periodic and random line gratings. Both components of the square wave are fully identified in the left column. (b) Same stimuli as (a) with a high T. No output response for the sine waves, but the outputs for the periodic and random lines and square wave gratings remain unchanged. (c) Opposite directions; in the sine waves output components cannot be separated. For the periodic and random lines and square waves, peaks corresponding to the two components were observed.

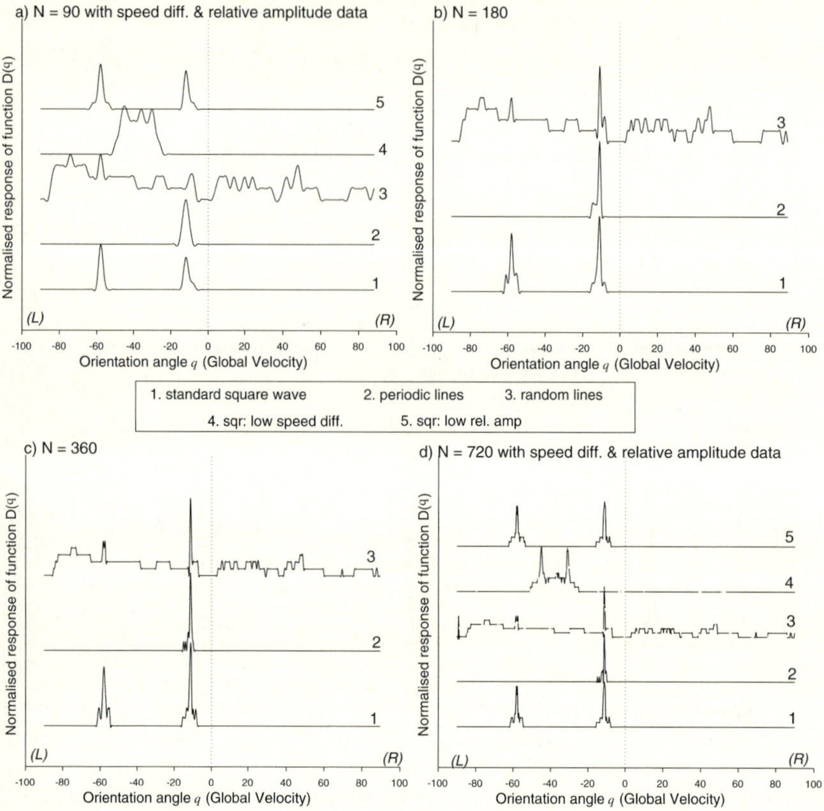

Fig. 4. Separation of transparently moving gratings in the same direction for different N. (a) For $N = 90$, the output for square gratings can be separated into two peaks (line 1). The random lines do not show any distinct peak (line 3), for periodic lines only the slower component (line 2) is identified. A smaller speed difference between the two square gratings returns a single wide peak (line 4). A lower amplitude of one of the superimposed gratings (line 5) makes no difference to separation. (b) For $N = 180$, the slower component is identified for all grating types, and for the square wave the faster moving component is also identified. (c) For $N = 360$ results similar to (b) but noise in the random line output is reduced and a small peak corresponding to the faster component appears. (d) At $N = 720$, all except the periodic lines lead to peaks corresponding to both transparent components. For the random line gratings, non-informative signals reduce in relative magnitude. The square wave grating with a lower speed difference between the components (line 4) shows two clear peaks close to each other.

than periodic line gratings. Because of model sensitivity to speed through θ (see eq. 1), there is no separate processing for the determination of direction and the same direction configuration should be no more difficult a computational task than the opposite direction gratings.

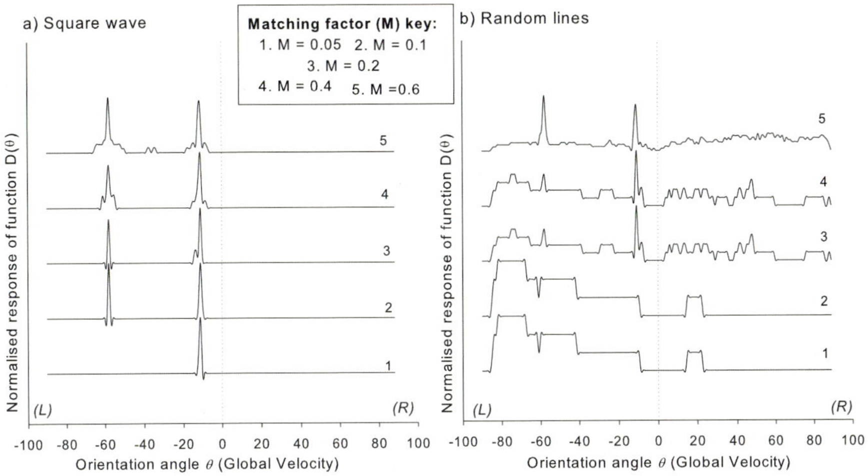

Fig. 5. Model performance for a range of M. (a) For square waves the low M only enables the detection of the slower moving component. (b) For Random lines, the responses are dominated by non-informative signals for low M but with increasing M, show the components.

The three grating types containing edges were used in further simulations with the number of iteration steps, N and the template matching parameter, M. As N is increased from 90 to 720, the output functions for square wave and periodic lines show little change in their appearance (compare the lines labelled 1 & 2 in fig. 4a to the same lines in 4d). Both components of the square wave are detected while only the slower component for the periodic line is detected. For the random lines, outputs contain speeds that are not present in the stimulus, but this effect reduces as N is increased. The effect of N on component separation for smaller speed differences is shown by line 4 in fig. 4a & d which can be separated at $N = 720$, but not $N = 90$. When one of the two transparently moving square wave gratings has lower amplitude (rel. amp = 0.1, line 5 fig. 4a & d), the output is unaffected by N. The results shown in fig. 3 & 4 suggest that the detectability is similar for periodic line and square wave gratings at low speeds, but lines become less detectable at higher speeds. The effect of a range of values of the template matching factor M is tested for these two stimuli. For square waves, reducing M to achieve a more stringent template matching leads to disappearance of the peak corresponding to the faster grating (line 1 in fig. 5a), similar to what is observed for the periodic lines in fig. 3 & 4. If edge detection were equally effective for both stimuli, there should be little difference between the periodic lines and the square wave outputs. From the results shown in fig. 5b, it appears that with appropriate values of M, separation is possible for random lines.

5 Discussion

In this paper, a phenomenological model with a set of operations to detect separate components of transparently moving gratings was developed and tested. Taking the problem of detecting motion as that of detecting orientation in x-t, it was observed that detecting the motion of a single moving pattern in x-t is easier than detecting transparently moving components. The difficulty arises when detecting overlapping component contours. Separation is impossible for superimposed sine waves in which continuous contours in x-t are lost.

Edges are critical for separating moving gratings. It was suggested by Marr that the human visual system extracts useful features during image processing [10]. Using this principle, the model initially extracts edges as a relevant "feature". With an image of edges the complexity of the input is reduced by removing the contrast dependence and other changes not immediately informative of motion [11]. This introduces the first parameter T which is a non linear threshold operating on the differentiated image. An iterative template matching then returns estimates of their orientation densities, introducing two additional parameters: N, which is the orientation resolution and M which sets the tolerance for a positive match as the line filter searches the image.

T is shown to have little effect on the outputs for the grating model outputs particularly those with edges above threshold following differentiation. Larger noise levels of the random line grating outputs may reveal inherent differences in detecting transparency for stimuli with random positions compared to those with periodic spatial arrangement. In contradiction, random lines have been reported to be more transparent than periodic lines in psychophysical experiments [3]. The responses to periodic lines are similar to those of square waves but square waves are more detectable particularly at higher speeds. The reduced detectability for periodic lines at higher speeds highlights the limitations of the computation used for detecting edges and its differential effects on the range of grating types. N determines the resolution of the transparency detection in x-t. Such a parameter exponentially relates to the number of calculations required to separate the components and must be as low as possible where real time processing is a fundamental requirement. M determines the tolerance when performing template matching and allows for the fact that edges may not be fully detected in x-t in the initial differentiation and imperfect detection is an inherent edge detection problem. Higher M makes it more responsive and less vulnerable to non-informative signals.

The phenomenological model has many limitations and is intended to show that, under a set of conditions, the motion transparency problem is tractable, even when components move in the same direction. The parameters introduced give an insight into the separation of motion signals, demostrating how the identification of features (using filter T), the number of calculations performed (N) and the noise tolerance levels in the processes (M) affect the outcome of the separation of transparent components.

References

1. Snowden, R.J., Verstraten, F.A.J.: Motion transparency: making models of motion transparency transparent. Trends in Cognitive Sciences 3(10), 369–377 (1999)
2. Qian, N., Andersen, R.A., Adelson, E.H.: Transparent Motion Perception as Detection of Unbalanced Motion Signals. III. Modeling. Journal of Neuroscience 14, 7381–7392 (1994)
3. Qian, N., Andersen, R.A., Adelson, E.H.: Transparent Motion Perception as Detection of Unbalanced Motion Signals. I. Psychophysics. Journal of Neuroscience 14, 7357–7366 (1994)
4. Qian, N., Andersen, R.A.: Transparent Motion Perception as Detection of Unbalanced Motion Signals. II. Physiology. Journal of Neuroscience 14, 7367–7380 (1994)
5. Treue, S., Hol, K., Rauber, H.-J.: Seeing multiple directions of motion - physiology and psychophysics. Nature Neuroscience 3, 270–276 (2000)
6. Nowlan, S.J., Sejnowski, T.J.: Filter selection model for motion segmentation and velocity integration. Journal of the Optical Society of America A11, 3177–3200 (1994)
7. Koechlin, E., Anton, J.L., Burnod, Y.: Bayesian interference in populations of cortical neurons: a model of motion integration and segmentation in area MT. Biological Cybernetics 80, 25–44 (1999)
8. Zanker, J.M.: A computational analysis of separating motion signals in transparent random dot kinematograms. Spatial Vision 18(4), 431–445 (2005)
9. Jasinschi, R., Rosenfeld, A., Sumi, K.: Perceptual motion transparency: the role of geometrical information. Journal of the Optical Society of America A 9, 1865–1879 (1992)
10. Marr, D.: Vision: A Computational Investigation into the Human Representation and Processing of Visual Information. Freeman & Co., San Francisco (1982)
11. Del Viva, M.M., Morrone, M.C.: Motion analysis by feature tracking. Vision Research 38, 3633–3653 (1998)

Lateral Excitation between Dissimilar Orientation Columns for Ongoing Subthreshold Membrane Oscillations in Primary Visual Cortex

Yuto Nakamura, Kazuhiro Tsuboi, and Osamu Hoshino*

Department of Intelligent Systems Engineering, Ibaraki University, 4-12-1 Nakanarusawa, Hitachi, Ibaraki, 316-8511 Japan
Tel.: +81-294-38-5194; Fax +81-294-38-5229
{07nm928s@hcs,ktsuboi@mx,hoshino@mx}ibaraki.ac.jp

Abstract. Primary visual cortex (V1) is the first stage of visual information processing; detection of particularly oriented bars. These elemental visual features are sent to higher visual regions including V2, in which their combinations such as "corners" and "junctions" are processed. A recent study has demonstrated that corners and T-junctions could be processed even at the early visual stage (V1), raising the question of why the binding of bars by V1 is necessary. We simulated a V1 neural network model, in which the so-called "orientation-columns" were connected through both excitatory and inhibitory synapses. The lateral excitatory connections contributed not only to binding paired bars constituting corners but also to making membrane oscillations near firing-threshold during ongoing (spontaneous) neuronal activity periods. This ongoing subthreshold neuronal state led to accelerating the reaction speed of neurons to paired bar-stimuli. The lateral inhibitory connections effectively enhanced the selective responsiveness of neurons to the stimuli. We suggest that coordinated lateral excitation and inhibition between orientation-columns in V1 could send angular information such as corners and junctions, presented to retina, rapidly to the next stage V2 for its full and precise analyses.

Keywords: Neural network model, Lateral excitation, Feature binding, Orientation map, Ongoing subthreshold membrane oscillation, Primary visual cortex.

1 Introduction

Primary visual cortex (V1) is the first stage of visual information processing, and V1 neurons have orientation specificity; a preference to a particular bar oriented in a narrow range of angles. It is also well known that V1 neurons that have the same orientation preference tend to be organized in a close neighbor and form the so-called "orientation column" [1]. These elemental visual features are sent to higher visual regions including V2. V2 neurons respond to more complex stimuli

* Corresponding author.

V. Kůrková et al. (Eds.): ICANN 2008, Part II, LNCS 5164, pp. 318–327, 2008.

such as angles and junctions, which are important visual attributes embedded within contours to represent the shape of objects [2].

Das and Gilbert [3] demonstrated that corners and T-junctions could be processed even at the early visual stage (V1). The researchers measured correlation between neuron pairs that have dissimilar orientation preferences; orthogonal bars. These neurons showed significant correlated activity, when presented with paired bar-stimuli constituting corners and T-junctions. It was suggested that local intracortical circuitry could endow V1 neurons with processing not only elemental but also their combinations. These findings raise the question of why the binding of orthogonal bars in V1 is necessary if these attributes are processed predominantly in V2.

The purpose of the present study is to understand the significance of the binding of paired bars expressing corners and T-junctions in V1. We propose a V1 neural network model, in which excitatory and inhibitory connections are made between columns with both similar and dissimilar orientation preferences [4]. The lateral excitation is employed for mutually exciting columnar neurons that are relevant to paired bar-stimuli, and the lateral inhibition is for suppressing columnar neurons that are irrelevant to the stimuli. To assess the cognitive performance of the V1 network, we measure the reaction time of neurons to the stimuli.

The assumed lateral intracortical (excitatory and inhibitory) connections between different orientation columns presumably influence not only stimulus-evoked neuronal activity but also ongoing-spontaneous neuronal activity, which is known to have a great impact on subsequent sensory information processing [5,6,7,8]. Statistically analyzing the membrane potentials of neurons, we investigate how the lateral intracortical circuitry influences ongoing neuronal behaviors, and how it affects the binding of paired bars expressing corners and T-junctions in V1.

2 Neural Network Model

Figure 1A is a schematic drawing of the neural network model functioning as an orientation preference map. Each orientation-column has a preference to one specific orientation of a bar-stimulus, ranging from 0 (θ_0) to $7\pi/8$ (θ_7), and consists of cell units (Figure 1B; "gray circle"). Each cell unit contains one principal cell ("P"), one feedback inhibitory cell ("F") and one lateral inhibitory cell ("L"). Each P-cell receives excitatory inputs from P-cells that belong to the same orientation-column, a feedback inhibitory input from its accompanying F-cell, and lateral inhibitory inputs from L-cells that receive excitatory inputs from P-cells belonging to other orientation-columns. Each F-cell receives excitatory input from its accompanying P-cell. The recurrent excitatory connections between P-cells within orientation-columns make the so-called "cell assemblies". P-cells between different orientation-columns (θ_n and $\theta_{n'}$) are mutually connected through lateral excitatory synapses. P-cells receive input signals from the lateral geniculate nucleus (LGN).

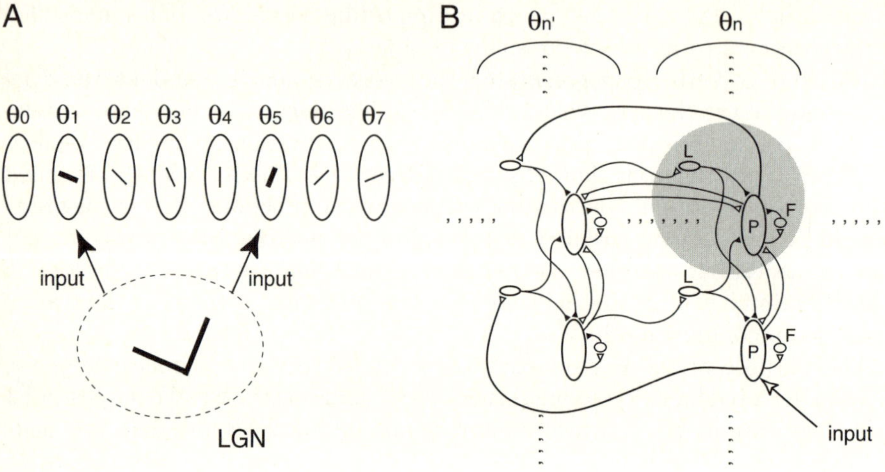

Fig. 1. (A) A schematic drawing of the neural network model for a primary visual cortical (V1) area. The network receives combinatorial input signals from LGN. (B) Orientation-columns consist of cell-units ("gray circle"); one principal cell ("P"), one feedback inhibitory cell ("F") and one lateral inhibitory cell ("L") constitute one cell-unit.

Dynamic evolution of the membrane potential of the ith P-cell that belongs to orientation-column θ is defined by

$$c_m^P \frac{du_i^P(\theta;t)}{dt} = -g_m^P(u_i^P(\theta;t) - u_{rest}^P) + I_{i,rec}^P(\theta;t) + I_{i,fed}^P(\theta;t) + I_{i,lat}^P(\theta;t) + \tilde{I}_{i,lat}^P(\theta;t) + I_{LGN}(\theta;\theta_{inp}), \quad (1)$$

where $I_{i,rec}^P(\theta;t)$ is a recurrent excitatory current, $I_{i,fed}^P(\theta;t)$ a feedback inhibitory current, $I_{i,lat}^P(\theta;t)$ a lateral inhibitory current, $\tilde{I}_{i,lat}^P(\theta;t)$ a lateral excitatory current, and $I_{LGN}(\theta;\theta_{inp})$ an excitatory input current triggered by an oriented bar-stimulus (θ_{inp}), where $\theta, \theta_{inp} \in \{\theta_0, \theta_1, \theta_2, \theta_3, \theta_4, \theta_5, \theta_6, \theta_7\}$. These currents are defined by

$$I_{i,rec}^P(\theta;t) = -\hat{g}_{AMPA}(u_i^P(\theta;t) - u_{rev}^{AMPA}) \sum_{j=1}^{N_\theta} w_{ij,rec}^P(\theta) r_j^P(\theta;t), \quad (2)$$

$$I_{i,fed}^P(\theta;t) = -\hat{g}_{GABA}(u_i^P(\theta;t) - u_{rev}^{GABA}) w_{i,fed}^P(\theta) r_i^F(\theta;t), \quad (3)$$

$$I_{i,lat}^P(\theta;t) = -\hat{g}_{GABA}(u_i^P(\theta;t) - u_{rev}^{GABA}) \sum_{j=1}^{N_\theta} w_{ij,lat}^P(\theta) r_j^L(\theta;t), \quad (4)$$

$$\tilde{I}_{i,lat}^P(\theta;t) = -\hat{g}_{AMPA}(u_i^P(\theta;t) - u_{rev}^{AMPA}) \sum_{\theta'=\theta_0(\theta'\neq\theta)}^{\theta_7} \sum_{j=1}^{N_\theta} \tilde{w}_{ij,lat}^P(\theta,\theta') r_j^P(\theta';t), \quad (5)$$

$$I_{LGN}(\theta;\theta_{inp}) = \alpha_P e^{-[\frac{(\theta-\theta_{inp})/(\pi/8)}{\tau_P}]^2}. \quad (6)$$

Dynamic evolution of the membrane potentials of the ith F-cell that belongs to orientation-column θ is defined by

$$c_m^F \frac{du_i^F(\theta;t)}{dt} = -g_m^F(u_i^F(\theta;t) - u_{rest}^F) + I_i^F(\theta;t), \quad (7)$$

where $I_i^F(\theta;t)$ is an excitatory current, and defined by

$$I_i^F(\theta;t) = -\hat{g}_{AMPA}(u_i^F(\theta;t) - u_{rev}^{AMPA})w_i^F(\theta)r_i^P(\theta;t). \quad (8)$$

Dynamic evolution of the membrane potentials of the ith L-cell that belongs to orientation-column θ is defined by

$$c_m^L \frac{du_i^L(\theta;t)}{dt} = -g_m^L(u_i^L(\theta;t) - u_{rest}^L) + I_i^L(\theta;t), \quad (9)$$

where $I_i^L(\theta;t)$ is an excitatory current, and defined by

$$I_i^L(\theta;t) = -\hat{g}_{AMPA}(u_i^L(\theta;t) - u_{rev}^{AMPA}) \sum_{\theta'=\theta_0(\theta'\neq\theta)}^{\theta_7} w_i^L(\theta,\theta')r_i^P(\theta';t), \quad (10)$$

$$w_i^L(\theta,\theta') = w_L e^{-[\frac{(\theta-\theta')/(\pi/8)}{\tau_{lat}}]^2}. \quad (11)$$

In these equations, c_m^Y is the membrane capacitance of Y (Y = P, F, L) cell, $u_i^Y(\theta;t)$ the membrane potential of the ith Y-cell at time t, g_m^Y the membrane conductance of Y-cell, and u_{rest}^Y resting potential. $\hat{g}_Z$ and u_{rev}^Z (Z = AMPA or GABA) are, respectively, the maximal conductance and the reversal potential for the current regulated by Z-type receptor. N_θ is the number of cell-units constituting each orientation-column.

$w_{ij,rec}^P(\theta)$ is the recurrent excitatory synaptic strength from jth to ith P-cell within orientation-column θ. $w_{i,fed}^P(\theta)$ is the feedback inhibitory synaptic strength from F-cell to P-cell within cell-unit i belonging to orientation-column θ. $w_{ij,lat}^P(\theta)$ is the inhibitory synaptic strength from the jth L-cell to the ith P-cell within orientation-column θ. $\tilde{w}_{ij,lat}^P(\theta,\theta')$ is the excitatory synaptic strength between different orientation-columns θ and θ'. $w_i^F(\theta)$ is the excitatory synaptic strength from P-cell to F-cell within cell-unit i belonging to orientation-column θ. $w_i^L(\theta,\theta')$ is the excitatory synaptic strength from the ith P-cell belonging to a different orientation-column (θ') to the ith L-cell belonging to orientation-column θ ($\theta' \neq \theta$), which weakens as the functional distance between orientation-columns θ' and θ increases (equation 11).

$r_j^P(\theta;t)$ is the fraction of AMPA-receptors in the open state induced by presynaptic action potentials of the jth P-cell belonging to orientation-column θ at time t. $r_j^F(\theta;t)$ and $r_j^L(\theta;t)$ are the fractions of GABAa-receptors in the open state induced by presynaptic action potentials of the jth F-cell and L-cell, respectively. Receptor dynamics is described by [9]

$$\frac{dr_j^P(\theta;t)}{dt} = \alpha_{AMPA}[Glut]_j^P(\theta;t)(1 - r_j^P(\theta;t)) - \beta_{AMPA} r_j^P(\theta;t), \quad (12)$$

$$\frac{dr_j^Y(\theta;t)}{dt} = \alpha_{GABA}[GABA]_j^Y(\theta;t)(1 - r_j^Y(\theta;t)) - \beta_{GABA} r_j^Y(\theta;t), \qquad (Y = F, L) \quad (13)$$

where α_z and β_z (z = AMPA or GABA) are positive constants. $[Glut]_j^P(\theta;t)$ and $[GABA]_j^Y(\theta;t)$ are concentrations of glutamate and GABA in the synaptic cleft, respectively. $[Glut]_j^P(\theta;t) = Glut_{max}^P$ and $[GABA]_j^Y(\theta;t) = GABA_{max}^Y$ for 1 ms when the presynaptic jth P-cell and Y-cell fire, respectively. Otherwise, $[Glut]_j^P(\theta;t) = 0$ and $[GABA]_j^Y(\theta;t) = 0$.

Probability of firing of the jth Y-cell belonging to orientation-column θ is defined by

$$Prob[Y_j(\theta;t); firing] = \frac{1}{1 + e^{-\eta_Y(u_j^Y(\theta;t) - \zeta_Y)}}, \qquad (Y = P, F, L) \quad (14)$$

where η_Y and ζ_Y are, respectively, the steepness and the threshold of the sigmoid function. After firing, the membrane potential is reset to the resting potential.

Unless otherwise stated elsewhere, c_m^P = 0.5 nF, c_m^F = 0.2 nF, c_m^L = 0.6 nF, g_m^P = 25 nS, g_m^F = 20 nS, g_m^L = 15 nS, u_{rest}^P = -65 mV and $u_{rest}^F = u_{rest}^L$ = -70 mV. $\hat{g}_{AMPA}$ = 0.5 nS, $\hat{g}_{GABA}$ = 0.7 nS, u_{rev}^{AMPA} = 0 mV and u_{rev}^{GABA} = -80 mV [10,11,12]. N_θ = 20, $w_{ij,rec}^P(\theta) = 8.0$, $w_{i,fed}^P(\theta) = 20.0$, $w_{ij,lat}^P(\theta) = 100.0$, $\tilde{w}_{ij,lat}^P(\theta,\theta') = 5.0$, $w_i^F(\theta) = 50.0$, $w_L = 7.0$, $\alpha_P = 5.0 \times 10^{-10}$, $\tau_P = 0.1$, $\tau_{lat} = 2.0$, $\alpha_{AMPA} = 1.1 \times 10^6$, $\alpha_{GABA} = 5.0 \times 10^5$, $\beta_{AMPA} = 190.0$, $\beta_{GABA} = 180.0$ and $Glut_{max}^X = GABA_{max}^Y$ = 1.0 mM. $\eta_P = \eta_F = \eta_L = 350.0$ and $\zeta_P = \zeta_F = \zeta_L$ = -45 mV.

3 Results

Figure 2A presents how neurons respond to a pair of orthogonally oriented bars, where "reaction time" denotes the time at which the P-cells corresponding to the paired stimuli begin to respond. If the lateral excitatory connections between orientation-columns are not considered in the model, reaction time tends to be prolonged (Figure 2B). Figure 2C presents a crosscorrelation function of membrane potential between input (θ_1, θ_5) relevant P-cells, indicating stronger correlated activity (see the "solid" line) if the lateral excitatory connections exist. This result implies that the lateral excitatory connections between orientation-columns effectively bind the information about paired bars.

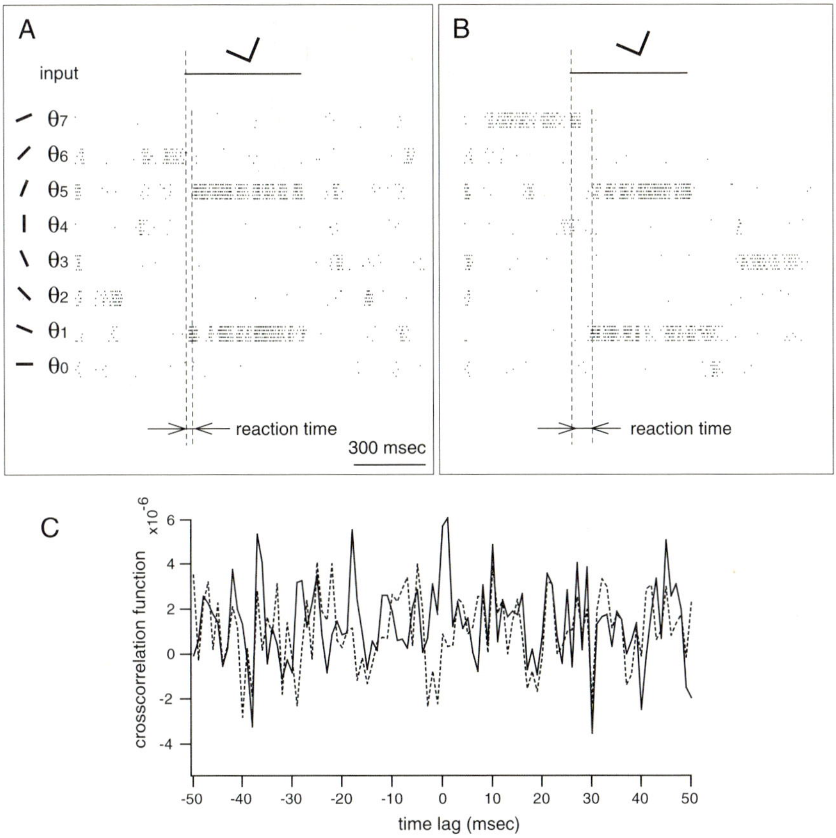

Fig. 2. Raster plots (P-cells) for orientation columns ($\theta_0 - \theta_7$) when stimulated with a corner consisting of paired bars (θ_1 and θ_5). "reaction time" denotes the time at which the populations of P-cells corresponding to the pair begin to respond, where lateral excitatory connections between orientation-columns (A) existed or (B) were cut off. (C) Crosscorrelation function of membrane potential between θ_1- and θ_5-responsive P-cells for the stimulation period with ("solid" line) or without ("dashed" line) lateral excitatory connections.

Figure 3A presents the membrane potential of a stimulus-relevant P-cell for these two cases; with (top) or without (bottom) lateral excitatory connections. An "arrow" indicates reaction time. Figure 3B presents the distribution of reaction time with ("open" rectangles) or without ("filled" rectangles) lateral excitatory connections. We stimulated the network at arbitrary times between 10 and 20 sec. The time-bin for the count was 2 msec. Figure 3C presents a relationship of reaction time between these two cases. The points deviating away from the diagonal line ("dashed line") imply that the reaction time can be reduced if the lateral excitatory connections exist. These results indicate that the lateral excitatory connections contribute to accelerating the reaction speed of V1 neurons to sensory input.

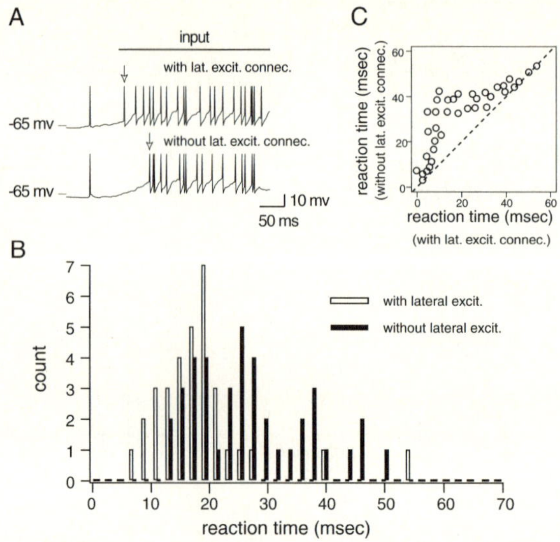

Fig. 3. (A) Reaction of a P-cell to sensory input with (top) or without (bottom) lateral excitatory connections between orientation-columns. (B) Histogram of reaction time with (”open” rectangles) or without (”filled” rectangles) lateral excitatory connections. (C) A relationship of reaction time between the two cases; with and without lateral excitatory connections. The points on the diagonal line (”dashed line”) indicate equal reaction times. For details, see the text.

To elucidate how the lateral excitation between different orientation columns enables the network to respond rapidly to the input, we recorded ongoing membrane potentials. Note that one of the possible mechanisms for the acceleration of reaction speed might be membrane potentials that oscillate near firing-threshold prior to sensory input. Figure 4A presents the time courses of membrane potentials of P-cells for an ongoing time period with (left) or without (right) lateral excitatory connections. Figure 4B presents the distribution of membrane potential for the ongoing period, where the lateral excitatory connections existed (”solid line”) or were cut off (”dashed line”).

Figure 4C presents ”cumulative hyperpolarization” during an ongoing period, which measures a degree of membrane hyperpolarization below firing-threshold and is defined by

$$H_c(t) = \frac{1}{8N_\theta} \sum_{\theta=\theta_0}^{\theta_7} \sum_{i=1}^{N_\theta} \int_0^t \tilde{u}_i^P(\theta;t')dt',$$

$$\tilde{u}_i^P(\theta;t') = \begin{cases} u_i^P(\theta;t') - u_{rest}^P & \text{if } u_i^P(\theta;t') < u_{rest}^P \\ 0 & \text{otherwise.} \end{cases} \quad (15)$$

The ongoing membrane potential tends to be less hyperpolarized as the strength of lateral excitatory connections $\tilde{w}_{ij,lat}^P(\theta,\theta')$ increases. Note that excessive lateral

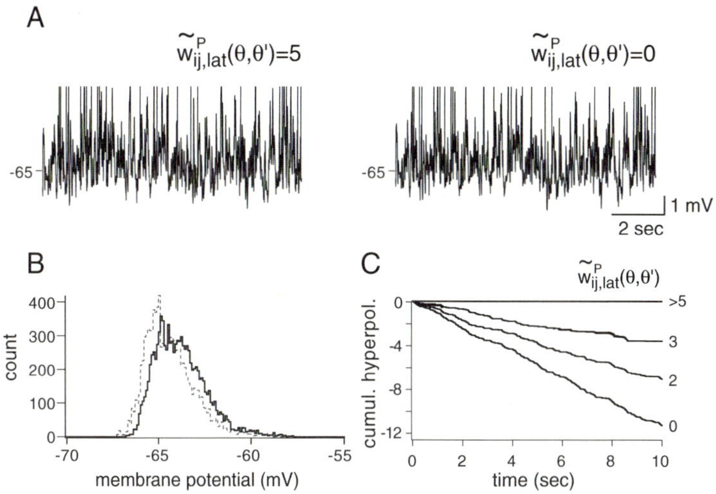

Fig. 4. (A) Membrane potentials of P-cells for an ongoing time period with (left) or without (right) lateral excitatory connections between orientation-columns. (B) Histogram of membrane potential with ("solid line") or without ("dashed line") lateral excitatory connections. (C) Dependence of membrane hyperpolarization on the strength of lateral excitatory connections. "Cumulative hyperpolarization" was calculated (see equation 15) during the ongoing neuronal activity period. For details, see the text.

excitation, e.g. $\tilde{w}^P_{ij,lat}(\theta, \theta') > 5$, all P-cells fire (not shown) and therefore the selective responsiveness is no longer possible. The less membrane hyperpolarization (or rather, depolarization) at $\tilde{w}^P_{ij,lat}(\theta, \theta') = 2 - 3$ allows the V1 network to be near firing-threshold during the ongoing period, whereby the P-cells can respond quickly to the input (see Figs. 2A and 3A). We call this network state an "ongoing subthreshold neuronal state", whose fundamental dynamic behavior and its significance in neural information processing have been suggested [5,6,7].

To elucidate how the lateral inhibitory connections contribute to processing the paired bars, we reduced its strength from the optimal value $w^P_{ij,lat}(\theta) = 100$ to 20. As shown in Figure 5, the selective responsiveness of orientation-columns (θ_1 and θ_5) is deteriorated, compared to that shown in Fig. 2A.

4 Discussion

We proposed a V1 neural network model of an orientation preference map, in which neuronal columns were connected through both excitatory and inhibitory synapses. Lateral excitatory connections were employed for binding pairs of bar-stimuli constituting corners. It was found that the lateral excitatory connections contribute not only to binding the paired bar-stimuli but also to making membrane oscillations near firing-threshold during ongoing (spontaneous) neuronal activity periods. This enables the V1 to respond rapidly to subsequent sensory input.

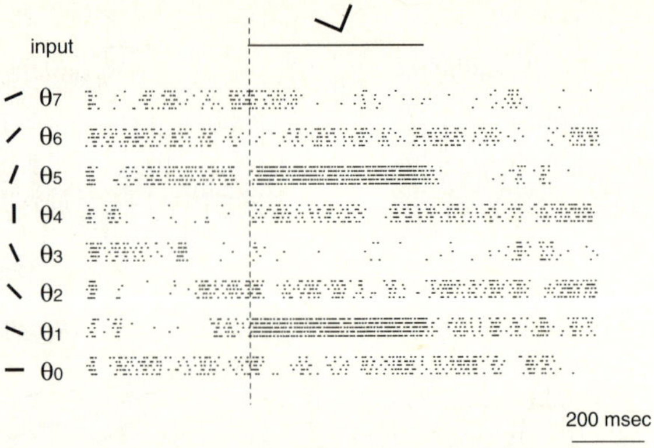

Fig. 5. Influences of lateral inhibition on selective responsiveness of P-cells. The strength of lateral inhibitory connections between orientation-columns was reduced. Raster plots (P-cells) for orientation columns ($\theta_0 - \theta_7$) are shown. The network was stimulated with a corner consisting of bars θ_1 and θ_5. For details, see the text.

Since synchronous common input, per se, was sufficient to some extent for binding paired bars as well (see Figure 2B), the lateral excitation between orientation-columns may have another important role; establishing an ongoing subthreshold neuronal state. This allows the V1 to send information about combinations of elemental features (bars) expressing corners and T-junctions, presented to retina, rapidly to the next stage V2 for its full and precise analyses.

Das and Gilbert [3] demonstrated that connections between dissimilar orientation-columns have suppressive property, and suggested that the suppressive function could be used to analyze combinatorial bar information including corners and T-junctions. The present study showed that the combinatorial visual information was blurred if the suppressive function declined (see Figure 5). We suggest that coordinated lateral excitation and inhibition among orientation-columns might be essential for effectively processing angular information at this early stage of vision.

In this study, we showed the importance of lateral excitation between orientation columns for establishing an ongoing subthreshold neuronal state. However, a balance of excitation and inhibition might be crucial for it. In a previous study [6], we proposed a cortical neural network model that has local excitatory and inhibitory circuitry. Simulating the model, we investigated how two distinct types of inhibitory interneurons (fast-spiking interneurons with narrow axonal arbors and slow-spiking interneurons with wide axonal arbors) have spatiotemporal influences on the ongoing activity of principal cells and subsequent neuronal information processing. In the model, dynamic cell assemblies expressed information about specific sensory features. Within cell assemblies, fast-spiking interneurons gave a feedback inhibitory effect on principal cells. Between cell assemblies, slow-spiking interneurons gave a lateral inhibitory effect on principal cells. These interneurons contributed to depolarizing principal cells below firing-threshold during ongoing neuronal activity periods, by which the network was kept at a subthreshold level for action potential generation. This led to accelerating the reaction speed of principal cells to sensory input. It might be inferred that a balance of lateral excitation and inhibition among orientation-columns is also important for achieving the ongoing subthreshold neuronal state in V1. This will be confirmed in our future studies.

References

1. Ts'o, D.Y., Roe, A.W.: Functional compartment in visual cortex: segregation and interaction. In: Gazzaniga Walter, M.S. (ed.) The cognitive neuroscience, pp. 325–337. MIT Press, Cambridge (1995)
2. Ito, M., Komatsu, H.: Representation of angles embedded within contour stimuli in area V2 of macaque monkeys. J. Neurosci. 24, 3313–3324 (2004)
3. Das, A., Gilbert, C.D.: Topography of contextual modulations mediated by short-range interactions in primary visual cortex. Nature 399, 655–661 (1999)
4. Eysel, U.T.: Turing a corner in vision research. Nature 399, 641–644 (1999)
5. Hoshino, O.: Cognitive enhancement mediated through postsynaptic actions of norepinephrine on ongoing cortical activity. Neural Comput. 17, 1739–1775 (2005)
6. Hoshino, O.: Coherent ongoing subthreshold state of a cortical neural network regulated by slow- and fast-spiking interneurons. Network: Comput. Neural Syst. 17, 351–371 (2006)
7. Hoshino, O.: Enhanced sound-perception by widespread onset neuronal responses in auditory cortex. Neural Comput. 19, 3310–3334 (2007)
8. Hoshino, O.: Extrasynaptic-GABA-mediated neuromodulation in a sensory cortical neural network. Network: Comput. Neural Syst. (in press)
9. Destexhe, A., Mainen, Z.F., Sejnowski, T.J.: Kinetic models of synaptic transmission. In: Koch, C., Segev, I. (eds.) Methods in Neuronal Modeling, pp. 1–25. MIT Press, Cambridge (1998)
10. Koch, C.: Biophysics of computation. Oxford Univ. Press, New York (1999)
11. McCormick, D.A., Connors, B.W., Lighthall, J.W., Prince, D.A.: Comparative electrophysiology of pyramidal and sparsely spiny stellate neurons of the neocortex. J. Neurophysiol. 54, 782–806 (1985)
12. Kawaguchi, Y., Shindou, T.: Noradrenergic excitation and inhibition of GABAergic cell types in rat frontal cortex. J. Neurosci. 18, 6963–6976 (1998)

A Computational Model of Cortico-Striato-Thalamic Circuits in Goal-Directed Behaviour

N. Serap Şengör[1,⋆], Özkan Karabacak[1,2], and Ulrich Steinmetz[3]

[1] Istanbul Technical University, Electrical and Electronics Engineering Faculty, Maslak 34469, Istanbul, Turkey
[2] School of Engineering, Computing and Mathematics, Harrison Building, University of Exeter, North Park Road, Exeter EX4 4QF, UK
[3] FGAN-FOM, Forschungsinstitut fr Optronik und Mustererkennung, Gutleuthausstr. 1, D-76275 Ettlingen, Germany
sengorn@itu.edu.tr, ok213k@ex.ac.uk, ulrich.steinmetz@fom.fgan.de

Abstract. A connectionist model of cortico-striato-thalamic loops unifying learning and action selection is proposed. The aim in proposing the connectionist model is to develop a simple model revealing the mechanisms behind the cognitive process of goal directed behaviour rather than merely obtaining a model of neural structures. In the proposed connectionist model, the action selection is realized by a non-linear dynamical system, while learning that modifies the action selection is realized similar to actor-critic model of reinforcement learning. The task of sequence learning is solved with the proposed model to make clear how the model can be implemented.

1 Introduction

From simplest actions we perform, as deciding which path to take on our way to home to complicated decisions as choosing a career, our evaluation is determined by goal-directed behavior. Its main feature that distinguishes it from the stimulus-response behaviour is that in the goal-directed behaviour presentations of stimuli depend on the prior occurrence of designated responses. Thus, the stimuli are direct result of the subject's behaviour.

Almost for two decades, neuroscientists are trying to comprehend the organization and functioning of neural substrates in goal-directed behaviour. In the meanwhile, the pioneering works on cortico-striato-thalamic (C-BG-TH) loops pointed out that different organizations of these neural substrates are needed to achieve different cognitive, emotional, and motor processes [1,2,3]. It is now well-known that even though these loops have different roles and are processing in parallel, they do have interactions and processing of more than one loop in cooperation with others is required in performing complex behaviors as goal-directed behavior, planning, etc.

⋆ Part of this work is completed during the first author's visit to LCE, Helsinki University of Technology, during Summer 2006.

V. Kůrková et al. (Eds.): ICANN 2008, Part II, LNCS 5164, pp. 328–337, 2008.

[4,5,6]. As shown in [4,5], loops containing ventral striatum take part in limbic system, whereas loops through the dorsal striatum have a role in generating the actions and these two loops together with motor loops progress to perform reward-guided behaviors while the neurotransmitter systems modulate this process.

Some of the computational models of goal-directed behaviour [11,12,13] utilizes the actor-critic models as they are developed in [14] but there are also computational models where the actor part is realized considering a biologically realistic model of basal ganglia-thalamo-cortical pathways as in [16]. The relation between the ventral and dorsal parts of striatum has been anticipated in [15,11]. In later works, as [17,18,19] the relation between dorsal and ventral parts of striatum amongst with other neuronal substrates during goal-directed learning are investigated. In [22], a heterarchical reinforcement learning model based on [4] is proposed and this model gives an explanation about how the information is shared and transferred between the loops rather than giving a computational model of goal-directed behaviour.

There are also computational models for the action selection function of C-BG-TH loop [9,10,20,21], and models for sequence storage and generation [7,8]. Even though none of these later mentioned works deal with goal-directed behaviour and their primary concern is to develop a biologically plausible model for action selection, they deserve to be mentioned in the context of modelling goal-directed behaviour as action selection is a part of goal-directed behaviour.

The aim of this work is to propose a connectionist model that reveals the effect of ventral stream on dorsal stream during goal directed behaviour. Furthermore the model incorporates the connection between premotor and motor loops. This incorporation is beneficial to explain how one action is selected from many possible actions which originate due to exploration. This unique choice emerges from the contribution of noise signal and from the effect of premotor loop on motor loop. While deriving the model, system theoretical tools, as bifurcation diagrams, domain of attractions, are used to understand the behaviour of nonlinear dynamical systems responsible for action selection. An analogy between the neural substrates, neurotransmitter systems and dynamical system's structure, parameters of the dynamical systems are drawn to explain the behavioral phenomenon. There are already some work using such tools in dealing with cognitive processes as [7,8,9,10,20,21].

In the next section, a short explanation on how the neural substrates organize and affect each other during goal-directed behaviour will be given. The computational model proposed is based on the known functions of these neural structures and their interrelations. In section 3, the proposed computational model will be introduced. A simple sequence-learning task given in [12] will be simulated with the proposed model to show its convenience.

2 Evidence from Neuroscience in Deriving the Model

The roles of C-BG-TH loops in different cognitive, motor processes and in emotional behaviours have been investigated since the papers of Alexander et.al.

[1,2]. In [1,2], it is stated that parallel C-BG-TH loops have been considered for different processes which can also work together and furthermore affect each other for some tasks. Besides the connections of related sub-structures for different processes, the modulatory effect of neurotransmitters, especially the role of dopamine on these circuits has to be considered [23].

For a goal-directed behaviour the motor, associative and limbic circuits have to work in harmony [4]. The modulatory effect of midbrain dopamine system would provide this harmony through mesocortical and nigrostriatal pathways. In [4,5], based on animal studies, it is stated that anterior cingulate, orbital cortices and ventral striatum process different aspects of reward evaluation, while the dorsolateral prefrontal cortex and dorsal striatum are involved in cognitive function during decision making. The relation between dorsal and ventral parts of striatum is initiated through midbrain dopamine pathways and striatonigrostriatal subcircuit directs information flow between limbic, cognitive and motor circuits through an ascending spiral formed by these circuits. A summary of these connections are depicted in Fig. 1.

Thus, for a goal-directed behaviour, these C-BG-TH loops have to operate together, first in order to learn to select the appropriate action for a goal and then to keep up with this knowledge till a new condition in the environment comes up. In order to realize this process, the ventral part has to evaluate the

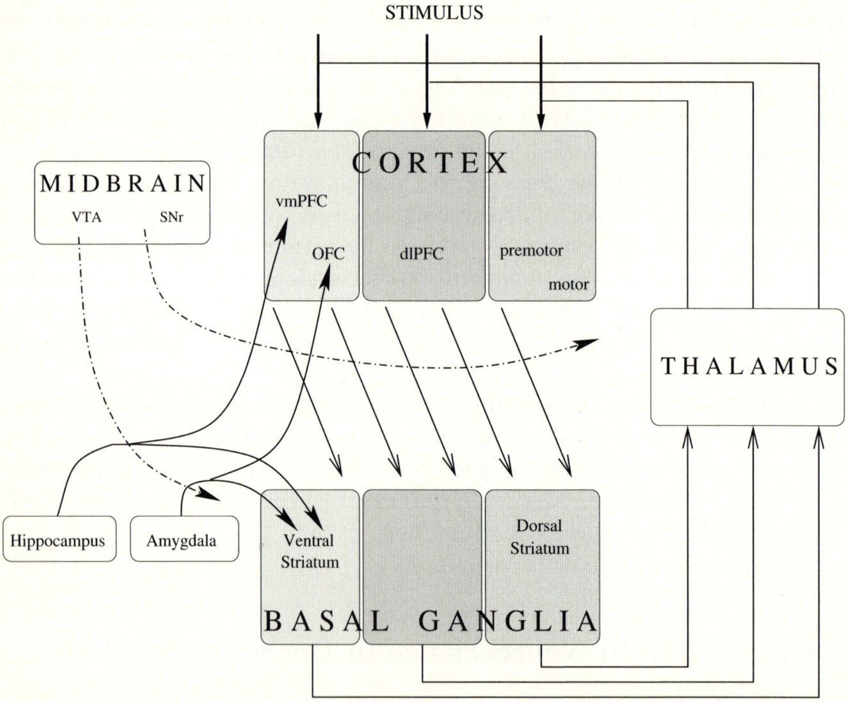

Fig. 1. The neural substrates considered in deriving the computational model

inputs from cortex, and the dorsal part being in connection with the premotor cortex has to select some actions out of many according to this evaluation. The motor-cortex will then consider the result of the action selection, and the ventral part will either continue evaluation or keep up with the previous decision in case the result of the selected action is satisfactory. While the mesolimbic pathway is responsible for dopamine modulation of ventral part, the dopamine modulation of dorsal part is carried out by nigrostriatal pathway. Following the literature on computational models of reinforcement learning, ventral part corresponds to critic and dorsal part corresponds to actor. In the next section, a computational model of goal-directed behaviour will be proposed which is originated from the above summarized interactions of neural substrates and the actor-critic method of reinforcement learning.

3 A Computational Model for Goal-Directed Behaviour

Two essential components of goal-directed behaviour are to select an action and evaluate the outcome of this action before determining the next action. As discussed in previous works [11,12,15,17,18], the actor-critic method, especially with temporal difference (TD) learning is appropriate to have a computational model of goal-directed behaviour. The computational model proposed in this work is also an actor-critic model.

However, the actor part which corresponds to action selection module in Fig. 2 is a discrete-time, nonlinear dynamical system model of C-BG-TH loops derived for action selection [9,10] and the critic part which corresponds to value assigner

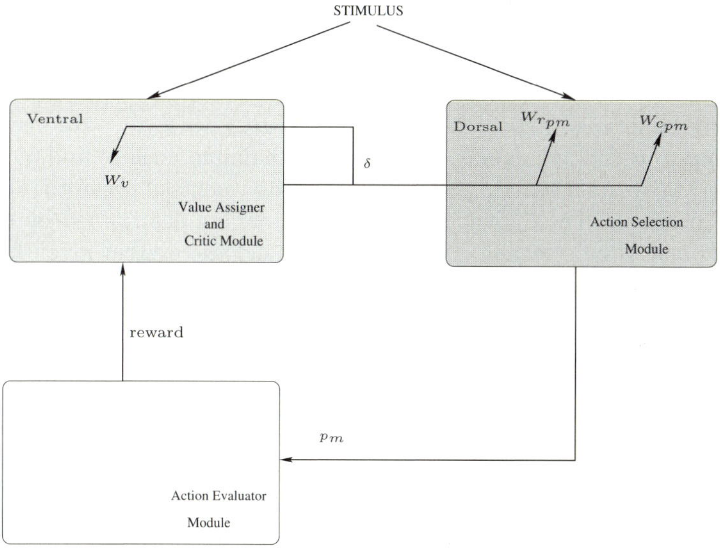

Fig. 2. A block diagram for the computational model

and critic module in Fig. 2. The value assigner and critic module adjusts the parameters of dynamical system via δ to set up the association of stimulus in cortex and to maintain explore/exploit phenomenon. δ also modifies the value function in the value assigner and critic module Thus, action selection module, value assigner and critic module, and δ correspond to dorsal, ventral pathways and modulatory neurotransmitter, respectively.

3.1 The Proposed Model

In this work, the nonlinear dynamical system proposed in [9,10] for action selection is expanded to include both the motor and premotor loops as given in Equations 2 and 3.

$$\begin{aligned}
p_{pm}(k+1) &= f(\lambda p_{pm}(k) + m_{pm}(k) + W_{c_{pm}} I(k)) \\
m_{pm}(k+1) &= f(p_{pm}(k) - d_{pm}(k)) \\
r_{pm}(k+1) &= W_{r_{pm}} f(p_{pm}(k)) \\
n_{pm}(k+1) &= f(p_{pm}(k)) \\
d_{pm}(k+1) &= f(W_{d_{pm}} n_{pm}(k) - r_{pm}(k))
\end{aligned} \tag{1}$$

$$\begin{aligned}
p_m(k+1) &= f(\lambda p_m(k) + m_m(k) + \beta p_{pm} + \text{noise}) \\
m_m(k+1) &= f(p_m(k) - d_m(k)) \\
r_m(k+1) &= W_{r_m} f(p_m(k)) \\
n_m(k+1) &= f(p_m(k)) \\
d_m(k+1) &= f(W_{d_m} n_m(k) - r_m(k))
\end{aligned} \tag{2}$$

The nonlinear function is a sigmoidal function and given as following:

$$f(x) = 0.5(\tanh(a(x - 0.45)) + 1) \tag{3}$$

In these equations, $p_{pm/m}$, $m_{pm/m}$, $r_{pm/m}$, $n_{pm/m}$, $d_{pm/m}$ are vectors corresponding to cortex, thalamus, striatum, subthalamic nucleus and globus pallidus interna/substantia nigra pars reticulate constituents of premotor and motor loops, respectively. The dimension of these vectors are determined by the number of actions to be selected. The need to expand the system given for action selection in [9,10] arised since the fine grain-coarse grain discrimination property of action selection module became important for goal-directed behaviour. The action selection module has to realize a selection based on evaluation of the value of the presented stimuli, but it is also expected to generate a random exploration in case the reward is disappointing. So, the premotor part completes the evaluation and determines possible actions and then the motor part acts as a fine discriminator and selects one action amongst the many determined by premotor part. Thus, while the sensory stimulus, denoted by I, affects the cortex consistituent of the premotor loop , the cortex consistituent of the motor loop is modulated by the cortex consistituent of the premotor loop. The noise signal enables the randomness of selection in the motor loop.

The diffusive effect of subthalamic nucleus is maintained by $W_{d_{pm/m}}$ which have identical entries. The diagonal matrix $W_{r_{pm}}$ represents the effect of of ventral striatum on dorsal striatum. Finally, the matrix $W_{c_{pm}}$ corresponds to the mapping with which the representation of sensory stimulus are formed.

The dynamical behaviour of very similar systems have been investigated in [9,10] and it has been shown that changing the parameters affects the action selection process especially its fine grain - coarse grain property. This effect of parameters has been demonstrated by figures showing the change in domain of attractions in [9,10]. Utilizing this knowledge,parameters of premotor and motor loops are either fixed to values to obtain fine grain properties of motor loops or modified in some range to get coarse grain effect.

The action selection module in Fig. 2 is composed of premotor and motor loops, but their behaviour is modulated by value assigner and critic module. This modification is accomplished by adapting the weights $W_{c_{pm}}$ and $W_{r_{pm}}$ as following:

$$W_{c_{pm}}(k+1) = W_{c_{pm}}(k) + \eta_c \delta(k) p_m(k) I(k)' \tag{4}$$

$$W_{r_{pm}}(k+1) = W_{r_{pm}}(k) + \eta_r \delta(k) f(p_m(k)) E r_m(k) \tag{5}$$

Here E is the identity matrix. The variable δ which represents the error in expectation is determined similar to traditional TD learning algorithm [14] as stated in Eq. 6.

$$\delta(k) = \text{reward}(k) + \mu V(k+1) - V(k) \tag{6}$$

The effect of the modification of $W_{r_{pm}}$ on the premotor system is demonstrated with the bifurcation diagram in Fig. 3. As the value of the $W_{r_{pm}}$'s elements are changing from 0.1 to 0.8, the first component of p_{pm} changes from small fixed values near zero to nonconvergent solutions (possbily quasi-periodic) and then again to fixed values near one.

The value assigner and critic module represents the role of ventral stream which evaluates the inputs from cortex considering the action selected and the reward received in turn. Then it generates an expectation signal based on the value V which it attains to stimuli. This is given by the equation where W_v is a row vector and the term base is a row vector with identical entries.

$$V(k) = (W_v + \text{base}) I(k) \tag{7}$$

This expectation signal, together with the reward, gives rise to the error δ which represents the modulatory effect of neurotransmitters and modulates the behaviour of the dorsal stream via $W_{r_{pm}}$. So, it strengths the corresponding representation of the sensory input via $W_{c_{pm}}$ as given in Eq. 4 and updates the value of stimuli via W_v as stated in Eq. 8

$$W_v(k+1) = W_v(k) + \eta_v \delta(k) I(k)' \tag{8}$$

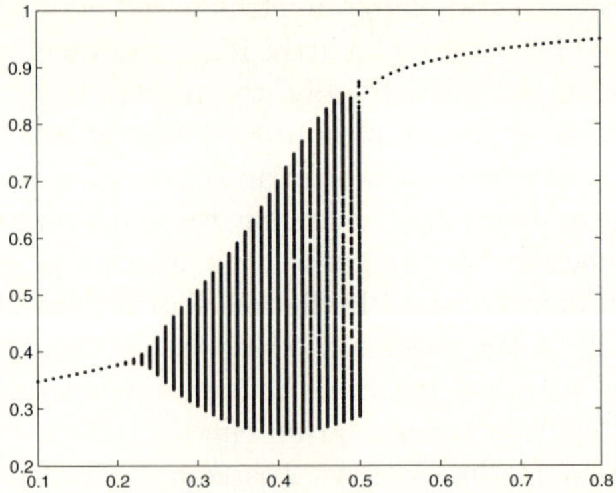

Fig. 3. The change in the behaviour of cortex component of premotor system with W_r

3.2 The Simulation Results for the Task of Sequence Learning

To investigate the appropriatness of the proposed model, the task of learning a sequence of stimulus-action pairs similar to one defined in [11] is considered. The task is to match the stimuli i.e., letters A, B, C to actions, i.e., numbers 1, 2, and 3, respectively. Reward is delivered at the end of the sequence, when each letter is correctly matched to numbers. The learning process begins with letter C as stimulus; once it is matched to correct action number 3 the reward is given. Then letters B and A are added to the sequence one at a time. Each time reward is given only when C is matched with 3.

During the simulation of task, first the stimulus C is presented, that is the vector I in Eq. 2 corresponding to letter C is applied. When the solution of non-linear discrete time system given by Eq. 2, 3 is settled to a stable equilibrium point, the action selected by the action selection module is determined considering the solution of p_m . If this action matches with the correct action that is with number 3, then reward is given, otherwise not given. The value function and the temporal difference error expressed with Eq.'s 6, 7 are calculated and the matrices $W_{c_{pm}}$, $W_{r_{pm}}$ and W_v are updated . If the action matches for 40 times then the stage of learning begins. This time the other stimulus B is given. If the action matches with 2, then stimulus C is given. If first B is matched with 2 and just following it C is matched with 3, then reward is given. The updating of matrices is made as above. This stage of the training ends if the reward is obtained for 40 times. In the last stage as the stimulus A is given and same process is continued.

The parameter values used in the simulation are $\lambda = 0.5$, $\beta = 0.03$, $a = 3$, $\mu = 0.95$, $\eta_c = 0.1$, $\eta_r = 0.2$, $\eta_v = 0.1$ and base$= 0.2$. The initial values of matrices $W_{c_{pm}}$, W_v are randomly generated small positive real numbers and initial value of the diagonal matrix $W_{r_{pm}}$ is ones. During updating the matrix values,$W_{c_{pm}}$ and $W_{r_{pm}}$ are normalized. The matrices $W_{d_{pm/m}}$ and W_{r_m} are composed of 0.5's and they are constant. The noise signal is created as a very small random number and the stimuli and are binary coded, as following $A = [1\,0\,0]$, $B = [0\,1\,0]$, $C = [0\,0\,1]$.

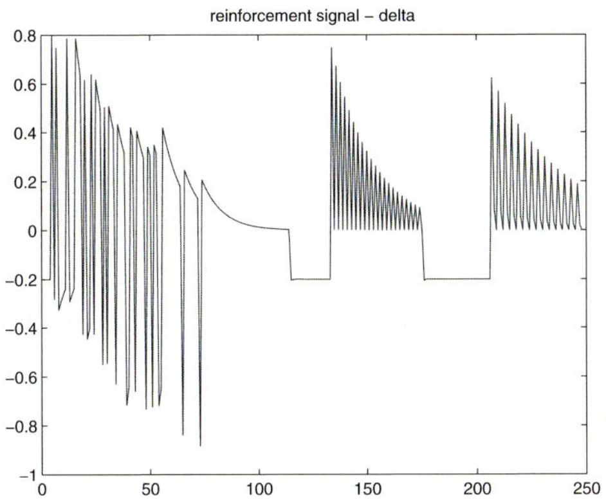

Fig. 4. The difference in δ during 255 trials

The reward signal is scalar and its value is one. The responses are the outputs of the first, second and third motor loops corresponding to numbers 1, 2, 3.

The model completed the task on average 346.35 ± 115.56 trials. The aveage of 20 different runs are taken and the difference between each run is due to randomness in exploration step and the initial values of matrices. During these 20 tryouts, for two times the trials lasted over 600 and stopped before task is completed. The final matrices for one such case and for one successful trial is given as following

$$W_{r_{pm}}^{(\text{success})} = \begin{bmatrix} 0.3862 & 0 & 0 \\ 0 & 0.5680 & 0 \\ 0 & 0 & 0.1589 \end{bmatrix}, W_{r_{pm}}^{(\text{fail})} = \begin{bmatrix} 0.0002 & 0 & 0 \\ 0 & 0.0010 & 0 \\ 0 & 0 & 0.0005 \end{bmatrix} \quad (9)$$

$$W_{c_{pm}}^{(\text{success})} = \begin{bmatrix} 0.6640 & 0.1189 & 0.0782 \\ 0.1052 & 0.8311 & 0.0667 \\ 0.1392 & 0.0891 & 2.4192 \end{bmatrix}, W_{c_{pm}}^{(\text{fail})} = \begin{bmatrix} 0.2017 & 0.1505 & 0.1023 \\ 0.1916 & 1.7053 & 0.0882 \\ 0.1936 & 0.1086 & 1.9519 \end{bmatrix} \quad (10)$$

As, it can be followed from the matrices obtained in the sucessful trial, the values of $W_{r_{pm}}$ and the values on the diagonal of $W_{c_{pm}}$ are bigger which reveals that the association is set up correctly.

In Fig. 4, the change in the value of δ is given for one run. Each time a stimulus is presented some time is spent for exploration to find out the adequate response, which can be followed from the Fig. 4 as big values of δ. The value of δ in other words, the error in expectation decreases as correct responses are given but then with a different stimuli the value of δ increases again as the old response is kept in the first trials following the presentation of new stimuli.

4 Discussion and Conclusion

In this work, a model of cortico-striato-thalamic circuits for goal-directed behaviour is proposed. This model is composed of two main components, one part corresponds to dorsal stream responsible for action selection while the other part corresponds to ventral stream which modulates the action selection While the action selection is realized as an interconnected nonlinear dynamical systems corresponding to premotor and motor loops, the part that affects the action selection is achieved similar to critic part of well-known actor-critic models of reinforcement learning. The simulation results of sequence learning task are given to demonstrate the effectiveness of the proposed model.

This work supports the idea that goal directed behaviour may arise from the interaction between cortico-striato-thalamic loops [4,3]. Modeling these loops as dynamical systems gives one possible mechanism of reinforcement learning in the neural structures.

References

1. Alexander, G.E., Crutcher, M.D.: Functional architecture of basal ganglia circuits:Neural substates of parallel processing. TINS 13, 266–270 (1990)
2. Crutcher, M.D., Alexander, G.E.: Basal ganglia-thalamacortical circuits:Parallel substrates for motor, oculomotor, prefrontal and limbic functions. Progress in Brain Research 85, 119–146 (1990)
3. Heimer, L.: A new anatomical frameworkfor neuropsychiatric disorders and drug abuse. Am. J. Psychiatry 160, 1726–1739 (2003)
4. Fudge, J.L., Haber, S.N., McFarland, N.R.: Striatonigrostriatal pathways in primates form an ascending spiral from the shell to the dorsolateral striatum. The Jour. Neuroscience 20, 2369–2382 (2000)
5. Haber, N.S., Kim, K.S., Mailly, P., Calzavara, R.: Reward-related cortical inputs definea large striatal region in primates that interface with associative cortical connections, providing a substrate for incentive-based learning. The Jour. Neuroscience 26, 8368–8376 (2006)
6. Ridderinkhof, K.R., van den Wildenberg, W.P.M., Segalowitz, S.J., Carter, C.S.: Neurocognitive Mechanisms of Cognitive Control: The Role of Prefrontal Cortex in Action Selection, Response Inhibition. Performance Monitoring and Reward-Based Learning, Brain and Cognition 56, 129–140 (2004)
7. Taylor, N.R., Taylor, J.G.: Hard-wired Models of Working Memory and Temporal Sequence Storage and Generation. Neural Networks 13, 201–224 (2000)
8. Taylor, J.G., Taylor, N.R.: Analysis of Recurrent Cortico-Basal Ganglia-Thalamic Loops for Working Memory. Biological Cybernetics 82, 415–432 (2000)
9. Karabacak, O., Sengor, N.S.: A Dynamical Model of a Cognitive Function: Action Selection. In: 16th IFAC World Congress, Prague (2005)
10. Karabacak, O., Sengor, N.S.: A computational model for the effect of dopamine on action selection during Stroop test. In: ICANN 2006, Athens (2006)
11. Schultz, W., Dayan, P., Montague, P.R.: A Neural Substrate of Prediction and Reward. Science 275, 1593–1599 (1997)
12. Suri, R.E., Schultz, W.: Learning of Sequential Movements by Neural Network Model with Dopamine-Like Reinforcement Signal. Exp. Brain Res. 121, 350–354 (1998)
13. Holyrod, C.B., Coles, M.G.H.: Neural basis of human error processing: reinforcement learning, dopamine, and error related negativity. Psychological Review 4, 679–709 (2002)

14. Sutton, R.S., Barto, A.G.: Reinforcement Learning (2nd printing), A Bradford Book. MIT Press, Cambridge (1998)
15. Berns, G.S., Sejnowski, T.J.: A Model of Basal Ganglia Function Unifying Reinforcement Learning and Action Selection. In: Proc. Joint Symp. Neural Comp., California (1994)
16. Suri, R.E., Bargas, J., Arbib, M.A.: Modeling functions of striatal dopamine modulation in learning and planning. Neuroscience 103, 65–85 (2001)
17. Dayan, P., Balleine, B.W.: Reward, Motivation and Reinforcement Learning. Neuron 36, 285–298 (2002)
18. Montague, P.R., Hyman, S.E., Cohen, J.D.: Computational Roles for Dopamine in Behavioural Control. Nature 431, 760–767 (2004)
19. Joel, D.A., Niv, Y., Ruppin, E.: Actor-Critic models of the basal ganglia: new anatomical and computational perspectives. Neural Networks 15, 535–547 (2002)
20. Gurney, K., Prescott, T.J., Redgrave, P.: A Computational Model of Action Selection in the Basal Ganglia I: A New Functional Anatomy. Biological Cybernetics 84, 401–410 (2001)
21. Gurney, K., Prescott, T.J., Redgrave, P.: A Computational Model of Action Selection in the Basal Ganglia II: Analysis and Simulation of Behaviour. Biological Cybernetics 84, 411–423 (2001)
22. Haruno, M., Kawato, M.: Heterarchical reinforcement learning model for integration of multiple cortico-striatal loops:fMRI examination in stimulus-action-reward association learning. Neural Networks 19, 1242–1254 (2006)
23. Graybiel, A.M.: Neurotransmitters and Neuromodulators in the Basal Ganglia. TINS 13, 244–254 (1990)

Firing Pattern Estimation of Synaptically Coupled Hindmarsh-Rose Neurons by Adaptive Observer

Yusuke Totoki[1], Kouichi Mitsunaga[2], Haruo Suemitsu[1], and Takami Matsuo[1]

[1] Department of Architecture and Mechatronics, Oita University, 700 Dannoharu, Oita, 870-1192, Japan
[2] Control Engineering Department, Oita Institute of Technology, Oita, Japan

Abstract. In this paper, we present adaptive observers for synaptically coupled Hindmarsh-Rose(HR) neurons with the membrane potential measurement under the assumption that some of parameters in an individual HR neuron are known. Using the adaptive observers for a single HR neuron, we propose a two-stage merging procedure to identify the firing pattern of a model of synaptically coupled HR neurons. The procedure allows us to recover the internal states and to distinguish the firing patterns of the synaptically coupled HR neurons, with early-time dynamic behaviors.

1 Introduction

In traditional artificial neural networks, the neuron behavior is described only in terms of firing rate, while most real neurons, commonly known as spiking neurons, transmit information by pulses, also called action potentials or spikes. Model studies of neuronal synchronization can be separated in those where models of the integrated-and-fire type are used and those where conductance-based spiking and bursting models are employed[1]. Bursting occurs when neuron activity alternates, on slow time scale, between a quiescent state and fast repetitive spiking. In any study of neural network dynamics, there are two crucial issues that are: 1) what model describes spiking dynamics of each neuron and 2) how the neurons are connected[2]. Izhikevich considered the first issue and compared various models of spiking neurons[3,2]. He reviewed the 20 types of real (cortical) neurons response, considering the injection of simple dc pulses such as tonic spiking, phasic spiking, tonic bursting, phasic bursting. Through out his simulations, he suggested that if the goal is to study how the neuronal behavior depends on measurable physiological parameters, such as the maximal conductance, steady-state (in)activation functions and time constants, then the Hodgkin-Huxley type model is the best. However, its computational cost is the highest in all models. He also pointed out that the Hindmarsh-Rose(HR) model is computationally simple and capable of producing rich firing patterns exhibited by real biological neurons. Nevertheless the HR model is a computational one of the neuronal bursting using three coupled first order differential equations[4,5],

V. Kůrková et al. (Eds.): ICANN 2008, Part II, LNCS 5164, pp. 338–347, 2008.

it can generate a tonic spiking, phasic spiking, and so on, for different parameters in the model equations. Charroll simulated that the additive noise shifts the neuron model into two-frequency region (*i.e.* bursting) and the slow part of the responses allows being robust to added noises using the HR model[6]. The parameters in the model equations are important to decide the dynamic behaviors in the neuron[7].

From the measurement theoretical point of view, it is important to estimate the states and parameters using measurement data, because extracellular recordings are a common practice in neuro-physiology and often represent the only way to measure the electrical activity of neurons[8]. Tokuda *et al.* applied an adaptive observer to estimate the parameters of the Hindmarsh-Rose neuron (HR neuron) by using membrane potential data recorded from a single lateral pyloric neuron synaptically isolated from other neurons[9]. However, their observer cannot guarantee the asymptotic stability of the error system. Steur[10] pointed out that the HR equations could not transformed into the adaptive observer canonical form and it is not possible to make use of the adaptive observer proposed by Marino[11]. He simplified the three dimensional HR equations and write as one-dimensional system with exogenous signal using contracting and the wandering dynamics technique. His adaptive observer with first-order differential equation cannot estimate the internal states of the HR neurons.

We have recently presented three adaptive observers for a single neuron with the membrane potential measurement under the assumption that some of parameters in the HR neuron are known[12]. Using the Kalman-Yakubovich lemma, we have shown the asymptotic stability of the error systems based on the standard adaptive control theory[13]. The estimators allow us to recover the internal states and to distinguish the firing patterns with early-time dynamic behaviors.

In this paper, we focus on synaptically coupled HR neurons. An individual HR neuron can generate a tonic bursting, tonic spiking, and so on. The firing pattern can be identified with the parameters of the individual neuron. We couple two model neurons both electronically and electrochemically in inhibitory and excitatory fashions. The model neurons synchronize in phase and out of phase depending on the strength of the coupling. Katayama *et al.* studied synchronous phenomena in neural network models with the HR neurons[14]. These neurons generate periodic spikes, quasiperiodic spikes and chaotic spikes in some range of bifurcation parameters. We adopt the synaptic coupling function modeled by the sigmoidal function[1]. First, we numerically confirm that the coupling of an intrinsic bursting neuron and a spiking neuron generate chaotic phenomena with strong coupling, and the strong coupling of two bursting neurons generates a spiking pattern. Next, using the adaptive observers for a single HR neuron[12], we propose a two-stage merging procedure to identify the firing pattern of a model of synaptically coupled HR neurons by estimating the parameter and the applied current of the membrane potential dynamics. Moreover, we propose an instantaneous Lyapunov exponent that is a real-time decay rate of time series data. The instantaneous Lyapunov exponent is one of the measures that estimate the decay rates of flows of nonlinear systems by assigning a criterion function.

We use the instantaneous Lyapunov exponent to check the firing patterns of the coupled HR neurons. The MATLAB simulations demonstrate the estimation performance of the proposed adaptive observers and the instantaneous Lyapunov exponent.

2 Review of Real (Cortical) Neuron Responses

There are many types of cortical neurons responses. Izhikevich reviewed 20 of the most prominent features of biological spiking neurons, considering the injection of simple dc pulses[2]. Typical responses are classified as follows[15]:

- Tonic Spiking (TS) : The neuron fires a spike train as long as the input current is on. This kind of behavior can be observed in the three types of cortical neurons: regular spiking excitatory neurons (RS), low-threshold spiking neurons (LTS), and first spiking inhibitory neurons (FS).
- Phasic Spiking (PS) : The neuron fires only a single spike at the onset of the input.
- Tonic Bursting : The neuron fires periodic bursts of spikes when stimulated. This behavior may be found in chattering neurons in cat neocortex.
- Phasic Bursting (PB) : The neuron fires only a single burst at the onset of the input.
- Mixed Mode (Bursting Then Spiking) (MM) : The neuron fires a phasic burst at the onset of stimulation and then switch to the tonic spiking mode. The intrinsically bursting excitatory neurons in mammalian neocortex may exhibit this behavior.
- Spike Frequency Adaptation (SFA) : The neuron fires tonic spikes with decreasing frequency. RS neurons usually exhibit adaptation of the interspike intervals, when these intervals increase until a steady state of periodic firing is reached, while FS neurons show no adaptation.

3 Single Model of HR Neuron

The Hindmarsh-Rose(HR) model is computationally simple and capable of producing rich firing patterns exhibited by real biological neurons.

3.1 Dynamical Equations

The single model of the HR neuron[1,4,5] is given by

$$\begin{aligned}\dot{x} &= ax^2 - x^3 - y - z + I \\ \dot{y} &= (a + \alpha)x^2 - y \\ \dot{z} &= \mu(bx + c - z)\end{aligned}$$

where x represents the membrane potential y and z are associated with fast and slow currents, respectively. I is an applied current, and a, α, μ, b and c are constant parameters. We rewrite the single H-R neuron as a vectorized form:

$$(S_0) : \dot{\boldsymbol{w}} = \boldsymbol{h}(\boldsymbol{w}) + \Xi(x, z)\boldsymbol{\theta}$$

where

$$\boldsymbol{w} = \begin{bmatrix} x \\ y \\ z \end{bmatrix}^T, \boldsymbol{h}(\boldsymbol{w}) = \begin{bmatrix} -(x^3+y+z) \\ -y \\ 0 \end{bmatrix}, \Xi(x,z) = \begin{bmatrix} x^2 & 1 & 0 & 0 & 0 & 0 \\ 0 & 0 & x^2 & 0 & 0 & 0 \\ 0 & 0 & 0 & x & 1 & -z \end{bmatrix},$$

$$\boldsymbol{\theta} = \left[\theta_1, \theta_2, \theta_3, \theta_4, \theta_5, \theta_6\right]^T = \left[a, I, a+\alpha, \mu b, \mu c, \mu\right]^T.$$

3.2 Intrinsic Bursting Neuron and Spiking Neuron

The HR model shows a large variety of behaviors with respect to the parameter values in the differential equations[7]. Thus, we can characterize the dynamic behaviors with respect to different values of the parameters. We focus on the parameter a and the applied current I. The parameter a is an internal parameter in the single neuron and I is an external depolarizing current. For the fixed $I = 0.05$, the HR model shows a tonic bursting with $a \in [1.8, 2.85]$ and a tonic spiking with $a \geq 2.9$. On the other hand, for the fixed $a = 2.8$, the HR model shows a tonic bursting with $I \in [0, 0.18]$ and a tonic spiking with $a \in [0.2, 5]$. For a small external current,I, the difference between the tonic bursting and the tonic spiking is only the value of the parameter a. We call the single neuron with bursting oscillation the intrinsic bursting neuron(IBN) and that with spiking oscillation the intrinsic spiking neuron(ISN), respectively.

4 Synaptically Coupled Model of HR Neurons

4.1 Dynamical Equations

Consider the following synaptically coupled HR neurons[1]:

$$\begin{aligned}
\dot{x}_1 &= a_1 x_1^2 - x_1^3 - y_1 - z_1 + I_1, \quad I_1 = -g_s(x_1 - V_{s1})\Gamma(x_2) \\
\dot{y}_1 &= (a_1 + \alpha_1)x_1^2 - y_1, \\
\dot{z}_1 &= \mu_1(b_1 x_1 + c_1 - z_1) \\
\dot{x}_2 &= a_2 x_2^2 - x_2^3 - y_2 - z_2 + I_2, \quad I_2 = -g_s(x_2 - V_{s2})\Gamma(x_1) \\
\dot{y}_2 &= (a_2 + \alpha_2)x_2^2 - y_2, \\
\dot{z}_2 &= \mu_2(b_2 x_2 + c_2 - z_2)
\end{aligned}$$

where $\Gamma(x)$ is the sigmoid function given by

$$\Gamma(x) = \frac{1}{1 + \exp(-\lambda(x - \theta_s))}$$

We call the neurons (x_1, y_1, z_1) and (x_2, y_2, z_2) the first and second neurons, respectively.

4.2 Synchronization of Coupled Neurons

Consider the IBN neuron with $a = 2.8$ and the ISN neuron with $a = 10.8$ whose other parameters are as follows:

$$\alpha_i = 1.6, c_i = 5, b_i = 9, \mu_i = 0.001, V_{si} = 2, \theta_s = -0.25, \lambda = 10.$$

The responses of the membrane potentials in the coupling of IBN neuron and ISN neuron with the coupling strength $g_s = 0.05$ behave as a pair of intrinsic single neurons. As increasing the coupling strength, however, the IBN neuron shows a chaotic behavior. The membrane potential of IBN neuron in the coupling of IBN neuron and ISN neuron with the coupling strength $g_s = 1$ behaves a chaotic like response. Though two IBN neurons synchronize as bursting neurons in the coupling of two same IBN neurons with the coupling strength $g_s = 0.05$, two IBN neurons synchronize as spiking neurons with the coupling strength $g_s = 1$.

5 Two-Stage Estimation Procedure with Adaptive Observers

The parameter a and I are key parameters that determine the firing pattern. When the full states, x, y, and z are measurable, the adaptive observer can estimate two parameters simultaneously with gradient-type adaptive update law[12]. However, it is difficult to measure the ion currents, y and z. We assume that the membrane potential x_1 and the external current I_1 of the first neuron are measurable, but the others are immeasurable.

In this case, we propose a two-stage merging procedure to estimate the states and the parameters of one of the synaptically coupled HR neurons:

- *First stage*: Estimate y_1, z_1, a_1 using the available signal x_1 and I_1 in a short time range. Distinguish the firing patterns by using early-time dynamic behaviors;
- *Second stage*: After the first stage, the measurement of the external current I is not required. Estimate y_1, z_1 and I_1 using the estimate of a_1 and the available signal x_1. We can monitor the firing pattern using the estimate of I_1.

5.1 Construction of Adaptive Observers

We focus on the first one of the synaptically coupled HR neurons. The parameters a_1 and I_1 are key parameters that determine the firing pattern. The HR model can be rewritten by the following two forms:

$$(S_{11}) : \dot{\boldsymbol{w}}_1 = A_1 \boldsymbol{w}_1 + \boldsymbol{h}_{11}(x_1) + \boldsymbol{b}_1(x_1^2 a_1) \tag{1}$$

$$(S_{21}) : \dot{\boldsymbol{w}}_1 = A_1 \boldsymbol{w}_1 + \boldsymbol{h}_{21}(x_1) + \boldsymbol{b}_2(I_1) \tag{2}$$

where

$$A_1 = \begin{bmatrix} 0 & -1 & -1 \\ 0 & -1 & 0 \\ \mu_1 b_1 & 0 & -\mu_1 \end{bmatrix}, \boldsymbol{h}_{11} = \begin{bmatrix} -x_1^3 + I_1 \\ \alpha_i x_1^2 \\ \mu_1 c_1 \end{bmatrix}, \boldsymbol{h}_{21} = \begin{bmatrix} -x_1^3 + a_1 x_1^2 \\ (a_1 + \alpha_1) x_1^2 \\ \mu_1 c_1 \end{bmatrix},$$

$$\boldsymbol{b}_1 = \left[1\,1\,0\right]^T, \boldsymbol{b}_2 = \left[1\,0\,0\right]^T.$$

In (S_{11}) and (S_{21}), the unknown parameters are assumed to be a_1 and I_1, respectively. We use the models (S_{11}) and (S_{21}) in the first stage and the second one, respectively.

Since the measurable signal is x_1, the output equation is given by

$$x_1 = \boldsymbol{c}\boldsymbol{w}_1 = \begin{bmatrix} 1\,0\,0 \end{bmatrix} \boldsymbol{w}_1.$$

In the first stage, we present an adaptive observer to estimate the parameter a_1 and the states $\boldsymbol{w}_1$ as follows:

$$(O_{11}) : \dot{\hat{\boldsymbol{w}}}_1 = A_1\hat{\boldsymbol{w}}_1 + \boldsymbol{h}_1(x_1) + \boldsymbol{b}_1(x_1^2\hat{a}) + \boldsymbol{g}(x_1 - \hat{x}_1), \dot{\hat{a}}_1 = \gamma_1 x_1^2(x_1 - \hat{x}_1) \quad (3)$$

where $\hat{\boldsymbol{w}}_1 = \begin{bmatrix} \hat{x}_1\; \hat{y}_1\; \hat{z}_1 \end{bmatrix}^T$ is the estimate of the states and $\hat{a}_1$ is the estimate of the parameter a. The vector $\boldsymbol{g}$ is selected such that $A_1 - \boldsymbol{g}\boldsymbol{c}$ is a stable matrix. Using the Kalman-Yakubovich (KY) lemma, we can show the asymptotic stability of the error system based on the standard adaptive control theory[13].

In the second stage, we present another adaptive observer to estimate the parameter I_1 and the states $\boldsymbol{w}_1$ as follows:

$$(O_{21}) : \dot{\hat{\boldsymbol{w}}}_2 = A_1\hat{\boldsymbol{w}}_2 + \boldsymbol{h}_2(x_1) + \boldsymbol{b}_2(\hat{I}) + \boldsymbol{g}(x_1 - \hat{x}_1), \dot{\hat{I}}_1 = \gamma_2(x_1 - \hat{x}_1). \quad (4)$$

where $\hat{I}$ is the estimate of the external current I_1. The vector $\boldsymbol{g}$ is also selected such that $A_1 - \boldsymbol{g}\boldsymbol{c}$ is a stable matrix.

5.2 Instantaneous Lyapunov Exponent

The Lyapunov exponent gives a measure of the mean decay/divergence rates of the flows of nonlinear systems and can be utilized to check chaotic behaviors. However, the Lyapunov exponent needs an infinite time interval of flows and the Jacobian matrix of system parameters. Yoden *et al.*[16] presented the finite-time Lyapunov exponent that does not require the limit of infinite time interval. However, it requires the knowledge of systems parameters. We propose an instantaneous Lyapunov exponent that is a real-time decay rate of time series data. The instantaneous Lyapunov exponent is one of the measures that estimate the decay rates of flows of nonlinear systems by assigning a criterion function.

Thus, we define an instantaneous Lyapunov exponent (ILE) as follows:

$$\lambda(t) = \frac{1}{t}\log\left[\frac{||\boldsymbol{x}(t)||}{\phi(t)} + \exp\left\{-\alpha\frac{||\boldsymbol{x}(t)||}{\phi(t)}\right\}\right] \quad (5)$$

where $\phi(t)$ is a designable function that assign the decay order and satisfies $\lim_{t\to\infty}\phi(t) = 0$. We apply the ILE to check the decay rate of the error states between the neuron states and its estimates. Selecting $\alpha = \infty$, the ILE is given by

$$\lambda_\infty(t) = \frac{1}{t}\log\left[\frac{||\boldsymbol{x}(t)||}{\phi(t)}\right]. \quad (6)$$

By choosing the signals as $\boldsymbol{x}(t) = x_1(t)$ and $\phi(t) = x_2(t)$, λ_∞ is one of the measures of the synchronization rate of two neurons.

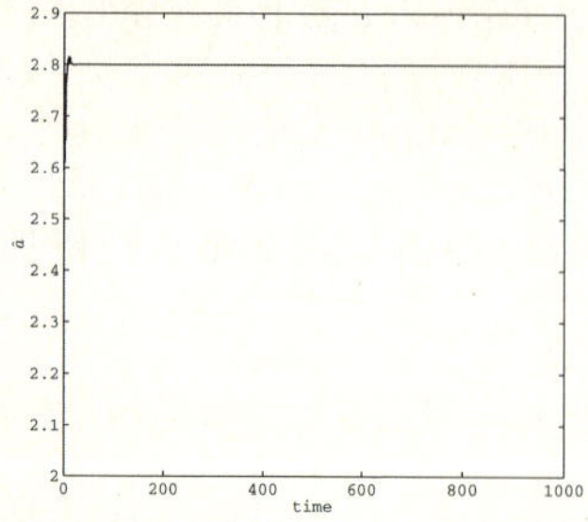

Fig. 1. The response of $\hat{a}_1$ by the adaptive observer (O_{11})(IBN-ISN coupling with $g_s = 0.05$)

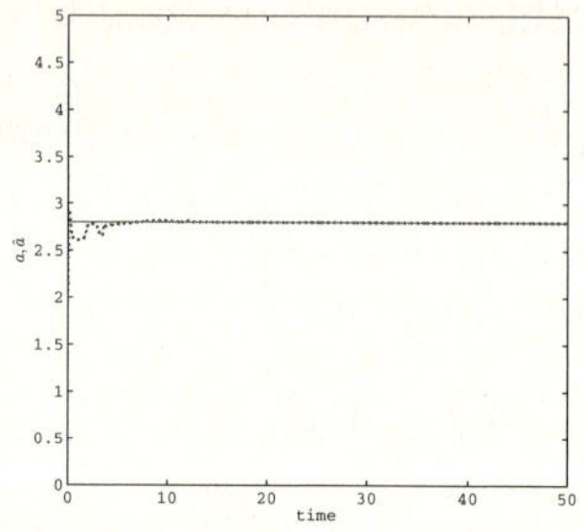

Fig. 2. The response of $\hat{a}_1$ in the short-time range(IBN-ISN coupling with $g_s = 0.05$)

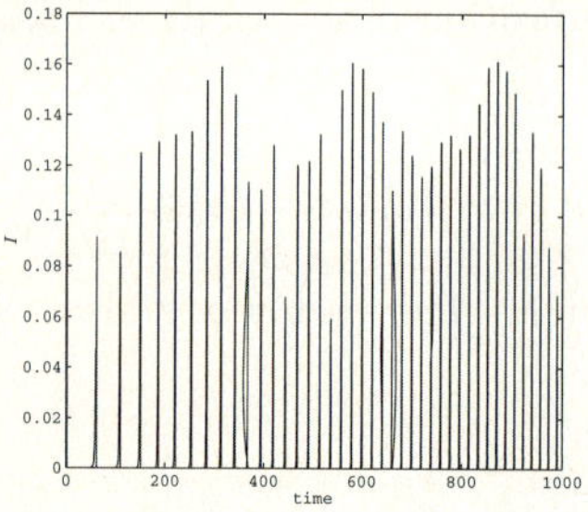

Fig. 3. The response of I_1(IBN-ISN coupling with $g_s = 0.05$)

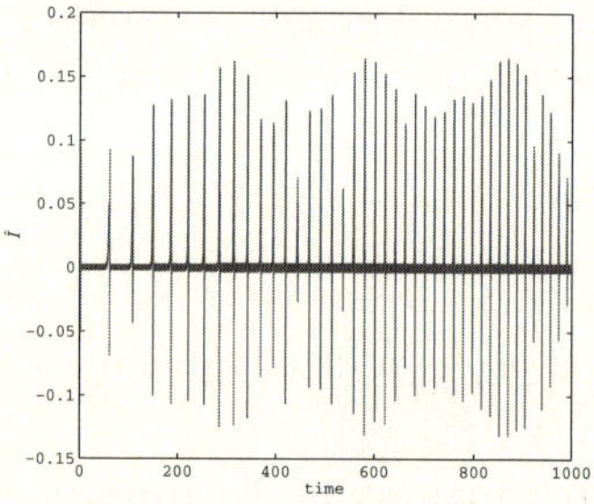

Fig. 4. The response of $\hat{I}_1$ by the adaptive observer (O_{21})(IBN-ISN coupling with $g_s = 0.05$)

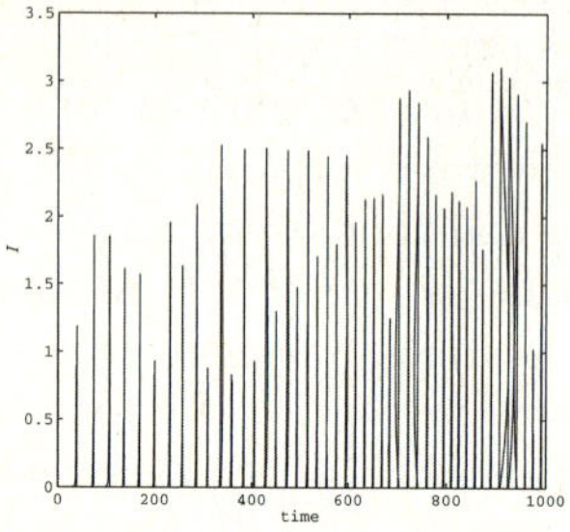

Fig. 5. The response of I_1 by the adaptive observer (O_{21})(IBN-ISN coupling with $g_s = 1$)

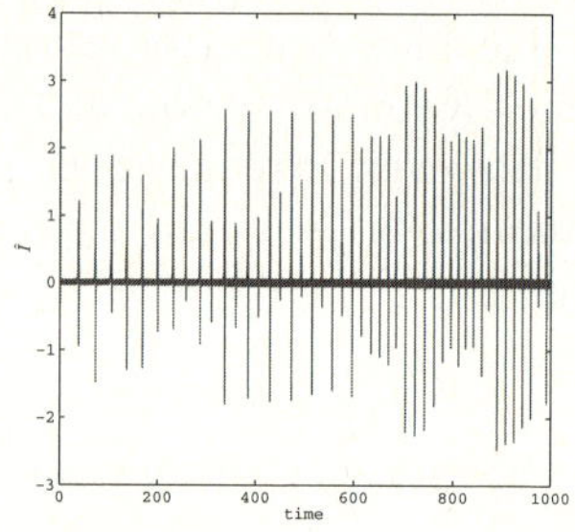

Fig. 6. The response of $\hat{I}_1$ by the adaptive observer (O_{21})(IBN-ISN coupling with $g_s = 1$)

5.3 Numerical Examples

We demonstrate the two-stage estimator using MATLAB/Simulink. Assuming that the output is x_1 in the synaptically coupled Hindmarsh-Rose neurons, we apply the proposed observers to estimate the parameters of the first neuron with the available signals x_1 and I_1.

The first case is where the IBN neuron with $a_1 = 2.8$(the first neuron) and the ISN neuron with $a_2 = 10.8$(the second neuron) are synaptically coupled with a weak coupling. The coupling strength $g_s = 0.05$. Each neuron behaves intrinsically. Figure 1 shows the estimate $\hat{a}_1$ by the adaptive observer (O_{11}). Figure 2 is its response in the short-time range. Figures 3 and 4 show I_1 and

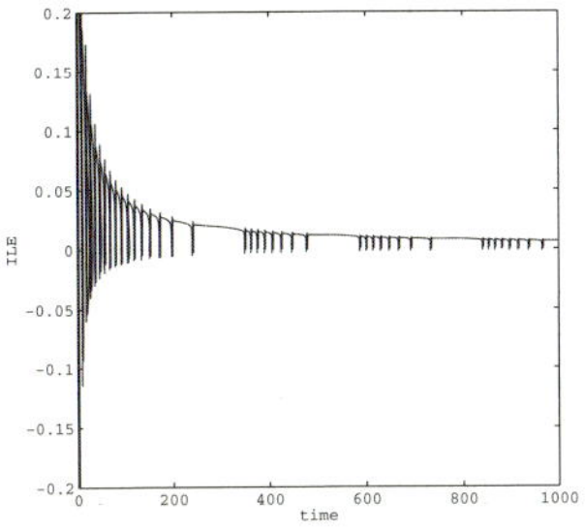

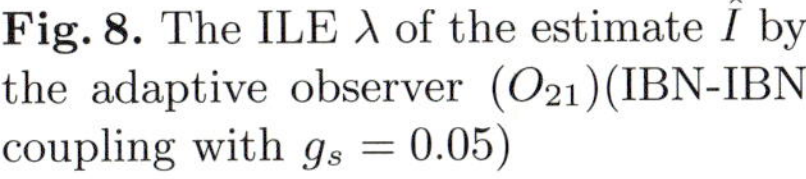

Fig. 7. The ILE λ of the membrane potential x_1(IBN-IBN coupling with $g_s = 0.05$)

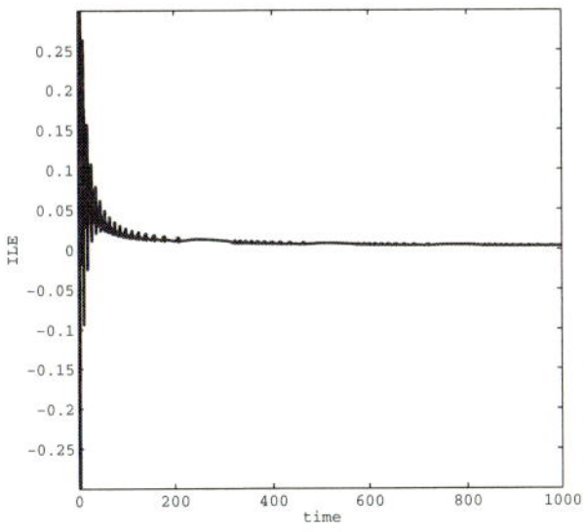

Fig. 8. The ILE λ of the estimate $\hat{I}$ by the adaptive observer (O_{21})(IBN-IBN coupling with $g_s = 0.05$)

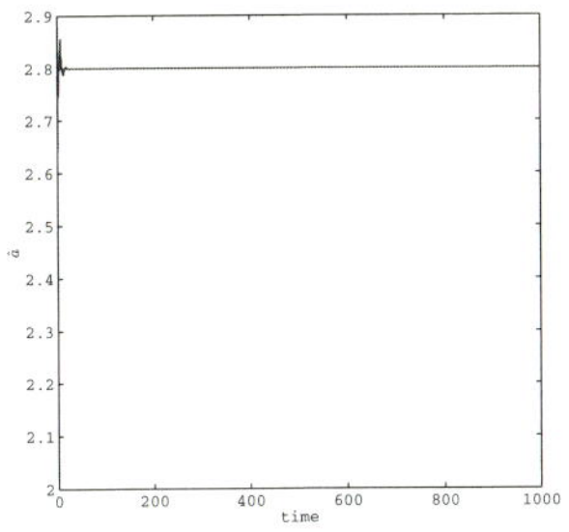

Fig. 9. The response of $\hat{a}_1$ by the adaptive observer (O_{11})(IBN-IBN coupling with $g_s = 1$)

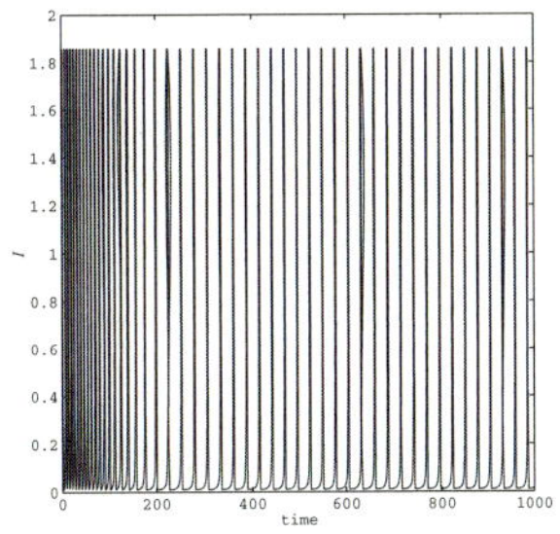

Fig. 10. The response of I_1 ($g_s = 0.05$)(IBN-IBN coupling with $g_s = 1$)

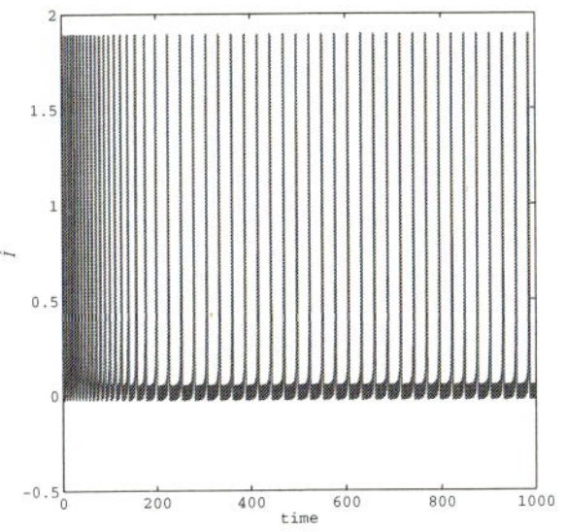

Fig. 11. The response of $\hat{I}_1$ by the adaptive observer (O_{21})(IBN-IBN coupling with $g_s = 1$)

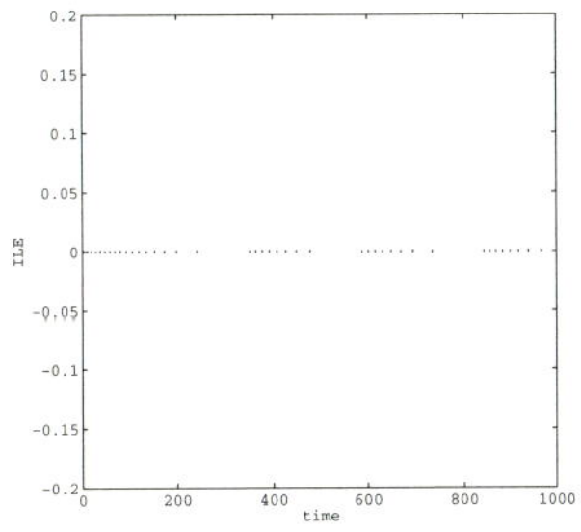

Fig. 12. The ILE $\lambda_\infty(t)$ between x_1 and x_2(IBN-IBN coupling with $g_s = 1$)

its the estimate $\hat{I}_1$ by the adaptive observer (O_{21}), respectively. The estimate $\hat{a}_1$ converges to the intrinsic value and $\hat{I}_1$ nearly equal to I except for the negative values. Though the adaptive estimation law is designed for an unknown constant parameter, it can follow time-varying parameters. The second case is where the IBN neuron with $a_1 = 2.8$(the first neuron) and the ISN neuron with $a_2 = 10.8$(the second neuron) are synaptically coupled with a strong coupling. The coupling strength $g_s = 1$. The estimate $\hat{a}_1$ converges to the intrinsic values and $\hat{I}_1$ nearly equal to I except for the negative values. Figures 5 and 6 show I_1 and its the estimate $\hat{I}_1$ by the adaptive observer (O_{21}). In this case, the first neuron generates a chaotic firing pattern. The third case is where two bursting neurons $(a_1 = a_2 = 2.8)$ are synaptically coupled with the weak coupling, $g_s = 0.05$. In this case, two neurons synchronize with the bursting firing pattern. Figure 7 shows the ILE, $\lambda(t)$, of the membrane potential x_1 selecting $\phi(t) = \frac{1}{t}, \alpha = 10$. Figure 8 shows the ILE, $\lambda(t)$, of $\hat{I}_1$. The last case is where two bursting neurons $(a_1 = a_2 = 2.8)$ are synaptically coupled with the strong coupling, $g_s = 1$. In this case, we can also identify the upper neuron as a bursting neuron by the estimated parameters, $\hat{a}$ and $\hat{I}$. Two intrinsic bursting neurons behave spurious spiking neurons. Figure 9 shows the estimate $\hat{a}_1$ by the adaptive observer (O_{11}). Figures 10 and 11 show I_1 and its the estimate $\hat{I}_1$ by the adaptive observer (O_{21}). The estimate $\hat{a}_1$ converge to the intrinsic values and $\hat{I}_1$ converge to I. Figure 12 shows $\lambda_\infty(t)$ between x_1 and x_2. It indicates that two signal x_1 and x_2 synchronize since $\lambda_\infty \approx 0$.

6 Conclusion

We presented the adaptive estimators of the parameters of the HR model using the adaptive observer technique with the output measurement data such as the membrane potential. The proposed observers allow us to distinguish the firing pattern in early time and to recover the immeasurable internal states in the case of a model of synaptically coupled Hindmarsh-Rose neurons.

References

1. Belykh, I., de Lange, E., Hasler, M.: Synchronization of bursting neurons: What matters in the network topology. Phys. Rev. Lett. 94, 188101 (2005)
2. Izhikevich, E.M.: Which model to use for cortical spiking neurons? IEEE Trans on Neural Networks 15(5), 1063–1070 (2004)
3. Izhikevich, E.M.: Simple model of spiking neurons. IEEE Trans on Neural Networks 14(6), 1569–1572 (2002)
4. Hindmarsh, J., Rose, R.: A model of the nerve impulse using two first order differential equations. Nature 296, 162–164 (1982)
5. Hindmarsh, J., Rose, R.: A model of neuronal bursting using three coupled first order differential equations. Proc. R. Soc. Lond. B. 221, 87–102 (1984)
6. Carroll, T.L.: Chaotic systems that are robust to added noise. CHAOS 15, 013901 (2005)

7. Arena, P., Fortuna, L., Frasca, M., Rosa, M.L.: Locally active Hindmarsh-Rose neurons. Chaos, Soliton and Fractals 27, 405–412 (2006)
8. Meunier, N., Narion-Poll, R., Lansky, P., Rospars, J.O.: Estimation of the individual firing frequencies of two neurons recorded with a single electrode. Chem. Senses 28, 671–679 (2003)
9. Tokuda, I., Parlitz, U., Illing, L., Kennel, M., Abarbanel, H.: Parameter estimation for neuron models. In: Proc. of the 7th Experimental Chaos Conference (2002)
10. Steur, E.: Parameter estimation in Hindmarsh-Rose neurons,traineeship report (2006), http://alexandria.tue.nl/repository/books/626834.pdf
11. Marino, R.: Adaptive observers for single output nonlinear systems. IEEE Trans. on Automatic Control 35(9), 1054–1058 (1990)
12. Mitsunaga, K., Totoki, Y., Matsuo, T.: Firing pattern estimation of biological neuron models by adaptive observer. In: Ishikawa, M., et al. (eds.) ICONIP 2007, Part I. LNCS, vol. 4984, pp. 83–92. Springer, Heidelberg (2008)
13. Narendra, K., Annaswamy, A.: Stable Adaptive Systems. Prentice Hall Inc., Englewood Cliffs (1989)
14. Katayama, K., Horiguchi, T.: Synchronous phenomena of neural network models using Hindmarsh-Rose equation. Interdisciplinary Information Sciences 11(1), 11–15 (2005)
15. Watts, L.: A tour of neuralog and spike - tools for simulating networks of spiking neurons (1993), http://www.lloydwatts.com/SpikeBrochure.pdf
16. Yoden, S., Nomura, M.: Finite-time Lyapunov stability analysis and its application to atmospheric predictability. Journal of the Atmospheric Sciences 51(11), 1531–1543 (1993)

Global Oscillations of Neural Fields in CA3

Francesco Ventriglia*

Istituto di Cibernetica "E.Caianiello" del CNR
Via Campi Flegrei 34, 80078 Pozzuoli (NA), Italy
franco@ulisse.cib.na.cnr.it

Abstract. The investigation on the conditions which cause global population oscillatory activities in neural fields, originated some years ago with reference to a kinetic theory of neural systems, as been further deepened in this paper. In particular, the genesis of sharp waves and of some rhythmic activities, such as theta and gamma rhythms, of the hippocampal CA3 field, behaviorally important for their links to learning and memory, has been analyzed with more details. To this aim, the modeling-computational framework previously devised for the study of activities in large neural fields, has been enhanced in such a way that a greater number of biological features, extended dendritic trees–in particular, could be taken into account. By using that methodology, a two-dimensional model of the entire CA3 field has been described and its activity, as it results from the several external inputs impinging on it, has been simulated. As a consequence of these investigations, some hypotheses have been elaborated about the possible function of global oscillatory activities of neural populations of Hippocampus in engram formation.

1 Introduction

The modality according which the information is coded within the activities of the neural fields of Hippocampus and the way with which they produce the memory traces, or *engrams*, specifically related to the sensory events of brain, constitute a very elusive problem and no clear elucidation about its nature exists still now.

Some hypotheses on the possible role of the sharp waves and the global oscillatory activities of neural populations of Hippocampus on engram formation are presented in this article. In fact, the synchronous oscillatory activity constitutes one of the most characteristic aspects of brain activity and is associated closely to fundamental behavioral states. In Hippocampus, rhythmic oscillations in the theta (4-15 Hz) and in the gamma (20-80 Hz) ranges are among the most prominent patterns of activity [4] and the neuroscientists believe that both rhythms, and sharp waves, reflect essential aspects of the hippocampal functions, mainly related to learning and memory [5,3,9].

The importance of CA3 field in this setting springs from several elements. CA3 is the main source of rhythmic activity in the Hippocampus. It has a pyramidal axonic structure with one of the highest levels of reentry among all the

* Corresponding author.

V. Kůrková et al. (Eds.): ICANN 2008, Part II, LNCS 5164, pp. 348–357, 2008.

other structures of brain (1.9% of probability of coupling between two pyramidal neurons) [1]. At CA3, not only mossy fibers from granular neurons of DG arrive, but also fibers directly from EC (III layer) and from sub-cortical nuclei, among which the most important are Medial Septum (MS) and Diagonal Band, Raphe and Corpus Coeruleus. They are linked to attention, to overt behavior, to stress.

We used the CA3 field of the Hippocampus to evaluate the new hypotheses about the global oscillatory activity of neural populations as base of engrams.

2 The Kinetic Model

The kinetic theory of neural systems, formulated several years ago [12,11] to describe the activity of large neural fields, has been utilized here for the description of the activity of the CA3 field. The most original aspect of such a theory is the statistical description of the neuronal interaction. In fact, the action potentials traveling along the axonic branches are represented as massless particles with a volume, the *impulses*, having only statistical links to the axonic structure. In general, the impulses are short-range with a span of about $300-400\mu m$, but the CA3 pyramidal neurons can emit also long-range impulses which can reach distant zones in CA3, also some millimeters far from the emitting neuron. After the firing the neuron goes in a refractoriness state, for a period of time τ. Vice versa, in the event that the subthreshold potential, due to an excessive inhibition, reaches values well under the resting value, the neuron passes in the so called hyperpolarized state, where it remains until the excitation changes sign. We associate the variables $\mathbf{r}, \mathbf{v}, e$ and t to position, velocity, subthreshold membrane potential and time, respectively. Since our model of CA3 is based on a projection to a two-dimensional surface, both vectors $\mathbf{r}$ and $\mathbf{v}$ have two dimensions. The different impulses traveling within CA3 are denoted by an index s and an index s' is associated to the different families of neurons. Moreover, s_{ex} denotes a generic impulse coming from sources external to CA3 and s_{in} denotes a generic impulse generated by neurons within CA3. The functions $f_s(\mathbf{r}, \mathbf{v}, t)$ and $g_{s'}(\mathbf{r}, e, t)$ describe, respectively, the velocity distribution of impulses and the distribution of the subthreshold neuronal excitation within the neural field. The function $\psi_{s'}(\mathbf{r})$ denotes the local density of neurons of a specific type s' and it is related to $g_{s'}(\mathbf{r}, e, t)$ via integration on $\mathbf{v}$.

It is necessary to note that the set of equations, reported in previous articles by the author, do not describe accurately the dynamics of the neuronal interaction. In fact, by considering only local absorptions (restricted to elemental space $(\mathbf{r}, d\mathbf{r})$ where impulses collide with neurons–somata) these equations handle too grossly the process of absorption and, as a consequence, the activity of the neural field could not be adequately described. In cerebral structures, indeed, the span of the mean dendritic arborization of the neurons (about $300-400\mu m$), is much larger then the diameter of the neuronal somata and probably this is an important element to take into account. In a new setting, still based on the methods of the kinetic theory of neural systems, the mean number of impulses of type s absorbed in $\mathbf{r}$ at time t from neurons s' was described by the following function:

$$I_{s's}(\mathbf{r},t) = \int\int_{D(\mathbf{r})} f_s(\mathbf{r}',\mathbf{v},t)d\mathbf{r}'\psi_{s'}(\mathbf{r})\sigma_{s's} \mid \mathbf{v} \mid d\mathbf{v} \tag{1}$$

where $\sigma_{s's}$—the absorption coefficient—corresponds to the collision kernel as defined in the transport theory, and $D(\mathbf{r})$ denotes the region of extension of the dendritic arborization of neurons located in $\mathbf{r}$. As it appear evident from this equation, the absorption of impulses is no more a local process, as in equations of earlier articles, but it interests a larger extension of the neural field from which the impulses can be absorbed. Accordingly, the previous set of equations will be denoted *local absorption model*, while the actual set will be denoted *non-local absorption model.*

By using this function, the mean number of impulses of type s absorbed from each neuron in $\mathbf{r}$ at time t is given by:

$$i_{s's}(\mathbf{r},t) = \frac{I_{s's}(\mathbf{r},t)}{\psi_{s'}(\mathbf{r})}. \tag{2}$$

To compute the time course of the net excitatory effect on subtreshold neurons s' in $\mathbf{r}$ at time t, induced by all the absorbed impulses, a new function was introduced. It is denoted by $E_s(t)$ and describes the time course of the mean Post Synaptic Potential associated to the absorption of an impulse of type s—a positive function for excitatory impulses, negative for inhibitory ones. Based on this function, the net excitatory effect, $\epsilon_{s'}(\mathbf{r},t)$, is given by the following convolution equation:

$$\epsilon_{s'}(\mathbf{r},t) = \Sigma_{s=1}^{n} \int_0^t E_s(t) \cdot i_{s's}(\mathbf{r},t-t')dt'. \tag{3}$$

Finally, this function allows the computation of the probable number of neurons in $\mathbf{r}$ which are in a firing state at time t:

$$N_{s'}(\mathbf{r},t) = \int_{1-\epsilon(\mathbf{r},t)}^{1} g_{s'}(\mathbf{r},e,t)de \tag{4}$$

While, the probable number of neurons in hyperpolarized state are described by the equation:

$$\nabla_t M_{s'}(\mathbf{r},t) = \int_0^{-\epsilon(\mathbf{r},t)} g_{s'}(\mathbf{r},e,t)de - M_{s'}(\mathbf{r},t)\theta(\epsilon) \tag{5}$$

where $M_{s'}(\mathbf{r},t)$ denotes the probable number of neurons in $\mathbf{r}$ which stay in maximum hyperpolarization level at time t. In the above equations, use is made of the conditions that the function $g_{s'}(\mathbf{r},e,t) = 0$ if $e \leq 0$ or if $e > 1$, and 0 and 1 represent the normalized maximum hyperpolarization level and neuronal threshold, respectively. The function $\theta(.)$ denotes the step function.

Based on these preliminary definitions and equations, the time evolution of the two distribution functions $f_s(\mathbf{r}, \mathbf{v}, t)$ and $g_{s'}(\mathbf{r}, e, t)$ is governed by the following set of coupled differential equations:

$$\nabla_t f_s(\mathbf{r}, \mathbf{v}, t) + \mathbf{v} \cdot \nabla_{\mathbf{r}} f_s(\mathbf{r}, \mathbf{v}, t) + f_s(\mathbf{r}, \mathbf{v}, t)(\Sigma_{s'} \int_{D(\mathbf{r})} \psi_{s'}(\mathbf{r}') \mid \mathbf{v} \mid \sigma_{s's} d\mathbf{r}') =$$

$$S_s(\mathbf{r}, \mathbf{v}, t)\delta(s - s_{ex}) + f_s^s(\mathbf{r}, \mathbf{v}) N_{s'}(\mathbf{r}, t)\delta(s - s_{in})$$

$$+ f_s^l(\mathbf{v}) \int_A \xi_{s's}(\mathbf{r}, \mathbf{r}')\, d\mathbf{r}' \int f^{'}(v) N_{s'}(\mathbf{r}', t - \frac{|\mathbf{r} - \mathbf{r}'|}{v})\, dv \delta(s - s_{in}) \quad (6)$$

$$\nabla_t g_{s'}(\mathbf{r}, e, t) + \mu(e_r - e)\nabla_e g_{s'}(\mathbf{r}, e, t) =$$

$$[g_{s'}(\mathbf{r}, e - \epsilon, t) - g_{s'}(\mathbf{r}, e, t)](1 - \delta(\epsilon))$$

$$+ N_{s'}(\mathbf{r}, t - \tau_{s'})\delta(e - e_r)$$

$$+ M_{s'}(\mathbf{r}, t)\theta(\epsilon)\delta(e - e_0) \quad (7)$$

where $\delta(.)$ denotes the Dirac function. A not standard use of the Dirac function is done in equation 6. In this case, the Dirac function has been utilized to obtain a more compact presentation of the set of equations related to different type of impulses propagating in CA3. In particular, the factor $\delta(s - s_{ex})$ in equation 6 means that the source term $S_s(\mathbf{r}, \mathbf{v}, t)$ is present only for the specific subset of the impulses coming from external sources. When this term is present, the other two terms on the right side of the equation, which are multiplied by the factor $\delta(s - s_{in})$ are lacking, and vice versa. Moreover, also the short-range and the long-range terms (the second and the third terms on the right member in equation 6) are mutually exclusive. In the above equations, ϵ stands for $\epsilon_{s'}(\mathbf{r}, t)$, and e_r and e_0 denote the resting potential and the maximum hyperpolarization level (this last was assumed as normalized to 0), μ is the decay constant of the subthreshold excitation, and A is the surface occupied by CA3.

The third term on left member side of equation 6 is the other distinguishing element of the non-local model. It describes the fact that the impulses present at time t in $(\mathbf{r}, d\mathbf{r})$ are absorbed not only by neurons lying there, but by the entire set of neurons which, having somata occupying the space $D(\mathbf{r})$, extend their dendrites till to the position $\mathbf{r}$.

The impulses conveyed by the Mossy fibers, which terminate on the base of the shaft of the dendritic tree (*stratum lucidum*), can be still described by the non-local equations. In this case, the absorption space is very close to the pyramidal soma, and strictly confined within the soma boundary. For the description of Mossy impulses absorption we use the equation:

$$I_{s's}(\mathbf{r}, t) = \int \int_{D(\mathbf{r})} f_s(\mathbf{r}', \mathbf{v}, t)\delta(\mathbf{r} - \mathbf{r}')d\mathbf{r}' \psi_{s'}(\mathbf{r})\sigma_{s's} \mid \mathbf{v} \mid d\mathbf{v}. \quad (8)$$

which is equivalent to the corresponding equation of the local model.

3 Discretized Model

The activity of the entire CA3 field of the rat, having dimensions of $8.4mm$ long (septo-temporal axis) and $3mm$ large (transverse axis) as reported in the Atlas of Rat Brain [10], has been simulated by using the non-local absorption equation set. The integro-differential equations of the non-local model were transformed in discrete-difference equations, according to a procedure based on the suggestions of [2], Chapt. 9, for equations related to the theory of invariant embedding. The space-time course of some macroscopic parameters (local frequency of spikes, local mean sub-threshold excitation, number of firing neurons), which have close analogy with the *in vivo* recorded activity of the hippocampal CA3 field (i.e., population spike trains, local field potentials–*LFPs*), has been analyzed to obtain information on their ability to simulate oscillating hippocampal activity.

Excitatory stimuli originated from Entorhinal Cortex via Dentate Gyrus (through the Mossy Fibers) and by a direct path have been simulated. Also, the effects of inhibitory afferences from Medial Septum have been investigated. In summary, seven different families of impulses and three families of neurons –pyramidal, inhibitory fast and inhibitory slow– have been used in the description.

Moreover, to compute $E_s(t)$, the time-course of the mean Post Synaptic Potential produced at the axon hillock by absorption of the different impulses impinging on pyramidal neurons, we considered that the synapses on pyramidal neurons in CA3 field of Hippocampus are parceled in four main layers: *Oriens*, *Lucidum*— formed by the mossy synapses of axons coming from granular neurons in *Dentate Gyrus*, *Radiatum*, and *Lacunosum-moleculare.* Another layer, the *stratum Pyramidal*, containing the somata of pyramidal neurons, is the place of arrival of several inhibitory terminals. We know also, from [8], that Entorhinal Cortex neurons send axons in *Lacunosum-moleculare* layer, some of the pyramidal Schaffer's collaterals from CA3 terminate in *Oriens* and *Radiatum* layers, and Medial Septum neurons excite synapses in *Oriens* and, at a minor extent, synapses in *Radiatum.*

Due to the passive propagation along the dendritic trees, the distance between the synapses and the axon hillock produces a widening and weakening of the synaptic effect at the axon hillock. By using the results of [13], the time-course of $E_s(t)$ produced by input in *Lacunosum-moleculare* layer (EC input) was considered maximally widened and weakened, whereas the input in *Oriens* and *Radiatum* layers (MS and pyramidal Short and Long Range impulses) had a mean widening and weakening time-course, and the input in *Lucidum* and *Pyramidal* layers was only slightly widened and weakened. Dentate Gyrus input, conveyed by layered mossy fibers in stratum lucidum, was assumed to be distributed along parallel strips of pyramidal neurons, each strip being $3mm$ long and $50\mu m$ large and containing about 6.250 mossy fibers and 87.500 mossy synapses. The amplitude of the strip ($50\mu m$) being imposed by the space step.

4 Ca3 Global Activity

The genesis of sharp waves and rhythmic oscillations in gamma and theta ranges has been investigated by computational experiments which simulated the reaction of the CA3 model to external stimuli. In general, each simulation had a duration of 2 seconds, requiring 16000 time steps ($\delta t = 0.125ms$).

In all the experiments, the main stimuli came from Dentate Gyrus, Entorhinal Cortex, and Medial Septum. In the simulations described here, an inhibitory input from MS has been simulated which inhibited selectively the inhibitory neural populations of CA3 [7]. The Dentate Gyrus input to each neural strip was constituted by random volleys of impulses, whose arrival times were distributed according to a Poisson distribution, while the amplitudes were Gaussian. Once a volley arrived to a strip, it traveled through all the transverse length of CA3 ($3mm$), impulses being absorbed along the route. In some cases the volleys along the different strips were correlated. The CA3 field was also reached by Poissonian inputs originating from a direct Entorhinal Cortex path. The Entorhinal Cortex inputs had Gaussian distributed amplitudes. A Poisson distribution was assumed also for the inhibitory input coming from MS, but the amplitude was modulated by a square wave at different frequencies in different simulations.

The main aim of the simulations was the investigation of the effect of the inhibitory input from Medial Septum on the activity of CA3. A meaningful parameter was considered to be the summed potentials at the axon hillock of neurons, averaged on all the modules (10080) of the CA3 model. This parameter is a nice representation of the *LFP* recorded by micro-electrodes in electrophysiological experiments. At first, to bring forth a CA3 global activity to use as basic reference, we considered an ineffective MS inhibition, lasting for all the duration of the simulation. Some results are reported in figure 1, where the *LFPs*, separated for the three different neural populations (pyramidal, inhibitory–fast and slow), are shown. Five volleys (or complex waves) are evident in the activity of the pyramidal population, with a global frequency of $2.5Hz$, each volley being composed by three or four waves at about $25Hz$, at the lower edge of the gamma range. But, differently from the results obtained in electrophysiological experiments, the

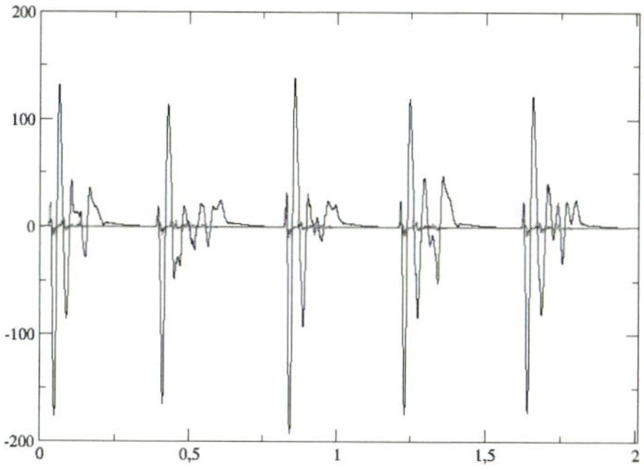

Fig. 1. Time course of the global input at axon hillock of the average neuron of the average modulus (*local field potentials–LFPs*) in CA3 under constant, but otherwise not effective, inhibitory MS input. The time course is shown for pyramidal, inhibitory fast and inhibitory slow neurons. The total time displayed is $2s$.

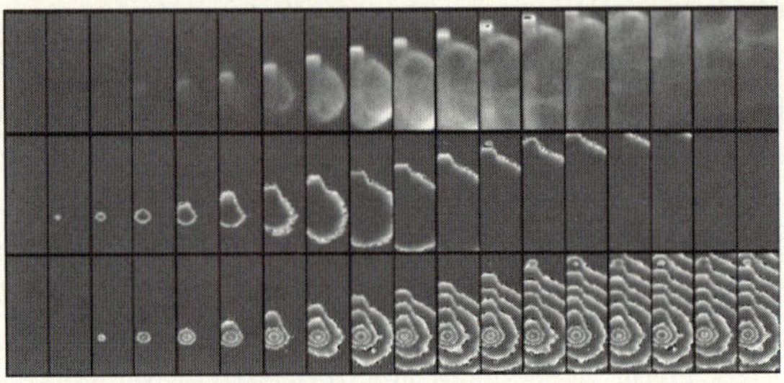

Fig. 2. CA3 global activity linked to the the first wave of *LFPs* shown in figure 1. Starting time at $27ms$ and ending time at $44ms$, the inter-frame time step is $1ms$. The upper frames show the course of long-range pyramidal impulses, the medial ones that of short-range pyramidal impulses, whereas the bottom frames are related to impulses emitted by fast inhibitory neurons.

model can provide not only this data, but a complete information about the CA3 activity. We can know both the firing space-time course of all the populations of neurons and the impulse density behavior, in space and time, for all the different impulses. Interesting aspects are shown by the propagation of short-range and long-range pyramidal impulses and of fast inhibitory impulses, in figure 2. From this figure we can describe a typical CA3 activity, when the inhibitory input from Medial Septum is not meaningful. At the starting time all the activity parameters are set to null (all the neurons are at the resting level and no impulse is flowing in the system), then the simulated input from DG and EC begin to influence the pyramidal neural population. At some times, depending on the characteristics of the Poisson distributions governing the DG input and the EC input, the first volleys of impulses begin to propagate in one or more strips, for DG input, or to hit some moduli, for the EC input. The absorption of impulses triggers some pyramidal neurons to fire action potentials. This induces firing in other pyramidal cells and also in fast and slow inhibitory neurons. In some milliseconds a patterned,

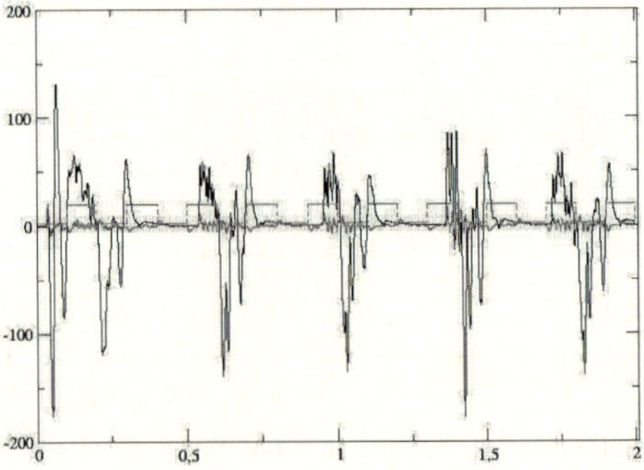

Fig. 3. Time course of *LFPs* of pyramidal, inhibitory fast and inhibitory slow neurons in CA3 under $5Hz$ inhibitory MS input. The total time displayed is $2s$.

self-organized activity begin to appear in the neuronal firing, which stabilizes and propagates throughout the entire CA3 field, involving both the pyramidal and the inhibitory neurons. The induced firing of fast and slow inhibitory neurons produces a level of inhibition sufficient to reduce to a silent state the pyramidal neurons. In such a way they remain unable to react to new inputs originating from the simulated external sources. The patterned activity of the inhibitory neural population remains active for a long period of time, as long as they have sufficient drive from their excitatory synapses. After, the decaying of the excitation on inhibitory neurons reduced gradually their firing. This permits the inputs from EC and DG to ignite again some of the neurons of the pyramidal population of CA3. A new cycle begins with a patterned activity that could be slightly or strongly different form the previous one (results not shown).

In a different simulation, a $20Hz$ inhibitory activity, coming from MS, and effectively inhibiting both fast and slow inhibitory populations of CA3 was investigated. The results of the global parameter showed that the CA3 field reacts to MS input producing an activity with the same frequency of the MS input (results not shown).

To study the reaction of CA3 to a MS inhibition at a lower frequency, a $5Hz$ input was utilized. The results, as expressed by the usual *LFP* parameter, are reported in figure 3. In this case CA3 is unable to follow the input with a $1:1$ regime. Vice versa, some complex waves are produced, each wave showing a time-course with different frequency contents. A two way course is shown in each wave, a sharp wave followed by some oscillations in the gamma range (from 21 to $35Hz$).

In figure 4 the local density of impulses are represented. In this case, the three families of impulses show different propagation behaviors during the gamma and the sharp waves courses.

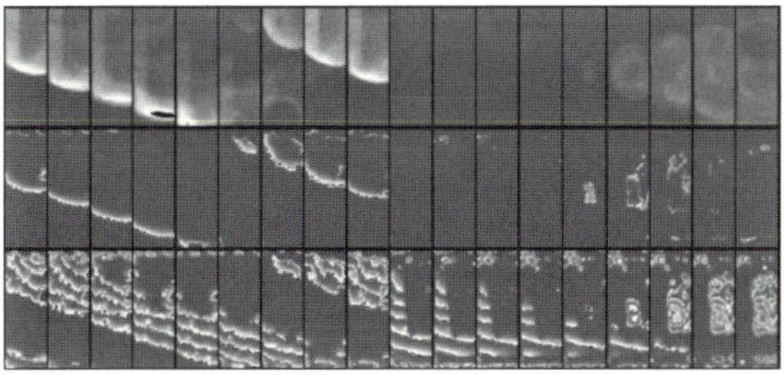

Fig. 4. CA3 global activity linked to the transition phase in the second wave of figure 3. Two time periods are shown. The first nine frames–from 0.560 to $0.576s$–frame step $2ms$, fall in a *sharp wave*. The last nine frames–from 0.648 to $0.664s$–frame step $2ms$, belong to a period of *gamma waves*.

5 Discussion

By simulating an oscillating, inhibitory input originating from Medial Septum, we investigated how the inhibition of fast and slow inhibitory neurons could modulate the CA3 activity. The following results were observed. When reached by an inhibitory input from Medial Septum at $6.6Hz$ frequency and higher, the CA3 field is able to produce oscillating activity at the same frequency of the driving input. If the frequency of the inhibitory input from MS decreases to $5Hz$ frequency or lower, a different behavior is manifested. In the simulation with $5Hz$ input, we observe five complex waves in 2 seconds. Leaving out the first one, each wave starts with a kind of *sharp wave* with a duration going from 80 to $50ms$, followed by 3-4 waves at about $30Hz$–falling in the range of the gamma waves.

The simulation suggests that the global processing structure of CA3 is organized in such a way to present specific time windows for the generation of excitatory activities by pyramidal neurons. These activities are very characteristic and depend strictly by the input from Medial Septum. In general, they are separated by short or long periods of patterned inhibition. A sort of temporal coding—with a meaning quite different from the common view—seems to be associated to the function of the entire CA3 field. Among all the inputs from cortical regions arriving to CA3, only those which reach it in appropriate time are able to trigger specific, global activities, and can produce effects on the brain regions driven by CA3. Other volleys, which arrive out of phase with the winning activity, cannot filter through the inhibitory barrage, and are unable to stimulate reactions. Hence, the information they convey is not allowed to pass to other brain stages. In such a way, some free periods of time, with a variable duration, are reserved to the successful inputs, during which they can drive specific activities in cortical regions without interferences by competing inputs. These activities may result in learning, memory and other cognitive effects.

Based on a long experimental activity also Vinogradova hypothesized a link between the theta rhythm and the attention mechanism [14]. A weaker, but similar hypothesis has been proposed in [6] on the inhibitory activity in Hippocampus. These authors suggest that oscillating inhibitory networks may provide temporal windows for single cells to suppress or facilitate their synaptic inputs in a coordinated manner.

References

1. Amaral, D.G., Ishizuka, N., Claiborne, B.: Neurons, number and the hippocampal network. In: Storm-Mathisen, J., Zimmer, J., Otterson, O.P. (eds.) Progress in Brain Research, vol. 83, pp. 1–11. Elsevier, Amsterdam (1990)
2. Angel, E., Bellman, R.: Dynamic Programming and Partial Differential Equations. Academic Press, New York (1972)
3. Buzsaki, G.: Hippocampal sharp waves: Their origin and significance. Brain Res. 398, 242–252 (1986)
4. Buzsaki, G., Chrobak, J.J.: Temporal structure in spatially organized neuronal ensembles: a role for interneuronal networks. Curr. Opin. Neurobiol. 5, 504–510 (1995)
5. Cohen, N.J., Eichenbaum, H.: Memory, amnesia, and the hippocampal system. MIT Press, Cambridge (1993)

6. Csicsvari, J., Hirase, H., Czurko, A., Mamiya, A., Buzsaki, G.: Oscillatory coupling of hippocampal pyramidal cells and interneurons in the behaving Rat. J. Neurosci. 19, 274–287 (1999)
7. Gulyas, A.I., Seress, L., Toth, K., Acsady, L., Antal, M., Freund, T.F.: Septal gabaergic neurons innervate inhibitory interneurons in the hippocampus of the macaque monkey. Neurosci. 41, 381–390 (1991)
8. Ishizuka, N., Weber, J., Amaral, D.G.: Organization of intra-hippocampal projections originating from CA3 pyramidal cells in the rat. J. Comp. Neurol. 295, 580–623 (1998); 20, 401–12 (1990)
9. Molter, C., Sato, N., Yamaguchi, Y.: Reactivation of behavioral activity during Sharp Waves: A computational model for two stage hippocampal dynamics. Hippocampus 17, 201–209 (2007)
10. Paxinos, G., Watson, C.: The rat brain in stereotaxic coordinates. Academic Press, San Diego (1986)
11. Ventriglia, F.: Kinetic approach to neural systems. I. Bull. Math. Biol. 36, 534–544 (1974)
12. Ventriglia, F.: Towards a kinetic theory of cortical-like neural fields. In: Ventriglia, F. (ed.) Neural Modeling and Neural Networks, pp. 217–249. Pergamon Press, Oxford (1994)
13. Ventriglia, F., di Maio, V.: Neural code and irregular spike trains. In: De Gregorio, M., Di Maio, V., Frucci, M., Musio, C. (eds.) BVAI 2005. LNCS, vol. 3704, pp. 89–98. Springer, Heidelberg (2005)
14. Vinogradova, O.S.: Hippocampus as comparator: role of the two input and two output systems of the hippocampus in selection and registration of information. Hippocampus 11, 578–598 (2001)

Selective Attention Model of Moving Objects

Roman Borisyuk[1], David Chik[1], and Yakov Kazanovich[2]

[1] University of Plymouth, Plymouth, PL4 8AA, UK
r.borisyuk@plymouth.ac.uk
http://www.tech.plym.ac.uk/soc/staff/roman/home2.htm
[2] Institute Mathematical Problems in Biology of the Russian Academy of Sciences
Pushchino, 240290 Russia

Abstract. Tracking moving objects is a vital visual task for the survival of an animal. We describe oscillatory neural network models of visual attention with a central element that can track a moving target among a set of distracters on the screen. At the initial stage, the model forms focus of attention on an arbitrary object that is considered as a target. Other objects are treated as distracters. We present here two models: 1) synchronisation based AMCO model of phase oscillators and 2) spiking neural model which is based on the idea of resource-limited parallel visual pointers. Selective attention and the tracking process are represented by partial synchronization between the central unit and subgroup of peripheral elements. The simulation results are in overall agreement with the findings from psychological experiments: overlapping between target and distractor is the main source of error. Future investigations include the dependence between tracking performance and neuron frequency.

1 Introduction

Selective visual attention is a mechanism that gives a living organism the possibility to extract from the incoming visual information a part that is most important at a given moment and that should be processed in more detail. This mechanism is necessary due to limited processing capabilities of the visual system which does not allow the rapid analysis of the whole visual scene.

The important property of attention is its metastability. This means that after being fixed, the focus of attention does not change for some time even when objects in a scene gradually vary their parameters (shape, brightness, position). In particular, the metastability of attention makes it possible to track a moving object. Special conditions should be fulfilled for attention to be switched from one object to another. These conditions include (a) an abrupt change of parameters of an object in the focus of attention, (b) appearance and disappearance of objects in the scene, (c) objects overlapping or hiding due to their movements, (d) termination or voluntary break of object processing.

In this paper we present two oscillatory neural networks with a central element for automatic object-oriented attention. The main advantage of these models is that they can work with non-stationary objects. It is assumed that objects in the input image can move and change their intensity and form. For simplicity we

V. Kůrková et al. (Eds.): ICANN 2008, Part II, LNCS 5164, pp. 358–367, 2008.

consider the case of greyscale images. The focus of attention is represented by those oscillators that work synchronously with the central oscillator. It is shown that a proper synchronization regime can be obtained by a suitable choice of parameters of oscillators and coupling strengths.

The first model is modification of our Attention Model with Central Oscillator (AMCO) comprising a layer of the so-called peripheral oscillators interacting with a special central oscillator [1,2,3]. Advanced phase oscillators have been used as the elements of this model. The state of such an oscillator is described by three variables: phase, amplitude, and natural frequency of oscillations. The functioning of the model has been based on the following main principles: 1) partial synchronization between the central oscillator and some subset of peripheral oscillators and 2) resonant increase of the amplitude during partial synchronization. The phase-locking mechanism used to synchronize oscillators allows them to achieve similar frequencies and the focus of attention is assumed to be formed by those peripheral oscillators whose activity is partially synchronous with the activity of the central oscillator. Functioning of the system is based on the principles of phase-locking, adaptation of the natural frequency of the central oscillator, and the resonance influence of the central oscillator on the assembly of oscillators that work in-phase with the central element.

The second model of object tracking is built from biologically derived Hodgkin-Huxley model neurons [4]. In this model we use the synchronization hypothesis [5,6] for implementation of a 'pointer' that identifies the neural assembly coding a specified object. Pylyshyn [7] postulated that pointers pick out and stay attached to individual objects in the visual field independent of the objects' features. The neural mechanism of visual pointers is unclear, but there are studies which provide some suggestions about its nature. In Intriligator's study [8], it was found that the spatial resolution of visual attention is different from (coarser than) that of the receptive field, suggesting the existence of a separate 'pointer field' in the parietal cortex. In the study by Egly et al. [9], it was shown that attention seems to start at one point, then propagate and fill the object. Another study by Alvarez and Scholl [10] showed that when tracking lines attention seems to concentrate at the middle of a line. This implies that the brain may be capable of quickly and roughly calculate the centre of an object in the receptive field.

Model descriptions and results are presented in the following sections.

2 Phase Oscillator Model of Object Tracking

2.1 Model Description

The network consists of a central oscillator (CO) that has feedforward and feedback connections to a rectangular grid of peripheral oscillators (POs). Each PO is coupled with its four nearest neighbours except on the boundaries where the mirror symmetry condition is used. An oscillator is described by three variables: the oscillation phase, the oscillation amplitude, and the natural frequency of the oscillator. We do not specify phase oscillator equations here, they can be found in our previous publications [3]. The values of these variables change in time according to prescribed rules of interaction between oscillators. The input to the network is an image on the plane that contains several connected greyscale objects on the white background. The objects are non-stationary, they can continuously change their form, brightness and position. Due to movements, objects can temporarily overlap with each other. While two (or several) objects are overlapping, it is always specified which object lies on the top. Therefore at any moment each pixel in the image has a single grey-level associated with it. The input to the network is an image on the plane grid of the same size as the grid of peripheral oscillators. So there is a one-to-one correspondence between the pixels in the image and the peripheral oscillators. Each PO receives an external input from the corresponding pixel. The darker is the pixel, the higher is the value assigned to the natural frequency of the corresponding oscillator. We call an assembly a connected set of POs that are stimulated by a single object. We say that an object is coded by the corresponding assembly of POs.

The coupling strength between oscillators is constant. The connections from POs to the CO and local connections between POs are synchronizing. The connections from the CO to POs are desynchronising. Due to synchronizing connections from POs to the CO, the latter can be phase-locked by an assembly of POs. Due to synchronizing connections between POs, the oscillators from an assembly of active POs become phase-locked and work nearly in-phase after they reach the resonant state. Desynchronising connections from the CO to POs are used to break the coherence between different assemblies of POs so that at each moment the system tends to include only one object in the attention focus.

Depending on the input signal and previous dynamics, a PO can be in one of three states: active, resonant, and silent. If a PO receives zero input (corresponding to the signal from the background), it is in the silent state. In this state the oscillator does not participate in the network dynamics and is not included in the dynamics equations. A PO becomes active as soon as it receives a non-zero signal from a pixel of an object. CO is synchronised by the assembly of POs that currently makes the greatest contribution into the controlling the phase dynamics of the CO. This assembly corresponds to the most salient object. Those POs that work synchronously with the CO significantly increase the amplitude of their oscillations. If the amplitude exceeds a certain threshold, the PO changes its state to resonant one. Being in the resonant state is interpreted as the fact that this oscillator is included in the focus of attention. If attention is switched to another object, the oscillators amplitude drops down and the oscillator returns to the silent state.

The attention focus is kept stable until the object in the focus of attention, say A, is crossed (overlapped) by another object B. If this event happens, the combination of two objects is considered by the system as a connected object C. The assembly of oscillators that codes C will become synchronous (due to local connections between POs) and will synchronize the CO. The next change of the attention focus happens when C again separates into two isolated object A and B. At that moment the focus of attention will move to A or B depending on their current saliency.

In biological terms, POs represent cortical columns and are constituted of locally interacting populations of excitatory and inhibitory neurons of the cortex. CO represents the septo-hippocampal system whose final position in the pyramid of cortical convergent zones and feedforward and feedback connections to the cortex give it a direct or indirect access to cortical structures.

2.2 Simulation Results

In the following examples,the frames show the state of the network at the integer moments of time. The frames are ordered from left to right and from top to bottom. The state of an oscillator is represented in the figures according to the following scheme: a pixel is white, grey, or black depending on whether the corresponding oscillator is silent, active, or resonant at the current moment of time. Thus, black pixels represent the focus of attention. The arrows show the direction of movement.

Example 1. Consider the case when the saliency of an object is solely determined by its size. Let a image contain two circles of a fixed size and intensity (Fig. 1). The circles move towards each other (the direction is fixed for each object). The circle of radius 5 moves to the right, the circle of radius 4 moves to the left. Denote by A and B the assemblies of POs coding the circles in the network. The natural frequencies of the oscillator in A and B are 5 and 4, respectively. This reflects the fact that the circles have different grey-levels (not shown in the figure), but in this example this has no influence on the strength of interaction between A and B and the CO (the difference in grey-level is below the sensitivity of the attention system).

As can be seen from Fig. 1, after a short transitional period (two time units) the attention is focused on the larger circle (note that this takes place despite

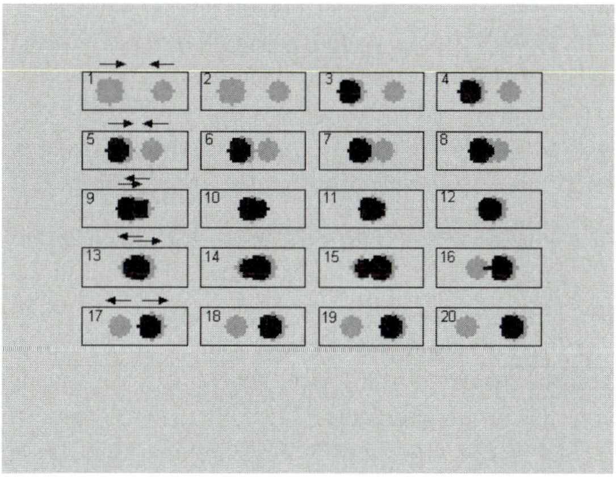

Fig. 1. Selection of moving objects that is based on the size of the objects. Two circles of different size move towards each other. Attention is focused on the larger circle all the time except a short initial transitional period and the time when both circles collide and form a complex object.

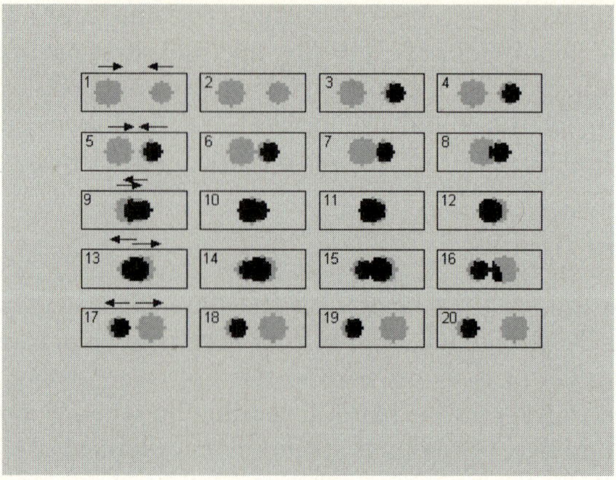

Fig. 2. Selection of moving objects that is based on the intensity. Two circles of different size and intensity move towards each other. Attention is focused on the smaller object since it is darker (and has a better contrast) than the larger one.

the fact that the initial frequency of the CO coincides with the natural frequency of oscillators in B). When the circles collide and overlap the focus of attention is spread on both circles as if they form a complex object (see frames 9-15). Lately, when the circles separate the attention is again focused on the larger and therefore more salient circle (frames 16-20).

Example 2. This example presents the case when the saliency of an object depends on its intensity (Fig. 2). The example differs from the previous one by the value of the natural frequency of oscillators in the assembly B, which is now equal to 6. This means that the smaller circle is darker than the lager one and, hence, more salient on the white background. Therefore the influence of the oscillators from B on the CO is increased in comparison to example 1. As a result, after a short transient period (two time unites) the attention is focused on the smaller circle (frame 3). After the collision (overlapping) of the circles, the focus of attention is temporarily spread on both circles (frames 10-15), but their separation results in attention returning to the smaller circle (frames 17-20).

Example 3. This example shows the case when the circles simultaneously move and change size (Fig. 3). It is presented to show that the focus of attention has the property of metastability. The image of size contains two circles whose radii periodically changes in the range between 3 and 8. The radii vary in antiphase, that is while the first radius is increasing the other one is decreasing. At the initial moment both radii are identical. The natural frequencies of oscillators in the assemblies A and B are equal to 5. This means that both circles have the same grey level and therefore their saliency is determined by their size only. At the moment when the focus of attention is formed, the circle that moves to the right has larger size (frame 3), therefore attention is attracted by this circles and is fixed on it during some time (frames 3-8) despite the fact that the circle in the attention focus becomes smaller than the other circle (frames 7-8). Frames

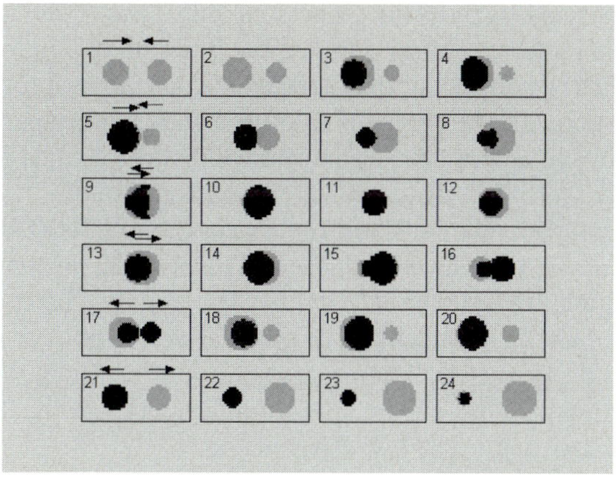

Fig. 3. Selection of moving objects of varying size. Two circles of the same grey-level move towards each other. Initially, attention is focused on the circle moving to the right but after collision and separation of circles it is switched to the circle moving to the left.

9-17 correspond to the time when both circles are included in the attention focus. Then the largest circle (that moves to the left) wins the competition for attention (frame 18) and holds it even when this circle becomes small (frames 22-24).

3 Spiking Element Model of Object Tracking

3.1 Model Description

The architecture of connections is similar to the first model. There is a central neuron (CN), which represents a neural ensemble in perhaps hippocampus or prefrontal cortex. It functions as a central executive which inhibits unwanted signals and promotes desirable signals. Second, there is a 2D grid of peripheral neurons (PNs), which represents the receptive field in visual cortex. Third, there is also a 2D grid of pointer neurons, which represents the attentional pointers stored in the parietal cortex as suggested by experiments [7,8,9,10]. All neurons are described using Hodgkin-Huxley equations with synaptic connections [11].

CN receives an external current and also some excitatory synaptic inputs from all PNs (peripheral neurons). The external current is subthreshold, therefore the activity of CN has nonlinear dependency with the activities of PNs. Each PN receives inputs from four sources: the visual image, pointer neurons, the central neuron, and neighboring PNs. Signals from the visual image together with the pointer neurons would determine which subgroup of PNs has a higher firing rate. CN provides inhibition so that those PNs with higher firing rates would be selected. Local connections between neighboring PNs ensure that a complete target object would be selected (because a physical object is usually locally connected in space). A partial synchronization between CN and a subgroup of PNs would be interpreted as attention being focused on the object represented by those PNs (which also means that that object is under tracking). Finally, the function of pointer neurons is trying to catch up with the objects under tracking. Experiments imply that the pointers always try to point to the centre of an object and then propagate outward [9]. In our simulations, an active pointer neuron will collect signals from the vicinity (9x9) of the current location. The signal is the existence of objects in the receptive field when PNs fire again. We call these non-background PNs as active pixels. The system then calculates the centre of the active pixels. This is done approximately by drawing a minimal rectangle that confines the object. The coordinates of this centre will be used as the new location of the next active pointer neuron (the previous one will be deactivated). The process of updating the pointers is assumed to be discrete and dependent on the operation frequency of the system.

3.2 Simulation Results

Before the tracking starts, the subject would be told which target to track. In our simulations, we assume that a top-down signal is provided for those PNs which are encoding the target object. Generally, a PN receives an external current from the visual field $I_{ext} = 5(1 + 0.5\xi)$ mA if the peripheral participates in representation of some object and the a smaller current $I_{ext} = 2(1+0.5\xi)$ mA for neurons represemting the background, where ξ is a random number uniformly distributed in $[-1, 1]$ representing the noise. A subgroup of PNs (which are encoding the target) receives an additional external current from the top-down process. Those PNs are therefore receiving a higher total current. We assume that $I_{ext} = 10(1 + 0.5\xi)$ mA for each of the 'attended PN'. The top-down signal is turned off once the tracking starts.

After the tracking task starts, the pointer neurons replace the role of the initial top-down signal. The coordinates of the active pointer keep updated by calculation approximate position of the centre of object in the vicinity of previous coordinates. A signal will be sent to the corresponding PN from the active pointer neuron. The PN at the same coordinates will become 'attended' and receives a higher external current ($I_{ext} = 10(1+0.5\xi)$ mA). This current spreads towards neighbouring PNs through local connections, so neighbouring PNs will also become 'attended'. Finally, the system will exhibit a partial synchronization between CN and a subgroup of PNs which receive higher external currents, representing selective attention on the object encoded by those PNs.

Fig. 4 shows the potential traces of CN and some of the PNs. The top panel shows the trace of CN. The next three panels show the potential traces of three of the attended PNs. The bottom three panels show the traces of three of the unattended PNs. The unattended PNs remains subthreshold (no spikes are produced by them), but the attended PNs form stochastic synchronization with CN. For individual PNs, the trace is quite irregular (due to noise ξ), but as a whole population the synchronized activities are prominent.

Our simulations show that the system performs reasonably well in the object tracking task. However, if a target and some distracter overlaps the confusion will occur and attention might mistakenly shift from target to the distracter.

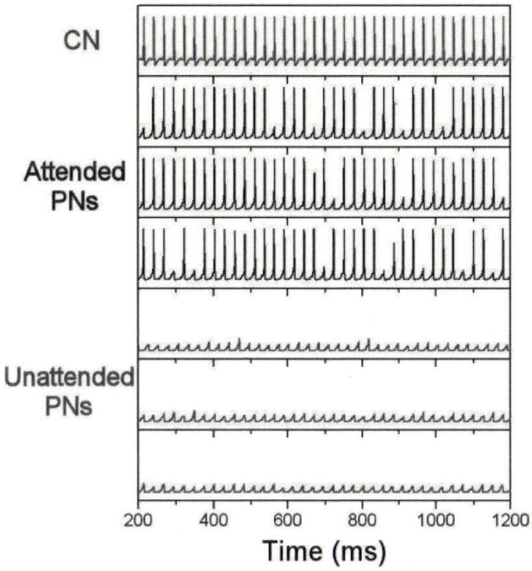

Fig. 4. Partial synchronization between attended PNs and CN. The traces in second to fourth panels are the membrane potentials of attended PNs. They stochastically synchronized with the CN (top panel). The bottom three traces are the potential traces of unattended PNs, which produce only sub-threshold fluctuations.

Tracking accuracy also depends on the operating frequency of the system. If the CN and PN operate at a high frequency, the update rate of pointer coordinates will also be higher, so the probability of error will be lower. In relation to experimental data [12,13] we propose that the system operates at gamma frequency band (around 30-40 Hz). It is important to note that this system can be easily adjusted to track more than one target by using additional active pointer neurons.

4 Discussion

Two main ideas are combined in the presented models: 1) key role of synchronization in formation of the attention focus and 2) connections are realised through the central unit, therefore the number of connections in the network is of the same order as the number of elements (i.e. linearly dependant on the size of the visual field). The functioning of our models is invariant relative to locations of objects in the image. This is why the processing of non-stationary objects does not evoke any additional difficulties. Note that the situation is quite different for winner-take-all models because in these models the modification of connection strengths is 'attached' to a definite position of objects.

The models of attention can be subdivided into two main categories. Traditional connectionist approach to attention modelling [14,15], is based on a winner-take-all procedure and is implemented through a proper modification of the weights of bottom-up and top-down connections in a hierarchical neural network. It works well in solving particular tasks but it is incompatible with modern hypothesis about distributed representation of visual information in the brain [5,6]. There are many experimental evidences showing the importance of neural synchronization and oscillation in various brain functions. Neural oscillations in the gamma range is found to be correlated to attention and memory (see for example, [13,16]). The coherence of spikes play a major role in the control of attention (see for example, [12,17]).

Therefore, an alternative approach to attention modelling is based on the synchronization principle realized in the frames of an oscillatory neural network. Usher and Niebur [18] suggested an oscillatory neural network where the competition between oscillator assemblies is biased by a top-down signal. Wang and Terman [19] developed a network (LEGION) of locally interacting Van-der-Pol type oscillators whose activity is globally controlled by an inhibitory neuron. Assemblies of oscillators representing different objects compete for the synchronization with the inhibitory neuron. A winning assembly increases its activity while the activity of other oscillators is temporarily shut down. A variant of LEGION developed in [20] is able to select a most salient object from the visual scene containing stationary objects. Kazanovich and Borisyuk [1], and Corchs and Deco [21] use a network with a central oscillator to work out a flexible mechanism of attention focus formation and switching. Comparing to these models, the important advantages of our models include: 1) The model can successfully function even the objects change brightness or size (Fig. 1–3). 2) A biologically realistic, stochastic neural synchronization can be seen (Fig. 4).

As a conclusion, we presented two models and show how the focus of attention is formed and can track one object or switch from one object to the other. The models capture many biological and psychological findings, such as neural synchronization, gamma oscillations, resource-limited parallel tracking mechanism, visual indexing, and so on. The simulation results are in overall agreement with the findings from psychological experiments: overlapping between target and distracter is the main source of error. Future investigations include the dependence between tracking performance and neuron frequency.

Acknowledgments. This research work was supported by the UK EPSRC (Grant EP/D036364/1) and by the Russian Foundation of Basic Research (Grant 07-01-00218).

References

1. Kazanovich, Y.B., Borisyuk, R.M.: Dynamics of neural networks with a central element. Neural Networks 12, 441–454 (1999)
2. Borisyuk, R., Kazanovich, Y.: Oscillatory neural network model of attention focus formation and control. BioSystems 71, 29–38 (2003)
3. Borisyuk, R., Kazanovich, Y.: Oscillatory Model of Attention-Guided Object Selection and Novelty Detection. Neural Networks 17, 899–910 (2004)

4. Hodgkin, A.L., Huxley, A.F.: A quantitative description of membrane current and its applications to conduction and excitation in nerve. J. Physiol. 117, 500–544 (1952)
5. Gray, C.M.: The temporal correlation hypothesis is still alive and well. Neuron 24, 31–47 (1999)
6. Singer, W.: Neuronal synchrony: A versatile code for the definition of relations. Neuron 24, 49–65 (1999)
7. Pylyshyn, Z.W.: Visual indexes, preconceptual objects, and situated vision. Cognition 80, 127–158 (2005)
8. Intriligator, J., Cavanagh, P.: The spatial resolution of visual attention. Cognitive Psychology 43, 171–216 (2001)
9. Egly, R., Driver, J., Rafal, R.D.: Shifting visual attention between objects and locations: Evidence from normal and parietal lesion subjects. Journal of Experimental Psychology, Genera 123, 161–177 (1994)
10. Alvarez, G.A., Scholl, B.J.: How does attention select and track spatially extended objects? New effects of attentional concentration and amplification. Journal of Experimental Psychology, General 134, 461–476 (2005)
11. Gerstner, W., Kistler, W.M.: Spiking Neuron Models: Single Neurons, Populations, Plasticity,Ch. 2. Cambridge University Press, Cambridge (2002)
12. Fries, P., Schroeder, J.-H., Roelfsema, P.R., Singer, W., Engel, A.K.: Oscillatory neural synchronization in primary visual cortex as a correlate of stimulus selection. Journal of Neuroscience 22, 3739–3754 (2002)
13. Jensen, O., Kaiser, J., Lachaux, J.-P.: Human gamma-frequency oscillations associated with attention and memory. Trends in Neurosciences 30, 317–324 (2007)
14. Olshausen, B.A., Anderson, C.H., Van Essen, D.C.: A neurobiological model of visual attention and invariant pattern recognition based on dynamic routing of information. J. Neuros. 13, 4700–4719 (1992)
15. Tsotsos, J.K., Culhane, S., Wai, W., Lai, Y., Davis, N., Nuflo, F.: Modeling visual attention via selective tuning. Artificial Intelligence 78, 507–547 (1995)
16. Tallon-Baudry, C., Bertrand, O.: Oscillatory gamma activity in humans and its role in object representation. Trends in Cognitive Sciences 3, 151–162 (1999)
17. Steinmetz, P.N., Roy, A., Fitzgerald, P., Hsiao, S.S., Johnson, K.O., Niebur, E.: Attention modulates synchronized neuronal firing in primate somatosensory cortex. Nature 404, 187–190 (2000)
18. Usher, M., Niebur, E.: Modeling the temporal dynamics of IT neurons in visual search: a mechanism for top-down selective attention. J. Cognitive Neurosci. 8, 311–327
19. Wang, D.-L., Terman, D.: Image segmentation based on oscillatory correlation. Neural Computation 9, 805–836 (1997)
20. Wang, D.L.: Object selection based on oscillatory correlation. Neural Networks 12, 579–592 (1999)
21. Corchs, S., Deco, G.: A neurodinamical model for selective visual attention using oscillators. Neural Networks 14, 981–990 (2001)

Tempotron-Like Learning with ReSuMe

Răzvan V. Florian[1,2]

[1] Center for Cognitive and Neural Studies (Coneural),
Str. Cireşilor nr. 29, 400487 Cluj-Napoca, Romania
florian@coneural.org
[2] Babeş-Bolyai University, Department of Computer Science,
Str. Al. Kogălniceanu nr. 1, 400084 Cluj-Napoca, Romania

Abstract. The tempotron is a model of supervised learning that allows a spiking neuron to discriminate between different categories of spike trains, by firing or not as function of the category. We show that tempotron learning is quasi-equivalent to an application for a specific problem of a previously proposed, more general and biologically plausible, supervised learning rule (ReSuMe). Moreover, we show through simulations that by using ReSuMe one can train neurons to categorize spike trains not only by firing or not, but also by firing given spike trains, in contrast to the original tempotron proposal.

1 Introduction

The tempotron has been recently proposed as a "new, biologically plausible supervised synaptic learning rule that enables neurons to efficiently learn a broad range of decision rules, even when information is embedded in the spatiotemporal structure of spike patterns rather than in mean firing rates" [1]. A few other supervised rules for spiking neurons have been previously proposed (for a review, see [2]). Here we show that a particularization of ReSuMe, one of those rules [3,4,5], is quasi-equivalent to the tempotron. Moreover, ReSuMe allows the training of tempotrons that are able to fire specific spike patterns in response to each input category.

2 The Tempotron

The tempotron learning rule [1] can be applied to a spiking neuron driven by synaptic afferents. The learning rule modifies the efficacies of the afferent synapses such that the trained neuron emits one spike when presented with inputs corresponding to one category and no spike when the inputs correspond to another category. The tempotron setup assumes that, before being presented with an input spike train, the neuron's potential is at rest, and that after the neuron emits a spike in response to an input pattern all other incoming spikes are shunted and have no effect on the neuron. Thus, even if the neuron would fire more than one spike, the spikes following the first one are artificially eliminated.

V. Kůrková et al. (Eds.): ICANN 2008, Part II, LNCS 5164, pp. 368–375, 2008.

The subthreshold membrane voltage u of the trained neuron is modeled as a sum of postsynaptic potentials:

$$u(t) = u_0 + \sum_i w_i \sum_{t_i^f < t} \epsilon(t - t_i^f), \tag{1}$$

where u_0 is the resting potential, w_i is the synaptic efficacy of synapse i, and $\epsilon(t - t_i^f)$ describes the form of the postsynaptic potential induced in the neuron by a spike at t_i^f received from neuron i. The first sum runs over all presynaptic neurons, and the second one runs over all spikes of neuron i prior to t. When u overcomes the firing threshold θ, the neuron emits a spike.

Tempotron learning minimizes the following cost function, for each input pattern [1]:

$$C = \begin{cases} \theta - u_{\max}, & \text{if } u_{\max} < \theta \text{ and the neuron should fire for this pattern,} \\ u_{\max} - \theta, & \text{if the neuron fired } (u_{\max} \geq \theta) \text{ and it should have been silent,} \\ 0, & \text{otherwise,} \end{cases} \tag{2}$$

where $u_{\max} = u(t_{\max})$ is the maximal value of the postsynaptic potential u, in the case that the neuron did not fire. In the case that the neuron fired, $u_{\max}$ is the maximal value that u would have been reached if the neuron would have not fired.

Applying the gradient descent method in the space of synaptic efficacies for minimizing the above cost function leads to the tempotron learning rule [1]:

$$\Delta w_i = \begin{cases} \lambda \sum_{t_i^f < t_{\max}} \epsilon(t_{\max} - t_i^f), & \text{if the neuron should fire but it did not,} \\ -\lambda \sum_{t_i^f < t_{\max}} \epsilon(t_{\max} - t_i^f), & \text{if the neuron should not fire but it did,} \\ 0, & \text{otherwise,} \end{cases} \tag{3}$$

where $\lambda > 0$ is the learning rate. During learning, synaptic changes Δw_i are applied after each presentation of an input pattern.

It can be seen that the dynamics of this learning rule has little biological plausibility, since: a) It requires the monitoring of the maximum of u; b) For trials when the neuron fires while it should not, it requires, for computing $t_{\max}$, simulating a dynamics of the neuron that is different from the real one (since it ignores that the neuron fired and the membrane potential was reset). The setup also has little biological plausibility, since: a) While it assumes that the precision of spike times in input patterns is important, it ignores the time when the trained neuron fires. If the coding of information in the brain depends on the precision of spike times, as experimental studies have suggested [6], then the timing of the firing of both the afferents and the trained neurons should matter; b) It is assumed that the trained neuron can fire either no spike or just one spike per input pattern; c) It is assumed that learned input patterns are isolated from other inputs and thus that the trained neuron is initially at rest, which is not the case in the brain.

3 ReSuMe

The ReSuMe learning rule [3,4,5,7,8] also allows the supervised training of a neuron and is defined by the following equation:

$$\frac{\mathrm{d}w_i(t)}{\mathrm{d}t} = \lambda \left[\tilde{\Phi}(t) - \Phi(t)\right] \left[a + \int_0^\infty W(s)\, \Phi_i(t-s)\, \mathrm{d}s\right], \tag{4}$$

where $\tilde{\Phi}(t) = \sum_{f=1}^{\tilde{n}} \delta(t - \tilde{t}^f)$ is the target spike train to be learned by the neuron, represented by a sum of Dirac pulses; $\Phi(t) = \sum_{f=1}^{n} \delta(t - t^f)$ is the actual output of the neuron; $\Phi_i(t)$ is the input spike train coming from synapse i, also a sum of Dirac pulses; a is a constant; $\tilde{t}^f$ are the moments of spikes in the target spike train and $\tilde{n}$ their number; t^f are the moments of spikes in the actual output spike train and n their number; and W is a learning window that was originally proposed to be $W(s) = E(s)$ with

$$E(s) = A\, \exp(-s/\tau_E) \tag{5}$$

where A and τ_E are positive constants. It can be seen that, after a learning trial, the synaptic change is

$$\Delta w_i = \lambda\, a\, (\tilde{n} - n) + \lambda \sum_{\tilde{t}^g} \sum_{t_i^f \le \tilde{t}^g} W(\tilde{t}^g - t_i^f) - \lambda \sum_{t^g} \sum_{t_i^f \le t^g} W(t^g - t_i^f). \tag{6}$$

4 Applying ReSuMe to the Tempotron Problem

By applying the ReSuMe rule to the tempotron setup (one target output spike or none, one output spike at t^1 or none), and if we note that for having one output spike, regardless of its timing, it is the easiest to have it at $\tilde{t}^1 = t_{\max}$, if the neuron did not fire, or at the actual time of firing $\tilde{t}^1 = t^1$, if it fires, we get:

$$\Delta w_i = \begin{cases} \lambda\, a + \lambda \sum_{t_i^f \le t_{\max}} W(t_{\max} - t_i^f), & \text{if } \tilde{n} = 1, n = 0, \\ -\lambda\, a - \lambda \sum_{t_i^f \le t^1} W(t^1 - t_i^f), & \text{if } \tilde{n} = 0, n = 1, \\ 0, & \text{if } \tilde{n} = n. \end{cases} \tag{7}$$

In the first and the lase case in the last equation, the ReSuMe learning rule is equivalent to the tempotron learning rule for $a = 0$ and $W(s) = \epsilon(s)$. In the second case, the learning rules can be considered equivalent if we note that, if the trained neuron fires, the maximum of u, θ, is actually reached at the firing time t^1, and this is the closest approximation to $t_{\max}$ that can be made with quantities locally available to the neuron. The postsynaptic potential ϵ can be well approximated by the exponential E that was originally used for ReSuMe; for example, for integrate-and-fire neurons with postsynaptic currents that are Dirac pulses, $\epsilon(s)$ is exactly an exponential. Thus, the tempotron learning rule can be considered a particular application of ReSuMe for a particular problem.

But ReSuMe is a more general and more biologically plausible learning rule, since, in contrast to the tempotron: a) If the trained neuron is assumed to fire in response to a pattern, one can not only teach it to fire, but also to fire at particular moments, which is more biologically plausible and also permits a finer control of the neuron's behavior; b) It allows discriminating between more than two input categories; c) It does not require monitoring the maximum of u; d) It allows not only episodic learning but also online learning. Of course, the biological plausibility of supervised learning rules for spiking neural networks is limited by the constraint of providing a teaching signal for each considered neuron.

Because the tempotron learning rule minimizes the cost function defined by Eq. 2 and of the quasi-equivalence of ReSuMe with the tempotron learning rule for the tempotron problem, ReSuMe, with $W(s) = \epsilon(s)$, is an optimal learning rule for the tempotron problem. In the case that, additionally to the tempotron setup, we would like to teach the neuron to fire at a particular moment in time, it is currently known that ReSuMe, with $W(s) = E(s)$ will converge to an optimal solution at least for the case of one input with one spike and one target output spike [8].

5 Simulations

In order to verify the assertions presented above, we performed several simulations. We implemented the setup for learning to classify latency patterns, which

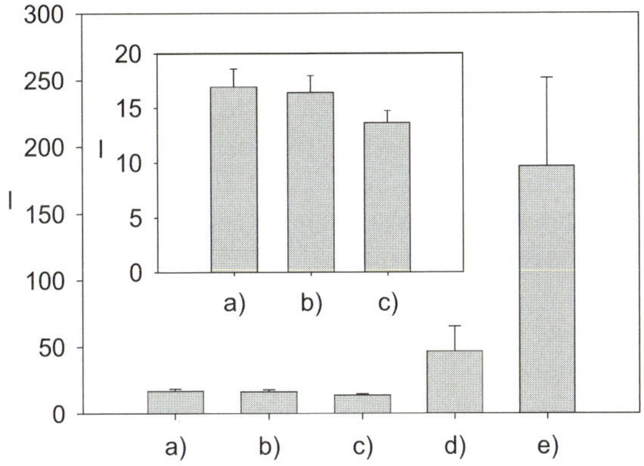

Fig. 1. The number l of learning trials needed for perfect learning of classifying latency patterns, for various implementation of the learning rule and various problems. For each rule / setup, averages and standard deviations are computed over 100 experiments with different, random initial conditions. a) Original tempotron learning rule. b) Modified tempotron learning rule ($t_{\max}$ replaced by t^1 when the output neuron fires), equivalent to the ReSuMe learning rule with $a = 0$ and $W(s) = \epsilon(s)$. c) ReSuMe learning rule for the tempotron setup (with $W(s) = E(s)$). d) Learning with ReSuMe of a classification task where, when the neuron fires, it has to fire at a particular moment in time. e) Learning with ReSuMe of a classification task where, for both categories, the neuron has to fire one spike at particular moments in time. See text for details. Inset: zoom over the results of a), b), c).

was originally used for demonstrating the efficacy of the tempotron learning rule [1]. The trained neuron must learn to separate p input patterns into two categories. For each category, the neuron has to have a distinct, characteristic output. The input patterns are generated randomly and are assigned randomly to one of the two categories. Each input pattern has a duration $T = 500$ ms and consists of one spike for each of the $N = 500$ afferent synapses of the trained neuron. The timing of each of these spikes is generated randomly with uniform distribution between 0 and T. Except where specified, the parameters of the simulation are as in [1]. The trained neuron is an integrate-and-fire neuron with the time constant of the membrane decay τ. Each input spike generates an exponentially decaying current with time constant τ_s. Thus, for the case that the neuron has not yet emitted a spike,

$$\epsilon(t - t_i^f) = \epsilon_0 \left[\exp\left(-\frac{t - t_i^f}{\tau} \right) - \exp\left(-\frac{t - t_i^f}{\tau_s} \right) \right] , \tag{8}$$

as in [1]. For the results presented here, we use $p = 50$ input patterns.

In the original tempotron setup, the two categories are named $\ominus$ and $\oplus$ and the trained neuron has to fire no spike for the $\ominus$ patterns and one spike, regardless of its timing, for the $\oplus$ patterns. Our first simulations implemented the original tempotron learning rule for this setup and checked whether the changes suggested by the application of ReSuMe to the tempotron setup affect the efficacy of learning. In the simulations, we trained the neuron until there was no error (all patterns were classified correctly) and recorded the number l of learning trials needed for learning. A trial consists of presentations of each of the p patterns, followed by applying the changes of w_i given by the learning rule.

We first reproduced learning with the original tempotron rule (Eq. 3; Fig. 1a). We then performed the simulation by replacing $t_{\max}$ in Eq. 3, for the cases where the output neuron spiked, with the actual timing of the output spike t^1. Thus, we effectively implemented the ReSuMe learning rule with $a = 0$ and $W(s) = \epsilon(s)$. The results are presented in Fig. 1b and show that there is no increase in learning time, for equal performance.

We then ran the simulation by using an exponential learning window, $W(s) = E(s)$ (Eq. 5), as originally proposed for the ReSuMe rule, with $A = 1$ and $\tau_E = \tau$. Again, we can still achieve learning with no errors, and learning time for this method is even better than for the tempotron (Fig. 1c). This is consistent with other results obtained in simulations, that have shown that ReSuMe has better performance with an exponential learning window (E, Eq. 5) than with a double-exponential learning window (like ϵ, Eq. 8) [9]. This is, however, somehow surprising, given current theoretical understanding of ReSuMe: it has been shown analytically that ReSuMe with $W(s) = E(s)$ converges to the solution only for one input [8] (but we use here multiple inputs); while the tempotron learning rule performs gradient descent towards the solution [1].

We then explored the more difficult, but more general and relevant problem where we remove the artificial shunting of the inputs after the output neuron fires (thus allowing it to fire more than one spike) but still require the neuron

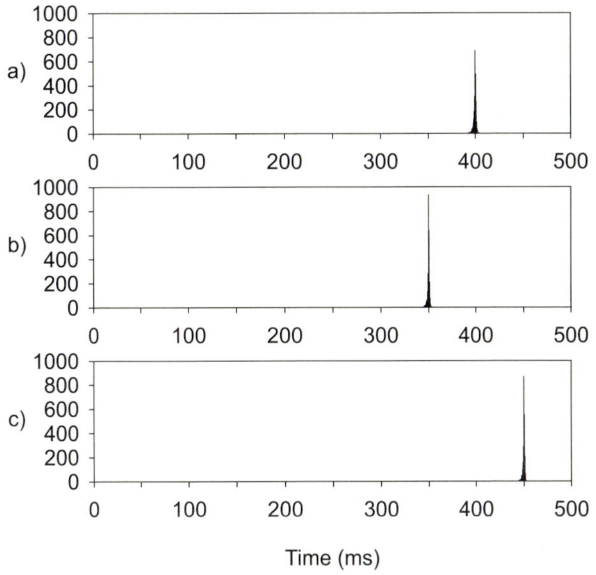

Fig. 2. Histograms of the distribution of spike timings of the output neuron, after learning, when responding to input spike pattern categories. The distribution contains the responses of the neuron to each pattern within a category (about $p/2 = 250$ patterns per category), for each trial, and for 100 trials with random initial conditions. Bin width is 0.5 ms. a) The neuron has to either fire a spike at $\tilde{t}^1 = 400$ ms in response to one category or not to fire at all. b), c) The neuron has to fire one spike at $\tilde{t}^1_\oplus = 350$ ms in response to a $\oplus$ pattern (b) and one spike at $\tilde{t}^1_\ominus = 450$ ms in response to a $\ominus$ pattern (c).

to learn to fire only one spike for the $\oplus$ patterns. Thus, there will be one output spike for the $\oplus$ patterns not because the number of spikes is restricted artificially but because the neuron has to adapt its synapses in order to do so. Moreover, we require the neuron to fire this spike at precisely $\tilde{t}^1 = 400$ ms after the beginning of an input pattern. As previously, the neuron has to fire no spike for $\ominus$ patterns. For this, we use the general ReSuMe learning rule, Eq. 6, with $a = 0$, $W(s) = E(s)$, $A = 1$, and $\tau_E = \tau$, and we keep training the neuron until the number of output spikes is the desired one for each input pattern. We now also have to consider the dynamics of the trained neuron after it emits a spike. We do this according to the standard integrate-and-fire model [10], and in this case we have, when the timing of a presynaptic spike t_i^f precedes the timing of the latest postsynaptic spike $\hat{t}$,

$$\epsilon(t - t_i^f) = \epsilon_0 \exp\left(-\frac{\hat{t} - t_i^f}{\tau_s}\right)\left[\exp\left(-\frac{t - \hat{t}}{\tau_m}\right) - \exp\left(-\frac{t - \hat{t}}{\tau_s}\right)\right], \quad (9)$$

while when $t_i^f > \hat{t}$ the form of ϵ from Eq. 8 is still valid. We used a reset potential equal to the rest potential, so there was no refractory kernel. Again, we achieve learning, and even if the problem is significantly more difficult the learning time is only about 3 times higher than for the simpler tempotron problem (Fig. 1d). As it can be seen in Fig. 2a, the time at which the neuron fires after learning in response to $\oplus$ patterns is very narrowly distributed around $\tilde{t}^1$.

Finally, we tackled the still more difficult problem of having the neuron fire one spike at $\tilde{t}^1_\oplus = 350$ ms in response to a $\oplus$ pattern and one spike at $\tilde{t}^1_\ominus = 450$ ms in response to a $\ominus$ pattern, in the same conditions as for the previous simulation. We still achieve learning, after a larger number of training epochs (Fig. 1e), and the distribution of the moments at which the neuron fires after learning is again narrowly distributed around the desired times (Fig. 2b,c).

6 Conclusions

We have demonstrated the equivalence between tempotron and ReSuMe, under certain conditions, and we have shown that ReSuMe is a more general and more biologically-plausible approach to supervised learning for spiking neurons than the tempotron. The tempotron learning rule is, in fact, a particular case of the ReSuMe learning rule for a specific, quite artificial problem. Moreover, we have shown in simulations that by using the ReSuMe learning rule one can train neurons to classify input patterns not only by indicating the class by firing or not firing in a given time interval, but also by firing spikes with precise timings.

If one considers that representing information in the precise spike timings is relevant, as it was considered for the input spike trains in the tempotron setup, then the output of spike train classifiers should also be capable of representing information temporally. Hence general learning rules that are capable of learning spike times, such as ReSuMe, should be used instead of the tempotron learning rule for such problems. Having the same type of coding for both input and output permits using the output of a classifier as the input of another similar classifier, thus forming networks with higher information processing capabilities. A spiking classifier with temporally coded output is also important for reservoir computing [11] when using a spiking reservoir. In this case, one may train a spiking readout and feed its output back into the reservoir (since information is coded as in the reservoir), thus improving the computational power of the network [12].

Despite its limitations, the tempotron has been proved quite efficient for spoken digit recognition, outperforming with only 15 spiking neurons complex state-of-the-art Hidden Markov Model word recognition systems [13]. This shows that spiking neural networks are quite powerful, and using appropriate learning methods for training them might reveal even more of their potential.

Acknowledgements. This work has been sponsored by grants of the Romanian National Authority for Scientific Research (PNCDI II, Parteneriate, contract no. 11-039/2007; CEEX, contract no. 1474/2006). I thank to Filip Ponulak and Raul C. Mureşan for useful suggestions.

References

1. Gütig, R., Sompolinsky, H.: The tempotron: a neuron that learns spike timing-based decisions. Nature Neuroscience 9(3), 420–428 (2006)
2. Kasiński, A., Ponulak, F.: Comparison of supervised learning methods for spike time coding in spiking neural networks. International Journal of Applied Mathematics and Computer Science 16(1), 101–113 (2006)
3. Ponulak, F.: ReSuMe — new supervised learning method for the spiking neural networks. Technical report, Institute of Control and Information Engineering, Poznań University of Technology, Poland (2005)
4. Kasiński, A., Ponulak, F.: Experimental demonstration of learning properties of a new supervised learning method for the spiking neural networks. In: Duch, W., Kacprzyk, J., Oja, E., Zadrożny, S. (eds.) ICANN 2005. LNCS, vol. 3696. Springer, Heidelberg (2005)
5. Ponulak, F.: Supervised learning in spiking neural networks with ReSuMe method. PhD thesis, Poznań University of Technology, Poland (2006)
6. Bohte, S.M.: The evidence for neural information processing with precise spike-times: A survey. Natural Computing 3(2), 195–206 (2004)
7. Ponulak, F., Kasiński, A.: Generalization properties of SNN trained with ReSuMe. In: Proceedings of the European Symposium on Artificial Neural Networks, ESANN 2006, Bruges, Belgium (2006)
8. Ponulak, F.: ReSuMe — proof of convergence. Technical report, Institute of Control and Information Engineering, Poznań University of Technology, Poland (2006)
9. Ponulak, F.: Analysis of ReSuMe learning process for spiking neural networks. International Journal of Applied Mathematics and Computer Science (in press)
10. Gerstner, W., Kistler, W.M.: Spiking neuron models. Cambridge University Press, Cambridge (2002)
11. Schrauwen, B., Verstraeten, D., Van Campenhout, J.: An overview of reservoir computing: theory, applications and implementations. In: Proceedings of the 15th European Symposium on Artificial Neural Networks, pp. 471–482 (2007)
12. Maass, W., Joshi, P., Sontag, E.D.: Computational aspects of feedback in neural circuits. PLoS Computational Biology 3(1), e165 (2007)
13. Gütig, R., Sompolinsky, H.: Neural mechanisms of speech processing: time warp invariants in adaptive integration time. In: Cosyne 2008: Computational and Systems Neuroscience, p. 337 (2008)

Neural Network Capable of Amodal Completion

Kunihiko Fukushima

Kansai University, Takatsuki, Osaka, Japan

Abstract. When some parts of a pattern are occluded by other objects, the visual system can often estimate the shape of missing portions from visible parts of the contours. This paper proposes a neural network model capable of such function, which is called *amodal completion.* The model is a hierarchical multi-layered network that has bottom-up and top-down signal paths. It contains cells of area V1, which respond selectively to edges of a particular orientation, and cells of area V2, which respond selectively to a particular angle of bend. Using the responses of bend-extracting cells, the model predicts the curvature and location of the occluded contours. Missing portions of the contours are gradually extrapolated and interpolated from the visible contours. Computer simulation demonstrates that the model performs amodal completion to various stimuli in a similar way as observed by psychological experiments.

1 Introduction

When parts of a pattern are occluded by other objects, the visual system can often estimate the shape of missing portions from visible parts of the contours. If we see a picture like the one in the upper left of Fig. 8, in which a circular disk is occluded by a square, we feel as though the contour of the disk is connected behind the square. Such process of perceptually filling in parts of objects that are hidden from view is called *amodal completion*, and a great deal of psychological research has been performed so far.

This paper proposes a neural network model that has an ability of amodal completion. In the primary visual cortex (area V1), there are cells that respond selectively to oriented edges, and these cells are classified into simple and complex cells. Ito, et al. reported that, in area V2 of the monkey, there are cells that show highly selective responses to a particular angle of bend of line stimuli [1]. We will call them *bend-extracting cells* in this paper. Our model is a hierarchical multi-layered neural network that has bottom-up and top-down signal paths, and contains simple, complex and bend-extracting cells in it. Using the response of bend-extracting cells, the model predicts the curvature and location of the occluded contours. The missing contours are then extrapolated a little. Using the extrapolated signals that are fed back through top-down path to edge-extracting cells, the model predicts the missing contours again. The process of amodal completion thus progresses gradually while the signals are circulating in the feedback loop, and finally the missing contours are interpolated completely.

We demonstrate by computer simulation how our model makes amodal completion for various stimuli.

V. Kůrková et al. (Eds.): ICANN 2008, Part II, LNCS 5164, pp. 376–385, 2008.

2 Neural Network Model

The model is a hierarchical multi-layered network as illustrated in Fig 1. It has bottom-up and top-down signal paths, by which a feedback loop is formed. Each layer of the network consists of a number of *cell-planes*, depending on the difference in the features to which cells respond selectively. Incidentally, a cell-plane is a group of cells that are arranged retinotopically and share the same set of input connections. All cells in a cell-plane have receptive fields of an identical characteristic, but the locations of the receptive fields differ from cell to cell.

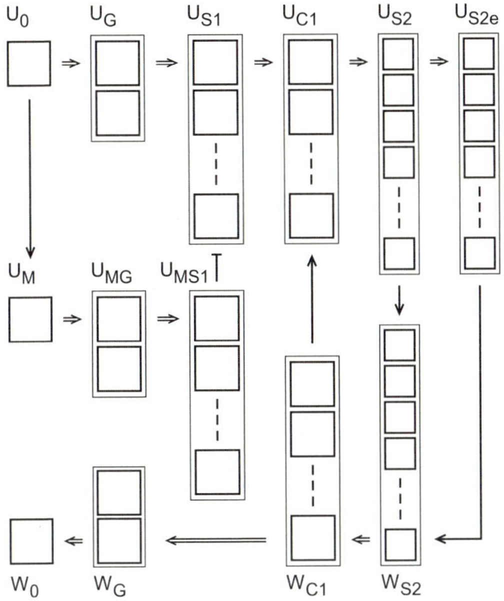

Fig. 1. Architecture of the model for amodal completion

2.1 Extraction of Oriented Edges

The stimulus pattern is presented to input layer U_0, which consists of a two-dimensional array of photoreceptors. The output of U_0 is fed to contrast-extracting layer U_G, whose cells resemble retinal ganglion cells or lateral geniculate nucleus cells. Layer U_G consists of two cell-planes: one with concentric on-center receptive fields, and the other with off-center receptive fields. The former cells extract positive contrast in brightness, whereas the latter extract negative contrast from the image presented to the input layer.

The output of U_G is fed to edge-extracting layer U_{S1}, which consists of S1-cells. S1-cells resemble simple cells in area V1, and respond selectively to edges of a particular orientation. Namely, layer U_{S1} consists of K_1 cell-planes, and all cells in the kth cell-plane respond selectively to edges of orientation $2\pi k/K_1$. As a result, the contours of input image are decomposed into edges of every orientation. We take $K_1 = 32$ in the computer simulation discussed later.

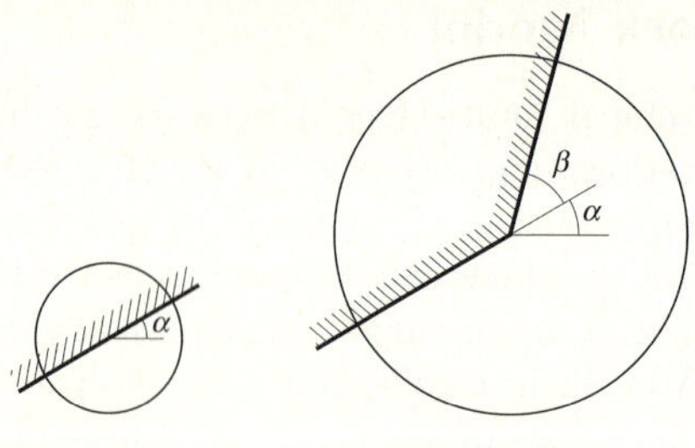

Fig. 2. Training patterns for S1- and S2-cells. These patterns become the preferred stimuli for S1- and S2-cells, respectively.

S1-cells, like S-cells in the neocognitron, are cells with shunting-inhibition and yield analog outputs. Their input connections are produced by a supervised learning, like the one used for the neocognitron [2]. When we train the kth cell-plane, for example, we first choose an arbitrary cell from the cell-plane. We present a training pattern like the one shown in Fig. 2(a) to input layer U_0. To be more specific, the training pattern is an edge of orientation $\alpha = 2\pi k/K_1$ and crosses the center of the receptive field of the chosen S1-cell. The input connection to the S1-cell, whose initial values are zero, are increased in proportion to the responses of the cells of layer U_G, from which the connections are leading.

The output of layer U_{S1} is fed to layer U_{C1}, which consists of C1-cells. C1-cells resemble complex cells in area V1. Similarly to the classical hypothesis by Hubel and Wiesel, a C1-cell receives excitatory signals from a group of S1-cells of the same preferred orientation whose receptive fields are spatially deviated. It responds if and only if at least one of these S1-cells is active. Hence its response is orientation selective like S1-cells but is more tolerant of shift in location. We can also express that the response of layer U_{C1} is a blurred version of the response of layer U_{S1}.

2.2 Masker Layer

The stimulus presented to layer U_0 usually contains the images of both occluding objects and a target pattern partly occluded by them. If the stimulus is processed directly, edges are extracted in layer U_{S1}, not only from the contours of the target pattern, but also from the contours of the occluding objects. The path, $U_M \rightarrow U_{MG} \rightarrow U_{MS1}$, works to eliminate features irrelevant to the target pattern.

Layer U_M, which is called *masker layer*, responds only to the occluding objects. We assume here a situation where the segmentation of occluding objects has already been finished: The shape of the occluding objects is detected and appears in U_M, in the same shape and at the same location as in input layer U_0. The connections from U_M through U_{MG} to U_{MS1} are identical to those from U_0 through U_G to U_{S1}. As a result, only the information on edges of the occluding objects appears in U_{MS1}. If the response of U_{S1} is inhibited by the response of U_{MS1}, only the information on edges of the occluded pattern remains in U_{S1}.

It should be noted here that, in the visual scene, the occluding objects can be either brighter or darker than the target pattern. Hence both positive and negative edges of the occluding objects have to be used to inhibit the response of U_{S1}. Incidentally, oriented edges of different polarities are extracted by cell-planes whose preferred orientations differ by π. In our model, a cell-plane of U_{S1} whose preferred orientation is α, for example, is inhibited by two cell-planes of U_{MS1} whose preferred orientations are α and $\alpha + \pi$.

2.3 Extraction of Bend

The output of layer U_{C1} is fed to bend-extracting layer U_{S2}, which consists of S2-cells. An S2-cell resembles the cells found in area V2 of the monkey [1]. It shows a highly selective response to a particular angle in the bend of a contour, as illustrated in Fig. 2(b). If the input pattern has a curved contour, the cell extracts the angle between adjacent tangential lines.

Input connections to S2-cells, like S1-cells, are produced by a supervised learning. The training pattern given to input layer U_0, however, is a bent edge, namely, a pattern in which two edges of different orientations are combined at the center. We will now represent a pattern shown in Fig. 2(b), in which one of the edge has orientation α $(0 \le \alpha < 2\pi)$ and the angle of the bend is β $(-\pi < \beta < \pi)$ (this means that the orientation of the other edge is $\alpha + \beta$), using two-dimensional variable (α, β). In the computer simulation, α and β are chosen with a step of $2\pi/K_1$, in the same way as for U_{S1}. As a result, α has K_1 different values, and β has $K_1 - 1$ different values because $\beta = \pm 2\pi$ is excluded. Hence the total number of the cell-planes, K_2, comes to be $K_1(K_1 - 1)$.

2.4 Extrapolation of a Curve

S2e-cells in the next layer U_{S2e} receive signals from S2-cells and extrapolate the curves of the occluded contours. In other words, S2e-cells predict the curvature to restore partly occluded contours.

Suppose a curved contour shown by the solid line in Fig. 3(a) elicits a response from a bend-extracting cell whose receptive field center is located at A. Let the bend extracted by this S2-cell be (α, β), using the notation defined by Fig. 2(b). We search for the largest-output cell from S2-cells that extract bend $(\alpha - \beta,\ \beta)$ and are located along the line extended to the direction of the left edge (i.e., the edge of orientation α). Let the location of the cell be B, and let the distance between A and B be p. If this contour extends to the right of A keeping the same curvature, an S2-cell that extracts bend $(\alpha + \beta,\ \beta)$ will responds strongly at location C, which is at distance p from A in the direction of $\alpha + \beta$.

Our model is so designed that cell C of layer U_{S2e} responds when cells A and B responds in layer U_{S2}. This can be realized with a simple network as illustrated in Fig. 3(b). In the figure, each cell on the extension from the left edge of the central cell A sends an excitatory signal to the cell at the same distance on the extension from the right edge, and this excitatory signal is gated by the output of cell A. The cells on the extension of the left edge compete each other,

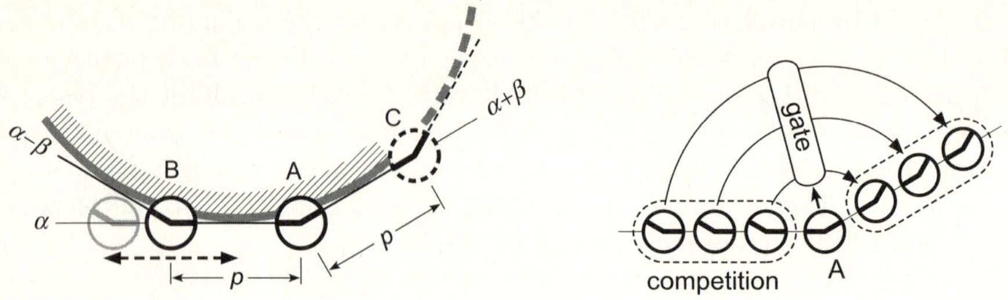

(a) The principle. (b) The network.

Fig. 3. Contour extrapolation from the response of bend-extracting cells

and only the winner can send the excitatory signal. Cell A and the cells on the extension of the left edge are the cells of layer U_{S2}, and the cells on the extension of the right edge are the cells of layer U_{S2e}. Although they are omitted in the illustration in the figure, there are also signal flows that are reversed bilaterally. This mechanism works only for bends in the range of $|\beta| \leq \pi/4$.

2.5 Top-Down Signal Path

The response of layer U_{S2}, which represent the bends of visible contours, and the response of U_{S2e}, which shows the bend of occluded contours extrapolated from the visible contours, are sent to layer W_{S2} in the top-down path and are added. We can say that the response of W_{S2} represents bends of restored contours.

The connections from W_{S2} to W_{C1} are identical to those from U_{C1} to U_{S2} in the bottom-up path, although the direction of signal flow is reversed. Thus oriented edges are generated from the information on bends of the contours. As a result, the edges of the restored contours appear in layer W_{C1}.

The response of W_{C1}, which represents the information on the restored edges, is fed back to U_{C1} and added to it. Since a feedback loop is thus produced in the hierarchical network, the missing contours are gradually restored while the signals circulate in the loop.

2.6 Effect of Blur by C1-Cells

Since the predicted curvature of occluded contours usually contains some amount of error, the location of the edges restored in W_{C1} also has some error. Hence some amount of discrepancy in location arises between two contours that have been extrapolated from both sides of an occluded area. If there is a discrepancy, the two contours often extend in different directions without merging into one, while the signals circulate in the feedback loop.

Layer U_{C1}, which is inserted between layers U_{S1} and U_{S2}, works to absorb this discrepancy. The response of U_{S1} is spatially blurred in U_{C1}. Since the input connections to S2-cells of U_{S2} have also been produced by the use of blurred response from U_{C1}, the feedback signals to W_{C1} from W_{S2} have also the same amount of blur. Even if there are small positional errors in the feedback signals

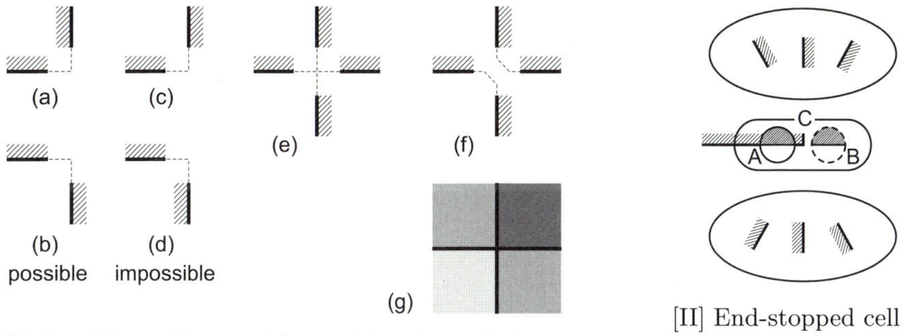

[I] Possible and impossible combinations of edges.

[II] End-stopped cell C inhibits contradictory edges.

Fig. 4. Suppressing contradictory contours

from W_{C1} to U_{C1}, the response to the two contours, whose widths have been made broader by the blur, can overlap in U_{C1}. The errors are thus absorbed, and the two contours that have come from both sides merge into a single contour.

2.7 Displaying the Result of Amodal Completion

Layer W_G on the top-down path, which corresponds to layer U_G on the bottom-up path, shows positive and negative components of the brightness contrast of the completed contours. The connections from W_{C1} to W_G are identical to those from U_G to U_{S1} in the bottom-up path, although the direction of signal flow is reversed. Corresponding to the oriented edges restored in W_{C1}, the information on brightness contrast thus appears in W_G.

Layer W_0 at the lowest stage in the top-down path is used only for the purpose of monitoring the result of amodal completion intuitively. It corresponds to input layer U_0 in the bottom-up path, and displays how the contours of the occluded patterns are restored. On-center cells of W_G, which represent positive components of the brightness contrast, send positive signals to W_0, and off-center cells send negative signals. These signals are diffused to neighboring cells in W_0, where the signal of opposite polarity works as a barrier to the diffusion [3].

2.8 Suppressing Contradictory Contours

In many cases, missing contours are correctly restored by the process discussed in section 2.4. If one of the two contours that have been extended from different directions arrives at their meeting point much earlier than the other, however, the former contour might extend far beyond the meeting point without being stopped there. This might occur, for example, for a pattern shown in the upper left of Fig. 10.

When two edges of different orientations meet, the combination of the polarities (namely, the direction of the brighter and darker sides) of the edges like (a) and (b) in Fig. 4[I] are possible, but combinations like (c) and (d) are impossible as a visual pattern. It might be thought that contradictory edges can be removed by a mechanism of lateral inhibition between edges that make a combination like (c) or (d), but this is not always the case. The case like (e) becomes possible as a visual pattern, although it contains an impossible combination like (c). If four edges are combined as illustrated in (f), both become possible combinations of edges. These combinations can actually occur, for example, in a pattern like (g).

The difference between (c) and (e) is caused by the condition whether an edge stops in one side of the meeting point or extends to both sides. Then the problem can be solved by introducing to W_{C1} a mechanism that inhibits contradictory edges, only when the end of an edge exists.

Suppose a horizontal edge, whose upper side is darker, has been extended from the left and is broken in the middle, as illustrated in Fig. 4[II]. The break of the edge is detected by *end-stopped cell* C, which receives excitatory signals from edge-extracting cells around point A and inhibitory signals from cells around B. When end-stopped cell C detects the break, it inhibits edge-extracting cells that are located in the upper ellipse and have selectivity to edges whose right sides are darker (to be more exact, edges whose orientation is in the range of $\pm\pi/2$ from the vertical). It also inhibits edge-extracting cells that are located in the lower ellipse and have selectivity to edges whose left sides are darker.

3 Computer Simulation

We simulated the model on a computer. Fig. 5 shows an example how the cells in the network respond to a partly occluded circular disk. The figure shows, however, not all layers, but only important layers. This is the response at the time when the signals have circulated 7 times through the feedback loop made by the bottom-up and top-down paths. Layer W_0 shows how the model perceives the contours from the occluded pattern presented to input layer U_0.

Fig. 6 shows how the response to this stimulus changes with time in layers U_0, U_G, W_0 and W_G. In other words, it shows how the amodal completion goes on. We can see that a disk covered by two rectangles is completed smoothly.

Figs. 7 and 8 show the response to occluded disks of different diameters. It can be seen from these figures that the contours are completed smoothly regardless of the diameters of the occluded disks.

For all stimuli in Figs. 8–10, the visible contours crosses the same occluding object at the same locations, but the missing contours are completed differently, depending on the curvature of the visible contours.

In Fig. 9, although two contours are perpendicular to each other, they stop extending beyond the crossing point, because they reach the crossing point almost at the same time. In these cases, the mechanism for suppressing contradictory contours, which was discussed in secion 2.8, is not necessarily required. We can have almost the same response without using the mechanism.

It is in the case like Fig. 10 that this mechanism becomes necessary. If this mechanism does not exist, the contour that has been extrapolated from the bottom will extend upward without being affected by its counterpart, which reaches the crossing point later.

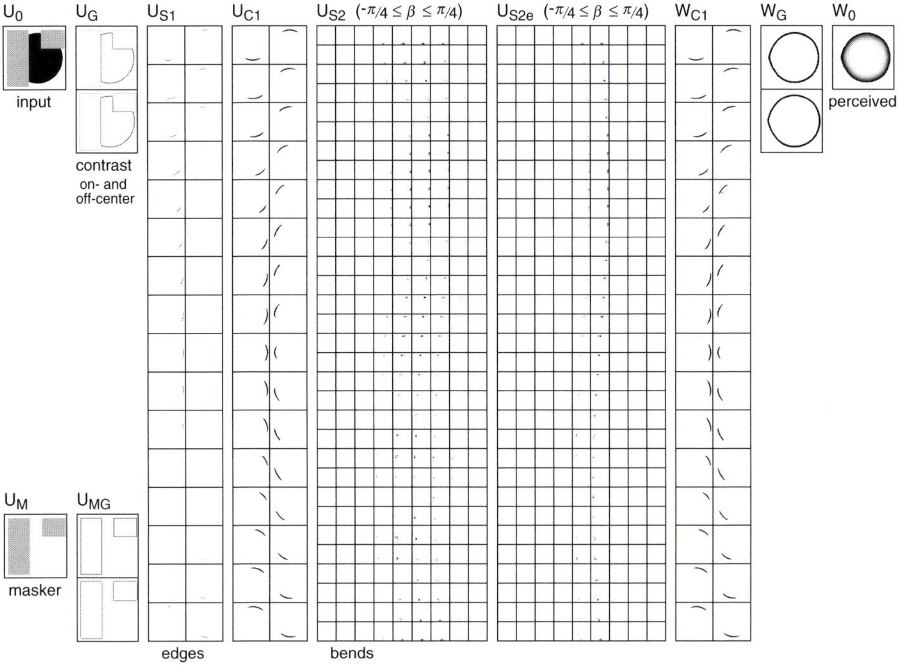

Fig. 5. An example of the response of the cells in the network to a partly occluded disk. This is the response after 7 times of circulation of signals in the feedback loop.

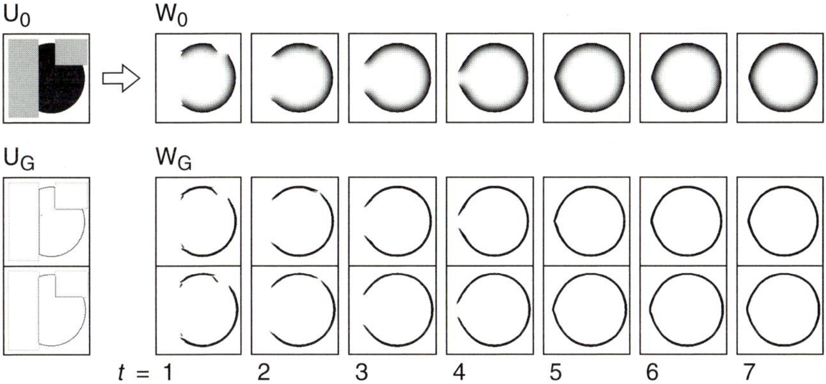

Fig. 6. The progress of amodal completion for the same stimulus as Fig. 5

The visible contours on both sides of an occluded area have different curvature in Fig. 11, but the extrapolated contours merge into a single contour of a natural shape. This is because the blur by C1-cells works effectively to absorb the discrepancy in location between two contours that have been extrapolated from both sides of an occluded area, as was discussed in section 2.6.

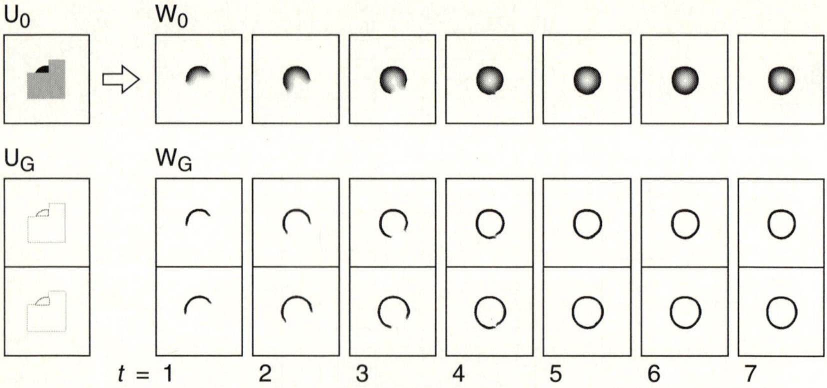

Fig. 7. The response to an occluded small disk

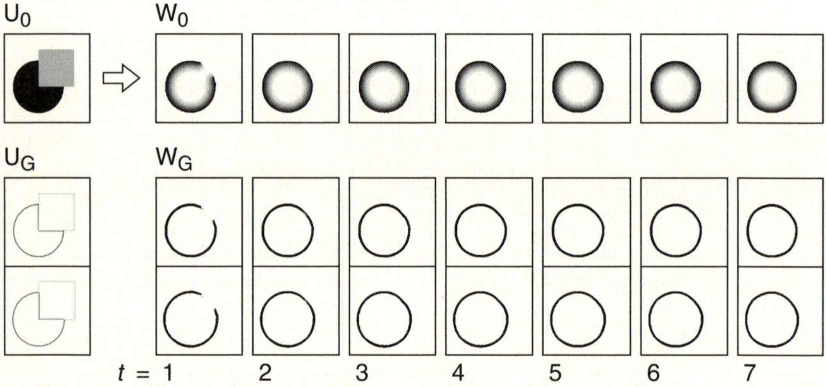

Fig. 8. The response to a disk partly occluded by a square

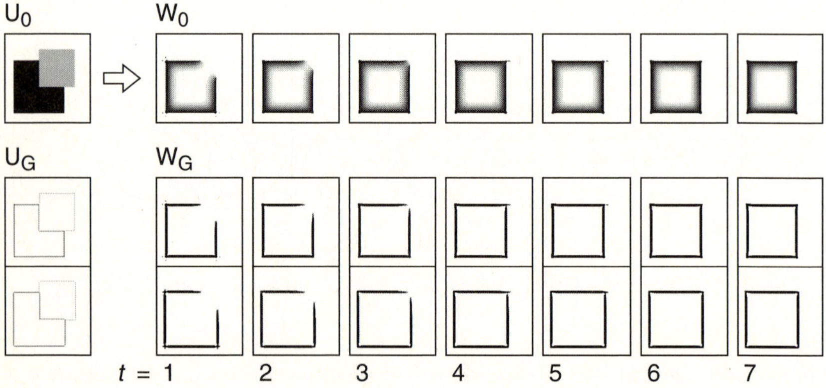

Fig. 9. The response to a square partly occluded by the same square as Fig. 8

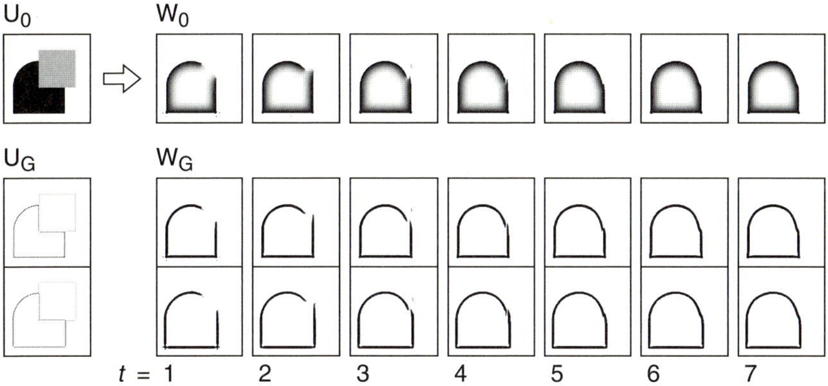

Fig. 10. The response to a half disk partly occluded by the same square as Fig. 8

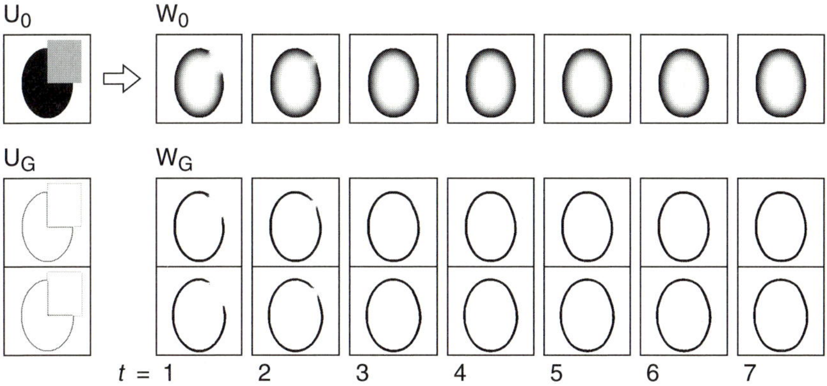

Fig. 11. Even if the visible contours on both sides of an occluding object differ slightly in curvature, a smooth interpolation can be performed

References

1. Ito, M., Komatsu, H.: Representation of angles embedded within contour stimuli in area V2 of macaque monkeys. J. Neuroscience 24(13), 3313–3324 (2004)
2. Fukushima, K.: Neocognitron for handwritten digit recognition. Neurocomputing 51, 161–180 (2003)
3. Fukushima, K.: Restoring partly occluded patterns: a neural network model. Neural Networks 18(1), 33–43 (2005)

Predictive Coding in Cortical Microcircuits

Andreea Lazar[1,*], Gordon Pipa[1,2,3], and Jochen Triesch[1]

[1] Frankfurt Institute for Advanced Studies, J.W. Goethe University, 60438, Frankfurt am Main, Germany
[2] Max-Planck Institute for Brain Research, 60528, Frankfurt am Main, Germany
[3] Massachusetts Institute of Technology, Department of Brain and Cognitive Sciences, MA 02139-4307, USA
{lazar,pipa,triesch}@fias.uni-frankfurt.de

Abstract. We investigate the influence of spike timing dependent plasticity (STDP) on the prediction properties of recurrent microcircuits. We use sparsely connected networks in which the synaptic modifications introduced by STDP are complemented by two homeostatic plasticity mechanisms: synaptic normalization and intrinsic plasticity. In the presence of structured external input, STDP changes the connectivity matrix of the network such that the recurrent connections capture the particularities of the input stimuli, allowing the network to anticipate future inputs. Network activation patterns reflect different input expectations and can be separated by an unsupervised clustering technique.

Keywords: STDP, intrinsic plasticity, prediction, recurrent networks.

1 Introduction

A fundamental task for any cognitive system is to find the causal relationships between stimuli and predict future events. The recurrent structure of cortical networks is endowed with memory which can provide context for prediction. Network architectures like the liquid state machine [1] and the echo state network [2] utilize this memory component of the system in which present responses influence subsequent ones. These random recurrent networks nonlinearly transform input streams into high-dimensional activation patterns partially constrained by the previous network activation states. The learning required by the computational task is performed entirely outside the recurrent network through trained projections to separate readout neurons. Since the recurrent connectivity stays fixed, the properties of these recurrent connections constrain the networks' performance.

In a more biologically plausible framework local synaptic modulations of the recurrent connections could reflect statistical properties of inputs and optimize the network for a required task. Especially in the context of dynamical integration of information over time a temporal plasticity rule like *spike timing dependent plasticity* (STDP) [3,4] could have a significant role in adjusting and enforcing the computational performance of recurrent networks.

* Corresponding author.

V. Kůrková et al. (Eds.): ICANN 2008, Part II, LNCS 5164, pp. 386–395, 2008.

Learning via STDP in recurrent networks [5,6] has been little studied mostly because of the instabilities that such a learning rule can bring about. Unless network parameters are carefully adjusted the level of activity in a neural circuit can grow or shrink in an uncontrolled manner. Narrow parameter settings for stable dynamics are implausible in a biological sense. A more appropriate approach is to complement synaptic learning with additional homeostatic mechanisms that are sensitive to the post-synaptic firing rate or to the total level of synaptic efficacy. We use synaptic normalization to keep the total synaptic input to a neuron approximately constant. We also use a model of intrinsic plasticity (IP), to maintain the activity of each neuron in a desired regime. Desai et al. showed that the excitability of a neuron increases under activity deprivation, which suggests that the mean firing rate of a neuron is constantly regulated [7]. We model this with a simple threshold adaptation to keep firing activity approximately constant.

Together the two homeostatic processes complement STDP such that the resulting network dynamics remains 'healthy' and computationally powerful. Our results show that, under these conditions, STDP embeds the likely transitions of inputs into the network's structure, forcing specific connectivity routes. Moreover STDP prunes down competing projections from afferent units and creates self-organized pathways that tend to separate from each other with respect to the time dependent transitions they reflect. We use an unsupervised clustering technique that groups activity patterns based on a distance measure and we show that the resulting clusters are not random but contain valuable information reflecting input expectancy. The network's dynamics incorporates valuable prediction properties which vanish in the absence of STDP.

2 Recurrent Network with Plasticity

2.1 Network Architecture

We use a discrete representation of time and a simple recurrent network of threshold excitatory (E) and inhibitory (I) units, obeying Dale's law (see Fig.1). Our cortical microcircuit has N^E excitatory and $N^I = 0.2 \times N^E$ inhibitory units. Neurons are connected through weighted synaptic connections, where W_{ij} is the connection strength from unit j to unit i. All possible E$\rightarrow$I and I$\rightarrow$E connections are present, while the E$\rightarrow$E connections are random and sparse. For simplicity we did not consider any I$\rightarrow$I connections. Self-connections are prohibited.

Inputs are time series of different symbols, each associated with a specific group of N^U input units, which all receive positive input drive when the associated symbol is active.

The firing activity of the network at the discrete time t is given by:

$$x_i(t+1) = \Theta\left(\sum_{j=1}^{N^E} W_{ij}^{EE}(t)x_j(t) - \sum_{k=1}^{N^I} W_{ik}^{EI} y_k(t) + U_i(t) - T_i^E(t)\right); \quad (1)$$

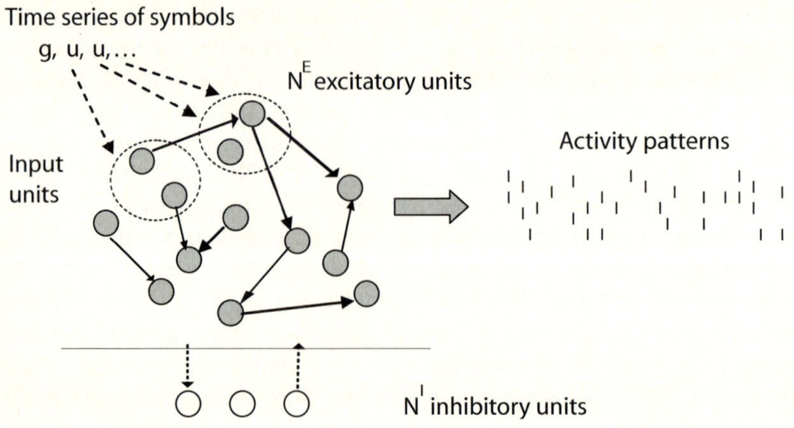

Fig. 1. Abstract representation of network model

$$y_i(t+1) = \Theta(\sum_{j=1}^{N^E} W_{ij}^{IE} x_j(t) - T_i^I); \tag{2}$$

where x and y are activations of excitatory and inhibitory units respectively, W^{EE}, W^{IE}, W^{EI} are synaptic weights, T^E and T^I are threshold values, and U is the additional input value received only by a predefined selection of input neurons. The heavyside step function Θ constrains the activation of the network at time t to a binary representation: the neuron fires if the total drive it receives is greater then its threshold ($x_i(t) = 1$) otherwise it stays silent ($x_i(t) = 0$).

2.2 Network Dynamics and Plasticity Rules

The network's activity patterns depend on intrinsic cellular and circuit properties. The predefined parameters of our model are few: the number of excitatory units N^E, the number of units corresponding to each input state N^U, the sparsity of the connectivity matrix W^{EE}, an expected rate value H_0, and η_{IP}, η_{STDP} which define the timescales of the two plasticity rules.

The initial threshold values T^E and T^I are drawn from a uniform distribution in the interval $[0, T^E_{max}]$ and $[0, T^I_{max}]$, respectively. The ratio T^E_{max}/T^I_{max} can be chosen such that the network exhibits on average H_o spikes per time step (see Appendix). We select a sufficiently strong input drive U so that for any input state the corresponding group of input neurons will fire, with or without help from the recurrent network activity, resulting in N^U out of $H_0 \times N^E$ input specific spikes.

We are interested in the properties of activity patterns produced by our recurrent architecture under the influence of synaptic learning via *spike timing dependent plasticity* (STDP). STDP refers to the sensitivity of synaptic plasticity to the relative timing of pre and post-synaptic action potentials. We use a simple model of STDP that strengthens the synaptic weight W_{ij}^{EE}, from unit j

to i, by a fixed amount η_{STDP} whenever unit i is active in the time step following activation of unit j. When unit i is active in the time step preceding activation of unit j, W_{ij}^{EE} is weakened by the same amount.

$$\Delta W_{ij}^{EE}(t) = \eta_{\text{STDP}}(x_i(t)x_j(t-1) - x_i(t-1)x_j(t)); \tag{3}$$

STDP changes the synaptic strength in a temporally asymmetric 'causal' fashion.

Recurrent networks are especially sensitive to neuron/circuit properties which if chosen wrongly will lead to breakdowns or explosions of activity. For a healthy dynamics we condition the architecture of the circuit in the following way:

- The initial connectivity matrix W^{EE} is sparse (the probability of an initial connection W_{ij}^{EE} is set to 0.2 for the following simulations). Learning with STDP is constrained to this set of initial W^{EE} synapses, with no synapses incorporated or lost.
- *Synaptic normalization* proportionally adjusts the values of incoming connections to a neuron so that they sum up to a constant value. The initial weights are adjusted such that $\sum_j W_{ij}^{IE} = 1$ and $\sum_j W_{ij}^{EI} = 1$. Since time dependent modulations driven by STDP learning affect W^{EE} connections, these are rescaled at every time step:

$$W_{ij}^{EE}(t) = W_{ij}^{EE}(t) / \sum_j W_{ij}^{EE}(t); \tag{4}$$

  The synaptic normalization rule does not change the relative strengths of synapses established by STDP but it does introduce an upper limit on the total incoming drive a neuron receives which has a stabilizing effect on the network behavior.
- An intrinsic plasticity rule regulates the activity level of each neuron. A unit that has just been active increases its threshold while an inactive unit lowers its threshold by a small amount. In our model this is given by:

$$T_i^E(t+1) = T_i^E(t) + \eta_{IP}(x_i(t) - H_0); \tag{5}$$

  where η_{IP} is a small learning rate. This rule encourages every excitatory unit to fire with a rate H_0.

In spite of the constant synaptic modulations introduced by STDP, the two homeostatic mechanisms, intrinsic plasticity and synaptic normalization along with the sparse connectivity between excitatory units ensure a computationally powerful network dynamics (see Results section).

3 Predictive Coding

3.1 Structure in Time Sequences

We have chosen input time series identical to the letter sequences used by Elman [8]. The sequences are formed out of six symbols (b, d, g, a, i, u) and designed

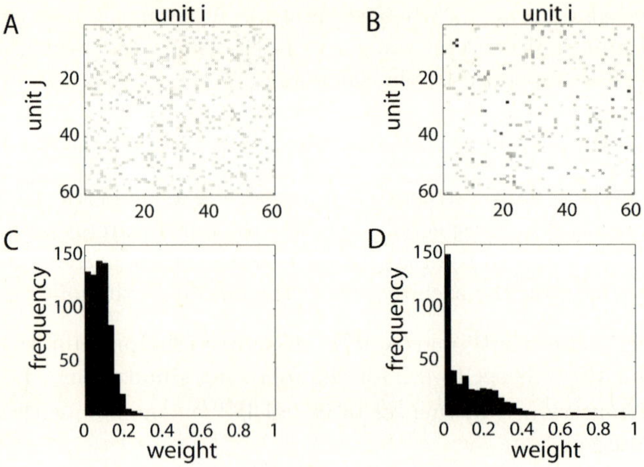

Fig. 2. Weight matrices (upper panels) and weight histograms (lower panels) for the E→E connections before (left) and after STDP training (right)

from the following building blocks ('words'): ba, dii, guuu. These words are fixed and alternate randomly. An example sequence is: b, a, g, u, u, u, b, a ...

Such a time varying input contains patterns of different durations, which means that the network has to build inventories of possible sequences with different temporal extents. It also has to deal with the intrinsic differences in the input statistics (e.g. there will be 3 times more u then a).

In the following experiments we consider networks with $N^E = 60$ ($N^I = 12$), from which five neurons are randomly selected as input population for each of the six symbols ($N^U = 5$). E→I and I→E connections are first uniformly distributed in the [0, 1] interval and then adjusted by synaptic normalization. The initial connectivity matrix between excitatory units W^{EE} is sparse with 0.2 probability of having a link between any two nodes. The distribution of W^{EE} is initially uniform in the interval [0, 0.2] and then rescaled following (4) and looks like the example figure below (Fig. 2 C). After initialization, each neuron will have on average 12 outgoing/incoming connections. Only these E-E connections are affected by STDP learning.

We set $T^E_{max} = 1$, $T^I_{max} = 2$ for rates of $H_0 = 0.6$ spikes/time step (see Appendix), or 10 population spikes/time step. An input drive of U = 1 makes sure that initially the input populations are responding independent of the recurrent network drive.

We run the network for 50000 steps while modulating E→E connections via STDP learning, with the homeostatic influence of synaptic normalization and intrinsic plasticity. As a result a number of connections grow stronger while others weaken (see Fig. 2). A training interval of 50000 time steps was enough for the configuration of the weight matrix to stabilize (with some small jitter, tested for up to 400000 time steps - not shown).

3.2 Predictive Learning Via STDP

The effect of time is implicit in the structural changes of the weight matrix since they are the result of a temporal learning rule under the influence of a time varying input. Under these conditions, we are interested in the aptitude of the recurrent network to capture the particularities of the input and make use of them for a computational task. Specifically we are interested in the capability of the network, at each moment in time, to anticipate the next symbol in the input sequence, by projecting drive towards the corresponding coding units.

To this end, we test the network's ability to generate similar activation patterns when inputs are missing. Using the last 20000 steps produced during STDP simulation, we group together the activity patterns corresponding to each of the input symbols. We differentiate between repeated instances of a letter within a word such that the sequence: b, a, d, i, i, g, u, u, u will count as nine different conditions. For each of these symbols we compute an average of all the networks states that resulted in response to that specific input stimulation and refer to it as a 'representative' pattern of activity for this input.

After STDP learning we freeze the trained network and use another random input sequence of the aforementioned building blocks for 40000 time steps (test phase). As we are interested in prediction, we analyze the patterns of activity produced by the network at a specific time step t in the absence of input. Basically, we record a second network state $x'(t+1)$, which is influenced by the complete history of the system but does not receive the present input ($U(t) = 0$).

For each moment in time we compute the euclidean distance between pattern $x'(t)$ and each of the nine representative patterns generated during training. Results are shown in Fig. 3 A for 20 consecutive time steps (out of 40000). On the horizontal axes we show the missing input at time t, on the vertical the degree of similarity of $x'(t)$ to each of the nine representative patterns. Darker colors display stronger similarity (smaller euclidean distance). It is worthwhile

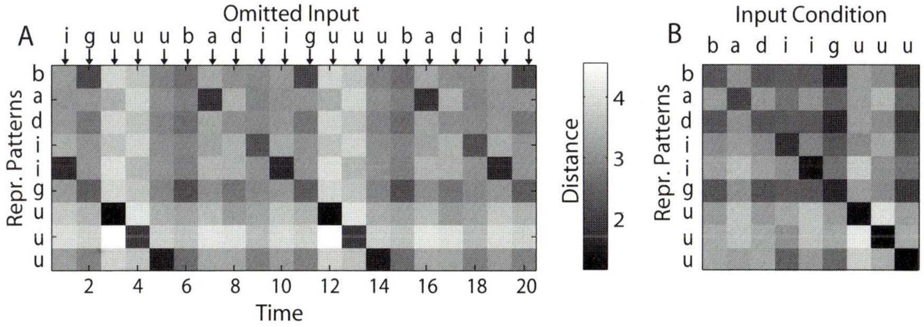

Fig. 3. (A) Sample window of 20 time steps (out of 40000) shows the distance between the test activity patterns ($x'(t)$) and each of the 9 representative patterns. Note that the input at present time is ignored ($U(t) = 0$) but is ilustrated in the figure for comparison with the expectancy of the network. (B) Average distance over all occurences of each input condition to each of the 9 representative patterns.

to remember that $x'(t)$ is fully determined by the propagation traces of previous inputs through the network. Thus, we can say that the more similar $x'(t)$ is to one of the nine patterns, the more prediction it carries.

Given the temporal structure of the input sequence, it is only sometimes possible to correctly predict the next input. At the end of each word the only prediction possible is that a new word will start, i.e. a consonant will follow. Time steps 2, 6, 8, 11, 15, 17, 20 correspond to such situations and the prediction is mainly distributed across the three consonants. In contrast, within each word, the predictions point towards the correct incoming symbol and the network state $x'(t)$ is very similar to the expected network state if the input had been present.

Fig. 3 B groups together the distances obtained for all the test patterns corresponding to the same missing input. Again, the network correctly anticipates the next state despite the missing input if it is a vowel. If a consonant input is missing, the network anticipates the occurrence of any consonant.

3.3 Self-organized Predictive Patterns

Activity patterns generated by the recurrent connections as prediction for the same input condition show strong similarity. We tested if these pattens are similar enough to cluster together in an unsupervised fashion.

We employed a hierarchical clustering algorithm, namely agglomerative clustering, to analyze the structure of the activity patterns (x') obtained during the test phase in the absence of input. Each of the 40000 patterns is a point in a 60 dimensional space. In agglomerative clustering, we start by considering each of these points to be centers of their own cluster. The distance between two clusters is computed as the euclidean distance between their centers. Repeatedly, the two closest clusters are merged into a single cluster, until the entire data is collapsed. In this way we can observe how the data distributes across any number of clusters without deciding on this number in advance.

Fig. 4 A shows an example hierarchy of the last 20 clusters. The data belonging to each of these 20 clusters is plotted in Fig. 4 B as a histogram with respect to the missed input condition. Cluster 18 is expanded for better visibility: it contains 99 patterns corresponding to input 'b', 1368 patterns corresponding to input 'd' and 504 corresponding to input 'g'. Interestingly clusters tend to correspond either to one input condition (for 'a', 'i', 'u') or they group together the input conditions 'b', 'd', 'g'. As discussed before, this behavior is sensible: the consonants (b, d, g) are always succeeded by specific vowels (a, i, u). The vowels can therefore be predicted properly making use of the activity history of the network. The position of the vowels inside the words is relevant, e.g. clusters 1, 2 and 3 correspond to the first 'u' in a 'guuu' sequence, cluster 16 to the second 'u' and cluster 10 to the third. At the end of each word no precise prediction is possible, only a guess towards a new start of a word which is apparent in the results. This implies that the network has assimilated the rules according to which the input sequences are structured.

In Fig. 4 C we contrast these results with networks that lack STDP training. In this situation no reliable structure is found, which is understandable since

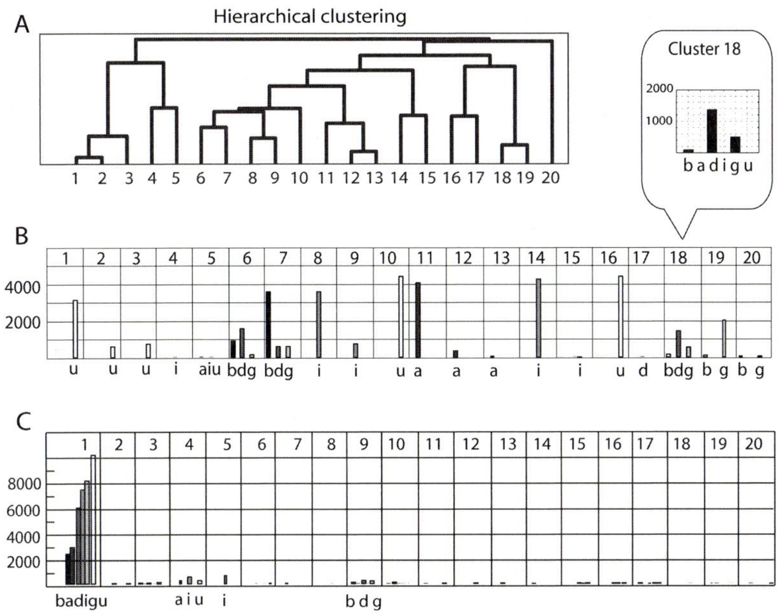

Fig. 4. (A, B) Predictive power of activity patterns produced by networks with STDP. Top: last 20 steps of the hierarchical clustering method. Middle: each cluster is plotted as a histogram of the number of patterns corresponding to each input condition (what the input would have been if present). Clusters show predictive power for each individual vowel and consonants as a group. (C) Networks in the absence of STDP: the last 20 clusters of the hierarchical clustering method don't show any structure with respect to the 'expected' inputs.

the connectivity matrix is random and lacks input specificity. It is worth mentioning that such randomly connected networks can have good performance in the case of a supervised readout projection for a prediction task. But even so, such a supervised mapping might be easier to perform for networks trained with STDP. In the presence of STDP the patterns of activity group together in space with respect to their input expectation and a readout is only required to assign the emergent clusters to correct output conditions. The advantages brought by the self-organization of the network, prior to any task demand, will be further addressed in a future study.

4 Conclusions

We have shown that a sparsely connected recurrent network, endowed with a combination of different forms of plasticity, can capture the temporal structure of its input stimuli and make use of it as context for prediction. The self-organization of the connectivity matrix was an outcome of local STDP modulations and thus unsupervised.

Learning via STDP was only possible in the presence of two homeostatic mechanisms: synaptic normalization and intrinsic plasticity. In the absence of synaptic normalization the changes introduced by STDP drive the network into an unstable regime; the same can happen if the connectivity matrix of the network is not sparse. Intrinsic plasticity spreads activity across all units in the network leading to distributed representations. In its absence only some of the units are involved in the activity while half of the network becomes silent.

Our results showed that the network activation after STDP learning was context dependent and carried strong predictive information with regard to the incoming inputs. The activity patterns became more similar when they reflected the same input expectation, and different otherwise. Because the size of the network was small and the patterns of activity were short we could make use of an agglomerative clustering technique to extract the regularities in the space of activity patterns.

It is important to emphasize that these internal representations were the outcome of the local modulations of recurrent weights under input stimulation and the prediction properties emerged naturally with no connection to a task. This is in sharp contrast to explicitly training a readout in a supervised fashion to make such predictions.

Acknowledgments

This work was supported by the Hertie Foundation and grant EC MEXT-CT-2006-042484 (PLICON).

References

1. Maass, W., Natschläger, T., Markram, H.: Real-time computing without stable states: A new framework for neural computation based on perturbations. Neural Computation 14, 2531–2560 (2002)
2. Jaeger, H.: Adaptive nonlinear system identification with echo state networks. NIPS 20, 593–600 (2003)
3. Markram, H., Lübke, J., Frotscher, M., Sakmann, B.: Regulation of synaptic efficacy by coincidence of postsynaptic APs and EPSPs. Science 275(5297), 213–215 (1997)
4. Bi, G.Q., Poo, M.M.: Synaptic modifications in cultured hippocampal neurons: Dependence on spike timing, synaptic strength, and postsynaptic cell type. Journal of Neuroscience 18, 10464–10472 (1998)
5. Henry, F., Daucé, E., Soula, H.: Temporal pattern identification using spike-timing dependent plasticity. Neurocomputing 70, 2009–2016 (2007)
6. Lazar, A., Pipa, G., Triesch, J.: Fading memory and time series prediction in recurrent networks with different forms of plasticity. Neural Networks 20, 312–322 (2007)
7. Desai, N.S., Rutherford, L.C., Turrigiano, G.G.: Plasticity in the intrinsic excitability of cortical pyramidal neurons. Nature Neuroscience 2, 515–520 (1999)
8. Elman, J.: Finding structure in time. Cognitive Science 14, 179–211 (1990)

Appendix

Let H_0 and I_0 be the desired rate values of excitatory and inhibitory units, respectively. For a network activity with small rate fluctuations there will be approx. $H_0 \times N^E$ excitatory spikes and approx. $I_0 \times N^I$ inhibitory spikes per time step. All possible W^{EI} and W^{IE} connections are present while W^{EE} are sparse with probability of a connection between any 2 units of $p_{con} = 0.2$. Both the incoming excitatory connections and the incoming inhibitory connections to a unit are scaled following Eq. 4 and sum up to 1. Thus the average synaptic strengths are $\overline{W}^{EE} = 1/(p_{con}N^E)$, $\overline{W}^{IE} = 1/N^E$, $\overline{W}^{EI} = 1/N^I$ The initial threshold values T^E and T^I are drawn from a uniform distribution in the interval $[0, T^E_{max}]$ and $[0, T^I_{max}]$, respectively.

From Eq. 2 the probability of an inhibitory unit y_i firing at time t+1 is given by:

$$\begin{aligned} P(y_i(t+1) = 1) &= P(\textstyle\sum_{j=1}^{N^E} W_{ij}^{IE} x_j(t) - T_i^I > 0) \\ &\approx P(\overline{W}^{IE} \times H_0 \times N^E - T_i^I > 0) \approx P(H_0 > T_i^I) = H_0/T^I_{max} \end{aligned}$$

For a sufficiently strong input drive U, all the excitatory units which receive input will be active. We will have N^U input specific spikes/ time step with $N^U < H_0 \times N^E$. Based on Eq. 1 an excitatory unit x_i which does not receive input will fire at time $t + 1$ with probability:

$$\begin{aligned} P(x_i(t+1) = 1) &= P(\textstyle\sum_{j=1}^{N^E} W_{ij}^{EE} x_j(t) - \sum_{k=1}^{N^I} W_{ik}^{EI} y_k(t) - T_i^E > 0) \\ &\approx P(0.2 \times H_0 \times N^E \times \overline{W}^{EE} - I_0 \times N^I \times \overline{W}^{EI} - T_i^E > 0) \\ &= (H_0 - I_0)/T^E_{max} \end{aligned}$$

For stable rate regimes: $P(y_i(t) = 1) \approx I_0$ and $P(x_i(t) = 1) \approx H_0 - N^U/N^E$. Let r_U be the ratio of input specific activations $r_U = N^U/N^E$ and for simplicity $T^E_{max} = 1$. The equation reduces to: $T^I_{max} = H_0/r_U$.

Thus, for a desired rate of $H_0 = 10/60$ when $N^U = 5$ and $N^E = 60$ we have to choose $T^I_{max} = (10/60)/(5/60) = 2$.

A Biologically Inspired Spiking Neural Network for Sound Localisation by the Inferior Colliculus

Jindong Liu, Harry Erwin, Stefan Wermter, and Mahmoud Elsaid

University of Sunderland, Sunderland, SR6 0DD, United Kingdom
jindong.liu@sunderland.ac.uk
http://www.his.sunderland.ac.uk

Abstract. We introduce a biologically inspired azimuthal sound localisation system, which simulates the functional organisation of the human auditory midbrain up to the inferior colliculus (IC). Supported by recent neurophysiological studies on the role of the IC and superior olivary complex (SOC) in sound processing, our system models two ascending pathways of the auditory midbrain: the ITD (Interaural Time Difference) pathway and ILD (Interaural Level Difference) pathway. In our approach to modelling the ITD pathway, we take into account Yin's finding that only a single delay line exists in the ITD processing from cochlea to SOC for the ipsilateral ear while multiple delay lines exists for the contralateral ear. The ILD pathway is modelled without varied delay lines because of neurophysiological evidence that indicates the delays along that pathway are minimal and constant. Level-locking auditory neurons are introduced for the ILD pathway network to encode sound amplitude into spike sequence, that are similar to the phase-locking auditory neurons which encode time information to the ITD pathway. A leaky integrate-and-fire spiking neural model is adapted to simulate the neurons in the SOC that process ITD and ILD. Experimental results show that our model performs sound localisation that approaches biological performance. Our approach brings not only new insight into the brain mechanism of the auditory system, but also demonstrates a practical application of sound localisation for mobile robots.

1 Introduction

The known performance of animals using two ears in the sound localisation task has inspired researchers to work on new computational auditory models to understand the mechanisms of auditory neural networks. During the last twenty-five years, the structure and function of a number of pathways in the auditory brainstem have been well studied and become better understood [1]. For example, multiple spectral representations [2] are known to exist both in the early stages of sound processing, in cochlea and cochlear nucleus, and in later stage, from SOC to IC. Excitatory and inhibitory connections in the neural network of the midbrain sound processing pathways have been clarified [3]. Modelling these networks can help us to understand the brain mechanisms and provide a robust approach of sound understanding to mobile robots.

V. Kůrková et al. (Eds.): ICANN 2008, Part II, LNCS 5164, pp. 396–405, 2008.

Binaural sound localisation systems take advantage of two important cues [4] of the arriving sound signals in two ears: (i) interaural time difference (ITD) or interaural phase difference (IPD), and (ii) interaural level difference (ILD). Using these two cues, the sound source position can be estimated in the horizontal or azimuthal plane. ILD is believed to be more efficient at localising low- and mid- frequency sounds (50 Hz ~3 kHz) while ILD is better for mid- and high-frequency sound (>1.2 kHz) [4].

Models of ITD and ILD processing have been developed by several researchers [4][5]. Jeffress [4] originally proposed a widely used model to detect ITDs, in which sound from each ear passes different delay lines before reaching a coincidence neuron which fires with a maximum rate when two specific delay times are present for the sound. Yin [5] improved Jeffress's model in response to biological evidence by introducing a single delay line for the ipsilateral ear while retaining multiple delay lines for the contralateral ear. For ILD, Jeffress suggested a so-called "latency hypothesis" to explain the processing mechanism, in which the latency of inhibitory input is delayed relative to the excitatory input from the ipsilateral ear. Evidence for this idea was provided by Hirsch [6], however, the mechanism of ILD processing remains unclear.

This paper presents a new auditory processing system designed to provide live sound source positions via a spiking neural network (SNN). In this SNN, we implement Yin's ITD model and additionally propose an ILD model. Then ITD and ILD are combined together to simulate the function of IC in the human. It is the first example of applying both ITD and ILD into a spiking neural network to cover a wide frequency range of sound inputs. The synaptic and soma models in the system are treated as independent of sound frequency, ITD or ILD as there is no current evidence for specialisation. This simplifies the system and is a potential advantage for future adaptive auditory systems.

The rest of this paper is organised as follows. Section 2 presents the neurophysiological data of human auditory pathway. In that section, we make an assumption of level locking auditory neurons. Section 3 proposes an pyramid model to calculate ILDs. Section 4 proposes a system model with combines the ITD pathway and the ILD pathway. In Section 5, experimental results are presented to show the feasibility and performance of the sound localisation system. Finally, conclusions and future work are given in Section 6.

2 Biological Fundamentals and Assumptions

When sound arrives at the external ear, it is directed through the ear drum and then propagates to the inner ear, i.e. cochlea. The inner hair cells along the cochlea respond to the sound pressure to generate spikes that are sent through the auditory nerve (AN) up to the central nervous system. The temporal and amplitude information of the sound wave are encoded by the hair cells up to auditory nerve [4]. Two properties of the hair cells are important for this encoding. First, the cochlea is tonotopically organised so that each inner hair cell is maximally excited by stimulation at a characteristic frequency (CF) [7]. In other words, each hair cell has a specific frequency with a highest sensitivity. Second, the hair cells are polarised so that their movement is excited during one specific phase of the sinusoidal sound wave while inhibited during other phases. This phase locking occurring at frequency of 50 $\sim$1.2 kHz is the basis for the encoding of timing information for sound.

As the sound pressure level (SPL) increases, auditory nerve fiber increase their rate of discharge with a sigmoidal relationship to the decibel of sound over a relative range of 30 dB. In order to cover a wide SPL range, e.g. 120 dB, the relative range is adaptively changed according to the background sound level. However, the mechanism of this adaptivity is still not clear. In this paper, we assume there are "level locking" auditory nerve fibers which generate spiking signals only if the sound signal level lies in a specific fixed range.

After encoding the temporal and amplitude information, AN fibers pass the spiking sequence through the superior olivary complex (SOC) up to the inferior colliculus (IC) to calculate the ITDs and ILDs. Two separate circuits, the ITD pathway and the ILD pathway, are involved in the calculation. The ITDs [7] have been proved to be processed in the medial superior olive (MSO), which is one part of the SOC, and the ILDs are processed in another part of the SOC, i.e. the lateral superior olive (LSO). The MSO in one side receives excitation from the AVCN (anteroventral cochlear nucleus) from both the ipsilateral and contralateral ears. Besides the excitation, recent neurophysiological studies have revealed that the MSO also receives inhibition shortly following the excitation [7]. After the ITD processing in the MSO, the results are projected to the IC. For the ILDs processing, cells in the LSO are excited by sounds that are greater in level at the ipsilateral ear than the contralateral ear and inhibited by sounds that are greater in level at the contralateral ear [7]. Therefore, the LSO receives excitation from the ipsilateral AVCN, but inhibition from the MNTB (medial nucleus of the trapezoid body) which converts the excitation from the contralateral AVCN to inhibition. Finally, the spiking output of the LSO is also tonotopically projected to IC.

3 Model of ILD Processing in LSO

In the LSO, the model of the ILD processing is unclear yet. In this paper, we propose an ILD model based on the neurophysiological evidence of single delay line and the assumption of the existing of level locking AN. This model, called pyramid ILD model, is illustrated in Figure 1. The left slope of the pyramid receives the inhibition inputs from the contralateral MNTB and the level order of the inhibition is arranged in the reversed way of the pyramid layer, i.e. the level number increases from the top to bottom layer. The inhibition of one specific sound level gets into the network along the dot lines in the figure. The right slope of the pyramid takes the excitation inputs from the ipsilateral AVCN and the level order of the excitation is the same way of the pyramid layer. The excitation of one sound level passes down the network along the dash line. The neuron at the cross point of the excitatory dash line and the inhibitory dot line is the ILD

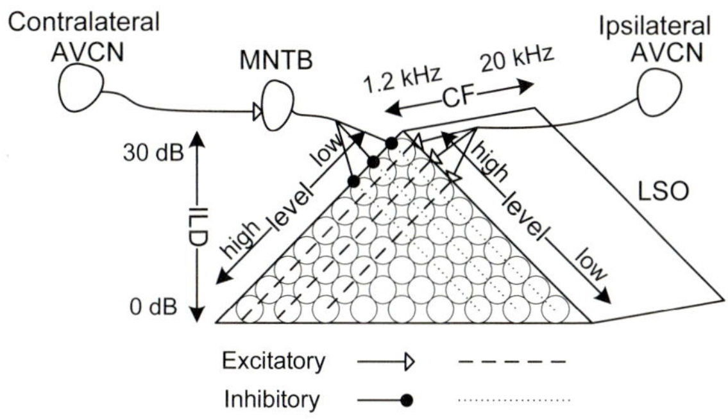

Fig. 1. Schematic diagram the ILD processing of the LSO in the human

coincidence neuron. It fires only if there are spikes at the excitatory line and no spike at the inhibitory line. The layer where the coincidence neuron belongs indicates the ILD, i.e. the lower layer represents the small level difference and vice versa.

4 System Model of Sound Localisation

Inspired by the neurophysiological data and the proposed ILD pathway assumption presented in Section 3, we designed a system model of sound localisation by using spiking neural networks (SNNs). The sound is first encoded into spikes as inputs to the SNN. The synaptic response $I(t)$ to a spike, occurred at $t = t_s$, is modelled as a constant square current with an amplitude (also called weight) of w_s, a latency l_s relative to the timing of the spike, and a lasting time τ_s. The sign of the the amplitude indicates whether the synapse inhibits (negative) or excitates (positive) the following soma. The soma response to $I(t)$ can be modelled based on the leaky integrate-and-fire model in [8].

$$u(t) = u_r \exp(-\frac{t - t_s}{\tau_m}) + \frac{1}{C}\int_0^{t-t_s} \exp(-\frac{s}{\tau_m}) I(t - s) ds \qquad (1)$$

where u_r is the initial membrane potential, and τ_m is a time constant. In this paper, a typical value for τ_m is 1.6 ms. C is the capacitor which is charged by $I(t)$, in order to simulate the procedure of the postsynaptic current charging the soma. The soma model has one more parameter, the action potential φ. When $u(t) = \varphi$, the soma will fire a spike, then $u(t)$ is reset to 0.

In contrast to other sound localisation systems, e.g. [9], which only applied an ITD pathway, our model utilises both the ITD and ILD pathways for both sides of ear. This feature provides a wide-frequency localisation processing ability to the model as we will see in Section 5. Furthermore, the SNN of the model simplifies the model of the synapse, and the parameters of the synapse and soma are independent to sound frequencies in contrast to the model in [9]. This feature enables our system to have a real-time computation ability and potential

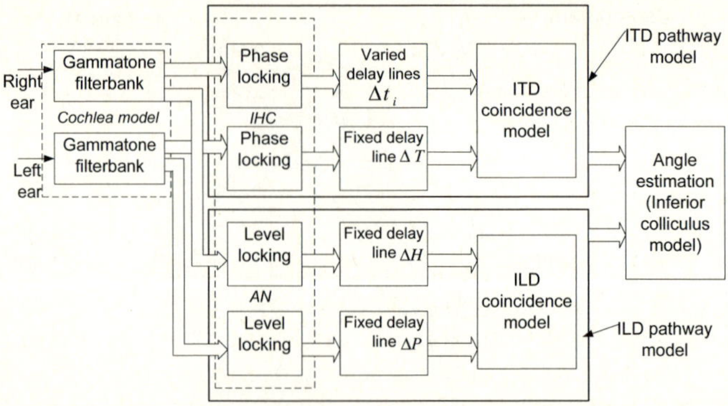

Fig. 2. Schematic structure of biologically inspired sound localisation system. This example assumes the left ear is the ipsilateral ear.

optimisation possibility. Figure 2 shows a schematic structure for the sound localisation procedure in the left ear side. Please note that there is another corresponding mode in the right ear side. In the figure, the tonotopical sound encoding feature of the cochlea is simulated by a bandpass filterbank consisting a series of Gammatone filters [10]. The filterbank decomposes the sound into several frequency channels in a similar way as the cochlea processing the sound.

After the Gammatone filterbank, the temporal information of the sound in each frequency channel is encoded into a spike sequence by the phase locking module in Figure 2, which simulates the phase locking character of the inner hair cell in the cochlea. At every positive zero-crossing of the incoming sound, a spike is triggered. At the same time, the amplitude information is encoded into several spike sequences by the level locking module.

Following the ITD pathway in Yin's model [5], the spike sequence of the contralateral ear, i.e. the right ear in Figure 2, passes variable delay lines Δt_i. We denote the delayed spike sequence as $S_{CP}(\Delta t_i, f_j)$, where C stands for the contralateral, P for phase locking, Δt_i for the delay time, f_j for the frequency channel j. Similarly, $S_{IP}(\Delta T, f_j)$ represents the delayed spike sequence of the ipsilateral ear with a fixed delay time ΔT. $S_{CP}(\Delta t_i, f_j)$ and $S_{IP}(\Delta T, f_j)$ are then input to the ITD coincidence model (please see Figure 3-(a) for details) to calculate the ITD. The calculated output of the ITD coincidence model is a new spike sequence represented as $S_{ITD}((\Delta_T - \Delta t_i), f_j)$. If there are spikes in $S_{ITD}((\Delta_T - \Delta t_i), f_j)$, it means the sound arriving to the ipsilateral ear is earlier than that to the contralateral ear by $ITD = \Delta_T - \Delta t_i$ second. Once the ITD calculation is implemented for all frequency channels, a three dimension ITD distribution map can be drawn, where the x-axis is for ITD, y-axis for the frequency channel and z-axis for the mean spike number in a unit time.

The ILD pathway is modelled in the bottom rectangular box in Figure 2. The level locking spike sequences from the contralateral and ipsilateral sides pass fixed delay lines, Δ_H and Δ_P, respectively. Then they go into the ILD coincidence model for calculating the level difference. Figure 3-(b) illustrates the coincidence model in detail. In the figure, $S_{CL}(l_i, f_j)$ represents the contralateral level locking spike sequence of frequency channel j and level l_i, while $S_{IL}(l_k, f_j)$ the ipsilateral level locking spike of level l_k. The output of ILD model is $S_{ILD}((l_k - l_i), f_j)$, which indicates the spike sequence when $ILD = l_k - l_i$. Once the ILD calculation is implemented for all frequency channels and for both sides, a ILD distribution map can be drawn in the similar way to the ITD distribution.

The calculation results of the ITD and ILD coincidence model are finally merged together as shown in the last module of Figure 2. Considering the complex head transfer function between the ILDs and source source angles [2], we use the ILD results to identify whether the sound came from the left or right side. Then we use this information to remove the ambiguity in the ITD results. For example, if the sum of the negative ILD spikes is larger than that of positive ILD spikes, we can conclude that the sound came from left side and then all the positive ITD spikes can be ignored in the following angle estimation of the sound source. After correcting ITD spikes, we can choose the significant ITD by using several methods, such as winner-take-all and the weighted mean method. Finally, the sound source azimuth angle can be calculated by:

$$\theta = \arcsin(\mathrm{ITD} \times \mathrm{V}_{sound}/\mathrm{d}_{\mathrm{ear}}) \tag{2}$$

where $\mathrm{V}_{\mathrm{sound}}$ is the sound speed, typically 344 m/s in 20°C. $\mathrm{d}_{\mathrm{ear}}$ is the distance between two ears, i.e. microphones in robot.

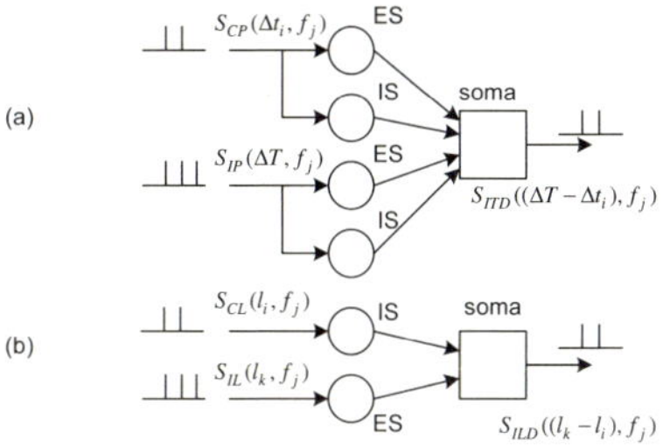

Fig. 3. (a) ITD coincidence model (b) ILD coincidence model. ES stands for excitatory synapse, and IS inhibitory synapse.

5 Experimental Results

To justify the feasibility and performance of our sound localisation model, we designed two groups of experiments: (i) artificial pure tone localisation, and (ii) real pure tone localisation. The sound in the former experiments is generated by a computer and it is manually added time difference and level difference on two sound channels according to the physical parameter of our mobile robot, MIRA. For example, to simulate a sound coming from the left 90 degree, the right channel sound is added by an extra time $\frac{\sin(90^\circ)\mathrm{d}_{ear}}{\mathrm{V}_{sound}}$ at the beginning and decreased in level by an arbitrary 50%. The ear distance $\mathrm{d}_{\mathrm{ear}}$ is 21 cm. The sound in real pure tone localisation is recorded in a general environment with 30 dB background noise. The distance of the sound source to the robot is 90 cm and the sound pressure at the speaker side is controlled to be 90±5 dB. The sample rate is 22050 Hz and the duration is 100 ms with 10 ms silence at the beginning. Pure tones with frequency 500, 1000 and 2000 Hz were recorded at the positions of -90, -60, -30, 0, 30, 60 and 90 degrees. The parameters of the spiking neural network are listed in Table 1. The Gammatone filterbank is set with 16 channels to cover from 200 to 3000 Hz.

Table 1. Parameters in experiments

	Excitatory Synapse			Inhibitory Synapse			Soma		
	l_s	τ_s	w_s	l_s	τ_s	w_s	φ	t_m	C
ITD	2.1	0.08	0.1	2.28	0.08	-0.1	8e-4	1.6	10
ILD	2.1	0.08	0.1	2.28	0.08	-0.1	8e-4	1.6	10
	channels		channels		Δt_i		ΔT	ΔH	ΔP
ITD	17		n/a		[1.4 2.6]		2	n/a	n/a
ILD	n/a		10		n/a		n/a	2	1.8

*Note: the unit of l_s, τ_s, w_s, t_m,Δ_{t_i}, Δ_T, ΔH and ΔP is ms and the unit of C is mF.

Figure 4 shows the spike distributions in the sound localisation of the artificial signal. In these distribution figures, the x-axis is the expected angle of sound localisation and the y-axis is the estimated angle after calculation. The size of square is proportional to the spiking number in the corresponding angle estimation. For example, in Figure 4(a), a big square at (0 0) means that when the artificial signal came from 0 degree the spiking number of estimated angle of 0 degree is the majority of spikes of all estimated angles. The localisation efficiency of the system is defined as a percentage of the spiking number at the correct estimation point, such as (-30 -30) and (60 60), to the total spiking number. In the figure, we compare the performance of two methods, i.e. the localisation using (i) the ITDs only and (ii) the ITDs with the ILDs.

In the top column of Figure 4, the localisation efficiency across all frequencies (500, 1000, 2000 Hz) and angles is about 70%. 85% of the sound signals across all frequencies were recognised correctly between -45 to 45 degree. The highest localisation efficiency occurred at the 0 degree. The efficiency decreases when the frequency goes high or the sound moves to the sides. This result matches the fact [7] that (i) the ITDs cue has the highest efficiency for sound localisation when the sound source is in front of the observer, and (ii) the ITD cue effect on sound localisation fades down over 1.2 kHz.

After adding the ILDs into our system, the experiments results shown in the bottom column of Figure 4 demonstrate that the spiking distribution is more concentrated, especially in the figure of 1000 Hz and 2000 Hz, than that of the results by only using the ITDs. This results match the fact [7] that the ILDs is the main cue for the high frequency sound localisation. The overall localisation

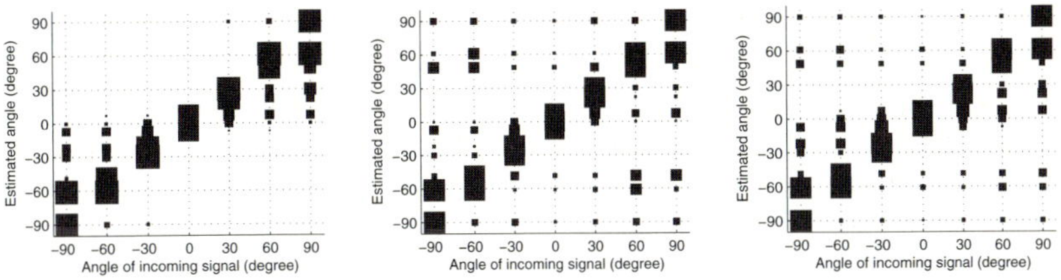

(a) 500 Hz with ITD only (b) 1000 Hz with ITD only (c) 2000 Hz with ITD only

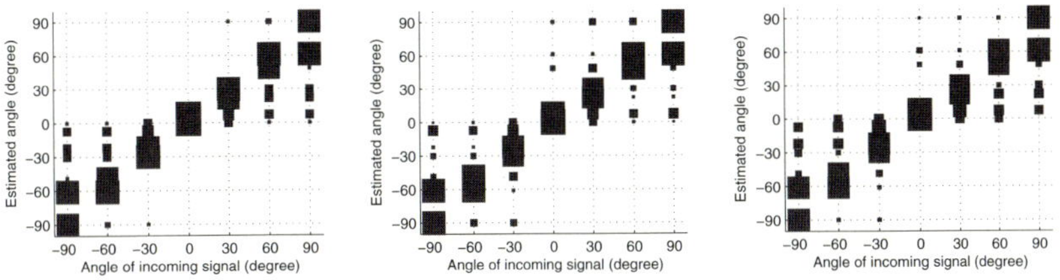

(d) 500 Hz with ITD & ILD (e) 1000 Hz with ITD & ILD (f) 2000 Hz with ITD & ILD

Fig. 4. The artificial sound localisation results for 500, 1000 and 2000 Hz. The size of square represents the spiking proportion in the corresponding angle estimation.

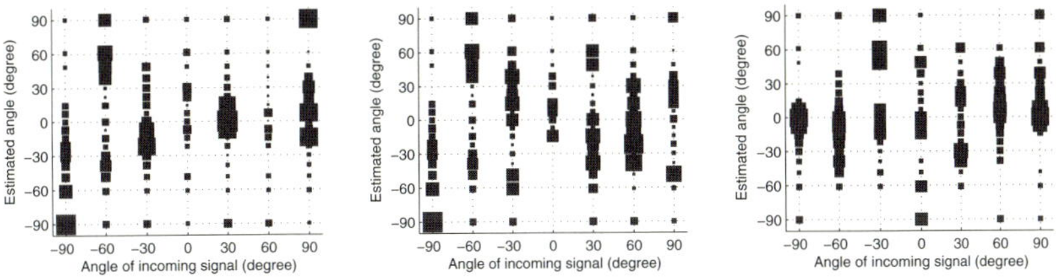

(a) 500 Hz with ITD only (b) 1000 Hz with ITD only (c) 2000 Hz with ITD only

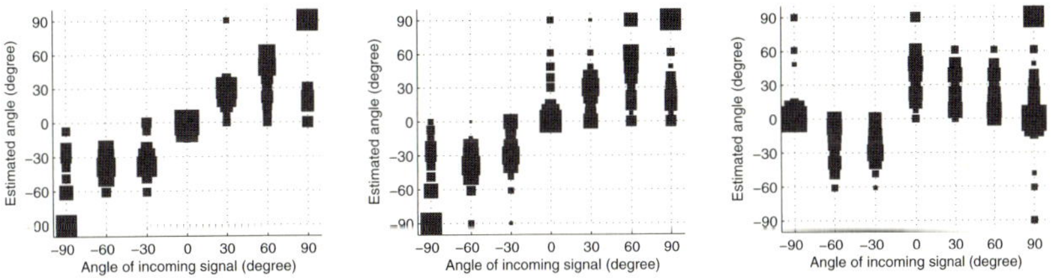

(d) 500 Hz with ITD & ILD (e) 1000 Hz with ITD & ILD (f) 2000 Hz with ITD & ILD

Fig. 5. The real sound localisation results for 500, 1000 and 2000 Hz. The size of square represents the spiking proportion in the corresponding angle estimation.

efficiency is increased to 80% and 95% of the sound signals across all frequencies were recognised correctly between -45 to 45 degree.

Figure 5 shows the spike distributions for the sound localisation of a real pure tone signal. The same experimental methods for the artificial sound localisation were applied. In these figures, the ambiguity of the estimated sound source angle is large when only the ITD cue is used. The overall localisation efficiency dropped down below 50% due to acoustic clatter which affects the ITD calculations because the phase-locking block in our system will generate more irrespective spike sequence in terms of these noise. However, after adding ILDs into the system, the ambiguity in the ITD calculation is improved and the overall localisation efficiency is raised to 65% because the noise level in the signal generally is not different in two microphones and therefore does not affect the ILD calculation much.

The time cost for processing a 100 ms sound signal is less than 50 ms on a 2.6 GHz CPU. Comparing the 9.1 s time cost when using the model in [9], our system performs much quicker without losing localisation efficiency. It is feasible to apply our system for real time sound localisation in real world.

6 Conclusion and Future Work

In this paper, we implemented a sound localisation system using spiking neural network inspired by mammalian auditory processing. In this system, both the ITD and ILD pathway were adopted and modelled based on recent neurophysiologic findings. In the ITD pathway, inhibitory inputs to the MSO are added together with traditional excitatory inputs in order to get a shape localisation results. In the ILD pathway, we proposed an assumption of the level-locking auditory nerve and built a pyramid model to calculate ILDs. The parameters of the SNN were set independent to the sound frequency and ITDs (ILDs) in contrast to similar work in [9]. The experimental results showed that our system can localise the sound source from the azimuth -90 to 90 degree. The sound frequency varied from 500 to 2000 Hz. The effect of frequency and sound source position to the localisation efficiency had a high correspondence with neurophysiologic data. It proved the reasonability of the proposed system.

In the future, active sound localisation, which can specify the feature frequencies of an interesting object, will be the next step of our research. In addition, an adaptive relative range of the level-locking encoding by using the feedback from SOC will increase the localisation efficiency of our system in a cluttered environment. For the application of our system to a mobile robot, we are planning to implement a self-calibration sound localisation system which can adaptively adjust the synapse and soma parameters according to the environment or electrical hardware change.

Acknowledgment

This work is supported by EPSRC (EP/D055466). Our thanks go to Dr. Adrian Rees and Dr. David Perez Gonzalez at University of Newcastle for their contribution on neurophysiologic support toward the project. This project, MiCRAM, is a cooperative project with them and they are mainly specialising in the neurophysiologic structure on the inferior colliculus for sound processing. We would also like to thank Chris Rowan for helping with building the MIRA robot head.

References

1. Oertel, D., Fay, R., Popper, A. (eds.): Integrative Functions in the Mammalian Auditory Pathway. Springer, New York (2002)
2. Young, E., Davis, K.: Circuitry and Function of the Dorsal Cochlear Nucleus. In: Integrative Functions in the Mammalian Auditory Pathway, pp. 160–206. Springer, New York (2002)
3. Fitzpatrick, D., Kuwada, S., Batra, R.: Transformations in processing interaural time differences between the superior olivary complex and inferior colliculus: Beyond the jeffress model. Hear. Res. 168(1-2), 79–89 (2002)
4. Jeffress, L.: A place theory of sound localization. J. Comp. Physiol. Psychol. 61, 468–486 (1948)
5. Smith, P., Joris, P., Yin, T.: Projections of physiologically characterized spherical bushy cell axons from the cochlear nucleus of the cat: evidence for delay lines to the medial suprior olive. J. Comp. Neurol. 331, 245–260 (1993)
6. Hirsch, J.A., Chan, J.C., Yin, T.C.: Responses of neurons in the cat's superior colliculus to acoustic stimuli. i. monaural and binaural response properties. J. Neurophysiol. 53, 726–745 (1985)
7. Yin, T.: Neural mechanisms of encoding binaural localization cues in the auditory brainstem. Integrative Functions in the Mammalian Auditory Pathway, 99–159 (2002)
8. Gerstner, W., Kistler, W.M.: Spiking Neuron Models, Single Neurons, Populations, Plasticity. Cambridge Unviersity Press, Cambridge (2002)
9. Voutsas, K., Adamy, J.: A biologically inspired spiking neural network for sound source lateralization. IEEE Trans Neural Networks 18(6), 1785–1799 (2007)
10. Meddis, R., Hewitt, M., Shackleton, T.: Implementation details of a computation model of the inner hair-cell/auditory-nerve synapse. J. Acoust. Soc. Am. 87(4), 1813–1816 (1990)

Learning Structurally Analogous Tasks

Paul W. Munro

School of Information Sciences, University of Pittsburgh
Pittsburgh PA 15260, USA
pwm@pitt.edu

Abstract. A method for training overlapping feedforward networks on analogous tasks is extended and analyzed. The learning dynamics of simultaneous (interlaced) training of similar tasks interact at the shared connections of the networks. The influence of one task on the other can be studied by examining the output of one network in response to a stimulus to the other network. Using backpropagation to train networks with shared hidden layers, a "crosstraining" mechanism for specifying corresponding components between structurally similar environments is introduced. Analysis of the resulting mappings reveals the potential for analogical inference.

1 Background

A fundamental operator for high-level cognition is the establishment of correspondences between analogous components of structurally similar tasks. Thorndike and Woodworth [1] put forward a theory that two different "mental functions" may share cognitive structures in their processing. A system that includes common representations of features in different domains can conceivably establish mappings between them [2].

The intrinsic pattern-matching properties of the connectionist framework can be brought to bear on feature mapping [3] [4] [5]. Approaches using hybrid symbolic-connectionist models have addressed the processing of structural mapping in the context of language processing [6]. Skill transfer between similar, if not strictly analogous, tasks has been the subject of connectionist models that transfer knowledge by reusing weights learned on one task on a second task [7]. Caruana [8] devised a multitask learning (MTL) architecture that exhibits faster convergence and better performance learning multiple tasks simultaneously than on one of the constituent tasks alone.

2 Overlapping Networks

A *partially shared network (PSN)* is defined here as a feedforward network that processes several input-output tasks with some input and output units dedicated to individual tasks, and other intermediate units and connections shared among the tasks.

The *strictly layered PSN* has an architecture in which the input vector is partitioned into K ($K > 1$) subvectors with the output vector is partitioned into K corresponding subvectors. There are 2 or more hidden layers. The connectivity from the input banks to the first hidden layer is *parallel*; hidden layers project from one to the next in *serial* fashion; the final step from the last hidden layer to the output banks is again in *parallel*.

V. Kůrková et al. (Eds.): ICANN 2008, Part II, LNCS 5164, pp. 406–412, 2008.

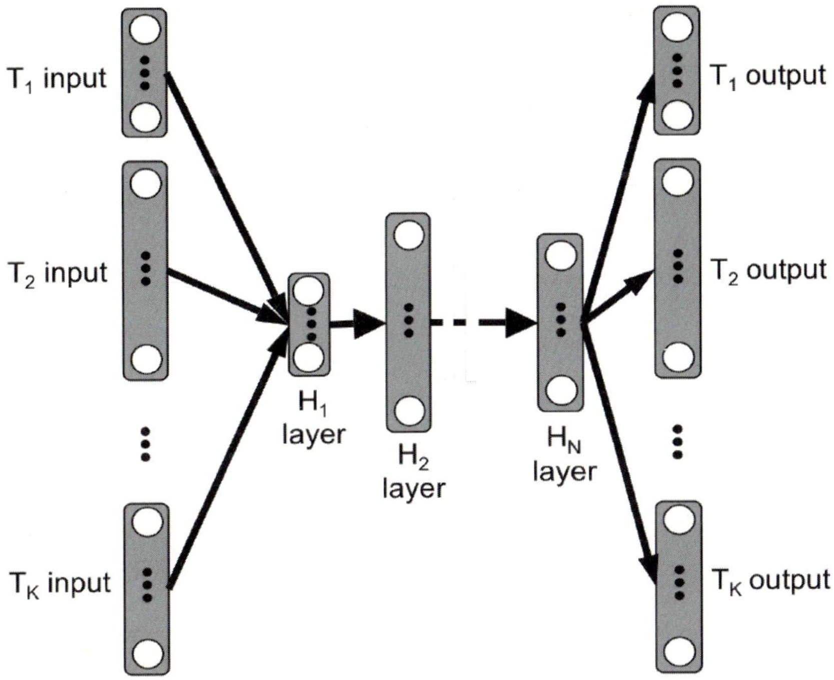

Fig. 1. A Strictly Layered PSN. Several banks of input units, T_j, project to a common hidden layer H_1, which in turn projects to a second hidden layer H_2, which projects to a subsequent layer H_3, and so on to H_N, $(N \geq 2)$. In the final stage, layer H_N projects to several output banks corresponding to the input banks. Arrows indicate full unit-to-unit connectivity between connected layers.

3 Methodology

3.1 Interlaced Training

Networks were initialized by setting the weights to small random values from a uniform distribution. The networks were trained with two similar tasks (T_1 and T_2) by presenting a random input-output pair from T_1 to input bank 1, training the net with backpropagation, then training on a pair from T_2 to input bank 2, and proceeding by alternating between T_1 and T_2.

3.2 Crosstesting

During training and testing, patterns are only presented on one input bank. This restriction maintains a consistent net input to the units of H_1. There is no reason for such a constraint among the output layers during testing. Thus output patterns across all banks can be examined in response to input on a single input bank. This makes possible a correspondence analysis between items in different tasks. That is, the output in T_2 generated by a T_1 input (A) can be compared to the outputs in T_2 generated by the set of

T_2 inputs. Any T_2 input (A') that generates a T_2 output that is sufficiently close to that generated by A is a candidate for the image of A in the input space of T_2.

Another approach to correspondence analysis is to compare representations of inputs from different input banks at the H_1 level. If a T_1 pattern and a T_2 pattern have identical (or nearly identical) H_1 representations, then they must give the same (or nearly the same) output patterns on all output banks. Hence they would be considered corresponding points in their respective tasks.

3.3 Crosstraining

The network can be explicitly trained to form a correspondence between a given pattern A in T_1 and a given pattern A' in T_2. A number of techniques were attempted:

1. Pattern A was presented at T_1. The resulting output at T_2 was compared with the target output for A' for training by backpropagation.
2. A more complex network architecture with additional output banks that were trained as autoencoders for each input bank was examined. With this network design, pattern A' was used as a target on its encoder bank for pattern A as input.
3. Patterns generated at H_1 by A' at T_2 were used as a target at H_1 to directly crosstrain input A at T_1.
4. The backpropagation algorithm is applied by first presenting A' at the T_2 input. The output at T_2 is then used as a training pattern for A on the T_1 input. This is very similar to technique 1 above except the actual output generated by A' is the target rather than the target of A' from the training set.

Of these techniques, only number 1 and number 4 gave good results. Technique 4 was ultimately chosen since it makes fewer assumptions about knowledge of the training stimulus.

3.4 Neighborhood Tasks

The networks in the study were trained on a class of tasks defined by Ghiselli-Crippa and Munro [9] called "neighborhood tasks", in which each input node represents a location on a graph. The set of output nodes is the same, each node representing a graph location. Each input-output pair in the training set consists of a single input node (a representation of the graph location L). The corresponding output is the set of nodes representing the locations that are neighborhood of L on the graph. Thus the neighborhood task is to map L to $N(L)$.

4 Results

All results presented here are results of single simulations. A more complete statistical analysis is anticipated in the near future. Results are depicted as an array of grids in which the location of the grid in the array represents he position of the referent of the input node in the graph. The grid is the corresponding output. That is, the lower left grid is the output generated by the input unit for the lower left corner.

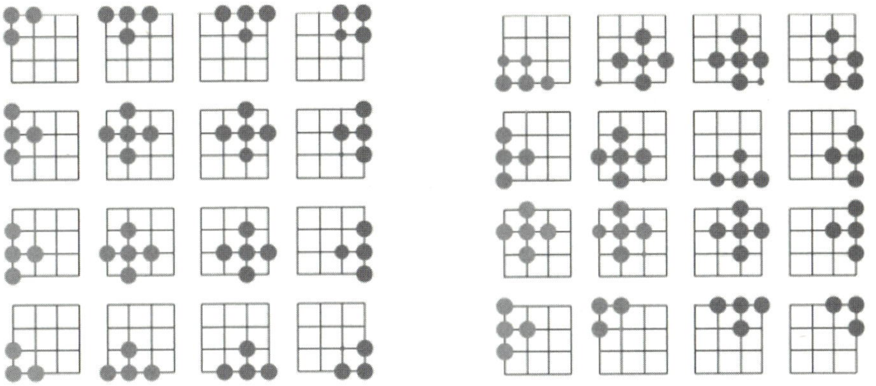

Fig. 2. Both tasks are neighborhood tasks on a 4-4 grid. Left: A 4x4 array of 4x4 grids. Each grid in the array represents the output grid during test mode of a single input. The position in the array indicates the grid position of the input. Right: The output from T_2 generated by a T_1 input (cross-testing).

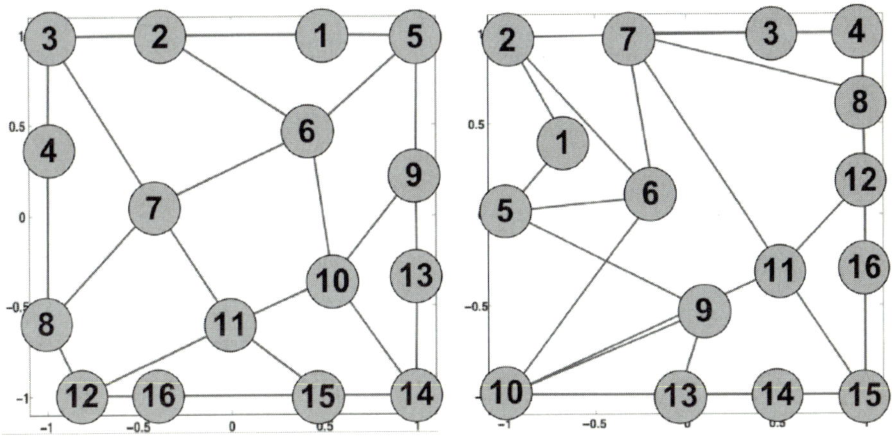

Fig. 3. H_1 representations. Left: H_1 representations of T_1. The lines indicate neighboring points from the task. For example the neighbors of 6 are 2, 5, 7, 10. Right: H_1 representations of T_2. Note that nodes 16 and 13 are in the same positions in the two plots. These points correspond to the lower right and uper right corners respectively in Fig 2.

4.1 No Crosstraining – The "Vertical" Condition

In all cases with sufficient resources (number of hidden units) and sufficient time, networks learned the separate tasks with low error. The left side of Fig. 2 depicts the network performance on T1 as an array of grids. Note each output grid shows the neighborhood of corresponding input. Visual inspection of the left array confirms that the network learns his task well (note the small error in the upper right grid). The network performs as well on T_2 (not shown). The right side of Fig 2 shows the T_2 output for a T_1 input. This example of cross-testing shows a mapping from the T_1 input space to T_2

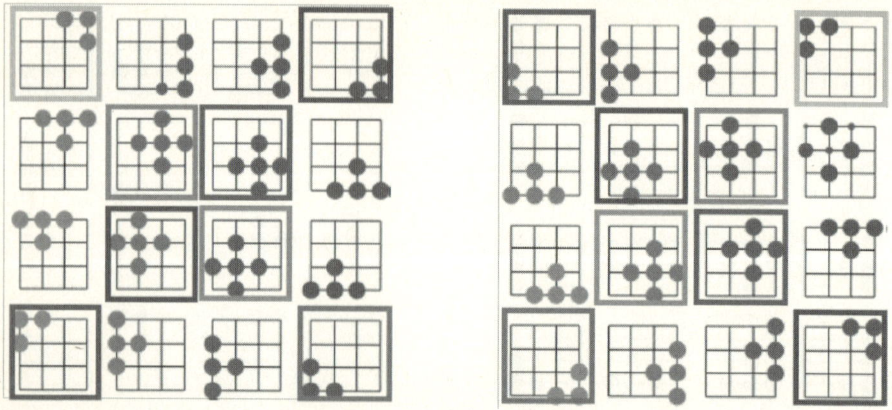

Fig. 4. The results of cross-training. Cross-training examples are indicated by boxes (corners and middle). The mapping was a 90 degree rotation (T_1-T_2 clockwise). Note hat the cross-trained items show low error, and he non cross-trained items are also correct for the most part.

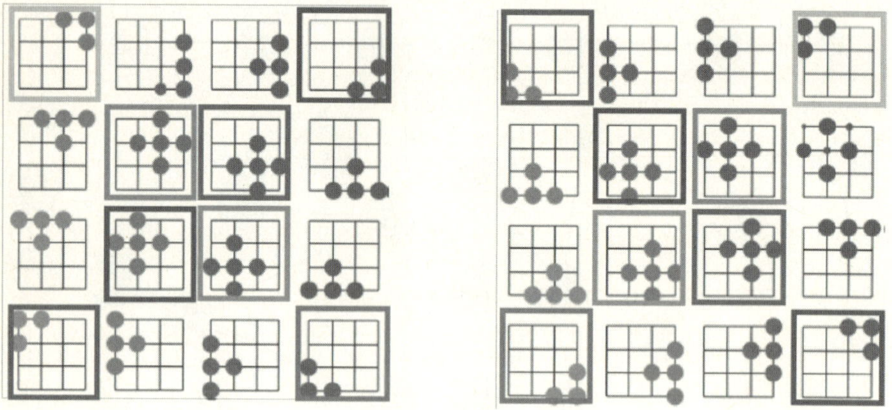

Fig. 5. The neigborhood task on grids of different sizes. All items in the 3x3 environment (left) are mapped to the corresponding corner nodes, edge center nodes, and central node of the 5x5 grid (right). Note that the cross-trained grids perform very well, and that the non-crosstrined items seem to map to the 3x3 output in a relatively consistent fasion.

that is roughly a vertical reflection. That is, the T_2 output for the T_1 lower right corner shows the neighborhood of the upper right corner. Layer H_1 has only two hidden units, which enables depiction of the hidden unit representations (Fig 3).

4.2 Assigning Correspondences with Cross-Training

In several simulations, cross-training was used to impose correspondences between the input spaces of the two tasks. Eight of the sixteen input patterns were mapped to points that were rotated 90 degrees. Cross-testing (Fig. 4) shows an example for which the training was successful (boxes indicate cross-trained items). Note that the items that were not cross-trained are, in general, consistent with the 90 degree rotation suggested by the cross-trained items. Other cases using fewer cross-trained items (for example, corners only) did not give such elegant results.

4.3 Correspondences between Non-identical Spaces

Since the tasks use dedicated input and output banks, they are not constrained to use the same input or output spaces. In Fig 5, the results of such an experiment are shown.

5 Discussion

The ability to establish correspondences between similar situations is fundamental to intelligent behavior. Here, a network has been introduced that can identify corresponding items in analogous spaces. A key feature of this approach is that here is no need for the tasks to use a common representation. Essentially the first hidden layer provides a common representational scheme for all the input spaces.

Simulations have been a bit uneven with respect to the tendency for items that are not cross-trained to generalize well. The factors that play a role in determining good generalization have not yet been identified. But candidates include the dimensionality of the hidden unit space, noise in the hidden units, and he learning rate.

It should be noted that the partially shared network architecture used here is virtually identical to the network used in Hinton's classic "family trees" example [10]. The network in that paper also had independent inputs and shared hidden units, but only briefly addresses the notion of generalization.

A possible application of this approach is for *ontology mapping*. Here, the tasks would be to learn the structure of various ontologies using a task like the neighborhood task. Explicit cross-training could be used for specific "known" correspondences.

References

1. Thorndike, E.L., Woodworth, R.S.: The influence of improvement in one mental function upon the efficiency of other functions. Psychological Review 8, 247–261 (1901)
2. Gentner, D.: Structure-mapping: A theoretical framework for analogy. Cognitive Science 7, 155–170 (1983)
3. Holyoak, K., Thagard, P.: Analogical mapping by constrant satisfaction. Cognitive Science 13, 295–355 (1989)
4. Halford, G., Wilson, W., Guo, J., Gayler, R., Wiles, J., Stewart, J.: Connectionist implications for processing capacity limitations in analogies. In: Holyoak, K., Barnden, J. (eds.) Advances in connectionist and neural computation theory. Analogical Connections, vol. 2, pp. 363–415. Ablex, Norwood (1994)
5. Hummel, J., Holyoak, K.: Distributed representations of structure: A theory of analogical access and mapping. Psychological Review 104, 427–466 (1997)
6. Mitchell, M.: Analogy-making as Perception: A computer model. MIT Press, Cambridge (1993)
7. Pratt, L.Y., Mostow, J., Kamm, C.A.: Direct transfer of learned information among neural networks. In: Proceedings of the Ninth National Conference on Artificial Intelligence (AAAI 1991), Anaheim CA (1991)

8. Caruana, R.: Multitask learning. Machine Learning 28, 41–75 (1997)
9. Ghiselli-Crippa, T., Munro, P.: Emergence of global structure from local associations. In: Cowan, J.D., Tesauro, G., Alspector, J. (eds.) Advances in Neural Information Processing Systems 6. Morgan Kaufmann Publishers, San Mateo (1994)
10. Hinton, G.: Learning distributed representations of concepts. In: Proceedings of the Eighth Annual Conference of the Cognitive Science Society, pp. 1–12. Amherst, Lawrence Erlbaum, Hillsdale (1986)

Auto-structure of Presynaptic Activity Defines Postsynaptic Firing Statistics and Can Modulate STDP-Based Structure Formation and Learning

Gordon Pipa[1,2,3,4,⋆], Raul Vicente[1,2], and Alexander Tikhonov[1]

[1] Max-Planck Institute for Brain Research, Department Neurophysiology, Deutschordenstr. 46, 60528 Frankfurt am Main, Germany
[2] Frankfurt Institute for Advanced Studies (FIAS), Ruth-Moufang-Str. 1, 60438 Frankfurt am Main, Germany
[3] Massachusetts Institute of Technology (MIT), Department of Brain and Cognitive Sciences, 77 Massachusetts Ave., Cambridge, MA 02139-4307, USA
[4] Massachusetts General Hospital, Dept. of Anesthesia and Critical Care, 55 Fruit Street, Boston, MA 02114, USA
{pipa,vicente,tikhonov}@mpih-frankfurt.mpg.de
http://www.mpih-frankfurt.mpg.de

Abstract. In this paper we study the impact of presynaptic activity that is deviating from Poissonian firing onto the postsynaptic firing of a conductance based integrate and fire neuron. We first show that the compound activity of a large group of neurons, e.g. presynaptic cells, cannot be described by a Poisson process in general. Then we demonstrate that the auto-structure of the presynaptic drive has strong impact onto the auto-structure of the postsynaptic spike-train. And finally, we discuss the potential impact of non-Poissonian presynaptic activity on the structure formation in recurrent networks based on Spike Timing Dependent Plasticity (STDP).

Keywords: STDP, self-organization, non-Poissonian spike activity, recurrent networks, *Gamma*-process, *Log-normal*-process.

1 Introduction

Neuronal self-organization and structure formation in recurrent networks have been proposed to be crucial elements in shaping the way the brain processes information [1,2,3,4,7]. However, most theoretical and simulation based approaches that investigate neuronal self-organization tacitly use the assumption that spiking activity can be modeled by Poisson point processes. Since it is now undoubted that real neuronal activity is often strongly deviating from Poisson processes these theoretical and simulation based approaches might lack a major component of real neuronal firing.

⋆ Corresponding author.

V. Kůrková et al. (Eds.): ICANN 2008, Part II, LNCS 5164, pp. 413–422, 2008.

2 Presynaptic Activity Deviating from Poisson

A typical cortical neuron receives input from many, up to more than 10^4, other neurons. Taking the presynaptic neurons as independent it was believed that the compound process of the activity of many input spike-trains can be described as a Poisson process [5,6]. In this case the auto-structure of individual pre-synaptic neurons could be ignored and a Poissonian statistics would have been a good approximation.

However, it has been shown analytically that this belief is wrong [8]. Lindner demonstrated that only the inter-spike interval (ISI) distribution and the ISI correlation at a finite time-lag approaches the ones of a Poisson process in the limit of large number of presynaptic spike-trains. Nevertheless, the power-spectrum of the compound process is identical to the power spectrum of the original spike-trains. This clearly demonstrates that presynaptic activity that is individually non-Poissonian will lead to a also non-Poissonian compound activity. For example, individual either regular or oscillatory presynaptic activity will lead to the same type of regular or oscillatory spiking in the compound process at the same time scale. This occurs even if many thousands spike-trains are added up. Hence, the impact of any presynaptic activity that is for example either regular or oscillatory is falsely ignored by present studies that model the compound activity as Poisson processes.

3 Impact of Non-poissonian Firing on Synchronous Drive

Neurons are very sensitive to the synchronous firing of sub-populations, especially if they are in a state of balanced excitation and inhibition. Under this

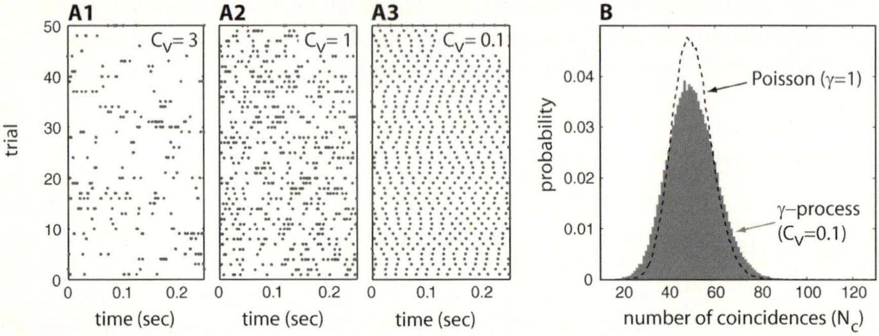

Fig. 1. (**A1-3**) Raster plot for 50 trials of *Gamma*-processes with 3 different coefficient of variations of the inter-spike interval distribution ($A1$: C_V=0.1; $A2$: C_V=1, $A3$: C_V=3). (**B**) Distribution of coincidence counts shared by pairs of mutually independent *Gamma*-processes of the same kind as shown in A. Dashed curves in (**B**) shows the coincidence count distribution for the cases of Poisson processes ($\gamma = 1$) and a *Gamma*-process with C_V=0.1. Coincidences were evaluated per trial based on a binned version (bin-length: $\Delta t = 4$ ms) of the original renewal process. Each distribution in B represents sample from $T = 100000$ trials of each 5 s length. The spike rate was chosen to be $R = 50$ ap/s.

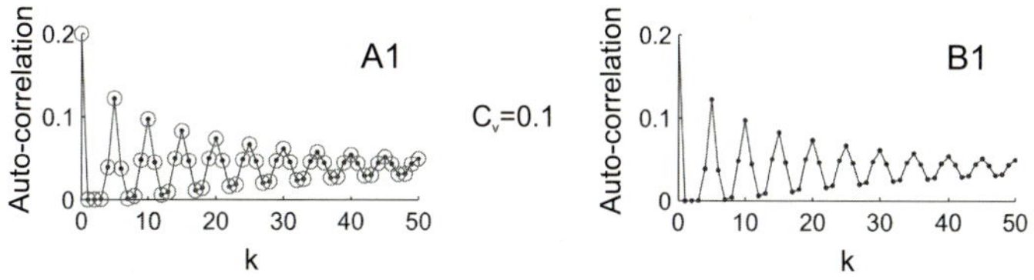

Fig. 2. Auto-correlation of the coincidences of *Gamma*-point processes (**A**), and a *log-normal*-point processes (**B**) with a C_V of 0.1 of the inter-spike interval distribution. A bin-length of $\Delta t = 4$ ms was used. The Auto-correlation is computed for time-lags τ with $\tau = k \cdot \Delta t$). The spike rate were chosen to $R = 50$ ap/s. Solid line indicate results derived from simulations, while circles locate an analytical evaluation.

regime the membrane potential of a neuron is typically fluctuating around its spiking threshold and a small number of coincident input can be particularly effective in driving the cell above threshold. It is therefore important for to determine the parameters affecting the synchrony of presynaptic spike-trains. Only recently the influence of the auto-structure of renewal processes on the likelihood of synchronous firing between mutually independent processes was studied in Ref.[9]. The authors described the influence of the auto-structure of different mutually independent renewal processes (Fig. 1, panels A1-A3) on the shape of the coincidence count distribution (Fig. 1, panel B) and on the auto-correlation of coincidences (Fig. 2) for individual and independent trials. To this end the authors described the shape of the distribution by the Fano factor FF_C of the coincidence distribution. Please note that the Fano factor FF_C is equal to a scaled variance in case of a renewal processes, since the expected frequency of coincidences is independent of the type of the process [9].

Neuronal activity is continuous stream in time. Therefore the concept of individual and independent trials might look inappropriate. However, if the auto-correlation time of the spiking activity is short compared to the length of trials, one can understand trials as subsequent and independent pieces taken form the same continuous stream of neuronal activity. In this case, the Fano factor expresses the variability of counts of coincident events for different periods of the neuronal activity. In other words, the Fano factor FF_c describes the degree of clustering of coincidences in time. Please note that since the expected value of coincidences is independent of the type of the renewal process, different auto-structures lead only to a redistribution of the same number coincidences and therefore to different clustering in time.

4 Clustering of Synchronous Activity in Time

We have studied then the Fano factor FF_C as a measure of clustering of synchronous activity for renewal processes with *Gamma*- and *log-normal*-distributed interspike intervals. The Fano factor of the coincidence count distribution for pairs of processes with identical spike-rates is represented in Fig. 3A for several coefficients of variations (C_V). High values of FF_C indicate strong clustering. To

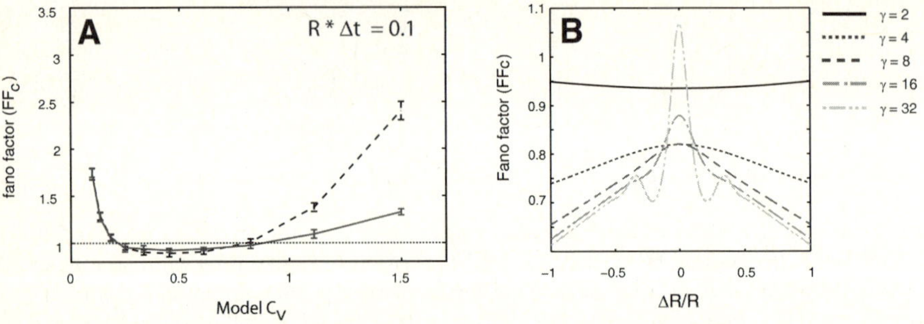

Fig. 3. (**A**) Fano factor of the coincidence count distribution as a function of the model C_V, for a *Gamma*-process and a *log-normal*-process. Coincidences were evaluated based on a binned version (bin-length: $\Delta t = 1$ ms) of the original renewal process. Both process types (*Gamma*- and *log-normal*) were parameterized by the product of the bin-length Δt in units of seconds and the firing rate R in units of [ap/s] (see legend). Each estimation of the C_V or Fano factor was based on $T = 2000$ trials of each 5 s length. The spike rate were chosen to $R = 50$ ap/s. (**B**) Analytically determined Fano Factor FF_C of the coincidence count distribution for two neurons with different rate $R_1 = R + \Delta R$ and $R_2 = R - \Delta R$ as function of $\Delta R/R$. *Gamma*-processes with shape parameter $\gamma = 2, 4, 8, 16, 32$ are explored. $R\Delta t = 0.2$.

describe the changes of FF_c we will use the Poisson process as a reference model since the Poisson process was commonly used in modeling studies. We have observed that only intermediate regularity in the processes ($0.2 < C_V < 0.8$) lead to smaller values of FF_c and consequently to less clustering than in the case of a Poisson process. For very low and very high C_Vs (corresponding to very regular and bursty spiking, respectively) the Fano factor can exceed the FF_C of a Poisson process by a factor larger than 2. However, if the two independent processes have different rates (Fig. 3B), the FF_c and therefore the clustering of synchrony for *Gamma*-processes is in general lower than in case of a Poisson process that exhibits a $FF_c = 1.4$ for the given parameters.

In summary, our first major finding is that deviations from individual Poissonian firing can induce clustering of coincidences between mutually independent point-processes in time. The amount of clustering depends on the detailed properties of the inter-spike interval distribution, and can be described by the Fano Factor FF_c of the coincidence count distribution. Our second major finding is that the clustering depends very critically on whether the rate across different point-processes it the same or not. We found that small differences of the the spike-rates of two processes in the order of a few percent can cause changes in the clustering measured by changes of FF_c of up to 50% in some cases (Fig. 3B). The third major finding is that both, bursty and regular firing, can induce clustering of coincidences. However, the underlying mechanism and cause of the clustering for bursty and regular processes is different. In the case of bursty processes clustering is due to a repeated firing of individual neurons in a period as short as the bin-width Δt that defines the time-scale of a coincidence. In contrast, in case of regular firing clustering is caused by a kind of periodic repetition of the same pattern on a time-scale defined by the expected inter-spike interval.

5 IF Neurons and Presynaptic *Gamma*-Processes

A prior step in demonstrating that deviations from Poissonian firing can modulate the structure formation in recurrent networks based on neuronal plasticity like STDP, is to show first that different variations of non-Poissonian presynaptic activity impact the postsynaptic spiking activity of a neuron.

To this end we simulated a conductance based Integrate and Fire neuron that receives input from one excitatory and one inhibitory population, with $N_E = 400$ and $N_I = 100$ cells, respectively. Exponentially decaying synaptic currents with time constants $\tau_{ampa} = 2$ ms and $\tau_{gaba} = 5.6$ ms were used respectively for the AMPA and GABA mediated receptors. The membrane time constant was set to $\tau_m = 20$ ms. The synaptic conductance strength g_{gaba} and g_{ampa} were chosen to be identical across all synapses of the same type. We model the postsynaptic neuron to be in a state close to balanced excitation and inhibition. To this end we defined g_{ampa} and g_{gaba} based on the balance between excitation and inhibition β as well as the total conductance in units of the leakage conductance g_{leak}. Motivated by studies that measured this relation in vivo we have chosen the total conductance to be four times higher than the leakage.

Panels A in Figure 4 shows the postsynaptic spike triggered average of the presynaptic spike activity, which describes the average presynaptic activity of the excitatory population (upper trace) and the inhibitory average activity (downer trace) in a window center around a postsynaptic spike. For the two point processes studied (Poissonian in A1 and *Gamma*-distributed in A2), on average an increase in the excitatory and/or a decrease in the inhibitory activity are necessary to drive the postsynaptic neuron to spike. Please note that both are by definition random fluctuation since the presynaptic activity has a constant expected spike count and is mutually independent across neurons. Remarkably, the the presynaptic spike triggered average for regular *Gamma*-processes (Fig. 4, A2) shows a damped periodic oscillation with the expected interspike interval of the presynaptic activity. This demonstrates that random fluctuations leading to increases of the excitatory or decreases of the inhibitory drive have the same auto-correlation structure as each of the original presynaptic point-processes.

This notable feature relies on the thresholding properties of the postsynaptic neuron and can be intuitively understood as follows. Since the neuron is close to a state of balanced excitation/inhibition, synchronous activity from a rather small subpopulation of neurons is sufficient to make the neuron firing. For simplicity we assume for now that all presynaptic processes of such driving subpopulation are identical. We know from the work of Linder [8] that the power-spectrum of the compound process is the same as for each individual neuron, and consequently we know that a subpopulation that is synchronous by chance

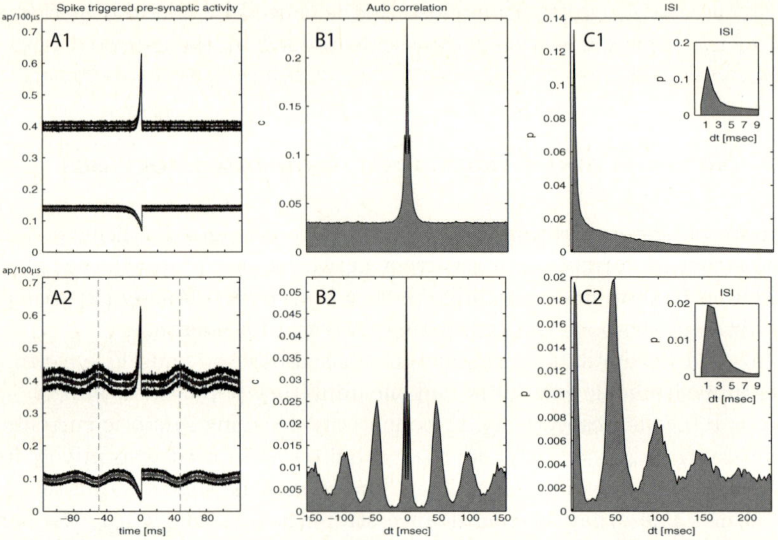

Fig. 4. (**A1,2**) Postsynaptic spike triggered average of the presynaptic activity in units of number of spikes per 100 μs of the whole presynaptic population of 400 excitatory (upper trace) and 100 inhibitory (downer trace) cells. (**B1,2**) Auto-correlation function of the postsynaptic spiking activity (bin-width Δt=1 ms). (**C1,2**) Inter-spike interval distribution of the postsynaptic spiking activity (bin-width Δt=2 ms). The indices 1 and 2 in the panels distinguishes between the individual Poissonian and *Gamma*-distributed presynaptic point process cases. The average postsynaptic spike rate was, 15.2 ap/s in the Poissonian case, and 3.36 ap/sec in the *Gamma* process. In both cases the presynaptic spike rate of the excitatory population was 28 ap/sec per synapse. The presynaptic spike rate of the inhibitory population was 20 ap/sec per synapse.

will have a cross-correlation with the same periodicity as the auto-correlation as each of the individual neurons. Hence, a chance synchronization of a subpopulation will lead to a package of postsynaptic spikes that follow on average the same auto-structure as each individual presynaptic process. This is illustrated in the auto-correlation function of the postsynaptic point-process shown in panel B2 of Figure 4. Interestingly, the ISI distribution shows multiple peaks. This is because the average postsynaptic spike rate is much lower than the compound presynaptic activity. Hence, the postsynaptic cell does not fire at the timing corresponding at each peak of the auto-correlation of the presynaptic drive. Such effect can be understood as a modulation of the firing probability of a stochastic point-process. Therefore, the postsynaptic neurons might do cycle skipping what induces presynaptic and postsynaptic spike-rates to be different.

6 Non-poissonian Input and Structure Formation in Recurrent Networks

Once we have evaluated the role of non-Poissonian presynaptic input in the firing properties of a postsynaptic neuron we are now in position to discuss the potential impact of non-Poissonian presynaptic activity on structure formation and learning via STDP. This form of plasticity has been applied to sequence learning and has been discussed to be involved in spontaneous and activity driven pattern formation [10,1,4]. STDP strengthens potentially causal relations between presynaptic drive and postsynaptic activity by increasing the synaptic strength of all synapses that have been activated immediately before a postsynaptic spike is generated. Since the clustering of coincidences in the presynaptic activity has been shown to impact postsynaptic events, it is also expected to modulate STDP-based pattern formation. Specifically, bursty presynaptic activity leads to clusters of presynaptic coincident events on a very short timescale and just a single postsynaptic potential can be sufficient to modulate the strength of neuronal plasticity.

In case of regular presynaptic activity the situation is different. In case of regular presynaptic activity the situation is different, such that regular presynaptic activity leads to periods postsynaptic activity with the same regularity as the presynaptic drive. In this scenario the temporal relation of presynaptic and postsynaptic activity needs to be maintained for a duration of the order of a few expected interspike intervals of the presynaptic activity to exert a significant influence. Exactly this maintenance of the temporal order between presynaptic drive and the postsynaptic activity has been observed in our simulation of an IF neuron subjected to structured presynaptic input. Hence, the clustering of coincident presynaptic activity, described by the Fano factor FF_C, is a critical parameter in studying the impact of different point-processes onto STDP.

7 Clustering and Repetition of Presynaptic Synchronous Spike Pattern in Dependence of the Fano Factor

Given our results concerning the temporal clustering of coincidences described by FF_C we can already distinguish between two scenarios. First, let us start by the high variance case where FF_C is large, and coincidences are more clustered than in case of Poissonian firing. This case occurs for very regular and bursty auto-structures (see Fig. 3A), and only in the case that the rates of the all spikes trains are identical or have a n:m relationship. In the second scenario FF_C is lower, and coincidences are less clustered than in case of Poissonian firing. This second scenario occurs predominately for non-rational rate relations of the processes and intermediate regularity.

Whether sets of individual neurons are either in the low FF_C, or high FF_C state, can have important implications on the neuronal dynamics and could be used by the neural system to modulate the clustering of synchronous neuronal activity that is occurring by chance. For example, the high FF_C regime seems to be well suited for spontaneous and activity driven structure formation due to neuronal plasticity, like STDP [10,1,4,11]. Spontaneous pattern formation requires a spontaneous symmetry breaking in the activity driving the network such that a randomly occurring pattern is strengthened and embedded as a reliable reoccurring sequence. A regime where the frequency of coincident neuronal activity is highly clustered in time is especially suited for such a spontaneous pattern formation since a pattern that occurs once by chance have a higher likelihood to be repeated in a short temporal window than in case of Poissonian activity. This repetition of individual chance patterns could boost the effect of STDP for short periods in time since the same patterns occur more than once. However, there is a critical period length that STDP requires to reinforce a given pattern. If STDP requires many reoccurrences of the same pattern for reinforcement the period length will grow larger than, first a couple of interspike intervals, and second the auto-correlation time of the process. For periods that are long, the advantage of the high variability regime vanishes since periods with high numbers of coincidences will be followed by periods with low numbers of coincidences such that the variability of the coincide count distribution is reduced.

More relevant is the case where individual neurons have different spike rates. In that case changing the auto-structure of the individual processes can also regulate STDP processes. Our findings based on *Gamma*-processes with rates that are not related to each other as n:m (with $n, m \in \mathbb{N}$) demonstrate that Poissonian firing yields the highest FF_C and therefore a higher clustering of coincidences than regular and bursty *Gamma*-processes. Hence, a Poissonian profile of spikes boost the effect of STDP compared to other statistics of spike-trains as long as the spike rates from individual neurons are different and does not express a n:m relationship.

8 Conclusions

We have described how auto-structure of presynaptic activity results in general in a compound activity on non-Poissonian statistics. We have also concluded that the driving of a postsynaptic neuron by spike-trains deviating from a Poissonian profile influences the statistics of firing the postsynaptic cell in a well defined relationship with respect to the presynaptic characteristics. Finally, and assuming that a random pattern occurring between mutually independent processes can be seen as a seed for the process of spontaneous pattern formation, we have discussed how changes of the auto-structure, such as regularity of presynaptic spike-trains, can modulate and boost the efficiency of STDP for short epochs of the length of a few interspike intervals. Consequently, the regularity and burstiness of neuronal spike-train can potentially act as a gating or boosting mechanism for the ability of spontaneous symmetry breaking and pattern formation.

Acknowledgements

The authors would like to thank Carl van Vreeswijk, and Larry Abbott for inspiring discussions and great support during the first stages of this project. This work was partially supported by the Hertie Foundation and the European Commission Project (FP6-NEST contract 043309).

References

1. Song, S., Abbott, L.F.: Cortical development and remapping through spike timing-dependent plasticity. Neuron 32(2), 339–350 (2001)
2. Izhikevich, E.M., Gally, J.A., Edelman, G.M.: Spike-timing Dynamics of Neuronal Groups. Cereb. Cortex 14, 933–944 (2004)
3. Xie, X., Seung, H.S.: Learning in neural networks by reinforcement of irregular spiking. Phys. Rev. E 69, 041909–041919 (2004)
4. Lazar, A., Pipa, G., Triesch, J.: Fading memory and time series prediction in recurrent networks with different forms of plasticity. Neural Netw. 20, 312–322 (2007)
5. Shimokawa, T., Rogel, A., Pakdaman, K., Sato, S.: Stochastic resonance and spike-timing precision in an ensemble of leaky integrate and fire neuron models. Phys. Rev. E 59, 3461–3470 (1999)
6. Hohn, N., Burkitt, A.N.: Shot noise in the leaky integrate-and-fire neuron. Phys. Rev. E 63, 031902 (2001)
7. Legenstein, R., Näger, C., Maass, W.: What can a neuron learn with spike-timing-dependent plasticity? Neural Comput. 17(11), 2337–2382 (2005)
8. Lindner, B.: Superposition of many independent spike trains is generally not a Poisson process. Phys. Rev. E 73, 022901 (2006)
9. Pipa, G., Vreeswijk, C., Grün, S.: Impact of spike-train autostructure on probability distribution of joint-spike events (submitted)
10. Markram, H., Lübke, J., Frotscher, M., Sakmann, B.: Regulation of synaptic efficacy by coincidence of postsynaptic APs and EPSPs. Science 275, 213–215 (1997)
11. Morrison, A., Aertsen, A., Diesmann, M.: Spike-timing-dependent plasticity in balanced random networks. Neural Comput. 19, 1437–1467 (2007)

Appendix

A: *Gamma*-Process and *log-normal*-Process

The ISI distribution for a *Gamma*-process with a given constant spike rate is described by:

$$p_\gamma(t) = t^{\gamma-1}\frac{(\gamma R)^\gamma \ e^{-\gamma R t}}{(\gamma-1)!} \qquad \text{for } t > 0 \quad . \tag{1}$$

This distribution is characterized by two variables, the spike rate, R, and the shape-parameter, γ. The *Poisson*-process is a special case of the *Gamma*-process for which $\gamma = 1$. Values of the shape-parameter γ smaller than 1 make short intervals more likely to occur than for a *Poisson*-process, and are used to model bursty firing, while large values of the shape parameter are used to model regular firing. We use the C_V of the ISI distribution to characterize the processes. The the shape-parameter, γ and the C_V value are related as follows:

$$\gamma = \frac{1}{C_V^2} \quad . \tag{2}$$

As a second kind of renewal process characterized by two variables we use the *log-normal*-process.ts interspike interval distribution $(p(t)_{log-normal})$ is defined by:

$$p(t)_{log-normal} = \frac{1}{k\sqrt{2\pi}} \frac{\exp\left(-\frac{(\ln(t)-a)^2}{2k^2}\right)}{t} \quad . \tag{3}$$

The spike rate, R, and coefficient of variation, C_V, can be expressed by a and k as follows:

$$a = -\ln R - \ln\left(C_V^2 + 1\right) \tag{4}$$

and

$$k = \sqrt{\ln\left(C_V^2 + 1\right)} \quad . \tag{5}$$

B: Coincidences

Given two parallel spike-train processes we define a coincidence based on binned versions of the original processes. The binned spike trains are obtained by segmenting the time axis into exclusive bins, each of length Δt, and counting the number of spikes per bin k. The number of coincidences in bin k shared by two spike trains for two simultaneous bins and of neuron 1 and neuron 2 is then defined by $N_c^k = n_k^1 * n_k^2$. This definition can be trivially extended to more than pairs.

Decision Making Logic of Visual Brain

Andrzej W. Przybyszewski[1,2]

[1] Dept of Neurology, University of Massachusetts Medical Center, Worcester, MA US
[2] Dept pf Psychology McGill University, Montreal, Canada
przy@ego.psych.mcgill.ca

Abstract. A popular view is that the brain is making fast decisions in temporal and frontal cortices predominantly on the bases of the feed-forward pathways (FF). In later stages iterations (reverberations) with delayed feedback connections (FB) may be helpful. We propose, an opposite concept, that decisions are made in single neurons from the retina to the cortex, and that FB is fast as FF, and from the beginning participates in making decisions. The main differences between FF and FB are their different logics: FF follows driver logical rules, but FB follows modulator logical rules. Driver logical rules are gathering all possible information together therefore and they are context dependent, FB pathways, however, using selective modulator logical rules extract only hypothetically important information. We say that FF FB interaction is prediction hypothesis testing system. Our psychophysical system is different than Turing Machine because we are often insensible to changes of some symbols but same symbols in different configuration may lead to different classification of the same object. In present work we are looking for the anatomical and neurophysiological basis of these perceptual effects. We describe interactions between parts and their configurations on the basis of a single cell electrophysiological activity in cortical area V4. This area is related to simple shape classification. We have divided area V4 cell responses into three categories and found equivalent classes of object attributes for each cell response category. On this basis, we found decision rules for different area V4 cells (rough set theory - Pawlak, 1992 [1]). Some of these rules are not consistent, which may suggest that the brain may use different, non-consistent strategies in parallel in order to classify significant attributes of the unknown object.

Keywords: Imprecise computation, bottom-up, top-down processes, neuronal activity.

1 Introduction

Primates outperform any AI system in such difficult tasks as complex objects (like faces) recognition even if they have never seen them in a particular context before. An important feature of primates brain is insensitive to the exact properties of an objects parts, as well as recognition based on only partial and variable information. Therefore, because objects attributes are not sharp and may have different unknown ahead meanings, to model them we are using rough set theory

V. Kůrková et al. (Eds.): ICANN 2008, Part II, LNCS 5164, pp. 423–432, 2008.

[1]. In this work, we have explored area V4 cells responses stimulated by pair of bars or light patches placed in variable parts of the receptive field (RF). Our analysis lead us to propose decision rules related to the neurophysiological basis of the interactions between parts.

Like Pawlak [1], we define an information system as $S = (U, A)$, where U is a set of objects and A is set of attributes. If $a \in A$, $u \in U$ the value $a(u)$ is a unique element of V_a (a value set of the attribute a). In agreement with the Leibniz principle we assume that objects are completely determined by their set of properties, meaning that attributes map objects to a set of attributes values. The indiscernibility relation of any subset B of A, or $IND(B)$, is defined [1] as the equivalence relation whose elements are the sets $u : b(u) = v$ as v varies in V_b, and $[u]_B$ - the equivalence class of u form *B-elementary granule*. The concept $X \subseteq U$ is *B-definable* if for each $u \in U$ either $[u]_B \subseteq X$ or $[u]_B \subseteq U - X$. $\underline{B} X = \{u \in U : [u]_B \subseteq X\}$ is a lower approximation of X. The concept $X \subseteq U$ is *B-indefinable* if exists such $u \in U$ that $[u]_B \cap X \neq \emptyset$. $\bar{B}X = \{u \in U : [u]_B \cap X \neq \emptyset\}$ is an upper approximation of X. The set $BN_B(X) = \bar{B}X - \underline{B}X$ will be referred to as the *B-boundary region* of X. If the boundary region of X is the empty set then X is exact (crisp) with respect to B; otherwise if $BN_B(X) \neq \emptyset$ X is not exact (rough) with respect to B. In this paper, the universe U is a set of simple visual patterns that we used in our experiments [2], which can be divided into equivalent indiscernibility classes related to their physically measured, computer generated attributes or *B-elementary granules*, where $B \subseteq A$.

The purpose of our research is to find how these objects are classified in the brain. W will therefore modify the definition of the information system as $S = (U, C, D)$ where C and D are condition and decision attributes respectively. Decision attributes will classify elementary granules in accordance with neurological responses from a specific area of the visual brain. From a cognitive perspective, the percept of the object is classified into different categories in different visual areas, leading to different decisions (actions). In this work we are looking into single cell responses only in one area - V4 that will divide all patterns into equivalent indiscernibility classes of *V4-elementary granules*. Neurons in V4 are sensitive only to the certain attributes of the stimulus, like for example space localization pattern must be in the receptive field, and most of them are insensitive to contrast changes. Different V4 cells have different receptive field properties, which means that one *B-elementary granule* can be classified in many ways by different *V4-elementary granules*.

2 Method

We will represent our experimental data [2]. in the following table. In the first column are neural measurements. Neurons are identified using numbers related to a collection of figures in [2]. concatenated with the cell number. Different measurements of the same cell are denoted by additional letters $(a, b,)$. For example, $11a$ denotes the first measurement of a neuron numbered 1 Fig. 1, 11b the second measurement, etc. Simple stimuli properties are as characterized as follows: Most of our analysis will be related to data from Pollen et al. [2].

1. orientation in degrees appears in the column labeled o, and orientation bandwidth is sob.
2. spatial frequency is denoted as sf, spatial frequency bandwidth is sfb
3. x-$axis$ position is denoted by xp and the range of x-positions is xpr
4. y-$axis$ position is denoted by yp and the range of y-positions is ypr
5. x-$axis$ stimulus size is denoted by xs
6. y-$axis$ stimulus size is denoted by ys
7. stimulus shape is denoted by s, values of s are following: for grating $s = 1$, for vertical bar $s = 2$, for horizontal bar $s = 3$, for disc $s = 4$, for annulus $s = 5$, for two stimuli $s = 22$ - two vertical bars, etc.

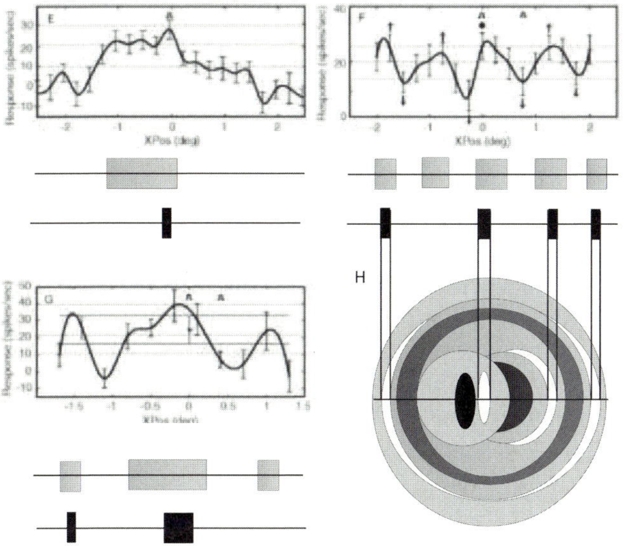

Fig. 1. Modified plots from [2]. Curves represent responses of two cells from area V4 to small single (E) and double (F, G) vertical bars. Bars change their position along x-axis (Xpos). Responses are measured in spikes/sec. Mean cell responses SE are marked in E, F, and G. Cell responses are divided into three ranges by thin horizontal lines. Below each plot are schematics showing bar positions giving r_1 (gray) and r_2 (black) responses; below (E) for a single bar, below (F and G) for double bars (one bar was always in position 0). (H) This schematic extends responses for horizontally placed bars (E) to the whole RF: white color shows excitatory related to r_2 responses, gray color is related to r_1 responses and black color inhibitory interactions between bars. Bars' interactions are asymmetric in the RF.

Decision attributes are divided into several classes determined by the strength of the neural responses. Small cell responses are classified as class 0, medium to strong responses are classified as classes 1 to n-1 (min(n)=2), and the strongest cell responses are classified as class n. Therefore each cell divides stimuli into its own family of equivalent objects Cell responses (r) are divided into n+1 ranges:

category 0 : activity below the threshold *10 - 20 sp/s* labeled by r_0;
category 1 : activity above the threshold labeled by r_1, ...
category n : activity above 30 - 40sp/s labeled by r_2.

Stimulus attributes can be express as: $B = \{o, ob, sf, sfb, xp, xpr, yp, ypr, xs, ys, s\}$.

3 Results

3.1 Analysis of the Interactions between Parts

We have used our model as a basis for an analysis of the experimental data from the neurons recorded in the monkey's area V4 [2]. One example of V4 cell responses to thin (0.25 deg) vertical bars in different horizontal - x positions is shown in the upper left part of Fig. 1 (Fig. 1E). Cell responses show a maximum for the middle (XPos = 0) bar position along the x-axis. Cell responses are not symmetrical around 0. In Fig. 1F the same cell (cell 61 in table 1) is tested with two bars. The first bar stays at the 0 position, while the second bar changes its position along the x-axis. Cell responses show several maxima dividing the receptive field into four areas. However, this is not always the case as responses to two bars in another cell (cell 62 in table 1) show only three maxima (Fig. 1G). Horizontal lines in plots of both figures divide cell responses into the three classes: r0, r1, r2, which are related to the response strength (see Methods). Stimuli attributes and cell responses divided into two: r1 and r2 classes are shown in table 1 for cells from Fig. 1 and in table 2 for cells from Fig. 2

We assign the narrow (xpr_n), medium (xpr_m), and wide (xpr_w) x position ranges as follows: xpr_n if ($xpr : 0 < xpr \leq 0.6$), medium xpr_m if ($xpr : 0.6 < xpr \leq 1.2$), wide xpr_w if ($xpr : xpr > 1.2$).
On the basis of Fig. 1 the decision table the two-bar horizontal interaction study can be presented as the following

Decision Rules of Two-bar (DRT):
DRT1:

$$o_{90} \wedge xpr_n \wedge (xp_{-1.9} \vee xp_{0.1} \vee xp_{1.5}) \wedge xs_{0.25} \wedge ys_4)_1 \wedge (o_{90} \wedge xp_0 \wedge xs_{0.25} \wedge ys_4)_0 \rightarrow r_2 \tag{1}$$

DRT2:

$$o_{90} \wedge xpr_m \wedge (xp_{-1.8} \vee xp_{-0.8} \vee xp_{0.4} \vee xp_{1.2}) \wedge xs_{0.25} \wedge ys_4)_1 \wedge$$

$$\wedge (o_{90} \wedge xp_0 \wedge xs_{0.25} \wedge ys_4)_0 \rightarrow r_1 \tag{2}$$

One-bar decision rules can be interpreted as follows: the narrow vertical bar evokes a strong response in the central positions, and wide size bars evoke medium responses in also near the central certain positions. Two-bar decision rules claim that: the cell responses to two bars are strong if one bar is in the middle of the RF (bar with index 0 in decision rules) and the second narrow bar (bar with index 1 in decision rules) is in the certain positions of the RF eq.

Table 1. Decision table for cells shown in Fig. 1. Attributes *o, ob, sf, sfb* were constant and are not presented in the table. In experiments where two stimuli were used, the shape value was following: for two bars s=22, for two discs s=44.

cell	*xp*	*xpr*	*xs*	*ys*	*s*	*r*
61e	-0.7	1.4	0.25	4	2	1
61f1	-1.9	0.2	0.25	4	22	2
61f2	0.1	0.2	0.25	4	22	2
61f3	1.5	0.1	0.25	4	22	2
61f4	-1.8	0.6	0.25	4	12	1
61f5	-0.4	0.8	0.25	4	22	1
61f6	0.4	0.8	0.2 5	4	22	1
61f7	1.2	0.8	0.25	4	22	1
62g1	-1.5	0.1	0.25	4	22	2
62g2	-0.15	0.5	0.25	4	22	2
62g3	-1.5	0.6	0.25	4	22	1
62g4	-0.25	1.3	0.25	4	22	1
62g5	1	0.6	0.25	4	22	1
63h1	-0.5	0	0.5	1	44	2
63h2	1	1	1	1	44	1
63h3	0.2	0.1	0.25	4	22	2

(1). But when the second bar has medium width the max cell responses became weaker eq. (2). Responses of other cells are sensitive to other bar positions (Fig. 1G). These differences could be correlated with anatomical variability of connections especially of the descending axons. As mentioned above V4 axons in V1 have distinct clusters or linear branches. Descending pathways are modulators and therefore their rules contain logical or which consequence is that not all excitatory areas become more active as a result of the feedback.

The decision table (Table 2) based on Fig. 2 describes cell responses to two patches placed in different positions along x-axis of the receptive field (RF). Figure 2 shows that adding the second patch reduced single patch cell responses. We have assumed that cell response to a single patch places in the middle of the RF is r_2. The second patch suppresses cell responses stronger when is more similar to the first patch (Fig. 2D)

Table 2. Decision table for one cell shown in Fig. 2. Attributes yp, ypr are constant and are not presented in the table. We introduce another parameter of the stimulus, difference in the direction of drifting grating of two patches: ddg = 0 when drifting are in the same directions, and ddg = 1 if drifting in two patches are in opposite directions.

cell	*xp*	*xpr*	*xs*	*ys*	*ddg*	*r*
64c	-4.5	3	1	1	1	2
64c1	-1.75	1.5	1	1	1	1
64c2	-0.5	1	1	1	1	2
64d	-6	0	1	8	0	2
64d1	-3.5	4.8	1	8	0	1

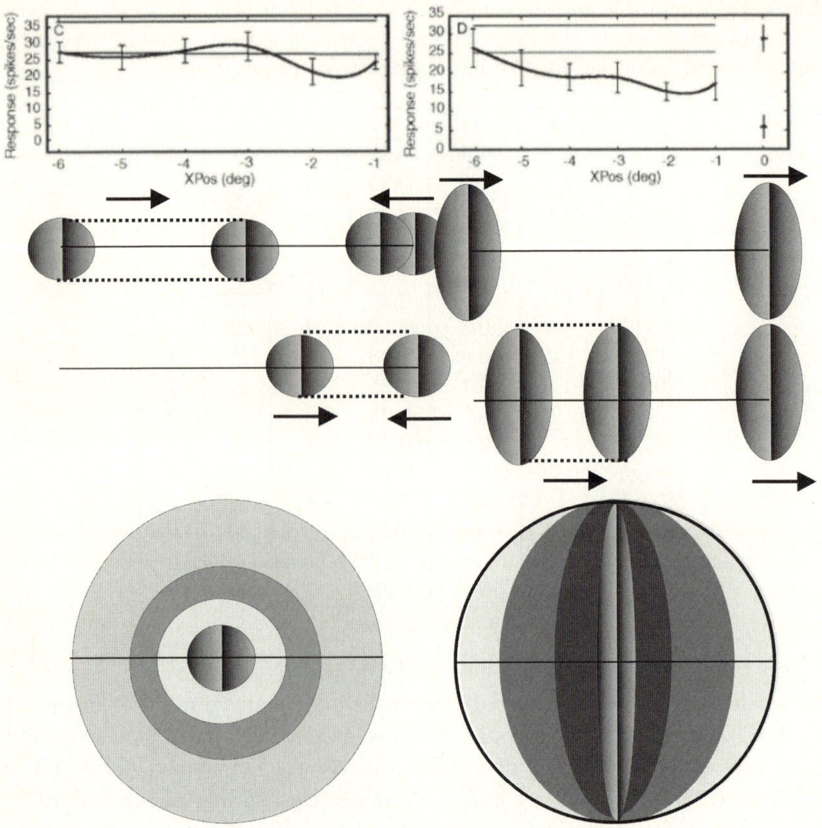

Fig. 2. Modified plots from [2]. Curves represent V4 cell responses to two patches with gratings moving in opposite direction - patch 1 deg diameter (C) and in the same (D) directions for patch 1 deg wide and 8 deg long. One patch is always at x-axis position 0 and the second patch changes its position as it is marked in *XPos* coordinates. The horizontal lines represent 95% confidence intervals for the response to single patch in position 0. Below C and D schematics showing positions of the patches and their influences on cell responses. Arrows are showing directions of moving gratings. Lower part of the figure shows two schematics of the excitatory (white) and inhibitory (black) interactions between patches in the RF. Patches with gratings moving in the same directions (right schematic) show larger inhibitory areas (more dark color) than patches moving in opposite directions (left schematic).

Two-patch horizontal interaction decision rules are as follows:

DRT3:

$$ddg_1 \wedge (o_0 \wedge xpr_3 \wedge xp_{4.5} \wedge xs_1 \wedge ys_1)_1 \wedge (o_0 \wedge xp_0 \wedge xs_1 \wedge ys_1)_0 \rightarrow r_2 \qquad (3)$$

DRT4:

$$ddg_1 \wedge (o_0 \wedge xpr_1 \wedge xp_{0.5} \wedge xs_1 \wedge ys_1)_1 \wedge (o_0 \wedge xp_0 \wedge xs_1 \wedge ys_1)_0 \rightarrow r_2 \qquad (4)$$

DRT5:

$$ddg_0 \wedge (o_0 \wedge xpr_{4.8} \wedge xp_{3.5} \wedge xs_1 \wedge ys_8)_1 \wedge (o_0 \wedge xp_0 \wedge xs_1 \wedge ys1)_0 \rightarrow r_1 \quad (5)$$

These decision rules can be interpreted as follows: patches with drifting in opposite directions gratings give strong responses when positioned very near (overlapping) or 150% of their width, apart one from the other eqs. (3, 4). Interaction of patches with a similar gratings evoked small responses in large extend of the RF eq. (5).

We propose following classes of the objects **Parts Interaction Rules:**

PIR1: facilitation when stimulus consists of multiple similar thin bars with small distances (about 0.5 deg) between them, and suppression when distance between bars is larger than 0.5 deg. Suppression/facilitation is very often a nonlinear function of the distance. In our experiments (Fig. 1) cells responses to two bars were periodic along the receptive field with dominating periods of about 30, 50, or 70% of the RF width. These nonlinear interactions were also observed along vertical and diagonals of the RF and often show strong asymmetries in relationship to the RF middle.

PIR2: strong inhibition when stimulus consists of multiple similar patches filled with gratings with distance between patch edges ranging from 0 deg (touching) to 2 deg, weak inhibition when distance is between 3 to 5 deg through the RF width.

PIR3: if bars or patches have different attributes like polarity or drifting directions than suppression is smaller and localized facilitation at the small distance between stimuli is present. As in bars interaction, suppression/facilitations between patches or bright/dark discs can be periodic along different RF axis and often asymmetric in the RF.

We have tested above rules in nine cells from area V4 by using discs or annuli filled stimuli with optimally oriented and variable in spatial frequencies drifting gratings (Pollen et al. [2] Figs. 9, 10). Our assumptions were that if it is a strong inhibitory mechanism as described in the rule PRI2 than responses to annulus with at least 2 deg inner diameter will be stronger than to disc. In addition by changing spatial frequencies of gratings inside the annulus, we have expected eventually to find other periodicities along the RF width as described by PIR3.

In summary, we wanted to find out what are relations between stimulus properties and area V4 cell responses or whether B-elementary granules have equivalence classes of the relation $IND\{r\}$ or V4-elementary granules, or whether $[u]_B \Rightarrow [u]_{B4}$. It was evident from the beginning that because different area V4 cells have different properties their responses to the same stimuli will be different, therefore we wanted to know if the rough set theory will help us in our data modeling. We can also ask what percentage of cells we fully classified. We have obtained consistent responses from 2 of 9 cells, which means that $\gamma = 0.22$. This is related to the fact that for some cells we have tested more than two stimuli. What is also important from an electrophysiological point of view is there are negative cases. There are many negative instances for the concept 0, which means that in many cases this brain area responds to our stimuli; however it seems that our concepts are still only roughly defined.

We have following decision rules:

DR_V4_7:

$$sf_l \wedge xo_7 \wedge xi_2 \wedge s_5 \rightarrow r_1 \tag{6}$$

DR_V4_8:

$$sf_l \wedge xo_7 \wedge xi_0 \wedge s_4 \rightarrow r_0 \tag{7}$$

DR_V4_9:

$$sf_l \wedge xo_8 \wedge xi_0 \wedge s_4 \rightarrow r_0 \tag{8}$$

DR_V4_10:

$$(sf_m \vee sf_h) \wedge xo_6 \wedge xi_2 \wedge s_5 \rightarrow r_2 \tag{9}$$

These can be interpreted as the statement that a large annulus (s_5) evokes a weak response, but a large disc (s_4) evokes no response when there is modulation with low spatial frequency gratings. However, somewhat smaller annulus containing medium or high spatial frequency objects evokes strong responses. It is unexpected that certain stimuli evoke inconsistent responses in different cells [3]:

103: $sf_l \wedge xo_6 \wedge xi_0 \wedge s_4 \rightarrow r_0$
106: $sf_l \wedge xo_6 \wedge xi_0 \wedge s_4 \rightarrow r_1$
107: $sf_l \wedge xo_6 \wedge xi_0 \wedge s_4 \rightarrow r_2$

A disc with not very large dimension containing a low spatial frequency grating can evoke no response (103), a small response (106), or a strong response (107).

4 Discussion

In this work we have concentrated on the decision rules in pre-attentive processes. These so-called early processes extract and integrate into many parallel channels the basic features of the environment. These processes are related to the human perceptions property of objects with unsharp boundaries of values of attributes put together by similarities [4]. These similarities may be related to synchronizations of the multi-resolution parallel computations that are difficult to simulate in the digital computer [5] On the basis of the objects physical properties we can define values of it attributes and in agreement to it classify or named the object. General problem appears when the same object in different conditions changes values of its attributes, or in other words its parts became unsharp. One solution is that the brain extracts as elementary parts so-called basic features [6].

Our eyes constantly perceive changes in light colors and intensities. From these sensations our brain extracts features related to different objects. The basic features were identified in psychophysical experiments as elementary features that can be extracted in parallel. Evidence of parallel extraction comes from the fact that their extraction time is independent of the number of objects.

Other features need serial search, so that the time needed to extract them is proportional to the number of objects. The high-level serial process is associated with the integration, and consolidation of items and with a conscious report. Other, low-level parallel processes are rapid, global, related to high efficiency categorization of items and largely unconscious [6] . Our work is related to the constitution of decision rules extracting basic features from the visual stream.

In conclusion, we have demonstrated previously [3] that the brain may use multi-valued logic in order to test learned predictions about object attributes by comparing them with actual stimulus-related hypotheses. Neurons in V4 integrate objects attributes from its parts in two ways: one is related to local excitatory-inhibitory interactions described here as PIR (parts interaction rules), and another way by changing possible part properties using feedback connections tuning lower visual areas. Different neurons have different PIRs watching objects by multiple unsharp windows (Fig. 1). If objects attributes fit to the unsharp window, neuron sends positive feedback [8] to lower areas filters which in end-effect sharpen the attribute-extracting window changing neuron response from class 1 to class 2 (Fig. 1).

We have suggested [8] that the role of FB pathways is related with extraction of the expected, significant information (hypothesis testing) from a large amount of incoming information (predictions). In this Separation Rule different logics are used: incoming information in related to AND logic (driver rules), whereas FB pathways are choosing (OR rules) certain information.

In summary, we have shown that using rough set theory we can divide stimulus attributes in relationships to neuronal responses into different concepts. Even if most of our concepts were very rough, they determine rules on whose basis we can predict decision of the brain to new, natural images.

Acknowledgement. The Parts Separation Theory (PST) was originally proposed and implemented in different system by George Sporzynski. Thanks to Diane Muhr for explaining to me the PST theory.

References

1. Pawlak, Z.: Rough Sets - Theoretical Aspects of Reasoning about Data. Kluwer Academic Publishers, Boston (1991)
2. Pollen, D.A., Przybyszewski, A.W., Rubin, M.A., Foote, W.: Spatial receptive field organization of macaque V4 neurons. Cereb Cortex 12, 601–616 (2002)
3. Przybyszewski, A.W.: Checking Brain Expertise Using Rough Set Theory. In: Kryszkiewicz, M., et al. (eds.) RSEISP 2007. LNCS (LNAI), vol. 4585, pp. 746–755. Springer, Heidelberg (2007)
4. Zadah, L.A.: Toward a perception-based theory of probabilistic reasoning with imprecise probabilities. Journal of Statistical Planning and Inference 105, 233–264 (2002)
5. Przybyszewski, A.W., Linsay, P.S., Gaudiano, P., Wilson, C.: Basic Difference Between Brain and Computer: Integration of Asynchronous Processes Implemented as Hardware Model of the Retina. IEEE Trans Neural Networks 18, 70–85 (2007)

6. Treisman, A.: Features and objects: the fourteenth Bartlett memorial lecture. Q. J. Exp. Psychol. A. 40, 201–237 (1988)
7. Przybyszewski, A.W., Gaska, J.P., Foote, W., Pollen, D.A.: Striate cortex increases contrast gain of macaque LGN neurons. Vis. Neurosci. 17, 485–494 (2000)
8. Przybyszewski, A.W., Kon., M.A.: Synchronization-based model of the visual system supports recognition. Program No. 718.11, 2003 Abstract Viewer/Itinerary Planner. Washington, DC, Society for Neuroscience, Online (2003)

A Computational Model of Saliency Map Read-Out during Visual Search

Mia Šetić and Dražen Domijan

Department of Psychology, Faculty of Arts and Sciences,
University of Rijeka, Omladinska 14, 51000 Rijeka, Rijeka, Croatia
mia-setic@ffri.hr, ddomijan@ffri.hr

Abstract. When searching for a target in a visual scene filled with distractors, the mechanism of inhibition of return prevents revisiting previously attended locations. We proposed a new computational model for the inhibition of return, which is able to examine priority or saliency map in a manner consistent with psychophysical findings. The basic elements of the model are two neural integrators connected with two inhibitory interneurons. The integrators keep the saliency value of the currently attended location in the working memory. The inhibitory inter-neurons modulate a feedforward flow of information between the saliency map and the output map which points to the location of interest. Computer simulations showed that the model is able to read-out the saliency map when the objects are moving or when eye movements are present. Also, it is able to simultaneously select more then one location, even when they are non-contiguous. The model can be considered as a neural implementation of the episodic theory of attention.

Keywords: Attention, Computational model, Neural integrators, Saliency, Target selection, Visual search.

1 Introduction

Visual search for targets of interest constitute an important function performed by the brain. Initial psychophysical investigations suggest that the search can be performed in two distinctive modes: 1) the fast, parallel mode when a target differs from distracters in a single feature and 2) the slow, effortful, serial mode when a target is distinguished from distracters by a conjunction of two or more features. Later research indicated that this distinction is not as simple and some conjunctions of features can be processes in a fast, parallel manner [19].

Computational models of visual search assume the existence of a saliency or a priority map, which represents the measure of the relative importance of spatial locations in the visual field [18]. The saliency map combines a registered feature contrast in feature maps for colours, brightness, orientations, etc. [10,11]. Different features are conjoined into a unique map over which serial process is issued to select the most active location. It is not yet clear where in the brain the saliency map is computed, but there are several candidate regions, including

V. Kůrková et al. (Eds.): ICANN 2008, Part II, LNCS 5164, pp. 433–442, 2008.

parts of the parietal cortex, the frontal cortex and the thalamus [18]. Lateral inhibition implementing winner-takes-all (WTA) behaviour is assumed to read-out points of interest in the saliency map. In order to move attention to a new location, the most active location is inhibited and, therefore, removed form the competition, which allows the selection of the next most salient location. This process is called the inhibition of return. Psychophysical investigation confirmed the existence of such a mechanism [12].

However, the proposed mechanism for the inhibition of return has several limitations; it can search a visual field only if all the objects are stationary and if eye movements are not allowed. That is, it does not operate on an object-level frame of reference because there is no way to move an inhibitory tag with the moving object [11]. Furthermore, it can select only one location at a time, while the psychophysical data indicates that humans are able to group elements based on their features, and to evaluate them through a single processing step [7]. Also there is evidence that the focus of attention could be split into more than one location [15]. All these facts point to a need for a more elaborate mechanism for spatial shifts of visual attention. The aim of the presented research is to introduce a new model of visual search which exhibits flexibility observed in psychophysical studies. It incorporates two neural integrators and two neural modulators, as a central part which keeps track of the history of the visited locations.

2 The Model

A new proposal for a neural model of serial visual search is based on the idea that there are two specialised neurons which modulate the feedforward flow of information between the saliency map and the output map which points to the location of interest (Fig 1.). This is a minimal neural architecture that can support sampling of neural activity in the saliency map by its magnitude [5]. It enables selective amplification of inputs with different magnitudes using simple operations such as neural integration and thresholding. Proposed neurons operate as neural integrators. They store the saliency value of the current spatial location in the working memory until a decision is made to move attention to a new location. The integrators are connected with two inhibitory neurons which contact dendrites of the neurons in the output map. These inhibitory interneurons can be called modulators or gating cells [3]. Neural modulators do not influence target neurons directly, rather, they modulate the amount of activation that a target neuron receives from other sources. Effectively, interneruons create sliding thresholds for the neurons in the output map. The neurons in the saliency map project excitatory and inhibitory connections to the output map which signals the current location of the focus of attention. Projections are one-to-one without the lateral spread. Inhibitory connections are mediated through a separate set of feedforward inhibitory interneurons. The neurons in the output map operate as a simple McCullogh-Pitts type of units. They are activated only if the excitatory input cross the threshold but the inhibitory input does not cross

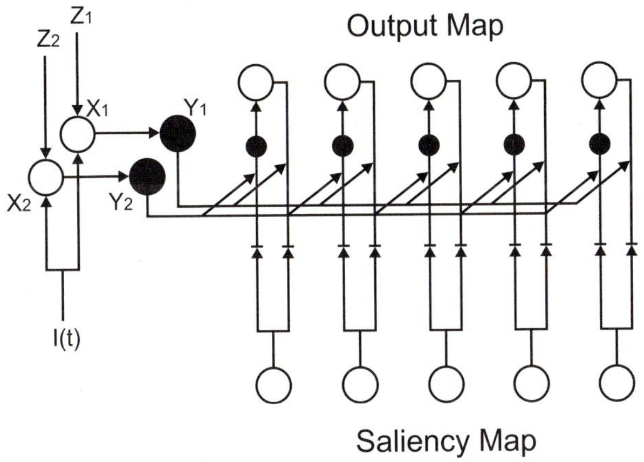

Fig. 1. A neural model for the inhibition of return which reads-out the saliency map and the result is presented in the output map. Open circles are excitatory neurons and filled circles are inhibitory neurons. Arrows indicates connections and the direction of the signal flow. Dendrites are depicted as lines with T endings. The feedforward signal flow between the saliency and the output map is dynamically modulated by two neural integrators (x_1 and x_2), which are connected to two inhibitory interneurons (y_1 and y_2). Interneurons are connected with dendrites of the neurons in the output map and their corresponding interneurons.

the threshold set up by the interneurons. Formally, the model is described as follows. The activity, O_i, in the output map at spatial location, {i}, is given by

$$\frac{dO_i}{dt} = -AO_i + (B - O_i)f[S_i - y_1] - (C + O_i)f[S_i - y_2] \qquad (1)$$

where A is a passive decay, B(C) is an excitatory (inhibitory) saturation point. The term, S_i, denotes activity in the saliency map at the corresponding location, {i}. The function, f[a], denotes the output from separate dendritic branches of the O_i neurons and it is modelled as a 0-1 function with the value 1 if $a > 0$ and 0 if $a \leq 0$. The total activity in the output map is denoted as

$$O_T = \sum O_i \qquad (2)$$

The activity of the inhibitory modulators y_1 and y_2 is given by

$$\frac{dy_1}{dt} = -y_1 + x_1 \quad and \quad \frac{dy_2}{dt} = -y_2 + x_2 \qquad (3)$$

Inhibitory modulators receive excitation from two neural integrators x_1 and x_2 described as

$$\frac{dx_1}{dt} = (-x_1 I(t)) * (I(t) + [z_1]^+) \qquad (4)$$

and

$$\frac{dx_2}{dt} = (-x_2 I(t)) * (I(t) + [z_2]^+) \qquad (5)$$

where

$$\frac{dz_1}{dt} = -Az_1 + (B - z_1)(R(t) * f[1 - O_T]) - (C + z_1)wf[1 - O_T] \quad (6)$$

and

$$\frac{dz_2}{dt} = -Az_2 + (B - z_2)(R(t) * f[O_T]) - (C + z_2)wf[O_T] \quad (7)$$

The term $-x_1$ and $-x_2$ describes the passive decay of the neural activity in the absence of the input I. However, the passive decay is multiplicatively gated by top-down signals z_1 and z_2, which enable or disable activity decay. Term w denotes synsaptic weight of inhibitory influence to top-down signals, which can assume values in the range between -1 and 2 due to the fact that we set $B = 2$ and $C = 1$ in the simulations. Top-down cells are driven by the reset signal $R(t)$ which provide information about mismatch between input pattern at currently attended location and target pattern which is searched [7]. When $R(t)$ is 0, there is no mismatch, z_1 and z_2 remains inacitve and the passive decay is disabled so the integrators x_1 and x_2 sustain a current activity value. In this way, the basic property of the neural integrator is achieved without a need to go deeper into a complicated network and cellular mechanisms responsible for the neural integration [2].

Prior to the start of the visual search, z_1 and z_2 are 0 while x_1 and x_2 are set to a certain high value by the non-specific input signal, $I(t)$. Non-specific input signals are above zero for an initial period of time $t < t_0$ after which they are set to $I(t_0) = 0$ for the rest of the visual search, $t > t_0$. When the mismatch between target pattern and input pattern occurs at $t > t_0$, the reset signal, R, is set to 1 (for a short period of time), which enables activity decay of the x_1 but not x_2 due to the fact that $O_T = 0$. The decay is stopped when the activity of the x_1 reaches the value of the location with the highest saliency in the saliency map. This moment is signalled by the presence of the activity in the output map. Therefore, $z_1 = 0$ when $O_T > 0$. The activity in the output map emerges as a result of the reduced inhibition in the dendrite which transmits excitation to the O_i. In this way, the location with the highest saliency is represented in the output map. When the template matching between input and target pattern is finished and if another mismathch occured, another reset signal, R, is issued, which enables activity decay of the x_2 becasue, now $O_T > 0$. The decay of the x_2 is stopped when it reaches the value of the x_1. This moment is signalled by the removal of the activity in the output map $O_T = 0$. That is, z_2 becomes 0 when $O_T = 0$ due to the fact that excitation and inhibition cancel each other out at a particular O_i. After the disengagement, the attention may be directed to a new location by next reset signal R, which enables x_1 activity to decay further to the next most salient value.

The saliency map, S_i, is assumed to be computed in a manner proposed previously [10]. Synaptic weights between the saliency map and the output map are all set to the unit strength and they are not explicitly represented in the model. Therefore, excitatory and inhibitory connections cancel each other and the output map can not be activated directly by S_i alone. However, the inhibition from modulators enables selective gating of feedforward signal flow between S_i and O_i. Modulators operate like a gate which opens or closes depending on the top-down signals. The gate opens when the first modulator y_1 decreases in its activity up to the level of the location with the highest salience. This location is selected in the output map and it can be evaluated by other visual areas. When the reset signal forces the network to move attention to a new location, the second modulator y_2 decreases its activity to the level of the first modulator. At that moment, the gate is closed and attention is disengaged form the old location. Closing of the gate is signalled by the absence of the activity in the output map. However, the first integrator (and modulator) is pushed to further decrease in its activity (by another R) due to the mismatch between target pattern and empty inputt. The new location is selected when its activity is bellow the second most active cell in the saliency map (the gate is open again). The same cycle of activity reduction can be repeated as long as the target is found or all the locations are visited. With the constraint that the integrator activity can only decrease during visual search, we implemented the mechanism of inhibition of return. A similar model has been proposed for the read-out from visual short-term memory [5]. The difference between these two models is the introduction of neural integrators in the present proposal which were not necessary for modelling a feature binding.

The proposed mechanism depends on an assumption that every object is represented in the saliency map with a single node or with several nodes of the same activity value. However, activity in the saliency map is distributed and can be described with The Gaussian Function of Distance from the object's centre-of-mass. This poses a problem for the model because neural integrators decrease their activity in a continuous fashion. Therefore, the visual search may be stuck on the same object because many of its points may have greater activity value than other objects due to the Gaussian spread of activity. A simple solution to this problem is to add a processing stage between the saliency map and the output map which sharpen the saliency representation by using lateral inhibition. This extra network layer should implement the local WTA behaviour where the locally strongest nodes remain active, while the other nodes are shut down.

3 Computer Simulations

Computer simulations illustrate several important characteristic of a new model. Fig 2 illustrates the network state at different points in time during the visual search. The simulations start with integrators set to a high activity above the saliency value of any point in the saliency map (Fig 2A). When the first integrator reduces its activity, the most salient location is selected as it is shown in the output map (Fig 2B). When the second modulator reduces its activity, the output map is silenced until the first modulator reduces its activity further and the next most salient location is found. Fig 2 demonstrates that attention jumps between nonadjacent locations without visiting intermediate locations because activity in the output map is silenced during the jump. Furthermore, the model

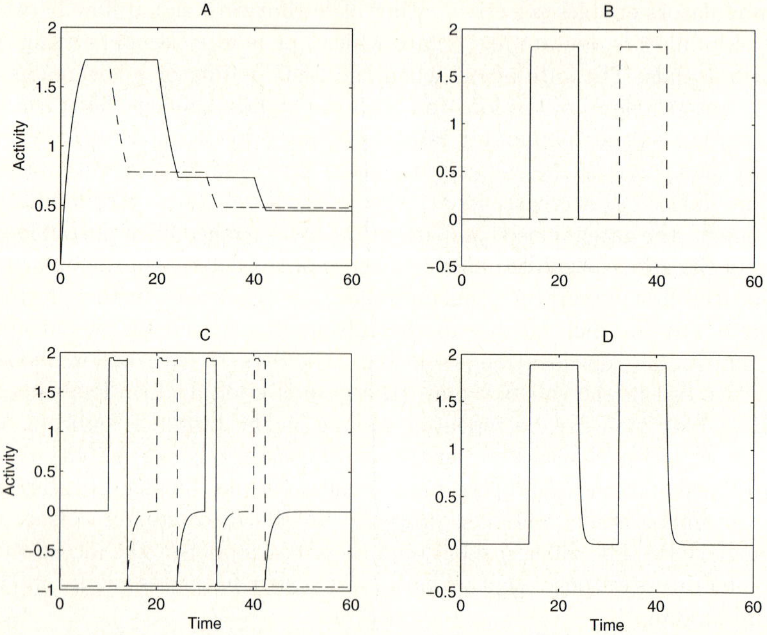

Fig. 2. Computer simulation illustrating the basic properties of the proposed model. A) Activity of neural integrators, x_1 and x_2. B) Activity of the cells in the output map positioned at the location of stimulus, O_i. C) Activity of the top-down signals, z_1 and z_2. D) Activity of the cell summing the total activity in the output map, O_T.

is also able to select more than one location at once due to the fact that the inhibitory modulation is global and there are no lateral interactions between nodes in the output map. The selection of multiple locations does not depend on whether they are connected or not. Therefore, the model can exhibit object based attentional selection and split attentional focus at the same time [11,15]. Moreover, the present model can select simultaneously all locations belonging to the object's representation. This is true under the assumption that saliency computation is able to assign the same activity amplitude to all locations that belong to the same object [6].

Figure 3 illustrates the model ability to select objects despite their movements or the movements of the eyes. This ability depends on the assumption that integrators (and modulators) can only decrease in their activity level during the search. When the attention is moved from the most salient object to another place, the object's saliency level will remain above both modulators so it cannot be selected again. The same issue arises with eye movements. They disrupt spatial memory and disable visual search in serial models. In the present model this problem is avoided by using the parametric working memory which keeps track of the saliency value of the object and not of its spatial location.

Fig 3 also showed that the model is able to perform object-based inhibition-of-return. This is true under the assumption that the object during motion will

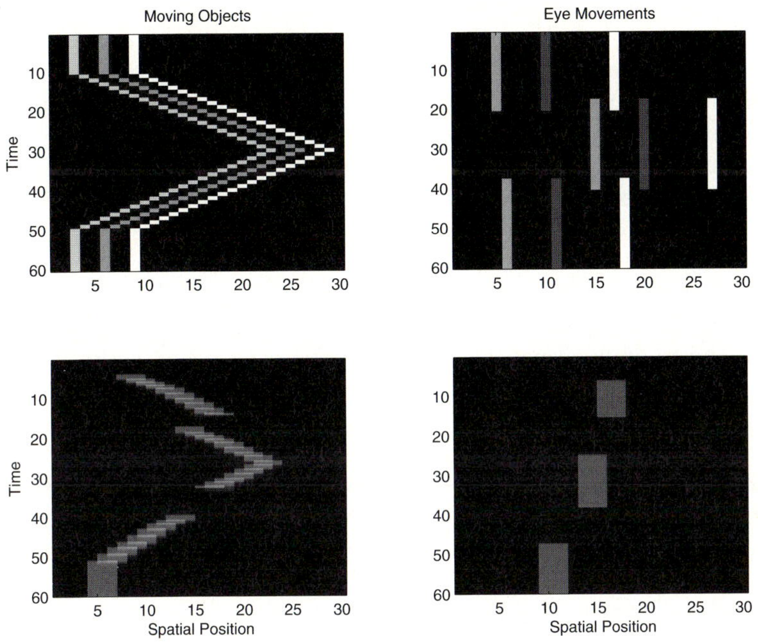

Fig. 3. Computer simulation illustrating the model ability to perform visual search even when the objects are moving or eye movements are present. Top row indicates stimulus movement and bottom row shows the response of the output map. A) Three objects with different saliency values moves to the right and then to the left. During the right movement they visited the same locations as during the left movement. Nevertheless, visual search proceeds from the more salient stimulus to less salient stimulus without interruption. B) Saccadic eye movement to the right and then second saccadic movement to the left do not disrupt the visual search either.

not change the level of saliency. If they come close to each other they may interfere with the process of saliency computation. However, in the experiments with object-based inhibition of return, objects are usually far apart and there is probably no violation of this assumption. Similar issue arises with eye movements which can introduce new objects in the field of view. Adding of new objects may disrupt the saliency of the old objects. Therefore, the visual search should be stopped and restated. In cases where eye movements do not introduce new objects in the saliency map, the visual search may proceed without interruptions.

4 Discussion

Computer simulations illustrate that the new model for the visual search has several advantages over previous proposals. Firstly, it can select more than one location simultaneously since it does not depend on the recurrent lateral inhibition. It is enough that two or more locations have the same level of activity in order to be selected together. Secondly, it is able to search the saliency map even if eye movements are allowed and if object can move. Finally, the model is a simple feedforward network without recurrent connections in the output map which enable it to perform fast searches. Recurrent connections are necessary only for integrators in order to be able to temporary sustain the current activity level. A recent research showed that the sustained neural activity arises even if network connections are disabled, suggesting that the intracellular mechanisms are responsible for maintaining the elevated firing rate [2]. Here, the specific instantiation of the neural integrator is not the issue. The important point is that neural integrators are able to store the saliency value of the currently attended location and to guide visual search by a gradual decrease in activity.

The new model can be considered as a neural implementation of the episodic theory of the dynamics of visual attention. Based on a paradigm of attentional gating, Sperling and Weichselgartner [17] suggest that attention moves across space in discrete episodes. Each episode starts with opening of the attentional gate in a particular location, extracting information at that location and closing the attentional gate, which enables attention to jump to a new location. In the proposed model, the opening of the attentional gate begins when the first integrator (x1) reduces the activity bellow the level of activity in the saliency map. In that moment, the location with the highest saliency is read-out in the output map and can be processed by other visual areas. When the attention should be directed to new location, the attentional gate is closed by reducing the activity in the second integrator (x2). The activity of the second integrator decays until the activity levels of both integrators are equal. At that moment, attention is disengaged from a previously selected location which is observed as a lack of activity in the output map. After that, the first integrator is again pushed towards the lower activity levels by a top-down signal. The activity of the first integrator decays until the next location is found. The implication of the episodic theory of attention is that the time course of attention shift does not depend on the spatial separation between objects, which is observed in the behavioural study [17]. This result suggests that attention does not traverse through intermediate locations between objects during the shift.

The proposed model shares some similarities with the model of the pointer map for attentional recruitment. Hahnloser et al. [8] proposed that the attentional influence on neural activity could be instantiated by pointer neurons which are connected with all neurons in a sensory (or saliency) map. However, the pointer map is a feedback model where pointer neurons modulate the recurrent signal flow between neurons in a spatial map. Attention can be redirected to a new location by moving the activity hill continuously over the map. On the other hand, in the model presented here, pointers (integrators and modulators) do not directly influence neurons in a saliency map but redirect the focus of attention due to the gradual decrease in their activity. The pointer's activity is not registered in the output map. Their influence on the signal flow is indirect.

In this way, the attentional shift between two nonadjacent points is achieved without visiting intermediate locations consistent with psychophysical findings.

In order to remember the current level of activity during the attention engagement at a certain location, we proposed that pointers should have the properties of neural integrators [13]. That is, they should be able to sustain a certain level of activity as long as the top-down signal is issued to reduce their activity. Neural integrators are a biophysically plausible mechanism for temporary storage of activity magnitude. They are discovered in various brain areas where they subserve short-term memory for analogue quantities such as current eye position, frequency of vibration or spatial position. Also, there are several models which explain how they may operate in real neurons [2,13]. Here, gated decay is used as a simple variant of neural integrators, just to illustrate the point that they may be responsible for visual search behaviour. Another feature of our model is the introduction of the special inhibitory neurons which have modulatory influence on the feedforward flow of signals between the saliency map and the output map. Anatomical studies suggest that calretinin-expressing interneurons may operate in this way. Callaway [3] hypothesized that these cells may dynamically route the signal flow by selectively inhibiting one pathway while at the same time disinhibiting another pathway. Inhibitory interactions are assumed to occur at dendrites of the neurons in the output map and their corresponding inhibitory interneurons. Dendrites are modelled as independent computational units with their own output function, f(a), consistent with recent theoretical and experimental work on dendritic computation [14].

Several models of visual search propose that the saliency map is not necessary and there is no need for serial searching over it. The search is assumed to occur in parallel within feature maps, and longer time of searching for targets defined by conjunction of features is ascribed to dynamic properties of competition within the maps [4,9]. However, recent brain imaging and electrophysiological studies reveal that there is a serial component in the visual search. Muller et al. [16] used fMRI to show that activity in the right intraparietal sulcus (IPS) is modulated in proportion to the difficulty of the search task. Woodman and Luck [20] showed that the attention shift between two objects requires about 100-150 ms and there is no temporal overlap in the allocation of processing resources between two locations. They used event-related potentials (N2pc) as a measure of the allocation of attention in the visual field during demanding the search task. Behavioural evidence also supports the view that there is a serial process in the visual search [1]. Here, we showed how the serial search could be implemented in the brain using biophysically realistic mechanisms such as neural integrators, modulators and dendritic computation. The model is able to read-out the saliency map and to move spatial attention in a manner consistent with several behavioural findings.

Acknowledgment. This work was supported by the Bial Foundation grant 80/06 and by the Croatian Ministry of Science, Education and Sport grant 009-0362214-0818.

References

1. Bricolo, E., Gianesini, T., Fanini, A., Bundasen, C., Chelazzi, L.: Serial attention mechanisms in visual search: A direct behavioural demonstration. J. Cogn. Neurosci. 14, 980–993 (2002)
2. Brody, C.D., Romo, R., Kepecs, A.: Basic mechanisms for graded persistent activity: discrete attractors, continuous attractors, and dynamic representations. Curr. Opin. Neurobiol. 13, 204–211 (2003)
3. Callaway, E.M.: Feedforward, feedback and inhibitory connections in primate visual cortex. Neural Netw. 17, 625–632 (2004)
4. Deco, G., Pollatos, O., Zihl, J.: The time course of selective visual attention: theory and experiments. Vision Res. 42, 2925–2945 (2002)
5. Domijan, D.: A mathematical model of persistent neural activity in human prefrontal cortex for visual feature binding. Neurosci. Lett. 350, 89–92 (2003)
6. Domijan, D.: Recurrent network with large representational capacity. Neural Comput. 16, 1917–1942 (2004)
7. Grossberg, S., Mingolla, E., Ross, W.D.: A neural theory of attentive visual search: Interactions of boundary, surface, spatial and object representations. Psychol. Rev. 101, 470–489 (1994)
8. Hahnloser, R., Douglas, R.J., Mahowald, M., Hepp, K.: Feedback interactions between neuronal pointers and maps for attentional processing. Nat. Neurosci. 2, 746–752 (1999)
9. Herd, S.A., O'Reilly, R.C.: Serial visual search from a parallel model. Vision Res. 45, 2987–2992 (2005)
10. Itti, L., Koch, C.: A saliency-based search mechanism for overt and covert shifts of visual attention. Vision Res. 40, 1489–1506 (2000)
11. Itti, L., Koch, C.: Computational modelling of visual attention. Nat. Rev. Neurosci. 2, 194–204 (2001)
12. Klein, R.M.: Inhibition of return. Trends Cogn. Sci. 4, 138–147 (2000)
13. Koulakov, A.A., Raghavachari, S., Kepecs, A., Lisman, J.E.: Model for a robust neural integrator. Nat. Neurosci. 5, 775–782 (2002)
14. London, M., Hausser, M.: Dendritic computation. Annu. Rev. Neurosci. 28, 503–532 (2005)
15. McMains, S.A., Somers, D.C.: Multiple spotlights of attentional selection in human visual cortex. Neuron 42, 677–686 (2004)
16. Muller, N.G., Donner, T.H., Bartelt, O.A., Brandt, S.A., Villringer, A., Kleinschmidt, A.: The functional neuroanatomy of visual conjunction search: a parametric fMRI study. NeuroImage 20, 1578–1590 (2003)
17. Sperling, G., Weichselgartner, E.: Episodic theory of the dynamics of spatial attention. Psychol. Rev. 102, 503–532 (1995)
18. Treue, S.: Visual attention: the where, how and why of saliency. Curr. Opin. Neurobiol. 13, 428–432 (2003)
19. Wolfe, J.M.: Moving towards solutions to some enduring controversies in visual search. Trends Cogn. Sci. 7, 70–76 (2003)
20. Woodman, G.F., Luck, S.J.: Serial deployment of attention during visual search. J. Exp. Psychol. Hum. Percept. Perform. 29, 121–138 (2003)

A Corpus-Based Computational Model of Metaphor Understanding Incorporating Dynamic Interaction

Asuka Terai and Masanori Nakagawa

Tokyo Institute of Technology, 2-12-1 Ookayama, Meguro, Tokyo 152-8550, Japan
{asuka,nakagawa}@nm.hum.titech.ac.jp

Abstract. The purpose of this study is to construct a computational model of metaphor understanding based on statistical corpora analysis. The constructed model consists of two processes: a categorization process and a dynamic interaction process. The model expresses features based not only on adjectives but also on verbs using adjective-noun and three types of verb-noun modification data. The dynamic interaction is realized based on a recurrent neural network employing differential equations. Generally, in recurrent neural networks, differential equations are converged using a sigmoid function. However, it is difficult to compare the estimated meaning of the metaphor to the estimated meaning of the target which is represented with conditional probabilities computed through statistical language analysis. In the present model, the differential equations converge over time, which makes it possible to compare the estimated meaning. Accordingly, the constructed model is able to highlight the emphasized features of a metaphorical expression. Finally, a psychological experiment is conducted in order to verify the psychological validity of the constructed model of metaphor understanding. The results from the psychological experiment support the constructed model.

1 Introduction

This paper constructs a computational model that realizes the understanding processes for metaphorical expressions, represented in the form of "A like B". Metaphor understanding basically requires a knowledge structure for concepts (a target, a vehicle and so on)[1]. However, it is not practically feasible to collect sufficient data to cover enough concepts by psychological methods alone, because participants cannot rate the entire range of concepts that are commonly used in metaphorical expressions within limited amounts of time. Thus, a model based only on psychological experimentation cannot be extended to computational systems (e.g. search engines). Accordingly, in this paper, a computational model is constructed based on a knowledge structure for concepts extracted from linguistic corpora.

One theory within psychology seeks to account for the understanding process of metaphors. It is the categorization theory which explains metaphor understanding in terms of class-inclusion statements, where a target is regarded as a

V. Kůrková et al. (Eds.): ICANN 2008, Part II, LNCS 5164, pp. 443–452, 2008.

member of an ad hoc category of which the vehicle is a prototypical member [2]. For example, in comprehending the metaphor "Socrates is like a midwife", the target "Socrates" can be regarded as belonging to a "helpful" category that could be typically represented by a vehicle such as "midwife".

Some computational models of metaphor understanding have been constructed using linguistic corpora based on categorization theory[3][4]. Kintsch's and Utsumi's models have employed knowledge structures estimated with Latent Semantic Analysis (LSA)[5]. Even though concept meanings are represented by vectors in LSA, the vector dimensions themselves do not have meaning. Thus, the meaning of a metaphor represented by a particular vector must generally be defined in terms of the cosines of the angles between other vectors according to the LSA method. This aspect of LSA makes it quite difficult to interpret metaphors represented by vectors.

It is worth noting that some studies have focused on the low-salient features of a target and a vehicle that are emphasized in the process of metaphor understanding; a phenomenon referred to as feature emergence[6][7]. Feature emergence has also been described in terms of an interaction among features[8][9][10]. However, one model[8] does not represent the dynamic interaction among attribute values. While another model[9] represents the dynamic interaction with a recurrent neural network, the model has a problem in that it is difficult to explain the role differences between targets and vehicles. In contrast, The other model[10] (Terai&Nakagawa's model) consists of two processes; the first is a categorization process and the second is a dynamic interaction process. The dynamic interaction process, realized with a recurrent neural network, represents feature emergence in terms of the dynamic interaction among features. However, Terai&Nakagawa's model[10] was constructed only using frequency data for adjective-noun modifications. Thus, the represented features for concepts in the model were limited to adjectives, and, consequently, the model cannot represent the understanding process for metaphors involving "verbs" (e.g. "a rumor like a virus" with the underlying meaning of "a rumor that spreads like a virus"). Moreover, while the meaning of a target is represented using the conditional probabilities of features given the target, the meaning of a metaphor is represented by the sigmoid function outputs through dynamic interaction. Thus, the output values that represent the meaning of the metaphor meaning cannot be compared directly to the output values that represent the meaning of the target. That is, the model cannot clearly indicate what kinds of features are emphasized by a metaphorical expression comparing to the meaning of the target.

Similar to Terai&Nakagawa's model[10], the present study also assumes that metaphor understanding is realized through two processes (categorization and dynamic interaction processes). In order to overcome problems with previous models, the present model is constructed as follows. First, concepts are represented using vectors that are estimated from a statistical language analysis[11] for four kinds of modification patterns (frequency data for adjective-noun modifications and three kinds of verb-noun modifications). The meanings of concepts estimated from the statistical language analysis are represented by conditional probabilities of the concepts given the features. This makes it easier to determine estimated meanings than with the LSA approach. In addition, features are expressed using not only adjectives but also for verbs for four modification patterns. Second, the categorization process model is constructed. The model assigns the meaning of a target to an ad-hoc category for the vehicle using estimated concept vectors. Third, the recurrent neural network model using differential equations to represent the dynamic interaction process estimates the meaning of the metaphor based on the assigned meaning of the target (the results of the categorization model). In order to indicate the emphasized features of the metaphor expression, the differential equations that converge over time are developed for the dynamic interaction process. Finally, a psychological experiment is conducted in order to verify the psychological validity of the constructed model.

2 Representation of Nouns Using Vectors

2.1 Statistical Language Analysis

This study applies Kameya and Sato's statistical method[11] using extracted frequency data for adjective-noun modifications and three types of verb-noun modifications. g The statistical method assumes that the terms n_i (noun) and a_j (adjective or verb) co-occur through latent classes and that the co-occurrence probabilities of these terms, $P(n_i, a_j)$, can be computed using the following formula(1):

$$P(n_i, a_j) = \Sigma_k P(n_i|c_k)P(a_j|c_k)P(c_k), \tag{1}$$

where c_k indicates the kth latent class assumed in the method. The parameters ($P(n_i|c_k)$, $P(a_j|c_k)$, and $P(c_k)$) are estimated as the value that maximizes the log likelihood of the co-occurence freaquency data between n_i and a_j using the EM algorithm.

$P(c_k|n_i)$ and $P(c_k|a_j)$ are computed using Bayes' theorem, as follows:

$$P(c_k|n_i) = \frac{P(c_k)P(n_i|c_k)}{\sum_{k'} P(c_{k'})P(n_i|c_{k'})}, \tag{2}$$

$$P(c_k|a_j) = \frac{P(c_k)P(a_j|c_k)}{\sum_{k'} P(c_{k'})P(a_j|c_{k'})}. \tag{3}$$

The meanings of the latent classes are identified from the conditional probability of the latent class c_k given the adjective a_j and the conditional probability of the latent class c_k given the noun n_i.

2.2 The Results of the Statistical Language Analysis

The present analysis employs four kinds of frequency data extracted for adjective-noun modifications (Adj) and three kinds of verb-noun modifications: noun (subject)-verb (S-V), verb-noun(modification) (M-V) and verb-noun(object) (O-V). As an example, let us take the sentence "A little child drops a letter into the post box", "little-child" is extracted as an adjective-noun, "child drop" as a noun(subject)-verb, "drop (into) the post box" as a verb-noun(modification) and "drop a letter" as a verb-noun(object). The data were extracted from the Japanese newspaper MAINICHI SHINBUN for the period 1993-2002 by using a modification analysis tool known as "CaboCha" [12]. The extracted data consists of 21,671 noun types and 3,403 adjective types for adjective-noun modifications, 29,745 noun types and 22,832 verb types for verb-noun(object), 26,113 noun types and 21,487 verb types for noun(subject)-verb, and 28,451 noun types 24,231 verb types for verb-noun(modification). The statistical language analysis is individually applied to each set of co-occurrence data fixing the number of latent classes at 200.

This model deals with the 18,142 noun types (n_h) that are common to all four data sets. The nouns are represented by vectors using four kinds of conditional probabilities of latent classes given the nouns ($P(c_k^{Adj}|n_h)$, $P(c_k^{S-V}|n_h)$, $P(c_k^{M-V}|n_h)$, $P(c_k^{O-V}|n_h)$), which are computed using the four data sets, as follows.

$$V_p(n_h) = P(c_k^r|n_h), \tag{4}$$

where $V_p(n_h)$ indicates the pth component of the vector that corresponds to the noun n_h. c_k^r indicates the latent classes extracted from the four data sets and r refers to the kind of data set (adjective-noun modification, noun(subject)-verb (S-V), verb-noun(modification) (M-V) and verb-noun(object) (O-V) data). p refers to the successive number of latent classes extracted from the four data sets. When $1 \leq p \leq 200$, r indicates the "Adj" modification and $k = p$, when $201 \leq p \leq 400$, r indicates the "S-V" modification and $k = p - 200$, when $401 \leq p \leq 600$, r indicates the "M-V" modification and $k = p - 400$, and when $601 \leq p \leq 800$, r indicates the "O-V" modification and $k = p - 600$. The dimensions of the vectors represent the latent classes.

3 The Metaphor Understanding Model

The model consists of two kinds of process. One is the categorization process and the other is the dynamic interaction process.

3.1 The Categorization Process

A vector, representing an assigned target as a member of an ad hoc category for a vehicle, is estimated based on categorization theory using the meaning vectors of concepts. The algorithm for the categorization process is as follows.

First, the semantic neighborhood ($N(n_h)$) of a vehicle of size s_1 is computed on the basis of similarity to the vehicle, which is represented by the cosine of the angles between meaning vectors using formula(5):

$$sim(n_h, n'_h) = \frac{V(n_h) \cdot V(n'_h)}{\|V(n_h)\| \|V(n'_h)\|}, \tag{5}$$

where $sim(n_h, n'_h)$ indicates the similarity between concept n_h and concept n'_h. Next, L concepts are selected from the semantic neighborhood ($N(n_h)$) of the vehicle on the basis of similarity to the target. Finally, a vector ($V(M)$) is computed as the centroid of the meaning vectors for the target, the vehicle and the selected L concepts (n_l: $l = 1, 2, .., L$). The computed vector ($V(M)$) indicates the assigned meaning of the target as a member of the ad-hoc category of the vehicle in the metaphor M. $V(M)$ is computed using formula(6):

$$V(M) = \frac{\sum_l V(n_l) + V(target) + V(vehicle)}{L+2}, \tag{6}$$

where n_l indicates the lth selected concepts and L denotes the number of the selected concepts. This algorithm is the same as Kintsch's [3] algorithm, that used in Utsumi1's categorization model [4] and the categorization process in Terai&Nakagawa's model[10]. The category consisting of the vehicle and the selected k concepts is regarded as an ad hoc category of which the vehicle is a prototypical member.

In order to estimate the meanings of $V(M)$, the conditional probability of an adjective or a verb given M, $P(a^r_j|M)$ is computed using the function(7) for $P(a^r_j|c^r_k)$ based on Bayes theory:

$$P(a^r_j|M) = \Sigma_k P(a^r_j|c^r_k)V_p(M), \tag{7}$$

when r indicates "Adj" modification, $p = k$, when r indicates "S-V" modification, $p = k + 200$, when r indicates "M-V" modification, $p = k + 400$ and when r indicates "O-V" modification, $p = k + 600$. Similarly, in order to estimate the meanings of a certain concept, the conditional probability of an adjective or a verb given n_h, $P(a^r_j|n_h)$ is computed using function(8):

$$P(a^r_j|n_h) = \Sigma_k P(a^r_j|c^r_k)P(c^r_k|n_h). \tag{8}$$

3.2 The Dynamic Interaction Process

The meaning of the metaphor is computed using the vector estimated by the categorization process ($P(a^r_j|M)$) by applying the dynamic interaction process model. The algorithm for the dynamic interaction process is as follows.

First, features are selected if $P(a^r_j|M)$ exceeds the threshold ζ^r. The selected features are related to metaphor understanding. Next, the recurrent neural network model is constructed using the selected features (Fig. 1). Each node corresponds to a selected feature. These nodes have both inputs and outputs.

The dynamics of the network are based on the following set of simultaneous differential equations(9):

$$\frac{dx_q(t)}{dt} = \frac{1}{\alpha^t}(-x_q(t) + \beta\Sigma_{q'} w_{qq'} x_{q'}(t) + I_q(M)), \tag{9}$$

where $x_q(t)$ represents the activation strength of the qth node at time t. The range is between 1 and 0. $\frac{1}{\alpha^t}$ is the coefficient for convergence. When $dx_q/dt = 0$,

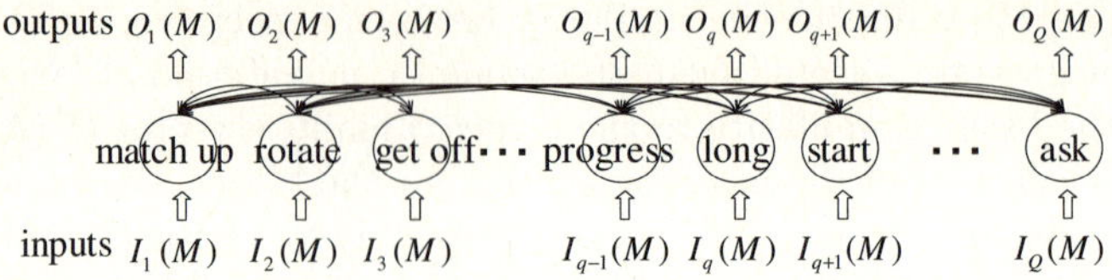

Fig. 1. Architecture of the model for "conversation like a gear (=M)". The nodes represent the selected features. These are both input and output nodes.

the node outputs $O_q(M) = x_q(t)$. The vector $(O(M))$, which is a set of $O_q(M)$, represents the meaning of the metaphor M. $I_q(M)$ represents the input value of the qth node related to the metaphor M. The value of $P(a^r_{j(q)}|M)$ is used as the input value $I_q(M)$ where a^r_j corresponds to the meaning of the qth node. In formula(9), β denotes the influences of the dynamic interaction among features. $w_{qq'}$ denotes the weight of the connection from the q'th to the qth node and is the correlation coefficient among the qth and q'th features related to the sibling concepts of the target and the vehicle. A sibling neighborhood $(N^s(vehicle))$ for a vehicle of size s_2 and a sibling neighborhood $(N^s(target))$ for a target of size s_2 are computed on the basis of similarity. The concepts included in $N^s(vehicle)$ and $N^s(target)$ are regarded as sibling concepts.

Thus, the mutual and symmetric connections among nodes $(w_{qq'})$ represent the interactions among features in metaphor understanding. If the metaphor is changed, then the weights for the connection between the same pair of features may change. For example, in the case of "a dog like a cloud", "white" and "puffy" should be connected strongly. On the other hand, in the case of "skin like snow", "white" and "puffy" should only be weakly connected. Therefore, each weight for mutual connections between nodes is estimated using the correlation coefficient between the two features.

The differential equations converged based on the sigmoid function in the previous model. In this model, the differential equations (formula(9)) converge over time. The output values of the present model include the influences of interaction among the features. Thus, the output values that represent a metaphor meaning could be directly compared to the meaning of the target which is computed using formula(8). Accordingly, the model clearly indicates what kinds of features are emphasized by the metaphor expression in comparison with the meaning of the target.

3.3 Model Simulation

In this study, the model is simulated using the parameters $s_1 = 250$, $L = 5$, $s_2 = 100$, $\alpha = 10$, $\beta = 0.15$, $\zeta^{Adj} = 0.0294$ $(= 100/\mathit{the\ number\ of\ adjectives})$, $\zeta^{S-V} = 0.0047$ $(= 100/\mathit{the\ number\ of\ verbs}(S-V))$, $\zeta^{M-V} = 0.0041$ $(= 100/\mathit{the\ number\ of\ verbs}(M-V))$, and $\zeta^{O-V} = 0.0044$ $(= 100/\mathit{the\ number\ of\ verbs}(O-V))$. The simulation results for the metaphors of "conversation like a gear" and "discussion like a war" are shown in Table1. Only the results for the categorization process model $(P(a^r_j|M))$ and for the dynamic interaction process model $(O_q(M))$ are presented.

Table 1. Metaphor meaning computed by the categorization process model (CPM) and the two-process model (TPM) ("conversation like a gear", "discussion like a war"). The output values are shown in parentheses.

	M = conversation like a gear		M = discussion like a war	
	CPM($P(a^r_{j(q)}\|M)$)	TPM (O_q(M))	CPM ($P(a_j(q)^r\|M)$)	TPM (O_q(M))
1	match up (0.1103)	match up (0.0922)	furious (0.2065)	furious (0.1755)
2	rotate (0.1008)	rotate (0.0868)	come up (0.0966)	expand (0.1029)
3	fun (0.0656)	get off (0.0468)	start (0.0843)	long (0.0959)
4	huge (0.0391)	capture (0.0422)	long (0.0806)	come up (0.0873)
5	get off (0.0283)	engage (0.0372)	expand (0.0620)	start (0.0805)
6	be sent (0.0234)	huge (0.0364)	new (0.0466)	unfurl (0.0795)
7	capture (0.0197)	fun (0.0361)	hot (0.0459)	be expanded (0.0764)
8	come up (0.0194)	film (0.0351)	strict (0.0450)	get furious (0.0763)
9	go (0.0176)	respond (0.0232)	end (0.0442)	be unfurled (0.0751)
10	appear (0.0162)	encounter (0.0228)	participate (0.0341)	be repeated (0.0739)

The results for both models (the categorization process model (CPM) and the two-process model (TPM)) seem to be appropriate. However, the following psychological experimentation is necessary in order to verify which model is more appropriate and to examine the validity of the two-process model.

4 Psychological Experiment

In order to examine the validity of the model, a psychological experiment was conducted. The participants were 85 undergraduates. The participants were divided into two groups (Group A: 42 undergraduates; Group B: 43 undergraduates). Two metaphorical expressions were used in the experiment ("conversation like a gear" and "discussion like a war"). First, the target, the vehicle and the metaphor were presented to the participants (Group A: "conversation", "gear" and "discussion like a war", Group B: "discussion", "war" and "conversation like a gear"). Next, they were asked to respond with the appropriate features of the target, of the vehicle and of the metaphor.

Table 2 lists features that were given by three or more participants when the vehicle, and the target, and the metaphor ("conversation like a gear" or "discussion like a war") were presented. The features of "conversation like a gear", such as "engage", "match up", "fun", "progress" and "stop" that were provided by two or more participants were also identified in both models. In particular, the most frequently responded features ("engage", "match up" and "fun") were estimated to be among the top 10 features by the two-process model, although "engage" was not estimated as one of the top 10 features by the categorization process model. Correlation coefficients were calculated for these features. The correlation coefficient is 0.414($n.s.$) for the output values for the two-process model and response frequency in the experiment, and is 0.079($n.s.$) for the output values for the categorization process model and experimental response frequencies. While both correlation coefficients are not significant, the results are supportive of

Table 2. Results of the psychological experiment ("conversation like a gear" or "discussion like a war"). The numbers of the participants who responded with a particular feature are shown in parentheses.

conversation like a gear			discussion like a war		
conversation	gear	conversation like a gear	discussion	war	discussion like a war
engage (18)	fun (21)	engage (19)	die (11)	quarrel (9)	furious (9)
rotate (14)	speak (8)	match up (10)	sad (8)	difficult (7)	quarrel (9)
get off (5)	get lively (5)	fun (9)	awfull (7)	say (6)	swirl around (5)
move (5)	know (5)	good (7)	kill (6)	come up (5)	noisy (3)
	difficult (3)	convey (5)	bitter (5)	hot (5)	hot (4)
	boring (3)	quick (4)	painful (4)	think (4)	do (3)
	warm up (3)	speak (4)	use (3)	discuss (4)	say (3)
	long (3)	stop (3)	come up (3)	brainstorm (3)	
	sad (3)	understand (3)	live (3)	agree on (3)	
	listen (3)			various (3)	

the two-process model. The experimental results suggest the existence of emergent features. In particular, the feature of "progress" was more appropriately estimated by the two-process model ($O_{progress}(M) = 0.0118$) than the categorization process model ($P(progress|M) = 0.0093$). In addition, the output values for the two-process model are compared with target meanings (Fig.2, Table 4). The shifts from the meaning of "conversation" ($P(feature|conversation)$) to the image of "a conversation like a gear" are consistent with the experimental results. Both sets of results indicate an emphasis on the images of "engage", "match up", "progress" and "stop" and a weakening of the image of "fun" (The value decreases from $P(fun|conversation) = 0.0978$ to $O_fun(M) = 0.0361$). While 21 of the participants responded with "fun" as an appropriate feature of "conversation", only 9 participants responded that it is an appropriate feature of "conversation like a gear". Thus, decline from 21 to 9 responses suggest that while "fun" is an image elicited from the noun "conversation" as a single entity, it is much less likely to be evoked for the metaphorical expression of "conversation like a gear".

Both the categorization process model and the two-process model successfully estimate the features of "discussion like a war", such as "furious" which was given by most of the participants in the experiment. Although the experimental results indicate that "furious" is an emergent feature, the categorization process model without interaction among features successfully estimated "furious". This can be account for by the fact that "furious" is strongly estimated as a feature for both "discussion" and "war" in the statistical language analysis. Actually, participants did not respond to "get furious". However, "get furious", as a verbal phrase from "furious", can be considered as an emergent feature of the metaphor. In the simulation results, the value for gget furioush in the context of the meaning of "discussion" ($P(getfurious|discussion) = 0.0294$) is emphasized in the output value for the two-process model ($O_{getfurious}(M) = 0.0763$), while it is weakened in the output value for the categorization model ($P(getfurious|M) = 0.0203$).

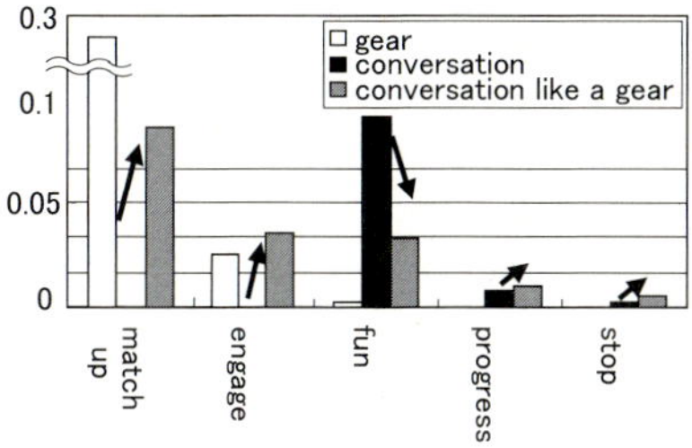

Fig. 2. The shifts from the meaning of "conversation" to the meaning of "conversation like a gearh in terms of the images of "engage", "match up", "fun", "progress" and "stop"

feature	gear	conversation	conversation like a gear
engage	0.0278 (18)	0.0004 (0)	0.0922 (19)
match up	0.2792 (1)	0.0000 (0)	0.0372 (10)
fun	0.0019 (1)	0.0978 (21)	0.0361 (9)
progress	0.0001 (1)	0.0075 (0)	0.0118 (5)
stop	0.0007 (1)	0.0019 (1)	0.0049 (3)

Fig. 3. Simulation results ("conversation like a gear) relating to "engage", "match up", "fun", "progress" and "stop". The numbers of participants responding with a particular feature are shown in parentheses.

This can be explained in terms of the value for gget furioush in the context of the meaning of "war" ($P(getfurious|war) = 0.0099$) which is relatively small, but "get furious" is emphasized by the interaction among the features. The two-process model can estimate not only the features estimated by the categorization process model but also emergent features that the categorization process model cannot estimate. Consequently, the experimental results support that the constructed model, which consists of two processes.

5 Discussion

In order to represent features not only from adjectives but also from verbs, this study has constructed a computational model of metaphor understanding based on statistical language analysis of corpora. Accordingly, the constructed model handles frequency data for adjective-noun modifications and three kinds of verb-noun modifications. The model consists of both a categorization process and a dynamic interaction process. In order to indicate the features emphasized by the metaphor expression, the dynamic interaction process was improved using a recurrent neural network model with differential equations that converge over time. A psychological experiment was conducted in order to verify the constructed model. The results from the experiment support the constructed model. Emergent features estimated by the model were also validated by the psychological experiment.

However, in the psychological experiment, the participants were asked to respond with appropriate features for the target, for the vehicle and for the metaphor. Therefore, the features provided by the participants could tend to rather limited. That is, it is possible that there are other appropriate features that were not provided by participants (e.g. "get furious" in the case of "discussion like a war"). Thus, it is necessary to evaluate the strength of relationships between the target, the vehicle and the metaphor and the features by the Semantic Differential (SD) method in order to elucidate the images evoked by the target, the vehicle and the metaphor. An experiment into the strengths of relationships between the target, the vehicle and the metaphor and the features as determined by the SD method could possibly provide further insights into the kinds of features that are emphasized by metaphorical expressions compared to target meanings.

Acknowledgement. This research is supported by the Tokyo Institute of Technology 21COE Program, "Framework for Systematization and Application of Large-scale Knowledge Resources" and MEXT's program "Promotion of Environmental Improvement for Independence of Young Researchers" under the Special Coordination Funds for Promoting Science and Technology.

References

1. Kusumi, T.: Hiyu no Syori Katei to Imikozo. Kazama Syobo (1995)
2. Glucksberg, S., Keysar, B.: Understanding Metaphorica Comparisons: Beyond Similarity. Psychological Review 97(1), 3–18 (1990)
3. Kintsch, W.: Metaphor comprehension: A computational theory. Psychonomic Bulletin & Review 7(2), 257–266 (2000)
4. Utsumi, A.: Computational exploration of metaphor comprehension processes. In: Proc. of the 28th Annual Meeting of the Cognitive Science Society, pp. 2281–2286 (2006)
5. Deerwester, S., Dumais, S., Furnas, G., Landauer, T., Harshman, R.: Indexing by Latent Semantic Analysis. Journal of the Society for Information Science 41(6), 391–407 (1990)
6. Nueckles, M., Janetzko, D.: The role of semantic similarity in the comprehension of metaphor. In: Proc. of the 19th Annual Meeting of the Cognitive Science Society, pp. 578–583 (1997)
7. Gineste, M., Indurkhya, B., Scart, V.: Emergence of features in metaphor comprehension. Metaphor and Symbol 15(3), 117–135 (2000)
8. Utsumi, A.: Hiyu no ninchi / Keisan Moderu. Computer Today 96(3), 34–39 (2000)
9. Terai, A., Nakagawa, M.: A Neural Network Model of Metaphor Understanding with Dynamic Interaction based on a Statistical Language Analysis; Targeting a Human-like Model. International Journal of Neural Systems 17(4), 265–274 (2007)
10. Terai, A., Nakagawa, M.: A Computational Model of Metaphor Understanding Consisting of Two Processes. In: de Sá, J.M., Alexandre, L.A., Duch, W., Mandic, D.P. (eds.) ICANN 2007. LNCS, vol. 4669, pp. 963–972. Springer, Heidelberg (2007)
11. Kameya, Y., Sato, T.: Computation of probabilistic relationship between concepts and their attributes using a statistical analysis of Japanese corpora. In: Proc. of Symposium on Large-scale Knowledge Resources: LKR 2005, pp. 65–68 (2005)
12. Kudoh, T., Matsumoto, Y.: Japanese Dependency Analysis using Cascaded Chunking. In: Proc. of the 6th Conference on Natural Language Learning, pp. 63–69 (2002)

Deterministic Coincidence Detection and Adaptation Via Delayed Inputs

Zhijun Yang[1,2], Alan Murray[2], and Juan Huo[2]

[1] School of Mathematics and Computer Science
Nanjing Normal University, Nanjing 210097, China
[2] School of Electronics and Engineering
The University of Edinburgh, Edinburgh, EH9 3JL, UK
{Zhijun.Yang,Alan.Murray,J.Huo}@ed.ac.uk

Abstract. A model of one integrate-and-firing (IF) neuron with two afferent excitatory synapses is studied analytically. This is to discuss the influence of different model parameters, i.e., synaptic efficacies, synaptic and membrane time constants, on the postsynaptic neuron activity. An activation window of the postsynaptic neuron, which is adjustable through spike-timing dependent synaptic adaptation rule, is shown to be associated with the coincidence level of the excitatory postsynaptic potentials (EPSPs) under several restrictions. This simplified model, which is intrinsically the deterministic coincidence detector, is hence capable of detecting the synchrony level between intercellular connections. A model based on the proposed coincidence detection is provided as an example to show its application on early vision processing.

Keywords: Coincidence detection, Leaky integrate and fire neuron, Model parameters, Time constant, Synchrony.

1 Introduction

Recent Biophysical and Neurobiological advances have revealed that temporal asymmetric Hebbian learning, a generalized version of the traditional Hebbian learning rule [1], may be the intrinsic mechanism underlying synaptic plasticity in various brain regions to coordinate pre- and postsynaptic neuronal activities within a critical time window to conduct information storage and processing [2,3,4]. Usually, such synaptic plasticity, termed spike timing dependent plasticity (STDP), involves a large amount of synaptic connections and, is stochastic in nature for either single- or multiple-compartment models of neuronal nets [5]. On the other hand, the physiological and anatomical data shows that nearby neurons usually fire in a correlated way and share similar functionalities [6], a property hence forms a basic principle in neuronal development [7].

By considering both theoretical formalism and application feasibility, we study the input-output relationship of a simplified model consisting of one postsynaptic neuron and two afferent excitatory synapses from the perspective of potential neuromorphic implementation. It has been shown that analog very large scale

V. Kůrková et al. (Eds.): ICANN 2008, Part II, LNCS 5164, pp. 453–461, 2008.

integrated (aVLSI) circuit design of as few as four synapses converging to a postsynaptic neuron is able to induce competition among synapses and hence STDP [8,9]. However, in our case the competition mechanism vanishes due to the fact that there are only two synaptic connections whilst each of their EPSPs from the spontaneous poisson current source is not sufficient alone to depolarize the resting postsynaptic neuron. A joint contribution from both synapses rather than competition between them is thus necessary to activate the postsynaptic neuron. We will show that this simplified model is fundamentally a leakage coincidence detector which can use either the synapse and membrane decay rates or the synaptic efficacies to associate the neuron response with input delays. Following the fact that EPSP decay rate adjustment has the same effect as adjusting the axonal conduction delay [10], we will further characterize the neuronal dynamics on delayed inputs analytically and numerically. An early vision model is then taken as an example to describe the effect of adjusting the synaptic efficacy. It is possible to generalize this simple, two-input coincidence detector by cascading them to formulate a multiple-input coincidence detector.

2 Model

In the model we have one leaky IF neuron fed by two spontaneous Poisson δ-function pulse streams through two identical excitatory synapses. For analytical simplicity we have the following assumptions.

1. No delay during input pulse propogation within the axons; Or in other words, since our purpose is to deal with the time difference of two events represented by the pulse inputs to an IF neuron, the axonal delays can be absorbed by the original event time difference.

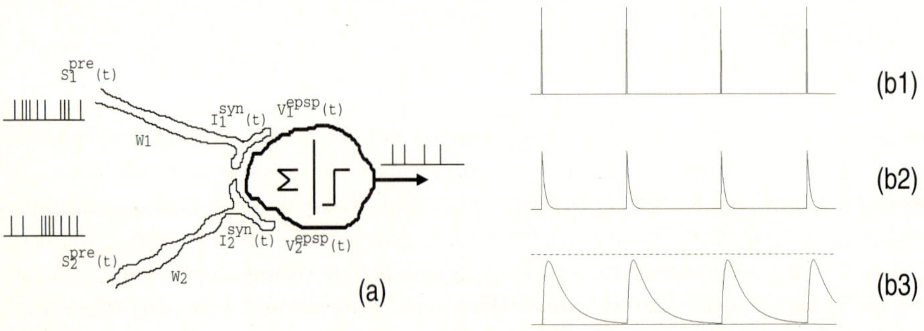

Fig. 1. A schematic circuit diagram and simulated pulse stream propogation in one afferent synapse. (a) two synapses converge to an IF neuron, with Poisson δ-function current pulse streams propogate from the dendritic input terminals $S_i^{pre}(t)$ (b1) to dendritic output terminals $I_i^{syn}(t)$ (b2), and induce the EPSPs $V_i^{epsp}(t)$ (b3), where dashed line represents the neuron threshold. The IF neuron adds up the EPSPs and has a threshold mechanism.

2. The pulses of either synapse have large interspike intervals due to spontaneous nature so that no cumulative effect of the consecutive input pulses from a same synapse can activate the neuron. One individual EPSP from either synaptic input does not, by itself, exceed the voltage threshold of the IF neuron (see Fig. 1).
3. Two pulses from two synapses are regarded as being synchronized (possibly with small tolerant delay) if and only if the postsynaptic IF neuron is depolarized to issue a spike.

Hence in this ideal, simplified model, the activation of the IF neuron only depends on the interaction of two synaptic inputs. A traditional input correlation detection has a multiplicative nature to use the convolution of the received signal contaminated by the noise in an information channel with the channel impulse response [11]. This approach has been widely used to detect signals in telecommunication. In biology, however, the subthreshold membrane activity summarizes the current charges before it can fire a spike due to the additive nature of the postsynaptic membrane activities [4].

2.1 Coincidence of Inputs

As shown in Fig. 1, the additive operation of two inputs is a basic 2-input coincidence detector. The dynamics of the postsynaptic membrane potential can be described by the following differential equation,

$$\tau_{mem} \frac{dv(t)}{dt} = v_m - v(t) + r_m \sum_i I_i^{syn} \tag{1}$$

where the afferent, post-dendrite current is,

$$\tau_{syn} \frac{dI_i^{syn}(t)}{dt} = -I_i^{syn}(t) + w_i \sum_j \delta(t - t_j) \tag{2}$$

here, τ_{mem} and τ_{syn} are the membrane and synaptic time contants, respectively, r_m is the membrane resistance, w_i is the synaptic efficacy, $\delta(t - t_j)$ is a δ-function dendritic input at time t_j, $I_i^{syn}(t)$ is the current pulse at the post-dendrite terminal, $v(t)$ is the EPSP induced by $I_i^{syn}(t)$, in our case $i = 1, 2$.

We are interested in any more restrictions for the firing of postsynaptic neuron, given the fact that the sum of two synchronized EPSPs will activate it. It is hence necessary to estimate, quantitatively, the synchrony level of two post-dendrite current pulses which will (or will not) result in a spike from IF neuron. Since the time constant reflects the basic activation property of a neuronal or synaptic circuit, the dynamic range of the IF neuron, in terms of the variation of its membrane and synaptic time constants, is measured to form a window, within which the sum of EPSPs reaches the neuron threshold and a spike is issued. A larger window allows more delay between two input pulses to depolarize the IF neuron.

By solving the current equation in (2) through Laplace transform we get,

$$I_i^{syn}(t) - \frac{w_i}{\tau_{syn}} exp(-\frac{t}{\tau_{syn}}) \tag{3}$$

Since two EPSPs induced by the input current pulses arise on the neuron membrane at different time, we set the time difference explicitly using t_1 for the first EPSP and t_2 the second with a delay Δt, i.e., $t_1 = t$ and $t_2 = t + \Delta t$. Use the current solution to the membrane voltage equation of (1), we get the neuron's membrane potential with Laplace tranform,

$$v(t, \Delta t) = \left(\frac{w_1}{\tau_{syn} - \tau_{mem}} + \frac{w_2 exp(-\frac{\Delta t}{\tau_{syn}})}{\tau_{syn} - \tau_{mem}}\right)\left(exp(-\frac{t}{\tau_{syn}}) - exp(-\frac{t}{\tau_{mem}})\right) \tag{4}$$

This membrane potential equation is the summation of two α-function caused by the dendritic propagation of afferent pulse waves, where τ_{syn} is the rising time and τ_{mem} the decay time of each α-function, usually $\tau_{mem} \gg \tau_{syn}$. The amplitude of the second α-function decays exponentially depending on its peak-peak delay with the first one. Since two afferent excitatory stimuli will lead to, if synchronized, only one postsynaptic spike due to the IF neuron's refractory period, by making the neuron's membrane potential equal to its threshold and gradually increasing the input delay Δt we can get a postsynaptic neuron's response window. The explicit analytical solution on the subthreshold activity of an IF neuron is often not easy while numerical analysis is a good alternative [12]. The numerical solution of the postsynaptic neuron's firing window width with respect to the change of synaptic and membrane time constants is shown in Fig. 2. It is obvious that larger synaptic time constant leads to more energy loss of input pulse on larger value of synaptic resistance or capacitance, hence less response window width; larger membrane time constant leads to more gradually decayed EPSP, hence the possibility that the neuron's membrane potential is greater than its threshold is increased.

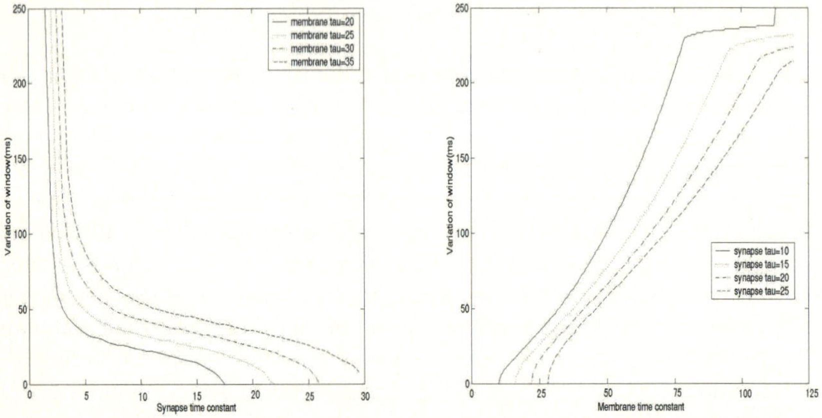

Fig. 2. Numerical solution of variation of the postsynaptic neuron's firing window size with respect to the change of synaptic (*top*) and membrane time constant (*bottom*), respectively. The synaptic efficacies $w_1 = w_2 = 20$, reversal potential $v_m = 0$, threshold $v_{thres} = -54mv$, resting potential $v_{rest} = -70mv$, refrectory period $t_{ref} = 3ms$.

2.2 Delay Adaptation Mechanism

The proposed model is not only a coincidence detector but also an efficient delay adaptor in the sense that it can adjust the activation window size of the postsynaptic neuron to adapt different input delays between two afferent synapses. The correlation between two pulses in two synapses can be displayed by depolarizing an IF neuron. The possible delay between the EPSPs of two pulses can be measured, in advance and offline, and then be saved in a memory to optimize the window size and to modify the circuit parameters through *a priori* adaptation. The initial random window width will thus converge to a deterministic value allowing the correlated stimuli to activate the postsynaptic neuron whilst rejecting the uncorrelated signals or noises out of the window.

Although both membrane and synaptic time constants are adjustable for the purpose of optimizing the activation window size, the adjustment of synaptic efficacies is usually preferred in neural computing because of the straightforward computational significance of adaptable synapses. The adaptation of window size is iteratively realized through depressing/potentiating a synaptic efficacy by a small amount according to whether the second (delayed) stimulus stays outside/inside the activation window of the last iteration. Thus we can have the adaptation as,

$$w_i^{syn}(t) = w_i^{syn}(t-1) \pm \Delta w_i^{syn} \tag{5}$$

where $w_i^{syn}(t)$ is the synaptic efficacy, Δw_i^{syn} is the change amount in an adaptation iteration. The activation window size of the IF neuron will be increased if its initial size is too small, otherwise it can be reduced.

3 The Retinotopic Application – A Case Study

When a three-dimensional object is mapped to a two-dimensional plane, its depth information is lost. Although binocular disparity is the overwhelming method used

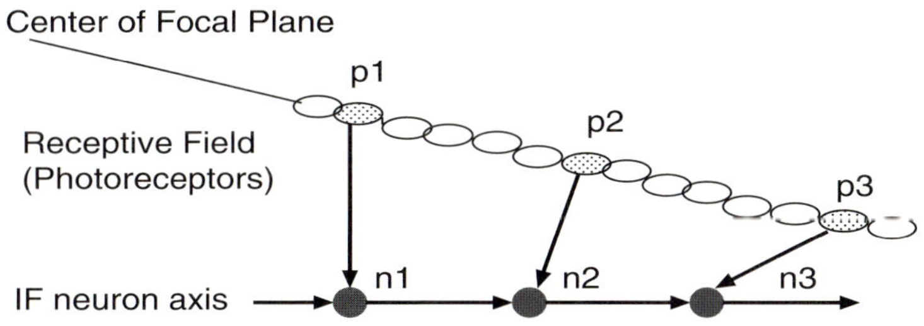

Fig. 3. Architecture of an axis of IF neurons and the corresponding photoreceptors. The photoreceptors align radially in the receptive field to transfer the irradiance edge inputs in the optical flow field to electrical pulse for the IF neurons. The dotted photoreceptors have actual connection with a corresponding neuron.

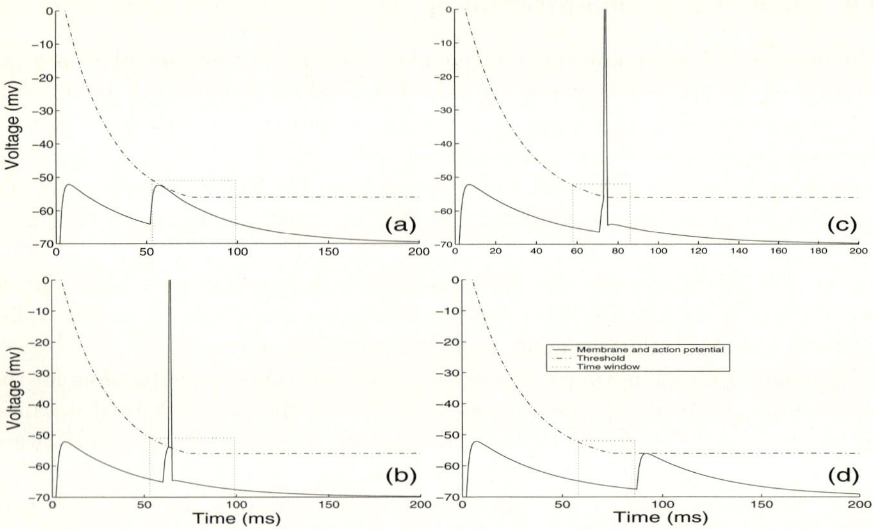

Fig. 4. Membrane dynamics of a neuron shows the temporal correlation of stimuli and response, and window adaptation. The first membrane depolarization is caused by the *flow synapse* spike, and the second by the edge from the *receptive synapse*. If the edge comes within a time window, then the neuron spikes. (a) An edge arrives at 52ms; (b) an edge arrives at 60ms; (c) an edge arrives at 71ms; and (d) an edge arrives at 87ms. A time window of [53, 99] is applied for (a), (b), and [58, 86] for (c), (d). $w_{recept} = 13.91$ for (a) and (b), $w_{recept} = 13.20$ for (c) and (d), the *flow synapse* $w_{flow} = 20$, membrane time constant $\tau_{mem} = 40$ms, synapse time constant $\tau_{syn} = 2$ms, neuron reset potential $v_{reset} = -65$mv, resting potential $v_{rest} = -70$mv.

to recover the depth [13], in this study we use a monocular depth perception algorithm which uses the optical flow in a dynamic scene to restore depth [14]. Our monocular vision model consists of a large amount of IF neurons. Since there is no vision comparison between two devices and each IF neuron is an independent computational unit which has only two inputs and one outputs, this model is highly scalable for different vision resolutions and amenable to aVLSI implementation.

We chose to use 400 identical radial axes arranged in a polar coordination plane to achieve a balance between image resolution and simulation time (and, ultimately, aVLSI hardware cost). There are 74 IF neurons in one axis, where the neurons have only nearest-neighbor, on axis connections (see Fig. 3). Besides this *flow synapse* connection, a neuron has another connection with a fixed pixel in the input layer, which is termed the *receptive synapse*. In the input layer we use 512×512 pixel array which records the input image sequence at the same resolutoin. An edge is detected if a pixel has more than 10% irradiance difference compared to its surrounding pixels.

If a neuron spikes due to the correlation of input stimuli from both synapses, it stimulates its neighboring neuron through the *flow synapse*. After a while when the corresponding edge comes, that neighboring neuron's membrane is

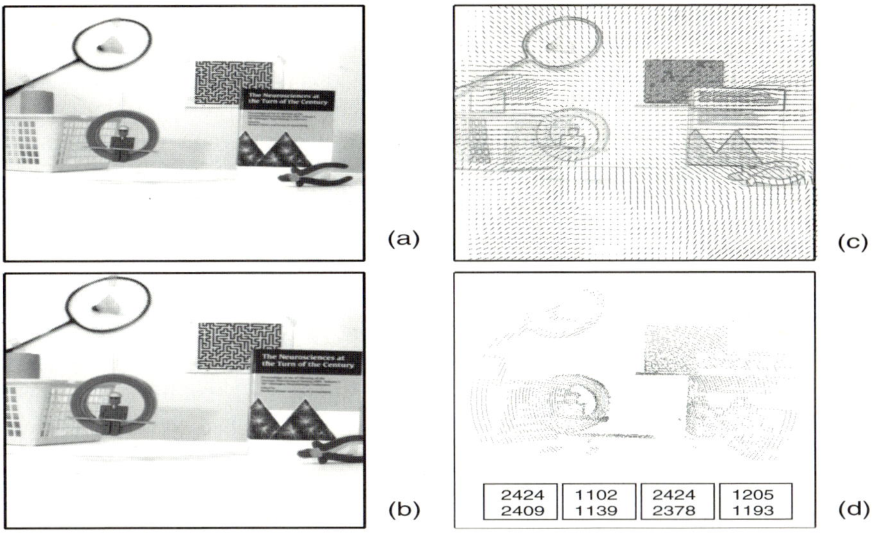

Fig. 5. Pixel maps of a real scene. (a) and (b) Snapshots of image frames at the start and 170 steps respectively. (c) Still contour of the first frame without depth detection processing, the size of the short lines signifies the strength of optical flow in the 2-D space. (d) A pixel map with the depth boxes corresponding to badminton racquet, toy pendulum, maze board and book, respectively, from left to right. Upper values in the boxes are the actual original depth of each object, lower values are estimated depth, both in millimeters.

further stimulated by its receptive synapse. If the EPSPs caused by the *flow* and *receptive* synapses are not correlated, then the neighboring neuron will not issue a spike (Fig. 4a). Otherwise a spike is issued (Fig. 4b). By adjusting the *flow synaptic* efficacy the activation window of a neuron can be changed (Fig. 4c,d), for more details of the model including the threshold mechanism see [15]).

The depth information of different objects in a real scene can be restored by feeding a sequence of recorded image frames to the model (Fig. 5). In the experiment we recorded 250 frames with a camera motion speed approximately 1.1 mm/frame. The image frames are saved in a bitmap forrmat with a resolutoin of 1276×1016 and 24 bits of color. The recording speed is 24 frames per second. The image frames are thereafter preprocessed to fit the final resolution of 512×512 and 8 bits of gray level, and transferred to a SUN Blade 100 workstation for processing.

The contours of the toy pendulum and book are easily recognizable from the pixel map in Fig. 5d. However, as the distance between edges in the maze board approximates that of the pixel layout density, significant detail cannot be seen. The less density of the badminton racquet pixels is due to its location at the periphery of the visual plane, where the photoreceptors and neurons are less dense. These results show that our model works well in a realistic setting with a reliable recovery of the depth information.

4 Conclusion

In this paper we have explored a simplified coincidence detection model in which the postsynaptic neuron's activity is dependent on the correlation of two synaptic inputs. We demonstrate that the intrinsic neuronal circuit parameters will determine a temporal tolerant window within which the effect of delayed inputs can result in the postsynaptic neuron's response. Afferent synchrony can thus be detected between the intercellular connections. The window size is adjustable through various model parameters such as synaptic efficacies, synaptic and membrane time constants which are directly related with the aVLSI circuit parameters, hence render the model applicable in hardware application. Our study thus provides an approach to the realization of transmission delay adaptation which has been recently suggested to coexist with traditional synaptic weight adaptation [16,17]. It is worth noting that coincidence detection can also be performed on multiple intercellular connections converging onto a postsynaptic cell by cascading many of this simple model together.

We have also proposed a vision application of this coincidence detection model in the optical flow field of an artificial retina algorithm. We suggest that it be possible to implement the neuromorphic information flow and noise/outlier rejection in an artificial retina by employing this biological synaptic plasticity mechanism.

Acknowledgement

This work is supported by a British EPSRC grant EP/E063322/1. Z.Yang is also supported by a Chinese NSF grant 60673102 and a Jiangsu Province grant BK2006218.

References

1. Hebb, D.: The organization of behavior. Wiley, New York (1949)
2. Markram, H., Lubke, J., Frotscher, M., Sakmann, B.: Regulation of synaptic efficacy by coincidence of postsynaptic aps and epsps. Science 275, 213–215 (1997)
3. Bi, G., Poo, M.: Activity-induced synaptic modifications in hippocampal culture, dependence on spike timing, synaptic strength and cell type. Journal of Neuroscience 18, 10464–10472 (1998)
4. Zhang, L., Tao, H., Holt, C., Harris, W., Poo, M.: A critical window for cooperation and competition among developing retinotectal synapses. Nature 395, 37–44 (1998)
5. Abbott, L., Nelson, S.: Synaptic plasticity: taming the beast. Nature Neuroscience 3, 1178–1183 (2000)
6. Zohary, E., Shadlen, M., Newsome, W.: Correlated neuronal discharge rate and its implications for psychophysical performance. Nature 370, 140–143 (1994)
7. Sheth, B., Sharma, J., Rao, S., Sur, M.: Orientation maps of subjective contours in visual cortex. Science 274, 2110–2115 (1996)
8. Bofill, A., Murray, A.: Circuits for vlsi implementation of temporally-asymmetric hebbian learning. Advances in Neural Information Processing Systems 14 (2001)

9. Indiveri, G.: Circuits for bistable spike-timing-dependent plasticity neuromorphic vlsi synapses. Advances in Neural Information Processing Systems 15 (2002)
10. Choe, Y., Miikkulainen, R.: The role of postsynaptic potential decay rate in neural synchrony. Neurocomputing 52-54, 707–712 (2003)
11. Kailath, T.: Correlation detection of signals perturbed by a random channel. IEEE Transaction on Information Theory 6(3), 361–366 (1960)
12. Tuckwell, H.: Introduction to theoretical neurobiology, vol. 1. Cambridge University Press, Cambridge (1988)
13. Qian, N.: Binocular disparity and the perception of depth. Neuron 18, 359–368 (1997)
14. Worgotter, F., Cozzi, A., Gerdes, V.: A parallel noise-robust algorithm to recover depth information from radial flow fields. Neural Computation 11, 381–416 (1999)
15. Yang, Z., Murray, A., Worgotter, F., et al.: A neuromorphic depth-from-motion vision model with stdp adaptation. IEEE Transactions on Neural Networks 17(2), 482–495 (2006)
16. Napp-Zinn, H., Jansen, M., Eckmiller, R.: Recognition and tracking of impulse patterns with delay adaptation in biology-inspired pulse processing neural net (bpn) hardware. Biological Cybernetics 74, 449–453 (1995)
17. Eurich, C.W., et al.: Dynamics of self-organized delay adaptation. Physical Review Letters 82(7), 1594–1597 (1999)

Synaptic Formation Rate as a Control Parameter in a Model for the Ontogenesis of Retinotopy

Junmei Zhu

Frankfurt Institute for Advanced Studies
Johann Wolfgang Goethe University, 60438 Frankfurt Germany
jzhu@fias.uni-frankfurt.de

Abstract. We study the shape of patterns formed under different values of a control parameter in a model system for the ontogenesis of retinotopy proposed by Häussler and von der Malsburg. Guided by linear modes, their eigenvalues and nonlinear interactions, a few deciding values of the synaptic formation rate α are chosen, under which final states are obtained using computer simulations of the full dynamics. We find that a precise topographic mapping can only be developed under a very narrow range of α close to its critical value. The allowed range of α is relaxed if the system is equipped with a proper structure, presumably by evolution.

1 Introduction

Mathematical models have made significant progress in understanding biological pattern formation and morphogenesis [1,2,3,4], with notable examples in cortical map formation [5,6,7,8]. Most studies focus on determining conditions under which an inhomogeneous pattern can emerge from a homogeneous state [1,9,8]. A less studied but very important question concerns the shape of the final pattern. Which pattern will actually appear not only determines how beautiful the coat of an animal is [10], or what hallucination patterns one sees [9], but, more importantly, can have functional consequences.

In this paper, we study the shape of patterns formed in a model system for the ontogenesis of retinotopy proposed by Häussler and von der Malsburg (here called the *Häussler system*) [7]. In particular, we are interested in conditions under which the desired patterns, which in retinotopy development are topographic mappings, can be developed. It has been shown that the Häussler system is able to develop a precise topological projection [7]. A spatial pattern is possible when the homogeneous state becomes unstable as the control parameter, the synaptic formation rate α, passes its critical value α_c ($\alpha < \alpha_c$). The proof in [7] is performed where α is slightly below α_c. For such a large α the formed mapping is rather broad, so α has to be reduced to 0 for a precise mapping to be formed. However, we have proven that when $\alpha = 0$, the system has exponentially many fixed points, including at least all permutation matrices [11]. A random sample showed that over half of the fixed points are in fact stable. As random permutation matrices rarely correspond to topographic mappings, it has thus been a

V. Kůrková et al. (Eds.): ICANN 2008, Part II, LNCS 5164, pp. 462–470, 2008.

puzzle why normal development does not end up at discontinuous mappings, which could lead to functional disaster. We attempt to answer this question by studying the range of α under which the desired topographic mappings can be formed. By studying this particular system, which is typical for brain organization and is simple enough to allow analytical treatment, we hope to gain insights on general self-organizing systems.

Our study uses a combination of analysis and simulation. Whereas a linear analysis can determine whether patterns are possible, it is a nonlinear problem to specify what patterns will emerge. A systematic approach to nonlinear systems is the weakly nonlinear analysis which is an analysis performed near the bifurcation point, where the system can be reduced into a low dimensional manifold of modes that become unstable [12,4]. When control parameters are far away from their critical values, however, a large number of modes could become unstable, and the weakly nonlinear analysis can at most provide a guidance. We therefore resort to using computer simulations of the full dynamics to obtain final patterns. This simulation, however, is no longer aimless as we will be able to choose a few deciding values of α, from the analysis of linear modes, their eigenvalues and nonlinear interactions.

2 The Häussler System for the Ontogenesis of Retinotopy

2.1 System Description

The Häussler system [7] addresses the establishment of ordered projections between two brain areas, retina and tectum. These are modelled as one-dimensional chains with N elements each (for generalization to general manifolds, see [13]). The projection between the two brain areas is represented by a set of links (τ, ρ), where τ and ρ are points in the tectum and retina, respectively. The weight $w_{\tau\rho}$ of link (τ, ρ) indicates the strength with which τ and ρ are connected, with a value zero representing the absence of a connection. The set of all links forms a mapping $W = (w_{\tau\rho})$. The Häussler system is described by the set of $n = N \times N$ differential equations:

$$\dot{w}_{\tau\rho} = f_{\tau\rho}(W) - \frac{1}{2N} w_{\tau\rho} \left(\sum_{\tau'} f_{\tau'\rho}(W) + \sum_{\rho'} f_{\tau\rho'}(W) \right) \qquad (1)$$

where the growth term $f_{\tau\rho}(W)$ of link $w_{\tau\rho}$ expresses the cooperation from all its neighbors, and α is a non-negative synaptic formation rate:

$$f_{\tau\rho}(W) = \alpha + w_{\tau\rho} \sum_{\tau',\rho'} C(\tau, \tau', \rho, \rho') w_{\tau'\rho'}. \qquad (2)$$

The coupling function $C(\tau, \tau', \rho, \rho')$, which in [7] was assumed to be separable and isotropic, describes the mutual cooperative support that link (τ, ρ) receives from its neighbor (τ', ρ'). Here we model the C function as a 2D Gaussian

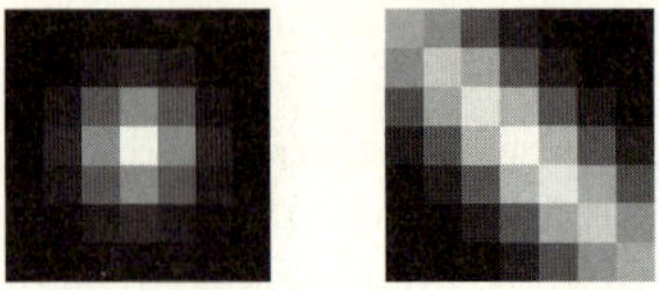

(a) C (b) T

Fig. 1. Coupling functions in the form of 2D Gaussian. (a): C function, isotropic; (b): T function, more structured.

$C(\tau,\tau',\rho,\rho') = C(\tau-\tau',\rho-\rho') = \frac{1}{2\pi\sigma_1\sigma_2}exp(-\frac{(\tau-\tau')^2}{2\sigma_1^2} - \frac{(\rho-\rho')^2}{2\sigma_2^2})$. An example of an isotropic C function is shown in Fig. 1(a), with $\sigma_1 = \sigma_2 = 1$ and support size 7×7. The coupling function can also be more structured, such as the T function shown in Fig. 1(b), which is a Gaussian with $\sigma_1 = 1, \sigma_2 = 5$ rotated by $\pi/4$. T is elongated toward one orientation, favoring mappings in that orientation.

Unless otherwise stated, we use $N = 32$, C coupling function as in Fig. 1(a) and periodic boundary conditions in all examples throughout this paper. A final pattern in simulation means the steady state solution obtained by simulating the full dynamics (1) using the fourth-order Runge-Kutta method.

2.2 System Analysis

The goal of analyzing the Häussler system is to prove that it develops a precise topographic projection. Specifically, starting from around a homogeneous state, the steady state solution of system (1) is a diagonal matrix. A complete proof is given in [7], and we briefly describe the method here.

Under periodic boundary conditions, the homogeneous state $W_0 = \mathbf{1}$ (where $\mathbf{1}$ is the unit matrix in which all entries equal 1) is a fixed point of the system. By introducing the deviation $V = W - W_0$ as a new variable, the system becomes, in a more general form,

$$\dot{V} = \mathbf{L}V + Q(V) + K(V), \tag{3}$$

where $V = (v_1, \cdots, v_n)^T$ (with ^{T} for transpose) is the n-dimensional state vector ($n = N^2$). $\mathbf{L}V, Q(V), K(V)$, indicating linear, quadratic and cubic terms, respectively, are functions of N, α and the coupling function, whose exact form can be found in [7].

The field V can be represented as the superposition of modes u^i which are eigenfunctions of the linear term: $\mathbf{L}u^i = \lambda_i u^i, i = 1\cdots n$. An example of modes, together with their respective eigenvalues (with $\alpha = 0$) is shown in Fig.2. The control parameter α shifts the eigenvalues of all modes uniformly (i.e., eigenvalue λ becomes $\lambda - \alpha$).

The temporal dynamics of the mode amplitude ξ_i ($V(t) = \sum_i \xi_i(t)u^i$) are

$$\dot{\xi}_i = \lambda_i\xi_i + \sum_{ll'} \xi_l\xi_{l'}\tilde{u}^i Q^{ll'} + \sum_{ll'l''} \xi_l\xi_{l'}\xi_{l''}\tilde{u}^i K^{ll'l''}. \tag{4}$$

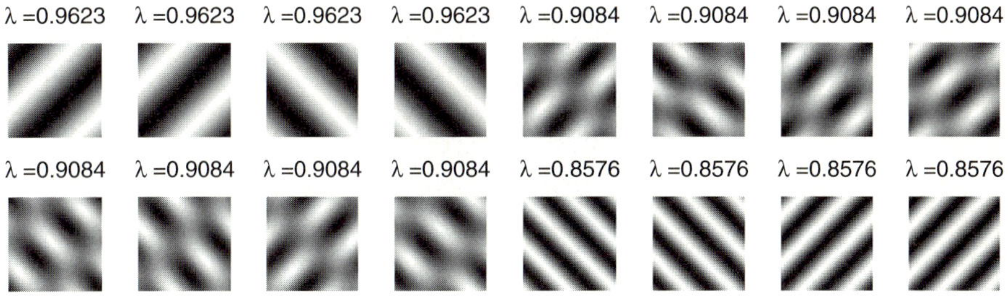

Fig. 2. Modes and their eigenvalues of the Häussler system with $N = 32$, C coupling function as in Fig. 1(a), calculated at $\alpha = 0$. Shown are the 16 modes with the largest eigenvalues.

where $\tilde{u}^i, i = 1, \cdots, n$ are the left eigenvectors of $\mathbf{L}$: $\tilde{u}^i\mathbf{L} = \lambda_i\tilde{u}^i$, which are orthonormal to the modes ($\tilde{u}^i u^j = \delta_{ij}$). $Q^{ll'}$ ($K^{ll'l''}$) are vectors computed from the quadratic (cubic) term with modes u^l, $u^{l'}$ (and $u^{l''}$). Definition of these terms and a derivation of (4) are given in [14]. These mode amplitude equations give the nonlinear interaction between modes. In weakly nonlinear analysis, they are reduced to those of the unstable modes only, by adiabatic elimination [14,12,7]. In the Häussler system the control parameter α is set to be slightly below its critical value, such that only the 4 modes with the maximum eigenvalue become unstable (see Fig. 2). The two diagonal orientations compete with each other, and the winning one activates modes of the same orientation, forming a narrow diagonal [7].

3 Control Parameter α

We study patterns formed in the Häussler system under different values of the control parameter α. Its critical value α_c is equal to the largest eigenvalue of the system calculated when $\alpha = 0$. When $\alpha > \alpha_c$, $W = \mathbf{1}$ is stable and therefore no pattern will be formed. With the specific values of N and C function we use, $\alpha_c = 0.9623$, which can be read out from the eigenvalues in Fig. 2.

An initial state is a random perturbation around the homogeneous state $W = \mathbf{1}$. In our experiments, perturbation is a number uniformly distributed in $[-0.1, 0.1]$, independent for each element of the matrix. An example is shown in Fig. 3. In the following experiments, we show final patterns formed starting from this initial state. Other such randomly generated initial states give qualitatively similar behavior. Especially we are interested in whether the final pattern is a precise topographic mapping, a term we reserve for the diagonal matrices. By precise we mean one-to-one.

3.1 α Close to Its Critical Value

The pattern formed at $\alpha = 0.95$ is shown in Fig. 4(a). This is a topographic mapping, as the analytical treatment implies. Note that the diagonal in the

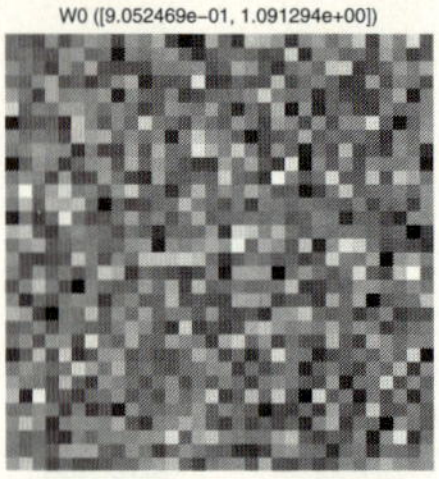

Fig. 3. Initial state, a random perturbation around $W = \mathbf{1}$. The numbers on top are the actual range of link weights.

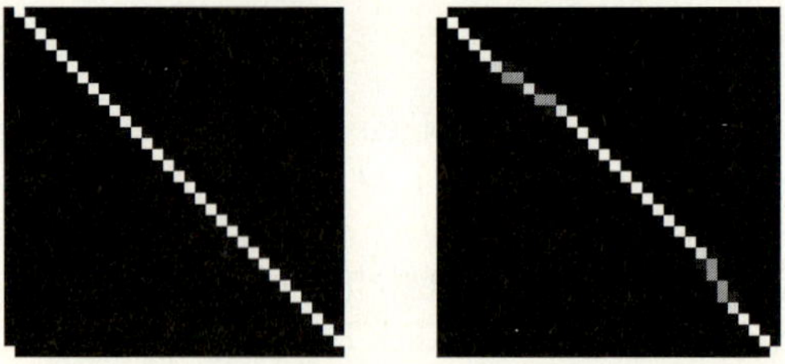

(a) $\alpha = 0.95$ (b) $\alpha = 0.90$

Fig. 4. Final state at different α values

figure is shifted, but it is still a topographic mapping because of the periodic boundary condition.

In the theoretical analysis, α needs to be very close to α_c. If $\alpha - \alpha_c$ is beyond a certain scale, the weakly nonlinear analysis no longer holds. Simulations can provide hint on this scale in real systems. Fig. 4(b) shows the finial pattern for $\alpha = 0.90$. It can be seen that this mapping, although still continuous, is distorted from the diagonal. From this example, we can see that only a narrow range of the control parameter α can lead to a perfect topographic mapping.

3.2 α Decreases at a Stable Pattern

A precise mapping needs $\alpha = 0$, as final patterns obtained when α is big could be rather broad. When a control parameter is changed, the stability of states needs to be reevaluated. Assume that a stable pattern with amplitudes ξ^* is formed under α. When the control parameter is decreased to $\alpha' < \alpha$, we have, at ξ^*, $\dot{\xi}_i = (\alpha - \alpha')\xi_i$, which is nonzero unless $\xi_i = 0$. This means that pattern ξ^* is no longer a fixed point of the system when α changes.

Linear stability analysis performed at the stable pattern formed under $\alpha = 0.95$ (Fig. 4(a)) shows that all eigenvalues are still negative when $\alpha = 0$. A plot of the first 150 largest eigenvalues, at both $\alpha = 0.95$ and $\alpha = 0$, is shown in Fig. 5. In this case the decrease of α does not lead to any bifurcation. The dynamics is simply that all nonzero amplitudes approach their respective equilibrium exponentially fast, which help to sharpen the mapping. The evolution of this stable

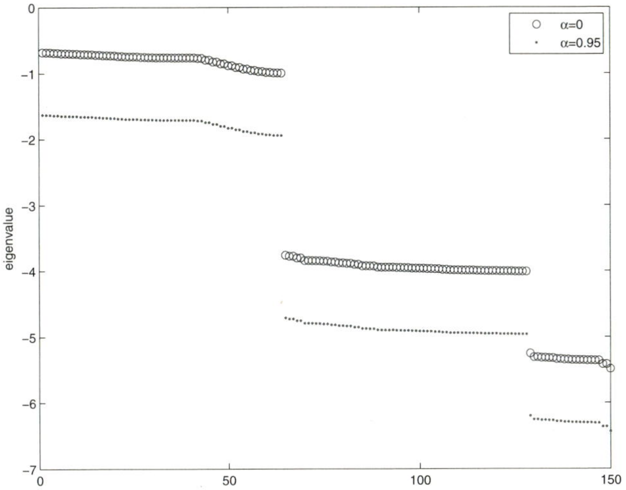

Fig. 5. Linear analysis at the final state formed under $\alpha = 0.95$ (Fig. 4(a)). The first 150 largest eigenvalues are shown, for $\alpha = 0.95$ and $\alpha = 0$.

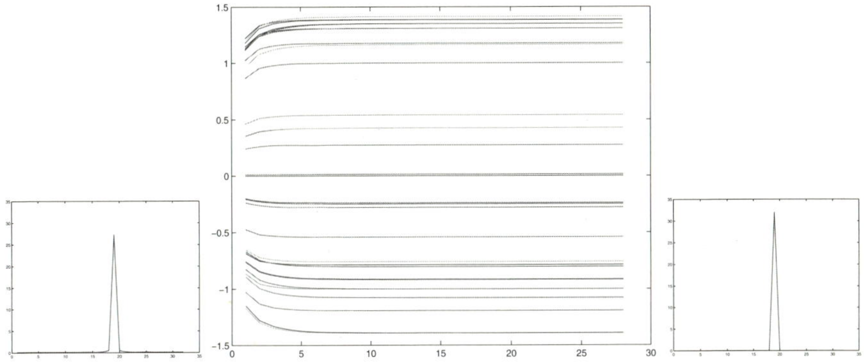

Fig. 6. Dynamics of mode amplitude when α is decreased to 0. The left figure shows one row of the final state with $\alpha = 0.95$ (Fig. 4(a)); When α is decreased to 0, this state settles down to another pattern, whose same row is shown in the right figure. The transition between the two states is shown in the middle figure, plotted as mode amplitudes versus the number of iterations. Each curve is for one mode.

pattern (Fig. 4(a)) when α is decreased to 0 is shown in Fig. 6. The amplitudes are based on modes calculated at $W = \mathbf{1}$.

3.3 $\alpha = 0$

The pattern formed at $\alpha = 0$ is shown in Fig. 7. It is not even a continuous mapping, but nonetheless a stable fixed point of the system. One hypothesis to explain why development does not end up at mappings like this was that these discontinuous stable fixed points have impractically small basins of attraction that do not include normal initial conditions. In light of the example here, however, we now believe that it is the proper value of the control parameter that defines the path to a precise topographic mapping.

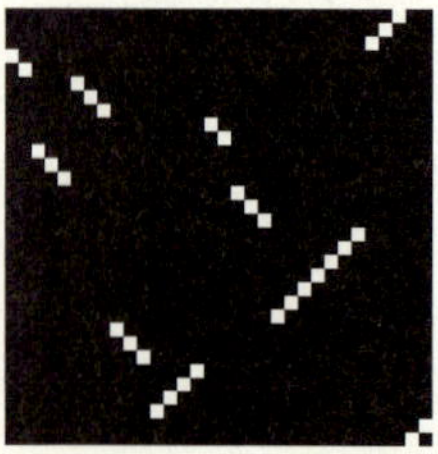

Fig. 7. Final state, $\alpha = 0$

3.4 *T* Coupling Function

In this section we show a few examples of pattern formation in the Häussler system with the more structured T coupling function. Presumably evolution provides biases like this.

The modes of the Häussler system with the T coupling function as in Fig. 1(b), together with their respective eigenvalues (with $\alpha = 0$) are shown in Fig.8. The critical value for α is now 0.9685, comparable to that of the C system.

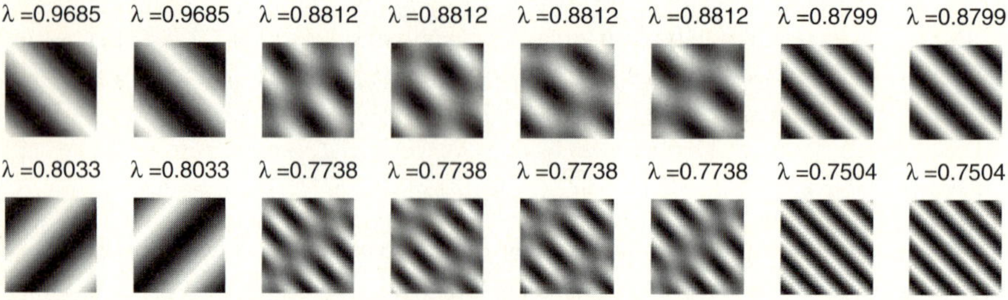

Fig. 8. Modes and their eigenvalues of the Häussler system with coupling function T. Shown are the 16 modes with the largest eigenvalues.

A few final patterns formed at different values of α are shown in Fig. 9. In this setting, a perfect topographic mapping fails to develop only when $\alpha < 0.81$. Compared to the C function, a larger range of α can give topographic mappings in the T system. (The choice of 0.81 is guided by the intuition to exclude modes of the other orientation, which first show as the first two modes on the second row in Fig. 8. Justification of the intuition is currently under way.)

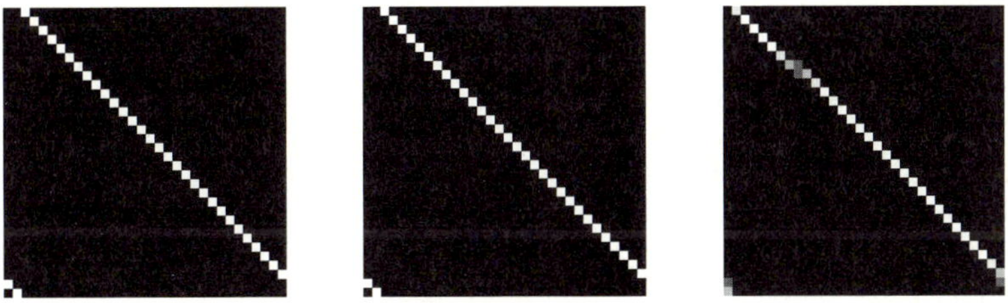

(a) $\alpha = 0.95$ (b) $\alpha = 0.81$ (c) $\alpha = 0.80$

Fig. 9. Final state of T coupling function at different α values

4 Conclusions and Discussions

We have studied the shape of patterns formed in the Häussler system at different values of a control parameter, the synaptic formation rate α. A better understanding of the system behavior has been obtained by following the system at a few deciding values of α using computer simulations of the full dynamics. We found that in order to develop a topographic mapping, α has to be first set to be close to its critical value. When α is subsequently reduced to zero, the broader topographic mapping formed under larger α is no longer a fixed point, and will approach exponentially to its final state, a precise mapping. Moreover, the interval of values that permit a topographic mapping is very narrow: α too large, no patterns will be formed; α too small, mappings become discontinuous. We also found that, if the system has more structure, as in the T coupling function, the range of α under which a topographic mapping can be developed is bigger. This less sensitivity on the control parameter gives a more robust system.

It would be interesting to measure α in biological systems at different stages of development. Note that α is a synaptic formation rate which is different from actual number of synapses. A related concept in biology is that during development there are more synapses formed which are selectively eliminated in later stage [15].

The methodology we use in this work shows an effective way to explore the behavior of a system. Guided by the concept of mode analysis, a few examples can be designed and solutions be obtained by simulating the full dynamics, from which intuitive understanding of the system can be obtained. Although not a rigorous proof, the results can be very revealing.

Acknowledgements. Supported by EU projects Daisy and SECO, the Hertie Foundation and the Volkswagen Foundation.

References

1. Turing, A.M.: The chemical basis of morphogenesis. Phil. Trans. Roy. Soc. London B, Biological Sciences 237(641), 37–72 (1952)
2. Meinhardt, H.: Models of biological pattern formation. Academic Press, London (1982)

3. Murray, J.D.: Mathematical Biology. Springer, Heidelberg (2003)
4. Cross, M.C., Hohenberg, P.C.: Pattern formation outside of equilibrium. Reviews of Modern Physics 65, 851–1112 (1993)
5. Willshaw, D.J., von der Malsburg, C.: A marker induction mechanism for the establishment of ordered neural mappings: its application to the retinotectal problem. Phil. Trans. Roy. Soc. London B B287, 203–243 (1979)
6. Swindale, N.V.: A model for the formation of orientation columns. Proc. R. Soc. Lond. B 215, 211–230 (1982)
7. Häussler, A.F., von der Malsburg, C.: Development of retinotopic projections — an analytical treatment. Journal of Theoretical Neurobiology 2, 47–73 (1983)
8. Obermayer, K., Blasdel, G.G., Schulten, K.: Statistical-mechanical analysis of self-organization and pattern formation during the development of visual maps. Phys. Rev. A 45, 7568–7589 (1992)
9. Ermentrout, G.B., Cowan, J.D.: A mathematical theory of visual hallucination patterns. Biological Cybernetics 34, 137–150 (1979)
10. Ermentrout, B.: Stripes or spots? Nonlinear effects in bifurcation of reaction-diffusion equations on the square. Proceedings: Mathematical and Physical Sciences 434(1891), 413–417 (1891)
11. Zhu, J.: Analysis of map formation in visual perception. In: Botelho, Hagen, Jamison (eds.) Fluids and Waves — Recent Trends in Applied Analysis. AMS Contemporary Mathematics, vol. 440, pp. 273–287. American Mathematical Society (2007)
12. Haken, H.: Synergetics: An Introduction: Nonequilibrium Phase Transitions and Self-Organization in Physics, Chemistry, and Biology. Springer, Berlin (1977)
13. Güssmann, M., Pelster, A., Wunner, G.: Synergetic analysis of the Häussler-von der Malsburg equations for manifolds of arbitrary geometry. Annalen der Physik (Leipzig) 16(5–6), 379–394 (2007)
14. Zhu, J., von der Malsburg, C.: Derivation of nonlinear amplitude equations for the normal modes of a self-organizing system. In: Proceedings European Symposium on Artificial Neural Networks (ESANN), pp. 409–414 (2007)
15. Luo, L., O'Leary, D.D.: Axon retraction and degeneration in development and disease. Annu. Rev. Neurosci. 28, 127–156 (2005)

Fuzzy Symbolic Dynamics for Neurodynamical Systems

Krzysztof Dobosz[1] and Włodzisław Duch[2]

[1] Faculty of Mathematics and Computer Science
[2] Department of Informatics, Nicolaus Copernicus University, Toruń, Poland
kdobosz@mat.umk.pl

Abstract. Neurodynamical systems are characterized by a large number of signal streams, measuring activity of individual neurons, local field potentials, aggregated electrical (EEG) or magnetic potentials (MEG), oxygen use (fMRI) or concentration of radioactive traces (PET) in different parts of the brain. Various basis set decomposition techniques try to discover components that carry meaningful information are used to analyze such signals, but these techniques tell us little about the activity of the whole system. Fuzzy Symbolic Dynamics (FSD) may be used for dimensionality reduction of high-dimensional signals, defining non-linear mapping for visualization of trajectories that shows various aspects of signals that are difficult to discover looking at individual components, or to notice observing dynamical visualizations. FSD can be applied to raw signals, transformed signals (for example, ICA components), or to signals defined in the time-frequency domain. Visualization of a model system with artificial radial oscillatory sources, and of the output layer (50 neurons) of a neural Respiratory Rhythm Generator model (RRG) that includes 300 spiking neural units, are presented to illustrate the method.

1 Introduction

Neuroimaging data and simulated neurodynamical systems are characterized by multiple streams of nonstationary data, and thus may be represented only in highly dimensional signal spaces. Understanding such signals is not easy because of high volume of data that quickly changes in time. Simulation of complex dynamics is usually described in terms of attractors, but precise characterization of their basins and possible transitions between them is never attempted. Popular signal processing techniques remove artifacts by various filtering techniques, involve waveform analysis, morphological analysis, decomposition of data streams into meaningful components using Fourier or Wavelet Transforms, Principal and Independent Component Analysis (ICA), etc [1,2]. Interesting events are then searched for using processed signal components, with time-frequency-intensity maps calculated for each component. Such techniques are very useful, but do not show global properties of the processes in the high-dimensional signal spaces. Global analysis is needed to see attractors that trap dynamics, characterize the type of system's behavior, notice partial desynchronization or high frequency noise that blurs the trajectories. For brain-computer interfaces and other applications a quasi-static snapshot of the whole trajectory, showing its main characteristics, could be very useful.

V. Kurková et al. (Eds.): ICANN 2008, Part II, LNCS 5164, pp. 471–478, 2008.

Most important properties of the dynamics may be uncovered if dimensionality of the problem is sufficiently reduced. This may be done with the help of fuzzy symbolic dynamics (FSD), introduced in this paper. To see the trajectories of the global system state "probes", or localized functions that are activated by the trajectories that pass near their center, are placed in the signal space. Using k such functions, strategically placed in important points of the signal space, a non-linear reduction of dimensionality suitable for visualization of trajectories is achieved. Inevitably a lot of details will be lost but with a proper choice of parameters the information that correlates with observed behavior or experimental task may be preserved, while irrelevant information will be suppressed.

In the next section FSD mapping that captures interesting properties of trajectories is described. To understand how to set up mapping parameters and how to interpret resulting images a model EEG situation is analyzed in Sec. 3, with several waves sources placed in a mesh, and sensors that record the amplitude of the incoming waves in different points of the mesh. As an example of real application in Sec. 4 trajectory visualizations for neural Respiratory Rhythm Generator model (RRG) are analyzed. The final section contains a brief discussion with a list of several open questions.

2 Fuzzy Symbolic Dynamics

Assume that some unknown sources create a multi-dimensional signal that is changing in time, for example an EEG signal measured by n electrodes:

$$x(t) = \{x_i(t)\} \quad i = 1, \dots, n \quad t = 0, 1, 2, \dots \ . \tag{1}$$

Vectors $x(t)$ represent the state of the dynamical system at time t, forming a trajectory in the signal space. Observing the system for a longer time should reveal the landscape created by this trajectory, areas of the signal space where the state of the system is found with the highest probability, and other areas where it never wonders. Recurrence maps and other techniques may be used to view this trajectory, but do not capture many important properties that it reflects.

In the symbolic dynamics [3] the signal space is partitioned into regions that are labeled with different symbols, emitted every time the trajectory is found in one of these regions. The sequence of symbols gives a coarse-grained description of dynamics that can be analyzed using statistical tools. Dale and Spivey [4] argue that symbolic dynamics gives an appropriate framework for cognitive representations, although discretization of continuous dynamical states looses the fluid nature of cognition. Symbols obviously reduce the complexity of dynamical description but partitioning of highly-dimensional signal spaces into regions with sharply defined boundaries is highly artificial.

The notion of symbolic dynamic is generalized in a natural way to a Fuzzy Symbolic Dynamics (FSD). Instead of discrete partitioning of the signal space leading to symbols, interesting regions are determined analyzing probability density $p(x)$ of finding the trajectory $x(t)$ in some point x, averaging over time. Local maxima of this probability define quasi-stable states around which trajectories tend to clusters. Such maxima may serve as centers μ_k of prototypes associated with fuzzy membership functions $y_k(x; \mu_k)$ that measure the degree to which the $x(t)$ state belongs to the prototype μ_k. Membership functions may be defined using knowledge-based clustering [5], or as prototype-based rules with context-based clustering techniques [6]. Context is defined by questions that are of interest, for example discrimination between different experimental conditions, or searching for invariants in one of these condition. For visualization Gaussian membership functions are quite useful:

$$y_k(x; \mu_k, \Sigma_k) = \exp\left(-\left(x-\mu_k\right)^T \Sigma_k^{-1} \left(x-\mu_k\right)\right) \tag{2}$$

Diagonal dispersions Σ_k are frequently sufficient, suppressing irrelevant signals, but general covariance matrices (used in Mahalanobis distance) may extract more meaningful combinations of signals that correlate with experimental conditions, or with qualities that may be estimated in a subjective way. Such brain-mind mapping will be closer to the idea of cognitive representations than symbolic dynamics [4]. Symbolic description may be easily generated by strongly activated prototypes, but other prototypes may correspond to sensorimotor actions that are not directly connected with symbolic labels.

Selecting only two prototypes trajectories $x(t)$ may be visualized in a two-dimensional space $\{y_i(t), y_j(t)\}$. If all Gaussian components have the same variance a single parameter will define dispersion. For visualization each pair of functions should have sufficiently large dispersions σ_i and σ_j to cover the space between them, for example $\sigma_i = \sigma_j = \frac{1}{2}\left\|\mu_i - \mu_j\right\|$. 3D visualization can also be done by plotting transformed points for three clusters, one for each dimension. Dispersions should then be set to the largest among the 3 pairs. Pairwise plots can be used to observe trajectory from different points of view. Normalization of vectors in the signal space is assumed. To distinguish several experimental conditions optimization of parameters of membership functions should be done using learning techniques, creating clear differences in corresponding maps. Adding more localized functions in some area where dynamics is complex will show fine structure of the trajectory.

An alternative to fuzzy membership functions is to define reference points in the signal space, and measure the distance between the trajectory and these points using some metric function. Non-linear metric functions should have some advantage in analysis of neurodynamics, as the influence of the trajectory on prototypes should sharply decrease to zero with the distance, reflecting non-linear properties of neurons. We shall not consider here adaptation of parameters or distance-based visualization, concentrating instead on the interpretation of global mappings.

3 Plane and Radial Waves on a Grid

To understand the structure of complex EEG and similar signals a very simple artificial model has been created. Sensors are placed on a quadratic grid with $n \times n$ points, where plane and radial waves generated by several sources are traveling, creating additive patterns and activating these sensors. Similar assumptions are made about electric potentials reflecting neuronal activity in the brain (for example, in the low resolution electromagnetic tomography, LORETA[1]).

---

[1] See http://www.unizh.ch/keyinst/loreta

The grid has equally spaced points $p_{ij} = (x_i, y_j)$ inside the square:

$$x_i, y_j \in \left\{0, \frac{1}{n-1}, \ldots, \frac{n-2}{n-1}, 1\right\} \quad i, j = 1, \ldots, n \ . \tag{3}$$

The activation of the sensor due to a plane wave $F^{(l)}(t, x)$ traveling through the square in the grid point p_{ij} at the time $t = 0, 1, 2, \ldots$ is given by the equation:

$$F^{(l)}\left(t, p_{ij}\right) = \cos\left(\omega_l t - \boldsymbol{k}_l \cdot \boldsymbol{p}_{ij}\right) \ , \tag{4}$$

where ω_l is the frequency of the wave (defining time intervals), the wave vector $\boldsymbol{k}_l$ defines the direction of the wave movement and its length is equal to the inverse of the wave length and $\boldsymbol{p}_{ij}$ is the vector pointing to the grid point p_{ij}. Thus, for horizontal plane wave $(\boldsymbol{k} = ||\boldsymbol{k}||[1, 0]^T)$ formula (4) becomes:

$$F\left(t, p_{ij}\right) = \cos\left(\omega t - k x_i\right) \ . \tag{5}$$

Radial wave reaching the sensor at grid point p_{ij} leads to an activation:

$$R^{(l)}\left(t, p_{ij}\right) = \cos\left(\omega_l t - k_l r^{(l)}\right) \ , \tag{6}$$

where

$$r^{(l)} = \sqrt{\left(x_i - x_0^{(l)}\right)^2 + \left(y_j - y_0^{(l)}\right)^2} \tag{7}$$

is the distance between point p_{ij} and the wave source (x_0, y_0).

The final activation $A\left(t, p_{ij}\right)$ of the sensor in point p_{ij} at time $t = 0, 1, 2, \ldots$ is obtained by summing and normalizing all wave values in every grid point:

$$A\left(t, p_{ij}\right) = \left(\sum_{l=1}^{Nf} F^{(l)}(t, p_{ij}) + \sum_{l=1}^{Nr} R^{(l)}(t, p_{ij})\right) / (Nf + Nr) \ . \tag{8}$$

Sensor activations form a $n \times n$ matrix $A(t)$ containing values for all sensors at time t. Elements of $A(t)$ are defined in n^2-dimensional signal space and are in the $[-1, 1]$ interval. Gaussian membership functions (2) may serve as probes (detectors of activity) in this space. Placing their centers in two opposite vertices of the hypercube $\mathcal{S} = [-1, 1]^{n^2}$:

$$\mu_1 = [-1, \ldots, -1]^T \qquad \mu_2 = [1, \ldots, 1]^T \tag{9}$$

the membership functions take all n^2 sensor activations $A(t)$ as their argument:

$$G_k(A(t); \mu_k, \sigma_k) = \exp\left(-\frac{\|A(t) - \mu_k\|}{2\sigma_k^2}\right) \ , \tag{10}$$

where σ_k is the dispersion.

A lot of experiments have been conducted using the 16×16 grid with 256 points (maximum number of electrodes used in real EEG experiments), and various number of stationary and non-stationary sources, frequencies and directions. For this grid $\sigma_1 = \sigma_2 = \|\mu_1 - \mu_2\| / 10$ gives relatively wide range of sensor activations. In Fig. 1 examples of trajectories for one and two radial waves are presented, using $\omega = 0.1$, which is sufficient for smooth trajectory change, and the wave vector length $\|k\| = 2\pi$. Specific position of sources and combinations of planar and radial waves may be identified with correct placement of centers and dispersions of the membership functions.

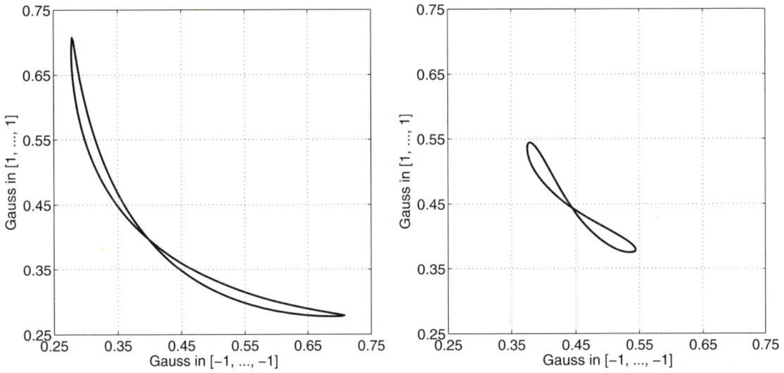

Fig. 1. Trajectories for one radial wave with the source at point $(\frac{1}{2}, \frac{1}{2})$ (left side), and two radial waves with the sources at $(\frac{1}{4}, \frac{1}{4})$ and $(\frac{3}{4}, \frac{3}{4})$ (right side)

4 Visualization of the Activity of Respiratory Rhythm Generator

FSD approach has been used to study behavior of the neural Respiratory Rhythm Generator model (RRG). The RRG is a parametric neural network model constructed from three populations of spiking neurons: beaters (200 in the model), bursters (50 units) and followers (50 units). The last population produce an output of model activity that is used for synaptic excitation of motoneurons and in consequence control of upper and lower lung muscles. Our implementation of RRG is based on the spiking neural network model described in [7].

Below visualization of the followers (output layer neurons) is examined. The first trajectory for time series corresponding to a single burst is presented in Fig. 2. The number of samples along these trajectory was 49090, each vector containing membrane potentials of 50 follower cells (normalized in every dimension). Clusterization was done with the k-means algorithm, for two clusters where Gaussian probe functions have been placed. Trajectories have been drawn with a thick pen to account for a jitter that blurs them when longer time sequences are taken.

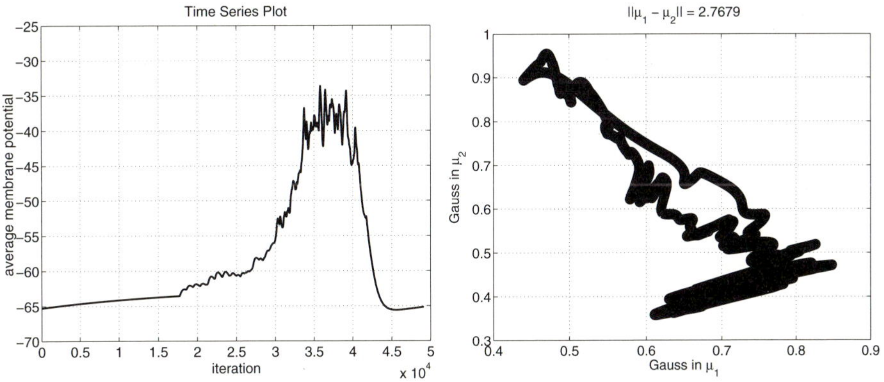

Fig. 2. The time series plot (left) representing average membrane potential vs. iteration number, and the mapping of the corresponding trajectory (right)

In Fig. 3 three cluster centers have been defined using the k-means algorithm ($k = 3$). Pairwise diagrams show trajectories for all three cluster pairs. Distances between cluster centers are printed above the graphs. The second pair is more sensitive to variability that appears during building of the discharge activity, showing quite a bit of variance in this process.

The RRG model may generate various rhythms that correspond to different breathing patterns. Trajectory examples in Fig. 4 compare two distinct cases, one for normal, regular burst generation, and one for pathological case with different burst strengths

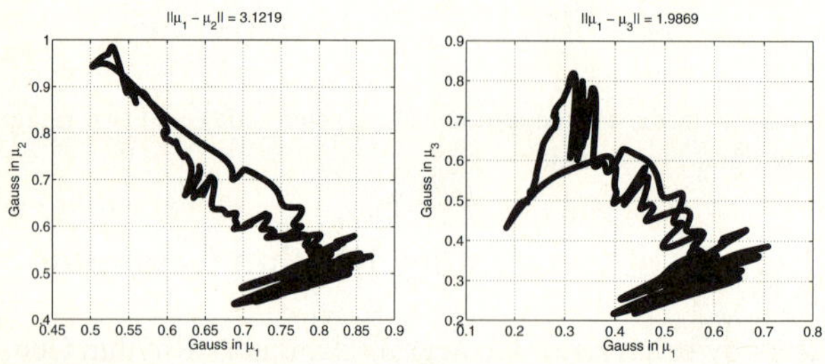

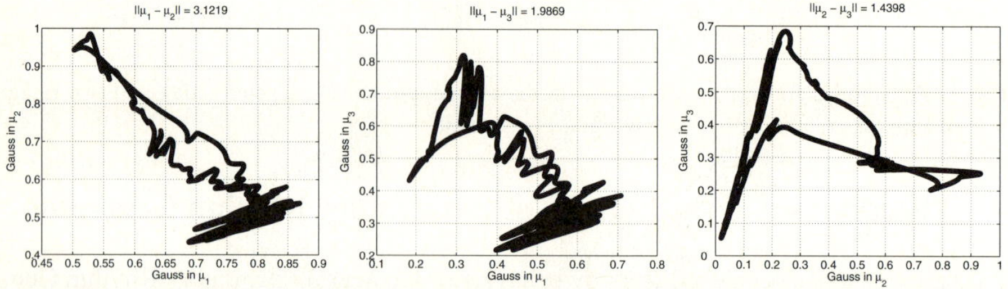

Fig. 3. Pairwise diagrams for 3 clusters found by k-means algorithm representing trajectories for time series with one burst

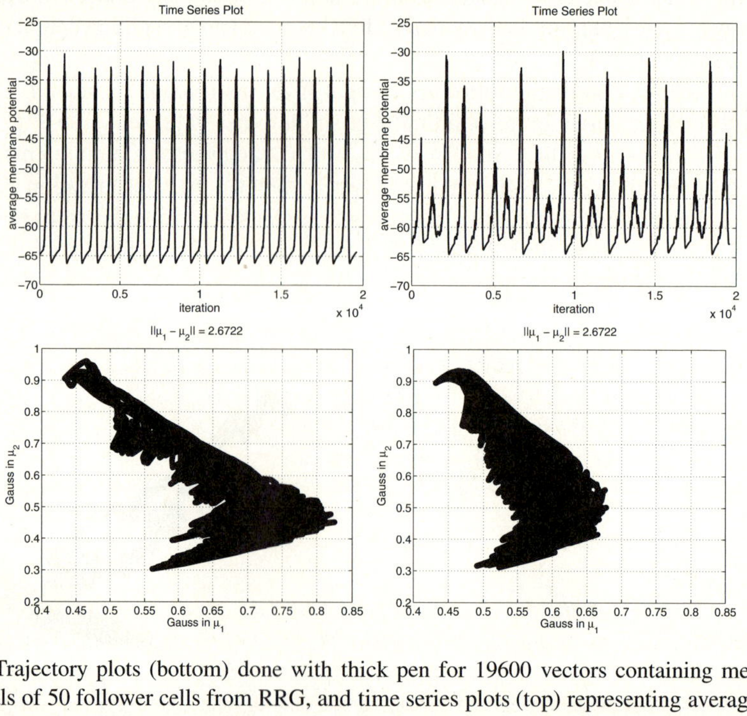

Fig. 4. Trajectory plots (bottom) done with thick pen for 19600 vectors containing membrane potentials of 50 follower cells from RRG, and time series plots (top) representing average membrane potential vs. iteration number. Graphs on the left correspond to a normal rhythm case, and on the right to a pathological one, both presented using the same membership functions.

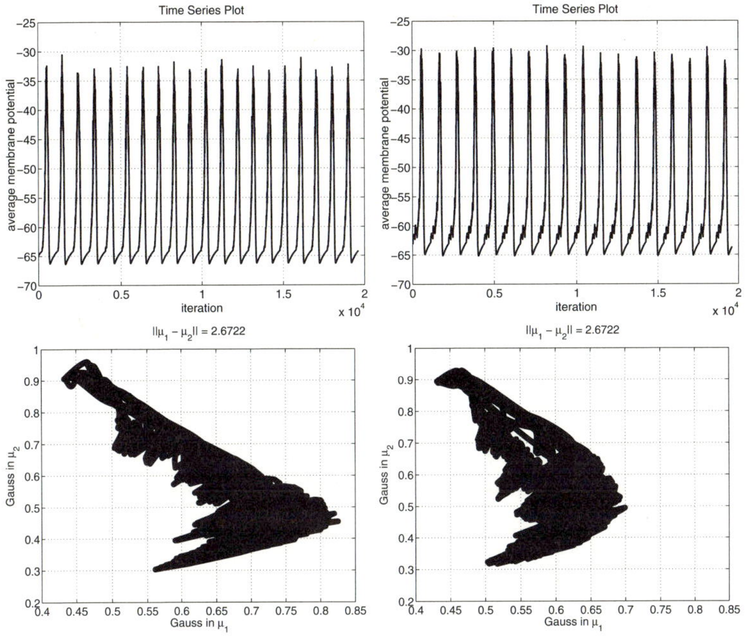

Fig. 5. Comparison of two similar normal rhythm cases. Time series plots (top) look very similar while trajectory plots (bottom) show noticeable differences (19600 points).

(i.e. different peak heights). The trajectories have been drawn using 19600 vectors, each containing values of membrane potentials of 50 follower cells, covering about 20 spikes. Two clusters have been found using the k-means algorithm, and the same parameters of membership functions used in both cases. Pathological case seems to reach the same amplitude but as a whole behaves quite differently, reaching much smaller values in first dimension, due to the lack of synchronization between different output neurons.

When two similar time series plots are compared small differences between them may not be noticeable. The FSD method is sensitive to small changes in the global dynamical state and consequently it allows for quite accurate comparison. Figure 5 compares two normal rhythms that differ only slightly. Time series plots looks very similar but global trajectories in FSD graphs show significant differences.

In all examples presented in this section dispersions of Gaussians were set to the half of the distance between centers ($\|\mu_1 - \mu_2\| / 2$).

5 Discussion

Symbolic dynamics has found many applications, while its fuzzy version has never been developed. It seems to be a very interesting method that should find many applications. In this paper it has been applied to visualization of high-dimensional neurodynamical systems. Many aspects of dynamics may be analyzed using this technique:

1. In which part of the signal space the state of the system spends most of its time?
2. How many attractors can be identified?
3. What are the properties of basins of attractors (size, depths)?
4. What are the probabilities of transition between them?
5. What type of oscillations occur around the attractors?

Quantitative measures to compare different dynamical systems should be introduced, for example:

- the number of attractors;
- percentage of time spent by the system in a given attractor basin;
- character of oscillations around attractors, including some measures of chaos;
- distances between attractors, measured by the time of transitions;
- probabilities of system transitions between attractors.

Such measures will give interesting characterization of dynamical systems. Application of FSD to recurrent networks should show transitions between their states. Applications to real EEG signals will require careful optimization of membership functions, with conditional clustering to remove irrelevant information by finding most informative center locations and weights for different signals. Visualization of highly-dimensional trajectories obviously depends on what aspects of the system behavior is of interest. Methods of parameter adaptation that include context [5,6] will soon be applied to visualization of real experimental data. For strongly non-stationary signals the whole landscape containing basins of attractors may slowly rotate, preserving relations between main attractors. For example, change in the level of neuromodulation may influence the landscape by increasing the overall activations in some regions of signal space. Parametrization of probes that should then change in time to counter this effect would be important. The great challenge is to find quantitative measures of the FSD representations that would be useful in brain computer interfaces, and to find meaningful combinations of signals correlated with inner experiences.

References

1. Rangayyan, R.M.: Biomedical Signal Analysis: A Case-Study Approach. IEEE Press Series on Biomedical Engineering. Wiley-IEEE Press (2001)
2. Sanei, S., Chambers, J.A.: EEG Signal Processing. Wiley, Chichester (2008)
3. Hao, B., Zheng, W. (eds.): Applied Symbolic Dynamics and Chaos. World Scientific, Singapore (1998)
4. Dale, R., Spivey, M.J.: From apples and oranges to symbolic dynamics: a framework for conciliating notions of cognitive representation. Journal of Experimental & Theoretical Artificial Intelligence 17(4), 317–342 (2005)
5. Pedrycz, W.: Knowledge-Based Clustering: From Data to Information Granules. Wiley-Interscience, Chichester (2005)
6. Blachnik, M., Duch, W., Wieczorek, T.: Selection of prototypes rules - context searching via clustering. In: Rutkowski, L., Tadeusiewicz, R., Zadeh, L.A., Żurada, J.M. (eds.) ICAISC 2006. LNCS (LNAI), vol. 4029, pp. 573–582. Springer, Heidelberg (2006)
7. Butera, R., Rinzel, J., Smith, J.: Models of respiratory rhythm generation in the pre-Bötzinger complex: I. Bursting pacemaker neurons; II. Populations of coupled pacemakers. Journal of Neurophysiology 82, 382–397, 398–415 (1999)

Towards Personalized Neural Networks for Epileptic Seizure Prediction

António Dourado, Ricardo Martins, João Duarte, and Bruno Direito

Centro de Informática e Sistemas da Universidade de Coimbra
Departament of Informatics Engineering
Pólo II Universidade 3030-290 Coimbra
{dourado,brunodireito}@dei.uc.pt,
{rfam,jvd}@student.dei.uc.pt

Abstract. Seizure prediction for untreatable epileptic patients, one of the major challenges of present neuroinformatics researchers, will allow a substantial improvement in their safety and quality of life. Neural networks, because of their plasticity and degrees of freedom, seem to be a good approach to consider the enormous variability of physiological systems. Several architectures and training algorithms are comparatively proposed in this work showing that it is possible to find an adequate network for one patient, but care must be taken to generalize to other patients. It is claimed that each patient will have his (her) own seizure prediction algorithms.

Keywords: Epilepsy, data mining, seizure prediction, classification, neural networks.

1 Introduction

About one third of epileptic people, meaning 0.4% of population, are not treatable by medication or ressective surgery [1]. At any time, anywhere, they can suffer from a seizure, *"like a bolt from the sky"*, during some seconds or some minutes, seriously affecting their motoricity, perception, language, memory and conscious. If they could predict the seizures, their life would change substantially.

Seizure prediction has been the object of extensive and intensive research for the last 20 years. For an excellent review see for example [2] and [3]. More recently computational intelligence techniques, such as neuron-fuzzy systems [4] or neuron-fuzzy systems associated with wavelets [5], have been identified as having a high potential for seizure identification. Seizure prediction, the object of the present work, is a different problem from seizure identification. Prediction is much harder than identification. However, from the clinical point of view, no significative practical advance has been verified: there is not any system usable by patients allowing them to predict a coming seizure and to take action to preserve his (her) safety and privacy, improving substantially his (her) social integration. This is probably because most of the researchers look for a general method and algorithm that would work for every patient. And although several

V. Kůrková et al. (Eds.): ICANN 2008, Part II, LNCS 5164, pp. 479–487, 2008.

authors propose methods to which they claim a high performance, the considered performance criteria is only partial, neglecting other parts of the problem that prevent them to be used in a clinical environment. Physiological systems, as every biological one, have a high variability, and, in the case of seizure prediction, it seems more advisable to look for a predictor well designed for each patient. Neural networks, by their diversity in architectures and training algorithms, have a high plasticity well suited for that purpose.

In the present study this problem is worked out as a classification task. The ElectroEncephaloGram (EEG) is the main electrical measure of the activity of the brain. It is supposed that epileptic seizures are an abrupt change in the electrical activity of the brain and that these changes are captured by the EEG. The challenge is then to process the EEG in such a way that four brain states can be identified: the normal state, the time interval preceding a seizure, the seizure itself, and the time interval for (re)normalization of the brain activity. This cannot be done directly with the EEG; instead some special features must be extracted from the EEG signal. These features must change as the brain evolves among these states and these changes, particularly during the pre-seizure period, may eventually lead the seizure prediction.

In the present work a set of features is extracted from the EEG signal. They quantify several characteristics about energy, time-frequency decomposition, nonlinear behavior, composing a 15 dimensional features space where classification is then to be done into the four classes (brain states): inter-ictal, pre-ictal, ictal, pos-ictal. These features have been considered by several authors with a high potential for the discrimination of the brain state with respect all or some of these four classes For example [6] and [7] used energy variation, [8] accumulated energy, [9] nonlinear systems concepts, [10] wavelet transform, [11] Fourier and wavelet transform, [12] wavelets and similarity analysis.

In this work two patients from Freiburg Database [13] are considered. They have been chosen by their different epileptic zones, one in frontal lobe, and the other in temporal lobe. Only one EEG channel is considered (the focus one) to test the possibility of prediction in such circumstances. Other authors (for example [14]) use more channels for other kind of approaches.

Several architectures and training algorithms are comparatively used for seizure prediction in one and in the other. The performance criterion has three facets: specificity, sensitivity, overall classification rate. The results show that there are several architectures adequate for a patient but they do not work properly for the other patient.

In the next Paragraph 2 the data and the features used for the classification stage are presented. Then in Paragraph 3 the results obtained with several network architectures are discussed. Conclusions and future work are set in the last Paragraph 4.

2 The Data and the Set of Features for Classification

The data used in this investigation has been collected from the epilepsy database of Freiburg Center for Data Analysis and Modeling (FDM) of Albert Ludwig

University of Freiburg [13]. Two patients have been selected, patient A with frontal lobe epilepsy and patient B with temporal lobe epilepsy. The intracranial recordings utilized were acquired using Neurofile NT digital video system with 128 channels, 256 Hz sampling rate, and a 16 bit analogue-to-digital converter.

Applying energy concepts, wavelet transform, nonlinear dynamics, 14 features have been extracted, listed in Table 1.

Intracranial EEG data is processed by the developed methods. The time interval between two consecutive computations of the 14 presented features is 5 seconds. One single channel of the EEG, the focal one, is used. Other studies use more channels [14] for an approach based on synchronization of neurons in different regions of the brain.

This section presents an overview of the methods which lead to this set of features. The methods were developed in Matlab and its toolboxes (including Neural Networks Toolbox) [15], and other freely available software, like the nonlinear time series analysis TSTOOL) [16].

Energy variation analysis is based on the algorithm presented in [7]. EEG signal is processed through two windows with different length to analyze energy patterns. The main objective is to confirm the increase of energy bursts in the periods that precede seizures. Accumulated energy was approximated by using moving averages of signal energy (using a short-term energy observation window versus a long-term energy observation window). A similar displacement was applied to both windows and both ended at the same time point. These features allow the observation of energy patterns before epileptic seizures.

Wavelet coefficients have been submitted to a similar energy analysis, allowing by this way the identification of variations in the different frequency bands that constitute the EEG signal. Based on the mechanism previously explained, the coefficients obtained by wavelet decomposition are processed and the

Table 1. The 14 extracted features from EEG to be used in classification of the brain state

Concept	**Feature**
Signal Energy	Accumulated energy
Signal Energy	Energy level
Signal Energy	Energy variation (short term energy)
Signal Energy	Energy variation (long term energy)
Wavelet Transform coeficient energy	short term energy band (0Hz – 12,5Hz)
Wavelet Transform coeficient energy	long term energy band (0Hz – 12,5Hz)
Wavelet Transform coeficient energy	short term energy band (12,5Hz – 25Hz)
Wavelet Transform coeficient energy	long term energy band (12,5Hz – 25Hz)
Wavelet Transform coeficient energy	short term energy band (25Hz – 50Hz)
Wavelet Transform coeficient energy	long term energy band (25Hz – 50Hz)
Wavelet Transform coeficient energy	short term energy band (50Hz – 100Hz)
Wavelet Transform coeficient energy	long term energy band (50Hz – 100Hz)
Nonlinear system dynamics	Correlation dimension
Nonlinear system dynamics	Max Lyapunov Exponent

accumulated energy of these series is determined. As before, accumulated energy was approximated by using moving averages of coefficients energy (using a short-term energy observation window versus a long-term energy observation window). The mother wavelet used in the presented study was daubechies-4; the decomposition was completed with four levels.

Nonlinear analysis faces the EEG as trajectories of a nonlinear system. Two nonlinear dynamic features, maximum Lyapunov exponent and correlation dimension through a sliding window, are computed using [15]. The construction of the attractor, after the determination of the parameters delay time and embedding dimension, allows the calculation of the maximum Lyapunov exponents and correlation dimension. The estimation of the maximum Lyapunov exponents consists in the quantification of the exponential growth of the average distance between two nearby trajectories of the attractor, through error approximation. Correlation dimension is determined by takens estimator method [15].

The joint analysis of the extracted features created a 14-dimension space which represents the EEG signal in several components (energy signal, frequency and system dynamics). The objective of the study is to investigate the eventual occurrence of hidden characteristics in data such that clusters can be discovered allowing an acceptable classification of EEG data into 4 classes:

- inter-ictal (normal EEG pattern)
- pre-ictal (two minutes prior to the seizure onset)
- ictal (the seizure onset)
- pos-ictal (two minutes subsequent to seizure end)

One cycle is composed of one series of these classes.
The overall approach is illustrated in Fig. 1.

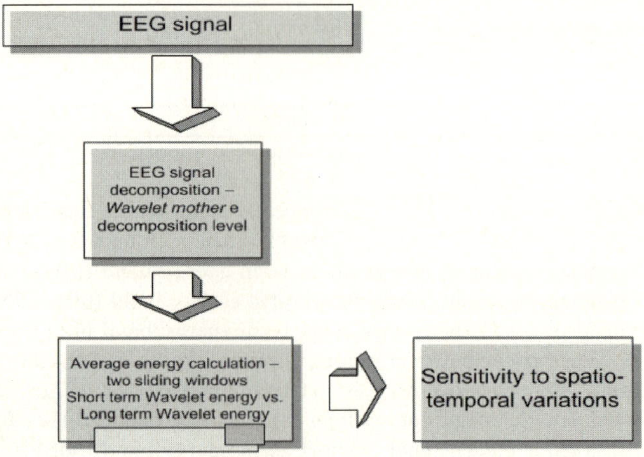

Fig. 1. The approach EEG features extraction-classification

3 The Applied Neural Network Architectures and Its Results

The data sets have the following characteristics: patient A- 2 cycles, 1366 points; patient B- 3 cycles, 1824 points. Data have been normalized feature by feature in [0 1].

3.1 The Best Found Architectures and Training Algorithms

After an extensive experimentation, the following neural network structures have been applied and compared, because they have been found to be the most promising:

(i) Three layer logsig feedforward (FFNN): 14 neurons in the first layer, 56 in the second and 4 in the output layer, Fig. 2. The output layer numerical values are rounded to integers and it has been trained to classify each class accordingly to the following coding:

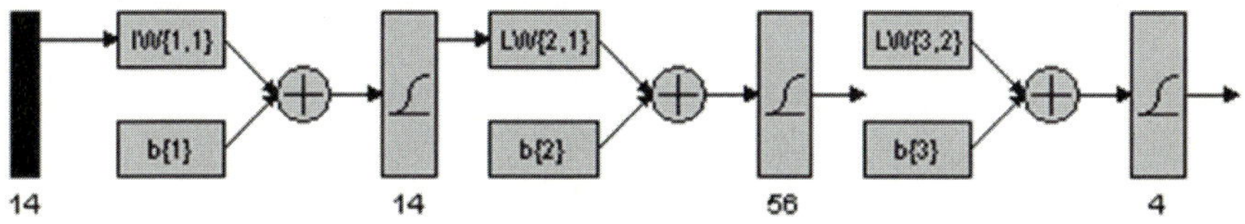

Fig. 2. The best architecture found patients A and B. Bias and weights are proper to each patient.

Class 1 (inter-ictal) : [1 0 0 0] Class 2 (pre-ictal): [0 1 0 0] Class 3 (ictal): [0 0 1 0] and Class 4 (pos-ictal) [0 0 0 1].Training was done using the Levenberg-Marquardt algorithm, better that the backpropagation one.

(ii) In order to catch the nonlinear dynamic nature of the signal, experiments have been made introducing a taped delay line in the network inputs (first layer), as implemented in the Matlab Neural Networks Toolbox. Delays of 1 and 2 have been experimented.

(iii) Radial Basis Function neural network (RBF) with variable size of the first (radial) layer and 4 linear neurons in the output layer. It was trained by the hybrid algorithm.

3.2 The Performance Criteria of the Classifier

In seizure prediction (as in the general problem of medical diagnosis) there are four possible outcomes to a diagnosis operation:

- positive true (PT), when the diagnose is positive and the event has been confirmcd,

– positive false (PF) when the diagnose is positive and the event has not been confirmed,
– negative true (NT), when the diagnose is negative and the event has been confirmed as not existing,
– negative false (NF) when the diagnose is negative abut the event has finally been confirmed as true.

For clinical applications any automatic diagnosis systems must give all the PT events and all the NT events. But it must also give zero PF and zero NF answers. Two performance indexes are defined:

Sensitivity: related to the Positive outcome, given by (1)

$$SENSIT = \frac{PT}{PT + NF}$$

Specificity: related to Negative outcome, given by (2)

$$SPECIF = \frac{NT}{NT + PF}$$

We can also define the overall index given by (3)

$$OVERALL = \frac{PT + NT}{PT + PF + NT + NF} = \frac{PT + NT}{ALL}$$

It is frequent that one author presents one of these indexes to measure the performance of a seizure prediction algorithm. However from a clinical judgment,

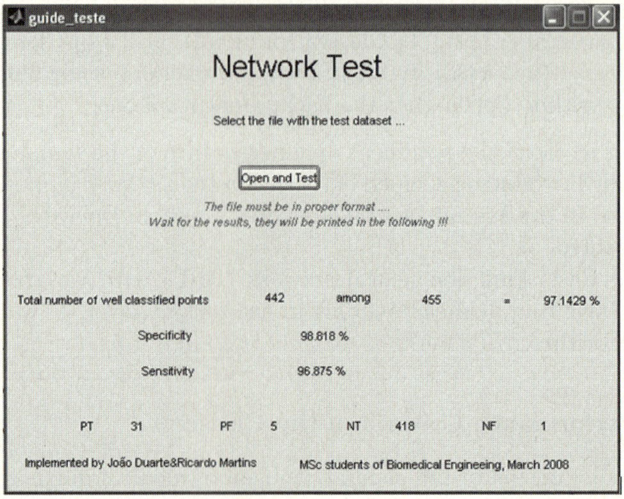

Fig. 3. The interface for testing the networks. It works under Matlab 2007b with NN Toolbox. The networks, data sets and interface are freely available at http://eden.dei.uc.pt/dourado/seizureprediction/ICANN2008BESTNNETS.zip.

only when the sensitivity and the specificity are both near to 1 can the algorithm be applied. A perfect system has both sensitivity and specificity equal to one. Probably that is why very few application of automatic diagnosis systems are really working today, although there is an extensive published literature on diagnosis algorithms with or high sensitivity or high specificity.

All the three indexes are used, as shown in the interface in Fig. 3.

Results. Table 2 shows some results for patient A. The FFNN has been used with and without input delays. But, although in theory a better result could be expected with delays (considering the brain as a dynamic system), in fact these two networks have a much worse performance. RBF shows also a poor performance. If one only cares about specificity, then all four nets are very good. The FFNN with 2 delays shows an absolutely good specificity of 1 and an absolutely bad sensitivity of 0. This illustrates the fact that only one of these parameters is not a proper performance index. The data set has 1366 instants (70% for training and 30% for testing). When the input delay isn't zero, the data is separated in two blocks: the first 70% instants of the data set are used for training, the remaining 30% are used for testing, avoiding the separation to be done in the midlle of a crisis. When the input delay is zero the data is separated selecting in each 3 successive instants of the data set, 2 for training and 1 for testing.

Table 2. Some results for patient A wth FFNN and RBF in test training set. The training criteria has been SSE (sum of Squared Error), with Levenberg-Marquard algorithm. Data is normalized. The last line is for RBF.

Input Delay (FFNN)	Size of test data set	N° well classified	PT	PF	NT	NF	SENSIT	SPECIF
0	455	442	31	5	418	1	0.9688	0.9812
1	409	252	0	2	348	59	0	0.9943
2	409	203	0	0	350	59	0	1
RBF	455	391	14	1	422	18	0.4375	0.9976

Table 3 shows similar results for patient B and the same comments can be done. The RBF in last line has been trained and tested, in this case, with original, non normalized data. It shows a slightly better specificity but a much worse sensitivity. The data set is bigger, with 1824 instants, and the training and testing data separation was done the same way as for patient A.

If the FFNN is trained simultouneously for the datasets of both patients, the training performance is rather poor. It is very hard, because of the different patients and different types of epilepsy, to find a network that, with the same weights and bias, works well for both. Of course one can always increase the dimension and improve training, until probably overtraining, loosing the generalization capability of the network. From a pratical clinic use, for example in ambulatory, where a patient transports with him some alarmig device forecasting the eminent coming of a seizure, the need for a personalized neural network is not a serious problem.

Table 3. Some results for patient B wth FFNN and RBF in test training set. The training criteria has been SSE (sum of Squared Error), with Levenberg-Marquard algorithm. Data is normalized. The last 2 lines are for RBF: normalized and original data.

Input Delay (FFNN)	Size of test data set	N° well classified	PT	PF	NT	NF	SENSIT	SPECIF
0	608	594	55	4	544	5	0.9167	0.9967
1	547	167	0	0	486	61	0	1
2	547	409	0	0	486	61	0	1
RBF	608	499	48	52	496	12	0.8	0.9051
RBF	608	465	1	0	548	59	0.0167	1

Table 4. Case of joining the data sets of both patients (training results) with Levenberg-Marquardt algorithm

Input Delay (FFNN)	Size of test data set	N° well classified	PT	PF	NT	NF	SENSIT	SPECIF
0	1064	912	18	12	958	75	0.20	0.98

Table 5. Testing patient A network into patient B and vice-versa

Case	Size of test data set	N° well classified	PT	PF	NT	NF	SENSIT	SPECIF
A in B	608	406	0	35	513	60	0	0.9361
B in A	455	255	2	69	354	30	0.625	0.8368

Testing the network A in patient B, or network B in patient A, gives the results presented in table 5. The degradation of performance is evident.

4 Conclusions

There is still a long way to set extensive guidelines for building seizure predictors for epileptic patients. However the shown results evidence two simple principles: (i) there is no general predictor good for all patients, and (ii) the predictor of one patient is not acceptable for other patient. This has as consequence that each patient must be the object of a personalized study, using as much data as possible, following its behavior and training it permanently. Neural networks have a high plasticity that can be profitably used for this purpose. However, other techniques should also be studied, such as support vector machines (SVM) that may have an important role in constructing nonlinear boundaries in the high dimensional features space, resulting eventually in better classification among the four classes in the context of seizure prediction.

Acknowledgments. This research is supported by European Union FP7 Framework Program, EPILEPSIAE Grant 211713 and by CISUC (FCT Unit 326). The authors express their gratitude to Freiburg Center for Data Analysis and Modeling (FDM) of Albert Ludwig University of Freiburg, for the access to the epilepsy database.

References

1. Browne, T., Holmes, G.: Handbook of epilepsy. Lippincott Williams & Wilkins (2000)
2. Mormann, F., Andrzejak, R.G., Elger, C.E., Lehnertz, K.: Seizure prediction: The long and winding road. Brain 130, 314–333 (2007)
3. Schelter, B., Winterhalder, M., Feldwisch, G., Drentrup, H., Wohlmuth, J., Nawrath, J., Brandt, A., Schulze-Bonhage, A., Timmer, J.: Seizure prediction: The impact of long prediction horizons. Epilepsy research 73, 213–217 (2007)
4. Guler, I., Ubeyli, E.D.: Application of adaptive neuro-fuzzy inference system for detection of electrocardiographic changes in patients with partial epilepsy using feature extraction. Expert Systems with Applications 27, 323–330 (2004)
5. Subasi, A.: Application of adaptive neuro-fuzzy inference system for epileptic seizure detection using wavelet feature extraction. Computers in Biology and Medicine 37, 227–244 (2007)
6. Litt, B., Esteller, R., Echauz, J., D'Alessandro, M., Shor, R., Henry, T., et al.: Epileptic seizures may begin hours in advance of clinical onset: a report of five patients. Neuron 30, 51–64 (2001)
7. Esteller, R., Echauz, J., D'Alessandro, M., Worrell, G., et al.: Continuous energy variation during the seizure cycle: towards an on-line accumulated energy. Clinical Neurophysiology 116, 517–526 (2005)
8. Gigola, S., Ortiz, F., D'Atellis, C., Silva, W., Kochen, S.: Prediction of epileptic seizures using accumulated energy in a multiresolution framework. Journal of Neuroscience Methods 138, 107–111 (2004)
9. Chaovalitwongse, W., Iasemidis, L.D., Pardalos, P.M., Carney, P.R., Shiau, D.S., Sackellares, J.C.: Performance of a seizure warning algorithm based on the dynamics of intracranial EEG. Epilepsy 64, 93–113 (2005)
10. Subasi: Epileptic seizure detection using dynamic wavelet network. Expert System with applications 29, 343–355 (2005)
11. Kemal Kiymik, M., Guler, Í., Dizibuyuk, A., Akin, M.: Comparison of STFT and Wavelet Transform Methods in Determining Epileptic Seizure Activity in EEG Signals for real-time application. Computers in Biology and Medicine 35, 603–616 (2005)
12. Ouyang, G., Li, X., Li, Y., Guan, X.: Application of wavelet-based similarity analysis to epileptic seizures prediction. Computers in Biology and medicine 37, 430–437 (2007)
13. Freiburger Zentrum fur Datenanalyse und mollbildung, `http://www.fdm.uni-freiburg.de/groups/timeseries/epi/EEGData/download/infos.txt`
14. Le Van Quyen, M., Amor, F., Rudrauf, D.: Exploring the dynamics of collective synchronizations in large ensembles of brain signals. J. Physiol. (in press, 2007)
15. The Mathworks, Inc.
16. Merkwirth, C., Parlitz, U., Wedekind, I., Lauterborn, W.: TSTOOL User Manual, Version 1.11, `http://www.dpi.physik.uni-goettingen.de/tstool/HTML/index.html`

Real and Modeled Spike Trains: Where Do They Meet?

Vasile V. Moca[1], Danko Nikolić[2,3], and Raul C. Mureşan[1,2]

[1] Center for Cognitive and Neural Studies (Coneural), Str. Cireşilor nr. 29, 400487 Cluj-Napoca, Romania
moca@coneural.org, muresan@coneural.org
http://www.coneural.org/muresanlab
[2] Max Planck Institute for Brain Research, Deutschordenstrasse 46, 60528 Frankfurt am Main, Germany
[3] Frankfurt Institute for Advanced Studies, Ruth-Moufang-Str. 1, 60438 Frankfurt am Main, Germany

Abstract. Spike train models are important for the development and calibration of data analysis methods and for the quantification of certain properties of the data. We study here the properties of a spike train model that can produce both oscillatory and non-oscillatory spike trains, faithfully reproducing the firing statistics of the original spiking data being modeled. Furthermore, using data recorded from cat visual cortex, we show that despite the fact that firing statistics are reproduced, the dynamics of the modeled spike trains are significantly different from their biological counterparts. We conclude that spike train models are difficult to use when studying collective dynamics of neurons and that there is no universal 'recipe' for modeling cortical firing, as the latter can be both very complex and highly variable.

1 Introduction

The activity of cortical neurons arises from complex firing patterns that are determined by the intricate brain architecture, the external stimuli and the interaction with subcortical structures [1]. Each cortical neuron receives on the order of $10^3 - 10^4$ inputs from other neurons, and hence it is prone to being exposed to a high input bombardment [2]. Indeed, it has been suggested that such a bombardment exists, and, in addition to a balanced excitation-inhibition state, it keeps the neuron in a so-called "high-conductance state" [3]. Moreover, under such heavy input, cortical neurons have a tendency to fire highly irregularly, such that the distribution of their inter-spike intervals (ISI) takes an exponential form [4]. As a result, numerous models of spike trains have been proposed [5,6], which assume an underlying homogenous (constant firing probability over time) or inhomogeneous (the instantaneous firing probability can fluctuate over time) Poisson process [7]. Recently however, the heavy input bombardment hypothesis has been challenged [8], new [9] and old [10] data suggesting that the assumption

V. Kůrková et al. (Eds.): ICANN 2008, Part II, LNCS 5164, pp. 488–497, 2008.

that neurons are simple Poisson spike generators is an extreme oversimplification. The firing properties of neurons depend critically on neuron type [11], brain structure [12], brain state [13], arousal [14], and other factors.

A particular case, where firing is non-Poissonian, is represented by the oscillatory discharge of cortical neurons. In such cases, neurons can engage into various rhythms in multiple frequency bands [15]. The oscillatory firing properties of neurons can be characterized by computing auto-correlation histograms (ACH), and we have previously introduced a measure, called Oscillation Score [16] that is useful in determining the degree to which a neuron oscillates. In order to precisely quantify the oscillation strength we needed to develop a model of oscillatory spike trains, in which the oscillatory behavior and the firing rate could be independently controlled [16]. Here, we explicitly study the properties of this model: the precision with which firing rate and oscillation strength can be controlled and the ability to independently express the two properties in the simulated spike trains. In addition, we are interested in the degree to which artificial spike trains, generated from statistical parameter distributions of recorded neuronal data, reflect the temporal structure of the latter.

2 The Model

The model produces artificial spike trains that retain basic properties of a recorded data set. The considered properties are: the firing rates, burst probability, oscillation strength and frequency, spike counts in bursts, refractory periods, and intra-burst inter-spike intervals, all of which are quantitatively determined from biological data. Two processes are used to generate the spike trains: at a coarse scale a discharge probability function $p_s(t)$ (Fig. 1A, B), and at a finer scale another process (Fig. 1C) that controls the exact spike timings (refractoriness, burst properties, etc). Since our model is not a biophysical one, most of the parameters that we used have no direct biological correspondents.

The spike discharge probability $p_s(t)$ should have the following properties: first, it should allow the spike train to exhibit a preferred oscillation frequency for transient periods of time; second, it should allow control over the strength and stability of the desired oscillation; and third, it should enable the control over the firing rates. To control the amount of oscillations, two discharge probabilities $p_o(t)$ and $p_b(t)$ corresponding respectively to an oscillatory and a background process are intermixed with a factor o (oscillation strength) (1). To obtain the transient oscillatory behavior we modulate the frequency of a sine probability function $p_o(t)$ (2) by a random process $f_o(t)$ (4). This random process takes into account past values and thus, it has memory. The history dependence is given by a decay constant, τ, while another factor, m, controls the amount of noise added to the random process. After a duration of 3τ, a value fades to less than 5% and its effect is very small. The interplay between τ and m controls how the oscillation frequency changes over time. The frequency range of the modulatory function, $f_o(t)$, can be bounded to increase the stability of the oscillation. It is assumed that prior to $t = 0$ the function $f_o(t)$ varies slightly around the desired oscillation frequency. The background probability $p_b(t)$ (3) is generated in the same way except that in this case, the frequency of the process $f_b(t)$ (which has the same form as $f_o(t)$) varies in a much broader frequency range (Fig. 1B). For a spike to be generated, $p_s(t)$ must be positive. The greater $p_s(t)$, the greater is the chance that a spike will be generated and thus, most of the spikes will be concentrated at the peaks of the probability functions (Fig. 1C). By manipulating the $offset$ (1) we can control how well the spikes are aligned to the desired oscillation (negative values of $p_s(t)$ mean that no spike is generated), and by manipulating the spike-probability-positive-integral ($SPPI$) (5) trough the amplitude parameter, A (1), we control the firing rates (Fig. 1A and Fig. 2 - left column) for the duration T of the spike train. Thus, the value of $p_s(t)$ controls the periodicity and firing rates of spike-trains in a manner that realistically mimics the oscillatory behavior of the recorded neurons.

$$p_s(t) = A \cdot [o \cdot p_o(t) + (1 - o) \cdot p_b(t) - offset] \tag{1}$$

$$p_o(t) = \frac{1}{2} + \frac{1}{2} \cdot \sin\left(2 \cdot \pi \cdot \int_0^t f_o(x)dx\right) \tag{2}$$

$$p_b(t) = \frac{1}{2} + \frac{1}{2} \cdot \sin\left(2 \cdot \pi \cdot \int_0^t f_b(x)dx\right) \tag{3}$$

$$f_o(t) = \lim_{\epsilon \to 0} \frac{\int_\epsilon^{3\tau} e^{-\frac{x}{\tau}} \cdot f_o(t - x)dx}{\int_\epsilon^{3\tau} e^{-\frac{x}{\tau}} dx} + m \cdot \mathrm{rand}_{[-1,1]}(t) \tag{4}$$

$$SPPI = \frac{1}{T} \int_0^T p_s^+(t)dt \tag{5}$$

$$\text{where: } p_s^+(t) = \begin{cases} p_s(t), & p_s(t) \geq 0 \\ 0, & p_s(t) < 0 \end{cases} \tag{6}$$

At smaller time scales, the model controls the timings between the spikes and the burstiness of the spike trains. The burst probability, p_{burst}, is modeled as a constant ratio between the count of bursts relative to the number of tonic spikes and burst occurrences altogether. These values can be extracted from a recorded data set. Once a discharge is initiated based on, $p_s(t)$, p_{burst} determines whether that discharge will be a burst or tonic spike. If a burst is generated, the number of spikes in the burst b_{SpkCnt} and the spacing between the spikes within the burst b_{ISI} (Fig. 1C) are set according to probabilities measured from real data. After a tonic spike a refractory period, r, prevents the occurrences of other discharges in a given period of time. The model makes a clear distinction between the tonic spikes and bursts. The bursts are defined as groups of spikes with successive ISIs smaller or equal to 8 ms [17]. Thus, the refractory period, r, is set to 8 ms such that tonic spikes can occur with an ISI of at least 9 ms. The control over these timings at a small scale produces a realistic center-peak shape in the ACH (Fig. 3) of artificial spike trains which model real spike trains.

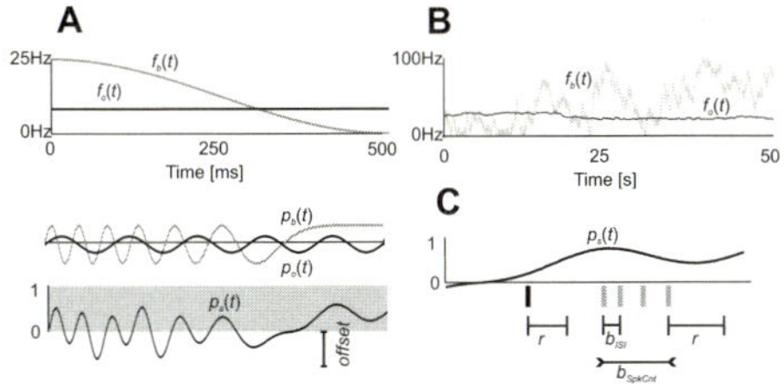

Fig. 1. Model for simulation of oscillatory spiking activity. **A:** Spike discharge probability. The spike discharge probability $p_s(t)$ is obtained by mixing two time-varying processes $p_b(t)$ and $p_o(t)$ with frequencies $f_b(t)$ and $f_o(t)$, respectively and with a mixing factor $o = 0.3$ (see Eq. 1). The amplitude and offset of the two processes are fixed for one run of the algorithm. A spike can be generated only where $p_s(t)$ is positive (grey band). **B:** Fluctuations of the modulation functions $f_b(t)$ and $f_o(t)$. While $f_b(t)$ varies between 0 and 100 Hz (gray) $f_o(t)$ varies slowly around 25 Hz due to its strong history dependence and boundaries (20 to 30 Hz). **C:** Spiking. A burst of spikes is represented by grey vertical bars, while a tonic spike is depicted in black. The spikes are generated taking into account the refractory period after a burst or a tonic spike r, the intra-burst inter-spike interval b_{ISI} and the burst spike count b_{SpkCnt}.

3 Results

We addressed here two important aspects related to the model: the relationship between firing rate and oscillation strength, and the degree to which the model can reproduce the fine temporal structure of recorded spike trains.

3.1 Firing Rate and Oscillation Strength

We computed the relationship between firing rate and oscillation strength by producing artificial spike trains, with a length of 30 s, and controlling independently the two parameters for rate and oscillation strength. The rate parameter ($SPPI$; see Methods) and the oscillation strength parameter (o) were varied in 24 steps. For each combination, two spike trains were produced, yielding a total of 1152 spike trains. In addition, we generated such spike trains for each oscillation frequency band separately: theta (4-8 Hz), alpha (8-12 Hz), beta-low (12-20 Hz), beta-high (20-30 Hz) and gamma (30-80 Hz) [16]. For the oscillation frequency in the model (f_o), we took the central (middle) frequency for each frequency band. The firing rate and the oscillation score [16] (the measured strength of oscillation) were measured for each spike train. We found that, the firing rate could be more precisely controlled for higher than for lower frequency bands (Fig. 2, left column, note the lower variance for higher frequency bands) while the oscillation score seemed to be controlled equally well across all frequency bands (Fig. 2, middle column). Moreover, the firing rate scaled as a power function ($R^2 > 0.93$) with respect to the $SPPI$ while the oscillation score scaled as a sigmoid with respect to the oscillation strength parameter. The noisy clouds in Fig. 2 (middle column), that deviate from the sigmoid shape are due to trials with very small firing rates, for which the oscillation score estimate becomes imprecise [16]. Finally, there was no correlation between the firing rate and oscillation score, as shown in Fig. 2, right column ($R^2 < 0.08$). Thus, the resulting firing rate and oscillation strength can be independently controlled by the two corresponding parameters in the model.

3.2 Temporal Structure of Spike Trains

Although our model is quite detailed, including statistics of bursting, oscillatory modulation, and other spike train properties, we wanted to estimate how closely some model spike trains resemble their corresponding, real, spike trains recorded from cat visual cortex. We considered data recorded from an anaesthetized cat on two experimental sessions: one without or poor oscillatory responses (col11b44; see Fig. 3) and one with strong oscillations (col11b68; see Fig. 3) in the beta-high frequency band. Since the experimental trials were rather short (6.5 s) we selected, from one single electrode, multi-unit activity (MUA) such as to yield sufficient number of spikes when computing single-trial statistics (see next). We also chose a stimulation condition (center-surround stimulus, with a small sinusoidal grating placed in the center and a larger one in the surround) for which the MUA showed strong oscillatory behavior in col11b68 session but not in col11b44. For this stimulation condition, we obtained 20 experimental trials in both sessions. Next, for each experimental trial, one corresponding model trial (spike train) was generated. We computed the ISI distributions of spikes within bursts (burst ISI), the burst spike-count distributions, and the firing rate on the experimental trial, and plugged these parameters into the model, producing an artificial spike train having the same statistics (see Methods). Importantly, we wanted to model each trial independently such that the temporal structure of firing is maximally similar to the corresponding recorded trial. Statistics are thus computed separately for each trial. For oscillatory spike trains, the oscillation frequency was computed on the whole stimulation condition and plugged into each model trial (because the oscillation frequency was very stable across trials in the analyzed dataset). Next, the autocorrelation histograms (ACH) per stimulation condition were computed by averaging the ACHs computed on each of the 20 trials. The inter-spike interval (ISI) distributions were also computed.

In Figs. 3A and 3B we show that the ACHs and ISI distributions of the real and model spike trains were remarkably similar, both for the non-oscillatory and oscillatory case. This indicates that statistically, the real and modeled spike trains had similar local structure. At a first look, one could say that the model spike train can successfully replace the real one. To get further insight into this hypothesis, we computed the cross-correlation histograms (CCH) between the model and their corresponding, real, spike trains. Ideally, if the model successfully replaces the real spike train, the CCH should be very similar to the ACHs of both the real and modeled spike trains. This was, however, not the case. For non-oscillatory spike trains (Fig. 3A) the CCH was flat, without a peak, indicating that there was no consistent relationship (fine temporal correlation) between real and modeled spike timings. The oscillatory case was somewhat better, such that weak oscillatory modulation could be seen in the CCH (Fig. 3B), but the central peak was missing here as well. These results suggest that, although the statistical properties of the original spike trains are faithfully reproduced, the temporal dynamics of the trials are not. We show in Fig. 3C that, for a given

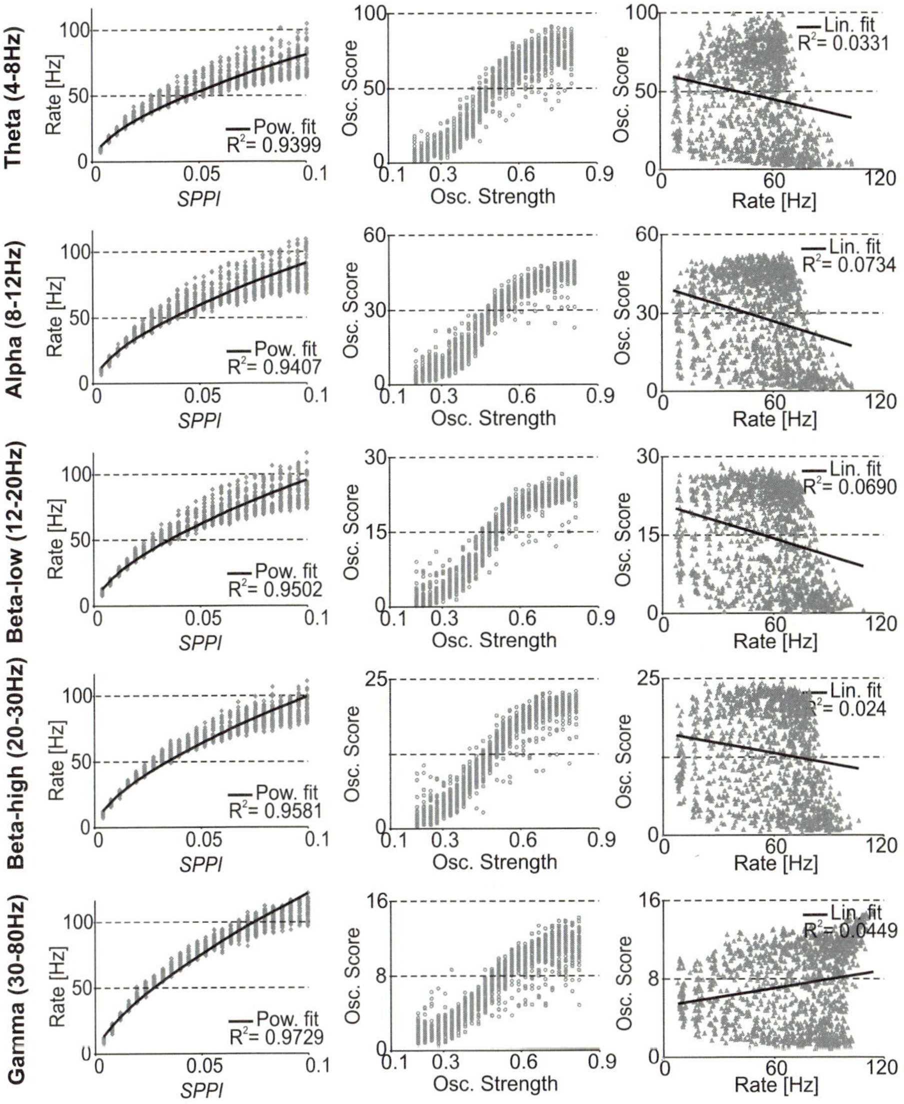

Fig. 2. Control of rate and oscillation strength. On the left column, scatter plots show the dependence of firing on the spike-probability-positive-integral: $SPPI$ (5). The black line shows a power fit. The center column depicts the dependency of the measured oscillation score on the oscillation strength parameter, o, in the model (1). In the right column, scatter plots of the rate and oscillation score show no dependency between these two, as indicated by linear fits (black lines).

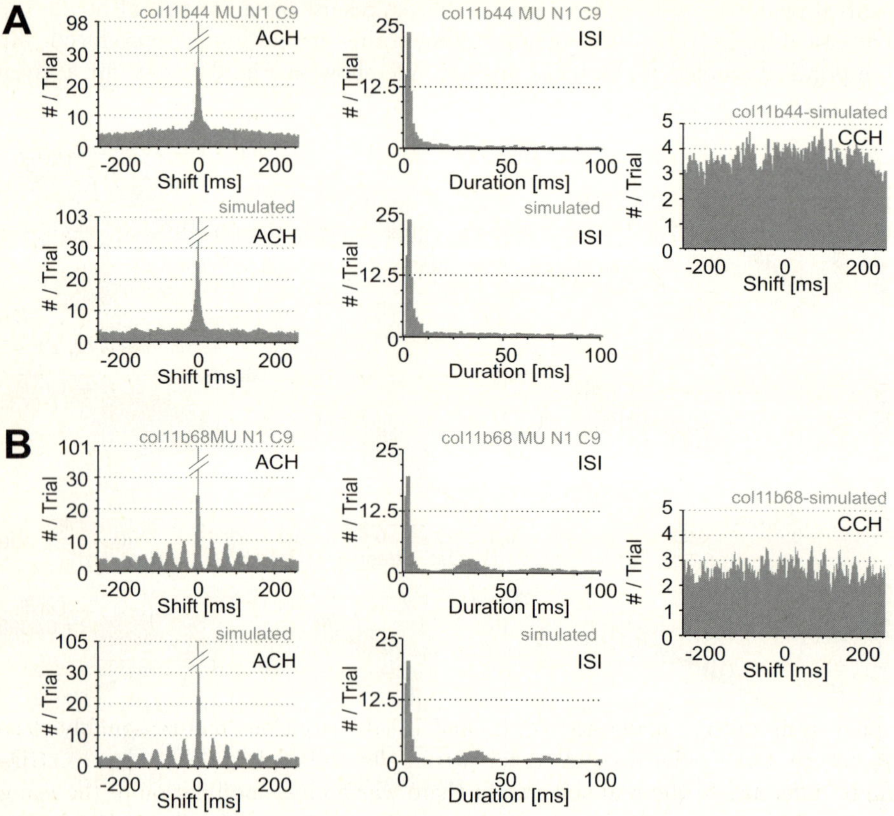

Fig. 3. Comparison between recorded and simulated spike trains. **A:** Non oscillatory data. The first row shows the ACH (left panel) and the ISI distribution (center panel) for non-oscillatory data. On the second row, the ACH (left panel) and ISI for simulated data (center panel) are shown. Note that the ACH and ISI plots are averages across 20 trials of the same condition. On the right panel the average cross-correlation between each recorded and artificial spike train is shown. **B:** The same plots as in A are shown for oscillatory data. The similarity of ACHs and ISI distributions between the recorded and simulated data is again clear. **C:** Differences between the recorded and simulated spike trains. The spikes are shown on the left: recorded data (upper traces, 20 consecutive trials) and simulated data (lower traces). For one recorded spike train and for its artificial model (both depicted in black) the ACHs (right panels) and ISI distributions (bottom panels) are shown. For each plot depicting recorded data the dataset, neuron, and condition labels are shown above the graph; for simulated data "simulation" is shown.

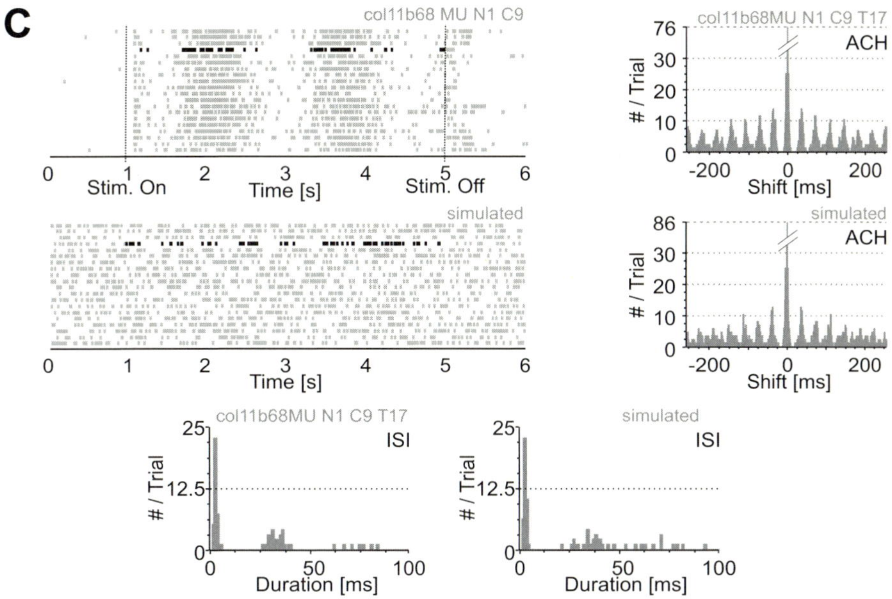

Fig. 3. (*continued*)

spike train and its model, the ACHs and ISI distributions had reasonably close structure, but the location of spikes within the real and model trials was strikingly different. In the real spike train there was a clear modulation of the firing by the drifting sinusoidal grating stimulus. In the modeled spike train, the firing events were more or less arbitrarily located, despite the fact that global ISI statistics and ACHs structure were accurately reproduced (Fig. 3C).

4 Discussion

We have shown that the proposed spike train model allows one to independently control firing rate and oscillation strength. Thus, the spike train model can isolate reasonably well the two processes such that one can use it to calibrate data analysis methods which need to separate the effects of the firing rate from those of the oscillation strength [16].

Moreover, our results indicate that the model can reproduce the statistical properties of the real data quite faithfully. The ACHs and the ISI distributions of the model were strikingly similar to those of the real data and thus, we validated the model as being appropriate for studying properties of spike trains in terms of their ACHs. On the other hand, we have found that reproducing the statistics of neuronal firing can not account for the dynamics of the spike trains. There are multiple conclusions that stem from the above. First, given the complexity required by our model in order to reproduce realistic spike trains, we infer that simple, homogenous or inhomogeneous Poisson processes are crude and largely inappropriate approximations of cortical firing. These two types of processes need to be complemented by other, more complex ones, as real spike trains are very hard to model. One needs to add many constraints to the model and these constraints vary across neurons, stimuli, cortical states, and so forth. There is no universal 'recipe' for generating spike-trains, since neuronal behavior can be very rich and highly variable.

Second, modeling the fine temporal structure of spike trains, including their temporal dynamics can become very difficult because one needs to know the underlying drive received by neurons from the sensory (thalamic) and cortical inputs. An important implication of this fact is that collective coding strategies (where multiple, simultaneous spike trains are analyzed) cannot be easily studied with model spike trains. The reason is that, for collective codes, the relative dynamics of different neurons plays a crucial role. Reproducing the statistical properties of each individual spike train is obviously not enough. One could use a hybrid approach and also measure some instantaneous firing probability over time, but it is doubtful that the exact spike times can be accurately reproduced. Some have been able to achieve accurate modeling of firing processes, but only when the visual stimulus is known and mostly for early processing stages, such as the retina or LGN [18].

Finally, we want to emphasize that it is always important to properly assess the usefulness of a given spike train model. Models can prove useful for developing and calibrating data analysis methods [16], or for studying and quantifying given properties of the data [18]. In general however, the spiking behavior of neurons is both complex and variable. One needs to judge carefully which model captures the interesting properties relevant for a given scientific question. Furthermore, one has to be aware that, even if statistical properties of neuronal firing are precisely reproduced, the exact spiking dynamics stem from the complex interactions with other neurons and the drive from external stimuli. When one needs to generate a spike train that is very close to the original, the best model for the spike train is probably the spike train itself.

Acknowledgments

The authors acknowledge three grants of the Romanian Government (Human Resources Program RP-5/2007 contract 1/01.10.2007 and Ideas Program ID_48/2007 contract 204/01.10.2007 both financed by MECT/UEFISCSU, and Partnerships Program contract 11039/18.09.2007 financed by MECT/ANCS), a grant for the "Max Planck - Coneural Partner Group", and a Deutsche Forschungsgemeinschaft grant (NI 708/2-1).

References

1. Braitenberg, V., Schz, A.: Cortex: Statistics and Geometry of Neuronal Connectivity, 2nd edn. Springer, Berlin (1998)
2. Churchland, P.S., Sejnowski, T.J.: The Computational Brain. MIT Press, Cambridge (1999)

3. Destexhe, A., Rudolph, M., Par, D.: The high-conductance state of neocortical neurons in vivo. Nat. Rev. Neurosci. 4, 739–751 (2003)
4. Tuckwell, H.C.: Stochastic Processes in the Neurosciences. Society for Industrial and Applied Mathematics, Philadelphia PA (1989)
5. Rieke, F., Warland, D., de Ruyter van Steveninck, R., Bialek, W.: Spikes: Exploring the Neural Code. MIT Press, Cambridge (1997)
6. Dayan, P., Abbott, L.F.: Theoretical Neuroscience. MIT Press, Cambridge (2001)
7. Heeger, D.: Poisson model of spike generation (2000), `http://www.cns.nyu.edu/david/handouts/poisson.pdf`
8. Waters, J., Helmchen, F.: Background synaptic activity is sparse in neocortex. J. Neurosci. 26(32), 8267–8277 (2006)
9. Tiesinga, P., Fellous, J.M., Sejnowski, T.J.: Regulation of spike timing in visual cortical circuits. Nat. Rev. Neurosci. 9, 97–109 (2008)
10. Evarts, E.V.: Temporal patterns of discharge of pyramidal tract neurons during sleep and waking in the monkey. J. Neurophysiol. 27, 152–171 (1964)
11. Soltesz, I.: Diversity in the neuronal machine. Oxford University Press, New York (2005)
12. Kara, P., Reinagel, P., Reid, R.C.: Low response variability in simultaneously recorded retinal, thalamic, and cortical neurons. Neuron 27, 635–646 (2000)
13. Steriade, M., McCormick, D.A., Sejnowski, T.J.: Thalamocortical oscillations in the sleeping and aroused brain. Science 262, 679–685 (1993)
14. Steriade, M., Timofeev, I., Dürmüller, N., Grenier, F.: Dynamic properties of corticothalamic neurons and local cortical interneurons generating fast rhythmic (30-40 Hz) spike bursts. J. Neurophysiol. 79, 483–490 (1998)
15. Buzsáki, G.: Rhythms of the brain. Oxford University Press, Oxford (2006)
16. Mureşan, R.C., Jurjuţ, O.F., Moca, V.V., Singer, W., Nikolić, D.: The Oscillation Score: An Efficient Method for Estimating Oscillation Strength in Neuronal Activity. J. Neurophysiol. 99, 1333–1353 (2008)
17. DeBusk, B.C., DeBruyn, E.J., Snider, R.K., Kabara, J.F., Bonds, A.B.: Stimulus-dependent modulation of spike burst length in cat striate cortical cells. J. Neurophysiol. 78, 199–213 (1997)
18. Keat, J., Reinagel, P., Reid, R.C., Meister, M.: Predicting Every Spike: A Model for the Responses of Visual Neurons. Neuron 30, 803–817 (2001)

Appendix

To generate the data for Fig. 2 the following parameters were used: for $f_o(t)$, $\tau = 0.5 * 1$, $m = 0.5$ and $f_o(t)$ was allowed to deviate from the target oscillation frequency with at most 1 Hz; for $f_b(t)$, $\tau = 0.1$, $m = 3$ and $f_b(t)$ was bounded between 0 and 100 Hz. Note that τ is given relative to the period of the target oscillation frequency. The oscillation strength, o, took values from 0.2 to 0.8 in 24 steps, $SPPI$ was varied between 0.004 to 0.1 in 24 steps and the $offset$ was set to 0.5. To generate the artificial data presented in Fig. 3, some of the parameters given above had to be adapted to each recorded spike train (trial) that was modeled.

The InfoPhase Method or How to Read Neurons with Neurons

Raul C. Mureşan[1,2], Wolf Singer[2,3], and Danko Nikolić[2,3]

[1] Center for Cognitive and Neural Studies (Coneural), Str. Cireşilor nr. 29, 400487 Cluj-Napoca, Romania
muresan@coneural.org
http://www.coneural.org/muresanlab
[2] Max Planck Institute for Brain Research, Deutschordenstrasse 46, 60528 Frankfurt am Main, Germany
[3] Frankfurt Institute for Advanced Studies, Ruth-Moufang-Str. 1, 60438 Frankfurt am Main, Germany

Abstract. We present a novel method that can be used to characterize the dynamics of a source neuronal population. A set of readout, regular spiking neurons, is connected to the population in such a way as to facilitate coding of information about the source in the relative firing phase of the readouts. We show that such a strategy is useful in revealing temporally structured processes in the firing of source neurons, which have been recorded from cat visual cortex. We also suggest extensions of the method to allow for the direct identification of temporal firing patterns in the source population.

Keywords: Spiking neurons, temporal coding, readouts, visual cortex.

1 Introduction

Information coding in the brain remains, to this date, the subject of lively debate. It has been proposed that in addition to the traditional firing-rate coding [1], the temporal structure of spike-trains also carries a significant amount of information [2]. A special case of temporal code is represented by neuronal synchronization, which has been recently shown to correlate with the perception of brightness [3], and also with conscious perception [4]. Beyond synchronization, temporal codes might assume more generalized forms, whereby spikes are not perfectly aligned, but arranged into temporal patterns. Polychronization, which relies on synaptic delays, is such an example [5]. A different putative mechanism organizing spikes into temporal patterns is represented by timed inhibition, precisely controlled by fast oscillatory rhythms in the gamma band [6]. In this latter case, temporal phase patterns may even be expressed within a single gamma cycle, with more excited neurons firing earlier relative to the onset of the cycle than less excited ones. Furthermore, we now know that there is a tight interplay between the expression of neuronal oscillations and temporal coding in the cortex [7], with rhythms expressed in different frequency bands [8]. Even further,

V. Kůrková et al. (Eds.): ICANN 2008, Part II, LNCS 5164, pp. 498–507, 2008.

temporal codes could be defined beyond single-spike timing. For example, bursts or very fast rate fluctuations could be timed relative to each other, as dictated either by internal network constraints, or by the temporal dynamics of the stimulus. It is very likely that multiple timescales play important roles [9] and the relative timing of processes evolving on these different timescales, is of crucial importance to the brain. Identifying these processes and their specific combinations, which carry information about stimuli, remains however a tremendous challenge. Various methods have been put forward, so far, for the detection and proper interpretation of synchronous spike patterns [10,11] or temporal structure of membrane potential fluctuations [12,13]. However, the difficulty remains to identify, in a flexible way, more general activity patterns (beyond synchrony) across a large population of neurons. There are two categories of problems: First, one cannot always know the timescale on which the relevant processes should be searched since these are likely to coexist on multiple timescales. Second, it is unclear how much history dependence one should take into account when detecting candidate correlation patterns. Nontrivial combinations of single spikes, bursts or even fast rate fluctuations, could be co-occurring in a stimulus specific manner and precisely timed relative to each other. The number of possible combinations far exceeds the analysis capabilities of most modern techniques. Thus, one needs to narrow down the search to a specific temporal window and a specific process/timescale. We wanted to break this limitation by considering that in the brain, neuronal information is read out by other neurons which integrate complex input patterns. Following a similar previous approach [14], we considered a readout set of simulated regular spiking (RS) neurons (observers) that preferentially sample subpopulations of a larger neuronal pool (the source neurons), the latter being recorded from the brain. We attempt to recover the information coded in the source neurons' dynamics by observing the relative firing phase of the readouts. At least to a first degree of approximation, interesting events can be assumed to span the timescale defined by the time constant of the neuronal membrane and the dynamics of synaptic currents. Except for the implicit time constants of readouts and their corresponding synapses, we did not restrict the information extraction to a fixed timescale, allowing for a broad range of temporal processes to be observed in a biologically relevant fashion. Moreover, the activation of readouts could reveal nontrivial combinations in the activity of source subpopulations, even when having source processes delayed with respect to each other on a broad temporal interval.

2 Methods

Since the method we describe here relies on information encoded in the relative firing phases of artificial readout neurons, we called the method "InfoPhase".

2.1 Experimental Procedures, Stimulation and Recording

The experiment was performed in the visual cortex of a lightly anesthetized, paralyzed cat. Anesthesia was induced with ketamine (Ketanest, Parke-Davis,

10 mg kg^{-1}, intramuscular) and xylazine (Rompun, Bayer, 2 mg kg^{-1}, intramuscular) and maintained with a mixture of 70% N_2O and 30% O_2 supplemented with halothane (0.5%-1.0%). The animal was paralyzed with pancuronium bromide (Pancuronium, Organon, 0.15 mg kg^{-1} h^{-1}). Visual stimuli were presented binocularly on a 21 inch computer screen (HITACHI CM813ET) with 100 Hz refresh rate. Data were recorded from area 17 by inserting multiple silicon-based multi-electrode probes (16 channels per electrode) supplied by the University of Michigan (Michigan probes). Signals were amplified 10,000× and filtered between 500 Hz and 3.5 kHz for extracting multi-unit activity. Offline spike-sorting techniques were next applied and 61 single units were extracted. The investigated neuronal activity was acquired in response to sinusoidal gratings moving in 12 directions in steps of 30°. For each condition, 20 trials have been recorded, yielding a total of 240 trials. Each trial had a length of 4800 ms. After a spontaneous activity lasting for 1000 ms, the stimulus was presented for 3500 ms, and subsequently, after the stimulus was removed, 300 ms additional OFF response activity was recorded.

2.2 The Readout Model

For the artificial readouts we used the two-dimensional phenomenological model of Izhikevich [15]. Each neuron is described by a set of differential equations:

$$dv/dt = 0.04v^2 + 5v + 140 - u + I \tag{1}$$

$$du/dt = a \cdot (bv - u) \tag{2}$$

where v - membrane potential, u - recovery variable, I - total post-synaptic current ($I = \sum psc_i$), a,b - parameters. When the membrane potential reaches a value larger than 30 mV, a spike is recorded and the membrane potential is reset to its resting value (-65 mV), while the recovery variable is increased by a given amount (8 for the RS neuron).

We used the parameter settings for the RS neurons, as they best represent the dynamics of pyramidal neurons in the cortex [15]. We next connected 10 RS readouts to the recorded population of cortical neurons (61 source neurons). This was the most important step, as we wanted to obtain the following: reduce the dimensionality of the data by projecting the activity of the source neurons onto the readouts, preserve as much temporal information as possible, and finally, reduce the influence of the firing rate. To this end, we created balanced synapses for each readout: Every readout was connected with all source neurons via excitatory synapses that produced exponentially decaying post-synaptic currents [16]:

$$psc_i = A \cdot W_i \cdot g_i \cdot (E_{syn} - v_{post}) \tag{3}$$

$$dg_i/dt = -g_i/\tau_{syn} \tag{4}$$

where, psc_i - post-synaptic current contributed by synapse i, A - an amplitude parameter determining the maximal amplitude of psc (global constant that allows scaling all synapses), W_i - synaptic strength, g_i - instantaneous synaptic conductance, E_{syn} - reversal potential of the synapse (taken 0 mV for excitatory synapses), v_{post} - membrane potential of the post-synaptic neuron, τ_{syn} - time constant for the decay of the synaptic conductance (here 30 ms).

The synapses of neurons are balanced in the sense that the total synaptic gain (sum of all input weights) is similar for all readouts. Each connection is instantiated such that it represents a relatively small random fluctuation ($<$ 28%) on top of a constant baseline:

$$W_i = 0.9 + rand_{[-0.25...+0.25]} \tag{5}$$

where $rand_{[]}$ represents the uniform random function in a given interval.

The instantiation scheme for the synapses is crucial. Each readout neuron is receiving roughly the same amount of excitation but samples slightly differently the source population of cortical neurons. By choosing a small enough value for A (here 0.05) the readouts will engage into quasi-periodic firing, with slight differences in their phases (Fig. 1). Because all synapses have very similar strengths, except for the small random fluctuation ($<$ 28%), when one source cortical neuron increases its firing rate, all readout neurons receive simultaneously increased excitation, and hence tend to shift their phase together. The exact amount of phase shift, for each readout, will depend on the particular sampling given by the synaptic distribution, and the activity of the underlying source subpopulations. Each readout can be considered as an independent observer that integrates, over time, the activity of preferred source subpopulations.

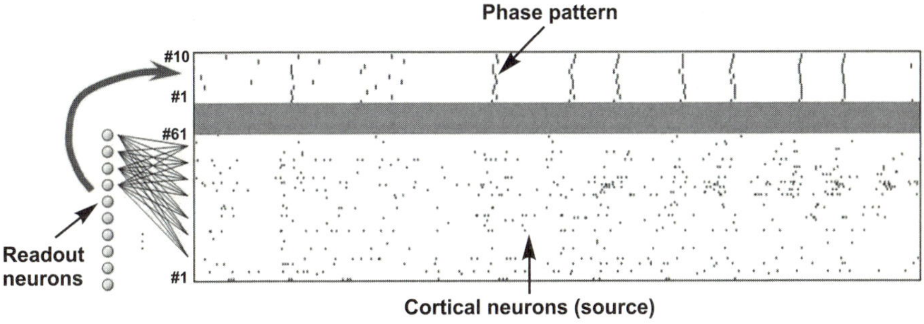

Fig. 1. Schematic representation of the InfoPhase readout method. Readout neurons are connected via conductance-based synapses to recorded cortical neurons, and produce quasi-periodic phase patterns.

We can now go one step further and extract the phase relationships among readout neurons (Fig. 2). These relative phases represent differences in the activation of various source subpopulations (randomly selected by the particular synaptic instantiation) that were integrated over time by the readouts and translated into small phase differences. Thus, we are observing non-trivial combinations of spatio-temporal patterns evolving in the source population (cortical neurons) and reflected in the phases of the readouts.

2.3 The Classifier

To assess the amount of information that can be extracted by the readout population from the cortical source, we next constructed a simple classifier, in three steps: extraction of phase vectors, clustering, and training. First, the phase patterns of the readouts are isolated using a sliding window for detection (Fig. 2A) and then converted into phase vectors by computing the relative phases with respect to neuron 1 (Fig. 2B). We obtained a sequence of phase vectors that represents the mapping of the cortical firing onto the activity of the readouts (Fig. 2C). Each experimental trial can then be represented as a sequence of phase vectors.

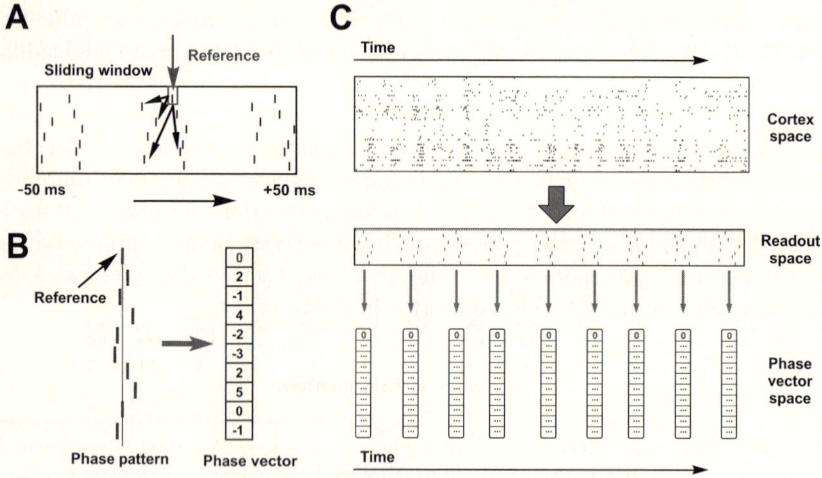

Fig. 2. Building of phase vectors. A. A sliding window is used to detect and isolate the firing phase patterns of the readouts. B. The isolated phase pattern is converted into a phase vector by computing the relative phases with respect to a reference neuron. C. The information from the cortical population is reflected in the readout phase patterns which are then converted into a sequence of phase vectors. Schematic illustration only.

Many phase vectors that represent a trial are similar to each other such that the vector space appears clustered. There are a few dense clouds and many scattered points. To have a good representation of this space we first applied a K-Means clustering with K being determined empirically (for this dataset the optimal K=40). Clustering, ensures that dense clouds are represented only by one phase vector (cluster center), and hence, increases robustness against noise. We should mention, however, that the method functions reasonably well without this step. For simplicity, we chose a classifier that memorizes, for each stimulus class, a set of representative, specific, phase vectors, called model vectors. The model vectors are computed during the training phase. We used a split-half procedure by randomly selecting, for each condition, half the trials (10 trials) for training and half for testing. From each training trial, the phase vectors are considered one by one. For a given vector, the closest phase vector (in Euclidian distance) is searched in the other training trials (of all conditions). If the closest vector belongs to the same stimulus condition, then the vector is marked as being stimulus specific and added to the list of model vectors for the corresponding stimulation condition. After training, each stimulation condition (class) will be represented by a variable number of model vectors, assigned by the previously described procedure. During testing, for each test trial, we first compute the predicted stimulus class by applying a scoring procedure. The score is computed as follows: For each phase vector from the trial to be classified, we search for the closest model vector among all models. For the matching condition, a value is added to the corresponding global scoring:

$$S_k = S_k + 1/NrModels(k) \quad (6)$$

where, S_k - global score corresponding to the matching condition k, $NrModels(k)$ - the number of model vectors describing stimulus class k.

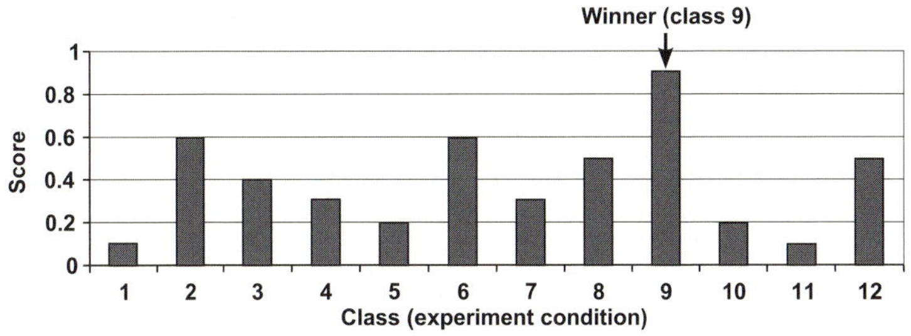

Fig. 3. A typical result of the scoring procedure for a trial. The trial is assigned the predicted class (stimulus) with the highest score.

The scoring protocol normalizes against the number of model vectors of each class such that classes having different number of models are treated equally. All phase vectors of the trial to be classified are scored in a similar manner. Finally, the predicted class of the trial is computed as the class with the maximum score (Fig. 3). After classifying all the test trials, we computed the classification performance as being the percent correctly classified trials from the testing set. Trials for which there were two or more classes sharing the same maximum score were considered as unclassifiable and were not included in the computation of the classification performance. This case appeared however very rarely.

3 Results

We applied the InfoPhase method to a dataset recorded from cat area 17, stimulated with drifting sinusoidal gratings moving in 12 different directions (30° steps, see Methods). We classified the test trials, computed the classification performance, and tried to identify the time scale of the information that has been extracted from the cortical population's activity.

3.1 Classification Performance

After training on 120 trials and classifying the other 120 trials we obtained a classification performance of 57%.

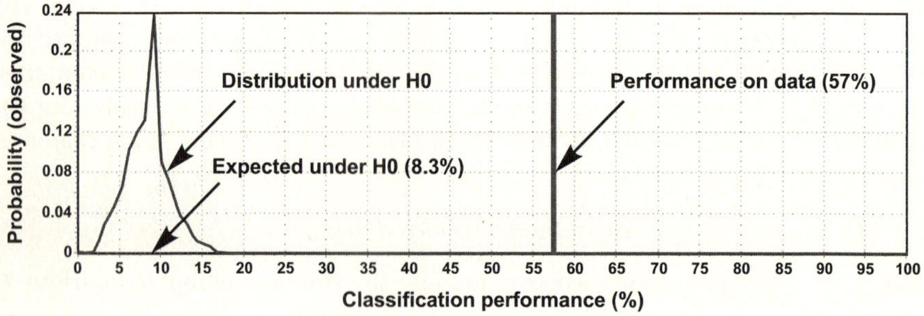

Fig. 4. Classification performance on the original data (57%) and distribution of classification performances when the labels (classes) are randomly permuted (H0)

We next wanted to determine whether this performance was obtained by chance or it reflected legitimate information about the stimuli that was extracted by the method from the cortical spikes. We hypothesized that the observed performance was obtained by chance (H0) and computed the distribution of the performance when the trial labels (classes) were randomly permuted. We used 5000 permutations to estimate the distribution. The expected performance under H0 was (as can also be computed theoretically) 8.3%. More importantly, the performances obtained by chance never exceeded 18% (Fig. 4). Thus, the difference between chance performance and the observed classification performance of 57% on the original, unpermuted data, was highly significant ($p \ll 0.001$).

3.2 Temporal Dynamics of Cortical Neurons

Sinusoidal gratings, which were used in the experimental protocol, are known to produce robust rate responses [17]. By integrating 3.5 seconds of activity across the 61 source neurons and computing the mean firing rate, one can classify the test trials 98.3% correct. The mean firing rate code is completely insensitive to the temporal structure of the spike trains. At the other extreme, there might be fast codes, where information is coded in up to tens of milliseconds of activity. We estimated the sensitivity of the readouts to the temporal structure of the source spike trains by progressively altering the original cortical spike times, in a way that preserves mean firing rates but destroys the fast temporal structure of the data.

We applied progressive jittering of spike times in the source dataset (Fig. 5A) by independently shifting each spike with a value drawn from a Gaussian distribution with 0 mean and standard deviations between 5 and 30 ms, changed in steps of 5 ms. The classification performance dropped rapidly with the amount of

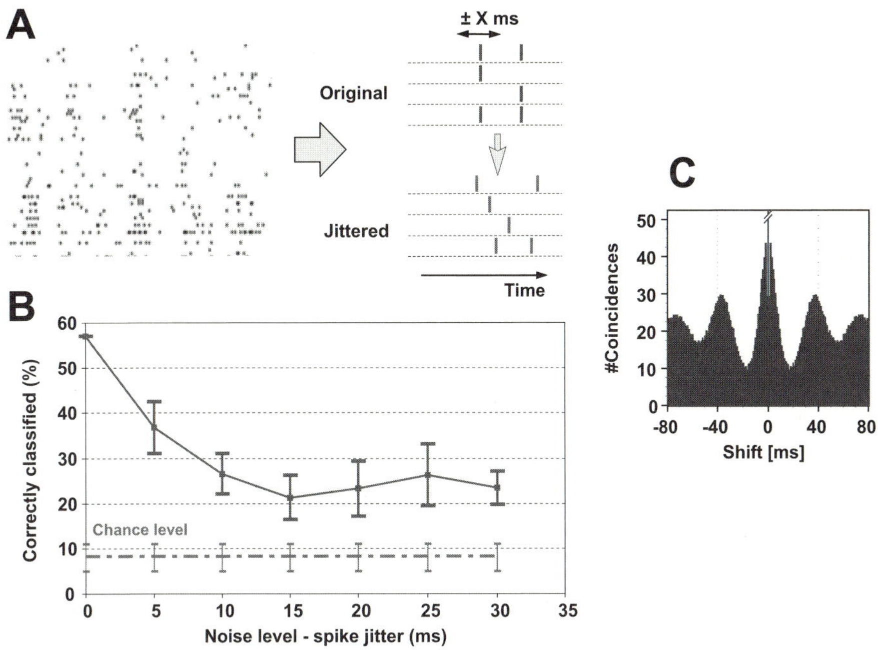

Fig. 5. Determining the temporal scale of the information extracted by the readout population. A. The temporal structure of the source spike trains can be progressively altered by jittering individual spike times. B. The classification performance drops steeply with the amount of jitter, indicating sensitivity of the method to the temporal structure of the source spike trains (error bars are SD). C. Autocorrelation histogram on the spike train of a source unit.

jitter applied to the source spike trains (Fig. 5B), hinting towards the possibility that the method extracts information that is encoded on fast temporal scales. Also note that the classification performance remained significantly better than chance, regardless of the amount of jitter. This suggests that, despite using phase readout patterns to classify, the readouts can also take advantage of the firing rate codes in the source dataset and thus the influence of the rate code cannot be completely eliminated. In addition, autocorrelation analysis of the source spike trains (Fig. 5C) revealed oscillatory activity with an oscillation period around 25-27 ms (beta-high band [8]) suggesting that the oscillation cycles might play an important role in enabling the coding of fine-grained temporal information in the source neurons [6].

4 Discussion

As we have seen, artificial spiking neurons can be successfully used to read out information from cortical spike trains. The method we presented here relies on regular spiking neurons that represent biased observers, sampling subpopulations of the cortical neuron set. The bias is minimal such that differences in the activation of the respective subpopulations are only enough to bring readouts out of phase but not enough to change their mean firing rate. The information can be thus extracted without imposing a rigid, fixed time scale for the observers. Moreover, the specificity of phase differences among the readouts seems to correlate with the fine temporal structure of the cortical spike trains. However, there are several aspects that need to be addressed in future studies.

First, the synapses of the readouts are biased in a random way. Thus, each readout represents an independent observer of the source population, transforming the input in a non-linear fashion, with a fixed amount of memory that is given by the membrane and synaptic time constants. However, the collective memory (phase relations between readouts) might go well beyond the individual time constants, as the advancement/delay of readouts relative to each other could last for much longer durations than the individual time constants. A future improvement of the method should consider some synapse training procedure allowing the readouts to optimally encode information in their relative firing phases.

Second, we need to address more in depth the exact signature of the information that is extracted by the readouts from the cortical spike trains. The high sensitivity of the classification performance with respect to the temporal structure of the cortical source suggests that the extracted information is coded on very fast time scales. Nonetheless, there is always the possibility that jittering affects only the capability of the method to correctly extract the information, even if the latter is coded on slow time scales and is thus insensitive to jittering. However, this possibility is remote since we always retrain the classifier after each jittering of the inputs. Retraining insures that the best possible model vectors are built, given the structure of the source spike trains. To find the exact source of the information and the relevant time scale, we need to go back to the original spike trains, guided by the occurrences of the phase patterns that were specific to each class (model vectors). A possibility would be to compute "phase pattern-triggered average" of the source spike trains. In this case, one computes an average of the source spike trains, around the time stamps where a specific pattern occurs, thus identifying the combination of source spikes that produced the pattern. If, indeed, a specific constellation of input spikes produced a specific readout pattern, then the next question is how fast can the readouts extract information? Or how much time is needed to integrate the relevant input and reach a decision? Even more so, what is the role of oscillatory activity in structuring these input patterns?

Finally, we conclude that using artificial spiking neurons to read out information from cortical neurons can be extremely fruitful and opens the path for a new generation of studies that might help us reveal the nature of the neuronal code.

Acknowledgments

The authors acknowledge two grants of the Romanian Government (Resurse Umane RP-5/2007 contract 1/01.10.2007 and Idei ID_48/2007 contract

204/01.10.2007 financed by MECT/UEFISCSU), a grant for the "Max Planck - Coneural Partner Group", a DFG grant (NI 708/2-1), and support from the Hertie Foundation.

References

1. Adrian, E.D.: The basis of sensation. W.W. Norton, New York (1928)
2. Singer, W.: Neuronal Synchrony: A Versatile Code for the Definition of Relations? Neuron 24, 49–65 (1999)
3. Biederlack, J., Castelo-Branco, M., Neuenschwander, S., Wheeler, D.W., Singer, W., Nikolić, D.: Brightness Induction: Rate Enhancement and Neuronal Synchronization as Complementary Codes. Neuron 52, 1073–1083 (2006)
4. Melloni, L., Molina, C., Pena, M., Torres, D., Singer, W., Rodriguez, E.: Synchronization of Neural Activity across Cortical Areas Correlates with Conscious Perception. Journal of Neuroscience 27, 2858–2865 (2007)
5. Izhikevich, E.M.: Polychronization: Computation With Spikes. Neural Computation 18, 245–282 (2006)
6. Fries, P., Nikolić, D., Singer, W.: The gamma cycle. Trends in Neuroscience 30, 309–316 (2007)
7. Buzsáki, G.: Rhythms of the brain. Oxford University Press, Oxford (2006)
8. Mureşan, R.C., Jurjuţ, O.F., Moca, V.V., Singer, W., Nikolić, D.: The Oscillation Score: An Efficient Method for Estimating Oscillation Strength in Neuronal Activity. Journal of Neurophysiology 99, 1333–1353 (2008)
9. Abbott, L.F., Regehr, W.G.: Synaptic Computation. Nature 431, 796–803 (2004)
10. Grün, S., Diesmann, M., Aertsen, A.: Unitary events in multiple single-neuron spiking activity: I. Detection and significance. Neural Computation 14, 43–80 (2002)
11. Pipa, G., Wheeler, D.W., Singer, W., Nikolić, D.: NeuroXidence: A Non-parametric Test on Excess or Deficiency of Joint-Spike Events. Journal of Computational Neuroscience (2008), doi:10.1007/s10827-007-0065-3
12. Ikegaya, Y., Aaron, G., Cossart, R., Aronov, D., Lampl, I., Ferster, D., Yuste, R.: Synfire Chains and Cortical Songs: Temporal Modules of Cortical Activity. Science 304, 559–564 (2004)
13. Mokeichev, A., Okun, M., Barak, O., Katz, Y., Ben-Shahar, O., Lampl, I.: Stochastic emergence of repeating cortical motifs in spontaneous membrane potential fluctuations in vivo. Neuron 53, 413–425 (2007)
14. Mureşan, R.C., Pipa, G., Florian, R.V., Wheeler, D.W.: Coherence, Memory and Conditioning. A Modern Viewpoint. NIP-LR 7(2), 19–28 (2005)
15. Izhikevich, E.M.: Simple Model of Spiking Neurons. IEEE Transactions on Neural Networks 14, 1569–1572 (2003)
16. Mureşan, R.C., Ignat, I.: The "Neocortex" Neural Simulator. A Modern Design. In: Proceedings of the International Conference on Intelligent Engineering Systems, Cluj-Napoca (2004)
17. Hubel, D., Wiesel, T.: Receptive fields, binocular interaction and functional architecture in the cat's visual cortex. Journal of Physiology 160, 106–154 (1962)

Artifact Processor for Neuronal Activity Analysis during Deep Brain Stimulation

Dimitri V. Nowicki[1], Brigitte Piallat[2], Alim-Louis Benabid[3], and Tatiana I. Aksenova[3,4]

[1] Institute of Mathematical Machines and Systems, 42 Glushkov ave., 03187 Kiev Ukraine
nowicki@nnteam.org.ua
[2] INSERM U 836-UJF-CEA-CHU CHU A. Michallon, BP 217, 38043 Grenoble, Cedex 09, France
brigitte.piallat@ujf-grenoble.fr
[3] Department of microtechnologies for Biology and Helth, CEA LETI MINATEC 17, Rue des Martyrs, 38054, Grenoble, France
alim-louis.benabid@cea.fr
[4] Institute of Applied System Analysis, Ukrainian Academy of Sciences, Prospekt Peremogy, 37, Kiev 03056, Ukraine
tetiana.aksenova@cea.fr

Abstract. The present paper is devoted to the suppression of spurious signals (artifacts) in records of neural activity during deep brain stimulation. An algorithm of adaptive filtering based on the signals synchronization in phase space is presented. The algorithm was implemented and tested using synthetic data and recordings collected from patients during the stimulation.

1 Introduction

High-frequency (100-300 Hz) DBS is an effective treatment of a variety of disabling neurological symptoms, in particular motor symptoms associated with Parkinson's disease. Studies of the appropriate signal of neuronal activity during the stimulation, namely the extracellular microelectrode recording of action potentials (spikes), are hampered by stimulation artifacts presence in the records (Fig. 1a). The artifacts are induced by the periodically repeated electrical pulses delivered to the target zone in the brain. The artifacts have a common waveform but are not identical due to sampling errors and irregularities of stimulus production. Stimulus artefacts have amplitudes 5 to 20 times lager then spikes of neuronal activity. Typical artifact duration in the timeline is 20%-50% of each stimulation period. Such artefacts hamper spike detection and sorting leading to the problems to studies of mechanism of DBS [1].

There are several recent studies focused on removal, subtraction, or filtering of stimulation artifacts. Subtraction techniques were developed to avoid the suppression of the high frequency components. However, most subtraction techniques suffer from an inability to adapt to the nonlinear dynamics of the artifacts and hence suffer from the residual artifacts. Estimates of representative stimulus artifact waveforms were studied by [2]. The template is generated by averaging a set of peri-stimulus segments adjusted

V. Kůrková et al. (Eds.): ICANN 2008, Part II, LNCS 5164, pp. 508–516, 2008.

a)

b)

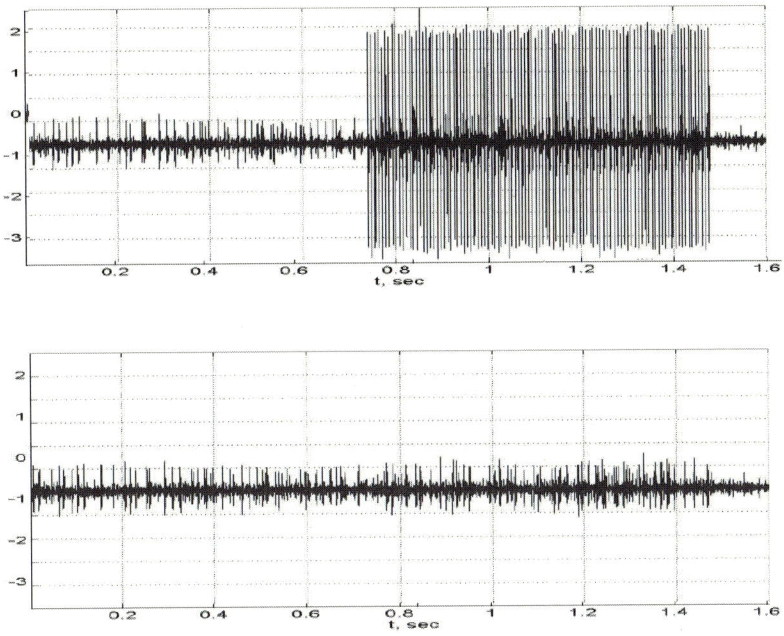

Fig. 1. Record S1529, the ratio of the amplitude of the artifacts to the averaged amplitude of spikes of neuronal activity equals 5.1; (a) signal of neuronal activity before, during and after DBS; (b) the same signal after filtering

by time shifting and scaling. [3] created several templates in order to account for artifact variability. Residual artifacts of 0.8-0.9 ms duration ([3], and [4]) are cut from the signal before further analysis. This part (about 11% of the recording time for DBS at 130Hz) is no longer available for spike sorting, etc.

In existing algorithms (solutions) only one-dimensional recordings are processed, although actual recordings are taken simultaneously from 3 to 16 microelectrodes located in the same brain nucleus. In this case spikes of most of neurons are not observed at least in one channel; but artifacts have exactly the same timing everywhere.

So, dealing with vector recordings helps to increase processing reliability and spike detection rate. Here we focus on this case of multi-channel data analysis.

This paper presents an algorithm and software implementation for filtering the stimulus artefacts from the signal of neuronal activity during DBS. It is based on the use of a multidimensional nonlinear oscillation model to explain variability of the artifacts of stimulation. After signal synchronization, templates are estimated and subtracted from signal in the phase space. Presented algorithm shows better results then previous approaches (2-5 times less residuals on the synthetic data).

2 Methods

2.1 Signal Characteristics

The following indices of original and processed signal are used. *Artifact level (AL)* is a mean of maximal (original or processed) signal absolute values, such maxima were taken across artifact peaks. *Spike level (SL)* is a mean of spikes amplitudes in the absence of stimulation. Confidence interval (at confidence level 0.95) in assumption of Gaussian distribution of noise was used to estimate *Noise level (NL)*. Then artifact-spike ratio ASR=AL/SL, artifact-noise ratio ANR=AL/NL, spike-noise ratio SNR= SL/NL.

2.2 Modelling

For the sake of simplicity we consider a scalar (single-channel) case. The general case differs only by phase space dimension which is nd for degree of model d and number of channels n.

Basic methods [3] treat artifacts as a periodic function with additive noise:

$$xst(t + kT) = xst^0(t) + \xi_k(t), 0 < tT \tag{1}$$

where T is the period of stimulation; k is the number of stimuli; $xst^0(t)$ is the mean stimulus; $\xi_k(t)$ are independent random variables with zero mean and finite variance $\sigma^2_\xi(t)$. In this case, the mean artifact can be estimated by averaging the observed signals of stimulation. Subtracting this mean from the signal can achieve a relatively rough signal cleaning (Fig. 2b).

To improve artefacts synchronisation the approach based on the use of oscillation models was recently proposed [5,6]. It is based on stimulus description as a solution of an ordinary differential equation with perturbation. Namely observed signal $x(t), t = 1, 2, \ldots$ is considered as a sum of the stimulation artifacts $xst(t)$ and the signal $xNr(t)$ of neuronal activity: $x(t) = xst(t) + xNr(t)$. Observations are available for discrete time moments. The signal of stimulation is assumed to be a solution of an ordinary differential equation with perturbation

$$\frac{d^n xst}{dt^n} = f(x, \ldots, \frac{d^{n-1}xst}{dt^{n-1}}) + F(x, \ldots, \frac{d^{n-1}xst}{dt^{n-1}, t}) \tag{2}$$

where n is the order of the equation, $F()$ is a perturbation function, and equation

$$\frac{d^n xst}{dt^n} = f(x, \ldots, \frac{d^{n-1}xst}{dt^{n-1}}) \tag{3}$$

describes a self-oscillating system with stable limit cycle $x^0(t) = (x^0_1(t), \ldots, x^0_n(t))$, $0 < t \leq T$, in phase space with coordinates $x_1 = xst, x_2 = \frac{dxst}{dt}, \ldots, x_n = \frac{dxst^{n-1}}{dt^{n-1}}$. Here T is a period of stable oscillations which is the period of stimulation. The perturbation function $F()$, bounded by a small value, is a random process with a zero mean and a correlation time Δt_{corr} which is small in comparison to the period of stable oscillations: $B(F(\cdot, t), F(\cdot, t + \Delta t)) \approx 0 if \Delta t > \Delta t_{corr}, \Delta t_{corr} << T$.

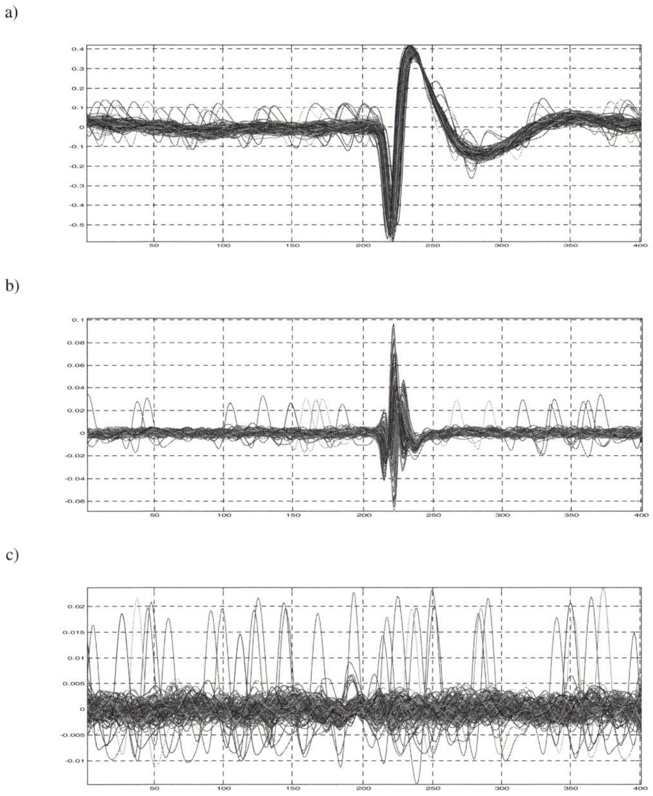

Fig. 2. Signal #S467, first channel ASR= 4.1 sliced into the stimulation-period windows (a) before filtering; (b) after filtering based on the model with additive noise; (c) after filtering according to phase-space algorithm

For the following parameterization the local coordinates in the neighbourhood of the limit cycle, phase deviation $\gamma(\theta)$ and normal deviations $n(\theta)$ are introduced. $n(\theta)$ is defined by orthogonal projection of current point on the limit cycle. Phase deviation $\gamma(\theta) = t(\theta) - \theta$, where $t(\theta)$ is a time shift along the trajectory from the initial point $\theta = 0$. New independent variable θ is a time shift from initial point through the limit cycle. Then Eq. (2) is rewritten in linear approximation in the deviations:

$$\frac{n(\theta)}{d\theta} + N(n) = F_n(\theta)\frac{d\gamma}{d\theta} + (\Theta n) = F_\gamma \Theta \tag{4}$$

Here $N(\theta)$ and $\Theta(\theta)$ are the functions of the parameters. As a result the signal trajectory in phase space is presented in linear approximation as a sum of periodic components of limit cycle and function of deviation $x(t(\theta)) = x_0(\theta) + n(\theta)$, where $n(\theta)$ and $t(\theta) = \gamma(\theta) - \theta$ are following Eq.4.

Thus, both models of the artifacts (Eq. 1 and Eq. 2) describe signals close to periodic ones. The model with additive noise (Eq. 1) explains the distortion of the amplitude of the signal only, while the model of nonlinear oscillations with perturbations (Eq. 2) explains the distortion of both amplitude and phase (Eq. 4).

For the following artifact filtering, the limit cycle that represents a periodic component of signal should be estimated and then subtracted from the signal.

2.3 Algorithm

The algorithm consists of the following stages: (1) Unsupervised learning procedure; (2) On-line filtering. More detailed description of this algorithm could be found in [5].

Unsupervised learning procedure.

- Approximate the signal trajectories in phase space and detect artifacts. For this purpose we compute the smoothed signal and the approximation of signal derivatives using convolution [7]. Artifacts are detected by the smoothed signal threshold crossing.
- Collect the set of periods of stimulation to the training set.
- Estimate a limit cycle. The element from training set that provides the maximum of probability density in the neighbourhood were used to estimate the limit cycle [8]. The time scale of the estimated limit cycle was considered as a phase.

On-line filtering has the following steps.

- Approximate the signal trajectories in phase space and detect i-th artifact xi(t).
- Synchronize artifact with the limit cycle by computing

$$t^i(\theta) = argmind(\theta) = argmin\|x^i(t) - x^0(\theta)\| \tag{5}$$

- Subtract the mean cycle from each artifact in phase space, according to the synchronization.
- Present the result in the time domain.

The parameters of algorithms were chosen as follows.

$Degree\ of\ the\ model,\ n = 3$ allows observing the limit cycles without self-intersection in the region of peak of artifact.

$Maximal\ phase\ deviation$ was estimated directly from a set of peak times.

2.4 Software Implementation

Artifact-processing algorithms were implemented in software package called Artifact Processor. The software consists of multi-platform library written in C++; and a Windows-based GUI. Implementation includes classes for reading data from text and binary files; there is an API that enables extension for direct reading from an electrophysiological device.

Then, the learning and subtraction procedures are implemented in the main class Signal. The output could be saved in text files readable by spike sorting systems like CED Spike2. The GUI (Fig. 3) has windows that visualize original and processing signal, as well as controls for navigating in the signal, setting up processing interval and parameters, loading and saving data etc.

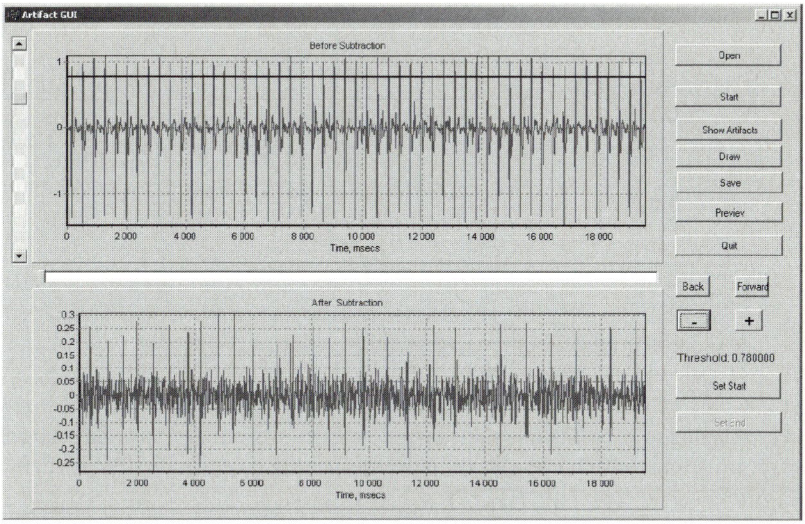

Fig. 3. Artifact Processor, synthetic data filtering, ASR=4

Speed test showed, that the training took about 30 sec for 1000 artifact training set (7 sec. recording at 130 Hz). Since the training is made, subtraction itself needs approximately same time as the recording time. These tests were mad for a standard PC with Intel Centrino 1400 MHz processor and 512MB RAM. So, we can conclude that the software might be used online during DBS surgery.

3 Results

3.1 Test of the Algorithm

Synthetic Dataset. In order to explore the algorithms behaviour depending on artifact amplitude, we constructed a set of artificial signals with predefined ASR. In the recordings (6-second segments) containing artifacts with no neuronal activity, another signal with single-neuron spikes and no stimulation was added. Appropriate constant K gives the desired value of ASR. So, for m-th channel we have:

$$X_{Syntethic}(t) = X_{Artifacts}(t) + KX_{Single-Neuron}(t). \quad (6)$$

Collection of such signals for ASR from 2 to 16 was generated. Here we mean ASR values for the first channel, the second and the third ones had ASR about 1.5 and 2 times bigger respectively.

Tests of the Algorithm with Synthetic data. The test intends to reveal what happens to spike trains in the presence of stimuli and cleaning procedures. Spike trains obtained from an original signal of neuronal activity $X_{Single-Neuron}(t)$ before mixing with artifacts and after full processing were compared. A simple threshold filter was used for

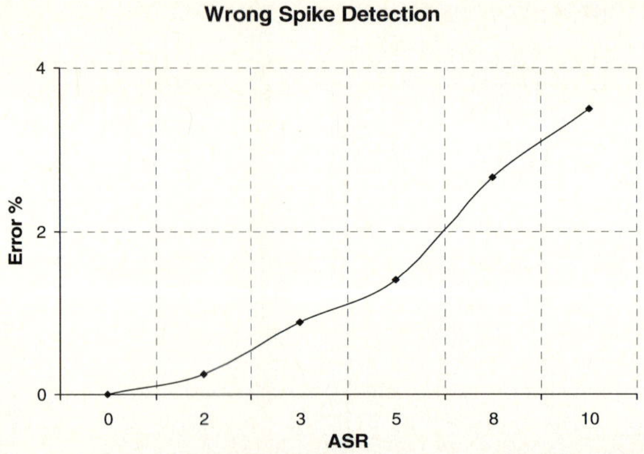

Fig. 4. Post-filtering spike quality for synthetic data: percentage of errors of spike detection in first channel after filtering. ASR=0 corresponds to the original neuronal activity with no stimulation added.

spike trains extraction. The percentages of spikes found at their correct positions in the processed signal were measured. Results are presented at Fig. 4. For the signals with $ASR \in [2, 10]$ (in the first channel) the percentage of wrong detection is from 1 to 3.8%. In particular there are 1.6% of errors for ASR=5 while in previous study [3], and [4] at least 11% of information is lost.

Real Data Description. Eight records collected from five human patients during DBS surgery were used to test the algorithm. Neuronal activity is recorded before, during, and after high-frequency stimulation within the subthalamic nucleus, Globus Pallidus or Substantia Nigra (for some records the post-stimulation segment is not available). Stimulation as produced using a Medtronic external stimulator, and the neuronal signals (measured in volts after amplification) were captured using the AlphaOmega Microguide TM recording system. The recordings, using a sampling rate of 48 kHz, had 9 to 100 seconds duration, the stimulation lasting for 7-80 seconds. Periodic stimuli were delivered through microelectrodes placed 2mm apart in the same brain nucleus, with a frequency of 130Hz (i.e. with period approx 370 sampling ticks) and a pulse width of 60 μs. The pulse current intensities were in the range of 1000 μA - 6000 μA.

The recordings described above were used to test the algorithm on real data. Actually, only 5 channels were related to microelectrodes and bore neuronal activity together with artifacts. For processing we selected 3 most suitable channels from each recording, containing both stimulation artifact and spikes.

The signals are characterized by ASR $\in$ [1.3, 14], ANR $\in$ [3, 160] and SNR $\in$ [1.4, 5.1]. Examples of original and filtered signals are shown in Fig. 1. Fig. 2 displays the signal sliced into sets of stimulation-period windows, before and after filtering. The figure demonstrates the shape of the original artifacts and of the residuals left after full processing. One can see also the neuronal spikes.

Dependences between Artifact Spike Ratio before and after signals filtering, both in phase space and time domain are presented in the Fig 5. For comparison, we have also shown such results for synthetic data. Results for eight real records and synthetic signals

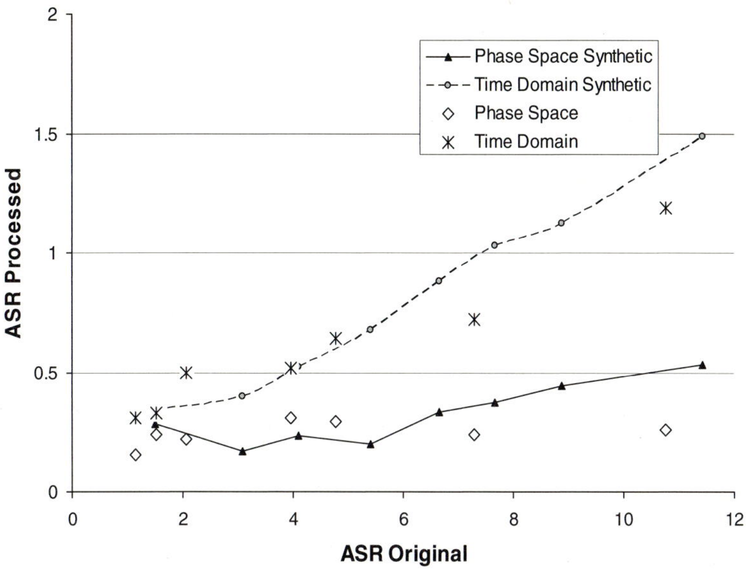

Fig. 5. Residual-spike ratio depending on pre-filtering ASR comparing time-domain and phase-space algorithms for synthetic and real data. ASR was measured by the first channel. Results for seven real records (markers) and synthetic signals (dash and continuous lines) are in a good fit.

are in a good fit. The use of phase space filtering produced smaller artifact residuals in all experiments: An advantage of phase space filtering augments with increasing of initial ASR of the signal.

4 Discussion

This paper presents algorithm and software for the filtering of the multi-channel signal of neuronal activity during DBS. The method is based on artifact template subtraction in phase space. Its implementation and the tests on the synthetic and real data sets are provided. The use of the proposed algorithm allows a 2-3 times reduction of the residual artifact of stimulation in comparison with the standard model with additive noise. The tests with 3-channeled artificial signals that are the combination of real signals of neuronal activity and stimulus artifacts showed that spike trains are less corrupted: we can increase spike detection reliability 7 times in comparison to existing methods (1.6% vs. 11% of spikes dropped). This promising approach will enable analysis of neuronal activity during DBS to more effectively study the mechanism and to therefore improve the DBS technique.

Acknowledgement

This study was partially supported by a Medtronic Grant to INSERM U318. The authors wish to express their appreciation to the Neurosurgical team for providing the recording used in the studies.

References

1. Benaid, A.L., Wallace, B., Mitrofanis, J., Xia, R., Piallat, B., Chabardes, S., Berger, F.: A putative generalized model of the effects and mechanism of action of high frequency electrical stimulation of the central nervous system. Acta neurol. 105, 149–157 (2005)
2. Winchman, T.: A digital averaging method for the removal of stimulus artifacts in neurophysiologic experiments. J. Neurosci. Meth 98, 57–62 (2000)
3. Hashimoto, T., Elder, C.M., Vitek, J.L.: A template subtraction method for stimulus artifact removal in high-frequency deep brain stimulation. J. Neurosci. Meth. 113, 181–186 (2002)
4. Tai, C.H., Boraud, T., Bezard, E., Bioulac, B., Gross, C., Benazzouz, A.: Electrophysiological and metabolic evidence that high-frequency stimulation of the subthalamic nucleus bridles neuronal activity in the subthalamic nucleus and the substantia nigra reticulata. FASEB 17, 1820–1830 (2003)
5. Aksenova, T.I., Novitsky, D.V., Benabid, A.-L.: Filtering out of artifacts of deep brain stimulation using nonlinear oscillatory model. Accepted to Neural Computation (to appear, 2008)
6. Nowicki, D.V., Benabid, A.L., Aksenova, T.I.: Adaptive model for filtering of stimulation artifacts in multichannel records of neuronal activity, international conference of numerical analysis and applied mathimatics. In: AIP Conference Processdings (2007)
7. Aksenova, T.I., Chibirova, O.K., Dryga, O.A., Tetko, I.V., Benabid, A.L., Villa, A.E.P.: An unsupervised automatic method for sorting neuronal spike waveforms in awake and freely moving animals. Methods 30, 178–187 (2003)
8. Aksenova, T.I., Shelekhova, V.Y.: Fast algorithms of derivative estimation on noisy observation. SAMS (1995)

Analysis of Human Brain NMR Spectra in Vivo Using Artificial Neural Networks

Erik Saudek[1], Daniel Novák[1], Dita Wagnerová[2], and Milan Hájek[2]

[1] Czech Technical University in Prague, Czech Republic
xnovakd1@labe.felk.cvut.cz
[2] Institute for Clinical and Experimental Medicine, Czech Republic
milan.hajek@ikem.cz

Abstract. Magnetic resonance has proven to be a successful method of in-vivo imaging. Although MRI can help detect various pathologies, its ability to classify the nature of the pathological tissue is limited. Magnetic resonance spectroscopy allows identifying metabolite content of the tissue and estimating the metabolite concentration. Map of metabolite concentration along with the MR image allows proper classification of many pathologies, for example progressive tumorous tissue identification in brain. Standard methods used to analyze nuclear magnetic resonance spectra such as singular value decomposition or curve fitting algorithms are very time consuming taking several minutes to analyze spectrum from a single voxel. To analyze the spectra from a chemical shift imagine sequence (CSI) in maximal resolution hundreds of spectra need to be processed. The suggested ANN framework proved to be much faster. Networks were trained on the outputs of LCModel curve fitting algorithm. Time needed to process a spectrum from a single voxel was reduced to the order of seconds. The total time needed to analyze a CSI in full resolution (hundreds of spectra) was significantly reduced to 5 minutes.

1 Introduction

Accurate quantification of cerebral metabolites by in vivo proton magnetic resonance spectroscopy is essential to the study of many brain disorders. Proton nuclear magnetic resonance (NMR) spectroscopy remains one of the few methods by which the chemical content of brain tissue may be probed non-invasively.

Several procedures have been applied in biomedical NMR spectroscopy to quantify areas of metabolite resonances from frequency domain spectra. With methods such as the lineshape fitting analysis [1] and the linear combination of model spectra [2], noisy 1H NMR spectra or overlapping peaks in spectra can be assessed. However, the reliable use of the lineshape fitting analysis methods needs spectroscopic expertise, so that a fully automated analyzer by these methods is rather difficult to develop. All these procedures involve relatively long computational times [3].

It has recently been shown that ANN analysis offers some important advantages for biomedical NMR data analysis [4],[5]. However, these techniques were limited to the use of simulated artificial inputs with mathematically estimated outputs for training. The proposed study shows the ability of an artificial neural network (ANN) to estimate concentrations of 3 main metabolites from human brain spectrum measured.

V. Kůrková et al. (Eds.): ICANN 2008, Part II, LNCS 5164, pp. 517–526, 2008.

In this work, absolute spectra of Free Induction Decay (FID) signals were used as input vectors. Matching metabolite concentrations acquired from the LCModel curve fitting algorithm [6] are set as targets.

Curve fitting algorithms, used to analyze NMR spectra in frequency domain, calculate the best possible combination of base signals (simulated or previously measured in-vitro) to fit the spectrum of the measured FID signal. Currently, the most commonly used quantification methods are based on curve fitting using a non linear least square optimization, such as Levenberg - Marquardt algorithm [2].

Once the network is trained the processing of a new (unknown) input is reduced to a set of matrix multiplications and can be done within seconds. The developed ANN gives comparable results in much shorter time than widely used curve fitting algorithms.

2 Methodology

The suggested methodological approach is summarized in Figure 1. The raw NMR spectrum is acquired by performing Fourier transform of nuclear spin echo of a free induction decay (FID) signal. The former case of NMR spectra derived from FID signal was used in this study. To increase a spatial resolution, the original 16 x 16 voxel matrix was interpolated to 32 x 32 voxel NMR spectra matrix by inverse 2D FFT. First, noise reduction was carried out by multiplying signal by a time damped exponential. Since the contribution of water is very significance, HLSVD (Singular Value Decomposition of Hankel matrix using Lanczos algorithm) method for water component removal was applied. Next, spectrum was calculated and frequency band between 0 ppm and 4 ppm was selected. The final vector representing the metabolite spectrum used as an input for the neural network was 800 points long. After ANN training the metabolic CSI map was reconstructed.

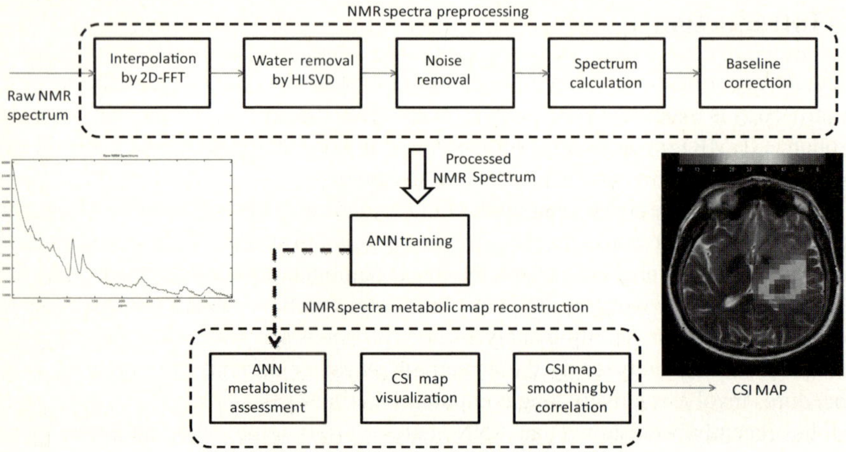

Fig. 1. Algorithm flow-chart diagram

2.1 Free Induction Decay

During the measurement a patient is placed in a strong magnetic field. Then a sequence of radio frequency pulses is applied for excitation. It consists of the signals from all the nuclei in the volume preselected by phase coding of the gradient fields added to the external field during the measurement. With a strong enough field and the right sequence of excitation pulses the sum of all the components of magnetization vectors perpendicular to the external field can be measured. The time evolution of the magnetic field in the plane perpendicular to the external field measured by a set of coils is called free induction decay (FID). It consists of the signals from all the nuclei in the patient. The frequency components of the FID correspond to the resonance frequencies of the nuclei in the patient.

In-vivo NMR spectra are measured from relatively large volumes (1.5 ml) of tissue enclosed within human body. Therefore homogeneity of the magnetic field within the measured volume is much harder to achieve. Moreover, in-vivo measurement has to be short with respect to the patient. As a result, in-vivo FID signal spectrum is noisy, the peaks representing various hydrogen nuclei are relatively wide and overlap - see Figure 2.

The FID signal of a single nucleus is a time damped complex exponential (complex plane allows expressing the orientation of the magnetization vector within the plane perpendicular to the external field). Singular value decomposition can be used to decompose the FID signal to a sum of such exponentials. These exponentials represent various groups of hydrogen nuclei.

2.2 Data Acquisition

Data were measured on a 3T Siemens Trio Tim system. Spin echo sequence with echo time of 30 ms and repetition time of 1510 ms for chemical shift imaging was used. Due to small SNR ratio, 4 measurement which were futher avereged were taken. The signal was sampled at 2048 samples per second. With the acquisition time of 1 second the length of the measured vector for each voxel was 2048 points. The signal was measured from a 15 mm thick matrix of 16 by 16 voxels of 10 mm on a side. The matrix is schematically represented in Figure 2(a). While the signal was acquired from the whole matrix the number of voxels excited by the selective radio frequency pulse (RF) during the measurement was patient specific.

Data for ANN modeling and testing were taken from clinical examinations of patients and healthy volunteers with written consent. The measurement procedures followed a protocol approved by the Ethic Committee. A set of 10 measurements in 9 subjects was used for training. Four of the subjects were healthy individuals, three were patients diagnosed with epilepsy, one patient diagnosed with a brain tumor, and one with a metabolic disease. The measured volume was from various locations within human brain. For each subject two measurements were taken, one with water suppression using a water saturation RF pulse, one with no water suppression. The mean frequency of the excitation RF pulse was set to -2.2 ppm for measurements with water suppression and to 0 ppm with no water suppression.

2.3 NMR Spectra Preprocessing

The signal of hydrogen bound in molecules of water is the strongest component of the FID signal from human brain tissue. During water saturation a RF pulse is used to keep the hydrogen in water molecules fully excited so that it can not contribute to the FID signal. With no water suppression the FID consists almost only of the water component. The signal of water is used as a scale and allows quantifying the concentration of metabolites from the signal with suppressed water by comparing it to the strength of the signal of unsuppressed water. The signal of unsuppressed water is so strong that only one acquisition needed to be taken compared to four averages without the suppression.

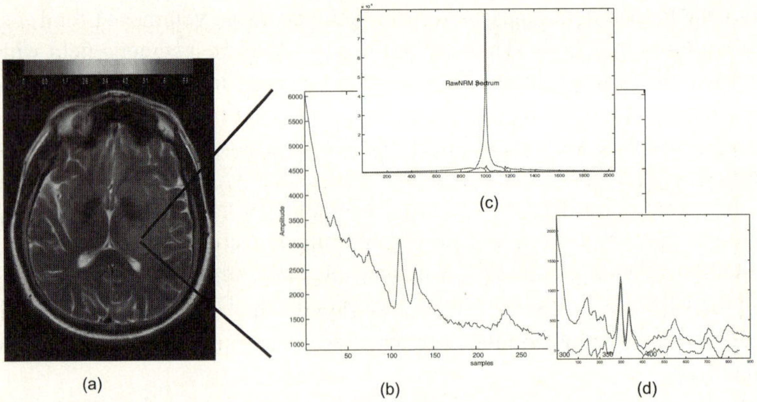

Fig. 2. Absolute NMR spectrum illustation. (a) Spatial configuration (16x16 voxel) of NMR recording. (b) example of NMR raw spectra. (c) after HLSVD water suppression apodisation and zero filling. (d) Same spectrum after baseline correction.

Data Interpolation. Before the spectra are calculated the signals are preprocessed. The measured signals were stored by the 3T Siemens Trio Tim system in spatial domain. To increase spatial resolution the data were interpolated. In order to interpolate a two dimensional inverse Fourier transform was preformed, then the data were padded by zeros to double resolution, then again transformed back to spatial domain by a 2D-FFT. That way the size of the matrix acquired from one subject was 32 by 32 voxels with a signal of 2048 time points each.

Noise Removal. To reduce noise the signal was multiplied by a time damped exponential $\exp({}^{-t}/500ms)$. The significant signal components are time damped as well. After multiplication by a damped exponential the significance of noise does not increase as rapidly with time. The same is done for the signals measured with no water suppression so that the amplitudes of both signals are processed equally. The FFT of the reference signal of water, measured with no water suppression, is then calculated and integrated across all frequencies. The integral corresponds to the water content in the voxel. For each voxel the signal with water suppression is divided by the integral of the water spectrum of that voxel.

Residual Water Removal. Even after application of water suppression a significant remainder of the signal of water persists in the signal. To remove the remaining water component HLSVD (Singular Value Decomposition of Hankel matrix using Lanczos algorithm) was used [7]. The most significant component of the signal after SVD shows to be a good approximation of the remaining water signal. In order to quickly calculate the first singular value and the corresponding vectors, the Lanczos algorithm was applied on a Hankel matrix composed for the signal from each voxel. Once the component corresponding to the first singular value is calculated it is subtracted from the signal. The example of water suppression is shown in Figure 2(b).

NMR Spectrum Calculation. After removing the remaining water components the signal is padded with zeros to double resolution, FFT is preformed and an absolute value is taken. The spectrum now consists of a few metabolite peaks and an underlying baseline. The points of the spectrum that are part of the peaks are usually a part of a steep slope. To identify the location of the peaks for each point of the spectrum a standard deviation of the points in its near neighborhood is calculated. If the standard deviation is above a certain threshold the point is classified as a part of a metabolite peak. The remaining points are classified as a part of the baseline. Through these a polynomial of the order of 6 is fitted and subtracted from the spectrum resulting in baseline correction. The baseline correction is depicted on Figure 2(d).

Finally only 800 points of the spectrum in frequency band from approximately 0 ppm to 4 ppm (all the metabolite peaks are within this band) is cut. Therefore, the final vector representing the metabolite spectrum used as an input for the neural network is 800 points long.

A spectrum before processing is depicted in Figure 3(b) and after processing is shown in Figure 2.

ANN Set-Up. The neural network has three layers. 800 points in the input layer are necessary to meet the length of the input vector. The number of neurons in the hidden layer was set to 100 experimentally (values between 50 and 300 were tested). The output layer consists of only one neuron. The best results were achieved when a separate network was trained for each metabolite. The output is a real number and is directly equal to the approximation of the concentration of the metabolite for which the network was trained. Concentrations are in laboratory units (no correction for T1, T2, white and gray matter and CSF or field inhomogeneity were done). The output function for the hidden layer is tansig. There is no output function in the output layer (it remains purely linear). Three networks were trained to calculate concentrations of three main metabolites in the NMR spectra of human brain (NAA, Choline, and Creatine). As a convergence criterion the sum of squared error was used. Scaled conjugate gradient backpropagation algorithm was used to train the network. The convergence criterion was set to total sum of error squared of 0.5 in all 700 inputs used for training. The time needed for the training algorithm to converge was approximately 10 minutes and 3000 to 5000 iterations were needed.

Because of bad homogeneity of the magnetic field, some of the signals were not good enough to be processed correctly. While such signals were present in the training set the training algorithm could not converge. To enable the convergence of the training

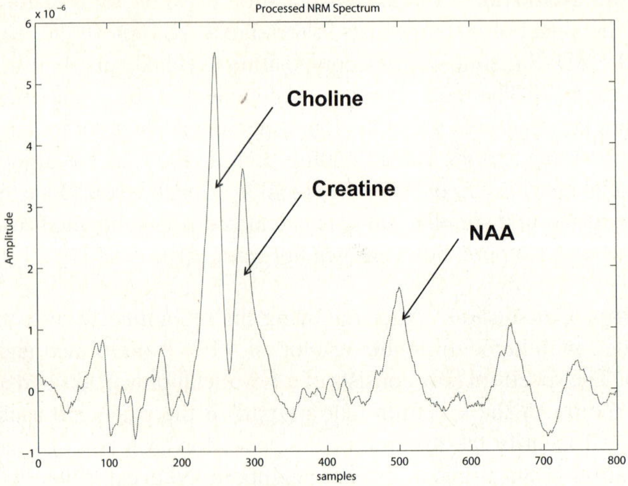

Fig. 3. Processed NMR spectrum with metabolites assessment. The concentrations are the followings: Choline: 4.39, Creatine: 7.32, NAA: 3.21.

algorithm the number of vectors in the training set was reduced to 700 spectra randomly selected from the spectra of the training subjects. Moreover, the spectra too different an average spectrum from the training set were restricted from the training. The similarity was determined by the correlation coefficient of the spectrum an average spectrum calculated from a large set of manually picked spectra used for training. The right threshold needed to detect useless spectra was found experimentally and set to 0.8.

For a CSI signal from a MR tomograph to be analyzed it needs to be preprocessed as were all the training signals. Once the set of 32 by 32 spectra is calculated it is fed one by one to the three neural networks to calculate the concentrations of NAA, Choline, and Creatine respectively for each voxel. Once the concentrations are calculated a metabolite map can be plotted over a corresponding MR image to demonstrate the spatial distribution of the metabolite.

3 Results

The performance of the system was tested on three healthy volunteers, one patient with a progressive brain tumor, one patient with a suspicion for a brain tumor and one patient with a suspicion for a metabolic disease. The resulting metabolic maps for Creatine Choline and NAA are displayed in Figures 4,5,6, respectively.

An MRI scan showed pathological tissue in the brain of the patient, possibly a brain tumor - see Figure 4. High concentration of Choline and low NAA points out low content of healthy neurons, rapid cell membrane decay along with high amount of Creatine showing intense metabolic processes. The pathological tissue is probably a progressive brain tumor. Both LCM and ANN models show the location and point out the nature of the pathology.

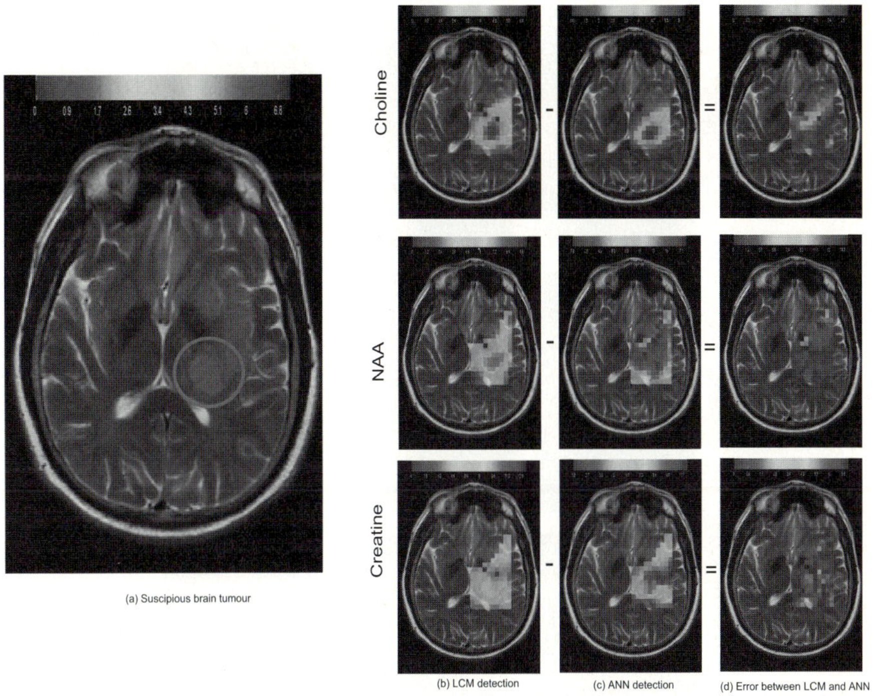

Fig. 4. Comparison of Choline, Creatine and NAA metabolite map calculated by LCM. (a) MR image of suspicious brain tumour. (b) and ANN model. (c) in case of patient with brain tumor. (d) The tumor was detected very precisely in both cases.

Analysis of 7 month old patient with a suspicion for a metabolic disorder is reported in Figure 5. In this examination it is vital to detect possible asymetry in metabolite distribution in the two hemispheres. Both LCM and ANN models showed good symetry and almost exactly the same space distribution. The diagnose would not differ no matter which model would be used. However, ANN approach takes 5 minutes compared to several hours in case of LCM.

In total over 1000 FID spectra were registered. The spatial distribution represented by the metabolite map according to LCModel curve fitting algorithm was visually very close to the approximation calculated by the neural network. For each subject the mean relative error and sum of square error was calculated as can be seen in Table 1. Also mean and variance of the targets and the calculated values were compared. The method did not show to be biased and the concentrations estimated by the neural networks showed to have a similar variance.

Performance results are summarized in Table 1. Subject 1 is an adult patient with a progressive brain tumor, subject 2 is a 7 month old patient with a suspicion for a metabolic disease, subject 3 is an adult healthy individual. High error and discrepancies in mean or variance are usually caused by a small number of voxels near the border of a measured volume with the poorest signal quality.

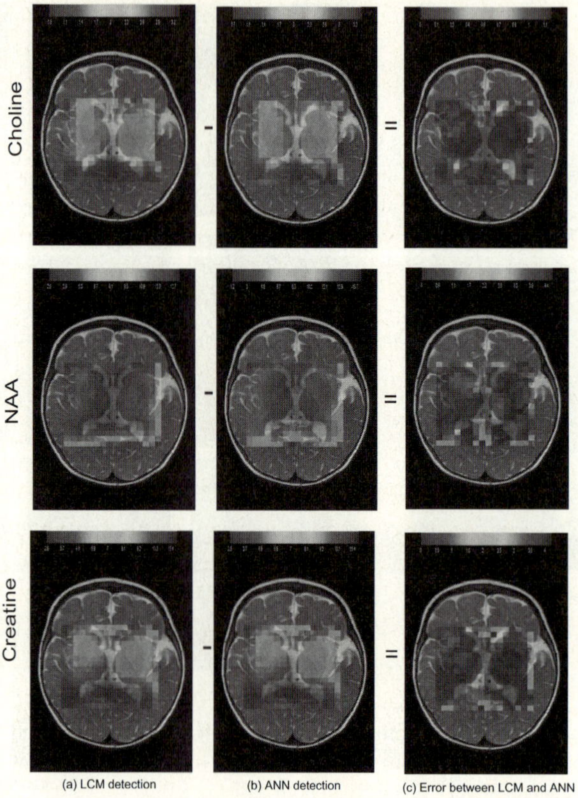

Fig. 5. Comparison of Choline, Creatine and NAA metabolite map calculated by LCM (a) and ANN model (b) in case of patient with a suspicion for a metabolic disorder

Table 1. Metabolite concentrate evaluation for LCM and ANN method

		Choline			Creatine			NAA		
Subject	Method	avg	std	er(%)	avg	std	er(%)	avg	std	er(%)
1	LCM	1.40	3		5.16	1.83		6.86	1.89	
	ANN	1.42	2.56	17.7	5.42	3.15	34	6.75	2.20	13.50
2	LCM	1.63	0.46		5.37	1.89		5.16	1.5	
	ANN	1.58	0.45	8.90	5.15	2.10	12.90	5.23	1.29	12.5
3	LCM	1.48	0.61		7.45	2.43		6.36	2.07	
	ANN	1.51	0.46	19.1	7.63	2.32	15.9	6.13	1.67	16.2

The data of adult healthy volunteer was acquired from the cerebellum as can be seen in Figure 6. The surrounding brain cavities and bone tissue causes magnetic field inhomogeneities and therefore the signal measured from this region is characteristic for its low quality. The results manifest the ability of the ANN model to follow the pattern of the curve fitting algorithm even in case of these low quality inputs. Although the error is significantly higher than in case of signals measured from the upper parts of the brain.

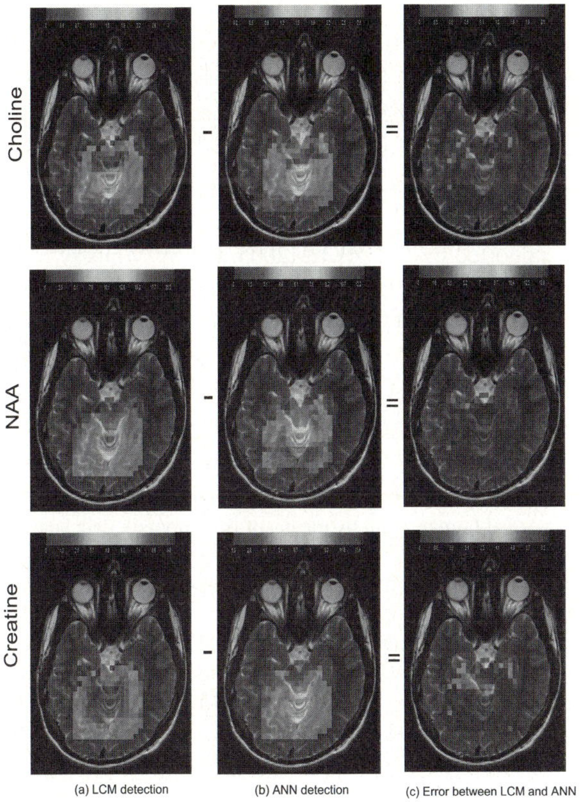

Fig. 6. Comparison of Choline,Creatine and NAA metabolite map calculated by LCM (a) and ANN model (b) in case of adult healthy volunteer

4 Conclusions

Even though the processing through the neural network is almost immediate the preprocessing of the signals takes up to a couple of seconds for each signal. Commercial software LCModel takes in order of minutes to process a signal from a single voxel. The processing of a full CSI signal takes several hours to run. The approximation using neural networks can give results with the mean error of under 15% within 5 minutes. Since the whole measurement is affected by a significant error due to high amount of noise and bad homogeneity of the magnetic field the error of 15% is acceptable.

The method is limited by the error of the LCModel algorithm. It is a black box algorithm and has limited generalization ability. It is impossible to ensure proper function for inputs significantly different from the training set. For best results the training set needs to be as large as possible with maximum possible variety. The optimal topology of the network can not be derived. Due to the size of the inputs the training process for a large training set is time and memory demanding. Testing the performance for more possible topologies as well as reducing the length of the input vector could improve the convergence of the training algorithm and that way enable larger training sets to be used. The preprocessing does not count for possible phase shift. Some spectra are improperly classified because of being out of phase. Phase correction as a part of the preprocessing might be considered.

Acknowledgment

The project was supported by MZOIKEM2005 and by Ministry of Education, Youth and Sport of the Czech Republic with the grant No. MSM6840770012 entitled "Transdisciplinary Research in Biomedical Engineering II". We thank Filip Jírů for technical discussion and remarks.

References

1. Ala-Korpela, M., Usenius, J.P., et al.: Quantification of metabolites from single-voxelin vivo 1h NMR data of normal human brain by means of time-domain data analysis. Magma 3(3–4), 129–136
2. Provencher, S.W.: Estimation of metabolite concentrations from localized in vivo proton nmr spectra. Magn. Reson. Med. 30(6), 672–679 (1993)
3. Poullet, J., Sima, D.M., Van Huffel, S.: Mrs signal quantitation: a review of time- and frequency-domain methods. Technical report, ESAT-SISTA, K.U.Leuven (2008)
4. Kaartinen, J., Mierisova, S., Oja, J.M.E., Usenius, J.P., Kauppinen, R., Hiltunen, Y.: Automated quantification of human brain metabolites by artificial neural network analysis from in vivo single-voxel 1h nmr spectra. Journal of Mag. Res. 134, 176–179 (1998)
5. Bhat, H., Sajja, B.R., Narayana, P.A.: Fast quantification of proton magnetic resonance spectroscopic imaging with artificial neural networks. Journal of Mag. Res. 183, 110–122
6. Provencher, S.: LCModel & LCMgui Users Manual (2007)
7. Cabanes, E., Confort-Gouny, S., Le Fur, Y., Simond, G., Cozzone, P.: Optimization of residual water signal removal by hlsvd on simulated short echo time proton mr spectra of the human brain. Journal of Magnetic Resonance 150, 116–125 (2001)

Multi-stage FCM-Based Intensity Inhomogeneity Correction for MR Brain Image Segmentation

László Szilágyi[1,2], Sándor M. Szilágyi[2], László Dávid[2], and Zoltán Benyó[1]

[1] Budapest University of Technology and Economics, Department of Control Engineering and Information Technology, Budapest, Hungary
[2] Sapientia - Hungarian Science University of Transylvania, Faculty of Technical and Human Science, Târgu-Mureş, Romania
{lalo,szs,ldavid}@ms.sapientia.ro, benyo@iit.bme.hu

Abstract. Intensity inhomogeneity or intensity non-uniformity (INU) is an undesired phenomenon that represents the main obstacle for MR image segmentation and registration methods. Various techniques have been proposed to eliminate or compensate the INU, most of which are embedded into clustering algorithms, and they generally have difficulties when INU reaches high amplitudes. This paper proposes a multiple stage fuzzy c-means (FCM) based algorithm for the estimation and compensation of INU, by modeling it as a slowly varying additive or multiplicative noise, supported by a pre-filtering technique for Gaussian and impulse noise elimination. The slowly varying behavior of the bias or gain field is assured by a smoothing filter that performs a context dependent averaging, based on a morphological criterion. The experiments using 2-D synthetic phantoms and real MR images show, that the proposed method provides accurate segmentation. The resulting segmentation and fuzzy membership values can serve as excellent support for 3-D registration and segmentation techniques.

Keywords: Intensity inhomogeneity, fuzzy c-means algorithm, image segmentation, context dependent filter, magnetic resonance imaging, morphological operations.

1 Introduction

Magnetic resonance imaging (MRI) is a very popular medical imaging technique, mainly because of its high resolution and contrast, which represent great advantages above other diagnostic imaging modalities. Besides all these good properties, MRI also suffers from three considerable obstacles: noises (mixture of Gaussian and impulse noises), partial volume effect (pixels containing at least two types of tissues), and intensity inhomogeneity [14]. This latter one, also known as intensity non-uniformity (or INU artefact), manifests as a spatially slowly varying function, that makes pixels belonging to the same tissue be observed having different intensities. In order to produce a correct segmentation

V. Kůrková et al. (Eds.): ICANN 2008, Part II, LNCS 5164, pp. 527–536, 2008.

or registration of MR images, the INU artefact needs to be modeled and compensated.

INU has two different types of sources: those related to the MRI device, and those related to the imaged patient's shape, position, structure and orientation. While the first type of source has efficient compensation and calibration methods, the second type of INU artefacts are much more difficult to handle [17]. A widely used technique, handling mostly the first type of INU, consists of the usage of a uniform phantom to produce prior information [2].

Homomorphic filtering represents a popular compensation method [4,7], built upon the theoretical assumption, that the frequency spectra of the image structures and of the INU artefact are not overlapping each other. The efficiency of such methods are limited because the initial assumption does not hold.

Although several INU compensation approaches exist [8,11,12,19], one of the most widely used methods is the adaptation of the fuzzy c-means (FCM) clustering algorithm to iteratively approximate the INU as a smooth varying bias or gain field. In this order, Pham and Prince introduced a modified objective function producing bias field estimation and containing extra terms that force this artefact vary smoothly [12,13]. They also provided a multigrid technique to speed up the computationally heavy algorithm, but even this way, the algorithm performs slowly. A probabilistic formulation leading to the same objective function was given in [9]. Liew and Hong created a log bias field estimation technique that models the INU with smoothing B-spline surfaces [10].

Further FCM-based bias field estimation techniques were introduced recently by Ahmed *et al.* [1] and Siyal and Yu [14]. The modification introduced by Ahmed *et al.* allows the labeling of a voxel to be influenced by its immediate neighbors. This approach has reduced some of the complexity of its ancestors, but the zero gradient condition that was used for bias field estimation leads to several misclassifications [14]. The other approach provided a mean spread filtering method to smoothen the estimated bias field in every cycle of the FCM algorithm. This approach reduces the amount of necessary computations, but the result of the segmentation is not deterministic due to the nature of the smoothing filter. All these clustering techniques suffer from sensitivity to high amplitude INU artefact: when the INU-contaminated intensities of different tissue types overlap, segmentation accuracy falls, as the clustering algorithm is unable to correctly compensate the disturbing phenomenon.

In order to handle this latter problem, in this paper we propose a multi-stage FCM-based technique for bias- or gain field estimation of the intensity inhomogeneity. Furthermore, we introduce two filtering techniques to improve the segmentation accuracy. The proposed methods are tested using real MR images and artificial phantoms.

The rest of the paper is organized as follows: section 2 describes the proposed filtering techniques and the multi-stage FCM-based bias- and gain field estimation approaches. Section 3 provides a qualitative analysis and short discussion of segmentation results. In Section 4 the conclusions are formulated.

2 Methods

2.1 Context Dependent Pre-filtering

Impulse and Gaussian noises are removed from the original MR image using a context dependent local filtering, which combines averaging and median filtering effects based on physical distances and gray level differences between neighbor pixels.

The proposed technique acts like a low-pass masking, but the mask weights are separately computed for each pixel, based on the distances and gray level differences encountered within a neighborhood. The weight of each neighboring pixel is the product of two terms: a distance term that decreases exponentially with Euclidean distance measured from the middle pixel; and a gray level term, which is high for pixels that have similar intensities with the middle pixel and low if they differ significantly. The middle pixel gets its weight based on the reliability of its gray level intensity. Details of such techniques are described in [5,16].

2.2 Conventional FCM Clustering and Derivations

The conventional FCM algorithm has been applied successfully in a wide variety of clustering problems [3]. This algorithm optimally partitions a set of object data into a previously set number of c clusters based on the iterative minimization of a quadratic objective function. When applied to segment grayscale images, FCM clusters the intensity value of each pixel x_k, $k = 1 \dots n$. The objective function to be minimized is given as follows:

$$J_{\mathrm{FCM}} = \sum_{i=1}^{c} \sum_{k=1}^{n} u_{ik}^{m} ||x_k - v_i||^2 \ , \tag{1}$$

where $m > 1$ is the fuzzy exponent or fuzzyfication parameter, v_i represents the centroid or prototype of the i-th cluster, and $u_{ik} \in [0, 1]$ is the fuzzy membership function indicating the degree to which pixel k belongs to cluster i. The definition of fuzzy logic implies that for any $k = 1 \dots n$, we have $\sum_{i=1}^{c} u_{ik} = 1$. The constrained optimization of the objective function is achieved using Lagrange multipliers. The minimization is reached by alternately applying the optimization of J_{FCM} over $\{u_{ik}\}$ with v_i fixed, $i = 1 \dots c$, and the optimization of J_{FCM} over $\{v_i\}$ with u_{ik} fixed, $i = 1 \dots c$, $k = 1 \dots n$ [3]. In each cycle, optimal fuzzy membership and optimal centroid values are computed using the formulas:

$$u_{ik}^{\star} = \frac{||x_k - v_i||^{-2/(m-1)}}{\sum_{j=1}^{c} ||x_k - v_j||^{-2/(m-1)}} \quad \forall i = 1 \dots c, \forall k = 1 \dots n \ , \tag{2}$$

and

$$v_i^{\star} = \frac{\sum_{k=1}^{n} u_{ik}^{m} x_k}{\sum_{k=1}^{n} u_{ik}^{m}} \quad \forall i = 1 \dots c \ . \tag{3}$$

After adequate initialization of cluster prototype values v_i, eqs. (2) and (3) are alternately applied until the norm of the variation of the vector $\mathbf{v}$, composed of the centroid values, stays within a previously set bound ε.

The above presented algorithm clusters the set of data $\{x_k\}$, which was recorded among ideal circumstances, containing no noise. However, in the real case, the observed data $\{y_k\}$ differs from the actual one $\{x_k\}$: there are impulse and Gaussian noises, which were treated in the previous subsection, and there is the intensity non-uniformity (INU) artefact, which will be handled here.

Literature recommends two different data variation models for intensity inhomogeneity: the bias and the gain field model. If we consider the INU as a bias field, for each pixel k we will have $y_k = x_k + b_k$, where b_k represents the bias value at pixel k. In case of gain field modeling, there will be a gain value g_k for each pixel k, such that $y_k = g_k x_k$. In case of both models, the variation of the intensity between neighbor pixels has to be slow. This is assured by the smoothing filter presented in the next section.

In case of modeling INU as a bias field, the objective function becomes:

$$J_{\mathrm{FCM-b}} = \sum_{i=1}^{c} \sum_{k=1}^{n} u_{ik}^{m} ||y_k - b_k - v_i||^2 \ . \tag{4}$$

Using the Lagrange multiplier technique, taking the derivatives of $J_{\mathrm{FCM-b}}$, with respect to u_{ik}, v_i and b_k, respectively, and equaling them to zero, we obtain the following optimization formulas:

$$u_{ik}^{\star} = \frac{||y_k - b_k - v_i||^{-2/(m-1)}}{\sum_{j=1}^{c} ||y_k - b_k - v_j||^{-2/(m-1)}} \quad \forall i = 1 \ldots c, \forall k = 1 \ldots n \ , \tag{5}$$

$$v_i^{\star} = \frac{\sum_{k=1}^{n} u_{ik}^{m} (y_k - b_k)}{\sum_{k=1}^{n} u_{ik}^{m}} \quad \forall i = 1 \ldots c \ , \tag{6}$$

and

$$b_k^{\star} = y_k - \frac{\sum_{i=1}^{c} u_{ik}^{m} v_i}{\sum_{i=1}^{c} u_{ik}^{m}} \quad \forall k = 1 \ldots n \ . \tag{7}$$

If we approximate the INU artefact as a gain field, the objective function should be:

$$\overline{J}_{\mathrm{FCM-g}} = \sum_{i=1}^{c} \sum_{k=1}^{n} u_{ik}^{m} ||y_k / g_k - v_i||^2 \ . \tag{8}$$

Because the derivatives of this function are hard to handle, we slightly modify this objective function the following way:

$$J_{\mathrm{FCM-g}} = \sum_{i=1}^{c} \sum_{k=1}^{n} u_{ik}^{m} ||y_k - g_k v_i||^2 \ . \tag{9}$$

This is an affordable modification, which distorts the objective function such a way, that it gives slightly higher impact to lighter pixels (as their gain field value will probably be over unity). Taking the derivatives of $J_{\mathrm{FCM-g}}$, with respect to u_{ik}, v_i and g_k, respectively, and equaling them to zero, we obtain the following optimization formulas:

$$u_{ik}^{\star} = \frac{||y_k - g_k v_i||^{-2/(m-1)}}{\sum_{j=1}^{c} ||y_k - g_k v_j||^{-2/(m-1)}} \quad \forall i = 1 \ldots c, \forall k = 1 \ldots n \ , \tag{10}$$

$$v_i^{\star} = \frac{\sum_{k=1}^{n} u_{ik}^m g_k y_k}{\sum_{k=1}^{n} u_{ik}^m g_k^2} \quad \forall i = 1 \ldots c \ , \tag{11}$$

and

$$g_k^{\star} = y_k \frac{\sum_{i=1}^{c} u_{ik}^m v_i}{\sum_{i=1}^{c} u_{ik}^m v_i^2} \quad \forall k = 1 \ldots n \ . \tag{12}$$

Similarly to the conventional FCM, these optimization formulas are applied alternatively in each iteration.

2.3 Smoothening Filter

The intensity inhomogeneity artefact varies slowly along the image. This property is ignored by both the bias or gain field estimation approaches presented above. To avoid this problem, a filtering technique is applied in each computation cycle, to smoothen the bias or gain field. This filtering introduces an extra step into each optimization cycle, after proceeding with eq. (7) or (12).

Several, not only FCM-based INU compensation approaches apply large sized, 11-31 pixels wide averaging filters performed once or several times in each cycle [18]. These filters efficiently hide tissue details, which may appear in the estimated b_k or g_k values, at the price of transferring bias or gain components to distantly situated pixels. Using larger averaging windows amplifies this latter undesired effect. In order to reduce the transfer of bias data to distant pixels, we need to check the necessity of averaging at all locations, and decide to proceed or skip the averaging accordingly. Averaging is declared necessary or not, based on the maximum intensity difference encountered within a small neighborhood of the pixel. The computation of the maximum difference is accomplished by a morphological gradient operation using a 3×3 square or slightly larger cross-shaped structuring element. Wherever the morphological gradient value exceeds the previously set threshold value θ, the averaged bias or gain value will be used; otherwise the estimated value is validated. The proposed filter can be easily implemented and efficiently performed by batch-type image processing operations.

2.4 Multi-stage Bias and Gain Field Estimation

Bias or gain field estimation using the previous FCM-based approaches [1,14] can only handle the INU artefact to a limited amplitude. For any pixel, the FCM algorithm assigns the highest fuzzy membership to the closest cluster. Consequently, when the INU amplitude is comparable with the distance between clusters, these pixels will be attracted by the wrong cluster, and the bias or gain field will be estimated accordingly. The smoothing of the bias and gain field may repair this kind of misclassifications, but the larger these wrongly labelled spots are, the harder will be to eliminate them via smoothing.

In order to deal with high-amplitude INU artefacts, we propose performing the bias or gain field estimation in multiple stages. When the FCM-based algorithm given by eqs. (5)-(7) or (10)-(12) has converged, we modify the input (observed) image according to the estimated bias or gain field:

$$y_k = y_k^{(old)} - b_k \qquad \text{or} \qquad y_k = y_k^{(old)} / g_k \ , \tag{13}$$

and then restart the algorithm from the beginning, using the modified input image.

2.5 Algorithm

The presented algorithm can be summarized as follows:

1. Remove the Gaussian and impulse noises from the MR image using the context dependent pre-filtering technique.
2. Initialize cluster prototypes v_i, $i = 1 \ldots c$, with random values differing from each other.
3. Initialize the bias (gain) field values with 0-mean (1-mean) random numbers having reduced variance, or simply set $b_k = 0$ $(g_k = 1)$ for all pixels.
4. Compute new fuzzy membership function values u_{ik}, $i = 1 \ldots c$, $k = 1 \ldots n$, using (5) or (10).
5. Compute new cluster prototype values v_i, $i = 1 \ldots c$, using (6) or (11).
6. Perform new bias or gain field estimation for each pixel k using (7) or (12).
7. Smoothen the bias or gain field using the proposed smoothing filter.
8. Repeat steps 4-7 until there is no relevant change in the cluster prototypes. This is tested by comparing any norm of the difference between the new and the old vector $\mathbf{v}$ with a preset small constant ε.
9. Modify the input image according to the estimated bias or gain field using (13), and repeat steps 2-8 until the INU artefact is compensated. The algorithm usually requires a single repetition.

3 Results and Discussion

We applied the presented filtering and segmentation techniques to several T1-weighted real MR images, artificially contaminated with different kinds of noises. Figure 1. demonstrates the efficiency of the pre-filtering technique on a real MR slice taken from Internet Brain Segmentation Repository [6].

The results of bias and gain field estimation performed on a phantom image are shown in Fig. 2. Conventional FCM is unable to compensate the INU artefact, but with the use of smoothened bias or gain field, this phenomenon is efficiently overcome.

In case of low-amplitude inhomogeneity, a single stage of bias or gain field estimation is sufficient. Figure 3. shows the accuracy and efficiency of the proposed segmentation technique, using a real T1-weighted MR brain slice. The presence of the smoothening filter supports the accurate segmentation, while the pre-filter has a regularizer effect on the final result.

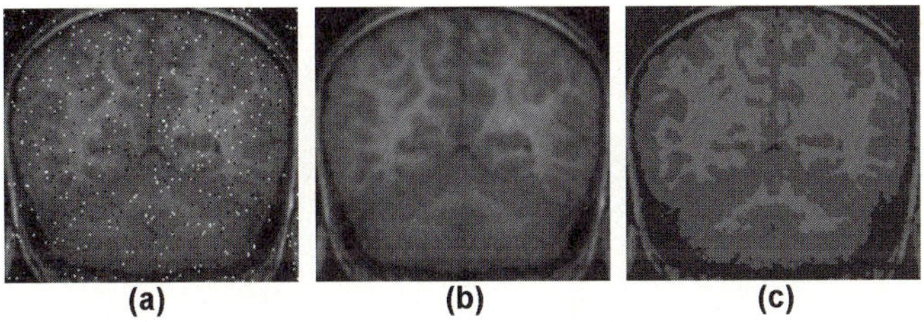

Fig. 1. Elimination of impulse and Gaussian noises demonstrated with a real T1-weighted MR brain image contaminated with artificial noise: (a) original, (b) filtered, (c) FCM segmentation after correction

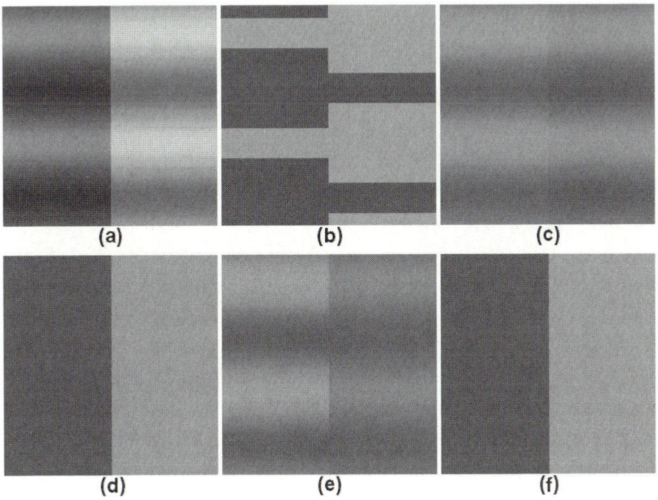

Fig. 2. Inhomogeneity correction using a phantom: (a) original, (b) FCM segmentation result without correction, (c) estimated bias field, (d) segmentation result with bias field estimation, (e) estimated gain field, (f) segmentation result with gain field estimation

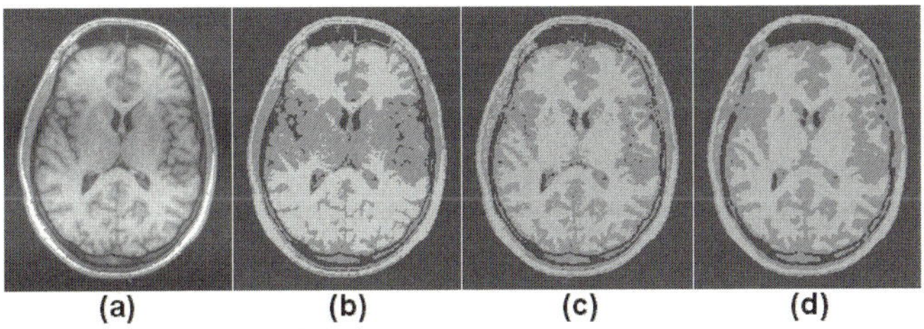

Fig. 3. Inhomogeneity correction demonstrated on an artificially contaminated real MR image: (a) original, (b) FCM segmentation without correction, (c) result of FCM-based segmentation with no pre-filtering, (d) result of FCM-based segmentation with context sensitive pre-filtering

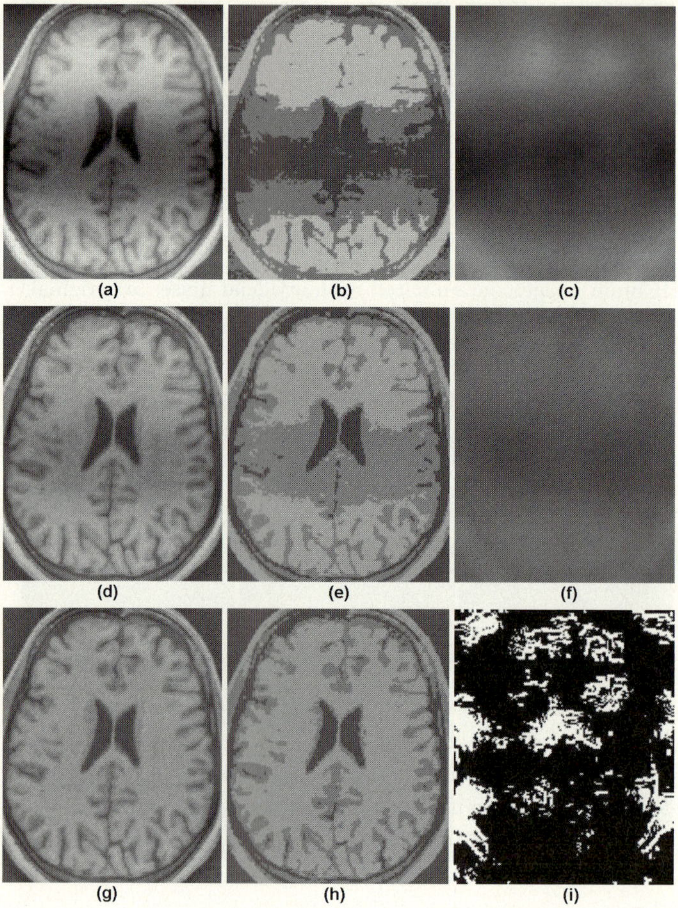

Fig. 4. Segmentation of a heavily inhomogeneous real MR image: (a) original, (b) segmentation without compensation, (c) bias field estimated in the first stage, (d) compensated MR image after first stage, (e) FCM-based segmentation after first stage, still unusable, (f) bias field estimated in the second stage, (g) final compensated image, (h) segmented image, (i) a smoothening mask computed by the proposed filter

Table 1. Misclassification percentages with various smoothening filters, in case of heavily INU-contaminated MR images

Structuring element	Window size 11 execution once	Window size 11 3 times	Window size 19 execution once	Window size 19 3 times
Averaging	7.357 %	5.766 %	4.368 %	7.141 %
3×3, square	6.281 %	6.201 %	2.852 %	3.379 %
5×5, cross	6.711 %	5.674 %	3.873 %	3.938 %
7×7, cross	6.254 %	5.518 %	3.470 %	5.029 %
11×11, cross	6.351 %	4.432 %	3.271 %	6.437 %

Figure 4. shows the intermediary and final results of a segmentation process, performed on a heavily INU-contaminated MR image. The inhomogeneity correction succeeds after two stages. Figure 4(i) shows the behavior of the proposed smoothing technique: white pixels indicate places which required averaging in a given computational cycle, while black ones signify those places where averaging was unnecessary.

Table 1. shows the misclassification percentages of the proposed INU compensation and MR image segmentation method, depending on the size of the averaging window expressed in pixels, the structuring element of the morphological criterion of the proposed filter, and the number of smoothening iterations performed in each cycle of the modified FCM algorithm. The experimental data reveal that the proposed filtering technique improves the segmentation quality assured by the averaging filter. The best segmentation was obtained using a 3×3 square shaped structuring element used by the morphological criterion, combined with averaging using a window size of 19×19 pixels.

Using several repetitive stages during INU compensation may reduce the intensity difference between tissue classes, which leads to misclassifications. That is why the estimation is limited to two steps, performing several stages is not recommendable.

4 Conclusions

A novel multi-stage approach has been proposed to produce accurate segmentation of INU-contaminated magnetic resonance images. A new smoothing filter has been proposed to assist bias or gain field estimation embedded into the FCM-based algorithmic scheme.

The proposed method proved to segment accurately and efficiently MR images in the presence of severe intensity non-uniformity. Although the proposed method segments 2-D MR brain slices, it gives a relevant contribution to the accurate volumetric segmentation of the brain, because the segmented images and the obtained fuzzy memberships can serve as excellent input data to any level set method that constructs 3-D cortical surfaces.

Further works aim at developing a context sensitive pre-filter for the elimination of INU artefacts, too, so that the segmentation can be performed using a histogram-based quick FCM algorithm [15].

Acknowledgments. This research was supported in part by the Hungarian National Research Funds (OTKA) under Grant No. T069055, the Hungarian National Office for Research and Technology (NKTH), and the Sapientia Institute for Research Programs (KPI).

References

1. Ahmed, M.N., Yamany, S.M., Mohamed, N., Farag, A.A., Moriarty, T.: A modified fuzzy c-means algorithm for bias field estimation and segmentation of MRI data. IEEE Trans. Med. Imag. 21, 193–199 (2002)

2. Axel, L., Costanini, J., Listerud, J.: Inhomogeneity correction in surface-coil MR imaging. Amer. J. Roentgenol. 148, 418–420 (1987)
3. Bezdek, J.C., Pal, S.K.: Fuzzy models for pattern recognition. IEEE Press, Piscataway (1991)
4. Brinkmann, B.H., Manduca, A., Robb, R.A.: Optimized homomorphic unsharp masking for MR grayscale inhomogeneity correction. IEEE Trans. Med. Imag. 17, 161–171 (1998)
5. Cai, W., Chen, S., Zhang, D.Q.: Fast and robust fuzzy c-means algorithms incorporating local information for image segmentation. Patt. Recogn. 40, 825–838 (2007)
6. Internet Brain Segmentation Repository, http://www.cma.mgh.harvard.edu/ibsr
7. Johnston, B., Atkins, M.S., Mackiewich, B., Anderson, M.: Segmentation of multiple sclerosis lesions in intensity corrected multispectral MRI. IEEE Trans. Med. Imag. 15, 154–169 (1996)
8. Leemput, K.V., Maes, F., Vandermeulen, D., Suetens, P.: Automated model-based bias field correction of MR images of the brain. IEEE Trans. Med. Imag. 18, 885–896 (1999)
9. Li, X., Li, L.H., Lu, H.B., Liang, Z.G.: Partial volume segmentation of brain magnetic resonace images based on maximum a posteriori probability. Med. Phys. 32, 2337–2345 (2005)
10. Liew, A.W.C., Hong, Y.: An adaptive spatial fuzzy clustering algorithm for 3-D MR image segmentation. IEEE Trans. Med. Imag. 22, 1063–1075 (2003)
11. Nagy, L., Benyó, B.: Filtering and contrast enhancement on subtracted direct digital angiograms. In: Ann. Int. Conf. IEEE Eng. Med. Biol. Soc., vol. 26, pp. 1533–1536. San Francisco (2004)
12. Pham, D.L., Prince, J.L.: An adaptive fuzzy C-means algorithm for image segmentation in the presence of intensity inhomogeneity. Patt. Recogn. Lett. 20, 57–68 (1999)
13. Pham, D.L., Prince, J.L.: Adaptive fuzzy segmentation of magnetic resonance images. IEEE Trans. Med. Imag. 18, 737–752 (1999)
14. Siyal, M.Y., Yu, L.: An intelligent modified fuzzy c-means based algorithm for bias field estimation and segmentation of brain MRI. Patt. Recogn. Lett. 26, 2052–2062 (2005)
15. Szilágyi, L.: Medical image processing methods for the development of a virtual endoscope. Period. Polytech. Ser. Electr. Eng. 50(1–2), 69–78 (2006)
16. Szilágyi, L., Szilágyi, S.M., Benyó, Z.: Efficient Feature Extraction for Fast Segmentation of MR Brain Images. In: Ersbøll, B.K., Pedersen, K.S. (eds.) SCIA 2007. LNCS, vol. 4522, pp. 611–620. Springer, Heidelberg (2007)
17. Vovk, U., Pernuš, F., Likar, B.: A review of methods for correction of intensity inhomogeneity in MRI. IEEE Trans. Med. Imag. 26, 405–421 (2007)
18. Wells, W.M., Grimson, W.E.L., Kikinis, R., Jolesz, F.A.: Adaptive segmentation of MRI data. IEEE Trans. Med. Imag. 15, 429–442 (1996)
19. Zhang, Y., Brady, M., Smith, S.: Segmentation of brain MR images through a hidden Markov random field model and the expectation-maximization algorithm. IEEE Trans. Med. Imag. 20, 45–57 (2001)

KCMAC: A Novel Fuzzy Cerebellar Model for Medical Decision Support

S.D. Teddy

Data Mining Departement, Institute for Infocomm Research,
21 Heng Mui Keng Terrace, Singapore 119613
sdteddy@i2r.a-star.edu.sg

Abstract. Most of the current advanced clinical decision support systems rely on some form of computational intelligence methodologies. As the machine intelligence paradigm shifted towards brain-inspired computing approach, it is interesting to investigate the performance of such a computing methodology in clinical data analysis. The human cerebellum constitutes a vital part of the brain system that possesses the capability to accurately model highly nonlinear physical dynamics. This paper presents a novel brain-inspired computational model of the human cerebellum named the *kernel density-based CMAC* (KCMAC) model for clinical decision support. The structure of the KCMAC model is inspired by the neurophysiological aspects of cerebellar learning and development process. The proposed KCMAC model is then applied to two medical case studies; namely, breast cancer diagnosis and the modeling of the human glucose metabolic process. The experimental results are encouraging.

1 Introduction

In recent years, medical information processing and decision support systems have become increasingly important with the need to analyze the huge amount of medical data that is generated through various clinical procedures. Such systems provide an objective and consistent approach to information handling that enable doctors and clinicians to extract valuable knowledge from existing data to enhance their diagnosis and clinical decisions. Computational intelligence and machine learning techniques have been the predominating approach to medical decision support systems, with the ultimate objective of developing an intelligent system that is capable of large-scale human-like analysis of clinical data.

The traditional approach to machine intelligence thus far has provided many achievements, but has generally fallen short of the vision of developing versatile human-like intelligent systems. Revolutionary advances may be possible by building upon new approaches inspired by cognitive psychology and neuroscience. Such approaches have the potential to help researchers understand and model significant aspects of intelligence thus far not attained by classic formal knowledge modeling techniques. With regard to this notion, *brain-inspired computing* (BIC) architectures are an actively pursued topical research [1,2], where the primary focus is to develop advanced computing structures and information

V. Kůrková et al. (Eds.): ICANN 2008, Part II, LNCS 5164, pp. 537–546, 2008.

processing systems that are often inspired by the neurophysiologic understanding of the underlying neural organisation, physiological functions and signal pathways of the various brain systems. Given that the most fundamental aspects of the human intelligence are arguably the capacity for learning and memory, the main objective of BIC research is therefore to emulate and replicate facets of the human intelligence via a functional description of the mechanisms and processes involved in human learning and memory formation. This is achieved through the use of computational intelligence and soft computing techniques to construct intelligent brain-inspired computing systems that possess human-like information processing capabilities.

One of the most extensively explored regions of the human brain is the cerebellum. As a motor movement calibrator [3], the human cerebellum possesses the capability to model highly complex and nonlinear physical dynamics to facilitate the precise and rapid executions of dexterous movements and fluid motor reflexes. The rapid and non-linear function learning capability of the human cerebellum has motivated the development of many computational models to solve ill-defined, complex problems. The *Cerebellar Model Articulation Controller* (CMAC) [4] is an established computational model of the human cerebellum. CMAC manifests as an associative memory network, and employs error correction learning to drive its memory formation process. This allows for advantages such as fast computation and ease of hardware implementation. However, CMAC employs a rigid and highly regularized grid-like computing structure that leads to poor memory utilization and a trade-off between the generalization capability and the modeling accuracy of the model [2]. Moreover, the multi-layered structure of CMAC often renders the network operation difficult to comprehend.

This paper proposes a brain-inspired cerebellar-based learning memory model named *kernel density-based CMAC* (KCMAC) as a functional model of the human cerebellum to facilitate knowledge construction for medical decision support. The proposed KCMAC model adopts a zero-ordered Takagi-Sugeno-Kang

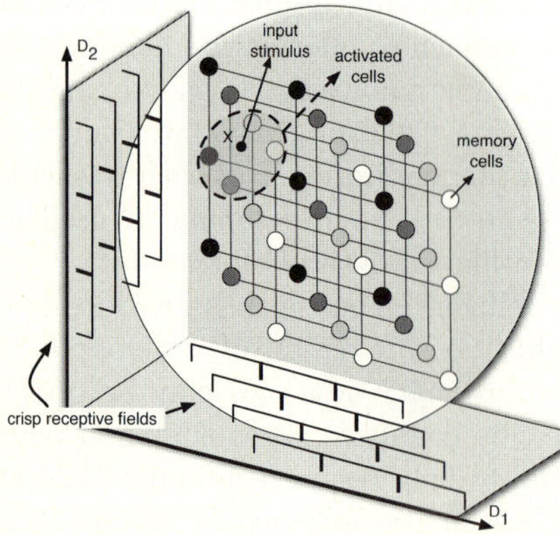

(a) CMAC

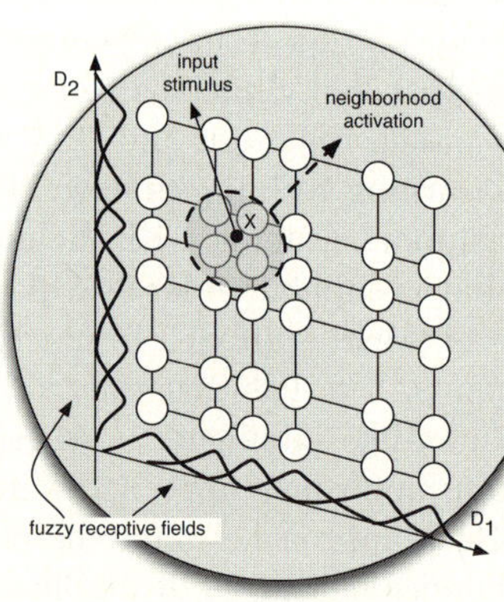

(b) KCMAC

Fig. 1. Comparison of CMAC and KCMAC architectures for a 2D input problem

(TSK) [5] fuzzy model to define its network computations. This novel architecture differs from the CMAC network in *two* aspects. Firstly, the KCMAC network employs only *one* layer of network cells but maintained the computational principles of the multi-layered CMAC network by adopting a fuzzy neighborhood activation of its computing cells. This facilitates: (1) smoothing of the computed output; (2) a distributed learning paradigm; and (3) enhanced data generalization ability. Secondly, instead of a uniform quantization of the memory cells, the KCMAC model employs *kernel density estimation* [6] to form an adaptive fuzzy partitioning of its network cells. Figure 1 illustrates this fundamental architectural distinction. The proposed KCMAC model is inspired by various neurophysiological studies on the cerebellar learning and development process, where it has been shown that motor skill learning and experience results in the multi-resolution characteristics of the cerebellar connections [7]. In addition, unlike CMAC which assigns equal weightage to all of its activated memory cells, KCMAC employs a fuzzy neighborhood activation scheme that modulates the activation level of each activated cell based on its computed fuzzy similarity to the input. Such a graded activation not only ehances the accuracy of the computed output, but also increases the interpretability of the computation.

The rest of the paper is organized as follows. Section 2 briefly describes the neurophysiological properties of the cerebellum that inspired the development of the KCMAC model and outlines the basic computational principles of the CMAC network. In Section 3, the architecture of the proposed KCMAC model is presented. Section 4 evaluates the performance of the KCMAC model on two medical applications; namely: (1) breast cancer diagnosis; and (2) modeling the human glucose metabolic process. Section 5 concludes this paper.

2 Human Cerebellum and the CMAC Network

The human cerebellum functions primarily as a movement regulator, where it is provided with an extensive repertoire of information about the objectives, motor commands and feedback signals associated with a physical movement. Studies in neuroscience have established that the cerebellum performs an associative mapping from the input sensory afferent and cerebral efferent signals to the cerebellar output, which is subsequently transmitted back to the cerebral cortex [3]. This physiological process of constructing an associative pattern map constitutes the underlying neuronal mechanism of learning in the cerebellum.

Learning in the human cerebellum is facilitated by the modifiable synaptic transmissions (cerebellar synaptic plasticity) and the synaptic reorganization capability (cerebellar structural plasticity) of the neuronal connections. Research into the physiology of the cerebellum has sufficiently demonstrated that *Long Term Depression* (LTD) modifies the synaptic weights of the cerebellar neuronal connections and it is the underlying cellular mechanism responsible for cerebellar synaptic plasticity [3]. However, further studies have suggested that synaptic depression alone may not be adequate for forming permanent, long term memories of motor programs [8]. Instead, there are evidences of morphological alterations of the cerebellar cortex following extensive cerebellar learning. Studies on the experience-driven cerebellar structural plasticity phenomenon have demonstrated that motor skill learning actually leads to an increased dendritic aborization of the activated cerebellar Purkinje cells [7]. This implies that during the motor learning process, the activated Purkinje cells grow to be more complex to encode more of the training episodes. Such an observation constitutes a learning-driven biological manifestation of the multi-resolution nature of the cerebellar circuitry. These neurophysiological properties of cerebellar learning and memory formation subsequently provides the neuroscience inspirations for the development of the proposed KCMAC model presented in this paper.

The CMAC network [4] is an established computational model of the human cerebellum. It is essentially a multi-dimensional memory array, where an input acts as the address decoder to access the respective memory (computing) cells containing the adjustable weight parameters that constitute the corresponding output. In the CMAC network, the memory cells are uniformly quantized to cover the entire learning space (see Figure 1(a)). The operation of the CMAC network is then characterized by the table lookup access of its memory cells. Each input vector to the CMAC network selects a set of active computing cells (one winner neuron from each layer) from which the output of the network is computed. CMAC employs equal activation for all its activated computing cells such that the network output is the aggregate of the weights of these activated cells. Following that, CMAC learns the correct output response to each input vector by modifying the contents of the selected memory locations.

Although the look-up table structure of the CMAC network allows for rapid algorithmic computations, the static uniform allocation of the CMAC memory cells saddles the network with several major limitations [2]. The uniform quantization of each input into crisp receptive fields results in a static output resolution for the entire input range. Its output granularity thus depends on the network size. There is also a trade-off between the generalization capability and the modeling fidelity of the network. That is, a small-sized CMAC is able to better generalize the characteristics of the training data, while a large-sized CMAC produces more accurate outputs. Finally, the layered structure of the CMAC computing cells often leads to limited comprehensibility of the computation.

3 The KCMAC Model

In this section, a novel cerebellar model that addresses the above-mentioned CMAC problems is presented. The proposed KCMAC model employs a two-phased learning process; namely: *structural learning* and *parameter tuning*. The objective of the structural learning phase is to construct the KCMAC's associative structure by performing adaptive fuzzy partitioning of its input dimensions. Subsequently, the input to output associative information of the training data samples is learnt by adapting the memory contents of the KCMAC model during the parameter tuning phase. The initial step in the KCMAC structural learning phase is to identify the data density profile of each input dimension.

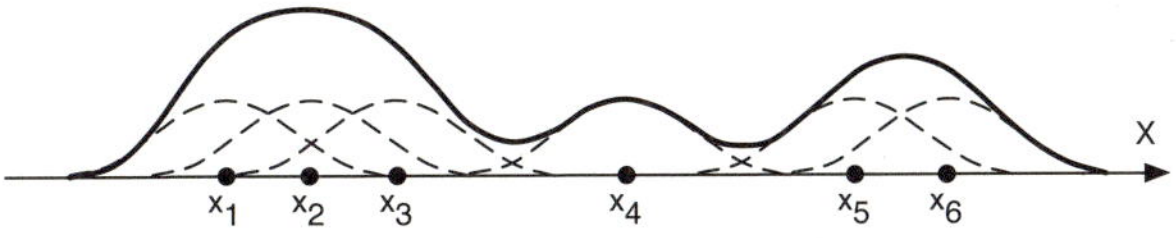

Fig. 2. Gaussian kernel density estimation for an input dimension

Subsequently, a fuzzy receptive field is assigned to each of the identified density clusters to emulate the biological experience-driven dendritic aborization phenomenon observed in the cerebellar learning process during skill acquisition.

The structural learning process of the KCMAC model proceeds as follows:

1. *Computation of the data density clusters*
   Figure 2 illustrates the computation for the density profile of an input dimension of the KCMAC model. To generate the density profile of a given dimension, KCMAC employs the kernel density estimation technique. Each training data point generates a stand alone gaussian-shaped density distribution. An overall density profile is then obtained by aggregating the density contributions of all the training points. Following that, the appropriate number of density clusters of a given dimension is determined by locating the local maximas (peaks) of the generated data density profile (see Figure 2).
2. *Definition of the KCMAC receptive fields*
   Each of the identified density clusters subsequently forms a fuzzy receptive field in the corresponding input dimension. KCMAC adopts a fuzzy pseudo-partition with asymmetric gaussian membership functions, where the fuzzy receptive fields are centered at the respective centroids of the density clusters.
3. *The KCMAC memory allocation*
   Finally, each KCMAC computing cell is formed from the multi-dimensional intersection of a receptive field from each of the input dimensions.

In the parameter tuning stage, KCMAC employs a fuzzy neighborhood activation of its computing cells to learn the correct output response to a given stimulus. For each input stimulus **X**, the computed output is derived as follows:

1. *Determine the fuzzy neighborhood activation*
   Each input stimulus $\mathbf{X} = [x_1, \cdots, x_d, \cdots, x_D]^T$ activates a fuzzy neighborhood of KCMAC computing cells, which consists of the cells that are defined by the intersections of the activated fuzzy receptive fields in each input dimension (see Fig 1(b)). Each fuzzy receptive field $Q_{d,j}$ in each input dimension d generates a membership value $\mu_{d,j}(x_d)$ with respect to the input stimulus for that dimension. A KCMAC input fuzzy receptive field $Q_{d,j}$ is activated if and only if $\mu_{d,j}(x_d) \geq \epsilon_\mu$, where ϵ_μ is the activation threshold.
2. *Compute the strength of the activated rules*
   Each activated rule (computing cell) has a varied degree of activation that is proportional to the membership values of its receptive fields. The rule activation strength of the kth activated computing cell is computed as:

$$\lambda_k = \min_{d \in [1,D]} \mu_{d,j_{d,k}}(x_d) \tag{1}$$

where $j_{d,k}$ is the index of the receptive field of the kth activated cell in the dth input dimension. These degrees of activation functioned as weighting factors to the stored contents of the active cells.

3. *Compute the KCMAC output*
The output Y is the weighted sum of the memory contents of the active cells:

$$Y = \sum_{k \in K} \frac{\lambda_k}{\sum_{k' \in K} \lambda'_k} W(k) \tag{2}$$

where K denotes the set of activated KCMAC computing cells and $W(k)$ is the stored weight value of the kth activated cell.

Following this, the network learning process is briefly described as follows:

1. *Computation of the network output at the ith training iteration*
The output $Y^{(i)}$ of the network corresponding to the input stimulus $\mathbf{X}^{(i)}$ is computed as described above.
2. *Computation of the learning error at the ith training iteration*
The learning error $Err^{(i)}$ is defined as the difference between the expected output $\hat{Y}^{(i)}$ and the current output of the KCMAC model $Y^{(i)}$.

$$Err^{(i)} = \hat{Y}^{(i)} - Y^{(i)} \tag{3}$$

3. *Update of the activated KCMAC cells*
The learning error is subsequently distributed to all of the activated cells based on their respective activation strength:

$$W^{(i+1)}(k) = W^{(i)}(k) + \alpha \frac{\lambda_k}{\sum_{k' \in K} \lambda'_k} Err^{(i)}, \qquad k \in K \tag{4}$$

where α is the network learning constant.

4 Case Studies

4.1 Breast Cancer Diagnosis

Breast cancer is a malignant tumor that develops when cells in the breast tissue divide and grow without the normal controls regulating cell death and cell division [9]. Although breast cancer is the second leading cause of cancer death in US women, the survival rate is high when treated early. With early diagnosis, 97% of the patients survive for 5 years or more [10]. The traditional method of surgical biopsy for breast cancer diagnosis is painful and highly invasive. Currently, the *Fine Needle Aspiration* (FNA) method offers the best alternative to biopsy as it is cost effective, non-traumatic and one of the most non-invasive methods available. The *Wisconsin Diagnostic Breast Cancer* (WDBC) dataset [11]

Table 1. Classification performance of the various benchmarked systems

System	Classification Rate (CR) [%]					Mean CR [%]	StdDev CR [%]
	CV1	**CV2**	**CV3**	**CV4**	**CV5**		
CMAC	92.43	88.64	91.20	91.95	89.63	90.77	1.5945
KCMAC	**95.28**	**96.15**	**95.70**	**95.98**	**96.52**	**95.93**	**0.4675**
GenSoFNN	94.52	92.73	93.40	94.36	94.57	93.916	0.8159
LR	95.67	93.52	95.93	94.06	96.55	95.15	1.2926
IDTM	89.61	91.57	92.17	87.5	92.83	90.74	2.1723
RBF	93.64	94.39	88.11	91.43	94.39	92.39	2.6834

is commonly used as a benchmarking data set to evaluate the performances of medical decision support systems. The data is based on the FNA test. There are two classes of tissue samples; namely *malignant* (denoted as "1") and *benign* (denoted as "0"). The dataset comprises of 569 samples with 357 and 212 instances of benign and malignant cases respectively. A total of thirty real-valued features of the cell nucleus were computed for each data sample, as described in [11]. Due to the large number of input features available, a feature selection algorithm named *Monte Carlo Evaluative Selection* (MCES) [12] is employed to identify the prominent features for the correct classification of the cancerous tissues. From the MCES algorithm, the top *five* features are selected for this classification study. These five features are: worse area, mean concave points, worse concave points, worse texture, and mean perimeter.

Based on the five selected features, the breast cancer diagnosis task is performed using KCMAC and several benchmarked architectures that include the basic CMAC, a well-established fuzzy rule-based system named *Generic Self-Organizing Fuzzy Neural Network* (GenSoFNN) [13], as well as the classical techniques of *Linear Regression* (LR), *Decision Table* (IDTM), and *Radial Basis Function* (RBF) network. The parameters of all the models have been empirically optimized for best performance. Five cross-validation (CV) groups are created to evaluate the classification performances of the various models. Each CV group consists of one training and one test set. The WDBC dataset is initially segregated into positive and negative data pools. For each CV, 50% of both pools are randomly selected as training data and the remaining 50% constitutes the test set. Table 1 tabulates the classification rates achieved by the various evaluated models. The classification rate is computed as the point where sensitivity equals specificity in the *receiver-operating-characteristic* (ROC) curve. As shown in Table 1, KCMAC achieved the best average classification rate (Mean CR) and the lowest CR variation across all the evaluated models. The multi-resolution structure and the fuzzy neighborhood computation of the KCMAC model results in a substantial improvement of 5.7% in the mean CR value as compared to that of the basic CMAC model. This set of results further reinforced the importance of judicious allocation of the computing cells in a cerebellar-based model. In addition, the adaptive memory allocation of the KCMAC model is able to derive an appropriate number of fuzzy receptive fields in each input dimension based on the characteristics of the training data. Finally, the implementation of the TSK

Table 2. Sample fuzzy rules extracted from the KCMAC model with training set CV1

Positive Rule
IF worse area is *medium* and mean concave points is *medium-large* and worse concave points is *large* and worse texture is *medium* and mean perimeter is *medium-large* THEN data sample is MALIGNANT
Negative Rule
IF worse area is *medium* and mean concave points is *medium-small* and worse concave points is *small* and worse texture is *medium* and mean perimeter is *medium-small* THEN data sample is BENIGN

fuzzy operations on the proposed KCMAC model also allows for the extraction of fuzzy rules for human analysis as tabulated in Table 2.

4.2 Modeling the Human Glucose Metabolic Process

Diabetes is a chronic disease where the body is unable to properly and efficiently regulate the use and storage of glucose in the blood. The key component to a successful management of diabetes is essentially to develop the ability to maintain long-term near-*normoglycaemia* state of the patient. With respect to this objective, accurate modeling of the human glucose metabolic process precedes the search for a suitable therapy regime.

In this study, the function approximation capability of the proposed KCMAC model is evaluated through the modeling of the dynamics of the healthy human blood glucose cycle. A well-known simulator named *GlucoSim* [14] is employed to simulate a healthy subject to generate the blood glucose data needed for the study. The simulated healthy person, referred to as Subject A, is a typical middle-aged Asian male. Based on the profile of Subject A, his recommended daily allowance (RDA) of carbohydrate intake from meals is computed and this RDA is used to design his simulated eating patterns. It is hypothesized that the instantaneous human blood glucose level is a non-linear function of prior food intakes and the historical traces of the insulin and blood glucose levels. To properly account for the effects of prior food ingestions to the fluctuations of the blood glucose level, three normalized weighting functions are introduced to compute the carbohydrate content of the meal(s) taken within the recent (i.e. previous 1 hour), intermediate past (i.e. previous 1 to 3 hours) or long ago (i.e. previous 3 to 6 hours) periods. The details on the data collection process is reported in [2]. There are a total of five inputs to the KCMAC glucose model.

A total of 100 days of glucose metabolic data for Subject A are generated using GlucoSim. The carbohydrate contents and the timings of the daily meals were varied on a daily basis during the data collection phase. The collected data set is partitioned into two non-overlapping groups: 20 days of data for training and the remaining 80 days for evaluation. Simulations to model the dynamics of the blood glucose level of Subject A using the KCMAC model were performed and the results were benchmarked against those of the basic CMAC, the *Hierarchically Clustered Adaptive Quantization CMAC* (HCAQ-CMAC) [2], the GenSoFNN

Table 3. Comparison of modeling performances for the various benchmarked models

	Recall		Generalization	
Models	RMSE [mg/ml]	Pearcorr	RMSE [mg/ml]	Pearcorr
CMAC	16.3755	0.9210	17.8086	0.9190
HCAQ-CMAC	9.9606	0.9715	10.4312	0.9748
GenSoFNN	13.7240	0.9600	13.8150	0.9400
KCMAC	**5.5115**	**0.9914**	**6.2974**	**0.9900**
MLP	4.9314	0.9962	5.2116	0.9962
RBF	10.6622	0.9673	11.8027	0.9638

model, as well as the classical MLP and RBF networks. The parameters of all the models have been empirically optimized for best performance, and the structure of the MLP network consists of five input, ten hidden and one output nodes respectively. Table 3 details the recall (training) and generalization (testing) performances of the various models. Two performance indicators are employed to quantify the modeling quality of the networks: the *root mean-squared error* (RMSE) and the *Pearson correlation coefficient* between the actual and the computed blood glucose level.

From Table 3, one can observe that MLP possesses the most accurate modeling performance as compared to the other benchmarked systems. However, it is a black-box model as its complex synaptic connections are hardly human interpretable. Moreover, the need to empirically determine the network structure often renders the MLP network hard to use. On the other hand, the proposed KCMAC model achieved a highly comparable modeling performance to the MLP network while at the same time offers interpretability of its network computations. Furthermore, the accuracy of the computed KCMAC glucose output far exceeded those of the other benchmarked models as evidenced by the simulation results tabulated in Table 3. To further analyze the modeling performance of the KCMAC model, a three-days modeling result of the KCMAC glucose model is compared against the actual observed blood glucose dynamics and the results

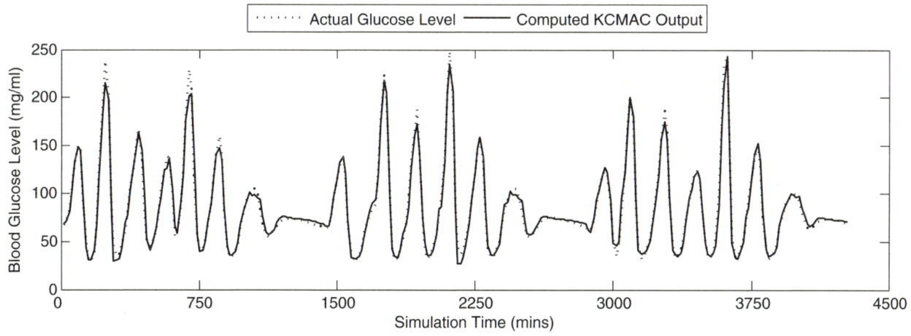

Fig. 3. Three days glucose modeling results of the KCMAC network

are depicted as Figure 3. The glucose plots in Figure 3 clearly demonstrated the superior modeling performance of the KCMAC model.

5 Conclusions

This paper proposes a novel neurophysiologically-inspired computational model of the human cerebellum named KCMAC for medical decision support. The proposed KCMAC model extends from the classical CMAC by employing a multi-resolution organization scheme of its computing cells and adopting the TSK fuzzy model to define its network computation. The proposed KCMAC model was subsequently evaluated on two medical problems; namely, breast cancer diagnosis and the modeling of the human glucose metabolic cycle. The simulation results presented in Section 5 of the paper have sufficiently demonstrated the strengths of the proposed KCMAC model as a medical decision support system when benchmarked to other established computational models.

References

1. Amari, S.: Part 1: Tutorial series on brain-inspired computing. New Generat. Comput. 23(4), 357–359 (2005)
2. Teddy, S.D., et al.: Hierarchically clustered adaptive quantization CMAC and its learning convergence. IEEE Trans. Neural Netw. 18(6), 1658–1682 (2007)
3. Kandel, E.R., et al.: Principles of Neural Science. McGraw-Hill, New York (2000)
4. Albus, J.S.: A new approach to manipulator control: The cerebellar model articulation controller (CMAC). J. Dyn. Syst., Meas., Cont., 220–227 (1975)
5. Takagi, T., Sugeno, M.: Fuzzy identification of systems and its application to modeling and control. IEEE Trans. Syst., Man, Cybern. 15(1), 116–132 (1985)
6. Duda, R.O., et al.: Pattern Classification. Wiley-Interscience, Chichester (2000)
7. Federmeier, K.D., et al.: Learning-induces multiple synapse formation in rat cerebellar cortex. Neurosci. Lett. 332, 180–184 (2002)
8. Houk, J.C., et al.: Models of the cerebellum and motor learning. Behav. Brain Sci. 19(3), 368–383 (1996)
9. Gomez-Ruiz, J.A., et al.: A combined neural network and decision trees model for prognosis of breast cancer relapse. Artif. Intell. Med. 27(1), 45–63 (2003)
10. O'Malley, C.D., et al.: Socio-economic status and breast carcinoma survival in four racial/ethnic groups. a population-based study. Cancer 97(5), 1303–1311 (2003)
11. The UCI Machine Learning Repository: Wisconsin breast cancer dataset, http://archive.ics.uci.edu/ml/datasets/Breast+Cancer+Wisconsin+(Diagnostic)
12. Quah, K.H., Quek, C.: MCES: A novel monte carlo evaluative selection approach for objective feature selections. IEEE Trans. Neural Netw. 18(2), 431–448 (2007)
13. Tung, W.L., Quek, C.: GenSoFNN: A generic self-organizing fuzzy neural network. IEEE Trans. Neural Netw. 13(5), 1075–1086 (2002)
14. IIT: GlucoSim [Online], `http://216.47.139.198/glucosim/gsimul.html`

Decoding Population Neuronal Responses by Topological Clustering

Hujun Yin[1], Stefano Panzeri[2,3], Zareen Mehboob[1], and Mathew Diamond[4]

[1] School of Electrical and Electronic Engineering, University of Manchester, UK
h.yin@manchester.ac.uk,
zareen.mehboob@postgrad.manchester.ac.uk
[2] School of Life Sciences, University of Manchester, Manchester, UK
s.panzeri@manchester.ac.uk
[3] IIT, Via Morego 30, 16163 Genova, Italy
[4] Italian Institute of Technology, SISSA unit, Trieste, Italy
diamond@sissa.it

Abstract. In this paper the use of topological clustering for decoding population neuronal responses and reducing stimulus features is described. The discrete spike trains, recorded in rat somatosensory cortex in response to sinusoidal vibrissal stimulations characterised by different frequencies and amplitudes, are first interpreted to continuous temporal activities by convolving with a decaying exponential filter. Then the self-organising map is utilised to cluster the continuous responses. The result is a topologically ordered clustering of the responses with respect to the stimuli. The clustering is formed mainly along the product of amplitude and frequency of the stimuli. Such grouping agrees with the energy coding result obtained previously based on spike counts and mutual information. To further investigate how the clustering preserves information, the mutual information between resulting stimulus grouping and responses has been calculated. The cumulative mutual information of the clustering resembles closely that of the energy grouping. It suggests that topological clustering can naturally find underlying stimulus-response patterns and preserve information among the clusters.

Keywords: Spike trains, clustering, self-organising maps, mutual information, barrel cortex.

1 Introduction

The stimulus information encoded in spike trains from multiple neurons or sites is a primary focus of research in neuroscience and is often examined in terms of various responses or features such as spike counts, firing rate, first spike latency, mean response time and interspike interval. Recent advances allow recordings up to hundreds of channels simultaneously [4][7][8]. However interpretation of these recordings and therefore interactions among neurons critically depends on accurately identifying activity of individual neurons. The massive response data arrays recorded in a large number of experiments pose great challenges

V. Kůrková et al. (Eds.): ICANN 2008, Part II, LNCS 5164, pp. 547–556, 2008.

for analysis and modelling of neuronal functions [3]. Not enough progress has yet been made on developing suitable data analysis methods that can efficiently analyse the wealth of the data and help determine how neurons encode external stimuli and exchange information to produce a reliable neuronal population code of sensory events.

Spike trains from multiple neurons are multi-dimensional point process [4]. These processes are stochastic in nature and their properties change over time in a manner that can often be characterised by a probability model describing the likelihood of spikes at a given time. Most of the standard signal processing methods are designed primarily for continuous-valued signals. It is difficult to directly evaluate how two spike trains synchronise or how they differ in their firing patterns. In addition, high dimensionality and massive response arrays of the spike trains in a growing number of experiments make systematic interpretation of the data and decoding of neuronal functions one of the biggest challenges facing the field.

Clustering is a classic technique of statistical data analysis and is used in many fields for discovering interesting patterns and underlying structures of the data [5]. Clustering can help discover intrinsic groups in the data and summarise the data. Many clustering methods have been proposed and studied [5]. The self-organising map (SOM) [6] has recently become a one of the widely used clustering algorithms (e.g. [11]) owing to its topology preserving property along the data manifold. The SOM uses an array of nodes, often arranged in 2-D, to quantise the data space, so forming a set of prototypes. In such a way the SOM can be and has been used to visualise high dimensional data sets. The SOM has been linked with the principal manifold as well as multidimensional scaling [15].

The aim of this paper is to use cluster analysis for population spike trains so that to prototype the neuronal responses and reduce the response space and stimulus features. It is also to study whether or how clustering will preserve the information among the responses with respect to stimuli. The binary point processes of spike trains are first interpreted as continuous synchronous processes by inserting a causal decaying exponential kernel at each spike. Then interpreted spike functions are grouped by the topology-preserving SOM. Clustering reveals the underlying response patterns and their relationship with the stimuli. The paper is organised as follows. In Section 2 the neuronal response data is described, along with the previous work on decoding such data sets. Preprocessing of binary spike train is described and training the SOM is presented in Section 3, together with the results and comparison of mutual information preservation of various groupings. Section 4 concludes the work.

2 Neuronal Responses and Mutual Information

This section describes the stimuli and neuronal responses studied in the paper. Spike count and spike timing are two principal media in conveying information in neuron population in existing studies. Mutual information is used to quantify the information between stimuli and responses.

2.1 Data

The data set used in this study consists of simultaneously recorded spike responses obtained from the whisker representation in the rat somatosensory ("barrel") cortex taken from a previously published series of experiments of Diamond and colleagues [1][2], where full details of the experiments are reported.

The stimuli were sinusoidal whisker vibrations characterised by different amplitudes (A) and frequency (F) and induced with high repeatability and temporal precision by a piezoelectric wafer. Seven A's and seven F's give a total of 49 combinations. Recorded signals of 100 channels (electrodes) were analysed using standard neurophysiological criteria [9], and only channels containing well detectable spiking responses from barrel cortex neurons were retained for further analysis [1]. In this study, a set of 24 simultaneously recorded responses (in spike format) were used. The data is in the form of bit-arrays (0 = no spike and 1= a spike in the considered time bin). In each trial, the 0-100 ms post-stimulus time was divided into 100 1-ms long time bins.

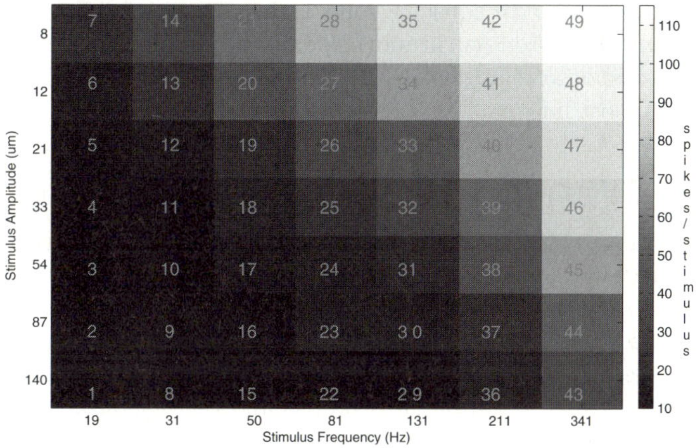

Fig. 1. Stimuli in terms of 7 frequencies (horizontal axis) and 7 amplitudes (vertical axis) of the sinusoidal whisker vibrations. In total there are 49 stimuli indicated by the number in each box. Colour map denotes the spike counts recorded in all channel.

Each stimulus is a combination of two features, frequency and amplitude, of the vibration. The frequency and amplitude on each trial were taken from the set of seven values (frequency = 19, 30, 50, 81, 131, 211 and 341 Hz; and amplitude = 8, 12, 21, 33, 54, 87 and 140 μm). The resulting 49 different frequency-amplitude combinations are represented in Fig. 1, with horizontal axis representing the frequency and vertical the amplitude. The number in each box represents the stimulus number (1-49). The stimuli 1 to 7 are the seven amplitudes combined with the lowest frequency. Similarly, stimuli 8 to 14 are the seven amplitudes combined with the second-lowest frequency, and so on. These stimuli were presented in the experiments in pseudo-random order 200 times per stimulus [1].

The previous studies have shown that the spike-count responses to these stimuli were not proportional to amplitude or frequency of the vibration alone, but were related to the product of the both, i.e. AF [1][2]. Since AF represents the vibration energy, this type of stimulus selectivity corresponds to the hypothesis that neurons respond to the energy and would lead to an idealised stimulus grouping by energy. The stimuli are grouped on the basis of their $\log(AF)$ values. The energy coding hypothesis can be approximately verified by considering the spike counts (on 0-500 ms post-stimulus [1]). Fig. 1 reports such spike counts, in which the colour in each box (A and F combination) denotes the total spike count of that stimulus in all channels. It can be appreciated that the spike count is approximately proportional to $\log(AF)$.

However, it has been demonstrated that, in addition to spike counts, the timing of individual spikes also carry information about the stimulus [9]. It is therefore important to investigate whether the neurons tend to respond primarily to energy even when we take into account the timing of the individual spikes and not only the spike counts. This issue is difficult to investigate with previously used algorithms [10]. We therefore developed a new approach to decoding population spike trains. We consider spike trains as temporal firing patterns or events and then we cluster synchronous spike trains in order to uncover underlying stimulus-response relationships. In this paper, we use the first 100 ms of the data set as most information is retained in this period [2].

2.2 Mutual Information Analysis of Population Responses

Information theoretical approach employs Shannon information theory (entropy) to quantify the "information" encoded in the responses with respect to the stimuli. Mutual information between a stimulus and evoked response tells the excess amount of code (information) produced or required by erroneously assuming that the two are independent, and thus is a natural way to quantify the deviation from independence between the stimulus and the response. In other words it shows how relevant the response is with respect to the stimulus.

In the vibrissal stimulation experiment on the "barrel" cortex, the mutual information between the stimuli, combinations of frequency and amplitude, $\{(A, F)\}$, and the response, r, can be expressed as [2],

$$MI(\{A,F\};R) = \left\langle \sum_r P(r|A,F) \log_2 \frac{P(r|A,F)}{P(r)} \right\rangle_{\{A,F\}}, \tag{1}$$

where $P(r|A,F)$ is the conditional probability of observing a neuronal response r given a stimulus, i.e. a combination of (A,F); $P(r)$ is the marginal probability of response r, i.e. the average of $P(r|A,F)$ across all stimuli; and $< \ldots >_{\{A,F\}}$ denotes the average over all stimuli. As all stimulus, $\{(A,F)\}$, were set equally probable, such an average is used. If different stimuli have different probabilities, then the probabilities $P(A,F)$ should be used as the weightings in the average.

To study which stimulus feature or what combination of features are encoded in the response. The mutual information between the response and individual features as well as various groupings of the features can be studied. It is widely believed that neurons encode few stimulus features. Previous studies [1][2] reveal that the grouping of AF carries the most information compared to other combinations. The mutual information between a (hypothetical) grouping of the stimulus features, g, and the response, r, can be expressed as [2],

$$MI(g;R) = \left\langle \sum_r P(r|g) \log_2 \frac{P(r|g)}{P(r)} \right\rangle_g \tag{2}$$

Comparing $MI(g;R)$ and $MI(\{A,F\};R)$ can tell whether the grouping encode the information conveyed in the response.

3 Topological Clustering of Population Responses

The use of clustering for spike trains can reveal underlying or natural grouping of the stimulus features encoded in the responses and reduce data quantity and feature dimension. To use any clustering algorithm on spike trains, a metric has to be defined. There exist a number of distance measures for spike trains [13]. Recently a novel spike distance has been proposed [12]. Spike trains are convolved with a causal decaying exponential kernel, so resulting continuous spike functions. Then the Euclidean distance or correlation function can be used to differentiate the spike functions. The SOM is chosen as the clustering algorithm due to its various advantages such as topology preservation among clusters and robust to noise and outliers. The SOM and its variants (e.g. ViSOM [14]) have been widely used to dimension-reduction mapping and data visualisation and have recently been associated with the principal manifold and multidimensional scaling [15].

3.1 Interpolating Spike Trains

There are several measures for spike trains such as edit distance and Euclidean distance (see [13] for a review). In [12], it has been proposed to apply a causal exponential kernel on each spike so that bit-arrays occupy a range of values or become a continuous function to enable comparison (such as correlation) between the spike trains - thus giving more appropriate interpretation.

Previous work on clustering of spike trains mainly relies on binning [7]. However, binning may not be effective when measuring synchronous neural (spiking) activities. In order to observe similarity between spike trains, one can convolve each spike with a decaying exponential kernel [12]. The parameter τ is carefully chosen so that if spike count is needed to be calculated, it remains very close to the spike count of the original bit-array.

Decaying exponential smoothed spike train can be expressed as,

$$f = \sum_{k=1}^{N} e^{-(t-t_k)/\tau} u(t-t_k), \tag{3}$$

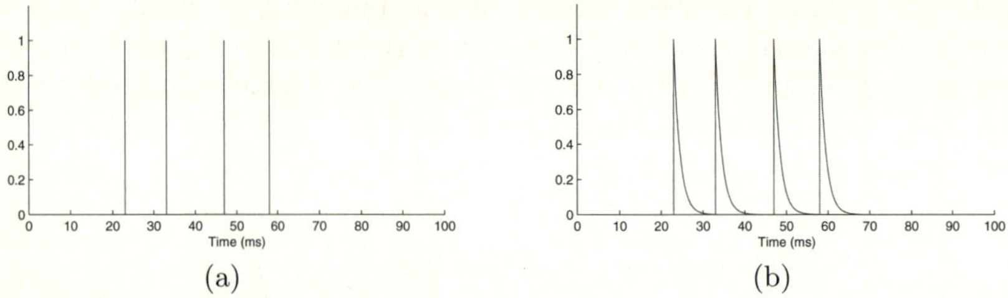

Fig. 2. (a) A typical spike train, (b) a (continuous) spike function after convolving a causal decaying exponential kernel with each spike

where N is the total number of spikes in the spike train, k denotes the k-th spike and t_k its time, τ is the decaying constant, and $u(.)$ is the unit step function.

The resulting response function of a typical spike train shown in Fig. 2(a) is plotted in Fig. 2(b). A decaying exponential kernel is inserted at each spike location. The decaying constant reflects the decaying speed. In calculation a window is used for each kernel, whose size covers its effective range for convolving the spike. It can be from 5-10 ms [12]. The window size is set to 10 ms in this study.

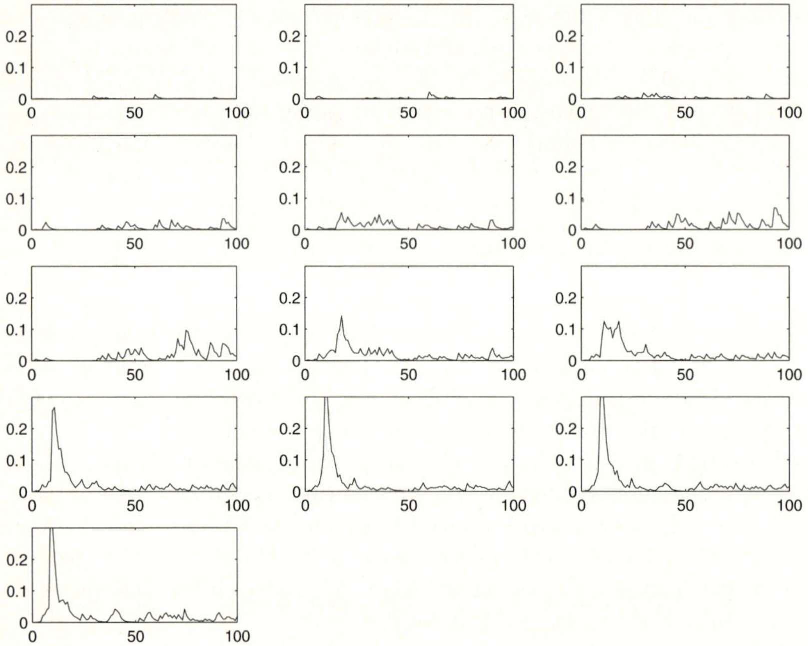

Fig. 3. Stimulus prototypes across the nodes (rearranged in rows) of 1×13 SOM trained on spike functions interpreted by causal decaying exponential kernel. Horizontal axis: time (ms).

Euclidean distance can then be used to measure the similarity between the kernel-smoothed spike trains. The distance between two spike trains $S_i(t)$ and $S_j(t)$ is calculated as,

$$d_{ij} = \left(\sum_{t=1}^{T} [S_i(t) - S_j(t)]^2 \right)^{1/2} , \tag{4}$$

where T is the total number of time points. T=100 in this study.

3.2 Topological Clustering of Population Responses

As it has been shown, the "barrel" cortex neuronal response to simulated vibration is a function of a single feature, i.e. product of A and F [2]. Thus we have chosen to use a 1-D SOM. The size of the SOM was chosen through a validation process. The best results were verified on the size of 1×13. The SOM was trained on randomly selected spike trains from all the trials. The neighbourhood size was varied slowly from 13 to 1 whereas the learning rate decreased slowly from 0.6 to 0.05. This seemed to ensure a good convergence of SOM [6].

Table 1. Main SOM clusters of stimuli and corresponding AF and spike count codes

Cluster	Stimuli	log(AF)	ln(spike count)
	35	4.26	4.47
	41	4.26	4.47
Cluster 1	42	4.47	4.63
	48	4.47	4.63
	49	4.68	4.53
	28	4.05	4.32
	34	4.06	4.23
Cluster 2	40	4.06	4.22
	46	4.05	4.43
	47	4.27	4.47
	21	3.85	4.04
	27	3.85	3.99
	33	3.85	3.89
Cluster 3	38	3.65	3.69
	39	3.84	3.99
	44	3.61	3.91
	45	3.85	4.23
	14	3.62	3.61
Cluster 4	20	3.64	3.71
	26	3.64	3.61
	32	3.64	3.53

One of the unique properties of SOM -compared to other clustering methods -is that the resulting clusters are topologically ordered on the map, at least partially if a global topological order is not achievable or does not exist. In other words, the mapping is to cluster the data and also to preserve the topology (the order) among the clusters. The SOM has been linked with the principal manifold as well as multidimensional scaling [14][15]. The SOM forms or extracts a manifold in the response space. In this study, the SOM is trained on the spike trains smoothed by the decaying kernel. The nodes of the SOM will present stimuli clusters or prototypes topologically formed on a 1-D manifold. The resulting prototypes of the SOM are shown in Fig. 3 (row by row). The stimulus-response relationship captured by the resulting manifold can be confirmed by mutual information preservation, as it will be seen in the next sub-section.

In total 7 significant clusters have been formed. The first four main clusters formed by the SOM are listed in Table 1. Stimuli and their corresponding AF and spike count codes are shown. The other clusters are mainly weak responses. From the table, it can be seen that the grouping yields similar codes to the energy code revealed in the previous studies [1][2].

3.3 Mutual Information Decoding

To further investigate how the clustering preserve the information encoded in the responses, the cumulative mutual information between the clusters and the

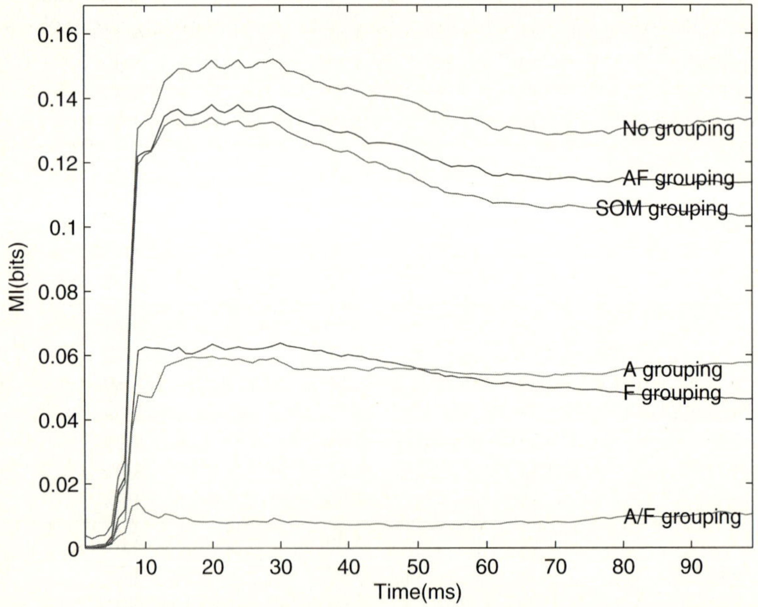

Fig. 4. Cumulative information of five different groupings: no grouping, AF grouping, SOM grouping, A grouping. F grouping, and A/F grouping.

responses is calculated using Eq. (2). The mutual information of the responses with respect to the *idealised* grouping (i.e. energy grouping, AF), the entire stimulus set (i.e. no grouping), individual features (F and A), and A/F grouping are also calculated for comparison. The results are shown in Fig. 4. It can be seen that the mutual information of the SOM clustering follows closely that of the energy grouping and that the clustering preserves the most information encoded in the response set. The small discrepancy may be due to the limited sample effect as well as the further reduction of the stimulus groups (from 13 to 7).

One advantage of using clustering for decoding spike trains information is that when idealised grouping cannot be inferred and is not obtainable in analytical forms, clustering can simply form response groups among key stimulus features. Clustering is often nonlinear and thus can detect naturally nonlinear relationships among stimuli and responses.

4 Conclusions

In this paper, clustering of decaying-exponential-kernel smoothed spike trains using the self-organising map is investigated. The resulting topological clustering can reveal the underlying manifold of the stimulus space and thus decode the stimulus-response relationship. This new approach has been applied to decoding the responses recorded from rat somatosensory cortex to whisker vibrissal stimulation. The result of this experiment supports the hypothesis that neurons encode mainly the energy - the product of amplitude and frequency of the vibration stimuli. Spike timing codes are implicitly considered here when converting spike trains into continuous spike trains and clustering the entire spike train functions. The mutual information of the clustering also confirms that the topologically formed clusters are analogous to the energy code as well as that the clustering preserves the information transfer between stimuli and responses.

Acknowledgment. The work was supported by the UK EPSRC under grant EP/E057101/1.

References

1. Arabzadeh, E., Petersen, R.S., Diamond, M.E.: Encoding of whisker vibration by rat barrel cortex neurons: implications for texture discrimination. Journal of Neuroscience 23, 9146–9154 (2003)
2. Arabzadeh, E., Panzeri, S., Diamond, M.E.: Whisker vibration information carried by rat barrel cortex neurons. Journal of Neuroscience 24, 6011–6020 (2004)
3. Brown, E.N., Kass, R.E., Mitra, P.P.: Multiple neural spike train data analysis: state-of-the-art and future challenges. Nature Neuroscience 7, 456–461 (2004)
4. Dayan, P., Abbott, L.F.: Theoretical Neuroscience: Computational and Mathematical Modeling of Neural Systems. MIT Press, Cambridge (2001)
5. Jain, A.K., Murty, M.N., Flynn, P.J.: Data clustering: a review. ACM Computing Surveys 31, 264–323 (1999)
6. Kohonen, T.: Self-Organizing Maps. Springer, Heidelberg (1997)

7. MacLeod, K., Bäcker, A., Laurent, G.: Who reads temporal information contained across synchronized and oscillatory spike trains? Nature 395, 693–698 (1998)
8. Nicolelis, M.A.L.: Methods for Neural Ensemble Recordings. CRC Press, Boca Raton (1999)
9. Panzeri, S., Petersen, R., Schultz, S., Lebedev, M., Diamond, M.: The role of spike timing in the coding of stimulus location in rat somatosensory cortex. Neuron 29, 769–777 (2001)
10. Panzeri, S., Senatore, R., Montemurro, M.A., Petersen, R.S.: Correcting for the sampling bias problem in spike train information measures. Journal of Neurophysiology 98, 1064–1072 (2007)
11. Tamayo, P., Slonim, D., Mesirov, D.J., Zhu, Q., Kitareewan, S., Dmitrovsky, E., Lander, E.S., Golub, T.R.: Interpreting patterns of gene expression with self-organizing maps: Methods and application to hematopoietic differentiation. Proc Natl Acad Sci. 96, 2907–2912 (1999)
12. van Rossum, M.C.W.: A novel spike distance. Neural Computation 13, 751–763 (2001)
13. Victor, J.D.: Spike train metrics. Current Opinion in Neurobiology 15, 585–592 (2005)
14. Yin, H.: Data visualization and manifold mapping using the ViSOM. Neural Networks 15, 1005–1016 (2002)
15. Yin, H.: On multidimensional scaling and the embedding of self-organizing maps. Neural Networks 21, 160–169 (2008)

Learning of Neural Information Routing for Correspondence Finding

Jan D. Bouecke[1,2] and Jörg Lücke[3,4]

[1] Institute of Neural Information Processing, Universität Ulm, 89081 Ulm, Germany
[2] Institute for Neuroinformatics, Ruhr-Universität Bochum, 44801 Bochum, Germany
[3] FIAS, Goethe-Universität Frankfurt, 60438 Frankfurt am Main, Germany
[4] Institute for Theoretical Physics, Goethe-Universität Frankfurt, 60438 Frankfurt am Main, Germany

Abstract. We present a neural network model that learns to find correspondences. The network uses control units that gate feature information from an input to a model layer of neural feature detectors. The control units communicate via a network of lateral connections to coordinate the gating of feature information such that information about spatial feature arrangements can be used for correspondence finding. Using synaptic plasticity to modify the connections amongst control units, we show that the network can learn to find the relevant neighborhood relationship of features in a given class of input patterns. In numerical simulations we show quantitative results on pairs of one-dimensional artificial inputs and preliminary results on two-dimensional natural images. In both cases the system gradually learns the structure of feature neighborhood relations and uses this information to gradually improve in the task of correspondence finding.

1 Introduction

Neural network models that use mechanisms to actively route information between stages of neural processing have repeatedly been suggested for invariant object recognition. Examples are early sketches, e.g., by Hinton [1] and prominent systems such as shifter circuits [2] and the dynamic link matching architecture (DLM) [3]. All these systems use an active control of information transfer between neural layers, which allows to establish correspondences between the presented input and a memory. Such systems are therefore sometimes referred to as *correspondence-based* [4]. In the technical domain correspondence-based systems represent state-of-the-art technology [5].

A common way to implement the control of information flow is to use neural *control units* [1,2,6,7] which, by their activities, gate the information flow between feature sensitive cells of different layers. Control units can up- or down-regulate the strengths of connections between two neural layers and are themselves driven by the input stimulus and the system's memory.

In all of the prominent neural systems so far suggested, the control of information flow has been fixed from the beginning using hand-crafted connections

V. Kůrková et al. (Eds.): ICANN 2008, Part II, LNCS 5164, pp. 557–566, 2008.

that excite or inhibit control unit activities. In this paper we present a system that can *learn* to control the flow of information from a set of examples.

2 The Correspondence Problem

Let us briefly illustrate the correspondence problem. Consider the sketch in Fig. 1A which shows two images of a woman, one on the input side and one on the model side. Both images are represented by a layer of feature detectors or *nodes* (black circles). The correspondence problem is now simply the problem to find those connections between input and model nodes that connect corresponding points, i.e., connections linking two feature detectors that represent the same part of an object (for instance head to head, neck to neck etc). In Fig. 1A the black connections or *links* show the correct correspondences as a subset of all possible ones.

To be able to find correspondences between images it is obvious that we need a mechanism to compute feature similarities. Unfortunately, such a mechanism alone is not sufficient to find correspondences in realistic applications. This is due to the fact that a pair of reaflistic inputs, e.g. different images of the same object, usually have a high variability, which frequently results in situations in which one feature is more similar to a feature at a location different from its corresponding one. Fig. 1B shows a cartoon of such a situation: the black lines connect the features with highest similarity, which results in a wrong correspondence mapping in this case. For realistic inputs such situations are the norm and the ambiguities increase the more feature detectors are used.

For the human observer it is, however, very easy to find the correct correspondences, e.g., between the images in Fig. 1B. The reason for this is that an object is defined by its features *and* their spatial arrangement rather than by an unordered collection of them. Correspondence-based systems therefore try to take

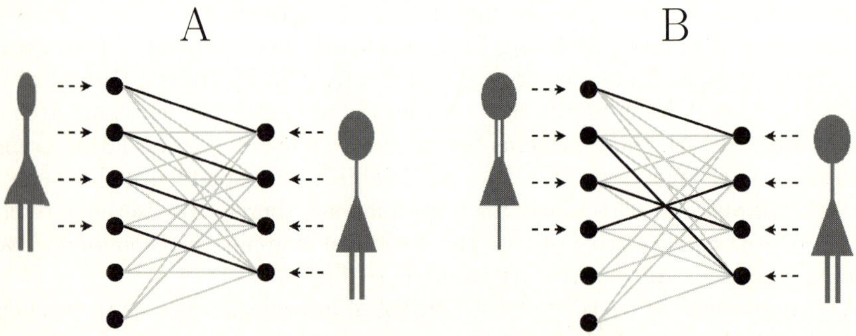

Fig. 1. In vision the correspondence problem is the task to find the corresponding points between two images. In **A** an input and a model images are represented by arrays of feature nodes (black circles). All potential corresponding points in this architecture are symbolized by links between the feature nodes. The real correspondences are the black links. **B** shows an example for wrong wrong correspondences.

both into account. Shifter circuits, for instance, relatively rigidly control subsets of links whereas DLM has a more flexible dynamic control that lets neighboring links communicate with each other. In this paper, we base our modeling on a recently suggested system [7] that uses neural control units similar to those in shifter circuits and a flexible recurrent dynamics to drive their activity reminiscent of the dynamics in DLM systems. This new dynamics is in the following shown to be readily extendable to a system that learns to structure the control of information routing.

3 Control Dynamics

The neural architecture used in this paper is essentially given in Fig. 1A, i.e., the system consists of an input layer and a model layer each of which contains an array of nodes that represent an input and a model datum by a set of abstract features. Input and model nodes are all-to-all connected (see Fig. 1A gray and black lines). To select the links that connect the corresponding points, each model node contains a set of control units to regulate the strength of each link of the node, i.e., there is one control unit per link. If the control unit is active, so is its link. To be consistent with the literature [7], we will refer to the set of all control units for one model node as its *control column*. Fig. 2 shows such a control column for model node $\mathcal{M}i$ and for an input layer of three feature nodes $\mathcal{I}1$, $\mathcal{I}2$ and $\mathcal{I}3$.

The activity of a control unit (and thus the link activity) is determined by its inputs which consist of two parts: (1) input from the pair of feature nodes associated with the controlled link and (2) input from other control units (see Fig. 2). Let us denote the activity of control unit j in model node $\mathcal{M}i$ by $w^{\mathcal{M}i\,\mathcal{I}j}$ and its input by $I^{\mathcal{M}i\,\mathcal{I}j}$. This control unit is responsible for the link from input node $\mathcal{I}j$ to model node $\mathcal{M}i$. The actual local features, which describe k different

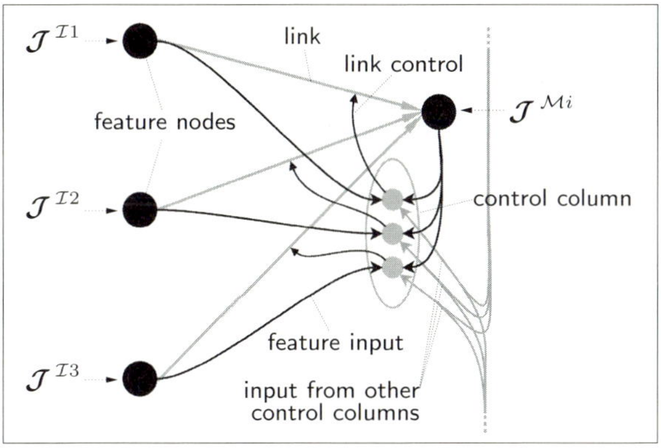

Fig. 2. Sketch of a model node with control units. Each control unit regulates the strength of one incoming link. The units themselves receive input from the pair of feature nodes whose link they control and from other control columns.

local properties (indexed by α or β) that are connected via such a link, are represented by a set of different feature node activities $p_\beta^{\mathcal{I}j}$ $p_\alpha^{\mathcal{M}i}$ within each node $\mathcal{I}j$ $\mathcal{M}i$. With this, the over-all input $I^{\mathcal{M}i\mathcal{I}j}$ to control unit $w^{\mathcal{M}i\mathcal{I}j}$ is given by:

$$I^{\mathcal{M}i\mathcal{I}j} = C_I \underbrace{\sum_{\alpha\beta} p_\alpha^{\mathcal{M}i} R_{\alpha\beta}\, p_\beta^{\mathcal{I}j}}_{\text{similarity term}} + (1 - C_I) \underbrace{\sum_{k \neq i,l} T_{k\,l}^{i\,j} w^{\mathcal{M}k\mathcal{I}l}}_{\text{topology term}} \tag{1}$$

The similarity term has the form of a dot product of the mean free[1] activities at two linked feature nodes. The topology term takes account of the spatial feature relations via the lateral interaction between control units. The matrix $T_{k\,l}^{i\,j}$ parameterizes this interaction. Without any input from the topologic term, the correspondence are computed only on the basis of similarity between linked feature nodes. Both types of input, the feature similarity and the preferred feature arrangement, drive the control units activities. To select a subset or just one of these links for each node, they compete through mutual inhibition. We use a kind of soft-max competition given by a set of non-linear differential equations as suggested in [7,8]:

$$f(w,h) := a\left(w^2 - w \cdot h - w^3\right) + \eta \tag{2}$$

$$\frac{d}{dt} w^{\mathcal{M}i\mathcal{I}j} = f(w^{\mathcal{M}i\mathcal{I}j}, \nu \cdot \max_{\tilde{j}}\{w^{\mathcal{M}i\mathcal{I}\tilde{j}}\}) + \kappa I^{\mathcal{M}i\mathcal{I}j} \tag{3}$$

where $w^{\mathcal{M}i\mathcal{I}j}$ denotes the activity of control unit j of model node $\mathcal{M}i$, η is a noise term[2], and ν denotes the strength of inhibitory coupling. If ν is increased, a control column is forced to successively deactivate its control unit until, for relatively high values of ν, just one unit survives. On the basis of this mechanism the routing or matching process is managed and within each so called ν-cycle one result of correspondence finding is established.

To neurally represent the features by activity distributions within feature nodes, the same dynamics can be used, e.g., for the feature unit $p_\alpha^{\mathcal{L}i}$: $\frac{d}{dt} p_\alpha^{\mathcal{L}i} = f(p_\alpha^{\mathcal{L}i}, \nu \cdot \max\{p_{\tilde{\alpha}}^{\mathcal{L}i}\}) + \kappa I_\alpha^{\mathcal{L}i}$ with $\mathcal{L} \in \{\mathcal{M}, \mathcal{I}\}$ (compare equation 3). The input to feature nodes within the input and model layer is given directly by the feature vectors they represent $I_\alpha^{\mathcal{I}j} = \mathcal{J}_\alpha^{\mathcal{I}j}$ and $I_\alpha^{\mathcal{M}i} = \mathcal{J}_\alpha^{\mathcal{M}i}$, where α again denotes the different types of features at a given position.

Much of the dynamics has been discussed in [9,7]. Important differences in this work are, however, the restriction of information flow to one direction (from input to model layer). As a consequence, control units are just needed in the model layer, which significant reduces the complexity of the dynamics and the computational cost for its simulation. However, the most important difference to earlier systems is the usage of learning.

---

[1] $R_{\alpha\beta} = \delta_{\alpha\beta} - \frac{1}{k}$.

[2] Normally distributed iid noise.

4 Learning

The learning rule used, modifies the lateral connections amongst the control units. The form is similar to the learning rule for component extraction as used in [8] but uses a slow decay of synaptic weights in addition.

$$\frac{d}{dt} T^{i\,j}_{k\,l} = \varepsilon \left(w^{i\,j}\, w^{k\,l} - \big(\sum_{\acute{k}\,\acute{l}} w^{i\,j} w^{\acute{k}\,\acute{l}}\big) T^{i\,j}_{k\,l} \right) \Theta(\chi - \sum_{\acute{j}}^{k} w^{i\,\acute{j}}) - \rho T^{i\,j}_{k\,l} \tag{4}$$

To improve readability we have dropped the index prefixes $\mathcal{I}$ $\mathcal{M}$ in equation 4. This is possible because for w and T the left index always refers to a link destination (model node) and the right to the link origin (input node).

The decay term is necessary to avoid learning of strong receptive fields for control units of links which are wrongly activated during correspondence finding. Such control units would, withought the slow decay, develop strong receptive fields similar to control units which are frequently used (even though much more slowly).

In general, the learning rule is of Hebbian type (positive term) with a normalization term to avoid divergence of the synaptic weights (first negative term). Additionally there is a Heaviside function Θ which prevents the synapses from learning when too many control units within a control column are active.

5 Simulation Results

Now we have to test the ability of our model system to learn regularities in feature arrangements from example data. If the system is able to find such regularities, the quality of correspondence finding should slowly improve and should result in topologic connections adapted to the presented kind of data.

At first the system is tested with sets of artificially created feature vectors. This input is usefull because the identified correspondences can easily be compared to ground truth and thus a mean error can be computed to observe the progress during learning. In a second set of trails preliminary results for natural images are shown.

5.1 Learning of Lateral Connection for 1-dimensional Data

The artificial feature vectors[3] used for this experiments have 10 entries, i.e., 10 single scalar feature values describe the state of any node. The network has to find corresponding positions for 5 model nodes within an input of 10 input nodes in this experiment. For every ν-cycle (within each ν-cycle the system tries to find corresponding points) a new set of model features is generated and noisily copied to the input layer. The noise, applied to the feature vectors is additive and

[3] We use randomly generated feature vectors that are composed of random numbers drawn from a uniform distribution in the range of $[0;1]$.

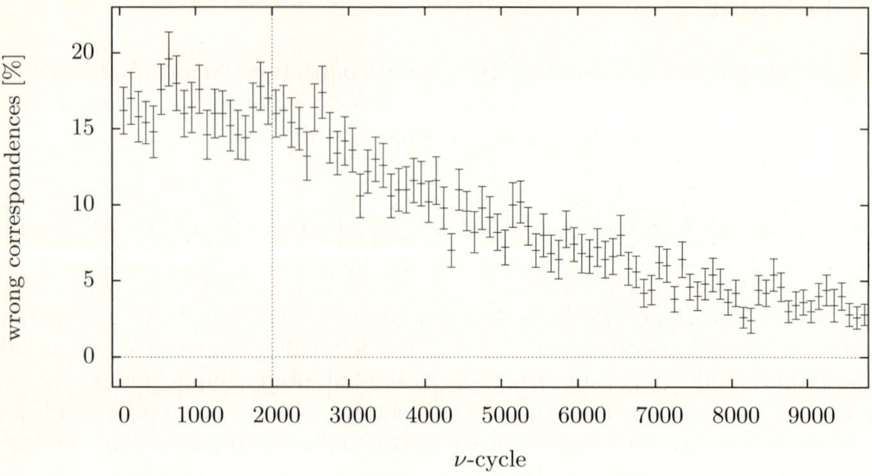

Fig. 3. Simulation results of routing quality before and during learning. The learning is enabled after 2000 ν-cycles. Each point shows the mean percentage of wrong correspondences of the respective past 100 ν-cycles together with its standard deviation.

uniformly distributed[4] in the range of $[-0.3, 0.3]$. Thereby the 5 model feature vectors keep their neighborhood relationship but are collectively shifted to a random position within the input layer. The remaining 5 input nodes of the input are filled with other random feature vectors.

We simulated the dynamics for control columns (equation 3), the feature nodes and the learning rule (equation 4) by using the explicit Euler method for stochastic differential equations using the following paremeters: $a = 200$, $\kappa = 5$, $C = 0.2$, $\epsilon = 0.01$,$\chi = 0.55$, $\rho = 0.01 \cdot \epsilon$. The results, which are calculated using a stepsize of $\triangle t = 0.02$ are shown in Fig. 3. The plotted values of wrong correspondences are a mean of 100 routing results and the error bars correspond to standard deviation of the mean value. For comparison, we run the system without learning for the first 2000 ν-cycles (lateral connections T are initialized to zero). From 2000 ν-cycles onwards, we use equation 4 to modify them. During the first 2000 ν-cycles, due to the very noisy feature vectors, about $\frac{1}{3}$ of the correspondences found are not between originally identical feature vectors (see Fig. 3). After enabling synaptic plasticity, the error rate decrease because of growing lateral connections. Over time connections adapt increasingly to the structure of the presented data.

Note that the presented arrangement of features and model nodes is initially not encoded in the network connectivity. The only connections which can account for a neighborhood structure are the lateral connections amongst control units, which we initialized to zero. The connectivity of the 10 control units (horizontally arranged) whithin each of the 5 control columns (vertically arranged) are shown

[4] Experiments with other types of artificially created input and other types of noise have not shown any significant changes.

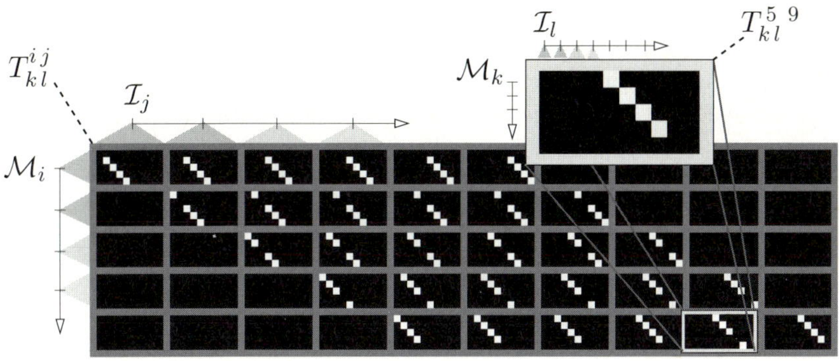

Fig. 4. Simulation results showing the receptive fields of all lateral connections after learning (7000 cycles) of 1-dimensional artificial input. Horizontally arranged are the receptive fields of control units for different link origins within the input layer. The destination position within the model layer changes vertically. The receptive field, for the Link from input node $\mathcal{I}9$ to model node $\mathcal{M}5$ is highlighted as an example. Each single value within this receptive filed is again ordered in the same way while the weight to itself is set to zero (black). The diagonal structure reflects the neighborhood preserving mapping of the presented 1-dimensional shifts.

in Fig. 4. Initially they were all set to zero but the more a special structure of common link probability emerge, they are bind together through growing lateral connections.

This improvement can be reproduced for a great variety of input structures. We have tested, e.g., input at always the same position, shifted input with cyclic boundary conditions, scaled/stretched positions or for shifts at which some feature positions disappear form the input.

5.2 Learning of Lateral Connection from Natural Images

The proposed network has shown to be able to learn the constraints of e.g. a 1-dimensional, position invariant mapping. In other words, it has learned the variations due to position invariance alongside with the restrictions due to the maintenance of neighborhoods. The thereby observed improvement in routing quality results from the learned lateral connections between control units. Due to all-to-all connectivity, a huge amount of different kinds of routing constraints and variations are possible and can potentially be learned from example data, e.g., from pairs of natural images.

To extract a set of local features from natural images, we use Gabor wavelets with different spatial frequencies and orientations (we use five spatial frequencies and eight orientations as, e.g., in [10,7]). The feature vectors presented to the network are taken from a regular square grid at a randomly selected position. The data to the input layer consists of $7 \cdot 7$ feature vectors and the smaller model layer receives its input from a square grid of $5 \cdot 5$ feature vectors. The images, used for this experiment are taken from three different objects (teapot and two different stapler). The $3 \cdot 10 = 30$ images were taken with a standard compact camera freehand out of approximately the same position.

For the simulations with this new kind of input the same network dynamics as before is used, just its size is adapted to the new data and the synamptic decay is drecreased to $\rho = 0.0005 \cdot \epsilon$. This is neccesary because the amount of links increases which results in less learning time per link. In comparison with the experiments in the previous secition, for example, the number of possible variations is increased by a factor of 49 : 5. To that, the amount of synaptic weights which have to be learned increses by the factor of $(49 \cdot 25)^2 : (5 \cdot 10)^2$. This results in simulation time, increased by a factor of about 6000 and is the price to pay, if no constraints are applied initially. For this reason, for such a large network, only some preliminary results in form of an example routing (see Fig. 5) and learned receptive field for one link (see Fig. 6) can be shown.

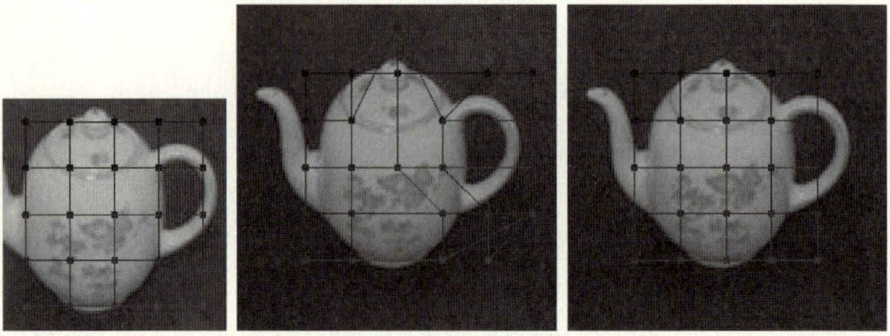

Fig. 5. Corresponding points, between grids (model gird: left and input grid: middle and right) for one example image. The middle result was produced withought any topological connecctions and the right one with the learned connections.

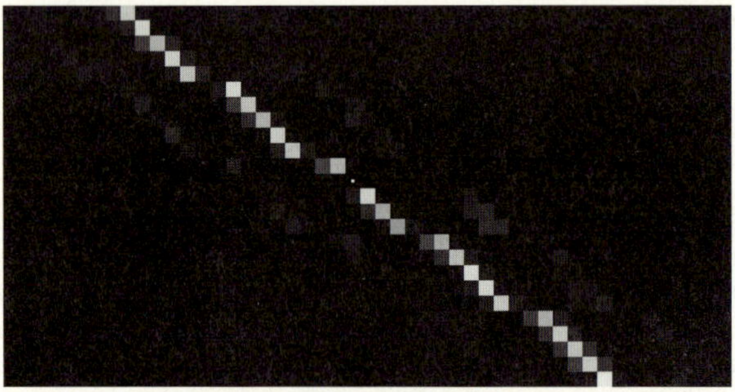

Fig. 6. Learned connectivity after 15000 ν-cycles for the link from the center of the input layer to the center of the model layer. Its structure matches what is expected from a 2-dimensional neighborhood relationship but has not jet (after 15000 ν-cycles) reached its final strength.

6 Discussion

In this paper we have presented a system that learns regularities in feature relations for correspondence finding. The system uses a set of control units to gate feature information between neural layers. Unlike dynamic routing systems such as [1,2,7], information routing in the system presented can be trained on examples. If these examples contain a typical neighborhood relationship of features, the system is able to learn it and to use feature relations for correspondence finding. For instance, we have seen that the system is able to learn the for images typical two-dimensional feature relationship. That is, the system has learned that the image of an object is characterized by its features and their two-dimensional relation. Note again, that the initial architecture of the system is not biased towards any neighborhood relationship (the lateral connections that define the model nodes' neighborhood relations are initially zero). For the one-dimensionally structured inputs, e.g. as used in Sec. 5.1, the system has learned a control unit connectivity that reflects one-dimensional feature relations. Likewise, inputs with other feature neighborhood relations result in other corresponding control unit connections. If the appropriate features are used, any neighborhood relationship typical for the used set of input images can in principle be learned in this way. These relationships are not necessarily restricted to be one-, two- or n-dimensional but can be more complex. An example would be to use training images of translated objects together with their mirror images.

Importantly, control unit connectivity captures information about feature relations and their *variations* in contrast to a fixed relation of specific features (as in a variety of other systems, e.g. [11,12]). This is illustrated by the fact that, for instance, training on images of different translations of an object can improve correspondence finding between images of translations of a different object. The mechanism investigated in this paper is thus addressing a problem complementary to the problem of learning input features, namely the problem of learning the typical variations in feature arrangements in a given class of inputs. While learning from training inputs is a well-established paradigm to obtain class specific features, the same can not be said about learning of feature arrangements which is usually just dealt with in hand-crafted ways (e.g., [12,2,7]; but see [13] for an exception on artificial data). Here we have shown how variations in feature arrangements can be learned from examples and how this knowledge can be used to improve correspondence finding. In future work such mechanisms hold some promise in improving more complex recognition systems, especially if combined with state-of-the-art systems for feature learning.

Acknowledgments. We thank Rolf Würtz and Christoph von der Malsburg for helpful discussions and acknowledge funding by the EU project FP6-2005-015803 (Daisy), the EU project FP7-2007-216593 (Seco), and the Hertie Foundation.

References

1. Hinton, G.E.: A parallel computation that assigns canonical object-based frames of reference. In: Proc. IJCAI, pp. 683–685 (1981)
2. Olshausen, B.A., Anderson, C.H., Van Essen, D.C.: A neurobiological model of visual attention and invariant pattern recognition based on dynamic routing of information. The Journal of Neuroscience 13(11), 4700–4719 (1993)
3. Wiskott, L., von der Malsburg, C.: Face recognition by dynamic link matching. In: Sirosh, J., Miikkulainen, R., Choe, Y. (eds.) Lateral Interactions in the Cortex: Structure and Function, ch. 4 (1995), `www.cs.utexas.edu/users/nn/book/bwdraft.html` ISBN 0-9647060-0-8
4. Zhu, J., von der Malsburg, C.: Maplets for correspondence-based object recognition. Neural Networks 17(8–9), 1311–1326 (2004)
5. Messer, K., Kittler, J., Sadeghi, M., Hamouz, M., Kostin, A., et al.: Face authentication test on the BANCA database. In: Proceedings of the International Conference on Pattern Recognition, Cambridge, vol. 4, pp. 523–532 (2004)
6. Weber, C., Wermter, S.: A self-organizing map of sigma-pi units. Neurocomputing 70(13-15), 2552–2560 (2007)
7. Lücke, J., Keck, C., von der Malsburg, C.: Rapid convergence to neural feature field correspondences. Neural Computation (page accepted, 2007)
8. Lücke, J., Bouecke, J.D.: Dynamics of cortical columns – self-organization of receptive fields. In: Duch, W., Kacprzyk, J., Oja, E., Zadrożny, S. (eds.) ICANN 2005. LNCS, vol. 3696, pp. 31–37. Springer, Heidelberg (2005)
9. Lücke, J., von der Malsburg, C.: Rapid correspondence finding in networks of cortical columns. In: Kollias, S.D., Stafylopatis, A., Duch, W., Oja, E. (eds.) ICANN 2006. LNCS, vol. 4131, pp. 668–677. Springer, Heidelberg (2006)
10. Lades, M., Vorbrüggen, J.C., Buhmann, J., Lange, J., von der Malsburg, C., Würtz, R.P., Konen, W.: Distortion invariant object recognition in the dynamic link architecture. IEEE Transactions on Computers 42(3), 300–311 (1993)
11. Mel, B.W.: SEEMORE: Combining color, shape, and texture histogramming in a neurally-inspired approach to visual object recognition. Neural Computation 9, 777–804 (1997)
12. Sali, E., Ullman, S.: Combining class-specific fragments for object classification. In: Pridmore, T.P., Elliman, D. (eds.) BMVC, British Machine Vision Association (1999)
13. Zhu, J., von der Malsburg, C.: Learning control units for invariant recognition. Neurocomputing 52–54, 447–453 (2003)

A Globally Asymptotically Stable Plasticity Rule for Firing Rate Homeostasis

Prashant Joshi[1,2,*] and Jochen Triesch[2]

[1] Computation and Neural Systems Program, Division of Biology, 216-76 California Institute of Technology, Pasadena, CA 91125 USA
joshi@fias.uni-frankfurt.de
[2] Frankfurt Institute for Advanced Studies, Johann Wolfgang Goethe University, Ruth-Moufang-Str. 1, 60438 Frankfurt am Main, Germany

Abstract. How can neural circuits maintain stable activity states when they are constantly being modified by Hebbian processes that are notorious for being unstable? A new synaptic plasticity mechanism is presented here that enables a neuron to obtain homeostasis of its firing rate over longer timescales while leaving the neuron free to exhibit fluctuating dynamics in response to external inputs. Mathematical results demonstrate that this rule is *globally asymptotically stable*. Performance of the rule is benchmarked through simulations from single neuron to network level, using sigmoidal neurons as well as spiking neurons with dynamic synapses.

1 Introduction

External environment and experiences modify the properties of cortical neurons using various synaptic plasticity mechanisms. Correlation-based Hebbian plasticity is accepted as the principal mechanism that modifies synaptic strengths, enabling the organism to adapt to a new environment and/or to learn new tasks [1]. Nevertheless, it is a well known fact that Hebbian plasticity by itself is a positive feedback process and will lead a circuit into unstable activity regimes in the absence of some kind of regulatory mechanism [2]. Recent neurobiological studies suggest that slow homeostatic plasticity mechanisms enable the neurons to maintain average firing rate levels by dynamically modifying the synaptic strengths in the direction that promotes stability [3,4]. For example, pharmacological blocking of activity in cultured networks causes an increase in the amplitudes of the miniature excitatory postsynaptic currents (mEPSCs), and increasing activity by blocking inhibition scales down the mEPSCs amplitude. This global regulatory mechanism known as 'synaptic scaling' has been observed in cultured networks of neocortical [5], hippocampal [6] and spinal-cord [7] neurons.

Several learning rules from the more classical ones such as BCM rule [8], Oja's rule [9] to more recent ones [10,11] have been proposed that provide some sort of homeostatic or normalizing mechanism. A new biologically plausible synaptic

* Corresponding author.

V. Kůrková et al. (Eds.): ICANN 2008, Part II, LNCS 5164, pp. 567–576, 2008.

modification mechanism is presented here that provides firing rate homeostasis. Rigorous mathematical results demonstrate that this learning rule is globally asymptotically stable. Results from computer simulations using spiking neurons and dynamic synapses are also provided showing the working of the rule from monosynaptic to network levels.

2 Computational Theory and Learning Rule

The simplest plasticity rule following Hebb's premise is given by:

$$\tau_w \frac{d\mathbf{w}}{dt} = \nu_{\mathbf{pre}}(t) \cdot \nu_{post}(t), \tag{1}$$

where $\nu_{\mathbf{pre}}(t)$ and $\nu_{post}(t)$ are the pre and postsynaptic firing rates and τ_w is the time constant that controls the rate of change of synaptic weights. A simple analysis[1] can show that the only stable points for the above equation are the trivial cases of $\nu_{\mathbf{pre}}(t) = 0$ and/or $\nu_{\mathbf{post}}(t) = 0$ when the whole system goes into a quiescent state. A new rule that enables the postsynaptic neuron to maintain an average firing rate of ν_{base} over time is obtained by adding a multiplicative term of $\nu_{base} - \nu_{post}(t)$ to equation 1, where ν_{base} is the steady state firing rate to which the postsynaptic neuron relaxes over time:

$$\tau_w \cdot \frac{d\mathbf{w}(t)}{dt} = \nu_{\mathbf{pre}}(t) \cdot \nu_{post}(t) \cdot (\nu_{base} - \nu_{post}(t)). \tag{2}$$

Theorem 1 shows that the plasticity mechanism described in equation 2 is *globally asymptotically stable.* Furthermore, Theorem 2 demonstrates that an analytical solution for the postsynaptic firing rate is possible.

For the sake of simplicity and intuitiveness, the learning rule is analyzed for the single input, single output (SISO) case, i.e. when a pre- and postsynaptic neuron are connected via a synapse with weight w. Both the pre- and postsynaptic neurons are assumed to be linear and excitatory in nature, so the output of the postsynaptic neuron is just the weighted value of its input. It is to be noted that the stability of the learning rule is not dependent on linearity, as is shown by simulation results in section 3.

Theorem 1. *For a SISO case, with the presynaptic input held constant at ν_{pre}, and the postsynaptic output having the value ν^0_{post} at time $t = 0$, and ν_{base} being the homeostatic firing rate of the postsynaptic neuron, the system describing the evolution of $\nu_{post}(.)$ is globally asymptotically stable. Further ν_{post} globally asymptotically converges to ν_{base}.*

Proof of Theorem 1. For the SISO case, with the pre- and postsynaptic neurons being excitatory, the weight update is given by:

$$\tau_w \cdot \frac{dw(t)}{dt} = -\nu_{pre} \cdot \nu_{post}(t) \cdot (\nu_{post}(t) - \nu_{base}). \tag{3}$$

[1] Set $\frac{dw}{dt} = 0$ in equation 1.

Since the neurons are linear, we have:

$$\nu_{pre} \cdot w(t) = \nu_{post}(t). \tag{4a}$$

On differentiating equation 4a we get:

$$\nu_{pre} \cdot \frac{dw(t)}{dt} = \frac{d\nu_{post}(t)}{dt}. \tag{4b}$$

By substitution in equation 3 we get:

$$\frac{d\nu_{post}(t)}{dt} = -\frac{\nu_{pre}^2}{\tau_w} \cdot \nu_{post}(t) \cdot (\nu_{post}(t) - \nu_{base}). \tag{5}$$

The temporal evolution of $\nu_{post}(.)$ is described by the system in equation 5. To prove that the system in equation 5 is globally asymptotically stable, the *global invariant set theorem* is used (see Appendix).

Consider as the Lyapunov function candidate (see Appendix), the scalar function $V(\nu_{post})$ defined as:

$$V(\nu_{post}) = \tau_w(\nu_{post} - \nu_{base} \cdot ln(\nu_{post})). \tag{6}$$

The function $V(\nu_{post})$ is radially unbounded, since $V(\nu_{post}) \to \infty$ as $\|\nu_{post}\| \to \infty$. Its derivative is:

$$V'(\nu_{post}) = \tau_w \cdot \left(1 - \frac{\nu_{base}}{\nu_{post}}\right) \cdot \frac{d\nu_{post}}{dt}. \tag{7a}$$

Substituting from equation 5 we get:

$$V'(\nu_{post}) = \tau_w \cdot \left(1 - \frac{\nu_{base}}{\nu_{post}}\right) \cdot \left(-\frac{\nu_{pre}^2}{\tau_w} \cdot \nu_{post} \cdot (\nu_{post} - \nu_{base})\right). \tag{7b}$$

or

$$V'(\nu_{post}) = -\nu_{pre}^2 \cdot (\nu_{post} - \nu_{base})^2. \tag{7c}$$

From equation 7c, it is evident that $V'(\nu_{post})$ is negative definite, i.e. $V'(\nu_{post}) \leq 0$ over the whole state-space.

The set **R** defined by $V'(\nu_{post}) = 0$ consists of the single point $\nu_{post} = \nu_{base}$. Furthermore, in this case the set **M**(the largest invariant set in **R**) equals the set **R**, i.e. **M** = **R**.

Thus by *global invariant set theorem* (see Appendix), all trajectories globally asymptotically converge to **M** as $t \to \infty$. Further since the set **M** contains the single point $\nu_{post} = \nu_{base}$, which is thus the attractor, the value of $\nu_{post}(t)$ asymptotically converges to ν_{base}

Theorem 2. *For a SISO case, with the presynaptic input held constant at* ν_{pre}*, and the postsynaptic output having the value* ν_{post}^0*, at time* $t = 0$*, and* ν_{base} *being* *the homeostatic firing rate of the postsynaptic neuron, the postsynaptic value at any time $t > 0$ is given by:*

$$\nu_{post}(t) = \frac{\nu_{post}^0 \cdot \nu_{base}}{\nu_{post}^0 + (\nu_{base} - \nu_{post}^0) \cdot e^{-\frac{\nu_{pre}^2 \cdot \nu_{base} \cdot t}{\tau_w}}}.$$

Proof of Theorem 2. To prove the statement in theorem 2, we have to solve the differential equation described by equation 5. To solve it, we first convert the above equation into a linear form. Dividing equation 5 by $\nu_{post}^2(t)$, we get:

$$\frac{1}{\nu_{post}^2(t)} \cdot \frac{d\nu_{post}(t)}{dt} - \frac{\nu_{pre}^2 \cdot \nu_{base}}{\tau_w} \cdot \frac{1}{\nu_{post}(t)} = -\frac{\nu_{pre}^2}{\tau_w}. \quad (8)$$

Now let:

$$-\frac{1}{\nu_{post}(t)} = y(t), \quad (9a)$$

On differentiating equation 9a we get:

$$\frac{1}{\nu_{post}^2(t)} \cdot \frac{d\nu_{post}(t)}{dt} = \frac{dy(t)}{dt}. \quad (9b)$$

Substituting in equation 8 we get:

$$\frac{dy(t)}{dt} + \frac{\nu_{pre}^2 \cdot \nu_{base}}{\tau_w} \cdot y(t) = -\frac{\nu_{pre}^2}{\tau_w}. \quad (10)$$

Equation 10 is a linear differential equation. Taking the Laplace transform on both sides of equation 10, we get:

$$s \cdot Y(s) - y(0) + \frac{\nu_{pre}^2 \cdot \nu_{base}}{\tau_w} \cdot Y(s) = -\frac{\nu_{pre}^2}{\tau_w} \cdot \frac{1}{s}, \quad (11a)$$

or

$$Y(s) \cdot \left(s + \frac{\nu_{pre}^2 \cdot \nu_{base}}{\tau_w}\right) = \left(y(0) - \frac{\nu_{pre}^2}{\tau_w} \cdot \frac{1}{s}\right) = \frac{y(0) \cdot \tau_w \cdot s - \nu_{pre}^2}{\tau_w \cdot s}, \quad (11b)$$

or

$$Y(s) = \frac{y(0) \cdot \tau_w \cdot s - \nu_{pre}^2}{s(\tau_w s + \nu_{pre^2} \cdot \nu_{base})} = \frac{y(0) \cdot s - \frac{\nu_{pre}^2}{\tau_w}}{s(s + \frac{\nu_{pre}^2 \cdot \nu_{base}}{\tau_w})}. \quad (11c)$$

Expanding the RHS of equation 11c into partial fractions we get:

$$Y(s) = \frac{1 + y(0) \cdot \nu_{base}}{\nu_{base}} \cdot \frac{1}{\left(s + \frac{\nu_{pre}^2 \cdot \nu_{base}}{\tau_w}\right)} - \frac{1}{\nu_{base} \cdot s}. \quad (11d)$$

Taking the inverse Laplace transform of equation 11d we get:

$$y(t) = \frac{1 + y(0) \cdot \nu_{base}}{\nu_{base}} \cdot e^{-\frac{\nu_{pre}^2 \cdot \nu_{base} \cdot t}{\tau_w}} - \frac{1}{\nu_{base}}. \quad (12)$$

Now substitute $y(t) = -\frac{1}{\nu_{post}(t)}$ and $y(0) = -\frac{1}{\nu^0_{post}}$ (from equation 9a) into equation 12:

$$-\frac{1}{\nu_{post}(t)} = \left(\frac{1 - \frac{\nu_{base}}{\nu^0_{post}}}{\nu_{base}}\right) e^{-\frac{\nu^2_{pre} \cdot \nu_{base} \cdot t}{\tau_w}} - \frac{1}{\nu_{base}}, \tag{13a}$$

or

$$-\frac{1}{\nu_{post}(t)} = \frac{(\nu^0_{post} - \nu_{base}) e^{-\frac{\nu^2_{pre} \cdot \nu_{base} \cdot t}{\tau_w}} - \nu^0_{post}}{\nu^0_{post} \cdot \nu_{base}}. \tag{13b}$$

From equation 13b we get:

$$\nu_{post}(t) = \frac{\nu^0_{post} \cdot \nu_{base}}{\nu^0_{post} + (\nu_{base} - \nu^0_{post}) \cdot e^{-\frac{\nu^2_{pre} \cdot \nu_{base} \cdot t}{\tau_w}}}, \tag{14}$$

which is the statement of theorem 2.

3 Results

The computational theory and learning rule introduced in this article were verified through simulations using sigmoidal neurons as well as integrate-and-fire neurons with dynamic synapses.

As a first test[2], a postsynaptic neuron with sigmoidal transfer function was used that was receiving presynaptic inputs from two different and independent Gaussian input streams[3]. Panel A of figure 1 shows the two input streams. Panel B shows the output of the postsynaptic neuron when the synaptic weights are being modified according to equation 2. Note the convergence of the output to ν_{base}. Panel C shows the corresponding evolution of synaptic weights. The postsynaptic rate converges to baseline value at a slower timescale from below than above due to the nature of equation 2, i.e. the magnitude of weight change is proportional to the value of presynaptic input. Panel D shows the temporal evolution of $\nu_{post}(n)$ when the postsynaptic neuron was receiving a single fixed presynaptic input of $\nu_{pre}(t) = 0.5$, for various initial values of ν^0_{post} and the initial value of synaptic weight $w^0 = \frac{1}{\nu_{pre}} ln(\frac{\nu^0_{post}}{1-\nu^0_{post}})$. Panel E shows similar results for various initial values of ν_{pre} and ν^0_{post}. Panel F shows the surface corresponding to the numerical solution of equation 14. Finally panel G shows the phase plane diagram. It is interesting to note that despite the inherent non-linearity in the neuron, the postsynaptic firing rate consistently converges to the baseline value in all the simulations.

---

[2] The simulation was run for $n = 5000$ steps. Initial weights $\langle w^0_1, w^0_2 \rangle$ uniformly drawn from $[0, 0.1]$, $\nu_{base} = 0.6$, $\tau_w = 30$.

[3] Input channel 1: mean = 0.3, SD = 0.01 for $n \leq 2500$, mean = 0.6, SD = 0.04 otherwise. Input channel 2: mean = 0.8, SD = 0.04, for $n \leq 2500$, mean = 1.6, SD = 0.01 otherwise.

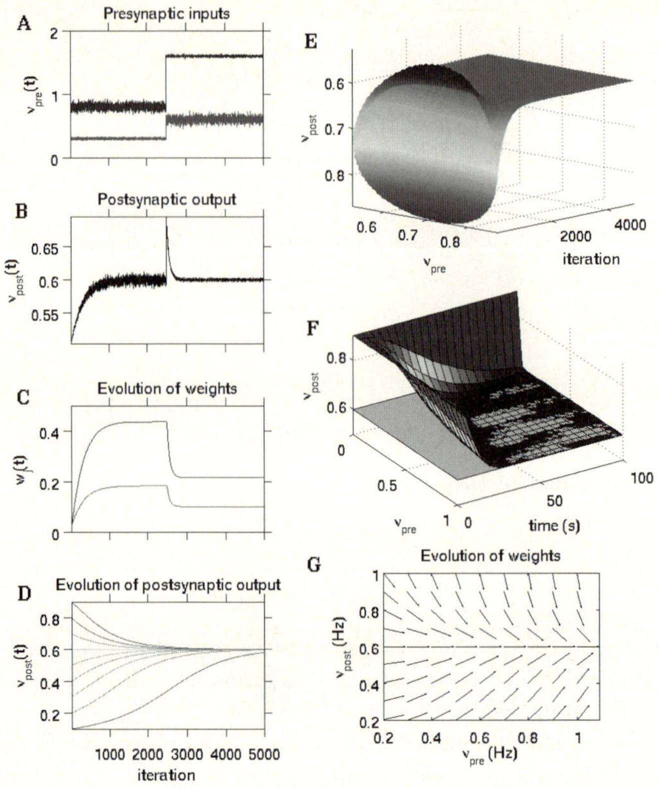

Fig. 1. For this experiment $\nu_{base} = 0.6, \tau_w = 30$ (A) The two presynaptic inputs (B) The output of the postsynaptic neuron (C) Evolution of synaptic weights (synaptic homeostasis) (D) Evolution of ν_{post} to the baseline rate ν_{base} over time, for a fixed value of ν_{pre} and different initial values of ν_{post}^0 (E) 3D plot showing system evolution for different initial conditions (F) Surface plot corresponding to a numerical solution of equation 14 with the value of $\nu_{post}^0 = 0.9$ and with different values of ν_{pre}. The plane marked with yellow color indicates the baseline rate (G) The weights evolve so that the postsynaptic rate reaches the baseline.

Another setup[4] studied the performance of the learning rule when a single postsynaptic integrate-and-fire neuron received presynaptic inputs from 100 Poisson spike trains[5] via dynamic synapses. Panel A of figure 2 plots the presynaptic input spike trains and panel B shows the membrane potential V_m and the spikes in the postsynaptic neuron while the synaptic weights were being modified according to the learning rule. In panel C, the mean-firing rate of the

[4] The simulation was run for $t = 10$ sec with a simulation time step of 1 msec. Initial weights $w_j^0 = 10^{-8}, j = 1, \ldots, 100$, $\nu_{base} = 40$ Hz, $\tau_w = 3600$.

[5] First 50 Poisson spike trains (shown in red in Pannel A of figure 2): drawn at 3 Hz for $0 < t \leq 5$ sec and 60 Hz for $t > 5$ sec. Remaining 50 Poisson inputs (shown in blue in Panel A of figure 2): drawn at 7 Hz for $0 < t \leq 5$ sec and 30 Hz for $t > 5$ sec.

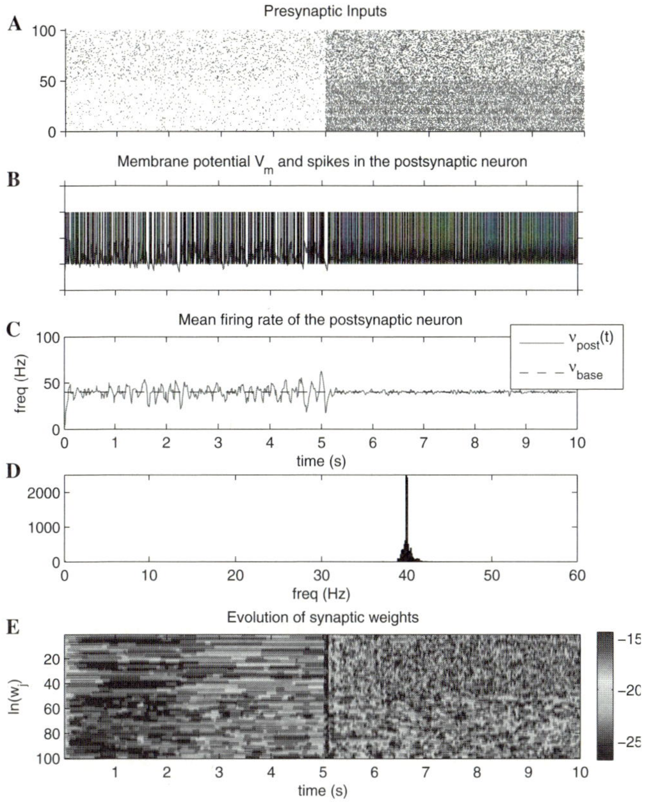

Fig. 2. Firing rate homeostasis in a single postsynaptic integrate-and-fire neuron receiving Poisson spiking inputs via 100 dynamic synapses. (A) The presynaptic input, and (B) the membrane potential and spikes, (C) the mean-firing rate, and (D) the histogram of firing rate of the postsynaptic neuron. (E) Temporal evolution of synaptic weights shown on a logarithmic scale.

postsynaptic neuron is illustrated and it is seen that $\nu_{post}(t)$ converges to baseline value[6]. Panel D shows the histogram of firing rate of the postsynaptic neuron over the course of simulation. The sharp peak at 40 Hz is attributed to the value of ν_{base}. In panel E, the evolution of synaptic weights on a logarithmic scale is shown and it is observed that the plasticity mechanism uses different weight update strategies to maintain the steady state firing rate during the first and second half respectively, depending on the strength of presynaptic input.

Can synaptic homeostasis mechanism be used to maintain stable ongoing activity in recurrent circuits? The next experiment[7] used a circuit made of 250 integrate-and-fire neurons, with 20% of these neurons chosen to be inhibitory.

[6] The synapses were modified using equation 2 by computing instantaneous mean firing rate (running average).

[7] Simulation was run for $t = 5$ sec with a simulation time step of 1 msec.

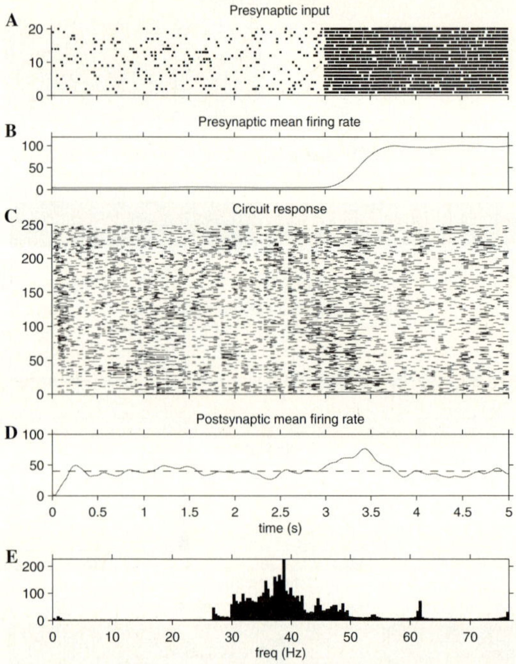

Fig. 3. Homeostasis of firing rate in a circuit of 250 integrate-and-fire neurons that received Poisson spiking input via 20 dynamic synapses. (A) The presynaptic input and (B) mean firing rate of the input. (C) The circuit response, (D) mean firing rate of circuit dynamics, and (E) histogram of mean-firing rate.

In addition models for dynamic synapses were used whose individual mixture of paired-pulse depression and facilitation is based on experimental data [12,13]. Sparse connectivity between neurons with a biologically realistic bias towards short connections was generated by a probabilistic rule, and synaptic parameters were randomly chosen, depending on the type of pre-and postsynaptic neurons, in accordance with empirical data (see [14] for details). Inputs to this circuit came from 20 Poisson spike trains that were drawn at the rate of 5 Hz for $0 < t \leq 3$ sec, and 100 Hz for $t > 3$ sec. Panel A of figure 3 shows the input spike trains while panel B shows the mean firing rate of the external input to the circuit. In panel C the circuit response is illustrated and panel D shows the mean firing rate of the neural circuit. The learning rule enables the circuit to maintain the mean firing rate of ν_{base} over time as is also evident from panel E which shows the histogram of firing rates. It should be noted that the time-scale of firing-rate homeostasis in biological systems is several orders of magnitude slower than what is observed in these simulations. Such fast time-constants were used to reduce computational time.

4 Discussion

This article describes a new synaptic plasticity mechanism that enables a neuron to maintain a stable firing rate over slower time scales while leaving it free to show moment-to-moment fluctuations in its activity based on variations in its presynaptic inputs. It should be noted that this plasticity mechanism is completely local. The rule is shown to be globally asymptotically stable through mathematical analysis. It is to be noted that the assumption of linearity used in the mathematical analysis is for the sake of simplicity only as the rule is able to obtain homeostasis of firing rate even in presence of non-linearity in the neuron and synaptic dynamics as is evident from the simulation results. Furthermore, computer simulations demonstrate that the rule is able to achieve firing rate homeostasis from single neuron to network level. A potential advantage of this learning rule is its inherent stability without the need for a sliding threshold as required by the BCM rule. Future work shall also explore the working of this rule in conjunction with other plasticity mechanisms, for example STDP.

Acknowledgments

We acknowledge financial support by the European Union through projects FP6-2005-015803 and MEXT-CT-2006-042484 and the Hertie Foundation. The work was also supported through the Broad Foundation. The authors would also like to thank Cornelius Weber, Andreea Lazar and Costas Anastassiou for helpful comments.

References

1. Hebb, D.O.: Organization of Behavior. Wiley, New York (1949)
2. Gerstner, W., Kistler, W.M.: Spiking Neuron Models. Cambridge University Press, Cambridge (2002)
3. Abbott, L.F., Nelson, S.B.: Synaptic plasticity: taming the beast. Nature Neurosci. 3, 1178–1183 (2000)
4. Turrigiano, G.G., Nelson, S.B.: Homeostatic plasticity in the developing nervous system. Nature Neuroscience 5, 97–107 (2004)
5. Turrigiano, G.G., Leslie, K.R., Desai, N.S., Rutherford, L.C., Nelson, S.B.: Activity-dependent scaling of quantal amplitude in neocortical neurons. Nature 391(6670), 892–896 (1998)
6. Lissin, D.V., Gomperts, S.N., Carroll, R.C., Christine, C.W., Kalman, D., Kitamura, M., Hardy, S., Nicoll, R.A., Malenka, R.C., von Zastrow, M.: Activity differentially regulates the surface expression of synaptic ampa and nmda glutamate receptors. Proc. Natl. Acad. Sci. USA 95(12), 7097–7102 (1998)
7. O'Brien, R.J., Kamboj, S., Ehlers, M.D., Rosen, K.R., Fischbach, G.D., Huganir, R.L.: Activity-dependent modulation of synaptic ampa receptor accumulation. Neuron 21(5), 1067–1078 (1998)
8. Bienenstock, E.L., Cooper, L.N., Munro, P.W.: Theory for the development of neuron selectivity: orientation specificity and binocular interaction in visual cortex. J. Neurosci. 2, 32–48 (1982)

9. Oja, E.: A simplified neuron model as a principal component analyzer. J. Math. Biol. 15, 267–273 (1982)
10. Triesch, J.: Synergies between intrinsic and synaptic plasticity in individual model neurons. In: Advances in Neural Information Processing Systems 2004 (NIPS 2004). MIT Press, Cambridge (2004)
11. Triesch, J.: A gradient rule for the plasticity of a neuron's intrinsic excitability. In: International Conference on Artificial Neural Networks 2005 (ICANN 2005) (2005)
12. Markram, H., Wang, Y., Tsodyks, M.: Differential signaling via the same axon of neocortical pyramidal neurons. PNAS 95, 5323–5328 (1998)
13. Gupta, A., Wang, Y., Markram, H.: Organizing principles for a diversity of GABAergic interneurons and synapses in the neocortex. Science 287, 273–278 (2000)
14. Maass, W., Joshi, P., Sontag, E.D.: Computational aspects of feedback in neural circuits. PLOS Computational Biology 3(1), e165, 1–20 (2007)
15. Slotine, J.-J.E., Li, W.: Applied Nonlinear Control. Prentice-Hall, Englewood Cliffs (1991)

Appendix

Mathematical Definitions

Lyapunov Function. Let V be a continuously differentiable function from $\Re^n$ to $\Re$. If G is any subset of $\Re^n$, we say that V is a Lyapunov function on G for the system $d\mathbf{a}/dt = \mathbf{g}(\mathbf{a})$ if

$$\frac{dV(\mathbf{a})}{dt} = (\nabla V(\mathbf{a}))^T \mathbf{g}(\mathbf{a}) \tag{15}$$

does not change sign on G.

Note that this is a generalization of the classical definition of Lyapunov function (c.f. definition 3.8 in [15] for example). More precisely, it is not required that the function V be positive-definite (just continuously differentiable). The only requirement is on the derivative of V, which can not change sign anywhere on the set G.

Global Invariant Set Theorem. Consider the autonomous system $d\mathbf{x}/dt = \mathbf{f}(\mathbf{x})$, with $\mathbf{f}$ continuous, and let V($\mathbf{x}$) be a scalar function with continuous first partial derivatives. Assume that

a) $V(\mathbf{x}) \to \infty$ as $\|\mathbf{x}\| \to \infty$
b) $V'(\mathbf{x}) \leq 0$ over the whole state space

Let $\mathbf{R}$ be the set of all points where $V'(\mathbf{x}) = 0$, and $\mathbf{M}$ be the largest invariant set in $\mathbf{R}$. Then all solutions of the system globally asymptotically converge to $\mathbf{M}$ as $t \to \infty$.

Analysis and Visualization of the Dynamics of Recurrent Neural Networks for Symbolic Sequences Processing

Matej Makula[1] and Ľubica Beňušková[2]

[1] Faculty of Informatics and Information Technologies, Slovak Technical University, Ilkovičova 3, 842 16 Bratislava, Slovakia
makula@fiit.stuba.sk
[2] Department of Computer Science, University of Otago, 9054 Dunedin, New Zealand
lubica@cs.otago.ac.nz
[3] Department of Applied Informatics, Faculty of Mathematics, Physics and Informatics, Comenius University, Mlynská dolina, 842 48 Bratislava, Slovakia
benus@ii.fmph.uniba.sk

Abstract. Recurrent neural networks unlike feed-forward networks are able to process inputs with time context. The key role in this process is played by the dynamics of the network, which transforms input data to the recurrent layer states. Several authors have described and analyzed dynamics of small sized recurrent neural networks with two or three hidden units. In our work we introduce techniques that allow to visualize and analyze the dynamics of large recurrent neural networks with dozens units, reveal both stable and unstable points (attractors and saddle points), which are important to understand the principles of successful task processing. As a practical example of this approach, dynamics of the simple recurrent network trained by two different training algorithms on context-free language $a^n b^n$ was studied.

1 Introduction

It is well-known fact that recurrent neural networks (RNNs) have universal approximation capability, although development of desired dynamics in training might be difficult or even unfeasible task. Various training sets on various architectures of RNNs were studied in many works which confirm that RNNs can learn to become perfect finite state recognizers of regular grammars [1,2]. Higher classes of grammars from Chomsky hierarchy have also been studied. For successful processing of context-free languages (CFL), a stack or counter in RNN memory is required [3]. On the other hand when RNN is used to process symbolic sequences, activations of hidden units show considerable amount of information about input sequence prior to training [4]. It was theoretically and experimentally shown that RNNs initialized with small weights are inherently biased towards Markov models [5,6,7].

V. Kůrková et al. (Eds.): ICANN 2008, Part II, LNCS 5164, pp. 577–586, 2008.

All these phenomena in discrete time RNNs can be characterized and studied as a discrete time dynamical systems [8] with inputs, outputs, and state variables. For example Markovian dynamics of randomly initialized RNN is quite simple and can be explained by the iterated function theory [9,10]. The state space of RNN initialized with small weights contains a set of attractors that are responsible for producing organization of state space closely related to variable length Markov models. When RNN is trained on regular grammars, attractive sets are rearranged to perform desired classification [11]. Analysis of RNNs processing context-free and complex context-sensitive languages demonstrated that their nonlinear state space can incorporate counting mechanism [12].

These analyses bring new insight into the internal mechanisms of the symbolic sequences processing by RNNs. Their performance should be studied by means of activities of hidden units (network state) and their changes in time (network dynamics). In our work, we focus on the RNN state space analysis by combination of visualisation and numerical techniques. Basic concepts were presented in previous works [12,13] on small recurrent networks with 2 dimensional state space, where so-called fixed points and their relation to network dynamics was analysed. We extended these techniques for analysis of networks with dozens of hidden units.

In following sections, method for visualisation and localization of unstable fixed points in its high dimensional state space of simple recurrent network (SRN) architecture will be described. As a practical example analysis of various sized SRNs trained on context-free language $a^n b^n$ and classification of counters developed in their nonlinear dynamics is performed. Bias of backpropagation through time (BPTT) [14] and extended Kalman filter (EKF) [15] algorithm to specific class of dynamics is demonstrated.

2 Dynamics of Simple Recurrent Network Processing Symbolic Sequences

As it was mentioned earlier RNN for symbolic sequences processing can be thought of as a set of dynamical systems. Consider activation of hidden units $\mathbf{s}(t)$ and output units $\mathbf{o}(t)$ of simple recurrent network with sigmoid activation function f:

$$\mathbf{s}(t) = f(\mathbf{M} \cdot \mathbf{s}(t-1) + \mathbf{W} \cdot \mathbf{i}(t) + \mathbf{b}), \tag{1}$$

$$\mathbf{o}(t) = f(\mathbf{V} \cdot \mathbf{s}(t) + \mathbf{d}), \tag{2}$$

where $\mathbf{M}$, $\mathbf{W}$, $\mathbf{V}$ are matrices of recurrent, input and output weights, respectively. Vectors $\mathbf{b}$ and $\mathbf{d}$ represent bias. When SRN process symbolic sequences, i.e. finite set of symbols from alphabet $A = \{a, b, \ldots\}$, each symbol $x \in A$ is encoded by one predefined input activation vector $\mathbf{i}_x$. Network dynamics (Eq. 1) thus can be reduced to:

$$\begin{aligned} \mathbf{F}_a:\ & \mathbf{s}(t) = f(\mathbf{M} \cdot \mathbf{s}(t-1) + \mathbf{h}_a) \\ \mathbf{F}_b:\ & \mathbf{s}(t) = f(\mathbf{M} \cdot \mathbf{s}(t-1) + \mathbf{h}_b) \\ & \ldots \\ \mathbf{F}_x:\ & \mathbf{s}(t) = f(\mathbf{M} \cdot \mathbf{s}(t-1) + \mathbf{h}_x) \end{aligned} \tag{3}$$

where $\mathbf{h}_x$ is a fixed vector assigned to input symbol x, i.e. $\mathbf{h}_x = \mathbf{W} \cdot \mathbf{i}_x + \mathbf{b}$. System of equations 3 is a set of autonomous discrete time dynamical systems $\{\mathbf{F}_a, \mathbf{F}_b \ldots\}$, each corresponds to one symbol of alphabet A. Although, these dynamical systems are autonomous, they are sharing network state - vector of hidden unit activities $\mathbf{s}(t)$. When a new symbol from input sequence is presented, corresponding transformation is applied to the network state. Network dynamics thus should be studied on two different levels:

1. Analysis of partial dynamical systems $\mathbf{F}_x$, i.e. exploration of network state trajectory while processing only one input symbol of A.
2. Analysis of whole system, i.e. exploration of network state trajectory while processing real input sequence.

Each particular dynamical system is mostly influenced by the location and character of so-called fixed points. A fixed point $\tilde{\mathbf{s}}$ of system $\mathbf{F}_x$ is a point in the state space which is mapped to itself:

$$\tilde{\mathbf{s}} = \mathbf{F}_x(\tilde{\mathbf{s}}) . \tag{4}$$

When network state space contracts to a fixed point, that point is an attractor (Fig. 1a), otherwise that point is repelling (Fig. 1b). In some cases, the repelling point may be contracting in one direction and expanding in another direction, so we call it a saddle point (Fig. 1c).

Although, dynamical systems from equation 3 are not linear (nonlinearity is introduced by activation function f), linearization can be performed to study character of fixed points. The eigenvalues λ_i and eigenvectors $\mathbf{v}_i$ of Jacobian $\mathbf{J}(\tilde{\mathbf{s}})$ (partial derivative matrix) calculated at fixed point $\tilde{\mathbf{s}}$ express how system changes in time [16]. They must satisfy condition $\mathbf{J}(\tilde{\mathbf{s}}) \cdot \mathbf{v}_i = \lambda_i \cdot \mathbf{v}_i$ and $\mathbf{v}_i \neq 0$, i.e. they are expressing direction and intensity of linear contraction/expansion of linearized system. If the complex eigenvalue λ_i lies in the unit circle, fixed point is contracting in the direction $\mathbf{v}_i$, otherwise is repelling. Moreover, the non-zero imaginary part of eigenvalue is a sign of rotation around the fixed point. The

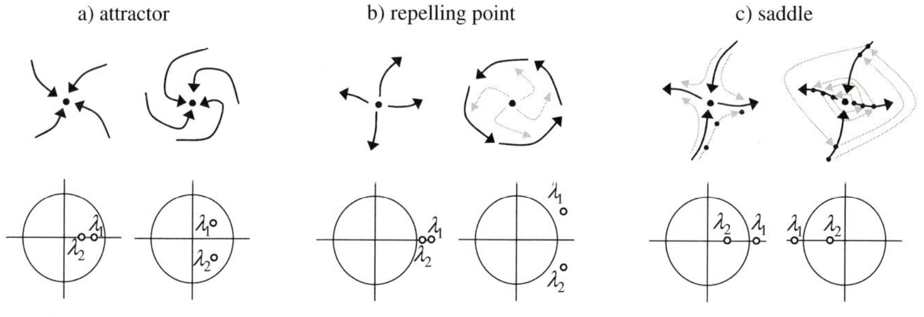

Fig. 1. State space topology near the fixed point $\tilde{\mathbf{s}}$ and its relation to the eigenvalues λ_i of partial derivative matrix $\mathbf{J}(\tilde{\mathbf{s}})$. a) attractor - state trajectories lead state to fixed point; b) repelling point; c) saddle - repelling in one direction and contracting in other.

negative value of λ_i indicates that the state is driven to/from fixed point by a 2-periodic oscillations (Fig. 1c, left).

3 Visualisation and Fixed-Point Analysis of SRN

3.1 Visualization of Network State Space

The vector 'flow' field plot represents popular approach for state space visualization. Direct insight into network dynamics is performed by arrows showing how particular transformation from equation 3 affects state (Fig. 2). Vector flow field thus can be used for both localization and analysis of the fixed-point character [12]. Since vector flow field of dynamical systems can be created only in networks

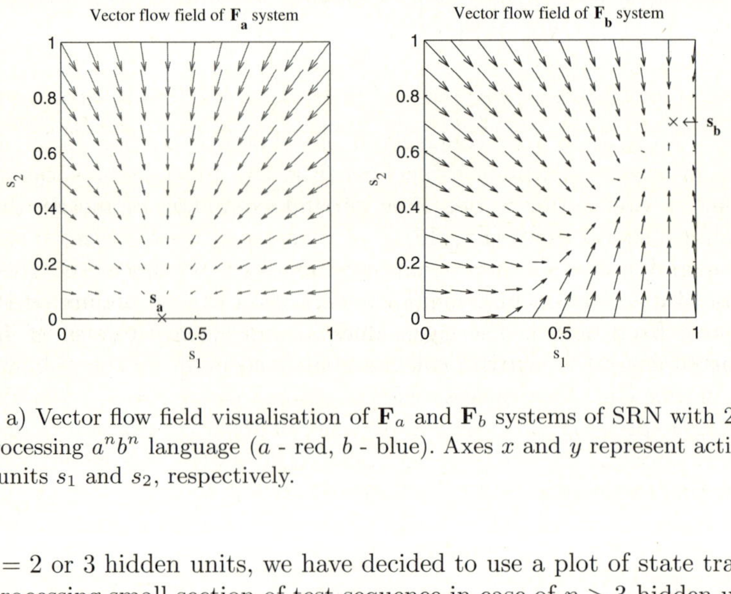

Fig. 2. a) Vector flow field visualisation of $\mathbf{F}_a$ and $\mathbf{F}_b$ systems of SRN with 2 hidden units processing $a^n b^n$ language (a - red, b - blue). Axes x and y represent activities of hidden units s_1 and s_2, respectively.

with $n = 2$ or 3 hidden units, we have decided to use a plot of state trajectory while processing small section of test sequence in case of $n > 3$ hidden units.

At first, projection of the state space trajectory into 2D or 3D subspace is performed. For this purpose well-known data analysis & reduction technique - principle component analysis (PCA) is used. PCA looks for orthogonal directions with maximal variance of input data, i.e. principal components. Coordinates $\mathbf{c}(t)$ in 2D or 3D subspace are calculated as:

$$\mathbf{c}(t) = \mathbf{P} \cdot \mathbf{s}(t), \tag{5}$$

where $\mathbf{s}(t)$ is a state vector at time step t and $\mathbf{P}$ is projection matrix composed of 2 or 3 principal components vectors of PCA analysis. In our work, we decided to use this simple linear visualisation technique, because more advanced nonlinear visualisation techniques will affect original nonlinearity of studied dynamical system. On the other hand, projection matrix $\mathbf{P}$ can be used several times in

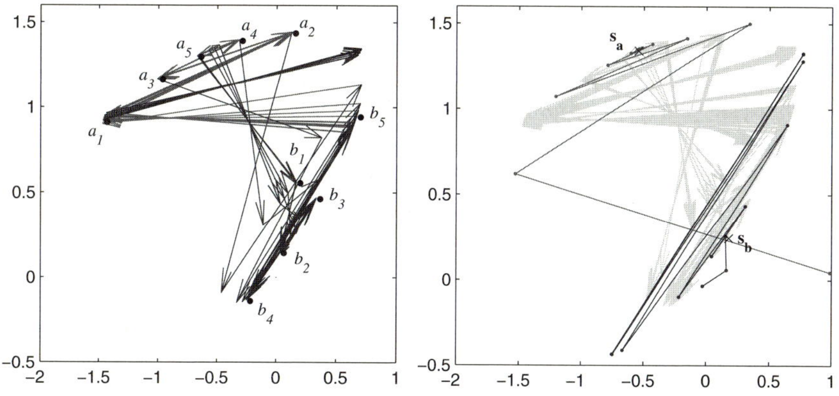

Fig. 3. a) Visualisation of state trajectory of SRN with 10 hidden units while processing $a^n b^n$ language (a - red, b - blue). Projection to 2D subspace is performed by PCA technique. b) Fixed points: attractor $\tilde{\mathbf{s}}_{\mathbf{a}}$ in red $\mathbf{F}_a$ dynamics and saddle $\tilde{\mathbf{s}}_{\mathbf{b}}$ in blue $\mathbf{F}_b$ dynamics. Sequence $aaaa\ldots$ leads network state from random position to $\tilde{\mathbf{s}}_{\mathbf{a}}$ attractor. Sequence $bbbb\ldots$ drives network state to 2-periodic oscillations around saddle $\tilde{\mathbf{s}}_{\mathbf{b}}$.

consequent visualisations of specific network state space areas, allowing us to view various parts of the state space from the same direction. Since state trajectory is influenced by set of autonomous dynamical systems, we distinguish them by colour of line connecting two states (Fig. 3a). In the black and white print out, unfortunately, one has to rely on direction of arrows, thickness of lines and the labels to deduce the trajectories.

3.2 Revealing Unstable Fixed Points

In previous works, localisation of fixed points was performed either visually by already mentioned vector flow field plot or numerically [12,17]. Numerical localization of fixed point works only for stable points, which attract network state. In figure 3b sequence $aaaa\ldots$ drives network state after few iterations to attractor $\tilde{\mathbf{s}}_{\mathbf{a}}$. If fixed point is unstable, i.e. saddle or repelling point, dominant expansive manifold drives network state away from it.

For localisation of unstable fixed points in high dimensional state space, we make use of property, which is common for both stable and unstable fixed points. According to equation 4, the Euclidean distance of two states:

$$||\mathbf{s}(t) - \mathbf{s}(t+1)|| \tag{6}$$

in neighbourhood of fixed point is close to zero. Network state space thus can be explored by hill climbing algorithm, searching for areas where value of Eq. 6 is minimal. In our experiments states located by hill climbing algorithm have Euclidean distance $||\mathbf{s}(t) - \mathbf{s}(t+1)|| < 10^{-5}$.

Then, each candidate area is explored more precisely by Newton's method. Denote by $\mathbf{J}(\mathbf{s}^j)$ the Jacobian matrix of examined system $\mathbf{F}$ evaluated at a point

$\mathbf{s}^j$. If $\mathbf{s}^j$ is close to fixed point $\tilde{\mathbf{s}}$ and matrix $\mathbf{J}(\mathbf{s}^j)$ is invertible then we could calculate $\mathbf{s}^{j+1}$ as:

$$\mathbf{s}^{j+1} = \mathbf{s}^j - \mathbf{J}(\mathbf{s}^j)^{-1} \cdot \mathbf{F}\left(\mathbf{s}^j\right), \tag{7}$$

which would be closer to $\tilde{\mathbf{s}}$ than $\mathbf{s}^j$. If this iteration converge to some $\mathbf{s}^j$ then $\mathbf{s}^j$ is a fixed point.

The use of Newton's method improved robustness of our localization method. Sometimes, even when Euclidean distance (Eq. 6) is small, fixed point was not in the area localized by hill climbing algorithm (Fig. 4c). To determine character of the fixed point, detailed analysis described in the next section is carried out.

3.3 Recognition of a Fixed Point Character

Two different methods for recognition of a fixed point character can be used. The first is visualisation of the network dynamics near the fixed point by PCA projection (Eq. 5). Figure 4 shows three time steps of 10 state trajectories in

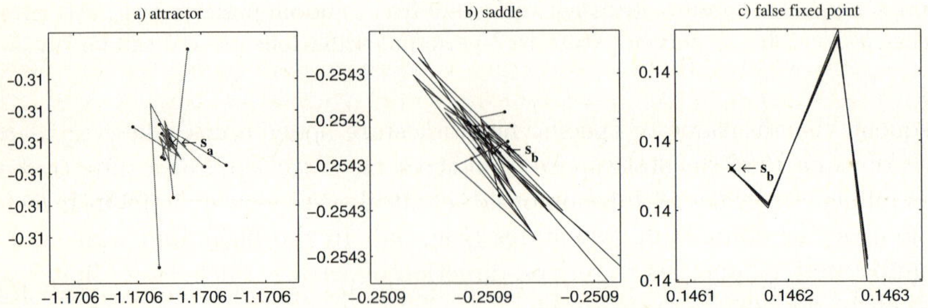

Fig. 4. Visualisation of 10 state trajectories near the fixed point - random starting positions are denoted as black dots: a) attractor $\mathbf{s}_a$; b) saddle point $\mathbf{s}_b$; c) false fixed point $\mathbf{s}_b$

small neighbourhood of fixed point. Visualisation demonstrates contracting behaviour near attractor, unstable direction (with 2–periodic oscillating behaviour) near saddle and uniform divergence of state trajectories from wrongly localized fixed point. The second method of analysis uses linearization near fixed point by Jacobian matrix calculation. As it was mentioned earlier, its eigenvalues allows us to recognize both stability and rotation/oscillations around fixed point.

4 Experiments and Results

To provide a practical example of our approach we performed experiments on context-free language $a^n b^n$ according to [12]. To our knowledge, no one has yet attempted to analyse dynamics of SRN with more than 3 hidden units on CFL language. All previous works were limited to small networks in which visualisation and fixed point analysis of network dynamics can be done in a straightforward way.

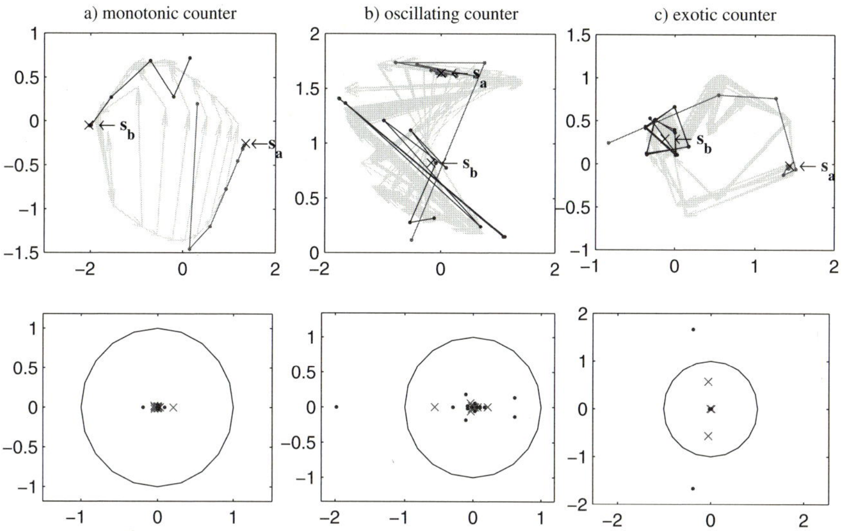

Fig. 5. Three typical solutions of counter in the SRN state space: a) monotonic counter - two attractive points; b) oscillating counter - attractor $\mathbf{s}_a$ and oscillating saddle $\mathbf{s}_b$; c) exotic counter - attractor $\mathbf{s}_a$ and unstable fixed point $\mathbf{s}_b$. First and second network have 15 hidden units. Third network has 5 hidden units. The eigenvalues of Jacobian matrix calculated at $\mathbf{s}_a$ and $\mathbf{s}_b$ points are shown as red crosses and blue dots, respectively.

We trained SRNs with 2, 5, 10, 15 hidden units to predict next symbol of the $a^n b^n$ sequence. There were more short strings in training sequence, i.e.: 1800 strings of $n = 1$, 1200 strings of $n = 2$, 800 strings of $n = 3$, 600 strings of $n = 4$ and 200 strings for $n > 5$. Sigmoid activation function $f(x) = 1/(1 + e^{-x})$ was used on hidden and output neurons. Network weights and biases were initialized to small random values from interval $\langle -0.3, 0.3 \rangle$. Input sequence was encoded by one-hot coding, i.e. $\mathbf{i}_a = (1, 0)$ and $\mathbf{i}_b = (0, 1)$.

We used both backpropagation through time (BPTT) and extended Kalman filter training (EKF) algorithms. EKF training was 100 times shorter due to fast convergence in terms of training epochs. Each network was trained 50 times by both BPTT and EKF, then weights were frozen and network dynamics was analyzed. For correct processing of the $a^n b^n$ sequence network is required to build a stack or a counter in its state space. According to previous results [12,18] on small networks with 2 and 3 hidden units there are three basic classes of SRN dynamics when trained on $a^n b^n$: monotonic, oscillating and exotic (Fig. 5).

The first solution contains two attracting points in opposite corners of the state space (Fig. 5a) All eigenvalues of Jacobian matrix in both fixed points $\mathbf{s_a}$ and $\mathbf{s_b}$ lay in the unit circle indicating attracting behaviour. While processing a and b inputs, the network state is moving between these two attractors, which can be interpreted as symbol counting. However, this type of counter has limited capacity, because once network state converges to attractor it stops counting.

The oscillating dynamics, which is known to achieve some kind of generalization, is performed by combination of attractive and saddle points (Fig. 5b). In this case counting is performed by oscillating towards the attractive point ($\mathbf{F}_a$ system) and from the saddle point ($\mathbf{F}_b$ system). Better capacity of this solution is provided by good symmetry of count-up and count-down operations. This is caused by inverse speed of contraction and expansion of $\mathbf{s}_a$ and $\mathbf{s}_b$ fixed points, respectively. In the bottom figure 5b the largest eigenvalue of $\mathbf{s_a}$ is inverse to the largest eigenvalue of $\mathbf{s_b}$.

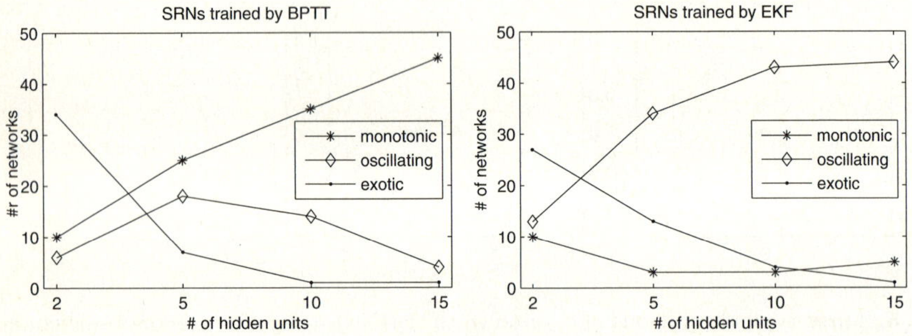

Fig. 6. Proportions of different dynamics which emerged in 50 training of the four SRNs with 2, 5, 10 and 15 hidden units: a) BPTT training; b) EKF training

The third exotic solution is mostly present in small networks. $\mathbf{F}_a$ system contains attractor and $\mathbf{F}_b$ system contains unstable fixed point. Complex eigenvalues outside unit circle of $\mathbf{F}_b$ indicate rotation around unstable fixed point. This can be interpreted as counting of b symbols (Fig. 5c).

The proportions of different dynamics which emerged in 50 training of the four SRNs (2, 5, 10 and 15 hidden units) are shown in Fig. 6. As you can see results for BPPT differs from EKF. In both cases, exotic dynamics is dominating only in small networks with 2 hidden units. This is due the fact that exotic dynamics appears only if network training stuck in local minima. Increasing number of hidden units results in larger ratio of two other dynamics. In the case of BPTT monotonic counter is dominant. As opposite, EKF training more often led to interesting oscillating dynamics. Analysis why this happens is intended for future work, but it may be related to a more robust convergence of the EKF training algorithm [19].

5 Conclusion

Our work describes visualisation and numerical analysis of the dynamics of recurrent neural networks. This approach allows us to perform a visualisation of multidimensional network state space, localize both stable and unstable fixed points, and analyze behaviour of system near these points. This information can be useful for classification of different types of network dynamics emerged during training. Another option is to visualize process of network dynamics evolution and study various bifurcation mechanisms that drive network to desired solution.

As a practical example four networks with 2, 5, 10 and 15 hidden units were trained by BPTT and EKF training algorithms on context-free language $a^n b^n$. Training was repeated for 50 times and proportions of three different types of counters were studied. Results showed bias of BPTT training towards simpler solution. Here counter is performed by combination of two attracting fixed points. On contrary, dynamics composed of both stable and unstable fixed point dominated in the case of EKF training. This difference points to EKF as being a better training algorithm than BPTT for the next-symbol prediction tasks. Why this type of dynamics develops more often in EKF trained RNNs deserves future analysis. Our simulations showed usefulness of proposed techniques for recurrent networks with dozens hidden units. In the future, various architectures and training dataset can be examined by these visualisation techniques.

Acknowledgments

This work was supported by the VEGA grants 1/0848/08 and 1/0822/08.

References

1. Cleeremans, A., Servan-Schreiber, D., McClelland, J.L.: Finite state automata and simple recurrent networks. Neural Computation 1(3), 372–381 (1989)
2. Tiňo, P., Šajda, J.: Laearning and extracting initial mealy automata with a modular neural network model. Neural Computation 7(4), 822–844 (1995)
3. Wiles, J., Elman, J.: Learning to count without a counter: A case study of dynamics and activation landscapes in recurrent networks. In: Proceedings of the Seventeenth Annual Conference of the Cognitive Science Society, pp. 482–487 (1995)
4. Christiansen, M.H., Chater, N.: Toward a connectionist model of recursion in human linguistic performance. Cognitive Science 23, 417–437 (1999)
5. Tiňo, P., Čerňanský, M., Beňušková, L.: Markovian architectural bias of recurrent neural networks. IEEE Transactions on Neural Networks 15(1), 6–15 (2004)
6. Hammer, B., Tiňo, P.: Recurrent neural networks with small weights implement definite memory machines. Neural Computation 15(8), 1897–1926 (2003)
7. Tiňo, P., Hammer, B.: Architectural bias in recurrent neural networks: Fractal analysis. Neural Computation 15(8), 1931–1957 (2003)
8. Tiňo, P., Horne, B.G., Giles, C.L., Collingwood, P.C.: Finite state machines and recurrent neural networks - automata and dynamical systems approaches. Neural Networks and Pattern Recognition, 171–220 (1998)
9. Kolen, J.F.: The origin of clusters in recurrent neural network state space. In: Proceedings from the Sixteenth Annual Conference of the Cognitive Science Society, pp. 508–513. Lawrence Erlbaum Associates, Hillsdale (1994)
10. Kolen, J.F.: Recurrent networks: state machines or iterated function systems? In: Mozer, M.C., Smolensky, P., Touretzky, D.S., Elman, J.L., Weigend, A. (eds.) Proceedings of the 1993 Connectionist Models Summer School, pp. 203–210. Erlbaum Associates, Hillsdale (1994)

11. Čerňanský, M., Makula, M., Beňušková, L.: Organization of the state space of a simple recurrent network before and after training on recursive linguistic structures. Neural Networks 20(2), 236–244 (2007)
12. Rodriguez, P., Wiles, J., Elman, J.L.: A recurrent neural network that learns to count. Connection Science 11, 5–40 (1999)
13. Boden, M., Wiles, J.: Context-free and context-sensitive dynamics in recurrent neural networks. Connection Science 12(3), 197–210 (2000)
14. Werbos, P.J.: Backpropagation through time; what it does and how to do it. Proceedings of the IEEE 78, 1550–1560 (1990)
15. Feldkamp, L., Prokhorov, D., Eagen, C., Yuan, F.: Enhanced multi-stream kalman filter training for recurrent networks. Nonlinear Modeling: Advanced Black-Box Techniques, 29–53 (1998)
16. Kuznetsov, Y.A.: Elements of applied bifurcation theory, 2nd edn. Springer, New York (1998)
17. Rodriguez, P.: Simple recurrent networks learn contex-free and contex-sensitive languages by counting. Neural Computation 13, 2093–2118 (2001)
18. Boden, M., Wiles, J.: On learning context-free and context-sensitive languages. IEEE Transactions on Neural Networks 13(2), 491–493 (2002)
19. Černaňský, M., Beňušková, L.: Simple recurrent network trained by rtrl and extended kalman filter algorithms. Neural Network World 13(2), 223–234 (2003)

Chaotic Search for Traveling Salesman Problems by Using 2-opt and Or-opt Algorithms

Takafumi Matsuura and Tohru Ikeguchi

Graduate School of Science and Engineering, Saitama University,
225 Shimo-Ohkubo, Sakura-ku, Saitama-city 338-8570, Japan
takafumi@nls.ics.saitama-u.ac.jp

Abstract. The traveling salesman problem (TSP) is one of the widely studied combinatorial optimization problems. Because, the TSP belongs to a class of $\mathcal{NP}$-hard, it is almost impossible to obtain an optimal solution in a reasonable time frame. To find the near optimum solutions of TSPs, a method with chaotic neurodynamics has already been proposed. In this paper, we propose a new method to solve TSP introducing chaotic neurodynamics, which uses not only the 2-opt algorithm but also the Or-opt algorithm, which is one of the powerful local searches. Namely, in the proposed method, the 2-opt and the Or-opt algorithms are adaptively driven by the chaotic neurodynamics. Thus, the local minimum problem in these algorithms is resolved by controlling executions of these local searches. As a result, the proposed method shows higher performance than the previous chaotic search methods.

1 Introduction

The traveling salesman problem (TSP) is one of the famous combinatorial optimization problems. Many real world problems can be formulated the TSP, for example, drilling problem, computer wiring, routing problem, VLSI design, job sequencing, and so on. Then, if we can develop an effective algorithm for solving TSPs, we can adapt the algorithm for these different problems.

The TSP is described as follows. First, a set of cities $C(c_1, c_2, \cdots, c_N)$ is given. Let $d(c_i, c_j)$ be distance from city c_i to city c_j. Then, the goal of TSP is to find a permutation σ of the cities that minimizes the following objective function $f(\sigma)$:

$$f(\sigma) = \sum_{i=1}^{N-1} d(c_{\sigma(i)}, c_{\sigma(i+1)}) + d(c_{\sigma(N)}, c_{\sigma(1)}). \quad (1)$$

If $d(c_i, c_j)$ equals to $d(c_j, c_i)$, this problem is a symmetric traveling salesman problem (STSP). For the other case, this problem is asymmetric. In this paper, we only deal with the STSP.

The TSP generally belongs to a class of $\mathcal{NP}$-hard. Therefore, it is believed to be almost impossible to obtain an optimal solution in a reasonable time frame. Thus, it is inevitable to develop an approximate algorithm for finding near optimum solutions or approximate solutions. To find the approximate solutions

V. Kůrková et al. (Eds.): ICANN 2008, Part II, LNCS 5164, pp. 587–596, 2008.

to TSPs, many local search algorithms, for example, the 2-opt, the 3-opt, the Or-opt [1], the k-opt and the Lin-Kernighan algorithm [2], have already been proposed. However, the local search algorithms cannot find an optimal solution, because the local search algorithms are usually trapped by local minima. To avoid getting trapped into the local minimum, it needs to jump from the local minimum to other regions in a searching state space. Several methods using different strategies for avoiding local minima have already been proposed. Stochastic approaches, for example, a simulated annealing [3] and a genetic algorithm, avoid local minima by changing a solution space stochastically. On the other hand, a tabu search method avoids local minima by a deterministic rule [4,5,6,7]. One of the essential idea of the tabu search is that a deterministic approach can be diversified by using a list of prohibited solutions known as a tabu list. A previous state is added to the tabu list and is not allowed to jump the searching states in the tabu list for a while.

In recent years, we have already proposed several effective methods for different type of combinatorial optimization problems such as TSP [8,9,10], quadratic assignment problems [11] and motif extraction problems [12,13,14,15]. In these methods, a chaotic neural network [16] model is used to realize a chaotic search. The chaotic neural network model, which can reproduce chaotic behavior observed in real nerve membrane, was proposed by K. Aihara, T. Takebe and M. Toyoda[16] in 1990. A chaotic neuron in the chaotic neural network is a one-dimensional map which can produce chaotic dynamics. The chaotic neuron realizes a refractory effect, which is one of the important characteristics of a real biological neuron: once a neuron fires, this neuron becomes hard to fire for a begin.

In the chaotic methods for solving TSPs [8,9,10], an execution of the 2-opt algorithm is driven by chaotic neurodynamics. The 2-opt algorithm is the simplest and the most common local search algorithm (Fig. 1(a)). It exchanges two paths for other two paths until no further improvement can be obtained (Fig. 1(a)), However, an improved tour by the 2-opt algorithm is not a global optimum but a local optimum. To escape from such a local optimum, we applied the chaotic neurodynamics to the 2-opt algorithm [8,9,10]. As a result, this method shows good results [8,9,10].

However, it is natural to expect that the chaotic search can explore much better solutions by modifying this algorithm, or introducing another powerful local search [8,9,10]. In general, performance of the k-opt algorithm becomes better, if k increases. However, calculating costs of the k-opt becomes larger as k increases. These facts indicate that it is very important to improve the conventional chaotic search by using a simpler algorithm.

Then, we proposed a new chaotic algorithm by introducing another simple local search, the Or-opt [1]. It attempts to improve the current tour by moving a partial tour of maximum three consecutive path in a different location until no further improvement can be obtained (Fig. 1(b)). The Or-opt algorithm is considered to obtain solutions that are comparable to the 3-opt in terms of quality of solutions and an amount of time is closer to that of the 2-opt algorithm.

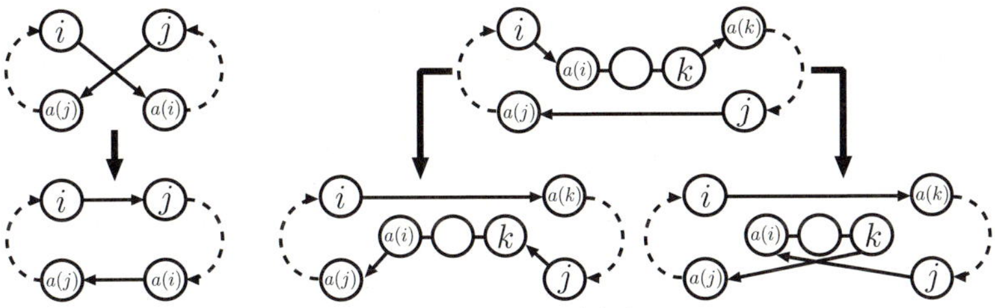

Fig. 1. Executions of the 2-opt algorithm and the Or-opt algorithm. In this example, $a(i)$ is the next city to i. (a) Two paths (i-$a(i)$ and j-$a(j)$) are deleted from the current tour, then new two paths, i-j and $a(i)$-$a(j)$, are connected to obtain a shorter tour. (b) A partial tour $a(i)$-$\cdots$-k is inserted into another path (j-$a(j)$).

In this paper, we propose a new method to solve TSP introducing both the 2-opt and the Or-opt algorithms which are driven by the chaotic dynamics. As a result, the proposed method obtains better solutions for the previous chaotic search methods [8,9,10].

2 The Proposed Method

In the proposed method, two local search algorithms, the 2-opt and the Or-opt [1], are driven by chaotic neurodynamics. To realize chaotic search, we use a chaotic neural network (CNN) composed of chaotic neurons [16]. Each chaotic neuron [16] is assigned to each city. If a neuron fires, the local searches related to the corresponding city are carried out. The firing of the ith chaotic neuron is defined by $x_i(t) = f(y_i(t)) > \frac{1}{2}$, where $f(y) = 1/(1 + \exp(-y/\epsilon))$, and $y_i(t)$ is the internal state of the ith chaotic neuron at time t. If $x_i(t) > \frac{1}{2}$, the ith chaotic neuron fires at the time t, otherwise resting. In the proposed method, $y_i(t)$ is decomposed into two parts, $\zeta_i(t)$ and $\xi_i(t)$. Each component represents a different factor to the dynamics of neurons, a gain effect and a refractory effect, respectively.

The gain effect is expressed as:

$$\xi_i(t+1) = \begin{cases} \max_j \{\beta_2(t)\Delta_{ij}(t) + \zeta_j(t)\} & \text{(2-opt)}, \\ \max_{j,k} \{\beta_{\mathrm{Or}}(t)\Delta_{ijk}(t) + \zeta_j(t)\} & \text{(Or-opt)}, \end{cases} \tag{2}$$

where $\Delta_{ij}(t)$ is a difference between the length of a current tour and that of a new tour when the city j is an adjacent city of the city i after applying the 2-opt algorithm to the city i (Fig.1(a)); $\zeta_j(t)$ is a refractory effect of the city j at time t which is defined by Eq. (3). Namely, the city corresponding to the smaller refractory effect $\zeta_j(t)$ and the larger gain Δ_{ij} is chosen. In Eq. (2), $\Delta_{ijk}(t)$ is a difference of the tour length in case of the Or-opt algorithm (Fig.1(b)). $\beta_2(t)$ and $\beta_{\mathrm{Or}}(t)$ are scaling parameters. Both of them increase in proportion to time

t: $\beta_2(t+1) = \beta_2(t) + \lambda$, and $\beta_{\mathrm{Or}}(t+1) = \beta_{\mathrm{Or}}(t) + \gamma$. If we use these functions, the searching space is gradually limited as the simulated annealing [3]. If the $\beta_2(t)$ (or $\beta_{\mathrm{Or}}(t)$) takes a small value, the proposed method can explore a large solution space. On the other hand, if $\beta_2(t)$ (or $\beta_{\mathrm{Or}}(t)$) takes a large value, the proposed method works like a greedy algorithm.

The second factor is a refractory effect which works to avoid the local minima. The refractory effect has a similar memory effect as the tabu search [4,5]: the same selection of a solution can be avoided by the refractory effect. In the tabu search, to avoid a local minimum, previous states are memorized by adding them to a tabu list and are not allowed for a certain temporal duration called a tabu tenure. In case of the chaotic search, past firings are memorized as previous states to decide strength of the refractory effect. The strength of the refractory effect increases just after corresponding neuron firings and recovers exponentially with time. Thus, while the tabu search perfectly inhibits to select the same solutions for a certain temporal period, the chaotic search might permit to select the same solutions if a corresponding neuron fires due to a larger gain than the refractory effect or an exponential decay of the refractory effect. The refractory effect is expressed as:

$$\zeta_i(t+1) = -\alpha \sum_{d=0}^{t} k_r^d x_i(t-d) + \theta \tag{3}$$

$$= k_r \zeta_i(t) - \alpha x_i(t) + \theta(1 - k_r), \tag{4}$$

where α controls the strength of the refractory effect after the firing ($\alpha > 0$); k_r is a decay parameter of the refractory effect ($0 < k_r < 1$); θ is a threshold value.

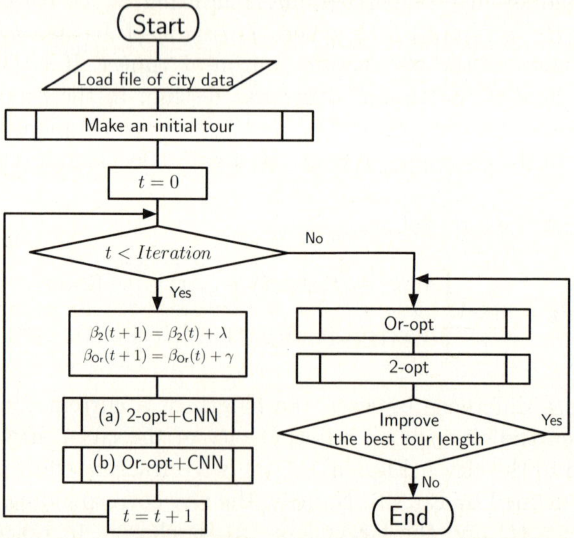

Fig. 2. Flow chart of the proposed method. Two main parts (a) 2-opt + CNN and (b) Or-opt + CNN are described in Fig. 3.

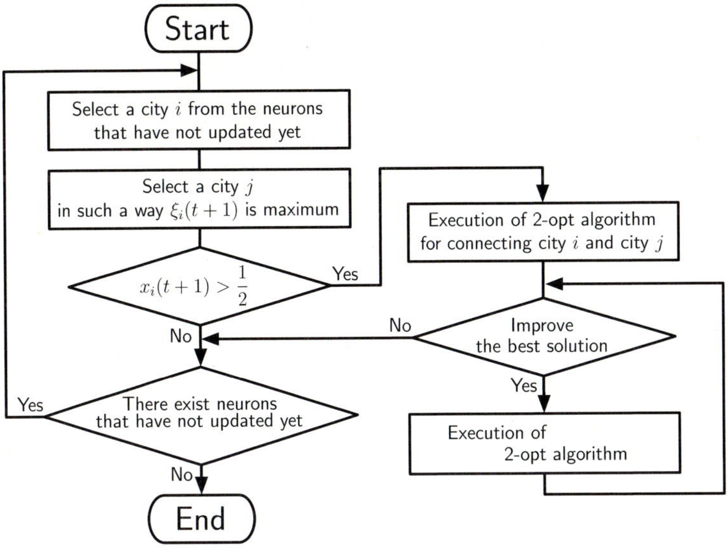

(a) 2-opt + CNN

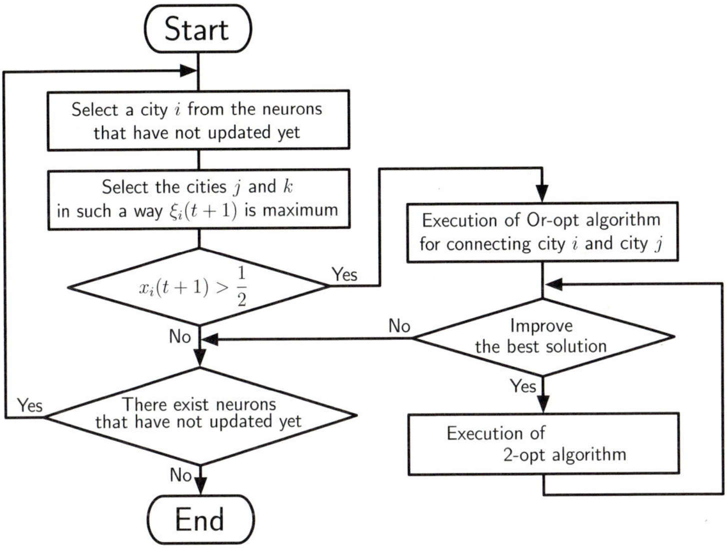

(b) Or-opt + CNN

Fig. 3. Detailed flow charts of (a) the 2-opt algorithm driven by chaotic dynamics and (b) the Or-opt algorithm driven by chaotic dynamics in the proposed method. In the both methods, if the best solution is updated, the 2-opt algorithm is carried out to the corresponding tour to obtain a local optimal solution.

Table 1. Percentages of how nearer neighbors are connected as a next city. For example, in case of kroA100, 43 cities are connected to their nearest neighbor (the 1st near neighbor) city in the optimal solutions. In case of pcb442, 178 cities out of 442 cities. Thus, 178/442 = 40.3%.

	near neighbor (clockwise / counterclockwise)				
Problem	1st	2nd	3rd	3rd∼ 10th	10th∼
kroA100	43.4 / 43.4	25.3 / 25.3	13.1 / 13.1	18.2 / 18.2	1.0 / 1.0
pcb442	40.3 / 40.3	47.7 / 34.4	7.2 / 16.1	4.3 / 8.6	0.5 / 0.6
pr2392	43.4 / 43.4	34.5 / 32.3	12.0 / 12.8	9.6 / 10.9	0.5 / 0.6
pla33810	38.3 / 38.3	56.0 / 35.0	4.2 / 17.4	1.3 / 9.1	0.2 / 0.2

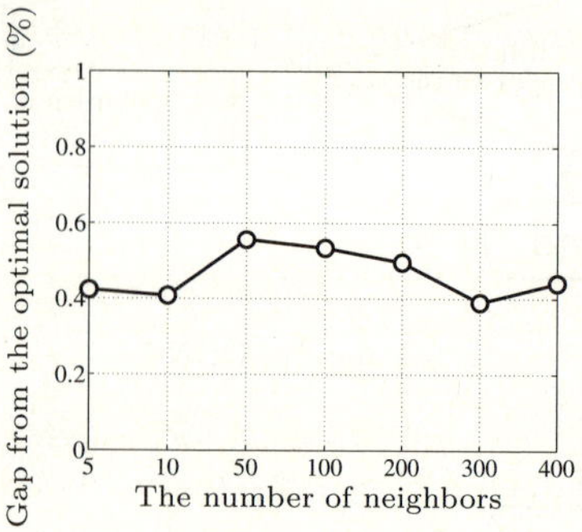

Fig. 4. Relationship between the number of neighbors and solving performance for pcb442

Then, $\zeta_i(t+1)$ expresses a refractory effect with a factor k_r, because the first term of the right hand side of Eq. (3) becomes negative, if the neuron frequently fires in its past history, which then depresses the value of $\zeta_i(t+1)$, and relatively leads the neuron state to a resting state.

Figs.2 and 3 show flow charts of the proposed method. The 2-opt and the Or-opt driven by the chaotic neurodynamics are carried out just one time in one iteration (Fig. 2). First, an initial tour is constructed by the nearest neighbor method (Fig. 2). This method constructs the initial tour by exploring nearest cities. Then, the initial tour is improved by the proposed method to find shorter tours. In the proposed method, if the best solution is updated, the corresponding tour is improved to obtain a local optimal solution by the 2-opt algorithm (Fig. 3).

Even if the simple local search algorithms, or the 2-opt and the Or-opt, are used in the proposed method, if we apply the large scale TSPs, such as 10^4 order problem, it requires heavy calculation. To reduce computational loads, we introduced an idea of a near neighbor. In the case of Euclidean TSPs, it is rare to connect longer paths. It means that connections between two cities can usually be limited to shorter paths. Then, we investigate how close city is visited as a next city in case of optimal solutions with several TSP instances. Depending on a direction on the tour, the next city of a city becomes different one, then, we searched both directions, clockwise and counterclockwise. Table 1 shows the results for several problems in TSPLIB [17]. From the results in Table 1, about 40 percents of all cities are connected to their 1st near neighbor (the nearest neighbor), and almost cities are connected to the near neighbor cities in the top 10. This result indicates that when we search a next city of a city i, it is possible to use the near neighbors of the city i only. In the proposed method, the next city is searched as a city j in Eq. (2). Therefore, when we search the best j in Eq. (2), we use the near neighbors of the city i only.

The near neighbors of a city i is decided by the distance between the city i and other cities. If the number of near neighbors is m, m cites are selected in the ascending order. If there exist many cities with the same distance, all the cities can be selected as near neighbors of the city i. Figure 4 shows relationship between the number of neighbors and solving performance for pcb442 in TSPLIB [17]. From this result, the performance does not depend on the number of neighbors. This result indicate that introduction of near neighbors is an effective idea for proposed method.

3 Results

To evaluate the performance of the proposed method, we solved benchmark problems in TSPLIB [17], and compared the performance between the proposed method and the conventional chaotic search methods: the 2-opt algorithm driven by chaotic dynamics [10], the 2-opt algorithm driven by chaotic dynamics which includes control and annealing of parameters [10,9], an adaptive k-opt algorithm is driven by chaotic dynamics which includes control and annealing of parameters [10,9]. The 2-opt algorithm is the simplest local search (Fig. 1(a)). The adaptive k-opt algorithm is almost the same as the Lin-Kernighan algorithm which is considered to be the best local search for the TSP. In this method, the value of k is not fixed but varied [9]. First, the 2-opt ($k = 2$) algorithm is applied to a current tour to improve the tour. Second, if a positive gain value is obtained by the 2-opt algorithm, the 3-opt ($k = 3$) algorithm is applied to improve the tour by a deterministic rule [9]. While the positive gain value is obtained, the k-opt algorithm is applied by increasing the value of k. Moreover, a double bridge (DB) algorithm is applied to change the solution space, when a better solution could not be obtained within 100 iteration. The double bridge algorithm is a special case of the 4-opt algorithm (Fig. 5).

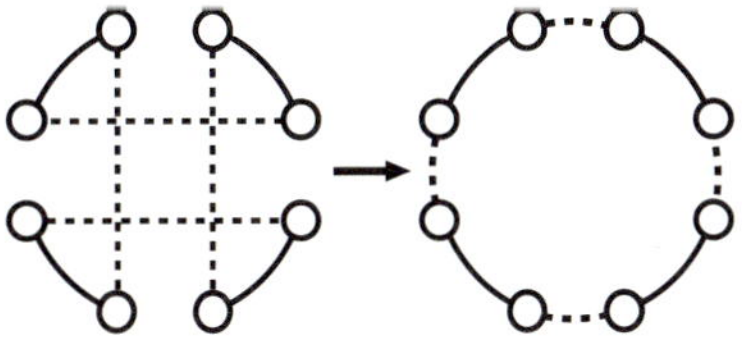

Fig. 5. Example of the double bridge algorithm

When we apply the proposed method to real life problems, one of the important issue to be solved is how to decide parameters. In fact, the proposed method has many parameters. To investigate how the parameters should be decided, we analyzed the proposed method in the following way. For the sake of simplicity, we fixed the values of α, k_r, θ and ϵ for all problems: $\alpha = 0.95$, $k_r = 0.30$, $\theta = 1.0$ and $\epsilon = 0.002$. Then, the scaling parameters in Eq. (2) are changed to several values to investigate the issue. The key idea we introduced in this paper is based on spatial ranges and spatial densities of city-distributions of TSP instance. The value of $\Delta_{ij}(t)$ (or $\Delta_{ijk}(t)$) in Eq.(2) represents a difference between the length of the current tour and that of a new tour. Then, if the spatial range becomes larger, the scaling parameters must be tuned to smaller values, because the value of $\Delta_{ij}(t)$ (or $\Delta_{ijk}(t)$) becomes larger. In addition to the spatial range, the spatial density of the cities also affects the scaling of $\Delta_{ij}(t)$ (or $\Delta_{ijk}(t)$), because even if the spatial range is same, the value of $\Delta_{ij}(t)$ with lower densities is larger than that with higher density. First, we decided a nominal problem instance. In this experiment, we used pcb1173. Next, using pcb1173, we searched good parameter values of $\beta_2(0)$, λ, $\beta_{\mathrm{Or}}(0)$ and γ manually. Then, we calculated the spatial density of city-distribution of pcb1173 (Table 2). Using the following equation:

$$\beta_2(0) = 0.008 \times \sqrt{\frac{\text{(the density for a problem)}}{2.1418}}, \tag{5}$$

Table 2. The values of the scaling parameters, and densities of each problem

Problem	$\beta_2(0)$	λ	$\beta_{\mathrm{Or}}(0)$	γ	Density
pcb442	0.0033988	0.000008497	0.0033988	0.000012745	0.3877
pcb1173	0.0080000	0.000020000	0.0080000	0.000030000	2.1481
pr2392	0.0022834	0.000005709	0.0022834	0.000008563	0.1750
rl5915	0.0028591	0.000007148	0.0028591	0.000010721	0.2744
rl11849	0.0040852	0.000010213	0.0040852	0.000015319	0.5601

Table 3. The results of the 2-opt based chaotic search [10], the results of the 2-opt based chaotic search with double bridge [9], the adaptive k-opt based chaotic search with double bridge [9] and, the proposed method. For each method, 5,000 iterations are used. Results are expressed by percentages of gaps between obtained solutions and the optimal solutions.

	2-opt [10]	2-opt+DB [9]	k-opt+DB [9]	The proposed method			
Problem	Ave.	Ave.	Ave.	Ave.	Min	Max	Std
pcb442	1.034	0.982	0.825	**0.409**	0.002	0.906	0.300
pcb1173	1.692	1.748	1.569	**0.804**	0.436	1.366	0.234
pr2392	1.952	2.000	1.839	**1.153**	0.716	1.614	0.239
rl5915	2.395	2.273	1.742	**1.291**	0.824	1.825	0.304
rl11849	2.223	1.730	1.186	**1.139**	0.690	1.496	0.200

where 0.1 of the right hand side of Eq.(5) is the manually decided value of $\beta_2(0)$ of pcb1173, and the denominator of the right hand side of Eq.(5), 2.1418, is the spatial density of pcb1173. Using Eq.(5), we set the parameters of $\beta_2(0)$, λ, $\beta_{\mathrm{Or}}(0)$ and γ for other problem instances. For example, for pcb442, $\beta_2(0) = 0.008 \times \sqrt{\frac{0.3877}{2.1481}} = 0.0033988$, $\lambda = 0.00002 \times \sqrt{\frac{0.3877}{2.1481}} = 0.000008497$, $\beta_{\mathrm{Or}}(0) = 0.008 \times \sqrt{\frac{0.3877}{2.1481}} = 0.0033988$ and $\gamma = 0.00003 \times \sqrt{\frac{0.3877}{2.1481}} = 0.000012745$. The scaling parameters for other cases are summarized in Table 2

In this experiment, we have applied only 5,000 iteration for each calculation, and the number of near neighbors m is set to 10. The results in Table 3 are shown by percentage of gaps between obtained best solutions and the optimal solutions. From Table 3, the proposed method obtains shorter tours than the other methods.

4 Conclusions

In this paper, we proposed a new method for solving the TSP using two local searches, the 2-opt algorithm and the Or-opt algorithm, driven by chaotic neurodynamics. From the computational results, although the proposed method is simple, it obtains good solutions comparing with the previous chaotic search methods it is important to clarify why the proposed method can improve performance with statistical methods [13]. It is also desirable to solve much larger problems, such as 10^5 order problems. To improve the proposed method, we also consider how to control different types of two local search algorithms. Moreover, we will apply more effective parameter adjustment methods to the proposed method [10]. The authors thank K. Aihara, Y. Horio, M. Adachi and M. Hasegawa for their valuable comments and discussions. The research of T.M. is partially supported by a Grant-in-Aid from the JSPS Fellows. The research of T.I. is partially supported by Grant-in-Aid for Scientific Research (B) (No.20300085) from the JSPS and research grant from the Mazda Foundation.

References

1. Or, I.: Traveling salesman-type combinatorial problems and their relation to the logistics of regional blood banking. Ph.D thesis. Department of Industrial Engineering and Management Science, Northwestern University, Evanston, Illinois (1967)
2. Lin, S., Kernighan, B.: An effective heuristic algorithm for the traveling salesman problem. Operations Research 21, 498–516 (1973)
3. Kirkpatrick, S., et al.: Optimization by simulated annealing. Science 220, 671–680 (1983)
4. Glover, F.: Tabu search i. ORSA Journal on Computing 1, 190–206 (1989)
5. Glover, F.: Tabu search ii. ORSA Journal on Computing 2, 4–32 (1989)
6. Hasegawa, M., et al.: Exponential and chaotic neurodynamical tabu searches for quadratic assignment problems. Control and Cybernetics 29(3), 773–788 (2000)
7. Matsuura, T., et al.: A tabu search for extracting motifs from dna sequences. In: Proc. of NCSP 2004, pp. 347–350 (2004)

8. Hasegawa, M., et al.: Combination of chaotic neurodynamics with the 2-opt algorithm to solve traveling salesman problems. Physical Review Letters 79(12), 2344–2347 (1997)
9. Hasegawa, M., et al.: On the effects of the k-opt method with chaotic neurodynamics. Technical Report of IEICE 101(229), 25–32 (2001)
10. Hasegawa, M., et al.: Solving large scale traveling salesman problems by chaotic neurodynamics. Neural Networks 15(2), 271–283 (2002)
11. Hasegawa, M., et al.: A novel chaotic search for quadratic assignment problems. European Journal of Operational Research 39(3), 543–556 (2002)
12. Matsuura, T., et al.: A tabu search and chaotic search for extracting motifs from dna sequences. In: Proc. of MIC 2005, pp. 677–682 (2005)
13. Matsuura, T., Ikeguchi, T.: Refractory effects of chaotic neurodynamics for finding motifs from dna sequences. In: Corchado, E., Yin, H., Botti, V., Fyfe, C. (eds.) IDEAL 2006. LNCS, vol. 4224, pp. 1103–1110. Springer, Heidelberg (2006)
14. Matsuura, T., Ikeguchi, T.: Analysis on memory effect of chaotic dynamics for combinatorial optimization problem. In: Proc. of MIC 2007 (2007)
15. Matsuura, T., Ikeguchi, T.: Statistical analysis of output spike time-series from chaotic motif sampler. In: Proc. of the 14th International Conference on Neural Information Processing (2007)
16. Aihara, K., et al.: Chaotic neural networks. Physics Letters A 144, 333–340 (1990)
17. TSPLIB, `http://elib.zib.de/pub/mp-testdata/tsp/tsplib/tsplib.html`

Comparison of Neural Networks Incorporating Partial Monotonicity by Structure

Alexey Minin[1] and Bernhard Lang[2]

[1] Saint-Petersburg State University
alexey.minin@siemens.com
[2] OOO Siemens, Fault Analysis and Prevention group, 191186, Russia
Saint-Petersburg, Volynskiy per. dom 3A liter A
bernhard.lang@siemens.com

Abstract. Neural networks applied in control loops and safety-critical domains have to meet more requirements than just the overall best function approximation. On the one hand, a small approximation error is required, on the other hand, the smoothness and the monotonicity of selected input-output relations have to be guaranteed. Otherwise the stability of most of the control laws is lost. Three approaches for partially monotonic models are compared in this article, namely Bounded Derivative Network (BDN) [1], Monotonic Multi-Layer Perceptron Network (MONMLP) [2], and Constrained Linear Regression (CLR). Authors investigated the advantages and disadvantages of these approaches related to approximation performance, training of the model and convergence.

1 Introduction

Incorporating prior knowledge in order to improve the modelling quality of neural networks is a popular approach. Neural networks are universal approximators but practical applications often show problems due to insufficient training data. Prior knowledge about monotonicity, smoothness, upper or lower bounds of inputs and/or targets can increase robustness of neural network's training procedures. We address in this article monotonicity. Simple rules like "if A increases then B will increase as well" can be transferred to a monotonicity constraint $dB/dA > 0$. Such behavior is often known so that we can extend the classical training approach which is minimization of cost functions consisting of target values and corresponding output values of a neural network. Early research in this area [5] recommends the extension of cost functions by penalty terms related to the derivatives. Monotonicity is evaluated at all training data. The two objectives minimization of target error and minimization of derivative error can compete with each other. To sum up, this approach improves the training procedure but cannot guarantee monotonic behavior for the complete input space. Other approaches provide topologies of function networks with guaranteed monotonicity [1,2]. Such monotonicity constraints do no longer depend on input values x but on corresponding weights only. In this paper we focus on

V. Kůrková et al. (Eds.): ICANN 2008, Part II, LNCS 5164, pp. 597–606, 2008.

such function networks, namely with partial monotonicity: monotonic behavior related to selected input-output relation but not necessarily for all inputs.

2 Monotonicity Issues

In order to induce monotonicity for some inputs of the input space one should mine monotonic behavior from the data. The acquisition of the monotonic behavior can be divided into two parts. First one is context acquisition which means that one should extract monotonicity from deep understanding of the process he or she is dealing with. Second possibility is the extraction of prior knowledge from the data using scatter plots. Using scatter plots one can analyze potential monotonic behavior in input-output relations. An example is given by figure 1.

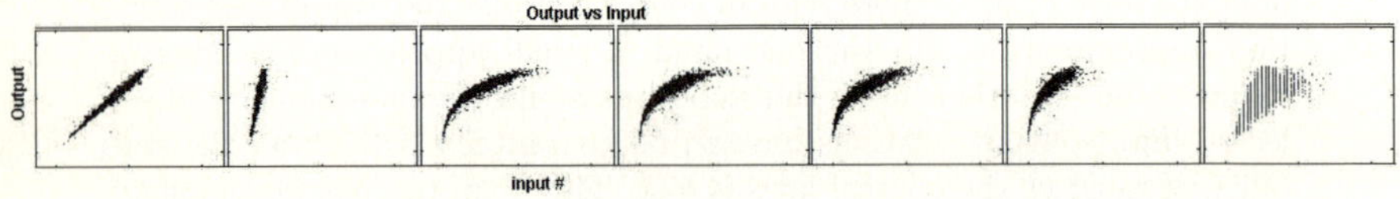

Fig. 1. Scatter plot of the data. Output vector against input vectors

Analyzing figure 1 one can assume monotonic (increasing) behavior between e.g. 3rd input and output. This information can be used for the introduction of an inequality constraint described in eq.(1):

$$\frac{dY}{dX_k} \geq 0, \tag{1}$$

where k is an input number, X is an input vector and Y is an output vector. Authors refer to [2,4,5,7] for more details about embedding monotonic behavior.

3 Constrained Linear Regression (CLR)

In statistics, linear regression is a regression method that models the relationship between a dependent variable Y, independent variables $X_{1..p}$ and a random term ε. The model can be written as:

$$Y = \beta_0 + \beta_1 X_1 + \ldots + \beta_p X_p + \varepsilon, \tag{2}$$

where β_0 is the intercept (the so-called bias parameter), β_i are the respective parameters of independent variables, and p is the number of parameters to be estimated in the linear regression. In order to induce prior knowledge about the data into training procedure one can use monotonic behavior in input-output relation. Constructing scatter plots (see "Monotonicity issues") one can extract monotonicity in the data. Using eq.(1) one can obtain a system of equations which consists of eq.(2) and several inequality constraints which describes Constrained Linear Regression, namely:

$$\begin{cases} Y = \beta_0 + \beta_1 X_1 + \ldots + \beta_p X_p + \varepsilon \\ \frac{dY}{dX_k} \geq 0 \end{cases} \tag{3}$$

4 Monotonic Multi-Layer Perceptron Network (MONMLP)

Multi-layer perceptron networks (MLP) consist of multiple layers of computational units, usually interconnected in a feed-forward way. Each neuron in one layer has directed connections to the neurons of the subsequent layer. In many applications the units of these networks apply a hyperbolic tangent function as an activation function (see fig.2).

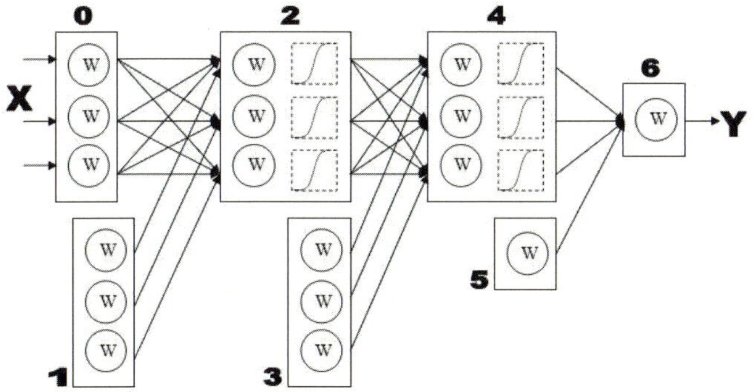

Fig. 2. Feed forward network, namely MLP is shown in the figure. Here X are inputs, W are weights between layers, Y are outputs. Connections between nodes are presented with the lines. One can see activation functions within dotted rectangulars. Bold numbers display the layer number.

Authors used nonlinear hyperbolic tangent function since it is differentiable and has two saturation limits. An MLP with four layers is described by eq.(4).

$$Y = \sum_{k=1}^{nh4} w_k^{4,6} \tanh \underbrace{\left(\sum_{l=1}^{nh2} w_{k,l}^{2,4} \tanh \overbrace{\left(\sum_{i=1}^{ni} w_i^{0,2} X_i + w_i^{0,1} \right)}^{R} + w_l^{3,4} \right)}_{Q} + w^{5,6} \tag{4}$$

where upper index of the w shows layer number and lower index shows node number inside the layer, i stands for input number, l is the node number in layer 2 (see fig. 2), k is the node number in layer 4. In [2] it was shown that two hidden layers are required for the universal approximation capability under monotonicity constraints. Let's say, that one knows that the input - output relation has monotonic increasing behavior. Then, related constraints for eq.(1) can be derived from eq.(5)):

$$\frac{dY}{dX_k} = \sum_{k=1}^{nh2} w_k^{4,6} \underbrace{\left(1 - \tanh^2(Q)\right)}_{>0} \sum_{l=1}^{nh4} w_{k,l}^{2,4} \underbrace{\left(1 - \tanh^2(R)\right)}_{>0} w_{i,m}^{0,2} \geq 0 \quad (5)$$

Simple constraints for signs of the weights $w_{i,m}^{0,2}$ related to input i can be derived over the equation $w_{i,m}^{0,2} w_{k,l}^{2,4} w_k^{4,6} \geq 0$ (see in detail in [2]).

5 Bounded Derivative Network (BDN)

Following the work [1] one should integrate a standard three-layer MLP (expression Q of eq.(4)) in order to obtain the architecture of the BDN – Bounded Derivative Network [1] described by eq.(6).

$$\begin{aligned} Y = {} & w_{1,1}^{6,1} + \sum_{i=1}^{ni} w_{1,i}^{6,2} w_{i,i}^{2,0} X_i + \sum_{j=1}^{nh} w_{1,j}^{6,5} w_{j,j}^{5,4} \times \\ & \times \left[\log \left(\cosh \left(w_{j,1}^{3,1} + \sum_{i=1}^{ni} w_{j,i}^{3,2} w_{i,i}^{2,0} X_i \right) \right) + \right. \\ & \left. + w_{j,j}^{5,3} \left(w_{j,1}^{3,1} + \sum_{i=1}^{ni} w_{j,i}^{3,2} w_{i,i}^{2,0} X_i \right) \right] \end{aligned} \quad (6)$$

Following the idea one should calculate the constraint, which can take into account monotonic behavior. To do so, one should use eq.(7):

$$\begin{aligned} \frac{\partial Y}{\partial X_k} = w_{k,k}^{2,0} \times \\ \times \left(w_{1,k}^{6,2} + \sum_{j=1}^{nh} w_{1,j}^{6,5} w_{j,k}^{3,2} \left(w_{j,j}^{5,3} + w_{j,j}^{5,4} \tanh \left[w_{j,1}^{3,1} + \sum_{i=1}^{ni} w_{j,i}^{3,2} w_{i,i}^{2,0} X_i \right] \right) \right) \end{aligned} \quad (7)$$

Then it is possible to see, that eq.(7) is bounded by its nature due to the limitation of the hyperbolic tangent function. Eq.(7) is very similar to the expression Q of eq.(4) - general equation for standard three-layer perceptron network. Taking into account $\lim_{x \to \pm\infty} \tanh(x) = \pm 1$ leads to eq.(8):

$$\begin{aligned} \frac{\partial Y}{\partial X_{k\,bound}} = w_{k,k}^{2,0} \times \\ \times \left(\sum_{j=1}^{nh} w_{1,j}^{6,5} w_{j,k}^{3,2} w_{j,j}^{5,3} \pm \sum_{j=1}^{nh} \mid w_{1,j}^{6,5} w_{j,k}^{3,2} w_{j,j}^{5,4} \mid + w_{1,k}^{6,2} \right) \end{aligned} \quad (8)$$

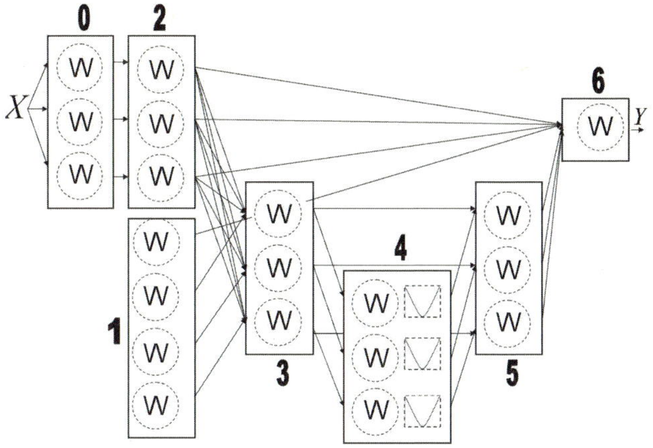

Fig. 3. Bounded Derivative Network. Here X are inputs, W are weights between layers, Y are outputs. Connections between nodes are presented with the lines. Activation functions are depicted within dotted rectangulars. Bold numbers display the layer numbers.

Depending on the sign of the $w^{2,0}_{k,k}$ derivative $\frac{\partial Y}{\partial X_k}_{bound}$ can be maximum (further max) or minimum (further min) bound. In the present paper assume $w^{2,0}_{k,k} > 0$ Then the derivative lies in some bounded region: $\min < \frac{\partial Y}{\partial X_k} < \max \Rightarrow \text{if} \frac{\partial Y}{\partial X_k} > 0$. Then taking $\min > 0$ it is sufficient that $\frac{\partial Y}{\partial X_k} > 0$. To constrain the model derivative minimal value for the derivative should be greater then zero. On the figure 4 one can find visualization of the activation function for the BDN network and the visualization for the activation function for the MONMLP network. From the figure 4 it is clear, that MONMLP has constrained activation function as well as constrained derivative. Activation function is limited due to its nature and allow to smooth outliers. In contrast BDN has constrained derivative and unconstrained activation function. In case BDN one can see that extrapolation (activation function) is linear in its major part and has very small nonlinear domain [1].

6 Comparison of MONMLP, BDN and CLR for the Benchmark Data

In order to show how powerful constrained networks are, authors decided to compare performance of each constrained network on benchmark datasets. To estimate the quality of the result one should use two measures, namely Root Mean Squared Error (RMS) and R^2 which is a statistic that provides information about the goodness of fit of a model. In regression, the R^2 coefficient of determination is a statistical measure of how well the regression line approximates the real data points. An R^2 of 1.0 indicates that the regression line perfectly fits the data (see eq.(9)).

$$R^2 = 1 - \frac{(x - \widetilde{x})^2}{(x - x)^2} \qquad (9)$$

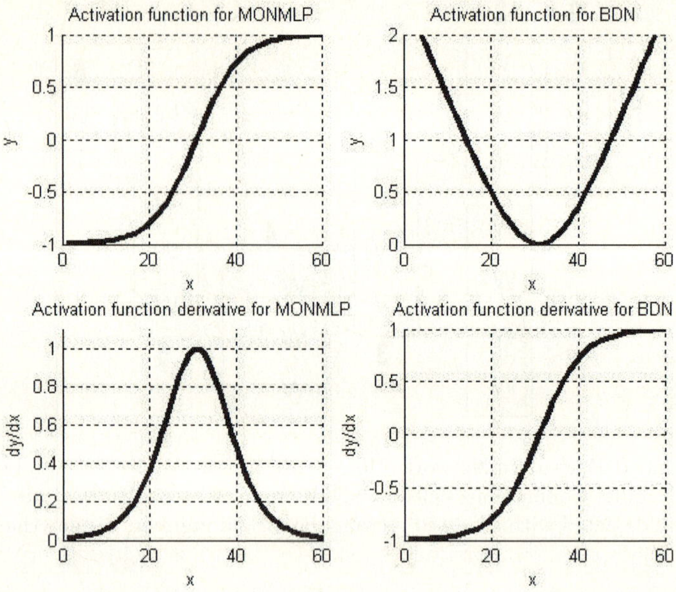

Fig. 4. Visualization of activation functions and their derivatives

where x is a real value,$\overline{x}$ - is an average value over all real values, $\widetilde{x}$ - is a value obtained in experiment. The results of the research work are presented below. All data except last two rows were taken from De Moor B.L.R. (ed.), DaISy: Database for the Identification of Systems, Department of Electrical Engineering, ESAT/SISTA, K. U. Leuven, Belgium.
URL http://www.esat.kuleuven.ac.be/sista/daisy/.

6.1 Description of the Data (Inputs, Outputs, Monotonicity)

Abalone dataset. Description: predicting the age of abalone from physical measurements. Monotonicity for this data set: the larger all inputs the larger the age is.

Ethane-Ethylene dataset. Description: Data of a simulation (not real) related to the identification of an ethane-ethylene distillation column. Monotonicity for this data set: the larger ratio between the distillate and the feed flow, the lower top ethane composition.

CD player Arm dataset. Description: Data from the mechanical construction of a CD player arm. Monotonicity: the larger the force the lower the tracking accuracy.

Boston housing problem. Description: Concerns housing values in suburbs of Boston. Monotonicity: The higher crime level, the lower median value, the higher nitric oxide concentration the lower median value, The higher average number of rooms per dwelling the higher the median value, the higher distances to center the lower median value, the higher tax rate the lower the median value, the higher lower status of the population the lower median value.

Steel dataset. Description: this is the real world dataset. Each strip of steel in a hot rolling mill must satisfy customer quality requirements. Monotonicity: the higher pressure of carbon the higher the quality, the higher the proportion of Mn the higher the quality, and the higher the temperature the lower the quality.

Dow Jones dataset. Description: this is real world dataset for DJIA index. This dataset is available at http://finance.yahoo.com. Monotonicity: there was no monotonicity found.

6.2 Training Procedure and Architecture Parameters

For the training of the MONMLP and BDN Sequential Quadratic Programming (further SQP) technique was used due to the constrained optimization procedure required. The implementation of the SQP was done by Optimization toolbox of MATLAB. Due to the fact, that all datasets (except DJIA) were not time series, for the cross validation randomly chosen points were used. Each network (CLR, MONMLP and BDN) was used 10 times for the same data set to find the range for the RMS and R2. The architecture was chosen to be like following: for the MONMLP one should use 8 to 4 units in first hidden layer and 4 to 3 units in second layer, for the BDN one should use 1 to 4 units in hidden layer. For each data set number of epochs was chosen to be 400. Moreover early stopping for the training procedure was used. Early stopping means that in case the error on the test set starts ascending (after it was descending) and at the same time error on the training set continue descending, one should stop training procedure. In the table below (see table 1) one can find results for each dataset.

7 Convergence Issues

Figure 5 depicts the convergence of training procedure based on the "Functional value" label. This is the functional constructed by MATLAB while solving the optimization problem. All figures were provided for the "Abalone" benchmark dataset. Both architectures (MONMLP and BDN) were used and compared. One should not pay attention to the initial functional values, since different number of hidden neurons was used in order to show possible outcomes.

As one can see from the figure 5 the convergence for MONMLP is not guaranteed in case of infeasible start point. The opposite story is with the BDN (see figure 5). For the BDN one can watch absence of the convergence for some regions (epochs $20-150$ at the figure 5) but at the end solution converges to some local optimum. The final error for the MONMLP as the final error for the BDN does not depend very much on the weight initialization (in case of feasible start point). Nevertheless, one should note that according to experiments presented

Table 1. Comparison of the results for each dataset

Data set	CLR_{RMS}	MLP_{RMS}	BDN_{RMS}	CLR_{R^2}	MLP_{R^2}	BDN_{R^2}
ABALONE	2.45± 0.05	2.3± 0.05	2.0 ± 0.7	0.4±0.1	0.53±0.8	0.37±0.1
ETHANE	1.26±0.2	1.40±0.1	1.30±0.1	-0.55±0.4	-1.0±0.1	-0.6±0.1
CD ARM	0.1±0.00	0.09±0.00	0.09±0.00	0.87±0.01	0.90±0.00	0.90±0.00
BOSTON	9.1±0.3	3.80±0.2	3.30±0.4	-1.9±0.2	0.45±0.15	0.4±0.1
DJIA	56.80±0.1	31.0±1.0	31.5±0.5	0.37±0.02	0.45±0.05	0.45±0.1
STEEL	35.45±0.65	6.75±0.5	6.80±0.5	-0.07±0.03	0.96±0.01	0.96±0.01
Av.Var	±0.22	±0.32	±0.38	±0.12	±0.18	±0.06

in the table 1 even in case of infeasible start point solution provided by BDN converge to some local optimum. For MONMLP the situation is worse, since in case of optimization start from infeasible start point, one can see absence of convergence. In order to start from a feasible start point, one can compute feasible start point for MONMLP and BDN. In case of MONMLP feasible start point can be computed like following: if $w_{i,m}^{0,2} w_{k,l}^{2,4} w_{k}^{4,6} \geq 0$ then any positive combination of weights will provide feasible start point and more over monotonicity can be fixed by the signs of weights in the initial layer ($w_{i,m}^{0,2}$) if other weights are fixed as positive ($w_{k,l}^{2,4} w_{k}^{4,6} \geq 0$). Therefore, fixing the sign of the input weights $w_{i,m}^{0,2} > 0$ or $w_{i,m}^{0,2} < 0$ leads to guaranteed monotonicity in I/O relation. In case of BDN the situation is a bit more difficult (see eq.(10)).

$$\begin{cases} \sum_{j=1}^{nh} w_{1,j}^{6,5} w_{j,k}^{3,2} \left(w_{j,j}^{5,3} + w_{j,j}^{5,4} \right) + w_{1,k}^{6,2} < 0, & if \frac{\partial Y}{\partial X_k} < 0 \\ \sum_{j=1}^{nh} w_{1,j}^{6,5} w_{j,k}^{3,2} \left(w_{j,j}^{5,3} - w_{j,j}^{5,4} \right) + w_{1,k}^{6,2} > 0, & if \frac{\partial Y}{\partial X_k} > 0 \\ w_{k,k}^{0,2} > 0 \\ w_{1,j}^{6,5} w_{j,k}^{3,2} w_{j,j}^{5,4} > 0, \forall j = 1 \ldots nh \end{cases} \tag{10}$$

In order to fulfill eq.(10) in case inputs have ascending behavior ($\frac{\partial Y}{\partial X_k} > 0$) one should initialize weights according to eq.(11):

$$\begin{cases} w_{j,j}^{5,3} = -\frac{w_{1,k}^{6,2}}{w_{1,j}^{6,5} w_{j,k}^{3,2}} \\ w_{k,k}^{0,2} > 0 & \forall j = 1 \ldots nh \\ w_{1,j}^{6,5} < 0, w_{j,j}^{5,4} < 0, w_{j,k}^{3,2} > 0 \end{cases} \tag{11}$$

eq.(11) clearly shows how to start from a feasible start point for BDN.

Authors used Pentium 4 DUO processor, 2.2 GHz, 2 GB RAM to train neural networks. Training time varies for different data sets. Nevertheless, a basic es-

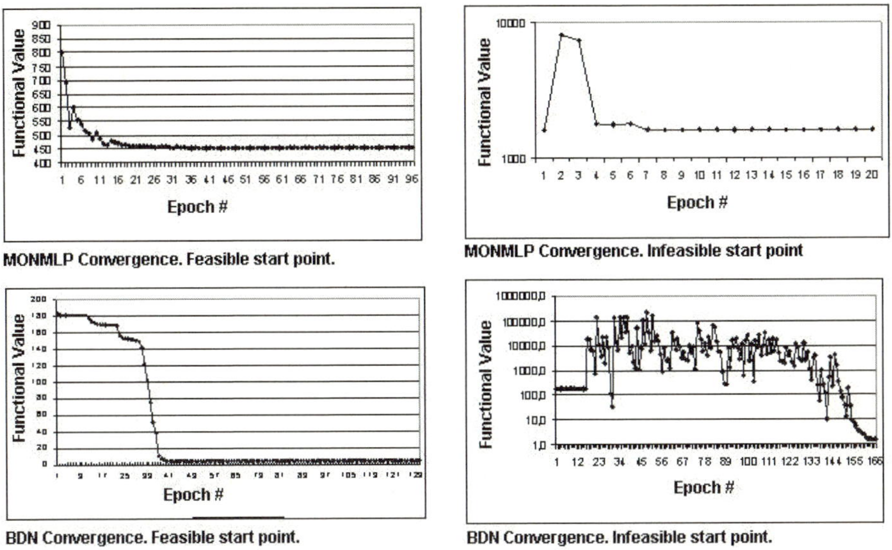

Fig. 5. Convergence of BDN and MONMLP for different start points

timation for all data sets is presented: one should spend 10-60 seconds to train CLR, 3-10 minutes for the MONMLP and 10-20 min for the BDN.

8 Conclusions

Present paper clearly shows that it is difficult to say what architecture to use. In any case, one has to mine for monotonicity in the input-output relation. But on the other hand, one can see that constrained neural networks are better in modelling quality than unconstrained ones, since such networks obey monotonicity rules induced during training into the structure. Compared architectures are more or less equal in the results. Obviously monotonicity constraints make the optimization problem more difficult for any architecture. SQP method should be used instead of other methods. Moreover, optimization to a reasonable minimum takes much more time (see Convergence issues).

Optimization time in case of the BDN should be considered, since the mathematical expressions of the BDN and its constraints are much more difficult than the corresponding expressions of the MONMLP. This makes usage of the MONMLP easier and faster in computational time. Convergence for the BDN is more robust with respect to the starting point but setting feasible starting points is easier for the MONMLP. The solution for BDN always converges to some optimum (fig. 5). In case of infeasible starting point for MONMLP training converges to an insufficient local optimum (fig.5).

Obtained results show the possibility to use monotonic neural networks in different applications. Note that wrong monotonicity rules can lead to very poor approximation results. Nevertheless, if monotonic rules were selected in a proper way and optimization procedure finished successfully according to some criteria, monotonic neural networks not only have the same approximation capability as neural networks without monotonicity constraints, but also guarantee monotonicity in input-output relation by their structure.

References

1. Turner, P., Guiver, J., Brian, L.: Introducing The State Space Bounded Derivative Network For Commercial Transition Control. In: Proceedings of the American Control Conference, Denver, Colorado, June 4-6 (2003)
2. Lang, B.: Monotonic Multi-layer Perceptron Networks as Universal Approximators. In: Duch, W., Kacprzyk, J., Oja, E., Zadrożny, S. (eds.) ICANN 2005. LNCS, vol. 3697, pp. 31–37. Springer, Heidelberg (2005)
3. Zhang, H., Zhang, Z.: Feed forward networks with monotone constraints. In: IEEE International Joint Conference on Neural Networks IJCNN 1999, Washington, DC, USA, vol. 3, pp. 1820–1823 (1999)
4. Sill, J.: Monotonic Networks, Advances in Neural Information Processing Systems, Cambridge, MA, vol. 10, pp. 661–667 (1998)
5. Sill, J., Abu-Mostafa, Y.S.: Monotonicity hints, Advances in Neural Information Processing Systems, Cambridge, MA, vol. 9, pp. 634–640 (1997)
6. Kay, H., Ungar, L.H.: Estimating monotonic functions and their bounds. AIChE J. 46, 2426
7. Tarca, L.A., Grandjean, B.P.A., Larachi, F.: Embedding monotonicity and concavity information in the training of multiphase flow neural network correlations by means of genetic algorithms. Computers and Chemical Engineering 28(9), 1701–1713 (2004)

Effect of the Background Activity on the Reconstruction of Spike Train by Spike Pattern Detection

Yoshiyuki Asai[1,2] and Alessandro E.P. Villa[2,3,4]

[1] The Center for Advanced Medical Engineering and Informatics. Osaka University, Toyonaka, Japan
asai@bpe.es.osaka-u.ac.jp
[2] NeuroHeuristic Research Group, Information Science Institute University of Lausanne, Switzerland
{oyasai,avilla}@neuroheuristic.org
http://www.neuroheuristic.org/
[3] Université Joseph Fourier, NeuroHeuristic Research Group, Institut des Neurosciences, Grenoble, F-38043, France
[4] INSERM, U836, Grenoble, F-38043, France
alessandro.villa@ujf-grenoble.fr

Abstract. Deterministic nonlinearity has been observed in experimental electrophysiological recordings performed in several areas of the brain. However, little is known about the ability to transmit a complex temporally organized activity through different types of spiking neurons. This study investigates the response of a spiking neuron model representing five archetypical types to input spike trains including deterministic information generated by a chaotic attractor. The comparison between input and output spike trains is carried out by the pattern grouping algorithm (PGA) as a function of the intensity of the background activity for each neuronal type. The results show that the thalamo-cortical, regular spiking and intrinsically busting model neurons can be good candidate in transmitting temporal information with different characteristics in a spatially organized neural network.

1 Introduction

The time series of the exact timing of the occurrences of neuronal action potentials is referred to as "spike train" and can be searched for detecting embedded temporal structures. Since a neuron can be considered as a nonlinear filter of spike trains, a neuronal network may be considered as a highly complex nonlinear dynamical system able to exhibit deterministic chaotic behavior, as suggested by experimental observations [1,2,3]. Previous studies [4,5] showed that deterministic nonlinear dynamics in noisy time series could be detected by applying algorithms aimed at finding preferred firing sequences with millisecond order time precision from simultaneously recorded neural activities. A neural network is also characterized by the presence of background activity of unspecified or

V. Kůrková et al. (Eds.): ICANN 2008, Part II, LNCS 5164, pp. 607–616, 2008.

unknown origin, usually referred to as "spontaneous activity", that is generally represented by stochastic inputs to each cell of the network. Then, a neuron belonging to a cell assembly within the network is expected to receive inputs potentially characterized by an embedded temporal structure as well as inputs corresponding to the stochastic background activity.

The purpose of the present work is to investigate how different types of archetypical neurons react to an input spike train generated by a deterministic dynamical system in presence of a stochastic background activity with several intensities. The assessment of the performance of the transmission of the temporal structure in the spike train is carried out by comparing various indices derived by applying the pattern grouping algorithm [6,7] to the input and output spike trains in presence and absence of background activity.

2 Method

2.1 Spiking Neuron Model

The simple spiking neuron model [8] used in this work is described as follows;

$$\frac{dv}{dt} = 0.04v^2 + 5v + 140 - u + I_{bg} + I_{ext} \tag{1}$$
$$\frac{du}{dt} = a(bv - u) ,$$

with the auxiliary after-spike resetting, $v \leftarrow c$ and $u \leftarrow u + d$ when $v \geq +30\ mV$. v represents the membrane potential $[mV]$, and u is a membrane recovery variable. The time unit is millisecond. a and b control the time scale of the recovery variable and its sensitivity to the subthreshold fluctuation of the membrane potential. Five neuron types were considered in this study according to the literature [8,9]: (*i*) a neo-cortical neuron of regular spiking (RS) type with parameters $a = 0.02$, $b = 0.2$, $c = -65$, $d = 8$; (*ii*) a thalamo-cortical (TC) neuron with parameters $a = 0.02$, $b = 0.25$, $c = -65$, $d = 2$; (*iii*) a neuron type characterized by its sustained subthreshold oscillation of the membrane potential called resonator (RZ) with $a = 0.1$, $b = 0.25$, $c = -65$, $d = 2$; (*iv*) an intrinsically bursting neuron (IB) that fires burst of spikes then generates single spikes with parameters $a = 0.02$, $b = 0.2$, $c = -55$, $d = 4$; (*v*) an chattering neuron (CH) making repetitive bursty spikes with $a = 0.02$, $b = 0.2$, $c = -50$, $d = 2$;

Let us denote I^{ext} the input synaptic current, defined as

$$I^{ext} = -A^{ext} g^{syn} (V - V^{syn}), \tag{2}$$

where V^{syn} is the synaptic reversal potential, set to 0 in this study; A^{ext} is an intensity of the synaptic transmission of the spike received as an external input. g^{syn} is the post synaptic conductance represented by

$$g^{syn} = C_0 \frac{e^{-\tilde{t}/\tau_1} - e^{-\tilde{t}/\tau_2}}{\tau_1 - \tau_2}, \tag{3}$$

where $\tilde{t}$ is interval between the last pre-synaptic neuron's discharge and current time; τ_1 and τ_2 are time constants given by 0.2 and 2 ms, respectively and C_0 is a coefficient used to normalize the maximum amplitude of g^{syn} to 1.

Let us assume that each neuron receives multiple inputs called *background activity* whose current is denoted I^{bg}. We assume that I^{bg} is defined by the same type of Eq. 2 with different intensity of the transmission denoted by A^{bg}. A^{ext} and A^{bg} were tuned for each cell type such that a single pulse of the pre-synaptic input can evoke a spike in the post-synaptic cell, if the membrane potential is at rest, but a single pulse of the background activity cannot. The values of (A^{ext}, A^{bg}) for TC were set to $(0.033, 0.032)$, for RS $(0.15, 0.14)$, for RZ $(0.018, 0.017)$, for IB $(0.15, 0.14)$, and for CH $(0.15, 0.14)$. We also used $A^{bg} = 0$ as a control case, meaning that there is no background activity.

2.2 Input Spike Train

We considered one deterministic and two stochastic processes as input spike trains. Ten thousand points ($N = 10,000$) were generated in each series. We used the Zaslavskii map [10] to generate the deterministic spike train, which is:

$$\begin{cases} x_{n+1} = x_n + v(1 + \mu y_n) + \varepsilon v \mu \cos x_n & (mod.\ 2\pi) \\ y_{n+1} = e^{-\gamma}(y_n + \varepsilon \cos x_n) \ , \end{cases} \tag{4}$$

where x, $y \in \mathbf{R}$, the parameters are real numbers with $\mu = \frac{1-e^{-\gamma}}{\gamma}$, $v = \frac{4}{3} \cdot 100$ and initial conditions set to $x_0 = y_0 = 0.3$. The system exhibits a chaotic behavior with this parameter set. Time series $\{x_n\}$ are generated by iterative calculation. A new time series $\{w_n\}$ corresponding to the sequence of the inter-spike-intervals is derived by $w_n = x_{n+1} - x_n + C$, where $C = \min\{(x_{n+1} - x_n)\} + 0.1$ is a constant to make sure $w_n > 0$. The dynamics was rescaled in milliseconds time units with an average rate of 3 $events/s$ (*i.e.*, 3 $spikes/s$) in order to let the *Zaslavskii spike train* be comparable to neurophysiological experimental data. Similarly *Poissonian* and *Gaussian* spike trains were generated according to the corresponding probability density function.

The background activity was simulated as the Poissonian spike train. To investigate the effect of the background intensity to the performance of the spike pattern detection from the neural outputs, we used the average firing rate of $0.5, 1, 2, 3, 6, 9, 12$ $events/s$. Return maps of the Zaslavskii, Poissonian and Gaussian spike trains are shown in Fig. 1.

2.3 Pattern Detection and Reconstruction of Time Series

Time series that are subset of neural spike trains referred to as "reconstructed time series" were obtained by using the Pattern Grouping Algorithm (PGA) as described in previous work [5]. The procedure can be briefly described as follows. At first, the PGA [11,6,7] detects a minimum number of spatiotemporal patterns of spikes (at least n times set to $n = 5$ in this study) that repeated above the chance level (significance set to $p = 0.05$ in this study), where the spike

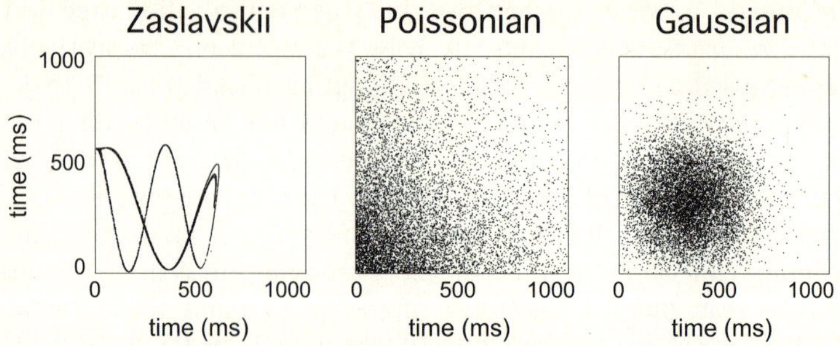

Fig. 1. Return maps of input spike trains. The n-th inter-spike-interval were plotted against the $(n-1)$-th inter-spike-interval. The axes are scaled in ms time units. Each panel shows Zaslavskii, Poissonian, and Gaussian spike trains from left to right. Each map includes $10,000$ spikes and the mean firing rate was 3 $spikes/s$.

patterns were defined as sequences composed of three or four spikes which were not necessarily separated by immediately successive intervals. The maximum duration of the pattern was set to $1,000$ ms. A clusterization procedure allows to group patterns whose difference in spike timing is below a threshold of accuracy (in this study the accuracy was set to 3 ms). The representative pattern of one such group is called a template pattern. For example if there are 10 triplets (*i.e.*, patterns formed by 3 spikes) in one cluster, then the group is formed by $10 \times 3 = 30$ spikes. All events belonging to all such template patterns detected by PGA are pooled together and form a subset of the original spike train referred to as "reconstructed time series" [5].

Let us term V_0 the set of spikes of the output spike train in the absence of any background activity and V the set of spikes of the output spike train, otherwise. The set of points reconstructed from V_0 is referred to as R_0 and from V as R. We evaluated the effect of the background activity by the number of spikes involved in both of R_0 and R denoted as $R_0 \cup R$ using the logical expression, *i.e.*, spikes in spike pattern groups associated to template patterns found in both of R_0 and R were counted as elements in $R_0 \cup R$.

3 Result

We investigated the response of the neuron model to the input spike trains, *i.e.* Zaslavskii, Poissonian and Gaussian spike trains, in the absence of the background activity. Return maps of spike trains of the neural outputs and corresponding reconstructed spike trains are shown in Fig. 2. For Zaslavskii spike train one can recognise the sets of points which reflect the original Zaslavskii attractor in the return map of the output spike train of all neuronal types (Fig. 2 panels in the top row) except CH neuronal type. Cell type CH generated doubled spikes for one input pulse with a short interval, and hence the points in the return map stayed near the ordinate and abscissa axes. Notice that the margins without points near both axes in each panel are related to the refractoriness of

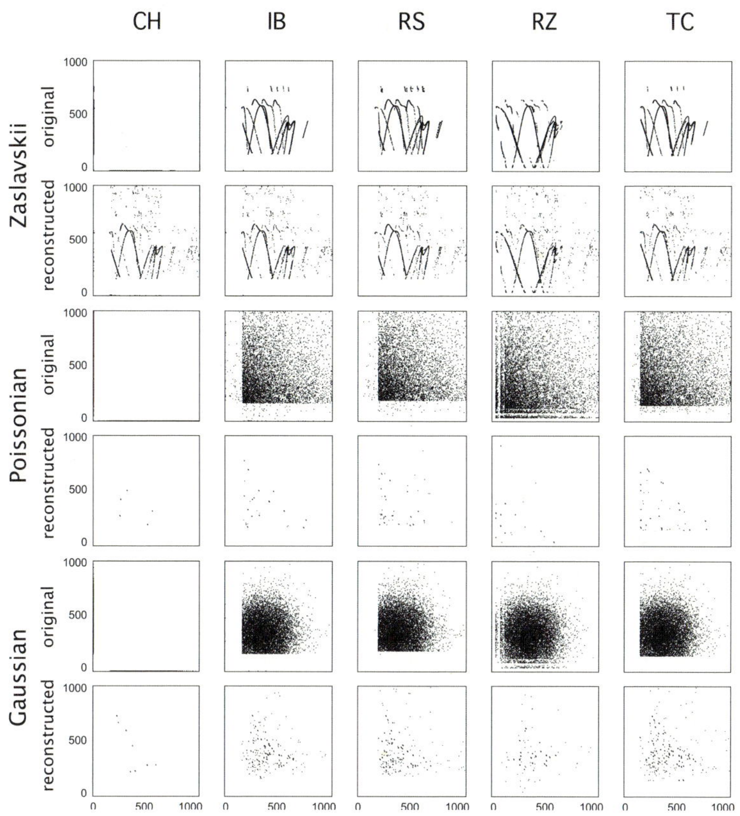

Fig. 2. Return maps of the neural outputs and corresponding reconstructed spike trains. Zaslavskii, Poissonian and Gaussian spike train were given to the neuron model of chattering (CH), intrinsically bursting (IB), regular spiking (RS), resonator (RZ) and thalamo-cortical (TC) neurons without the background activity. The axes are scaled in *ms* time units.

the neuronal dynamics. Panels in the second row of Fig. 2 show the return maps of the reconstructed spike trains which include principal features of the shape of the attractor. Even in the case of CH neuron, the shade of the attractor which was invisible in the return map of the neural output was revealed by applying the reconstruction method. An example of a preferred spike pattern found by PGA in the RS neural output without (*i.e.*, with background activity set to 0 *spikes/s*) and in presence of background activity of 6 *spikes/s* were shown in Fig. 3.

We analyzed at first the responses to stochastic inputs. In case of a Poissonian input spike train the time series is entirely stochastic and PGA could not detect any preferred spike pattern in it. However, a few patterns were found in the corresponding output spike train due to the intrinsic dynamics of the neuron model.

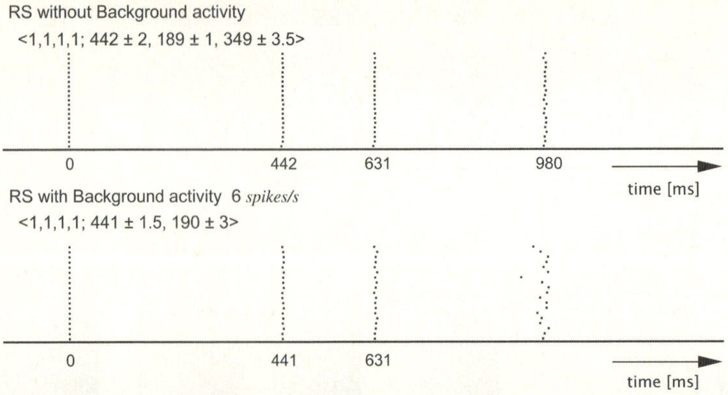

Fig. 3. Raster plot of an example spike patterns aligned by the first spike in the pattern. The upper panel shows a quadruplet found in the output of RS neuron receiving the Zaslavskii spike train at the absence of the background. The lower panel shows a triplet corresponding to a subpattern shown in the upper panel found in presence of the background with mean firing rate 6 *spike/s*.

In case of a Gaussian input spike train all neuron types were characterized by nearly 30% more spikes in the output train compared to the Poissonian input. In the corresponding reconstructed time series, about 3 times more spikes were observed for neuronal types IB, RS and TC, and about twice for RZ and CH. The stripe pattern near the axes was clearly observed in RZ following its typical subthreshold oscillatory dynamics.

The number of spikes in the output spike trains and in the corresponding reconstructed time series as a function of several intensities of background activity were summarized in Table 1. With a Poissonian input and for all cell types except RZ the number of spikes in the output time series tended to decrease when the background activity increased up to the range $[3, 6]$ *spikes/s*. Beyond this range of background intensity the number of spikes increased in parallel with an increase of background activity. For the reconstructed time series the number of events tended to decrease with an increase in background activity. The dynamics of cell type RZ was such that the number of events in the output trains increased monotonously with an increase in background activity. With a Gaussian input the cell types responses were similar to those observed with a Poissonian input.

Figure 4 shows that an increase in background activity tended to blur the contour of the Zaslavskii attractor, especially in RZ neuron, Notice that for all cell types the background activity up to a mean rate near 6 *spikes/s* tended to reduce the number of spikes in the output train. Beyond this intensity higher levels of background activity provoked an increase in the firing rate of the output spike train. The number of events belonging to the reconstructed spike trains decreased almost linearly with an increase in background activity except for RZ cell type.

Table 1. Summary of the spike number in the spike train of the neural output (V), and corresponding reconstructed time series (R). V_0 and R_0 represent the spike set at the absence of the background activity, respectively. BG indicated by the mean firing rate (*spikes*/*s*) of the background activity. $R_0 \cup R$ represents number of spikes composing the spike patterns found in both of R_0 and R. The Detected Ratio: $DR = \frac{R_0 \cup R}{R_0}$, and the Affected Pattern ratio: $AP = \frac{R-(R_0 \cup R)}{R_0}$ are shown for the case of the Zaslavskii input spike train.

		Zaslavskii					Poissonian		Gaussian	
	BG	V	R	$R_0 \cup R$	DR	AP	V	R	V	R
	0	15,676	5,690				13,573	102	17,824	180
	0.5	15,402	5,239	4,441	0.78	0.14	13,378	39	17,509	167
	1	15,189	4,821	3,950	0.69	0.15	13,321	90	17,361	143
CH	2	14,906	4,055	3,119	0.55	0.16	13,057	15	16,801	172
	3	14,628	3,358	2,357	0.41	0.18	12,965	0	16,421	144
	6	14,612	1,823	841	0.15	0.17	13,320	21	16,019	21
	9	15,411	767	355	0.06	0.07	14,610	0	16,665	15
	12	17,061	96	0	0	0.02	16,226	15	17,902	36
	0	7,861	5,887				6,778	290	8,989	1,163
	0.5	7,689	5,408	4,507	0.76	0.15	6,651	172	8,786	1,051
	1	7,539	5,031	3,793	0.64	0.21	6,581	140	8,664	717
IB	2	7,323	4,119	3,286	0.55	0.14	6,384	111	8,312	646
	3	7,148	3,634	2,705	0.45	0.15	6,283	76	8,046	531
	6	6,888	1,990	1,322	0.22	0.11	6,247	51	7,589	321
	9	7,098	1,007	467	0.07	0.09	6,683	69	7,697	225
	12	7,739	355	118	0.02	0.04	7,306	45	8,110	117
	0	7,587	5,543				6,422	335	8,670	1,093
	0.5	7,427	5,096	4,174	0.75	0.16	6,306	228	8,483	1,091
	1	7,287	4,663	3,739	0.67	0.16	6,237	226	8,366	925
RS	2	7,055	3,808	2,775	0.50	0.18	6,057	149	8,022	737
	3	6,874	3,090	2,227	0.40	0.15	5,943	174	7,777	712
	6	6,579	1,656	994	0.17	0.11	5,911	75	7,279	402
	9	6,736	1,109	474	0.08	0.11	6,243	62	7,359	232
	12	7,266	378	0	0	0.07	6,772	93	7,607	203
	0	8,840	6,914				8,503	293	9,763	630
	0.5	8,860	6,463	5,224	0.75	0.18	8,512	187	9,776	345
	1	8,906	5,825	4,786	0.69	0.15	8,613	226	9,842	407
RZ	2	9,186	5,132	3,717	0.53	0.27	8,892	602	10,075	487
	3	9,710	4,811	3,058	0.44	0.25	9,390	1,155	10,422	900
	6	11,432	3,239	799	0.11	0.35	11,318	1,916	12,176	2,106
	9	14,495	4,592	196	0.02	0.67	14,141	3,568	15,017	3,798
	12	17,528	6,072	652	0.09	0.78	17,386	6,852	18,052	6,870
	0	8,005	6,232				7,052	395	9,192	1,350
	0.5	7,758	5,375	4,525	0.72	0.13	6,855	272	8,896	886
	1	7,580	4,739	3,850	0.61	0.14	6,752	260	8,687	781
TC	2	7,300	3,580	2,600	0.41	0.15	6,573	256	8,336	721
	3	7,111	2,661	1,809	0.29	0.13	6,493	150	8,032	434
	6	7,270	1,239	563	0.09	0.10	6,841	162	7,892	170
	9	7,963	472	140	0.02	0.05	7,756	182	8,446	232
	12	9,138	603	21	0	0.09	8,834	90	9,492	131

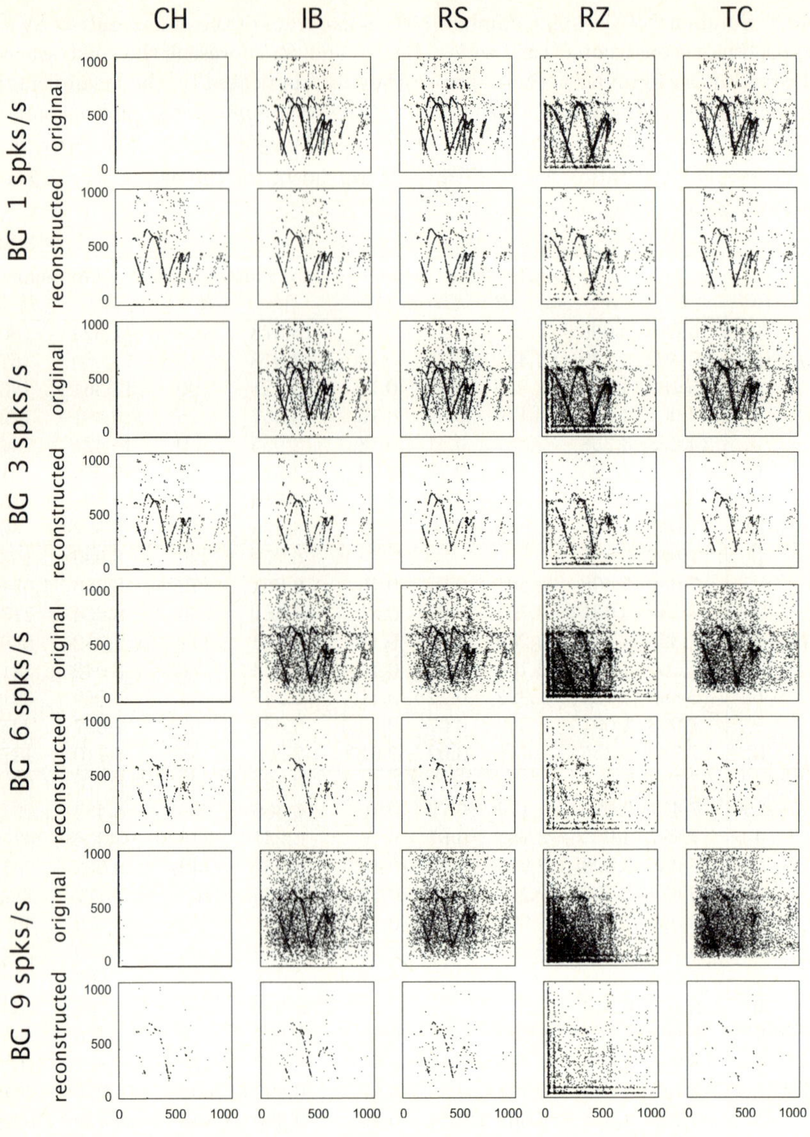

Fig. 4. Return maps of raw output spike trains and reconstructed output spike trains by PGA for a Zaslavskii input with several intensities of background activity

The similarity between the reconstructed spike trains in absence (R_0) and presence (R) of background activity was analyzed for each cell type. The events found in both R_0 and R are denoted by $R_0 \cup R$ and their number is summarized in Table 1. For all neuron types but RZ, the number of such events decreased as the background activity increased. We defined two indexes, DR and AP, to assess the performance of the reconstruction method (Table 1). The detection ratio (DR) was defined as the ratio of the number of spikes in $R_0 \cup R$ to R_0. Thus, large values of DR correspond to good reconstructions of the attractor dynamics. The affected pattern ratio (AP) was defined as the ratio of the number of spikes in R but not in R_0 to R_0. Thus, large values of AP mean that many spike patterns were spurious, *i.e.*, many patterns were associated to the background noise and were formed by events not belonging to R_0. According to DR, the performance of TC neurons (for BG=3 $DR = 0.29$) appeared somehow slightly lower than for the other cell types ($DR \geq 0.4$). According to AP the RZ neurons showed more spurious patterns than other cell types (Fig. 4). In the case of TC neurons AP values were lower, thus suggesting that TC cell type might be better suited to transmit some temporal information embedded in the input spike train despite increasing levels of background activity.

4 Discussion

We investigated the responses of five types of modeled neurons, *i.e.* regular spiking (RS), thalamo-cortical (TC), resonator (RZ), intrinsically busting (IB) and chattering (CH) neurons [8,9] to inputs characterized by deterministic temporal structure or stochastic properties in presence of several intensities of background activity. RS and IB types are characterized by a rather long refractory period and RZ is characterized by subthreshold oscillations of the membrane potential. In the case of RZ its peculiar dynamics provoked the appearance of stripe patterns in the return maps of the output spike trains and increased the number of spurious patterns which are irrelevant to deterministic temporal information embedded in the inputs.

It is interesting to notice the nonlinear effect of the background intensity to the output firing rate. The background activity near 6 $spikes/s$ provoked the minimal output rates for CH, IB and RS neurons. Low values of AP indicate less spurious patterns detected by PGA in the neural output. According to this index the TC cell type appeared to be the best model investigated here to preserve the input deterministic dynamics. The fact that TC neurons appear as a good candidate for the transmission of temporal information may raise several hypotheses on the role of the thalamo-cortical loop with respect to feed-forward networks formed by cortical regular spiking neurons [12]. According to the detection rate RD IB and RS cell types could also transmit information associated to the temporal structure embedded in the input spike train even in presence of background intensity larger than the input firing rate.

The current work is only at an incipient state but it clearly address the need to investigate in detail the ability of transmitting detailed temporal information as a function of neuronal internal dynamics and the background activity level in the network composed of these neurons. Furthermore, the nature of additional stochastic inputs might act as additional controlling parameters on the transmission of temporal information through chains of different types of neurons.

Acknowledgements

This study was partially funded by the bi-national JSPS/INSERM grant SYRNAN and Japan-France Research Cooperative Program.

References

1. Mpitsos, G.J.: Chaos in brain function and the problem of nonstationarity: a commentary. In: Basar, E., Bullock, T.H. (eds.) Dynamics of sensory and cognitive processing by the brain, pp. 521–535. Springer, Heidelberg (1989)
2. Celletti, A., Villa, A.E.P.: Low dimensional chaotic attractors in the rat brain. Biological Cybernetics 74, 387–394 (1996)
3. Villa, A.E.P., Tetko, I.V., Celletti, A., Riehle, A.: Chaotic dynamics in the primate motor cortex depend on motor preparation in a reaction-time task. Current Psychology of Cognition 17, 763–780 (1998)
4. Tetko, I.V., Villa, A.E.: A comparative study of pattern detection algorithm and dynamical system approach using simulated spike trains. In: ICANN 1997. LNCS, vol. 1327, pp. 37–42. Springer, Heidelberg (1997)
5. Asai, Y., Yokoi, T., Villa, A.E.P.: Detection of a dynamical system attractor from spike train analysis. In: Kollias, S.D., Stafylopatis, A., Duch, W., Oja, E. (eds.) ICANN 2006. LNCS, vol. 4131, pp. 623–631. Springer, Heidelberg (2006)
6. Tetko, I.V., Villa, A.E.P.: A pattern grouping algorithm for analysis of spatiotemporal patterns in neuronal spike trains. 1. detection of repeated patterns. J. Neurosci. Meth. 105, 1–14 (2001)
7. Abeles, M., Gat, I.: Detecting precise firing sequences in experimental data. Journal of Neuroscience Methods 107, 141–154 (2001)
8. Izhikevich, E.M.: Simple model of spiking neurons. IEEE Transactions on Neural Networks 14, 1569–1572 (2003)
9. Izhikevich, E.M.: Which model to use for cortical spiking neurons? IEEE Transactions on Neural Networks 15, 1063–1070 (2004)
10. Zaslavskii, G.M.: The simplest case of a strange attractor. Phys. Let. 69A, 145–147 (1978)
11. Villa, A.E.P., Tetko, I.V.: Spatiotemporal activity patterns detected from single cell measurements from behaving animals. Proceedings SPIE 3728, 20–34 (1999)
12. Villa, A.E.P.: Cortical modulation of auditory processing in the thalamus. In: Lomber, S.G., Galuske, R.A.W. (eds.) Virtual lesions: Examining Cortical Function with reversible Deactivation, pp. 83–119. Oxford University Press, Oxford (2002)

Assemblies as Phase-Locked Pattern Sets That Collectively Win the Competition for Coherence

Thomas Burwick

Frankfurt Institute for Advanced Studies (FIAS)
Johann Wolfgang Goethe-Universität
Ruth-Moufang-Str. 1, 60438 Frankfurt am Main, Germany
and
Thinking Networks AG, Markt 45-47, 52062 Aachen, Germany
burwick@fias.uni-frankfurt.de

Abstract. The formation of temporal coding assemblies requires synchronizing mechanisms to establish the coherence of the corresponding patterns, as well as desynchronizing mechanisms that separate it from other possibly overlapping patterns. Here, we use a recently proposed desynchronizating mechanism, denoted as acceleration. This mechanism lets each unit of the network oscillate with higher frequency in case of stronger and/or more coherent input from the other units. In the context of Hebbian memory, it implies a competition for coherence among the stored patterns. The profound effect on the segmentation of patterns is demonstrated. Going beyond earlier discussions, we also illustrate the concept of an assembly as a collection of phase-locked coherent patterns. This points to a role of synchronization strength as controlling the degree of association, that is, determining the number of patterns that are collected in the assembly.

Keywords: Associative memory, complex-valued neural networks, oscillatory networks, pattern recognition, synchronization, temporal coding.

1 Introduction

Neural information processing has to combine cooperative with competitive mechanisms. In the context of temporal coding, the cooperative mechanisms should be based on excitation and some form of temporal correlation, commonly identified with synchronization; see [1,2] for introductions to temporal coding and the related concepts of assemblies and binding. Without desynchronizing interactions present, however, neural systems may reach global coherence and the superposition catastrophe would be present. In fact, it was this superposition catastrophe that motivated the temporal correlation hypothesis in [3]. In consequence, merely adding synchronization mechanisms to the classical neural couplings does not realize temporal coding. Instead, identifying the appropriate competitive mechanisms, resulting in meaningful desynchronization, is of utmost importance for realizing the benefits that temporal coding may hold.

V. Kůrková et al. (Eds.): ICANN 2008, Part II, LNCS 5164, pp. 617–626, 2008.

As a counterpart of excitation, inhibition is a natural first candidate to implement competition for activity among the units of the network. Nevertheless, in particular with a view on the workings of the brain, it should be noticed that inhibition rather may serve to implement competition on a local scale; see [4] for a recent review where the local nature of the competition resulting from the cortical inhibitory (interneuron) system is emphasized. Cooperation among the neural units serves to establish assemblies that are distributed and, in the context of temporal coding, bound together through synchronization among its different parts. Given the distributed nature of assemblies versus the local nature of the competition based on inhibition, another competition may be needed that complements the local one by acting in a distributed manner.

Recently, it was demonstrated that temporal coding may provide such a competition mechanism based on surprisingly simple assumptions. Assuming that a number of patterns are stored in a network according to the rules of Hebbian memory, a competition for coherence among the patterns may be established through complementing synchronization with another interaction, denoted as acceleration, that lets the units of the network tend to oscillate with higher frequency in case of stronger and/or more coherent (synchronous) input from the connected units [5,6,7,8]; see also [9].

In section 2, the model is briefly reviewed. Moreover, following a proposal made in [8, section 6] in the context of hierarchical architectures, an assembly is defined as a set of phase-locked patterns (here, we discuss only auto-associative memory). In section 3, first examples are given to demonstrate the formation of such configurations. Section 4 contains the summary and discussion.

2 Neural Dynamics with Amplitudes and Phases

2.1 The Model in Real Coordinates

The model that is used in the following arises from an appropriate complex-valued generalization of the Cohen-Grossberg-Hopfield model; see [5]. Thereby, it implements temporal coding through an oscillatory network model that shares important features with the classical models, for example, the saturation property of activities and the presence of on- and off-states.

Consider a network with N units, where each unit k is described in terms of amplitude u_k and phase θ_k, $k = 1, \ldots, N$. Then, in terms of these real coordinates, the model takes the form

$$\tau(u_k)\frac{du_k}{dt} = I_k - u_k + \frac{\alpha}{N}\sum_{l=1}^{N} h_{kl}c(\theta_l - \theta_k)V_l \tag{1a}$$

$$\tau\frac{d\theta_k}{dt} = \tau\omega_k\,(u,\theta) + \underbrace{\frac{\sigma}{N}\sum_{l=1}^{N} h_{kl}s(\theta_l - \theta_k)V_l}_{\text{synchronization terms}}, \tag{1b}$$

with

$$\omega_k(u,\theta) = \omega_{1,k} + \omega_{2,k} V_k + \underbrace{\frac{\omega_3}{N} \sum_{l=1}^{N} h_{kl} \Delta(\theta_l - \theta_k) V_l}_{\text{acceleration terms}} . \tag{2}$$

Here, t is the time, τ is a time-scale, I_k is an external input, the $\omega_{1,k}$ are the eigenfrequencies of the oscillators, and the $\omega_{2,k}$ parameterize the shear terms. The activity V_k is related to u_k via the activation function g, chosen as:

$$V_k = g(u_k) = \frac{1}{2}(1 + \tanh(u_k)). \tag{3}$$

Equation 1 may be described as complex-valued gradient system [5]. This form implies the phase-dependent couplings

$$c(\theta) = 1 + \frac{\sigma}{2\alpha}\cos\theta - \frac{\tau\omega_3}{2\alpha}\sin\theta = 1 + \frac{\beta}{2\alpha}\cos(\theta + \phi), \tag{4}$$

$$\tau\omega_3 \Delta(\theta) + \sigma s(\theta) = \tau\omega_3 \cos\theta + \sigma \sin\theta = \beta \sin(\theta + \phi), \tag{5}$$

and the scaling factor

$$\tau(u_k) = (1 - V_k)\tau \ , \tag{6}$$

where $\alpha > 0$ is the classical (that is, phase-independent) coupling strength, $\sigma > 0$ is the strength of synchronization, and $\omega_3 > 0$ is the acceleration parameter. Equations 4, 5 also introduce β, ϕ with $\beta\cos\phi = \sigma$, $\beta\sin\phi = \tau\omega_3$, where ϕ may be referred to as acceleration phase, since $\phi = 0$ implies $\omega_3 = 0$.

The fact that acceleration and synchronization terms may be combined into single terms in equations 4 and 5 points to their common origin as imaginary and real part of the coupling terms in the complex-valued formulation.

See [6] for a list of references to implementations of temporal coding through oscillatory networks. What makes the approach of equations 1 to 6 special is the phase- and amplitude-dependency of the ω_k as given by the acceleration terms. It implies that a more coherent and/or stronger input to unit k will lead to a higher phase velocity through increasing the value of ω_k (the contributions from $\tau\omega_3 h_{kl}\cos(\theta_l - \theta_k)$ increase as $\theta_k \to \theta_l$).

2.2 Hebbian Memory and Competition for Coherence

The storage of P patterns ξ_k^p, with $p = 1, \ldots, P$, and $k = 1, \ldots, N$, enters equation 1 through the couplings h_{kl}. In this paper, it will be sufficient to assume that $\xi_k^p \in \{0, 1\}$, where 1 (0) corresponds to an on-state (off-state). We refer to the units k with $\xi_k^p = 1$ as the units of pattern p. In equations 1 with 2, Hebbian memory may be used that is defined by

$$h_{kl} = \sum_{p,q=1}^{P} \lambda_p\, \xi_k^p \xi_l^p , \tag{7}$$

with $\lambda_p > 0$. The λ_p give the weights for patterns p.

In the following, the collective dynamics of the network is described in terms of pattern activities A_p, coherences C_p, and phases Ψ_p, given by

$$A_p = \frac{1}{N_p} \sum_{k=1}^{N} \xi_k^p V_k \tag{8}$$

$$Z_p = C_p \exp(i\Psi_p) = \frac{1}{N_p A_p} \sum_{k=1}^{N} \xi_k^p V_k \exp(i\theta_k), \tag{9}$$

where N_p denotes the number of units of patterns p, that is, the number of units k with $\xi_k^p = 1$. The quantities A_p,C_p,Ψ_p are real and $0 \leq A_p, C_p < 1$.

These quantities allow to understand that the interplay of synchronization and aceleration terms introduces a competition for coherence among the stored patterns. A pattern p may win this competition (it is then said to be dominating) through having a larger value of $\lambda_p n_p$ than the other patterns, where λ_p is the pattern weight (see equation 7) and n_p is the number of active units of pattern p, that is, units k with $\xi_k^p = 1$ and $V_k \simeq 1$; see [6, section 2.3] or [8, section 3].

2.3 Assemblies as Phase-Locked Pattern Sets

The earlier discussions of equation 1, given in [5,6,7,8], concentrated on the case that the competition for coherence results in a single dominating pattern. This pattern gets segmented from the other patterns through taking a coherent state. Such a case will be reviewed with example 1 in section 3.2.

Here, we want to take a more general point of view. While the case of only one pattern being retrieved corresponds to identifying a neural assembly with a single pattern, we now allow for assemblies made up of several patterns. Given certain values for the acceleration strength ω_3, it should be obvious that increasing values of synchronization strength σ lead to global synchronization of the connected units of the network when σ is sufficiently large. Examples 2 and 3 in section 3.3 demonstrate two steps on this way to global coherence. These examples illustrate an interesting aspect: increasing the synchronization strength does not imply a gradual increase in synchronization of the complete network. Instead, a stability hierarchy among the patterns is observed, that is, subsets of the stored patterns form well-defined sets of phase-locked groups. The cardinality of this subsets increases with synchronization strength.

The units of the network may be interpreted as representing features. Accordingly, phase-locking of the units may be interpreted as association of features. Then, given the described behavior in the context of associative memory, synchronization strength may be seen as controlling the degree of association.

More will be said on this phenomenon in the context of the corresponding examples 2 and 3 in section 3.3 and the discussion of section 4.

3 Examples

3.1 Network Architecture and Parameters

In this section, examples are given that use a network with $N = 30$ units and $P = 6$ overlapping patterns ξ_k^p, where $k = 1, ..., N$ and $p = 1, ..., P$. The overlap between patterns p and q may be described through

$$\mathcal{O}^{pq} = \sum_{k=1}^{N} \xi_k^p \xi_k^q \, . \tag{10}$$

The patterns of the following examples have

$$(\mathcal{O}^{pq}) = \begin{pmatrix} 19 & 4 & 5 & 3 & 3 & 2 \\ . & 7 & 2 & 0 & 1 & 1 \\ . & . & 7 & 1 & 1 & 1 \\ . & . & . & 6 & 0 & 0 \\ . & . & . & . & 4 & 0 \\ . & . & . & . & . & 3 \end{pmatrix} , \tag{11}$$

where the symmetric elements are not repeated. For example, patterns $p = 1$ and $p = 3$ have $N_1 = 19$ and $N_3 = 7$ units, respectively, and they are overlapping at five units. We ordered the patterns so that $N_p \geq N_{p+1}$. With our examples, we assume that the patterns enter the Hebbian couplings with equal weight, $\lambda_p = 1/P$ for every p.

The examples use acceleration strength $\omega_3/N = 2\pi/\tau$. Eigenfrequency and shear parameters are chosen to vanish, $\omega_{1,k} = \omega_{2,k} = 0$ for any k. This assures that any observed desynchronization (and the corresponding segmentation) may be traced back to the presence of acceleration. The inputs are chosen as $I_k = -(\alpha/(2N)) \sum_{l=1}^{N} h_{kl} + J_k$, with $J_k = 0$, thereby describing a kind of "neutral" input (see [10, subsection 4.1]).

As with the foregoing studies of equation 1, the discretized dynamics is obtained by introducing a parameter $1 \gg \epsilon > 0$, replacing the scaling factor of equation 6 with $\tau_\epsilon(u_k) = (1 - V_k + \epsilon)\tau$, and applying a simple Euler discretization with time step $dt = \epsilon\tau$ [5]. Here, we choose $\epsilon = 0.01$.

Initial values are chosen that let each unit get active, $V_k \simeq 1$, thereby reducing the system to the phase dynamics. (Examples for pattern retrieval, where only part of the network gets active may be found in [6,7]).

Each of the following example uses the same initial values and parameters, except for an increasing value of synchronization: example 1 uses $\sigma = \sigma_1 = \tau\omega_3$, while examples 2 and 3 use $\sigma = 3\sigma_1$ and $\sigma = 4\sigma_1$, respectively. We also mention (without demonstration) the behavior for even larger values.

3.2 Example 1: Retrieval of Dominating Pattern

Example 1 uses $\sigma = \sigma_0 = \tau\omega_3$. It may serve to illustrate the competition for coherence that results from equation 1, as mentioned in section 2.2. Thereby,

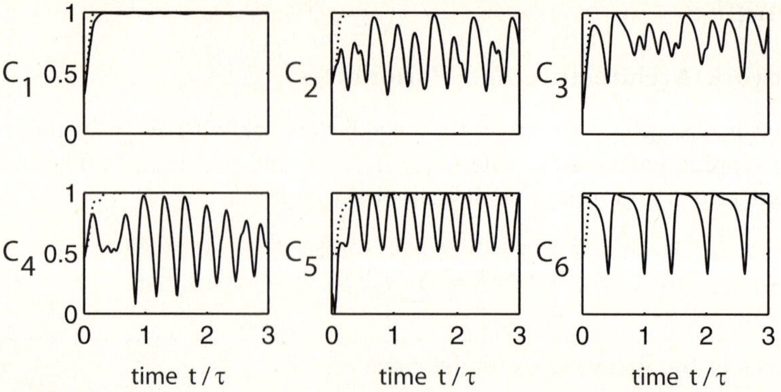

Fig. 1. Example 1: Synchronization strength is $\sigma = \sigma_1 = \tau\omega_3$, where ω_3 gives the strength of acceleration and τ is the time scale; see equations 1 and 2. The solid curves show the pattern coherences C_p, while the dotted curves give the pattern activities A_p. The superposition problem is present, since all units get active, due to the chosen inputs and initial values. Temporal coding resolves this problem by making the dominating pattern $p = 1$ coherent, while the other patterns return to states of decoherence. This segmentation is a result of the interplay of synchronization and acceleration. Pattern $p = 1$ is dominating, that is, it takes a state of enduring coherence, because it has the highest number of active units (given that each pattern enters Hebbian memory of equation 7 with same weight λ_p), as mentioned in section 2.2. See also the additional remarks in section 3.2.

we review the case of a dominating pattern that is unambigously retrieved as coherent part of the network [5,6,7,8].

As mentioned in section 3.1, the initial values are chosen such that the network gets completely active, that is, each of the pattern activities approaches $A_p \simeq 1$. Thus, from a classical point of view, the superposition problem is present. Temporal coding has to resolve this problem with respect to coherences C_p by identifying the retrieved patterns as the ones that take coherent states: the retrieved patterns p show a dynamic that approaches and remains close to $C_p \simeq 1$.

The patterns compete with each other for coherence with an effective weight given by the product $\lambda_p n_p$, as mentioned in section 2.2. Since the $\lambda_p = 1/P$ are identical for all patterns and each pattern gets completely active, that is, $n_p = N_p$, we may expect from equation 11 that pattern $p = 1$ is the dominating pattern. Indeed, we find that this pattern wins the competition for coherence; see figure 1.

In order to value the described behavior, it is essential to notice that winning the competition for coherence is a dynamical process that is determined by choosing external inputs I_k and initial values. Here, to illustrate a most simple situation, we chose external inputs and initial values such that each unit of the networks gets active: $V_k \simeq 1$ for each k. However, other inputs and initial values may cause only a subset of the units to get active. Then, the n_p may differ from the N_p and the competition will decide for the pattern with the most active units (or the pattern with highest value of $\lambda_p n_p$ in case of different Hebbian pattern

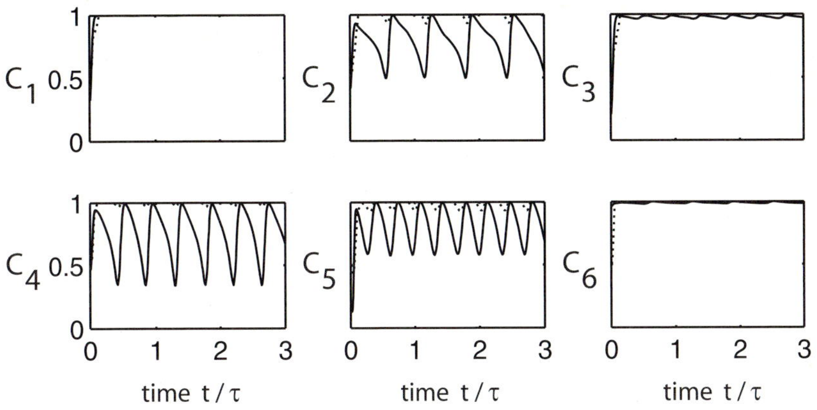

Fig. 2. Example 2. The synchronization is stronger than with example 1: $\sigma = 3\sigma_1$, while the other parameters and initial values are identical. Now, three patterns, $p = 1, 3, 6$, are phase-locked, with $C_p \simeq 1$. Simulations confirmed that for an intermediate range of synchronization strength values σ (for example, $\sigma = 3\sigma_1/2$) only patterns $p = 1$ and $p = 6$ are phase locked. See the additional remarks in section 3.3.

weights). Obviously, this is a remarkably reasonable behavior that realizes a useful complementary behavior of classical (amplitude) and temporal coding (phase) dynamics. Examples illustrating this behavior were given in [6,7].

3.3 Examples 2 and 3: Retrieval of Phase-Locked Pattern Sets. Synchronization Strength as Degree of Association

The following examples result from increasing the synchronization strength σ. Other parameters, including the acceleration strength ω_3, as well as initial values remain the same as with example 1. It should be obvious that a sufficiently large value of σ leads to global coherence among the active and connected units. The examples will illustrate that the road to global coherence is not through gradually increasing the coherence of each pattern but through assembling an increasing set of patterns that are phase locked to each other in a coherent state.

Example 2 uses $\sigma = 3\sigma_1$, while example 3 uses $\sigma = 4\sigma_1$; see figures 2 and 3 and read the captions for remarks on results of other simulations that are not displayed here. Notice, the phase locking follows from the fact that the overlapping patterns could not reach $C_p \simeq 1$ without phase-locking. Substantial phase differences would make one of the patterns decoherent.

The examples confirm the stability hierarchy among patterns that was described in section 2.3. Here, we cannot give a detailed discussion of this phenomenon. At least, we want to mention one systematic behavior that may be observed when going from example 2 to example 3. This observation concerns the fact that the pattern with lowest rate of transitions to decoherent states is phase-locked when the synchronization is strengthened: compare the coherence of pattern $p = 2$ in figure 2 with the other coherences. This may be understood

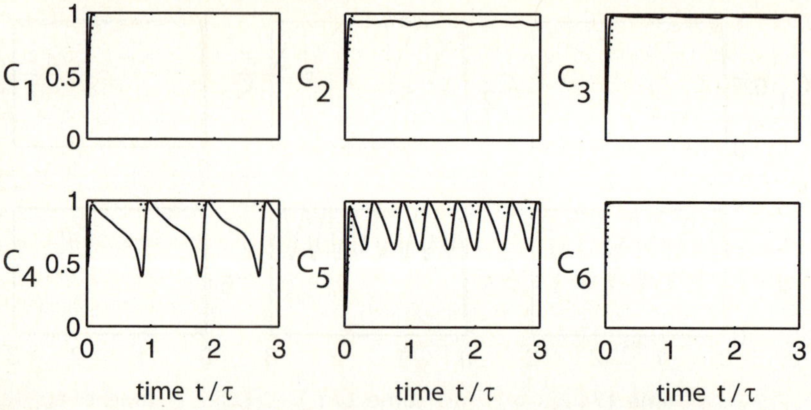

Fig. 3. Example 3 uses synchronization that is stronger than the value in example 2: $\sigma = 4\sigma_1$. Again, the other parameters and initial values are identical with examples 1 (that is, they also are the same as with example 2). Now, an additional pattern, $p = 2$, is phase-locked to the set that is observed with example 2. Simulations confirmed that for even higher values the assembly is extended to pattern $p = 4$ (for example, with $\sigma = 5\sigma_1$), and additionally to $p = 5$ (for example, with $\sigma = 7\sigma_1$). The latter state does then correspond to global coherence. See the additional remarks in section 3.3.

through realizing that the behavior of the drifting patterns is related to interference phenomena; see [6]. Roughly speaking, this mentioned rate is given by the difference of frequencies between the part of pattern $p = 2$ that is overlapping with the phase-locked pattern set (patterns $p = 1, 3, 6$) and the non-overlapping part. Thus, the lowest rate of transitions to decoherent states indicates that this frequency difference is smaller for pattern $p = 2$ than for the other drifting patterns. With example 2, the synchronization strength is not large enough to balance the frequency difference. However, with example 3, due to larger synchronization strength, small phase differences may establish (approximately) the same phase velocities for the overlapping and non-overlapping part of pattern $p = 2$, thereby making the phase locking of pattern $p = 2$ possible.

With example 1, the drifting pattern with the lowest rate of transitions to decoherence is $p = 6$. Correspondingly, a simulation showed that with $\sigma = 3\sigma_1/2$ (larger than the value of example 1 but smaller than the value of example 2), pattern $p = 6$ is phase locked to pattern $p = 1$, while the other patterns remain drifting. This confirms the above argument. Moreover, consider the situation of example 3. There, the drifting pattern with lowest rate is pattern $p = 4$. Correspondingly, a simulation confirmed that this pattern is phase-locked when σ is increased, for example, to $\sigma = 5\sigma_1$. To complete the road to global coherence, let us also mentioned that even larger values, for example, $\sigma = 7\sigma_1$, let all $P = 6$ patterns get phase-locked.

4 Summary and Discussion

Here, we continued the discussion of a recently proposed mechanism that serves to segment patterns in the context of temporal coding. The segmentation is achieved through establishing a competition for coherence among the patterns that are assumed to be stored according to Hebbian memory. This competition results from the interplay of synchronizing and accelerating interactions, where the accelerating interaction imply desynchronization of the patterns that loose the competition, as discussed in [5,6,7,8].

The present discussion dealt with the changes that arise through increasing the strength of synchronization while keeping the strength of acceleration constant. Considering the extremal case, it should be evident that strong enough synchronization causes global coherence of the network. Therefore, our consideration deals with the road towards such global coherence. The interesting observation reported here is that this road reveals a stability hierarchy among the stored patterns.

The discussions in [5,6,7,8] concentrated on the case that the synchronization strength is in a range where exactly one of the overlapping patterns is winning the competition for coherence (or none, in case of undecided situations gestalt-switching like phenomena were observed). Here, we find that strengthening the synchronization versus acceleration leads to the phase locking of more than one pattern to each other, thereby letting a set of patterns win the competition rather than only a single pattern. The number of patterns that get coherent and phase-locked to each other increases with increased synchronization while keeping acceleration strength constant (it may be expected that the same effect occurs through lowering the acceleration strength while keeping the synchronization strength constant).

This observation is interesting in at least two respects. First, the observed stability hierarchy may point to some interesting bifurcation structure that arises with Hebbian memory in the context of oscillatory networks. As such, it may deserve some further studies as a dynamical phenomenon.

Second, one may wonder whether the observed phase-locking has functional relevance with respect to information processing. There is an argument that pattern storage through Hebbian memory may rather be a local mechanism, in a sense that is explained in the following. Hebbian coupling strengths are supposed to be the result of repeated experience. However, the notion of an assembly (see [1,2] for reviews) is hardly compatible with the necessity to repeat. After all, intelligent systems have to deal with complex situations that may have never occurred to them before. For example, the system may look at a visual scene with a composition that it sees the first time. Nevertheless, on a local scale, for example, with respect to orientations in a small part of the visual field, patterns may be present that occurred in several situations. Therefore, these patterns may be memorized through Hebbian couplings. However, the complete scene or particular objects of this scene, realized through the formation of assemblies, may need a more flexible representation.

Following these arguments, one could speculate that patterns should rather be building blocks for the more complex and flexible assemblies. Temporal coding is understood as binding features through temporal correlation. Correspondingly, one may also consider temporal coding as binding patterns, thereby providing a flexible mechanism to collect the patterns into an assembly. The examples given here may illustrate a particular form of this binding of patterns.

Evidently, more studies are needed to apply the described mechanisms to complex applications as well as understanding of brain dynamics phenomena. In particular, the competition for coherence that arises through the mechanisms described here should be embedded into some hierarchical architecture and be combined with competition for local activity through inhibitory couplings; see, correspondingly, [8, section 6] and [9] for some steps into these directions.

Acknowledgements. It is a pleasure to thank Christoph von der Malsburg for valuable discussions. This work is supported by Hertie Foundation and Volkswagen Foundation.

References

1. von der Malsburg, C.: The what and why of binding: The modeler's perspective. Neuron 24, 95–104 (1999)
2. Singer, W.: Synchronization, binding and expectancy. In: Arbib, M. (ed.) Brain Theory and Neural Networks, 2nd edn., pp. 1136–1143. MIT Press, Cambridge (2003)
3. von der Malsburg, C.: The correlation theory of brain function. Internal Report 81-2, Max-Planck Institute for Biophysical Chemistry (1981)
4. Fries, P., Nicolić, D., Singer, W.: The gamma cycle. Trends in Neurosciences 30(7), 309–316 (2007)
5. Burwick, T.: Oscillatory neural networks with self-organized segmentation of overlapping patterns. Neural Computation 19, 2093–2123 (2007)
6. Burwick, T.: Temporal coding: Assembly formation through constructive interference. Neural Computation (in press, 2008)
7. Burwick, T.: Neural assembly formation with complementary roles of classical and temporal coding. In: Proceedings of the 7th International Workshop on Information Processing in Cells and Tissues (IPCAT 2007), Oxford, United Kingdom, 29-31 August 2007, pp. 215–226 (2007)
8. Burwick, T.: Temporal coding with synchronization and acceleration as complementary mechanisms. Neurocomputing 71, 1121–1133 (2008)
9. Burwick, T.: The gamma cycle and its role in the formation of assemblies. In: Proceedings of the 16th European Symposium on Artificial Neural Networks (ESANN 2008), Bruges, Belgium, 23-25 April (2008) (accepted for publication)
10. Burwick, T.: Oscillatory networks: Pattern recognition without a superposition problem. Neural Computation 18, 356–380 (2006)

A CA^{2+} Dynamics Model of the STDP Symmetry-to-Asymmetry Transition in the CA1 Pyramidal Cell of the Hippocampus

Vassilis Cutsuridis[1,⋆], Stuart Cobb[2], and Bruce P. Graham[1]

[1] Department of Computing Science and Mathematics, University of Stirling, Stirling, FK9 4LA, U.K.
{vcu,b.graham}@cs.stir.ac.uk
[2] Division of Neuroscience and Biomedical Systems, University of Glasgow, Glasgow, G12 8QQ, U.K.
s.cobb@bio.gla.ac.uk

Abstract. Recent experimental evidence has reported that the profiles of spike-timing-dependent plasticity (STDP) in the CA1 pyramidal neuron can be classified into two types depending on the location along the stratum radiatum (SR) dendrite: (1) A symmetric STDP profile centered at 0 ms (largest LTP value) with two distinct LTD windows at about ±20ms in the proximal SR dendrite, and (2) an asymmetric one in the distal SR dendrite. Bicuculline application revealed that GABAA is responsible for the symmetry of the STDP curve. We investigate via computer simulations the STDP symmetry-to-asymmetry transition in the proximal SR dendrite. Our findings indicate the transition from symmetry-to-asymmetry is indeed due to decrease of GABAA, but the simulated symmetrical STDP profile is centered at +10ms (and not at 0ms) with two distinct LTD tails at -10ms and +40ms (and not at ±20ms). The simulated LTD tails are strongly dependent on the GABAA conductance.

Keywords: Hippocampus, CA1 pyramidal neuron, computer model, STDP, GABA, LTP, LTD, calcium.

1 Introduction

In 1949, Hebb [1] postulated that a synapse is strengthened only if the pre- and postsynaptic neurons are activated simultaneously. Recently, Hebb's law has been refined ever further with STDP, where the precise timing of presynaptic and postsynaptic action potentials (spikes) determines the sign and magnitude of synaptic modifications [5]. Bi and Poo [7] showed that the profile of the STDP curve in the in-vitro hippocampal network has an asymmetrical shape with the largest LTP/LTD value at $\Delta\tau = t_{post} - t_{pre} = \pm 10$ms, respectively.

Recently, a study by Nishiyama and colleagues [4] reported that "the profile of STDP induced in the hippocampal CA1 network with inhibitory interneurons

⋆ Corresponding author.

V. Kůrková et al. (Eds.): ICANN 2008, Part II, LNCS 5164, pp. 627–635, 2008.

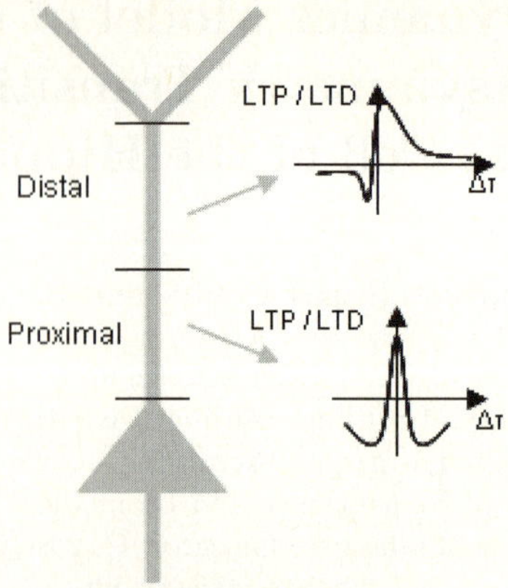

Fig. 1. Pyramidal cell. Asymmetric STDP learning curve at distal dendrite. Symmetric STDP learning curve at proximal dendrite.

is *symmetrical* for the relative timing of pre- and post-synaptic activation". Two long-term depression (LTD) windows at ±20ms and a central long-term potentiation (LTP) peak at 0 ms have been observed (see figure 1) [4]. Further optical imaging studies revealed that the shape of the STDP profile depended on the location on the SR dendrite. A symmetric STDP profile was observed in the proximal SR dendrite and an asymmetric STDP profile in the distal one [2], [3]. It was reported that the transition from symmetry-to-asymmetry is due to the presence of GABAA inhibition in the proximal SR dendrites.

In this study, we investigate via computer simulations the validity of the reported GABAA effects on the symmetry-to-asymmetry transition. To do so, we employed and extended a well established Ca^{2+} dynamics model of the CA1 pyramidal neuron [6] by incorporating the effects of inhibitory interneurons in the SR dendrites. In support of the experimentally observed evidence, we report that the symmetry-to-asymmetry transition is indeed due to GABAA depletion. However, in the model, the simulated symmetrical STDP curve is centered at +10ms ($\Delta\tau = t_{post} - t_{pre} > 0$) and not at 0ms [2],[3]. Two distinct LTD tails are present at -10ms and +40ms and not at ±20ms [2],[3]. Finally, the presence/absence of LTD tails strongly depends on the conductance value of GABAA, with high GABAA conductance ($g_{GABA} = 0.5$) leading to the lift-off of both LTD tails (and not only the positive LTD tail [2],[3]) towards the LTP region.

2 The Model

Rubin and colleagues [6] have recently advanced a Ca^{2+} dynamics model for the CA1 pyramidal cell. Briefly, their model neuron had two compartments: a soma and a dendrite. The generation of action potentials was due to the interplay of a wealth of Na^+, K^+, Ca^{2+}-activated K^+ and Ca^{2+} currents as well as synaptic currents (AMPA and NMDA) [8],[9],[10]. Two excitatory transient inputs to the soma and SR dendrite were used to simulate the experimental STDP protocol.

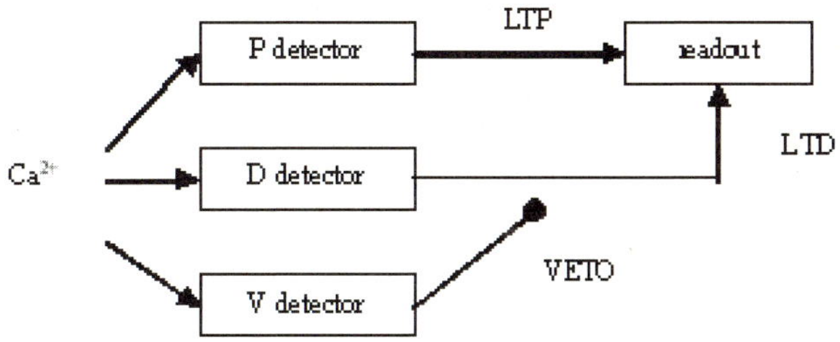

Fig. 2. Rubin and colleagues [6] calcium detection system. P detector: potentiation detector, D detector: depression detector, V detector: veto detector, LTP: long-term potentiation, LTD: long-term depression.

Their mechanism for plasticity had a modular structure consisting of three biochemical detectors, which responded to the instantaneous calcium level in the SR dendrite. More specifically, their detection system (see figure 2) consisted of: (1) a potentiation detector which detected calcium levels above a high-threshold (4 μM) and triggered LTP, (2) a depression detector which detected calcium levels that exceeded a low threshold level (0.6 μM) and remained above it for a minimum time period and triggered LTD, and (3) a veto detector which detected levels exceeding a mid-level threshold (2 μM) and triggered a veto of the model's depression components. Their detector system was inspired by the molecular pathways of protein phosphorylation and dephosphorylation. Their potentiation detector was an abstraction of the phosphorylation cascade of CaMKII, whereas the depression detector represented the kinetics of dephosphorylation agents such as PP1. The veto system represented the competition of kinases and phosphatases such as the inhibitory action of PKA.

We extended the Rubin et al. model [6] by incorporating the GABAA effects of inhibitory interneurons in the form of a second transient input to the SR dendrite.

3 Experiments

To investigate the transition of the STDP curve from symmetry to asymmetry in the SR dendrite, we designed the following experimental paradigms (figure 3):

1. Excitatory spike pairs repeatedly applied to the SR dendrite and soma for 2s (7 times at about 3 Hz) in the absence of GABA for various interspike intervals $\Delta\tau$.
2. Excitatory spike pairs repeatedly applied to the SR dendrite and soma for 2s (7 times at about 3 Hz) in the presence of a single pre-synaptic GABA spike slid between the interspike interval $\Delta\tau$.
3. Spike pairs repeatedly applied to the SR dendrite and soma for 2s (7 times at about 3 Hz) in the presence of a GABA inhibitory spike train presented at 100 Hz (gamma frequency) between the interspike interval $\Delta\tau$.

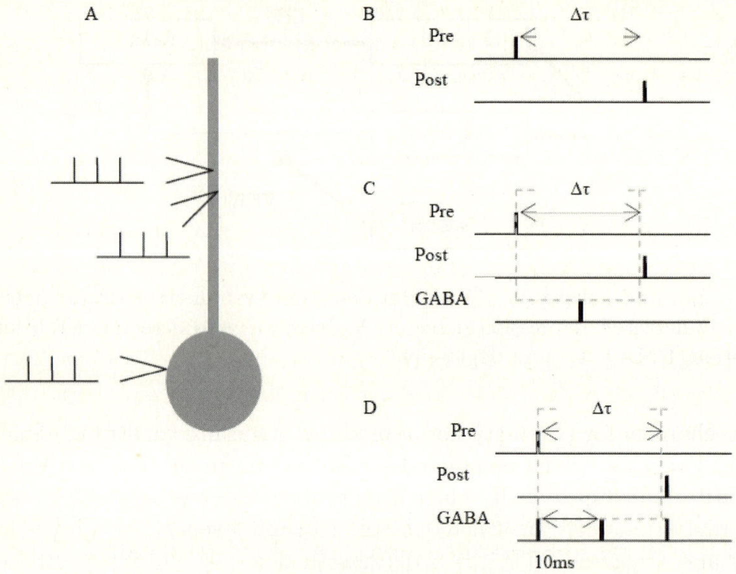

Fig. 3. (A) Our model CA1 neuron with its three transient inputs to the soma and SR dendrite. (B) Experimental paradigm 1: spike doublets in the absence of GABA. $\Delta\tau$ is the relative interval between the pre- and post-synaptic pairing. $\Delta\tau$ is positive (> 0) if pre-activation precedes the post-activation and negative (< 0) otherwise. (C) Experimental paradigm 2: spike doublets in the presence of a single GABA spike occurring during the interspike interval (gray dashed square). (D) Experimental paradigm 3: spike doublets in the presence of a GABA spike train at 100 Hz (individual spikes presented every 10ms) during the interspike interval (gray dashed square).

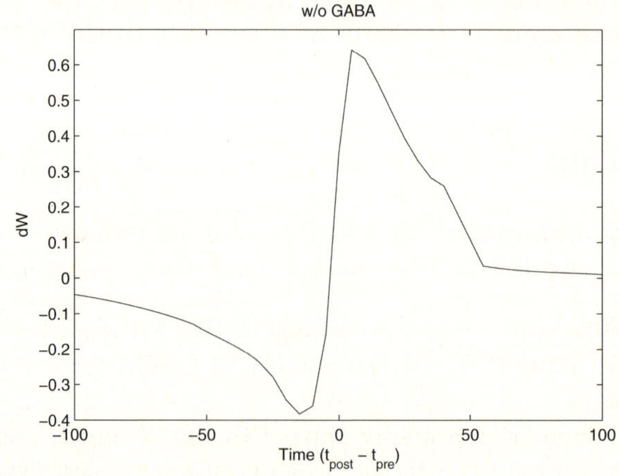

Fig. 4. Simulated asymmetric STDP profile in the absence of GABA. $\Delta\tau$ (t_{post} - t_{pre}) ranges from -100 to 100 in increments of 5ms.

During experimental paradigms 2 and 3, we varied the conductance of GABA and observed its effects on the amplitude of the proximal SR Ca^{2+} spike and the STDP curve. These results are reported in the next section.

4 Results

4.1 Spike Doublets in the Absence of GABA

Figure 4 depicts the saturated synaptic weight values (W_{∞}) as a function of the interspike interval, $\Delta\tau = t_{post} - t_{pre}$. Simulations were performed with $\Delta\tau$ ranging from -100 to 100 in increments of 5ms. An asymmetrical STDP curve is shown with the largest LTP value at +10ms and the largest LTD value at -15ms [7].

4.2 Spike Doublets in the Presence of a Single GABA Event Introduced in the Interspike Interval

Figure 5 is a composite graph of W_{∞} as a function of $\Delta\tau$ (figures 5A and 5D) and the time course of the calcium spike for different values of the GABA conductance ($g_{GABA} = 0.05$ mS/cm^2 and $g_{GABA} = 0.1$ mS/cm^2) (figures 5B, C, E and F) in the SR dendrite in the presence of a single GABA transient input spike occuring at different times in the interspike interval.

More specifically, figures 5A (GABA spike is introduced in the middle of the pre-post interval) and 5D (GABA spike is introduced at 15ms after the pre- activation) depict that as the conductance of GABA is increased, the STDP curve amplitude is reduced, but the asymmetry is preserved. As we further increased the conductance of GABA ($g_{GABA} = 0.2$ mS/cm^2 and $g_{GABA} = 1$ mS/cm^2), the STDP curve amplitude decreased even further (data not shown). The amplitude reduction is more pronounced with the GABA spike occurring 15ms after pre-activation, and also there is a shift in the peak LTP (LTD) changes to more positive (negative) time intervals.

Figures 5B, C, E and F show that even a single GABA spike has significant effects on the amplitude of the calcium spike. The effects are more pronounced for higher values of GABA as well as for shorter $\Delta\tau$ intervals and earlier GABA spike onset times (compare the [Ca^{2+}] values in figures 5B and 5C and in figures 5E and 5F).

4.3 Spike Doublets in the Presence of an 100Hz GABA Spike Train Introduced in the Inter-spike Interval

In this experimental paradigm, we tested the hypothesis of whether the frequency of GABA input presentation has any effect on the STDP profile. A GABA spike input train was presented at a frequency of 100Hz (i.e. a spike event every 10ms; see figure 3D). The GABA spike train was bounded by the onsets of the pre- and post-synaptic activations.

Direct comparison of figures 4 and 6A shows that the absence or presence of low conductance ($g_{GABA} = 0.05$ mS/cm^2) of GABA has no effect on the STDP

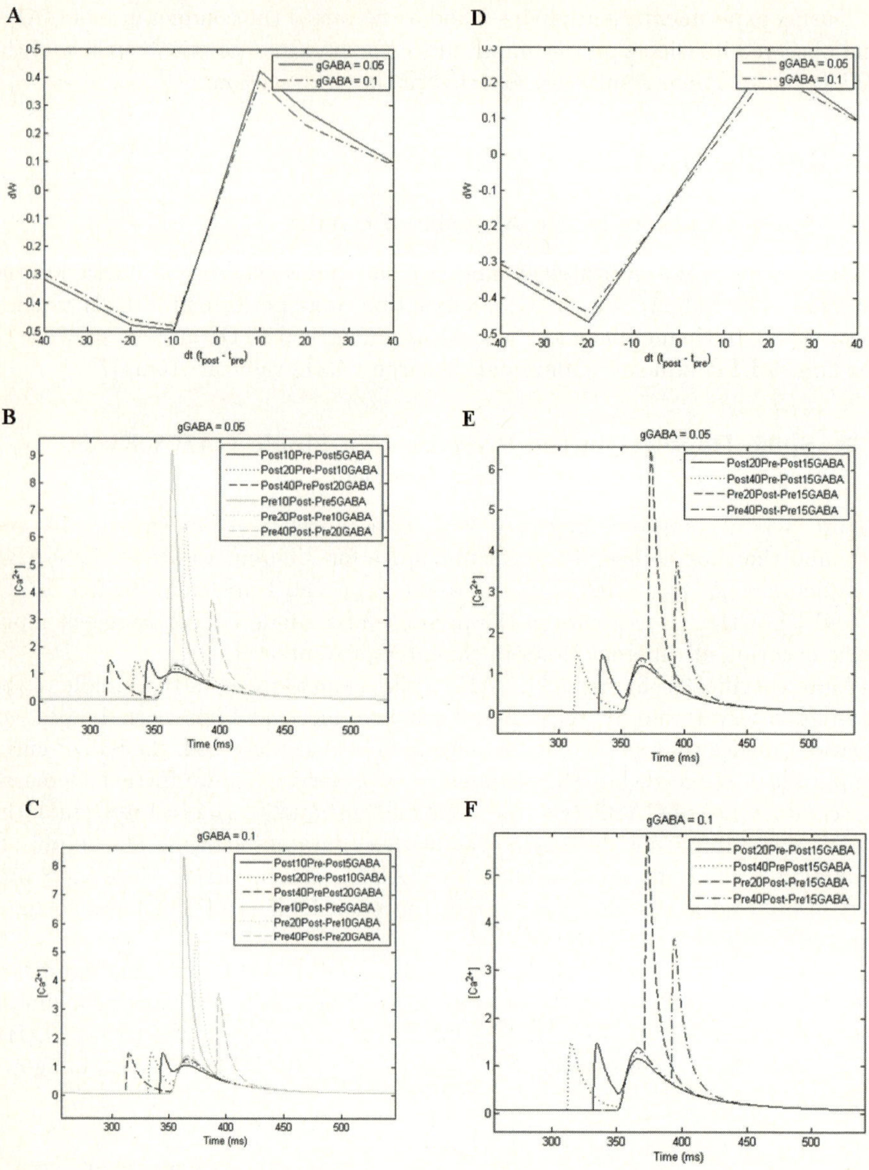

Fig. 5. Composite graph of the saturated synaptic weight values (W_∞) as a function of interspike interval, $\Delta\tau = t_{post} - t_{pre}$ (5A and 5D) and time course of the calcium spike in the SR dendrite (5B, C, E and F) when no GABA is present or a single GABA spike is introduced in the middle of the interspike interval $\Delta\tau$ (5A, B and C) and after 15ms from the pre-synaptic activation (5D, E and F) and for $g_{GABA} = 0.05$ mS/cm^2 (5B and E) and 0.1 mS/cm^2 (5C and F)

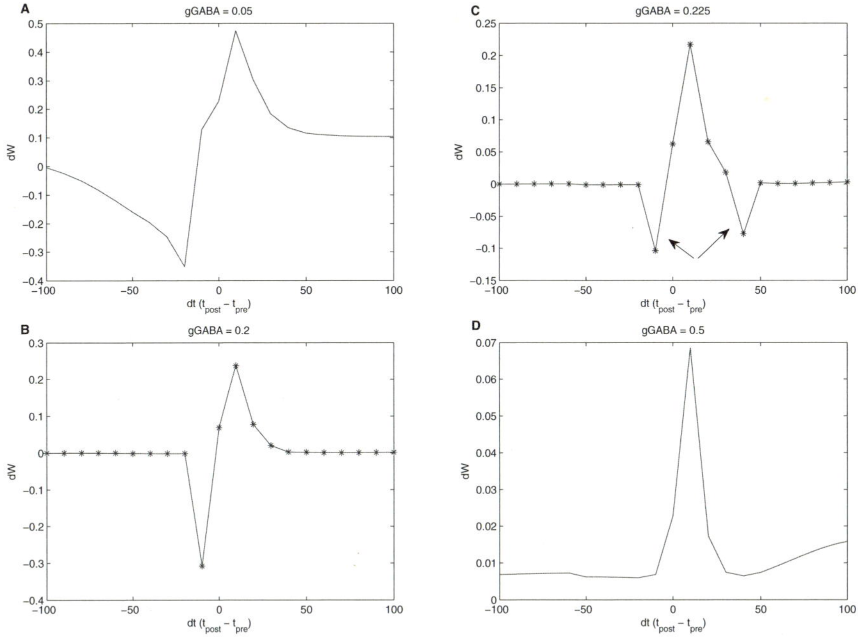

Fig. 6. Plot of the STDP profile in the presence of 100Hz GABAA input during the interspike interval, as a function of the interspike interval $\Delta\tau$ for different levels of GABAA conductance

asymmetry. At g_{GABA} = 0.2 mS/cm^2, the width of the negative LTD window decreases considerably and it is present only at +10ms (compare Figs. 6A and 6B). At g_{GABA} = 0.225 mS/cm^2 (0.2 < g_{GABA} ≤0.3), the asymmetrical STDP curve becomes symmetric centered at +10ms (and not at 0ms [2], [3]) with two distinct LTD windows. In contrast to experimental evidence [2], [3], which indicate the presence of two distinct LTD windows at ±20ms, our simulated negative LTD window is centered at -10ms, whereas the simulated positive one is centered at +40ms. At high GABAA conductance values (see fig. 6D), the two LTD windows are lifted-off towards the LTP regime. This finding supports and extends recent experimental evidence, which showed that in the presence of bicuculline, a GABA blocker, only the positive LTD tail is blocked [2], [3].

5 Conclusion

A well established Ca^{2+} dynamics model of the CA1 pyramidal neuron with three calcium amplitude detectors was extended by incorporating the effects of GABAergic interneurons to simulate the symmetry-to-asymmetry transition of the STDP profile in the proximal SR dendrite. In support of the experimental evidence [2], [3], [4], the transition was found to be due to the GABAA depletion. From our simulation results, we conclude that the transition is strongly dependent on the conductance value of GABAA, where (1) at low GABAA conductance, the asymmetry is preserved, but the width of the negative LTD tail is reduced as GABAA is increased, (2) at intermediate conductance values, the symmetry appears with the positive LTP window centered at +10ms and two equal in height LTD tails at -10ms and +40ms, and (3) at high GABAA conductance values, the two distinct LTD tails disappear and a symmetrical LTP curve centered at +10ms becomes evident. Furthermore, an inverse relationship was found to exist between the Ca^{2+} spike amplitude in SR dendrite and the GABAA conductance. As the GABA conductance is increased, the amplitude of Ca^{2+} is decreased.

While GABA-A inhibition is sufficient to achieve a symmetric STDP curve, it does not provide the peak of LTP at 0ms, as found experimentally. So the question remains, is GABA-A inhibition sufficient, or is there another timing mechanism, perhaps outside the interspike interval, or another synaptic mechanism (e.g. GABA-B) that produces this shift? Several computational models [11], [12] have been published over the years that have modelled in detail the biochemical intracellular mechanisms of synaptic (meta)plasticity. To gain perhaps a better understanding of the symmetry-to-asymmetry transition in the CA1 SR dendrite, we need to incorporate some of this knowledge into our Ca^{2+} dynamics model.

Moreover, a more detailed compartmental model of the CA1 pyramidal neuron needs to be constructed to model the conditions under which the STDP asymmetry in the distal SR dendrites appears. Experimental evidence has shown that both distal and proximal SR dendrites receive excitatory inputs from CA3 cells as well as inhibitory inputs from local CA1 interneurons. An additional excitatory input drives the lacunosum-moleculare (LM) dendrites of the CA1 pyramidal neuron. Pairings of the SR and LM presynaptic excitatory and inhibitory inputs with the postsynaptic somatic activation will provide us with a more realistic picture of STDP in the SR dendrites.

Acknowledgement. This work was funded by an EPSRC project grant to B. Graham and S. Cobb.

References

[1] Hebb, D.O.: The organization of behavior. John Wiley, New York (1949)
[2] Tsukada, M., Aihara, T., Kobayashi, Y., Shimazaki, H.: Spatial Analysis of Spike-Timing-Dependent LTP and LTD in the CA1 Area of Hippocampal Slices Using Optical Imaging. Hippocampus 15, 104–109 (2005)
[3] Aihara, T., Abiru, Y., Yamazaki, Y., Watanabe, H., Fukushima, Y., Tsukada, M.: The relation between spike-timing dependent plasticity and Ca2+ dynamics in the hippocampal CA1 network. Neuroscience 145, 80–87 (2007)
[4] Nishiyama, M., Hong, K., Mikoshiba, K., Poo, M., Kato, K.: Calcium stores regulate the polarity and input specificity of synaptic modification. Nature 408, 584–589 (2000)
[5] Bi, G.Q., Rubin, J.: Timing in synaptic plasticity: From detection to integration. TINS 8(5), 222–228 (2005)

[6] Rubin, J.E., Gerkin, R.C., Bi, G.Q., Chow, C.C.: Calcium time course as signal for spike-timing-dependent plasticity. J. Neurophysiol. 93, 2600–2613 (2005)
[7] Bi, G.Q., Poo, M.M.: Synaptic modifications in cultured hippocampal neurons: dependence on spike timing, synaptic strength and postsynaptic cell type. J. Neurosci. 18, 10464–10472 (1998)
[8] Poirazzi, P., Brannon, T., Mel, B.W.: Arithmetic of subthreshold synaptic summation in a model CA1 pyramidal cell. Neuron 37, 977–987 (2003)
[9] Poirazzi, P., Brannon, T., Mel, B.W.: Pyramidal neuron as a 2-layer neural network. Neuron 37, 989–999 (2003)
[10] Traub, R., Jeffreys, J.G., Miles, R., Whittington, M.A., Toth, K.: A branching dendritic model of a rodent CA3 pyramidal neurone. J. Physiol. 481, 79–95 (1994)
[11] Lindskog, M., Kim, M.S., Wikstrom, M.A., Blackwell, K.T., Kotaleski, J.H.: Transient calcium and dopamine increase of PKA activity and DARPP-32 phosphorylation. PLOS 2(9), 1045–1060 (2006)
[12] Castellani, G.C., Quinlan, E.M., Bersani, F., Cooper, L.N., Shouval, H.Z.: A model of bidirectional synaptic plasticity: From signalling network to channel conductance. Learning & Memory 12, 423–432 (2005)

Improving Associative Memory in a Network of Spiking Neurons

Russell Hunter[1,⋆], Stuart Cobb[2], and Bruce P. Graham[1]

[1] Department of Computing Science and Mathematics, University of Stirling, Stirling, FK9 4LA, U.K.
{rhu,b.graham}@cs.stir.ac.uk
[2] Division of Neuroscience and Biomedical Systems, University of Glasgow, Glasgow, G12 8QQ, U.K.
s.cobb@bio.gla.ac.uk

Abstract. Associative neural network models are a commonly used methodology when investigating the theory of associative memory in the brain. Comparisons between the mammalian hippocampus and neural network models of associative memory have been investigated [7]. Biologically based networks are complex systems built of neurons with a variety of properties. Here we compare and contrast associative memory function in a network of biologically-based spiking neurons [14] with previously published results for a simple artificial neural network model [6]. We investigate biologically plausible implementations of methods for improving recall under biologically realistic conditions, such as a sparsely connected network.

Keywords: Associative memory, mammalian hippocampus, neural networks, pattern recall, inhibition.

1 Introduction

Graham and Willshaw [6] tested the pattern recall performance of an artificial neural network model of associative memory comprised of binary units with 10% partial connectivity, with the intention of improving the quality of the pattern recall. They used a method known as the winners-take-all (WTA) approach to decide which output units should fire based on the weighted (dendritic) sum of their inputs [16]. The WTA approach simply chooses the required number of units with the highest dendritic sum to fire during pattern recall. Various mathematical transforms of the dendritic sum compensated for the partial connectivity and noise due to pattern overlap. We investigate whether biologically-plausible implementations of these transforms can be found.

Sommer and Wennekers [14] studied associative memory using a pool of Pinsky-Rinzel two-compartment model CA3 pyramidal cells [12] connected by a structured connectivity matrix. They found that this biologically realistic network provided robust associative storage of sparse patterns at a capacity close to

⋆ Corresponding author.

V. Kůrková et al. (Eds.): ICANN 2008, Part II, LNCS 5164, pp. 636–645, 2008.

that of theoretical neural networks [14]. Starting with the Sommers and Wenneker's model, we test how well global inhibition implements standard WTA recall of a stored pattern. Then further inhibitory circuitry and a change in the membrane properties of the pyramidal cells is implemented in an attempt to improve recall quality and match the methods of Graham and Willshaw [6].

2 The Model

2.1 The Cell

The Pinsky-Rinzel two-compartment, eight-variable reduced model of a CA3 pyramidal cell [12] has soma-like and dendrite-like compartments, which are coupled electrotonically using parameters g_c, which represents the strength of coupling and p, the percentage of total area in the soma-like compartment (fig. 1).

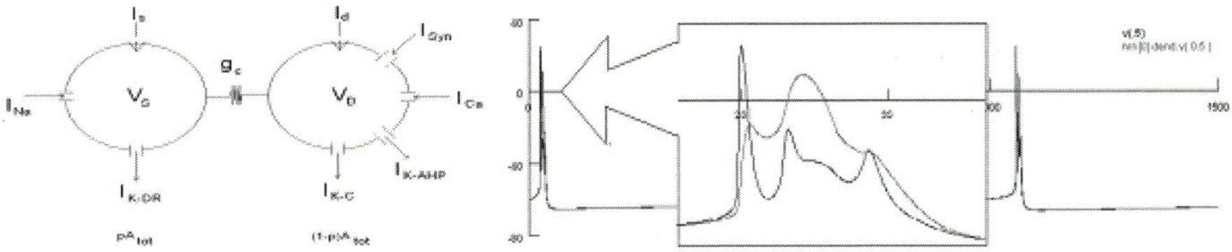

Fig. 1. Pinsky and Rinzel [1] two-compartment model of a CA3 pyramidal cell. Voltage traces show the bursting response of the model in the soma and dendrite.

The soma-like compartment has fast sodium (I_{Na}) and potassium (I_{Kdr}) currents that can generate action potentials (spikes). The dendrite-like compartment contains slower calcium and calcium-modulated currents.

A distinctive characteristic of this two-compartment neuron is the effect of the interactions between the soma and the dendrite mediated by the coupling parameters. Compared to simple integrate-and-fire neurons, the two-compartment cell gives a variety of modes due to the dynamics between the soma and the dendrite such as spiking, bursting or joined spiking/bursting sequences (fig. 1). The exact dynamic equations and parameter settings can be found in [12].

2.2 The Associative Net

The network model contains 100 Pinsky-Rinzel pyramidal cells.The model was created and simulated using the Neuron computer simulation package [1]. Each pyramidal cell is connected to every other pyramidal cell (non-reciprocal) with a certain probability, ranging from 0.1 (for 10% physical connectivty) up to 1 (for full connectivity). Connections use an AMPA synapse which generates a fast excitatory post-synaptic potential. These excitatory connections were tested with varying values of synaptic delay from 0.3 to 1 ms, and with similarly varied peak conductances up to $G_{AMPA} = 0.0154\ \mu S$. Higher conductance is required at lower levels of connectivity to maintain synaptic drive onto each cell.

The actual connectivity is dependent upon the number of patterns stored, in combination with the physical connectivity, which is determined randomly with a minimum connection probability of 10%, but tested at various intermediate levels between 10% and 100%. A pattern consists of 10 randomly-chosen active neurons out of the population of 100. Patterns are stored by clipped Hebbian synaptic modification, resulting in a structured connectivity matrix. For a given pattern, the Hebbian rule specifies a weight of 1 for a connection between two neurons that are both active in the pattern, with all other weights being 0. The connectivity matrix, possibly containing multiple patterns that may overlap, where one cell may be active in one or more patterns, is imposed upon the actual network by randomly choosing connections with a weight of 1 up to the percentage of the physical connectivity.

Within the model there is also global inhibition, similar to the inhibition provided by basket cells. This inhibition acts as a threshold control that moderates the activity level in the network and keeps that activity from unphysiological states where all cells fire at very high rates [14]. It restricts firing rates to approximately gamma frequencies [2]. In the model, the inhibitory dynamics are not induced by interneurons individually, but it is assumed that action potentials of pyramidal cells evoke not only EPSPs on their target cells, but also IPSPs on all cells in the network via inhibitory connections [14]. Accordingly, any spike of a principal cell evokes equally weighted IPSCs onto all principal cells. These inhibitory synapses employ a fast GABA-ergic conductance change with reversal potential V_{CL} = -75 mV and a fast rise-time and slow decay. The connection delay was around 2 ms. The inhibitory peak conductance was fixed at G_{GABA} = 0.00017 μS.

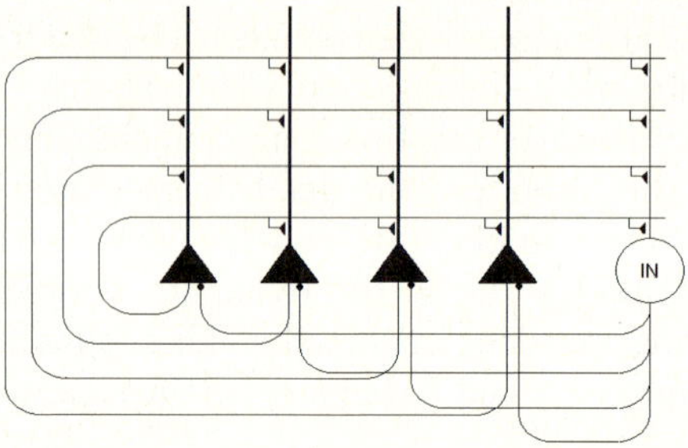

Fig. 2. Circuit diagram of the network. Pyramidal cells have an apical dendrite and soma with excitatory connections between cells but with no connections onto the same cell. Network is fully connected in this example. The IN cell represents the global inhibition mediated by activity from spiking pyramidal cells.

Recall from the network was tested by tonically stimulating 5 from a known pattern of 10 pyramidal cells using current injection to either the soma or dendrite with a strength ranging between 0.00075 and 0.0075 nA. Successful recall would result in the remaining 5 pyramidal cells in the pattern becoming active, but no other cells.

3 Thresholding Strategies for Recall

3.1 Standard Winners-Take-All (WTA)

As described above, the connectivity matrix is determined both by the physical connectivity between cells and the learnt connection weights during pattern storage. Pattern recall proceeds by only those pyramidal cells that receive the greatest excitatory input (dendritic sum) becoming active. For a network with full physical connectivity and not too many patterns stored, this should result in accurate recall of a stored pattern when starting from an initial partial pattern cue [6]. An interesting problem is how well the network can recall a pattern when there is partial connectivity or corruption due to noise (possibly by overlap in pattern storage). In this network of spiking neurons, the standard winners-take-all (WTA) method recalls a pattern where the threshold of PC's firing is set by the intrinsic threshold of the PC itself and the pseudo-basket-cell inhibition. The intrinsic threshold of a PC is largely determined by membrane resistance and sodium channel density.

3.2 Normalised WTA Network (Localized Inhibition)

Partial connectivity complicates recall as a neuron cannot distinguish between missing physical connections, and connections that have not been modified during storage (and consequently have a weight of 0 and so cannot contribute to the cell's dendritic sum). This adds to the variance of dendritic sums across the network. The dendritic sums of cells that belong to a pattern and should be active (high cells) and the sums of cells that should be silent (low cells) may overlap, leading to errors in recall. The overlap between the dendritic sums of high cells and low cells can be reduced by using a normalised winners-take-all approach [6]. The normalised WTA uses the fact that all dendritic sums lie between a range 0 and some maximal level of input activity, which equates with the number of physical connections onto a cell that are active, irrespective of the learnt synaptic weight. Thus this input activity is the amount of excitation each cell could receive, whereas the dendritic sum is the amount of excitation the cell actually receives. Graham and Willshaw [6] found that by normalising a cell's dendritic sum by its input activity, giving an input value between 0 and 1, reduces the error/overlap during recall. This technique is not transferable to a network of realistic pyramidal cells in a direct form but by using a method of localized inhibition proportional to the excitation a cell could receive, the range of EPSPs and thus the dendritic sums produced, are better separated between high and low cells.

The local inhibition is implemented by having inhibitory connections between pyramidal cells corresponding to all possible modified excitatory connections in a partially connected net. Thus the local inhibition inhibits a PC in proportion to the excitation it could receive. This inhibition could be considered as part of a disynaptic inhibitory drive with a fast acting GABAa type synapse [4]. The actual model implements a form of synaptic circuitry that allows two pyramidal cells to rapidly inhibit one another [13]. There is experimental data which implies

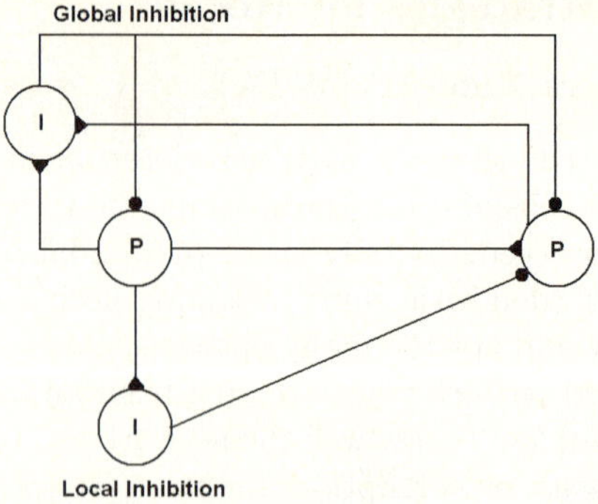

Fig. 3. Normalised WTA with two pyramidal cells: basic global inhibition and a local inhibitory connection

that glutamatergic synapses from pyramidal cells directly excite presynaptic GABAergic terminals, which then inhibit pyramidal cell somata [3]. Fast GABAa mediated events through basket cells perhaps are a more realistic possibility. Such connections have very fast kinetics, short synaptic delays and are very robust. Basket cells are endowed with ionic conductances and specific glutamate receptors to enable very fast forward activation.

3.3 Amplified WTA Method

Another source of noise in the recall process arises from the number of stored patterns each cell belongs to (unit usage). The average excitation a cell receives during recall increases with cell's unit usage, leading to increasing overlap between the dendritic sums of high and low cells. Graham and Willshaw [6] found that for cells with a given unit usage, the variations/overlap due to unit usage can be reduced by a suitable transformation of the dendritic sum as a function of a cell's unit usage. It is not clear how this transformation could be implemented in a biological network, so an alternative method to increase separation between the dendritic sums of low and high cells was tried.

Graham [5] used a method of signal (EPSP) amplification to help discriminate between low and high cells and therefore improve pattern recognition. The summation of EPSPs has a near linear distribution in our PC model. We want to create a non-linear increase in this summation so that cells, after reaching a certain membrane potential, increase the summed amplitude of EPSPs. This has the presumption that cells in a pattern (high cells) will receive slightly more excitation than cells out-with a pattern (low cells). Adding a persistent sodium channel to the soma with a low voltage activation range and appropriate maximum conductance should amplify the signal (summed EPSPs). Testing on a single cell shows a non-linear increase in dendritic summation above a given threshold (fig. 4). We call this the amplified WTA method when incorporated into the network model.

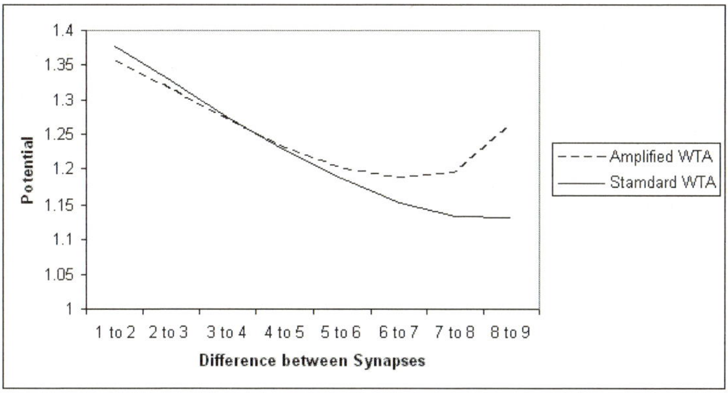

Fig. 4. The difference in membrane potential between the summed EPSPs from n and (n + 1) synapses, where n = 1, 2...8. Summation is slightly sublinear, but it becomes supralinear at high n in the amplified case.

4 Results

Recall performance was tested by storing 50 random patterns, each consisting of 10 active cells, in the network, and then using 5 of the 10 cells of a stored pattern as a recall cue. Physical connectivity was set at 100% or 10%.

Figure 5 shows the spiking activity in a fully connected network resulting from cued input activity when global inhibition is absent (fig. 5a) and present (fig. 5b). Clearly, without inhibition all cells in the network become active, but only cells belonging to the cued pattern largely are active when PC activity is thresholded by global inhibition. The nature of the network determines that the

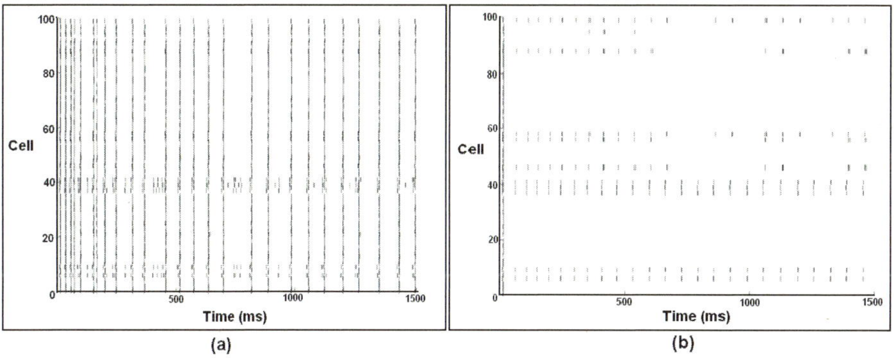

Fig. 5. Spiking activity from a fully connected network. Excitatory conductances are $G_{AMPA} = 0.003\ \mu S$. There are 50 stored patterns in the network and $I_d = 0.0075$ nA is applied to the input cue (5/10 active cells in a pattern) and all hyperpolarizing currents in the cells not associated with the input cue are relaxed i.e. set to 0 nA. (a) no inhibition present. (b)network with global inhibition, $G_{GABA(g)} = 0.00017\ \mu S$.

recall process is synchronous in which cell activation out with the input cue is dependent on action potentials from the cued cells. Thus the recalled cells are in synchrony with the activity of the input cue. The banks of excitation become further apart over time and show no noticeable pattern recall. The introduction of a global inhibitory circuit synchronises the activity to more even time steps approaching the gamma frequency range (fig. 5b). Now the stored pattern is approximately recalled.

Quality of recall was measured by examining spiking activity over 1500ms and calculating a quality measure as:

$$C = \frac{\sum_{i=1}^{N}(B_i - \bar{B})(B^* - \alpha_B)}{\left(\sum_{i=1}^{N}(B_i - \bar{B})^2 \sum_{i=1}^{N}(B^* - \alpha_B)^2\right)^{1/2}}$$

where C is the recall quality, B is a recalled output vector, B^* is the required output, N is the number of cells, α_B is the mean output activity and $\bar{B}$ is the mean activity of the recalled pattern. The required output vector is the selected memory pattern stored in the structured connectivity matrix. The actual output vector is determined by the action potentials from any cell occurring within a given sliding time window of 16 ms. This time was selected on the basis of spiking frequency during recall, so that at most a single spike from a cell would occur in the window.

Recall quality was tested in a 10% partially connected net. With standard WTA recall the mean pattern recall quality over 1500ms is approximately 61% (fig. 6a). Using the normalised WTA approach (fig. 6b) the addition of localised inhibition improves the mean pattern recall quality over 1500ms to approximately 64%. Also a significant improvement can be measured using the amplified WTA where the addition of a persistent Na channel [5] to the PCs gives a mean pattern recall quality over 1500ms of approximately 65%.

Pattern recall is not perfect for any of the recall methods, suggesting confusion from the noise due to overlap in patterns during the pattern storage procedure and partial physical connectivity. The global inhibition tends to synchronize the

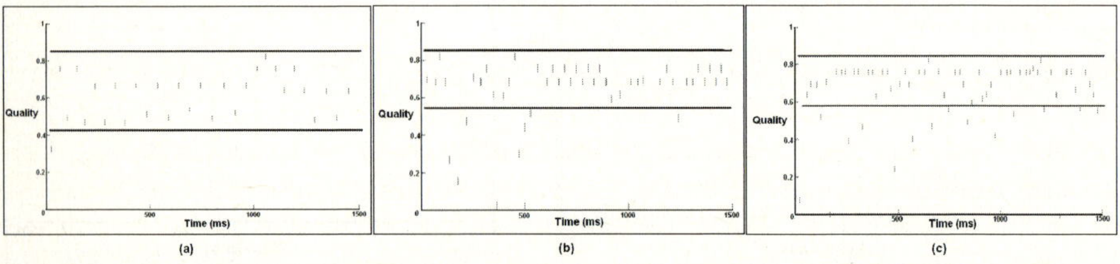

Fig. 6. Recall quality over time in a 10% partially connected network when recalling a single pattern using (a) standard WTA, (b) normalised WTA and (c) amplified WTA. Throughout, $I_d = 0.0075$ nA and $G_{GABA(g)} = 0.00017\ \mu S$. (a) $G_{AMPA} = 0.0154\ \mu S$; (b) $G_{AMPA} = 0.0154\ \mu S$, $G_{GABA(l)} = 0.00748\ \mu S$; (c) $G_{AMPA} = 0.008\ \mu S$, $G_{pNa} = 0.000165\ \mu S$. The horizontal lines are qualitative indicators of the main spread of recall performance in each case.

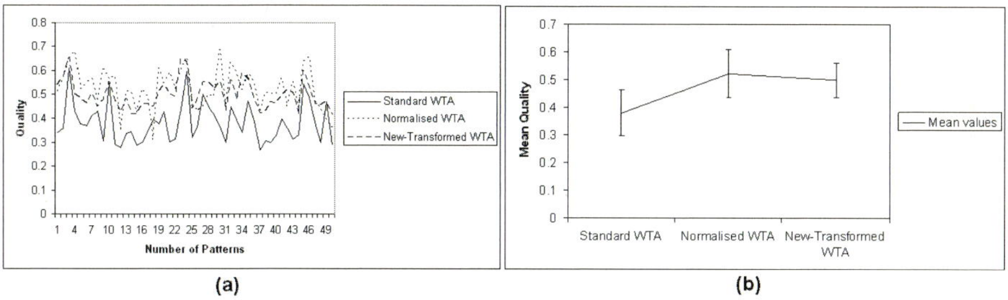

Fig. 7. (a) Mean recall quality of each of the 50 stored patterns over a 1500 ms trial, and (b) mean and standard deviation of recall quality averaged over all stored patterns, for the three recall methods

activity of pyramidal cells in the network. The standard WTA approach (fig. 6a) shows an oscillation between high and low values of recall and a wide variation in the quality of pattern recall over time. The normalised WTA (fig. 6b) has a faster rate of cell spiking due to the localised inhibitory circuit. Also, the variation in recall quality is greatly reduced, with a range of 60% to 80% (excluding some outliers), compared to the standard WTA (fig. 6a) at approximately 40% to 80%. Similarly, the amplified WTA approach (fig. 6c) shows less variation in quality of recall per iteration with a range of 60% to 80% and fewer outliers. Inherent outliers can be attributed to increased iterations from the extra inhibition in the normalised WTA method and the persistent Na channel increasing the likeliness of an action potential.

The quality of recall varies across the 50 stored patterns due to the noise from pattern overlap and partial connectivity (fig. 7a). The mean quality over all stored patterns shows a statistically significant (95% confidence interval) increase when using the normalised and amplified methods compared to the standard WTA method (fig. 7b).

5 Conclusion

Our modelling experiments demonstrate that the results from methods to improve recall in a network of spiking neurons show significant correlations with the results found in artificial neural networks of associative memory [6]. Biologically plausible methodology using inhibitory microcircuits, based on linear algebraic functions (see [6]), to improve recall in a network of cells has resulted in an overall improvement in the quality of pattern recall from a partially connected network with high pattern loading.

Inhibitory microcircuits can play a variety of roles within autoassociative memory spiking neural networks. As shown experimentally [2, 10], global feedback inhibition acts to synchronize principal cell activity to regular firing within the gamma frequency range (40-100Hz). Reduction in GABA synthesis in a subpopulation of inhibitory GABA-ergic neurons results in a diminished capacity for the gamma-frequency synchronized neuronal activity [9]. Such network activity is seen in our model and other similar models [4, 8, 11, 14, 15]. These models also show that, within the context of associative memory, such global inhibition provides a simple thresholding of PC activity that can lead to pattern recall. Here we show that an additional localised inhibitory circuit can improve pattern recall in partially connected networks. Another use of inhibition in associative memory models is for setting the negative weights that result from the general Hebbian learning rule for pattern storage [4, 11]. This also requires a specifically structured inhibition in which the negative affect of one PC on another is generated by a disynaptic pathway involving an interposed inhibitory cell.

In our model, adding a persistent Na channel to the cell to amplify large EPSPs also improved the quality of pattern recall. This result suggests that the membrane properties of pyramidal cells may be able to reduce noise in patterns of synaptic input. The added persistent Na channel confirms the methods explored in [5], where it was found that voltage-gated ion channels act to boost synaptic input and thus improve recall in a model of associative memory.

Acknowledgement. This work was funded by an EPSRC project grant to B. Graham and S. Cobb.

References

[1] Carnevale, N.T., Hines, M.L.: The Neuron Book. Cambridge University Press, Cambridge (2005)

[2] Cobb, S.R., Buhl, E.H., Halasy, K., Paulsen, O., Somogyi, P.: Synchronization of neuronal activity in hippocampus by individual GABAergic interneurons. Nature 378, 75–78 (1995)

[3] Connors, B.W., Cruickshank, S.J.: Bypassing interneurons: inhibition in neocortex. Nature Neuroscience 10(7), 808–810 (2007)

[4] Fransen, E., Lasner, A.: A model of cortical associative memory based on a horizontal network of connected columns. Network Comput Neural Syst. 9, 235–264 (1998)

[5] Graham, B.P.: Pattern recognition in a compartmental model of a CA1 pyramidal neuron. Network Comput Neural Syst. 12, 473–492 (2001)

[6] Graham, B., Willshaw, D.: Improving recall from an associative memory. Bio Cybernetics 72, 337–346 (1995)

[7] Graham, B., Willshaw, D.: Capacity and information efficiency of the associative net. Network 8, 35–54 (1997)

[8] Jensen, O., Idiart, M.A., Lisman, J.E.: Physiologcally realistic formation of autoassociataive memory in networks with theta/gamma oscillations: role of fast NMDA channels. Learn. Mem. 3, 243–256 (1996)

[9] Lewis, D.A., Hashimoto, T., Volk, D.W.: Cortical Inhibitory Neurons and Schizophrenia. Nature 6, 312–324 (2005)

[10] Mann, E.O., Catrin, A.R., Paulsen, O.: Hippocampal gamma-frequency oscillations: from interneurones to pyramidal cells, and back. J. Physiol. 562, 55–63 (2005)

[11] Menschik, E.D., Finkel, L.H.: Neuromodulatory control of hippocampal function: towards a model of Alzheimer's disease. Artificial Intelligence in Medicine 13, 99–121 (1998)
[12] Pinsky, P., Rinzel, J.: Intrinsic and Network Rhythmogenesis in a Reduced Traub Model for CA3 Neurons. J. Comput. Neuroscience 1, 3960 (1994)
[13] Rolls, E.T., Kesner, R.P.: A computational theory of hippocampal function, and empirical tests of the theory. Progress in Neurobiology 79, 1–48 (2006)
[14] Sommer, F.T., Wennekers, T.: Associative memory in networks of spiking neurons. Neural Networks 14, 825–834 (2001)
[15] Wennekers, T., Sommer, F.T., Palm, G.: Iterative retrieval in associative memories by threshold control of different neural models. In: Herrmann, H.J., Wolf, D.E., Poppel, E. (eds.) Supercomputing in Brain Research: From Tomography to Neural Networks, pp. 301–319. World Scientific, Singapore (1995)
[16] Willshaw, D.: Ph. D. Thesis. J. University of Edinburgh (1971)

Effect of Feedback Strength in Coupled Spiking Neural Networks

Javier Iglesias[1,2], Jordi García-Ojalvo[2], and Alessandro E.P. Villa[1]

[1] Grenoble Institut des Neurosciences-GIN, NeuroHeuristic Research Group
Université Joseph Fourier, Grenoble, France
{Javier.Iglesias,Alessandro.Villa}@ujf-grenoble.fr
http://www.neuroheuristic.org/
[2] Departament de Física i Enginyeria Nuclear
Universitat Politècnica de Catalunya, Terrassa, Spain
{Javier.Iglesias,Jordi.G.Ojalvo}@upc.edu
http://www-fen.upc.es/

Abstract. We simulated the coupling of two large spiking neural networks (10^4 units each) composed by 80% of excitatory units and 20% of inhibitory units, randomly connected by projections featuring spike-timing dependent plasticity, locality preference and synaptic pruning. Only the first network received a complex spatiotemporal stimulus and projected on the second network, in a setup akin to coupled semiconductor lasers. In a series of simulations, the strength of the feedback from the second network to the first was modified to evaluate the effect of the bidirectional coupling on the firing dynamics of the two networks. We observed that, unexpectedly, the number of neurons which activity is altered by the introduction of feedback increases in the second network more than in the first network, suggesting a qualitative change in the dynamics of the first network when feedback is increased.

1 Introduction

Recent studies have shown analogies between the dynamics of neurons and lasers [1]. In a series of experiments and simulations [2], collective dynamics were investigated in the case of two lasers coupled by mutual injection of their emitted light beams in presence of a transmission delay. The experimental setup featured a first semiconductor laser L_1 operating in the low-frequency fluctuation (LFF) chaotic regime [3] in presence of optical feedback from an external mirror. The LFF regime is characterized by sudden drops in the emitted light at irregular times with periods on the order of tens of nanoseconds, followed by a gradual recovery. When the L_1 output beam was injected into a second semiconductor laser L_2 exhibiting similar physical properties but operating in a non-chaotic mode, synchronized power dropouts were induced in L_2 with a lag corresponding to the transmission delay. Then, by injection of variable amounts of L_2 light into L_1, the leader-laggard dynamics could be controlled in a reversible way such that, given enough feedback, L_1 dynamics could be synchronized with L_2 with

V. Kůrková et al. (Eds.): ICANN 2008, Part II, LNCS 5164, pp. 646–654, 2008.

a lag equal to the transmission delay, thus exchanging at will the leader and laggard roles between the lasers.

For this study, a similar setup was prepared where two simulated large spiking neural networks substituted the semiconductor lasers chaotic dynamics. The simulations reported here investigate the effect of different injection/feedback strength on the networks spiking dynamics. The effect of the number of projecting units, the relative sizes of the injection and the feedback, as well as the number of projections involved in the injection/feedback were tested.

2 Neural Network Model

The complete neural network model is described in details elsewhere [4]. A short description of the model with specific model parameters related to the current study follows below.

As sketched in Figure 1, we assume that at time $t = 0$ of the simulation each of the two networks N_1 and N_2 are characterized by two types of integrate-and-fire units and by its maximum over growth in terms of connectivity. The amount of units in each network is 10,000 (8,000 excitatory and 2,000 inhibitory) laid down on a 100×100 2D lattice according to a space-filling quasi-random Sobol distribution, summing to a total of 20,000 units.

The network model features cell death mechanisms that may be provoked by: (i) an excessive firing rate and (ii) the loss of all excitatory inputs. An excessive firing rate is assumed to correspond to the biological effect known as glutamate neurotoxicity [5]. During an initial phase called "early developmental

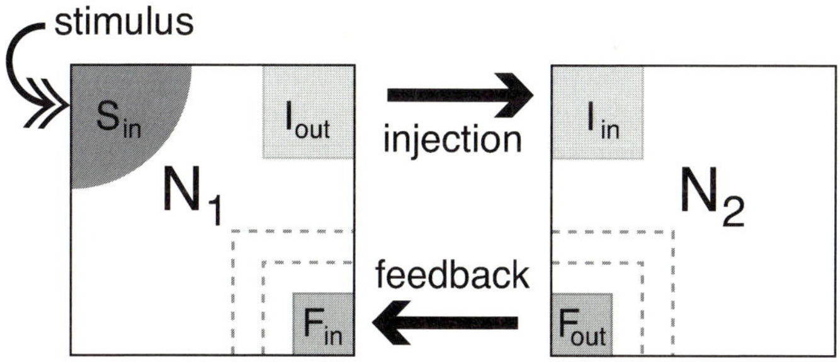

Fig. 1. A sketch of the experimental layout. Two networks N_1 and N_2 with identical connectivity properties are synchronously simulated. A group S_{in} of N_1 composed by a fix number of units (800) receives a spatio-temporally organized stimulus. A group I_{out} composed by a number of N_1 units projects to I_{in}, an equal number of N_2 units. In a similar way, a group F_{out} composed by a variable number of N_2 units projects to F_{in}, an equally variable number of N_1 units. Two procedures are used for the wiring of the injection I and the feedback F: procedure ψ_t in which the number of projections is fixed to 10^5 for I and F; and procedure ψ_u in which the number of projections is fixed to 100 per unit of I_{out} and F_{out}. The size of the F group is defined as a ratio of the size of the corresponding I group. All units of S_{in}, I_{in}, I_{out}, F_{in} and F_{out} are excitatory.

phase", at each time step and for each unit a maximum firing rate was arbitrarily determined following a parameter search procedure described in [6]. Whenever the rate of discharges of a unit exceeds the maximum allowed rate the cell had a probability to die according to a "genetically" determined probability function. A dead unit is characterized by the absence of any spiking activity.

From the network N_1, two sets of 400 "sensory units" (S_{in}) are randomly selected among the 8,000 excitatory units at time $t = 0$. The units belonging to these sets are the "sensory units" of the network, meaning that in addition to sending and receiving connections from the other units of both types they are stimulated by patterned activity organized both in time and space described elsewhere [6]. This stimulus is assumed to correspond to content-related activity generated elsewhere in the brain. The overall stimulus duration is set to 100 ms, followed by a period without stimulation that lasted 1900 ms. Thus, the rate of stimulation was 0.5 $stim/s$. Such a stimulus was not delivered to the network N_2.

Some randomly selected excitatory units of network N_1 (I_{out}) project on an equal number of randomly selected excitatory units of network N_2 (I_{in}), such that the activity of I_{out} units is transmitted with a fixed 1 time step delay to some units of I_{in}. The exact number of projections depends on the wiring procedure. Two wiring procedures were used in this study: (ψ_t) a fixed total number of 10^5 projections from I_{out} to I_{in} and (ψ_u) a fixed number of 100 projections per projecting units from I_{out} to I_{in}. In a similar way, some units of network N_2 (F_{out}) project feedback on an equal number of units in network N_1 (F_{in}). The exact number of projections is determined with the corresponding wiring procedures ψ_t and ψ_u described above.

At the end of the early developmental phase, the synaptic plasticity [7] is enabled. It is assumed a priori that modifiable synapses are characterized by discrete activation levels [8] that could be interpreted as a combination of two factors: the number of synaptic *boutons* between the pre- and post-synaptic units and the changes in synaptic conductance. In the current study we attributed a fixed activation level (meaning no synaptic modification) $A_{ji}(t) = 1$, to *(inh, exc)* and *(inh, inh)* synapses while activation levels were allowed to take one of $A_{ji}(t) = \{0, 1, 2, 4\}$ for *(exc, exc)* and *(exc, inh)*, $A_{ji}(t) = 0$ meaning that the projection was permanently pruned out. For $A_{ji}(t) = 1$, the post-synaptic potentials were set to 0.84 mV and -0.8 mV for excitatory and inhibitory units, respectively. According to the wiring procedure $[\psi_t, \psi_u]$ and to I and F values, the number of projections varied between 6.1 and $6.9 \cdot 10^6$ at time $t = 0$. Between 0.1 and 13.2% of these were related to the injection and feedback. The projections from and to "dead" units undergo a decay of their synapses leading eventually to their pruning when $A_{ji}(t) = 0$. Other projections may be pruned due to synaptic depression driven by STDP and also leading to $A_{ji}(t) = 0$. Thus, some units that survived the early phase can also remain without any excitatory input. The loss of all excitatory inputs also provokes the cell death and these units stop firing (even in presence of background activity) immediately after the pruning of the last excitatory afference from within the network.

3 Simulations

The values of the set of parameters specified above were kept constant throughout all this study. Each simulation run lasted $5 \cdot 10^5$ discrete time steps (T_{end}), with 1 time step corresponding to 1 ms in the model, that means $500,000$ ms as total duration of a simulation run. The states (spiking/not spiking) of all N_1 and N_2 units were updated synchronously and recorded for further investigation. After spiking, the membrane potential was reset to its resting potential, and the unit entered an absolute refractory period lasting 3 and 2 time steps for excitatory and inhibitory units, respectively. Starting at time $t = 0$ and throughout all the simulation run each unit received a background activity following an independent Poisson process of 5 $spikes/s$ on average.

The early developmental phase, characterized by cell death provoked by excessive firing rate, begins at time $t = 0$ and lasts until $t = T_{edp}$, that was fixed at 700 ms for this study. The spike timing dependent plasticity is enabled at $t = T_{edp} + 1$ for all projections, including those involved in the injection and the feedback. At time $t = 1001$ ms the first stimulation is applied to S_{in} units of N_1, lasting 100 ms until $t = 1100$ ms. Between $t = 1101$ ms and $t = 3000$ ms only the background activity is getting into the two networks. At time $t = 3001$ ms another stimulation is applied and so forth until the end of the simulation run. Overall this corresponds to 250 presentations of the stimulus along one simulation run.

The events belonging to the univariate time series formed by the spike train may have different origins. Spikes may be simply associated to the Poisson background noise whenever the cell excitability is high and close enough to the threshold of activation. Other spikes can be produced by the convergence of synchronous activity (i.e., temporal summation of excitatory post-synaptic potentials) generated within the network or through the injection/feedback projections.

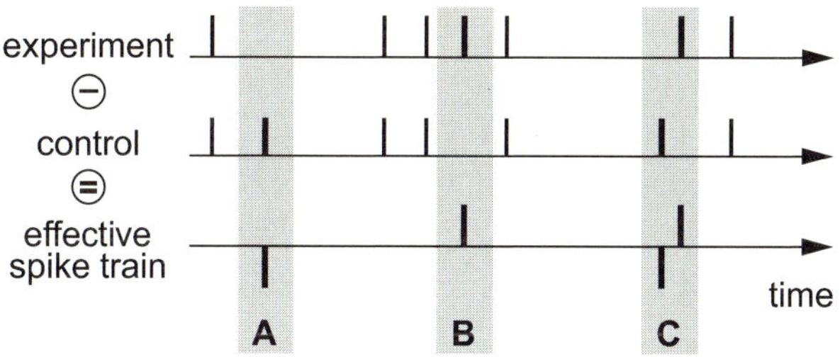

Fig. 2. The experimental conditions alter the spike timing of the units with respect to a control condition. An effective spike train is computed by subtracting the spikes recorded in the control simulation (including the background activity) to the spikes recorded in the experiment in order to quantify these changes. Three types of differences can be accounted for **(A)** deleted episodes relative to the control condition; **(B)** inserted episodes relative to the control condition; and **(C)** drifted (anticipated or delayed) episodes which result in both a deleted and an inserted episode in the effective spike train.

In order to study the spikes associated to the network activity in presence of injection and feedback for $1000 < t < T_{end}$ (i.e. during the STDP and synaptic pruning phase of the simulation), those spikes associated to the background noise and generated within each network were discarded from the spike trains and the so-called "effective spike trains" were extracted [9] as sketched in Figure 2. In this study, for each value of I and for each procedure (fixed total number of projections ψ_t or fixed number of projections per unit ψ_u), the control condition was defined as being the simulation where the corresponding size of $F = 0$, i.e. in presence of injection but in absence of feedback. This method permits to subtract the noisy background activity, the intrinsic network activity and the change of activity in N_2 induced by the direct injection $N_1 \rightarrow I_{out} \rightarrow I_{in} \rightarrow N_2$, as those are also recorded in the control condition in absence of feedback. The changes in activity induced in N_1 by the feedback $N_2 \rightarrow F_{out} \rightarrow F_{in} \rightarrow N_1$, and those induced in N_2 through one (or multiple) complete loops $N_2 \rightarrow F_{out} \rightarrow F_{in} \rightarrow N_1 \rightarrow I_{out} \rightarrow I_{in} \rightarrow N_2 \rightarrow \ldots$ are isolated and studied for each simulation in the respective effective spike train.

4 Results

For each I and F values and for each ψ_t, ψ_u procedure, we computed from the effective spike train previously described the fraction of units of $N_1 \notin F_{in}$ and $N_2 \notin I_{in}$ which spiking activity was altered by the introduction of the feedback with respect to the control condition. In order to compare the results obtained with the different values of I and F, the data was normalized by using the relative size F/I. Therefore, in the following figures, a value of $F/I < 1$ means that there are more units projecting from N_1 to N_2 than units projecting from N_2 to N_1 while a value of $F/I > 1$ means that there are more units projecting from N_2 to N_1 than units projecting from N_1 to N_2. The number of injection and feedback units is balanced when $F/I = 1$. We investigated the network activity for F/I between $\frac{1}{4}$ and 3.

Figure 3 shows the results for the fixed total number of projections ψ_t procedure. The ratio $\rho_{\psi_t}^{N_1}$ of units in $N_1 \notin F_{in}$ which activity is modified by the presence of the feedback from N_2 to N_1 is plotted in Fig. 3A for different values of I and F, while the corresponding ratio $\rho_{\psi_t}^{N_2}$ in $N_2 \notin I_{in}$ is plotted in Fig. 3B. The effect of the increasing I and F sizes is visible at $F/I = 1$ where, despite the fact that with the ψ_t procedure all have the same total number of projections (10^5), larger I and F lead to larger $\rho_{\psi_t}^{N_1}$ and $\rho_{\psi_t}^{N_2}$. It is noticeable that this group size effect in N_1 is canceled for $F/I > 1$ while it is amplified in N_2, suggesting that, when the total number of projections is kept constant, recruiting more N_2 units to send feedback to N_1 counterintuitively leads to relatively more changes in the dynamics of N_2 (through $F_{in} \rightarrow N_1 \rightarrow I_{out} \rightarrow I_{in}$) without recruiting more units from N_1 where $\rho_{\psi_t}^{N_1}$ is constant and below 10%.

The situation is different for the fixed 100 projections per I_{out} and F_{out} units ψ_u procedure depicted in Figure 4. Again, the ratio $\rho_{\psi_u}^{N_1}$ of units in $N_1 \notin F_{in}$

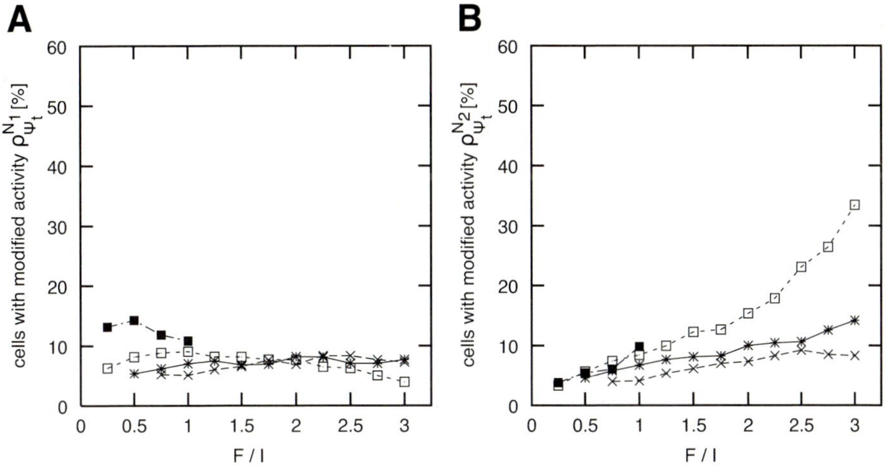

Fig. 3. The effective spike trains were computed between $1000 < t < T_{end}$ for all the units of N_1 and N_2 for different values of I and F group sizes in the ψ_t procedure, maintaining a fix total number of projections (10^5) from I_{out} to I_{in} and from F_{out} to F_{in}, respectively. **(A)** ratio $\rho_{\psi_t}^{N1}$ of excitatory units in $N_1 \notin F_{in}$ which activity is modified by the presence of feedback; **(B)** ratio $\rho_{\psi_t}^{N2}$ of excitatory units in $N_2 \notin I_{in}$ which activity is modified by the presence of feedback. $\times$ $I = 500$; $*$ $I = 1000$; □ $I = 2000$; ■ $I = 4000$.

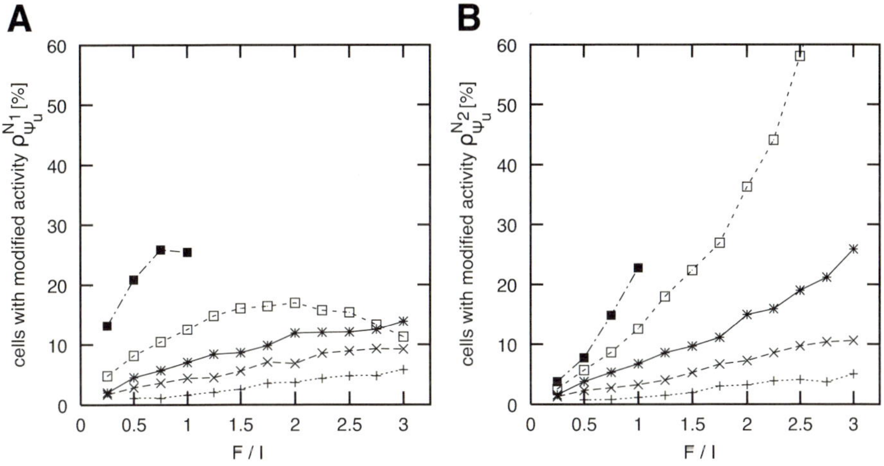

Fig. 4. The effective spike trains were computed between $1000 < t < T_{end}$ for all the units of N_1 and N_2 for different values of I and F group sizes in the ψ_u procedure, maintaining 100 projections from each unit of I_{out} to I_{in} and from each unit of F_{out} to F_{in}, respectively. Note that this results in an increasing number of projections for increasingly larger I and F. **(A)** ratio $\rho_{\psi_u}^{N1}$ of excitatory units in $N_1 \notin F_{in}$ which activity is modified by the presence of feedback; **(B)** ratio $\rho_{\psi_u}^{N2}$ of excitatory units in $N_2 \notin I_{in}$ which activity is modified by the presence of feedback. $+$ $I = 250$; $\times$ $I = 500$; $*$ $I = 1000$; □ $I = 2000$; ■ $I = 4000$.

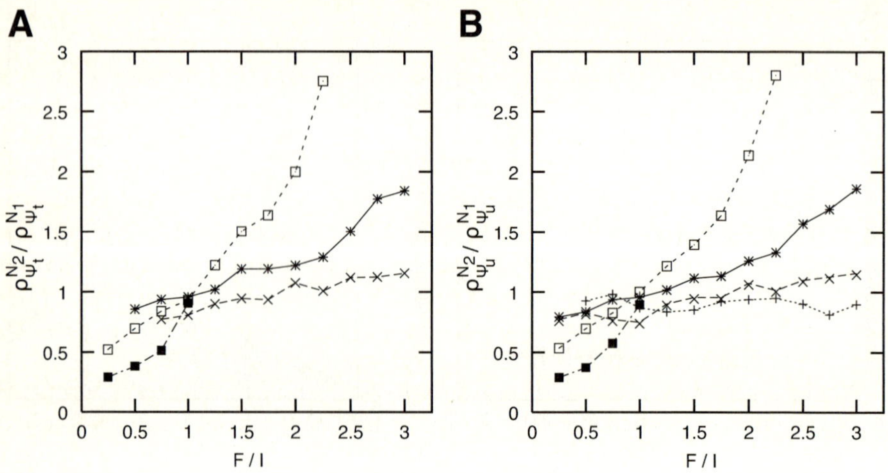

Fig. 5. Comparing the relative effect of the feedback in N_1 and N_2: **(A)** ratio $\rho^{N_2}_{\psi_t}/\rho^{N_1}_{\psi_t}$ between the effect in N_2 and N_1 in the procedure ψ_t with fixed total number of projections from I_{out} to I_{in} and from F_{out} to F_{in}, as presented in Figure 3 ; **(B)** ratio $\rho^{N_2}_{\psi_u}/\rho^{N_1}_{\psi_u}$ between the effect in N_2 and N_1 in the ψ_u procedure with 100 projections from each unit of I_{out} to I_{in} and from each unit of F_{out} to F_{in}, as presented in Figure 4 ; $+$ $I = 250$; $\times$ $I = 500$; $*$ $I = 1000$; $\square$ $I = 2000$; $\blacksquare$ $I = 4000$.

which activity is modified by the presence of the feedback is plotted in Figure 4A for different values of I and F, while the corresponding ratio $\rho^{N_2}_{\psi_u}$ in $N_2 \notin I_{in}$ is plotted in Figure 4B. With this ψ_u procedure, the total number of projections increases with the size of I and F, resulting in a stronger group size effect in both N_1 (Fig. 4A) and N_2 (Fig. 4B) as can be seen at $F/I = 1$. Note that the number of projections is equal for Fig. 3 and 4 in the specific case of $F/I = 1$ and $I = 1000$, providing a direct comparison point. With this wiring procedure ψ_u, increasing F/I values result in almost linear increase of ρ^{N_1} and above linear increase of ρ^{N_2}.

We investigated the relative effect on the dynamics of N_2 and N_1 across the different simulations. For that purpose, for each values of I and F, we plotted in Figure 5A (Figure 5B) the ratio computed between the values of $\rho^{N_2}_{\psi_t}$ and $\rho^{N_1}_{\psi_t}$ shown in Fig. 3B and Fig. 3A ($\rho^{N_2}_{\psi_u}$ and $\rho^{N_1}_{\psi_u}$ from Fig. 4B and Fig. 4A, respectively). A value above 1 for this ratio results from a relatively larger effect of the presence of feedback in N_2 than N_1. With both procedures, this ratio is roughly above 1 for $F/I > 1$ and, most importantly, increases with F/I in every condition, confirming that the addition of feedback results in increasingly more changes in the activity of N_2 than in N_1. It is also interesting to notice that Fig. 5A and Fig. 5B can be almost perfectly superimposed, suggesting that this ratio ρ^{N_2}/ρ^{N_1} cancels the differences between the two procedures ψ_t and ψ_u used for the simulations.

5 Discussion

In an experimental setup inspired by semiconductor laser experiments [2], we simulated two coupled large scale spiking neural networks, with the time resolution of 1 ms, characterized by a brief initial phase of cell death [6]. During this phase the units that exceeded a certain threshold of firing had an increasing probability to die with the passing of time until 700 time units. After the stop of the massive cell death, spike timing dependent plasticity (STDP) and synaptic pruning were made active. Selected sets of units of only one of the networks were activated by regular repetitions of a spatiotemporal pattern of stimulation. During the STDP phase, the cell death could occur only if a unit became deafferented, i.e. it lost all its excitatory afferences because of synaptic pruning.

We recorded the spike trains of all excitatory units that were not directly stimulated either by the external stimulus or the injection/feedback projections for $1000 < t < T_{end}$. For different injection/feedback strength and wiring procedures, we extracted the effective spike train by subtracting the activity recorded in absence of feedback from network N_2 to network N_1, removing the background activity, the intrinsic networks activity and the unidirectional injection-induced activity. We searched the effective spike trains for the units which activity was modified by the introduction of feedback, either through insertion or deletion of episodes.

We observed that the relative number of units affected by the injection tended to increase with an increase in the size of the injection/feedback groups, with both constant and increasing number of projections wiring protocols. As expected, the group size effect is amplified by the presence of an increased number of projections, but the main difference resides in the modified response to the change in the balance of injection/feedback.

A less expected feature of the system is that recruiting more feedback units (i.e. increasing F/I) leads to more alteration of the dynamics of network N_2 than of N_1. In the case of the ψ_t wiring procedure, in which the total number of injection/feedback projections is not altered by changes of F/I, the ratio $\rho_{\psi_t}^{N_1}$ of N_1 units remains constant for all $F/I > 1$, while this ratio increases in N_2. This might be explained by qualitative changes of the dynamics of N_1 which are not taken into account by the ρ^{N_1} ratio. Understanding this aspect will require a precise analysis of the effective spike trains recorded during the simulations. Indeed, similar networks have been shown to produce precise firing sequences when stimulated with spatially and temporally organized stimuli [10]. In future work, the recorded effective time series will be studied to characterize the dynamical changes involved in their alteration in order to determine if the activity organized in time is exchanged and reinforced through the injection/feedback loop.

Acknowledgments. This work was partially funded by the European Community New and Emerging Science and Technology NEST program, grant #043309 (GABA), and the Swiss National Science Foundation SNF, grant #PA002–115330/1.

References

1. Meucci, R., di Garbo, A., Allaria, E., Arecchi, F.T.: Autonomous bursting in a homoclinic system. Phys. Rev. Lett. 88(14), 144101 (2002)
2. González, C.M., Torrent, M.C., García-Ojalvo, J.: Controlling the leader-laggard dynamics in delay-synchronized lasers. Chaos 17 (33122), 1–8 (2007)
3. Fischer, I., van Tartwijk, G.H., Levine, A.M., Elsässer, W., Gbel, E., Lenstra, D.: Fast pulsing and chaotic itinerancy with a drift in the coherence collapse of semiconductor lasers. Physical Review Letters 76 (2), 220–223 (1996)
4. Iglesias, J., Eriksson, J., Grize, F., Tomassini, M., Villa, A.E.: Dynamics of pruning in simulated large-scale spiking neural networks. BioSystems 79(1), 11–20 (2005)
5. Choi, D.W.: Glutamate neurotoxicity and diseases of the nervous system. Neuron 1(8), 623–634 (1988)
6. Iglesias, J., Villa, A.: Neuronal cell death and synaptic pruning driven by spike-timing dependent plasticity. In: Kollias, S.D., Stafylopatis, A., Duch, W., Oja, E. (eds.) ICANN 2006. LNCS, vol. 4131, pp. 953–962. Springer, Heidelberg (2006)
7. Iglesias, J., Eriksson, J., Pardo, B., Tomassini, M., Villa, A.E.: Emergence of oriented cell assemblies associated with spike-timing-dependent plasticity. In: Duch, W., Kacprzyk, J., Oja, E., Zadrożny, S. (eds.) ICANN 2005. LNCS, vol. 3696, pp. 127–132. Springer, Heidelberg (2005)
8. Montgomery, J.M., Madison, D.V.: Discrete synaptic states define a major mechanism of synapse plasticity. Trends in Neurosciences 27(12), 744–750 (2004)
9. Hill, S., Villa, A.E.: Dynamic transitions in global network activity influenced by the balance of excitation and inhibtion. Network: computational neural networks 8, 165–184 (1997)
10. Iglesias, J., Chibirova, O., Villa, A.E.: Nonlinear dynamics emerging in large scale neural networks with ontogenetic and epigenetic processes. In: de Sá, J.M., Alexandre, L.A., Duch, W., Mandic, D.P. (eds.) ICANN 2007. LNCS, vol. 4668, pp. 579–588. Springer, Heidelberg (2007)

Bifurcations in Discrete-Time Delayed Hopfield Neural Networks of Two Neurons*

Eva Kaslik and Stefan Balint

Department of Mathematics and Computer Science
West University of Timisoara, Romania
kaslik@info.uvt.ro

Abstract. This paper is devoted to the bifurcation analysis of a two-dimensional discrete-time delayed Hopfield-type neural network. In the most general framework considered so far in the known literature, the stability domain of the null solution and the bifurcations occurring at its boundary are described in terms of two characteristic parameters. By applying the center manifold theorem and the normal form theory, the direction and stability of the existing bifurcations are analyzed.

1 Introduction

Since the pioneering work of [1], the dynamics of continuous-time Hopfield neural networks have been thoroughly analyzed. However, discrete-time counterparts of continuous-type neural networks have only been in the spotlight since 2000, although they are essential when implementing continuous-time neural networks for practical problems such as image processing, pattern recognition and computer simulation.

One of the first problems that needed to be clarified, concerned the discretization technique which should be applied in order to obtain a discrete-time system which preserves certain dynamic characteristics of the continuous-time system. In [2] a semi-discretization technique has been presented for continuous-time Hopfield neural networks, which leads to discrete-time neural networks which faithfully preserve some characteristics of the continuous-time network, such as the steady states and their stability properties.

In recent years, the theory of discrete-time dynamic systems has assumed a greater importance as a well deserved discipline. Many results in the theory of difference equations have been obtained as natural discrete analogs of corresponding results from the theory of differential equations. Nevertheless, the theory of difference equations is a lot richer than the corresponding theory of differential equations. For example, a simple difference equation resulting from a first order differential equation may exhibit chaotic behavior which can only happen for higher order differential equations. This is the reason why, when

* This work has been supported by the Romanian National Authority for Research under the contract PN-II-11028/14.09.2007 (NatComp - New Natural Computing Models in the Study of Complexity).

V. Kůrková et al. (Eds.): ICANN 2008, Part II, LNCS 5164, pp. 655–664, 2008.

studying discrete-time counterparts of continuous neural networks, important differences and more complicated behavior may also be revealed.

The analysis of the dynamics of neural networks focuses on three directions: discovering equilibrium states and periodic or quasi-periodic solutions (of fundamental importance in biological and artificial systems, as they are associated with central pattern generators [3]), establishing stability properties and bifurcations (leading to the discovery of periodic solutions), and identifying chaotic behavior (with valuable applications to practical problems such as optimization [4,5,6,7], associative memory [8] and cryptography [9]).

We refer to [10,11] for the study of the existence of periodic solutions of discrete-time Hopfield neural networks with delays and the investigation of exponential stability properties.

In [12,13] and in the most general case, in [14], a bifurcation analysis of two dimensional discrete neural networks without delays has been undertaken. In [15,16], the bifurcation phenomena have been studied, for the case of two- and n-dimensional discrete neural network models with multi-delays obtained by applying the Euler method to a continuous-time Hopfield neural network with no self-connections. In [17,18], a bifurcation analysis for discrete-time Hopfield neural networks of two neurons with self-connections has been presented, in the case of a single delay and of two delays. In [19], a generalization of these results was attempted, considering three delays; however, only two delays were considered independent (the third one is a linear combination of the first two) and the analysis can be reduced to the one presented in [17].

The latest results concerning chaotic dynamics in discrete-time delayed neural networks can be found in [20] and [21].

In this paper, we study the discrete-time Hopfield-type neural network of two neurons with finite delays defined by:

$$\begin{cases} x_{n+1} = a_1 x_n + T_{11} g_1(x_{n-k_{11}}) + T_{12} g_2(y_{n-k_{12}}) \\ y_{n+1} = a_2 y_n + T_{21} g_1(x_{n-k_{21}}) + T_{22} g_2(y_{n-k_{22}}) \end{cases} \forall n \geq \max(k_{11}, k_{12}, k_{21}, k_{22}) \quad (1)$$

In this system $a_i \in (0,1)$ are the internal decays of the neurons, $T = (T_{ij})_{2\times 2}$ is the interconnection matrix, $g_i : \mathbb{R} \to \mathbb{R}$ represent the neuron input-output activations and $k_{ij} \in \mathbb{N}$ represent the delays. In practice, due to the finite speeds of the switching and transmission of signals in a network, time delays unavoidably exist in a working network, therefore, they should be incorporated into the model equations of the network.

In order to insure that delays are present, we consider $\max(k_{11}, k_{12}, k_{21}, k_{22}) > 0$. The non-delayed case was extensively studied in [14]. In the followings, we will denote $k_1 = \max(k_{11}, k_{21})$ and $k_2 = \max(k_{12}, k_{22})$.

We will suppose that the activation functions g_i are of class C^3 in a neighborhood of 0 and that $g_i(0) = 0$. In the followings, let $g : \mathbb{R}^2 \to \mathbb{R}^2$ be the function given by $g(x,y) = (g_1(x), g_2(y))^T$ and

$$B = TDg(0) = \begin{pmatrix} T_{11} g_1'(0) & T_{12} g_2'(0) \\ T_{21} g_1'(0) & T_{22} g_2'(0) \end{pmatrix} = \begin{pmatrix} b_{11} & b_{12} \\ b_{21} & b_{22} \end{pmatrix}$$

Considering equal internal decays $a_1 = a_2 = a$, delays satisfying $k_{11} + k_{22} = k_{12} + k_{21}$ and supposing that $b_{11} = b_{22}$, we will present a complete stability and bifurcation analysis in a neighborhood of the null solution of (1). This analysis allows the description of the stability domain of the null solution and the types of bifurcation occurring at its boundary, in terms of the characteristic parameters $\beta = b_{11} = b_{22}$ and $\delta = b_{11}b_{22} - b_{12}b_{21}$. By applying the center manifold theorem and the normal form theory, the direction and stability of the existing bifurcations will be analyzed. A numerical example will be presented to substantiate the theoretical findings.

2 Stability and Bifurcation Analysis

We transform system (1) into the following system of $k_1 + k_2 + 2$ equations without delays:

$$\begin{cases} x_{n+1}^{(0)} = a_1 x_n^{(0)} + T_{11} g_1(x_n^{(k_{11})}) + T_{12} g_2(y_n^{(k_{12})}) \\ x_{n+1}^{(j)} = x_n^{(j-1)} \qquad \forall j = \overline{1, k_1} \\ y_{n+1}^{(0)} = a_2 y_n^{(0)} + T_{21} g_1(x_n^{(k_{21})}) + T_{22} g_2(y_n^{(k_{22})}) \\ y_{n+1}^{(j)} = y_n^{(j-1)} \qquad \forall j = \overline{1, k_2} \end{cases} \qquad \forall n \in \mathbb{N} \tag{2}$$

where $x^{(j)} \in \mathbb{R}$, $j = \overline{0, k_1}$ and $y^{(j)} \in \mathbb{R}$, $j = \overline{0, k_2}$.

Let be the function $F : \mathbb{R}^{k_1+k_2+2} \to \mathbb{R}^{k_1+k_2+2}$ given by the right hand side of system (2). The jacobian matrix of system (2) at the fixed point $\bar{0} \in \mathbb{R}^{k_1+k_2+2}$ is $\hat{A} = DF(\bar{0})$.

The following characteristic equation is obtained:

$$(z - a_1 - b_{11} z^{-k_{11}})(z - a_2 - b_{22} z^{-k_{22}}) - b_{12} b_{21} z^{-(k_{12}+k_{21})} = 0 \tag{3}$$

Studying the stability and bifurcations occurring at the origin in system (1) reduces to the analysis of the distribution of the roots of the characteristic equation (3) with respect to the unit circle. The difficulty of this analysis is due to the large number of parameters appearing in the characteristic equation.

In the followings, we suppose that the following hypotheses are satisfied:

- the internal decays are equal: $a_1 = a_2 = a$;
- the delays satisfy $k_{11} + k_{22} = k_{12} + k_{21}$;
- $b_{11} = b_{22}$.

To the best of our knowledge, this framework is the most general one considered so far in the existing literature. It includes the cases studied in [17] - where all four delays were considered equal, [18] - where it was assumed that $k_{11} = k_{21}$ and $k_{12} = k_{22}$, and [19] - where it was additionally assumed that $k_{11} = k_{22}$.

The characteristic equation becomes

$$z^{k_{11}+k_{22}}(z - a)^2 - \beta z^{k_{22}}(z - a) - \beta z^{k_{11}}(z - a) + \delta = 0 \tag{4}$$

where $\beta = b_{11} = b_{22}$ and $\delta = b_{11}b_{22} - b_{12}b_{21}$. This equation is the same as the one obtained and analyzed in [17], for $k_{11} = k_{22}$ and in [18], for $k_{11} \neq k_{22}$. The results from [17,18] provide us with the conclusions presented below.

First, a list of notations will be introduced and some mathematical results will be presented, which can be proved using basic mathematical tools:

- $m = \frac{1}{2}(k_{11} + k_{22})$ and $l = \frac{1}{2}|k_{11} - k_{22}|$;
- $[x]$ denotes the integer part of x;
- the function $c : [0, \pi] \to \mathbb{R}$, $c(\theta) = \cos(m+1)\theta - a\cos m\theta$;
- the function $s : [0, \pi] \to \mathbb{R}$, $s(\theta) = \sin(m+1)\theta - a\sin m\theta$;
- $S_1 = \{\phi_0 = 0, \phi_1, \phi_2, ..., \phi_{[m]+1}\}$ the set of all solutions of the equation $s(\theta) = 0$ from the interval $[0, \pi]$ (see [18] for details);
- $S_2 = \{\psi_j = \frac{(2j-1)\pi}{2l} / j \in \{1, 2, ..., [\frac{2l+1}{2}]\}\}$ if $l \neq 0$ and $S_2 = \emptyset$ if $l = 0$;
- $\theta_1 = \min(\phi_1, \psi_1)$ if $l \neq 0$ and $\theta_1 = \phi_1$ if $l = 0$;
- the strictly decreasing function $h : [0, \theta_1) \to \mathbb{R}$, $h(\theta) = c(\theta)\sec(l\theta)$;
- $\alpha = \lim_{\theta \to \theta_1} h(\theta) = \begin{cases} c(\phi_1)\sec(l\phi_1) < 0 & \text{if } l \neq 0 \text{ and } \phi_1 < \psi_1 \text{ or if } l = 0 \\ -\infty & \text{if } l \neq 0 \text{ and } \phi_1 \geq \psi_1 \end{cases}$
- $h^{-1} : (\alpha, 1-a] \to [0, \theta_1)$ the inverse of the function h;
- the function $U : (\alpha, 1-a] \to (0, \infty)$, $U(\beta) = 1 + a^2 - 2a\cos(h^{-1}(\beta))$; remark: U is strictly decreasing on the interval $(\alpha, 1-a]$;
- the functions $\lambda_j : \mathbb{R} \to \mathbb{R}$, $\lambda_j(\beta) = 2c(\phi_j)\cos(l\phi_j)\beta - c(\phi_j)^2$, $j \in \{0, 1, ..., [m]+1\}$;
- the function $L : (\alpha, 1-a] \to \mathbb{R}$, $L(\beta) = \max(\lambda_j(\beta) / j \in \{0, 1, ..., [m]+1\})$;
- β_{ij} the solution of the equation $\lambda_i(\beta) = \lambda_j(\beta)$, $i \neq j$;
- $\beta_0 = \max(\beta_{0j} / j \in \{1, 2, ..., [m]+1\}, \beta_{0j} < 0)$; remark: $L(\beta) = \lambda_0(\beta) = 2(1-a)\beta - (1-a)^2$ for any $\beta \in [\beta_0, 1-a]$;
- if the equation $U(\beta) = L(\beta)$ has some roots in the interval (α, β_0), then β_1 is the largest of these roots; otherwise, $\beta_1 = \alpha$.

The following three cases will be referred to:

(C1) $k_{11} = k_{22} = 0$ (i.e $m = l = 0$);
(C2) $k_{11} = k_{22} > 0$ (i.e. $m = k_{11} = k_{22} \geq 1$ and $l = 0$);
(C3) $k_{11} \neq k_{22}$ (i.e. $m \geq l \geq \frac{1}{2}$) with the subcases:
  (C3.1) At least one of the delays k_{11} or k_{22} is odd;
  (C3.2) Both delays k_{11} and k_{22} are even.

Theorem 1. *The null solution of (1) is asymptotically stable if β and δ satisfy the following inequalities:*

$$\beta_1 < \beta < 1 - a \quad \text{and} \quad L(\beta) < \delta < U(\beta). \tag{5}$$

On the boundary of the set $D_S = \{(\beta, \delta) \in \mathbb{R}^2 : \beta_1 < \beta < 1 - a \text{ and } L(\beta) < \delta < U(\beta)\}$ the following bifurcation phenomena causing the loss of asymptotical stability of the null solution of (1) take place:

i. For $\beta \in (\beta_1, 1-a)$ and $\delta = U(\beta)$ system (1) has a Neimark-Sacker bifurcation with multipliers $z = e^{\pm i h^{-1}(\beta)}$, except the following situations:

(C1) *if* $k_{11} = k_{22} = 0$ *then*
- *for* $\beta = -a$ *and* $\delta = U(\beta) = 1 + a^2$ *the system (1) has a strong* $1:4$ *resonant bifurcation;*
- *for* $\beta = -a - \frac{1}{2}$ *and* $\delta = U(\beta) = 1 + a + a^2$ *the system (1) has a strong* $1:3$ *resonant bifurcation;*

(C3) *if* $k_{11} + k_{22} = 1$ *then*
- *for* $\beta = -a - 1$ *and* $\delta = U(\beta) = 1 + a^2$ *the system (1) has a strong* $1:4$ *resonant bifurcation;*

ii. *For* $\beta \in (\beta_1, \beta_0)$ *such that the function* L *is differentiable at* β*, and* $\delta = L(\beta)$*, system (1) exhibits, depending on the case, the following type of bifurcation:*
(C1) *a Flip bifurcation.*
(C2) and (C3.1) *a Neimark-Sacker bifurcation, with multipliers* $z = e^{\pm i\phi_j}$*, where* $j \in \{1, 2, ..., [m] + 1\}$ *such that* $L(\beta) = \lambda_j(\beta)$*.*
(C3.2) *a Flip or a Neimark-Sacker bifurcation, with multipliers* $z = e^{\pm i\phi_j}$*, where* $j \in \{1, 2, ..., [m] + 1\}$ *such that* $L(\beta) = \lambda_j(\beta)$*.*

iii. *For* $\beta \in (\beta_0, 1 - a)$ *and* $\delta = L(\beta) = 2(1 - a)\beta - (1 - a)^2$ *system (1) has a Fold bifurcation.*

iv. *For* $\beta = (1 - a)$ *and* $\delta = (1 - a)^2$*, system (1) has a strong* $1:1$ *resonant bifurcation.*

v. *For* $\beta = \beta_0$ *and* $\delta = L(\beta_0) = 2(1 - a)\beta_0 - (1 - a)^2$*, system (1) exhibits, depending on the case, the following type of bifurcation:*
(C1) *a Fold-Flip bifurcation.*
(C2) and (C3) *a Fold-Neimark-Sacker bifurcation.*

vi. *For* $\beta = \beta_1$ *and* $\delta = U(\beta_1)$ *system (1) exhibits, depending on the case, the following type of bifurcation:*
(C1) *a strong* $1:2$ *bifurcation.*
(C2) and (C3.1) *a double Neimark-Sacker bifurcation.*
(C3.2) *a double Neimark-Sacker or a Flip-Neimark-Sacker bifurcation.*

vii. *(applies only to the case* ***(C3)****) If there exists* $\beta^\star \in (\beta_1, \beta_0)$ *such that the function* L *is not differentiable at* $\beta^\star$*, then for* $\beta = \beta^\star$ *and* $\delta = L(\beta^\star)$*, system (1) exhibits, in the case:*
(C3.1) *a double Neimark-Sacker bifurcation.*
(C3.2) *a double Neimark-Sacker or a Flip-Neimark-Sacker bifurcation.*

We underline that Theorem 1 completely characterizes the stability domain (in the (β, δ)-plane) of the null solution of (1) and the bifurcations occurring at its boundary, if the given hypotheses hold.

3 Direction and Stability of Codimension 1 Bifurcations

Let be the function $F : \mathbb{R}^{k_1+k_2+2} \to \mathbb{R}^{k_1+k_2+2}$ given by the right hand side of system (2). Let be the operators $\hat{A} = DF(\bar{0})$, $\hat{B} = D^2F(\bar{0})$ and $\hat{C} = D^3F(\bar{0})$.

In this section, we will describe a method [22] for analyzing the direction and stability of Neimark-Sacker, Fold and Flip bifurcations that occur at the origin in system (1), according to Theorem 1.

The following result plays an important role in studying the direction and stability of codimension 1 bifurcations.

Proposition 1. *Let $z \in \mathbb{C}$ be an eigenvalue of modulus 1 of the matrix $\hat{A}$. The vectors q and p of $\mathbb{C}^{k_1+k_2+2}$ which verify*

$$\hat{A}q = zq \quad ; \quad \hat{A}^T p = \bar{z}p \quad ; \quad \langle p, q \rangle = 1$$

are given by:

$$q = (z^{k_1}q_1, z^{k_1-1}q_1, ..., zq_1, q_1, z^{k_2}q_2, z^{k_2-1}q_2, ..., zq_2, q_2)^T$$

$$p = (p_1, (\bar{z}-a)p_1, \bar{z}(\bar{z}-a)p_1, ..., \bar{z}^{k_1-1}(\bar{z}-a)p_1, p_2, (\bar{z}-a)p_2, \bar{z}(\bar{z}-a)p_2, ..., \bar{z}^{k_2-1}(\bar{z}-a)p_2)^T$$

where $q_1 = z^{k_{22}}(z-a) - \beta$; $q_2 = b_{21}$; $\bar{p}_1 = \frac{1}{P'(z)}$; $\bar{p}_2 = \frac{z^{k_{11}}(z-a)-\beta}{b_{21}P'(z)}$ and $P(z)$ is the polynomial defined by $P(z) = [z^{k_{11}}(z-a) - \beta][z^{k_{22}}(z-a) - \beta]$.

In the case of a Neimark-Sacker bifurcation, matrix $\hat{A}$ has a simple pair $(z, \bar{z})$ of eigenvalues on the unit circle, such that z is not a root of order $1, 2, 3, 4$ of the unity. The restriction of system (2) to its two dimensional center manifold at the critical parameter value can be transformed into the normal form written in complex coordinates:

$$w \mapsto zw(1 + \frac{1}{2}c_1|w|^2) + O(|w|^4), \qquad w \in \mathbb{C} \tag{6}$$

where

$$c_1 = \bar{z}\langle p, \hat{C}(q, q, \bar{q}) + 2\hat{B}(q, (I - \hat{A})^{-1}\hat{B}(q, \bar{q})) + \hat{B}(\bar{q}, (z^2 I - \hat{A})^{-1}\hat{B}(q, q))\rangle$$

and the eigenvectors p and q are given by Proposition 1. The direction and stability of the Neimark-Sacker bifurcation is determined by the sign of $Re(c_1)$. If $Re(c_1) < 0$ then the bifurcation is supercritical, i.e. the closed invariant curve bifurcating from the origin is asymptotically stable. If $Re(c_1) > 0$, the bifurcation is subcritical, i.e. the closed invariant curve bifurcating from the origin is unstable.

In the case of a Fold bifurcation, matrix $\hat{A}$ has a simple root $z = 1$. The restriction of system (2) to its one dimensional center manifold at the critical parameter value can be transformed into the normal form:

$$w \mapsto w + \frac{1}{2}c_2 w^2 + \frac{1}{6}c_2' w^3 + O(w^4), \qquad w \in \mathbb{R} \tag{7}$$

where $c_2 = \langle p, \hat{B}(q, q)\rangle$ and the eigenvectors p and q are given by Proposition 1 (considering $z = 1$). If $c_2 \neq 0$, then a Fold bifurcation occurs at the origin. If the activation functions satisfy $g_i''(0) = 0$, then $c_2 = 0$, and a Cusp (degenerate Fold) bifurcation occurs at the origin. In this case, the parameter c_2' appearing in the normal form (7), which characterizes the Cusp bifurcation is given by $c_2' = \langle p, \hat{C}(q, q, q)\rangle$.

In the case of a Flip bifurcation, matrix $\hat{A}$ has a simple root $z = -1$ on the unit circle. The restriction of system (2) to its one dimensional center manifold at the critical parameter value can be transformed into the normal form:

$$w \mapsto -w + \frac{1}{6}c_3 w^3 + O(w^4), \qquad w \in \mathbb{R} \tag{8}$$

where

$$c_3 = \langle p, \hat{C}(q,q,q) + 3\hat{B}(q,(I-\hat{A})^{-1}\hat{B}(q,q))\rangle$$

and the eigenvectors p and q are given by Proposition 1 (considering $z = -1$). The direction and stability of the Flip bifurcation is determined by the sign of c_3. If $c_3 < 0$ then the bifurcation is subcritical, i.e. the period-2 orbit bifurcating from the origin is unstable. If $c_3 > 0$, the bifurcation is supercritical, i.e. the period-2 orbit bifurcating from the origin is asymptotically stable.

4 Example

In the following example, we will consider the delays $k_{11} = 2$, $k_{22} = 6$, $k_{12} = 5$ and $k_{21} = 3$. We will also choose $a = 0.5$ and $b_{11} = b_{22} = \beta$. In this case, using Mathematica, we compute:

- case **(C3.2)**, $m = 4$, $l = 2$;
- $S_1 = \{0, 0.543422, 1.15252, 1.80265, 2.46923, \pi\}$ (rad), $S_2 = \{\frac{\pi}{4}, \frac{3\pi}{4}\}$;
- $\theta_1 = \phi_1 = 0.667561$ (rad), $\alpha = -1.34916$, $\beta_1 = -0.867628$, $\beta_0 = -0.0909356$;
- $\beta^\star = -0.794888$, $\beta^{\star\star} = -0.695344$.

Bifurcations at the boundary of D_S (according to Theorem 1):

- For $\beta \in (\beta_1, \beta_0)$ and $\delta = U(\beta)$ a Neimark-Sacker bifurcation occurs, with the multipliers $e^{\pm i h^{-1}(\beta)}$;
- For $\beta \in (\beta_0, 1-a)$ and $\delta = \lambda_0(\beta) = 2(1-a)\beta - (1-a)^2$ a Fold bifurcation occurs;
- For $\beta \in (\beta_1, \beta^\star)$ and $\delta = L(\beta) = \lambda_5(\beta)$ a Flip bifurcation occurs;
- For $\beta \in (\beta^\star, \beta^{\star\star})$ and $\delta = L(\beta) = \lambda_2(\beta)$ a Neimark-Sacker bifurcation occurs, with the multipliers $e^{\pm i\phi_2}$;
- For $\beta \in (\beta^{\star\star}, \beta_0)$ and $\delta = L(\beta) = \lambda_1(\beta)$ a Neimark-Sacker bifurcation occurs, with the multipliers $e^{\pm i\phi_1}$;
- For $\beta = 1-a$ and $\delta = (1-a)^2$ a $1:1$ resonant bifurcation occurs;
- For $\beta = \beta_1$ and $\delta = U(\beta_1)$ a Flip-Neimark-Sacker bifurcation occurs;
- For $\beta = \beta^\star$ and $\delta = L(\beta^\star)$ a Flip-Neimark-Sacker bifurcation occurs;
- For $\beta = \beta^{\star\star}$ and $\delta = L(\beta^{\star\star})$ a double Neimark-Sacker bifurcation occurs;
- For $\beta = \beta_0$ and $\delta = L(\beta_0) = \lambda_0(\beta_0)$ a Fold-Neimark-Sacker bifurcation occurs.

More precisely, we consider the delayed discrete-time Hopfield neural network:

$$\begin{cases} x_{n+1} = 0.5x_n + \beta\tanh(x_{n-2}) - \sin(y_{n-5}) \\ y_{n+1} = 0.5y_n + (\delta - \beta^2)\tanh(x_{n-3}) + \beta\sin(y_{n-6}) \end{cases} \quad \forall n \geq 6 \qquad (9)$$

The stability domain in the (β, δ)-plane for this network is the one presented in Fig. 1.

For $\beta = -0.82$, a supercritical Flip bifurcation occurs at $\delta = L(\beta) = 0.21$, and a supercritical Neimark-Sacker bifurcation occurs at $\delta = U(\beta) = 0.348902$. The bifurcation diagram presented in Fig. 2 shows the existence of asymptotically

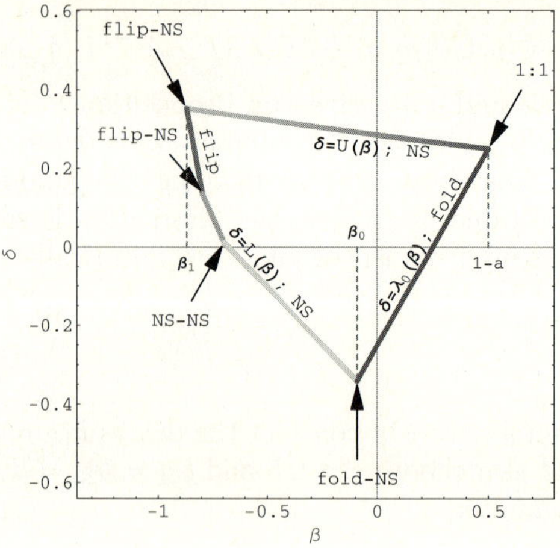

Fig. 1. Stability domain of (9) in the (β, δ)-plain

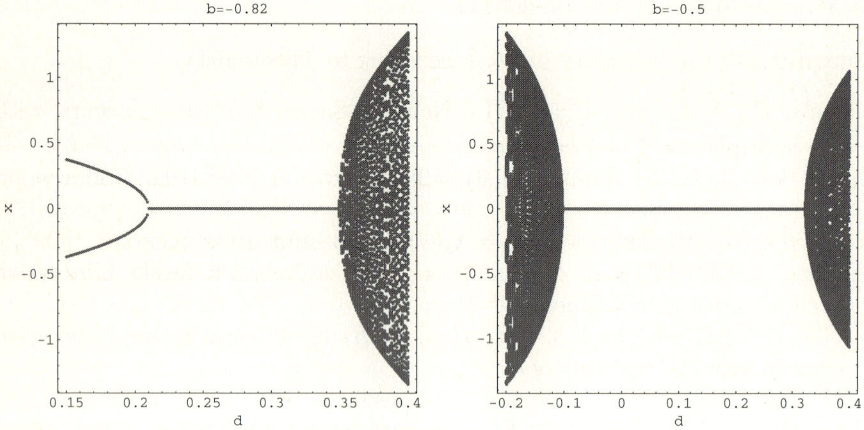

Fig. 2. Bifurcation diagrams in the (δ, x)-plain for system (9) considering $\beta = -0.82$ and $\beta = -0.5$ respectively. For each β value (step size of 0.001), the initial conditions were reset to $(x_0, y_0) = (0.01, 0.01)$ and 10^5 time steps were iterated before plotting the data (which consists of 10^2 points per β value).

stable period-2 orbits for $\delta \in (0.15, 0.21)$ and asymptotically stable limit cycles for $\delta \in (0.348902, 0.4)$.

For $\beta = -0.5$, a supercritical Neimark-Sacker bifurcation takes place at $\delta = L(\beta) = -0.101981$ and another supercritical Neimark-Sacker bifurcation takes place at $\delta = U(\beta) = 0.322838$. The bifurcation diagram presented in Fig. 2 shows the existence of asymptotically stable limit cycles for $\delta \in (-0.2, -0.101981) \cup (0.322838, 0.4)$.

For $\beta = 0.25$, a Fold bifurcation takes place $\delta = L(\beta) = 0$ and a supercritical Neimark-Sacker bifurcation takes place at $\delta = U(\beta) = 0.267028$ (see Fig. 3).

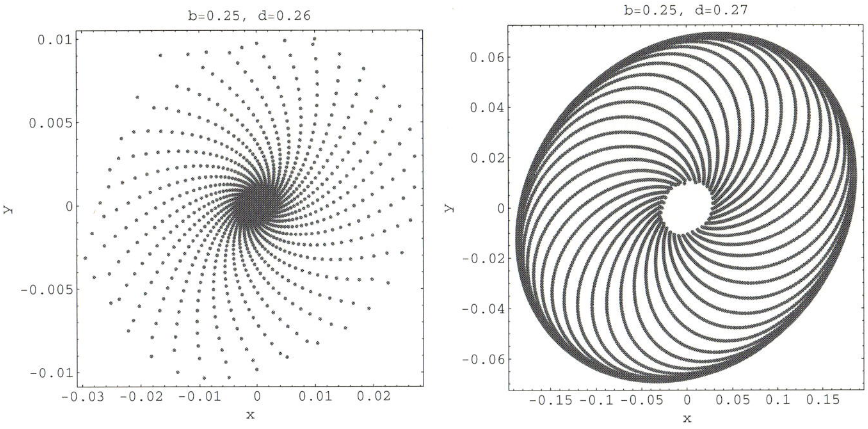

Fig. 3. When $b = 0.25$, a supercritical Neimark-Sacker bifurcation at $\delta = U(\beta) = 0.267028$. For $\delta = 0.26$, the null solution is asymptotically stable, and the trajectory converges to the origin. For $b = 0.27$, an asymptotically stable cycle (1-torus) is present, and the trajectory converges to this cycle.

5 Conclusions

In the most general framework considered so far in the existing scientific literature, a complete bifurcation analysis has been presented for a discrete-time Hopfield-type neural network of two neurons with several delays, uncovering the structure of the stability domain of the null solution, as well as the types of bifurcations occurring at its boundary. A numerical example has been provided which illustrates the theoretical findings. A generalization of these results to more complicated networks of two or more neurons may constitute a direction for future research.

References

1. Hopfield, J.: Neural networks and physical systems with emergent collective computational abilities. Proc. Nat. Acad. Sci. 79, 2554–2558 (1982)
2. Mohamad, S., Gopalsamy, K.: Dynamics of a class of discrete-time neural networks and their continuous-time counterparts. Mathematics and Computers in Simulation 53(1-2), 1–39 (2000)
3. Pasemann, F., Hild, M., Zahedi, K.: S0(2)-networks as neural oscillators. In: Proceedings IWANN 2003. LNCS, vol. 2686, pp. 144–151. Springer, Heidelberg (2003)
4. Chen, L., Aihara, K.: Chaos and asymptotical stability in discrete-time neural networks. Physica D: Nonlinear Phenomena 104(3-4), 286–325 (1997)
5. Chen, L., Aihara, K.: Chaotic simulated annealing by a neural network model with transient chaos. Neural Networks 8, 915–930 (1995)
6. Chen, L., Aihara, K.: Chaotic dynamics of neural networks ans its application to combinatorial optimization. Journal of Dynamical Systems and Differential Equations 9(3), 139–168 (2001)

7. Chen, S., Shih, C.: Transversal homoclinic orbits in a transiently chaotic neural network. Chaos 12, 654–671 (2002)
8. Adachi, M., Aihara, K.: Associative dynamics in a chaotic neural network. Neural Networks 10, 83–98 (1997)
9. Yu, W., Cao, J.: Cryptography based on delayed chaotic neural networks. Physics Letters A 356(4-5), 333–338 (2006)
10. Guo, S., Huang, L., Wang, L.: Exponential stability of discrete-time Hopfield neural networks. Computers and Mathematics with Applications 47, 1249–1256 (2004)
11. Guo, S., Huang, L.: Periodic oscillation for discrete-time Hopfield neural networks. Physics Letters A 329(3), 199–206 (2004)
12. Yuan, Z., Hu, D., Huang, L.: Stability and bifurcation analysis on a discrete-time system of two neurons. Applied Mathematical Letters 17, 1239–1245 (2004)
13. Yuan, Z., Hu, D., Huang, L.: Stability and bifurcation analysis on a discrete-time neural network. Journal of Computational and Applied Mathematics 177, 89–100 (2005)
14. He, W., Cao, J.: Stability and bifurcation of a class of discrete-time neural networks. Applied Mathematical Modelling 31(10), 2111–2122 (2007)
15. Zhang, C., Zheng, B.: Hopf bifurcation in numerical approximation of a n-dimension neural network model with multi-delays. Chaos, Solitons & Fractals 25(1), 129–146 (2005)
16. Zhang, C., Zheng, B.: Stability and bifurcation of a two-dimension discrete neural network model with multi-delays. Chaos, Solitons & Fractals 31(5), 1232–1242 (2007)
17. Kaslik, E., Balint, S.: Bifurcation analysis for a two-dimensional delayed discrete-time Hopfield neural network. Chaos, Solitons and Fractals 34(4), 1245–1253 (2007)
18. Kaslik, E., Balint, S.: Bifurcation analysis for a discrete-time Hopfield neural network of two neurons with two delays and self-connections. Chaos, Solitons and Fractals (in press, 2007), doi:10.1016/j.chaos.2007.01.126
19. Guo, S., Tang, X., Huang, L.: Stability and bifurcation in a discrete system of two neurons with delays. Nonlinear Analysis: Real World Applications (in press, 2007), doi:10.1016/j.nonrwa.2007.03.002
20. Huang, Y., Zou, X.: Co-existence of chaos and stable periodic orbits in a simple discrete neural network. Journal of Nonlinear Science 15, 291–303 (2005)
21. Kaslik, E., Balint, S.: Chaotic dynamics of a delayed discrete-time Hopfield network of two nonidentical neurons with no self-connections. Journal of Nonlinear Science (in press, 2007), doi:10.1007/s00332-007-9015-5
22. Kuznetsov, Y.A.: Elements of applied bifurcation theory. Springer, Heidelberg (1998)

EEG Switching: Three Views from Dynamical Systems

Carlos Lourenço

[1] Faculty of Sciences of the University of Lisbon - Informatics Department, Campo Grande, 1749-016 Lisboa - Portugal
csl@di.fc.ul.pt
http://www.di.fc.ul.pt/~csl
[2] Instituto de Telecomunicações - Security and Quantum Information Group, Av. Rovisco Pais, 1, 1049-001 Lisboa - Portugal

Abstract. There is evidence that the same neural substrate may support different dynamical regimes and hence give rise to different EEG signals. However, the literature lacks successful attempts to systematically explain the different regimes —and the switching between them— within a coherent setting. We explore a mathematical model of neural tissue and call upon concepts from dynamical systems to propose a possible explanation of such processes. The model does not aim to capture a high degree of neurophysiological detail. It rather provides an opportunity to discuss the change in the signals from a dynamical perspective. Notwithstanding, realistic values are adopted for the model parameters, and the resulting EEG also shows typical frequencies in a realistic range. We identify three mechanisms accounting for change: external forcing, bifurcation, and small perturbations of a chaotic attractor.

Keywords: EEG, neurodynamics, spatiotemporal dynamics, chaos.

1 Introduction

Laboratory experiments indicate that, for a given animal, a common neural substrate may display different spatiotemporal electrical activities corresponding e.g. to different behaviors or to the processing of different input patterns [1,2,3]. The switching between dynamical states occurs much faster than the typical time-scales of synaptic adaptation associated with learning. These findings do not contradict the well-known fact that the brain is organized in a modular fashion, with each module responsible for some type of task. However, it can be said that there continues to be a bias toward trying to identify different sub-populations that are responsible for different behaviors, rather than identifying a single population capable of a high degree of multitasking. For instance, in a study of the mouse sleep-wake behavior [4], the Authors tend to emphasize the role of the neural sub-populations that are active exclusively during either the wake stage or the sleep stages —considering also the necessary distinction between REM and non-REM sleep. These Authors attribute a minor role to the sub-populations that show activity overlapping wake and any of the sleep stages.

V. Kůrková et al. (Eds.): ICANN 2008, Part II, LNCS 5164, pp. 665–674, 2008.

In the present paper we aim to unveil some of the possible dynamical mechanisms that might explain multitasking by a common neural population. This exploration is done within the framework of dynamical systems and requires a not-too-complicated model of neural tissue in order to be effective. We hope that the evidence gathered in the following numerical simulations may provide further arguments in support of the multitasking view. We have to emphasize once more that this approach does not preclude a —complementary— modular view of the brain.

A 'single-channel' model electroencephalogram (EEG) is used as a tag to confirm the switching between dynamical regimes, whenever it occurs. However, we will assess that the underlying neural dynamics is spatiotemporal. We agree with other authors in that single-channel EEG provides an incomplete account of the dynamics [5]. Yet, the type of dynamical tagging that it allows is considered sufficient for the present purposes. Furthermore, generalizing to the multi-channel case would be trivially accomplished via minor modifications of the model discussed below.

In summary, this paper tries to model those specific EEG transitions that result from some neural population changing its dynamics. It does not address the case where different neural populations are alternatingly responsible for the observed EEG.

2 Mathematical Model

We adopt a 'minimal' model in what concerns biological realism. We wish to retain the main properties of neurons which may influence the dynamical behavior both at the single cell and at the network level. At least at an abstract level, these dynamics should be comparable to those of biological neurons. On the other hand, we assume that the intricacies of multi-compartment neurons and different types of real neuron connections are not necessary at this level of abstraction, as they should not invalidate the main dynamical mechanisms that we wish to unveil. When considering the full network, a range of diverse spatiotemporal dynamical regimes is our target for study. Some of these regimes may be denoted as complex, or even chaotic.

Neural oscillations in electrical activity, which are ultimately responsible for the observed EEG, result from an interplay between neural excitation and inhibition. Hence we consider two different neural populations, respectively, of excitatory and inhibitory neurons. Our model is based on an original proposal of Kaczmarek to explain neural firing patterns during epileptic seizures [6], but we believe it is general enough to be of use in modeling different brain states, notably in the cases of comparing different awareness states or states supporting cognitive actions. See also the model developments in [7]. The model of the individual cell can be described as a leaky integrator. Passive as well as active membrane properties are incorporated, but neural connectivity is simplified to the point where only first-neighbor coupling is considered. This elimination of distant connections provides a means to study those particular dynamical processes where the spatial dimension plays a crucial role, such as the development of certain complex spatiotemporal dynamics. This includes wave propagation and related phenomena. Biological plausibility is enhanced by a highly nonlinear neural coupling, as well as the consideration of delays in signal transmission between neurons. These model features account for the nonlinear transformation between membrane potential and firing rate, as well as for the several synaptic and axonal processes contributing to a non-instantaneous information propagation. Realistic values are chosen for the parameters wherever possible.

2.1 A Model of Neural Tissue

For lack of space in the present publication, we refer the reader to the detailed derivation of our model in [8]. Here we summarize only the main features.

The instantaneous values of each neuron's membrane potential are taken as network state variables. Since delays are considered, the dynamical evolution also depends on past membrane potential values; hence the latter also define the state vector, which is thus infinite-dimensional. The neurons communicate their state to each other trough a firing rate code. Therefore, there is no explicit spiking in the model. In spite of this, it can be noted that a stationary state corresponds to the firing of action potentials with a frequency constant in time. This frequency can be close to zero, or have some finite value. An oscillatory state, on the other hand, implies a time-modulation of the firing frequency.

The derivation starts with a resistive-capacitive equation for the electrical equivalent of each neuron's membrane potential, and ends up with a coupled set of delay differential equations for the neural population. The network comprises N_{ex} excitatory neurons X_i, and N_{in} inhibitory neurons Y_j. Their dynamics is given by $\mathrm{d}X_i/\mathrm{d}t = -\gamma(X_i - V_L) - (X_i - E_1)\sum_{k\neq i}\omega_{ik}^{(1)} F_X[X_k(t - \tau_{ik})] - (X_i - E_2)\sum_{l\neq i}\omega_{il}^{(2)} F_Y[Y_l(t - \tau_{il})]$ and $\mathrm{d}Y_j/\mathrm{d}t = -\gamma(Y_j - V_L) - (Y_j - E_1)\sum_{k\neq j}\omega_{jk}^{(3)} F_X[X_k(t - \tau_{jk})]$, respectively, where $i, k = 1, \ldots, N_{\mathrm{ex}}$ and $j, l = 1, \ldots, N_{\mathrm{in}}$. The inverse of the membrane's time-constant takes the value $\gamma = 0.25\ \mathrm{msec}^{-1}$. The propagation delay τ_{ik} between neurons k and i would generally depend on the distance between the two neurons. Since we consider only first-neighbor connections, a fixed value $\tau = 1.8$ msec is adopted. Other parameter values are $V_L = -60$ mV, $E_1 = 50$ mV, and $E_2 = -80$ mV, these being the different equilibrium potentials. The sigmoidal transfer function is defined by $F(V) = 1/(1 + e^{-\alpha(V - V_c)})$, with parameter values $V_c = -25$ mV, $\alpha_X = 0.09\ \mathrm{mV}^{-1}$, and $\alpha_Y = 0.2\ \mathrm{mV}^{-1}$. Notice the different α slopes for excitatory and inhibitory neurons, respectively. The synaptic weights $\omega_{ik}^{(1)}$, $\omega_{il}^{(2)}$, and $\omega_{jk}^{(3)}$ refer, respectively, to excitatory-to-excitatory, inhibitory-to-excitatory, and excitatory-to-inhibitory connections. No inhibitory-to-inhibitory connections are considered. We study a spatially homogeneous network, where all synaptic weights of the same type have the same value. Furthermore, the ω values are constant in time, thus no adaptation or learning takes place. However, we will consider deviations from a set of adopted nominal ω values, for instance in a bifurcation setting. The

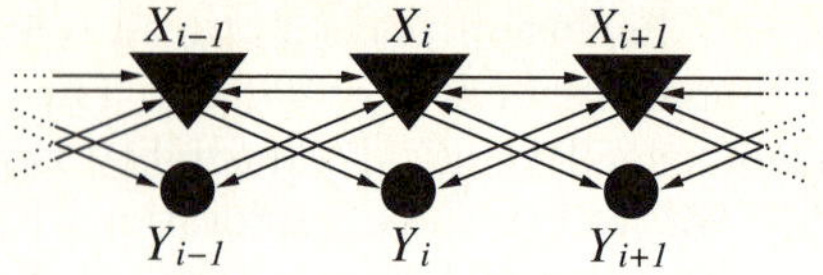

Fig. 1. Neural network where the dynamics takes place. Excitatory neurons and inhibitory neurons are represented as triangles and circles, respectively. Connections $X_i \rightarrow X_{i\pm1}$ and $X_i \rightarrow Y_{i\pm1}$ are excitatory, whereas connections $Y_i \rightarrow X_{i\pm1}$ are inhibitory.

weights of types (1) and (3) are fixed at $\omega^{(1)} = 3.15$ and $\omega^{(3)} = 2.5$ respectively. Parameter $\omega^{(2)}$ can have different values; a reference value is $\omega^{(2)} = 1.68$.

In the isolated model neuron, the response of the membrane to some finite-duration perturbation from the resting state would be a simple relaxation. However, through the coupling of the neurons, new dynamical regimes can be observed depending on details of the coupling and on the initial conditions of the network. One can observe multiple steady states, including global quiescence and global saturation, as well as a variety of oscillatory regimes for the electrical activity of the neurons. A form of spatiotemporal chaos is one of the observed complex regimes, as discussed in more detail in [8].

Different connectivity configurations have been explored for this general neural arrangement, including the case of bi-dimensional networks with many excitatory and inhibitory neurons. However, here we consider only a 1-D spatial arrangement. The network features 16 neurons, equally divided into an excitatory and an inhibitory population. Hence $N_{\rm ex} = N_{\rm in} = 8$. The network topology is depicted in Fig. 1.

The previous dynamical equations are simplified such that only first-neighbor connections are kept. The boundary conditions are of the zero-flux type.

2.2 A Model of EEG

Our model EEG measurement is best regarded as a signal denoting a field potential. We follow a few basic assumptions in defining this potential, or the resulting EEG. We avoid the complicated practical issues that involve the actual measurement of the EEG in a laboratory. The EEG results from extracellular current flow associated with integrated synaptic potentials in activated neurons. The intracellular potentials are not directly accessible via EEG measurements. Pyramidal cells, which are excitatory, are the major source of the EEG. They are oriented parallel to one another, and most of their dendrites are oriented perpendicularly to the surface of the cortex [9]. Therefore, current flows are mostly oriented in the same fashion. In contrast, other types of cells do not share any common orientation and thus their individual contributions do not sum up.

In the context of our model, the synaptic current sources are those corresponding to excitatory neurons X_i at spatial locations $\mathbf{r}_i$ and are given by

$$I_{\rm m}(\mathbf{r}_i,t) = C_{\rm m}\left[-(X_i - E_1)\sum_{j=i\pm1}\omega_{ji}^{(1)}\,F_X[X_j(t-\tau_{ji})] - (X_i - E_2)\sum_{j=i\pm1}\omega_{ji}^{(2)}\,F_Y[Y_j(t-\tau_{ji})]\right],$$

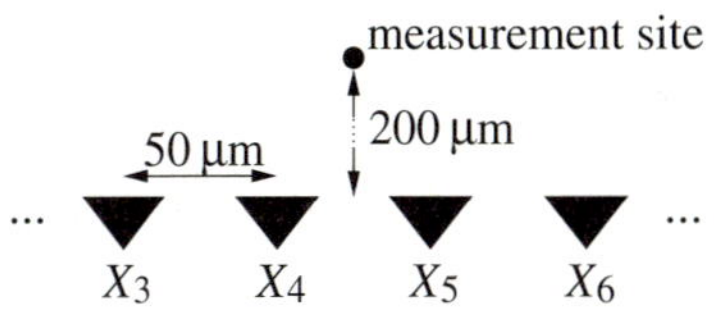

Fig. 2. Relative positions of the current sources and of the EEG voltage measurement site

where $i = 1, \ldots, N_{\text{ex}}$ and C_{m} is the membrane capacitance. The field potential at a point $\mathbf{r}_{\text{e}}$ of the extracellular space incorporates the contributions from all excitatory neurons: $V(\mathbf{r}_{\text{e}}, t) = (1/(4\pi\sigma)) \sum_i (I_m(\mathbf{r}_i, t)/|\mathbf{r}_i - \mathbf{r}_{\text{e}}|)$, where σ represents the electrical conductivity of the extracellular medium. The measurement of the model EEG is performed according to the scheme of Fig. 2.

3 Change in the Dynamics and the Resulting EEG

We take, as the network's typical dynamical state, one where spatiotemporal oscillations of the membrane potential are observed. This is the case for the parameter region considered [8]. Traveling waves are mathematically possible. However, due to the relation between wavelength, system size, and boundary conditions, the waves are in this case geometrically constrained. Details of the full dynamics are out of the scope of this paper. We focus on the fact that the underlying neural network may undergo specific changes in its dynamical state through any of the mechanisms identified in the next sections. The EEG follows these changes and may be used as a tag thereof.

3.1 Change in Response to External Forcing

Let us take $\omega^{(2)} = 1.68$ and all other parameters as indicated in Section 2.1. The dynamics of the unperturbed network is spatiotemporal. This means that, at any time, different neurons generally display different values of the membrane potential. However, the dynamics of the full network is periodic as revealed by a EEG signal which has a period $T = 25.62$ msec.

From this reference dynamical condition, we perturb a fraction of the neurons with an external stimulus of finite magnitude. Namely, we perturb excitatory neurons X_1 through X_4 with a simulated injected electric current. In the mathematical model, this is actually equivalent to momentarily shifting the value of the resting potential V_L. The external perturbation signal could also have its own intricate dynamics, and the network's response to such a signal might be viewed as an analog computation performed over an input pattern. Such ideas are discussed e.g. in [10], but they are not the focus of the present paper.

Here we use a static perturbation with a simple form. Having allowed the network to evolve autonomously for some time, we choose a certain instant to activate the perturbation, and mark this instant as $t = 0$. The perturbation then

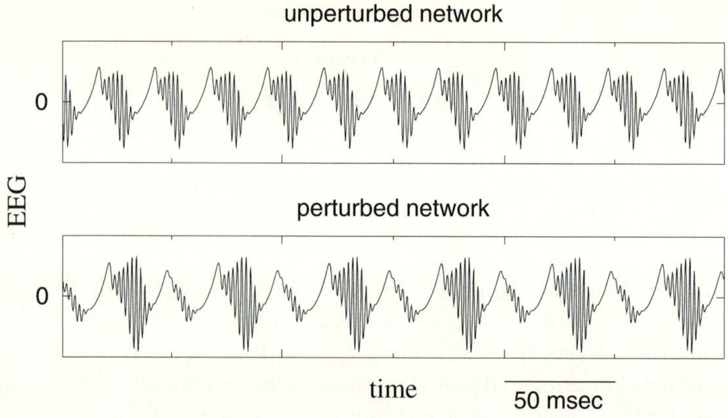

Fig. 3. Comparison between the EEG measured from a network which suffers a finite-size perturbation, and the EEG from an unperturbed network. An arbitrary 300 msec interval is shown. The transient dynamics right after the onset of perturbation is not depicted since we choose to make it occur prior to the plotted interval. See text for details of the perturbation. $\omega^{(2)} = 1.68$ and the remaining parameters are as indicated in Section 2.1.

remains activated for the duration of the numerical experiment. The perturbation consists in shifting V_L by $+5$ mV for neurons X_1 through X_4.

As a result of the perturbation, the dynamics remains periodic but the EEG period changes to $T = 50.13$ msec. Figure 3 displays the EEG signal, respectively, without and with the described perturbation activated.

3.2 Change by Following a Bifurcation Path

Let us now consider that no external perturbation is present. Instead, an internal system parameter is available for change. We select the value of the synaptic weight $\omega^{(2)}$ to be varied.

We take the values $\omega^{(2)} = 3.00$ and $\omega^{(2)} = 1.68$, one at a time, and inspect the dynamics that occurs for each of these values. Again, all other parameters are as indicated in Section 2.1. Also in this case, the EEG is used as a probe of the dynamics. Lowering $\omega^{(2)}$ from 3.00 to 1.68 is actually just walking a small path in the complex parameter space of the neural network. However, it is enough to elicit a visible transition in the dynamics. This parameter variation is part of a larger bifurcation path. A larger $\omega^{(2)}$ range than the one shown here would produce a full period-doubling bifurcation scenario [8]. Figure 4 displays the EEG signal for $\omega^{(2)} = 3.00$ and $\omega^{(2)} = 1.68$, respectively. The EEG period is $T = 17.28$ msec when the inhibition is higher and $T = 25.62$ msec when the inhibition is lower. This numerical experiment, if taken isolatedly and resorting only to the EEG signal, would not suffice to suggest the occurrence of period-doubling bifurcations.

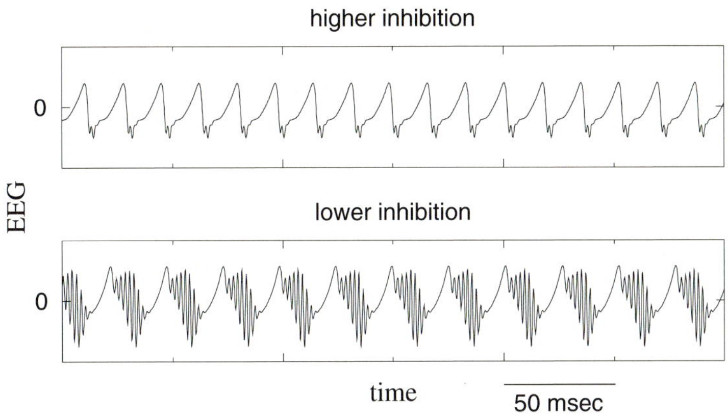

Fig. 4. Comparison of the EEG measured for two different values of synaptic inhibition in the network. As in Fig. 3, an arbitrary 300 msec interval is shown. The vertical scale is also the same as in Fig. 3. The higher inhibition corresponds to $\omega^{(2)} = 3.00$ and the lower inhibition to $\omega^{(2)} = 1.68$. The remaining parameters are as indicated in Section 2.1.

3.3 Switching within a Chaotic Attractor

We now take $\omega^{(2)} = 1.64$ and let all other parameters have the usual values as indicated in Section 2.1. Finite perturbation of the neural system is excluded, be it through some external influence or through parameter change as in the bifurcation scenario. However, 'infinitesimal' perturbations are allowed. The latter can consist in occasional minute changes to system variables, or in occasional minute shifts in a system parameter. Any of the the latter types of perturbations are to be distinguished from the finite perturbations of sections 3.1 and 3.2, since they correspond to different orders of magnitude.

As reported in [8], for these parameter values the system displays a form of low-dimensional spatiotemporal chaos. The dynamics is not periodic, but moderate coherence is observed in the neurons' activities. The change in electrical activity across the network tends to follow a wavy pattern. This is quite far from, say, a regime of fully developed turbulence. Furthermore, the dynamics is coherent enough that it can be switched in a controlled way via an infinitesimal perturbation such as the ones referred above. For this to be possible, two essential properties of chaos come into play: 1) the chaotic dynamics is very flexible and extremely sensitive to small perturbations; 2) under appropriate conditions (verified here), the chaotic attractor contains an infinite number of unstable periodic orbits (UPOs). Infinitesimal perturbations, if adequately tuned, may switch the chaotic dynamics into one of those UPOs, or switch the dynamics between different UPOs. These transitions can be very fast.

In [11,8,10], it is argued that this mechanism may place the chaotic system in an appropriate state for some type of information processing. This UPO selection need not last longer than the time required for a particular computational

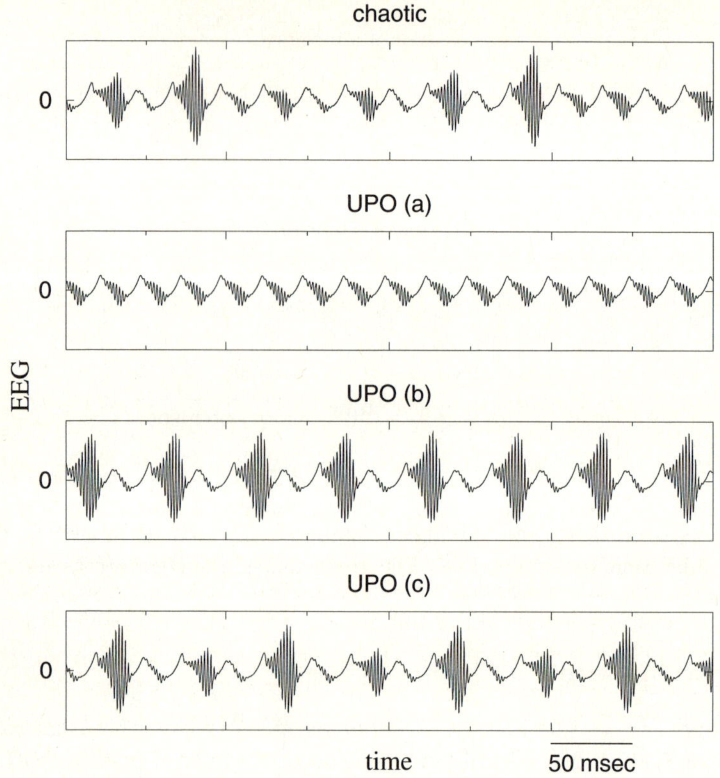

Fig. 5. With $\omega^{(2)} = 1.64$ and the remaining parameters as indicated in Section 2.1, the unperturbed network displays a chaotic regime. Its corresponding EEG is show at the top of the figure. Via infinitesimal perturbations, the dynamics can be switched into a number of different UPOs. In contrast to the broadband frequency spectrum of chaos, the UPOs possess well-defined natural frequencies. Their respective values are: **(a):** 39.79 Hz; **(b):** 18.96 Hz; **(c):** 9.61 Hz. Notice the different time-scale as compared to the one in Figs. 3 and 4. In the present figure an arbitrary 400 msec interval is shown. Also, the vertical scale is compressed by a factor of 1.875 with respect to the one in Figs. 3 and 4.

task. We do not intend to further re-state those ideas here, but rather put forth the EEG signal as a side-effect of, or a tag for, some of those UPOs that can be selected in practice. Since the UPOs are unstable by nature, they cannot be sustained in time unless some control mechanism performs occasional infinitesimal perturbations obeying a so-called control algorithm. This too is out of the scope of the present paper.

Figure 5 allows to compare the EEG signal, respectively, for the unperturbed chaotic attractor, and for three particular UPOs that are embedded in the chaotic attractor. allows to compare the EEG signal, respectively, for the unperturbed chaotic attractor, and for three particular UPOs that are embedded in the chaotic attractor. Thus a multitude of possible dynamical regimes coexist within the chaotic attractor. They are in principle accessible through tiny perturbations of the attractor. The EEG of the dynamical regimes illustrated in Fig. 5 turns out to possess natural frequencies in a realistic range. It should be noted that these EEG signals are all produced by the same neural population, which is (infinitesimally) perturbed in different manners starting from a common chaotic regime.

4 Discussion

We identified three different mechanisms that may account for changes in the spatiotemporal neural dynamics. These changes imply transitions in the EEG, which is a scalar observable, or tag.

The first type of change, through external forcing, requires that a finite-size perturbation reach the neurons. 'External' here means external to the neural population we are modeling, not necessarily external to the organism. As it comes, a primary cue, eventually one with limited time-span, may be used by some distinct neural population for it to enter a state in which it provides afferent input to our particular neural population —the one that does the dynamical multitasking.

The second type of change depends on an internal parameter being available for change. If the parameter range so allows, bifurcations may be observed. In the latter case, the transitions in the dynamics will be clearly noticeable. However, an adequate parameter is not always available, or the time-scales for the parameter change might be slower than required. Furthermore, such type of change might be metabolically costly. Notwithstanding, certain neuromodulators might play a role in these processes. Acetylcholine, for instance, has a confirmed action in the modulation of attentional processes. Interestingly, attentional processes also provide distinct EEG frequencies directly associated with behavior. We also note that the variation of certain internal parameters may have an effect equivalent to that of an external forcing.

The third type of dynamical switching is the fastest and most flexible, since the dynamics is not constrained to some limit-cycle behavior. Rather, the spatiotemporal chaotic regime allows that multiple behaviors be simultaneously available in the form of UPOs. Yet, this generally requires that an adequate mechanism be available to select the most favorable UPO for a given task.

Of course, a combination of more than one of the scenarios is possible. For instance, in separate articles we discuss the consequences of perturbing a chaotic regime with a finite-size external perturbation [10,11], apart from the infinitesimal perturbations that provide UPO switching. This is done from a computational standpoint, and the aim is to assess the pattern processing capabilities of chaotic neural networks.

The model that we discussed in this article supports all three scenarios in a homogeneous way. The chaotic 'mode' could be the most promising. However, the dispute over the very existence of low-dimensional chaos in the brain has not been settled even if more than 20 years have passed since the seminal claims [12,13]. Here we keep an open perspective, and try to determine what actual use chaos may have for any system —natural or artificial— where it may occur [11].

Our last comment is on the particular set of EEG frequencies that are somewhat 'hidden' in our chaotic model, and that can be elicited trough minute perturbations. The 9.61 Hz falls into the alpha range, whereas the 39.79 Hz is in the gamma range. The latter range contains the famous "40 Hz" oscillations. These frequencies have been obtained without any purposeful tuning of the system's parameters. They can be regarded as an emergent phenomenon.

Acknowledgments. The author acknowledges the partial support of Fundação para a Ciência e a Tecnologia and EU FEDER via Instituto de Telecomunicações and via the project PDCT/MAT/57976/2004.

References

1. Wu, J.-Y., Cohen, L., Falk, C.: Neuronal activity during different behaviors in *Aplysia*: A distributed organization? Science 263, 820–823 (1994)
2. Freeman, W.: The physiology of perception. Scientific American 264, 78–85 (1991)
3. Freeman, W.: Chaos in the CNS: Theory and practice. In: Ventriglia, F. (ed.) Neural Modeling and Neural Networks, pp. 185–216. Pergamon Press, New York (1994)
4. Diniz Behn, C., Brown, E., Scammell, T., Kopell, N.: Mathematical model of network dynamics governing mouse sleep-wake behavior. Journal of Neurophysiology 97, 3828–3840 (2007)
5. Lachaux, J.-P., Pezard, L., Garnero, L., Pelte, C., Renault, B., Varela, F., Martinerie, J.: Spatial extension of brain activity fools the single-channel reconstruction of EEG dynamics. Human Brain Mapping 5, 26–47 (1997)
6. Kaczmarek, L.: A model of cell firing patterns during epileptic seizures. Biological Cybernetics 22, 229–234 (1976)
7. Destexhe, A.: Ph.D. Dissertation, Université Libre de Bruxelles (March 1992)
8. Lourenço, C.: Dynamical computation reservoir emerging within a biological model network. Neurocomputing 70, 1177–1185 (2007)
9. Kandel, E., Schwartz, J., Jessel, T. (eds.): Principles of Neural Science, 3rd edn. Prentice-Hall, Englewood Cliffs (1991)
10. Lourenço, C.: Structured reservoir computing with spatiotemporal chaotic attractors. In: Verleysen, M. (ed.) 15th European Symposium on Artificial Neural Networks (ESANN 2007), April 25-27, pp. 501–506. d-side publications (2007)
11. Lourenço, C.: Attention-locked computation with chaotic neural nets. International Journal of Bifurcation and Chaos 14, 737–760 (2004)
12. Babloyantz, A., Salazar, J., Nicolis, C.: Evidence of chaotic dynamics of brain activity during the sleep cycle. Physics Letters A 111, 152–156 (1985)
13. Rapp, P., Zimmerman, I., Albano, A., de Guzman, G., Greenbaun, N.: Dynamics of spontaneous neural activity in the simian motor cortex: The dimension of chaotic neurons. Physics Letters A 110, 335–338 (1985)

Modeling Synchronization Loss in Large-Scale Brain Dynamics

Antonio J. Pons Rivero[1], Jose Luis Cantero[2], Mercedes Atienza[2], and Jordi García-Ojalvo[1]

[1] Departament de Física i Enginyeria Nuclear, Universitat Politècnica de Catalunya, Colom 11, 08222 Terrassa, Spain

[2] Laboratory of Functional Neuroscience, Universidad Pablo de Olavide, Ctra. de Utrera km. 1, 41013 Sevilla, Spain

Abstract. We implement a model of the large-scale dynamics of the brain, and analyze the effect of both short- and long-range connectivity degradation on its coordinated activity, with the ultimate goal of comparing the structural and functional characteristics of neurodegenerative diseases such as Alzheimer's. A preliminary comparison between the results obtained with the model and the activity measured in patients diagnosed with mild cognitive impairment (a precursor of Alzheimer's disease) and healthy elderly controls is shown.

1 Introduction

Optimal brain function requires coordinated activity at a wide range of length scales, from the microscopic, at the single-neuron level, all the way to the macroscopic, at the level of the whole brain. But modeling full-brain dynamics with detailed neuronal network models is currently unfeasible, due to both limited computational capacity and lack of experimental understanding of brain tissue at the microscopic level (see, however, recent progress in this direction, in [1]). Therefore, mesoscopic models have been developed in order to reproduce the dynamics of cortical columns from the perspective of large neuronal populations. Among these, special interest has been devoted to neural mass models, which were designed to simulate electrical brain activity (both spontaneous and evoked) as measured by the electroencephalogram (EEG).

Originally, neural mass models were aimed at modeling the nonlinear dynamics of a single cortical column [2]. In such models, the cortical column is described by a population of pyramidal neurons receiving feedback from local populations of excitatory and inhibitory neurons. Additional excitatory input from neighboring or more distant columns is represented by an external input signal, not coupled dynamically to the column itself. The model reproduces the dynamical behavior of the average excitatory and inhibitory post-synaptic potentials for the neural populations in the cortical column. Following these initial steps, neural mass models for two coupled cortical columns were introduced [3,4]. No variations in the average external input levels with respect to the single-column case

V. Kůrková et al. (Eds.): ICANN 2008, Part II, LNCS 5164, pp. 675–684, 2008.

were considered in those works, but in Ref. [4] the coupling coefficient between columns was tuned such that the standard deviation of the total presynaptic input in each area is conserved. More recently, a neural mass model was used to infer effective connectivity among a small number of cortical regions from high-resolution EEG data [5].

To our knowledge, the only attempt so far to use a massively coupled neural mass model to describe the large-scale activity of the brain is the one by Sotero *et al* [6]. In that work, the authors generalized Jansen's model [3], valid at the level of cortical columns, to voxels taking into account both local connectivity among neighboring cortical areas and long-range connectivity among more distant brain regions. Here we build upon that approach to study in a systematic way the effect of local and long-range connectivity on the collective dynamics of the brain, with the final goal of relating the synchronization loss observed in patients with Alzheimer's disease [7,8] to the cortical thinning reported in radiological studies [9].

2 The Model

As explained above, we use a neural mass model that describes three neuronal populations within a voxel: pyramidal cells and excitatory and inhibitory interneurons. The average postsynaptic potentials (PSPs) of each of these populations can be described by:

$$\frac{d^2y_1^{ni}}{dt^2} + 2a\frac{dy_1^{ni}}{dt} + a^2y_1^{ni} = Aa\left\{c_5S(y_1^{ni} - y_2^{ni}) + c_2S(y_3^{ni}) + p_{\text{lr}}^n + p_{\text{sr,exc}}^{ni} + p_{\text{th}}^n(x_4) + p_{\text{ext}}^n\right\} \tag{1}$$

$$\frac{d^2y_2^{ni}}{dt^2} + 2b\frac{dy_2^{ni}}{dt} + b^2y_2^{ni} = Bb\left\{c_4S(y_4^{ni}) + p_{\text{sr,inh}}^{ni}\right\} \tag{2}$$

$$\frac{d^2y_3^{ni}}{dt^2} + 2a\frac{dy_3^{ni}}{dt} + a^2y_3^{ni} = Aa\left\{c_1S(y_1^{ni} - y_2^{ni}) + p_{th}^n(x_5)\right\} \tag{3}$$

$$\frac{d^2y_4^{ni}}{dt^2} + 2a\frac{dy_4^{ni}}{dt} + a^2y_4^{ni} = Aa\left\{c_3S(y_1^{ni} - y_2^{ni}) + p_{th}^n(x_6)\right\}, \tag{4}$$

where y_1^{ni} (y_2^{ni}) represents the average excitatory (inhibitory) PSP acting on the pyramidal population in voxel i located in area n. y_3^{ni} (y_4^{ni}) represents the average excitatory PSP acting on the excitatory (inhibitory) interneurons. The left-hand side of the differential equations implement a linear filter that transforms incoming pulse trains into average postsynaptic potentials with amplitudes A, B and inverse time constants a and b. The corresponding impulse response functions are $h_e(t) = Aat\exp(-at)$ and $h_i(t) = Bbt\exp(-bt)$ for the excitatory and inhibitory PSPs, respectively. The functions $S(y)$ are sigmoidals that transform average PSPs into an average pulse density of action potentials: $S(y) = 2e_0/(1 + e^{r(v_0-v)})$, with $2e_0$ representing the maximum pulse density and v_0 the PSP corresponding to e_0. The constants c_i are average number of synaptic connections between the corresponding neuronal populations.

Besides the terms described above, which correspond to synaptic connections of populations within a voxel, other inputs coming from outside the voxel are considered in the model. First, input from distant brain areas is represented by the term

$$p_{\mathrm{lr}}^{n} = \sum_{\substack{m=1 \\ m \neq n}}^{N} K^{mn} \sum_{j=1}^{M_m} c_6 S(y_5^{mj}), \tag{5}$$

where the first sum runs over areas and the second one over voxels in area m, which has a total of M_m voxels, and $S(y_5^{mj})$ represents afferent input from the pyramidal population of distant voxels. Pyramidal neurons also receive excitatory input from neighboring voxels of the same area, given by

$$p_{\mathrm{sr,exc}}^{ni} = \sum_{\substack{j=1 \\ j \neq i}}^{M_n} \left(k_{\mathrm{exc1}}^{ij} c_7 S(y_6^{nj}) + k_{\mathrm{exc2}}^{ij} c_8 S(y_7^{nj}) \right) \tag{6}$$

Here the coupling strengths depend on the distance between voxels:

$$k_{\mathrm{exc1}}^{ij} = \frac{1}{2\sigma_{\mathrm{exc1}}} \mathrm{e}^{-\frac{|x_i - x_j|}{\sigma_{\mathrm{exc1}}}}, \qquad k_{\mathrm{exc2}}^{ij} = \frac{1}{2\sigma_{\mathrm{exc2}}} \mathrm{e}^{-\frac{|x_i - x_j|}{\sigma_{\mathrm{exc2}}}}. \tag{7}$$

Inhibitory input from neighboring voxels of the same area into the interneurons is given by

$$p_{\mathrm{sr,inh}}^{ni} = \sum_{\substack{j=1 \\ j \neq i}}^{M_n} k_{\mathrm{inh}}^{ij} c_9 S(y_8^{nj}), \tag{8}$$

where the coupling strength k_{inh}^{ij} is given by an expression analogous to Eq. (7). Input from the thalamus into voxels of area n is described by $p_{\mathrm{th}}^{n} = K^{th,n} c_3^t S(x_4)$, where x_4 is described below. Finally, we consider an additional input coming from other sources besides the cortex and the thalamus, to be given by the external signal p_{ext}^{n}, which acts equally upon each pyramidal population in all voxels of area n.

The incoming PSPs from voxels in the same area and other areas are described by the following relations:

$$\frac{d^2 y_5^{ni}}{dt^2} + 2a_{d1} \frac{dy_5^{ni}}{dt} + a_{d1}^2 y_5^{ni} = A a_{d1} S(y_1^{ni} - y_2^{ni}) \tag{9}$$

$$\frac{d^2 y_6^{ni}}{dt^2} + 2a_{d2} \frac{dy_6^{ni}}{dt} + a_{d2}^2 y_6^{ni} = A a_{d2} S(y_1^{ni} - y_2^{ni}) \tag{10}$$

$$\frac{d^2 y_7^{ni}}{dt^2} + 2a_{d3} \frac{dy_7^{ni}}{dt} + a_{d3}^2 y_7^{ni} = A a_{d3} S(y_3^{ni}) \tag{11}$$

$$\frac{d^2 y_8^{ni}}{dt^2} + 2b_{d4} \frac{dy_8^{ni}}{dt} + b_{d4}^2 y_8^{ni} = B b_{d4} S(y_4^{ni}) \tag{12}$$

where the inverse time constants a_{di} and b_{d4} are chosen now to account for delays in the different transmission mechanisms.

Finally, given the thalamus has strong reciprocal connections with the cerebral cortex, it is represented in the current model as a single voxel obeying the following equations:

$$\frac{d^2x_1}{dt^2} + 2a_t\frac{dx_1}{dt} + a_t^2x_1 = A_ta_t\left\{\sum_{n=1}^{N} K^{th,n}\sum_{i=1}^{M_n} c_6S(y_5^{ni}) + p_{\text{ext,th}}\right\} \tag{13}$$

$$\frac{d^2x_2}{dt^2} + 2b_t\frac{dx_2}{dt} + b_t^2x_2 = B_tb_tc_{2t}S(c_{1t}x_3) \tag{14}$$

$$\frac{d^2x_3}{dt^2} + 2a_t\frac{dx_3}{dt} + a_t^2x_3 = A_ta_tS(x_1 - x_2) \tag{15}$$

$$\frac{d^2x_4}{dt^2} + 2a_{d1t}\frac{dx_4}{dt} + a_{d1t}^2x_4 = A_ta_{d1t}S(x_1 - x_2) \tag{16}$$

$$\frac{d^2x_5}{dt^2} + 2a_{d2t}\frac{dx_5}{dt} + a_{d2t}^2x_5 = A_ta_{d2t}S(x_1 - x_2) \tag{17}$$

$$\frac{d^2x_6}{dt^2} + 2a_{d3t}\frac{dx_6}{dt} + a_{d3t}^2x_6 = A_ta_{d3t}S(x_1 - x_2) \tag{18}$$

where all constants have meanings analogous to those of the cortical voxels equations above.

3 From Single Voxels to Cortical Areas

We begin by reviewing the influence of the external input p_{ext} on the dynamics of a single voxel. A detailed bifurcation analysis of the classic Jansen's model was performed by Grimbert and Faugeras [10] showing that a cortical column becomes unstable for an intermediate range of external input values. For the specific model and parameter values used here, the bifurcation diagram for a single voxel is shown in Fig. 1. The figure represents the extrema of $y = y_1 - y_2$ during its time evolution at each p_{ext}. The system exhibits a constant state for low enough and high enough input, while for intermediate input levels y oscillates for the parameter values used (given in the figure caption). In the stability analysis performed by Grimbert and Faugeras [10] two types of dynamical activity are observed when the system becomes unstable: a "sinusoidal" type of activity, limited by two Hopf bifurcations, and a spike-like type of activity, limited by a homoclinic and a Hopf bifurcation and observed in a smaller region

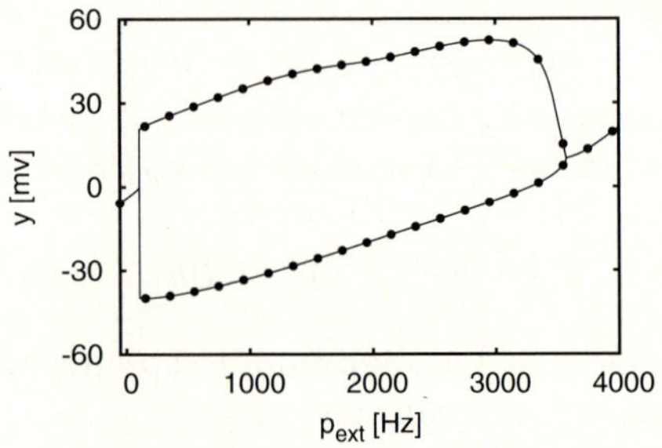

Fig. 1. Stability diagram for a single voxel. Parameters are those for an alpha voxel defined in Tab. 1 of [6], with $c_5 = 0.25\ c_1$.

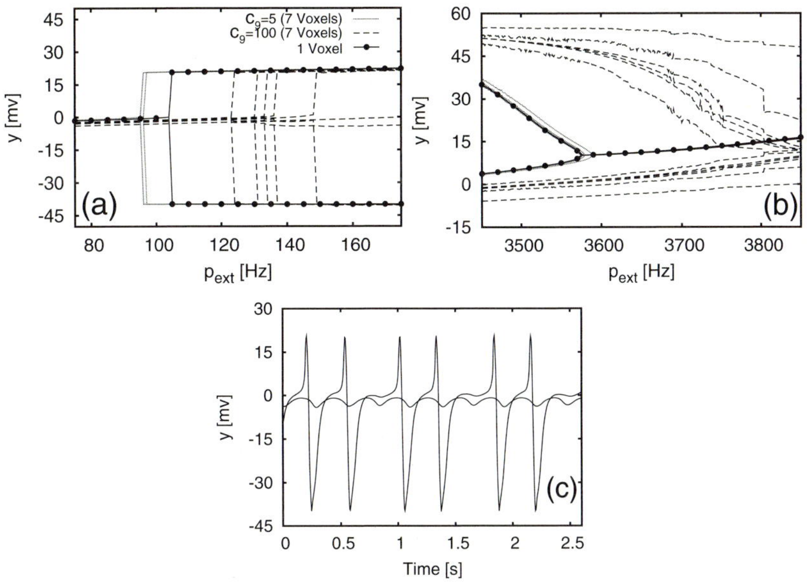

Fig. 2. (a) and (b) Stability diagram for voxels showing the effect of inhibition in local coupling for an area with 7 voxels. Parameters are those for alpha voxels defined in Tab. 1 of [6], with $c_5 = 0.25\ c_1$ and $c_7 = 100$, $c_8 = 100$, $c_9 = 5, 100$. (c) Examples of spike-like and "sinusoidal" activity exhibited by two voxels for $c_9 = 100$ and $p_{\text{ext}} = 160$.

of parameter space, coexisting with the "sinusoidal" activity. For the parameter values used in this paper, we obtain spike-like activity for isolated voxels, but we will observe coexistence with "sinusoidal" activity when voxels are coupled (see Fig. 2(c)). The previous results show that a single cortical column requires a minimum input level to become active (i.e. to exhibit oscillatory behavior), but if the input becomes too large its activity becomes saturated at a high y state.

When the model is scaled up by incorporating neighboring voxels to form a cortical area, one would expect that signals coming into a given column from its neighboring voxels could replace, at least partially, the external input needed to produce activity in the column. To verify this possibility, we simulated the activity of an area with 7 voxels, coupled via the short-range connectivity functions given in Eqs. (6) and (8) (and therefore in the absence of any long-range connectivity). As shown in Fig. 2(a), the influence of short-range coupling on the onset of the first activating bifurcation depends on the amount of the inhibitory short-range connections, measured by parameter c_9. For weak enough inhibition (small c_9, dotted line) the bifurcation point moves towards the left, and thus less external input is indeed necessary for the column to be active. This does not happen, however, for large values of c_9 (dashed line), in which case the bifurcation point moves towards the right, and stronger external input is required to overcome the excess of inhibition and thus activate the column. Note the strong variability exhibited by the different voxels in both cases, with the bifurcation point being shifted differently for the different voxels.

The effect of short-range coupling on the higher limit of the bifurcation diagram, at high p_{ext} values, is shown in Fig. 2(b). In contrast with the case of the first bifurcation, here the short-range coupling shifts the bifurcation point towards the right, thus increasing the region of activity, for both low and high values of c_9 (even though the shift is certainly stronger for higher c_9).

It is interesting to note that although most of the voxels exhibit spike-like type of activity when they become unstable (as the single voxel does) there are examples, for the same parameters, in which the "sinusoidal" type of activity is also observed. Examples of both activities are presented in Fig. 2(c). Notice also that the period of the spike-like oscillations is more irregular than that of the "sinusoidal" oscillations. The amplitude is, however, much larger for spikes than for the "sinusoidal" activity.

4 System-Size Effects on the Dynamics of Cortical Areas

The strength of the short-range connectivity terms, Eqs. (6) and (8), also changes when the number of voxels in the cortical area varies. We now study these effects in the case of a weak inhibitory short-range coupling, in order to determine whether the size of a cortical area might be a relevant factor in the relationship between the level of activity and the strength of the external input. In the same spirit of the previous Section, Fig. 3 shows the variation in the bifurcation points limiting the active region of p_{ext} values for three cortical areas with different number of voxels. The simulations show that, in the case where excitatory short-range coupling dominates over inhibitory interactions, larger areas requires a weaker external input in order to become active Fig. 3(a). Note again the large variability among voxels regarding the location of their activation threshold. The second bifurcation, on the other hand, is not very much affected by the number of voxels in the area (Fig. 3(b)).

Together, the results of these previous two Sections show that intra-area short-range coupling among voxels enhances the activity of a cortical area when such coupling has a weak inhibitory component ($p^{ni}_{\text{sr,inh}}$) with respect to the excitatory

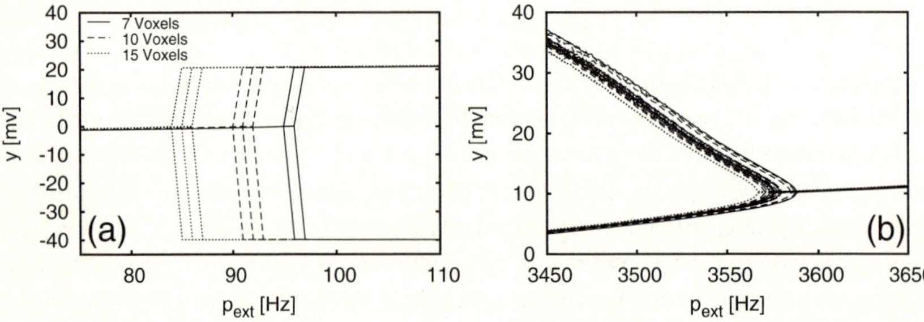

Fig. 3. Stability diagram for areas of increasing number of voxels. Only 7 voxels are represented in each case. Parameters are the same as those shown in Fig. 2 with $c_9 = 5$.

one ($p^{ni}_{\mathrm{sr,exc}}$), and reduces it when the former dominates over the latter. This fact will have implications when building a large scale model in Sec. 5.

5 Effects of Connectivity Loss in Large-Scale Brain Dynamics

Now, areas are connected, through the term defined in Eq. (5), to form the whole brain. The intensity of this coupling is controlled by the connectivity matrix $K^{m,n}$. Thalamus dynamics is also introduced as explained in Section 2. Thus, the model defines a system composed by 22 cortical areas (see Table II in Sotero et al [6]) and the thalamus. The signal obtained at each voxel ($y_1^{ni} - y_2^{ni}$) contributes with a different weight to the signal obtained at an electrode located in the scalp. Distant voxels contribute much less than closer ones. So, the voltage measured at each electrode is obtained by averaging the signals of all voxels with weights coming from a kernel called "electric lead field" [6]. As a result of this convolution, we obtain the voltage that would be measured in 120 electrodes distributed in the scalp for the system considered. See examples of time traces for several of these electrodes in Fig. 4(a). These signals are equivalent to those measured experimentally, as the ones considered below.

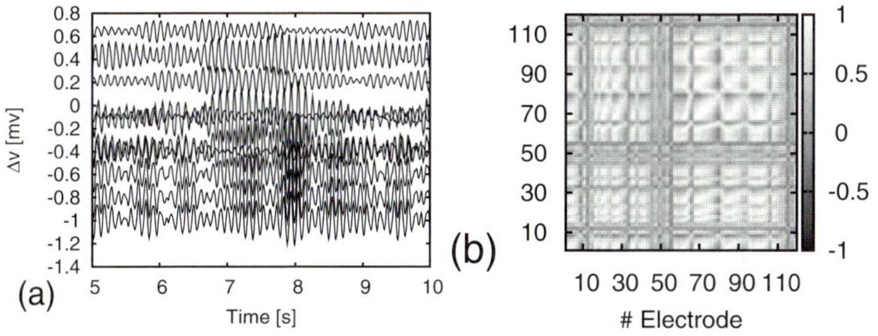

Fig. 4. Example of time traces of the electrode signals from simulations (left), and the corresponding correlation matrix (right). Parameters for voxels are those of Fig. 2 with $c_9 = 5$ and the connectivity matrix of [6] scaled such that Max($K^{m,n}$)= 10^{-1}. Parameters for thalamus are those defined in Tab. 1 of [6]. $p_{\mathrm{ext}}=0$ and $p_{\mathrm{ext,th}}=110$. For simplicity, $p^n_{th}(x_i)$ are mantained constant in each area and proportional to Max($K^{m,n}$).

To analyze the degree of coordinated activity produced at different coupling strengths we will calculate the correlation matrix among the 120 electrodes considered, plus the thalamus. As a result we obtain, from each realization of the model evolution, a correlation matrix as that one shown in Fig. 4(b). From these matrices we want to extract a quantitative measure of the synchronization produced in the system, to be compared with the behavior observed experimentally. Other characterizations of the signal could be considered (e.g. the power spectrum), but we will only examine here the histogram of correlations calculated from the matrix.

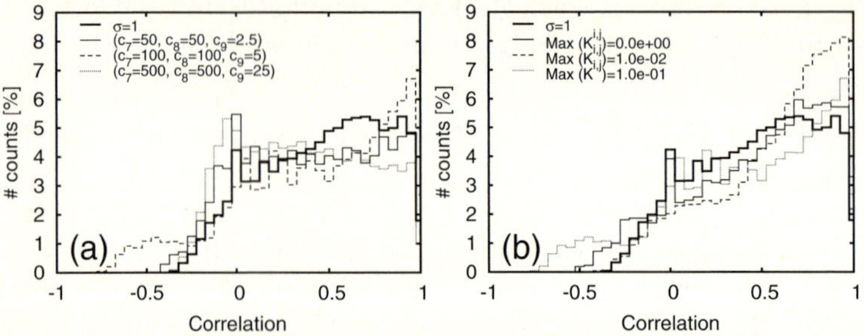

Fig. 5. (a) Histogram of correlations obtained for different short-range coupling intensities (given in the legend) and the rest of parameters as in Fig. 4. (b) Histogram of correlations obtained for different long-range coupling intensities (given in the legend) and the rest of parameters as in Fig. 4.

Note that when calculating the voltage at each electrode from the signal, y, of all voxels, we introduce artificial correlations between the different electrodes (termed background correlation in what follows). Therefore, before studying in detail the effect of increasing short- and long-range coupling on the correlation histograms, we analyze the signal that would produce a model with a random value of y (gaussian white noise, with zero mean and standard deviation equal to 1) in each voxel. The average histogram of correlations obtained from 25 noise realizations appears in Figs. 5(a) and 5(b) as a thick line. Observe that the histogram increases almost monotonically when correlation increases. A peak is observed for zero correlation (which presumably corresponds to correlations between distant electrodes) and a decrease at correlations close to 1, showing that full correlation is almost never attained.

Figure 5(a) shows the correlation histograms for increasing amounts of short-range coupling (maintaining the excitatory-to-inhibitory strength ratio). For low couplings (thin solid line), the correlation profile is very close to the background correlation induced by the lead-field convolution. For intermediate values of the short-range coupling (dashed line), on the other hand, deviations from the background correlations appear. Specifically, correlation peaks emerge around 1 and -0.5. Finally, for even larger coupling intensities (dotted line), differences with the background correlations decrease again.

In Fig. 5(b) we show the histograms produced when scaling up the connectivity matrix by a constant factor. For a system without long-range coupling (thin solid line) the correlations are again similar to their background level (although certainly, some increase of the number of correlations close to 1 and close to -0.5 is observed). A further increase of the connectivity matrix (dashed line) enhances very significantly the number of correlations close to 1. Finally, for the largest value of the connectivity matrix considered (dotted line), a substantial increase in correlations close to -0.5 is produced, together with a corresponding reduction of the correlation peak close to 1. The behavior of both panels of Fig. 5 seems to suggest that there are optimal strengths of the short- and long-range coupling that produce the strongest coordinated activity in our simulated model.

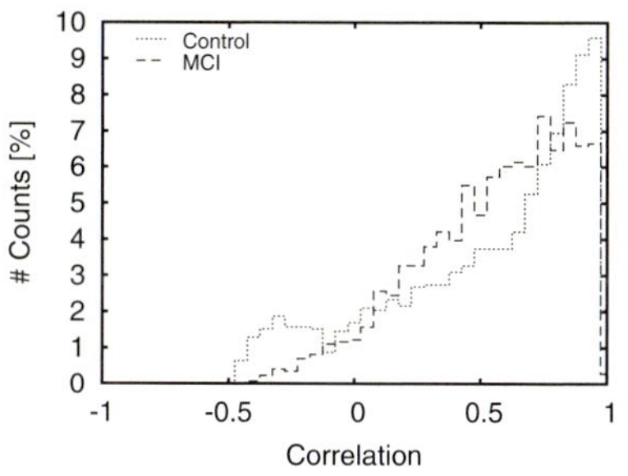

Fig. 6. Comparison of experimental correlation histograms of EEG measurements of the brain activity in healthy elderly subjects and MCI patients

To conclude this section, we show in Fig. 6 the same type of analysis for data obtained from EEG recordings for MCI patients and healthy elderly subjects [11]. Notice that control subjects exhibit a correlation histogram very similar to those shown in Figs. 6(a) and (b). The histogram of MCI patients shows a decrease in the correlation peak at 1 and a smaller anticorrelation. This behavior is consistent with that shown in cases with less connectivity, both in short and long range.

6 Conclusion

We have studied the effect of short- and long-range coupling on the dynamical behavior of a neural mass model, with the goal of modeling the dynamical consequences of a loss of synchronization at the level of the whole brain. Our results show that while isolated cortical columns require a minimum amount of input signal to become active (i.e. to exhibit non-stationary, dynamical behavior such as oscillations), short-range coupling between voxels in a given cortical area may elicit active behavior for weaker inputs, or even in the absence of input altogether, provided the excitatory component of the coupling dominates over the inhibition. Using this fact to establish meaningful parameter values, we have built a full-brain neural mass model in order to examine the effects of varying the strengths of short- and long-range coupling on the synchronization level of the brain, and compared it with experimental measurements taken from MCI patients and healthy elderly subjects. The results show that synchronization in our model is enhanced for optimal levels of short- and long-range coupling strengths. Thus both types of coupling seem to work together in establishing functional connectivity in the brain.

Acknowledgments

We thank R. C. Sotero for kindly providing us with very useful information regarding their numerical model and the experiments associated to it. This work was supported by the Ministerio de Educaci6n y Ciencia (Spain) under projects

FIS2006-11452 and SAF2005-00398, and by the GABA project (European Commission, FP6-NEST contract 043309). A.J. P. acknowledges financial support from the Juan de la Cierva program (Ministerio de Educación y Ciencia, Spain).

References

1. Markram, H.: The blue brain project. Nat. Rev. Neurosci. 7(2), 153–160 (2006)
2. Jansen, B., Zouridakis, G., Brandt, M.: A neurophysiologically-based mathematical model of flash visual evoked potentials. Biological Cybernetics 68(3), 275–283 (1993)
3. Jansen, B., Rit, V.: Electroencephalogram and visual evoked potential generation in a mathematical model of coupled cortical columns. Biological Cybernetics 73(4), 357–366 (1995)
4. David, O., Friston, K.J.: A neural mass model for MEG/EEG: coupling and neuronal dynamics. NeuroImage 20(3), 1743–1755 (2003)
5. Ursino, M., Zavaglia, M., Astolfi, L., Babiloni, F.: Use of a neural mass model for the analysis of effective connectivity among cortical regions based on high resolution EEG recordings. Biological Cybernetics 96(3), 351–365 (2007)
6. Sotero, R.C., Trujillo-Barreto, N.J., Iturria-Medina, Y., Carbonell, F., Jimenez, J.C.: Realistically coupled neural mass models can generate EEG rhythms. Neural Comp. 19(2), 478–512 (2007)
7. Stam, C., Jones, B., Nolte, G., Breakspear, M., Scheltens, P.: Small-world networks and functional connectivity in Alzheimer's disease. Cerebral Cortex 17(1), 92–99 (2007)
8. Stam, C.J., Nolte, G., Daffertshofer, A.: Phase lag index: Assessment of functional connectivity from multi channel EEG and MEG with diminished bias from common sources with diminished bias from common sources. Human Brain Mapping 28(11), 1178–1193 (2007)
9. Thompson, P.M., Hayashi, K.M., de Zubicaray, G., Janke, A.L., Rose, S.E., Semple, J., Herman, D., Hong, M.S., Dittmer, S.S., Doddrell, D.M., Toga, A.W.: Dynamics of gray matter loss in Alzheimer's disease. Journal of Neuroscience 23(3), 994–1005 (2003)
10. Grimbert, F., Faugeras, O.: Bifurcation analysis of Jansen's neural mass model. Neural Computation 18(12), 3052–3068 (2006)
11. Cruz-Vadell, A., Atienza, M., Gil-Neciga, E., Gomez-Herrero, G., Rodriguez-Romero, R., Garcia-Solis, D., Cantero, J.L.: Functional integrity of thalamocortical circuits differentiates normal aging from mild cognitive impairment (sumbitted, 2008)

Spatio-temporal Dynamics during Perceptual Processing in an Oscillatory Neural Network

A. Ravishankar Rao and Guillermo Cecchi

IBM T.J. Watson Research Center
Yorktown Heights, NY 10598, USA
ravirao@us.ibm.com, gcecchi@us.ibm.com

Abstract. We examine the dynamics of object recognition in a multi-layer network of oscillatory elements. The derivation of network dynamics is based on principles of sparse representation, and results in system behavior that achieves binding through phase synchronization. We examine the behavior of the network during recognition of objects with missing contours. We observe that certain network units respond to missing contours with reduced amplitude and temporal delay, similar to neuroscientific findings. Furthermore, these units maintain synchronization with a high-level object representation only in the presence of feedback.

Our results suggest that the illusory contour phenomena are formal consequences of a system that dynamically solves the binding problem, and highlight the functional relevance of feedback connections.

1 Introduction

An important goal of building *computational* models of neural function is to be able to explain the behavior of neural mechanisms as measured in biological organisms. This is a highly challenging task, as models do not generalize well. For instance, conventional neural network models ignore the temporal aspects of neural signaling [1]. Thus, a major issue in the area of computational neural models is appropriate model validation.

In an effort to address this issue, we investigate a recent model of oscillatory neural networks [2]. The model dynamics were derived using an optimization framework based on a sparse coding principle, which states that the goal of the cortex is to create a sparse and faithful representation of the inputs it receives. Desirable properties of the derived model include its ability to separate superposed objects in a simulated visual field and to associate multiple high-level object categories with their lower-level constituents. This is achieved through phase synchronization in oscillatory elements. This behavior is learnt in an unsupervised manner through the modification of synaptic weights. The model was initially developed for a two-layer system, and has been extended to multi-layer systems [3].

Using the network in [2,3], we performed simulations that mimic the biological experiments [4] in probing the neural responses during illusory contour perception. The model in [3] corroborates many of the biological findings, and

V. Kůrková et al. (Eds.): ICANN 2008, Part II, LNCS 5164, pp. 685–694, 2008.

thus could serve as a viable building block for more complex neural systems. We emphasize that our model was not explicitly created to explain illusory contour perception, as others in the literature have [5]. Rather the illusory contour responses arise as a natural consequence in a network that achieves a sparse representation.

Furthermore, from our experiments, we derive a testable hypothesis for the neuroscientific field, which states that disabling feedback connections should impair the phase (or timing) relation between higher and lower level units that represent a common percept, while having relatively little effect on the formation of the percept. The pursuit of this hypothesis should shed further light on the role of feedback in object perception.

2 Background

Lee and Mumford [4]. measured the response in visual cortical area V1 to a missing contour such as the side of a square. A neuron that responds vigorously to a complete contour still responds to a missing contour, with reduced amplitude and a temporal delay. This is depicted in Figure 1(A). We would like to observe whether the computational model in [2,3] can reproduce such a phenomenon. To the best of our knowledge, such a validation has not been performed for an oscillatory neural network model.

Engel *et al* [6] review the neuroscientific evidence regarding oscillatory phenomena observed in neural recordings and establish the importance of synchrony in the grouping of of neuronal populations. Bar *et al* [7] investigated the role of top-down processing in visual perception, and propose that low-spatial frequency information allows rapid formation of high-level representations.

Li [8] has investigated the phenomenon of contour formation in the primary visual cortex. The units in his model do not learn their synaptic connection

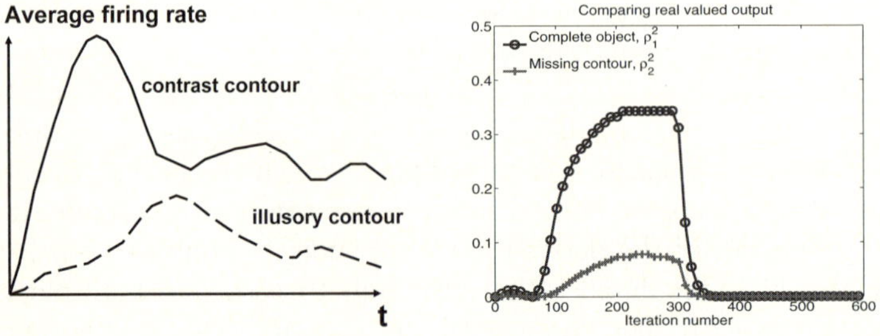

(A) (B)

Fig. 1. (A) Stylized depiction of actual neural responses to contrast contours, and illusory contours. Derived from Lee and Mumford [4]. (B) The result derived from the computational model in Section 4.7, which compares the real valued output ρ_1^2 when the entire square is presented, with the output ρ_2^2 when the square with missing contour is presented. Feedback is present. The input is presented for 300 iterations, and then removed.

weights. In contrast, our model learns weights in an unsupervised manner, and the overall system achieves the desirable properties of separation of inputs and segmentation.

Fukushima created the Neocognitron to model neural processing in the visual pathway [9]. Our proposed system is similar in terms of the hierarachical nature of processing. However, the precise computations carried out in our model are totally different, and have been derived from a sparse coding perspective.

3 Experimental Methods

Our model uses oscillatory units, whose state is defined by an amplitude, a frequency and a phase, which is represented by a phasor $Ae^{i\theta}$ where A is the amplitude, θ the phase. Rao *et al* [2,3] used an optimization approach based on sparse coding to derive the following state update equations.

Consider a network with three layers, a lower input layer $\{x\}$, a middle layer $\{y\}$ and an upper layer $\{z\}$, as shown in Figure 2. The amplitudes of units in these layers are denoted by x, y and z respectively. Phases are represented by ϕ in the lower layer, θ in the middle layer, and ξ in the upper layer.

The dynamics of the middle layer evolve as follows.

$$\begin{aligned}\Delta y_n &\sim \sum_j W_{nj} x_j [1+\cos(\phi_j-\theta_n)] + \kappa_1 \sum_k W_{nk} z_k \sin(\xi_k-\theta_n) \\ &- \alpha y_n - \gamma \sum_m W_{nm} y_m [1+\beta\cos(\theta_m-\theta_n)] \\ \Delta\theta_n &\sim \beta \sum_j W_{nj} x_j \sin(\phi_j-\theta_n) + \kappa_2\beta\gamma \sum_m W_{nm} z_m \sin(\xi_m-\theta_n) \\ &- \beta\gamma \sum_k W_{nk} y_k \sin(\theta_k-\theta_n) \qquad (1)\end{aligned}$$

The dynamics of the upper layer evolve as follows.

$$\begin{aligned}\Delta z_n &\sim \sum_j W_{nj} y_j [1+\cos(\theta_j-\xi_n)] - \alpha z_n - \gamma \sum_k W_{nk} z_k [1+\beta\cos(\xi_k-\xi_n)] \\ \Delta\xi_n &\sim \beta \sum_j W_{nj} y_j \sin(\theta_j-\xi_n) - \beta\gamma \sum_k W_{nk} z_k \sin(\xi_k-\xi_n) \qquad (2)\end{aligned}$$

The dynamics of the lower layer evolve as follows.

$$\Delta\phi_n \sim \sum_j W_{jn} y_j \sin(\theta_j-\phi_n) \qquad (3)$$

A phase-dependent Hebbian learning rule is used to learn synaptic weights between a source node with index j and a target node with index i as follows

$$\Delta W_{ij} \sim S_j T_i [1+\beta\cos(\Delta)] \qquad (4)$$

where S_j is amplitude of the source node, T_i is the amplitude of the target node, and Δ is the phase difference between the source and target node.

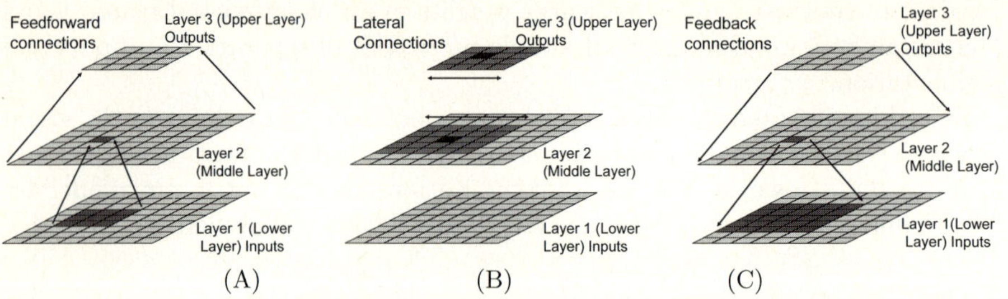

Fig. 2. Illustrating the network connectivity. (A) Feedforward connections. (B) Lateral connections. (C) Feedback connections.

4 Experimental Results

The system undergoes *unsupervised* learning to discriminate between 16 simple objects shown in [2]. This is done in accordance with the learning rules in Section 3 that achieve separation and segmentation, which are required to solve the binding problem. When the trained system is presented with an individual object, the response in the upper z layer consists of a single winner unit that uniquely represents the object. When presented with a superposition of objects that it has learnt, the system recognizes the individual objects, and also segments the lower layer into components that represent the superposed objects. This segmentation is exhibited through phase synchronization between the upper z layer unit that constitutes a high-level representation of an object and lower x layer units that form the object pixels.

The response of such a trained network to a missing contour is measured with the following visual stimuli. Figure 3(a) shows the original stimulus consisting of one of the 16 input objects, a square. Figure 3(b) shows the modified stimulus consisting of this square with its left edge missing.

4.1 Experimental Protocol

We examine the network behavior by varying the type of stimulus used, and the presence of feedback. For the stimulus type, we either use the full object in

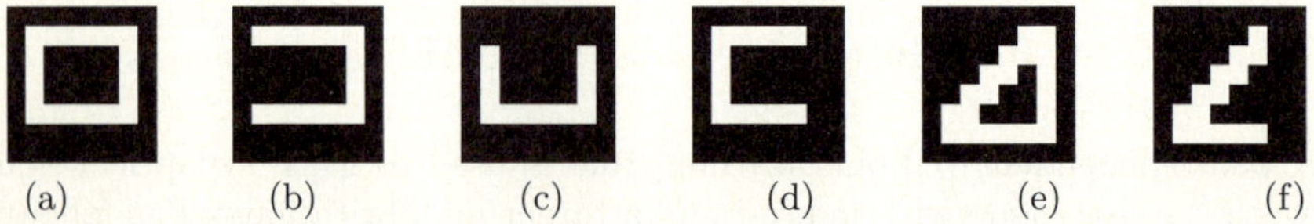

Fig. 3. (a) Original stimulus consisting of a complete square. (b)-(d) Modified stimulus consisting of the square with missing contours. (e) Original stimulus consisting of a complete triangle. (f) Modified stimulus consisting of the triangle with a missing contour.

Figure 3, or the object with missing contour. For the feedback we either use, or disable feedback connections from the z to y layer. This is explored as follows.

4.2 Spatial Phasor Maps in the Presence of Feedback

The system dynamics is visualized through a spatial phasor map. The first two columns in Figure 4 are for the full square stimulus and square with missing contour with feedback. Note that the winner in the z layer, the encircled unit[1] number 4, z_4, stays the same even when the square stimulus contains a missing contour. In the y layer, the encircled unit y_{35} shows an attenuated response to the missing contour. This unit receives feedforward connections from the region surrounding the missing contour.

4.3 Temporal Evolution of Amplitudes in the Presence of Feedback

Figures 5(A) and 6(A) show the amplitude evolution in the z layer and y_{35}.

Note that in Figure 5, the winner in the z layer, z_4, stays the same even when the square stimulus contains a misssing contour. In Figure 6(A), y_{35} shows an attenuated response to the missing contour, similar to the biological finding reported in [4]. In the first two columns of Figure 4, observe that y_{35} is synchronized in phase with the winner in the z layer, z_4. Since the winner in the z layer represents the higher-level percept 'square', this phase synchronization demonstrates that y_{35} is indeed responding to the square.

4.4 Spatial Phasor Maps in the Absence of Feedback

In order to demonstrate the importance of feedback, after training, we disabled the feedback connections from the upper z layer to the middle y layer.

The third and fourth columns in Figure 4 show the spatial phasor map for the presentation of the full square stimulus and square with missing contour with no feedback. Note that the encircled winner in the z layer, z_4, stays the same even when the square stimulus contains a misssing contour, *and* feedback is disabled. Also, y_{35} still responds to the missing contour.

4.5 Temporal Evolution of Amplitudes in the Absence of Feedback

Figures 5(B) and 6(B) show the amplitude evolution in the z layer and y_{35} respectively, when feedback from the z to y layer is disabled. Again, the winner in the z layer, the encircled z_4 stays the same in all cases. In Figure 6(B), y_{35} shows an attenuated response to the missing contour. Very interestingly, y_{35} is *not* synchronized with the winner z_4.

Figures 6(A) and 6(B) show that the amplitude response of y_{35} is essentially similar. Hence the amplitude response cannot be used to distinguish whether feedback from the z to y layer is present or absent. This leads us to examine whether the phase response is sensitive to the presence of feedback.

[1] Unit numbers are measured in sequential raster order starting from the upper left corner. They will be indicated with subscripts.

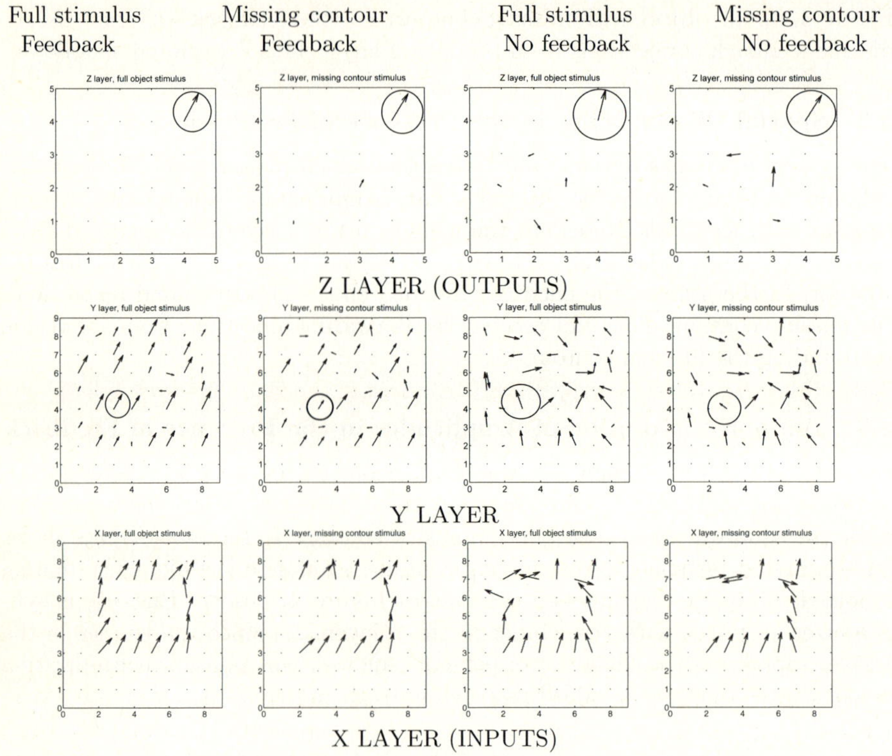

Fig. 4. This figure shows the phasor plot for units in the three layers at iteration 300. Each column represents a different condition as indicated at the top, such as full stimulus (square) with feedback, or stimulus with missing contour with feedback etc.

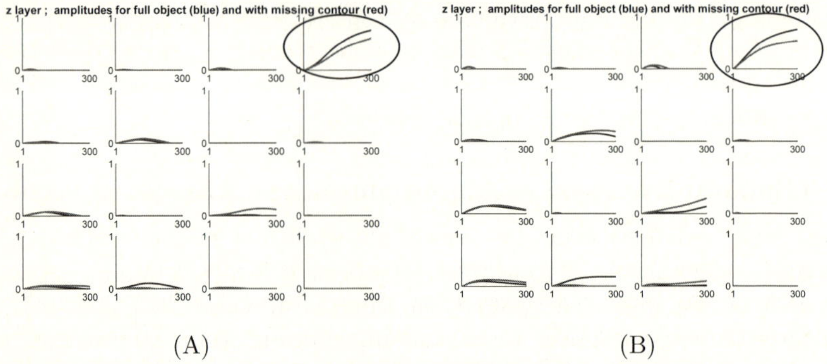

Fig. 5. The amplitude response in the z layer as displayed as a function of the iteration number. Each panel represents the amplitude response of a given z layer unit. In all figures, the response to the complete square is indicated in blue, and the response to the square with a missing contour is indicated in red. The encircled unit, number 4, is the winner in the z layer, and can be considered to represent the object 'square'. (A) Responses in the presence of feedback connections from the z to the y layer. (B) Responses in the absence of feedback connections from the z to the y layer.

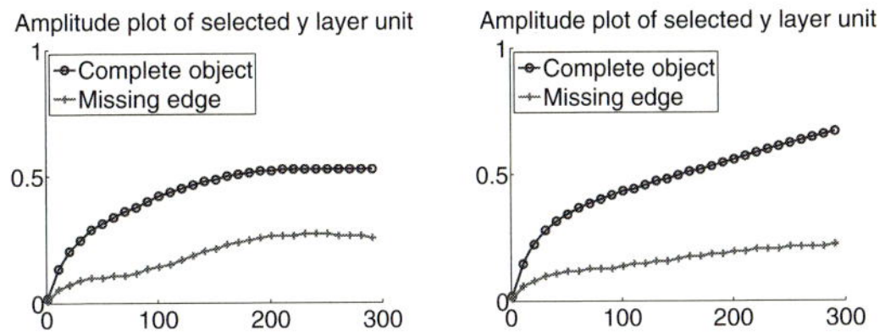

(A) (B)

Fig. 6. The evolution of amplitude activity in y_{35} as a function of the iteration number. (A) Feedback is present from the z to y layer. (B) Feedback is absent.

4.6 Temporal Evolution of Phases with and without Feedback

From Figure 6(A) we see that y_{35} shows an attenuated response to the missing edge of the square. This occurs when feedback connections from the z to y layer are present.

When the same set of stimuli are used after the feedback connections from the z to y layer are *disabled*, we get the activity pattern shown in Figure 6(B) Again, y_{35} shows an attenuated response to the missing edge of the square. However, it's phase response is very different in the case when feedback is removed.

This is shown in Figure 7. When feedback connections are present, Figure 7(A), the y layer units, even those responding in an attenuated fashion to missing contours, synchronize rapidly in phase with the winner in the z layer that represents the input object ($cos\Delta \rightarrow 1$ as $\Delta \rightarrow 0$). Note that synchronization is fastest when the full object is presented, such as the complete square. When feedback connections are *removed*, Figure 7(B), the same y layer unit that responded to the missing contour, y_{35} fails to synchronize in phase with the winner in the z layer that represents the input object. The measure of synchrony is poor even when the full object is presented, and is worse when there is a missing contour.

However, as can be seen in the phasor plots of third and fourth columns of Figure 4, the y layer units that receive direct input from existing features of the object are fairly synchronized with the z layer winner. This causes the units in the x layer to also be fairly synchronized with the winner in the z layer. This indicates that a behavioral task such as 'pick up the square' can still be carried out, but with less certainty, as the synchronization is not perfect.

Though the phase evolution of only one pair of units is presented here due to space constraints, we observed the same qualitative behavior when deleting different contours of the objects such as the square and triangle.

4.7 Real Valued Output of the Oscillating Units

As presented in Section 3, an oscillating unit is described by a phasor, $Ae^{i\theta}$. The real part of this is given by $\rho = A\cos(\theta)$. We plot the real valued output

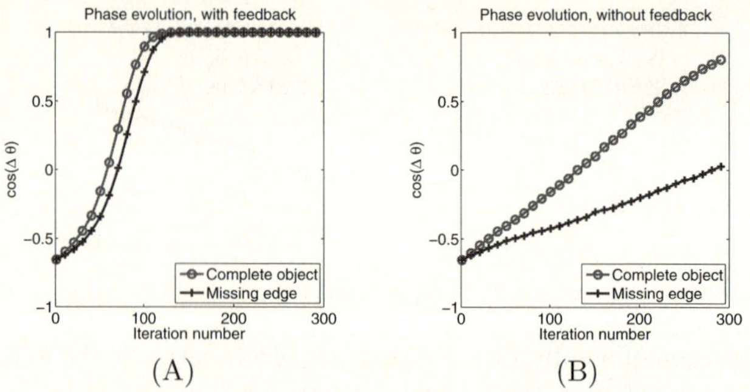

Fig. 7. (A) Shows the phase difference between z_4 and y_{35}. The phase difference is plotted as $cos(\Delta)$ where $\Delta = \xi_z - \theta_y$, and ξ_z and θ_y are phases of z_4 and y_{35}. Feedback is present. The two cases shown are for the presentation of the complete square and for the square with the missing left edge as in Figure 3(B). (B) In this plot, the evolution of phase differences is shown with no feedback.

of y_{35} when the square with and without the left contour is presented. Instead of using the absolute phase of y_{35} we compute a relative phase Δ, with respect to z_4. Let $\rho_1(t) = y_{35}(t)\cos(\Delta(t))$, which is the real valued output when the full square is presented. In Figure 1(B), we compare ρ_1^2 with ρ_2^2, when the missing contour is present. Note the resemblance of the plots in Figure 1. Specifically, the real-valued output for y_{35} shows an attenuated amplitude, as well as a temporal delay.

4.8 Further Results

Figure 8 shows abbreviated results when a triangle with and without a missing contour is presented. This type of behavior was also observed when other edges of the square were removed. They are not shown here for lack of space.

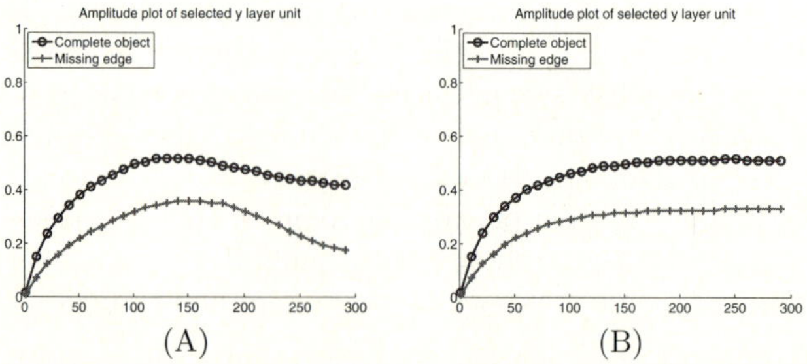

Fig. 8. (A) Shows the evolution of amplitude response of unit 55 in the y layer, which shows an attenuated response to a missing in the triangle. Feedback is present in this case. (B) Shows the response of the same unit when there is no feedback.

5 Discussion

The results in Figure 4 indicate the following: (1) The winner in the z layer, z_4, stays the same even when the square stimulus contains a misssing contour. (2) This holds true whether or not there is feedback from layer z to layer y. This strongly suggests that the formation of the percept 'square' at the higher level is driven primarily through feedfoward connections. Unit y_{35} in Figure 4 responds with reduced amplitude to a missing contour. However, phase synchronization with the winner in the z layer occurs only when feedback present.

The reduced amplitude response in the y layer in Figures 6, and Figure 1(B) are similar to Figure 1(A). Lee and Mumford [4] report that neurons in V1 respond to illusory contours, and that the V2 response precedes the V1 response. In Figure 5(A), the winner representing the square emerges early. Also, the unit in the y layer responding to the missing contour is not initially synchronized with the winner in the z layer, but eventually becomes so, as in Figure 7. The qualitative behavior of our model is in agreement with the findings in [4].

The results in Figures 4 and 7 show that when the full square is presented as stimulus, the elimination of feedback disrupts the synchronization in the middle y layer. Note that the feedback connections from the z to y layer affect both amplitude and phase of the units in the y layer (Equation 1). Furthermore, the feedforward connections also have an effect on the phase, and hence on the synchronization process, so synchronization cannot be assumed to depend on feedback connections alone.

This is an interesting result as it generates a testable hypothesis. The suggested biological experiment is to temporarily disable feedback connections between V2 and V1, and examine its effect on the response of V1 neurons. Our prediction is that there will be reduced activity in V1 neurons, and that they will not be synchronized with the activity of V2 neurons. So we show that feedback has a greater effect on synchronization, ie the phase than on the amplitude. A generalization of our hypothesis states that disabling feedback connections should impair the phase (or timing) relation between higher and lower level units that represent a common percept, while having relatively little effect on the formation of the percept. The pursuit of this hypothesis should shed further light on the role of feedback in object perception.

Such experimentation is possible, as Galuske *et al* [10] selectively deactivated feedback connections, leading to a disruption in the direction response for motion encoding direction in V1. Thus, our prediction seems reasonable, and could lead to further investigation in neuroscience.

Bar [7] showed that the feedback signal specifically accentuates the most likely input interpretation. This is explicitly exhibited in our model, as the feedback signals have a synchronizing effect on the target unit, thus accentuating likely interpretations.

Bar [7] states that a low spatial frequency representation of the input image may suffice to generate a high-level percept, which occurs rapidly. Indeed, our results support this observation, as the removal of specific object contours does not change the object's low frequency representation. In Figure 5, the high level object percept does not change. Also, the temporal dynamics show that the correct high-level representation (of a 'square') emerges rapidly.

Lamme [11] observed a fast feedforward sweep of information processing. The behavior of the z layer in our system rapidly determines the winner that represents the input object, even when the object contains missing contours.

Engel *et al* [6] report that modulatory top-down effects may influence average neuronal firing rates, and also the temporal structure of neuronal responses. Our experiments validate this observation, as Figure 7 clearly demonstrates the importance of feedback in determining the phase response, which is homologous to the temporal structure of the neural response.

6 Conclusions

By exploring recent biological findings about feedback processing [7], and illusory contour perception [4], with a computational model [2,3], we offer a unique, integrated perspective on object recognition and feedback processing. Illusory contour phenomena naturally arise in a system that dynamically solves the binding problem. Our results highlight the functional relevance of feedback connections, and show that they have a greater effect on synchronization than on the amplitude. For illusory contour perception, our model predicts reduced activity in V1 neurons when feedback from V2 is disabled, and a lack of synchronization with the activity of V2 neurons.

References

1. Wang, D.: The time dimension for scene analysis. IEEE Trans. Neural Networks 16(6), 1401–1426 (2005)
2. Rao, A.R., Cecchi, G.A., Peck, C.C., Kozloski, J.R.: Unsupervised segmentation with dynamical units. IEEE Trans. Neural Networks (January 2008)
3. Rao, A.R., Cecchi, G.A., Peck, C.C., Kozloski, J.R.: Efficient segmentation in multi-layer oscillatory networks. In: IJCNN (accepted, 2008)
4. Lee, T.S., Mumford, D.: Hierarchical bayesian inference in the visual cortex. J. Optical Soc. America A 20(7), 1434–1448 (2003)
5. Grossberg, S., Mingolla, E.: Neural dynamics of perceptual grouping: textures, boundaries and emergent segmentations. Perception and Psychophysics 38, 141–171
6. Engel, A., Pascal, F., Singer, W.: Dynamic predictions: oscillations and synchrony in top-down processing. Nature Reviews Neuroscience 2, 704–716 (2001)
7. Bar, M., et al.: Top-down facilitation of visual recognition. PNAS 103(2), 449–454 (2006)
8. Li, Z.: A neural model of contour integration in the primary visual cortex. Neural Compuation 10, 903–940 (1998)
9. Fukushima, K., Miyake, S., Ito, T.: Neocognitron: a neural network model for a mechanism of visual pattern recognition. IEEE Trans. Systems, Man, and Cybernetics 13(3), 826–834 (1983)
10. Galuske, R., et al.: The role of feedback representations in cat visual cortex. PNAS 99(26), 17083–17088 (2002)
11. Lamme, V., Roelfsema, P.: The distinct modes of vision offered by feedforward and recurrent processing. Trends in Neuroscience 23, 571–579 (2000)

Resonant Spike Propagation in Coupled Neurons with Subthreshold Activity

Belén Sancristóbal[1], José M. Sancho[2], and Jordi García-Ojalvo[1]

[1] Departament de Física i Enginyeria Nuclear, Universitat Politècnica de Catalunya, Colom 11, 08222 Terrassa, Spain
[2] Departament d'Estructura i Constituents de la Matèria, Universitat de Barcelona, Diagonal 647, 08028 Barcelona, Spain

Abstract. Chemical coupling between neurons is only active when the pre-synaptic neuron is firing, and thus it does not allow for the propagation of subthreshold activity. Electrical coupling via gap junctions, on the other hand, is also ubiquitous and, due to its diffusive nature, transmits both subthreshold and suprathreshold activity between neurons. We study theoretically the propagation of spikes between two neurons that exhibit strong subthreshold oscillations, and which are coupled via both chemical synapses and gap junctions. Due to the electrical coupling, the periodic subthreshold activity is synchronized in the two neurons, and affects propagation of spikes in such a way that for certain values of the delay in the synaptic coupling, propagation is not possible. This effect could provide a mechanism for the modulation of information transmission in neuronal networks.

1 Introduction

Information transmission in the form of spike propagation plays a vital role in the functioning of the nervous system [1]. The excitable nature of the neuronal response to perturbations allows for the untarnished propagation of action potentials along chains of neurons, coupled chemically to one another via unidirectional synaptic connections. But synapses are only activated when the pre-synaptic neuron undergoes an action potential, which elicits a constant-shaped post-synaptic potential (PSP) at the receiving neuron. Therefore, the only kind of information that is transmitted between neurons due to synaptic coupling is the timing at which spikes occur. However, an increasing amount of evidence shows that subthreshold oscillations constitute an important part of the dynamical activity of neurons [2,3], and thus the question arises as to what is the functional role of subthreshold activity. Here we study a potential effect of subthreshold oscillations in modulating the propagation of spikes along a chain of neurons.

Indeed, since successful propagation of a spike requires that the post-synaptic neuron overcomes its excitation threshold upon receipt of the synaptic current, the state of the post-synaptic neuron at the time at which it receives the synaptic pulse determines strongly whether a post-synaptic spike will be produced.

V. Kůrková et al. (Eds.): ICANN 2008, Part II, LNCS 5164, pp. 695–702, 2008.

This is specially relevant for neurons with subthreshold oscillations: if the pulse arrives to the neuron at around a minimum of a subthreshold oscillation, the effective distance to the excitation threshold will be large, and one could expect that producing a spike would become more difficult. Oppositely, when the pulse arrives near a maximum of a subthreshold oscillation, excitation should be easier. Consequently, one would expect that the propagation efficiency of a neuronal system would depend on the relationship between the period of the subthreshold oscillations and the delay incurred in the propagation due to the time required by the synaptic mechanism to operate. A coherent modulation of the propagation efficiency along a chain of neurons should be expected when the subthreshold activity between all neurons in the chain is synchronized. This can be accomplished by means of diffusive coupling due to gap junctions, which is also ubiquitous in neural tissue [1]. In this paper, we examine this possibility by studying the propagation of a spike train between two neurons, coupled via both synapses (with a delay) and gap junctions (instantaneously). Our results show that propagation appears resonantly, only for certain values of the synaptic delay such that the pre-synaptic pulse arrives at the receiving neuron at the right time to elicit a spike in it.

2 Model Description

We consider two coupled neurons whose dynamical behavior is described by a FitzHugh-Nagumo model, modified in order to obtain relatively strong subthreshold oscillations [4]. In dimensionless form, the model reads

$$\epsilon \dot{u}_i = u_i(u_i - a)(1 - u_i) - v_i + I_i^{\text{elec}} + I_i^{\text{chem}} + I_i^{\text{app}}, \tag{1}$$

$$\dot{v}_i = g(u_i - b) \tag{2}$$

where $i = 1, 2$ index the neurons, u_i is the voltage variable, and v_i is the recovery variable, which represents the effective membrane conductivity. The parameter ϵ is the ratio between the characteristic times of u and v, and I_i^{elec}, I_i^{chem} and I_i^{app} are the electrical, chemical and applied currents, respectively. In the absence of noise, an isolated FitzHugh-Nagumo neuron can exhibit different types of dynamical behavior. We consider in what follows the excitable regime, by setting $a = 0.9$, $\epsilon = 0.005$ and $b = 0.316$. The function $g(x)$ determines the dynamics of the recovery variable; for certain nonlinearities in $g(x)$ the neuron exhibits subthreshold oscillations. Here we use

$$g(u) = k_1 u^2 + k_2 \left(1 - \exp\left(-\frac{u}{k_2}\right)\right) \tag{3}$$

where k_1 and k_2 are chosen to give a period of the subthreshold oscillations larger than the duration of a spike. In what follows, we choose $k_1 = 7.0$ and $k_2 = 0.08$.

The current applied externally to each neuron I_i^{app} is a train of pulses of 2 units of amplitude with a period of 10 units of time, while the other current terms provide two mechanisms of coupling between the two neurons: a linear electrical coupling and a nonlinear pulsed coupling through chemical synapses. The electrical coupling term, which arises at gap junctions between adjacent neurons, depends on the membrane-potential difference between the two neurons:

$$I_i^{\text{elec}} = g^{\text{elec}}(u_i - u_j) \tag{4}$$

where g^{elec} is the effective conductance of the gap junction, whose value will be tuned in Sec. 3 in order to maximize the synchronization between the neurons without destroying the subthreshold activity.

Chemical coupling can be modeled by the following current term [5]:

$$I_i^{\text{chem}} = g^{\text{chem}} r_j (u_i - E_s) \tag{5}$$

where g^{chem} is the conductance of the synaptic channel, r_j represents the fraction of bound receptors, and E_s is fixed to 0.7 in order to make the synapse excitatory. The fraction of bound receptors has the following dynamical behavior [6]

$$\dot{r_j} = \alpha C_j (1 - r_j) - \beta r_j \tag{6}$$

where α and β are rise and decay time constants, respectively, and

$$C_j = C_{max}\theta\left((T_0^j - \tau) + \tau_{\text{syn}} - t\right)\theta\left(t - (T_0^j - \tau)\right)$$

is the concentration of neurotransmitter released into the synaptic cleft. T_0^j is the time at which the presynaptic neuron j fires, which happens whenever the pre-synaptic membrane potential exceeds a predetermined threshold value, in our case chosen to be 0.7. The time during which the synaptic connection is active is given by τ_{syn}. We add a delay time τ into the chemical current in order to model a more realistic situation in which some time is needed to get the signal of the pre-synaptic cleft, and then the neurotransmitters take a time to reach the post-synaptic neuron. The values of the coupling parameters that we use in what follows are $\alpha = 2.0$, $\beta = 1.0$, $C_{max} = 1.0$, and $\tau_{\text{syn}} = 0.006$. The term I^{chem} only affects the postsynaptic neuron, since the chemical current flows in only one direction. The coupling strength g^{chem} will be tuned in Sec. 4 so that activation of spikes in the post-synaptic neuron is modulated by the subthreshold oscillations.

3 Electrical Coupling: Resynchronization *Versus* Damping

Synchronization of subthreshold activity takes place only via gap junctions. Figure 1 shows the time evolution of the membrane potential u_i of the two neurons for increasing values of the electrical coupling strength g^{elec} and in the absence of chemical coupling ($g^{\text{chem}} = 0$). For small enough g^{elec}, a spike in neuron 1 (represented by a solid line in the figure) does not excite a spike in neuron 2, but perturbs sufficiently the dynamics of both neurons such that

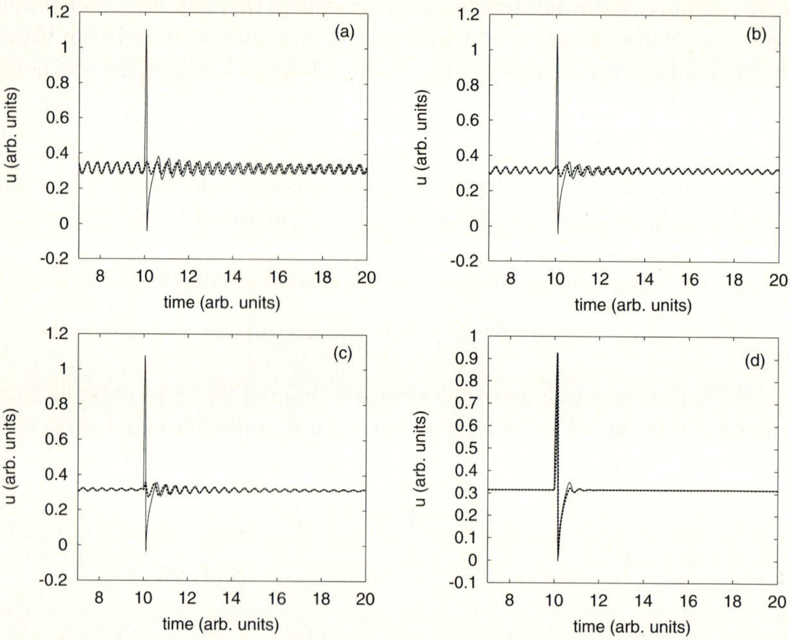

Fig. 1. Time traces for the two coupled neurons in the absence of chemical coupling and for increasing strength of the electrical coupling: (a) $g^{elec} = 0.001$, (b) $g^{elec} = 0.005$, (c) $g^{elec} = 0.01$, (d) $g^{elec} = 0.1$

synchronization of their subthreshold activity is temporarily lost [Fig. 1(a)]. The resynchronization time decreases as g^{elec} is increased [Fig. 1(b)]. For larger strengths of the electrical coupling, the subthreshold oscillations become heavily damped [Fig. 1(c)], and eventually the second neuron pulses as a response to the first [Fig. 1(d)]. For an intermediate level of electrical coupling [Fig. 1(b)], the resynchronization time is relatively small, the subthreshold oscillations are not heavily damped, and the second neuron does not pulse as a response to the first. We will use $g^{elec} = 0.005$ in what follows.

4 Chemical Coupling: Modulating Spike Excitation *Via* Subthreshold Oscillations

As mentioned above, chemical coupling is only able to propagate suprathreshold activity. If g^{chem} is too high the post-synaptic neuron will fire whenever the pre-synaptic neuron does, and if it is too low it will never fire. In both such cases, the subthreshold activity of the post-synaptic neuron does not play a relevant role. For intermediate values of g^{chem}, on the other hand, the receiving neuron will fire depending on its state at the time at which the input pulse from the first neuron is received. We examined the behavior of the neurons for different values of g^{chem} in the absence of electrical coupling in order to determine whether subthreshold oscillations have an effect on the propagation of a spike. Figure 2(a) shows the case of a low value of g^{chem}, for which the second neuron never fires irrespective of the instant at which the pulse from neuron 1 is received with respect to the subthreshold oscillation state (which can be varied by changing the delay τ; results not shown, see next Section for results in the presence of electrical coupling). For higher coupling strength g^{chem} [Fig. 2(b)],

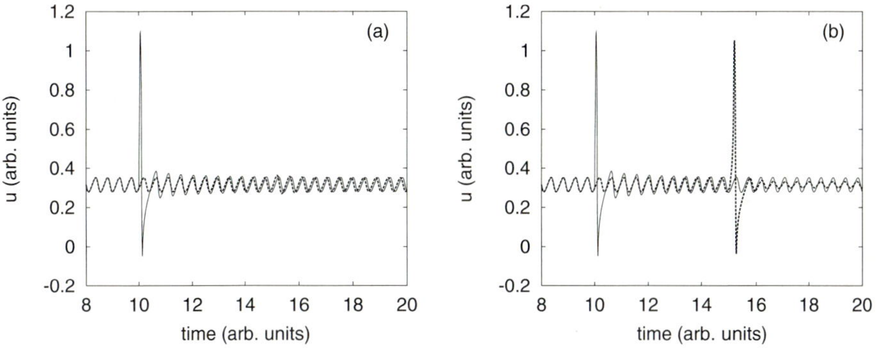

Fig. 2. Time traces for the two coupled neurons in the absence of electrical coupling and for increasing strength of the chemical coupling. A delay $\tau = 5$ is externally added in order to clearly separate the spikes of the two neurons. (a) $g^{\text{chem}} = 0.1$, (b) $g^{\text{chem}} = 1.0$.

on the other hand, the receiving neuron fires in response to the emitting one for the case presented in the figure, while it does not fire for other delay values (results not shown). We choose that coupling value, $g^{\text{chem}} = 1.0$, in what follows when adding the electrical coupling, in order to synchronize the background subthreshold oscillations.

5 Effect of the Delay in Synaptic Transmission

We now wish to compare the behavior of the two coupled neurons for different values of the delay introduced in the chemical current, in the presence of both synaptic and electrical couplings. To that end, we vary the delay and check the behavior of the receiving neuron for different values of g^{chem}. As mentioned in the previous section, it turns out that $g^{\text{chem}} = 0.1$ is too small and the spike of the pre-synaptic neuron does not propagate for any value of the delay. For larger values of g^{chem}, such as $g^{\text{chem}} = 1.0$ suggested above in the case of no electrical coupling, we obtain the desired behavior: spike transmission depends on the delay.

Figure 3 shows how the spiking behavior of the postsynaptic neuron changes in relation to the delay introduced. By varying this delay we can look at different situations. For example, the post-synaptic neuron can be at the maximum of its subthreshold oscillations, or at the minimum, or somewhere in between, at the moment at which it receives an external signal from the pre-synaptic neuron.

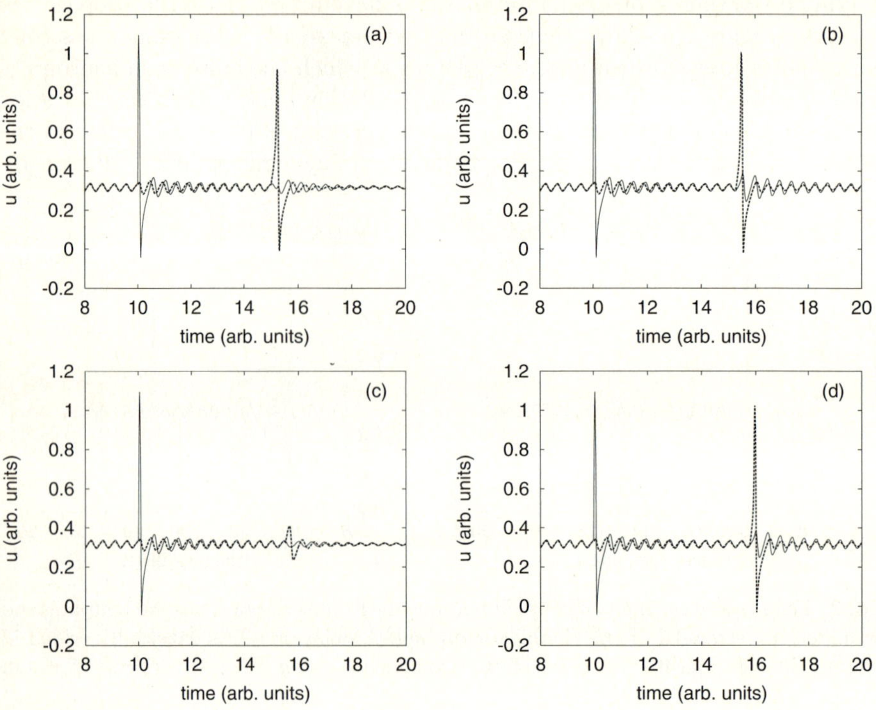

Fig. 3. Time traces for the two coupled neurons with $g^{\text{elec}} = 0.005$ and $g^{\text{chem}} = 1.0$ for different values of the delay introduced. (a) $\tau = 5.0$, (b) $\tau = 5.25$, (c) $\tau = 5.5$, (d) $\tau = 5.75$.

We are interested on how the subthreshold oscillations of the post-synaptic neuron can increase the probability to propagate a spike when it is excited by the previous neuron. As shown in Fig. 3, at different situations the response of the receiving neuron is different, it can either fire or not.

The results of Fig. 3 allow us to infer a non-monotonic dependence of the spike propagation efficiency on the timing at which the post-synaptic neuron receives the stimulus. That timing is controlled by the synaptic delay in our case, with certain delay values at which propagation is optimal and others at which it is absent. In order to quantify this observation, we vary the delay from $\tau = 5$ to $\tau = 6$ in units of 0.05, stimulate the pre-synaptic neurons in multiple realizations (100 in the results presented here), and calculate the percentage of successful spike transmission events for increasing delay. As shown in Fig. 4, some delays lead to a 100% success rate, while for some others the postsynaptic neuron never fires. The figure also includes a time trace with the subthreshold oscillations that underlie the activity of the post-synaptic neurons. Comparing the top plot with the bottom time trace, one can see that the post-synaptic neuron fires when it receives the input pulse while the value of its membrane potential is growing. This can also be seen by looking carefully at Fig. 3.

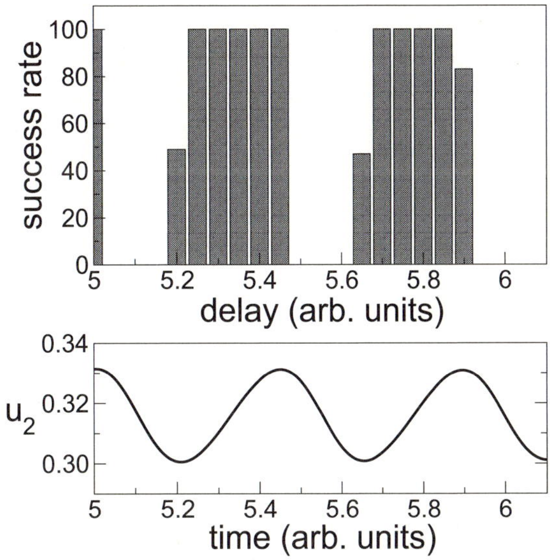

Fig. 4. Success rate for increasing delay in the chemical synaptic coupling, for $g^{\text{elec}} = 0.005$ and $g^{\text{chem}} = 1.0$. The bottom time trace shows the underlying subthreshold oscillation of the post-synaptic neuron.

6 Conclusions

We have seen that subthreshold oscillations play a relevant role in the propagation of spikes through two neurons coupled via chemical synapses. Electrical coupling via gap junctions leads to a synchronization of the background subthreshold activity, and would thus allow to scale up the phenomenon in a coherent manner to an array of coupled neurons, something which we are currently investigating. In the present case, when the membrane potential of the post-synaptic neuron is increasing the success rate of spike propagation is at its maximum, and spike propagation is achieved. When the membrane potential is decreasing, on the other hand, the success rate drops to zero and no propagation is carried out. We expect such type of mechanism to be of general importance, given the ubiquity of gap-junction coupling and subthreshold activity in neural tissue.

Acknowledgments

This work was supported by the Ministerio de Educación y Ciencia (Spain) under project FIS2006-11452 and by the GABA project (European Commission, FP6-NEST contract 043309).

References

1. Kandel, E.R., Schwartz, J.H., Jessell, T.M.: Principles of Neural Science, 4th edn. McGraw-Hill, New York (2000)
2. Llinas, R.R., Grace, A.A., Yarom, Y.: In vitro Neurons in Mammalian Cortical Layer 4 Exhibit Intrinsic Oscillatory Activity in the 10- to 50-Hz Frequency Range. Proceedings of the National Academy of Science, 88, 897–901 (1991)
3. Giocomo, L.M., Zilli, E.A., Fransen, E., Hasselmo, M.E.: Temporal frequency of subthreshold oscillations scales with entorhinal grid cell field spacing. Science 315(5819), 1719–1722 (2007)
4. Makarov, V.A., Nekorkin, V.I., Velarde, M.G.: Spiking behavior in a noise-driven system combining oscillatory and excitatory properties. Physical Review Letters 86(15), 3431–3434 (2001)
5. Balenzuela, P., Garcia-Ojalvo, J.: Role of chemical synapses in coupled neurons with noise. Physical Review E (Statistical, Nonlinear, and Soft Matter Physics) 72(2), 021901–7 (2005)
6. Destexhe, A., Mainen, Z.F., Sejnowski, T.J.: An efficient method for computing synaptic conductances based on a kinetic model of receptor binding. Neural Comput. 6(1), 14–18 (1994)

Contour Integration and Synchronization in Neuronal Networks of the Visual Cortex

Ekkehard Ullner[1], Raúl Vicente[2,3], Gordon Pipa[2,3,4], and Jordi García-Ojalvo[1]

[1] Departament de Física i Enginyeria Nuclear, Universitat Politècnica de Catalunya, Colom 11, E–08222 Terrassa, Spain
[2] Max-Planck Institute for Brain Research, Deutschordenstr. 46, D-60528 Frankfurt/Main, Germany
[3] Frankfurt Institute for Advanced Studies, Ruth-Moufang-Str. 1, D-60438 Frankfurt/Main, Germany
[4] Dep. of Brain and Cognitive Sciences, Massachusetts Inst. of Technology, and Dep. of Anesthesia and Critical Care, Massachusetts General Hospital, 77 Massachusetts Ave., Cambridge, MA 02139-4307, USA

Abstract. The visual perception of contours by the brain is selective. When embedded within a noisy background, closed contours are detected faster, and with higher certainty, than open contours. We investigate this phenomenon theoretically with the paradigmatic excitable FitzHugh-Nagumo model, by considering a set of locally coupled oscillators subject to local uncorrelated noise. Noise is needed to overcome the excitation threshold and evoke spikes. We model one-dimensional structures and consider the synchronization throughout them as a mechanism for contour perception, for various system sizes and local noise intensities. The model with a closed ring structure shows a significantly higher synchronization than the one with the open structure. Interestingly, the effect is most pronounced for intermediate system sizes and noise intensities.

1 The Introduction

Object representation in the brain relies on two properties: first, the detection of specific features (such as location and orientation) by individual neurons, and second, the integration of features of extended objects through dynamic association of neuronal assemblies [1]. Seminal experimental studies showed, for instance, that assembly coding allows for scene segmentation in the cat visual cortex [2,3]: the response to individual objects in a scene, in that case a single moving bar, was characterized by synchronization among neuronal assemblies corresponding to the different parts of the scene, while correlations were absent in response to different objects, such as two moving bars with different orientations. When the stimulus is a static contour, the different sections of the contour map to different neurons in the visual cortex. It is hypothesized that these neurons fire synchronously when the contour is perceived. The potential beneficial role of synchrony in that context lies in the possibility that at the following stages of cortical processing the receiving or downstream neurons will

V. Kůrková et al. (Eds.): ICANN 2008, Part II, LNCS 5164, pp. 703–712, 2008.

be much more reactive to synchronous input than to temporally dispersed input. In this way synchrony can effectively impact the dynamics of further stages of information processing, and neurons that synchronize can have a higher saliency than the ones that do not. Experimental observations show that contours embedded within a noisy background are detected more efficiently if they are closed than if they are open. Here we study this phenomenon theoretically in a one-dimensional array of model neurons subject to noise, considering the effect of boundary conditions in the synchronization efficiency. We also show that noise has a beneficial role in this process by leading to a near zero-lag synchronization in the firing of all neurons in the array.

The effect of noise in brain activity has evoked a large interest in recent years. Stochasticity in neurons originates from different sources, including random synaptic input from other neurons, stochastic switching of ion channels, and quasi-random release of neurotransmitter by synapses. Despite (or maybe because of) the many noise sources in neuronal networks, the brain acts very reliably and needs only a very small amount of energy (about 12 W according to Ref. [4]). A growing number of scientific results suggests that noise plays a constructive role in brain activity. For instance, a noise-induced effect has been demonstrated in the visual processing area of the human brain [5]. In that experiment a periodic light signal was sent to one eye, whereas the other eye was subjected by noise, represented by light with fluctuating intensity. The result was that noise improved the processing of the periodic signal sent to the first eye.

Coherence resonance (CR), also known as stochastic coherence, is a noise-induced effect through which periodic oscillatory behavior arises and its coherece is optimized by noise. It has been found that at a certain noise intensity the system responds with a maximal periodicity, i.e. with an enhanced coherence. Both an increase and a decrease of the noise amplitude away from this optimal value lead to a decreasing of the coherence. CR has been observed in excitable systems like the Hodgkin-Huxley model [6], the FitzHugh-Nagumo systems [7], leaky integrate-and-fire models [8], the Plant/Hindmarsh-Rose neural model [9], and in dynamic systems which besides show jumps between several attractors [10]. Besides the neural context, CR can be found in climate [11] and laser models [12,13].

Array-enhanced coherence resonance (AECR) [14] is an extension of the described noise-induced rhythm generation to an ensemble of many coupled excitable oscillators. Interestingly, the quality of the coherence in a large ensemble with diverse oscillators can be larger than in a single one with the same mean properties. The results presented below show that AECR provides a potential mechanism for contour perception, whose efficiency depends on the network topology, which should explain why a closed contour is better perceived than an open one.

2 The FitzHugh-Nagumo Model

The FitzHugh-Nagumo (FHN) model is a paradigmatic model describing the behavior of firing spikes in neural activity [15]. The model was proposed in

Refs. [16,17] as a simplification of the famous model by Hodgkin and Huxley [16] and is a simple example of two-dimensional excitable dynamics. It describes qualitatively the response of an excitable nerve membrane to external stimuli. Important features are the inclusion of a refractory mechanism and the existence of different refractory states, as well as states of enhanced and depressed excitability depending on the external stimulation. Beside the paradigmatic description of the firing spikes of neural activity [15], the FHN model is representative for activator-inhibitor dynamics of excitable media in general [18]. The model reads:

$$\varepsilon \dot{x}_i = x_i - \frac{x_i^3}{3} - y_i + \xi_i(t) + D_{i-1}(x_{i-1} - x_i) + D_i(x_{i+1} - x_i) \tag{1}$$

$$\dot{y}_i = a - x_i \tag{2}$$

In a neural context, $x(t)$ represents the membrane potential of the neuron and $y(t)$ is related to the time-dependent conductance of the potassium channels in the membrane [15]. The dynamics of the activator variable x is much faster than that of the inhibitor y, as indicated by the small time-scale-ratio parameter ε and is fixed to $\varepsilon = 0.01$ throughout the following calculations. It is well known that for $|a| > 1$ the only attractor is a stable fixed point. For $|a| < 1$, the limit cycle generates a periodic sequence of spikes. The parameter a is the bifurcation parameter and is fixed below to $a = 1.05$, in order to tune the system to the excitable regime. The index i distinguishes the separate oscillators and runs from 1 to N, the total number of coupled elements. The Gaussian (white) noise sources $\xi_i(t)$ satisfy $\langle \xi_i(t)\xi_j(t')\rangle = \sigma_a^2 \delta(t - t')\delta_{i,j}$ with the noise intensity σ_a^2.

We model the neurons in the visual cortex by diffusively coupled FHN and consider in the present model only those neurons that map the visual contour. We assume a constant and equal external stimulus to all neurons mapping the contour, which leads to a constant reduction of the excitation threshold for all involved neurons. This assumption results in a permanent reduction of the parameter a in the model closer to the bifurcation point for all neurons, so that the external stimulus need not appear explicitly in the model equations. The neurons not involved in the contour mapping remain with high threshold of excitation and are ignored. All FHN elements are thus in a excitable state and remain in their rest state without spiking activity if no noisy stimulus is present. The noise is absolutely needed to overcome the excitation threshold and evoke spikes. We assume local Gaussian white noise subject to every neuron with the same intensity σ_a^2. The spikes are noise-evoked and mediated by the local nearest neighbor coupling with diffusive property and the coupling strength D_i. Note that the coupling is instantaneous, i.e. it does not include any explicit delay. Effective delay in the signal transmission will appear, however, due to the natural inertia of each neuron in the chain to react to an input (see sec. 4 below). We consider three spatial architectures: the uncoupled situation ($D_i = 0.0$), a linear chain ($D_{1,\ldots,N-1} = 0.02$, $D_N = 0.0$, Fig 1 bottom), and a closed loop ($D_i = 0.02$, Fig 1 top).

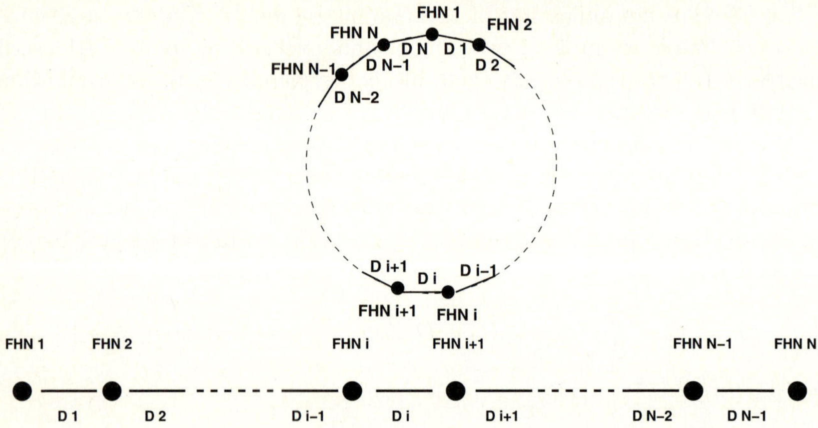

Fig. 1. Schemes of a closed loop and a linear open chain

3 Comparison of the Synchronization of an Open and a Closed Contour

This section should shed light on the question of whether synchronization is better in a closed contour than in an open one. We measure the degree of synchronization R_{syn} as the ratio of the variance of the mean field to the mean variance of the individual elements [19]:

$$R_{syn} = \frac{\langle \bar{x}^2 \rangle - \langle \bar{x} \rangle^2}{\frac{1}{N}\sum_{i=1}^{N}(\langle x_i^2 \rangle - \langle x_i \rangle^2)} = \frac{\mathrm{Var}(\bar{x})}{\mathrm{Mean}_i(\mathrm{Var}(x_i))} \tag{3}$$

with the mean membrane potential $\bar{x}$

$$\bar{x} = \frac{1}{N}\sum_{i=1}^{N} x_i \,. \tag{4}$$

The fully desynchronized state results in a synchronization measure $R_{syn} = 0$, whereas the complete synchronization amongst all oscillators is becoming manifest by $R_{syn} = 1$. Values between 0 and 1 describe states of partial synchronization. The measure R_{syn} detect only zero-lag synchronization, i.e. delay free synchronization and express the average difference between the mean field and the dynamics of the individual oscillators. For small ensembles R_{syn} results in values larger than zero also if one compares completely independent elements, e.g. for only two oscillators each of them influences the mean field to 50%. To eliminate this effect of the system size, we considere the fully uncoupled ensemble as the baseline to compare open and closed structures.

We consider two main influences on R_{syn}: the noise intensity σ_a^2 and the system size N. First we discuss the synchronization measure R_{syn} as a function of the system size N for different fixed noise intensities σ_a^2 (Fig. 2). In the first case

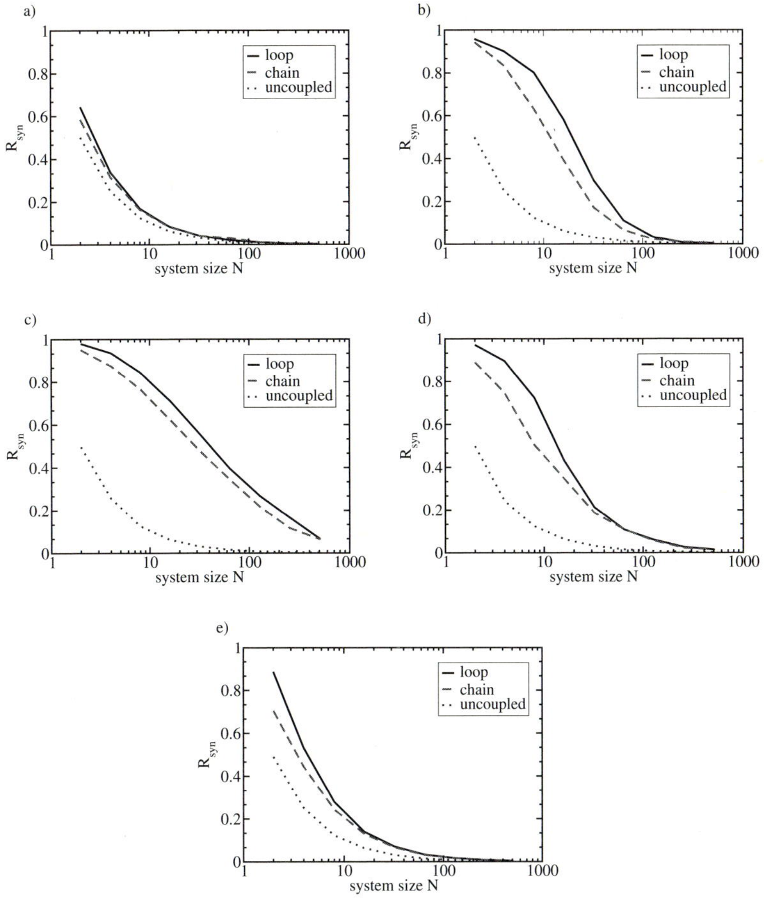

Fig. 2. Synchronization measure R_{syn} versus system size N. The plots differ by the noise intensities: a) $\sigma_a^2 = 0.01$, b) $\sigma_a^2 = 0.025$, c) $\sigma_a^2 = 0.2$, d) $\sigma_a^2 = 1.0$, e) $\sigma_a^2 = 5.0$. Each plot compares the synchronization in a closed loop, an open chain and in the fully uncoupled case as the reference.

($\sigma_a^2 = 0.01$) the noise intensity is sub-threshold and only few and irregular spikes were evoked. The dynamics is determined by the small noisy and uncorrelated sub-threshold fluctuations. Hence the coupling does not play a significant role and the loop and chain configuration are close to the reference of the uncoupled chain. An increased noise intensity leads to a spiking behavior and the coupling contributes to the dynamics. The next plot ($\sigma_a^2 = 0.025$) shows a clear difference between the loop, the chain and the uncoupled case. For intermediate system sizes the loop synchronizes the ensemble more effectively than the open chain. Further increase of the noise to $\sigma_a^2 = 0.2$ optimizes the synchronization. The difference between loop and chain is less pronounced, but the absolute synchronization quality is enhanced, especially for large system sizes, i.e. the zero lag

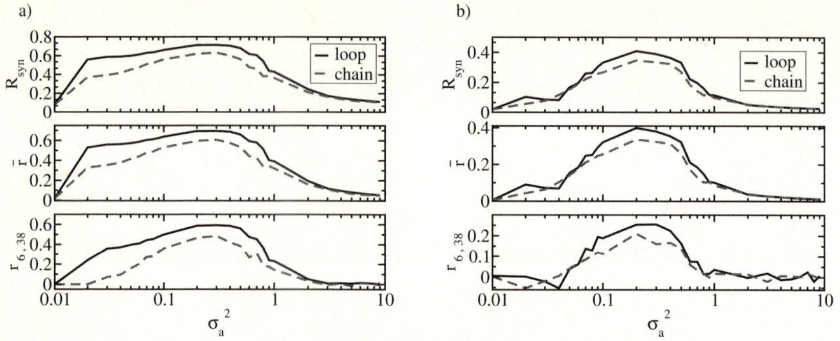

Fig. 3. Synchronization measure R_{syn} and correlation coefficient versus noise intensity σ_a^2. The left plot depicts a small ensemble ($N = 16$) and the right one shows the behavior of a larger system ($N = 64$).

synchronization becomes stronger. $\sigma_a^2 = 1.0$ is beyond the optimal noise intensity and the synchronization is less than in the optimal case over the complete range of N. Interestingly, in the non-optimal noise case but close to it, the closed loop configuration shows a significant better performance than the open chain for intermediate system sizes. Further increase of the noise intensity ($\sigma_a^2 = 5.0$) destroys the synchronization. The system is determined by the random fluctuations and the coupling is too small to smoothen the irregular spikes.

The previous plots of the synchronization measure R_{syn} as a function of the system size (Fig. 2) have shown a strong dependence of the performance on the noise intensity. The noise evoking a maximal difference between open and close contour differs from the overall optimal noise, and we see that intermediate system sizes demand a noise strength for optimization different than larger sizes. To investigate this issue further, we calculated R_{syn} as a function of the noise intensity σ_a^2 for two fixed system sizes ($N = 16$ and $N = 64$) and the same coupling, in the spirit of stochastic and coherence resonance as a noise-induced effect with a resonance like response curve. Besides the synchronization measure R_{syn}, we also computed the correlation coefficient $r_{i,j}$ and the mean correlation coefficient $\bar{r}$. The correlation coefficient $r_{i,j}$ is the value of the cross-correlation function at time delay $\tau = 0.0$ and reads:

$$r_{i,j} = \frac{\langle x_i x_j \rangle - \langle x_i \rangle \langle x_j \rangle}{\sqrt{\langle x_i^2 \rangle - \langle x_i \rangle^2} \sqrt{\langle x_j^2 \rangle - \langle x_j \rangle^2}} \tag{5}$$

and $\bar{r}$ is:

$$\bar{r} = \frac{2}{N(N-1)} \sum_{i=1}^{N-1} \sum_{j=i+1}^{N} r_{i,j} \tag{6}$$

All these measures (R_{syn}, $r_{i,j}$, and $\bar{r}$), plotted in Fig. 3, quantify the level of zero-lag synchronization, and express the same behavior.

Figure 3 reveals a typical resonance-like behavior. Too small noise leaves the oscillators near the stable fixed point without crossing the threshold. Too large noise dominates the overall dynamics and the coupling can not provoke synchronization. Only intermediate noise is able to improve the synchronization amongst the neurons in combination with the coupling. For the two system sizes considered, the closed contour of the loop surpasses the synchronization results of the open chain. In the case of intermediate size $N = 16$, the relative advantage of the loop to the chain is most striking before and after the absolute maximum located at $\sigma_a^2 \approx 0.3$, and the bell-shaped resonance curve is relatively broad compared to the narrower one for the large ensemble. For $N = 64$ the overall maximal synchronization is reached at $\sigma_a^2 \approx 0.1$, i.e. for a smaller noise intensity.

4 Near Zero-Lag Synchronization by Additive Uncorrelated Noise

The measures used above take into account only zero-lag synchronization. Local coupling and the limited transmission velocity of the signal leads normally to a time lag along the signal chain, as can be seen in Fig. 4(a). In contrast to this, experimental measurements show that synchronization with cell assemblies does not exhibit a significant time lag [2,3]. We therefore study in what follows the time-lag in the cross-correlation function as a function of the noise intensity.

Specifically, we investigated an ensemble of $N = 64$ FHN oscillators in the loop configuration, for varying additive noise intensity σ_a^2. The results for the

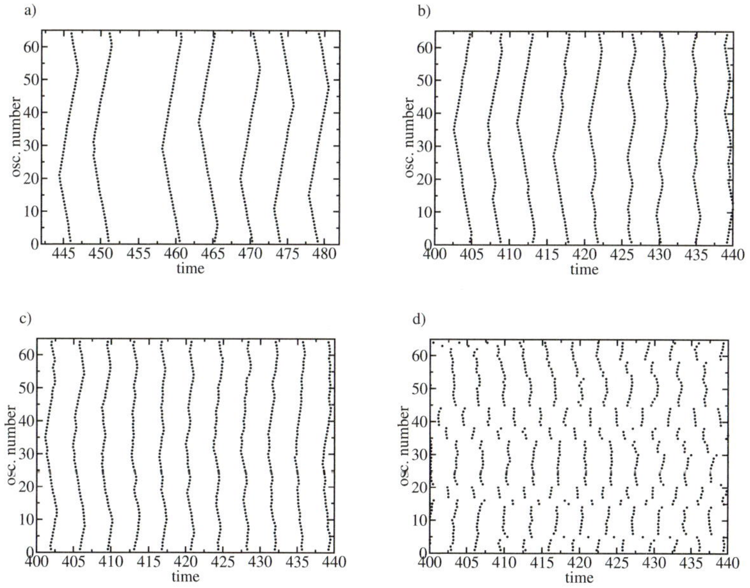

Fig. 4. Raster plots to illustrate the behaviour of the spiking dynamics. The noise intensity increases from top to bottom and left to right: a) $\sigma_a^2 = 0.04$, b) $\sigma_a^2 = 0.06$, c) $\sigma_a^2 = 0.2$, and d) $\sigma_a^2 = 1.0$. Compare with the cross-correlation functions in Fig. 5.

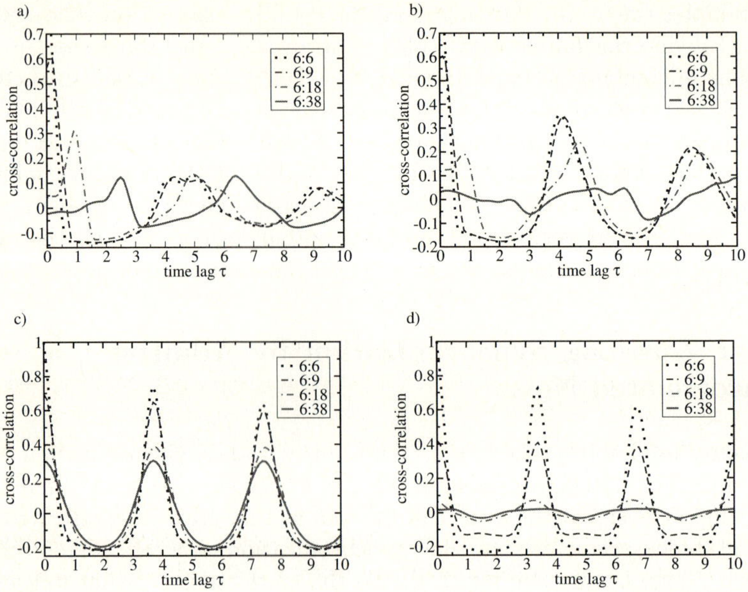

Fig. 5. Cross-correlation functions for different noise intensities and oscillator pairs of the closed contour of $N = 64$ FHNs. The noise intensity increases from top to bottom and left to right: a) $\sigma_a^2 = 0.04$, b) $\sigma_a^2 = 0.06$, c) $\sigma_a^2 = 0.2$, and d) $\sigma_a^2 = 1.0$.

open chain are not shown but they are comparable. We calculate the cross-correlation functions of various pairs of elements, noted in the legend of each diagram (Fig. 5), choosing oscillator #6 as a reference, without loss of generality. The pair 6-6 is the auto-correlation function of the time series of oscillator #6, and the pair 6-38 depicts the cross-correlation function for two oscillators separated the maximal spatial distance in the system of 64 coupled FHNs. The rasterplots corresponding to the cross-correlation functions Fig. 5 are plotted in Fig. 4, and give a snapshot of the spiking activity in the three situations: insufficient noise, optimal noise and too large noise intensity.

In the case of small noise nucleation occurs rarely and each excitation travels from its nucleating oscillator to all the others. The locations of the noise-induced nucleation events are fully random, The finite traveling time of the signal produces a time delay of the spike time of elements far from the nucleation site. Hence a growing spatial distance between neurons increases the time delay of the signal response, as one can see in the raster plot Fig. 4(a), and in the corresponding cross-correlation functions in Fig. 5(a). Small noise results in a non-zero-lag synchronization, with the lag time depending on the spatial distance.

In the situation of optimal and intermediate noise intensity ($\sigma_a^2 \approx 0.2$), multiple nucleation points appear almost simultaneously in the rasterplot [Fig. 4(c)]. The cross-correlation functions for all pairs, independently of their spatial separation, show maximal correlation at zero lag. Thus an optimal noise intensity in combination with local coupling leads to a zero-lag synchronization of all neurons in the ensemble [Fig. 5(c)]. As we noted there is no input signal in the model, thus the effect is fully noise driven.

Finally, too strong noise destroys the effect. In that case, noise is strong enough to evoke spikes everywhere at random times. The formation of small local synchronized cluster destroys the long range correlation completely, i.e. the zero-lag and the non-zero lag synchronization. Only very adjacent pairs like 6:9, 6:6 (trivially) and 6:3 show correlations [Fig. 5(d)].

5 Conclusion

We compared the synchronizability of open and closed structures of noisy neuronal networks, and relate it with the efficiency of contour perception in the visual cortex. For small and intermediate sized neuronal networks, a closed contour can be recognized better than a equally sized open contour. The effect shows a resonance-like relation between synchronization and the noise intensity. Furthermore, noise compensates the time lag in the signal transduction caused by the finite value of the signaling velocity. Note that the coupling in our model is instantaneous, so that the signaling velocity is not caused by axonal conduction delays, but by the inertia of each neuron to react to an input. In the case of insufficient noise, a clear time-lag can be seen in the cross-correlation function, increasing with the spatial distance between the neurons. An optimal non-zero noise enhances the correlation and synchronization amongst the neuron ensemble, and shifts the maximal of the cross-correlation function to zero-lag as one can see in experimental measurements [2,3].

Acknowledgments

E.U. acknowledges financial support from the Alexander von Humboldt Foundation. This research was supported by the Ministerio de Educacion y Ciencia (Spain) under project FIS2006-11452 and by the GABA project (European Commission, FP6-NEST contract 043309).

References

1. Singer, W.: Neuronal synchrony: A versatile code for the definition of relations? Neuron 24(1), 49–65 (1999)
2. Gray, C.M., König, P., Engel, A.K., Singer, W.: Oscillatory responses in cat visual cortex exhibit inter-columnar synchronization which reflects global stimulus properties. Nature 338, 334–337 (1989)
3. Castelo-Branco, M., Goebel, R., Neuenschwander, S., Singer, W.: Neural synchrony correlates with surface segregation rules. Nature 405, 685–689 (2000)
4. Sarpeshkar, R.: Analog versus digital: Extrapolating from electronics to neurobiology. Neural Comput. 10(7), 1601–1638 (1998)
5. Mori, T., Kai, S.: Noise-induced entrainment and stochastic resonance in human brain waves. Phys. Rev. Lett. 88, 218101 (2002)

6. Lee, S., Neiman, A., Kim, S.: Coherence resonance in a hodgkin-huxley neuron. Phys. Rev. E 57, 3292 (1998)
7. Pikovsky, A., Kurths, J.: Coherence resonance in a noise-driven excitable system. Phys. Rev. Lett. 78, 775 (1997)
8. Lindner, B., Schimansky-Geier, L., Longtin, A.: Maximizing spike train coherence or incoherence in the leaky integrate-and-fire model. Phys. Rev. E 66, 31916 (2002)
9. Longtin, A.: Autonomous stochastic resonance in bursting neurons. Phys. Rev. E 55, 868 (1997)
10. Palenzuela, C., Toral, R., Mirasso, C., Calvo, O., Gunton, J.: Coherence resonance in chaotic systems. Europhys. Lett. 56(3), 347–353 (2001)
11. Ganopolski, A., Rahmstorf, S.: Abrupt glacial climate changes due to stochastic resonance. Phys. Rev. Lett. 88, 038501 (2002)
12. Dubbeldam, J.L.A., Krauskopf, B., Lenstra, D.: Excitability and coherence resonance in lasers with saturable absorber. Phys. Rev. E 60, 6580–6588 (1999)
13. Buldú, J.M., García-Ojalvo, J., Mirasso, C.R., Torrent, M.C., Sancho, J.M.: Effect of external noise correlation in optical coherence resonance. Phys. Rev. E 64, 051109 (2001)
14. Hu, B., Zhou, C.: Phase synchronization in coupled nonidentical excitable systems and array-enhanced coherence resonance. Phys. Rev. E 61(2), R1001–R1004 (2000)
15. Keener, J.P., Sneyd, J.: Mathematical Physiology. Springer, New York (1998)
16. FitzHugh, R.A.: Impulses and physiological states in models of nerve membrane. Biophys. J. 1, 445–466 (1961)
17. Nagumo, J., Arimoto, S., Yoshitzawa, S.: An active pulse transmission line simulating nerve axon. Proc. IRE 50, 2061 (1962)
18. Mikhailov, A.S.: Foundations of Synergetics, 2nd edn. Springer, Berlin (1994)
19. García-Ojalvo, J., Elowitz, M.B., Strogatz, S.H.: Modeling a synthetic multicellular clock: Repressilators coupled by quorum sensing. Proc. Natl. Acad. Sci. U.S.A 101(30), 10955–10960 (2004)

Fuzzy Growing Hierarchical Self-Organizing Networks

Miguel Barreto-Sanz[1,2,3], Andrés Pérez-Uribe[2], Carlos-Andres Peña-Reyes[2], and Marco Tomassini[1]

[1] Université de Lausanne, Hautes Etudes Commerciales (HEC), Institut des Systèmes d'Information (ISI)
Miguel-Arturo.Barreto-Sanz@heig-vd.ch, Marco.Tomassini@unil.ch
[2] University of Applied Sciences of Western Switzerland (HEIG-VD)(REDS)
andres.perez-uribe@heig-vd.ch, Carlos.Pena@heig-vd.ch
[3] Corporación BIOTEC

Abstract. Hierarchical Self-Organizing Networks are used to reveal the topology and structure of datasets. Those structures create crisp partitions of the dataset producing branches or prototype vectors that represent groups of data with similar characteristics. However, when observations can be represented by several prototypes with similar accuracy, crisp partitions are forced to classify it in just one group, so crisp divisions usually lose information about the real dataset structure. To deal with this challenge we propose the Fuzzy Growing Hierarchical Self-Organizing Networks (FGHSON). FGHSON are adaptive networks which are able to reflect the underlying structure of the dataset, in a hierarchical fuzzy way. These networks grow by using three variables which govern the membership degree of data observations to its prototype vectors and the quality of the network representation. The resulting structure allows to represent heterogeneous groups and those that present similar membership degree to several clusters.

1 Introduction

It is usual to create representations of data relationships to improve the analysis of datasets, and hence permitting to reveal its topology and structure. These representations are used in different ways, some of their applications include: partitioning (i.e. clustering), classification and prototypes selection. However, producing an optimal dataset representation is a subjective process, since it depends on the application and on the complexity of the analysis to which it will be applied. Nevertheless, several methodologies are used in order to reduce the degree of subjectiveness and to improve the dataset representations.

On the one hand, hierarchical methods are used to help explain the inner organization of datasets, since the hierarchical structure imposed on the data produces a separation of clusters that is mapped onto different branches. Therefore, this hierarchy enables to analyze complicated structures as well as it allows the exploration of the dataset at multiple levels of detail. On the other hand,

V. Kůrková et al. (Eds.): ICANN 2008, Part II, LNCS 5164, pp. 713–722, 2008.

algorithms for learning topology-preserving mappings are employed to discover the topology of datasets in a self-organizing way (using fixed [7] or growing networks [9]). These algorithms attempt to distribute a certain number of vectors (prototype vectors) in a possibly low-dimensional space. The distribution of these vectors should reflect the probability distribution of the input signals.

Different approaches have been introduced in order to combine the capabilities of hierarchical and learning topology-preserving mapping methods [10], [5], [3], [12]. Hence, obtaining networks that permit representing the topology and the structure of datasets in a hierarchical self-organizing way. These networks are able to grow and adapt their structure in order to represent the characteristics of the datasets in the most accurate manner. Although these hybrid models provide satisfactory results, they generate crisp partitions of the datasets. The crisp segmentations tend to split the dataset in clusters which are represented in the network by one branch of the hierarchy or by unique prototype vector. Nevertheless, in many applications crisp partitions are not optimal dataset representations, since would be useful to have a degree of membership of each sample data to different branches or vector prototypes. With the purpose of representing degrees of membership, fuzzy logic is a feature that could be added to the aforementioned characteristics of hierarchical self-organized hybrid models.

We propose, thus Fuzzy Growing Hierarchical Self-Organizing Networks (FGHSON), which intends to synergistically combine the advantages of Self-Organizing Networks, hierarchical structures, and fuzzy logic. FGHSON are designed to improve the analysis of datasets where it is desirable to obtain a fuzzy representation of a dataset in a hierarchical way, then discovering its structure and topology. This new model will be able to obtain a growing hierarchical structure of the dataset in a self-organizing fuzzy manner. These kind of networks are based on the Fuzzy Kohonen Clustering Networks (FKCN) [1] and Hierarchical Self-Organizing Structures (HSS) [8], [11], [10], [12].

This paper is organized as follows: In the next section the Hierarchical Self-Organizing Structures and the FKCN will be explained serving as a base, to subsequently, introduce our model. Section three focuses on the application of the methodology using the Iris benchmark an a toy dataset. Finally, in Section four are presented some conclusions and future extensions of this work.

2 Methods

2.1 Hierarchical Self-Organizing Structures

The ability of obtaining hierarchically structured knowledge from a dataset using autonomous learning has been widely used in many areas. This is due to the fact that the Hierarchical Self-Organizing Structures (HSS) allow to the unevenly distributed real-world data to be represented in a suitable network structure, during an unsupervised training process. These networks capture the unknown data topology in terms of hierarchical relationships and cluster structures.

Different methodologies have been presented in this area with various approaches. Therefore, it is possible to classify the HSS in two classes taking into account the algorithm of self-organization used. The first family of models is based on Kohonen Self-Organizing Maps (SOM) [7], and the second on Growing Cell Structures (GCS) [4].

The algorithms derived on GCS [2], [5], [3] are based on periodic node deletion based on node activity and on the volume of the input space classified by the node. This approach tends to represent mainly the examples with high occurrence rates, and therefore takes as outliers or noise low frequency examples. As a result, examples with low presence rates are not represented in the model. Nevertheless, in many cases it is desirable to discover novelties in the dataset, so taking into account the observations with low occurrence rates could permit to find out those exceptional behaviors.

For the aforementioned reason, we focused our research in approaches based on SOM [8], [11], [10] particularly in the Growing Hierarchical Self-Organizing Map (GHSOM)[12] due to its ability to take into account the observations with low presence rates as part of the model. This is possible since the hierarchical structure of the GHSOM is adapted according to the requirements of the input space. Therefore, areas in the input space that require more units for appropriate data representation create deeper branches than others, this process is done without eliminating nodes that represent examples with low occurrence rates.

2.2 Fuzzy Kohonen Clustering Networks

FKCN [1] integrate the idea of fuzzy membership from Fuzzy c-Means (FCM) with the updating rules from SOM. Thus, creating a self-organizing algorithm that automatically adjust the size of the updated neighborhood during a learning process, which usually, terminates when the FCM objective function is minimized. The update rule for the FKCN algorithm can be given as:

$$W_{i,t} = W_{i,t-1} + \alpha_{ik,t}(Z_k - W_{i,t-1});\ for\ k = 1, 2, ..., n;\ for\ i = 1, 2, ..., c \quad (1)$$

where $W_{i,t}$ represents the centroid[1] of the ith cluster at iteration t , Z_k is the k_{th} vector example from the dataset and α_{ik} is the only parameter of the algorithm and according to [6]:

$$\alpha_{ik,t} = (U_{ik,t})^{m(t)} \quad (2)$$

Where $m(t)$ is an exponent like the fuzzification index in FCM and $U_{ik,t}$ is the membership value of the compound Z_k to be part of cluster i. Both of these constants vary at each iteration t according to:

$$U_{ik} = \left(\sum_{j=1}^{c} \left(\frac{\|Z_k - W_i\|}{\|Z_k - W_j\|} \right)^{2/(m-1)} \right)^{-1} \quad ;\ 1 \le k \le n\ ;\ 1 \le i \le c \quad (3)$$

$$m(t) = m0 - m\Delta \cdot t \quad ; \quad m\Delta = (m0 - m_f)/iterate\ limit \quad (4)$$

---

[1] In the perspective of neural networks it represents a neuron or a prototype vector. So the number of neurons or prototype vectors will be equal to the number of clusters.

Where $m0$ is a constant value greater than the final value (m_f) of the fuzzification parameter m. The final value m_f should not be less than 1.1, in order to avoid the divide by zero error in equation (3). The iterative process will stop if $\left\|W_{i,(t)} - W_{(i,t-1)}\right\|^2 < \epsilon$, where ϵ is a termination criterion or after a given number of iterations. At the end of the process, a matrix U is obtained, where U_{ik} is the degree of membership of the Z_k element of the dataset to the cluster i. In addition, the centroid of each cluster will form the matrix W where W_i is the centroid of the ith cluster. The FKCN algorithm is given below:

1. Fix c, and $\epsilon > 0$ to some small positive constant.
2. Initialize $W_0 = (W_{1,0}, W_{2,0}, \cdots, W_{c,0}) \in \Re^c$.
   Choose $m_0 > 1$ and $t_{max} = max.\ number\ of\ iterations$.
3. For $t = 1, 2, \cdots, t_{max}$
   a. Compute all cn learning rates $\alpha_{ik,t}$ with equations (2) and (3).
   b. Update all c weight vectors $W_{i,t}$ with
   $W_{i,t} = W_{i,t-1} + [\sum_{k=1}^{n} \alpha_{ik,t}(Z_k - W_{i,t-1})] / \sum_{j=1}^{n} \alpha_{ij,t}$
   c. Compute $E_t = \left\|W_{i,(t)} - W_{(i,t-1)}\right\|^2 = \sum_{i=1}^{c} \left\|W_{i,(t)} - W_{(i,t-1)}\right\|^2$
   d. If $E_t < \epsilon$ stop.

2.3 Fuzzy Growing Hierarchical Self-Organizing Networks

Fuzzy Growing Hierarchical Self-Organizing Networks (FGHSON) are based on a hierarchical fuzzy structure of multiple layers, where each layer consists of several independent growing FKCNs. This structure can grow by means of an unsupervised self-organizing process in two manners (inspired by [12]): individually, in order to find the more suitable number of FKCN units that may represent, in an accurate manner, the input dataset. On the other hand on groups of FKCNs in a hierarchical mode, permitting to the hierarchy to explain a particular set of characteristics of data. Both growing processes are modulated by three parameters that regulate the so-called breadth (growth of the layers), depth (hierarchical growth) and membership degree of data to the prototype vectors. The FGHSON works as follows:

1) Initial Setup and Global Network Control:
The main motivation of the FGHSON algorithm is to properly represent a given dataset. The quality of this representation is measured in terms of the difference among a prototype vector and the example vectors represented by this. The *quantization error* qe is used to reach this aim. The qe measures the dissimilarity of all input data mapped onto a particular prototype vector, hence it can be used to guide a growth process with the purpose of achieving an accurate representation of the dataset reducing the qe. The qe of a prototype vector W_i is calculated according to (5) as the mean Euclidean distance between its prototype and the input vectors Z_c that are part of the set of vectors C_i mapped onto this prototype.

$$qe_i = \sum_{Z_c \in Ci} \|W_i - Z_c\|; \; C_i \neq \phi \tag{5}$$

The first step of the algorithm is focused on obtaining a global measure that allows to know the nature of the whole dataset. For this purpose the training process begins with the computation of a global measure error qe_0. qe_0 represents the qe of the single prototype vector W_0 that form the layer 0, see Fig. 1(a), calculated as shown in (6). Where, Z_k represents the input vectors from the whole data set Z and W_0 is defined as a prototype vector $W_0 = [\mu_{0_1}, \mu_{0_2}, \ldots, \mu_{0_n}]$, where μ_{0_i} for $i = 1, 2, \ldots, n$; is computed as the average of μ_{0_i} in the complete input dataset, in other words W_0 is a vector that corresponds to the mean of the input variables.

$$qe_0 = \sum_{Z_k \in Z} \|W_0 - Z_k\| \tag{6}$$

The value of qe_0 will help to measure the minimum quality of data representation of the prototype vectors in the subsequent layers, therefore the next prototypes have the task of reducing the global representation error qe_0.

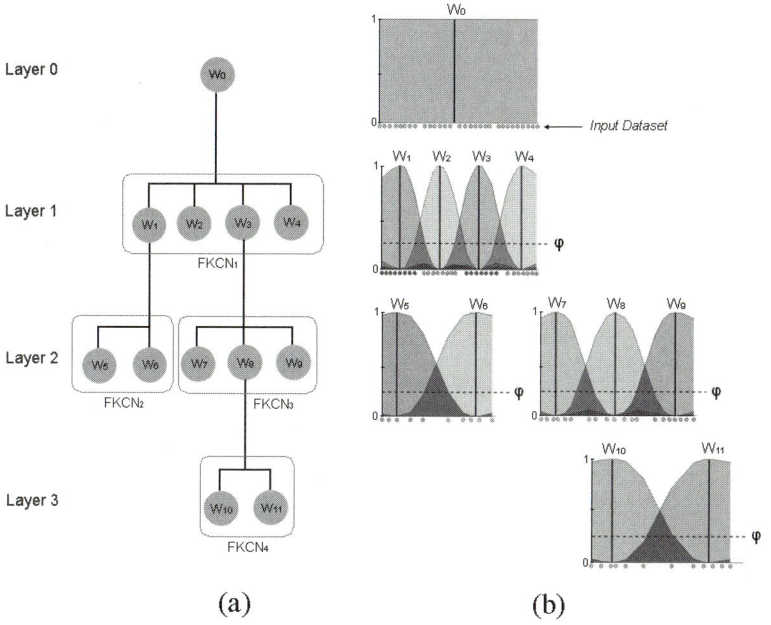

Fig. 1. (a) Hierarchical structure showing the prototype vectors and FKNC in the layers (b) Membership degrees in each layer

2) Breadth growth process:

The construction of the first layer starts after the calculation of qe_0. This first layer consists of a FKCN ($FKCN_1$) with two initial prototype vectors. Hence, the growth process of $FKCN_1$ begins by adding a new prototype vector and training it until a suitable representation of the dataset is reached. Each of these prototype vectors is an n-dimensional vector W_i (with the same dimensionality as the input patterns), which is initialized with random values. $FKCN_1$ is trained as is shown in section 2.2, taking as input (in the exceptional case of the first layer) the whole dataset. More precisely, $FKCN_1$ is allowed to grow until the qe present on the prototype of its preceding layer (qe_0 in the case of layer 1) is reduced to at least a fixed percentage τ_1. Continuing with the creation of the first layer, the number of prototypes in the $FKCN_1$ will be adapted. To achieve this, the *mean quantization error of the map* (MQE) is computed according to expression (7), where d refers to the number of prototype vectors contained in the FKCN, and qe_i represents the quantization error of the prototype W_i.

$$MQE_m = \frac{1}{d} \cdot \sum_i qe_i \tag{7}$$

The MQE is evaluated using (8) in an attempt to measure the quality of data representation, and is used also as stopping criterion for the growing process of the FKCN. In (8) qe_u represents the qe of the corresponding prototype u in the upper layer. In the specific case of the first-layer, the stopping criterion is shown in (9).

$$MQE < \tau_1 \cdot qe_u \tag{8}$$

$$MQE_{layer1} < \tau_1 \cdot qe_0 \tag{9}$$

If the stopping criterion (8) is not fulfilled, it is necessary to aggregate more prototypes for a more accurate representation. For this aim, the prototype with the highest qe is selected and is denoted as the error prototype e. A new prototype is inserted in the place where e was computed. After the insertion, all the FKCN parameters are reset to the initial values (except for the values of the prototype vectors) and the training begins according to the standard training process of FKCN. Note that the same value of the parameter τ_1 is used in each layer of the FGHSON. Thus, at the end of the process, a layer 1 is obtained with a $FKCN_1$ formed by a set of prototype vectors W, see Fig. 1(a). In addition, a membership matrix U is obtained. This matrix contains the membership degree of the dataset elements to the prototype vectors, as explained in section 2.2.

3) Depth growth process:
As soon as the breadth process of the first layer is finished, its prototypes are examined for further growth (depth growth or hierarchical growth). In particular, those prototypes with a large quantization error will indicate us which clusters need a better representation by means of new FKCNs. The new FKCNs thus form a second layer, namely W_1 and W_3 in Fig. 1(a). The selection of these prototypes is regulated by qe_0 (calculated previously in step 1) and a parameter τ_2 which is used to describe the desired level of granularity in the data representation. More precisely, each prototype W_i in the first layer that does not fulfill the criterion given in expression (10) will be subject to hierarchical expansion.

$$qe_i < \tau_2 \cdot qe_0 \tag{10}$$

After the expansion process and creation of the new FKCNs, the breadth process described in stage 2 begins with the newly established FKCNs, for instance, $FKCN_2$ and $FKCN_3$ in Fig. 1(a). The methodology for adding new prototypes, as well as the termination criterion of the breadth process is essentially the same as used in the first layer. Nevertheless, the difference among the training processes of the FKCNs in second the layer and the subsequent layers in comparison with the first, is that only a fraction of the whole input data is selected for training. This portion of data will be selected according to a minimal membership degree (φ). This parameter φ (the well known $\alpha - cut$) represents the minimal degree membership of an observation to be part of the dataset represented by a prototype vector. Hence, φ is used as selection parameter, so all the observations represented by W_i have to fulfill expression (11), where U_{ik} is the degree of membership of the Z_k element of the dataset to the cluster i. As an example, Fig. 1(b) shows the membership functions of the FKCNs in each layer, and how φ is used as a selection criteria to divide the dataset.

$$\varphi < U_{ik} \tag{11}$$

At the end of the creation of layer two, the same procedure described in step 2 is applied in order to build the layer 3 and so forth. The training process of the FGHSON is terminated when no more prototypes require further expansion. Note that this training process does not necessarily lead to a balanced hierarchy, i.e. a hierarchy with equal depth in each branch. Rather, the specific distribution of the input data is modeled by a hierarchical structure, where some clusters require deeper branching that others.

3 Experimental Testing

3.1 Iris Data Set

In this experiment the Iris dataset[2] is used in order to show the adaptation of the FGHSON to those areas where an absolute membership to a single prototype is not obvious. Therefore, FGHSON must (in an unsupervised manner) look at the representation of the dataset on the areas where observations of the same category share similar zones. For instance in the middle of the data cloud formed by the Virginica and Versicolor observations, see Fig. 2(a). The parameters of the algorithm were set to $\tau_1 = 0.2$, $\tau_2 = 0.03$, and $\varphi = 0.2$. After training, a structure of four layers was obtained. The zero layer is used to measure the whole deviation of dataset (see section 2.3). The first layer consist of a FKCN with three prototype vectors as is shown in Fig. 2(b), this distribution of prototypes attempt to represent three Iris categories.

The second layer as shows Fig. 2(c) reach a more fine-grain description of the dataset, placing prototypes in almost all the data distribution, adding prototypes in the zones where more representation was needed. Finally in Fig. 2(d), it is possible to observe an over population of prototypes in the middle of the cloud of Virginica and Versicolor observations. This occurs because this part of the dataset present observations with ambiguous membership in the previous

[2] There are three categories in the data set : Iris setosa, Iris versicolor and Iris virginical. Each having 50 observations with four features: sepal length (SL), sepal width (SW), petal length (PL), and petal width (PW).

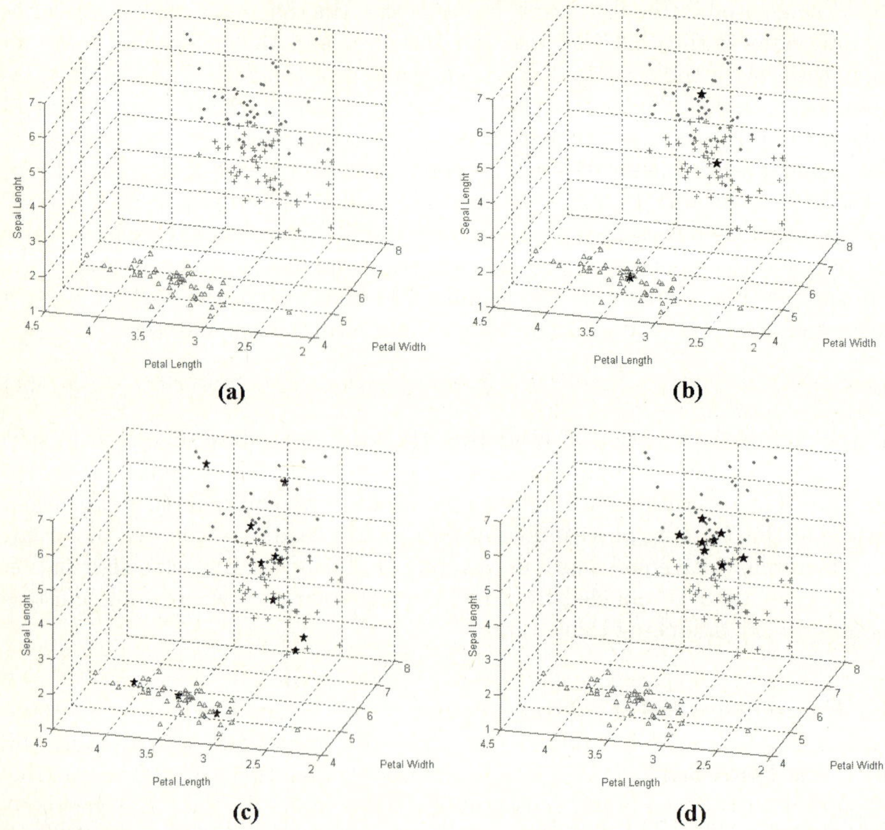

Fig. 2. Distribution of the prototype vectors, represented by stars, in each layer of the hierarchy. (a) Iris data sets. There are three Iris categories: Setosa, Versicolor, and Virginica represented respectively by triangles, plus symbols, and dots. Each having 50 samples with 4 features. Here, only three features are used: PL, PW, and SL. (b) First layer (c) Second layer and (d) Third layer of the FGHSON, in this layer prototypes are presented only in the zone where observations of Virginica and Vesicolor share the same area, so the new prototypes represent each category in a more accurate manner.

layer, then in this new layer, several prototypes are placed in this zone for a proper representation. Hence, permitting to those observations to obtain a higher membership to its new prototypes. Finally, to obtain at the end of the process a more accurate representation of this zone.

3.2 Toy Set

A toy set, as the one presented by Martinez et al [9] is used in order to show the capabilities of the FGHSON to represent a dataset that has multiple dimensionalities. In addition, it is possible to illustrate how the model stops the

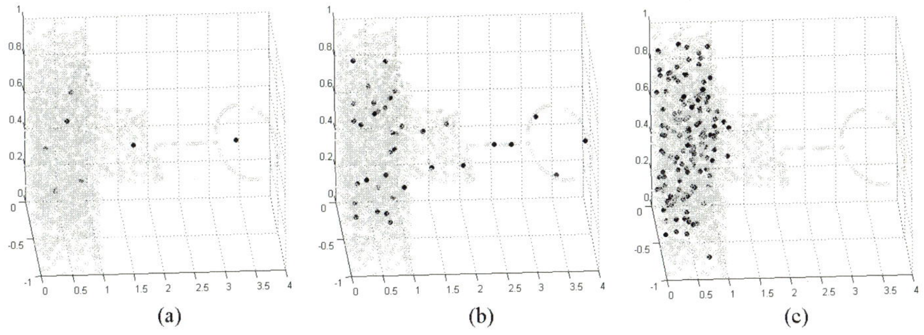

Fig. 3. Distribution of the prototype vectors (represented by black points) (a) First layer (b) Second layer (c) Third layer

growing process in those parts where the desired representation is reached and keep on growing where a low membership or poor representation is present. The parameters of the algorithm were set to $\tau_1 = 0.3$, $\tau_2 = 0.065$, and $\varphi = 0.2$ for φ. Four layers were created after training the network. In Fig. 3(a) the fist layer is shown, in this case seven prototypes were necessary to represent the dataset at this level, one for the 1D oval, one for the 2D plane and five for the 3D parallelepiped (note that there are no prototypes clearly associated to the line).

In the second layer (see Fig. 3(b)), a more accurate distribution of prototypes is reached, so it is possible to observe prototypes adopting the form of the dataset. Additionally, in regions where the quantization error was large, the new prototypes allow a better representation (e.g., along the line). In layer three (see Fig. 3(c)), no more prototypes are needed to represent the circle, the line and the plane; but a third hierarchical expansion was necessary to represent the parallelepiped. In addition, due to the data density in the parallelepiped, many points share memberships to different prototypes, so a several prototypes at this level were created.

4 Conclusion

The Fuzzy Growing Hierarchical Self-organizing networks are fully adaptive networks able to hierarchically represent complex datasets. Moreover, it allows for a fuzzy clustering of the data, allocating more prototype vectors or branches to heterogeneous areas and where there is presented similar membership degree to several clusters, this can help to better describing the dataset structure and the inner data relationships. Future work will be focused on a more accurate way to find the parameters used to tune the algorithm, more specifically φ. In some cases this value can change in order to find better fuzzy sets to represent the structure of the dataset.

Acknowledgments

This work is part of a cooperation project between BIOTEC, CIAT, CENICAA (Colombia) and HEIG-VD (Switzerland) named Precision Agriculture and the Construction of Field-crop Models for Tropical Fruits. The economical support is given by several institutions in Colombia (MADR, COLCIENCIAS, ACCI) and the State Secretariat for Education and Research (SER) in Switzerland.

References

1. Bezdek, J., Tsao, K., Pal, R.: Fuzzy Kohonen clustering networks. Fuzzy Systems, IEEE Int. Conf. on, 1035–1043 (1992)
2. Burzevski, V., Mohan, C.: Hierarchical Growing Cell Structures. Tech Report: Syracuse University (1996)
3. Doherty, J., Adams, G., Davey, N.: TreeGNG Hierarchical topological clustering. In: Proc. Euro. Symp. Artificial Neural Networks, pp. 19–24 (2005)
4. Fritzke, B.: Growing cell structures: a self-organizing network for unsupervised and supervised learning. Neural Networks 7(9), 1441–1460 (1994)
5. Hodge, V., Austin, J.: Hierarchical growing cell structures: TreeGCS. IEEE Transactions on Knowledge and Data Engineering 13(2), 207–218 (2001)
6. Huntsberger, T., Ajjimarangsee, P.: Parallel Self-organizing Feature Maps for Unsupervised Pattern Recognition. Int. Jo. General Sys. 16, 357–372 (1989)
7. Kohonen, T.: Self-organized formation of topologically correct feature maps. Biological Cybernetics 43(1), 59–69 (1982)
8. Lampinen, J., Oja, E.: Clustering properties of hierarchical self-organizing maps. J. Math. Imag. Vis. 2(2–3), 261–272 (1992)
9. Martinez, T., Schulten, J.: Topology representing networks. Neural Networks 7(3), 507–522 (1994)
10. Merkl, D., He, H., Dittenbach, M., Rauber, A.: Adaptive hierarchical incremental grid growing: An architecture for high-dimensional data visualization. In: Proc. Workshop on SOM, Advances in SOM, pp. 293–298 (2003)
11. Miikkulainen, R.: Script recognition with hierarchical feature maps. Connection Science 2, 83–101 (1990)
12. Rauber, A., Merkl, D., Dittenbach: The growing hierarchical self-organizing map: Exploratory analysis of high-dimensional data. IEEE Transactions on Neural Networks 13(6), 1331–1341 (2002)

MBabCoNN – A Multiclass Version of a Constructive Neural Network Algorithm Based on Linear Separability and Convex Hull

João Roberto Bertini Jr.[1] and Maria do Carmo Nicoletti[2]

[1] Institute of Mathematics and Computer Science, University of São Paulo, Av. Trabalhador São Carlense 400, São Carlos, Brazil
bertini@icmc.usp.br

[2] Department of Computer Science, Federal University of São Carlos, Via Washington Luiz, km. 238, São Carlos, Brazil
carmo@dc.ufscar.br

Abstract. The essential characteristic of constructive neural network (CoNN) algorithms is the incremental construction of the neural network architecture along with the training process. The BabCoNN (Barycentric-based constructive neural network) algorithm is a new neural network constructive algorithm suitable for two-class problems that relies on the BCP (Barycentric Correction Procedure) for training its individual TLU (Threshold Logic Unit). Motivated by the good results obtained with the two-class BabCoNN, this paper proposes its extension to multiclass domains as a new CoNN algorithm named MBabCoNN. Besides describing the main concepts involved in the MBabCoNN proposal, the paper also presents a comparative analysis of its performance versus the multiclass versions of five well known constructive algorithms, in four knowledge domains as an empirical evidence of the MBabCoNN suitability and efficiency for multiclass classification tasks.

Keywords: Constructive neural network algorithm, LS-discriminant learning, Barycentric Correction Procedure, Multiclass classification.

1 Introduction

There are many algorithms that allow the automatic learning of concepts, as can be seen in [1], [2]. A particular class of very relevant learning algorithms is based on the concept of linear separability (LS). The concept of linear separability permeates many areas of knowledge and based on the definition given in [3] it can be stated as: Let E be a finite set of N distinct patterns $\{E_1, E_2, ..., E_N\}$, each E_i $(1 \leq i \leq N)$ described as a set of K pairs *attribute - attribute value*. Let the patterns of E be classified in such a way that each pattern in E belongs to only one of C_j classes $(1 \leq j \leq M)$. This classification divides the set of patterns E into the subsets EC_1, EC_2, ..., EC_M. If a linear machine can classify the patterns in E into the proper class, the classification of E is a linear classification and the subsets EC_1, EC_2, ..., EC_M are linearly separable. Stated

V. Kůrková et al. (Eds.): ICANN 2008, Part II, LNCS 5164, pp. 723–733, 2008.

another way, a classification of E is linear and the subsets EC_1, EC_2, ..., EC_M, are linearly separable if and only if linear discriminant functions g_1, g_2, ..., g_n exist such that,

$$g_i(E) > g_j(E) \text{ for all } E \in EC_i \qquad (1)$$
$$j = 1, \ldots, M, j \neq i \text{ for all } i = 1, \ldots, M.$$

Since the decision regions of a linear machine are convex, if the subsets EC_1, EC_2, ..., EC_M are linearly separable, then each pair of subsets EC_i, EC_j, i, j = 1, ..., M, $i \neq j$, is also linearly separable. As proposed in [4], linearly separable based methods can be divided into four groups, depending on their main focus be based on linear programming, computational geometry, neural networks (NN) and quadratic programming.

This paper describes a linearly separable based NN method that is a multiclass version of the two-class CoNN named BabCoNN (Baricenter-based Constructive Neural Network) proposed in [5] and is organized as follows. Section 2 describes the basics of the two-class BabCoNN algorithm and briefly presents the main concepts and strategies used by the algorithm. Section 3 highlights the main characteristics of the four well-known CoNN multiclass algorithms used in the empirical experiments, described in Section 4, and also details the new algorithm i.e., the multiclass MBabCoNN. Section 5 highlights some conclusions and tasks to be accomplished for continuing the research work.

2 Constructive NN, the BCP and the BabCoNN Algorithms

Constructive neural network (CoNN) algorithms enable the network architecture to be constructed simultaneously with the learning process; both sub-processes i.e., learning and constructing the network, depend on each others performance. There are many CoNN algorithms suitable for two-class learning tasks such as the Tower and Pyramid [6], Tiling [7], Upstart [8], Pti and Shift [9], etc. A description of a few well-known CoNN algorithms can be found in [6] and [10].

2.1 Training Individual TLUs

Usually the basic step performed by a CoNN algorithm is the addition to the network architecture of a new TLU and its subsequent training. For training a TLU a constructive algorithm generally employs the Perceptron or any of its variants, such as Pocket or Pocket with Ratchet Modification (PRM) [6]. Considering that CoNN algorithms are very dependable on an efficient TLU training algorithm, there is still a need for finding new and better methods, although some of the Perceptron variants (especially the PRM) have been broadly used with good results. The Barycentric Correction Procedure (BCP) algorithm [11] [12], although not widely employed, has shown very good performance when used for training individual TLUs [12] and has established itself as a good competitor compared to the PRM when used by CoNN algorithms [13].

The BCP is an efficient TLU training algorithm that is based on the geometric concept of the barycenter of a convex hull and the algorithm iteratively calculates the barycenter of the regions defined by the positive and the negative training patterns. Unlike Perceptron based algorithms, the BCP calculates the weight vector and the bias separately. The algorithm defines the weight vector as the vector that connects two points: the barycenter of the convex hull of positive patterns and the barycenter of the convex hull defined by negative patterns. Further details of the BCP algorithm can be found in [11], [12] and [14].

2.2 The Two-Class BabCoNN Algorithm

The BabCoNN [5] is a new two-class CoNN algorithm based on some of the concepts used by the geometric-based algorithm (BCP) algorithm. As typical of constructive neural network algorithms, BabCoNN constructs the neural network at the same time as the learning process takes place. The network construction starts by training the output neuron, using the BCP. Next, the algorithm identifies all the misclassified training patterns; if there are none, the algorithm stops and outputs a BCP network, otherwise it starts adding neurons (one at a time) to the unique hidden layer of the network, aiming at not having misclassified patterns. A hidden neuron added to the hidden layer will be trained with the training patterns that were misclassified by the last added neuron. The process goes up to the point that no training patterns remain or all the remaining patterns belong to the same class. The process of building the network architecture is described in Fig. 1, where $E = \{E_1, E_2, \ldots, E_n\}$ represents the training set.

```
function void babconnNetworkConstruction(E)
begin
   output ← bcp(E);
   h ← 0;
   nE ← removeClassifiedPatterns(E);
   while bothClasses(nE) do
   begin
      h ← h + 1;
      hiddenLayer[h] ← bcp(nE);
      nE ← removeClassifiedPatterns(nE);
   end while
end function
```

Fig. 1. The BabCoNN algorithm for constructing a Neural Network

As explained in [5], the output and the hiddenLayer variables from Fig. 1 define the neural network. The output variable represents a single neuron, and the hiddenLayer is a vector representing all the hidden neurons added to the single hidden layer of the network. The function *bcp*() stands for the BCP algorithm, used for training individual neurons. The function *removeClassifiedPatterns*() removes the patterns that were correctly classified by the last neuron added to the hidden layer and *bothClasses*() is a Boolean function that returns 'true' if the current training set still has patterns belonging to both classes and 'false' otherwise.

Due to the way the learning phase is conducted by BabCoNN, each hidden neuron of the network is trained using patterns belonging to a region of the training space (i.e., that defined by the patterns that were misclassified by the previous hidden neuron added to the network). Given an input pattern to be classified, each hidden neuron has three possible outputs: 1, when the input pattern is classified as positive; −1, when the pattern is classified as negative and 0, when the pattern is classified as undetermined.

3 Multiclass Classification Using CoNN Algorithms

Multiclass classification tasks are common in pattern recognition. Frequently a classification task with $M(> 2)$ classes is treated as M two-class tasks. Although this approach may be suitable for some applications, there is still a need for more effective ways of dealing with multiclass problems. CoNNs have proved to be a good alternative for two-class tasks and have the potential to become good alternatives for multiclass domains as well. Multiclass CoNNs start by training as many output neurons as there are classes in the training set. For training neurons there are mainly two strategies, the individual (I) and the winner-takes-all (WTA) [6]. The main goal of this section is to provide a brief overview of the five well-known multiclass CoNN used in the experiments described in Section 4 and to describe the new multiclass algorithm MBabCoNN.

3.1 MTower, MPyramid, MUpstart, MTiling and MPerceptron-Cascade

The MTower algorithm [15] can be considered a direct extension of the two-class Tower algorithm [6]. The Tower creates a NN with only one TLU per hidden layer; each hidden neuron is connected to all the input neurons and to the hidden neuron previously created. Similarly to the two-class Tower, the MTower adds TLUs to the network; instead of adding one at a time, however, for an M-class problem the MTower adds M hidden neurons per hidden layer. Each one of the M neurons in a certain hidden layer has connections with all the neurons of the input layer as well as with all the M neurons of the previously added hidden layer. The multiclass MPyramid [15] is also a direct extension of its two-class algorithm, Pyramid [6]. The difference between the Tower and Pyramid algorithms (and consequently between their multiclass versions) relies on the connections. In a Pyramid network each newly added hidden neuron has connections with all the previously added hidden neurons as well as with the input neurons.

The two-class Upstart algorithm [8] constructs the neural network as a binary tree of TLUs starting with the output neuron and is governed by the addition of new hidden neurons specialized in correcting *wrongly-on* or *wrongly-off* errors made by the previously added neurons. The MUpstart [16] creates a single hidden layer. Every hidden neuron is directly connected to every neuron in the output layer. The input layer is connected to the hidden neurons as well as to the output neurons. The Tiling algorithm [7] constructs a neural network in successive layers where each hidden layer has a *master neuron* and a few *auxiliary neurons*. The output layer has only one master neuron. Similarly to the Tiling, the MTiling [17] constructs a multi layer neural network where the first hidden layer has connections to the input layer and each subsequent hidden layer has connections only to the previous hidden layer. Each layer has master and auxiliary neurons with the same functions they perform in a Tiling network i.e., the master neurons are responsible for classifying the training patternss and the auxiliary are responsible for turning the corresponding layer faithful. While the Tiling algorithm adds one master neuron per layer, the MTiling adds M master neurons (where M is the number of different classes in the training set).

The Perceptron-Cascade algorithm [18] constructs a neural network with an architecture that resembles the one constructed by the Cascade Correlation algorithm [19] and uses the same approach for correcting the errors adopted by the Upstart algorithm [8]. Differently from the Cascade Correlation, however, the Perceptron-Cascade uses the Perceptron (or any of its variant) for training individual TLUs. Similarly to the Upstart, the Perceptron-Cascade starts the construction of the network by training the output neuron. Hidden neurons are added to the network similarly to the process adopted by the Cascade Correlation. The MPerceptron-Cascade [15] is similar to the MUpstart. Their main difference is the neural network architecture induced. The MPerceptron-Cascade adds new hidden neurons in new layers while the MUpstart adds them in a single layer.

3.2 The Multiclass MBabCoNN Algorithm

The MBabCoNN constructs a network beginning with the output layer containing as many neurons as there are classes in the training set (each output neuron is associated to a class); the algorithm is flexible enough to allow the neurons to be trained using any TLU algorithm combined with either strategy, individual or WTA. After adding the M output neurons, the algorithm starts to add neurons to its single hidden layer in order to correct the classification mistakes made by the output neurons. Each hidden neuron can be considered a two-class BabCoNN-like neuron, i.e. it only outputs 1, -1 or 0 values. The constructive process continues by finding which output neuron (class) is responsible for the greatest number of misclassifications in relation to patterns belonging to all the other classes. A hidden neuron is then added to the hidden layer and is trained with a set containing patterns of two classes only: those belonging to the class the output neuron represents (which are labeled 1) and those belonging to the misclassified class (which are labeled -1). Each newly added hidden neuron is then connected only to the two output neurons whose classes it separates; the connection to the neuron responsible for the misclassifications has weight 1 and the other -1. The process of adding hidden neurons continues until the network converges (i.e., makes no mistakes). Figure 2(a) shows a simple example (six

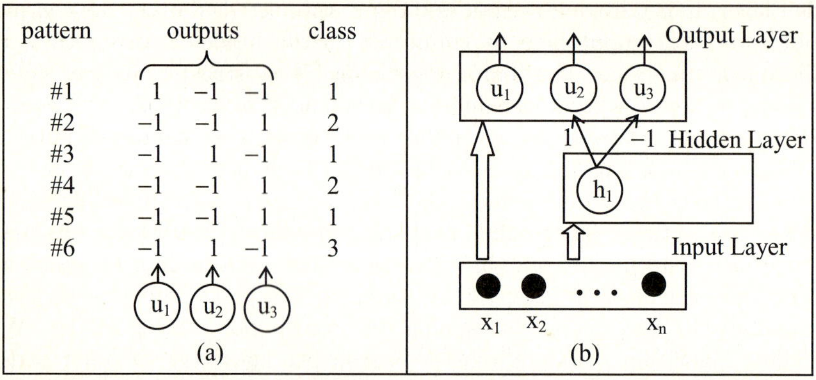

Fig. 2. The MBabCoNN architecture

training patterns identified by numbers 1 to 6 describing three classes identified by numbers 1 to 3) of the process for identifying the output neuron responsible for the greatest number of misclassifications.

In Fig. 2 the output neuron u_1 is associated with class 1, u_2 with class 2 and u_3 with class 3 and suppose that the three output neurons have been trained. As it can be seen on the left side of the figure, the pattern #1 was correctly classified and all the others were misclassified by the current network which up to this point only has output neurons. Neuron u_3 is responsible for the greatest number of misclassifications (it misclassified two patterns belonging to class 2 (i.e., #2 and #4) as class 3). All patterns belonging to classes 2 and 3 are then selected to compose the training set for the first hidden (h_1) neuron. Patterns from class 2 are relabeled as class 1 and those from class 3 as class −1; a connection is then established between h_1 and u_2, with weight 1 and between h_1 and u_3 with weight −1. This is justified by the fact that hidden neurons are two-class neurons and the weights allow the neuron to identify either two classes from the others. When h_1 fires 1 it favors class 2 and neglects class 3 and when h_1 fires −1, the opposite happens. As mentioned before, in situations of uncertainty BabCoNN neurons fire 0; this is convenient in a multiclass situation because it will not interfere with the patterns that do not belong to either of the two classes responsible for the hidden neuron creation. After a hidden neuron is added, the whole training set is input to the network constructed so far and training resumes. Depending on the misclassification results, new hidden neurons may be added to the network in a similar fashion as the one previously described.

For the classification process an output neuron that has any connections to hidden neurons is said to have dependencies. In Fig. 2(b) the neuron u_1 has no dependencies, while both u_2 and u_3 have dependencies. The classification process benefits the lack of dependency; if an output neuron fires 1 and has no dependencies then the class given to the pattern being classified is the class the neuron represents. Intuitively, a neuron that does not induce dependencies reflects the fact that it has not been associated with misclassifications during training. This can be an indication that the class represented by this particular neuron is reasonably ease to identify from the others (i.e., it is linearly separable from the others). If, however, the output neuron that classifies the pattern has dependencies, the classification result will be given by the sum of the outputs of all hidden neurons; if the sum is 0 the input pattern is classified as the class the output neuron represents.

4 Experimental Results and Discussion

This section presents and comparatively discusses the results of using MBabCoNN and the five other multiclass CoNN algorithms previously described, when learning from four multiclass knowledge domain data. Each of the five algorithms was implemented in Java using two different algorithms for training TLUs, PRM and BCP. All the multiclass CoNN algorithms use the respective WTA versions, PRMWTA and BCPWTA, for training the output neuron. The MBabCoNN algorithm, as have already explained, always uses the BCP for training its hidden neurons. For comparison purposes, results from running a multiclass version of the PRM and of BCP, each using the two training strategies (WTA and individual(I)) are presented. The four knowledge domains used in the experiments have been downloaded from the UCI-Repository [20]: Iris (150 patterns, 4 attributes and 3 classes), E.coli (336 patterns, 7 attributes and 8 classes), Balance (625 patterns, 4 attributes and 3 classes) and Glass (214 patterns, 9 attributes and 6 classes). Taking into consideration the MBabCoNN proposal and five algorithms used for comparison (MTower, MPyramid, MUpstart, MTiling and MPerceptron-Cascade) each implemented in two versions and the two different strategies employed for implementing the MPRM and the MBCP, a total of 16 different algorithms have been implemented and evaluated. In the experiments the accuracy of each neural network is based on the percentage of successful predictions on test sets for each domain. For each of the four datasets the experiments consisted of performing a ten-fold cross-validation process. The results are given by the average of the ten runs followed by their standard deviation.

Runs with the various learning procedures were carried out on the same training sets and evaluated on the same test sets. The cross-validation folds were the same for all the experiments on each domain. For each domain, each learning procedure was run considering one, ten, a hundred and a thousand iterations. All the results obtained with MBabCoNN and the seven other algorithms (and their variants) are presented in Table 1 to 4, organized by knowledge domain. Each algorithm version is identified after the TLU training algorithm it employ, PRM or BCP, by the suffixes **P** and **B** added to their names respectively. The following abbreviations were adopted for presenting the tables: #I: number of iterations, TR training set, TE testing set. The accuracy (Acc) is given in percentage followed by the standard deviation value. The 'Absolute Best' (AB) column gives the best performance of the learning procedure (in TE) over the ten runs and the 'Absolute Worst' (AW) column gives the worst performance of the learning

Table 1. Iris

Algorithm	# I	AccTR	AccTE	# HN	ABTE	AWTE	ABHN	AWHN
MBabCoNNP	10^3	98.7∼0.4	98.0∼3.2	3.1∼0.3	100.0	93.3	3.0	4.0
MBabCoNNB	10^2	94.2∼3.1	94.0∼4.9	4.9∼0.6	100.0	86.7	4.0	6.0
PRMWTA	10^2	98.7∼0.4	95.3∼8.9	-	100.0	73.3	-	-
PRMI	10^2	89.3∼2.6	86.7∼18.1	-	100.0	46.7	-	-
BCPWTA	10^2	87.9∼1.4	84.0∼13.4	-	100.0	53.3	-	-
BCPI	1	85.0∼7.3	72.7∼41.3	-	100.0	6.7	-	-
MTowerP	10^2	98.9∼0.4	96.7∼6.5	3.6∼1.3	100.0	80.0	3.0	6.0
MTowerB	10^2	87.8∼1.6	83.3∼14.5	3.3∼0.9	100.0	53.3	3.0	6.0
MPyramidP	10^2	98.8∼0.4	96.0∼8.4	3.3∼0.9	100.0	73.3	3.0	6.0
MPyramidB	10^2	88.3∼1.4	82.7∼13.8	3.6∼1.3	100.0	53.3	3.0	6.0
MUpstartP	10^2	98.9∼0.4	93.3∼12.9	3.3∼0.0	100.0	60.0	3.0	3.0
MUpstartB	10^2	88.6∼1.6	80.7∼13.9	3.6∼0.5	100.0	53.3	3.0	4.0
MTilingP	10	98.2∼0.8	95.3∼8.9	3.0∼0.0	100.0	73.3	3.0	3.0
MTilingB	10^2	89.6∼4.7	84.0∼14.5	7.0∼8.8	100.0	53.3	3.0	28.0
MPCascadeP	10^2	98.8∼0.4	95.3∼8.9	3.0∼0.0	100.0	73.3	3.0	3.0
MPCasdadeB	10^2	88.5∼1.4	80.7∼13.5	3.3∼0.5	100.0	53.3	3.0	4.0

Table 2. E. Coli

Algorithm	# I	AccTR	AccTE	# HN	ABTE	AWTE	ABHN	AWHN
MBabCoNNP	10^2	90.6∼0.6	83.9∼5.3	8.2∼0.6	91.2	75.8	8.0	9.0
MBabCoNNB	10^2	85.1∼1.5	81.0∼5.2	9.1∼0.7	88.2	73.5	8.0	10.0
PRMWTA	10^2	90.8∼1.3	77.8∼18.9	-	100.0	52.9	-	-
PRMI	10^2	87.2∼2.6	73.8∼24.0	-	100.0	42.4	-	-
BCPWTA	10^2	76.5∼3.0	69.1∼15.0	-	97.1	42.4	-	-
BCPI	10	85.1∼3.3	72.0∼28.2	-	100.0	27.3	-	-
MTowerP	10	90.2∼1.5	78.4∼22.2	27.8∼14.1	100.0	30.3	6.0	56.0
MTowerB	10^2	76.6∼2.9	69.2∼14.4	9.2∼3.2	97.1	48.5	8.0	16.0
MPyramidP	10	90.1∼1.3	79.6∼18.3	29.1∼10.3	100.0	42.4	14.0	48.0
MPyramidB	10^2	76.6∼2.8	69.5∼16.2	8.4∼2.1	97.1	36.4	6.0	14.0
MUpstartP	10^2	90.7∼1.6	76.9∼19.7	8.1∼1.3	100.0	50.0	6.0	10.0
MUpstartB	10^2	80.5∼2.8	75.2∼10.4	8.9∼1.3	97.1	60.7	6.0	11.0
MTilingP	10	87.9∼2.2	76.3∼20.7	7.7∼0.7	100.0	38.2	6.0	8.0
MTilingB	10^2	76.3∼3.3	67.3∼18.0	25.3∼55.4	97.1	33.3	6.0	183.0
MPCascadeP	10	88.8∼1.2	82.9∼16.2	8.6∼1.3	100.0	57.8	8.0	11.0
MPCasdadeB	10^2	79.6∼2.7	70.1∼15.3	9.0∼1.7	97.0	42.4	6.0	12.0

procedure (in TE) over the ten runs; #HN represents the number of hidden nodes; ABHN gives the smallest number of hidden nodes created and AWHN gives the highest number of hidden nodes created. Obviously the PRMWTA, PRMI, BCPWTA and BCPI do not have values for #HN, ABHN and AWHN because the networks they create do not have hidden layer.

Considering the results obtained in the experiments shown in Table 1 to 4, it can be said that as far as accuracy in test sets is concerned, MBabCoNNP outperformed all the others in three out of four domains, namely the Iris, E. Coli and Glass. In the domain Balance, although its result is very close to the best result (due to MTilingP), it is worth noticing that MTilingP created 28.1 hidden neurons, on average while MBabCoNN created only 3.5. The number of hidden neurons constructed particularly by the MBabCoNNP version (in comparison to the other multiclass CoNN algorithms) is much smaller - this is more evident

Table 3. Glass

Algorithm	# I	AccTR	AccTE	# HN	ABTE	AWTE	ABHN	AWHN
MBabCoNNP	10^2	100.0∼3.2	99.1∼2.0	6.2∼0.4	100.0	95.2	6.0	7.0
MBabCoNNB	10	99.3∼0.6	98.6∼3.2	6.3∼0.7	100.0	90.5	6.0	8.0
PRMWTA	10^2	99.9∼0.2	90.6∼15.2	-	100.0	57.1	-	-
PRMI	10^2	77.4∼3.7	49.9∼40.9	-	100.0	0.0	-	-
BCPWTA	10	98.5∼0.5	87.7∼22.1	-	100.0	33.3	-	-
BCPI	1	72.1∼3.7	68.3∼47.1	-	100.0	0.0	-	-
MTowerP	10^2	100.0∼0.0	92.9∼16.9	6.0∼0.0	100.0	47.6	6.0	6.0
MTowerB	10^2	99.6∼0.6	85.8∼25.8	9.4∼2.9	100.0	23.8	6.0	12.0
MPyramidP	10^2	100.0∼0.0	95.4∼10.6	7.7∼2.9	100.0	68.2	5.0	12.0
MPyramidB	10	98.8∼0.7	84.8∼24.7	7.7∼2.9	100.0	23.8	5.0	12.0
MUpstartP	10^2	97.3∼1.8	88.3∼18.6	6.6∼0.8	100.0	54.5	5.0	8.0
MUpstartB	10^2	99.7∼0.4	86.3∼24.2	6.5∼0.9	100.0	23.8	5.0	8.0
MTilingP	10^2	100.0∼0.0	94.4∼9.2	7.5∼3.4	100.0	77.3	5.0	14.0
MTilingB	10^2	100.0∼0.0	82.5∼26.2	12.7∼5.1	100.0	23.8	5.0	19.0
MPCascadeP	10^2	96.6∼1.4	79.2∼32.1	6.7∼0.8	100.0	9.5	5.0	8.0
MPCasdadeB	10	98.7∼0.7	82.9∼26.2	6.0∼0.5	100.0	28.6	5.0	7.0

Table 4. Balance

Algorithm	# I	AccTR	AccTE	# HN	ABTE	AWTE	ABHN	AWHN
MBabCoNNP	10	92.1∼0.8	91.4∼2.3	3.5∼0.7	93.7	87.1	3.0	5.0
MBabCoNNB	10	92.1∼0.9	89.3∼2.5	5.3∼0.5	93.5	85.5	5.0	6.0
PRMWTA	10^2	92.2∼0.5	90.1∼3.7	-	96.8	84.1	-	-
PRMI	10	89.1∼1.6	89.3∼4.8	-	98.4	84.1	-	-
BCPWTA	10^2	80.1∼4.4	77.9∼9.9	-	91.9	65.1	-	-
BCPI	10	89.5∼1.7	88.0∼3.5	-	93.5	82.3	-	-
MTowerP	10	94.8∼1.1	90.6∼6.2	20.4∼5.8	98.4	79.4	12.0	30.0
MTowerB	10^2	83.6∼4.8	80.8∼8.0	9.0∼3.5	91.9	65.1	6.0	15.0
MPyramidP	10	95.1∼0.9	90.1∼6.3	24.0∼6.2	96.8	76.2	15.0	33.0
MPyramidB	10^2	83.3∼4.1	83.1∼5.8	6.2∼2.2	93.5	76.2	3.0	9.0
MUpstartP	10	91.8∼0.4	91.2∼4.7	3.4∼0.9	98.4	85.5	3.0	6.0
MUpstartB	10^2	82.1∼4.9	81.1∼6.9	4.0∼0.8	91.9	69.8	3.0	5.0
MTilingP	10^2	95.5∼2.9	92.3∼3.3	28.1∼21.8	96.8	88.9	3.0	49.0
MTilingB	10^2	79.8∼4.5	76.4∼8.4	3.0∼0.0	91.9	66.7	3.0	3.0
MPCascadeP	10^2	92.1∼0.6	90.0∼4.5	3.2∼0.4	98.4	84.1	3.0	4.0
MPCasdadeB	10^2	81.6∼4.7	78.6∼9.9	4.0∼0.8	91.9	66.7	3.0	5.0

in domains E. coli and Balance where some of the algorithms induced over 20 hidden neurons.

5 Conclusions

This paper proposes a multiclass version of a recently proposed constructive neural network algorithm, named BabCoNN, based on the geometric concept of convex hull, which uses the BCP algorithm for training the individual TLU added to the network during learning. The paper presents the accuracy results of learning experiments conducted in four multiclass knowledge domains, using the proposed algorithm MBabCoNN implemented in two different versions: MBab-CoNNP and MBabCoNNB, *versus* five well-known multiclass algorithms (each implemented in two versions as well). Both versions of the MBabCoNN use the BCP for training the hidden neurons and differ from each other in relation to the algorithm used for training their output neurons (PMRWTA and BCPWTA respectively). As far as results in four knowledge domains are concerned, the new MBabCoNN proposal, particularly in its MBabCoNNP version has shown superior performance in relation to both, accuracy in test sets and size of the induced neural network. The work will continue by extending the comparative evaluation to a broader set of knowledge domains.

Acknowledgments. To CAPES and FAPESP for the project financial help granted to the first and to the second author, respectively.

References

1. Duda, R.O., Hart, P.E., Stork, D.G.: Pattern Classification. John Wiley, Chichester (2001)
2. Mitchell, T.M.: Machine Learning. McGraw-Hill, New York (1997)
3. Nilsson, N.J.: Learning Machines. McGrall-Hill Systems Science Series (1965)
4. Elizondo, D.: The linear separability problem: Some testing methods. IEEE Transactions on Neural Network 17(2), 330–344 (2006)
5. Bertini Jr., J.R., Nicoletti, M.C.: A constructive neural network algorithm based on the geometric concept of barycenter of convex hull. In: ICAISC (June 2008) (accepted for presentation)
6. Gallant, S.I.: Neural Network Learning & Expert Systems. MIT Press, Cambridge (1994)
7. Mézard, M., Nadal, J.: Learning feedforward networks: The tiling algorithm. J. Phys. A: Math Gen. 22, 2191–2203 (1989)
8. Frean, M.: The upstart algorithm: a method for constructing and training feedforward neural networks. Neural Computation 2, 198–209 (1990)
9. Amaldi, E., Guenin, B.: Two constructive methods for designing compact feedforward networks of threshold units. Int. Journal of Neural System 8(5), 629–645 (1997)
10. Nicoletti, M.C., Bertini Jr., J.R.: An empirical evaluation of constructive neural network algorithms in classification tasks. Int. Journal of Innovative Computing and Applications (IJICA) 1, 2–13 (2007)
11. Poulard, H.: Baricentric correction procedure: A fast method for learning threshold unit. In: WCNN, vol. 1, pp. 710–713 (1995)
12. Poulard, H., Labreche, S.: A new threshold unit learning algorithm. Technical Report 95504, LAAS (1995)
13. Bertini Jr., J.R., Nicoletti, M.C., Hruschka Jr., E.R.: A comparative evaluation of constructive neural networks methods using prm and bcp as tlu training algorithms. In: Proc. of the IEEE Int. Conf. Systems, Man, and Cybernetics, Taiwan, pp. 3497–3502 (2006)
14. Poulard, H., Estèves, D.: A convergence theorem for barycentric correction procedure. Technical Report 95180, LAAS-CNRS (1995)
15. Parekh, R.G., Yang, J., Honovar, V.: Constructive neural network learning algorithm for multi-category classification. Technical Report TR ISU-CS-TR95-15a, Iowa State University (1995)

16. Parekh, R.G., Yang, J., Honovar, V.: Mupstart - a constructive neural network learning algorithm for multi-category pattern classification. In: ICNN, vol. 3, pp. 1920–1924(1997)
17. Yang, J., Parekh, R., Honovar, V.: Mtiling - a constructive network learning algorithm for multi-category pattern classification. In: World Congress on Neural Networks, 182–187 (1996)
18. Burgess, N.: A constructive algorithm that converges for real-valued input patterns. International Journal of Neural Systems 5(1), 59–66 (1994)
19. Fahlman, S., Lebiere, C.T.: The Cascade Correlation Architecture. In: Advances in Neural Information Processing Systems 2, pp. 524–532. Morgan Kaufman, San Francisco (1990)
20. Asuncion, A., Newman, D.J.: Uci machine learning repository. University of California, School of Information and Computer Science, Irvine, CA (2007)

On the Generalization of the m-Class RDP Neural Network

David A. Elizondo[1], Juan M. Ortiz-de-Lazcano-Lobato[2], and Ralph Birkenhead[1]

[1] School of Computing
De Montfort University
The Gateway
Leicester, LE1 9BH
United Kingdom
{elizondo,rab}@dmu.ac.uk
[2] School of Computing
University of Málaga
Málaga, Spain
jmortiz@lcc.uma.es

Abstract. The Recursive Deterministic Perceptron (RDP) feed-forward multilayer neural network is a generalisation of the single layer perceptron topology. This model is capable of solving any two-class classification problem as opposed to the single layer perceptron which can only solve classification problems dealing with linearly separable sets. For all classification problems, the construction of an RDP is done automatically and convergence is always guaranteed. A generalization of the 2-class Recursive Deterministic Perceptron (RDP) exists. This generalization allows to always separate, in a deterministic way, m-classes. It is based on a new notion of linear separability and it falls naturally from the 2 valued RDP. The methods for building 2-class RDP neural networks have been extensively tested. However, no testing has been done before on the m-class RDP method. For the first time, a study on the performance of the m-class method is presented. This study will allow the highlighting of the main advantages and disadvantages of this method by comparing the results obtained while building m-class RDP neural networks with other more classical methods such as Backpropagation and Cascade Correlation. The networks were trained and tested using the following standard benchmark classification datasets: IRIS, SOYBEAN, and Wisconsin Breast Cancer.

1 Introduction

The RDP for 2-class classification problems was introduced in [1]. This topology is a generalization of the single layer perceptron topology (SLPT) developed by Rosenblatt [2]. This generalization is capable of transforming any non-linearly separable (NLS) 2-class classification problem into a linearly separable (LS) one, thus making it possible for the SLPT to find a solution to the problem. An extension of the RDP algorithm to m-class problems (with $m \geq 2$) was introduced in [3]. This extension is based on, a new notion of linear separability and, it falls naturally from the 2 valued RDP.

V. Kůrková et al. (Eds.): ICANN 2008, Part II, LNCS 5164, pp. 734–743, 2008.

1.1 Preliminaries

We use the following standard notions:

• $\mathcal{S}_m$ stands for the set of permutations of $\{1, ..., m\}$.
• If $\boldsymbol{u} = (u_1, ..., u_d), \boldsymbol{v} = (v_1, ..., v_d) \in I\!R^d$, then $\boldsymbol{u}^T\boldsymbol{v}$ stands for $u_1v_1 + ... + u_dv_d$; and $\boldsymbol{u}(j) = u_j$ *(i.e. $\boldsymbol{u}(j)$ is the jth component of $\boldsymbol{u}$).*
• $\Pi_{\{i_1,...,i_k\}}\ \boldsymbol{u} = (u_{i_1}, ..., u_{i_k})$ and by extension, if $S \subset I\!R^d$ then $\Pi_{\{i_1,...,i_k\}}\ S = \{\Pi_{\{i_1,...,i_k\}}\ \boldsymbol{x} \mid \boldsymbol{x} \in S\}$.
• Let $r \in I\!R$, $Adj(\boldsymbol{u}, r) = (u_1, ..., u_d, r)$ and by extension, if $S \subset I\!R^d$ $Adj(S, r) = \{Adj(\boldsymbol{x}, r) \mid \boldsymbol{x} \in S\}$.
• We define $Im(E, F) = \{(x_1, ..., x_d, x_{d+1}) \in F \mid (x_1, ..., x_d) \in E\}$ for $E \subset I\!R^d$ and $F \subset I\!R^{d+1}$.
• $\mathcal{P}(\boldsymbol{w}, t)$ stands for the hyperplane $\{\boldsymbol{x} \in I\!R^d \mid \boldsymbol{w}^T\boldsymbol{x} + t = 0\}$ of $I\!R^d$.

1.2 Some Definitions and Properties

In this section, we introduce the notions of convex hull (CH) and of linear separability.

Definition 1. *[4] Let S be a sub-set of $I\!R^d$*
$CH(S) = \{t_1\boldsymbol{x}_1 + \ldots + t_k\boldsymbol{x}_k \mid \boldsymbol{x}_1, \ldots, \boldsymbol{x}_k \in S,\ t_1, \ldots, t_k \in [0,1]\ \text{and}\ t_1 + \ldots + t_k = 1\}$. *Thus, if S is finite, then there exists $\boldsymbol{a}_1, ..., \boldsymbol{a}_k \in I\!R^d$ and $b_1, ..., b_k \in I\!R$ such that $CH(S) = \{\boldsymbol{x} \in I\!R^d \mid \boldsymbol{a}_i^T\boldsymbol{x} \geq b_i\ for\ 1 \leq i \leq k\}$.*

Definition 2. *Two subsets X and Y of $I\!R^d$ are said to be linearly separable if there exists a hyperplane $\mathcal{P}(\boldsymbol{w}, t)$ of $I\!R^d$, such that ($\forall \boldsymbol{x} \in X,\ \boldsymbol{w}^T\boldsymbol{x} + t > 0$ and $\forall \boldsymbol{y} \in Y,\ \boldsymbol{w}^T\boldsymbol{y} + t < 0$). In the following we will denote the fact that X and Y are LS by $X \mid\mid Y$ or $X \mid\mid Y\ (\mathcal{P}(\boldsymbol{w}, t))$ if we want to specify the hyperplane which linearly separates X and Y. A discussion on the different methods for testing linear separability can be found in [1].*

This paper is divided into four sections. The m-class generalisation of the RDP neural network, based on a notion of linear separability for m classes, is presented in section two. In this section also, some of the notions used throughout this paper are introduced. In section three, the procedure used to evaluated the generalisation of the m-class RDP model is presented. Three machine learning benchmarks (Iris, Soybean, and Wisconsin Breast Cancer) were used [5] and datasets were generated using cross validation. The method is compared with Backpropagation and Cascade Correlation in terms of their level of generalisation. A summary and some conclusions are presented in section four.

2 The m-Class RDP Algorithm

The m-class RDP algorithm is an adaptation of the 2-class RDP based on the following notion of linear separability for m classes ($m > 2$):

Definition 3. *Let $X_1, ..., X_m \subset I\!R^d$ and $a_0 < a_1 < ... < a_m$, $X_1, ..., X_m$ are said to be linearly separable relatively to the ascending sequence of real numbers $a_0, ..., a_m$ if $\exists \sigma \in \mathcal{S}_m, \exists \boldsymbol{w} \in I\!R^d, \exists t \in I\!R$ such that $\forall i, \forall \boldsymbol{x} \subset X_{\sigma(i)},\ a_{i-1} < \boldsymbol{w}^T\boldsymbol{x} + t < a_i$.*

Remarks
Let $X_1, ..., X_m \subset I\!R^d$ and $a_0 < a_1 < a_2 < ... < a_m$,

• $X_1, ..., X_m$ are linearly separable relatively to $a_0, ..., a_m$ iff $CH(X_1), ..., CH(X_m)$ are linearly separable relatively to $a_0, ..., a_m$.
• let $\sigma \in \mathcal{S}_m$. Put :
$X^\sigma = Adj(X_{\sigma(1)}, -a_0) \cup Adj(X_{\sigma(2)}, -a_1) ... \cup Adj(X_{\sigma(m)}, -a_{m-1})$, $Y^\sigma = Adj(X_{\sigma(1)}, -a_1) \cup Adj(X_{\sigma(2)}, -a_2) ... \cup Adj(X_{\sigma(m)}, -a_m)$, then, $X_1, ..., X_m$ are linearly separable relatively to $a_0, ..., a_m$ by using σ iff $X^\sigma \parallel Y^\sigma$. In other words, we reduce the problem of linear separability for m classes into the problem of linear separability for 2 classes. We do this by augmenting the dimension of the input vectors with the ascending sequence $a_0, ..., a_m$.
• If $X_1 \parallel X_2$ $(\mathcal{P}(\boldsymbol{w}, t))$ and $\alpha = Max(\{|\boldsymbol{w}^T\boldsymbol{x} + t| \; ; \; \boldsymbol{x} \in (X_1 \cup X_2)\}$, then X_1, X_2 are linearly separable relatively to $-\alpha, 0, \alpha$.

Proposition 1. *Let $X_1, ..., X_m \subset I\!R^d$, $a, b \in I\!R, h, k > 0$ and let $a_i = a + ih, b_i = b + ik$, for $0 \leq i \leq m$, then $X_1, ..., X_m$ are linearly separable relatively to $a_0, ..., a_m$ iff they are linearly separable relatively to $b_0, ..., b_m$. In other words, the linear separability between m classes is independent of the arithmetic sequence.*

Proof. Let $\sigma \in \mathcal{S}_m$ represent a class, and let $\boldsymbol{w} \in I\!R^d, t \in I\!R$ such that $\forall i$, $\forall \boldsymbol{x} \in X_{\sigma(i)}$, $a_{i-1} < \boldsymbol{w}^T\boldsymbol{x} + t < a_i$. Thus, $\forall i, \forall \boldsymbol{x} \in X_{\sigma(i)}$, $b_{i-1} < \frac{k}{h}\boldsymbol{w}^T\boldsymbol{x} + \frac{k}{h}(t - a) + b < b_i$ □

Definition 4. *$X_1, ..., X_m \subset I\!R^d$ are said to be linearly separable if there exists $a \in I\!R, h > 0$ such that $X_1, ..., X_m$ are linearly separable relatively to $a, a+h, ..., a+mh$.*

Definition 5. *A m-SLPT with the weight $\boldsymbol{w} \in I\!R^d$, the threshold $t \in I\!R$, the values $v_1, v_2, ..., v_m \in I\!R$ and the characteristic $(c, h) \in I\!R \times I\!R^+$ (c represents the value corresponding to the starting hyperplane, and h a chosen distance between a hyperplane which we will call the step size), has the same topology as the 2-class SLPT. The only difference is that the function corresponding to a m-SLPT is a $m-$valued function f defined by : $\forall \boldsymbol{y} \in I\!R^d$*

$$f(\boldsymbol{y}) = \begin{cases} v_1 & \text{if } \boldsymbol{w}^T\boldsymbol{y} + t < c + h \\ v_i & \text{if } c + (i-1)h < \boldsymbol{w}^T\boldsymbol{y} + t < c + ih, \text{for } 1 < i < m \\ v_m & \text{if } \boldsymbol{w}^T\boldsymbol{y} + t > c + (m-1)h \end{cases} \quad (1)$$

2.1 The Specialized NLS to LS Transformation Algorithm for m Classes

A specialized version of the transformation algorithm , from two to m classes, was proposed in [1]. This extension is based on the notion of linear separability for m classes described above.

Let $c \in I\!R, h > 0, b = -(m - \frac{3}{2})h$ and let $b_i = c + (m - i)b + (\frac{(m-1)(m-2)}{2} - \frac{(i-1)(i-2)}{2})h$ for $1 \leq i < m$ and $b_m = c$.

Table 1 shows the specialized NLS to LS transformation algorithm for m classes. We proceed as in the 2-class specialized transformation algorithm. That is to say, at each step we select a LS sub-set which belongs to a single class and add an artificial variable to the entire input data set. To this artificial variable we assign as value b_i for all the input vectors belonging to the selected LS sub-set and a value b_j to the rest of the set of input vectors, where $b_i \neq b_j$. Two cases for assigning the values to the artificial inputs are possible depending on the class to which the LS sub-set belongs:

1. If the selected LS sub-set belongs to the jth class, with $j < m$, we add to its input vector a new component with value b_j and we add to the rest of the input vector a new component with value b_{j+1}.
2. If the selected LS sub-set belongs to the last class *(mth class)*, we add to its input vector a new component with value b_m and we add to the rest of the input vector a new component with value b_{m-1}.

In the following theorem we prove the correctness and the termination of the algorithm presented in table 1 which allows to construct an $m-$RDP for linearly separating any given m classes.

Theorem 1. *If $X_1^i, ..., X_m^i$ are not linearly separable, then there exists Z_i such that ($Z_i \subset X_1'^i$ or ... or $Z_i \subset X_m'^i$), $Z_i \neq \emptyset$ and $Z_i \mid\mid (S_i \setminus Z_i)$.*

Proof. We will prove that, there exists $\boldsymbol{x} \in X_1'^i \cup ... \cup X_m'^i$ such that $\{\boldsymbol{x}\} \mid\mid (S_i \setminus \{\boldsymbol{x}\})$.
Assume that $\forall \boldsymbol{x} \in X_1'^i \cup ... \cup X_m'^i$, $\{\boldsymbol{x}\}$ and $(S_i \setminus \{\boldsymbol{x}\})$ are not linearly separable, then $X_1'^i \cup ... \cup X_m'^i \subset CH(S_i \setminus (X_1'^i \cup ... \cup X_m'^i))$.
So, if $S_i = \{\boldsymbol{v}_1, ..., \boldsymbol{v}_k, \boldsymbol{v}_{k+1}, ..., \boldsymbol{v}_{k+r_1}, \boldsymbol{v}_{k+r_1+1}, ..., \boldsymbol{v}_{k+r_1+r_2}, ..., \boldsymbol{v}_{k+r_1+...+r_m}\}$ where $X_1'^i \cup ... \cup X_m'^i = \{\boldsymbol{v}_1, ..., \boldsymbol{v}_k\}$ and
$X_j^i \setminus X_j'^i = \{\boldsymbol{v}_{k+r_1+...+r_{j-1}+1}, ..., \boldsymbol{v}_{k+r_1+...+r_j}\}$ for $1 \leq j \leq m$.
Let $\boldsymbol{x} \in X_1'^i \cup ... \cup X_m'^i$, then $\boldsymbol{x} = t_1\boldsymbol{v}_{k+1} + ... + t_{r_1+...+r_m}\boldsymbol{v}_{k+r_1+...+r_m}$
$t_1, ..., t_{r_1+...+r_m} \geq 0$ and $t_1 + ... + t_{r_1+...+r_m} = 1$
Let $j < m$, $1 \leq l \leq r_j$ and e_l such that $\boldsymbol{v}_{k+r_1+...+r_{j-1}+l}(e_l) = b_j$ and $\boldsymbol{v}_f(e_l) = b_{j+1}$ for $\boldsymbol{v}_f \notin X_j^i \setminus X_j'^i$ ($b_j < b_{j+1}$, $\boldsymbol{x}(e_l) = b_{j+1}$) . If $t_{r_1+...+r_{j-1}+l} > 0$ then $b_{j+1} = \boldsymbol{x}(e_j) < (t_1 + ... + t_{k+r_1+...+r_m})b_{j+1} = b_{j+1}$, which is absurd; thus, $\forall j \leq r_1 + ... + r_{m-1}$, $t_j = 0$,
Let $j < r_m$ and $1 \leq l \leq i$ such that $\boldsymbol{v}_{k+r_1+...+r_{m-1}+j}(l) = b_m$ and $\boldsymbol{v}_f(l) = b_{m-1}$ for $\boldsymbol{v}_f \notin X_m^i \setminus X_m'^i$ ($b_{m-1} < b_m$, $\boldsymbol{x}(l) = b_{m-1}$). If $t_{r_1+...+r_{m-1}+j} > 0$ then $b_{m-1} = \boldsymbol{x}(l) > (t_{r_1+...+r_{m-1}+1} + ... + t_{r_1+...+r_{m-1}+r_m})b_{m-1} = b_{m-1}$, which is absurd;
then, $\forall j \leq r_m$, $t_{r_1+...+r_{m-1}+j} = 0$. Thus $\boldsymbol{x} \notin CH((S_i \setminus (X_1'^i \cup ... \cup X_m'^i))$.
So, there exists $\boldsymbol{x} \in X_1'^i \cup ... \cup X_m'^i$ such that $\{\boldsymbol{x}\} \mid\mid (S_i \setminus \{\boldsymbol{x}\})$ □

Theorem 2. *if $X_1'^i \cup ... \cup X_m'^i = \emptyset$ then $X_1'^i, ..., X_m'^i$ are linearly separable. Thus, the algorithm stops, in the worse case, after $Card(X_1 \cup, ..., \cup X_m) - 1$ steps, and the result $[(\boldsymbol{w}_0, t_0, a_0, h_0, b_{0,1}, b_{0,2}), ..., (\boldsymbol{w}_{i-1}, t_{i-1}, a_{i-1}, h_{i-1}, b_{i-1,1}, b_{i-1,2}), (\boldsymbol{w}_i, t_i, a_i, h_i, b_1, ..., b_m)]$ is a m-RDP separating $X_1, ..., X_m$ where $b_{j,1}, b_{j,2}$ are b_k, b_{k+1} if at step j, $Z_j \subset X_k'^j$ and $k \leq m - 1$ or b_{m-1}, b_m if at step j, $Z_j \subset X_m'^j$.*

Table 1. Specialized NLS to LS transformation algorithm for m classes

---

SNLS2LS$(X_1, .., X_m, X_{0i}, .., X_{mi})$
– data: m data set vectors, $X_0, .., X_m$ representing m NLS classes
– result: A m-RDP $[(\boldsymbol{w}_0, t_0, a_0, h_0, b_{0,1}, b_{0,2}), ..., (\boldsymbol{w}_{i-1}, t_{i-1}, a_{i-1}, h_{i-1}, b_{i-1,1}, b_{i-1,2}),$
$(\boldsymbol{w}_i, t_i, a_i, h_i, b_1, ..., b_m)]$ which linearly separates $X_1, ..., X_m$.
INITIALIZE : Let $i := 0$; $X_1^0 := X_1; ...; X_m^0 := X_m$; ${X'_1}^0 := X_1; ...; {X'_m}^0 := X_m$;
$S_0 = X_1 \cup ... \cup X_m$;
WHILE $(X_1^i, ..., X_m^i)$ are not linearly separable
BEGIN
SELECT : Select a non-empty sub-set Z_i from ${X'_1}^i$ or ... or from ${X'_m}^i$
(if it exists) such that Z_i, $(S_i \setminus Z_i)$ are linearly separable
(i.e. $(Z_i \subset {X'_1}^i$ *or ... or* $Z_i \subset {X'_m}^i)$ *and* $Z_i \,||\, (S_i \setminus Z_i))$ $(\mathcal{P}(\boldsymbol{w}_i, t_i))$;
CASE :

Case $Z_i \subset {X'_1}^i$:
$S_{i+1} := Adj(Z_i, b_1) \cup Adj(S_i \setminus Z_i, b_2)$;
${X'_1}^{i+1} := Im({X'_1}^i, S_{i+1}) \setminus Im(Z_i, S_{i+1})$;
${X'_2}^{i+1} := Im({X'_2}^i, S_{i+1}); ...;$
${X'_2}^{i+1} := Im({X'_m}^i, S_{i+1})$;
$X_1^{i+1} := Im(X_1^i, S_{i+1})$;
$X_2^{i+1} := Im(X_2^i, S_{i+1}); ...;$
$X_m^{i+1} := Im(X_m^i, S_{i+1})$;
$i := i + 1$;
....

Case $Z_i \subset {X'_j}^i$:
$S_{i+1} := Adj(Z_i, b_j) \cup Adj(S_i \setminus Z_i, b_{j+1})$;
${X'_1}^{i+1} := Im({X'_1}^i, S_{i+1}); ...;$
${X'_{j-1}}^{i+1} := Im({X'_{j-1}}^i, S_{i+1})$;
${X'_j}^{i+1} := Im({X'_j}^i, S_{i+1}) \setminus Im(Z_i, S_{i+1})$;
${X'_{j+1}}^{i+1} := Im({X'_{j+1}}^i, S_{i+1}); ...;$
${X'_m}^{i+1} := Im({X'_m}^i, S_{i+1})$;
$X_1^{i+1} := Im(X_1^i, S_{i+1})$;
$X_2^{i+1} := Im(X_2^i, S_{i+1}); ...;$
$X_m^{i+1} := Im(X_m^i, S_{i+1})$;
$i := i + 1$;
....

Case $Z_i \subset {X'_m}^i$:
$S_{i+1} := Adj(Z_i, b_m) \cup Adj(S_i \setminus Z_i, b_{m-1})$;
${X'_1}^{i+1} := Im({X'_1}^i, S_{i+1}); ...;$
${X'_{m-1}}^{i+1} := Im({X'_{m-1}}^i, S_{i+1})$;
${X'_m}^{i+1} := Im({X'_m}^i, S_{i+1}) \setminus Im(Z_i, S_{i+1})$;
$X_1^{i+1} := Im(X_1^i, S_{i+1})$;
$X_2^{i+1} := Im(X_2^i, S_{i+1}); ...;$
$X_m^{i+1} := Im(X_m^i, S_{i+1})$;
$i := i + 1$;

END

Proof. Let $\boldsymbol{w} = (0, ..., 0, 1, ..., 1) \in \mathbb{R}^{d+i}$ with d times 0 and i times 1, and let $t = -(k_{X_1}(i)b_2 + ... + k_{X_{m-1}}(i)b_m + k_{X_m}(i)b_{m-1})$. Thus, $\forall j < m, \forall \boldsymbol{x} \in X_j^i,\ \boldsymbol{w}^T\boldsymbol{x} + t = b_j - b_{j+1} = b + (j-1)h$, and $\forall \boldsymbol{x} \in X_m^i,\ \boldsymbol{w}^T\boldsymbol{x} + t = b_m - b_{m-1} = b + (m-1)h$. Let $a_0 = b - \frac{h}{2}, a_i = a_0 + ih$ for $1 \leq i \leq m$, thus $\forall j \leq m,\ \forall \boldsymbol{x} \in X_j^i,\ a_{j-1} \leq \boldsymbol{w}^T\boldsymbol{x} + t \leq a_j$.

So, $X_1^i, ..., X_m^i$ are linearly separables by the hyperplane $\mathcal{P}(\boldsymbol{w}, t)$ □

3 Comparison Procedure

The three machine learning benchmark data sets used in the comparison study were identified in section 1.

The IRIS dataset classifies a plant as being an Iris Setosa, Iris Versicolour or Iris Virginica. The dataset describes every iris plant using four input parameters (Table 2). The dataset contains a total of 150 samples with 50 samples for each of the three classes. All the samples of the Iris Setosa class are linearly separable from the rest of the samples (Iris Versicolour and Iris Virginica). Some of the publications that used this benchmark include: [6] [7] [8] and [9].

The SOYBEAN classification problem contains data for the disease diagnosis of the Soybean crop. The dataset describes the different diseases using symptoms. The original dataset contains 19 diseases and 35 attributes. The attribute list was limited to those attributes that had non trivial values in them (Table 3). Thus there were only 20 out of the 35 attributes that were included in the tests. Only 15 of the 19 have no missing values. Therefore, only these 15 classes were used for the comparisons.

Table 2. Inputs and outputs used on the IRIS classification problem

Attributes (In cm)	Output	Output Classes
Sepal Length	Iris plant type	Iris Setosa
Sepal Width		Iris Versicolour
Petal Length		Iris Virginica
Petal Width		

Table 3. Inputs and outputs used in the SOYBEAN classification problem

Attributes		Output	Output classes
Date	leaf-shred	Disease type	brown-spot
plant-stand	stem		alternarialeaf-spot
precipitation	stem-cankers		frog-eye-leaf-spot
temperature	canker-lesion		
hail	fruiting-bodies		
crop-hist	external decay		
area-damaged	fruit-pods		
severity	fruit spots		
seed-tmt	seed		
germination	plant-growth		

Table 4. Inputs and outputs used on the Wisconsin Breast Cancer classification problem

Attributes (1 - 10)	Output	Output Classes
Clump Thickness	Class	Benign
Uniformity of Cell Size		Malignant
Uniformity of Cell Shape		
Marginal Adhesion		
Single Epithelial Cell Size		
Bare Nuclei		
Bland Chromatin		
Normal Nucleoli		
Mitoses		

The Wisconsin Breast Cancer dataset [10,11,12] consists of a binary classification problem to distinguish between benign and malignant breast cancer. The data set contains 699 instances and 9 attributes (Table 4). The class distribution is: Benign 458 instances (65.5 %), and Malignant 241 instances (34.5 %).

The technique of cross validation was applied to split the benchmarks into training and testing data sets. The datasets were randomly divided into 'n' equal sized testing sets that were mutually exclusive [13]. The remaining samples were used to train the networks. In this study, the classification benchmark data sets were divided into ten equally sized data sets. Sixty percent of the samples were used for training the networks and the remaining forty percent were used for testing purposes.

The simplex algorithm was used on this study for testing for linear separability. This algorithm was remarkably faster than the Perceptron one when searching for LS subsets. Other algorithms for testing linear separability include the Class of Linear Separability [14] and the Fisher method (see [15] for a survey on methods for testing linear separability).

These results provide a good basis to further develop this study and to compare the topology size (number of intermediate neurons), and convergence time obtained with the m-class RDP method and Backpropagation and Cascade Correlation. After describing the experimental setup, some conclusions are presented in the next section.

4 Results and Discussion

We now give a comparison between the m-class RDP construction method, Backpropagation and Cascade Correlation based on their level of generalisation on previously unseen data. As specified before, the m-class RDP uses a single output neuron for multiple classes.

Table 5 shows the results obtained with the Iris data set. A single output unit was used for the three neural network models. As it can be seen, the generalisation figures obtained with the m-class RDP are slightly lower but comparable with those obtained with the Backpropagation and the Cascade Correlation neural network models.

Table 6 summarizes the results obtained with the Soybean data set. Topologies for the backpropagation and the cascade correlation methods included both 1 and 15 output

Table 5. Results obtained with the m-class RDP learning method, backpropagation, and cascade correlation using the Iris benchmark data set in terms of the level of generalisation. The topologies for all three methods used a single output neuron.

Data Set	m-class RDP	Back Propagation	Cascade Correlation
1	98.33	98.33	98.33
2	98.33	98.33	98.33
3	95.00	96.67	96.67
4	95.00	98.33	100
5	91.67	98.33	98.33
6	91.67	96.67	98.33
7	96.67	96.67	96.67
8	96.67	98.33	100
9	95.00	96.67	98.33
10	93.33	95.00	96.67

Table 6. Results obtained with the m-class RDP learning method, backpropagation, and cascade correlation using the Soybean benchmark data set in terms of the level of generalisation. Topologies with 1 and 15 output neurons (one per class) were tried.

	m-class RDP	Back Propagation		Cascade Correlation	
Data Set		1 output neurons	15 output neurons	1 output neurons	15 output neurons
1	83.18	49.53	87.85	54.21	87.85
2	73.83	54.21	87.85	50.47	89.71
3	74.77	45.79	84.11	47.66	85.98
4	61.68	37.38	80.37	54.21	83.17
5	75.70	45.79	84.11	51.4	84.11
6	70.09	39.25	81.31	51.4	85.05
7	75.70	43.93	84.11	52.34	85.5
8	74.77	47.66	78.50	52.34	82.24
9	73.83	45.79	84.11	52.34	85.98
10	71.03	48.60	85.05	57.94	91.58

neurons. This problem is more difficult than the Iris one. Therefore, as can be seen from these results, both the backpropagation and the cascade correlation methods did not produce good results using a single output neuron. In order to improve results, it was necessary to use 15 output neurons (one per class).

The generalisation levels obtained with the Wisconsin breast cancer data set are shown in table 7. Contrary to the previous two classificatio problems, this is a 2 class classification problem. This is why results obtained with the 2-class RDP neural network are also provided. As can be seen, the generalisation results obtained with the m-class RDP are ovefrall slightly lower than the ones obtained with the backpropagation and the cascade correlation methods. In some of the data subsets, the m-RDP

Table 7. Results obtained with the m-class and 2-class RDP learning methods, backpropagation, and cascade correlation using the Wisconsin Breast Cancer data set benchmark in terms of the level of generalisation. Since this is a two class classification problem a single output neuron was used on the topologies.

Data Set	2-class RDP	m-class RDP	Back Propagation	Cascade Correlation
1	90	94.16	96.35	98.17
2	97	95.62	95.26	97.08
3	92.7	94.53	95.99	96
4	94.2	95.62	95.62	97
5	95.7	93.07	97.45	96.35
6	90.0	91.61	96.72	97
7	97.0	94.16	96.35	96.7
8	92.7	89.78	97.45	97
9	94.2	91.61	97.08	97.45
10	95.7	94.53	97.81	98.54

provides a slightly higher level of gereralisation than the ones obtained witht the 2-class RDP method. This is probably linked to the extra degrees of freedom provided by the m-class RDP.

Overall, results in terms of generalisation obtained with the m-class RDP method are slightly lower than those obtained with the backpropagation and the cascade correlation methods. Results however are higher for the m-class RDP than for the two other when considering topologies with a single output neuron. This can be interested when considering reduced topologies. An extension of this work could include a comparison of the three methods in terms of topology size (number of intermediate neurons) and convergence time. This is of special interest since, contrary to the backpropagation method, the topology of an RDP neural network is automatically produce with no used intervention.

References

1. Tajine, M., Elizondo, D.: Enhancing the perceptron neural network by using functional composition. Technical Report 96-07, Computer Science Department, Université Louis Pasteur, Strasbourg, France (1996)
2. Rosenblatt, F.: Principles of Neurodynamics. Spartan, Washington D.C (1962)
3. Tajine, M., Elizondo, D., Fiesler, E., Korczak, J.: Adapting the 2−class recursive deterministic perceptron neural network to m−classes. In: The International Conference on Neural Networks (ICNN). IEEE, Los Alamitos (1997)
4. Preparata, F.P., Shamos, M.: Computational Geometry. An Introduction. Springer, New York (1985)
5. Newman, D.J., Hettich, S., Blake, C.L., Merz, C.J.: UCI repository of machine learning databases (1998)
6. Fisher, R.A.: The use of multiple measurements in taxonomic problems. Annual Eugenics 7(II), 179–188 (1936)

7. Gates, G.W.: The reduced nearest neighbor rule. IEEE Transactions on Information Theory, 431–433 (May 1972)
8. Dasarathy, B.W.: Nosing around the neighborhood: A new system structure and classification rule for recognition in partially exposed environments. IEEE Transactions on Pattern Analysis and Machine Intelligence 2(1), 67–71 (1980)
9. Elizondo, D.: The Recursive Determinist Perceptron (RDP) and Topology Reduction Strategies for Neural Networks. PhD thesis, Université Louis Pasteur, Strasbourg, France (January 1997)
10. Mangasarian, O.L., Wolberg, W.H.: Cancer diagnosis via linear programming. SIAM News 23(5), 1–18 (1990)
11. Bennett, K.P., Mangasarian, O.L.: Robust linear programming discrimination of two linearly inseparable sets. Optimization Methods and Software 1, 23–34 (1992)
12. Wolberg, W.H., Mangasarian, O.: Multisurface method of pattern separation for medical diagnosis applied to breast cytology. In: Proceedings of the National Academy of Sciences, December 1990, vol. 87, pp. 9193–9196 (1990)
13. Weiss, S.M., Kulikowski, C.A.: Computer Systems That Learn. Morgan Kaufmann Publishers, San Mateo (1991)
14. Elizondo, D.: Searching for linearly separable subsets using the class of linear separability method. In: Proceedings of the IEEE-IJCNN, pp. 955–960 (April 2004)
15. Elizondo, D.: The linear separability problem: Some testing methods. In: IEEE TNN (2006) (accepted for Publication)

A Constructive Technique Based on Linear Programming for Training Switching Neural Networks

Enrico Ferrari and Marco Muselli

Institute of Electronics, Computer and Telecommunication Engineering
Italian National Research Council
via De Marini, 6 - 16149 Genoa, Italy
{ferrari,muselli}@ieiit.cnr.it

Abstract. A general constructive approach for training neural networks in classification problems is presented. This approach is used to construct a particular connectionist model, named Switching Neural Network (SNN), based on the conversion of the original problem in a Boolean lattice domain. The training of an SNN can be performed through a constructive algorithm, called *Switch Programming* (*SP*), based on the solution of a proper linear programming problem. Simulation results obtained on the StatLog benchmark show the good quality of the SNNs trained with SP.

Keywords: Switching Neural Network, constructive technique, positive Boolean function, Switch Programming.

1 Introduction

Backpropagation algorithms [1] have scored excellent results in the training of neural networks for classification. In general any technique for the solution of classification problems consists of two steps: at first a class Γ of functions is selected (*model definition*), then the best classifier $g \in \Gamma$ is retrieved (*training phase*).

The choice of Γ must take into account two considerations. If the set Γ is too large, it is likely to incur in the problem of *overfitting*: the optimal classifier $g \in \Gamma$ has a good behavior on the examples of the training set, but scores a high number of misclassifications on the other points of the input domain. On the contrary, the choice of a small set Γ prevents from retrieving a function with a sufficient level of accuracy on the training set.

In a multilayer perceptron the complexity of the set Γ depends on some topological properties of the network, such as the number of hidden layers and neurons. The drawback of this approach is that the architecture of the neural network must be chosen by the user before the training phase, often without any prior information.

In order to avoid the drawbacks related to backpropagation algorithms, two different approaches can be introduced: pruning methods and constructive techniques [2]. Pruning methods consider an initial trained neural network with a

V. Kůrková et al. (Eds.): ICANN 2008, Part II, LNCS 5164, pp. 744–753, 2008.

high number of neurons and adopt smart techniques to find and eliminate those connections and units which have a negligible influence on the accuracy of the classifier.

The complementary approach is followed by constructive methods. At the beginning of the training process the neural network only includes the input and the output layers. Then, hidden neurons are added one at a time until the input-output relations in the training set are satisfied within the desired level of accuracy. Generally, the connections between neurons are decided *a priori* so that only a small part of the weight matrix has to be updated at each iteration. Constructive methods usually present a rapid convergence to a well-generalizing solution and allow also the treatment of complex training sets. Nevertheless, since the inclusion of a new hidden unit involves only a limited number of weights, it is possible that some correlations between the data in the training set are missed.

In the following sections we will introduce a constructive algorithm, based on the solution of a linear programming problem, for training Switching Neural Networks (SNN), a recent connectionist model for the solution of classification problems [3].

SNN includes a first layer containing a particular kind of A/D converters, called *latticizers*, that suitably transform input vectors into binary strings. Then, the subsequent two layers of an SNN realize a positive Boolean function that solves in a lattice domain the original classification problem.

Preliminary results show that, with this constructive technique, SNN can achieve generalization errors comparable with those of the best machine learning techniques.

2 A General Structure for a Class of Constructive Methods

The following classification problem will be considered: d-dimensional vectors $\boldsymbol{x} \in X$ have to be assigned to one of two classes labeled by a Boolean value $y \in \{0, 1\}$. According to possible real world situations, the type of the components x_i, $i = 1, \ldots, d$, can be either *continuous ordered*, when x_i belongs to a subset of $\mathbb{R}$, or *discrete ordered*, when x_i assumes values inside a finite ordered set, or *nominal*, when x_i can assume values inside a finite set, where no ordering is defined.

The decision function $\hat{g}$ realizing such classification must be inferred from a set S of examples $\{\boldsymbol{x}_k, y_k\}$, $k = 1, \ldots, s$. Denote with $S_1 = \{\boldsymbol{x}_k \mid y_k = 1\}$ the set of positive examples and with S_0 the set of negative examples. Moreover, let $s_1 = |S_1|$ and $s_0 = |S_0|$, where $|\cdot|$ denotes the number of elements in the set.

In many constructive methods the function $\hat{g}$ is realized by a two layer neural network; the hidden layer is built incrementally by adding a neuron at each iteration of the training procedure. In order to characterize the hidden neurons consider the following

Definition 1. *A collection* $\{\{L_h, \hat{y}_h\}\ ,\ h = 1, \ldots, r+1\}$, *where* $L_h \subset X$ *and* $\hat{y}_h \in \{0, 1\}$ *for each* $h = 1, \ldots, r$, *will be called a* decision list *for a two class problem if* $L_{r+1} = X$.

In [2] the decision list is used hierarchically: a pattern $\boldsymbol{x}$ is assigned to the class y_h, where h is the lower index such that $\boldsymbol{x} \in L_h$. It is possible to consider more general criteria in the output assignment: for example a weight $w_h > 0$ can be associated with each domain L_h, measuring the reliability of assigning the output value $\hat{y}_h$ to every point in L_h.

It is thus possible to associate with every pattern $\boldsymbol{x}$ a weight vector $\boldsymbol{u}$, whose h-th component is defined by

$$u_h = \begin{cases} w_h & \text{if } \boldsymbol{x} \in L_h \\ 0 & \text{otherwise} \end{cases} \tag{1}$$

for $h = 1, \ldots, r$. The weight u_h can be used to choose the output for the pattern $\boldsymbol{x}$. Without loss of generality suppose that the decision list is ordered so that $\hat{y}_h = 0$ for $h = 1, \ldots, r_0$, whereas $\hat{y}_h = 1$ for $h = r_0 + 1, \ldots, r_0 + r_1$, where $r_0 + r_1 = r$. The value of $\hat{y}_{r+1}$ is the *default* decision, i.e. the output assigned to $\boldsymbol{x}$ if $\boldsymbol{x} \notin L_h$ for each $h = 1, \ldots, r$.

In order to fix a criterion in the output assignment for an input vector $\boldsymbol{x}$ let us introduce the following

Definition 2. *A function $\sigma(\boldsymbol{u}) \in \{0, 1\}$ is called an* output decider *if*

$$\sigma(\boldsymbol{u}) = \begin{cases} y_{r+1} & \text{if } u_1 = \ldots = u_{r_0} = u_{r_0+1} = \ldots = u_r = 0 \\ 1 & \text{if } u_1 = \ldots = u_{r_0} = 0 \text{ and some } u_h > 0 \text{ with } r_0 < h \le r \\ 0 & \text{if } u_{r_0+1} = \ldots = u_r = 0 \text{ and some } u_h > 0 \text{ with } 0 < h \le r_0 \end{cases}$$

This classifier can then be implemented in a two layer neural network: the first layer retrieves the weights u_h for $h = 1, \ldots, r$, whereas the second one realizes the output decider σ. The behavior of σ is usually chosen *a priori* so that the training phase consists in finding a proper decision list and the relative weight vector $\boldsymbol{w}$. For example, σ can be made equivalent to a comparison between the sum of the weights of the two classes:

$$\sigma(\boldsymbol{u}) = \begin{cases} 0 & \text{if } \sum_{h=1}^{r_0} u_h > \sum_{h=r_0+1}^{r} u_h \\ 1 & \text{if } \sum_{h=1}^{r_0} u_h < \sum_{h=r_0+1}^{r} u_h \\ \hat{y}_{r+1} & \text{otherwise} \end{cases}$$

The determination of the decision list $\{L_h, \hat{y}_h\}$, $h = 1, \ldots, r$, can be performed in a constructive way, by adding at each iteration h the best pair $\{L_h, \hat{y}_h\}$ according to a smart criterion. Each domain L_h corresponds to a neuron characterized through the function introduced by the following

Definition 3. *Consider a subset $Q \subset S_y$, $y \in \{0, 1\}$. The function*

$$\hat{g}_h(\boldsymbol{x}) = \begin{cases} 1 & \text{if } \boldsymbol{x} \in L_h \\ 0 & \text{otherwise} \end{cases}$$

is called a partial classifier *for Q if $Q \cap L_h$ is not empty whereas $L_h \cap S_{1-y} = \emptyset$. If $\hat{g}_h(\boldsymbol{x}) = 1$ the h-th neuron will be said to* cover *$\boldsymbol{x}$.*

Since the target of the training phase is to find the simplest network satisfying the input-output relations in the training set, the patterns already covered by

Constructive training for a two layer perceptron

For $y \in \{0,1\}$ do

1. Set $Q = S_y$ and $h = 1$.
2. Find a partial classifier $\hat{g}_h$ for Q.
3. Let $R = \{\{\boldsymbol{x}_k, y_k\} \in Q \mid \hat{g}_h(\boldsymbol{x}_k) = 1\}$.
4. Set $Q = Q \setminus R$ and $h = h + 1$.
5. If Q is nonempty go to step 2.
6. Prune redundant neurons and set $r_y = h$.

Fig. 1. General constructive procedure followed for neural network training

at least a neuron can be ignored when training further neurons having the same value of $\hat{y}_h$.

Fig. 1 shows a general constructive procedure for training a neural network in case of binary output. At each iteration the set Q contains the patterns belonging to the current output value not covered by the neurons already included in the network. Notice that removing elements from Q allows a high reduction of the training time for each neuron since a lower number of examples has to be processed at each iteration.

A pruning phase is performed at the end of the training process in order to eliminate redundant overlaps among the sets L_h, $h = 1, \ldots, r$.

Without entering in details about the general theoretical properties of constructive techniques, which can be found in [2], in the following sections we will present the architecture of Switching Neural Networks and an appropriate training algorithm.

3 Switching Neural Networks

A promising connectionist model, called Switching Neural Network (SNN), has been recently developed [3]. According to this model the input variables are transformed into m-dimensional Boolean strings by means of a particular mapping $\varphi : X \to \{0,1\}^m$ and a positive Boolean function f (such that $f(\boldsymbol{v}) \leq f(\boldsymbol{v}')$ for $\boldsymbol{v} \leq \boldsymbol{v}'$) is built starting from a converted training set S'. The mapping $\boldsymbol{\varphi}$, called *latticizer*, is obtained by discretizing the ordered inputs and by applying a proper binary coding that preserves ordering and distance [4].

If $\boldsymbol{a} \in \{0,1\}^m$, let $P(\boldsymbol{a})$ be the subset of $I_m = \{1, \ldots, m\}$ including the indexes i for which $a_i = 1$. It can be shown [4] that a positive Boolean function can always be written as

$$f(\boldsymbol{z}) = \bigvee_{\boldsymbol{a} \in A} \bigwedge_{j \in P(\boldsymbol{a})} z_j \tag{2}$$

where A is an antichain of the Boolean lattice $\{0,1\}^m$, i.e. a set of Boolean strings such that neither $\boldsymbol{a} < \boldsymbol{a}'$ nor $\boldsymbol{a}' < \boldsymbol{a}$ holds for each $\boldsymbol{a}$, $\boldsymbol{a}' \in A$. The symbol $\bigvee$ (resp. $\bigwedge$) in (2) denotes a logical sum (resp. product) among the terms identified by the subscript. In particular, $\bigwedge_{j\in P(\boldsymbol{a})} z_j$ is an *implicant* for the function f; however, when no confusion arises, the term implicant will also be used to denote the corresponding binary string $\boldsymbol{a} \in A$.

A drawback of the model described in [3] is that it treats in an asymmetrical way positive and negative patterns. In fact the SNN is built to assign positive labels whereas negative ones are associated by default. A possible way to recover this drawback consists in using a separate positive Boolean function f_y (i.e. an antichain A_y to be employed in (2)) for each value assumed by the output y.

In order to combine the functions relative to the different output values, each generated implicant can be characterized by a weight $w_h > 0$, which measures its significance level for the examples in the training set. Thus, to each Boolean string $\boldsymbol{z}$ can be assigned a weight vector $\boldsymbol{u}$ whose h-th component is

$$u_h = F_h(\boldsymbol{z}) = \begin{cases} w_h & \text{if } \bigwedge_{j\in P(\boldsymbol{a}_h)} z_j = 1 \\ 0 & \text{otherwise} \end{cases}$$

where $\boldsymbol{a}_h \in A_0 \cup A_1$ for $h = 1, \ldots, r$.

At the final step of the classification process an output decider $\sigma(\boldsymbol{u})$ assigns the correct class to the pattern $\boldsymbol{z}$ according to a comparison between the weights u_h of the different classes. If no h exists such that $u_h > 0$, the default output is assigned to $\boldsymbol{z}$.

The device implementing the function $\hat{g}(\boldsymbol{x}) = \sigma(\boldsymbol{F}(\varphi(\boldsymbol{x})))$ is shown in Fig. 2. It can be considered as a three layer feedforward neural network. The first layer

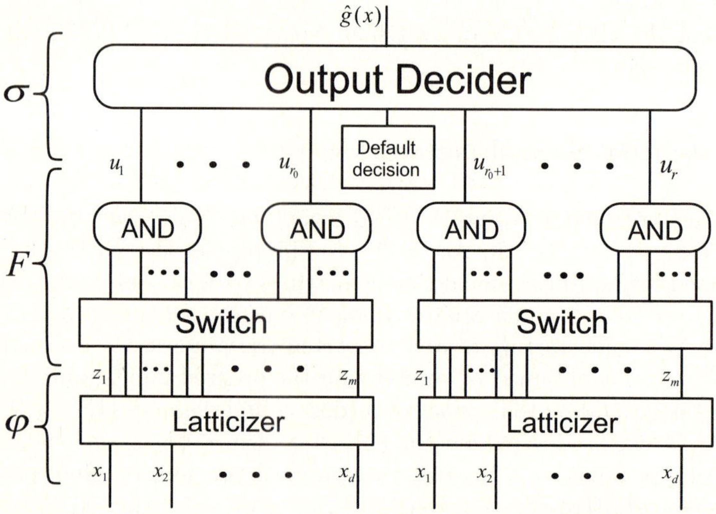

Fig. 2. The schema of a Switching Neural Network

is responsible of the latticization mapping φ; the second realizes the function $\boldsymbol{F}$ assigning a weight to each implicant. Finally, the third layer uses the weight vector $\boldsymbol{u} = \boldsymbol{F}(\boldsymbol{z})$ to decide the output value for the pattern $\boldsymbol{x}$.

Every AND port in the second layer is connected only to some of the outputs leaving the latticizers; they correspond to values 1 in the associated implicant. The choice of such values is performed by a *switch* port. For this reason the connectionist model shown in Fig. 2 is called *Switching Neural Network.*

Notice that the device can be subdivided into two parts: the left part includes the r_0 neurons characterizing the examples having output $y = 0$, whereas the right part involves the $r_1 = r - r_0$ implicants relative to the output $y = 1$. For this reason, the generalization of an SNN to a multiclass problem (where $y \in \{1, \ldots, c\}$, $c > 2$) is immediate: it is sufficient to create a set of implicants for each output value.

It is interesting to observe that, unlike standard neural networks, SNNs do not involve floating point operations. In fact the weights $\boldsymbol{w}$ are provided by the training process and can be chosen in a set of integer values. Moreover, the antichain A can be converted into a set of intelligible rules in the form

$$\textbf{if} \; < premise > \; \textbf{then} \; < consequence >$$

via the application of a smart inverse operator [3] of φ to the elements of A.

4 Training Algorithm

In this section the constructive algorithm for building a single f_y (denoted only by f for simplicity) will be described. The procedure must be repeated for each value of the output y in order to find an optimal classifier for the problem at hand. In particular, if the function f_y is built, the Boolean output 1 will be assigned to the examples belonging to the class y, whereas the Boolean output 0 will be assigned to all the remaining examples.

The architecture of the SNN has to be constructed starting from a set S' containing s_1 positive examples and s_0 negative examples. Suppose, without loss of generality, that the set S' is ordered so that the first s_1 examples are positive. Since the training algorithm sets up, for each output value, the switches in the second layer of the SNN, the constructive procedure of adding neurons step by step will be called *Switch Programming* (*SP*).

4.1 Implicants Generation

When a Boolean string $\boldsymbol{z}$ is presented as input, the output of the logical product $\bigwedge_{j \in P(\boldsymbol{a})} z_j$ at a neuron is positive if and only if $\boldsymbol{a} \leq \boldsymbol{z}$ according to the standard ordering in the Boolean lattice. In this case $\boldsymbol{a}$ will be said to *cover* $\boldsymbol{z}$.

The target of a training algorithm for an SNN is to find the simplest antichain A covering all the positive examples and no negative examples in the training set. A constructive approach for this target consists in generating implicants one at a time according to a smart criterion of choice determined by an objective function $\Phi(\boldsymbol{a})$.

In particular $\Phi(\boldsymbol{a})$ must take into account the number of examples in S' covered by $\boldsymbol{a}$ and the degree of complexity of $\boldsymbol{a}$, usually defined as the number of elements in $P(\boldsymbol{a})$ or, equivalently, as the sum $\sum_{i=1}^{m} a_i$. These parameters will be balanced in the objective function through the definition of two weights λ and μ.

In order to define the constraints to the problem, define, for each example $\boldsymbol{z}_k$, the number ξ_k of indexes i for which $a_i = 1$ and $z_{ki} = 0$. It is easy to show that $\boldsymbol{a}$ covers $\boldsymbol{z}_k$ if and only if $\xi_k = 0$. Then, the quantity $\sum_{k=1}^{s_1} \theta(\xi_k)$, where θ represents the usual Heaviside function (defined by $\theta(u) = 1$ if $u > 0$, $\theta(u) = 0$ otherwise), is the number of positive patterns not covered by $\boldsymbol{a}$. On the contrary, it is necessary that $\xi_k > 0$ for each $k = s_1+1, \ldots, s$, so that any negative pattern is not covered by $\boldsymbol{a}$.

Starting from these considerations, the best implicant can be retrieved by solving the following optimization problem:

$$\begin{aligned}
&\min_{\boldsymbol{\xi},\boldsymbol{a}} \ \frac{\lambda}{s_1}\sum_{k=1}^{s_1} \xi_k + \frac{\mu}{m}\sum_{i=1}^{m} a_i \\
&\text{subj to} \sum_{i=1}^{m} a_i(a_i - z_{ki}) = \xi_k \quad \text{for } k = 1, \ldots, s_1 \\
&\sum_{i=1}^{m} a_i(a_i - z_{ki}) \geq 1 \quad \text{for } k = s_1 + 1, \ldots, s \\
&\xi_k \geq 0 \quad \text{for } k = 1, \ldots, s_1\,, \quad a_i \in \{0, 1\} \quad \text{for } i = 1, \ldots, d
\end{aligned} \tag{3}$$

where the Heaviside function has been substituted by its argument in order to avoid nonlinearity in the cost function.

Since the determination of a sufficiently great collection of implicants, from which the antichain A is selected, requires the repeated solution of problem (3), it must be avoided at any extraction the generation of an already found implicant. This can be obtained by adding the following constraint

$$\sum_{i=1}^{m} a_{ji}(1 - a_i) \geq 1 \quad \text{for } j = 1, \ldots, q - 1 \tag{4}$$

where $\boldsymbol{a}$ is the implicant to be constructed and $\boldsymbol{a}_1, \ldots, \boldsymbol{a}_{q-1}$ are the already found $q - 1$ implicants.

Additional requirements can be added to problem (3) in order to improve the quality of the implicant $\boldsymbol{a}$ and the convergence speed. For example, in order to better differentiate implicants and to cover all the patterns in a fewer number of steps, the set S_1'' of positive patterns not yet covered can be considered separately and weighted by a different factor $\nu \neq \lambda$.

Moreover, in presence of noise it should be useful to avoid excessive adherence of $\boldsymbol{a}$ with the training set by accepting a small fraction ε of errors.

In this case a further term is added to the objective function, measuring the level of misclassification, and constraints in (3) have to be modified in order to allow at most εs_0 patterns to be misclassified by the implicant $\boldsymbol{a}$. If the training set is noisy, the optimal implicant can be found by solving the following LP problem, where it is supposed that the first s_1' positive patterns are not yet covered:

$$\begin{aligned}
&\min_{\boldsymbol{\xi},\boldsymbol{a}} \ \frac{\nu}{s_1'}\sum_{k=1}^{s_1'} \xi_k + \frac{\lambda}{s_1 - s_1'}\sum_{k=s_1'+1}^{s_1} \xi_k + \frac{\mu}{m}\sum_{i=1}^{m} a_i + \frac{\omega}{s_0}\sum_{k=s_1+1}^{s} \xi_k \\
&\text{subj to} \sum_{i=1}^{m} a_i(1 - z_{ki}) = \xi_k \quad \text{for } k = 1, \ldots, s_1 \\
&\sum_{i=1}^{m} a_i(1 - z_{ki}) \geq 1 - \xi_k \quad \text{for } k = s_1 + 1, \ldots, s \\
&\sum_{k=s_1+1}^{s} \xi_k \leq \varepsilon_0 s_0 \,, \quad a_i \in \{0,1\} \ \text{for } i = 1, \ldots, m \\
&\xi_k \geq 0 \ \text{for } k = 1, \ldots, s_1 \,, \quad \xi_k \in \{0,1\} \ \text{for } k = s_1 + 1, \ldots, s
\end{aligned} \tag{5}$$

4.2 Implicant Selection

Once the antichain A' has been generated, it should be useful to look for a subset A of A' which is able to describe the data in the training set with sufficient accuracy. To this aim both the number of implicants included in A and the number N_k of nonnull components in each element $\boldsymbol{a}_k \in A$ must be minimized.

Denote with $\boldsymbol{a}_1$, $\boldsymbol{a}_2$, ..., $\boldsymbol{a}_q$ the q implicants obtained in the generation step and with c_{kj} a binary variable asserting if the input vector $\boldsymbol{z}_k$, $k = 1, \ldots, s_1$, is covered by $\boldsymbol{a}_j$:

$$c_{kj} = \begin{cases} 1 & \text{if } \boldsymbol{z}_k \text{ is covered by } \boldsymbol{a}_j \\ 0 & \text{otherwise} \end{cases}$$

In addition, consider the binary vector $\boldsymbol{\zeta}$ having as jth component the value $\zeta_j = 1$ if the corresponding implicant $\boldsymbol{a}_j$ will be included in the final collection A. Then, an optimal subset $A \subset A'$ can be found by solving the following constrained optimization problem:

$$\begin{aligned}
&\min_{\boldsymbol{\zeta}} \ \sum_{j=1}^{q} \zeta_j(\alpha + \beta N_j)) \\
&\text{subj to} \sum_{j=1}^{q} c_{kj}\zeta_j \geq 1 \quad \text{for } k = 1, \ldots, s_1 \\
&\zeta_j \in \{0,1\} \quad \text{for } j = 1, \ldots, q
\end{aligned} \tag{6}$$

where α and β are constants.

Additional requirements can be added to the problem (6) in order to improve the generalization ability of A. For example, if the presence of noise has to be taken into account, the antichain A can be allowed not to cover a small fraction of positive examples.

5 Simulation Results

To obtain a preliminary evaluation of performances achieved by SNNs trained with SP, the classification problems included in the well-known StatLog benchmark [5] have been considered. In this way the generalization ability and the complexity of resulting SNNs can be compared with those of other machine learning methods, among which backpropagation algorithm (BP) and rule generation techniques based on decision trees, such as C4.5 [6].

The complexity of an SNN is measured through the number of AND ports in the second layer (corresponding to the number of intelligible rules) and the average number of conditions in the **if** part of a rule. Table 1 presents the obtained results. Performances of resulting SNNs are compared with those of models produced by C4.5 and BP. In the same table is also shown the best generalization error included in the StatLog report [5] for each problem, together with the rank scored by SNN when its generalization error is inserted into the list of available results.

Table 1. Generalization error and complexity of SNN, compared with C4.5, BP and other methods, on the StatLog benchmark

Test Problem	Generalization error					# Rules		# Conditions	
	SNN	C4.5	BP	Best	Rank	SNN	C4.5	SNN	C4.5
HEART	0.554	0.781	0.574	0.374	15	1.22	11.4	5.09	2.68
AUSTRALIAN	0.138	0.155	0.154	0.131	3	31.2	11.5	5.72	2.76
DIABETES	0.246	0.270	0.248	0.223	7	180.4	9.4	4.59	2.58
VEHICLE	0.299	0.266	0.207	0.150	18	143.7	26.1	5.19	4.03
GERMAN	0.696	0.985	0.772	0.535	10	66.1	21.1	6.45	2.77
SEGMENT	0.0424	0.040	0.054	0.030	8	105.5	28.0	6.50	3.94
DNA	0.0658	0.076	0.088	0.041	7	30.0	34	11.00	4.47
SATIMAGE	0.168	0.150	0.139	0.094	19	607	80.0	6.42	5.41
SHUTTLE	0.0001	0.001	0.43	0.0001	1	18.0	20.0	3.00	3.14

Notice that in one case (Shuttle), SP achieves the best result among the methods in the StatLog archive, whereas in other five problems SP achieves one of the first ten positions. Moreover, the classification accuracy achieved by C4.5 and BP is significantly lower than that obtained by an SNN trained with the SP algorithm in all the considered datasets, except for VEHICLE and SATIMAGE. This points out the good quality of the solutions offered by the SNNs.

Nevertheless, the performances obtained by SP are conditioned by the number of examples s in the training set and by the number of input variables d. Since the number of constrains in (3) or (5) depends linearly on s, SP becomes slower and less efficient when dealing with complex training sets. In addition, increasing the number of input bits may cause approximation errors while converting the optimal solution of the LP problem into a $\{0, 1\}$ domain. For these reasons the performances of SP get worse when the dimensionality of the training set increases.

In particular, the number of implicants generated by SP is in many cases is higher than that of the rules obtained by C4.5, causing an increasing in the training time.

Notice that the minimization of (3) or (5) is obtained using the package *Gnu Linear Programming Kit* (*GLPK*) [7], a free library for the solution of linear programming problems. It is thus possible to improve the above results by adopting more efficient tools to solve the LP problem for the generation of implicants.

Acknowledgement. This work was partially supported by the Italian MIUR project "Laboratory of Interdisciplinary Technologies in Bioinformatics (LITBIO)".

References

1. Rumelhart, D.E., Hinton, G.E., Williams, R.J.: Learning representations by back-propagating errors. Nature 323, 533–536 (1988)
2. Muselli, M.: Sequential constructive techniques. In: Leondes, C. (ed.) Optimization Techniques. Neural Network Systems Techniques and Applications, vol. 2, pp. 81–144. Academic Press, San Diego (1998)
3. Muselli, M.: Switching neural networks: A new connectionist model for classification. In: Apolloni, B., Marinaro, M., Nicosia, G., Tagliaferri, R. (eds.) WIRN 2005 and NAIS 2005. LNCS, vol. 3931, pp. 23–30. Springer, Berlin (2006)
4. Muselli, M., Quarati, A.: Reconstructing positive Boolean functions with Shadow Clustering. In: Proceedings of the 17th European Conference on Circuit Theory and Design (ECCTD 2005), Cork, Ireland (2005)
5. Michie, D., Spiegelhalter, D., Taylor, C. (eds.): Machine Learning, Neural, and Statistical Classification. Ellis-Horwood, London (1994)
6. Quinlan, J.R.: C4.5: Programs for Machine Learning. Morgan Kaufmann, San Francisco (1994)
7. Makhorin, A.: GNU Linear Programming Kit - Reference Manual (2008), `http://www.gnu.org/software/glpk/`

Projection Pursuit Constructive Neural Networks Based on Quality of Projected Clusters

Marek Grochowski and Włodzisław Duch

Department of Informatics, Nicolaus Copernicus University, Toruń, Poland
grochu@is.umk.pl

Abstract. Linear projection pursuit index measuring quality of projected clusters (QPC) is used to discover non-local clusters in high-dimensional multiclass data, reduce dimensionality, select features, visualize and classify data. Constructive neural networks that optimize the QPC index are able to discover simplest models of complex data, solving problems that standard networks based on error minimization are not able to handle. Tests on problems with complex Boolean logic, and tests on real world datasets show high efficiency of this approach.

1 Introduction

Theoretical analysis of non-separable classification problems, introduced in [1], shows that complexity of data classification is proportional to the minimum number of intervals that are needed to separate pure clusters of data in a single linear projection. Problems that require at least k such intervals are called k-separable. For example, n-bit parity problems are $n+1$-separable [2], because linear projection of binary strings exists that forms $n+1$ separated alternant groups of vectors for odd and even cases. Neural networks based on basis set expansions, such as the Radial Basis Function (RBF) networks, Multi-Layer Perceptrons (MLPs), other standard neural models [3] that use error minimization techniques, Support Vector Machines (SVMs) and almost all other classifiers are not able to discover simple data models when the k index is high. Yet many problems in bioinformatics or text analysis may have inherent complex logic that needs to be discovered.

Transformations of original features should help to find interesting low dimensional representations, revealing structures that are hard to anticipate looking at the original dataset. The simplest transformations with easy interpretations are linear projections. For many data distributions projection on a single direction creates two or more pure clusters. Even if such k-separable solution exist clusters may be small and close to each other. More reliable predictions are possible if all data vectors are projected into large pure clusters. Some additional projections may provide different useful solutions giving large clusters too. For example, in parity problem projection on a $[1, 1, 1..1]$ direction and $[1, -1, 1, -1...]$ direction creates pure clusters of different size allowing for high confidence predictions for all data vectors.

A lot of methods that search for optimal and most informative linear transformations have been developed. A general projection pursuit (PP) framework to find interesting data transformations by maximizing some "index of interest" has been proposed

V. Kůrková et al. (Eds.): ICANN 2008, Part II, LNCS 5164, pp. 754–762, 2008.

by Friedman [4,5]. PP index assigns numerical values to data projections (or general transformations). For example, the Fisher Discriminant Analysis (FDA) algorithm [6] defines an index that measures the ratio of between-class scatter to within-class scatter. Local FDA (LFDA) can deal with multimodal class distributions [7], trying to preserve local structure of data. The approach presented below also belongs to the general projection pursuit framework, and may be implemented as a constructive neural network. In the next section a new index based on the Quality of Projected Clusters (QPC) is proposed. It allows to find compact and pure (from the same class) groups of vectors, separated from other clusters. In contrast to most similarity-based methods [8,9] that optimize metric functions to capture local clusters, projection pursuit may discover non-local structures. A few practical issues related to the application of the QPC index are presented in the next section. Searching for projections into two or more-dimensional spaces allows for visualization of data, as shown in section three. The use of QPC index as a basis for constructive neural network is described in section four. The final section contains discussion and future perspectives.

2 The QPC Projection Index

The celebrated MLP backpropagation of errors training algorithm does not define specific target for hidden layers, trying instead to adapt all weights in a way that contributes to the overall reduction of some error function. It has been quite successful for problems that have relatively low complexity, but fails most of the time for Boolean functions that are 4 or 5-separable. The PP index should help to discover interesting linear projections of multiclass data, and localize vectors from the same class in compact clusters separated from other clusters. Consider a dataset $\mathcal{X} = \{\boldsymbol{x}_1, \ldots, \boldsymbol{x}_n\} \subset \mathcal{R}^d$, where each vector $\boldsymbol{x}_i$ belongs to one of k different classes. For a given vector $\boldsymbol{x} \in \mathcal{X}$ with a label C QPC index is defined as:

$$Q(\boldsymbol{x};\boldsymbol{w}) = A^+ \sum_{\boldsymbol{x}_k \in C} G\left(\boldsymbol{w}^T(\boldsymbol{x}-\boldsymbol{x}_k)\right) - A^- \sum_{\boldsymbol{x}_k \notin C} G\left(\boldsymbol{w}^T(\boldsymbol{x}-\boldsymbol{x}_k)\right) \tag{1}$$

where $G(x)$ is a localized function achieves maximum for $x = 0$ and should have compact ϵ-support for all $x \in \mathcal{R}$. The first term in Eq. (1) is large if projection on a line, defined by weight vector $\boldsymbol{w}$, groups many vectors from class C close to $\boldsymbol{x}$. The second term estimates separation between a given vector $\boldsymbol{x}$ and vectors from classes other than C. It introduces penalty for placing projected $\boldsymbol{x}$ too close to projected vectors from other classes. The average of the $Q(\boldsymbol{x};\boldsymbol{w})$ indices for all vectors:

$$QPC(\boldsymbol{w}) = \frac{1}{n} \sum_{\boldsymbol{x} \in \mathcal{X}} Q(\boldsymbol{x};\boldsymbol{w}) \,, \tag{2}$$

provides a leave-one-out estimator measuring quality of projection on $\boldsymbol{w}$. This index is large if projected clusters are pure, compact and well separated from clusters of vectors with other labels. For linearly separable problems function $QPC(\boldsymbol{w})$ achieves maximum if projection $\boldsymbol{w}\boldsymbol{x}$ creates two well-separated pure clusters. If dataset is k-separable then maximization of this index should find a projection with k separated clusters, that may then be easily classified defining simple intervals or using a special neural architecture [2].

Parameters A^+, A^- control influence of each term in Eq. (1), and may simply be fixed to balance and normalize the value of projection index, for example at $A^+ = 1/p(C)$ and $A^- = 1/(1 - p(C))$ (where $p(C)$ is the *a priori* class probability). If A^- is large strong separation between classes is enforced, while large A^+ impacts mostly compactness and purity of clusters. Influence of each vector on projection index is determined by properties of $G(x)$ function. This function should be localized, achieving a maximum value for $x = 0$. If $G(x)$ is continuous then gradient-based methods may by used to find maximum of the QPC index. All bell-shaped functions are suitable for $G(x)$, including Gaussian and an inverse quartic function:

$$G(x) = \frac{1}{1 + (bx)^4} \tag{3}$$

where parameter b controls the width of $G(x)$, and thus determines influence of the neighboring vectors on the index value. Another useful function is constructed from a combination of two sigmoidal functions (a bicentral function [10,11]):

$$G(x) = \sigma(x + b) - \sigma(x - b) \tag{4}$$

Constructive MLP networks with special architecture could be used to implement approximations to the $QPC(\boldsymbol{w})$ index using several hidden layers. Neurobiological justification for constructive models of networks calculating PP indices is given in the final section.

Multistart gradient approach has been quite effective in searching for interesting projections. Although solutions are not always unique they may sometimes provide additional insight into the structure of data. Calculation of function (2) requires $O(n^2)$ operations. Various "editing techniques" used for the nearest neighbor methods with very large number of vectors [12] may decrease this complexity to $O(n \log n)$. This may be done by sorting projected vectors and restricting computations of the sum in Eq. (1) only to vectors $\boldsymbol{x}_i$ with $G(\boldsymbol{w}(\boldsymbol{x} - \boldsymbol{x}_i)) > \epsilon$. The cost is further decreased if centers of projected clusters are defined and a single sum $G(\boldsymbol{w}(\boldsymbol{x} - t))$ is used. Gradient descent methods may be replaced by more sophisticated approaches [3,13]). Technical improvements are important, but here the potential of the model based on the localized projected clusters is explored.

Figure 1 presents examples of projections that give maximum value of the $QPC(\boldsymbol{w})$ index for 4 very different datasets. All projections were obtained taking Eq. (3) for $G(x)$, with $b = 3$, simple gradient descent maximization initialized 10 times, selecting after a few iterations the most promising solution that is trained until convergence. Values of weights and the value of $QPC(\boldsymbol{w})$ are shown in the corresponding figures. Positions of vectors from each class are shown below the projection line. Smoothed histograms may be normalized and taken as estimations of class conditionals $p(x|C)$, from which posterior probabilities $p(C|x) = p(x|C)p(C)/p(x)$ are easily calculated.

The top left figure shows the Wine dataset from the UCI repository [14], with 13 features and 3 classes. It can be classified using a single linear projection that gives 3 groups of vectors (one for each class). The weight for "flavanoids" dominates and is almost sufficient to separate all 3 classes.

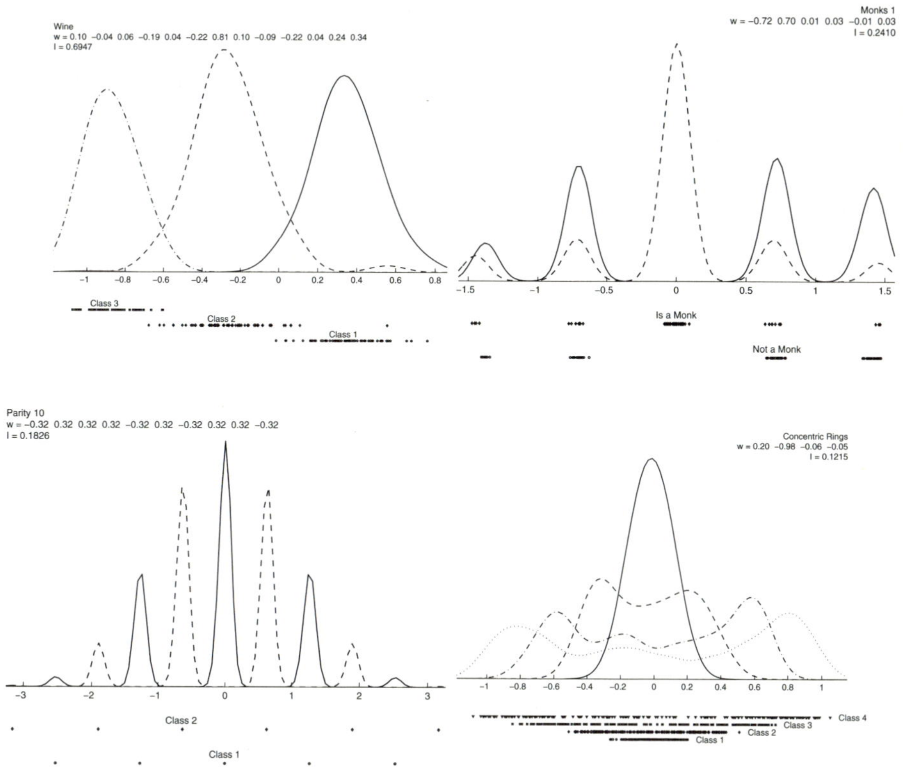

Fig. 1. Examples of projections found by maximization of the QPC index using gradient descent for Wine data (top-left), Monk's 1 problem (top-right), 10-bit Parity (bottom-left) and Concentric Rings with noise (bottom-right)

The top right figure shows artificial Monk 1 datasets [14], with 6 symbolic features and two classes. All vectors of the Monk 1 problem can be classified correctly with two simple rules. Large cluster of vectors in the middle of the projection presented in Fig. 1 (first two coefficients are equal, others are essentially zero) corresponds to a rule: if head shape = body shape then object is called a Monk. To separate the remaining cases a second projection is needed (see below). These logical rules may also be extracted from an MLP networks [15].

The lower left figure shows parity problem in 10 dimensions, with 512 even and 512 odd binary strings. This problem is 11-separable, with a maximum value of projection index obtained for diagonal direction in the 10 dimensional hypercube, therefore all weights have the same value. Although a perfect solution using a single projection has been found clusters at the extreme left and right of the line are quite small and finding another direction that puts these vectors in larger clusters is useful. Convergence of MLP or RBF networks for such complex data is quite unlikely.

The final dataset represents concentric rings, with 4 classes, each with 200 samples defined by 4 features. The first two features define points inside 4 concentric rings, and the last two are uniformly distributed random numbers. For this dataset best projection that maximizes the QPC index reduces influence of noisy features, with weights for dimensions 3 and 4 close to zero. This shows that the QPC index may be used for feature selection, but also that linear projections have limited power, and in this case leads to complicated solution requiring many projections at different angles, while much simpler network using localized functions is sufficient. The need for networks with different types of transfer functions [10,11] has been stressed some time ago, but still there are no programs capable of finding the simplest data models in all cases.

3 Visualization

Additional directions for interesting projections can be found in several ways. New direction may be generated by the same procedure used in the subspace orthogonal to all directions found earlier. This warrants that different directions are found at each iteration. Second, one can focus on clusters of vectors that overlap in the first projection, and use only subset of these vectors to maximize the PP index to find the second direction. The third possibility is to search for the next linear projection with additional penalty term that will punish solutions similar to those found earlier:

$$QPC(\boldsymbol{w};\boldsymbol{w}_1) = QPC(\boldsymbol{w}) - \lambda f(\boldsymbol{w},\boldsymbol{w}_1)\ . \tag{5}$$

The value of $f(\boldsymbol{w},\boldsymbol{w}_1)$ should be large if the current $\boldsymbol{w}$ is close to previous direction $\boldsymbol{w}_1$. For example, some power of the scalar product between these directions may be used: $f(\boldsymbol{w},\boldsymbol{w}1) = (\boldsymbol{w}_1{}^T \cdot \boldsymbol{w})^2$. Parameter λ controls the influence of additional term on the optimization process. Scatterplots of data vectors projected on two directions may be used for visualization. Figure 2 presents such scatterplots for the four datasets used in the previous section. The second direction $\boldsymbol{w}$, found by gradient descent optimization of function (5) with $\lambda = 0.5$, is used for the horizontal axis. The final weights of the second direction, value of the projection index $QPC(\boldsymbol{w})$ and the inner product of $\boldsymbol{w}_1$ and $\boldsymbol{w}$ are shown in the corresponding graphs.

For the Wine problem first projection was able to separate almost perfectly all three classes. Second projection (Fig. 2) gives additional insight at data structure leading to better separation of vectors placed near decision boundary. Two-dimensional projection of Monk's 1 data shows separate and compact clusters. The 5th feature (which forms the second rule describing this dataset: if it is 1 then object is a Monk) has significant value, and all unimportant features have weights equal almost to zero, allowing for simple extraction of correct logical rules. In case of the 10-bit parity problem each diagonal direction of a hypercube representing Boolean function gives a good solution with large cluster in the center. Two such orthogonal directions have been found, projecting each data vector into large pure cluster, either in the first or in the second dimension. Results for the noisy Concentric Rings dataset shows that maximization of the QPC index has caused vanishing of noisy and uninformative features, and has been able to discover two-dimensional relations hidden inside this data. Although linear projections in two directions cannot separate this data, such dimensionality reduction is sufficient for any similarity-based method, for example the nearest neighbor method, to perfectly solve this problem.

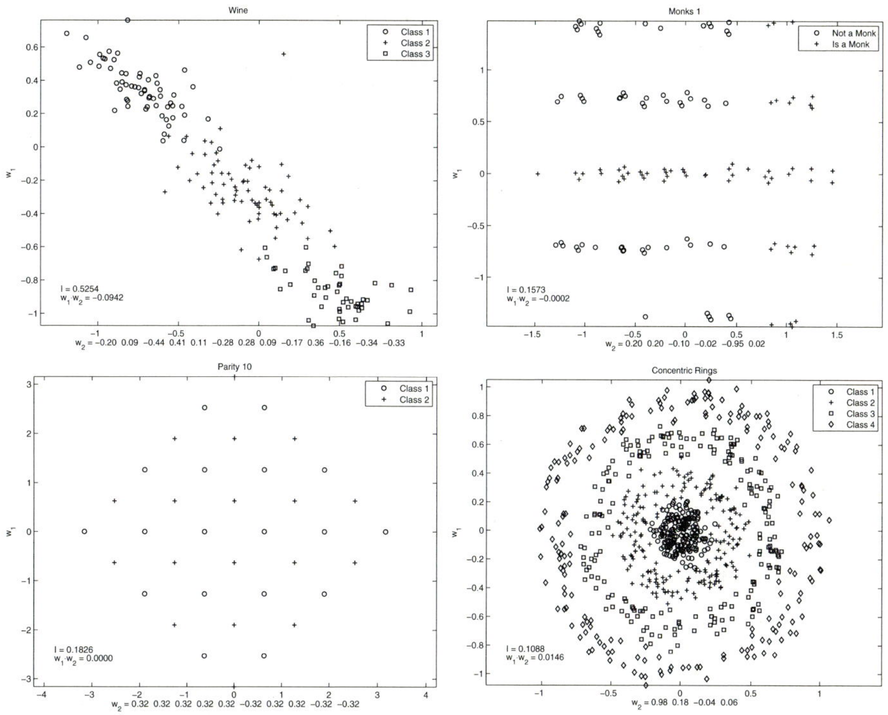

Fig. 2. Scatterplots created by projection on two directions for Wine and Monk 1 data (top-left/right), 10-bit parity and the noisy Concentric Rings data (bottom-left/right)

4 Constructive Neural Network

A single projection allows for estimation and drawing class-conditional and posterior probabilities, but may be not sufficient for optimal classification. Projection on 2 or 3 dimensions allows for visualization of scatterograms, showing data structures hidden in the high-dimensional distributions, suggesting how to handle the problem in a simplest way: adding linear output layer (Wine), localized functions, using intervals (parity), or nearest neighbor rule (Concentric Rings). Reduction to higher number of dimensions will be useful as a pre-processing for final classification. Coefficients of the projection vectors may be used directly for feature ranking and feature selection models, because maximization of the QPC index gives negligible weights for noisy or insignificant features, while important attributes have distinctly larger values. This method might be used to improve learning for many machine learning models sensitive to feature weighting, such as kNN. Interesting projections may also be used to initialize weights in various neural network architectures.

The QPC index Eq. (1) defines specific representation for the hidden layer of a constructive neural network. This may be used in several ways to construct nodes of the network. In most cases, projections that maximize $QPC(\boldsymbol{w})$ contain at least one large cluster. The center of this cluster can be directly estimated during maximization of the $Q(\boldsymbol{x}; \boldsymbol{w})$ index, because it is associated with some vector $\boldsymbol{x}_i \in \mathcal{X}$ that gives the maximum value of $Q(\boldsymbol{x}_i; \boldsymbol{w})$ in Eq. (1). Consider the node M implementing the following function:

$$M(\boldsymbol{x}) = \begin{cases} 1 & \text{if} \quad |G(\boldsymbol{w}(\boldsymbol{x}-\boldsymbol{t})) - \theta| \geq 0 \\ 0 & \text{otherwise} \end{cases} \tag{6}$$

where the weights $\boldsymbol{w}$ are obtained by maximization of the QPC index, and $\boldsymbol{t}$ is the center of cluster associated with maximum $Q(\boldsymbol{x}; \boldsymbol{w})$. This node splits input space into two disjoint subspaces, with output 1 for each vector that belongs to the cluster and 0 for all other vectors. It is fairly easy to solve the parity-like problems with such nodes, summing the output of all nodes that belong to the same class.

Further adjustments of weights and center of the cluster can enlarge the cluster and give better separation between classes. This can be done by maximizing $Q(\boldsymbol{t}; \boldsymbol{w})$ with respect to weights $\boldsymbol{w}$ and cluster center $\boldsymbol{t}$, or by minimization of an error function:

$$E(\boldsymbol{x}) = E_{\boldsymbol{x}} || G\left(\boldsymbol{w}(\boldsymbol{x}-\boldsymbol{t})\right) - \delta(c_{\boldsymbol{x}}, c_{\boldsymbol{t}}) || \tag{7}$$

where $\delta(c_{\boldsymbol{x}}, c_{\boldsymbol{t}})$ is equal to 1 when $\boldsymbol{x}$ belongs to the class associated with cluster with center $\boldsymbol{t}$, and 0 if not (error functions that are more sensitive to a number of separated vectors and purity of solution have been considered [2]).

This method has twice as many parameters to optimize (weights and center), but computational cost of calculation of function $Q(\boldsymbol{t}; \boldsymbol{w})$ is linear in the number of vectors $O(n)$, and since only a few iterations are needed this part of learning is quite fast. Final neuron should give good separation between the largest possible group of vectors with the same labels, and the rest of the dataset. In the general sequential constructive method [16] next node is trained only on vectors that have not been correctly handled by previous nodes. This leads in a finite number of steps to a neural network which classifies all samples of a given multiclass dataset, where each neuron derived by this method is placed in hidden layer, and weights in the output layer are determined from a simple algebraic equation (for details see [16]). Although we do not have space here to report detailed results they are an improvement over already excellent results obtained in [2] and similar to [7].

5 Discussion

Projection pursuit networks that reduce dimensionality and use clustering, such as the QPC networks described in this paper, are able to find the simplest data models (including logical rules) in case of quite diverse and rather complex data. They create interesting features, allowing for visualization and classification of data. Should such networks be called neural? Inspirations for the projection pursuit networks should be searched at a higher level than single neurons. Non-local projections that form low-dimensional clusters may be realized by neural columns that are activated by linear combinations of incoming signals, learn to remember vectors that give similar projections, and learn better weights to increase their excitations, inhibiting at the same time competing columns.

Different mechanisms may then be used to extract interesting transformation from such columns, reducing noise in data, selecting relevant information, learning to estimate similarity of responses. A column may learn to react to inputs of specific intensity, solving complex logical problem by clustering data in low-dimensional projections that are not linearly separable.

Linear separability is not the best goal of learning. The QPC index helps to solve problems that go well beyond capabilities of standard neural networks, such as the parity or the noisy Boolean functions. The class of PP networks is quite broad. One can implement many transformations in the hidden layer, explicitly creating hidden representations that are used as new inputs for further network layers, or used for initialization of standard networks. Brains are capable of deep learning, with many specific transformations that lead from simple contour detection to final invariant object recognition. Studying PP networks should bring us a bit closer to powerful methods for deep learning.

References

1. Duch, W.: k-separability. In: Kollias, S.D., Stafylopatis, A., Duch, W., Oja, E. (eds.) ICANN 2006. LNCS, vol. 4131, pp. 188–197. Springer, Heidelberg (2006)
2. Grochowski, M., Duch, W.: Learning highly non-separable Boolean functions using Constructive Feedforward Neural Network. In: de Sá, J.M., Alexandre, L.A., Duch, W., Mandic, D.P. (eds.) ICANN 2007. LNCS, vol. 4668, pp. 180–189. Springer, Heidelberg (2007)
3. Haykin, S.: Neural Networks - A Comprehensive Foundation. Maxwell MacMillian Int., New York (1994)
4. Friedman, J.: Exploratory projection pursuit. Journal of the American Statistical Association 82, 249–266 (1987)
5. Jones, C., Sibson, R.: What is projection pursuit. Journal of the Royal Statistical Society A 150, 1–36 (1987)
6. Fisher, R.: The use of multiple measurements in taxonomic problems. Annals of Eugenics 7, 179–188 (1936)
7. Sugiyama, M.: Dimensionality reduction of multimodal labeled data by local fisher discriminant analysis. Journal of Machine Learning Research 8, 1027–1061 (2007)
8. Duch, W.: Similarity based methods: a general framework for classification, approximation and association. Control and Cybernetics 29, 937–968 (2000)
9. Duch, W., Adamczak, R., Diercksen, G.: Classification, association and pattern completion using neural similarity based methods. Applied Mathemathics and Computer Science 10, 101–120 (2000)
10. Duch, W., Jankowski, N.: Survey of neural transfer functions. Neural Computing Surveys 2, 163–213 (1999)
11. Duch, W., Jankowski, N.: Transfer functions: hidden possibilities for better neural networks. In: 9th European Symposium on Artificial Neural Networks, pp. 81–94. De-facto publications, Brusells (2001)
12. Shakhnarovish, G., Darrell, T., Indyk, P. (eds.): Nearest-Neighbor Methods in Learning and Vision. MIT Press, Cambridge (2005)
13. Kordos, M., Duch, W.: Variable Step Search MLP Training Method. International Journal of Information Technology and Intelligent Computing 1, 45–56 (2006)
14. Asuncion, A., Newman, D.: UCI machine learning repository (2007)

15. Duch, W., Adamczak, R., Grąbczewski, K.: A new methodology of extraction, optimization and application of crisp and fuzzy logical rules. IEEE Transactions on Neural Networks 12, 277–306 (2001)
16. Muselli, M.: Sequential constructive techniques. In: Leondes, C. (ed.) Optimization Techniques. Neural Network Systems, Techniques and Applications, vol. 2, pp. 81–144. Academic Press, San Diego (1998)

Introduction to Constructive and Optimization Aspects of SONN-3

Adrian Horzyk

University of Science and Technology, Department of Automatics
Mickiewicza Av. 30, 30-059 Cracow, Poland
horzyk@agh.edu.pl
http://home.agh.edu.pl/~horzyk

Abstract. The paper introduces new valuable improvements of performance, a construction and a topology optimization of the Self-Optimizing Neural Networks (SONNs). In contrast to the previous version (SONN-2), the described SONN-3 integrates the very effective solutions used in the SONN-2 with the very effective ADFA algorithms for an automatic conversion of real inputs into binary inputs. The SONN-3 is a fully constructive ontogenic neural network classificator based on a sophisticated training data analysis that quickly estimates values of individual real, integer or binary input features. This method carries out all computation fully automatically from a data analysis and a data input dimension reduction to a computation of a neural network topology and its weight parameters. Moreover, the SONN-3 computational cost is equal $O(nlog^2n)$, where n is a sum of a data quantity, a data input dimension and a data output dimension. The results of the SONN-3 construction and optimization are illustrated and compared by means of some examples.

Keywords: Binary factorization, automatic conversion, discrimination, artificial intelligence, factorization neural networks.

1 Introduction

Nowadays, various types of constructive neural networks and other incremental learning algorithms play even more important role in neural computations. These algorithms usually provide an incremental way of building neural networks with reduced topologies for classification problems. Furthermore, these neural networks produce a multilayer topology, which together with the weights, are determined automatically by the constructing algorithm and thus avoid the search for finding a proper neural network architecture. Another advantage of these algorithms is that convergence is guaranteed by the method [1],[3],[4],[8],[9]. A growing amount of current research in neural networks is oriented towards this important topic. Providing constructive methods for building neural networks can potentially create more compact models which can easily be implemented in hardware and used on various embedded systems.

This paper introduces a new extended and generally improved version (SONN-3) of the constructive ontogenic SONNs, that are not only able to develop a network architecture and compute weights fully automatically but also to optimize

V. Kůrková et al. (Eds.): ICANN 2008, Part II, LNCS 5164, pp. 763–772, 2008.

an input data dimension, automatically convert any integer and real data into the discriminative bipolar binary vectors necessary for further SONN-3 computations. The previous version (SONN-2) was not be able to use integer and real input vectors at all. The SONN-3 can automatically use various inputs (boolean, integer, real) thanks to the ADFA conversion algorithms. This automatic conversion makes possible to optimize a final SONN-3 network architecture even more than it was possible using the SONN-2. The manual conversion necessary for the SONN-2 was usually not optimal nor it can assure discrimination properties of the converted data. Moreover, the simple manual factorization usually does not cover input data space sufficiently, spoils a quality of generalization and wastes potential of the further SONN-2 construction process. The described SONN-3 is free from these limitations. Many neural methods compete for a better generalization using various training strategies and parameters, neuron functions, various quantities of layers, neurons and interconnections. This paper confirms that generalization potential is also hidden in data preprocessing and a wideness of input data space that neurons can cover and suitably represent. The introduced SONN-3 can be successfully used to various classification tasks. The mentioned thesis is illustrated by means of some training data samples from MLRepository (e.g. fig. 2.-3).

2 The SONN-3 Construction Process and Elements

The SONN-3 construction process starts from a sophisticated training data (TD) analysis. First, the SONN-3 examines all input features of all training samples and then automatically factorizes integer and real inputs into bipolar binary vectors not loosing discrimination properties of input data. Next, it counts and estimates all factorized input features and starts a process of a neural network architecture development and an optimal weights calculation. The SONN-3 tries to compress input data features with the same discrimination coefficients values (7) for a selected subgroup of training samples by grouping them together and by transforming same features into single connections. The compression process is associated with the process of training samples divisions into subgroups that are used to create neurons. The described algorithm creates still smaller subgroups of training samples unless all training samples are correctly discriminated.

The SONN-3 consists of three types of neurons (fig. 1) that play a very important role in an input data transformation into a final classification achieved on network outputs. These neurons are arranged in layers. The number of layers is dependent of correlation between training samples. If TD of various classes are more correlated then a neural network architecture is also more complicated and has got more layers and neurons and *vice versa.* If there are little correlations between same class training samples there are more neurons in layers and *vice versa.* The SONN-3 outputs determine the similarity of given inputs to classes defined in TD. Many times some input features turn out to be redundant. The SONN-3 can automatically reduce the minor input features and a data input dimension as well (fig. 2-3). As a result, the SONN-3 qualifies some groups of

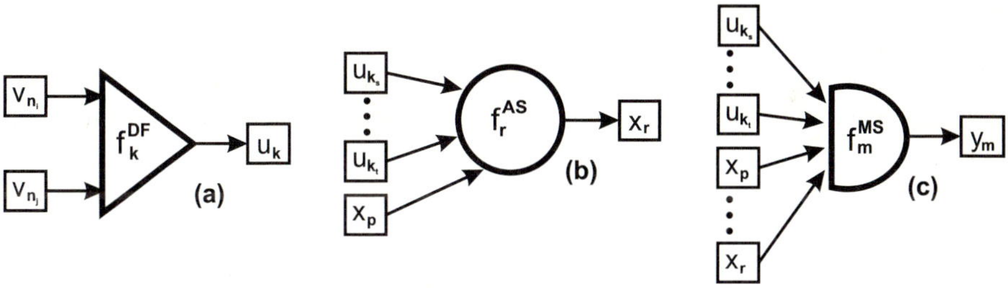

Fig. 1. Three types of SONN-3 neurons: (a) Discriminatively-Factorization Neuron (DFN) (b) Aggregation-Strengthening Neuron (ASN), (c) Maximum-Selection Neuron (MSN)

inputs to be more important and pays no attention to the inputs which are less or by no means important for a classification.

A very important part of a data transformation is processed by Aggregation-Strengthening Neurons (ASNs) (fig. 1b) placed in the middle part of the SONN-3 architecture (fig. 2-3). These neurons demand bipolar binary inputs during their adaptation process. They aggregate same inputs together (without loosing any information) and strengthen these of them that better discriminate training samples in-between various classes. The strengthening factors are called global discrimination coefficients (7). The ASNs always produce their outputs in the range of $[-1;+1]$ and are interconnected to next ASNs in such a way to propagate the sum of discrimination coefficients values of all previous connections to next layers (2). The discrimination coefficients values directly influence weights values (3)-(4) of the network. In this way, the proper strengthening is appropriately promoted and propagated through a network without a discrimination information lost.

$$x_r = f_r^{AS}(u_{k_s}, ..., u_{k_t}, x_p) = w_0^{x_p} x_p + \sum_{j \in \{k_s,...,k_t\}} w_j^{x_r} u_j \quad (1)$$

where

$$d_0^{AS_p} = \sum_{j \in J} d_j \quad (2)$$

$$w_0^{AS_p} = \frac{d_0^{AS_p}}{\sum_{j \subset \{k_s,...,k_t\}} d_j} \quad (3)$$

$$w_j^{AS_r} = \begin{cases} \frac{u_i^n d_i^+}{\sum_{j \in \{k_s,...,k_t\}} d_j} & if\ u_i^n \geq 0 \\ \frac{u_i^n d_i^-}{\sum_{j \in \{k_s,...,k_t\}} d_j} & if\ u_i^n < 0 \end{cases} \quad (4)$$

The bipolar binary inputs necessary for an ASNs adaptation are computed by Discriminatively-Factorization Neurons (DFNs) (fig. 1a) that can automatically factorize integer and real TD inputs in such a way that existing discriminative TD properties are not lost after this transformation. After the conversion, the input space of raw TD data is represented enough to discriminate all TD from various classes if only TD are not contrary. If simultaneously all associated inputs v_t lie in appropriate factorization ranges $[L_m^t; P_m^t]$ of a considered DFN then it produces the value +1 on its output and the value −1 otherwise (5).

$$f_k^{DF}(v_{n_i}, ..., v_{n_j}) = \begin{cases} 1 & if \bigwedge_{t=n_i}^{n_j} v_t^{L_m^t} \leq v_t \leq v_t^{P_m^t} \\ -1 & otherwise \end{cases} \quad (5)$$

The maximum similarity to each class is computed by Maximum-Selection Neurons (MSNs) (fig. 1c) that are used to select a maximal value from all connected ASNs that represent aggregated most important input features of training samples of appropriate same classes (6). There could be sometimes connections between factorized inputs and these neurons (fig. 3) for some trivial or very correlated training samples of same classes.

$$y_m = f_m^{MS}(u_{k_s}, ..., u_{k_t}, x_{n_i}, ..., x_{n_j}) = max\{ u_{k_s}, ..., u_{k_t}, x_{n_i}, ..., x_{n_j}\} \quad (6)$$

The above described three types of neurons are used to construct a partially connected multilayer neural network topology that can be precisely adjusted to whatever training data set and a classification task. Such adjustment is similar to the plasticity [5] processes that take place in natural nervous systems [10].

The SONN-3 is an extended and generalized version of the SONN-2 and the SONN-1 widely described and discussed in many papers (e.g. [4],[8]). The most important aspect of the SONN-3 is the ability to automatically factorize integer and real training data and thanks this to achieve a more optimized and better representative neural network topology and better generalization results. These are possible because the automatic factorization algorithm (ADFA) better covers an input data space and makes possible to represent a more important TD information in a neural network in a more consistent form.

The main goal of the ADFA-1 is to find out a possibly minimal set of factorization ranges for all integer and real data input features and to convert the values from these ranges into binary unipolar $\{0; 1\}$ or bipolar $\{-1; +1\}$ values. First, the real data inputs have to be sorted separately for each data input feature in order to start the computation of these ranges. There is preferred the heapsort algorithm which computational cost is always $O(nlogn)$. The stability of the sorting algorithm does not matter.

After all input data features are sorted, the ADFA-1 starts to search for optimal factorization ranges taking into account the following two criteria:

1. the optimal range should contain as many samples of the same class as possible and no samples of other classes (i.e. a maximum quantity of samples of the same class is looked for),
2. the optimal range containing maximum samples of the same class (there can be more than one) should be as wide as possible and be placed as far as possible from data samples of other classes (i.e. a maximum size of the range and its maximum distances from values of samples of other classes are looked for).

These two criteria are important to satisfy the requirements mentioned at the beginning of this paper, especially in a view of good generalization properties. The ADFA-1 is looking for those optimal ranges in the following way:

1. All data samples are marked as not discriminated at the beginning.
2. The ADFA-1 is looking through all input features of all yet not discriminated data samples in the sorted order and is looking for the range that contains a maximum quantity of training samples of the same class (to meet the criterion 1 described above). If there are two or more ranges that meet this criterion, there is selected this one which meets the criterion 2 described above. Each optimal range is described by the minimal and maximal values and the feature for which this range has been found. This input feature is called the winning input feature here.
3. All not discriminated samples which winning input feature is in the chosen optimal range are marked as discriminated.
4. Next, all input data features except the winning one are looked through in order to remove the indexes from the sorted index tables if only these indexes point out the already discriminated data samples.
5. The steps 2, 3 and 4 are repeated until all training data samples are discriminated and the sorted index tables for all input data features are empty or there can not be fixed any new range of not discriminated samples of the same class.

If there can not be found any new range and some training samples left not to be discriminated it means that TD are contrary or they can be discriminated only by some combination of various feature ranges. The combinations can be constructed by the ADFA-M (that will be introduced in a not too distant future). On the other hand, the ADFA-1 can successfully process a large majority of TD and is much faster than the ADFA-M.

Moreover, the SONNs require all data available at the beginning of the adaptation process, so they can process various global optimization processes. In a result, the SONNs can globally estimate importance of each input feature for all training samples and adapt the network structure and weights reflecting these estimations. The importance of each k-th input feature for each n-th training sample u_n of the m-th class C^m is globally measured by the use of the defined so called global discrimination coefficients (7):

$$\forall_{m\in\{1,...,M\}}\forall_{u^n\in C^m}\forall_{n\in\{1,...Q\}}\forall_{k\in\{1,...K\}}:$$
$$d^n_{k+} = \begin{cases} \frac{P^m_k}{(M-1)\cdot Q^m}\sum_{h=1\wedge h\neq m}^{M}\left(1-\frac{P^h_k}{Q^h}\right) & if\ u^n_k \geq 0 \wedge u^n \in C^m \\ 0 & if\ u^n_k < 0 \wedge u^n \in C^m \end{cases}$$
$$d^n_{k-} = \begin{cases} \frac{N^m_k}{(M-1)\cdot Q^m}\sum_{h=1\wedge h\neq m}^{M}\left(1-\frac{N^h_k}{Q^h}\right) & if\ u^n_k \leq 0 \wedge u^n \in C^m \\ 0 & if\ u^n_k > 0 \wedge u^n \in C^m \end{cases} \quad (7)$$

where M denotes a number of trained classes, Q is a number of all training samples, K is a number of factorized input features, u^n_k is the k-th feature value for the n-th training sample and P_k^m, N_k^m, Q^m are defined by the following formulas:

$$\forall_{m\in\{1,..,M\}}\ Q^m = \left\| \left\{ u^n \in U \bigcap C^m : n \in \{1,...Q\} \right\} \right\| \tag{8}$$

$$\forall_{m\in\{1,...,M\}} \forall_{k\in\{1,...,K\}} : P_k^m = \sum_{u_k^n \in \left\{ u\in U \vee C^m : u_k^n > 0,\ n\in\{1,...,Q\} \right\}} u_k^n \tag{9}$$

$$\forall_{m\in\{1,...,M\}} \forall_{k\in\{1,...,K\}} : N_k^m = \sum_{u_k^n \in \left\{ u\in U \vee C^m : u_k^n < 0,\ n\in\{1,...,Q\} \right\}} -u_k^n \tag{10}$$

Those discrimination coefficients (7) qualify the discrimination property of each k-th feature for the n-th training sample of the m-th class and have the following properties:

- they are insensitive for a different quantity of training samples that represent various classes,
- a discrimination property of each feature of each training sample is globally estimated,

The SONN topology optimization process is so designed to compute a minimal network structure using only these features that have maximal discrimination properties for each class. The SONN-2 topology and the middle part of the SONN-3 topology are created in a specific process of a TD division and a features aggregation. Each division produces a single neuron that represents a subgroup of training samples that meets a division criterion. The selected training samples represented by a neuron always deal at least one same discrimination coefficient value. They are computed in such a way to maximize a quantity of aggregated same obligatory discrimination coefficients by single connections. Neurons produce a simple weighted sum output (1) but the weight of inter-neuron connection (3) is so computed to reflect the sum of all discrimination coefficients values represented by all input connections of this neuron (2). This makes possible to keep the influence on a classification of all inputs at a level determined by the values of their corresponding discrimination coefficients. This assures appropriate influence on a resultant classification and provides a very good generalization.

The SONN-3 topology optimization process is based on discrimination coefficients (7) computed for all training data samples as well as for some subgroups of them. In order to find an optimal SONN-3 topology there is necessary to create only these connections which are necessary to classify correctly and unambiguously all training data using only the input features with maximal discrimination properties [5],[8]. Weights parameters and a topology are computed simultaneously during the SONN-3 construction process [4],[8]. Such strategy makes possible to precisely assign each feature representing a subgroup of training samples to an accurate optimal weight value (3)-(4) arising from its discrimination properties. It is fundamental in decision processes and a classification. The SONN classification results have been compared with the other AI methods (tab. 1-2) for the Wine and Iris data from MLRepository. The comparison results show that the SONN-3 is not only very fast [4],[8] but also it achieves very good generalization properties in the comparison with the other popular AI classification methods. There is not many other neural network training algorithms that can directly and optimally fit a topology and all weights to all training samples after a global information about them. Moreover, the SONNs have many interesting features that can be compared with biological neural networks and various neural processes in biological brains [5].

Moreover, the SONN-3 uses two kinds of information for the discrimination: an existence and a non-existence of some input values for some classes (7). The second kind of information is rarely used by other AI models. The majority of AI models and methods use only the information about an existence of some input values in some classes. The SONN methodology expands these abilities and supplies better possibilities for a generalization.

TD and relations between trainings samples of a same class can be very various. The described SONN methodology does not find dependencies between data different features but uses their given values to probabilistically estimate their discriminative properties (7) individually. The SONN-3 topology puts together the most important information about discriminative properties of different features constructing a uniform classification model for any given TD. The SONN-3 is always built up after the most important, well-differentiating and discriminating features of all training samples. The SONN-3 also automatically excludes artifacts of data bacause it focuses on the most discriminative features which artifacts are not. The SONN generalization property reflects the most important and characteristic information of individual classes for any given TD set.

3 Experiments and Comparisons

The SONN-3 has been applied to many classification problems in order to compare this method with other AI methods. The figures 2-3 and the tables 1-2 illustrate the sample architectures and compares various methods applied to the same TD sets from MLRepository.

The Wine data have been divided into disjunctive 101 training samples and 77 test samples and the Iris data into 75 training samples and 75 test samples in order to compare classification results of the various AI classification methods. GhostMiner 3.0 was used for developing and evaluation some AI models with the default training parameters. First, all classification methods have been trained and adapted to fit the training samples and then all methods have been tested. The comparisons of the classification results is shown in the tables 1-2. Moreover, the figure 3 shows the comparison of the SONN-3 network architectures automatically created for all and for 101 selected training Wine samples.

The presented samples from the figures 2-3 illustrate that:

- The original input data dimension have been really reduced in the both examples: There have been used only 3 from 4 available inputs for the Iris data and only 5 from 13 available inputs for the Wine data.

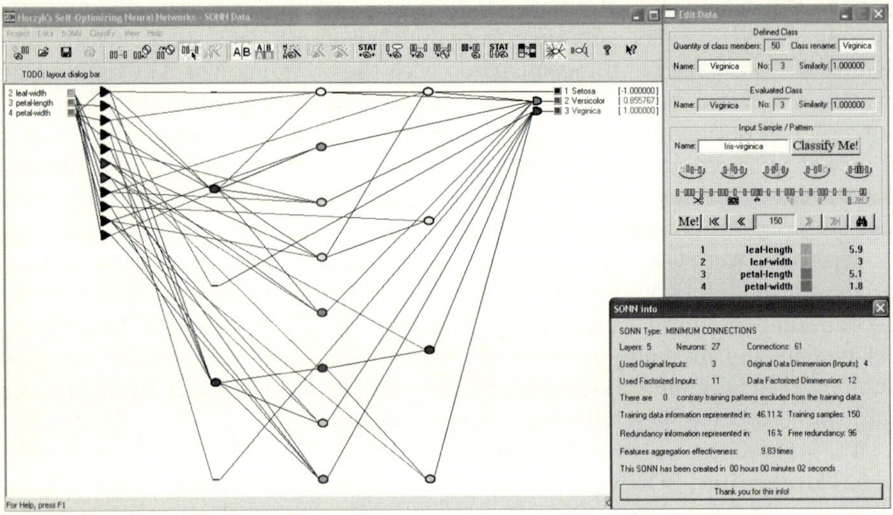

Fig. 2. The SONN-3 constructed automatically for the Iris data

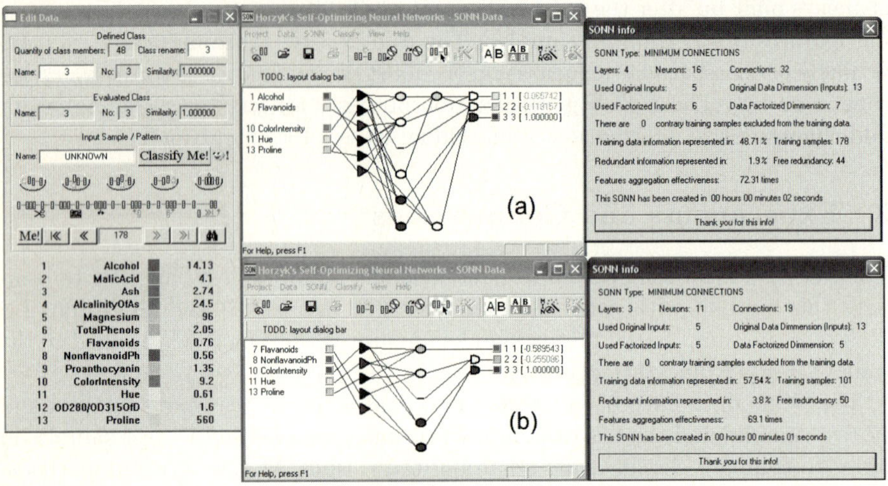

Fig. 3. The comparison of SONNs-3 constructed automatically for (a) all Wine samples and (b) for 101 selected training Wine samples from MLRepository

– The factorized input dimension for the ASNs have been also reduced in the both examples: There have been used only 11 from 12 available inputs for the Iris data and only 6 from 7 available inputs for the Wine data.
– The selected reduction of the original input dimensions and the factorized input dimensions have not spoil the discrimination properties of the processed data and have enabled the correct classification of them.

– In spite of the creation of only a few neurons and connections for the both examples all training data and almost all test data have been correctly classified.
– The whole process of the neural networks development and the optimization and the weights computation have taken 1-2 seconds.
– The generalization properties of the created neural networks are at the top level in the comparison with the other AI methods (Table 1).

Table 1. The comparison of the Wine data classification results

DATA	TRAIN RESULTS		TEST RESULTS		TOTAL	
Wine: 101 : 77	correct	errors	correct	errors	correct	errors
FSM	101	0	76	1	177	1
SONN-3	101	0	72	5	173	5
SSV Tree	98	3	61	16	159	19
k-Nearest Neighbors	101	0	57	20	158	20
IncNet	73	28	34	43	107	71
SVM	101	0	1	76	102	76

Table 2. The comparison of the Iris data classification results

DATA	TRAIN RESULTS		TEST RESULTS		TOTAL	
Iris: 75 : 75	correct	errors	correct	errors	correct	errors
k-Nearest Neighbors	75	0	72	3	147	3
SONN-3	75	0	71	4	146	4
IncNet	74	1	72	3	146	4
FSM	74	1	71	4	145	5
SSV Tree	73	2	71	4	144	6
SVM	72	3	71	4	143	7

4 Conclusions

The paper deals with the fully automatic construction of the universal ontogenic neural network classificator SONN-3 that is able to automatically adapt itself using binary, integer and real data without any limitations. It analyses and processes TD very fast and finds out the optimal SONN-3 architecture and weights for any given TD set. It can also optimize an input data dimension - automatically limiting a number of necessary inputs - what can have a special value in many practical uses. Not many methods can effectively select most discriminative inputs, so many times classification results are influenced under minor or not important parameters that can also spoil classification results and generalization properties of a finally adapted neural network. The SONN-3 reduces original real inputs to a set of the best discriminating inputs and then reduces binary bipolar inputs in the second phase of computations as well what can be noticed in the neural networks created for the Wine and Iris data shown in the fig. 2-3. Finally, the SONN-3 is very fast, cost-effective and fully automatic. On the other hand a computer implementation of this method is not easy because of a huge number of optimization algorithms that gradually analyze TD, transform them, develop a final neural network topology and compute its weights. The interactive web with SONN-3 will be publish soon at the site http://home.agh.edu.pl/˜horzyk.

References

1. Duch, W., Korbicz, J., Rutkowski, L., Tadeusiewicz, R. (eds.): Biocybernetics and Biomedical Engineering, EXIT, Warszawa (2000)
2. Dudek-Dyduch, E., Horzyk, A.: Analytical Synthesis of Neural Networks for Selected Classes of Problems. In: Bubnicki, Z., Grzech, A. (eds.) Knowledge Engineering and Experts Systems, pp. 194–206. OWPN, Wroclaw (2003)
3. Fiesler, E., Beale, R.: Handbook of Neural Computation. IOP Publishing Ltd. and Oxford University Press, Bristol (1997)
4. Horzyk, A.: A New Extension of Self-Optimizing Neural Networks for Topology Optimization. In: Duch, W., Kacprzyk, J., Oja, E., Zadrożny, S. (eds.) ICANN 2005. LNCS, vol. 3696, pp. 415–420. Springer, Heidelberg (2005)
5. Horzyk, A., Tadeusiewicz, R.: Comparison of Plasticity of Self-Optimizing Neural Networks and Natural Neural Networks. In: Mira, J., Alvarez, J.R. (eds.) Proc. of ICANN 2005, pp. 156–165. Springer, Heidelberg (2005)
6. Horzyk, A.: Interval Basis Neural Networks. In: Ribeiro, B., Albrecht, R.F., Dobnikar, A., Pearson, D.W., Steele, N.C. (eds.) Adaptive and Natural Computing Algorithms, pp. 50–53. Springer, New York (2005)
7. Horzyk, A.: Interval Basis Neural Networks as a New Classification Tool. In: Proc. of Int. Conf. Artificial Intelligence and Applications, pp. 871–876. ACTA Press, Anaheim Calgary Zurich (2005)
8. Horzyk, A., Tadeusiewicz, R.: Self-Optimizing Neural Networks. In: Yin, F., Wang, J., Guo, C. (eds.) Advances in Neural Networks - ISNN 2004, pp. 150–155. Springer, Berlin Heidelberg (2004)
9. Jankowski, N.: Ontogenic neural networks. EXIT, Warszawa (2003)
10. Kalat, J.: Biological Psychology. Thomson Learning Inc., Wadsworth (2004)

A Reward-Value Based Constructive Method for the Autonomous Creation of Machine Controllers

Andreas Huemer[1], David Elizondo[2], and Mario Gongora[2]

[1] Institute Of Creative Technologies, De Montfort University, Leicester, UK
[2] Centre for Computational Intelligence, De Montfort University, Leicester, UK
{ahuemer,elizondo,mgongora}@dmu.ac.uk

Abstract. A novel method for the creation of machine controllers autonomously is presented. The method is based on reward values which can represent the internal state of the machine or the rating of its task performance. The method consists of a biologically sound constructive neural network model with a minimum number of neurons and no connections initially. New connections and neurons are added when feedback is fed into the neural network using positive and negative reward values. This way the topology and the level of connectivity of the network are kept to a minimum. The method will be applied to a controller for an autonomous mobile robot.

Keywords: Constructive Neural Network, Spiking Neural Network, Reinforcement Learning, Growing Machine Controllers.

1 Introduction

Machines make decisions based on sensory input that bring them closer to certain goals. It has been shown that artificial neural networks can learn how to select adequate actions [1]. Neural networks can also be used for classification tasks [9]. In many cases the classes themselves are wanted as a result. In a machine controller classification those classes can also be used to reduce the size of the network that learns the action selection task.

A novel method, which does not only include the classification task and the action selection task into a single neural network is presented. It is also capable of defining the classes autonomously and it can grow new neurons and connections to learn to select correct actions from experience based on reward values. Positive and negative reward values build the feedback, which can be generated internally, without human intervention. Alternatively it is possible that a human feeds the controller with feedback during runtime acting as a teacher.

The initial neural network consists of a minimum number of neurons and no connections. The designer of the controller does not need to take care of the topology of the network. Only the interface from the controller to the machine has to be defined, which reduces the design effort considerably.

V. Kůrková et al. (Eds.): ICANN 2008, Part II, LNCS 5164, pp. 773–782, 2008.

The interface includes:

- *Input neurons*, which convert sensory input into values that can be sent to other neurons. For our experiments we used spiking neurons, so the sensory values are converted into spikes by the input neurons.
- *Output neurons*, which convert the output of the neural network into values that can be interpreted by the actuators.
- *A reward function*, which feeds feedback into the machine controller.

This paper first shows the characteristics of the neural network that is used for the novel methods to grow machine controllers autonomously in section 2. Section 3 discusses the crucial issue of how feedback can be defined, how it is fed into the neural network and how it is used for basic learning strategies. Section 4 shows how to use the feedback for the more sophisticated methods of growing new neurons and connections. A simulation of a mobile robot was used to test the novel methodology. The robot should learn to wander around in a simulated environment avoiding obstacles, such as walls. The results are presented in section 5. Section 6 draws some conclusions and future prospects.

2 Controller Characteristics

Our growing machine controller consists of a layered spiking neural network as illustrated in Fig. 1.a, in which neurons send Boolean signals that transport the information depending on whether or not the presynaptic neuron (the first of two connected neurons) was activated and has fired (a spike). The hidden layer B contains two neurons that store input combinations. One of them excites dendrite a via axon d, which itself excites an output neuron. The neuron in layer C inhibits the same output neuron by axon c. Also axon b inhibits that neuron but implements local inhibition. A neuron is activated when a certain threshold potential is exceeded. The neuron potential is increased by spikes arriving at excitatory connections and decreased by spikes arriving at inhibitory connections. A basic explanation of spiking neural networks can be found in [12].

The use of sparse neural networks was has been discussed in [3] and [5]. These models result in less computation requirement and better development of the network as well as smaller topologies. The main reason for obtaining smaller topologies when using classification is obvious:

If there are no neurons, which represent a certain class, all neurons of this class have to connect to the next layer in the network separately and not with a single connection. There is no problem if one neuron connects only to one neuron. An additional connection in this case is required. However, when a neuron connects to a second neuron, only one additional connection is made instead of more connections from all neurons of a class. Also the total number of neurons is reduced if the neurons are representing the possible combinations of neurons in the previous layer, because in a succeeding layer only the class neurons have to be considered.

To achieve an efficient topology along with action selection and classification, in a single neural network, we separate the connections into two parts: artificial dendrites and axons. An axon of the presynaptic neuron is connected to

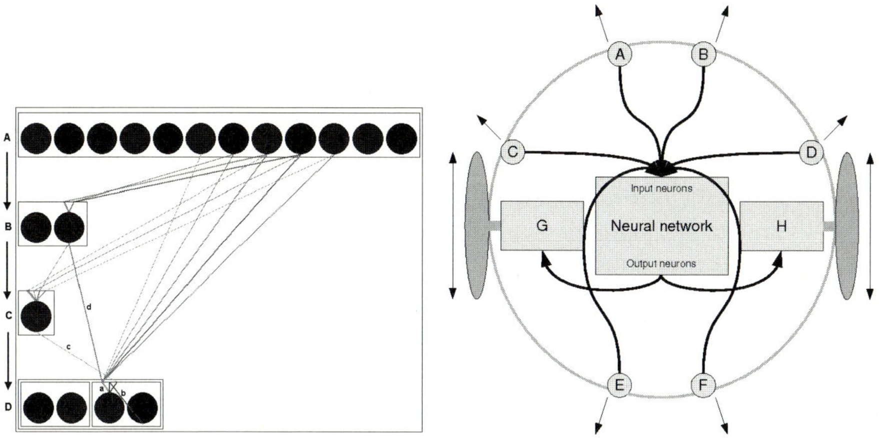

(a) (b)

Fig. 1. (a) Neural network topology after a short period of a simulation run. For clarity not all connections are shown. Layer A consists of the input neurons which are connected to sensors A to F from the robot shown in (b). Layer D contains the motor neurons, connected to G and H in (b).

a dendrite of the postsynaptic neuron. Excitatory connections have to be used for operations that are similar to the logical AND and OR. For inhibitory connections this separation is not necessary, because they represent an operation similar to the logical NOT.

A dendrite has got a low threshold potential and is activated when only one presynaptic neuron (or a few neurons) have fired a spike via their axons. All presynaptic neurons are handled equally at this point (logical OR) and represent neurons which are combined into one class. An axon weight defines how much one presynaptic neuron belongs to the class.

In the following we want to show the computations when signals travel from one neuron to another.

Input of a dendrite:

$$I_d = \sum O_{a+} \cdot w_{a+} \tag{1}$$

where I_d is the dendrite's input. O_{a+} is the output of an excitatory axon, which is 1, if the presynaptic neuron has fired and 0 otherwise. w_{a+} is the weight of the same excitatory axon, which has to be a value between 0 and 1.

Output of a dendrite:

$$O_d = \frac{1}{1 + e^{-b \cdot (I_d - \theta_d)}} \tag{2}$$

where O_d is the dendrite's output and I_d is its input. θ_d is a threshold value for the dendrite. b is an activation constant and defines the abruptness of activation.

Input of a neuron:

$$I_j = \sum O_d \cdot w_d - \sum O_{a-} \cdot w_{a-} \tag{3}$$

where I_j is the input of the postsynaptic neuron j, O_d is the output of a dendrite, w_d is the weight of this dendrite, O_{a-} is the output of an inhibitory axon and w_{a-} is the weight of this inhibitory axon. Dendrite weights and axon weights are in the range $[0, 1]$ and all dendrite weights sum up to 1.

Change of neuron potential:

$$P_j(t+1) = \delta \cdot P_j(t) + I_j \tag{4}$$

where the new neuron potential $P_j(t+1)$ is calculated from the potential of the last time step t, $P_j(t)$, and the current contribution by the neuron input I_j. δ is a constant between 0 and 1 for recovering to the resting potential (which is 0 in this case) with time.

The postsynaptic neuron is activated when its potential reaches the threshold θ_j and becomes a presynaptic neuron itself for neurons which its own axons are connected to. After firing the neuron resets its potential to the resting state. In contrast to similar neuron models that are for example summarised by Katic [8], a refractory period is not implemented here.

3 Feedback and Reward Values

The main challenge for a designer of a machine controller that uses the methods described in this paper to evolve itself, is to define an appropriate reward function. Positive and negative reward values are fed into the neural network as explained below and are used for all adaptation processes, like basic learning (adaptation of connection weights), creating new connections and creating new neurons (growing mechanisms).

In our experiments we have used a single global reward value, which represents a positive “rewarding“ or negative “aversive” feedback relative to the machine's performance in the task that has been assigned. The objective is to create a network which maximises positive feedback.

Depending on current measurements like fast movement, crashes, recovering from crashes, or the energy level the reward value is set from -1, very bad, to 1, very good. The challenge is to provide a useful value in all situations. For example, as experiments have shown (see Sect. 5), a reward function that is not capable of providing enough positive feedback may result in a hang-up of the machine, because all of its effort to find a good action is not evaluated properly. Also uniform positive feedback may result in a similar situation because of lacking contrast.

The reward value $\Pi(t)$, which is the result of the reward function at time t, has to be back-propagated to all neurons, where it is analysed and used for the adaptation mechanisms. To do so, the neurons of each layer, starting with the output layer and going back to the input layer, calculate their own reward value.

The value of the output neurons is equivalent to the global reward value. All other neurons calculate their reward as follows:

$$\Pi_i(t) = \frac{\sum \Pi_{j+}(t) - \sum \Pi_{j-}(t)}{N_+ + N_-} \tag{5}$$

where $\Pi_i(t)$ is the reward value of the presynaptic neuron i at time step t. $\Pi_{j+}(t)$ is the reward value of a postsynaptic neuron that has got an excitatory connection from neuron i, while $j-$ refers to a postsynaptic neuron that has got an inhibitory connection from neuron i. N_+ is the number of the postsynaptic neurons of the first kind and N_- is the number of the other postsynaptic neurons.

First the reward value of a neuron is used to adapt the connection weights. This is done after the basic processes of Sect. 2 for each neuron. All calculations for all neurons are done once at each time step t.

The reward value can be added to a learning rule as an additional factor. Different authors, all of them using different neuron functions and learning functions, have shown that this surprisingly simple method can successfully be used to implement reinforcement learning in a neural network [2,4,7]. They do not need an external module that evaluates and changes the connections of the network after each processing step any more.

Activation Dependent Plasticity is used to adapt connection weights in the experiments. Activation Dependent Plasticity (ADP) is based on Hebb's ideas of strengthening connections that fire together [6]. As shown by [4,7] reward can also be integrated into the more sophisticated Spike Time Dependent Plasticity (STDP) learning model.

Adaptation of an excitatory axon weight (axon connected to dendrite):

$$w_{a+}(t+1) = w_{a+}(t) + \eta_a \cdot \Pi_j(t) \cdot \phi_i \cdot \phi_j \tag{6}$$

where $w_{a+}(t)$ and $w_{a+}(t+1)$ are the axon weights before and after the adaptation. η_a is the learning factor for axons and $\Pi_j(t)$ is the current reward value of the postsynaptic neuron. ϕ_i and ϕ_j represent the recent activity of the presynaptic and the postsynaptic neuron. In our experiments ϕ_i was kept between -1 and 0 for very little activity and from 0 to 1 for more activity, ϕ_j is kept between 0 and 1. For positive reward much activity in the presynaptic neuron strengthens the axon weight if also the postsynaptic neuron was active but little presynaptic activity weakens the axon weight. A negative reward value turns around the direction of change.

Adaptation of a dendrite weight (always excitatory):

$$w_d(t+1) = w_d(t) + \eta_d \cdot \Pi_j(t) \cdot \phi_d \cdot \phi_j \tag{7}$$

where $w_d(t)$ and $w_d(t+1)$ are the dendrite weights before and after the adaptation. η_d is the learning factor for dendrites. ϕ_d represents the recent activity of the dendrite (which joins the activity of the connected axons) and is kept between 0 and 1. $\Pi_j(t)$ and ϕ_j are discussed with Equ. 6. When all dendrite weights of a neuron are adapted they are normalised to sum up to 1 again because of the dependencies between the weights (see Sect. 2).

Adaptation of an inhibitory axon weight (axon connected to neuron):

$$w_{a-}(t+1) = w_{a-}(t) - \eta_a \cdot \Pi_j(t) \cdot \phi_i \cdot \phi_j \qquad (8)$$

where $w_{a-}(t)$ and $w_{a-}(t+1)$ are the axon weights before and after the adaptation. η_a, $\Pi_j(t)$, ϕ_i and ϕ_j are discussed in Equ. 6. Axons that are part of local inhibition, were not changed in our experiments. Also if ϕ_i is negative, the weight was kept equal. An inhibitory axon is strengthened, if it was not able to prevent from bad reward, and it is weakened, if it tried to prevent good reward.

4 Autonomous Creation of the Neural Network

The methods that were discussed in the previous section tune a neural network. They are necessary to reinforce neural paths that were responsible for good actions and to weaken and finally remove connections that made a neuron output to classify incorrectly.

There are different methods that result in new connections or even new neurons.

If a neuron has got no input connections, it can connect to a random new predecessor in the previous layer. In our experiments the randomness was reduced by looking for neurons that have similar relative positions. Such a new connection is always excitatory and consists of a single axon and a single dendrite. This makes the postsynaptic neuron (or better: its single dendrite) representing a new class and having the potential to carry out a new action when activated. Other neurons can be added to the class by creating new axons occasionally. An axon that does not fit into the class will be weakened by the mechanisms of Sect. 3.

New neurons are created to remember activation patterns that were responsible for good or bad actions. Liu, Buller and Joachimczak have already shown that correlations between certain input patterns and a certain reward can be stored by creating new neurons [10] [11]. For this task we add a new potential value to each neuron. The common "neuron potential" function defines when a neuron fires a spike. Our new "reward potential" function defines when a neuron has got enough feedback to create a new neuron:

$$R_j(t+1) = \delta_R \cdot R_j(t) + \phi_j \cdot \Pi_j(t) \qquad (9)$$

where $R_j(t+1)$ is the new reward potential of neuron j while $R_j(t)$ is the old one. ϕ_j is the value for the recent activity that was also used in Sect. 3 and $\Pi_j(t)$ is the current reward value of neuron j.

When $|R_j|$ reaches a certain threshold θ_R (different thresholds for positive and negative feedback are possible) all dendrites (excitatory connections) that were active recently are evaluated. Young dendrites are ignored, because they have not proved themselves already. For a dendrite with more than one axon it may be worth to remember an activation combination. A single axon may still be interesting if there was bad feedback, because the axon should have been inhibitory in this case. A list of axons, that had influence on the activation of the neuron, is kept. If this combination is not yet stored in an existing preceding neuron, a new neuron is created and each axon of the list is copied. However, each of the new axons is connected to its own dendrite to record the combination. The new neuron will only be activated when the same combination is activated again. When the reward potential was positive, the new neuron is connected to the existing one by a new axon to the currently evaluated dendrite (neurons in layer B in Fig. 1.a). A negative reward potential results in a new inhibiting axon to the existing neuron (neuron in layer C in Fig. 1.a).

The new neuron is then inserted into the previous layer. However, if there is a neuron in that layer, that is connected to the new one, a new layer is created between the layers of the presynaptic and the postsynaptic neuron. The relative position of the new neuron will be similar to the relative position of the existing one.

Once the new neuron is inserted, local inhibition will be generated. Experiments have shown that local inhibition makes learning much faster and much more reliable (see Sect. 5). Hence, new inhibiting axons are created to and from each new neuron. This inhibitory web has been automatically created within one layer in our experiments. In more sophisticated future developments this web should maybe be limited to certain areas within a layer. Nested layers with neuron containers, which are basically supported by our implementation but are not used for the growing mechanisms yet, could help with this task.

5 Experiments

5.1 Setup

Our novel method for the creation of machine controllers autonomously was tested in a simulation of a mobile robot which moves using differential steering, as shown in figure Fig. 1.b. The initial neural network consists of 12 input neurons (2 for each sensor) and 4 output neurons (2 for each motor, see Fig. 1.a).

The input neurons are fed by values from 6 sonar sensors as shown in Fig. 1.b, each sensor feeds the input of 2 neurons. The sonar sensors are arranged so that 4 scan the front of the robot and 2 scan the rear as shown in the figure. The distance value is processed so that one input neuron fires more frequently as the measured distance increases and the other neuron connected to the same sensor fires more frequently as the distance decreases.

For the actuator control, the output connections are configured so that the more frequently one of the output neurons connected to each motor fires, the faster this motor will try to drive forward. The more frequently the other output neuron connected to the same motor fires, the faster that motor will try to turn backward. The final speed that each motor will drive is calculated by the difference between both neurons.

With this experimental setup the robot should learn to wander around in a simulated environment while avoiding obstacles.

The original robot's bumpers are included in the simulation and are used to detect collisions with obstacles, and are used to penalise significantly the reward values when such a collision occurs. The reward is increased continuously as the robot travels farther during its wandering behaviour. Backward movement is only acceptable when recovering from a collision, therefore it will only be used to increase the robot's reward value in that case, while it is used to decrease this value for all other cases. As time increases, linear forward movement will receive higher positive reward and this will discourage circular movement.

5.2 Results

The reward function that we have used in our experiments delivers -1 in the case the robot crashes into an obstacle. Backward movement is punished (negative value). There are two features that are not implemented in all three simulation runs of Fig. 2.a: First, no or small forward movement is punished; second, backward movement is rewarded (positive value), if the robot received bad feedback for a while, to keep the robot active.

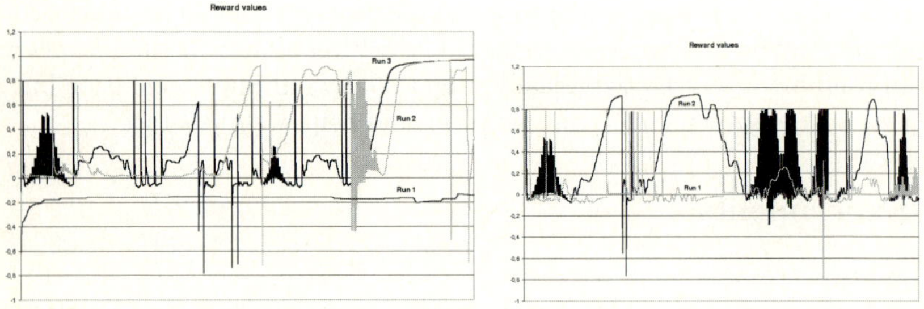

(a) (b)

Fig. 2. (a) The three curves show the reward values the robot received in simulation runs of the same duration and starting at the same position. In "Run 1" only forward movement is rewarded. The robot learns to hold its position to minimise negative feedback. In "Run 2" and "Run 3", the same reward algorithm is used, but in "Run 3" no or small forward movement is punished. The robot learns to receive positive feedback with time which makes it more stable. (b) "Run 1" shows the development of the reward value without local inhibition. This contrasting method increases the produced reward values significantly in "Run 2".

Figure 2.b shows the importance of local inhibition. Without local inhibition the simulation run did not produce a single phase in which significant positive feedback is received. Only short spots of positive reward can be identified where the robot acted in a good way by chance. Local inhibition increases the contrast of spiking patterns, which makes single neurons and hence single motor actions more powerful and the assignment of reward to a certain spiking pattern more reliable.

Table 1 shows the results of a test of 50 simulation runs. In many cases the robot was able to learn the wandering task with the ability to avoid obstacles. Each run of the test sample was stopped after 20000 control cycles (processing the whole neural network in each cycle).

Table 1. Results of 50 simulation runs

Performance of the controller			
	Min.	Max.	Avg.
Total reward	664.82	7486.24	4014.02
Average reward	0.03	0.37	0.20
Maximum speed	657.00	1056.00	1015.92
Average speed	102.74	971.63	260.37
Crashes	0.00	13.00	4.50

Topology of the controller			
	Min.	Max.	Avg.
Neurons	16.00	38.00	22.46
Excitatory axons	14.00	178.00	45.28
Excitatory dendrites	4.00	139.00	27.88
Inhibitory axons	4.00	7.00	4.36

The neural network contained no hidden neurons and no connections at the beginning. Connections for local inhibition were created at the very beginning. The speed values of the table are given in internal units of the simulation.

6 Conclusions and Further Work

We have shown that a neural network can be grown based on the reward measured by a feedback function which analyses the performance of a task in real time. The neural network can control a machine such as a mobile robot in an unpredictable or unstructured environment.

Since the controller constructs itself, only the input layer, the output layer and a feedback function that measures the task performance of the machine and rewards the controller have to be defined. This means that the task of the designer involves only the definition of these elements, no effort is required for the actual design of the network.

Because controlling the machine and learning from experience in a continuous way when running is integrated into a single and robust stage, the system can adapt to completely new situations without changing the complete control structure manually. This implicates three advantages additional to the possibility of evolving a control system from scratch:

1. The machine can learn to react appropriately in situations that did not occur before but can use the experience it has gathered so far.
2. The machine can learn to handle changes to its hardware, for example if a sensor breaks.
3. When new machines are developed, it may be possible to use the neural network of established machines to reduce the time necessary for training them by starting from a semi-optimal state rather than from scratch.

Further analysis and improvements of the growing methodology are necessary to gain more results of the potential of the growing methods for the neural network. For example it will be necessary to investigate the behaviour of the system with concurrent tasks and conflicting or noisy sensory data.

References

1. Alnajjar, F., Murase, K.: Self-organization of Spiking Neural Network Generating Autonomous Behavior in a Real Mobile Robot. In: Proceedings of the International Conference on Computational Intelligence for Modelling, Control and Automation 2005, vol. 1, pp. 1134–1139 (2005)
2. Daucé, E., Henry, F.: Hebbian Learning in Large Recurrent Neural Networks. Movement and Perception Lab, Marseille (2006)
3. Elizondo, D., Fiesler, E., Korczak, J.: Non-ontogenetic Sparse Neural Networks. In: International Conference on Neural Networks 1995, vol. 26, pp. 290–295. IEEE, Los Alamitos (1995)
4. Florian, R.V.: Reinforcement Learning Through Modulation of Spike-timing-dependent Synaptic Plasticity. Neural Computation 19(6), 1468–1502 (2007)
5. Gómez, G., Lungarella, M., Hotz, P.E., Matsushita, K., Pfeifer, R.: Simulating Development in a Real Robot: On the Concurrent Increase of Sensory, Motor, and Neural Complexity. In: Proceedings of the Fourth International Workshop on Epigenetic Robotics, pp. 119–122 (2004)
6. Hebb, D.O.: The Organization of Behaviour: A Neuropsychological Approach. John Wiley & Sons, New York (1949)
7. Izhikevich, E.M.: Solving the Distal Reward Problem through Linkage of STDP and Dopamine Signaling. Cerebral Cortex 10, 1093–1102 (2007)
8. Katic, D.: Leaky-integrate-and-fire und Spike Response Modell. Institut für Technische Informatik, Universität Karlsruhe (2006)
9. Kohonen, T.: Self-organization and Associative Memory. 3rd printing. Springer, Heidelberg (1989)
10. Liu, J., Buller, A.: Self-development of Motor Abilities Resulting from the Growth of a Neural Network Reinforced by Pleasure and Tension. In: Proceedings of the 4th International Conference on Development and Learning 2005, pp. 121–125 (2005)
11. Liu, J., Buller, A., Joachimczak, M.: Self-motivated Learning Agent: Skill-development in a Growing Network Mediated by Pleasure and Tensions. Transactions of the Institute of Systems, Control and Information Engineers 19(5), 169–176 (2006)
12. Vreeken, J.: Spiking Neural Networks, an Introduction. Intelligent Systems Group, Institute for Information and Computing Sciences, Utrecht University (2003)

A Brief Review and Comparison of Feedforward Morphological Neural Networks with Applications to Classification

Alexandre Monteiro da Silva and Peter Sussner

Institute of Math., Stat. and Sci. Comp,
University of Campinas
Campinas, SP 13081 − 970
{ra007954,sussner}@ime.unicamp.br

Abstract. The mathematical background of MNNs can be found in mathematical morphology (MM). Since MM can be conducted very generally in the complete lattice setting, MNNs are closely related to other lattice-based neurocomputing models.

This paper reviews some important types of feedforward morphological neural networks including their mathematical background. In addition, we analyze and compare the performance of feedforward morphological models and conventional multi-layer perceptrons in some classification problems.

Keywords: Mathematical Morphology, Complete Lattice, Morphological Neural Network, Pattern Classification.

1 Introduction

The theory of *morphological neural networks* (MNNs) and its applications has experienced a steady and consistent growth in the last few years [1]. An artificial neural network is said to be morphological if every neuron performs an elementary operation of MM [17,3], which is generally non-linear, possibly followed by the application of an activation function. Thus, MMNs are drastically different from conventional semi-linear artificial NNs which are endowed with linear aggregation functions. Morphological and hybrid morphological/linear neural networks [9] have been successfully applied to a variety of problems such as automatic target recognition, handwritten character recognition, and time series prediction [13,9,19]. MNNs include morphological perceptrons [6], dendritic morphological perceptrons [7], (fuzzy) morphological associative memories [17,19,5], modular morphological neural networks [8], and morphological shared-weight neural networks [13]. This paper clarifies that fuzzy lattice neural networks [10] can also be viewed as MNNs.

Although the theory of morphological neural networks (MNNs) and its applications has experienced a steady and consistent growth in the last few years [1], an overview and a comparison of MNNs such as the one presented in this paper has not yet appeared in the literature.

V. Kůrková et al. (Eds.): ICANN 2008, Part II, LNCS 5164, pp. 783–792, 2008.

The paper is organized as follows. After presenting the lattice background of MNNs, we review some of the most important types of MNNs. Section 3 compares the performances of morphological models and MLPs in some classification problems. We finish the paper with some concluding remarks.

1.1 Lattice Background for Morphological Neural Networks

Morphological neural networks consist of morphological neurons, which perform elementary operations of mathematical morphology. Complete lattices provide for a general framework in which MM can be conducted [18].

A *complete lattice* is a partially ordered set $\mathbb{L}$ in which all subsets (finite or infinite) have both a *supremum* and an *infimum* in $\mathbb{L}$ [14]. For any $Y \subseteq \mathbb{L}$, we denote the infimum of Y by the symbol $\bigwedge Y$ and we write $\bigwedge_{j \in J} y_j$ instead of $\bigwedge Y$ if $Y = \{y_j, j \in J\}$ for a index set J. We use similar notations to denote the *supremum* of Y. The extended real numbers $\bar{\mathbb{R}}$ and the unit interval $[0,1]$ represent specific examples of complete lattices. We say that an operator $\nu_{\mathbb{L}} : \mathbb{L} \longrightarrow \mathbb{L}$ is a *negation* on $\mathbb{L}$ if $\nu_{\mathbb{L}}$ is a involutive bijection that inverts the partial order relation of $\mathbb{L}$.

A central point in mathematical morphology is the decomposition of mappings between complete lattices in terms of elementary operations [4].

Definition 1. *Let $\varepsilon, \delta, \bar{\varepsilon}, \bar{\delta}$ be operators from the complete lattice $\mathbb{L}$ to the complete lattice $\mathbb{M}$, and let $Y \subseteq \mathbb{L}$.*

$$\varepsilon \quad \text{is called erosion} \Leftrightarrow \varepsilon(\bigwedge Y) = \bigwedge_{y \in Y} \varepsilon(y); \tag{1}$$

$$\delta \quad \text{is called dilation} \Leftrightarrow \delta(\bigvee Y) = \bigvee_{y \in Y} \delta(y); \tag{2}$$

$$\bar{\varepsilon} \quad \text{is called anti-erosion} \Leftrightarrow \bar{\varepsilon}(\bigwedge Y) = \bigvee_{y \in Y} \bar{\varepsilon}(y); \tag{3}$$

$$\bar{\delta} \quad \text{is called anti-dilation} \Leftrightarrow \bar{\delta}(\bigvee Y) = \bigwedge_{y \in Y} \bar{\delta}(y). \tag{4}$$

The following theorem establishes representations of anti-dilations and anti-erosions in terms of erosions, dilations, and negations [17].

Theorem 1. *Let $\mathbb{L}$ and $\mathbb{M}$ be complete lattices with negations $\nu_{\mathbb{L}}$ and $\nu_{\mathbb{M}}$, respectively.*

- *An operator $\bar{\delta} : \mathbb{L} \rightarrow \mathbb{M}$ is an anti-dilation $\Leftrightarrow \bar{\delta} = \varepsilon \circ \nu_{\mathbb{L}}$ or $\bar{\delta} = \nu_{\mathbb{M}} \circ \delta$, where δ is a dilation and ε is a erosion.*
- *An operator $\bar{\varepsilon} : \mathbb{M} \rightarrow \mathbb{L}$ is an anti-erosion $\Leftrightarrow \bar{\varepsilon} = \delta \circ \nu_{\mathbb{M}}$ or $\bar{\varepsilon} = \nu_{\mathbb{L}} \circ \varepsilon$, where ε is an erosion and δ is a dilation.*

Banon and Barrera [4] showed that for every mapping $\psi : \mathbb{L} \longrightarrow \mathbb{M}$ there exist erosions ε^i and anti-dilations $\bar{\delta}^i$ for some index set I such that

$$\psi = \bigvee_{i \in I} (\varepsilon^i \wedge \bar{\delta}^i) . \tag{5}$$

Similarly, the mapping ψ can be written as an infimum of supremums of pairs of *dilations* and *anti-erosions*. In the special case that ψ is increasing, ψ can be represented as a supremum of erosions or as an infimum of dilations.

2 Some Models of Feedforward Morphological Neural Networks

Morphological neural networks are equipped with morphological neurons. We speak of a morphological neuron if its aggregation function corresponds to an elementary morphological operation. In the following subsections, we describe some types of feedforward morphological neural networks.

2.1 Morphological Perceptron (MP)

Given a vector of inputs $\mathbf{x} \in \mathbb{R}^n$, a vector of synaptic weights $\mathbf{w} \in \bar{\mathbb{R}}^{1\times n}$ and an activation function f, a neuron of the morphological perceptron calculates the output y according to one of the following rules:

$$y = f(\varepsilon_{\mathbf{w}}(\mathbf{x})), \text{ where } \varepsilon_{\mathbf{w}}(\mathbf{x}) = \wedge_{i=1}^{n}(x_i + w_i);$$
$$y = f(\delta_{\mathbf{w}}(\mathbf{x})), \text{ where } \delta_{\mathbf{w}}(\mathbf{x}) = \vee_{i=1}^{n}(x_i + w_i);$$
$$y = f(\bar{\varepsilon}_{\mathbf{w}}(\mathbf{x})), \text{ where } \bar{\varepsilon}_{\mathbf{w}}(\mathbf{x}) = \vee_{i=1}^{n}(-x_i + w_i);$$
$$y = f(\bar{\delta}_{\mathbf{w}}(\mathbf{x})), \text{ where } \bar{\delta}_{\mathbf{w}}(\mathbf{x}) = \wedge_{i=1}^{n}(-x_i + w_i).$$

The values of the morphological perceptron's weights must be determined before it can act as a classifier. Actually, the weights are determined in a learning stage where the supervised learning algorithm [6] constructs n-dimensional boxes around sets of points which share the same class value. Convergence occurs in a finite number of steps.

Originally, a training algorithm for MPs was proposed to solve two-class classification problems [6]. This training algorithm automatically produces the architecture illustrated in Figure 1(**a**).

After training, the MP calculates the following output pattern $\mathbf{y}$ for an input pattern $\mathbf{x}$:

$$y = f(\vee_{j=1}^{m}(\varepsilon_{\mathbf{v}_j}(\mathbf{x}) \wedge \bar{\delta}_{\mathbf{w}_j}(\mathbf{x}))) \tag{6}$$

According to Equation 6, MP calculates a supremum of erosions and anti-dilations which is a decomposition suggested by Banon and Barrera [4] followed by the application of a hard-limiting function f. Equation 6 implies the set of points that are classified as belonging to class 1 ($y = 1$) is given by a union of hyperboxes. Note that the operators $\varepsilon_{\mathbf{w}}, \delta_{\mathbf{w}}, \bar{\varepsilon}_{\mathbf{w}}, \bar{\delta}_{\mathbf{w}}$ constitute examples of the elementary operators introduced in Definition 1.

2.2 Morphological Perceptrons with Dendrites (MPD)

Based on recent research in neuroscience, Ritter and Urcid [7] developed a new paradigm for computing with morphological neurons where the process occurs

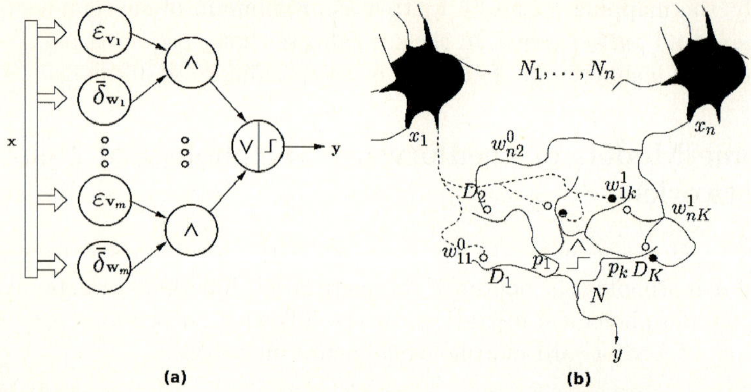

Fig. 1. Architectures of a morphological perceptron (**a**) and of an MPD (**b**), respectively

in the dendrites. The MPD training algorithm [7] resembles the one for MPs [6]. As is the case for MPs, the MPD training algorithm is guaranteed to converge in a finite number of steps and, after convergence, all training patterns will be classified correctly. Figure 1(**b**) provides a graphical representation of an MPD. The architecture of an MPD is not determined beforehand. During the training phase, the MPD grows new dendrites while the input neurons expand their axonal branches to synapse on the new dendrites. The output of an MPD is defined by the following equation:

$$y(\mathbf{x}) = f(p_k \bigwedge_{i=1}^{n} \bigwedge_{l \in L} (-1)^{l+1}(x_i + w_{ki}^l)), \tag{7}$$

where f is a hard limiting function, $L = \{0, 1\}$ and $p_k = \{-1, 1\}$ denotes the excitatory or inhibitory response of the kth dentrite.

2.3 Morphological/Rank/Linear Neural Network (MRL-NN)

MRL-NNs [9] employ a linear and a rank/morphological aggregation function at each node followed by the application of an activation function. The MRL-NN aggregation function generalizes the morphological operators $\delta_{\mathbf{w}}(\mathbf{x}) = \bigvee_{i=1}^{n}(x_i + w_i)$ and $\varepsilon_{\mathbf{w}}(\mathbf{x}) = \bigwedge_{i=1}^{n}(x_i + w_i)$. By sorting the components of the list $\mathbf{t} = (x_1 + w_1, \ldots, x_n + w_n)$ in decreasing order, which yields $t_{(1)} \geqslant t_{(2)} \geqslant \ldots \geqslant t_{(n)}$, we pick the rth element of the sorted list. In this way, we define the rth rank function of t by

$$\mathcal{R}_r(\mathbf{t}) \equiv t_{(r)}, \quad r = 1, \ldots, n. \tag{8}$$

The equations that formally define the lth layer of the MRL are given by:

$$x_k^{(l)} = \lambda_k^{(l)} \alpha_k^{(l)} + (1 - \lambda_k^{(l)}) \beta_k^{(l)}, \tag{9}$$

$$\alpha_k^{(l)} = \mathcal{R}_{r_k^{(l)}}(\mathbf{y}^{(l-1)} + \mathrm{a}_k^{(l)}), \tag{10}$$

$$\beta_k^{(l)} = \mathbf{y}^{(l-1)^{\top}} \mathrm{b}_k^{(l)} + \theta_k^{(l)}, \tag{11}$$

where $r_k^{(l)} \in \mathbb{Z}$, $\lambda_k^{(l)}, \theta_k^{(l)} \in \mathbb{R}$ and $\mathbf{a}_k^{(l)}, \mathbf{b}_k^{(l)} \in \mathbb{R}^{N_{l-1}}$ represents the parameters needed to be estimated by the MRL-NN. Thus, we denote the synaptic weight associated with a MRL-NN neuron by $\mathbf{w}_k^{(l)} = (\mathbf{a}_k^{(l)}, \rho_k^{(l)}, \mathbf{b}_k^l, \theta_k^{(l)}, \lambda_k^{(l)})$. Figure 2(**a**) illustrates the lth layer of a MRL-NN.

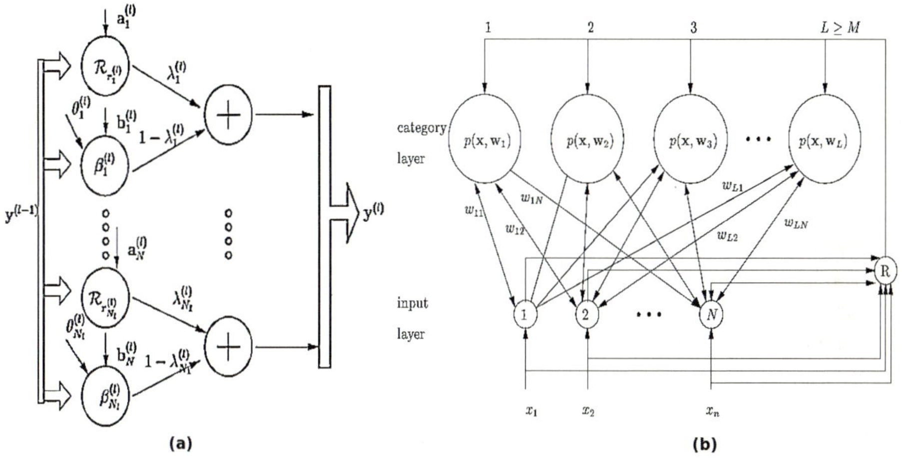

Fig. 2. lth layer of the MRL-NN (**a**), architecture of the FLNN (**b**), respectively

During the training phase, the non-differentiability of rank functions is overcome by using rank indicator vectors [9] which simply mark the locations in $\mathbf{t}$ where $\mathcal{R}_r(\mathbf{t})$ occur [9].

2.4 Modular Morphological Neural Network (MMNN)

The decompositions of Banon and Barrera [4] have provided the basis for the development of modular morphological neural networks (MMNNs) by Sousa *et al.* [12]. The architecture of a modular morphological neural network resembles the MP's architecture of Figure 1(**a**).

MMNN training [8] can be achieved using a simple genetic algorithm (SGA) or a modified backpropagation (BP) algorithm, which uses the methodology of Pessoa and Maragos [9] for estimating the derivatives of the training equation.

2.5 Fuzzy Lattice Neural Network-FLNN

The theoretical framework of FLNN constitutes a successful combination of fuzzy sets [15], lattice theory [2] and adaptive resonance theory [16]. Let $\mathbb{L}$ be a complete lattice. Given a vector of inputs $\mathbf{x} \subset \mathbb{L}$ and a vector of synaptic weights

$\mathbf{w} \in \mathbb{L}$, a neuron of an FLNN [10] computes $p(\mathbf{x}, \mathbf{w})$ which is a fuzzy partial order relation that calculates the degree of inclusion of $\mathbf{x}$ in $\mathbf{w}$. The function $p : \mathbb{L} \times \mathbb{L} \to [0, 1]$ is such that $p(\mathbf{x}, \mathbf{y}) = 1 \Leftrightarrow \mathbf{x} \leq \mathbf{y}$ and p represents both an anti-erosion and an anti-dilation. In general, $\mathbb{L}$ is given by the complete lattice of the so called *generalized intervals*.

Figure 2(**b**) illustrates the architecture of the FLNN that consists of an *input layer* and a *category layer*. The input layer has N artificial neurons used for storing and comparing input data. The category layer has L artificial neurons that define M classes.

FLNNs can be trained in supervised or unsupervised fashion. Both versions generate hyperboxes that determine the output of the FLNN. In this paper, we focus on the supervised learning algorithm [10].

3 Experiments Results

In order to assess the classification performance of our feedforward morphological models, we have conducted a series of experiments on two well known datasets: Ripley's synthetic problem [22] and the Pima Indian dataset [20].

3.1 Ripley's Synthetic Experiment

Ripley's synthetic dataset [22] consists of data samples characterized by two features in two classes. The training and test sets, respectively, consist of 250 and 1000 samples with the same number of samples belonging to each of the two classes. We employed 25-fold cross-validation to the training set.

As a result from the application of 25-fold cross-validation, we obtained four MPs that exhibited the same least validation error. After testing these four MPs, the overall error of testing arises as the average of the corresponding four testing errors. The training algorithms of the MPD and the FLNN generated 16 and 47 neurons. With respect to the other morphological models and the MLP, we used the same training parameters and network topologies as in the Pima Indians experiment that is described below.

Based on the results in Table 1, the FLNN achieved the best performance and the MP, MPD and MLP also exhibited favorable results whereas the MRL and MMNN did not perform well for the same reasons that we discussed in the preceding subsection. Figures 3 and 4 visualize the decision surfaces produced by the MP, MPD, FLNN, and MLP models after completion of the respective training phases.

3.2 Pima Indian Diabetes Experiment

The Pima Indian dataset [20] is not already partitioned into training and testing sets and it consists of 768 samples taken from patients who show signs of diabetes. The task is to diagnose whether a patient is diabetic or not. Each sample is described by 8 numerical variables representing clinical findings. There are 376

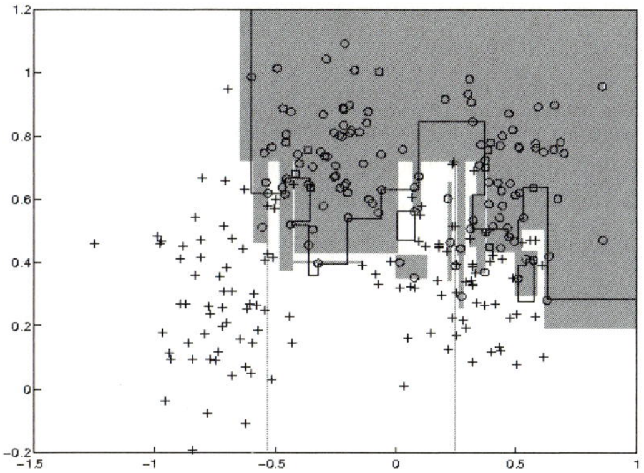

Fig. 3. Decision surfaces of an MP (*represented by the difference in shading*) and of an MPD (*represented by the continuous line*)

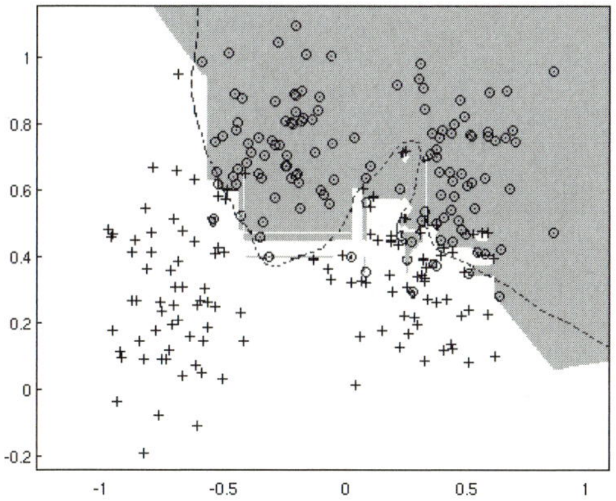

Fig. 4. Decision surfaces of an FLNN (*represented by the difference in shading*) and of an MLP (*represented by the dashed line*)

incomplete samples, mainly in variable 5. For this reason, we used the partition suggested by Ripley [22] where the 532 complete samples, excluding variable 5, are divided into 200 samples for training and 332 samples for testing. We applied 20-fold cross-validation to the training set.

In conjunction with 20-fold cross-validation, the MP's training algorithm [6] automatically generated 50 and 25 neurons in the first and the second hidden layer, respectively. In a similar manner, training an MPD using the constructive algorithm of Ritter and Urcid [7] yielded 27 dendrites. Furthermore, we employed an MRL-NN with one hidden layer consisting of 5 neurons and applied a modified backpropagation (BP) algorithm [9], using a step size $\mu = 0.01$ and a smoothing parameter $\sigma = 0.05$. We also used a simple genetic algorithm (SGA) to train an MMNN, considering an initial population of 50 elements (*synaptic weights*), maximum of 100 generations, crossover weight $w = 0.9$ and mutation probability of 0.1. The FLNN generated 72 neurons during the training phase. Moreover, we compared the morphological models with an MLP with ten hidden nodes that was trained using *gradient descent with momentum and adaptive step backpropagation* rule (learning rate $\eta = 10^{-4}$, increase and decrease parameters 1.05 and 0.5 respectively, momentum factor $\alpha = 0.9$).

We report our experimental results in Table 1. As Table 1 indicates, the FLNN outperformed all the other models in this experiment. The MP, MPD and MLP models exhibited favorable results. The MRL-NN and the MMNN presented the worst testing error rate. The poor generalization obtained by MRL-NN is probably due to the discontinuity of the rank variable $r_k^{(l)}$. The performance of the MMNN model in classification problems strongly depends on the initial feature and the corresponding shapes of the decision regions. Hence, it would be appropriate to employ pre-processing of the data. Table 1 also compares the computational effort that arises in the training phase of various models on the Pima Indian Diabetes experiment.

Table 1 shows that the morphological models converge much faster than the MLP. Table 1 also reveals that more neurons are needed in morphological models compared to an MLP. However, the artificial neurons of MNNs are much less complicated than those of MLPs since their calculations only involve operations of maximum, minimum, addition and subtraction.

Table 1. Percentage of misclassified patterns for training (T_r) and testing (T_e) in the experiments, # of epochs (N_e) and # of neurons (N_a)

	Ripley's synthetic dataset				Pima Indian dataset			
Model	T_r(%)	T_e(%)	N_e	N_a	T_r(%)	T_e(%)	N_e	N_a
MP	0.0	11.2	17	52	0.0	20	25	76
MPD	0.0	12.9	16	49	0.0	28	26	79
MRL	13.1	26.4	250	41	33.4	36.4	250	41
MMNN(BP)	15.0	26.8	250	31	32.0	39.2	250	31
MMNN(SGA)	14.6	33.0	150	31	24.0	33.3	150	31
FLNN	0.0	10.0	141	47	0.0	15.1	234	72
MLP	5.1	11.6	25000	11	23.6	21	3500	11

4 Conclusions

Several models of feedforward morphological neural networks were presented in this paper. Morphological models employ elementary morphological operations in every node. Many times, these elementary operations can be expressed in terms of maximums (or minimums) of sums, which lead to fast neural computational [7] and easy hardware implementation [21]. Unlike traditional feedforward neural networks, the MPD, MP and FLNN update their network topologies during learning, which eliminates the problem of choosing a network topology including a fixed number of neurons in advance. Their training algorithms require only a finite number of epochs to converge, resulting in a decision surface that perfectly separates the training data.

The effectiveness of feedforward morphological models has been demonstrated empirically using Ripley's synthetic dataset [22] and the Pima Indian Diabetes dataset [20]. The results in terms of classification error rate and computational effort have been analyzed and compared. Cross-validation has been used to provide a fair performance comparison.

In general, the FLNN outperformed all other models in the experiments. The MLP, MP and MPD exhibit similar percentages of misclassified patterns, whereas the MRL and the MMNN exhibit the worst performances among the models we tested.

Acknowledgements

This work was supported by Capes and by CNPq under grant no. 306040/2006 − 9.

References

1. Sussner, P., Graña, M.: Guest Editorial: Special Issue on Morphological Neural Networks. Journal of Mathematical Imaging and Vision 19(2), 79–80 (2003)
2. Heijmans, H.J.A.M.: Morphological Image Operators. Academic Press, New York (1994)
3. Serra, J.: Image Analysis and Mathematical Morphology. Academic Press, London (1994)
4. Banon, G., Barrera, J.: Decomposition of Mappings between Complete Lattices by Mathematical Morphology, Part 1. General Lattices. Signal Processing 30 3, 299–327 (1993)
5. Sussner, P., Valle, M.E.: Gray Scale Morphological Associative Memories. IEEE Transactions on Neural Networks 17(3), 559–570 (2006)
6. Sussner, P.: Morphological Perceptron Learning. In: Proceedings of IEEE ISIC/CIRA/ISAS Joint Conference, Gaithersburg, MD, September 1998, pp. 477–482 (1998)
7. Ritter, G.X., Urcid, G.: Lattice Algebra Approach to Single-Neuron Computation. IEEE Transactions on Neural Networks 14(2), 282–295 (2003)
8. Araújo, R.A., Madeiro, F., Pessoa, L.F.C.: Modular morphological neural network training via adaptive genetic algorithm for design translation invariant operators. In: Proceedings of the IEEE International Conference on Acoustics, Speech and Signal Processing, Toulouse, France (May 2006)
9. Pessoa, L.F.C., Maragos, P.: Neural networks with hybrid morphological/rank/linear nodes: a unifying framework with applications to handwritten character recognition. Pattern Recognition 33, 945–960 (2000)
10. Kaburlasos, V.G., Petridis, V.: Fuzzy lattice neurocomputing (FLN) models. Neural Networks 13, 1145–1170 (2000)
11. Raducanu, B., Graña, M., Albizuri, X.F.: Morphological Scale Spaces and Associative Morphological Memories: Results on Robustness and Practical Applications. Journal of Mathematical Imaging and Vision 19(2), 113–131 (2003)

12. Sousa, R.P., Pessoa, L.F.C., Carvalho, J.M.: Designing translation invariant operations via neural network training. In: Proceedings of the IEE International Conference on Image Processing, Vancouver, Canada, pp. 908–911 (2000)
13. Khabou, M.A., Gader, P.D.: Automatic target detection using entropy optimized shared-weight neural networks. IEEE Transactions on Neural Networks 11(1), 186–193 (2000)
14. Birkhoff, G.: Lattice Theory, 3rd edn. American Mathematical Society, Providence (1993)
15. Zadeh, L.A.: Fuzzy Sets. Information and Control 8(3), 338–353 (1965)
16. Carpenter, G.A., Grossberg, S.: ART 3: Hierarchical search using chemical transmitters in self-organizing pattern recognition architectures. Neural Networks 3, 129–152 (1990)
17. Valle, M.E., Sussner, P.: A General Framework for Fuzzy Morphological Associative Memories. Fuzzy Sets and Systems, 159(7), 747-768 (April 2008) (September 2007) (accepted for publication)
18. Ronse, C.: Why Mathematical Morphology Needs Complete Lattices. Signal Processing 21(2), 129–154 (1990)
19. Sussner, P., Valle, M.E.: Morphological and Certain Fuzzy Morphological Associative Memories with Applications in Classification and Prediction. In: Kaburlasos, V.G., Ritter, G.X. (eds.) Computational Intelligence Based on Lattice Theory. Studies in Computational Intelligence, vol. 67, pp. 149–173. Springer, Heidelberg (2007)
20. Blake, C.L., Hettich, S., Newman, D.J., Merz, C.J.: UCI Repository of machine learning databases. University of California, Irvine, Dept. of Information and Computer Sciences (1998), `http://www.ics.uci.edu/mlearn/MLRepository.html`
21. Porter, R., Harvey, N., Perkins, S., Theiler, J., Brumby, S., Bloch, J., Gokhale, M., Szymanski, J.: Optimizing digital hardware perceptrons for multispectral image classification. Journal of Mathematical Imaging and Vision 19(2), 133–150 (2003)
22. Ripley, B.D.: Pattern Recognition and Neural Networks. Cambridge University Press, Cambridge, United Kingdom (1996)

Prototype Proliferation in the Growing Neural Gas Algorithm

Héctor F. Satizábal[1,2,3], Andres Pérez-Uribe[2], and Marco Tomassini[1]

[1] Université de Lausanne, Hautes Etudes Commerciales (HEC),
Institut des Systèmes d'Information (ISI)
Marco.Tomassini@unil.ch
[2] University of Applied Sciences of Western Switzerland (HEIG-VD)(REDS)
hector-fabio.satizabal-mejia@heig-vd.ch, andres.perez-uribe@heig-vd.ch
[3] Corporación BIOTEC

Abstract. Growing Neural Gas is an incremental vector quantization algorithm with the capabilities of topology-preserving and distribution-matching. Distribution matching can produce overpopulation of prototypes in zones with high density of data. In order to tackle this drawback, we introduce some modifications to the original Growing Neural Gas algorithm by adding three new parameters, one of them controlling the distribution of the codebook and the other two controlling the quantization error and the amount of units in the network. The resulting learning algorithm is capable of efficiently quantizing datasets presenting high and low density regions while solving the prototype proliferation problem.

Keywords: Large Datasets, Vector Quantization, Topology-Preserving Networks, Distribution-matching, Growing Neural Gas.

1 Introduction

Processing information from large databases has become an important issue since the emergence of the new large scale and complex information systems (e.g., satellital images, bank transactions databases, marketing databases, internet). Extracting knowledge from such databases is not an easy task due to the execution time and memory constraints of current actual systems. Nonetheless, the need for using this information to guide decision-making processes is imperative.

Classical *data mining* algorithms exploit several approaches in order to deal with this kind of datasets [3, 7]. Sampling, partitioning or hashing the dataset drives the process to a *split and merge*, *hierarchical* or *constructive* framework, giving the possibility of building large models by assembling (or adding) smaller individual parts. Another possibility to deal with large datasets is incremental learning [8]. In this case, the main idea is to transform the modelling task in an *incremental task*[1] by means of a sampling or partitioning procedure, and the use

[1] A learning task is incremental if the training examples used to solve it become available over time, usually one at a time [8].

V. Kůrková et al. (Eds.): ICANN 2008, Part II, LNCS 5164, pp. 793–802, 2008.

of an incremental learner that builds a model from the single samples of data (one at a time).

Moreover, large databases contain a lot of redundant information. Thus, having the complete set of observations is not mandatory. Instead, selecting a small set of prototypes containing as much information as possible would give a more feasible approach to tackle the knowledge extraction problem. One well known approach to do so is *Vector Quantization* (VQ). VQ is a classical quantization technique that allows the modelling of a distribution of points by the distribution of prototype or reference vectors. Using this approach, data points are represented by the index of their closest prototype. The codebook, i.e., the collection of prototypes, typically has many entries in high density regions, and discards regions where there is no data [1].

A widely used algorithm implementing VQ in an incremental manner is Growing Neural Gas (GNG) [6]. This neural network is part of the group of *topology-representing networks* which are unsupervised neural network models intended to reflect the topology (i.e., dimensionality, distribution) of an input dataset [11]. GNG generates a graph structure that reflects the topology of the input data manifold (topology learning). This data structure has a dimensionality that varies with the dimensionality of the input data. The generated graph can be used to identify clusters in the input data, and the nodes by themselves could serve as a codebook for vector quantization [5].

In summary, building a model from a large dataset could be done by splitting the dataset in order to make the problem an *incremental task*, then applying an *incremental learning* algorithm performing *vector quantization* in order to obtain a reduced set of *prototypes* representing the whole set of data, and then using the resulting codebook to build the desired model.

Growing Neural Gas suffers from prototype proliferation in regions with high density due to the absence of a parameter stopping the insertion of units in sufficiently-represented[2] areas. This stopping criterion could be based on a local measure of performance. One alternative that exploits this approach to overcome the aforementioned drawback was proposed by Cselenyi [4]. In this case, the proposed method introduces eight new parameters to the GNG algorithm proposed by Fritzke [6]. In our case, we propose a modification that adds three new parameters to the original GNG algorithm in order to restrict the insertion of new units due to points belonging to already covered areas. Thus, promoting the insertion of new units in areas with higher quantization error in order to produce network structures covering a higher volume of data using the same number of units.

The rest of the article is structured as follows. In section 2 we make a brief description of the original GNG algorithm. Section 3 describes the proposed modifications made to the algorithm. Section 4 describes some of the capabilities of the resulting method using some "toy" datasets. Section 5 shows one application using a real large size dataset and finally, in section 6 we give some conclusions and insights about prototype proliferation and the exploitation of large datasets.

[2] Areas with low quantization error.

2 Growing Neural Gas

Growing Neural Gas (GNG) [6] is an incremental *point-based network* [2] that performs vector quantization and topology learning. The algorithm builds a neural network by incrementally adding units using a competitive hebbian learning strategy. The resulting structure is a graph of neurons that reproduces the topology of the dataset by keeping the distribution and the dimensionality of the training data [5]. The classification performance of GNG is comparable to conventional approaches [9] but has the advantage of being incremental. Hence, giving the possibility of training the network even if the dataset is not completely available all the time while avoiding the risk of catastrophic interference.

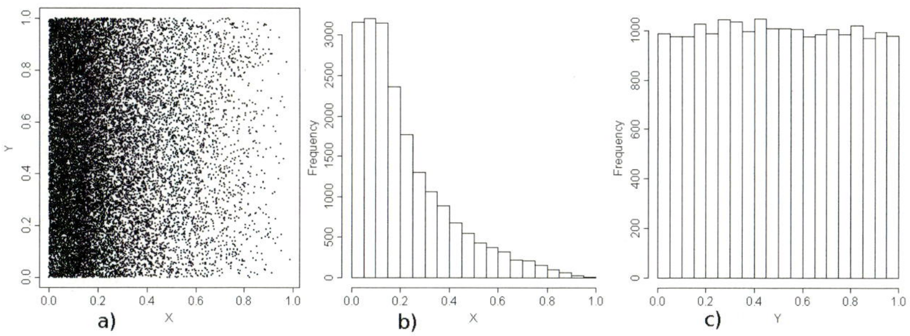

Fig. 1. a) Two dimensional non-uniform data distribution. b) Histogram of variable X. c) Histogram of variable Y.

The algorithm proposed by Fritzke is shown in table 1. In such approach, every λ iterations (step 8), one unit is inserted halfway between the unit q having the highest error and its neighbour f having also the highest error. Carrying out this insertion without any other consideration makes the network to converge to a structure where each cell is the prototype for approximately the same number of data points and hence, keeping the original data distribution. As an example, a GNG network was trained using the dataset shown in figure 1, and the training parameters shown in table 2. These values were selected after several runnings of the algorithm.

Figure 2 shows the position and distribution of the 200 cells of the resulting structure. As we can see, the distribution of each one of the variables is reproduced by the group of prototypes.

If we have a huge dataset where there is a lot of redundant information, and we want to keep only a relatively small number of prototypes describing the data with some distortion, then we are not interested in reproducing the data distribution. Instead, we want to distribute the prototypes over the whole volume of data without exceeding a maximum quantization error or distortion. In this case some modifications to the original algorithm are needed.

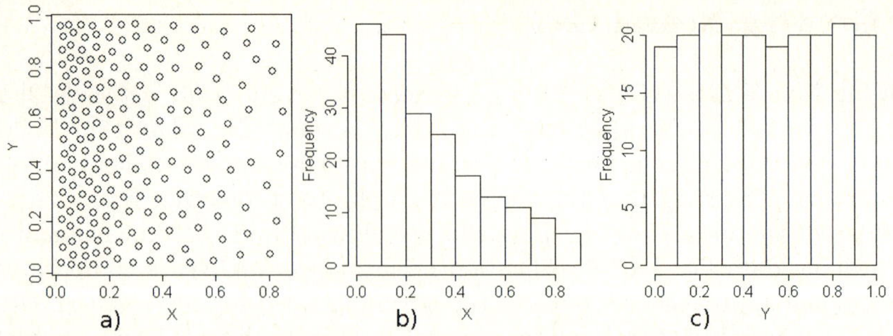

Fig. 2. Positions of neurons of the GNG model. a) Position of the neuron units. b) Distribution of X. c) Distribution of Y.

Table 1. Original growing neural gas algorithm proposed by Fritzke

Step 0:	Start with two units a and b at random positions w_a and w_b in $\Re^n$
Step 1:	Generate an input signal ξ according to a (unknown) probability density function $P(\xi)$
Step 2:	Find the nearest unit s_1 and the second-nearest unit s_2
Step 3:	Increment the age of all edges emanating from s_1
Step 4:	Add the squared distance between the input signal and the nearest unit in input space to a local counter variable: $$\Delta error(s_1) = \lVert w_{s_1} - \xi \rVert^2$$
Step 5:	Move s_1 and its direct topological neighbours towards ξ by fractions ϵ_b and ϵ_n, respectively, of the total distance: $$\Delta w_{s_1} = \epsilon_b (\xi - w_{s_1})$$ $$\Delta w_n = \epsilon_n (\xi - w_n) \qquad \text{for all direct neighbours } n \text{ of } s_1$$
Step 6:	If s_1 and s_2 are connected by an edge, set the age of this edge to zero. If such an edge does not exist, create it
Step 7:	Remove edges with an age larger than a_{max}. If the remaining units have no emanating edges, remove them as well
Step 8:	If the number of input signals generated so far is an integer multiple of a parameter λ, insert a new unit as follows: • Determine the unit q with the maximum accumulated error. • Insert a new unit r halfway between q and its neighbour f with the largest error variable: $w_r = 0.5 (w_q + w_f)$ • Insert edges connecting the new unit r with units q and f, and remove the original edge between q and f. • Decrease the error variables of q and f by multiplying them with a constant α. Initialize the error variable of r with the new value of the error variable of q.
Step 9:	Decrease all error variables by multiplying them with a constant d
Step 10:	If a stopping criterion (e.g., net size or some performance measure) is not yet fulfilled go to step 1

Table 2. Parameters for the Growing Neural Gas algorithm

Parameter	ϵ_b	ϵ_n	λ	a_{max}	α	d
value	0.05	0.005	100	100	0.5	0.9

3 Including a Fixed Quantization Error

The criterion driving the insertion of units in the GNG algorithm is the accumulated error of each unit (step 4). This local error measure grows each time a cell becomes the winner unit (i.e., the closest unit to the current data point), producing the insertion of more cells in zones with higher densities of data.

In order to attenuate that effect, we propose to modulate the error signal $\Delta error$ when it is produced by a data point having a quantization error smaller than a threshold qE.

$$accumulatedError = \begin{cases} accumulatedError + \Delta error & \text{if } \Delta error \geq qE \\ accumulatedError + (h * \Delta error) & \text{if } \Delta error < qE \end{cases}$$
$$0 \leq h \leq 1$$

The proliferation of prototypes could also be due to neuron movement. Neuron units located in zones with higher densities are chosen as winners with higher probability, attracting their neighbours belonging to less populated zones. As in the previous case, parameter h can be used to modulate the change of position of the units in each iteration (step 5).

$$\Delta w_{s_1} = \begin{cases} \epsilon_b\,(\xi - w_{s_1}) & \text{if } \Delta error \geq qE \\ h * \epsilon_b\,(\xi - w_{s_1}) & \text{if } \Delta error < qE \end{cases}$$
$$\Delta w_n = \begin{cases} \epsilon_n\,(\xi - w_n) & \text{if } \Delta error \geq qE \\ h * \epsilon_n\,(\xi - w_n) & \text{if } \Delta error < qE \end{cases} \quad \text{for all direct neighbours } n \text{ of } s_1$$

Parameter qE is the radius of an hypersphere determining the region of influence of each prototype. These regions of influence could overlap to a certain extent. In our approach this amount of superposition can be controlled by a parameter sp. Every λ iterations (step 8), one unit is inserted halfway between the unit q having the highest error and its neighbour f having also the highest error. Knowing the distance between unit q and unit f, and taking the quantization error qE as the radius of each unit, one could change the step 8 of the original algorithm proposed by Fritzke as shown in table 3.

In summary, the proposed modifications add three parameters, qE, sp and h, to the original GNG algorithm. Parameters *quantization error* qE and *superposition percent* sp depend on the application and are strongly related. Both control the amount of units in the resulting neural network structure, the former by controlling the region of influence of each unit, and the latter by controlling the superposition of units. In a less obvious sense, parameter h controls the distribution of units between high and low density areas. Examples using these parameters are given in section 4.

Table 3. Proposed modification to the original algorithm

Step 8: If the number of input signals generated so far is an integer multiple of a parameter λ, insert a new unit as follows:

- Determine the unit q with the maximum accumulated error.
- Determine the unit f in the neighbourhood of q with the maximum accumulated error.
- Calculate the distance $dist$ between units q and f.

$$dist = \|q - f\|$$

- Calculate the available space between the two units as follows:

$$available = dist - qE$$

If $(available > (sp * qE))$ go to step 9. $0 \leq sp \leq 1$
Else,

- Insert a new unit r halfway between q and its neighbour f with the largest error variable: $w_r = 0.5\,(w_q + w_f)$
- Insert edges connecting the new unit r with units q and f, and remove the original edge between q and f.
- Decrease the error variables of q and f by multiplying them with a constant α. Initialize the error variable of r with the new value of the error variable of q.

4 Toyset Experiments

In section 2, a non-uniform distribution of points in two dimensions was used to train a GNG network. Figure 2 shows a high concentration of prototypes in the zone with higher density due to the property of density matching of the model. This is an excellent result if we do not have any constraint on the amount of prototypes. Having more prototypes also increases the execution time of the algorithm and this is not desirable if we have a very large dataset. Adding the parameters that control prototype proliferation by introducing the measure of quantization error allows us to overcome this situation.

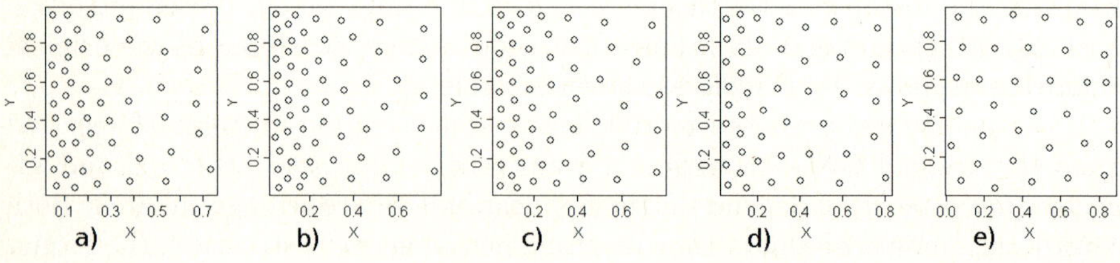

Fig. 3. Results of the modified algorithm varying parameter h ($qE = 0.1$ and $sp = 0.75$). a)Using $h = 1.00$ b)Using $h = 0.75$ c)Using $h = 0.50$ d)Using $h = 0.25$ e)Using $h = 0.00$.

Figure 3 shows the results of the modified algorithm when trained with the data distribution showed in figure 1. Figure 3 shows how parameter h effectively controls the proportion of units assigned to regions with high and low densities. In this case the parameters qE and sp were kept constants ($qE = 0.1$ and $sp = 0.75$) because their effects are more global and depend less on the data distribution. The rest of parameters were set as shown in table 2.

Another interesting test consists on using a dataset similar to the one proposed by Martinetz [12] in the very beginnings of this kind of networks (neural gas, growing cell structures, growing neural gas) [5]. This distribution of data has been used by several researchers [4, 5, 6, 11, 12] in order to show the ability of the *topology-preserving* networks in modelling the distribution and dimensionality of data. The generated dataset shown in figure 4.a) presents two different levels of densities for points situated in three, two and one dimension, and has points describing a circle.

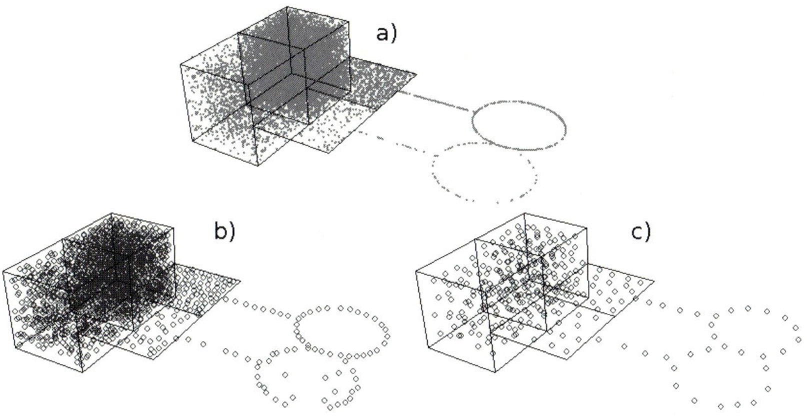

Fig. 4. The original and the modified version of GNG trained with a dataset like the used by Cselenyi [4]. a) Training data. b) Algorithm of GNG by Fritzke. c) Modified GNG algorithm, $qE = 0.1$, $h = 0.1$, $sp = 0.5$.

When this dataset is used, the model has to deal with data having different dimensionalities, different densities and different topologies. Figures 4.b) and 4.c) show the position of the units of two GNG networks, one of them using the original algorithm and the other one using the modified version. Both structures preserve the topology of the data in terms of dimensionality by placing and connecting units depending of local conditions. Conversely, the two models behave differently in terms of the distribution of the data. The codebook of the original GNG algorithm reproduces the distribution of the training data by assigning almost the same quantity of data points to each vector prototype. In the case of the modified version, the parameter h set to 0.1 makes the distribution of prototypes more uniform due to the fact that the insertion of new units is conditioned with the quantization error. Other parameters were set as shown in table 2.

5 Obtaining a Codebook from a Large Dataset

This section summarizes a series of experiments using a large database of climate. The database contains information of the average temperature over approximately the last fifty years in Colombia, with a spatial resolution of 30 seconds (≈900m) (WORLDCLIM) [10]. There are 1'336.025 data points corresponding to the amount of pixels covering the region, and each one has twelve dimensions corresponding to the months of the year (i.e., one vector of twelve dimensions per pixel). The main objective of modelling this dataset is to find zones with similar environmental conditions (i.e, temperature) by means of some measure of distance.

Processing the whole dataset implies the use of a lot of memory resources and takes hours of calculation. Moreover, the situation could get even worse if we consider the addition of other variables (e.g., precipitation). Instead of processing every pixel in the dataset, we could use vector quantization to extract a codebook representing the data, and then process this set of prototypes finding the zones having similar properties. Figure 5 shows the resulting quantization errors using both algorithms.

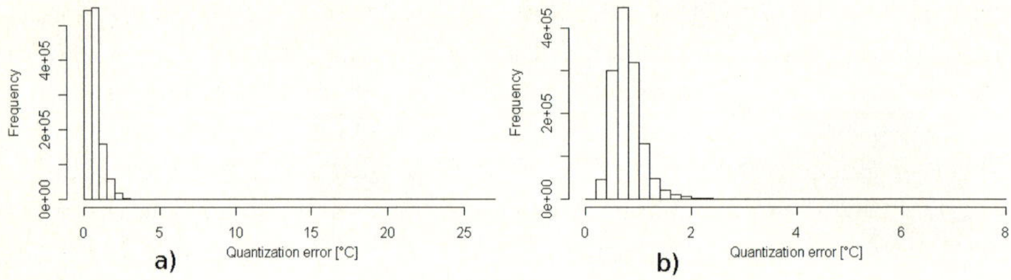

Fig. 5. Histogram of the quantization error for a large dataset. a) Fritzke's original GNG algorithm. b) Modified GNG algorithm, $qE = 1\,°\text{C}$, $h = 0.1$, $sp = 0.5$.

Both neural networks have only 89 neuron units, which means having a codebook with only 0.007% of the original size of the dataset. Nonetheless, the quantization error is astonishing low. This reduction is possible due to the low local dimensionality of the data, and the low range of the variables. Figure 5 shows that the modified algorithm presents quantization error values that are comparable to those from the original version, but with a sligthtly different distribution.

Having a dataset that allows a representation over two dimensions has some advantages. In this case, we can draw some information from the geographic distribution of the prototypes. Figure 6 shows the geographic representation of the boundaries (white lines) of the voronoi region of each prototype. The region delimited with a circle is a wide plain at low altitudes which presents homogeneous conditions in terms of temperature. Therefore, this large amount of pixels belong to a zone with high density in the space of twelve dimensions of our data. In this case, this high density zone does not mean more information to quantize. However, the original GNG algorithm is "forced" to proliferate prototypes due to its

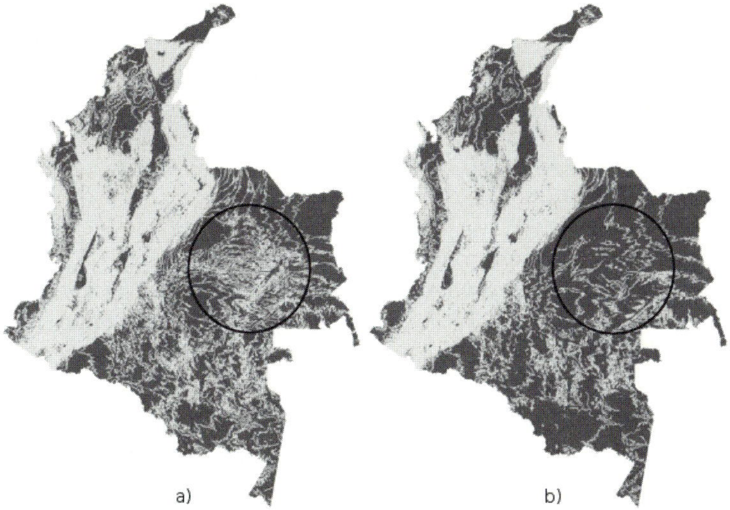

Fig. 6. Prototype boundaries. a) Original algorithm. b) Modified algorithm.

property of distribution matching. Thus incurring in a high computational cost when one uses the resulting model. Instead of representing better these areas, our approach is to avoid prototype proliferation in regions with regular conditions in order to better represent heterogeneous zones (e.g., mountains). Figure 6.b) shows that the modified version of GNG places less prototypes in flat areas (i.e., high density regions) than the original version (Figure 6.a), and assigns more prototypes (i.e., cluster centres) to the lower density points belonging to mountain areas (i.e., low density regions).

6 Conclusions

There is a increasing need of dealing with large datasets nowadays. A large dataset can be split or sampled in order to divide the modelling task into smaller subtasks that can be merged in a single model by means of an incremental learning technique performing vector quantization. In our case, we chose the Growing Neural Gas (GNG) algorithm as the vector quantization technique. GNG allows us to get a reduced codebook to analyse, instead of analysing the whole dataset. Growing Neural Gas is an excellent incremental vector quantization technique, allowing to preserve the topology and the distribution of a set of data.

However, in our specific application, we found the necessity of modulating the topology matching property of the GNG algorithm in order to control the distribution of units between zones with high and low density. To achieve this, we modified the original algorithm proposed by Fritzke by adding three new parameters, two controlling the quantization error and the amount of neuron units in the network, and one controlling the distribution of these units. The modified version continues to perform topology preserving, but contrary to the original version it permits the modulation of the distribution matching capabilities of the original algorithm. These changes allows the quantization of datasets having high contrasts in density while keeping the information of low density areas using a limited number of prototypes.

Acknowledgements

This work is part of a cooperation project between BIOTEC, CIAT, CENICAÑA (Colombia) and HEIG-VD (Switzerland) named *"Precision agriculture and the construction of field-crop models for tropical fruits"*. The economical support is given by several institutions in Colombia (MADR, COLCIENCIAS, ACCI) and the State Secretariat for Education and Research (SER) in Switzerland.

References

1. Alpaydin, E.: Introduction to Machine Learning. MIT Press, Cambridge (2004)
2. Bouchachia, A., Gabrys, B., Sahel, Z.: Overview of Some Incremental Learning Algorithms. In: Fuzzy Systems Conference. FUZZ-IEEE 2007, vol. 23-26, pp. 1–6 (2007)
3. Bradley, P., Gehrke, J., Ramakrishnan, R., Srikant, R.: Scaling mining algorithms to large databases. Commun. ACM 45, 38–43 (2002)
4. Cselenyi, Z.: Mapping the dimensionality, density and topology of data: The growing adaptive neural gas. Computer Methods and Programs in Biomedicine 78, 141–156 (2005)
5. Fritzke, B.: Unsupervised ontogenic networks. Handbook of Neural Computation, ch. c2.4. Institute of Physics and Oxford University Press (1997)
6. Fritzke, B.: A Growing Neural Gas Learns Topologies. Advances in Neural Information Processing Systems, vol. 7. MIT Press, Cambridge (1995)
7. Ganti, V., Gehrke, J., Ramakrishnan, R.: Mining very large databases. Computer 32, 38–45 (1999)
8. Giraud-Carrier, C.: A note on the utility of incremental learning. AI Commun. 13, 215–223 (2000)
9. Heinke, D., Hamker, F.H.: Comparing neural networks: a benchmark on growing neural gas, growing cell structures, and fuzzy ARTMAP. IEEE Transactions on Neural Networks 9, 1279–1291 (1998)
10. Hijmans, R., Cameron, S., Parra, J., Jones, P., Jarvis, A.: Very High Resolution Interpolated Climate Surfaces for Global Land Areas. Int. J. Climatol. 25, 1965–1978 (2005)
11. Martinetz, T., Schulten, K.: Topology representing networks. Neural Networks 7, 507–522 (1994)
12. Martinetz, T., Schulten, K.: A "neural gas" network learns topologies. Artificial Neural Networks, 397–402 (1991)

Active Learning Using a Constructive Neural Network Algorithm

José Luis Subirats[1], Leonardo Franco[1], Ignacio Molina Conde[2], and José M. Jerez[1]

[1] Departamento de Lenguajes y Ciencias de la Computación
Escuela Técnica Superior de Ingeniería en Informática
Universidad de Málaga,
Campus de Teatinos S/N, 29071 Málaga, Spain
[2] Departamento de Tecnología Electrónica
Escuela Técnica Superior de Ingeniería en Telecomunicación
Universidad de Málaga,
Campus de Teatinos S/N, 29071 Málaga, Spain

Abstract. Constructive neural network algorithms suffer severely from overfitting noisy datasets as, in general, they learn the set of examples until zero error is achieved. We introduce in this work a method for detect and filter noisy examples using a recently proposed constructive neural network algorithm. The method works by exploiting the fact that noisy examples are harder to be learnt, needing a larger number of synaptic weight modifications than normal examples. Different tests are carried out, both with controlled experiments and real benchmark datasets, showing the effectiveness of the approach.

1 Introduction

A main issue at the time of implementing feed-forward neural networks in classification or prediction problems is the selection of an adequate architecture [1,2,3]. Feed-forward neural networks trained by back-propagation have been widely used in several problems but yet the standard approach for selecting the number of layers and number of hidden units of the neural architecture is the inefficient trial-by-error method. Several constructive methods and pruning techniques [1] have been proposed as an alternative for the architecture selection process but it is a research issue whether these methods can achieve the same level of prediction accuracy. Constructive algorithms start with a very small network, normally comprising a single neuron, and work by adding extra units until some convergence condition is met [4,5,6,7]. On the other hand, pruning techniques start with a very large architecture and work by eliminating unnecessary weights and units [8].

Despite the existence of many different constructive algorithms, they have not been extensively applied in real problems. This fact is relatively surprising, given that they offer a systematic and controlled way of obtaining an architecture and also because they offer the possibility of an easier rule extraction procedure. In a 1993 work, Smieja [9] argued that constructive algorithms might be more

V. Kůrková et al. (Eds.): ICANN 2008, Part II, LNCS 5164, pp. 803–811, 2008.

efficient in terms of the learning process but cannot achieve a generalization ability comparable to back-propagation neural networks. Smieja arguments were a bit speculative more than based on obtained results, but nevertheless might explain the fact that constructive methods have not been widely applied to real problems. In recent years new constructive algorithms have been proposed and analyzed, and the present picture might have changed [7,10].

One of the problems that affects predictive methods in general, is the problem of overfitting [11,12]. In particular, overfitting affects severely neural network constructive algorithms as they, in general, learn towards zero error. The overall strategy in constructive algorithms for avoiding overfitting is by creating very compact architectures. Unfortunately, this approach is not enough when the input data is noisy as it is normally the case of real data. A solution to this overfitting problem might be the implementation of methods that exclude noisy instances from the training dataset [13,14,15,16,17]. In this work, we use a recently introduced constructive neural network algorithm named C-Mantec [18] for detecting noisy examples. The method can detect and filter noisy instances leading to an improvement in the generalization ability of the algorithm and permitting to obtain more compact neural network architectures.

2 The C-Mantec Algorithm

The C-Mantec algorithm is a constructive neural network algorithm that creates architectures with a single layer of hidden nodes with threshold activation functions. For functions with 2 output classes, the constructed networks have a single output neuron computing the majority function of the responses of the hidden nodes (i.e., if more than half of the hidden neurons are activated the output neuron will be active). The learning procedure starts with an architecture comprising a single neuron in the hidden layer and adds more neurons every time the present ones are not able to learn the whole set of training examples. The neurons learn according to the thermal perceptron learning rule proposed by Frean [5], for which the synaptic weights are modified according to Eq. 1.

$$\delta w_i = (t - o)\, \psi_i \,\frac{T}{T_0}\, \exp\{-\frac{|\phi|}{T}\}\,, \tag{1}$$

where t is the target value of the example being considered, o represent the actual output of the neuron and ψ is the value of the input unit i. T is an introduced temperature, T_0 the starting temperature value and ϕ is a measure of how far is the presented example from the actual synaptic vector. The thermal perceptron is a modification of the perceptron rule that incorporates a modulation factor forcing the neurons to learn only target examples close to the already learnt ones, in order to avoid forgetting the stored knowledge. For a deeper analysis of the thermal perceptron rule, see the original paper [5].

At the single neuron level the C-Mantec algorithm uses the thermal perceptron rule, but at a global level the C-Mantec algorithm incorporates competition between the neurons, that makes the learning procedure more efficient and permitting to obtain more compact architectures [18]. Competition between neurons is implemented as follows: for a given input example, the neuron with the smallest value of the parameter ϕ is picked as the neuron that will learn the presented input (note that the weights are modified only for wrongly classified inputs). The temperature, T, controlling each individual perceptrons, is lowered every time the neuron gets an update of its weights. When a new unit is added to the network, the temperature of all neurons is reset to the initial value T_0. The learning procedure continues in this way until enough neurons are present in the architecture, and the network is able to learn the whole sets of inputs. Regarding the role of the parameters, an initial high temperature T_0 ensures a certain number of learning iterations and an initial phase of global exploration for the weights values, as for high temperature values changes are easier to be accepted. The parameter setting for the algorithm is relatively simple as C-Mantec has been shown to be very robust to changes. The convergence of the algorithm is ensured because the learning rule is very conservative in their changes, preserving the acquired knowledge of the neurons and given by the fact that new introduced units learn at least one input example. Tests performed with noise-free Boolean functions using the C-Mantec algorithm show that it generates very compact architectures with less number of neurons than existing constructive algorithms [18]. However, when the algorithm was tested on real datasets, it was observed that a larger number of neurons was needed because the algorithm overfit noisy examples. To avoid this overfitting problem the method introduced in the next section is developed in this work.

3 The "Resonance" Effect for Detecting Noisy Examples

In Fig. 1, an schematic drawing shows the "resonance" effect that is produced when a thermal perceptron tries to learn a set of instances containing a

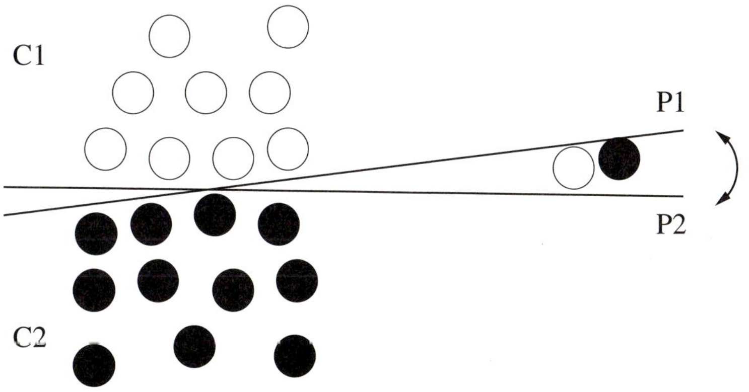

Fig. 1. Schematic drawing of the "resonance effect" that occurs when noisy examples are present in the training set. A thermal perceptron will learn the "good" examples, represented at the left of the figure, but will classify rightly only one of the noisy samples. Further learning iterations in which the neuron tries to learn the wrongly classified example will produce an oscillation of the separating hyperplane. The number of times the synaptic weights are adjusted upon presentation of an example can be used to detect noisy inputs.

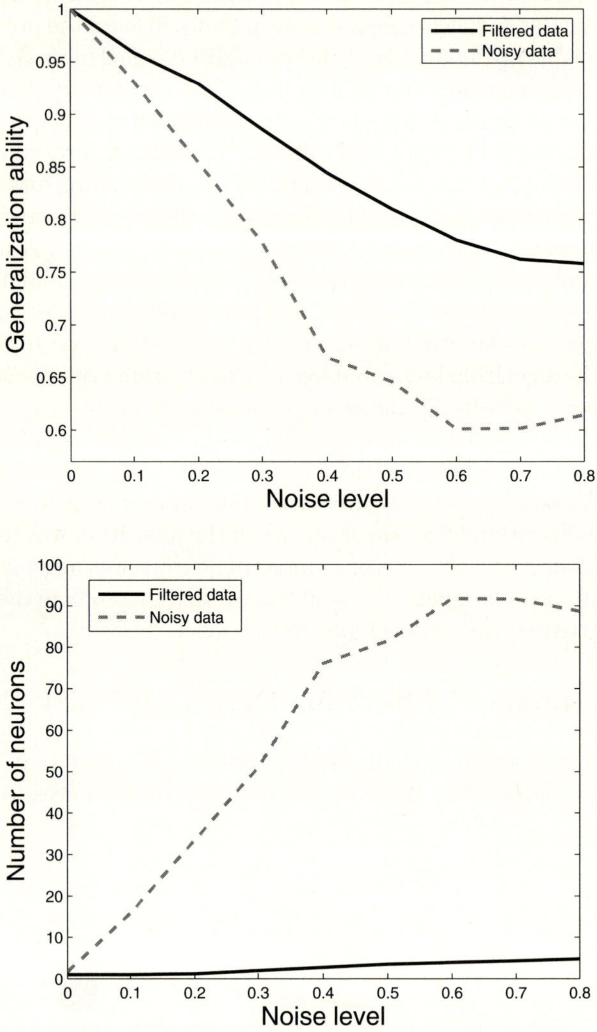

Fig. 2. The effect of attribute noise. Top: Generalization ability as a function of the level of attribute noise for the "modified" Pima indians diabetes dataset for the C-Mantec algorithm applied with and without the filtering stage. Bottom: The number of neurons of the generated architectures as a function of the level of noise. The maximum number of neurons was set to 101.

contradictory pair of examples. In the figure, the set of "good" examples is depicted in the left part of the figure, while the contradictory pair is on the right. When a single neuron tries to learn this set, the algorithm will find an hyperplane from a beam of the possible ones (indicated in the figure) that classifies correctly the whole set except for 1 of the noisy examples. Further learning iterations produce a resonant behavior, as the dividing hyperplane oscillates trying

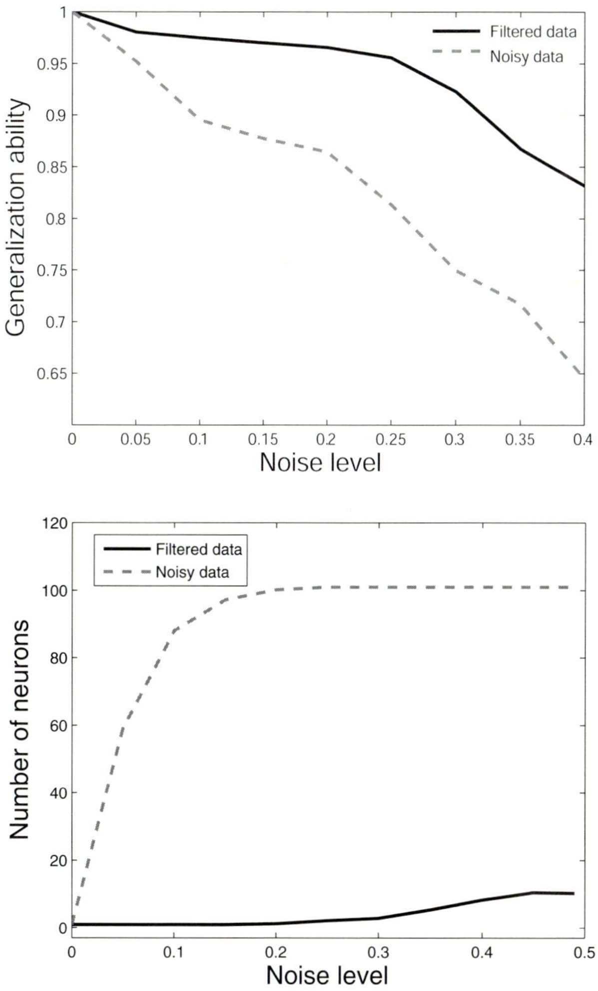

Fig. 3. The effect of class noise. Top: Generalization ability as a function of the level of class noise for the modified Pima indians diabetes dataset for the cases of implementing the filtering stage and for the case of using the whole raw dataset. Bottom: The number of neurons of the generated architectures for the two mentioned cases of the implementation of the C-Mantec algorithm.

to classify correctly the wrong example. Eventually, the iterations will end and as the whole set cannot be learnt, a new neuron will be added to the network. It was observed that these noisy examples make the network to grow excessively and degrade the generalization ability, and thus a method for removing them is quite useful. The method is based on counting the number of times each training example is presented to the network; and if the number of presentations for an example is larger by two standard deviations from the mean, it is removed from the training set. The removal of examples is made on-line as the architecture is constructed but a second learning phase with the selected set of examples was implemented as better results can be obtained. In this second phase, the training set is fix as no removal of examples is permitted.

To test the new method for removal of noisy examples a “noise-free” dataset is created from a real dataset, and then controlled noise was added on the attributes (input variables) and on the class (output), separately to see if there was any evident difference between the two cases [19]. The dataset chosen for this analysis is the Pima Indians Diabetes dataset, selected because it has been widely studied and also because it is considered a difficult set with an average generalization ability around 75%. To generate the “noise-free” dataset, the C-Mantec algorithm was run with a single neuron, that classify correctly approximately 70% of the dataset, and then the “noise-free” dataset was constructed by presenting the whole set of inputs through this network to obtain the “noise-free” output. Two different experiments were carried out: in the first one, noise was added to the attributes of the dataset and the performance of the C-Mantec algorithm was analyzed with and without the procedure for noisy examples removal. In Fig. 2 (top) the generalization ability for both mentioned cases is shown for a level of noise between 0 and 0.8 and the results are the average over 100 independent runs. For a certain value of added noise, x, the input values were modified by a random uniform value between $-x$ and x. The bottom graph shows the number of neurons in the generated architectures when the filtering process was and was not applied as a function of the added attribute noise. It can be clearly seen that the removal of the noisy examples help to obtain much more compact architectures and a better generalization ability. The second experiment consisted in adding noise to the output values and the results are shown on Fig. 3. In this case the noise level indicate the probability of modifying the class value to a random value. The results in this case also confirm the effectiveness of the filtering approach in comparison to the case of using the whole “noisy” dataset.

4 Experiments and Results

We tested the noise filtering method introduced in this work using the C-Mantec constructive algorithm on a set of 11 well known benchmark functions. The set of functions contains 6 functions with 2 classes and 5 multi-class problems with a number of classes up to 19. The C-Mantec algorithm was run a maximum number of iterations of 50.000 and an initial temperature (T_0) equals to the number of inputs of the analyzed functions, but it is worth noting that the algorithm is quite robust to changes on these parameters. The results are shown in Table 1, where it is shown the number of neurons of the obtained architectures and the generalization ability obtained, including the standard deviation values, computed over 100 independent runs. The last column of Table 1 shows, as a comparison, the generalization ability values obtained by Prechelt [20] in a work where he analyzed in a systematic way the prediction capabilities of different topologies neural networks, and thus we believe that the reported values are highly optimized. The number and the set of training and test examples were chosen identically in both compared cases. The results shows that the C-Mantec algorithm outperforms the ones obtained by Prechelt in 6 out of 11 problems and on average the generalization ability is 2.1% larger. Regarding the size of the networks obtained using the method introduced in this work, the architectures are very small for all problems with 2 or 3 classes, for which the architectures contain less than 4 neurons (on average) for all these cases. For the multi-class problems the algorithm generates networks with a larger number of hidden neurons but this is because of the method used to treat multiclass problems that will be reported in [18].

Table 1. Results for the number of neurons and the generalization ability obtained with the C-Mantec algorithm using the data filtering method introduced in this work. The last column shows the results from [20] (See text for more details).

Function	Inputs	Classes	Neurons	Generalization C-Mantec	Generalization NN [20]
Diab1	8	2	3.34 ± 1.11	76.62 ± 2.69	74.17 ± 0.56
Cancer1	9	2	1 ± 0.0	96.86 ± 1.19	97.07 ± 0.18
Heart1	35	2	2.66 ± 0.74	82.63 ± 2.52	79.35 ± 0.31
Heartc1	35	2	1.28 ± 0.57	82.48 ± 3.3	80.27 ± 0.56
Card1	51	2	1.78 ± 0.87	85.16 ± 2.48	86.63 ± 0.67
Mushroom	125	2	1 ± 0.0	99.98 ± 0.04	100.00 ± 0.0
Thyroid	21	3	3 ± 0.0	91.91 ± 0.59	93.44 ± 0.0
H orse1	58	3	3 ± 0.0	66.56 ± 5.08	73.3 ± 1.87
Gene1	120	3	3.03 ± 0.22	88.75 ± 1.07	86.36 ± 0.1
Glass	9	6	17.84 ± 1.19	63.75 ± 6.38	53.96 ± 2.21
Soybean	82	19	171 ± 0.0	91.63 ± 1.89	90.53 ± 0.51
Average	50.27	4.18	18.99 ± 0.43	84.21 ± 2.03	82.50 ± 0.63

5 Discussion

In this work we have introduced a new method for filtering noisy examples using a recently developed constructive neural network algorithm C-Mantec. The filtering method is based on the fact that noisy examples are more difficult to be learnt, and this fact is evident during the learning process in which the constructive algorithm tries to classify correctly the examples using the minimum number of neurons. Noisy examples need more learning updates of the synaptic weights, and this fact permits its identification and further removal. Simulations performed on benchmark datasets show that the generalization ability and size of the resulting network are very much improved after the removal of the noisy examples and a comparison, done against previous reported results [20], shows that the generalization ability was on average a 2.1% larger, indicating the effectiveness of the C-Mantec algorithm implemented with the new filtering stage. It has to be noted that the introduced method of data selection can be used as a pre-processing stage for its use with other prediction algorithms. We have also analyzed the performance of the filtering stage on datasets contaminated by only attribute or class noise, but did not find any clear difference between these two cases for which the filtering process worked equally well.

Acknowledgements

The authors acknowledge support from CICYT (Spain) through grant TIN2005-02984 (including FEDER funds) and from Junta de Andalucía through grant P06-TIC-01615. Leonardo Franco acknowledges support from the Spanish Ministry of Education and Science through a Ramón y Cajal fellowship.

References

1. Haykin, S.: Neural Networks: A Comprehensive Foundation. Macmillan/IEEE Press (1994)
2. Lawrence, S., Giles, C.L., Tsoi, A.C.: What Size Neural Network Gives Optimal Generalization? Convergence Properties of Backpropagation. In Technical Report UMIACS-TR-96-22 and CS-TR-3617, Institute for Advanced Computer Studies, Univ. of Maryland (1996)
3. Gòmez, I., Franco, L., Subirats, J.L., Jerez, J.M.: Neural Networks Architecture Selection: Size Depends on Function Complexity. In: Kollias, S.D., Stafylopatis, A., Duch, W., Oja, E. (eds.) ICANN 2006. LNCS, vol. 4131, pp. 122–129. Springer, Heidelberg (2006)
4. Mezard, M., Nadal, J.P.: Learning in feedforward layered networks: The tiling algorithm. J. Physics A. 22, 2191–2204 (1989)
5. Frean, M.: The upstart algorithm: A method for constructing and training feed-forward neural networks. Neural Computation 2, 198–209 (1990)
6. Parekh, R., Yang, J., Honavar, V.: Constructive Neural-Network Learning Algorithms for Pattern Classification. IEEE Transactions on Neural Networks 11, 436–451 (2000)
7. Subirats, J.L., Jerez, J.M., Franco, L.: A New Decomposition Algorithm for Threshold Synthesis and Generalization of Boolean Functions. IEEE Transactions on Circuits and Systems I (in press, 2008)
8. Reed, R.: Pruning algorithms - a survey. IEEE Transactions on Neural Networks 4, 740–747 (1993)
9. Smieja, F.J.: Neural network constructive algorithms: trading generalization for learning efficiency? Circuits, systems, and signal processing 12, 331–374 (1993)
10. do Carmo Nicoletti, M.C., Bertini, J.R.: An empirical evaluation of constructive neural network algorithms in classification tasks. International Journal of Innovative Computing and Applications 1, 2–13 (2007)
11. Bramer, M.A.: Pre-pruning Classification Trees to Reduce Overfitting in Noisy Domains. In: Yin, H., Allinson, N.M., Freeman, R., Keane, J.A., Hubbard, S. (eds.) IDEAL 2002. LNCS, vol. 2412, pp. 7–12. Springer, Heidelberg (2002)
12. Hawkins, D.M.: The problem of Overfitting. Journal of Chemical Information and Computer Sciences 44, 1–12 (2004)
13. Angelova, A., Abu-Mostafa, Y., Perona, P.: Pruning training sets for learning of object categories. In: IEEE Computer Society Conference on Computer Vision and Pattern Recognition, CVPR 2005, vol. 1, pp. 494–501 (2005)

14. Cohn, D., Atlas, L., Ladner, R.: Improving Generalization with Active Learning. Mach. Learn. 15, 201–221 (1994)
15. Cachin, C.: Pedagogical pattern selection strategies. Neural Networks 7, 175–181 (1994)
16. Kinzel, W., Rujan, P.: Improving a network generalization ability by selecting examples. Europhys. Lett. 13, 473–477 (1990)
17. Franco, L., Cannas, S.A.: Generalization and Selection of Examples in Feedforward Neural Networks. Neural Computation 12(10), 2405–2426 (2000)
18. Subirats, J.L., Franco, L., Jerez, J.M.: Competition and Stable Learning for Growing Compact Neural Architectures with Good Generalization Abilities: The C-Mantec Algorithm (in preparation, 2008)
19. Zhu, X., Wu, X.: Class noise vs. attribute noise: a quantitative study of their impacts. Artif. Intell. Rev. 22, 177–210 (2004)
20. Prechelt, L.: Proben 1 – A Set of Benchmarks and Benchmarking Rules for Neural Network Training Algorithms. Technical Report (1994)

M-CLANN: Multi-class Concept Lattice-Based Artificial Neural Network for Supervised Classification

Engelbert Mephu Nguifo[1], Norbert Tsopzé[1,2], and Gilbert Tindo[2]

[1] Université de Lille-Nord, Artois
CRIL-CNRS UMR 8188, IUT de Lens
SP 16, Rue de l'Université 62307 Lens Cedex, France
[2] Université de Yaoundé I, Faculté des Sciences,
Département d'Informatique BP 812
Yaoundé - Cameroun

Abstract. Multi-layer neural networks have been successfully applied in a wide range of supervised and unsupervised learning applications. However defining its architecture is a difficult task, and might make their usage very complicated. To solve this problem, a rule-based model, KBANN, was previously introduced making use of an apriori knowledge to build the network architecture. Neithertheless this apriori knowledge is not always available when dealing with real world applications. Other methods presented in the literature propose to find directly the neural network architecture by incrementally adding new hidden neurons (or layers) to the existing network, network which initially has no hidden layer. Recently, a novel neural network approach CLANN based on concept lattices was proposed with the advantage to be suitable for finding the architecture of the neural network when the apriori knowledge is not available. However CLANN is limited to application with only two-class data, which is not often the case in practice. In this paper we propose a novel approach M-CLANN in order to treat multi-class data. Carried out experiments showed the soundness and efficiency of our approach on different UCI datasets compared to standard machine learning systems. It also comes out that M-CLANN model considerably improved CLANN model when dealing with two-class data.

1 Introduction

An artificial neural network (ANN) is an adaptive system that changes its structure based on external or internal information that flows through the network during the learning phase. ANN are useful especially when there is no a priori knowledge about the analyzed data. They offer a powerful and distributed computing architecture, with significant learning abilities and they are able to represent highly nonlinear and multivariable relationships. ANN have been successfully applied to problems in pattern classification, function approximation, optimization, pattern matching and associative memories [13]. Different types

V. Kůrková et al. (Eds.): ICANN 2008, Part II, LNCS 5164, pp. 812–821, 2008.

of ANN have been reported in the literature, among which the multilayer feed-forward network, also called multi-layer perceptron (MLP), was the first and arguably simplest type of ANN devised. MLP networks trained using the back-propagation learning algorithm are limited to searching for a suitable set of weights in an a priori fixed network topology. The selection of a network architecture for a specific problem has to be done carefully. In fact there are no known efficient methods for determining the optimal network topology of a given problem. Too small networks are unable to adequately learn the problem well while overly large networks tend to overfill the training data and consequently result in poor generalization performance. In practice, a variety of architectures are tried out and the one that appears best suited to the given problem is picked. Such a trial-and-error approach is not only computationally expensive but also does not guarantee that the selected network architecture will be close to optimal or will generalize well. An ad-hoc and simple manner deriving from this approach is to use one hidden layer with a number of neurons equal to the average of the number of input neurons and the number of output neurons. In the literature, different automatic approaches have been reported to dynamically build the network topology. These works could be divided into two groups:

1. Search an optimal network to minimize the number of units in the hidden layers [13]. These techniques bring out a dynamic solution to the ANN topology problem when a priori knowledge is not available. One technique suggests to construct the model by incrementally adding hidden neurons or hidden layers to the network until the obtained network becomes able to better classify the training data set. Another technique is network pruning which begins by training a larger than necessary network and then eliminate weights and neurons that are deemed redundant. An alternative approach consists of using the genetic approach [4], which is computationally expensive. All these (incremental, pruning, genetic) techniques results to neural network that can be seen as black box system, since no semantic is associated to each hidden neuron. Their main limitation is the intelligibility of the resulting network (black-box prediction is not satisfactory [1,5]).
2. Use a set of an a priori knowledge (set of implicative rules) on the problem domain and derive the neural network from this knowledge [15]. The a priori knowledge is provided by an expert of the domain. The main advantage here is that each node in the network represents one variable in the rules set and each connection between two nodes represents one dependence between variables. The obtained neural network, KBANN (Knowledge-Based ANN), is a comprehensible ANN since each node is semantically meaningful, and the ANN decision is not viewed as deriving from a black-box system, but could be easily explain using a subset of rules from the initial apriori knowledge. But this solution is limited while the apriori knowledge is not available as it might be the case in practice.

There are also in the literature many works which help user to optimise [19] or prune networks by pruning some connections [10] or by selecting some variables [3] among the entire set example variables. But these works do not propose an efficient method to build neural network. We propose here a novel solution, M-CLANN (Multi-class Concept Lattices- based Artificial Neural Networks), to build a network topology where each node has an associated semantic without using an a priori knowledge. M-CLANN is an extended version of the CLANN approach [17]. Both approaches uses formal concept analysis (FCA) theory to build a semi-lattice from which the NN topology is derived and trained by error backpropagation. The main difference between M-CLANN and CLANN are two-folds. First M-CLANN can deal with multi-class classification problem, while CLANN is limited to two-class. Second, the derived topology from the semi-lattice is different in both systems. Our proposed approach presents many advantages: (1) the proposed architecture is a multi-layer feed-forward neural network, such that the use of learning algorithm such as back propagation is obvious; (2) each neuron has a semantic as it corresponds to a formal concept in the semi-lattice, which is a way to justify the presence of a neuron; (3) each connection (between input neuron and hidden neuron, and between hidden neurons) in the derived ANN has also a semantic as it is associated to a link in the Hasse diagram of the semi-lattice; (4) the knowledge for other systems (such as expert system) could be extracted from the training data through the model; (5) It better classifies the dataset after training, as observed during experimentations on some datasets of UCI repository [11].

The paper is organized as follows: Section 2 gives preliminary definitions. Section 3 presents the CLANN model. Section 4 is dedicated to our novel approach M-CLANN. The empirical evidences about the utility of the proposed approach are presented in Section 5.

2 Definitions – Preliminaries

A **formal context** is a triplet $K = (O, A, I)$ where O is a not empty finite set of objects, A a not empty finite set of attributes (or items) and I is a binary relation between elements of O and elements of A (formally $I \subseteq O \times A$).

Let f and g be two applications defined as follows: $f : 2^O \longrightarrow 2^A$, such that $f(O_1) = O_1' = \{a \in A \ / \ \forall o \in O_1 \ , \ (o, a) \in I\}$, $O_1 \subseteq O$ and $g : 2^A \longrightarrow 2^O$, such that $g(A_1) = A_1' = \{o \in O \ / \ \forall a \in A_1 \ , \ (o, a) \in I\}$, $A_1 \subseteq A$; a pair (O_1, A_1) is called **formal concept** iff $O_1 = A_1'$ *and* $A_1 = O_1'$. O_1 (resp. A_1) is the extension (resp. intension) of the concept.

Let L be the entire set of concepts extracted from the context K and $\leq$ a relation defined as $(O_1, A_1) \leq (O_2, A_2) \Rightarrow (O_1 \subset O_2)$ (or $A_1 \supset A_2$). The relation $\leq$ defines the order relation on L [7]. If $(O_1, A_1) \leq (O_2, A_2)$ is verified (without intermediated concept) then the concept (O_1, A_1) is called the successor of the concept (O_2, A_2) and (O_2, A_2) the predecessor of (O_1, A_1). The **Hasse diagram** is the graphical representation of the relation *successor/predecessor* on the entire set of concepts L. More details on FCA could be found in [7]. Different works in the literature (e.g. [2]) have shown how to derive implication rules or associations rules [8] from this join semi-lattice, sometimes named iceberg concept lattice [16]; or to use the lattice structure in classification [12]. Neithertheless CLANN only treats problem with binary-class data, which is not often the case in practice.

3 The CLANN Model

We describe in this section the different steps of our new approach as shown by figure 1. The process of finding the architecture of neural networks are three-folds: (1) build a join semi-lattice of formal concepts by applying constraints to select relevant concepts; (2) translate the join semi-lattice into a topology of the neural network, and set the initial connections weights; (3) train the neural network.

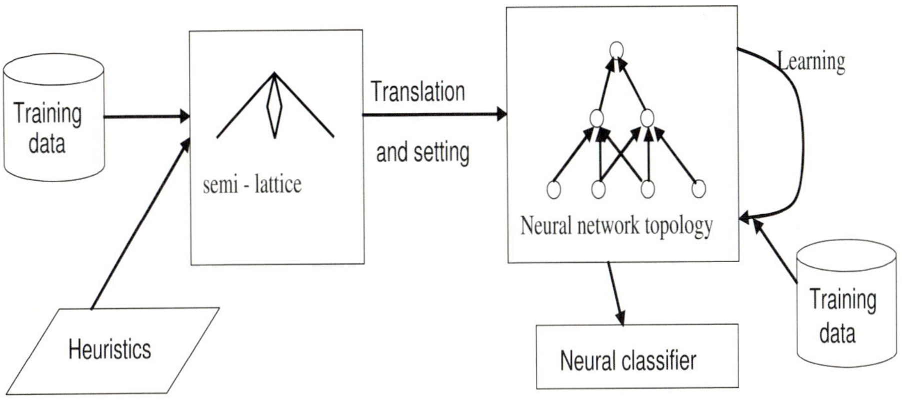

Fig. 1. Neural network topology definition

3.1 Semi-lattice Construction

There are many algorithms [6] which can be used to construct concept lattices; few of them build the Hasse diagram. Lattice could be processed using top-down or bottom-up techniques. In our case, a levelwise approach presents advantage to successively generate concepts of the join semi-lattice and the Hasse diagram. For this reason, we choose to implement the Bordat's algorithm [6] which is suitable here. Concepts included in the lattice are only those which satisfy the defined constraints.

3.2 Constraints

In order to reduce the size of lattice and then the time complexity, we present some constraints regularly used to select concepts during the learning process.

Frequency of concept. A concept is frequent if it contains at least α (also called minsupp is specified by the user) objects. The support s of a concept (X, Y) is the ratio between the cardinality of the set X and the total number of objects ($|O|$) ($s = \frac{100 \times |X|}{|O|}\%$). Frequency is an anti-monotone constraint which helps in pruning the lattice and reduce it computational complexity. Support could be seen as the minimal number of objects that the intention of one concept must verified before being taken in the semi-lattice.

Algorithm 1. Modified Bordat's algorithm

Require: Binary context K
Ensure: concept lattices (concepts extracted from K) and the Hasse diagram of the order relation between concepts.
Init the list L of the concepts $(G, \{\})$ $(L \leftarrow (G, \{\}))$
repeat
  for concept $c \in L$ such that his successors are not yet been calculated **do**
    Calculate the successors c' of c.
    if the specified constraint is verified by c' **then**
      add c' in L as successor of c if c' does not exist in L else connect c' as successor of c.
    end if
  end for
until no concept is added in L.
derive the neural network architecture as described in section **??** from the concept semi-lattice.

Validity of concept. Many techniques are used to reduce the size of lattice. The following notions are used in order to select concepts: a concept (X, Y) is **complete** if Y recognize all examples in dataset. A concept (X, Y) is **consistent** if Y throws back all counterexamples (formally, the set of consistent concept is $\{(X,Y)/Y \cap O^- = \{\}\}$ where $O = O^+ \cup O^-$). To reduce the restriction imposed by these two constraints, other notions are used:

1. **Validity**. A concept (X, Y) is valid if its description recognizes most examples; a valid concept is a frequent concept on the set of examples O^+; formally the set of valid concepts is defined as $\{(X, Y) \,/\, |X^+| \geq \alpha\}$ where $0 < \alpha \leq |O^+|$.
2. **Quasi-consistency**. A concept (X, Y) is quasi-consistent is if it is valid and its extension contains few counterexamples. Formally the set of quasi-consistent concepts is defined as $\{(X, Y) \,/\, |X^+| \geq \alpha \text{ } and \text{ } |X^-| \leq \beta\}$.

Height of semi-lattice. The level of a concept c is defined as the minimal number of connexions from the supremum concept to c. The height of the lattice is the greatest value of the level of concepts. Using levelwise approach to generate the join semi-lattice, a given constraint can be set to stop concept generation at a fixed level. The height of the lattice could be performed as the depth without considering the cardinality of concepts extension (or intention). In fact at each level, concept extensions (or intentions) do not have the same cardinality. The number of layers of the semi-lattice is a parameter corresponding to the maximum level (height) of the semi-lattice.

4 The M-CLANN Model

As CLANN, M-CLANN defines the topology of a neural network in two phases: in the first phase, a join semi-lattice structure is constructed and in the second one, the join semi-lattice structure is translated into a neural network architecture.

M-CLANN builds the join semi-lattice using a modified version of Bordat's algorithm [6]. The modified algorithm uses monotonic constraints to prune the lattice and then reduce its complexity. The semi-lattice construction process starts by finding the supremum element. The process continues by generating the successors of the concepts that belong to the existing set until there are no concept which satisfies the specified constraints. Algorithm 2 presents the M-CLANN method to translate the semi-lattice into ANN.

Objects used in this algorithm are defined as follows: K is a formal context (dataset); L is the semi-lattice built from the training dataset K; c and c' are formal concepts; n is the number of attributes in each training pattern; m is the number of output classes in the training dataset; c a formal concept, element of L; NN is the comprehensive neural network build to classify the data.

Algorithm 2. Translation of semi-lattice into ANN topology

Require: L a semi-lattice structure built using specified constraints.
Ensure: NN initial topology obtained from the semi-lattice L

```
for each concept c ∈ L do
    if the set of predecessor of c is empty, mark its successor as "last hidden neuron";
    Else c becomes neurons and add to NN with the successor and predecessor as
    in L; if the set of successor of c is empty then mark c as "first hidden neuron".
    Endif
end for
Create a new layer of n neurons and connect each neuron of this layer to the
  neurons marked as "first hidden neuron" in NN.
Create a new layer of m neurons and connect each neuron of this layer to the
  neurons marked as "last hidden neuron" in NN.
Initialize connection weights and train them.
```

Among the constraints used in CLANN, the validity of concept is not use in M-CLANN since M-CLANN does not consider the class of each object during the semi-lattice building process. The constraints used in M-CLANN are frequency of concept and the height of the semi-lattice.

Threshold is zero for all units and the connection weights are initialized as follows:

- Connection weights between neurons derived directly from the lattice is initialized to 1. This implies that when the neuron is active, all its predecessors are active too.
- Connection weights between the input layer and hidden layer is initialized as follows: 1 if the attribute represented by the input appears in the intention Y of the concept associated to the ANN node and -1 otherwise. This implies that the hidden unit connected to the input unit will be active only if the majority of its input (attributes including in its intention) is 1.

5 Experimentations

To examine the practical aspect of the approach presented above, we run the experiments on the data available on UCI repository [11]. The characteristics of these data are collected in table 1 which contains the name of the dataset (dataset label), the number of training patterns (#Train), the number of test patterns (#Test), the number of output classes (#Class), the initial number of (nominal) attributes in each pattern (#Nom), the number of binary attributes obtained after binarization (#Bin). Those attributes were binarized by the Weka [18] binarization procedure "Filters.NominalToBinary". The diversity of these data (from 24 to 3196 training patterns; from 2 to 19 output classes) helps in revealing the behaviour of each model in many situations. There is no missing values in these datasets.

Table 1. Experimental data sets

Dataset	#Train	#Test	#Class	#Nom	#Bin
Balance-scale (Bal)	625	0	3	4	20
Chess	3 196	0	2	36	38
Hayes-roth (Hayes)	132	28	3	5	15
Tic-tac-toe (Tic)	958	0	2	9	26
Spect	80	187	2	22	22
Monsk1	124	432	2	6	15
Monsk2	169	432	2	6	15
Monsk3	122	432	2	6	15
Lymphography (lympho)	148	0	3	18	51
Solar-flare1 (Solar1)	323	0	7	12	40
Solar-flare2 (Solar2)	1066	0	7	12	40
Soybean-backup (Soyb)	307	376	19	35	151
Lenses	24	0	3	4	12

The two constraints presented above have been applied in selecting concepts during experimentation. In the first step we separately use each constraint and we combine them in the second step during the join semi-lattice construction process.

The experiment results are obtained from the model trained by backpropagation algorithm [14] and validated by 10-cross validation or holdout [9]. The parameters of the learning algorithm are the following: as activation function, we use the sigmoid ($f(x) = \frac{1}{1+\exp x}$), 500 iterations in the weight modification process and 1 as learning rate. In the result table, the symbol "-" indicates that no concept satisfies the constraint and the process has not converged. Table 2 presents the accuracy rate (percentage) obtained with data in table 1 and those obtained using other classifiers. These classifiers are MLP (a WEKA implementation of the multilayer perceptron classifier), C4.5 (a decision tree based classifier), IB1 (a case based learning classifier model). The result presented is the accuracy rate. In this table, MCL1 is M-CLANN built using only one level of the semi-lattice; MCL30 and MCL20 are M-CLANN built using respectively 30 and 20 percent as frequency of the concepts. CLANN column represents the accuracy rate obtained using the original version of CLANN (with a constraint of one level; x indicates that it is not possible to use CLANN since it is a multi-class dataset). Finally MC1-30 and MC1-20 are respectively M-CLANN built with a combination of one level lattice and 30% minimum support and M-CLANN built with a combination of one level semi-lattice and 20% minimum support as constraints. The best results (accuracy rate) of M-CLANN are obtained with the α value equal to 20% (MCL20). With high minimum support values, the semi-lattice does sometimes not contain sufficient concepts to better classify the data. For instance, with the minimum support value set to 35%, the semi-lattice built from Balance-scale is empty.

M-CLANN was not compared with KBANN because we haven't an apriori knowledge about these data. The goal of this comparison is to see the behaviour (on the supervised classification problems) of M-CLANN regarding those of other learning models. From the average of accuracy rate presented in the table 2, MCL20 is the best classifier in average. Using different parameters settings, M-CLANN is still better than other classifiers. MLP of Weka platform is the best one compared to C4.5 and IB1. Another advantage of M-CLANN over MLP is that each neuron is meaningful and this can be used to explain its decision. During the experimentations, the running times of MLP and M-CLANN are much more greater than that of C4.5 and IB1. The difference between the running times of M-CLANN and those of MLP is not significative.

Table 2. Accuracy rate of used classifiers with data of table 1

Dataset	CLANN	MCL1	MCL30	MCL20	MC1-30	MC1-20	MLP	C4.5	IB1
Bal	x	99,76	-	99,89	-	99,89	98,40	77,92	66,72
Chess	93,60	99,87	93,60	93,78	99,87	99,87	99,30	98,30	89,9
Hayes	x	75,72	78,58	85,72	78,57	85,71	82,15	89,28	75,00
Tic	94,45	89,64	99,67	99,86	99,32	100	96,86	93,21	81,63
Spect	93,90	72,74	92,56	96,73	73,66	77,57	65,77	66,70	66,31
Monsk1	82,70	91,67	91,17	91,17	91,67	91,71	100	100	89,35
Monsk2	78,91	100	100	100	100	99,67	100	70,37	66,89
Monsk3	83,61	93,51	91,17	93,52	92,59	93,52	93,52	100	81,63
Lympho	x	80,78	84,67	88,91	85,71	92,56	81,76	74,32	80,41
Solar1	x	79,42	78,67	69,58	71,10	71,10	72,79	74,30	68,39
Solar2	x	75,00	76,71	70,91	75,34	78,95	68,11	69,97	66,56
Soyb	x	81,33	89,34	86,95	83,11	84,04	92,02	88,83	89,89
Lenses	x	98,67	100	99,87	98,67	99,87	95,83	91,67	100
Average	86,05	86,62	89,67	90,53	87,57	90,37	88,57	84,22	78,67

6 Conclusion

This paper shows how concept lattices can help to build a comprehensible ANN for a given task, by providing clear significance to each neuron. In this work, we use some monotonic constraints to reduce the size of the lattice and also the training time. Particularly for the classification tasks, looking at the experiment results, our model is better in average than many other classifiers in different test cases. M-CLANN model also uses binary data set. Our ongoing research firstly consists of extending M-CLANN to nominal and numeric data. Many attempts exist in the literature to build concept lattices for such data sets. We will also compare extracted rules from our model to those of other models as KBANN, or decision trees.

References

1. Andrews, R., Diederich, J., Tickle, A.B.: Survey and critique of techniques for extracting rules from trained artificial neural networks. Knowledge-Based Systems 8(6), 373–389 (1995)
2. Bastide, Y., Pasquier, N., Taouil, R., Stumme, G., Lakhal, L.: Mining minimal non-redundant association rules using frequent closed itemsets. In: CL 2000. LNCS (LNAI), vol. 1861, pp. 972–986. Springer, Heidelberg (2000)
3. Cibas, T., Fogelman, F., Gallinari, P., Raudys, S.: Variable Selection with Optimal Cell Damage. In: International conference on Artificial Neural Network (ICANN 1994), vol. 1, pp. 727–730 (1994)
4. Curran, D., O'Riordan, C.: Applying Evolutionary Computation to designing Neural Networks: A study of the State of the art. Technical report NUIG-IT-111002, department of Information Technology, NUI of Galway (2002)
5. Duch, W., Setiono, R., Zurada, J.M.: Computational intelligence methods for understanding of data. Proceedings of the IEEE 92(5), 771–805 (2004)
6. Kuznetsov, S., Obiedkov, S.: Comparing Performance of Algorithms for Generating Concept Lattices. JETAI 14(2/3), 189–216 (2002)
7. Ganter, B., Wille, R.: Formal Concepts Analysis: Mathematical foundations. Springer, Heidelberg (1999)
8. Gasmi, G., Ben Yahia, S., Mephu Nguifo, E., Slimani, Y.: A new informative generic base of association rules. In: Advances in Knowledge Discovery and Data Mining (PAKDD), vol. 3518, pp. 81–90 (2005)
9. Han, J., Hamber, M.: Data Mining: Concepts and Techniques. Morgan Kauffman Publishers, San Francisco (2001)
10. Le Cun, Y., Denker, J.S., Solla, S.A.: Optimal Brain Damage. In: Advances in Neural Information Processing Systems 2, pp. 598–605. Morgan Kaufmann Publishers, San Francisco (1990)
11. Newmann, D.J., Hettich, S., Blake, C.L., Merz, C.J.: (UCI)Repository of machine learning databases. Dept. Inform. Comput. Sci., Univ. California, Irvine, CA (1998), `http://www.ics.uci.edu/AI/ML/MLDBRepository.html`
12. Fu, H., Fu, H., Njiwoua, P., Mephu Nguifo, E.: Comparative study of FCA-based supervised classification algorithms. In: The proc of 2nd ICFCA, Sydney, pp. 313–320 (2004)

13. Parekh, R., Yang, J., Honavar, V.: Constructive Neural-Network Learning Algorithms for Pattern Classification. IEEE Transactions on Neural Networks 11, 436–451 (2000)
14. Rumelhart, D.E., Hinton, G.E., Williams, R.J.: Learning representations by back-propagating errors. Nature 323, 318–362 (1986)
15. Shavlik, J.W., Towell, G.G.: Kbann: Knowledge based articial neural networks. Artificial Intelligence 70, 119–165 (1994)
16. Stumme, G., Taouil, R., Bastide, Y., Pasquier, N., Lakhal, L.: Computing Iceberg concept lattices with TITANIC. Journal on Knowledge and Data Engineering (KDE) 2(42), 189–222 (2002)
17. Tsopze, N., Mephu Nguifo, E., Tindo, G.: CLANN: Concept-Lattices-based Artificial Neural Networks. In: Diatta, J., Eklund, P., Liquiére, M. (eds.) Proceedings of fifth Intl. Conf. on Concept Lattices and Applications (CLA 2007), Montpellier, France, October 24-26, 2007, pp. 157–168 (2007)
18. Witten, I.H., Frank, E.: Data Mining: Practical machine learning tools and techniques. Morgan Kaufmann, San Francisco (2005)
19. Yacoub, M., Bennani, Y.: Architecture Optimisation in Feedforward Connectionist Models. In: 8th International Conference on Artificial Neural Networks (ICANN 1998), Skövde, Sweden (1998)

A Classification Method of Children with Developmental Dysphasia Based on Disorder Speech Analysis

Marek Bártů and Jana Tučková

Czech Technical University, Faculty of Electrical Engineering,
Laboratory of Artificial Neural Networks Application,
Dept. of Circuit Theory,
Technická 6, 169 00, Prague 6

Abstract. This paper focuses on method developed for classifications of the speech with disorders. We are interested in children's neurological disorder called developmental dysphasia and its influence on the speech. Described classification method is based on children's speech signal analysis and allows observing the trend of the speech disorder during therapy. The classification is based on the fact that the disorder has an direct impact on speech production (i.e. movement of vocal tract). Thus, we can measure the trend of the disorders comparing patterns obtained from speech of healthy children to the patterns obtained from children with disorder. Described method is based on cluster analysis of Kohonen Self-Organizing Maps (KSOMs) trained on speech signals. The main advantage of using artificial neural network is adaptability to specific attributes of the signal and tolerance for the noise contained in recordings.

1 Introduction

The aim of described method is to distinguish between healthy and ill children and describe the trend of the disorder during therapy.

The method is based on a comparison of the differences in the parametrized speech. Purpose of the parametrization is not to give perfect representation for recognition, but describe the differences. Therefore, the MFCC parametrization is not convenient, as will be shown in the following experiment. We intentionally utilize LPC-based parametrization, that is not suitable for recognition. However, LPC coefficients are suitable for our analysis because it describing the differences in the speech better than MFCC coefficients.

The analysis itself is complicated with the fact that the regularity in children's speech evolution must be considered. Therefore our team created children's speech database. In the database are stored both the recordings of healthy and ill children from 4 to 10 years old. The recording utterances are divided into classes based on their phonetics properties (vowels, monosyllables, etc.).

The process of analysis is divided into the two phases. In the first phase, the samples from selected subset of healthy children are taken and the patterns are

V. Kůrková et al. (Eds.): ICANN 2008, Part II, LNCS 5164, pp. 822–828, 2008.

worked out. Input set is divided in two disjunctive parts. The first part is used for training KSOMs. The second part is utilized in computing the patterns using these trained KSOMs. There is one map for each class of samples from database.

In the second phase, patterns for selected ill children are calculated and compared with the patterns for training set. Whereas in the first part utilized only samples from healthy children, the computation in the second part is done using the samples from only one (ill) child acquired during one session. Comparing these two results we can observe the trend of the disorder. Important conditions are sufficient size of the training set and using different samples for computing the patterns. Satisfy these conditions ensures generalization of children's speech signal and avoid adaptation on individual speakers.

2 Description of Method

The patterns are estimated from the subset consisting only samples (speech) from healthy children. The speech is processed common way, firstly, the signal is divided into segments and weighted by the Hamming window. Then, every segment is represented in the terms of standard parameters: MFCC, PLP [2] and LPC coefficients [4] and [5]. For each segment, three different vectors are created one for each type of parameters. Whole speech is then represented by the series of such a vectors. These series are then processed using artificial neural network, namely by KSOM. Overview of the method is in Figure 1.

MFCC and PLP coefficients are utilized generally. With the respect to the results of previous experiments with processing of speech signals with KSOMs

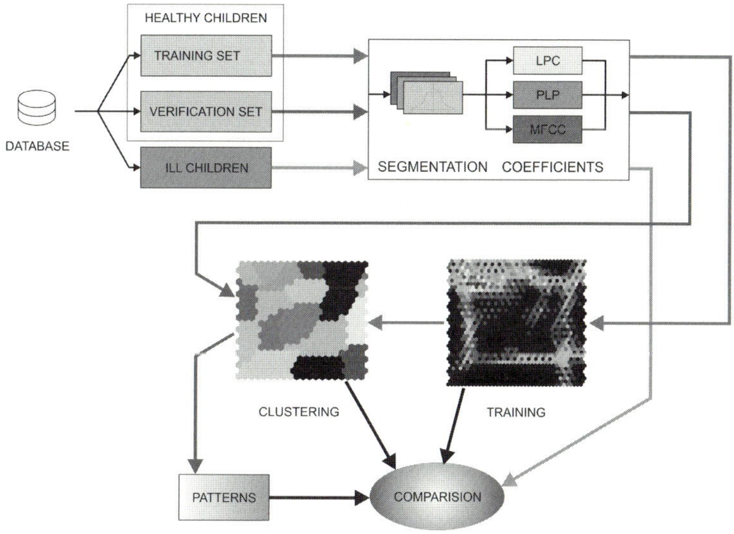

Fig. 1. Overview of the method

[6], our analysis involving also LPC coefficients. Whereas MFCC and PLP coefficients were created for recognition and thus, they tend to generalize, LPC coefficients clearly describes parameters of vocal tract with respect for these differences. Generated vectors are used to train KSOMs. There is three independent networks, one for each type of representation (i.e. one KSOM for MFCC, one for PLP and another for LPC). The greater part of the input data (training set) is used as the input for the training. The rest is for calculation of the patterns.

After training, the cluster analysis of each KSOM [3] is performed. For this purpose we utilize k-means algorithm. The analysis is utilized to obtain higher level of abstraction of the speech - every cluster represents specific features in the speech.

Using the clusters, patterns are generated. Patterns are vectors, which dimension is equal to the number of clusters in particular KSOM. Each component of the pattern represents the percent occurrence of the input vectors in the corresponding cluster. The patterns are calculated using the rest of the samples from the input set (verification set). These vectors are important for following comparison. The patterns are derived from map trained only on healthy children's speech.

Then we compute patterns for the child with disorder. After comparison with patterns estimated from healthy children's speech, we get observable differences. The measure of those differences (euclidean distance of these two vectors) qualifies differences between the ill child's speech and the speech of healthy children. The euclidean distance between representative vectors and the particular vector serves as a main criterion. Observing distances on the various classes of speech, we can approximate the trend of disease.

3 Experiment

The experiment with classification of vowels A, E, I, O and U is described in the following section. For the training set consisted of recording (samples) taken from twenty-one children (seven boys and fourteen girls). As was described above, whole training set was divided into two parts, the first part consisted of samples from that were used for training KSOMs, while the second part of the training set was utilized to calculate the patterns.

There was also verification set consisting of seven healthy children (two boys and five girls) and three children with speech disorder (two boys and one girl).

Table 1. Overview of the number of units

sets	boys	girls	together	samples
training	7	14	21	90
verification	2	5	7	35
children with disorder	2	1	3	43
total	11	20	31	168

The samples in this set were intended to confirm the results obtained from the described method. For every ill children, three recordings from various sessions were used. The overview of the division of the children taken to the experiment is in Table 1.

Speech signal was processed in the following way: whole recordings of the vowels were segmented by the 30ms Hamming window with 10ms overlay. The MFCC, PLP and LPC coefficients were calculated from these segments. For MFCC and PLP coefficients, the basic band (22kHz) was divided into twenty sub bands. The LPC coefficients were of the 8th order, the MFCC and PLP coefficients were counted for 20 bands.

We used three different sizes of the maps in the experiment 10×10 neurons, 20×20 neurons and 30×30 neurons. All the computation was done using Matlab and SOM Toolbox [7].

The maps of 10×10 neurons were too small. The exceedingly generalisation was performed and therefore the distribution function has not been captured in details. Better results were obtained using maps with 20×20 neurons, where the best results were observed for map trained on LPC coefficients. The approximation in the map trained for MFCC was worse compared to the maps trained on PLP or LPC.

Fig. 2. Clusters in LPC-trained map

There is completely different situation with the maps containing 30×30 neurons. Maps trained on MFCC and PLP coefficients generalized excessively. This is suitable for recognition, but not for our purposes. In the analyses of the differences in the speech maps trained for LPC gave better results. Nevertheless, the maps with dimension 30 neurons are not suitable for described training group because of the effect of limited generalization. For classification for described purposes, the 20×20 maps are appropriate.

The k-means algorithm was then utilized for clustering maps. The calculations were randomly initialized. The lowest number of clusters was extracted from maps trained on MFCC coefficients - from 10 to 14 clusters. For the maps trained on LPC and PLP the k-means sensitivity was better. For LPC coefficients there were extracted from 19 to 24 clusters, for PLP coefficients algorithm found between 20 and 24 clusters.

Fig. 3. Clusters in MFCC-trained map

Examples of clustered maps are in Figure 2 and in Figure 3. Each area filled in different shade of grey represents one cluster. The cluster is a group of the one or more neurons, which consist of neurons that represents similar features in the speech. Dividing the trained network into cluster bring a higher level of abstraction into analysis - instead 400 (almost) similar prototypes represented by the neurons, there will be only about 20 distinctive features represented by the clusters. As could be seen in Figure 3, MFCC parametrization gives fewer clusters. It means that this parametrization does not describing so much details as LPC based parametrization (Fig. 2). The reasons of such a differences is mentioned in previous section.

Then the patterns were calculated using the second part of the training set. Then the samples from the verification set and also the samples obtained from children with disease was compared against these vectors. In Table 2 and Table 3, there are results of comparison for the vowel A. In Table 2, each row represents results from different speaker from verification set. In Table 3, each row represents one recording session form one of the three children with disorder.

As could be seen, there is difference between result obtained from various parametrizations. For MFCC coefficients, there is not satisfactory variability and therefore these coefficients are not suitable for classification of children with

Table 2. Results for healthy children (vowel **A**)

speaker	LPC	PLP	MFCC
H1	1.4878	1.1767	1.1939
H2	1.1741	0.4844	0.6514
H3	0.8275	1.3865	1.1705
H4	0.9739	0.5795	0.7437
H5	0.7824	0.4418	0.4904
H6	1.3211	1.3218	1.1320
H7	0.6218	0.6435	0.7373
average	**1.0269**	**0.8620**	**0.8742**

Table 3. Results for the children with disorder (vowel **A**)

speaker	LPC	PLP	MFCC
D1R1	1.1273	1.1983	1.7149
D1R2	1.3461	1.2415	1.7149
D1R3	1.4625	1.1907	1.7149
D2R1	1.5789	1.2703	1.7151
D2R2	1.4989	1.2553	1.7149
D2R3	1.5127	1.2208	1.7149
D3R1	1.5613	1.3012	1.7149
D3R2	1.5663	1.3012	1.7149
average	**1.4568**	**1.2474**	**1.7149**

Table 4. Results for both groups of children – all parametrizations

vowel	health condition	LPC	PLP	MFCC
A	healthy	1.0269	0.8620	0.8742
	disorder	1.4568	1.2474	1.7149
E	healthy	1.3924	1.2464	1.2868
	disorder	1.7652	1.2795	2.0644
I	healthy	1.9056	0.4446	0.7204
	disorder	2.0020	0.9438	1.4936
O	healthy	1.1604	0.9426	1.0637
	disorder	1.4143	1.2510	1.4122
U	healthy	0.7170	0.6567	0.6683
	disorder	0.5177	1.3335	1.6071

disorders using KSOMs. For this task, the LPC and PLP parametrization giving better results. It is in accordance with the reasons mentioned above.

The average results obtained for both groups of children and all parametrisations are in Table 3. The result are distinguished by the method used to obtain coefficients from segments. As one could be seen, the LPC parametrization gives the best result in distinguish between healthy children and children with disorder. The PLP is also possibly useful, but MFCC parametrization giving unsatisfactory results that are not suitable for described purposes.

The result are strongly depended on the type of speech units. We suppose that vowels are the simplest units to analyse and in the more complicated cases, the result could be only worse. The method well detect ordinary problems in speech of the children with Developmental Dysphasia - interchange of the high wovels in the vocalic triangular. The performance of described method depends on using proper KSOMs size according to the cardinality of a training set.

4 Conclusion

Utilization of the KSOM allows modifications of the process that distinguish between speakers, to the process where common attributes are extracted and allows to distinguish between healthy children and children with disorder [1]. We plan to extend described method in order to get more information about children's speech and to reliable describe trend of the disorder.

We plan to validate method on larger, representative base and statistically prove the results, including comparison with the records of specialists. We hope that described method or its modification will bring new opportunity in diagnosis of the children with developmental dysphasia and other developmental disorders.

Acknowledgment. This work was supported by the research program Transdisciplinary Research in Biomedical Engineering No. II MSM 6840770012 of the Czech Technical University in Prague (experimental part) and by the grant of Czech Grant Agency No. 102/08/H008 (theoretical part).

References

1. Bártů, M.: The Patterns Extraction from Speech Signal of Children with Neurological Disorder. Czech Technical University in Prague, report - specialised study (in Czech) (2008)
2. Hermansky, H.: Perceptual linear predictive (PLP) analysis of speech. J. Acoust. Soc. Am. 87(4), 1738–1752 (1990)
3. Kohonen, T.: Self-Organizing Maps, 3rd edn. Springer, Heidelberg (2001)
4. Rabiner, L., Shafer, R.W.: Digital Processing of Speech Signals. Prentice-Hall, Englewood Cliffs (1978)
5. Rabiner, L., Juang, B.H.: Fundamentals of Speech Recognition. Prentice-Hall, Englewood Cliffs (1993)
6. Tvrdík, V., Žák, V., Tučková, J.: Parametrizations of a Children Speech and their Impact on Classification of Vowels. In: Digital Technologies 2007, Technical Unviversity of Žilina, 2007, Proc. on CD (2007) ISBN 978-80-8070-786-6
7. Vesanto, J., Himberg, J., Alhoniemi, E., Parhankangas, J.: SOM Toolbox for Matlab 5. Helsinky University of Technology, report A57 (2000) ISBN 951-22-4951-0

Nature Inspired Methods in the Radial Basis Function Network Learning Process

Miroslav Burša and Lenka Lhotská

Czech Technical University in Prague,
Gerstner Laboratory, BioDat Research Group
bursam@fel.cvut.cz, lhotska@fel.cvut.cz
http://bio.felk.cvut.cz

Abstract. In present we benefit from the use of nature processes which provide us with highly effective heuristics for solving various problems. Their advantages are mainly prominent in hybrid approach. This paper evaluates several approaches for learning neural network based on Radial Basis Function (RBF) for distinguishing different sets in $\mathcal{R}^L$. RBF networks use one layer of hidden RBF units and the number of RBF units is kept constatnt. In the paper we evaluate the $ACO_{\mathcal{R}}$ (Ant Colony Approach for Real domain) approach inspired by ant behavior and the PSO (Particle Swarm Optimization) algorithm inspired by behavior of flock of birds or fish in the nature. Nature inspired and classical algorithms are compared and evaluated.

Keywords: RBF, RBF Neural Network Learning, Radial Basis Function, Particle Swarm Optimization, PSO, Ant Colony Optimization, ACO, $ACO_{\mathcal{R}}$.

1 Introduction

In the last two decades, many advances in computer sciences have been based on the observation and emulation of processes of the natural world. The origins of *bioinspired informatics* can be traced up to the development of perceptrons and artificial life, which tried to reproduce the mental processes of the brain and biogenesis respectively, in a computer environment [1]. Bioinspired informatics also focuses on observing how the nature solves situations that are similar to engineering problems we face.

Radial basis function (RBF) neural networks (RBF-NN) have been introduced by Broomhead and Lowe in 1998 [2]. Their model is inspired by the locally tuned response observed in biologic neurons that can be found in several parts of the nervous system, for example cells in the auditory systems which are selective to small bands of frequencies, in visual cortex, etc. Such locally tuned neurons show response characteristics bounded to a small range of the input space.

Ant Colony Optimization (ACO) is an optimization technique that is inspired by the foraging behavior of real ant colonies. Originally, the method was introduced for the application to discrete and combinatorial problems. K. Socha presented variant for optimization in continuous domain: $ACO_{\mathcal{R}}$ [3] which is closest to the ACO approach.

V. Kůrková et al. (Eds.): ICANN 2008, Part II, LNCS 5164, pp. 829–838, 2008.

Particle Swarm Optimization (PSO) is an optimization technique inspired by the behavior of the flocks of birds, schools of fish, or swarms of bees, and even human social behavior. PSO is a population-based optimization tool, which could be implemented and applied easily to solve various function optimization problems, or the problems that can be transformed to function optimization problems.

In this paper we apply and evaluate the $ACO_{\mathcal{R}}$ and PSO methods in the RBF neural network training process. First, an overview of the RBF neural networks and the learning methods are presented. Next, the learning algorithms are evaluated over the freely available datasets (UCI database [4]) in the experimental section.

2 RBF Neural Network

Radial basis function networks consist of a layer of units with radial basis activation function (RBF) and an output layer of linear summation unit(s). As the RBF, often Gaussian activation functions are used (2), therefore the corresponding units are called *Gaussian (kernel) units*. The number of the units has to be known in advance, however techniques for dynamic growth of the structures have been introduced. See [5] for more information.

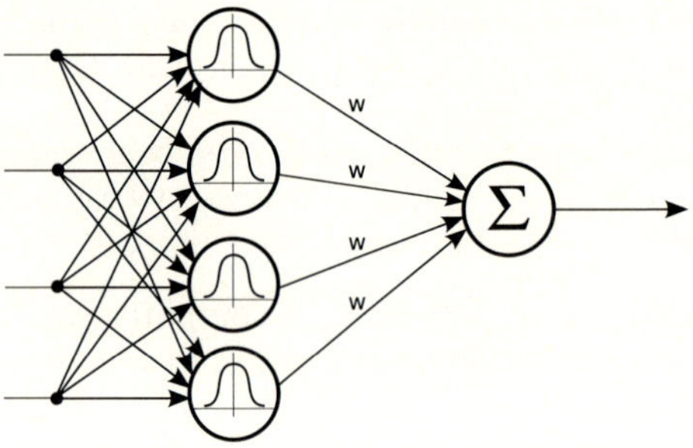

Fig. 1. An illustration of the RBF (Radial Basis Function) network (1). The input vector passes through a hidden layer of RBF (Gaussian) units. Then a weighted summation is performed. It can be easily extended for multiple outputs.

Each Gaussian unit c has an associated vector $\mathbf{m}_c \in \mathcal{R}^n$, indicating the position of the Gaussian in the n-dimensional input vector space, and a standard deviation σ_c. The response of the network $R(\circ)$ for the data vector $\mathbf{x}$ is given by the following equation:

$$R(\mathbf{x}) = \sum_{i=0}^{c} w_i \Phi_i(\mathbf{x}) \tag{1}$$

where $\Phi(\circ)$ stands for the RBF function (2), w_i represents the weight for linear summation output of the kernel functions. The equation can be simply generalized for multiple summation units. The response of the RBF unit in the case Gaussian function is used is given by the following equation:

$$\Phi_i(\mathbf{x}) = \exp\left(-\frac{\|\mathbf{x} - \mathbf{m}_i\|}{2\sigma_i^2}\right) \tag{2}$$

where $\mathbf{m}$ is the vector determining the center (basis) of the RBF function $\Phi_i(\mathbf{x})$. The parameter σ stands for the standard deviation and determines the width of the RBF function. Each kernel function returns higher output when the input is closer to its center, and the value decreases monotonically as the distance from the center increases. Note that normalized and non-normalized versions exist.

3 Learning Method Description

The goal of the learning algorithm is to set the free parameters of the network such that the output units produce suitable values for given input data. We can define an error function which measures the difference between the correct and predicted response on the given dataset. The task of the learning algorithm is to minimize such function.

The free parameters in the case of RBF networks are the centers $\mathbf{m}_i$ of kernel units $i \in c$, their widths σ_i and the weights w_i to the output unit(s). For each RBF unit and n-dimensional input we therefore obtain $n + 2$ free parameters, total number of parameters for the whole network is then $|c| \cdot n + 2$.

The learning phase of RBF NN basically consists of two consecutive phases, an unsupervised one and a supervised one:

1. Training of the first layer weights – determining the kernel function centers – by means of a clustering procedure. Moody and Darken [6] propose the use of k-means clustering algorithm. For the width parameter (standard deviation in the Gaussian function) they report good results for using the distance to the other nearest RBF kernel unit.
2. Calculation of the weights between the hidden layer and the output layer by solving a system of linear equations. The delta rule (least mean square rule) is commonly used.

Another approach for learning the network introduce growing cell structures, however such approach is beyond the scope of this paper. See [5] for more information. Many approaches for supervised training have been proposed using k-means [7], orthogonal least squares (using the Gram-Schmidt algorithm) [8], expectation maximization [9], etc.

In this paper we evaluate three different learning methods: The simplest one reduces the network to behave like the k-means algorithm: the weight and the width of each unit are kept constant, only the centers are updated. In such way we minimaze the k-means objective function (sum of squared distances of each datum to it's relevant center). Other approaches update all free parameters according to the error function evaluation over the training data set. After the learning phase is finished, the learned network is evaluated on a different (unknown) data set (testing data set).

3.1 Ant Colony Optimization

Ant Colony Optimization [10] is a metaheuristic approach, inspired by the ability of ants to discover shortest path between nest and food source. The process is guided by deposition of a chemical substance (pheromone). As the ants move, they deposit the pheromone on the ground (amount of the pheromone deposited is proportional to the quality of the food source discovered). The pheromone is sensed by other ants and the amount of pheromone changes the decision behavior of the ant individual. The ant will more likely follow a path with more pheromone.

The ants communicate indirectly through their environment. Such autocatalytic communication is called *stigmergy* (communication through the environment) (introduced by Grassé [11]).

The basic idea of ACO can be described as follows:

```
Repeat until stopping criterion is reached
  Create ants
  Construct solutions
  Evaporate Pheromone
  Daemon Actions (pheromone deposit)
End Repeat
```

New solutions are constructed continuously. It is a stochastic decision process based on the pheromone amount sensed by the ants. In the same way as it does in nature, the pheromone slowly evaporates over time (over iterations) in order to avoid getting stuck in local minimum and to adapt to dynamically changing environment. Daemon actions represent *background* actions which consist mainly of pheromone deposition. The amount is proportional to the quality of solution (and appropriate adaptive steps).

Several versions of the ACO strategy have been proposed, but they all follow the same basic ideas:

- search performed by a population of individuals
- incremental construction of solutions
- probabilistic choice of solution components based on stigmergic information
- no direct communication between the individuals

Many approaches to perform an optimization in continuous domain exist: CACO [12], API [13], CIAC [14]. The most related to ant colony metaphor is the $ACO_{\mathcal{R}}$ method presented by K. Socha [3]. In his work, a probabilistic density function (PDF) is used to estimate the distribution of Gaussian function.

A review of ACO-related methods can be found in [15] and [16].

Ant Colony Optimization for Continuous domain. Ant Colony Optimization (ACO) algorithms were originally introduced to solve discrete optimization (i.e., combinatorial) tasks. The ACO variants achieved interesting results in the field of NP-problem solving (traveling salesman, vehicle routing problem, etc.), therefore the methods for optimization in continuous domain have been introduced. Method that follows the original ACO framework as much as possible is the $ACO_{\mathcal{R}}$ algorithm [3].

$ACO_{\mathcal{R}}$ uses a solution archive T for the derivation of a probability distribution over the search space. While a pheromone model can be seen as an implicit memory of the search history, a solution archive is an explicit memory. Given an n-dimensional optimization problem and a solution s_l, $ACO_{\mathcal{R}}$ stores in T the solutions together with their objective function denoted hereby by $f(s_l)$ (also called error function, fitness). The capacity of the solution archive is denoted by k. The value of ith variable of the lth solution can be denoted as s_l^i.

At the beginning the archive is initialized with k random solutions. At each algorithm iteration, a set of m solutions is probabilistically generated and added to T. The same number of the worst solutions is removed. This biases the search process towards the best solution.

The probabilistic generation of solutions uses *probability density functions* (PDF). PDF may be any function $P(x) : \mathcal{R} \in x \rightarrow P(X) \in \mathcal{R}$ such that:

$$\int_{-\infty}^{\infty} P(x)\mathrm{d}x = 1 \tag{3}$$

For a given probability density function $P(x)$, an associated *cumulative distribution function* (CDF) $D(x)$ may be defined, which is often useful when sampling the corresponding PDF. The CDF $D(x)$ associated with PDF $P(x)$ is defined as follows:

$$D(X) = \int_{-\infty}^{x} P(t)\mathrm{d}t \tag{4}$$

The general approach to sampling the PDF $P(x)$ is to use the inverse of its CDF, $D^{-1}(x)$. Then it is sufficient to use a pseudo-random number generator that produces uniformly distributed real numbers. More information can be found in [3].

As the PDF function, Gaussian kernel PDF are used. They are formed from the weighted sum of Gaussian functions $g_l^i(x)$ (2):

$$G^i(x) = \sum_{l=1}^{k} \omega_l g_l^i(x) = \sum_{l=1}^{k} \omega_l \frac{1}{\sigma_l^i \sqrt{2\pi}} \exp\left(- \frac{(x - \mu_l^i)^2}{2{\sigma_l^i}^2} \right) \tag{5}$$

where $i = 1, \ldots, n$, that is, for each dimension of the given continuous optimization problem we define different Gaussian kernel PDF. Each kernel is parametrized by three vectors of parameters: ω is the weight vector associated with the individual Gaussian function, μ^i and σ^i are the vectors of means and standard deviations respectively. See Fig. 2 for illustration.

The construction of the solutions is performed as follows: At each construction step, the ant chooses a value for decision variable. This is done by sampling a Gaussian kernel PDF $G^i(x)$, which is derived from the k solutions stored in the solution archive T. The values of the i-th variable of all solutions in the archive become the elements of the vector μ^i and the vector of weights ω is created in

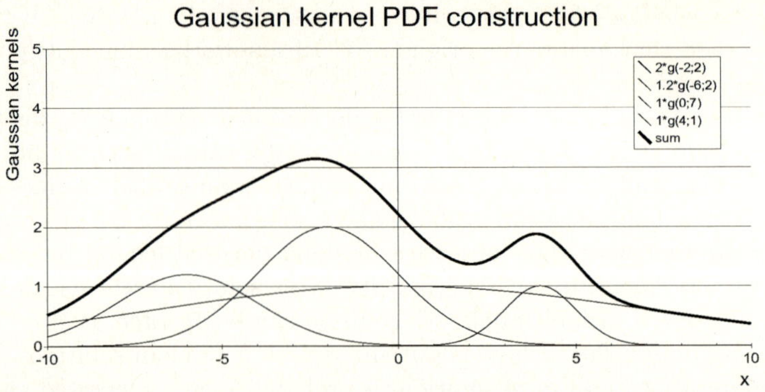

Fig. 2. Figure shows an example of four Gaussian functions and their superposition – the resulting Gaussian kernel PDF. Note that the x axis is limited to the range $\langle -10; 10\rangle$ in this illustration.

the following way. The solutions in the solution archive T are sorted according to their rank and the weight of the l-th solution is computed as follows:

$$\omega_l = \frac{1}{qk\sqrt{2\pi}} exp\left(\frac{(l-1)^2}{2q^2k^2}\right) \tag{6}$$

which defines the weight to be a value of the Gaussian function with argument l, mean value 1.0 and standard deviation qk where q is a parameter of the algorithm and k is the number of solutions. The q parameter causes the strong preference of the best-ranked solutions when it has a small value; when it has larger value, the probability becomes more uniform.

Only one Gaussian function to compose the Gaussian kernel PDF is selected. The probability p_l of choosing the l-th Gaussian functions is linearly proportional to the value of the correspondent weight ω_l. The standard deviation σ_l^i of the correspondent Gaussian function $g_l^i(x)$ is then computed (there is no need to recompute the whole vector σ^i):

$$\sigma_l^i = \rho \sum_{e=1}^{k} \frac{|x_e^i - x_l^i|}{k-1} \tag{7}$$

where the parameter $\rho > 0$ has an effect similar to the pheromone evaporation in ACO for combinatorial optimization. The higher the value of ρ, the lower the convergence speed of the algorithm.

The whole process is repeated for each dimension $i = 1, \ldots, n$ and each time the average distance σ_l^i is calculated only with the use of a single dimension i. This ensures that the algorithm is able to adapt to linear transformation of the considered problem.

3.2 Particle Swarm Optimization

Particle Swarm Optimization (PSO) technique [17] is inspired by the behavior observed in flocks of birds, schools of fish, or swarms of bees. PSO model consists of a swarm of k particles which are initialized with a population of random candidate solutions. During the iterative process, they move through the n-dimensional problem space and search for new solutions. The method needs an objective (fitness) function to be defined in the search space $\mathcal{R}^n$. Each i-th particle has its own position (in the solution space) – solution vector $\mathbf{x}_i$, and a velocity represented by a velocity vector $\mathbf{v}_i$.

In the adaptive process, each particle uses the local and global information: Each particle remembers its own best so far solution achieved $lBest$ (local information) and is provided with the information about the best so far solution of the algorithm $gBest$ (global information). During the iteration process, the position $\mathbf{x}_i$ of the i-th solution is updated according to the following rule:

$$\mathbf{x}_i(t+1) = \mathbf{x}(t) + \mathbf{v}_i(t+1) \tag{8}$$

where t denotes time (iteration number) and the velocity vector $\mathbf{v}_i$ is computed as follows:

$$\mathbf{v}_i(t+1) = w\mathbf{v}_i(t) + c_1 r_1 \|\mathbf{x}_{lBest}(t) - \mathbf{x}_i(t)\| + c_2 r_2 \|\mathbf{x}_{gBest}(t) - \mathbf{x}_i(t)\| \tag{9}$$

where $gBest$ and $lBest$ denote the global and local best so far solution respectively. Parameter w is called the inertia factor, $r_1 \in \langle 0; 1\rangle$ and $r_2 \in \langle 0; 1\rangle$ are uniformly distributed random numbers which serve for maintaining the diversity of the population. Parameter $c_1 > 0$ is called *coefficient of the self-recognition component* and the parameter $c_2 > 0$ is called *coefficient of the social component.*

Usually a maximum and minimum velocity constraint is used in order to increase the effectiveness of the algorithm.

4 Experimental Part

The experiments have been performed using the *training* and *validation* set with classification known in advance (a priori). These sets are used during the learning process of the classifier and to validate the model (avoid over-learning). After successful evolutionary process of the classifier finishes, the testing is performed (where the classification is not needed a priori, but should be known in order to be able to evaluate the performance of the classifier) on *testing* (unknown) data. If the classification is not known a priori, cluster validation techniques as described in [18] and [19] can be used to measure the classification (or clustering) efficiency of the clustering obtained.

4.1 Preliminary Tests

Numerous tests have been performed to estimate the optimal parameters and measure the dependency of error ratio on relevant parameters. As an objective

Table 1. Dataset characteristics

Data set	Examples	Categorical attributes	Continuous attributes	Classes
Ljubljana breast cancer	282	9	0	2
Wisconsin breast cancer	683	0	9	2
Cleveland heart disease	303	8	5	5
Hepatitis	155	13	6	2
Tic-tac-toe	958	9	0	2
Iris	150	0	4	3

function, the ratio of incorrectly classified data to the whole (training) data set has been used ($e \in \langle 0; 1 \rangle$). The minimization of this function is desired. First, the tests on artificial data set with Gaussian distribution have been performed. Then the iris data set [4] with 150 data items and 3 classes has been used to measure the performance and for preliminary parameter estimation.

4.2 Tests

Experimental evaluation of the test has been performed similarly to [20], where the description of the datasets can be found. However, we use the predefined number of RBF units which is equal to the number of classes in the data. Characteristics of the datasets are summarized in Table 1. Categorical attributes have been assigned integer number from linear sequence, starting with the value of 1.

4.3 Parameter Estimation

Based on the results of the preliminary experiments, the population size has been set to 20 for both $ACO_{\mathcal{R}}$ and PSO algorithm. The stopping criterion has been set to 300 iterations which allows the algorithms to stabilize.

For the $ACO_{\mathcal{R}}$, the values of $\rho = 1.0$ and $q = 0.5$ have been used, for the PSO, the values $c_1 = 1.0$, $c_2 = 1.0$ and $w = 0.5$ have been used. These values have been determined experimentally. Adaptive methods can be also used.

Table 2. Table presents mean accuracy (%) in classification of the UCI datasets selected. Different learning algorithms are compared. The result is a mean average over one hundred independent runs. For the evaluation, the testing data set has been used.

Data Set	K-means RBF	PSO RBF	$ACO_{\mathcal{R}}$ RBF
Ljubljana breast cancer	69.84±3.75	71.25±7.58	**72.58**±4.86
Wisconsin breast cancer	96.16±1.08	96.11±1.10	**96.72**±1.12
Cleveland heart disease	55.11±4.18	49.99±15.36	**57.14**±4.27
Hepatitis	82.50±5.92	81.73±7.20	**83.46**±6.65
Tic-tac-toe	65.44±2.10	**70.38**±12.24	68.36±2.36
Iris	86.75±8.78	90.33±7.60	**90.75**±4.40

5 Results

The results are summarized in the Table 2. Every result represents an average over one hundred of independent runs. The testing data set has been selected randomly for each test run. A standard deviation of the results is also presented. First column of the results table (*K-means RBF*) provides results for the basic algorithm where only the RBF mean vector has been updated. The next colums provide results for the PSO and $ACO_{\mathcal{R}}$ learning methods respectively.

It is obvious that the $ACO_{\mathcal{R}}$ method outperformed the other approaches in the datasets containing continuous attributes (both Breast cancer, Cleveland, Hepatitis and Iris data sets). In the case of more categorical attributes (Tic-tac-toe dataset), the PSO method performed better, however it often introduced higher standard deviation.

6 Conclusion

In this paper we have presented, applied and evaluated two nature inspired methods in the task of RBF neural network learning process: $ACO_{\mathcal{R}}$ and PSO algorithms for continuous optimization and compared them with deterministic approach. The networks have been used in the process of classification. The use of nature inspired methods increases robustness of the learning process.

The nature inspired methods have been compared with a simple algorithm for RBF neural network learning which effectively reduces the network to k-means algorithm. We have shown that the nature inspired methods increased the accuracy of the basic algorithm, and have proved to be an efficient tool for optimalization in the continuous domain and for ANN learning.

Acknowledgments

This research project has been supported by the research program number MSM 6840770012 "Transdisciplinary Research in the Area of Biomedical Engineering II" of the CTU in Prague, sponsored by the Ministry of Education, Youth and Sports of the Czech Republic. This work has been developed in the BioDat research group `http://bio.felk.cvut.cz`

The breast cancer data have been provided from the University Medical Centre, Institute of Oncology, Ljubljana, Yugoslavia. Thanks go to M. Zwitter and M. Soklic for providing the data. The Cleveland heart disease data have been provided by V. A. Medical Center, Long Beach and Cleveland Clinic Foundation. Thanks go to Robert Detrano, M.D., Ph.D. The breast cancer database was obtained from the University of Wisconsin Hospitals, Madison from Dr. William H. Wolberg.

References

1. Adami, C.: Introduction to Artificial Life. Springer, Heidelberg (1998)
2. Broomhead, D., Lowe, D.: Multivariable functional interpolation and adaptive networks. Complex Systems 2, 321–355 (1988)

3. Socha, K., Dorigo, M.: Ant colony optimization for continuous domains. European Journal of Operational Research 185(3), 1155–1173 (2008)
4. Asuncion, A., Newman, D.: UCI machine learning repository (2007)
5. Fritzke, B.: Growing cell structures – a self-organizing network for unsupervised and supervised learning. Neural Networks 7(9), 1441–1460 (1994)
6. Moody, J., Darken, C.: Learning with localized receptive fields. In: Rouretzky, D., Sejnowski, T. (eds.) Proceedings of the 1988 Connectionist Models Summer School, pp. 133–143. Morgan Kaufmann, San Mateo (1989)
7. Moody, J.: Fast learning in networks of locally-tuned processing units. Neural Computation 1, 281–294 (1989)
8. Chen, S., Cowan, C.F.N., Grant, P.M.: Orthogonal least squares learning algorithm for radial basis function nertworks. IEEE Transactions on Neural Networks 2(2), 302–309 (1991)
9. Cha, I., Kassam, S.A.: Rbfn restoration of nonlinearly degraded images. IEEE Transactions on Image Processing 5(6), 964–975 (1990)
10. Dorigo, M., Stutzle, T.: Ant Colony Optimization. MIT Press, Cambridge (2004)
11. Grasse, P.P.: La reconstruction du nid et les coordinations inter-individuelles chez bellicositermes natalensis et cubitermes sp. la thorie de la stigmergie: Essai d'interprtation des termites constructeurs. Insectes Sociaux 6, 41–81 (1959)
12. Bilchev, G., Parmee, I.C.: The ant colony metaphor for searchin continuous design spaces. In: Proc. of AISB Workshop on Evolutionary Computing. LNCS, pp. 25–39. Springer, Heidelberg (1993)
13. Monmarche, N., Venturini, G., Slimane, M.: On how pachycondyla apicalis ants suggest a new search algorithm. Future generation comput. syst. 16, 937–946 (2000)
14. Dreo, J., Siarry, P.: A new ant colony algorithm using the heterarchical concept aimed at optimization of multiminima continuous functions. In: Dorigo, M., Di Caro, G., Sampels, M. (eds.) Ant Algorithms 2002. LNCS, vol. 2463, Springer, Heidelberg (2002)
15. Blum, C.: Ant colony optimization: Introduction and recent trends. Physics of Life Reviews 2(4), 353–373 (2005)
16. Dorigo, M., Blum, C.: Ant colony optimization theory: A survey. Theoretical Computer Science 344(2–3), 243–278 (2005)
17. Kennedy, J., Eberhart, R.C.: Particle swarm optimization. In: Proceedings IEEE International Conference on Neural Networks IV, pp. 1942–1948 (1995)
18. Bezdek, J.C., Li, W.Q., Attikiouzel, Y.A., Windham, M.P.: A geometric approach to cluster validity for normal mixtures. Soft Computing 1, 166–179 (1997)
19. Davies, D.L., Bouldin, D.W.: A cluster separation measure. IEEE Transactions on Pattern Recognition and Machine Intelligence 1(2), 224–227 (1979)
20. Smaldon, J., Freitas, A.A.: A new version of the ant-miner algorithm discovering unordered rule sets. In: GECCO 2006: Proceedings of the 8th annual conference on Genetic and evolutionary computation, pp. 43–50. ACM, New York (2006)

Tree-Based Indirect Encodings for Evolutionary Development of Neural Networks

Jan Drchal and Miroslav Šnorek

Czech Technical University, Faculty of Electrical Engineering,
Department of Computer Science and Engineering
Karlovo náměstí 13, 121 35, Prague, Czech Republic

Abstract. This paper focuses on TWEANN (Topology and Weight Evolving Artificial Neural Network) methods based on indirect developmental encodings. TWEANNs are Evolutionary Algorithms (EAs) which evolve both topology and parameters (weights) of neural networks. Indirect developmental encoding is an approach inspired by multi-cellular organisms' development from a single cell (zygote) known from Nature. The possible benefits of such encoding can be seen in Nature: for example, human genome consists of roughly 30 000 genes, which describe more than 20 billion neurons, each linked to as many as 10 000 others. In this work we examine properties of known tree-based indirect developmental encodings: Cellular Encoding and Edge Encoding. Well known Genetic Programming is usualy used to evolve tree structures. We have employed its successors: Gene Expression Programming (GEP) and Grammatical Evolution (GE) to optimize the trees. The combination of well designed developmental encoding and proper optimization method should bring compact genomes able to describe large-scale, modular neural networks. We have compared GE and GEP using a benchmark and found that GE was able to find solution about 7 times faster then GEP. On the other hand GEP solutions were more compact.

1 Introduction

Using Evolutionary Algorithms (EAs) [1] to learn Artificial Neural Networks (ANNs) [2] is a well examined approach as EAs are very robust. Even harder optimization problems must be solved for TWEANNs (Topology and Weight Evolving Artificial Neural Networks) – these algorithms are not limited to search for proper weight settings only but they also optimize the topologies of neural networks. TWEANN approach is useful for ANN user, who does not have to experiment with different topologies. Also finding (at least near) optimal topology leads to better data modeling results.

The problem of most recent TWEANN algorithms is in their inability to evolve large-scale modular ANNs. This is mainly caused by the so-called *curse of dimensionality*, where optimization methods fail because of high-dimensional space to search through. Most current TWEANN methods use *direct encoding* approaches to represent ANNs. In direct encoding a single gene describes a single feature (e.g. a connection or a neuron) of an ANN. This is unlike in Nature

V. Kůrková et al. (Eds.): ICANN 2008, Part II, LNCS 5164, pp. 839–848, 2008.

where genome rather represents a "program" to build the target organism. Human genome consists of roughly 30 000 genes and is able to encode more than 20 billion neurons, each linked to as many as 10 000 others, plus, of course, the rest of the organism. This efficient information storage can be seen as a kind of compression. Hence, the translation of a genotype to phenotype is a decompression process. Of course, such a high level of compression is only possible, if the information is highly regular. This is true for Nature, as organisms and their brains are known to be highly modular, hierarchical systems. Artificial encodings which are trying to possess such attributes are known as *indirect encodings.*

In Nature, phenotypes are created by process of development (embryogeny). Multi-cellular organisms are grown from a single cell. The natural development, according to [3], involves 5 main processes: *cleavage division*, *pattern formation* (e.g. location of axes), *morphogenesis* (cell movement), *cellular differentiation* and *cellular growth.* The whole process of a cellular development is controlled by a complex interplay of protein diffusion and detection by binding on genes regulatory sites. It is often referred to a Gene Regulatory Network (GRN).

Reisinger [4] and Mattiussi [5] developed TWEANN systems which employ GRN models similar (although very simplified) to those in Nature. However, the target ANNs are not created by a development. Rather ANNs are directly associated with evolved GRNs – the number of genes in a GRN exactly corresponds to neurons of a final network. These approaches do not use *generative* representations. In [6,7] Eggenberger used very complex artificial embryogeny system to grow ANNs. Such system is computationally very demanding. The question is, if it is possible to model the involved Natural phenomena on a higher level of abstraction.

In [8,9] Stanley et al. presented very different approach – a HyperNEAT algorithm. It evolves neural networks in a two stage process: the standard NEAT [10] TWEANN is used to create ANNs with special transfer functions. These networks are called the Compositional Pattern Producing Networks (CPPNs). In the second step, planned neurons are given coordinates on an n-dimensional grid. The previously evolved CPPN is then used to determine synaptic weights between all pairs of neurons. The spatial coordinates of both neurons are fed into the CPPN's inputs and the CPPN then outputs their connection weight. The HyperNEAT itself is not a typical TWEANN method as the number of neurons and their spatial distribution has to be known in advance. Interesting feature of HyperNEAT is that its developmental process is not simulated in time – complex phenotypes are built just by questioning the relevant CPPN.

This work particularly focuses on experiments with tree-based indirect encodings of ANNs. In [11] Gruau introduced an indirect encoding called Cellular Encoding (CE), where the development of ANNs is controlled by evolved program trees. Trees are well examined data structures and there was already a great amount of interest in development of tree structure optimization Evolutionary Algorithms. The most widely known, Koza's Genetic Programming (GP) [12], which was originally used to evolve LISP program trees, was adopted by Gruau and used to optimize CE development trees. Here, we employ GP's successors Gene Expression

Programming and Grammatical Evolution. Along with Cellular Encoding we make experiments using a similar approach – the Edge Encoding (EE)[13].

The paper is organized as follows: Section 2 contains a brief decription of Gene Expression Programming and Grammatical Evolution. Section 3 describes Cellular and Edge encodings. In Section 4, experiments are performed. The last section concludes.

2 Evolution of Tree Structures

Koza's Genetic Programming (GP) [12] is a well-known Evolutionary Algorithm for optimization of tree structures. GP trees can represent mathematical models of data or programs. Figure 1 shows a simple function of two variables described by an evolved GP tree. The tree nodes are labeled by symbols which can represent either operations, variables or constants. Each node has a predefined arity (number of child nodes). Constants and variables are localized in leaf nodes. GP was already used for evolution of ANNs [14]. However, the encoding was direct.

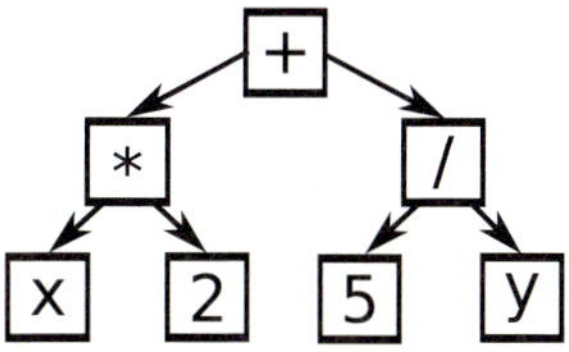

Fig. 1. Genetic Programming evolved tree. This tree is assembled of 7 symbols, it uses three operations (+, ×, /), two variables (x, y) and two constants (2, 5). It represents a function $f(x, y) = (x \times 2) + (5/y)$.

2.1 Gene Expression Programming (GEP)

Gene Expression Programming(GEP) is a method developed by Ferreira [15]. GEP is based on GP and it was reported to surpass the traditional GP by a factor 100 to 10 000 on many benchmarks. Unlike GP, where a genome is the evolved tree itself, GEP works with linear representation. Ferreira already used GEP for evolution of ANNs. The approach used similar direct encoding to the one mentioned in the GP section above.

Figure 2 clarifies the situation – it shows a two stage process of genotype to phenotype transformation. A genotype represented by a linear genome is translated into a development tree which is later used to grow the target ANN by simulated embryogenesis. Trees are efficiently encoded into a linear form by use of the Karva language, such encoded tree is then called a K-expression. In Karva, trees are read layer by layer starting by the root node The K-expression for the program tree shown in Figure 1 would be "`+*/x25y`". Karva facilitates efficient mutation and crossover by keeping newly created individuals valid. GEP allows multigenic genomes, where multiple trees are evolved simultaneously and

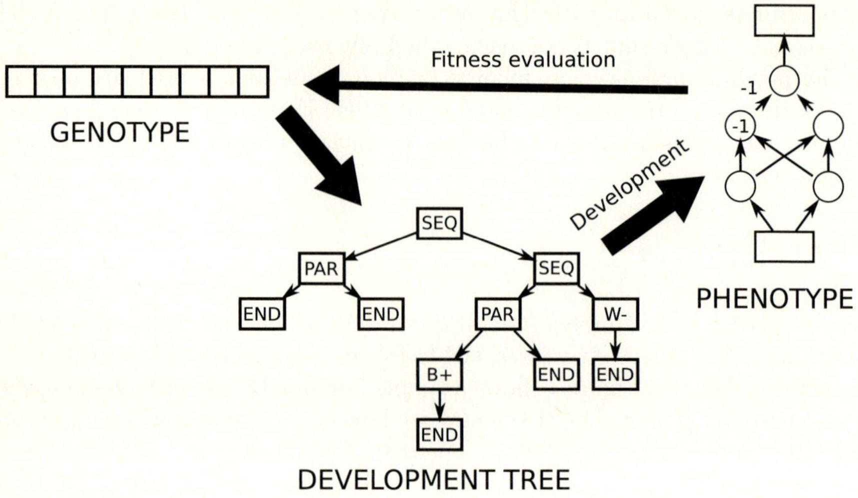

Fig. 2. The figure shows a two stage process of genotype to phenotype transformation for both Grammatical Evolution (GE) and Gene Expression Programming (GEP) Evolutionary Algorithms. Both approaches work with population of linear genomes. In the first stage, these genotypes are translated into development trees. In the second stage the simulated development starts by progressively traversing the tree from its root node, effectively growing the phenotype – a target ANN. The ANN is then evaluated in the domain. Its fitness is used by selection mechanisms during evolution.

later combined by a *linking function*. GEP also includes a possibility of Random Numerical Constants (RNCs) where an array of numerical constants is coevolved together with the tree. These constants can parametrize tree symbols which are connected to a special '?' terminal.

2.2 Grammatical Evolution

In [16] Ryan, Collins and O'Neil introduced a Grammatical Evolution (GE) which is an Evolutionary Algorithm able to evolve solutions according to a user-specified grammar. Unlike GP or GEP, GE brings more flexibility as user is able to constrain the way in which the program symbols are assembled together. In both GP and GEP it is impossible to control their order – any symbol can become a child or a parent of any other symbol. The GE approach is therefore able to radically cut down the search space. The grammar in GE is specified using Backus Naur Form (BNF). GE uses linear genome of integer numbers.

3 Tree-Based Indirect Encoding

3.1 Cellular Encoding

Cellular Encoding (CE) was introduced by Gruau in [11], where more detailed description can be found. In CE, ANN development starts with a single cell (neuron). A development tree is traversed using breath-first search starting from the root node. CE uses symbols for cell division (`PAR`, `SEQ`) and updating of cell registers like biases or input register (`B+`, `INC` etc.). The original Cellular Encoding was limited to Boolean Neural Networks where network's input and output is binary. Boolean Neural Networks use neurons which fire 1 for activities above given threshold and 0 for lower. Weights can be either 1 or -1. The following list contains short description of symbols used in CE development trees:

- `SEQ` sequential division – a new cell inherits mother cell's outputs, input of a new cell is connected to the mother cell's output using a connection of weight 1. Development instructions for a mother cell continues in a left subtree, while for a new cell in the right.
- `PAR` parallel division – create a new cell which has the same set of inputs and outputs as the mother cell.
- `INC`, `DEC` increase/decrease internal input register.
- `W+`, `W-` set weight of an input link designated by an internal input register to 1 or -1.
- `B+`, `B-` increase/decrease bias by 1.
- `CUT` cut incoming connection given by the internal input register.
- `WAIT` do nothing – continue with the next step (needed for synchronization).
- `END` end processing of this cell.

Gruau has later developed a modified method which is capable to encode real-valued weights [17].

3.2 Edge Encoding

In [13] Luke and Spector pointed out that Gruau's Cellular Encoding bias the evolutionary search towards highly connected networks. This is mostly due to its Cell-centric approach and its rather inefficient internal input register operators (`INC` and `DEC`). They developed a new technique called Edge Encoding (EE), which in contrast to Cellular Encoding grows trees rather by modification of edges then nodes. Edge Encoding was designed to be a tool for evolution of general networks. Hornby in [18] evolved ANNs with a slightly modified EE. EE is able to create general recurrent networks, the development starts with a single cell with a self-loop (the initial edge). In our work we have used the following symbols which can be applied to edges (from cell A to cell B):

- `SPLIT` creates a new cell C. New edges $A \rightarrow C$ and $C \rightarrow B$ are added.
- `REVERSE` creates a new edge $B \rightarrow A$.
- `CONNECT` creates a new edge $A \rightarrow B$.
- `LOOP` creates a self loop $B \rightarrow B$.
- `BIAS(n)` sets B's bias to n.
- `INPUT(n)`, `OUTPUT(n)` connects to the n-th input/output.
- `WAIT` do nothing – continue with the next step (needed for synchronization).
- `END` end processing of this cell.

Here, we have used a simplified version of Hornby's Parametric Edge Encoding – symbols `dest_to_next`, `source_to_next` and `source_to_parent` were not used as the above mentioned set of symbols was sufficient for our experiments.

4 Experiments

In this section we describe our experiments concerning Cellular and Edge encodings. The optimization methods chosen were the Grammatical Evolution and Gene Expression Programming described above. For our experiments we have chosen a well known XOR benchmark where ANN is evolved to approximate the XOR Boolean function. XOR was often used as a preliminary benchmark for TWEANN algorithms [11,10]. XOR is simple enough but it still requires hidden nodes in phenotype.

At first the Boolean Neural Network model as described in 3.1 was employed. Results of all experiments are summed up in a Table 1.

4.1 Grammatical Evolution

Cellular Encoding. To use a Grammatical Evolution for optimization of development trees one has to define a grammar in the BNF. For a basic set of Cellular Encoding symbols as described in 3.1 the grammar is straightforward:

```
<tree> ::= SEQ<tree><tree>|PAR<tree><tree>
           |INC<tree>|DEC<tree>
           |W+<tree>|W-<tree>|B+<tree>|B-<tree>
       |CUT<tree>|WAIT<tree>|END
```

Note that END is the only terminal symbol. One of possible handcoded solutions for ANN can be: `SEQ(PAR(B-(END),END),W-(END))` (for better readability the tree is written in the infix notation).

Results of this (and all following experiments) are summarized in Table 1. GE was able to evolve solution on average in 240 generations. It never achieved as compact development as the presented handcoded one. One of shorter solutions found was the following:

```
SEQ(PAR(END,END),SEQ(PAR(B-(PAR(DEC(END),W-(B-(B+(W+(W-(END))))))),
END),DEC(W+(W-(W-(END))))))
```

The average symbol count was 25 which is much worse result than the 7 symbols of the handcoded one. It is obvious, that the solution uses unnecessary combinations of alternating symbols (`W+`, `W-` and `B+`, `B-`).

Edge Encoding. For Edge Encoding as described in 3.2 we have used the following grammar:

```
<tree> ::= SPLIT<tree><tree><tree>|REVERSE<tree><tree>
           |CONNECT<tree><tree>|LOOP<tree><tree>
           |WAIT<tree>|END
       |INPUT<input><tree><tree>|OUTPUT<output><tree><tree>
       |B+<tree>|B-<tree>|W+|W-
<input> ::= 0|1
<output> ::= 0
```

Table 1. A comparison of GE and GEP vs. Cellular Encoding and Edge Encoding on a XOR problem. Average number and standard deviation of generations and symbols needed are given. The values were measured on 50 runs. If solution was not found within 5000 the run was considered a failure. Edge Encoding was measured only for GE. All experiments were done using population of 1500 individuals.

	GE					GEP				
	Failures	Generations		Symbols		Failures	Generations		Symbols	
		avg	stdv	avg	stdv		avg	stdv	avg	stdv
Cellular Encoding	4	240	173	25	17	4	1509	1256	15.2	2.7
Edge Encoding	3	173	97	135	135	-	-	-	-	-

Here, the grammar is adapted for two inputs and one output given by the solved XOR problem. EE in its basic form can evolve general recurrent neural networks in order to keep the results at least somewhat comparable to CE we removed `REVERSE` and `LOOP` symbols, which greatly reduced number of recurrent links in developed ANNs. Additionally each developed ANN was pruned off all recurrent links. One of possible handcoded solutions can be:

```
SPLIT(END,INPUT(0,INPUT(8,OUTPUT(0,END,W+),W+),W+),
INPUT(0,INPUT(8,OUTPUT(0,B-(END),W-),W+),W+))
```

A mean number of generations needed to evolve XOR was 173. The number of solution symbols was 135, which is extremely high in comparison with the handcoded solution (23 symbols). An example of a shorter evolved solution follows:

```
OUTPUT(6,INPUT(1,INPUT(3,OUTPUT(1,INPUT(8,W-,WAIT(W+)),
WAIT(SPLIT(B+(OUTPUT(1,W-,END)),WAIT(W-),
INPUT(3,END,B-(SPLIT(WAIT(W-),INPUT(8,W+,WAIT(W-)),B+(W+)))))))
B+(SPLIT(W-,W+,W-))),B-(W+)),WAIT(W+))
```

Interestingly, when recurrent ANNs were allowed, GE was able do develop shorter solutions as the recurrent version of the XOR network is topologically simpler.

4.2 Gene Expression Programming

In case of Gene Expression Programming we have performed experiments with Cellular Encoding only. Combination of Edge Encoding and GEP is currently being subject of research.

Cellular Encoding. For GEP we have used a classical set of Gruau's Cellular Encoding symbols as described in 3.1. We have evolved single gene genomes with `END` being the only one terminal symbol. An average number of generations was 1509 which is approximately 6.3 times more than in the case of GE. On the other hand GEP produced smaller solutions – only 15.2 symbols were needed (compared to 25 for GE). An example of the evolved solution:

```
SEQ(PAR(END,WAIT(WAIT(END))),WAIT(SEQ(PAR(PAR(END,B-(END)),
PAR(END,END)),W-(INC(END)))))
```

4.3 Real-Valued Weights

After experiments with Boolean Neural Networks we performed experiments with real-valued weights using GE and GEP. The GE grammar was modified in the following way:

```
<tree> ::= SPLIT<bias><tree><tree><tree>|REVERSE<tree><tree>
           |CONNECT<tree><tree>|LOOP<tree><tree>
           |WEIGHT<weight>|WAIT<tree>|END
       |INPUT<input><tree><tree>|OUTPUT<output><tree><tree>
       |BIAS<bias><tree>
<bias> ::= <number>
<weight> ::= <number>
<neuron> ::= <number>
<number> ::= <digit><digit><digit><digit>
<digit> ::= 0|1|2|3|4|5|6|7|8|9
<input> ::= 0|1
<output> ::= 0
```

Note that `<number>` describes an integer from interval 0..9999. This integer was transformed to $[-1, 1]$ for weights and $[-10, 10]$ for thresholds. The activation function used in the developed ANNs was a logistic sigmoid.

In the case of GEP, we used of its RNCs (see 2.1). The result of experiments for both GE and GEP were unsatisfactory as none approach was able to solve the XOR problem. In fact, even evolving simple functions OR and AND needed thousands of generations.

5 Conclusion

We performed experiments with Cellular and Edge indirect encodings. We employed Grammatical Evolution (GE) and Gene Expression Programming (GEP) for evolution of development trees. The comparison of GE and GEP was done using Cellular Encoding. We found that GE outperforms GEP in a smaller number of generations needed to optimize development trees. On the other hand, GEP tends to produce smaller solutions. A comparison of Cellular and Edge Encoding has shown that although Edge Encoding produced very large development trees (135 symbols on average), the solutions were found faster.

Still, looking at the results, we can see than even in the best case of GE and Edge Encoding $1500 \times 173 = 259\,500$ evaluations were needed on average, which is not satisfactory for a such simple problem as XOR. The experiments with real-weight ANNs were complete failures. The following observations might be helpful for improvement of current developmental TWEANNs:

- The use of a 2-stage phenotype construction as shown in Figure 2 brings complications: even very small change (mutation) of a development tree may cause massive changes in the final ANN. This effect must be minimized as much as possible.

- In [19] Stanley has shown advantages of a complexification – a gradual growth from a simple form. The complexification makes the traversal through the search space more efficient because its dimension grows as needed. The implementation of this approach might be very useful in the case of GE and GEP.
- As we have seen in the case of GE and Edge Encoding, development trees used 135 symbols on average which is unacceptable. We claim that developmental indirect encodings need mechanisms which will prevent such bloat (uncontrolled growth).

Acknowledgments

This research is partially supported by the grant Automated Knowledge Extraction (KJB201210701) of the Grant Agency of the Academy of Sciences of the Czech Republic and the research program "Transdisciplinary Research in the Area of Biomedical Engineering II" (MSM6840770012) sponsored by the Ministry of Education, Youth and Sports of the Czech Republic.

References

1. Michalewicz, Z.: Genetic algorithms + data structures = evolution programs, 2nd edn. Springer, New York (1994)
2. Tsoukalas, L.H., Uhrig, R.E.: Fuzzy and Neural Approaches in Engineering. John Wiley & Sons, Inc., New York (1996)
3. Kumar, S., Bentley, P.J. (eds.): On Growth, Form, and Computers. Elsevier Academic Press, London (2003)
4. Reisinger, J., Miikkulainen, R.: Acquiring evolvability through adaptive representations. In: GECCO 2007: Proceedings of the 9th annual conference on Genetic and evolutionary computation, pp. 1045–1052. ACM, New York (2007)
5. Mattiussi, C.: Evolutionary synthesis of analog networks. PhD thesis, Ecole polytechnique fédérale de Lausanne EPFL, Lausanne (2005)
6. Eggenberger-Hotz, P.: Creation of neural networks based on developmental and evolutionary principles. In: Gerstner, W., Hasler, M., Germond, A., Nicoud, J.-D. (eds.) ICANN 1997. LNCS, vol. 1327, pp. 337–342. Springer, Heidelberg (1997)
7. Eggenberger-Hotz, P., Gómez, G., Pfeifer, R.: Evolving the morphology of a neural network for controlling a foveating retina and its test on a real robot. In: Artificial Life VIII. The 8th International Conference on the Simulation and Synthesis of Living Systems (2003)
8. D'Ambrosio, D.B., Stanley, K.O.: A novel generative encoding for exploiting neural network sensor and output geometry. In: GECCO 2007: Proceedings of the 9th annual conference on Genetic and evolutionary computation, pp. 974–981. ACM, New York (2007)
9. Gauci, J., Stanley, K.: Generating large-scale neural networks through discovering geometric regularities. In: GECCO 2007: Proceedings of the 9th annual conference on Genetic and evolutionary computation, pp. 997–1004. ACM, New York (2007)
10. Stanley, K.O., Miikkulainen, R.: Evolving neural networks through augmenting topologies. Technical report, University of Texas at Austin, Austin, TX, USA (2001)

11. Gruau, F.: Neural Network Synthesis using Cellular Encoding and the Genetic Algorithm. PhD thesis, France (1994)
12. Koza, J.R.: Genetic programming: on the programming of computers by means of natural selection. MIT Press, Cambridge (1992)
13. Luke, S., Spector, L.: Evolving graphs and networks with edge encoding: Preliminary report. In: Koza, J.R. (ed.) Late Breaking Papers at the Genetic Programming 1996 Conference, July 1996, pp. 117–124. Stanford University, USA (1996)
14. Koza, J.R., Rice, J.P.: Genetic generation of both the weights and architecture for a neural network. In: International Joint Conference on Neural Networks, IJCNN 1991, Washington State Convention and Trade Center, Seattle, WA, USA, 8-12, 1991, vol. II, pp. 397–404. IEEE Computer Society Press, Los Alamitos (1991)
15. Ferreira, C.: Gene Expression Programming: Mathematical Modeling by an Artificial Intelligence (Studies in Computational Intelligence), 2nd edn. Springer, New York (2006)
16. Ryan, C., Collins, J.J., O'Neill, M.: Grammatical evolution: Evolving programs for an arbitrary language. In: EuroGP, pp. 83–96 (1998)
17. Gruau, F., Whitley, D., Pyeatt, L.: A comparison between cellular encoding and direct encoding for genetic neural networks. In: Koza, J.R., Goldberg, D.E., Fogel, D.B., Riolo, R.L. (eds.) Proceedings of the First Annual Conference on Genetic Programming 1996, Stanford University, CA, USA, July 1996, pp. 81–89. MIT Press, Cambridge (1996)
18. Hornby, G.S.: Shortcomings with using edge encodings to represent graph structures. Genetic Programming and Evolvable Machines 7(3), 231–252 (2006)
19. Stanley, K.O., Miikkulainen, R.: Competitive coevolution through evolutionary complexification. Journal of Artificial Intelligence Research (21), 63–100 (2004)

Generating Complex Connectivity Structures for Large-Scale Neural Models

Martin Hülse

Department of Computer Science, Aberystwyth University
Aberystwyth, Ceredigion, Wales, SY23 3DB, UK
msh@aber.ac.uk
http://users.aber.ac.uk/msh

Abstract. Biological neural systems and the majority of other real-world networks have topologies significant different from fully or randomly connected structures, which are frequently applied for the definition of artificial neural networks (ANN). In this work we introduce a deterministic process generating strongly connected directed graphs of fractal dimension having connectivity structures very distinct compared with random or fully connected graphs. A sufficient criterion for the generation of strongly connected directed graphs is given and we indicate how the degree-distribution is determined. This allows a targeted generation of strongly connected directed graphs. Two methods for transforming directed graphs into ANN are introduced. A discussion on the importance of strongly connected digraphs and their fractal dimension in the context of artificial adaptive neural systems concludes this work.

1 Introduction

Within the context of neural computation and cognitive science artificial neural networks (ANN) are frequently utilized as the basic building blocks for large-scale neural models in order to explore the nature of complex information processing exploited in animals and human beings. The majority of such neural models are based on connectivity structures which match with the classical types of ANN, such as, Multi-layered-perceptrons, Hopfield- or Elman-networks [5,7]. All these network types establish a connectivity structure close to fully connected networks. The application of fully connected networks, however, might become crucial with respect to plausibility if they are intended to model biological systems. Fully connected ANN can hardly represent brain-like neural structures, if, as only one example, approximately 10^{11} neurons in the human brain are coordinated by "only" 10^{15} synapses [10].

An alternative, in particular for large-scale neural models, to overcome fully connected neural networks is the creation of random graph structures [1]. Nevertheless, random graph models do not well describe some essential properties of real-world or biological networks, such as degree-distribution [13]. Therefore, we argue, while modeling large-scale neural networks one must consider alternatives for the projections between neural assemblies; alternatives which go beyond random graphs and fully connected structures.

V. Kůrková et al. (Eds.): ICANN 2008, Part II, LNCS 5164, pp. 849–858, 2008.

Furthermore, large-scale neural models are applied for robot control more and more [8,16]. Such implementations on autonomous robots might be motivated as a proof of concept as well as for targeting specific issues of embodiment [14]. However, autonomous robots have usually very limited computational resources CPU and memory. Hence, for performance reasons it becomes important to utilize highly connected networks established by as less connections as possible.

The objective of this paper is to introduce a deterministic method which enables us to create *strongly connected directed graphs* [15] established by a number of edges magnitudes smaller than in fully connected graphs / networks. The generation process is inspired by fractal sets, namely Sierpiński carpets [12]. This makes the resulting structures very distinct compared with random and fully connected graphs. Due to the simplicity and deterministic character of the generation process, this method seems to be a promising alternative for the targeted generation of directed graphs in general as well as it opens a wide field for applications in many areas of neural modeling. We will introduce two strategies which allow alternative definitions of connectivity structures for feedforward and recurrent neural networks.

2 Directed Graphs, Sierpiński Carpets, and Strongly Connected Digraphs of Fractal Dimension

In this section we demonstrate how Sierpiński carpets motivate a process for a targeted generation of strongly connected directed graphs (digraphs), which will lead us to the definition of *digraphs of fractal dimension.*

2.1 Directed Graphs and Sierpiński Carpets

A directed graph or a digraph is a pair $G(V, E)$, where V is a set of vertices (sometime also called nodes) and E is a subset of $V \times V$. An edge $e_{ij} \in E$ represents an edge from node v_i to v_j, where e_{ij} is the in-coming edge for v_j and the out-going for v_i. In the following we also allow edges e_{ii} and therefore, edges can be in-coming as well as out-going edge for one and the same node. We call the number of in-coming edges of node v_i in-degree $d_{in}(v_i)$ and the number of out-going edges the out-degree $d_{out}(v_i)$. In a digraph the number of in-coming edges is equal to the number of the out-going: $\sum_i^n d_{in}(v_i) = \sum_i^n d_{out}(v_i)$, where n is the number of nodes in G. A directed path between node v_s and v_t in G is a sequence of edges $e_1, e_2, \ldots e_k$ such that the end node of edge e_i is the start node of $e_{(i+1)}$, $i = 1, 2 \ldots k$. If there exists a directed path between each pair of nodes in a digraph, we call it *strongly connected.*

The structure of a digraph $G(V, E)$ can be represented by an adjacency matrix M. Each matrix element m_{ij} of an adjacency matrix can either be zero or one. The element m_{ij} is one, if and only if $e_{ij} \in E$.

In the following section we describe how adjacency matrixes can build a bridge between Sierpiński carpets and digraphs as well as they give us a process for the deterministic development of strongly connected digraphs.

Sierpiński carpets result from an iterative process where a pattern is successively used to replace specific regions in an evolving pattern. In Fig. 1 an example is shown which illustrates this process. One starts with a given pattern, here a square divided into 3×3 equal sub-squares. The sub-squares are labeled either black or white, which creates a specific pattern. In each iteration the original pattern is used to replace all the black labeled regions, by this pattern again. This leads to a finer partition of the originally given pattern. Thus, after the first iteration, when each black labeled square is replaced by the pattern, we have a square subdivided into $9 \times 9 = 81$ regions, instead of the 9 regions given in the original pattern. As we see, after 5 iterations we have a "fractal set" represented on a square, regularly subdivided into 729×729 equal sub-squares. For infinity iterations we get the Sierpiński carpet, that is a set of fractal dimension.

In this process towards a fractal set we have to distinguish between the pattern $\mathcal{P}$ which is transformed into a new pattern $\mathcal{P}'$ and the pattern which determines this transformation. The latter we call *mask* $\mathcal{M}$. In the following we only allow masks and patterns with dimension $n \times n$ $(n > 1)$, where n indicates the segmentation. Obviously, a mask of segmentation S, written as $\mathcal{M}_S$, applied to a pattern $\mathcal{P}$ of segmentation m results in a new pattern $\mathcal{P}'$ of segmentation $S \cdot m$.

We utilize this transformation process to generate digraphs simply by interpreting the resulting patterns as an adjacency matrix. Namely, the black labeled sub-squares are interpreted as edges, i.e. black color represents value 1 in the corresponding adjacency matrix, while white squares indicate the zero entries, i.e. no edge. In this way a pattern or a mask of segmentation S is transformed into an adjacency matrix representing a digraph of S nodes and k edges, where k is the number of black labeled sub-squares in the pattern. Examples of 3×3 patterns transformed into digraphs are given in Figure 1.

With respect to adjacency matrixes we see that the Kronecker product [6] can be applied in order the define an algorithm which is isomorph to the process generating Sierpiński carpets:

$$D_0 := M_S$$
$$D_{n+1} := D_n \otimes M_S,$$

where M_S is an adjacency matrix $(M(i,j) \in \{0;1\})$ of dimension $S \times S$. In the following we refer to this algorithm as the *digraph generating process* DGP.

Due to the direct interpretation of patterns and masks as adjacency matrixes and digraphs we make use of these terms synonymously. The only important thing here is that a mask, either written as pattern $\mathcal{M}_S$ or adjacency matrix M_S, is the only seed which initializes the DGP and therefore completely determines the resulting connectivity structure.

2.2 Masks Creating Strongly Connected Digraphs

We now ask which masks of a given segmentation S create strongly connected digraphs. For the investigation of this question we start with the simplest form of strongly connected digraphs: cycles, also called rings. Each strongly connected

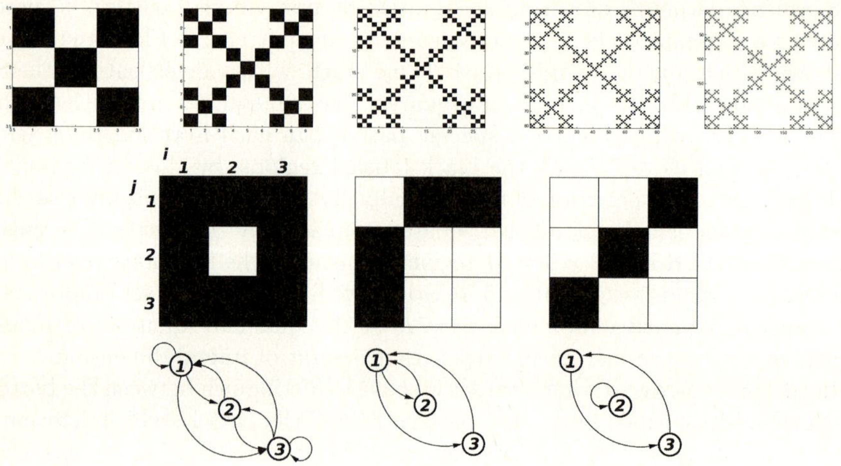

Fig. 1. Top: An example of the first five iterations towards Sierpiński carpets. After 5 iterations the original given 3×3 partition is transformed into a 729×729 partition, which can be interpreted as adjacency matrix for a digraph containing 729 nodes. See text for explanation. **Bottom:** Examples of digraphs derived from 3×3 patterns.

digraph $G(V, E)$ containing as many edges as nodes forms a cycle. Consequently such a digraphs has no loops (also called self-connections), i.e. $e_{ii} \notin E$. Further more, each node has only one in-coming and one out-going edge.

Assume the general case where we have a mask M_S and a digraph of segmentation T represented by the adjacency matrix D_0. Both, M_S and D_0, are digraphs forming cycles. Due to the digraph generation process we can deduce, that $D_1 := D_0 \otimes M_S$:

1. has as many nodes as edges $(T \cdot S)$,
2. only contains nodes with one out-going and one-incoming edge, and
3. has no node with a self-connection (loop).

Therefore, D_1 forms either a cycle or is fallen apart into "sub-cycles". In the latter case D_1 wouldn't be strongly connected anymore.

Whether D_1 is a cycle or not is actually determined by the number of nodes in M_S *and* D_0, in other words it depends on their segmentation. The cycles which M_S and D_0 are forming can be represented by the corresponding sequence of nodes. Taking the example shown in Fig. 2 we get for the following sequence for D_0:

$$(1^*) \rightarrow (3^*) \rightarrow (2^*) \rightarrow (4^*) \rightarrow (1^*),$$

for M_S we have:

$$(1^+) \rightarrow (3^+) \rightarrow (2^+) \rightarrow (1^+),$$

and D_1 is represented by:

$$(1^*, 2^+) \rightarrow (3^*, 1^+) \rightarrow (2^*, 3^+) \rightarrow (4^*, 2^+) \rightarrow (1^*, 1^+) \rightarrow (3^*, 3^+) \rightarrow \ldots$$

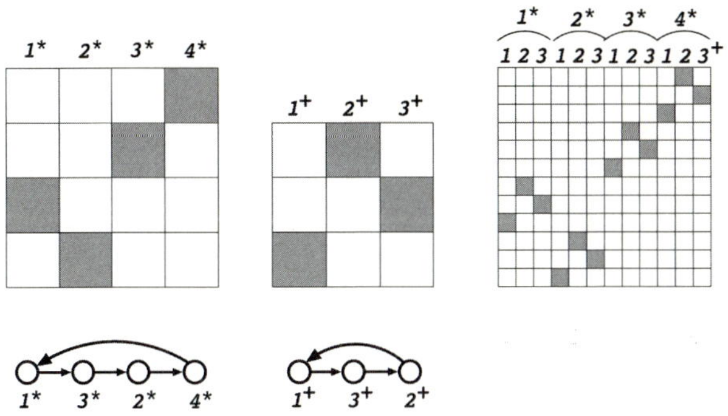

Fig. 2. Example of a process where a cycle with 4 nodes (D_0) is transformed by a mask (M_S, cycle with 3 nodes) into a new cycle of 12 nodes, D_1. See text for details.

In order to relate these two sequences to the sequence of the resulting graph D_1 we apply a numbering for D_1 which somehow preserves the numbering of D_0 and M_S. Fig. 2 shows how this can be achieved. Each node in D_1 is now represented by a number of two components (a^*, b^+). Starting with an arbitrary node (a_1^*, b_1^+) we can now write down the sequence of nodes forming a directed path in the following general form:

$$(a_1^*, b_1^+) \rightarrow (a_2^*, b_2^+) \rightarrow \ldots \rightarrow (a_k^*, b_k^+) \rightarrow (a_1^*, b_1^+).$$

Note, due to the definition of DGP each node in D_1 can have only one successor, which is not the node itself. Therefore, we can conclude: if $k = S \cdot T$ then we have a cycle formed by $S \cdot T$ nodes, which is the number of nodes in D_1. Thus, it would follow: D_1 is still strongly connected. On the other hand, if $k < S \cdot T$ then D_1 consist of sub-graphs forming cycles and therefore D_1 isn't neither strongly connected nor connected at all.

We see, it is the value of k which indicates whether or not the result is strongly connected. However, the crucial point for calculating k is to understand how the two sequences a_i^* and b_i^+ are determined by D_0 and M_S. In fact, the sequence for a_i^* (first component of the D_1 numbering) is exactly the same of D_0. No matter how the sequence of D_1 actually looks like, considering only the a^*-sequence, we see the same sequence as for D_0:

$$(1^*, b_t^+) \rightarrow (3^*, b_{t+1}^+) \rightarrow (2^*, b_{t+2}^+) \rightarrow (4^*, b_{t+3}^+) \rightarrow (1^*, b_{t+4}^+) \rightarrow \ldots$$

The same holds for the b_i^+-sequence (second component of the D_1 numbering). It is determined by the "cycle sequence" of M_S:

$$(a_t^*, 1^+) \rightarrow (a_{t+1}^*, 3^+) \rightarrow (a_{t+2}^*, 2^+) \rightarrow (a_{t+3}^*, 1^+) \rightarrow \ldots$$

Combining these two observation we get: $k = S \cdot T$ if, and only if S is not divisor of T or vice versa. In other words D_1 remains strongly connected. While for the other case, i.e. S is divisor of T or vice versa, then we get $k = max(S, T)$, meaning D_1 consist of $min(S,T)$ sub-graphs with k nodes forming separated cycles and therefore D_1 isn't connected at all.

This result tells us that DGP doesn't create strongly connected digraphs if it is initialized with a cycle, because D_0 and M_S have the same segmentation per definition.

However, examining the same argumentation we see that by extending M_S with a single loop (i.e. one self-connection) all separated sub-cycles will be connected. An additional self-connection operates like a junction connecting all separated sub-cycles. In this case the resulting digraph D_1 remains strongly connected. Further more, all resulting digraphs $D_n \quad (n > 1)$ would also be strongly connected, because, due to the definition of DGP, the given mask (with its self-connection) is always applied to a strongly connected digraph. Speaking precisely, the mask is always applied to a digraph containing a sub-graph forming a cycle which involves all nodes.

At this point we are able to formulate a sufficient criterion which guarantees strongly connected graphs for the DGP: *If M_S represents a cycle with at least one self-connection then all the resulting digraphs D_n are strongly connected.* In the following we call such masks *complete.*

We also see that each mask containing a complete mask as sub-structure will generate strongly connected digraphs as well.

2.3 Digraphs of Fractal Dimension and Their Degree Distribution

The generation of strongly connected digraphs therefore has always to start with a mask $\mathcal{M}_S$ containing at least $S+1$ edges. The maximal number of edges in M_S is S^2 and it is easy to see that such a mask generates only fully connected digraphs. Hence, the non-trivial cases of connected graphs are generated by masks with n edges, where $S < n < S^2$. Interestingly enough, masks with this number of labeled segments generate Sierpiński carpets of fractal dimensions d_f between 1 and 2 [12], since: $d_f = \frac{log(n)}{log(S)}$, from which follows: $1 < d_f < 2$. Therefore, we say a *fractal digraph* or a *digraph of fractal dimension* d_f is defined as a strongly connected digraph resulting from a mask of fractal dimension d_f.

The degree distribution is an important property in order to classify networks. Due to the deterministic nature of the DGP the mask determines this distribution in the following way. Be M_S the adjacency matrix. Out- and in-degree for each node in the digraph are directly given by the sums over the entries in the column or row of M_S:

$$d_{in}(v_n) = \sum_{j=1}^{S} M_S(n,j), \quad d_{out}(v_n) = \sum_{j=1}^{S} M_S(j,n), \qquad 1 \leq n \leq S.$$

Considering only the in-degree we can calculate the degree for each node in digraph D_i (i.e. resulting digraph after i iterations of DGP initialized with M_S and $i \geq 0$) as follows:

$$\left(d_{in}(v_1) + d_{in}(v_2) + \ldots + d_{in}(v_S)\right)^{i+1}$$

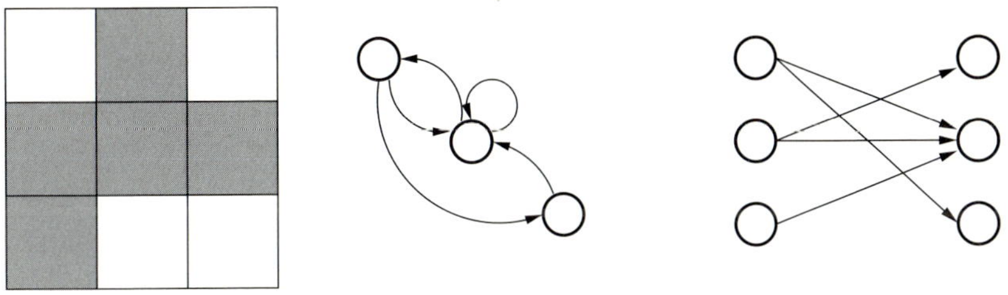

Fig. 3. Two ways of transforming a given digraph of fractal dimension (left) into a artificial neural networks. First, the digraph is directly interpreted as ANN with recurrent neural connections (middle). Second, the adjacency matrix as a description of a feed-forward network (right).

Solving this equation in a symbolic manner we get S^{i+1} products each representing the in-degree of one node in D_i. The distribution of these product values is the actual distribution of the in-degrees in D_i.

As an example let us assume we have a complete mask where only one node has an in- and out-degree greater 1, for all the other nodes in- and out-degree is one. The resulting digraphs will have a degree distribution characteristic of scale-free networks. Thus, the majority of nodes have very less in-coming and out-going edges but a few nodes (usually called hubs) have degrees magnitudes larger then the average [2]. On the other hand we can create digraphs out of masks, where each node has the same number of in- and out-degree. The resulting digraphs are going to have equal in- and out-degree as well. Due to this relation of the degree distribution between mask and resulting digraphs one is able to generate networks of specific degree-distribution.

3 From Fractal Digraphs to Artificial Neural Networks

There are in principle two strategies to turn a digraph of fractal dimension into an artificial neural network (see Figure 3). First, the adjacency matrix / the digraph can directly be interpreted as a neural network containing recurrences of any kind. Second, the adjacency matrix can purely be seen as the connections between two separated layers of neurons: input and output layer. Both layers contain the same number of neurons. In this way a feed-forward structure between two neuron layers is created. One can also think about a chain of feedforward connections where each projection layer might be based on different digraphs of fractal dimension.

The recurrent case might be interesting as method for the generation of reservoirs of non-linear dynamics. Based on random graphs, this has been done in the echo-state [9] and liquid-state-machine [11] approach. The intention of using digraphs of fractal dimension as dynamical reservoir is one reason for us to aim for strongly connected digraphs only. The dynamics of echo-state and liquid-state-machines rely on the recurrences. If an underlying digraph would be only

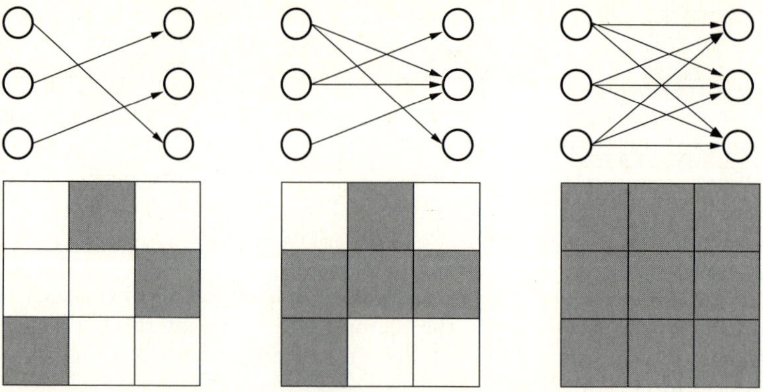

Fig. 4. Three examples of feedforward connections between two neural layers of an ANN. The left shows a non-degenerated matching between input and output signal. Each neuron in the left layer does only activate one neuron in the right layer. Due to the introduced representation this can be described as a cycle. On the right a completely degenerated matching. Each neuron on the left activates each neuron on the right. This is represented by a fully connected digraph. In between a degenerated matching formed by a digraph of fractal dimension.

connected (i.e. not strongly connected) neurons can emerge which only project signals out of the reservoir or which would feed constant signals into it. This is obviously not the intention of a dynamical reservoir for both approaches. In the worst case a connected digraph could have no recurrences at all and therefore no complex dynamics would emerge. Only strongly connected digraphs guarantee recurrent neural structures involving all components of the network.

For feedforward structures the use of strongly connected digraphs is also essential. Strongly connected digraphs guarantee that a signal feed into an arbitrary node can be propagated through several layers to any other node. Assume a multi-layered network structure and each projection between the layers is based on the same digraph of fractal dimension. In this case we know there must exist a finite number of layers between the input and output layer which guarantees that each neuron of the input layer has at least one path to each neuron in the output layer. In theory the number of layers cannot be larger than the number of nodes in the underlying strongly connected digraph. It is not shown here, but simulations indicate that the mean value of the shortest path scales with $log(n)$ (n number of nodes). Hence, the number for layers supporting a signal flow through all network components scales with log either.

ANN with feedforward structures based on fractal digraphs might become an object of investigation within the Neural Darwinism approach to the function of the brain introduced, developed and promoted by Edelman [3]. According to this approach, an essential element for the brain-function is the matching between specific signal configurations and neural groups, which respond in a specific manner. Obviously, this matching must be sufficient specific in order to allow distinction among different signals, called recognition. However, more important within the Neural Darwinism approach is the argumentation, that such a matching must be degenerated. The assumption is, that there is more than one way to recognize a signal, that is, one signal configuration activates different neural groups as well as one neural group can be activated by different signal configurations. Two extremes of degeneration can be distinguished: a non-degenerated (unique) matching on one side and the completely degenerated matching on the other side. The Neural Darwinism approach claims that the variability of brain functions occurs within a neural organization is, somehow, located between these two extremes of non- and complete degeneration.

It is interesting to see that the introduced digraphs of fractal dimension create feedforward networks between these two extremes. The examples shown in Figure 4 are only simple schemas. However, it is not hard to imagine that the fractal dimension and degree distribution of a digraph determine the grad of degeneration. Therefore, we argue, that within the Neural Darwinism approach the introduced digraphs of fractal dimension might be a promising substrate for future research in order to model brain-like mechanisms of adaptation which take into account not only weight dynamics but also specific neural connectivity structures.

4 Conclusion

In this work we have introduced a process, called DGP, which allows the deterministic generation of strongly connected digraphs. This process is inspired by Sierpiński carpets. In order to apply this process for the development of connectivity structures for recurrent and feedforward structures of ANN we have formulated a sufficient criterion which guarantees the generation of strongly connected digraphs. The resulting digraphs are called digraphs of fractal dimension.

Furthermore, we have shown how the degree-distribution of the resulting digraphs is determined by the initial structure. This allows us a targeted generation of strongly connected digraphs with respect to the size (number of nodes) and the degree distribution of the network. We have indicated that strongly connected digraphs with scale-free network topologies can be expected to emerge for specific initializations. Thus, the introduced process is an efficient tool for the generation of a wide variety of connectivity structures for ANN establishing interactions beyond those provided by fully connected or random graphs.

Within the context of Neural Darwinism, we have highlighted the importance of ANN based on digraphs of fractal dimensions for the implementation of adaptive processes. Due to the fractal dimension, the resulting strongly connected networks are established by a number of edges magnitudes smaller then given in fully connected topologies. Hence, graphs of fractal dimension support the instantiation of highly connected ANN with much less computational costs. Therefore, the DGP might be an efficient tool for the instantiation of ANN on autonomous robot systems having limited computational resources. In addition to this, fractal digraphs within the context of cognitive robotics and "brain-based devices" might be become a promising method for a systematic investigation and modelling of biological neural systems and "their combinations of interactions that we don't fully understand yet" [4]. Our future work will be focused on this issue.

Acknowledgement

The author would like to thank Prof. Dr. Vassilis C. Mavron and Prof. Dr. Frank Pasemann for fruitful comments on the topic of this paper and gratefully acknowledges the support of EPSRC through grant EP/C516303/1.

References

1. Bollobás, B.: Random Graphs. Academic Press Inc., London (1985)
2. Dorogovtsev, S.N., Mendes, J.F.F.: Evolution of Networks: From Biological Nets to the Internet and WWW. Oxford University Press, Oxford (2003)
3. Edelman, G.M.: Neural Darwinism. Oxford University Press, Oxford (1989)
4. Edelman, G.M.: Learning in and from brain-based devices. Science 318(5853), 1103–1105 (2007)
5. Elman, J.L.: Finding structure in time. Cognitive Science 14, 179–211 (1990)
6. Eves, H.: Elementary Matrix Theory. Dover publications (1980)
7. Hertz, J., Krogh, A., Palmer, R.G.: Introduction to the theory of neural computation. Perseus Books, USA (1991)
8. Ito, M., Noda, K., Hoshino, Y., Tani, J.: Dynamic and interactive generation of object handling behaviors by a small humanoid robot using a dynamic neural network model. Neural Networks 19, 323–337 (2006)
9. Jaeger, H., Haas, H.: Harnessing Nonlinearity: Predicting Chaotic Systems and Saving Energy in Wireless Communication. Science 304(5667), 78–80 (2004)
10. Kandel, E.R., Schwartz, J.H., Jessell, T.M.: Principles of Neural Science, 4th edn. McGraw-Hill Medical, New York (2000)
11. Maass, W., Natschläger, T., Markram, H.: Real-Time Computing Without Stable States: A New Framework for Neural Computation Based on Perturbations. Computation 14, 2531–2560 (2002)
12. Mandelbrot, B.B.: The Fractal Geometry of Nature. Freeman, San Francisco (1982)
13. Newman, M.E.J.: The Structure and Function of Complex Networks. SIAM Review 45, 167–256 (2003)
14. Pfeifer, R., Scheier, C.: Understanding Intelligence. MIT Press, Cambridge (2000)
15. PlanetMath.org, `http://PlanetMath.org`
16. Wermter, S., Palm, G., Elshaw, M.: Biomimetic Neural Learning for Intelligent Robots. Springer, Heidelberg (2005)

The GAME Algorithm Applied to Complex Fractionated Atrial Electrograms Data Set

Pavel Kordík[1], Václav Křemen[2], and Lenka Lhotská[2]

[1] Department of Computer Science and Engineering, FEE,
Czech Technical University, Prague, Czech Republic
[2] Department of Cybernetics, FEE,
Czech Technical University, Prague, Czech Republic
kordikp@fel.cvut.cz, xkremen@fel.cvut.cz

Abstract. Complex fractionated atrial electrograms (CFAEs) represent the electrophysiologic substrate for atrial fibrillation (AF). Individual signal complexes in CFAEs reflect electrical activity of electrophysiologic substrate at given time. To identify CFAEs sites, we developed algorithm based on wavelet transform allowing automated feature extraction from source signals. Signals were ranked by three experts into four classes. We compiled a representative data set of 113 instances with extracted features as inputs and average of expert ranking as the output. In this paper, we present results of our GAME data mining algorithm, that was used to (a) predict average ranking of experts, (b) classify into three classes. The performance of the GAME algorithm was compared to well known data mining techniques using robust ten times tenfold cross validation. Results indicate that wavelet signal decomposition could carry high level of predictive information about the state of electrophysiologic substrate and that the GAME algorithm outperforms other data mining techniques (such as decision trees, linear regression, neural networks, Support Vector Machines, etc.) in both prediction and classification accuracy.

1 Introduction

Atrial fibrillation (AF) is a cardiac arrhythmia characterized by very rapid and uncoordinated atrial activation with a completely irregular ventricular response [8]. Radiofrequency ablation of atrial areas that triggers or sustains AF is a nonfarmacological treatment available recently [4].

During AF, multiple wavefronts propagate continuously through the right and left atria, separated by anatomical and functional barriers [10]. This can be electrophysiologically manifested as hierarchical distribution of dominant frequency [19] or complex fractionated electrograms (CFAEs) [17] during endocardial mapping. Local dominant frequency analysis of AF is burdened by many methodological problems of spectral analysis [18]. Therefore the software support for electroanatomical mapping system is focused on objective description and space representation of CFAEs distribution most recently. Till now there is no single automated classifier of A-EGMs, that would enable independently on operator

V. Kůrková et al. (Eds.): ICANN 2008, Part II, LNCS 5164, pp. 859–868, 2008.

classify A-EGMs and suggest its level of fractionation of CFAEs during mapping procedure.

Our approach is based on the idea, that there are signal complexes (Fig. 2) in every AEGM signal, which are related to electrical activation of electropathological substrate during AF. These signal complexes (SCs) can be found automatically and then used for several features extraction (degrees of freedom of the signal), which could be used for automatic evaluation of electrogram complexity (or level of fractionation).

In this paper we used these and other features to construct an automated classifier using the FAKE GAME approach [13]. The GAME algorithm constructs hybrid feedforward neural networks that enable to give an operator independent look on A-EGM signal and classify its degree of fractionation. The FAKE methods allow to extract useful knowledge from data.

In this contribution we studied optimal settings if the GAME algorithm and compared its performance to several well established methods available in the WEKA data mining environment [1]. For the comparison, we needed to repeat ten-fold cross validation 10 times to get reliable results. Results are presented in form of box plot charts.

2 Methodology

We used a representative data set ($n = 113$) of atrial electrograms (A-EGMs), which were pre-selected by an expert from large database of A-EGMs. This database was recorded during AF mapping procedures. Signals were sampled by frequency $977Hz$ during AF procedure and resampled to $1kHz$ after that. Each pre-selected A-EGM signal in this data set is $1500ms$ long. The expert signal selection was tailored to get a good quality signals with respect to low noise and high information value of signal for later evaluation of degree of A-EGMs fractionation by an expert. Although the degree of fractionation is supposed to be naturally continuous we decided to make a four degree set of classes (Fig. 2). Three experts used these four classes for ranking (1 – organized atrial activity, $n = 24$; 2 – mild, $n = 40$; 3 – intermediate, $n = 36$; 4 – high degree of fractionation, $n = 13$.).

Individual points of interest were found automatically after filtering of the signal. Both steps were performed by novel algorithm based on wavelet transform [15] for every A-EGM in data set (Fig. 2) using following parameters of the algorithm: discrete wavelet decomposition into 5 levels [5] was performed. Detail coefficients were thresholded by soft-tresholding [6] with these settings of thresholds (level 1 to level 5): 0.02, 0.04, 0.008, 0.008 and 0.008. Reconstruction of the signal was computed by wavelet reconstruction based on the original approximation coefficients and the modified detail coefficients of levels from 1 to 5. Thresholding of the signal with value of threshold $0.003mV$ was performed.

The filtered signal was decomposed again into 5 levels using Coiflet wavelet of order four. The reconstruction of the detailed coefficients of a signal (L3) of given wavelet decomposition structure was performed at level 3 (L3). Normalization of

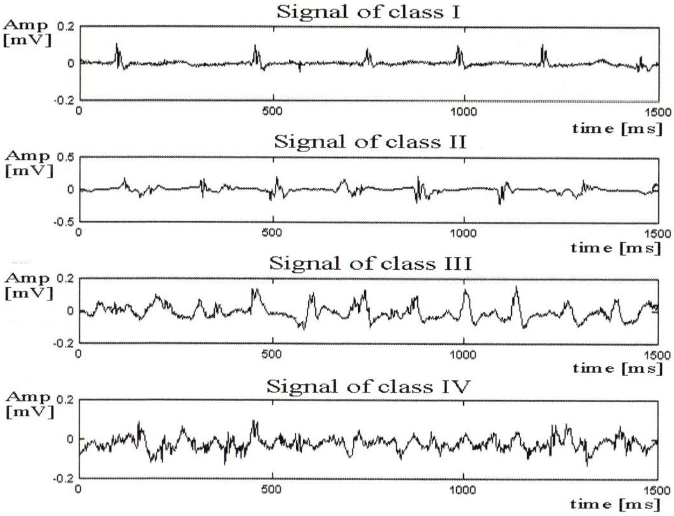

Fig. 1. Four complex fractionated electrograms are shown. These are representatives of each ranking class of degree of fractionation ranked by three independent experts. From the top to bottom: 1 – organized atrial activity; 2 – mild, 3 – intermediate; 4 – high degree of fractionation.

L3 was performed with respect to maximal absolute value of given L3 values to obtain uniform signals across the data set for next stages of SCs detection. Thresholding of normalized L3 signal values was performed with value of threshold 0.014. Then all parts of the signal, where absolute value of amplitude was higher than 0, and whose inter-distance was closer than threshold $5ms$ were joined together and they were marked as one individual SC.

Several feature extraction was performed to prepare input of classifier.

1. Number of inflection points in particular A-EGM signal (IP).
2. Mean value of number of inflexion points in the found SCs in particular A-EGM signal (MIPSC).
3. Variance of number of inflexion points in the found SCs in particular A-EGM signal (VIPSC).
4. Mean value of width of found SCs in particular A-EGM signal (MWSC).
5. Number of inflection points in found SCs in particular A-EGM signal (IPSC).
6. IPSC normalized per number of found SCs in particular A-EGM signal (NIPSC).
7. IP + MIPSC (IPMIPSC).
8. $\sqrt{IPSC^2 + TDM^2}$ (IPSCTDM).
9. Number of zero-level crossing points in found SCs in particular A-EGM signal (ZCP).
10. Maximum of IPSC in particular A-EGM signal (MIPSC).

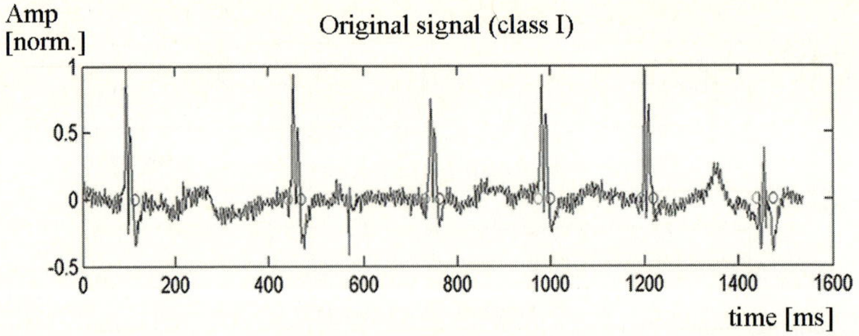

Fig. 2. Original CFAE signal recorded during AF mapping procedure. All expert rankings of the signal were into class I. Depicted amplitude is normalized with respect to maximal absolute value of this particular CFAE signal. Green circles denote the beginnings of SCs and red circles the ends of SCs found automatically.

11. Time domain method (see below) with rough (unfiltered) input A-EGM signal (TDM).
12. Time domain method using input A-EGM signal filtered by above described wavelet filter (FTDM).
13. Correlation power (number of correlated SCs in the A-EGM signal/number of found SCs in the A-EGM signal)(CP).

TDM extraction method used here was based on an algorithm which worked with A-EGM signals only in time domain. TDM described signal as mean of the intervals between discrete peaks of A-EGMs, which were detected using input parameters of TDM method: peak-to-peak sensitivity, signal width and refractory period. Input parameters were optimized by exhaustive search with respect to the used data set (sensitivity $0.02mV$, signal width $8ms$, refractory period $14ms$). Spearman's correlation coefficient between indexes of fractionation of TDM method and categories ranked by an experts was used as optimization criteria for exhaustive search of optimal input parameters.

3 Group of Adaptive Model Evolution

Group of Adaptive Models Evolution (GAME) [12] proceeds from the Group Model Data Handling (GMDH) theory [11]. GMDH was designed to automatically generate model of the system in the form of polynomial equations. An example of inductive model created by GAME algorithm is depicted on the Figure 3. Similarly to Multi-Layered Perceptron (MLP) neural networks, GAME units (neurons) are connected in a feedforward network (model). The structure of the model is evolved by special niching genetic algorithm, layer by layer. Parameters of the model (coefficients of units' transfer functions) are optimized independently [14]. Model can be composed from units of different transfer function type (e.g sigmoid, polynomial, sine, linear, exponential, rational, etc). Units with transfer function performing

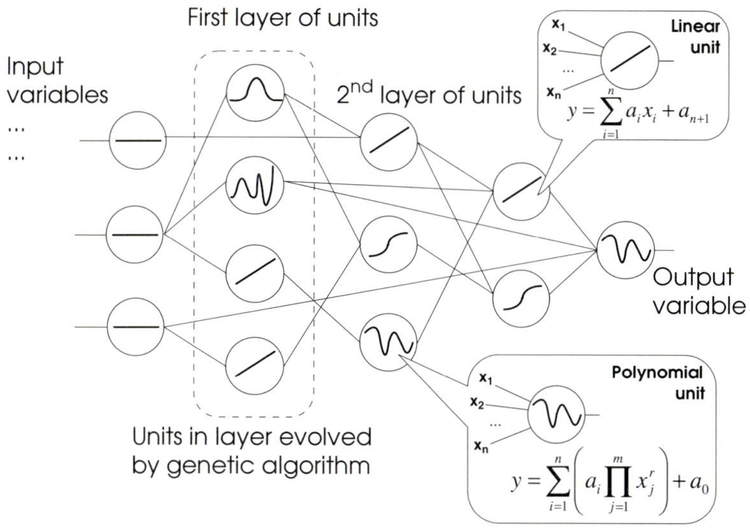

Fig. 3. The structure of the GAME model is evolved layer by layer using special genetic algorithm. While training an unit, selected optimization method adjusts coefficients $(a_1, ..., a_n)$ in the transfer function of the unit.

well on given data set survive the evolution process and form the model. Often, units of several different types survive in one model, making it hybrid.

In this paper, we use GAME models as predictors (continuous output variable) and also as classifiers (four output classes). The evolutionary design of the GAME algorithm makes this possible without need to change the learning strategy. For classification purposes, the GAME model consists mainly of units with sigmoid transfer function and for regression, polynomial, linear or exponential units are often selected to fit the input-output relationship.

4 Data Analysis

Each A-EGM signal in the data set ($n = 113$) was ranked by three independent experts. This resulted in a total of 339 rankings distributed among four classes of fractionation. Table 1 shows the distribution of the rankings assigned to given classes of fractionation by the three experts. The rankings of the three experts were averaged for each A-EGM signal and used as a class of fractionation of the signal for comparing the classifiers studied here. Table 2 shows the number of signals in the averaged classes of fractionation. The observation of particular importance is that expert rankings differ mostly for two inner classes (2 and 3). Several times happened that one two experts ranked the signal as class 2 and the third expert included the signal to class 3. This case was observed 19 times (class 2.3 in Table 2).

Table 1. Number of A-EGM signals in the data set assigned by three independent experts to the four classes of fractionation

CF[1]	1	2	3	4
NS[2]	76	116	102	45

[1]Four classes of fractionation used in the study.
[2]Number of A-EGM signals assigned by three independent experts to each CF.

Table 2. Number of A-EGM signals in the data set assigned to averaged classes of fractionation and used for comparison of classifiers

ACF[1]	1	1.3	1.7	2	2.3	2.7	3	3.3	3.7	4
NS[2]	19	5	9	17	19	4	21	4	4	11

[1] Classes of fractionation, determined as the average of the rankings assigned by three independent experts.
[2]Number of A-EGM signals in the data set assigned to the ACF.

The character of the data set, particulary the uncertain boundary of middle classes excludes the possibility to classify into 10 classes of fractionation with reasonable error. Even for classification into four classes, we got perfect (95%) accuracy for class 1 and 4 only, the accuracy for other two classes was bad (around 65%).

Reflecting these preliminary experiments we projected the data into two dimensions (using [7]) and found boundaries better separating the data vectors. We prepared two versions of the data set. The first was A-EGM-regression data set, where the output was continuous number - the average experts ranking (AER). The second was A-EGM-classification data set with three output classes: (1 – organized atrial activity; 2 – intermediate; 3 – high degree of fractionation), where for class one the AER was below 1.9, for class two the AER in $\langle 1.9, 3\rangle$ interval and for the class three the AER was above 3.

5 Results

At first, we studied the regression performance of GAME models produced by different configurations of the algorithm. The target variable was the average A-EGM signal ranking by three experts (the A-EGM-regression data set). We found out, and it is also apparent in the boxplot charts, that comparison of the 10-fold cross validation error is not stable enough to decide, which configuration is better. Therefore we repeated the 10-fold cross validation ten times, each time with different fold splitting. For each box plot it was necessary to generate and validate hundred models.

For all experiments we used three default configurations of the GAME algorithm available in FAKE GAME environment [2]. The *std* configuration uses just subset of units (those with implemented analytic gradient for faster optimization).

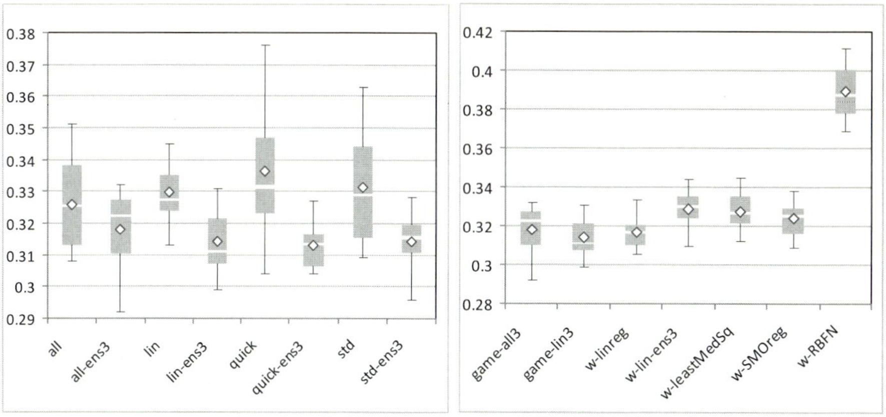

Fig. 4. The comparison of RMS cross validation errors for several configuration configuration of the GAME engine(left). Selected GAME models compared with models generated in Weka environment(right).

It evolves 15 units for 30 epochs in each layer. The *quick* configuration is the same as *std* except that it do not use the niching genetic algorithm (just 15 units in the initial population). The *linear* restricts type of units that can be used to linear transfer function units. The *all* configuration is the same as *std*, in addition it uses all units available in the FAKE GAME environment. This configuration is more computationally expensive, because it also optimizes complex units such as BPNetwork containing standard MLP neural network with the back-propagation of error [16].

The GAME algorithm also allows to generate ensemble of models [9,3]. Ensemble configurations contain digit (number of models) in their name.

The Figure 4 shows that the regression of the AER output is not too difficult task. All basic GAME configurations performed similarly (left chart) and ensembling of three models further improved their accuracy. The ensemble of three linear models performed best in average, but the difference from $all - ens3$ configuration is not significant.

In Weka data mining environment, *LinearRegression* with embedded feature selection algorithm was the best performing algorithm. Ensembling (bagging) did not improved results of generated model, quite the contrary. The Radial Basis Function Network (RBFN) failed to deliver satisfactory results in spite of experiments with its optimal setting (number of clusters).

Secondly, our experiments were performed on the A-EGM-classification data set. The methodology remained the same as for regression data. Additionally we tested classification performance of 5 models ensembles. Figure 5 left shows that the classes are not linearly separable - *linear* configuration generates poor classifiers and ensembling does not help. Combining models in case of all other configurations improve the accuracy. For *all* configuration the dispersion of cross validation errors is quite high. The problem is in the configuration of the genetic

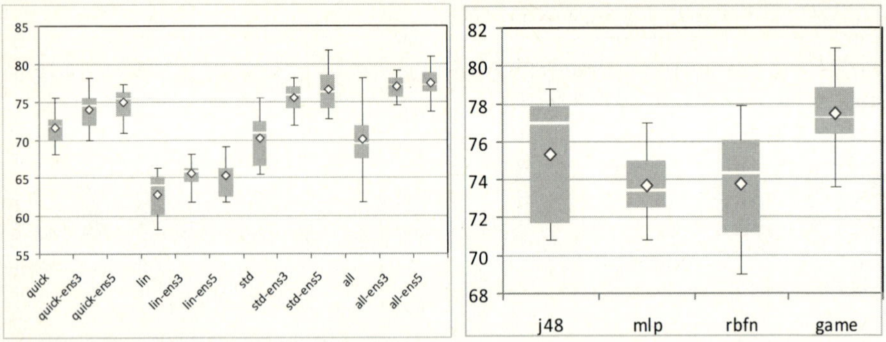

Fig. 5. Classification accuracy in percent for several GAME configurations (left) and comparison with Weka classifiers (right)

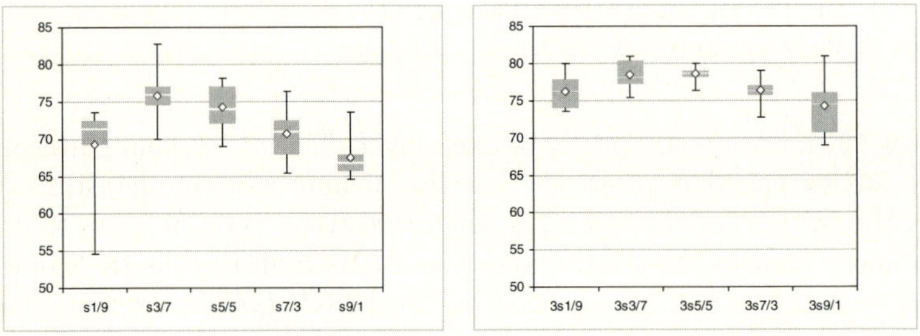

Fig. 6. Classification performance for different ratios of training/validation data split. Left - results for single game models generated by *std* configuration. Right - results for GAME ensemble ($std - ens3$).

algorithm - with 15 individuals in the population some “potentially useful” types of units do not have chance to be instantiated. Ensembling models generated by this configuration improves their accuracy significantly.

Comparison with Weka classifiers (Figure 5 right) shows that GAME ensemble significantly outperforms Decision Trees (j48), MultiLayered Perceptron (mlp) and Radial Basis Function network (rbfn) implemented in Weka data mining environment.

The last experiment (Figure 6) showed that the best split of the training and validation data set is 40%/60% (training data are used by optimization method to adjust parameters of GAME units transfer functions, whereas from validation part, the fitness of units is computed).

Implicitly, and in all previous experiments, training and validation data set was divided 70%/30% in the GAME algorithm. Changing the implicit setting to 40%/60% however involves additional experiments on different data sets.

6 Conclusion

Our results show, that extracted features for Complex Fractionated Atrial Electrograms bear significant information allowing us to classify signals into three classes with 80% accuracy. That is good result with respect to 60% of consistent assignments into four classes performed by experts. For this data set, the GAME algorithm outperforms well established methods in both classification and regression accuracy. What is even more important, both winning configurations were identical $all - ens$. Natural selection evolved optimal models for very different tasks - that is in accordance with our previous experiments and with our aim to develop automated data mining engine.

Acknowledgement

This research is partially supported by the grant Automated Knowledge Extraction (KJB201210701) of the Grant Agency of the Academy of Science of the Czech Republic and the research program "Transdisciplinary Research in the Area of Biomedical Engineering II" (MSM6840770012) sponsored by the Ministry of Education, Youth and Sports of the Czech Republic.

References

1. Weka open source data mining software (September 2006), D `http://www.cs.waikato.ac.nz/ml/weka/`
2. The fake game environment for the automatic knowledge extraction (March 2008), `http://www.sourceforge.net/projects/fakegame`
3. Brown, G.: Diversity in Neural Network Ensembles. PhD thesis, The University of Birmingham, School of Computer Science, Birmingham B15 2TT, United Kingdom (January 2004)
4. Calkins, H., Brugada, J., Packer, D., et al.: Hrs/ehra/ecas expert consensus statement on catheter and surgical ablation of atrial fibrillation: recommendations for personnel, policy, procedures and follow-up. Heart Rhythm 4, 816–861 (2007)
5. Daubechies, I.: Ten lectures on Wavelts, Philadelphia, Pennsylvania, USA (1992)
6. Donoho, D.: De-noising by soft-thresholding. IEEE Trans. on Inf. Theory 41(3), 613–662 (1995)
7. Drchal, J., Kordík, P., Šnorek, M.: Dataset Visualization Based on a Simulation of Intermolecular Forces. In: IWIM 2007 - International Workshop on Inductive Modelling, Praha, vol. 1, pp. 246–253. Czech Technical University in Prague (2007)
8. Fuster, V., Rydn, L., Cannom, D.S., et al.: Acc/aha/esc 2006 guidelines for the management of patients with atrial fibrillation: a report of the american college of cardiology/american heart association task force on practice guidelines and the european society of cardiology committee for practice guidelines (writing committee to revise the 2001 guidelines for the management of patients with atrial fibrillation): developed in collaboration with the european heart rhythm association and the heart rhythm society. Circulation 114(7), 257–354 (2006)
9. Hansen, L., Salamon, P.: Neural network ensembles. IEEE Trans. Pattern Anal. Machine Intelligence 12(10), 993–1001 (1990)

10. Allessie, M.A., Houben, R.P.M.: Processing of intracardiac electrograms in atrial fibrillation. diagnosis of electropathological substrate of af. IEEE Eng. Med. Biol. Mag. 25, 40–51 (2006)
11. Ivakhnenko, A.G.: Polynomial theory of complex systems. IEEE Transactions on Systems, Man, and Cybernetics SMC-1(1), 364–378 (1971)
12. Kordík, P.: Game - group of adaptive models evolution. Technical Report DCSE-DTP-2005-07, Czech Technical University in Prague, FEE, CTU Prague, Czech Republic (2005)
13. Kordík, P.: Fully Automated Knowledge Extraction using Group of Adaptive Models Evolution. PhD thesis, Czech Technical University in Prague, FEE, Dep. of Comp. Sci. and Computers, FEE, CTU Prague, Czech Republic (September 2006)
14. Kord'ık, P., Kovářík, O., Šnorek, M.: Optimization of Models: Looking For The Best Strategy. In: Proceedings of the 6th EUROSIM Congress on Modelling and Simulation, Vienna, vol. 2, pp. 314–320. ARGESIM (2007)
15. Kremen, V., Lhotska, L.: Evaluation of novel algorithm for search of signal complexes to describe complex fractionated atrial electrogram. In: Proceedings of Biosignal, Biosignals (2008)
16. Mandischer, M.: A comparison of evolution strategies and backpropagation for neural network training. Neurocomputing (42), 87–117 (2002)
17. Nademanee, K., McKenzie, J., Kosar, E., et al.: A new approach for catheter ablation of atrial fibrillation: mapping of the electrophysiologic substrate. J. Am. Coll. Cardiol. 43, 2044–2053 (2004)
18. Ng, J., Kadish, A., Goldberger, J.: Effect of electrogram characteristics on the relationship of dominant frequency to atrial activation rate in atrial fibrillation. Heart Rhythm 3, 1295–1305 (2006)
19. Sanders, P., Berenfeld, O., Hocini, M., et al.: Spectral analysis identifies sites of high-frequency activity maintaining atrial fibrillation in humans. Circulation 112, 789–797 (2005)

Geometrical Perspective on Hairy Memory*

Cheng-Yuan Liou

Department of Computer Science and Information Engineering
National Taiwan University
Republic of China
cyliou@csie.ntu.edu.tw

Abstract. This paper constructs a balanced training used in the hairy network [1] to balance the vulnerable memory parts and improve the memory. It provides a perspective view on the geometrical structure of memory patterns. This training fixes many drawbacks of the Hopfield network, such as loading capacity, limit cycle and error tolerance.

Keywords: Hairy network, hologram, conscious, neural network, associative memory, Hopfield network, Elman network, homeostasis, gene regulatory network.

1 Introduction

This paper presents a new training used in the hairy network [1][2] to balance the vulnerable parts of memory patterns. It enlarges the memory basins evenly in order to resist corruptions [3]. We briefly review the associative memory (AM) and notations.

Associative memory
Hopfield model (HM) has drawbacks in loading capacity, limit cycles, and error tolerance with respect to noisy patterns. Various designs aimed at remedying these drawbacks have achieved varying degrees of success [1][2][4][5]. Auto-AM is a connected network with N neurons. Each neuron i has N neural weights connecting it to all the neurons j, including itself, a threshold θ_i, and a state value v_i. The state value is evolved according to the formula

$$v_i(t+1) = \text{sgn}\left[\sum_{j=1}^{N} w_{ij} v_j(t) - \theta_i\right], \tag{1}$$

or in matrix form,

$$V(t+1) = \text{sgn}\left[W^T V(t) - \theta\right], \tag{2}$$

where W is an $N \times N$ neural weight matrix, θ is an $N \times 1$ threshold vector, $V(t)$ is an $N \times 1$ state vector representing the state at iteration (or evolution) time

* Supported by National Science Council.

V. Kůrková et al. (Eds.): ICANN 2008, Part II, LNCS 5164, pp. 869–878, 2008.

t, and sgn() is the signum function returning 1 with input greater than or equal to zero and -1 with negative input. During the training phase, the network is trained by P memory patterns $X^k, k = 1, \ldots, P$, using any training algorithm. During the retrieving phase, the input is presented to the network as $V(0)$. The network operates repeatedly according to Eq.(1) or Eq.(2) until the evolution converges to a stable state or falls into a limit cycle. A stable state (memory pattern) meets the requirement

$$V(t) = \operatorname{sgn}\left[W^T V(t) - \theta\right]. \tag{3}$$

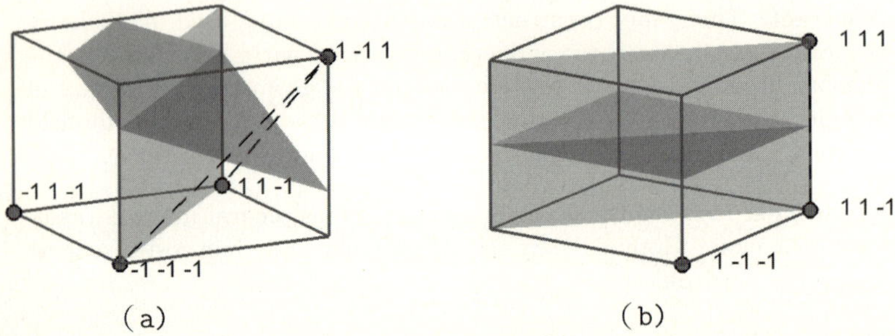

Fig. 1. 3-D cube network with three neurons. Black dots denote memory patterns. The patterns connected by dashed lines are used to determine the position of the decision (shaded) plane and they will be discussed in this paper.

Each neuron has a bipolar state (a bit), and there are 2^N states in total. Therefore, we view the network as an N-dimensional (N-D) hypercube with each state located at a cube corner [6]. The current state is located at a corner and serves as the next input to the network. After evolution proceeds according to Eq.(1) or Eq.(2), the current state either moves to another corner or stays at the original corner. Corners that remain unchanged are stable states. The memory patterns we intend to save are located at certain stable corners. The goal of AM is to evolve an arbitrary initial state to a nearby stable corner where a pattern is stored. Figure 1 shows a 3-D cube corresponding to a network with three neurons (three bits). In this figure, a neuron, the third bit, represents a plane (the dark shaded plane (blue)) dividing the cube into positive and negative regions (sides).

The state of each neuron i is determined by an $(N-1)$-D decision hyperplane (the shaded surfaces in Fig. 1) with the following equation:

$$w_{i1}v_1 + w_{i2}v_2 + w_{i3}v_3 + \cdots + w_{iN}v_N - \theta_i = 0, \quad i = 1, \ldots, N. \tag{4}$$

The $N \times 1$ weight vector $W_i = (w_{i1}, w_{i2}, \ldots, w_{iN})^T$ of neuron i is the normal vector of the corresponding hyperplane, and this hyperplane divides the hypercube into a positive region (side) to which the normal vector points and a negative region. The length of vector W_i, $|W_i|$, is normalized to one. The hairy network adjusts the hyperplane to make all the P patterns stable. That is, when the ith bit of a pattern is equal to 1, this stable pattern should be located in the positive region of the ith hyperplane; on the other hand, if its ith bit is equal to -1, it should be located in the negative region.

Since each decision hyperplane can be trained separately in the hairy network, we will discuss the hyperplane for neuron i only. For neuron i, we have P equations, $k = 1, \ldots, P$:

$$\begin{cases} w_{i1}X_1^k + w_{i2}X_2^k + \cdots + w_{iN}X_N^k - \theta_i = W_i^T X^k - \theta_i > 0, & if \quad X_i^k = 1 \\ w_{i1}X_1^k + w_{i2}X_2^k + \cdots + w_{iN}X_N^k - \theta_i = W_i^T X^k - \theta_i < 0, & if \quad X_i^k = -1 \end{cases}. \tag{5}$$

We discard the case in Eq.(5), where $W_i^T X^k - \theta_i = 0$, because it rarely happens in analog operation. For computational convenience, we multiply every element in Eq.(5) by X_i^k and obtain a compact formula:

$$\begin{cases} s_i^k = \{w_{i2}X_2^k + \cdots + w_{iN}X_N^k - \theta_i\}X_i^k \\ \quad = \{W_i^T X^k - \theta_i\}X_i^k > 0, \text{ for both cases } X_i^k = 1 \text{ and } X_i^k = -1, \end{cases} k = 1, \ldots, P. \tag{6}$$

Whenever we say that the ith hyperplane is stable, we mean that all P patterns in the correct (stable) region of the hyperplane satisfy Eq.(6). This hyperplane is not stable whenever there is a pattern in the wrong region.

The basin of a stable pattern is defined as the collection of all states, see [1], which eventually evolve to this stable pattern according to the dynamic equation, that is, Eq.(2). We further subdivide a basin according to the number of iterations, λ. Therefore, basin-λ of a stable pattern contains all the states that will evolve to this stable pattern in λ iterations. Basin-0 contains only the stable pattern. All such basins, {basin-λ; $\lambda = 0, 1, 2, \ldots$ }, constitute the overall basin of a stable pattern. The states in basin-0 and basin-1 of a memory pattern are contained in the cube region enclosed by the N decision hyperplanes and this region has a shape like an open polyhedral cone[7]. The hairy network increases the number of states in basin-1 and increases the size of the overall basin for a stable pattern indirectly.

The basin radius of a stable pattern is defined as the maximal number of errors for which recovery is guaranteed, no matter which bits are wrong [1][5]. The value of this radius is the distance (with a unit of one bit or a Hamming distance of 2 for each bipolar bit) between the stable pattern and a nearest state right on basin-1's border. This is the most rigorous definition. Hence, '0' radius means that there exists at least one bit or pixel that is vulnerable to noise contamination no matter how large the basin is.

To calculate the radius of each pattern, we calculate s_i^k for all the neurons according to Eq.(6) for each memory pattern, X^k, after the learning process is completed. If a bit j of the pattern X^k is reverted, s_i^k will increase or decrease by $2w_{ij}$, depending on whether the signs of $X_j^k X_i^k$ and w_{ij} are identical or not. If s_i^k is just larger than the sum of the r_i^k largest weight terms multiplied by two, then r_i^k is the radius of neuron i for pattern X^k. That is,

$$2\sum_{r=1}^{r_i^k+1} \{w_{ij}X_j^k X_i^k\}^r > s_i^k = |s_i^k| > 2\sum_{r=1}^{r_i^k} \{w_{ij}X_j^k X_i^k\}^r, \tag{7}$$

where $\{w_{ij}X_j^kX_i^k\}^r$ is the rth largest weight term sorted in descending order according to the values of the terms in the set $\{w_{ij}X_j^kX_i^k|\ j = 1, 2, ..., N\}$. The radius of the pattern X^k, r^k, is the smallest one among the radii of all the neurons, $r^k = \min\{r_i^k; i = 1, ..., P\}$.

We expect that increasing s_i^k will enlarge the size of basin-1 and will enlarge the overall basin size and improve tolerance. To achieve this goal, et-AM [5] and e-AM [1] maximize the minimum of $\{s_i^k, k = 1, \ldots, P\}$ to improve the hyperplane's tolerance for noisy neighbors as much as possible. The et-AM is designed with a distance s_i^k and the e-AM is designed with a quadratic distance $(s_i^k)^2$. They provide biologically plausible solutions in order to increase the memory basin without using any hidden neurons or an annealing process. They explore the flexible space in between the two pattern sets, $U_{i,1} = \{X^k; k = 1, \ldots, P,$ and $X_i^k = 1\}$ and $U_{i,-1} = \{X^k; k = 1, \ldots, P,$ and $X_i^k = -1\}$, in order to locate the decision hyperplane under the stability conditions in Eq.(5). Partial differentiation of the distance, s_i^k, gives

$$\frac{\partial s_i^k}{\partial w_{ij}} = X_j^k X_i^k \qquad \text{and} \qquad \frac{\partial s_i^k}{\partial \theta_i} = -X_i^k. \tag{8}$$

This equation is *consistent* with Hopfield model and Hebb's postulate [8].

The idea of the proposed balanced training comes from a selection proposed in (Eq.(14) in [1]; Eq.(9) in [5]). This selection says that one selects a nearest pair of patterns $\{c_{i,p}, c_{i,n}\}$ from each of the two sets, $U_{i,1}$ and $U_{i,-1}$, where $c_{i,p}$ is a pattern in $U_{i,1}$ and $c_{i,n}$ is a pattern in $U_{i,-1}$, $|c_{i,p} - c_{i,n}| = \min\{|X^k - X^l|, X^k \in U_{i,1}, X^l \in U_{i,-1}\}$. We suppose that both sets, $U_{i,1}$ and $U_{i,-1}$, are not empty. This pair is the nearest pair (in terms of the Euclidean distance) among all the pairs across the two sets (see Fig. 1(b)). These two nearest patterns are most vulnerable to each other's corrupted patterns. The weights are set as

$$w_{ij} = c_{i,p,j} - c_{i,n,j}, \text{ normalize } W_i\ , \tag{9}$$
$$\theta_i = \sum_{j=1}^{N} w_{ij} \left(\frac{c_{i,p,j} + c_{i,n,j}}{2} \right).$$

This hyperplane is right in the middle between the two nearest patterns. It is perpendicular to the line section connecting the two patterns $\{c_{i,p}, c_{i,n}\}$ and passes through the center of this section. $c_{i,p} - c_{i,n}$ is the direction of the normal vector of this hyperplane. From our evidence, this sclection may not be useful when there are multiple minima (see Fig. 1(a)) [2]. In other words, many pattern pairs, $\{c_{i,p}, c_{i,n}\}$, across the two sets will share the same minimum distance. There are many vulnerable pair patterns. To keep the balance of such multiple minima, we devised a constrained training to balance the hyperplane among all the minimal pair patterns. We expect that this training can effectively reduce the number of training iterations and achieve the tolerance.

2 Balanced Training

For the multiple minima, we save those patterns which have the same minimum pair distance in two sets, $U_{i,p}$ and $U_{i,n}$, where $U_{i,p} = \{X^k$; pattern X^k with the minimum distance, $k = 1, \ldots, P$, and $X_i^k = 1\}$ and $U_{i,n} = \{X^k$; pattern X^k with the minimum distance, $k = 1, \ldots, P$, and $X_i^k = -1\}$. Suppose there are $N_{i,p}$ patterns in set $U_{i,p}$ and $N_{i,n}$ patterns in $U_{i,n}$. In Fig. 1(a), there are two nearest pairs: $N_{3,p} = 1$, $N_{3,n} = 2$, $U_{3,p} = U_{3,1} = \{(1,1,1)^T\}$, $U_{3,n} = \{(1,-1,-1)^T, (-1,1,-1)^T\}$, $U_{3,-1} = \{(1,-1,-1)^T, (-1,1,-1)^T, (-1,-1,-1)^T\}$.

Since we prefer a hyperplane which evenly balances the minimum distance patterns. We formulate the balance in a set of constraints. This constraint sets all s_i^k equal for those patterns in $U_{i,p}$ and $U_{i,n}$. This is, in some sense, analogous to the mean-field theory for the extensive memory loading case [4], where the basin size is equal for all neurons and all patterns. We expect that this constraint can evenly distribute the neighboring corrupted patterns for each memory pattern as much as possible and alleviate competition for basin borders.

For each neuron i, we require that all s_i^k be equal for the patterns in sets $U_{i,p}$ and $U_{i,n}$. All patterns, $\{X_i^k, k = 1, \ldots, P\}$, must contain both -1 and 1 elements in their i^{th} elements.

According to the constraint, we have $s_i^1 = s_i^2 = \cdots = s_i^{N_{i,p}+N_{i,n}} = S_i$ and, hence, $N_{i,p} + N_{i,n} - 1$ homogeneous equations. The setting of these equations is discussed in Appendix. Note that we rearrange the pattern numbers in sets $U_{i,p}$ and $U_{i,n}$ to ease the expression. After performing Gaussian elimination and back substitution in these $N_{i,p} + N_{i,n} - 1$ equations, we obtain a simple row echelon form of these equations with free variables. To increase S_i, we proceed in the ascending direction with respect to these free variables $w_{iq_n}, w_{iq_{n+1}}, \ldots, w_{iq_N}$ and θ_i, $\frac{\partial S_i}{\partial w_{ij}}$, $j = q_n, \ldots, q_N$. That is,

$$w_{ij}(t+1) = w_{ij}(t) + \eta_1 \frac{\partial S_i}{\partial w_{ij}}$$

$$\theta_i(t+1) = \theta_i(t) + \eta_2 \frac{\partial S_i}{\partial \theta_i},$$

where η_1 and η_2 are training rates.

As an example, there is a six-neuron associative memory storing three patterns [9]: $X^1 = (1,1,1,-1,-1,-1)^T$, $X^2 = (1,-1,1,1,-1,1)^T$, and $X^3 = (1,1,-1,1,-1,-1)^T$. There are six hyperplanes to be set in a 6-D hypercube. Since the first bits of all the patterns are equal to 1, we simply put the first hyperplane outside the hypercube and allocate all the corners to the positive region of this hyperplane. We may set W_1 to $(1,0,0,0,0,0)^T$ and θ_1 to $-\sqrt{6}$. Similarly, since all the fifth bits are equal to -1, we may set W_5 to $(0,0,0,0,1,0)^T$ and set θ_5 to $\sqrt{6}$. The second hyperplane is set using the proposed method. In this case, $N = 6$, $P = 3$, $N_{2,p} = 2$, $N_{2,n} = 1$, $U_{2,p} = \{X^1, X^3\}$, $U_{2,n} = \{X^2\}$. We have the following equations:

$$\begin{bmatrix} s_2^1 \\ s_2^2 \\ s_2^3 \end{bmatrix} = \begin{bmatrix} 1 & 1 & 1 & -1 & -1 & -1 & -1 \\ -1 & 1 & -1 & -1 & 1 & -1 & 1 \\ 1 & 1 & -1 & 1 & -1 & -1 & -1 \end{bmatrix} \cdot \begin{bmatrix} w_{21} \\ w_{22} \\ w_{23} \\ w_{24} \\ w_{25} \\ w_{26} \\ \theta_2 \end{bmatrix} = \begin{bmatrix} >0 \\ >0 \\ >0 \end{bmatrix}.$$

Let $s_2^1 = s_2^2 = s_2^3 = S_2$. After performing Gaussian elimination and back substitution of the two equations $s_2^1 = s_2^2$ and $s_2^2 = s_2^3$, we obtain a row echelon form with $P - 1$ equations:

$$\begin{bmatrix} 1 & 0 & 0 & 1 & -1 & 0 & -1 \\ 0 & 0 & 1 & -1 & 0 & 0 & 0 \end{bmatrix} \cdot \begin{bmatrix} w_{21} \\ w_{22} \\ w_{23} \\ w_{24} \\ w_{25} \\ w_{26} \\ \theta_2 \end{bmatrix} = \begin{bmatrix} 0 \\ 0 \end{bmatrix}.$$

Now, we have $w_{21} = -w_{24} + w_{25} + \theta_2$ and $w_{23} = w_{24}$. Consider $X_2^1 = 1$; we calculate S_2 as follows:

$$\begin{aligned} S_2 = s_2^1 &= w_{21} + w_{22} + w_{23} - w_{24} - w_{25} - w_{26} - \theta_2 = w_{22} - w_{24} - w_{26}, \\ S_2 = s_2^2 &= -w_{21} + w_{22} - w_{23} - w_{24} + w_{25} - w_{26} + \theta_2 = w_{22} - w_{24} - w_{26}, \\ S_2 = s_2^3 &= w_{21} + w_{22} - w_{23} + w_{24} - w_{25} - w_{26} - \theta_2 = w_{22} - w_{24} - w_{26}, \end{aligned}$$

w_{22}, w_{24}, w_{25}, w_{26}, and θ_2 are free variables and can be assigned any values. Then, we update these free variables to increase the distance S_2.

Initial condition

The initial weights are set as follows:

$$w_{ij}(0) = \begin{cases} 1, & if \quad i = j \\ 0, & if \quad i \neq j \end{cases}, \qquad (10)$$
$$\theta_i(0) = 0.$$

The distance S_i is linearly proportional to the weights w_{ij} and threshold θ_i in the training. This distance will increase linearly following the updating of w_{ij} and θ_i, reach a limit, and then decrease linearly after further updating.

Experiments on associative memory

We performed simulations to compare the performance of the Little model (LM) [10], the error-correction rule (ECR) [11] and the proposed balanced training (bal-AM). The light shaded (green) planes in Fig. 1 are obtained by using the Hopfield model. In all the simulations, we set $\eta_1 = \eta_2 = 0.00011$ in the balanced training.

Table 1. Preformances of the bal-AM with different (N,P) values

$N = 10$; 10 sets of patterns				$N = 20$; 30 sets of patterns		
	$\alpha = 0.5$			$\alpha = 1$	$\alpha = 1.25\ (> 1)$	$\alpha = 2.25\ (> 2)$
	$P = 5$	$P = 5$	$P = 5$	$P = 20$	$P = 25$	$P = 45$
	LM	ECR	bal-AM	bal-AM	bal-AM	bal-AM
SP ($/P$)	1.9	5	5	20	25	45
SS ($/2^N$)	5.0	43.9	219	5.4×10^5	6.9×10^5	8.8×10^5
TS ($/2^N$)	744.8	978.4	805	5.0×10^5	3.5×10^5	1.6×10^5
C	44.3	0.2	0	0	0	0
IC ($/2^N$)	88.6	0.4	0	0	0	0
TC ($/2^N$)	185.6	1.3	0	0	0	0
R ($/(P \times N)$)	12.1	13.5	20.8	16.5	13.0	5.9
BR	N.A.	N.A.	0.44	0.045	0.028	0.0071

Experimental Results

Table 1 lists experimental results for ($N = 10, P = 5$). In this case the storage parameter $\alpha = \frac{P}{N} = 0.5$ [12]. In this experiment, we presented 10 sets of randomly produced patterns to the networks and then got the averaged results. The labels of rows in this table are: 'SP', given P patterns, the number of patterns successfully stored; 'SS', the number of stable states; 'TS', the number of transient states converging to all stable states; 'C', the number of limit cycles; 'IC', ($\geq$ 2C) the number of states involved in all limit cycles; 'TC', the number of transient states falling into limit cycles; 'R', given NP 1-bit-error patterns, the number of patterns converging to the original stored patterns; 'BR', average basin radius of each pattern. Table 1 lists the performances of a network with 20 neurons. In this table, we presented 30 sets of randomly produced patterns to the networks and then got the averaged results for $P = 20$, $P = 25$ and $P = 45$ separately. Note that the loading capacity α under various conditions has been studied [12], $\alpha = 0.14$ [13]; $\alpha < (1/[2(lnN)])$ [14]; $\alpha = 0.16$ [15]; $\alpha = 1$ [16]. The bound $\alpha < 2$ has been derived in [17] based on thermal statistic average. All of them are derived for *imperfect* recalls. The methods based on the basin-λ, et-AM, e-AM and bal-AM are designed for *perfect* recalls.

From Table 1, LM has rather limited capacity; it could not even store five patterns in a ten-neuron network, and neither could the Hopfield model. ECR produces cycles. The given five patterns can be successfully memorized using the bal-AM. All bal-AM's spurious stable states have very small basins and cannot withstand thermal pertubation. The bal-AM produced no limit cycles, as we expected. The performance of bal-AM in recovery from noisy patterns was excellent. This is because we tuned the decision hyperplane so as to enlarge the local (polyhedral) region, basin-1, for each memory pattern and include as many its noisy patterns as possible in that region. The total number of states in each basin for the bal-AM case with $P = 5$ is listed in Table 2. Note that one of the traits of the basin-λ is: $674 \geq 59 \geq 3 \geq 1 \geq 0 \geq 0$; $674 \geq (59 + 3 + 1 + 0 + 0)$; $59 \geq (3 + 1 + 0 + 0)$; $3 \geq (1 + 0 + 0)$. The number of overall states in these ten sets is $10240 = 2^{10} \times 10$.

Table 2. Total number of states in basin-λ in the all ten sets

	basin-o	basin-1	basin-2	basin-3	basin-4	basin-5	basin-6
states	50	674	59	3	1	0	0

3 Summary

In almost all of our simulations, many states evolved in a single iteration to their memory patterns during recall, for example, 674 in basin-1 in Table 2. This is very different from the evolutionary recall process in many other models. The bal-AM operates in one shift. Each iteration improves the location of a hyperplane. Each hyperplane is independent of all others during training. This is beneficial for training in asynchronous mode. This is very different from correlation matrix designs, such as the pseudo-inverse method [16]. The basins other than basin-1, for example, basin-2, can be enlarged similarly by backward tracing and stabilization. This means that the proposed bal-AM can be used to enlarge the local regions whose states will evolve to the memory patterns in exactly two evolutions.

The hairy network can be used in the study of the hologram hypothesis [18] and in the study of conscious [19]. The spatial temporal mode of the network provides a basis to study the remembering process, a reconstructive process -the assembly of dismembered events [20], see the website on the perception and generation of similar melodies after learning http://red.csie.ntu.edu.tw/MG/index.htm

Since the network possesses collective memories that are held by a group of populations, these stubborn memories can highly resist local corruptions and tolerate containmination. The states in basin-λ may resemble those serious mutations that can be self-restored automatically after λ generations [3]. There is no need to save a prototype copy for the restoration in the back mutation mode or reverse mode. Heritable traits can be well preserved by a group of dense canalized sites in the cell.

References

1. Liou, C.Y., Lin, S.L.: Finite memory loading in hairy neurons. Natural Computing 5, 15–42 (2006)
2. Liou, C.-Y.: Backbone structure of hairy memory. In: Kollias, S.D., Stafylopatis, A., Duch, W., Oja, E. (eds.) ICANN 2006. LNCS, vol. 4131, pp. 688–697. Springer, Heidelberg (2006)
3. Kauffman, S.: Antichaos and adaptation, August 1991, pp. 64–70. Scientific American (1991)
4. Gardner, E.: Optimal basins of attraction in randomly sparse neural network models. Journal of Physics A 22, 1969–1974 (1989)
5. Liou, C.Y., Yuan, S.K.: Error tolerant associative memory. Biological Cybernetics 81, 331–342 (1999)

6. Li, J., Michel, A., Porod, W.: Analysis and synthesis of a class of neural networks: linear systems operating on a closed hypercube. IEEE Transactions on Circuits and Systems 36(11), 1405–1422 (1989)
7. Mirchandani, G., Cao, W.: On hidden nodes for neural nets. IEEE Transactions on Circuits and Systems 36(5), 661–664 (1989)
8. Hebb, D.: The Organization of Behavior: A Neuropsychological Theory. Wiley, New York (1949)
9. Tank, D., Hopfield, J.: Collective computation in neuronlike circuits. Scientific American 257, 104–114 (1987)
10. Little, W.: The existence of persistent states in the brain. Mathematical Biosciences 19, 101–120 (1974)
11. Widrow, B., Hoff, M.: Adaptive switching circuits. Ire Wescon Convention Record, 96–104 (1960)
12. Amit, D.: Modeling brain function: The world of attractor neural networks. Cambridge University Press, Cambridge (1989)
13. Hopfield, J.: Neural networks and physical systems with emergent collective computational ability. Proceedings of the National Academy of Sciences of the United States of America 79, 2554–2558 (1982)
14. Weisbuch, G., Fogelman-Soulie, F.: Scaling laws for the attractors of hopfield networks. Journal De Physique Lett. 46, 623–630 (1985)
15. Amari, S., Maginu, K.: Statistical neurodynamics of associative memory. Neural Networks 1(1), 63–73 (1988)
16. Kanter, I., Sompolinsky, H.: Associative recall of memory without errors. Physical Review A 35, 380–392 (1987)
17. Gardner, E., Derrida, B.: Optimal storages properties of neural network models. Journal of Physics A 21, 271–284 (1988)
18. Pribram, K.: The neurophysiology of remembering. Scientific American 220, 73–86 (1969)
19. Seth, A., Reeke, E.I.G., Edelman, G.: Theories and measures of consciousness: An extended framework. Proceedings of the National Academy of Sciences of the United States of America (PNAS) 103(28), 10799–10804 (2006)
20. Cooper, L.: A possible organization of animal memory and learning. In: Lundquist, B., Lundquist, S. (eds.) Proceedings of the Nobel Symposium on Collective Properties of Physical Systems, pp. 252–264. Academic Presss, New York (1973)

Appendix

This appendix discusses the setting and several intuitive arguments in manipulating the balanced constraint. These arguments also reveal the geometrical operations of the method. Here we have $N_{i,p} \geq 1$, $N_{i,n} \geq 1$, and $P \geq N_{i,p} + N_{i,n}$. Let D_P, $D_{p,n}$, D_p, and D_n be the dimensionality of the spaces spanned and sustained by all P patterns, $U_{i,p}$ and $U_{i,n}$ patterns, $U_{i,p}$ patterns, and $U_{i,n}$ patterns, respectively. We have $N \geq D_{p,n}$, $P \geq D_{p,n}$, $D_{p,n} > D_p$, and $D_{p,n} > D_n$. In cases where $D_p = N-1$ (or $D_n = N-1$), the all stable hyperplane in Eq.(10) satisfies the constraint with distances $\{s_i^k = 1 |\ k = 1, ..., P\}$. This hyperplane has dimension $N-1$ and parallels the hyperplane spanned by all patterns in $U_{i,p}$ (or $U_{i,n}$). If both $D_p < N-1$ and $D_n < N-1$, a solution hyperplane parallels (and is in between) the two hyperlines D_p, and D_n. This parallel hyperplane guarantees the constraint. When $D_{p,n} = N$, the hyperplane in Eq.(10)

satisfies the constraint with distances $\{s_i^k = 1 | \ k = 1, \ldots, P\}$. In all cases, we may start with the hyperplane in Eq.(10) and move it according to the constraint to improve the basin-1 solution. In summary, when $\{D_p = N - 1\}$, or $\{D_n = N - 1\}$, or $\{D_{p,n} = N\}$, we can construct a hyperplane directly by setting $W_i = (w_{i1} = 0, w_{i2} = 0, \ldots, w_{i(i-1)} = 0, w_{ii} = 1, w_{i(i+1)} = 0, \ldots, w_{iN} = 0)^T$ and $\theta_i = 0$ as in Eq.(10).

For the six-neuron associative memory in [9], these parameters are $D_p = 1$, $D_n = 0$, $D_{p,n} = 2$, and $D_P = D_{p,n}$. As for the five cases in Fig.1 in the reference [2] (Fig.1, [13]), (a) $N = 3, P = 2, D_P = D_{p,n} = 1$, $D_p = D_n = 0$; (b) $N = 3, P = 4, D_P = 3, D_{p,n} = 2$, $D_p = 0$, $D_n = 1$; (c) $N = 3, P = 4, D_P = D_{p,n} = 3$, $D_p = 1$, $D_n = 1$; (d) $N = 3, P = 4, D_P = D_{p,n} = 2$, $D_p = 1$, $D_n = 1$; (e) $N = 3, P = 3, D_P = 2, D_{p,n} = 1$, $D_p = 0$, $D_n = 0$.

When $\{D_p < N - 1$, and $D_n < N - 1$, and $D_{p,n} < N\}$, we apply the following procedure. We arbitrarily select $D_{p,n} + 1$ patterns from both the $U_{i,p}$ and $U_{i,n}$ sets. These $D_{p,n} + 1$ patterns must span and sustain the whole $D_{p,n}$ space (and contain at least one pattern from each set). The solution hyperplane is parallel to (and is in between) the two hyperlines, D_p and D_n.

For each neuron i, there are $N + 1$ variables, N weights, and one threshold. The selected $D_{p,n} + 1$ patterns are used to set the $D_{p,n}$ homogeneous equations. Because $N + 1 > D_{p,n}$, there may exist an infinite number of solutions for the $N + 1$ variables. Then the method searches for a set of weights to make all of the identical $\{s_i^k | \ X_i^k \in U_{i,p}$ or $U_{i,n}\}$ as large as possible. The remaining $P-$ $(N_{i,p} + N_{i,n})$ memory patterns, $\{ \ X_i^k | X_i^k \notin \{ \ U_{i,p}, U_{i,n}\}, k = 1, \ldots P\}$, must be kept stable with distances larger than $\{s_i^k | \ X_i^k \in U_{i,p}$ or $U_{i,n}\}$ during learning. In practice, when the $D_{p,n} + 1$ patterns are not prepared in advance, we may use all the patterns in $\{ \ U_{i,p}, U_{i,n}\}$ to solve the hyperplane. This is a rather straightforward approach.

Let $\{s_i^k | \ X_i^k \in U_{i,p}$ or $U_{i,n}$, $k = 1, .., N_{i,p} + N_{i,n}\}$ be the distance for the patterns in $U_{i,p}$ and $U_{i,n}$. According to the constraint, we have $s_i^1 = s_i^2 = \cdots = s_i^{N_{i,p} + N_{i,n}}$ and, hence, $N_{i,p} + N_{i,n} - 1$ homogeneous equations. The number of free variables is large when both D_p and D_n are small values.

Neural Network Based BCI by Using Orthogonal Components of Multi-channel Brain Waves and Generalization

Kenji Nakayama, Hiroki Horita, and Akihiro Hirano

Graduate School of Natural Science and Technology, Kanazawa University
Kanazawa, 920-1192, Japan
nakayama@t.kanazawa-u.ac.jp
http://leo.ec.t.kanazawa-u.ac.jp/~nakayama

Abstract. FFT and Multilayer neural networks (MLNN) have been applied to 'Brain Computer Interface' (BCI). In this paper, in order to extract features of mental tasks, individual feature of brain waves of each channel is emphasized. Since the brain wave in some interval can be regarded as a vector, Gram-Schmidt orthogonalization is applied for this purpose. There exists degree of freedom in the channel order to be orthogonalized. Effect of the channel order on classification accuracy is investigated. Next, two channel orders are used for generating the MLNN input data. Two kinds of methods using a single NN and double NNs are examined. Furthermore, a generalization method, adding small random numbers to the MLNN input data, is applied. Simulations are carried out by using the brain waves, available from the Colorado State University website. By using the orthogonal components, a correct classification rate P_c can be improved from 70% to 78%, an incorrect classification rate P_e can be suppressed from 10% to 8%. As a result, a rate $R_c = P_c/(P_c + P_e)$ can be improved from 0.875 to 0.907. When two different channel orders are used, P_e can be drastically suppressed from 10% to 2%, and R_c can be improved up to 0.973. The generalization method is useful especially for using a sigle channel order. P_c can be increased up to $84 \sim 88\%$ and P_e can be suppressed down to $2 \sim 4\%$, resulting in $R_c = 0.957 \sim 0.977$.

Keywords: BCI, Brain waves, Neural network, Mental task, Orthogonal components, Gram-Schmidt, Generalization.

1 Introduction

Among the interfaces developed for the handicapped persons, Brain Computer Interface (BCI) has been attractive recently [1], [2]. Approaches to the BCI technology includes nonlinear classification by using spectrum power, adaptive auto-regressive model and linear classification, space patterns and linear classification, hidden Markov models, and so on [3],[4]. Furthermore, application of neural networks have been also discussed [5],[6],[7], [8], [9], [10]. In our works, FFT of the brain waves and a multilayer neural network (MLNN) have been

V. Kůrková et al. (Eds.): ICANN 2008, Part II, LNCS 5164, pp. 879–888, 2008.

applied to the BCI. Efficient pre-processing techniques have been also employed in order to achieve a high score of correct classification of the mental tasks [16]. Furthermore, the generalization methods have been applied to the neural network based BCI. The method of adding small random noise to the MLNN inputs can improve classification performance [17].

The multi-channel brain waves have some common features, which make features vague, and make mental task classification difficult. We will try to remove the common and vague features and emphasize individual features of each mental task. For this purpose, essential features of the multi-channel brain waves are extracted. Conventional methods have employed Independent Component Analysis (ICA), Blind Source Separation (BSS) and so on [11],[12]. However, these methods have an essential problem, that is 'Permutation'. Order of the components, which are extracted, is not fixed. It can be changed depends on data sets. Thus, these methods are difficult to be combined with the MLNN.

In this paper, the BCI using the FFT amplitude of the brain waves and the MLNN is employed. The brain wave in some time interval, that is a frame, can be regarded as vectors. Letting M be the number of channels, M vectors are obtained for one mental task and one measuring trial. Let M vectors be $\{\boldsymbol{x}_1, \boldsymbol{x}_2, \cdots, \boldsymbol{x}_M\}$. This vector set is transferred to the orthogonal vector set $\{\boldsymbol{v}_1, \boldsymbol{v}_2, \cdots, \boldsymbol{v}_M\}$. This vector set is further pre-processed and is used for the MLNN input data.

2 Brain Waves and Mental Tasks

2.1 Mental Tasks

In this paper, the brain waves, which are available from the web site of Colorado State University [13], are used. The following five kinds of mental tasks are used.

- Baseline - Relaxed situation - (B)
- Multiplication (M)
- Letter-composing (L)
- Rotation of a 3-D object (R)
- Counting numbers (C)

2.2 Brain Wave Measurement

Location of the electrodes to measure brain waves is shown in Fig.1. Seven channels including C3, C4, P3, P4, O1, O2, EOG, are used. EOG, which does not appear in this figure, is used for measuring movement of the eyeballs. In this paper, channel numbers Ch1 through Ch7 are assigned to C3, C4, P3, P4, O1, O2 and EOG, respectively, for convenience.

The brain waves are measured for a 10sec interval and sampled by 250Hz for each mental task. Therefore, 10sec $\times$ 250Hz $= 2,500$ samples are obtained for one channel and one mental task. One data set includes 2,500 samples for each channel and each mental task. Five mental tasks and seven channels are included in one data set.

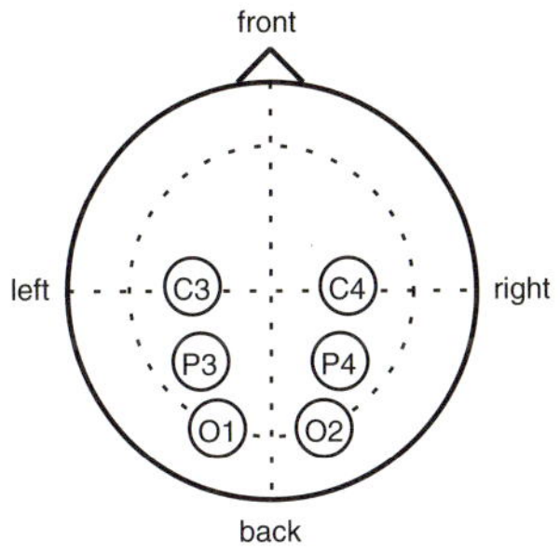

Fig. 1. Location of electrodes measuring brain waves

2.3 Mental Task Classification by Using Multilayer Neural Network

An MLNN having a single hidden layer is used. Activation functions used in the hidden layer and the output layer are a hyperbolic tangent and a sigmoid function, respectively. The number of input nodes is 10 samples×7 channels=70. Five output neurons are used for five mental tasks. The target for the output has only one non-zero element, such as $(1, 0, 0, 0, 0)$. In the testing phase, the maximum output becomes the winner and the corresponding mental task is assigned. However, when the winner have small value, estimation becomes incorrect. Therefore, the answer of the neural network is rejected, that is any mental task cannot be estimated. The error back-propagation algorithm is employed for adjusting the connection weights.

3 Pre-processing of Wave Forms

Several techniques for pre-processing proposed in [16] are also employed in this paper, and are briefly described here.

3.1 Amplitude of FFT of Brain Waves

In order to avoid effects of brain wave shifting along the time axis, which is not essential, the brain wave is first Fourier transformed and its amplitude is used.

3.2 Reduction of Samples by Averaging

In order to make the neural network size to be compact and to reduce effects of the noises added to the brain waves, the FFT samples in some interval are averaged. By this averaging, the number of samples is reduced from 2,500 to 20. Since the brain waves are real values and their FFT amplitude are symmetrical, a half of the 20 samples can represents all information. Finally, 10 samples are used.

3.3 Nonlinear Normalization

The amplitude of the FFT is widely distributed. Small samples also contain important information for classifying the mental tasks. However, in the neural networks, large inputs play an important role. If large samples do not include important information, correct classification will be difficult. For this reason, the nonlinear normalization as shown in Eq.(1) has been introduced [17]. x is the FFT amplitude before normalization and $f(x)$ is the normalized amplitude. In Eq.(1), x_{min} and x_{max} mean the minimum and the maximum values of x in all channels. The small samples are expanded and the large samples are compressed. In this paper, usefulness this nonlinear normalization method for the orthogonal components of the brain waves will be also investigated.

$$f(x) = \frac{\log(x - x_{min})}{\log(x_{max} - x_{min})} \tag{1}$$

The linear normalization given by $f_{linear}(x) = (x - x_{min})/(x_{max} - x_{min})$ will be examined for comparison.

4 Generalization by Adding Small Random Numbers

The brain waves are very sensitive, which easily change depending on health conditions of the subjects and the measuring environment. The data sets measured for the same subject, have different features. Therefore, generalization is very important for the BCIs. In our previous work, two kinds of generalization techniques, which are adding small random numbers to the MLNN input data [14] and a weight decay technique [15], have been applied. The former method can provide good classification performance [17].

In this paper, the method of adding small and different random numbers to the MLNN input data at each epoch of the learning process is applied, and its usefulness for the orthogonal components of the multi-channel brain waves will be investigated.

5 Orthogonal Components of Multi-channel Brain Waves

5.1 Orthogonal Component Analysis

There are several kinds of methods for analyzing orthogonal components, including blind source separation (BSS), independent component analysis (ICA), principal component analysis (PCA) and so on. They have an essential problem, that is 'Permutation'. In these methods, it is not guaranteed that the same component is analyzed in the same order. This point is described in detail here.

Letting M be the number of channels, a set of M vectors is measured for one mental task and one measuring trial. Let a whole vector be $\boldsymbol{X} = [\boldsymbol{x}_1^T, \boldsymbol{x}_2^T, \cdots, \boldsymbol{x}_M^T]^T$. $\boldsymbol{x}_i$ corresponds to the brain wave measured at the ith channel. Let the corresponding orthogonalized vector be $\boldsymbol{V} = [\boldsymbol{v}_1^T, \boldsymbol{v}_2^T, \cdots, \boldsymbol{v}_M^T]^T$. The whole vector $\boldsymbol{V}$ is used as the MLNN input data after the pre-processing. Therefore, the order of $\boldsymbol{v}_i$ in $\boldsymbol{V}$ is very important. Let consider two sets of the vectors $\boldsymbol{V}_1$ and $\boldsymbol{V}_2$ for the same mental task and for different measuring trials. Let $\boldsymbol{v}_{1i}$ and $\boldsymbol{v}_{2j}$ be the ith and jth vectors in $\boldsymbol{V}_1$ and $\boldsymbol{V}_2$, respectively. If $\boldsymbol{v}_{1i}$ is most similar to $\boldsymbol{v}_{2j}, i \neq j$, for instanse $\boldsymbol{v}_{1i}^T \boldsymbol{v}_{2j} / \parallel \boldsymbol{v}_{1i} \parallel\parallel \boldsymbol{v}_{2j} \parallel$ is most close to 1, then $\boldsymbol{V}_1$ and $\boldsymbol{V}_2$ cannot express the same feature even though they belong to the same mental task. BSS, ICA and PCA cannot guarantee the same order of the orthogonal components in the different measuring trials due to 'Permutation' problem [11],[12]. This is an essential problem to use the orthogonal components as the MLNN input data.

For this reason, in this paper, Gram-Schmidt orthogonalization is applied to the vector set $\{\boldsymbol{x}_1, \boldsymbol{x}_2, \cdots, \boldsymbol{x}_M\}$. The order of the orthogonal components can be controlled by selecting the channels to be orthogonalized in the specified order. 'Permutation' of the orthogonal components does not occur.

5.2 Gram-Schmidt Orthogonalization

The vectors $\{\boldsymbol{x}_1, \boldsymbol{x}_2, \cdots, \boldsymbol{x}_M\}$, which express the brain waves at M-channels, are usually linearly independent. This set can be transferred into the orthogonal vector set $\{\boldsymbol{v}_1, \boldsymbol{v}_2, \cdots, \boldsymbol{v}_M\}$ by Gram-Schmidt orthogonalization [18]. $\boldsymbol{x}_1$ is used for $\boldsymbol{v}_1$. $\boldsymbol{v}_2$ is a part of $\boldsymbol{x}_2$, which is orthogonal to $\boldsymbol{v}_1$. In the same way, $\boldsymbol{v}_k$ is a component of $\boldsymbol{x}_k$, which is orthogonal to all the previous orthogonal vectors $\boldsymbol{v}_1, \boldsymbol{v}_2, \cdots, \boldsymbol{v}_{k-1}$.

$\{\boldsymbol{v}_i\}$ are Fourier transformed and their amplitude are pre-processed as described in Sec.3, and are used as the MLNN input data.

5.3 Order of Orthogonalization

There exists degree of freedom of selecting the channel order, in which the brain waves are orthogonalized by the Gram-Schmidt method. The first channel can hold a whole information, and the following channels provide only a part of the vector, which is orthogonal to the previous orthogonal vectors. Therefore, the order of the channels will affect accuracy of classifying the mental tasks. We will investigate effects of the channel order through simulation.

5.4 Input Data Sets by Using Two Channel Orders

As described in the previous sections, we can use a plural number of the channel orders for generating the MLNN input data. Let the input data sets, which are generated from the same brain waves by using two kinds of the channel orders, be I_1 and I_2. Thus, the input data are equivalently doubled for each mental task.

Single MLNN Method. A single MLNN, denoted NN, is used for classifying the mental tasks. In the training phase, both I_1 and I_2, generated from the training brain waves, are used to train NN. In the testing phase, I_1, generated from the test brain waves, is applied to NN, and the output O_1 is obtained. Separately, I_2, generated from the same brain waves, is applied to NN, resulting in the output O_2. The final output O_t is given by $O_t = (O_1 + O_2)/2$. The mental task is estimated based on the maximum value of O_t.

Double MLNN Method. Two independent MLNNs are used, denoted NN_a and NN_b. I_1 of the training brain waves is used to train NN_a, and I_2 of the same brain waves is used to train NN_b, respectively. In the testing phase, I_1 of the test brain waves is applied to NN_a and the output O_{a1} is obtained. In the same way, I_2 of the same brain waves is applied to NN_b and O_{b2} is obtained. The final output O_t is evaluated by $O_t = (O_{a1} + O_{b2})/2$. The mental task is classified based on the maximum value of O_t.

The threshold of rejection, that is 'No estimation', is also employed in all methods. If all the outputs are less than the threshold, then the MLNN answers 'any mental task cannot be estimated'.

6 Simulations and Discussions

6.1 Simulation Setup

Training and Testing Brain Waves
The brain waves with a 10 sec length for five mental tasks were measured 10 times. Therefore, 10 data sets are available. Among them, 9 data sets are used for training and the remaining one data set is used for testing. Five different combinations of 9 data sets are used for the training. As a result, five different data sets are used for testing. Thus, five independent trials are carried out. Classification accuracy is evaluated based on the average over five trials [3].

Score of Correct and Error Classifications
Estimation of the mental tasks is evaluated based on a correct classification rate (P_c) and an error classification rate (P_e), and a rate of correct and error classification (R_c) as follows:

$$P_c = \frac{N_c}{N_t} \times 100\%, \quad P_e = \frac{N_e}{N_t} \times 100\% \tag{2}$$

$$R_c = \frac{N_c}{N_c + N_e}, \quad N_t = N_c + N_e + N_r \tag{3}$$

N_c, N_e and N_r are the numbers of correct and incorrect classifications and rejections, respectively. N_t is the total number of the testing data. R_c is used to evaluate a correct classification rate except for 'Rejection'.

Parameters in Neural Network Learning
A hyperbolic tangent function and a sigmoid function are used fin the hidden layer and the output layer, respectively. The number of hidden units is 20. The threshold for rejection is 0.7. A learning rate is 0.02.

6.2 Brain Waves before and after Orthogonalization

Figure 2 shows the brain waves before (gray) and after (black) orthogonalization. The horizontal axis shows the sample number in the time domain. The channel

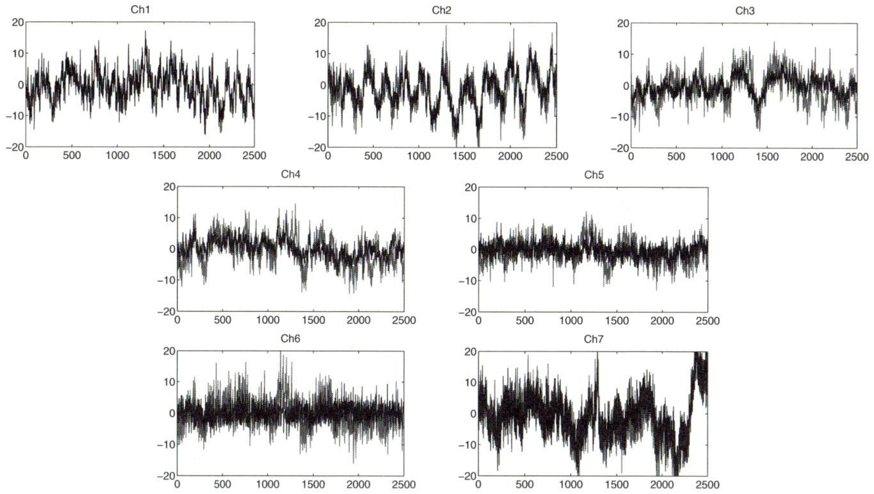

Fig. 2. Brain waves of 7 channels before (gray) and after (black) orthogonalization

order of orthogonalization is Ch1, 2, 3, 4, 5, 6, 7. The orthogonalized brain waves are gradually decreased along the channel order. Since Ch7 is used for detecting blinking, which is not the mental task, its brain wave is not correlated with that of the other channels, resulting in a large orthogonal component.

Figure 3 shows the MLNN input data before (dashed line) and after (solid line) orthogonalization. They are normalized by using the nonlinear function Eq.(1) [16]. The FFT amplitude responses are arranged from Ch1 through Ch7 along the horizontal axis from the left side to the right side. One channel includes 10 samples.

6.3 Classification by Using Orthogonal Components

Table 1 shows classification rates by using the orthogonal components of the brain waves. 7 kinds of channel orders are used, which are selected by circular shifting. Although they do not include all permutations, effects of the channel order can be investigated. 'Conventional' means our method, which employs the original brain waves and the pre-processing techniques [16]. By using the orthogonal components, P_c can be improved from 70% to 78%, and P_e can be suppressed from 10% to 8%, and R_c is increased from 0.875 to 0.907.

As expected in the previous section, also from Table 1, the classification accuracy depends on the channel order to be orthogonalization. The channel orders Ch2, 3, 4, 5, 6, 7, 1 and Ch3, 4, 5, 6, 7, 1, 2 can provide good classification accuracy. The optimum channel order can be searched for in advance, and can be fixed for an individual subject.

Furthermore, the generalization method of adding small random numbers to the MLNN input data was carried out for the best channel order Ch2, 3, 4, 5, 6,

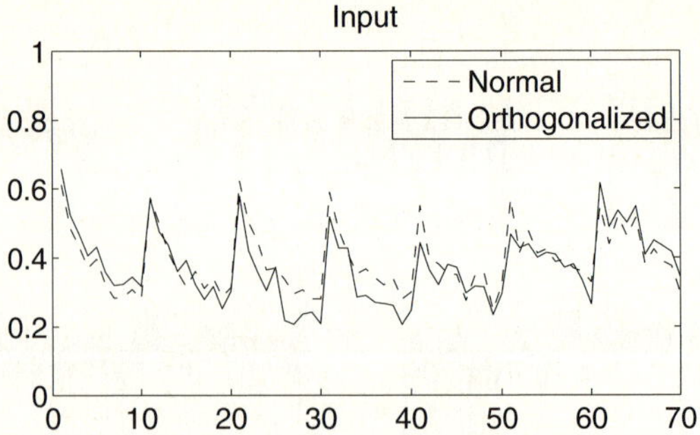

Fig. 3. Input data of MLNN. Dashed line and solid line indicate input data before and after orthogonalization, respectively.

Table 1. Score of classification by using orthogonal components

	P_c	P_e	R_c
Conventional	70	10	0.875
Ch1, 2, 3, 4, 5, 6, 7	70	12	0.854
Ch2, 3, 4, 5, 6, 7, 1	78	8	0.907
Ch3, 4, 5, 6, 7, 1, 2	74	8	0.902
Ch4, 5, 6, 7, 1, 2, 3	70	12	0.854
Ch5, 6, 7, 1, 2, 3, 4	68	12	0.85
Ch6, 7, 1, 2, 3, 4, 5	66	24	0.733
Ch7, 1, 2, 3, 4, 5, 6	70	12	0.854
Generalization ± 0.1 Ch2, 3, 4, 5, 6, 7, 1	88	4	0.957
Generalization ± 0.05 Ch2, 3, 4, 5, 6, 7, 1	84	2	0.977

7, 1. Random numbers uniformly distributed during ± 0.1 and ± 0.05 are used. As shown in the same table, P_c is well improved from 78% up to 84 $\sim$ 88%, P_e is well suppressed from 8% to 2 $\sim$ 4%, resulting in $R = 0.957 \sim 0.977$.

The linear normalization $f_{linear}(x)$ described in Sec.3.3 is also examined. The best channel order is Ch5, 6, 7, 1, 2, 3, 4, and $P_c = 62\%$, $P_e = 14\%$ and $R_c = 0.816$, which are not good compared with the nonlinear normalization.

6.4 Classification by Using Two Channel Orders

Two channel orders, Ch2, 3, 4, 5, 6, 7, 1 and Ch3, 4, 5, 6, 7, 1, 2, which provide good classification accuracy in Table 1, are used to generate the MLNN input data I_1 and I_2, respectively.

Table 2 shows simulation results. A method of using a single NN is not good. By using double NN, P_e is well suppressed and R_c can be well increased before the generalization. After the generalization, its performances $P_c = 82\%$, $P_e = 2\%$

Table 2. Score of classification by using two channel orders

	P_c	P_e	R_c
Conventional	70	10	0.875
Single NN	66	8	0.892
Generalization ±0.1	76	6	0.927
±0.05	78	8	0.907
Double NN	72	2	0.973
Generalization ±0.1	82	2	0.976
±0.05	72	4	0.947

and $R_c = 0.976$ are almost same as those of the method using a single channel order Ch2, 3, 4, 5, 6, 7, 1.

When the generalization method, adding random numbers to the MLNN input data, is embedded in the learning process, a single channel order can provide good performance by optimizing the channel order.

6.5 Dependence on Individual Subjects

Since brain waves are dependent on the subjects, the MLNN is needed to be optimized or tuned up for individual subjects. It is not useful to apply the same MLNN to different subjects. From our experiences, the best channel order of orthogonalization also depends on both the subjects and mental tasks. However, it can be searched for by using the training data in advance. The proposed method, using the orthogonal components, has been applied to three subjects, and almost the same improvement on the classification rates have been achieved.

7 Conclusion

In this paper, the BCI based on the FFT amplitude and the MLNN is dealt with. Especially, the orthogonal components of the multi-channel brain waves are used to generate the MLNN input data. Gram-Schmidt orthogonalization is applied. The proposed approach can improve P_c from 70% to 78%, P_e from 10% to 8%, and R_c from 0.875 to 0.907. When two channel orders are used, P_e can be well suppressed from 10% to 2%, and R_c can be well improved up to 0.973. The generalization method is also useful, which can improve P_c up to 88% and P_e down to 2%, resulting in $R_c = 0.977$.

References

1. Pfurtscheller, G., Neuper, C., Guger, C., Harkam, W., Ramoser, H., Schlögl, A., Obermaier, B., Pregenzer, M.: Current trends in Graz braincomputer interface (BCI) research. IEEE Trans. Rehab. Eng. 8, 216–219 (2000)

2. Obermaier, B., Muller, G.R., Pfurtscheller, G.: Virtual keyboard controlled by spontaneous EEG activity. IEEE Trans. Neural Sys. Rehab. Eng. 11(4), 422–426 (2003)
3. Anderson, C., Sijercic, Z.: Classification of EEG signals from four subjects during five mental tasks. In: Bulsari, A.B., Kallio, S., Tsaptsinos, D. (eds.) EANN 1996, Systems Engineering Association, PL 34, FIN-20111 Turku 11, Finland, pp. 407–414 (1996)
4. Pfurtscheller, G., Neuper, C.: Motor imagery and direct brain-computer communication. Proc. IEEE 89(7), 1123–1134 (2001)
5. Millan, J.R., Mourino, J., Babiloni, F., Cincotti, F., Varsta, M., Heikkonen, J.: Local neural classifier for EEG-based recognition of metal tasks. In: IEEE-INNS-ENNS Int. Joint Conf. Neural Networks (July 2000)
6. Muller, K.R., Anderson, C.W., Birch, G.E.: Linear and non-linear methods for brain-computer interfaces. IEEE Trans. Neural Sys. Rehab. Eng. 11(2), 165–169 (2003)
7. Millan, J.R.: On the need for on-line learning in brain-computer interfaces. In: Proc. IJCNN, pp. 2877–2882 (2004)
8. Fabiani, G.E., McFarland, D.J., Wolpaw, J.R., Pfurtscheller, G.: Conversion of EEG activity into cursor movement by a brain-computer interface (BCI). IEEE Trans. Neural Sys. Rehab. Eng. 12(3), 331–338 (2004)
9. Obermaier, B., Neuper, C., Guger, C., Pfurtscheller, G.: Information transfer rate in a five-classes brain-computer interface. IEEE Trans. Neural Sys. Rehab. Eng. 9(3), 283–288 (2001)
10. Anderson, C.W., Devulapalli, S.V., Stolz, E.A.: Determining mental state from EEG signals using neural networks. Scientific Programming, Special Issue on Applications Analysis 4(3), 171–183 (1995)
11. Kachenoura, A., Albera, L., Senhadji, L., Comon, P.: ICA: A potential tool for BCI systems. IEEE Signal Processing Magazine, 57–68 (January 2008)
12. Wang, S., James, C.J.: Enhancing evoked responses for BCI through advanced ICA techniques. In: Proc., 3rd Int. Conf. on Advances in Medical, Signal and Information Processing (MEDSIP 2006), Stevenage, UK (2006)
13. Colorado State University, `http://www.cs.colostate.edu/eeg/`
14. Robert, J., Burton, M., Mpitsos, G.J.: Event-dependent control of noise enhances learning in neurla networks. Neural Networks 5(4), 627–637 (1992)
15. Treadgold, N.K., Gedeon, T.D.: Simulated annealing and weight decay in adaptive learning: The SARPROP algorithm. IEEE Trans. Neural Networks 9(4), 662–668 (1998)
16. Nakayama, K., Inagaki, K.: A brain computer interface based on neural network with efficient pre-processing. In: Proc. IEEE, ISPACS 2006, Yonago, Japan, pp. 673–676 (December 2006)
17. Nakayama, K., Kaneda, Y., Hirano, A.: A brain computer interface based on FFT and multilayer neural network-Feature extraction and generalization. In: Proc. IEEE, ISPACS 2007, Xiamen, China, pp. 101–104 (December 2007)
18. Strang, G.: Linear Algebra and its Applications. Academic Press, Inc., New York (1976)

Feature Ranking Derived from Data Mining Process

Aleš Pilný, Pavel Kordík, and Miroslav Šnorek

Department of Computer Science and Engineering, FEE,
Czech Technical University, Prague, Czech Republic
pilnyal@fel.cvut.cz, kordikp@fel.cvut.cz, snorek@fel.cvut.cz

Abstract. Most common feature ranking methods are based on the statistical approach. This paper compare several statistical methods with new method for feature ranking derived from data mining process. This method ranks features depending on percentage of child units that survived the selection process. A child unit is a processing element transforming the parent input features to the output. After training, units are interconnected in the feedforward hybrid neural network called GAME. The selection process is realized by means of niching genetic algorithm, where units connected to least significant features starve and fade from population. Parameters of new FR algorithm are investigated and comparison among different methods is presented on well known real world and artificial data sets.

1 Introduction

Nowadays data with few input features is the exception. Each feature adds one dimension to the dimensionality of data vectors. For effective, more accurate data mining, it is necessary to use preprocessing methods which reduce dimensionality of input data or describe the relevance of each feature of data. Set of methods to reduce data dimension, so called Feature Selection(FS) [12], search for subset of relevant features from an initial set of features while Feature Extraction(FE) methods [14] create subset of new features containing information extracted from original set of features. Relaxed setting for FS are methods known as Feature Ranking [5], ranking of all original features in correspondence to their relevance.

Feature Selection algorithms may be divided into three categories. Algorithms in the first category are based on filters [2], where the significance of features is computed outside from classification algorithm. On the other side Wrapper methods [6], from the second category, depends on classifier to evaluate quality of selected features. Finally Embedded methods [3] selects relevant features within learning process of internal parameters (e.g. weights between layers of neural networks). The goal of feature selection is to avoid selecting too many or too few variables than necessary. In practical applications, it is impossible to obtain complete set of relevant features. Therefore, the modelled system is open system, and all important features that are not included in the data set (for what reason ever) are summarised as noise [11].

V. Kůrková et al. (Eds.): ICANN 2008, Part II, LNCS 5164, pp. 889–898, 2008.

On the other hand, it is not recommended to select as much features as possible. In fact, even if theoretically more variables should provide one with better modelling accuracy, in real cases it has been observed many times that this is not the case. This depends on the limited availability of data in real problems: successful models seem to be in good balance of model complexity and available information. In facts, variables selection tends to produce models that are simpler, clearer, computationally less expensive and, moreover, providing often better prediction accuracy [12].

In statistical analysis, forward and backward stepwise multiple regression (SMR) are widely used [12]. The resulting subset of features generated by adding features until the addition of a new feature no longer results in a significant increment in an R^2 (correlation coefficient) value.

Siedlecki and Sklansky [13] use genetic algorithms for variable selection by encoding the initial set of n variables to a chromosome, where 1 and 0 represents presence and absence respectively of variables in the final subset. They used classification accuracy, as the fitness function and obtained good neural network results. Mutual information (MI) [3] between features can be computed by integrating the probability density functions of input and output variables. MI is very often used in FS algorithms to distinguish between useful and irrelevant features. Several FS algorithms for the WEKA [15] data mining environment are based on measuring the mutual information of attributes. In this paper, we compare the performance of FS algorithms available in WEKA on synthetic data set generated by Tesmer and Estevez to measure the performance of the AMIFS method [14]. Then we introduce new methods for FR which are byproducts of the GAME data mining algorithm [7]. We adjust their performance to improve results on syntectic data sets. Finally, we applied all FS algorithms to real-world data set.

2 Feature Ranking Methods in WEKA

Seven Weka FR methods will be used for comparison of ranking performance. ChiSquared method evaluates the worth of an attribute by computing the value of the chi-squared statistic with respect to the class, GainRatio by measuring the gain ratio with respect to the class, InfoGain by measuring the information gain with respect to the class, OneR by using the OneR classifier. ReliefF evaluates the worth of an attribute by repeatedly sampling an instance and considering the value of the given attribute for the nearest instance of the same and different class. SVM evaluates the worth of an attribute by using an SVM classifier and SymmetricalUncert(SU) evaluates the worth of an attribute by measuring the symmetrical uncertainty with respect to the class.

3 FeRaNGA: Novel Method for Feature Ranking

In this section, we propose a method for FR derived from information gained during the data mining process. The GAME data mining engine needs to be briefly described.

3.1 Group Method of Data Handling

There are several algorithms for inductive models construction commonly known as Group Method of Data Handling (GMDH) introduced by Ivachknenko in 1966 [8]. The Multilayered Iterative Algorithm (MIA) uses a data set to construct a model of a complex system. Layers of units transfer input variables to the output of the network. The coefficients of units transfer functions are estimated using the data set describing the modeled system. Networks are constructed layer by layer during the learning stage. the Group of Adaptive Models Evolution (GAME) algorithm proceeds from the MIA algorithm. Modifications of the original algorithm are the following: maximal number of unit inputs equals to the number of layer the unit belongs to, interlayer connections are allowed, transfer function and learning algorithm of units can be of several types, niching genetic algorithm is used to select surviving units and an ensemble of models is generated. The more detailed description can be found in [7].

The niching genetic algorithm used in GAME is a cornerstone of feature ranking algorithm and needs to be described.

3.2 Niching Genetic Algorithm

Niching methods [10] extend genetic algorithms to domains that require the location of multiple solutions. They promote the formation and maintenance of stable subpopulations in genetic algorithms (GAs). One of these methods is deterministic crowding [9]. The basic idea of deterministic crowding is that offspring is often most similar to parents. We replace the parent who is most

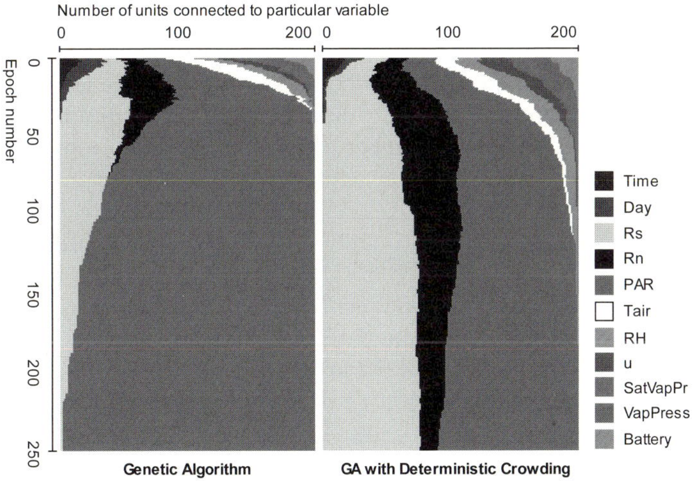

Fig. 1. GA versus niching GA with DC: first layer of the GAME network (units with single input)

similar to the offspring with higher fitness. The reason why we employ deterministic crowding instead of using just simple GA is the ability to maintain multiple subpopulations (niches) in the population. When the model is being constructed units connected to the most important input would soon dominate in the population of the first layer if we have used traditional GA(see Fig.1). All other units connected to least important inputs would show worse performance on the validation set and disappear from the population with exponential speed.

In inductive modeling we need also to extract and use information from least important features and therefore we prefer maintaining various niches in the population. The distance of genes is based on the phenotypic difference of units (to which inputs are connected). Each niche is thus formed by units connected to similar set of inputs. In the first layer just one input is allowed, therefore niches are formed by units connected to the same feature. After several epochs of GA with deterministic crowding the best individual (unit) from each niche is selected to survive in the layer of the model. The construction of the model goes on with the next layers, where niching is also important.

Proposed method of feature ranking takes into account two factors. First is the significance of feature for modelling the output variable. The second is the amount of additional information to the information carried by already selected variables. This resembles to state of the art methods based on mutual information analysis. These methods select set of features of the highest mutual information with the output variable while minimizing mutual information among the selected features.

We found out that by monitoring which genes exist in the population we can estimate the significance of each feature.

3.3 Significance Estimation

In the initial population of the first layer units are randomly generated. Connection to certain feature is represented as "1" in corresponding gene locus. Numbers of ones in locus are therefore uniformly distributed at the beginning of GA. After several epochs of GA with DC, numbers of ones in gene loci representing more important features increases whereas numbers in loci of least significant features decreases (see Fig. 1.).

This fact can be used for the estimation of features significance. In each layer of the network, after the last epoch of GA with DC, before the best gene from each niche is selected, we count how many genes (units) are connected to each input variable. This number is accumulated for each feature and when divided by sum of accumulated numbers for all features, we get the proportional significance of each feature.

3.4 FeRaNGA Algorihtm for Feature Ranking

The ranking of features can be easily extracted from their proportional significance estimated by the above described procedure. This is what we call Feature

Ranking utilizing information from Niching Genetic Algorithm (FeRaNGA) algorithm. The configuration of the GAME engine, particulary size of the population and number of epochs of Niching Genetic Algorithm (NGA) has considerable influence on results of the FeRaNGA algorithm. For very low number of epochs, significance of features is close to random number, because the niching genetic algorithm is unable to eliminate units connected to irrelevant features from the population. The GAME engine typically constructs ensemble of models [4]. We found out, that by applying FeRaNGA to all ensemble models and computing the median from estimated significance of features greatly improves the results. We will refer to this method as FeRaNGA-n where n is the number of ensemble models.

4 Data Sets Overview

Two artificial and one real data sets were used in experiments. Artificial data sets were generated to measure the performance of the AMFIS feature selection algorithm [14].

4.1 Gaussian Multivariate Data Set

This artificial data set consists of two clusters of points generated from two different 10th-dimensional normal Gaussian distributions. Class 1 corresponds to points generated from N(0, 1) for each dimension and Class 2 to points generated from N(4, 1). This data set consists of 50 features and 500 samples per class. By construction, features 1-10 are equally relevant, features 11-20 are completely irrelevant and features 21-50 are highly redundant with the first ten features. Ideally, the order of selection should be: at first relevant features 1-10, then the redundant features 21-50, and finally the irrelevant features 11-20.

4.2 Uniform Hypercube Data Set

Second artificial data set consists of two clusters of points generated from two different 10th-dimensional hypercube $[0,1]^1 0$, with uniform distribution. The relevant feature vector (f1, f2, . . . , f10) was generated from this hypercube in decreasing order of relevance from feature 1 to 10. A parameter $\alpha = 0.5$ was defined for the relevance of the first feature and a factor $\alpha = 0.8$ for decreasing the relevance of each feature. A pattern belongs to Class 1 if $(f_i < \gamma^{i-1} * \alpha \ / \ i - 1, \ldots, 10)$, and to Class 2 otherwise. This data set consists of 50 features and 500 samples per class. By construction, features 1-10 are relevant, features 11-20 are completely irrelevant, and features 21-50 are highly redundant with first 10 features. Ideally, the order of selection should be: at first relevant features 1-10 (starting with feature 1 until feature 10 in the last position), then the redundant features 21-50, and finally the irrelevant features 11-20.

4.3 Housing Data Set

This real-world multivariate data set consist of 506 instances, 13 continuous attributes [1]. The last attribute, MEDV, was for experiments in Weka discretized into 4 classes, because the nominal character of the output variable is expected.

5 Experimental Analysis

The first experiment compares the performance of feature ranking methods in their implicit configuration on three different data sets. The GAME used standard configuration with 15 individuals in the NGA and 30 epochs. We run the algorithm severel times to show the unstable behavior. Second part of analysis describe experiments with making the FeRaNGA algorithm more restrictive to prevent irrelevant features being selected by a chance. Finally the results of the FeRaNGA-n algorithm with well configured GAME engine are presented.

5.1 FeRaNGA Algorithm vs. WEKA FR Methods in Default Settings

Most of FR methods in Weka are giving exact results corresponding to artificial data sets characteristic, except SVM(as is shown in Table 1 and 2). In these tables only the interesting columns are displayed. The reference ranking of features is displayed in first rows. A light gray background cells with black coloured numbers represent features which have zero significance (no units connected to them were able to survive the evolutionary process) and dark gray with white coloured numbers redundant features. Black background cells with white numbers are irrelevant features and cells with white background are relevant. This colour separation is valid for all tables except number 3.

Results of the FeRaNGA algorithm are presented under label GAME followed by number of model used for significance extraction. For the Gaussian data set,

Table 1. Ranks of features for Gaussian Multivariate Data Set used in WEKA and GAME with default settings

Method	Relevant(1-10)										Redundant(21–50) and Irrelevant(11–20)																		
ChiSq	7	6	9	10	8	2	1	3	5	4	38	26	21	48	50	33	29	40	49	27	22	42	32	34	30	17	16	20	15
GainRatio	10	6	5	9	7	1	3	2	4	8	38	49	27	47	29	36	33	22	37	48	24	39	43	28	34	18	20	19	17
InfoGain	7	6	8	9	2	10	1	3	5	4	38	26	29	48	50	40	21	33	46	27	24	22	31	32	43	17	16	20	15
OneR	7	9	10	8	2	3	1	6	4	5	38	26	25	40	29	28	50	21	24	33	35	43	30	34	27	16	11	12	14
ReliefF	6	9	7	4	10	3	5	2	1	8	26	36	27	50	21	46	30	49	28	31	23	45	41	47	48	18	15	16	13
SVM	2	4	6	3	7	9	1	10	8	23	47	39	30	37	42	22	34	21	38	44	5	17	41	14	28	20	45	18	25
SU	10	6	7	9	5	1	3	2	4	8	38	47	49	27	29	33	36	48	37	22	24	44	39	40	30	18	20	19	17
GAME 1	7	8	2	6	26	23	9	35	28	4	1	33	38	40	43	20	42	47	3	5	15	21	22	24	34	36	37	39	48
GAME 2	10	7	3	9	37	27	24	50	8	5	23	46	44	1	2	4	6	11	12	13	19	28	29	30	36	38	39	40	47
GAME 3	10	2	5	8	7	6	3	29	26	4	37	47	36	46	16	34	43	1	9	11	18	23	24	25	32	33	35	38	48
GAME 4	3	9	10	41	6	7	5	8	27	25	31	48	16	39	20	34	44	28	35	1	14	21	22	23	32	33	36	37	46
GAME 5	9	6	3	10	28	50	4	38	29	8	13	36	48	27	20	15	18	23	42	1	14	22	24	25	34	37	39	40	46

Table 2. Uniform Hypercube Data Set analysed with default settings of WEKA and GAME methods

Method	1	2	3	4	5	6	7	8	9	10	Redundant(21-50) and Irrelevant(11-20)																					
ChiSquare	1	2	3	4	5	6	7	8	9	10	22	37	48	35	39	43	47	26	45	27	30	49	29	32	28	33	46	31	42	17	18	14
GainRatio	1	2	3	4	5	6	7	8	9	10	23	38	27	35	37	30	34	39	28	25	22	26	50	32	36	21	31	46	33	17	18	14
InfoGain	1	2	3	4	5	6	7	8	9	10	22	37	48	43	35	39	26	47	44	30	27	49	36	29	33	28	23	46	42	17	18	14
OneR	1	2	3	4	5	6	7	8	9	10	48	37	43	50	40	30	25	44	39	45	22	49	38	34	36	33	23	28	24	18	17	14
ReliefF	1	2	3	4	8	7	9	10	5	6	29	48	47	25	31	30	49	26	27	28	21	50	24	43	45	42	23	20	44	15	41	17
SVM	1	2	48	3	29	43	4	9	35	7	36	45	41	44	34	25	8	11	12	30	37	19	15	16	6	38	24	20	5	13	40	28
SU	1	2	3	4	5	6	7	8	9	10	23	38	35	37	27	48	39	34	26	25	50	45	49	29	32	21	31	46	24	17	18	14
GAME 1	1	2	4	3	5	27	49	45	6	47	32	7	8	33	35	38	24	43	13	16	21	23	37	14	28	40	9	12	18	20	29	50
GAME 2	1	2	3	4	44	38	5	48	25	7	22	40	32	19	47	50	20	46	31	35	43	18	37	8	9	10	12	14	16	21	30	49
GAME 3	1	2	3	5	7	47	37	4	48	6	8	26	30	43	45	13	14	18	42	9	10	11	12	17	19	20	23	25	29	32	38	50
GAME 4	1	2	3	33	4	30	41	7	44	29	46	47	8	5	9	25	34	36	48	10	35	40	6	13	14	16	18	20	22	24	31	50
GAME 5	1	6	3	2	27	8	50	4	5	48	7	22	25	28	39	47	9	16	29	30	31	11	13	40	46	49	10	15	20	26	35	45

Table 3. Housing data set - resutls from GAME vs Weka. Every column represents ranks of one feature by all compared methods.

Method	Features Significance											Method	Features Significance											
ChiSquare	1	6	4	7	5	2	8	10	11	3	9	12	GAME 1	1	6	7	4	5	10	9	2	3	11	8	12
GainRatio	1	10	3	2	7	12	5	11	9	8	4	6	GAME 2	1	7	6	4	2	8	9	10	11	3	12	5
InfoGain	1	4	6	7	2	5	8	10	11	3	9	12	GAME 3	1	7	4	6	3	9	8	2	10	5	12	11
OneR	1	7	6	5	4	2	11	8	3	10	9	12	GAME 4	1	6	7	5	2	12	10	9	4	8	11	3
ReliefF	1	7	6	4	11	5	9	3	12	10	8	2	GAME 5	1	7	6	8	5	9	10	2	4	3	11	12
SVM	1	6	2	7	4	12	3	8	5	11	9	10	-	-	-	-	-	-	-	-	-	-	-	-	-
SU	1	4	2	6	7	5	8	10	3	11	9	12	-	-	-	-	-	-	-	-	-	-	-	-	-
Average	1	7	6	4	2	5	3	10	8	11	12	9	**Average**	1	7	6	4	2	9	5	10	8	3	12	11

the order in which first ten features were selected is not important, because all 10 features are equally significant. WEKA's methods (except the SVM) ranked features correctly (see Table 1). The FeRaNGA algorithm demonstrated worse results in comparison with WEKA's methods. Due to randomness of niching genetic algorithm used and insufficient number of epochs, ranks are different for each GAME model. Table 2 shows more or less similar results. All methods from WEKA ranked first ten features correctly, except ReliefF and SVM methods. Results of the FeRaNGA method are unstable, except the first position. The most significant feature was identified correctly for all 5 models.

In the Table 3 we can see results for real-world data, Housing Data Set. This time, even methods from WEKA differ in their ranking. It indicates that this problem is more complex than synthetic data. All methods found first feature (Criminality in the area) as the most significant for the value of the value of housing in Boston. When the average ranking is computed from WEKA methods and from several GAME models, results are very similar.

5.2 Restricted FeRaNGA Algorithm – More Selective Ranking

The results of the FeRaNGA algorithm applied to synthetic data (Table 1) show that some redundant features received higher ranking then expected. This

Table 4. Restricted FeRaNGA algorithm on the Hypercube data set. The bigger number of UNC causes the bigger number of used features.

UNC	1	2	3	4	5	6	7	8	9	10	Redundant(21–50) and Irrelevant(11–20)														
All	1	2	3	4	5	6	46	15	24	37	48	13	17	21	28	29	30	41	43	45	49	50	7	8	9
1 / 2	1	2	3	4	26	12	35	46	5	6	7	8	9	10	11	13	14	15	16	17	18	19	20	21	22
1 / 3	1	2	37	21	3	4	5	6	7	8	9	10	11	12	13	14	15	16	17	18	19	20	22	23	24
1 / 4	1	2	41	3	4	5	6	7	8	9	10	11	12	13	14	15	16	17	18	19	20	21	22	23	24
3	1	2	3	4	5	6	7	8	9	10	11	12	13	14	15	16	17	18	19	20	21	22	23	24	25
2	1	2	3	4	5	6	7	8	9	10	11	12	13	14	15	16	17	18	19	20	21	22	23	24	25
1	1	2	3	4	5	6	7	8	9	10	11	12	13	14	15	16	17	18	19	20	21	22	23	24	25

behavior is caused by insufficient selection pressure in the niching genetic algorithm - number of epochs was to low to eliminate all redundant features.

In this section, we experiment with the restriction of the FeRaNGA algorithm. The ranking is computed just from best chromosomes (feature lists) from the population. Number of unique chromosomes (UNC) used for the ranking varies from 1 to all (Table 4). Results were measured on Hypercube Data Set with NGA number of epochs 150 and size of initial population 150. It can be observed, that when UNC is lower, number of used features is also reduced. For very low values of UNC only relevant features are ranked, while for all UNC redundant and irrelevant features are ranked as well. The FeRaNGA algorithm restricted to one UNC (the best chromosome from the population) can be used to find and rank just few most important features of the data set. In the next section we present results of the FeRaNGA-n algorithm which is powered by ensemble methods.

5.3 Results for FeRaNGA-n Algorithm

The GAME algorithm usually generate more models for one purpose (ensemble of models). The idea of FeRaNGA-n algorithm is to improve results of ranking by combining unstable FeRaNGA ranks from n GAME models. Final ranks of features are computed as median ranking of features from n models. In the Table 5 we show results of the FeRaNGA-5 algorithm on Hypercube data set. Restrictive trend corresponding with number of UNC is again evident. NGA

Table 5. Results of FeRaNGA-5 algorithm on Hypercube data set. All selected features are relevant and have correct ranks.

UNC	1	2	3	4	5	6	7	8	9	10	Redundant (21 – 50) and Irrelevant (11 – 20)													
2	1	2	3	4	5	6	7	8	9	10	11	12	13	14	15	16	17	18	19	20	21	22	23	24
3	1	2	3	4	5	6	7	8	9	10	11	12	13	14	15	16	17	18	19	20	21	22	23	24
1 / 4	1	2	3	4	5	6	7	8	9	10	11	12	13	14	15	16	17	18	19	20	21	22	23	24
1 / 3	1	2	3	4	5	6	7	8	9	10	11	12	13	14	15	16	17	18	19	20	21	22	23	24
1 / 2	1	2	3	4	5	6	7	8	9	10	11	12	13	14	15	16	17	18	19	20	21	22	23	24
2 / 3	1	2	3	4	5	6	7	8	9	10	11	12	13	14	15	16	17	18	19	20	21	22	23	24
All	1	2	3	4	5	6	7	8	9	10	11	12	13	14	15	16	17	18	19	20	21	22	23	24

configuration was 150 epochs and size of initial population as well. Results of the FeRaNGA-n are very accurate. All selected features are relevant and have correct ranks. For full nr. of UNC only 7 from 10 relevant features is selected(last row in Table 5), but their ranks are accordant with their real ranks. To change the number of selected features, we can reconfigure the NGA.

5.4 Parameters of NGA for FR

The performance of the FeRaNGA-n algorithm on Hypercube data can be improved by reconfiguration of NGA parameters. The table 6 presents results of FeRaNGA-7 algorithm with all UNC and NGA's configurations displayed in the first column. First is size of initial population and second is number of epochs. When the ranks for more (or for all) features are needed, one can easily reconfigure parameters of NGA. Results in table 6 are not quite accurate, but it can be improved by increasing number of epochs or number of GAME models from which medians are chosen.

Table 6. Different configuration of FeRaNGA-7 on Hypercube data set

Config.	1	2	3	4	5	6	7	8	9	10	Redundant(21-50) and Irrelevant(11-20)														
15-30	1	2	3	4	6	7	8	5	9	10	11	12	13	14	15	16	17	18	19	20	21	22	23	24	25
30-30	1	2	3	4	5	6	7	8	10	39	48	9	11	12	13	14	15	16	17	18	19	20	21	22	23
50-30	1	2	4	3	5	7	6	8	9	22	37	10	11	12	13	14	15	16	17	18	19	20	21	23	24
100-30	1	2	3	4	5	7	6	8	9	10	22	26	34	35	37	38	40	44	45	47	48	11	12	13	14
150-30	1	2	4	3	5	6	8	7	22	9	34	35	26	39	43	44	10	30	40	45	47	48	11	12	13
300-30	1	2	3	4	5	7	6	8	9	43	37	22	39	48	44	26	27	38	50	34	35	36	10	11	12

6 Conclusion

FeRaNGA-n algorithm for FR was presented and fine tuned. The comparison of FeRaNGA-n algorithm and FR methods from WEKA showed that on Hypercube data set results are equivalent, but the number of selected features depends on the configuration of NGA and on number of GAME models from which the median ranks are chosen. FeRaNGA-n algorithm can be used for FR as well as for feature selection. The advantage of the algorithm is that it is primary designed for data with continuous output variable, but it can be also used for categorical variables. It does not require any additional computation, all information for ranking are extracted from process of data mining (GAME) models evolution.

Acknowledgement

This research is partially supported by the grant Automated Knowledge Extraction (KJB201210701) of the Grant Agency of the Academy of Science of the Czech Republic and the research program "Transdisciplinary Research in the Area of Biomedical Engineering II" ($MSM6840770012$) sponsored by the Ministry of Education, Youth and Sports of the Czech Republic.

References

1. Uci machine learning repository (September 2006), `http://www.ics.uci.edu/mlearn/MLSummary.html`
2. Almuallim., T.G., Dietterich, H.: Learning with many irrelevant features (1991)
3. Biesiada, J., Duch, W., Kachel, A., Maczka, K., Palucha, S.: Feature ranking methods based on information entropy with parzen windows., 109–119 (2005)
4. Brown, G.: Diversity in Neural Network Ensembles. PhD thesis, The University of Birmingham, School of Computer Science, Birmingham B15 2TT, United Kingdom (January 2004)
5. Guyon, I., Elisseeff, A.: An introduction to variable and feature selection. Journal of Machine Learning Research 3, 1157–1182 (2003)
6. Kohavi, R.: Wrappers for Performance Enhancement and Oblivious Decision Graphs. PhD thesis, Stanford University (1995)
7. Kordík, P.: Fully Automated Knowledge Extraction using Group of Adaptive Models Evolution. PhD thesis, Czech Technical University in Prague, FEE, Dep. of Comp. Sci. and Computers, FEE, CTU Prague, Czech Republic (September 2006)
8. Madala, H., Ivakhnenko, A.: Inductive Learning Algorithm for Complex System Modelling. CRC Press, Boca Raton (1994)
9. Mahfoud, S.W.: A comparison of parallel and sequential niching methods. In: Sixth International Conference on Genetic Algorithms, pp. 136–143 (1995)
10. Mahfoud, S.W.: Niching methods for genetic algorithms. Technical Report 95001, Illinois Genetic Algorithms Laboratory (IlliGaL), University of Ilinios at Urbana-Champaign (May 1995)
11. Muller, J.A., Lemke, F.: Self-Organising Data Mining, Berlin (2000) ISBN 3-89811-861-4
12. Piramuthu, S.: Evaluating feature selection methods for learning in data mining applications. European Journal of Operational Research 156, 483–494 (2004)
13. Siedlecki, W., Sklansky, J.: On automatic feature selection. International Journal of Pattern Recognition 2, 197–220 (1988)
14. Tesmer, M., Estevez, P.: Amifs: adaptive feature selection by using mutual information. In: Proceedings of the 2004 IEEE International Joint Conference on Neural Networks, July 2004, vol. 1, page 308. Dept. of Electr. Eng., Chile Univ, Santiago (2004)
15. Witten, I., Frank, E.: Data Mining – Practical Machine Learning Tools and Techniques, 2nd edn. Elsevier, Amsterdam (2005)

A Neural Network Approach for Learning Object Ranking

Leonardo Rigutini, Tiziano Papini, Marco Maggini, and Monica Bianchini

Dipartimento di Ingegneria dell'Informazione
Università degli Studi di Siena
Via Roma 56, Siena, Italy
{rigutini,papinit,maggini,monica}@dii.unisi.it

Abstract. In this paper, we present a connectionist approach to preference learning. In particular, a neural network is trained to realize a comparison function, expressing the preference between two objects. Such a "comparator" can be subsequently integrated into a general ranking algorithm to provide a total ordering on some collection of objects. We evaluate the accuracy of the proposed approach using the LETOR benchmark, with promising preliminary results.

1 Introduction

Recently, the topic of *Preference Learning* received considerable attention in Artificial Intelligence (AI). For instance, in many tasks related to "Intelligent Agents" and "Planning", the best action to be taken at each step can not be unique, and could be appropriately chosen by exploiting an underlying "preference model". Similarly, the goal of "Object Ranking" can be cast in a framework where the criterion used to order a set of given objects is not predefined but inferred from users' preferences. In all these cases, the system behaviour should be adapted based on a set of feedbacks, provided by an user or by sensors. The approaches proposed in the literature vary from approximating the utility function of a single agent on the basis of a question-answer process (often referred to as "preference elicitation") to "collaborative filtering", where the preferences of a given user are estimated from the preferences of the other users. In this scenario, it is not surprising that, in recent years, automatic methods for learning and predicting preferences have been widely investigated in disciplines such as machine learning, knowledge discovery, and recommender systems. In the AI literature, preference learning problems have been formalized in various different settings, depending on the underlying preference model or the type of information provided to the learning system. However, two formalisms mainly appear to be central: the "Learning Labels Preference (LLP)" and the "Learning Objects Preference (LOP)".

This paper presents a neural model for Learning Objects Preferences based on a pairwise approach. In particular, a comparison function, referred to as a "comparator", is learned using a set of examples. Given an ordered input pair of examples,

V. Kůrková et al. (Eds.): ICANN 2008, Part II, LNCS 5164, pp. 899–908, 2008.

the comparator decides which of the two examples is preferable. The performances of the proposed approach are firstly tested in a classification framework by evaluating the accuracy of the preference assignment for each input pair, and then the comparator is applied to a ranking task by using three different schemes. These tests are performed using the LETOR (LEarning TO Rank) dataset [1], which is a standard benchmark for the "Learning to Rank" task. The comparison considers several state-of-the-art ranking algorithms for this task, like RankSVM [2], RankBoost [3], FRank [4], ListNet [5], and AdaRank [6].

The paper is organized as follows. In the next section, we introduce the model and provide a brief mathematical description, whereas, in Section 3 we describe the three ranking algorithms used to test the comparator. Subsequently, in Section 4, we present the experimental setup, the LETOR dataset, the related evaluation measures, and some comparative experimental results. Finally, in Section 5, same conclusions are drawn.

2 The Comparator Model

The connectionist method proposed in this paper is based on a pairwise preference learning approach. In particular, let S be a set of objects described by a vector of features, and let $< x, y > \in S \times S$ be the pair of objects to be compared. The model is trained to decide if x has to be preferred to y or vice-versa. Formally, a pairwise preference function realizes a function f such that:

$$f(x, y) = \begin{cases} > 0 \; if \; x \succ y \\ < 0 \; if \; x \prec y \\ = 0 \; if \; x \sim y \end{cases} \tag{1}$$

where $x \succ y$ means that x is preferred to y, $x \prec y$ if y is preferred to x, and $x \sim y$ if there is no preference between x and y (i.e. x and y are considered equivalent). In many cases, the equivalence relationship can be ignored. To induce a correct ordering, the function f has to meet the following constraints:

- anti-simmetry: if $x \succ y$ then $y \prec x$. It means that: $f(x, y) = -f(y, x)$.
- self-equivalence: if $x = y$, then $f(x, x) = 0$.
- transitivity: if $x \succ y$ and $y \succ q$, then $x \succ q$. It means that if $f(x, y) > 0$ and $f(y, q) > 0$, then $f(x, q) > 0$.

The transitivity property is not guaranteed by the proposed model, even if it can be learnt from data.

2.1 The Comparator Neural Network

The comparator neural network is a feed-forward neural network with a single output neuron such that:

- all neurons (hiddens and output) have an anti-symmetric transfer function;
- all neurons have no biases;
- the input pattern is the difference between the two exampels in the pair: $z = (x - y)$.

This particular architecture realizes an anti-symmetric function $f(z): f(\overrightarrow{z}) = -f(\overleftarrow{z})$, where $\overrightarrow{z}$ indicates the ordered pair $< x, y >$, while $\overleftarrow{z}$ indicates $< y, x >$. This can be easily shown as follows. The output, for the inputs $\overrightarrow{z}$ and $\overleftarrow{z}$ can be expressed as

$$f(\overrightarrow{z}) = s(\sum_i w_i \overrightarrow{h_i} + b)$$
$$f(\overleftarrow{z}) = s(\sum_i w_i \overleftarrow{h_i} + b)$$

where s is the transfer function of the output neuron, and h_i and w_i are the hidden neurons' outputs and the hidden-output weights, respectively . By imposing the anti-symmetry of f, we can write:

$$s\left(\sum_i w_i \overrightarrow{h_i} + b\right) = -s\left(\sum_i w_i \overleftarrow{h_i} + b\right) .$$

If s is an anti-symmetric function, it follows that

$$\sum_i w_i \overrightarrow{h_i} + b = -\sum_i w_i \overleftarrow{h_i} - b \Longrightarrow \begin{cases} \overrightarrow{h_i} = -\overleftarrow{h_i} \\ b = -b = 0 \end{cases}$$

where $\overrightarrow{h_i}$ and $\overleftarrow{h_i}$ represent the output of the ith hidden neuron for the inputs $\overrightarrow{z}$ and $\overleftarrow{z}$, respectively. Similarly, $\overrightarrow{h_i}$ and $\overleftarrow{h_i}$ can be expressed as:

$$\overrightarrow{h_i} = g(\sum_j v_{x_j,i} x_j + \sum_j v_{y_j,i} y_j + b_i)$$
$$\overleftarrow{h_i} = g(\sum_j v_{x_j,i} y_j + \sum_j v_{y_j,i} x_j + b_i)$$

where $v_{x_j,i}$ and $v_{y_j,i}$ are the weights between the jth feature of x and y, and the hidden neuron i. By imposing $\overrightarrow{h_i} = -\overleftarrow{h_i}$, we can write:

$$g(\sum_j v_{x_j,i} x_j + \sum_j v_{y_j,i} y_j + b_i) = -g(\sum_j v_{x_j,i} y_j + \sum_j v_{y_j,i} x_j + b_i) .$$

If the function g is anti-symmetric, it follows that

$$\sum_j v_{x_j,i} x_j + \sum_j v_{y_j,i} y_j + b_i = -\sum_j v_{x_j,i} y_j - \sum_j v_{y_j,i} x_j - b_i$$
$$\Longrightarrow \sum_j v_{x_j,i}(x_j + y_j) + b_i = \sum_j -v_{y_j,i}(x_j + y_j) - b_i \Longrightarrow \begin{cases} v_{x_j,i} = -v_{y_j,i} \\ b_i = -b_i = 0 \end{cases} .$$

Therefore, the weight between the ith hidden neuron and the jth feature of x has to be the opposite of that between the ith neuron and the jth feature of y, which simply corresponds to consider $z_j = x_j - y_j$ as the jth input to the network. Thus, the function implemented by each hidden neuron i is

$$h_i = g\left(\sum_j v_{x_j,i}\,(x_j - y_j)\right) .$$

This neural architecture also satisfies the self-equivalence property, since the network output is zero when the input pair is $< x, x >$. In fact, since g and s are anti-symmetric functions,

$$h_i = g\left(\sum_j v_{x_j,i}\,(x_j - x_j)\right) = 0$$

$$f(z) = s(\sum_i w_i h_i) = 0 .$$

The hyperbolic tangent can be used as transfer function both for the hidden and output neurons. When training the model, for each pair of inputs $< e_1, e_2 >$, the target is assigned as

$$\begin{cases} t = 1 & if\ e_1 \succ e_2 \\ t = 0 & if\ e_1 \sim e_2 \\ t = -1 & if\ e_1 \prec e_2 \end{cases} .$$

However, in the experimental phase, we noticed that the model obtains better performances when trained without using the equivalence relationship. In fact, the network is still able to learn the "no-preference" relationship hidden in the training set even without using explicit examples.

3 The Ranking Algorithm

A trained neural network comparator can be exploited to rank objects by means of a sorting algorithm, an one-vs-all scoring scheme, or a page-rank like approach.

Ranking by a Sorting Algorithm

In this approach, the trained preference function is used as the comparison function in a sorting algorithm. The set of objects can be ordered with $O(n \log n)$ time complexity, according to the ordering defined by the preference function. It must be noticed that the proposed model does not realize a perfect comparison function since the transitivity property is not guaranteed. This could result in the definition of a partial ordering for which different rankings are admissible. However, the experimental results show that, when the transitivity is implicit in the training set, it is learnt quite well by the proposed model. In these experiments, the same sorting algorithm was applied to different shufflings of the same objects, and different sorting algorithms were used on the same set of objects. The obtained final orderings were the same in most of the cases and seldom they differed only by a very small number of object positions ($1 - 5$ over 1000).

Ranking by One-vs-All Scoring
The objects are ranked by exploiting a sort of "championship" consisting in all the "matches" between pairs of competitors. The ith object is compared with the $n-1$ remaining objects using the trained preference model. If $o_i \succ o_j$, o_i wins the match and its total score is incremented by 1, otherwise the score of o_i is unchanged. When $o_i \sim o_j$, there is a draw and the corresponding scores are not modified. This algorithm has an $O(n^2)$ time complexity, since it requires $n(n-1)/2$ matches.

Ranking by Page-Rank Scoring
This ranking algorithm exploits the PageRank algorithm [7]. In particular, the neural network comparator is used to build the preference graph on the objects: the sign of the network output is used to determine the direction of the arc between the nodes corresponding to the compared objects. The graph adjacency matrix is normalized, such that the elements on each row sum to 1, and the PageRank vector $x \in \mathbf{R}^N$ is computed by the following iterative computation

$$\begin{cases} x^{t+1} = d \cdot W \cdot x^t + \frac{(1-d)}{N} \cdot \mathbf{1} \\ x^0 = \frac{1}{N} \cdot \mathbf{1} \end{cases}$$

where $W \in \mathbf{R}^{N \times N}$ is the transpose of the normalized adiacency matrix, N is the number of nodes in the graph (the number of objects to rank), and $d \in [0, 1]$ is the dumping factor, typically set to 0.85. The resulting score vector x is then used to order the objects.

4 Experimental Results

The performances of the proposed schemes have been evaluated on the LETOR dataset [1], a package of benchmark datasets[1] for LEarning TO Rank, released by Microsoft Research Asia. This dataset collects objects which correspond to query-document pairs on the OHSUMED and TREC document collections. The documents are represented using several classical information retrieval (IR) features, such as term frequency, inverse document frequency, BM25, language models, and other features proposed in the recent literature, such as HostRank, Feature propagation and Topical PageRank. For the experiments reported in this paper, we considered only the TREC collections, since the OHSUMED dataset considers three relevance degrees. LETOR contains two TREC datasets, TD2003 and TD2004. TD2003 is composed by 50 sets of documents from TREC 2003, each one containing the 1000 most relevant documents returned to one out of 50 different queries. Similarly, TD2004 contains documents from the TREC 2004 returned to 75 different queries. Each query-document pair is represented by 44 features described in the LETOR technical report. A label is provided for each

[1] `http://research.microsoft.com/users/LETOR/`

document to define its actual relevance with respect to the corresponding query, defining the set R of relevant documents and the set NR of the not-relevant ones. For all queries, the number of relevant documents is roughly the 1% with respect to the whole set of documents. Each dataset is partitioned into five subsets in order to apply a 5-fold cross-validation. Thus, for TD2003 there are 10 queries for each subset, whereas for TD2004 each subset collects the results of 15 queries. For each fold, three subsets are used for training, one subset for validation, and one for testing.

The feature vectors have been normalized to $[-1,1]$. In particular, if N^i documents d^i_j, $j=1,...,N^i$, are returned as result to the i-th query, and $x^i_{j,r}$, with $j=1,...,N^i$, indicates the r-th feature of the document d^i_j, the corresponding normalized feature $\hat{x}^i_{j,r}$ is

$$\hat{x}^i_{j,r} = \frac{x^i_{j,r} - \bar{x}^i_r}{max_{s=1,\dots,N^i}|x^i_{s,r}|} \quad (2)$$

where $\bar{x}^i_r = \frac{\sum_{s=1}^{N^i} x^i_{s,r}}{N^i}$ represents the average of $x^i_{j,r}$. The learning set is built by selecting a pair $< x,y >$ of documents, each from one of the two relevance classes. The corresponding target value for the network output is 1 if $x \in R$ and $y \in NR$, -1 otherwise. The number of hidden neurons was chosen by a trial-and-error procedure using the validation set.

To evaluate the accuracy of the comparison function learnt by the neural network, the network output is compared with respect to the actual preference order for each pair of documents in the test set, as defined by the relevance classes to which the two documents belong. The evaluation considers only pairs of documents belonging to different classes. In fact, the available labeling allows only a clear preference ordering when the two documents belong to different classes, and it is too restrictive to assume a strict equivalence relationship when they belong to the same class. Documents inside the same class can have any ordering, since no information is provided to prefer one ordering to the other. Finally, it should be noticed that if the network provides an erroneous value for the comparison of the pair $< x,y >$, then, due to its anti-symmetry, also the comparison of $< y,x >$ is wrong.

For this task, the best performances were obtained using a network with 20 hidden neurons. The total accuracy, averaged on the 5 folds, is quite different on the two datasets. On TD2003, the neural network was able to compare correctly $83,53\% \pm 0.1$ of the test pairs, while on TD2004 the performances were significantly better, yielding an accuracy of $96,51\% \pm 0.05$. This is probably due to the different characteristics of the two document sets and the related relevance criterion. However, the results show that, in the best case, the neural network comparator was able to approximate the pairwise preference ordering with an error at most of 3.5% on the considered pairs.

Since the final goal is to rank the results sets, the neural network comparator was exploited to order the documents in each set using the three algorithms described in Section 3. In this setting, a good ordering is characterized by the

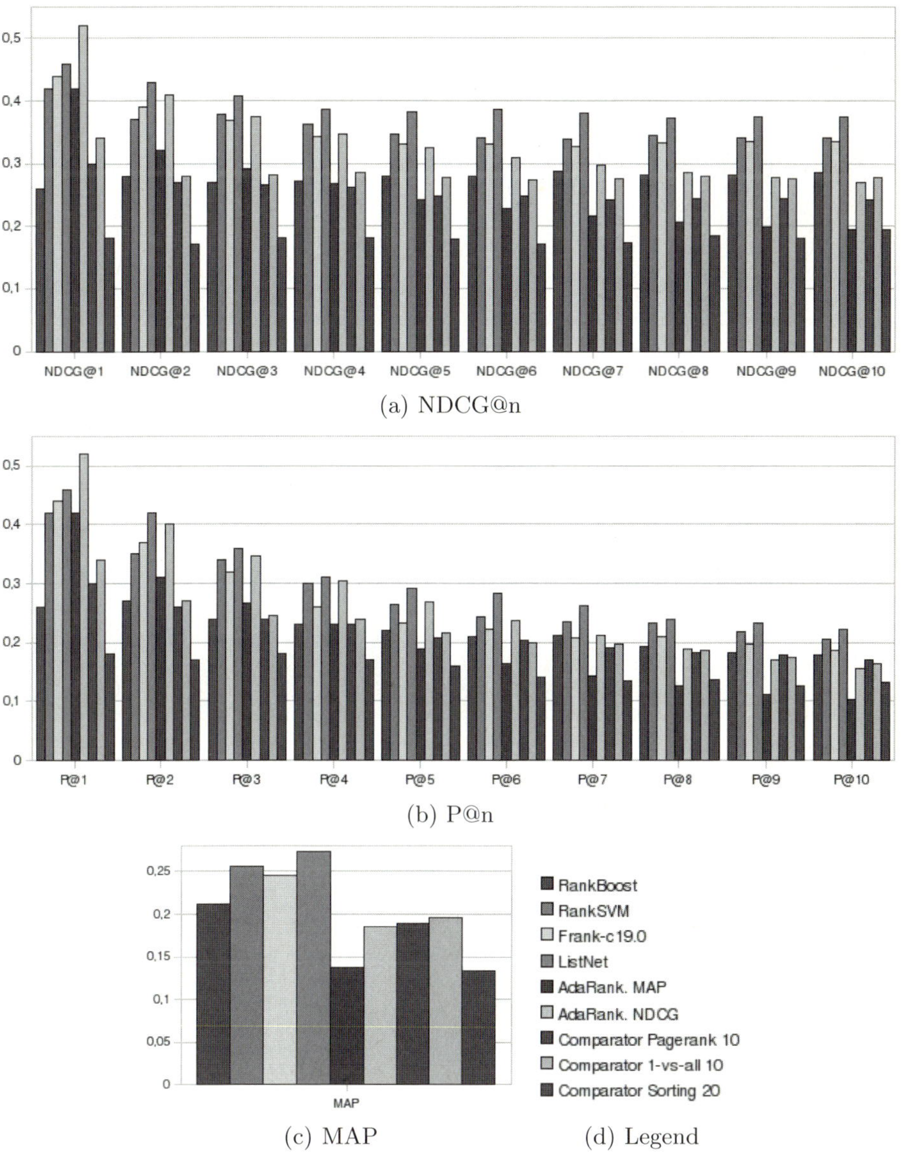

(a) NDCG@n

(b) P@n

(c) MAP (d) Legend

Fig. 1. Results on the test set of TREC2003

presence of the documents belonging to the relevant class in the top positions. The obtained results are compared to those reported in [1]: RankSVM [2], RankBoost [3], FRank [4], ListNet [5], and AdaRank [6].

The performances of these algorithms are compared using the ranking measures proposed in the LETOR technical report, defined as follows.

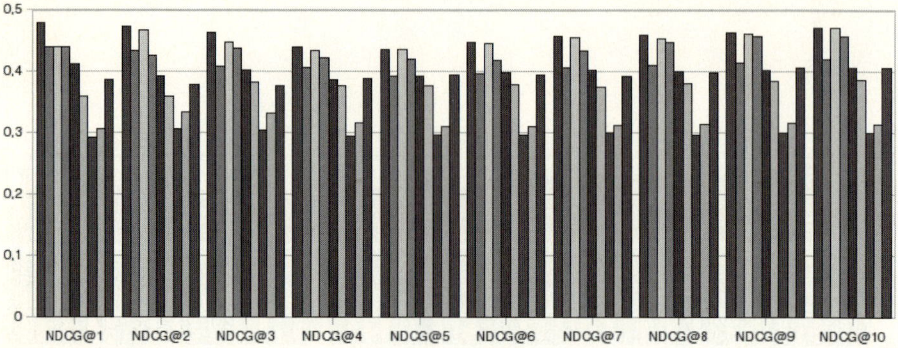

(a) NDCG@n

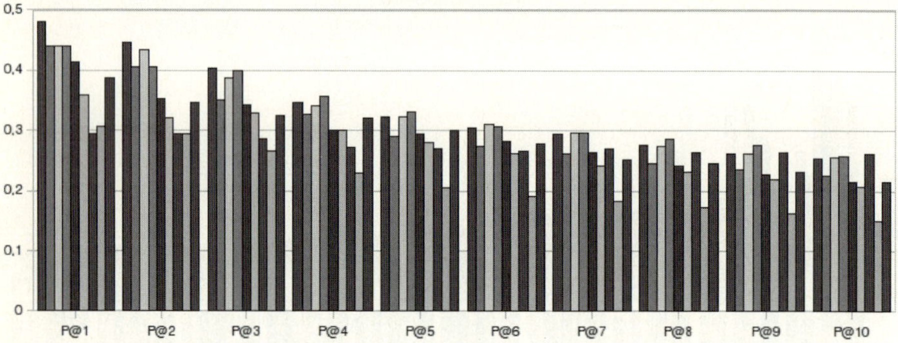

(b) P@n

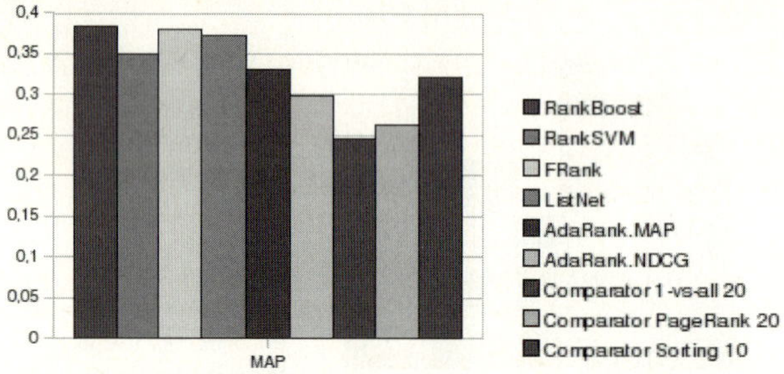

(c) MAP (d) Legend

Fig. 2. Results on the test set of TREC2004

- **Precision at position n (P@n)** — This value measures the relevance of the top n results of the ranking list with respect to a given query.

$$P@n = \frac{\text{relevant docs in top } n \text{ results}}{n}$$

- **Mean average precision (MAP)** — Given a query q, the average precision is evaluated as

$$AP_q = \frac{\sum_{n=1}^{N} P@n \cdot rel(n)}{\text{total relevant docs for } q}$$

where N is the number of documents in the result set of q and $rel(n)$ is 1 if the nth document in the ordering is relevant, 0 otherwise. Thus, AP_q averages the values of the $P@n$ when n equals the positions of each relevant document. Finally, the MAP value is computed as the mean of AP_q over a set of queries.
- **Normalized discount cumulative gain (NDCG)** — This measure is able to handle multiple levels of relevance. The NDCG value of a ranking list at position n is calculated as

$$NDCG@n \equiv Z_n \sum_{j=1}^{n} \frac{2^{r(j)} - 1}{log(1+j)}$$

where $r(j)$ is the rating of the jth document in the list (0 is the worst), and Z_n is chosen such that the ideal ordering (the DGC-maximizing one) gets a NDCG score of 1.

For each ranking algorithm, we report the results for the best neural network architecture. In particular, for the experiments using the TD2003 dataset, the best performances were obtained with 10 hidden neurons for the one-vs-all and the PageRank algorithms, whereas 20 hidden neurons were used for the neural network exploited in the sorting algorithm. The results on TD2003 dataset are shown in Figure 1. All the three algorithms show values of $P@n$ and $NDCG@n$ that are comparable to the RankBoost, AdaRank-MAP, and AdaRank-NDCG algorithms. In particular, the results improve when increasing the parameter n.

In the experiments on the TD2004 dataset, 20 hidden neurons were used for the neural network exploited in the one-vs-all and PageRank algorithms, whereas the neural network used as comparison function in the sorting algorithm had 10 hidden neurons. The results are shown in Figure 2. We can notice that the NDCG of the sorting algorithm is better than the values of the AdaRank-MAP and AdaRank-NDCG algorithms. Moreover, for increasing values of n, the algorithm reaches the performances of RankSVM.

5 Conclusions

In this paper a connectionist approach to the preference learning task has been proposed. A neural network is trained using patterns formed by pairs of examples. Given an ordered input pair, the comparator decides which of the two objects is preferable. The model is then exploited in a ranking algorithm to obtain a total ordering over a set of objects. Experiments were carried out to evaluate the performance of the proposed algorithm using the datasets TD2003 and TD2004, available among the LETOR benchmarks. RankingSVM, RankBoost,

FRank, ListNet and AdaRank were compared with the proposed approach and the results show that the neural comparator combined with different ranking algorithm obtains comparable performances with respect to all the methods, except AdaRank, that is outperformed.

References

1. Liu, T.Y., Xu, J., Qin, T., Xiong, W., Li, H.: LETOR: Benchmarking learning to rank for information retrieval. In: SIGIR 2007 – Workshop on Learning to Rank for Information Retrieval, Amsterdam, The Netherlands (2007)
2. Cao, Y., Xu, J., Liu, T.Y., Li, H., Huang, Y., Hon, H.W.: Adapting ranking SVM to document retrieval. In: Proceedings of ACM SIGIR 2006, pp. 186–193. ACM, New York (2006)
3. Freund, Y., Iyer, R., Schapire, R.E., Singer, Y.: An efficient boosting algorithm for combining preferences. In: Shavlik, J.W. (ed.) Proceedings of ICML 1998, pp. 170–178. Morgan Kaufmann Publishers, San Francisco (1998)
4. Tsai, M.F., Liu, T.Y., Qin, T., Chen, H.H., Ma, W.Y.: FRank: a ranking method with fidelity loss. In: Proceedings of the ACM SIGIR, pp. 383–390. ACM, New York (2007)
5. Cao, Z., Qin, T., Liu, T.Y., Tsai, M.F., Li, H.: Learning to rank: from pairwise approach to listwise approach. In: Proceedings of ICML 2007, pp. 129–136. ACM, New York (2007)
6. Xu, J., Li, H.: AdaRank: a boosting algorithm for information retrieval. In: Proceedings of ACM SIGIR 2007, pp. 391–398. ACM, New York (2007)
7. Page, L., Brin, S., Motwani, R., Winograd, T.: The pagerank citation ranking: bringing order to the web (1998)

Evolving Efficient Connection for the Design of Artificial Neural Networks

Min Shi[1] and Haifeng Wu[2]

[1] Department of Computer and Information Science,
Norwegian University of Science and Technology, Norway
minshi@idi.ntnu.no
[2] Department of Engineering Cybernetics,
Norwegian University of Science and Technology, Norway
haifeng@stud.ntnu.no

Abstract. Most of the recent neuroevolution (NE) approaches explore new network topologies based on a neuron-centered design principle. So far evolving connections has been poorly explored. In this paper, we propose a novel NE algorithm called Evolving Efficient Connections (EEC), where the connection weights and the connection paths of networks are evolved separately. We compare our new method with standard NE and several popular NE algorithms, SANE, ESP and NEAT. The experimental results indicate evolving connection weights along with connection paths can significantly enhance the performance of standard NE. Moreover the performances of cooperative coevolutionary algorithms are superior to non-cooperative evolutionary algorithms.

1 Introduction

Neuroevolution (NE) techniques evolve artificial neural networks (ANNs) by using evolutionary algorithms; it has been shown to have a promising efficacy in reinforcement learning problems [1] [2] [3]. As summarized by Yao, these kinds of techniques can be roughly classified into three levels: evolving connection weights; evolving architectures and evolving learning rules [4]. In the past decade, more and more approaches have focused on simultaneously evolving connection weights and architectures of ANNs [5] [6] [3] [7] [8] [9].

A major challenge in NE is to find both optimal weights and network topologies. Several recent approaches have been proposed to achieve this purpose. Moriarty and Miikkulainen [5] developed a cooperative coevolutionary model, SANE, that coevolves neurons with predefined numbers of connections in one population and evolves blueprints in another population to form functioning neural networks. Potter and De Jong [6] proposed an architecture for evolving coadapted subcomponents and applied this architecture to cascade networks, in which each hidden neuron was evolved in an independent population. In their model, a subcomponents can be removed or added based on the current performance of cooperative subcomponents. ESP [7], developed by Gomez and Miikkulainen, had some similarities with the model proposed by Potter and De

V. Kůrková et al. (Eds.): ICANN 2008, Part II, LNCS 5164, pp. 909–918, 2008.

Jong. More precisely, ESP evolves three-layer networks as SANE does, but replaces the blueprints with a fixed sequence of neurons and coevolves neurons in independent sub-populations. A Hierarchical Cooperative CoEvolution (HCCE) model was introduced by Maniadakis and Trahanias [10]. The model also coevolves subcomponents. In addition, a Coevolved agent Group (CG), similar to blueprint, is employed to explore effective cooperative subcomponents. A high level CG can enforce the cooperation of the lower ones. NEAT [3] is a new NE technique that works differently from all of the above. NEAT evolves both the connection weights and connection topologies of ANNs in one population. The topologies of networks are changed by using mutation operators. This mutation could change weights values, or insert a new node between an existing connection, or create a connection between two unconnected nodes, or delete a connection between two connected nodes. Garcia-Pedrajas et al. [8] developed a new cooperative coevolutionary model for evolving ANNs, called COVNET, where each subcomponent is not a neuron but a subnetwork called a nodule. The design of COVNET integrates the ideas of the above methods. Each nodule is evolved in an independent sub-population. A network population works similar to blueprints to evolve efficient combination of nodules. The mutation operators can change the network topologies as NEAT does.

Most of the recent NE approaches explore new network topologies based on a neuron-centered design principle [5] [7] [8] [6]: 1) networks are decomposed into and combined with neuron-centered subcomponents, and 2) network topologies are altered by evolving cooperative subcomponents. Few authors have devoted their attention to evolve connection topologies of ANNs. SANE defines evolvable connection paths in the representation scheme of a neuron population, but the number of connections is predefined and fixed. NEAT has very limited possibilities to exploit the connection topologies, since they are changed merely through mutation operators.

In this paper, we propose a new NE algorithm, called Evolving Efficient Connections (EEC). We break through the neuron-centered design principle, and introduce a novel connection-centered concept. EEC evolves connection weights and connection paths that are respectively represented in their own populations, which cooperatively coevolve. The most important contributions of EEC are twofold. First, it builds a new cooperative coevolutionary model for evolving ANNs. Second, it corrects the one-sided understanding that the learning capacity of an ANN is mainly dependent on the efficient number of hidden neurons. Our model demonstrates that efficient connections also play an important role in the performance of ANNs.

2 EEC: Evolving Efficient Connections

EEC is a cooperative coevolutionary model that separates the search space of a network into two sub-spaces: connection weights space and connection paths space. Thus, it contains two populations.

The population of connection weights evolves weight vectors for fully connected networks as does standard NE. Each individual represents connection weights of a network, in which each neuron hypothetically connects with all neurons in the next layer. The population of connection paths evolves switches of these connections; each individual represents a series of binary bits to specify the status of the connections: connected (1) and disconnected (0). Figure 1 shows how the connection-weights individuals, and the connection-paths individuals are related.

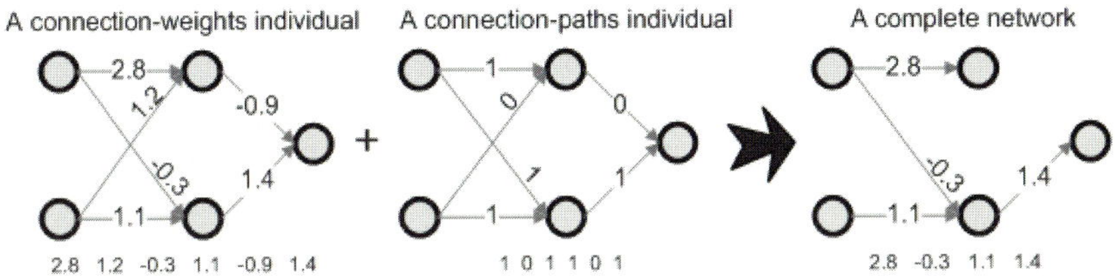

Fig. 1. In EEC, connection-weights individuals represent weight vectors of fully connected networks, complete networks switch on a partial connections specified by connection-paths individuals

EEC borrows many ideas from both SANE and ESP. Below we provide a briefly description of the EEC algorithm.

1. Initialization. The number of hidden neurons of the networks has to be specified at the beginning. Each chromosome of connection weights encodes fully-connected weight vectors with random real numbers. The chromosome of connection paths encodes connection status with a random string of binary bits, corresponding to the connection weights.
2. Evaluation. A complete network is formed through sampling paired individuals from the connection-weights population and the connection-paths population. To achieve the best performance, we suggest evaluating every individual of both populations. The top n individuals are regarded as elite based on their previous evaluation. An elite-rate percent of the cooperative individuals are selected from the elite to cooperative with the current evaluated individuals, while the rest are selected randomly from the cooperative population. The resulting network is evaluated on the task and assigned a fitness score. Each individual is awarded an average of the cumulative fitness of the networks in which it participated during one generation.
3. Recombination. The same recombination process of ESP is employed in our algorithm. All the individuals will be ranked according to their fitness within each population. The top 1/4 individuals are breeding members. The lowest ranking half of the individuals will be replaced by offspring of the breeding pairs. One-point crossover and one-point mutation are employed.

4. Iteration. During the generation process, the stagnation-handling strategy of ESP is also used in our method. A similar delta-coding technique of ESP will be employed when the best fitness of networks has not improved after a specified number of generations. In our delta-coding process, the connection-weights individual and the connection-paths individual from the current best network are regarded as seeds of each population. We replace all the individuals of each population with new individuals generated by perturbations of the selected seed. The generation is iterated until an optimal solution is found or the maximum generation is reached.

Besides evolving weights, EEC evolves efficient connection topologies of networks. Although we evolve a three-layer feedforward network using EEC in our experiments, this algorithm is applicable to other types of networks, such as recurrent networks, multi-hidden layer networks and so on.

3 Performance Evaluation

In order to get a clear impression of the performance of the proposed model, this section empirically evaluates it through comparisons with standard NE and several current popular NE techniques, SANE, ESP and NEAT[1].

Two board games, Tic-Tac-Toe (TTT) and Gobang, were selected as our benchmark tasks. The reasons that these two games are chosen as our test domains are twofold. First, games are widely used for evaluating NE methods, especially board games, and are easily implemented. Second, TTT can be regarded as an abbreviated version of Gobang since the two games have similar game rules but different board sizes. By performing simple tasks and then processing to more difficult problems, we may be able to analyze if EEC is able to evolve optimal structures and keep its learning capacities better than other methods.

All the networks were evolved to play with hand-coded strategies in both games in our experiments.

3.1 Board Games

Tic-Tac-Toe. TTT is a classic game that is commonly used to evaluate NE algorithms [11] [12]. In TTT two players alternately mark their symbols on a board. The one who first obtains three in a row horizontally, vertically, or diagonally wins. The game ties if all grids are filled with symbols and no one wins.

James and Tucker [12] evolved networks with three NEAT dynamics, simplification, complexification and blended, to play with five hand-coded strategies of TTT, called BEST, FORKABLE, CENTER, RANDOM and BAD. The same five strategies were employed in our TTT experiments.

[1] SANE, ESP and NEAT (ANJI) experiments of our work were implemented based on the relative software developed by Neural Networks Research Group at the University of Texas, see http://nn.cs.utexas.edu/

Gobang. Gobang is also known as GoMoku, or 5-in-a-row. This game has some similar game rules to TTT, but plays on a bigger size board with the goal of achieving five in a row. Gobang is as easy to learn as TTT but much more difficult to master even by humans, because the size of the board is bigger than the goal number of symbols in a row. To both reduce the time consumption for evolving networks and simplify the structure of networks, we chose to do the experiments of Gobang on a 7×7 size board.

The hand-coded strategy of Gobang used in our experiments was developed by Vladimir Shashin[2]. 106 patterns composed of four symbols, "*", "-", "o" and "x", with a length of five each (such as "-*ooo", "-x*x-") were created in his program. These patterns were used to make a decision of the best move for each next step after scanning the board state. We divided the hand-coded strategy into three increasingly difficult levels through disabling partial patterns:

1. BEST strategy employed all the 106 patterns.
2. MIDDLE strategy disabled six patterns that include four same symbols, such as "o*ooo" and "*xxxx". It contained 100 patterns.
3. BAD strategy disabled 30 patterns that included four or three same symbols. It employed 76 patterns.

3.2 Networks Representation and Evaluation

In both games, the board states are represented to the networks through mapping each grid to both an input neuron and an output neuron. An input neuron is: 1) 1 if play-1 marks its symbol in the corresponding grid of the board, 2) -1 if the corresponding grid is occupied by player-2, and 3) 0 denoting a blank grid that is one of legal moves for both players. After performing the activation computation, a move decision is made by the output neurons. The one with the highest output value corresponding to a legal move is chosen as the move decision of the network player.

To evaluate the performances of a network, 100 matches are played between the network player and a predefined hand-coded player. Each player takes turns going first. The network player is awarded 5 points for a win, 2 points for a tie and 0 for a loss. The fitness of the network, of course, is the cumulation of the points from the 100 matches. Due to the BEST hand-coded strategy of Gobang is too strong, no one network is able to evolve the ability to force a win or even a tie using the above fitness evaluation. So a bonus is awarded additionally to the network player. The bonus is the number of filled board grids divided by the total number of the board grids.

3.3 Parameter Setting

For the purpose of implementing the evaluation as fairly as possible, we evolved the solutions with the same number of generations and built the same number of

[2] The Java game of Gobang developed by Vladimir Shashin can be download at http://down1.tech.sina.com.cn/download/down Content/2004-03-16/9313.shtml

Table 1. Parameter settings for TTT and Gobang experiments

Methods	Parameters	Value	Parameters	Value
All	Num. of Gen.	200	Num. of Evaluation per Gen.	200
NE	Pop. size	200	Hidden neurons	10
	Mutation rate	0.2	Breeding rate	0.25
NEAT	Pop. size	200	Initialized hidden neurons	0
	Weight mutation rate	0.75	Survival rate	0.2
	Add connection mutation rate	0.2	Add neuron mutation rate	0.2
	Excess gene coefficient	1.0	Disjoint gene coefficient	1.0
	Common weight coefficient	0.4	Speciation threshold	0.9
	Delete connection mutation	0.02		
SANE	Pop. size of neurons	1000	Pop. size of blueprints	200
	Num. of connection for TTT	36	Num. of connection for Gobang	196
	Hidden neurons	10	Breeding neurons	250
	Elite neurons	250	Mutation rate	0.02
	Num. of top network	100	Num. of top network breedings	20
ESP	Sub-population size	100	Num. of trial networks	200
	Initialized hidden neurons	10	Mutation rate	0.2
	Gen. of stagnation	20	Breeding rate	0.25
EEC	Connection-weights Pop. size	200	Connection-paths Pop. size	200
	Num. of evaluated networks	200	Hidden neurons	10
	Mutation rate	0.2	Gen. of stagnation	20
	Elite of connection-weights	10	Elite rate of weights	0.2
	Elite of connection-paths	10	Elite rate of paths	0.2
	Breeding rate	0.25		

networks in each generation for all the experiments. Besides using some related work [2] [13] [12] in the same or similar problem domains, a number of experiments were carried out at the beginning to search effective parameters. Table 1 summarizes the parameters chosen for both TTT and Gobang experiments.

As we have presented, EEC performs full-evaluation on all the individuals. For the comparison purpose, however, only 200 evaluations were carried out in each generation. So a half-evaluated ECC was performed in our experiment, in which only every individual in the connection-weights population was evaluated actively. The individuals of connection-paths population were evaluated passively only if they were selected to participate in the cooperation. Using the half-evaluation, thus, some very good connection paths could be lost due to a failure to participate in the cooperation.

3.4 Experimental Results

We carried out 20 runs for each type of evaluation. These methods were compared based on the average performances of our experiments. Table 2 and table 3 lists the average results of wins, ties, losses and fitness for the 20 best solutions from each type of performance for TTT and Gobang tasks.

Table 2. Average performances of different networks playing with five hand-coded strategies of TTT (20 runs each)

		NE	NEAT	SANE	ESP	EEC
BAD	Wins	100	100	100	100	100
	Ties	0	0	0	0	0
	Losses	0	0	0	0	0
	Fitness	500	500	500	500	500
RANDOM	Wins	94.35	95.8	93.45	95.95	95.65
	Ties	2	1.05	1.85	1.8	1.2
	Losses	3.65	3.15	4.7	2.25	3.15
	Fitness	475.75	481.1	470.95	483.35	480.65
CENTER	Wins	95.3	96.95	96.55	97.6	96.55
	Ties	2.3	1.65	1.7	1.15	1.75
	Losses	2.4	1.4	1.75	1.25	1.7
	Fitness	481.1	488.05	486.15	490.3	490.45
FORKABLE	Wins	39.6	56.15	54.7	47.6	50.45
	Ties	38.95	25.65	28.15	38.3	30.3
	Losses	21.45	18.2	17.15	14.1	19.25
	Fitness	275.9	332.05	329.8	314.6	312.85
BEST	Wins	0	0	0	0	0
	Ties	91.05	98.1	97.7	97.7	96.9
	Losses	8.95	1.9	2.3	2.3	3.1
	Fitness	182.1	196.2	195.4	195.4	193.8

Table 3. Average performances of different networks playing with three hand-coded strategies of Gobang (20 runs each)

		NE	NEAT	SANE	ESP	EEC
BAD	Wins	96.3	97.95	100	100	100
	Ties	0.3	0.1	0	0	0
	Losses	3.4	1.95	0	0	0
	Fitness	482.1	489.95	500	500	500
MIDDLE	Wins	77.6	61.4	97.4	99.5	96.1
	Ties	0.05	0.1	0.05	0	0
	Losses	22.35	38.5	2.55	0.05	3.9
	Fitness	388.3	307.2	487.1	499.75	480.5
BEST	Wins	0	0	1.65	0	0
	Ties	9.4	2.42	60.5	69.3	56.05
	Losses	90.6	97.58	37.85	30.7	43.95
	Fitness	65.33	45.51	210.25	218.04	181.73

From these results we can see that, after 200 generations the average final results came out from half-evaluated EEC were quite close to NEAT, SANE, ESP and appreciably better than standard NE for TTT task. While for Gobang

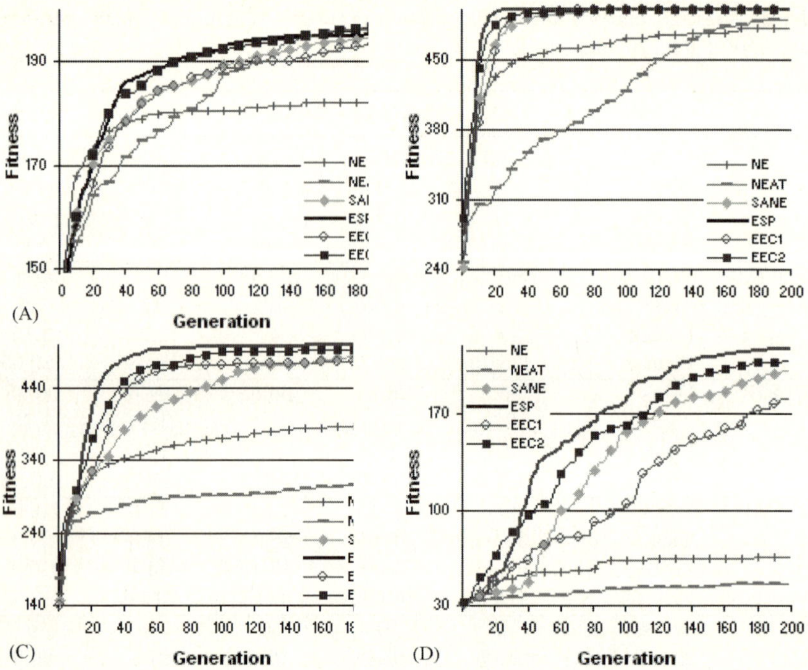

Fig. 2. Comparison of average learning speeds between EEC and some other methods. EEC1 is half evaluated EEC and EEC2 is full-evaluated EEC. (A) is the performances for the BEST strategy of TTT. (B) is the performances for the BAD strategy of Gobang. (C) is the performances for the MIDDLE strategy of Gobang. (D) is the performances for the BEST strategy of Gobang. (20 runs each).

task the performances of half-evaluated EEC were obviously better than that of standard NE and NEAT, but slightly inferior to SANE and ESP.

To facilitate a more explicit comparison, figure 2 shows the average learning curve for the BEST strategy of TTT and all the strategies of Gobang from 20 runs. In this figure we also show the performances of full-evaluated EEC (line EEC2). As a whole, ESP learned faster than all of the others. Half-evaluated EEC performed as well as SANE for the simple TTT task, the two curves of SANE and half-evaluated EEC almost overlapped each other. For the difficult Gobang task half-evaluated EEC achieved much better results than standard NE and NEAT. The performances of full-evaluated EEC were obviously superior to that of half-evaluated EEC, and even outperformed SANE.

It's not surprising that the learning speed of NEAT was slower than all of other methods, since NEAT started its search from zero hidden neurons. New neurons and new connections are added only when beneficial. For the tasks of MIDDLE and BEST strategies of Gobang, however, the evolutionary process of NEAT nearly stagnated after a few generations. Almost no hidden neurons were added

Table 4. Average complexity of best networks found by different methods (20 runs each)

	NE	NEAT	SANE	ESP	EEC
Hidden neurons	10	3.35	10	7.25	10
connections	180	61.3	180	130.5	90.3

during 200 generations. A significant advantage of NEAT could be in evolving minimal complexity of networks rather than in finding optimal solutions.

Table 4 lists the average complexity of the best networks evolved by different methods for the BEST strategy of TTT. NEAT found the most compact networks that, on average, contained 3.35 hidden neurons and 61.3 connections. ESP removed neurons that did not contribute to the solutions, and eventually, found simpler networks than the initialized ones, which, on average, contained 7.25 hidden neurons and 130.5 connections. The networks evolved by EEC, on average, contained 90.3 efficient connections. SANE evolved fully-connected networks in our experiments, so the networks of SANE contained the same number neurons and connections as those of standard NE.

EEC has achieved remarkable results for both domains, especially, for the hardest task, the BEST strategy of Gobang. Our model was able to keep the growth of learning even when NE and NEAT encountered premature convergence for the hardest problem.

4 Conclusion

Evolving connections has been poorly explored in ANNs so far. In this paper a novel cooperative coevolutionary algorithm, EEC, was presented. EEC, was developed based on standard NE model. An additional connection-paths population is evolved simultaneously in order to cooperate with connection weights to build complete networks with efficient connections. The experimental results have demonstrated that evolving connection weights along with connection paths can significantly enhance the performance of standard NE.

A fully-connected network could generate noise. As we have known, a neuron will be activated when its input signal reaches a threshold value, where that signal is the sum of weighted output signals from upstream neighbor neurons. Redundant products that come from inefficient connection could result in incorrect activation of neurons. Standard NE restricts inefficient connections by evolving their connection weights toward 0. However, a holistic search served by standard NE will be inefficient when the search space is large. Our experiments also indicate that decomposition will be helpful for complex domains. SANE, ESP and EEC are three cooperative coevolutionary algorithms, which search optimal solutions through decomposition and combination processes. They have shown highly robust performance compared to standard NE and NEAT.

Acknowledgments

The author would like to thank Keith Downing and all the members of the SOS group for their support.

References

1. Moriarty, D.E., Miikkulainen, R.: Efficient reinforcement learning through symbiotic evolution. Machine Learning 22, 11–32 (1996)
2. Richards, N., Moriarty, D., Mcquesten, P., Miikkulainen, R.: Evolving neural networks to play Go. In: Bäck, T. (ed.) Proceedings of the Seventh International Conference on Genetic Algorithms (ICGA 1997, East Lansing, pp. 768–775. Morgan Kaufmann, San Francisco (1997)
3. Stanley, K.O., Miikkulainen, R.: Evolving neural networks through augmenting topologies. Evolutionary Computation 10(2), 99–127 (2002)
4. Yao, X.: Evolving artificial neural networks. Proceedings of the IEEE 87(9), 1423–1447 (1999)
5. Moriarty, D.E., Miikkulainen, R.: Forming neural networks through efficient and adaptive co-evolution. Evolutionary Computation 5, 373–399 (1997)
6. Potter, M.A., De Jong, K.A.: Cooperative coevolution: An architecture for evolving coadapted subcomponents. Evolutionary Computation 8(1), 1–29 (2000)
7. Gomez, F., Miikkulainen, R.: Robust non-linear control through neuroevolution. Technical Report AI02-292, Department of Computer Sciences, The University of Texas at Austin (2002)
8. Garcia-Pedrajas, N., Hervas-Martinez, C., Munoz-Perez, J.: Covnet: a cooperative coevolutionary model for evolving artificial neural networks. IEEE Transactions on Neural Networks 14(3), 575–596 (2003)
9. Gomez, F., Schmidhuber, J., Miikkulainen, R.: Efficient non-linear control through neuroevolution. In: Fürnkranz, J., Scheffer, T., Spiliopoulou, M. (eds.) ECML 2006. LNCS (LNAI), vol. 4212. Springer, Heidelberg (2006)
10. Maniadakis, M., Trahanias, P.E.: Modelling robotic cognitive mechanisms by hierarchical cooperative coevolution. In: Antoniou, G., Potamias, G., Spyropoulos, C., Plexousakis, D. (eds.) SETN 2006. LNCS (LNAI), vol. 3955, pp. 224–234. Springer, Heidelberg (2006)
11. Rosin, C.D., Belew, R.K.: Methods for competitive co-evolution: Finding opponents worth beating. In: Eshelman, L. (ed.) Proceedings of the Sixth International Conference on Genetic Algorithms, pp. 373–380. Morgan Kaufmann, San Francisco (1995)
12. James, D., Tucker, P.: A comparative analysis of simplification and complexification in the evolution of neural network topologies. In: Proceedings of the Genetic and Evolutionary Computation Conference (GECCO) (2004)
13. Andres: Applying ESP and region specialists to neuro-evolution for Go. Technical Report CSTR01-24, Department of Computer Sciences, The University of Texas at Austin (2001)

The Extreme Energy Ratio Criterion for EEG Feature Extraction

Shiliang Sun

Department of Computer Science and Technology,
East China Normal University, Shanghai, 200241, China
shiliangsun@gmail.com

Abstract. Energy is an important feature for electroencephalogram (EEG) signal classification in brain computer interfaces (BCIs). It is not only physiologically rational but also empirically effective. This paper proposes extreme energy ratio (EER), a discriminative objective function to guide the process of spatially filtering EEG signals. The energy of the filtered EEG signals has the optimal discriminative capability under the EER criterion, and hence EER can as well be regarded as a feature extractor for distilling energy. The paper derives the solutions which optimize the EER criterion, shows the theoretical equivalence of EER to the existing method of common spatial patterns (CSP), and gives the computational savings EER makes in comparison with CSP. Two paradigms extending EER from binary classification to multi-class classification are also provided.

Keywords: Brain computer interface (BCI), common spatial patterns (CSP), EEG signal classification, feature extraction, optimal spatial filter.

1 Introduction

The development of recently emerging brain computer interface (BCI) technology has both theoretical and practical significance. Theoretically, a BCI provides a platform through which the biological brain (e.g., human brain) and the computer can communicate with each other in terms of information exchange. This makes it possible not only to decode the nature of life and cognition of the biological brain, but also to reconstruct the computer according to the information processing mechanism of the biological brain, although the realization of this process is definitely long-term and full of obstacles [1]. Practically, a BCI can serve as a communication and control channel for motion-disabled people, because it is a direct connection between the brain and external devices and thus independent of peripheral nerves and muscles [2,3]. Other potential applications include alarming paroxysmal diseases for neuropaths (e.g., epilepsy), and manipulating robot control in dangerous environments for healthy people.

Electroencephalogram (EEG) signal-based BCIs, our underlying concern in this paper, adopt electroencephalography as the recording technique for brain

V. Kůrková et al. (Eds.): ICANN 2008, Part II, LNCS 5164, pp. 919–928, 2008.

activities. Feature extraction of EEG signals is one of the important components for the operation of this kind of BCIs. Generally, the features used in current BCIs can be divided into three categories: spatial, temporal, and spatio-temporal. Most of these features represent some form of brain events that can be physiologically distinguished [3]. In this paper, we attempt to extract one sort of such features, namely energy features for the sake of subsequent EEG signal classification in BCIs. The specific form of energy features can be power spectral amplitudes, such as μ (8-13 Hz) and β (14-26 Hz) rhythm amplitudes, or equivalently the variances of temporally and/or spatially filtered signals. Here we propose a new criterion to spatially filter EEG signals for energy feature extraction.

Spatially filtering methods intend to transform EEG signals by combining recordings from different and often adjacent electrodes with examples such as bipolar derivation, Laplacian derivation, common average reference and common spatial patterns (CSP) [3]. The bipolar derivation calculates the first spatial derivative and thus reflects the voltage difference in some chosen direction. The Laplacian derivation derives the second spatial derivative and can be quantified by manipulating the voltages of the current site and its surrounding sites [4]. As the item itself tells, the method of common average reference converts voltages employing the mean voltage of all the recording electrodes as a reference.

The spatially filtering method of CSP led to signals which discriminate optimally between two classes of brain activities [5]. The point is to linearly project EEG signals to the most discriminative directions found after the simultaneous diagonalization of two covariance matrices respectively belonging to two different categories. Later it was extended to the multi-class scenario [6]. As reported in the literature, using the same classifiers the recognition rates obtained by the CSP method are usually as good as, or higher than, those obtained previously by other filtering methods [5]. Recently, to overcome the restriction of its linearity and make it suitable for feature extraction of time-varying signals, Sun and Zhang have proposed the nonlinear CSP and adaptive CSP methods [7,8].

The above mentioned spatial filters have been widely adopted in feature extraction of EEG signals. However, from a general review, we find that they lack an explicit discriminative objective function to help people understand the intuitive signification and conceivable efficacy of the corresponding method. To overcome this shortcoming, this paper devises a discriminative criterion for energy feature extraction of EEG signals. It also shows the computational savings of the new approach compared to the popular CSP method, and proves the theoretical equivalence relation between these two methods.

The remainder of this paper is organized as follows. In Section 2, the discriminative extreme energy ratio (EER) criterion for energy feature extraction in the case of two different classes is proposed, and the optimal solutions for extracting one or multiple sources from each class are also provided. In Section 3, after briefly summarizing the computational details of the CSP method, we show the theoretical equivalence of EER to CSP. Section 4 compares the computational complexities of CSP and EER, and gives the computational savings

EER generates. Section 5 proposes two possible paradigms for extending EER to the multi-class case. Finally, concluding remarks are made in Section 6.

2 The EER Criterion for Energy Feature Extraction

It is acknowledged that observed EEG waves are smeared brain activities generated by some inherent signal sources underneath the surface of the brain cortex [9,10]. We assume the process of spatial filtering to recover these latent sources. The energy features to be pursued can consequently be taken as the variances of these sources.

2.1 Feature Extraction of One Source

The proposed EER criterion is based on the EEG covariances of two different classes of brain activity patterns denoted by class A and class B, e.g., two different mental imagery tasks: imagination of repetitive left or right hand movements. Given some EEG samples (an EEG sample can correspond to the EEG signal of a trial or just a segment), each of which belongs to one of these two classes, below we provide the computation of covariances C_A and C_B. The covariances are applicable whether for feature extraction of one source or multiple sources.

Denote an EEG sample as an $N \times T$ matrix X^{raw}, where N is the number of recording electrodes and T is the number of total points in the recording period. Consequently, the observation from a snapshot can be represented as a vector in the N-dimensional Euclidean space, and an EEG sample is thus a distribution of T such vectors. For EEG signal analysis, bandpass filters are usually employed to remove the constant components. Therefore, we can take it for granted that the mean value of each EEG sample is zero. In addition, to eliminate the energy difference resulting from the varying recording time, a normalization is often executed in advance of further processing. The normalized EEG sample X is defined as

$$X \triangleq X^{raw}/\|X^{raw}\|_F \,, \tag{1}$$

where $\|X^{raw}\|_F = (\sum_{i=1}^N \sum_{j=1}^T |X_{ij}^{raw}|^2)^{1/2}$ is the Frobenius norm [5,8].

Without loss of generality we can ignore the multiplicative factor $\frac{1}{T}$ in the following calculation of covariance, and thus write the estimation for the covariance of one EEG sample as

$$C \triangleq XX^\top \,. \tag{2}$$

Usually there are a number of samples which belong to the same class. Thus the covariance of this specific class can be computed as the average of all these single covariances. One merit of this averaging operation is that it could enhance the accuracy and stability of the estimated covariance. In this way, the covariances C_A and C_B respectively for class A and class B are obtained.

Assume only one latent signal source from each class is to be recovered. For the EEG sample X, the spatially filtered signal with a spatial filter denoted by $\phi_{(N\times 1)}$ will be $\phi^\top X$. The signal energy after filtering can be represented by the sample variance as $\phi^\top XX^\top\phi = \phi^\top C\phi$, where similarly to (2) the multiplicative factor is also omitted.

Keeping in mind feature extraction for binary classification, we define the discriminative criterion of EER as

$$R(\phi) \triangleq \frac{\phi^\top C_A\phi}{\phi^\top C_B\phi}, \tag{3}$$

where $R(\phi)$ indicates the energy ratio after spatial filtering for two classes A and B. To make it well defined, we regularize C_A and C_B as

$$C_A = C_A + \sigma^2 I, \quad C_B = C_B + \sigma^2 I, \tag{4}$$

where I is an identity matrix, σ^2 is a positive number made small enough so that the value of $R(\phi)$ is basically not affected. So far, matrices C_A and C_B are both positive definite. For the purpose of succeeding classification, we can optimize (3) to find the filter ϕ^* which maximizes or minimizes the ratio. Therefore, about the energy feature extraction of one latent source from each class, there are in fact two optimal spatial filters ϕ^*_{max} and ϕ^*_{min} to be sought which satisfy

$$\begin{cases} \phi^*_{max} = \arg\max\limits_{\phi} R(\phi) = \arg\max\limits_{\phi} \dfrac{\phi^\top C_A\phi}{\phi^\top C_B\phi}, \\ \phi^*_{min} = \arg\min\limits_{\phi} R(\phi) = \arg\min\limits_{\phi} \dfrac{\phi^\top C_A\phi}{\phi^\top C_B\phi}. \end{cases} \tag{5}$$

The EER objective function $R(\phi)$ shown in (3) is a generalized Rayleigh quotient whose maximum denoted by $R_{max}(\phi)$ and minimum denoted by $R_{min}(\phi)$ have the following conclusions [11,12]

$$\begin{cases} R_{max}(\phi) = \lambda_{max}, & \text{if } C_A\phi = \lambda_{max} C_B\phi, \\ R_{min}(\phi) = \lambda_{min}, & \text{if } C_A\phi = \lambda_{min} C_B\phi, \end{cases} \tag{6}$$

where $\lambda_{min} = \lambda_1 \le \lambda_2 \le \ldots \le \lambda_{N-1} \le \lambda_N = \lambda_{max}$ are the eigenvalues of matrix $C_B^{-1}C_A$, that is, the generalized eigenvalues of the matrix pair (C_A, C_B).

Thus far, we have completed the process of deriving the two optimal spatial filters ϕ^*_{max} and ϕ^*_{min} in (5), which are two eigenvectors respectively corresponding to the maximal and minimal eigenvalues for the matrix pair (C_A, C_B). For a new EEG sample, its energy feature will be a vector consisting of two entries which are the energy values of the sample respectively filtered by ϕ^*_{max} and ϕ^*_{min}.

2.2 Feature Extraction of Multiple Sources

On energy feature extraction for classifying two classes of brain activities, as shown above, a pair of spatial filters can be obtained by optimizing the EER criterion if we assume one latent source from one class. Nevertheless, sometimes one source is insufficient to describe the inherent brain activities. Therefore, the demand of extracting multiple sources from one class is put forward.

The key to extending the former EER criterion to suit the case of feature extraction of multiple sources is in grasping the meanings of energy. When only one source is defined, EER criterion searches for a direction so that the variance of the EEG signal projected to this direction is maximized or minimized. For 1-dimensional signals, the variance actually reflects the energy value. For multi-dimensional signals, the following lemma indicates that the determinant is closely related to the concept of energy.

Lemma 1 ([13]). *For any matrix $U_{(n\times n)}$, denote its eigenvalues as $\{\gamma_i\}$ $(i = 1, ..., n)$. The determinant of U is equal to the product of its all eigenvalues, that is*

$$|U| = \prod_{i=1}^{n} \gamma_i \,. \tag{7}$$

Since eigenvalues of a covariance matrix can be explained as the variances on principal directions of data distribution, the equality of (7) manifests that the determinant of a covariance matrix represents the product of signal energy from all the principal directions. As a result, the concept of determinant can be used to extend the former EER criterion to feature extraction of multiple sources.

Suppose there are m sources to be extracted from one class of brain activity pattern, then EER will seek totally $2m$ sources, half of which maximize the objective function while the other half minimize the objective function. The m spatial filters for extracting m sources constitute a spatial filter bank $\Phi \triangleq [\phi_1, \phi_2, \ldots, \phi_m]$. Now the discriminative criterion of EER can be rewritten as

$$R(\Phi) \triangleq \frac{|\Phi^\top C_A \Phi|}{|\Phi^\top C_B \Phi|} \,, \tag{8}$$

where like $R(\phi)$ in (3), $R(\Phi)$ shows the energy ratio after spatial filtering for two classes A and B, and C_A and C_B are also regularized in a similar way as in (4). The optimization process is to find the filter bank Φ^* which maximizes or minimizes the ratio. On the analogy of (5), the two optimal spatial filter banks Φ^*_{max} and Φ^*_{min} should satisfy

$$\begin{cases} \Phi^*_{max} = \arg\max_{\Phi} R(\Phi) = \arg\max_{\Phi} \dfrac{|\Phi^\top C_A \Phi|}{|\Phi^\top C_B \Phi|}, \\ \Phi^*_{min} = \arg\min_{\Phi} R(\Phi) = \arg\min_{\Phi} \dfrac{|\Phi^\top C_A \Phi|}{|\Phi^\top C_B \Phi|}. \end{cases} \tag{9}$$

The process of raveling out the equalities in (9) is a bit tricky. Fukunaga has already given the details for solving problems of this type [14]. That is, Φ^*_{max} consists of m generalized eigenvectors of the matrix pair (C_A, C_B) which correspond to the m maximal eigenvalues, while Φ^*_{min} consists of m generalized eigenvectors whose eigenvalues are minimal. We can see that when $m = 1$, (9) degenerates to (5) with the same solutions, and therefore (9) is actually a superset of (5). For a new EEG sample, its energy feature will be a vector consisting of $2m$ entries which are the energy values of the sample respectively filtered by $2m$ spatial filters coming from two filter banks Φ^*_{max} and Φ^*_{min}.

3 Theoretical Equivalence to CSP

3.1 The Computational Details of CSP

To facilitate further statements, we first briefly summarize the spatially filtering method of CSP, mainly on its computational aspect. Readers can refer to [5] for a comprehensive description. The summary here is only for facilitating the latter comparison between EER and CSP.

The CSP method was first introduced in the field of EEG analysis in 1990 [15], and recently Müller-Gerking et al. developed this method and used it for single-trial EEG signal classification in movement tasks [5,16]. It decomposes the raw multi-channel signals into new spatial signals based on a kind of transformation that is obtained from the data of two populations of EEG signals in a manner that maximizes their differences. The transformation actually provides a weighting of the electrodes [5].

Computationally, the CSP method is implemented by the simultaneous diagonalization of two covariance matrices, and specifically whitening transformation and projection transformation are included. With the conventions of C_A and C_B discussed in Section 2.1, CSP defines a composite covariance matrix C_{Com} as

$$C_{Com} \triangleq C_A + C_B \, . \tag{10}$$

Being the summation of two covariance matrices, C_{Com} is positive semidefinite. And it has the following eigenvalue decomposition

$$C_{Com} = U_C \Sigma_C U_C^\top \, , \tag{11}$$

where U_C is the eigenvector matrix and Σ_C is the diagonal matrix of eigenvalues.

The whitening transformation matrix for C_{Com} can be defined as

$$W \triangleq \Sigma_C^{-\frac{1}{2}} U_C^\top \, . \tag{12}$$

In case of Σ_C being singular, the above equation can be replaced by the subspace whitening transformation. If C_A and C_B are transformed by W as

$$S_A = W C_A W^\top , \tag{13}$$
$$S_B = W C_B W^\top , \tag{14}$$

then S_A and S_B would share common eigenvectors, i.e., there exists an eigenvector matrix U and two diagonal eigenvalue matrices Σ_A and Σ_B, which satisfy

$$S_A = U \Sigma_A U^\top , \tag{15}$$
$$S_B = U \Sigma_B U^\top , \tag{16}$$
$$\Sigma_A + \Sigma_B = I \, , \tag{17}$$

where the diagonal elements of Σ_A are assumed to be sorted in descending order.

Since the sum of two corresponding eigenvalues in Σ_A and Σ_B is always equal to one as shown in (17), the eigenvector with largest eigenvalue for S_A has the smallest eigenvalue for S_B and vice versa. Consequently, the projection of the whitened EEG onto the first and last eigenvectors in U (i.e., projection transformation matrix) will give waveforms optimal for discriminating two populations of EEG in the least squares sense [16]. To sum up, the total transformation matrix of CSP is

$$P_{CSP} = W^\top U \,, \tag{18}$$

which will transform the EEG sample X to a new sample as

$$X' = P_{CSP}^\top X \,. \tag{19}$$

Based on the new sample, energy features can be extracted which are the variances of some of its row vectors.

3.2 Theoretical Equivalence

Lemma 2 ([14]). *We can diagonalize two symmetric matrices Q and Q_1 as*

$$\Phi^T Q \Phi = I \ \text{ and } \ \Phi^T Q_1 \Phi = \Lambda, \tag{20}$$

where Λ and Φ are respectively the generalized eigenvalue and eigenvector matrices of the matrix pair (Q_1, Q), i.e.,

$$Q^{-1} Q_1 \Phi = \Phi \Lambda. \tag{21}$$

According to Lemma 2, we see that to implement the simultaneous diagonalization of two symmetric matrices Q and Q_1, as in (20), we only need to decompose $Q^{-1}Q_1$ as in (21), and then normalize the eigenvector matrix Φ to meet $\Phi^T Q \Phi = I$.

In fact, the spatial filter of CSP is an instantiation of Lemma 2 if we take

$$Q = C_{Com},\ Q_1 = C_A,\ \Phi = W^\top U,\ \Lambda = \Sigma_A. \tag{22}$$

Therefore, the final transformation directions found by CSP are identical to the directions of the eigenvector matrix of $C_{Com}^{-1} C_A$.

Based on the discussions Fukunaga gave on linear discriminative analysis [14], we present the following theorem.

Theorem 1. *For two symmetric matrices Q_1 and Q_2, if defining $Q = Q_1 + Q_2$ then $Q^{-1}Q_1$ and $Q_2^{-1}Q_1$ have the same eigenvectors.*

Proof. From Lemma 2, we can have $Q^{-1}Q_1\Phi = \Phi\Lambda$. That is

$$Q_1 \Phi = (Q_1 + Q_2)\Phi\Lambda$$

and thus

$$Q_1 \Phi (I - \Lambda) = Q_2 \Phi \Lambda.$$

The last equality can be converted to

$$Q_2^{-1} Q_1 \Phi = \Phi \Lambda (I - \Lambda)^{-1},$$

which means that Φ is the eigenvector matrix of $Q_2^{-1}Q_1$ and $\Lambda(I - \Lambda)^{-1}$ is the corresponding eigenvalue matrix. Considering Lemma 2, $Q^{-1}Q_1$ and $Q_2^{-1}Q_1$ have the same eigenvector matrix Φ.

Naturally, we can let Q_2 be the covariance of the second class of EEG pattern, i.e., C_B. Thus, matrices $C_B^{-1}C_A$ and $C_{Com}^{-1}C_A$ have identical eigenvectors which are respectively the results derived by EER and CSP. The theoretical equivalence of EER and CSP is therefore proved.

4 Computational Savings

EER is a discriminative objective function which helps to show the explicit nature of energy feature extraction. Although theoretically the EER criterion is equivalent to CSP, these two methods for learning spatial filters may have different computational burdens. In this section we derive the computational complexity of EER and give a comparison with that of CSP.

CSP mainly involves two kind of operations: whitening transformation and projection transformation. In order to get the total transformation matrix P_{CSP}, given the class covariances C_A and C_B, the least amount of computation required is: 1 (*matrix addition* (10)) + 1 (*eigenvalue decomposition* (11)) + 1 (*matrix inversion* (12)) + 1 (*matrix multiplication* (12)) + 2 (*matrix multiplication* (13)) + 1 (*eigenvalue decomposition* (15)) + 1 (*matrix multiplication* (18)) = 1 (*matrix addition*) + 4 (*matrix multiplication*) + 1 (*matrix inversion*) + 2 (*eigenvalue decomposition*). The numbers in brackets indicate the equations where the corresponding operations appear.

Given a matrix $U_{(n \times n)}$, we know the fact that the computational complexity for matrix addition is $O(n^2)$, while the computational complexities for matrix multiplication, matrix inversion, and eigenvalue decomposition are all $O(n^3)$. Therefore, the computational complexity of CSP can be represented as $O(n^3)$.

To obtain the final filtering matrix or transformation matrix, EER only needs one kind of operation: the generalized eigenvalue decomposition for (C_A, C_B) after the regularization as in (4). In order to compare the computational burden with CSP, suppose we use the eigenvalue decomposition for $C_B^{-1}C_A$ to calculate the generalized eigenvalue decomposition, though this is of course not an efficient way. Given the preregularized class covariances C_A and C_B, the computation required is: 2 (*matrix addition*) + 1 (*matrix inversion*) + 1 (*matrix multiplication*) + 1 (*eigenvalue decomposition*). At the moment, we can see that the computational complexity of EER can also be represented as $O(n^3)$.

Let us carry out a more fine comparison. Compared to the computational burden of CSP, EER brings forth several savings: 3 (*matrix multiplication*) + 1 (*eigenvalue decomposition*), and one more *matrix addition*. Remember that the computational complexity of matrix addition is $O(n^2)$, therefore it is computationally neglectable to other matrix operations. Thus we can draw the conclusion that EER has lower computational cost than CSP does. The low computational cost can make the derivation of optimal spatial filters in EER more economical and feasible than in CSP, and facilitates the utilization of the EER criterion for EEG signal feature extraction. For those applications requiring fast responses, e.g., on-line analysis, low computational cost would be more desirable.

5 EER for Multi-class EEG Signal Classification

Although the EER criterion is initially presented for feature extraction in binary EEG signal classification, it can, like the CSP method, be easily extended to the case of feature extraction in multi-class EEG signal classification. Using the idea of extending the CSP method to the multi-class case for reference [5,6], we give two possible paradigms of extending the EER criterion. For each paradigm, the number of total spatial filters to be learned, which is equal to the number of features for describing an EEG sample, is also listed.

One class versus another class. The multi-class problem, e.g., L classes, can be divided into C_L^2 groups where each group corresponds to a binary classification problem. For each binary classification problem, the EER criterion can apply naturally and get $2m$ spatial filters to recover m sources from each class. Therefore, totally $2m \times C_L^2$ spatial filters are obtained for extracting features of a new EEG sample. The resultant feature vector for each EEG sample would have $2m \times C_L^2$ entries.

One class versus the remaining classes. This paradigm divides the L classes into two parts, with one part containing only one class and the other part containing the remaining $L-1$ classes. L such pairs would be generated where each pair is treated as a binary classification problem. The EER criterion can apply and get $2m$ spatial filters for each pair. As a result, totally $2m \times L$ spatial filters are obtained for feature extraction of a new EEG sample. The feature vector for each EEG sample would have $2m \times L$ entries.

6 Conclusion

In this paper, a discriminative and intuitive criterion EER for learning optimal spatial filters with the intent of energy feature extraction for subsequent EEG signal classification is proposed. In binary classification, we give the optimal spatial filters derived from the criterion, whether for extracting one source or multiple source from one class. Besides proving the theoretical equivalence to the CSP method, we also show its computational superiority. The computational savings are very important for the application of BCIs, especially for those scenarios requiring fast responses, such as on-line analysis. Further to extend the feature extractor to cope with the case of multi-class classification, we also give two possible paradigms based on the existing extensions for CSP.

Since the effectiveness of CSP has already be confirmed, e.g., empirical evaluations in the literature [5,6,16], and EER is equivalent to CSP in principle, it is redundant for the current paper to further conduct experiments to validate the effectiveness of EER. Hence, this paper is mainly about the theoretical aspects of the related methods. Future research can consider generalizing the EER criterion to cope with the case of nonlinear feature extraction by means of the kernel trick [17], and comparing the performance of different kernel functions.

Acknowledgments. This work was supported in part by the National Natural Science Foundation of China under Project 60703005, and in part by Shanghai Educational Development Foundation under Project 2007CG30.

References

1. Sun, S.: Research on EEG Signal Classification for Brain-Computer Interfaces Based on Machine Learning Methodologies. Ph.D. Thesis, Tsinghua University, Beijing (2006)
2. Nicolelis, M.A.L.: Actions from thoughts. Nature 409, 403–407 (2001)
3. Wolpaw, J.R., Birbaumer, N., McFarland, D.J., Pfurtscheller, G., Vaughan, T.M.: Brain-computer interfaces for communication and control. Clin. Neurophysiol. 113, 767–791 (2002)
4. Nunez, P.L., Srinivasan, R., Westdorp, A.F., Wijesinghe, R.S., Tucker, D.M., Silberstein, R.B., Cadusch, P.J.: EEG coherency I: Statistics, reference electrode, volume conduction, laplacians, cortical imaging, and interpretation at multiple scale. Electroenceph. Clin. Neurophysiol. 103, 499–515 (1997)
5. Müller-Gerking, J., Pfurtscheller, G., Flyvbjerg, H.: Designing optimal spatial filters for single-trial EEG classification in a movement task. Clin. Neurophysiol. 110, 787–798 (1999)
6. Dornhege, G., Blankertz, B., Curio, G., Müller, K.R.: Boosing bit rates in noninvasive EEG single-trial classifications by feature combination and multiclass pardigms. IEEE Trans. Biomed. Eng. 51, 993–1002 (2004)
7. Sun, S., Zhang, C.: Adaptive feature extraction for EEG signal classification. Med. Biol. Eng. Comput. 44, 931–935 (2006)
8. Sun, S., Zhang, C.: An optimal kernel feature extractor and its application to EEG signal classification. Neurocomputing 69, 1743–1748 (2006)
9. Curran, E.A., Stokes, M.J.: Learning to control brain activity: A review of the production and control of EEG components for driving brain-computer interface (BCI) systems. Brain Cogn. 51, 326–336 (2003)
10. Kamousi, B., Liu, Z., He, B.: Classification of motor imagery tasks for brain-computer interface applications by means of two equivalent dipoles analysis. IEEE Trans. Neural Syst. Rehabil. Eng. 13, 166–171 (2005)
11. Golub, G.H., Van Loan, C.F.: Matrix Computation, 2nd edn. The John Hopkins University Press, Baltimore (1989)
12. Zhang, X.: Matrix Analysis and Applications. Tsinghua University Press, Beijing (2004)
13. Searle, S.R.: Matrix Algebra Useful for Statistics. John Wiley & Sons, New York (1982)
14. Fukunaga, K.: Introduction to Statistical Pattern Recognition, 2nd edn. Academic Press, San Diego (1990)
15. Koles, Z.J., Lazar, M.S., Zhou, S.Z.: Spatial patterns underlying population differences in the background EEG. Brain Topogr. 2, 275–284 (1990)
16. Ramoser, H., Müller-Gerking, J., Pfurtscheller, G.: Optimal spatial filtering of single trial EEG during imagined hand movement. IEEE Trans. Rehabil. Eng. 8, 441–446 (2000)
17. Müller, K.R., Mika, S., Rätsch, G., Tsuda, K., Schölkopf, B.: An introduction to kernel-based learning algorithms. IEEE Trans. Neural Netw. 12, 181–201 (2001)

The Schizophrenic Brain: A Broken Hermeneutic Circle

Péter Érdi[1], Vaibhav Diwadkar[2], and Balázs Ujfalussy[3,*]

[1] Center for Complex Systems Studies, Kalamazoo College, Kalamazoo, MI 49006, U.S.A. and Department of Biophysics, Research Institute for Particle and Nuclear Physics of the Hungarian Academy of Sciences
perdi@kzoo.edu

[2] Dept. Psychiatry and Behavioral Neurosciences, Wayne State Univ. School of Medicine, Detroit, MI, USA
vdiwadka@med.wayne.edu

[3] Department of Biophysics, Research Institute for Particle and Nuclear Physics of the Hungarian Academy of Sciences
ubi@rmki.kfki.hu

Abstract. A unifying picture to the hermeneutical approach to schizophrenia is given by combining the philosophical and the experimental/computational approaches.

1 Introduction

Hermeneutics is a branch of continental philosophy which treats the understanding and interpretation of texts. For an introduction for non-philosophers please see [1]. One of the most important concepts in hermeneutics is the hermeneutic circle. This notion means that the definition or understanding of something employs attributes which already presuppose a definition or understanding of that thing. The method is in strong opposition to the classical methods of science, which do not allow such circular explanations.

Motivated by Ichiro Tsuda [2,3] who applied the principles of hermeneutics to the brain by using chaos as a mechanism of interpretation, one of us (PE) played with the idea [4] of how, if at all, two extreme approaches, the "device approach" and the "philosophical approach" could be reconciled. It was cautiously suggested by turning to the philosophical tradition that hermeneutics, i.e., the "art of interpretation", which is neither monist nor dualist a priori, can be applied to the brain. Further, we stated that the brain is both the "object" of interpretation as well as the interpreter: therefore the brain is itself a hermeneutic device. For our dialog with Tsuda see [5].

Recently new initiatives for applying hermeneutics in the context of neuroscience and cognitive science have emerged. Chris Firth [6] uses neural hermeneutics as the neural basis of social interaction, and explains psychiatric disorders, such as schizophrenia, as failure in the ability to interpret (represent and model) the world. Shaun

* Thanks to Brad Flaugher and Trevor Jones for participating in the model building and testing. PE thanks the Henry R. Luce Foundation the general support.

V. Kurková et al. (Eds.): ICANN 2008, Part II, LNCS 5164, pp. 929–938, 2008.

Gallagher's analysis points out that hermeneutics and cognitive science have an overlapping interest [7].

Independently from our interest in hermeneutics we have started to work on combined behavioral, brain imaging and computational approaches to associative learning in healthy and schizophrenia patients to explain their normal and reduced performance. The working hypothesis we adopt is that schizophrenia is a "disconnection syndrome", as was suggested among others by Friston and Frith [8] and our aim is to qualitatively and quantitatively understand the functional bases of these disconnections.

Rethinking these studies from the perspective of the hermeneutic approach together with the preliminary results of our combined experimental and computational studies [9,10] leads us to believe that the hermeneutic circle necessary to associative learning is broken for schizophrenic patients, and that therapeutic strategies should act to repair this circle.

In this paper we provide a unifying picture by combining the philosophical and the experimental/computational approaches. First, we briefly review both our own old perspective and newer developments on neural hermeneutics (Sec. 2, Sec. 3). Then we provide a sketch of the associative learning paradigm (Sec. 4), specific neural implementations (Sec. 5), and the results of simple and a more complex computational models (Sec. 6). Simulation results suggest that reduced performance of schizophrenia patients may be due to reduced cognitive capacity and learning rate related to impairment of functional connectivities between brain regions.

2 The Brain as a Hermeneutic Device

The brain can be considered as different types of device. Among these: the brain can be seen as a thermodynamic device, a control device, a computational device, an information storing, processing and creating device, or a self-organizing device.

The device approach is strongly related to the *dynamic metaphor* of the brain [11]. Dynamic systems theory offers a conceptual and mathematical framework to analyze spatiotemporal neural phenomena occurring at different levels of organization. These include oscillatory and chaotic activity both in single neurons and in (often synchronized) neural networks, the self-organizing development and plasticity of ordered neural structures, and learning and memory phenomena associated with synaptic modification. Systems exhibiting high structural and dynamic complexity may be candidates of being thought of as hermeneutic devices. The human brain, which is structurally and dynamically complex thus qualifies as a hermeneutic device. One of the characteristic features of a hermeneutic device is that its operation is determined by circular causality. Circular causality was analyzed to establish self-organized neural patterns related to intentional behavior [12].

The world of systems determined by linear (and only linear) causal relationships belongs to the class of "simple systems" or mechanisms. The alternative is not a "subjective" world, immune to science, but a world of complex systems, i.e., one which contains closed causal loops.

Systems with feedback connections and connected loops can be understood based on the concepts of circular and network causality. Leaving aside the clear and well-organized world of linear causal domains characterizing "simple systems" we find ourselves in the jungle of the complex systems [13].

As we know from engineering control theory, large systems consist of both controller and controlled units. The controller discharges control signals towards the controlled system. The output of the controlled system is often sent back to the controller ("feedback control") forming a closed loop. Negative feedback control mechanisms serve to reduce the difference between the actual and the desired behavior of the system. In many cases, specific neural circuits implement feedback control loops which regulate specific functions.

Analyzing the question of whether the technical or "device approach" to the brain and the "philosophical approach" can be reconciled, it was concluded that the brain is a physical structure which is controlled and also controls, learns and teaches, process and creates information, recognizes and generates patterns, organizes its environment and is organized by it. It is an "object" of interpretation, but also it is itself an interpreter. The brain not only perceives but also creates new reality: it as a hermeneutic device [4].

3 Hermeneutics, Cognitive Science, Schizophrenia

Frith's research group is working on establishing a scientific discipline they call neural hermeneutics dealing with the neural basis of social interaction. The key elements of their approach is the assumption that there representations of the external world can be shared with others, and this share representation may be the basis of predicting others actions during interactions. They use combined behavioral and brain imaging studies to uncover both the normal neural mechanisms, and pathological ones leading to schizophrenia.

Gallagher's analysis implies: (i) Hermeneutics and cognitive science is in agreement on a number of things. An example is the way we know objects. The interpretation of objects needs "schema theory" (a modern version is given by Michael Arbib [14]); (ii) Hermeneutics can contribute to cognitive science. The basis of the argument is that understanding situations needs hermeneutic interpretation. The usual critique is that logic, rule-based algorithms, and other similar computational methods are too rigid to interpret ill-defined situations, but hermeneutics "the art of interpretation" can do it. ("Mental models", which also helps to analyze situations also should have mentioned. Mental models have played a fundamental role in thinking and reasoning, and were proposed in a revolutionary suggestion by Kenneth Craik (1914 - 1945) [15]. The idea that people rely on mental models can be traced back to Craik's suggestion that the mind constructs "small-scale models" of reality that it uses to predict events.) (iii) Cognitive science also has something to offer to hermeneutics, particularly to understand other minds. The most popular notion today is the *theory of mind* or more precisely "theory of other's minds". The most effective method of cognitive science to understand other minds, i.e. to show empathy is to simulate other minds by using analogical thinking [16]. The neural basis of theory of mind now seems to be related to mirror neurons, which is the key structure of imitation, and possibly language evolution [17]. A failure of attributing self-generated action generated by the patient himself (what we may label as the lack of ability to close the hermeneutic circle) can be characteristic for schizophrenic patients [18].

4 Associative Learning

Our own studies investigate the associative learning performance of healthy control and schizophrenic patients. The learning procedure does not happen during ”one shot”, but over a series of encoding/consolidation and retrieval epochs: learning has an iterative characteristic. Associative learning relies on the consolidation and retrieval of associations between diverse memoranda, sensory inputs and streams of neural activity, particularly by hippocampal and medial temporal lobe neurons. During the task subjects alternate between blocks of consolidation, rest/rehearsal and retrieval. During consolidation, nine equi-similiar objects with monosyllabic object names are presented in sequential random order (3s/object) in grid locations for naming (e.g. “bed” and “book”). Following a brief rest/rehearsal interval, memory for object-location associations is tested by cuing grid locations for retrieving objects associated with them (3s/cue). Object names are monosyllabic to minimize head motion. Eight blocks (each cycling between consolidation, rest and retrieval) are employed. Learning dynamics in controls and schizophrenia patients are shown on Figure 1.

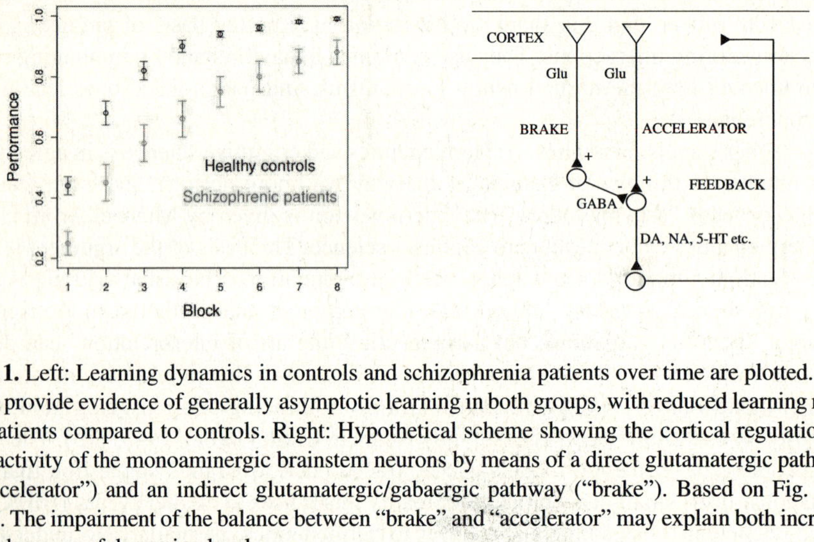

Fig. 1. Left: Learning dynamics in controls and schizophrenia patients over time are plotted. The data provide evidence of generally asymptotic learning in both groups, with reduced learning rates in patients compared to controls. Right: Hypothetical scheme showing the cortical regulation of the activity of the monoaminergic brainstem neurons by means of a direct glutamatergic pathway (“accelerator”) and an indirect glutamatergic/gabaergic pathway (“brake”). Based on Fig. 1 of [21]. The impairment of the balance between “brake” and “accelerator” may explain both increase and decrease of dopamine level.

5 Specific Implementation of Neural Circular Causality

There are several neural implementations of circular causality for our system under studies. Three levels of connections - an anatomical, a functional and a neurochemical - will be mentioned.

5.1 Cortico-Hippocampal Loop

It is generally agreed that the hippocampal formation has a crucial role in learning and memory processes. The hippocampus is reciprocally connected to many neural centers and is thought to prepare information for long term storage. The cortico-hippocampal-cortex loop might be considered as the structural basis of a circular causal chain, where information can be stored, circulated, recalled and even created. Such kinds of loops have *control* functions.

5.2 The Functional Macro-network for Associative Memory

Based on the available data on the activity of five interconnected regions (superior parietal cortex, inferio-temporal cortex, prefrontal cortex, primary visual cortex and the hippocampus) are supposed to form the functional macro-network (Fig. 2). In accordance with the spirit of the "disconnection syndrome" a question to be answered is which connections are impaired during schizophrenia, and what is the measure of functional reduction of the information flow?

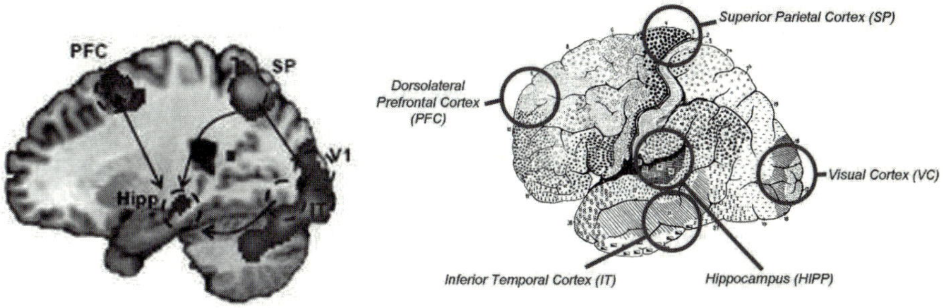

Fig. 2. Left: Information flow during object-location associative memory (based on [19].). The cortical pathway is overlaid on a medial slice depicting brain activity ($p<.001$) during memory consolidation during object-location associative learning (see Preliminary data for further details). Regions labeled are: V1: Primary Visual Cortex; IT: Inferior Temporal Cortex; SP: Superior Parietal Cortex; Hipp: Hippocampus; PFC: Dorso-lateral prefrontal cortex. Right: An enlarged view of the brain areas involved.

5.3 The Glutamate – Dopamine Interplay

There is another type of loop (characterizing the neurochemical machinery), failures of which may be related to schizophrenia. The dopamine hypothesis, which has been the predominant hypothesis, postulates that symptoms of schizophrenia may result from failure of the dopaminergic control system. Both increases (mostly in striatum), and decreases (mostly in prefrontal regions) in dopaminergic levels have been found. Glutamatergic mechanisms also seem to have a major role. Drugs by blocking neurotransmission at NMDA-type glutamate receptors cause symptoms similar to those of schizophrenia. The understanding of dopamine-glutamate interaction may lead to new therapeutic strategies [20]. A simple control loop for glutamatergic regulation of dopamine release [21] is illustratcd on Fig. 1.

6 Computational Modeling

6.1 A Simple Learning Model

Our initial modeling efforts were directed toward a simple model to simulate behavioral performance on the associative learning task, with model output as learning curves depicting performance over each iteration of recall [9].The model incorporates the separation between encoding/consolidation and cued recall while also retaining biologically plausible relationships between model architecture and neural systems, as well as known learning parameters in the brain. In particular, the model accounts for (i) the separation between "where" and "what" regions (ii) reduced synaptic plasticity in schizophrenia and reduced cognitive capacity in schizophrenia.

Separate neural systems are represented by two separate nine (nine objects and nine locations) dimensional binary vector inputs supplied to the model representing the object shown to the subject and the location of the object named a_L and a_O respectively. Nine unique object-location vector pairs for each trial represent the nine unique object-location relationships in the task. Background neural activity is simulated by the addition of a normally distributed noise term (mean= .5) to each element. Each vector pair is dyadically multiplied to form a single object-location association matrix A, which has elements that fall into three categories. The element that results from multiplying the active signal of a_L and a_O contains the strongest association. Every other element in the row and column that it occupies is the product of active signal and a noise term. The remaining elements are the product of only noise terms and contain no meaningful signal.

The learning rule adopted was motivated by the Rescorla-Wagner rule [22]. Each A is added to form $W(t)$ with nine elements that hold correct multiplicative associations with noise, and 72 elements that hold incorrect additive associations. $W(t)$ is multiplied by a learning rate $r(t)$ modulating the strength of associations on a trial-wise basis to form $W(t+1)$ as in 1. Synaptic plasticity can be represented by the rate parameter r_max, modulating the learning rate $r(t)$:

$$r(t) = r_{max}(S_{max} - S(t-1)) \tag{1}$$

Cognitive capacity is indexed by S_{max} and $S(t-1)$ is performance on the previous iteration of the task. During encoding, the learning rate $r(t)$ functions as a supervisory parameter by modulating the strength of the encoding matrix W at any instant during learning, depending on its own parameterization. Crucially, at any given time "t" during learning the association matrix $W(t)$ represents the strengths between associations to be learned during the task. During recall, the model is given a noiseless input a_L which represents the location cue. a_L is multiplied with the encoding matrix $W(t+1)$ to select the column of $W(t)$ that contains the information of which object is associated with the chosen a_L. This recalled column is a vector y:

$$y = a_L W(t) \tag{2}$$

Each element in y is tested by a threshold function (τ to determine if it is an active recall or a noise induced value:

$$\tau := 1/81 \sum_{i=1}^{9} \sum_{j=1}^{9} W_{ij}(t+1) + \pi[maxW(t+1))], \tag{3}$$

where i is a column, and j is a row of $W(t)$ and π is a chosen multiplier between 0 and 1. The threshold is determined by averaging the elements of $W(t)$ and adding the largest element in $W(t+1)$ multiplied by π. This premium controls the sensitivity of the model to noise. As seen in Figure 3, model performance for "controls" ($S_max = 1$ and $r_max = 0.4$) and "patient" behavior ($S_max = 0.7$ and $r_max = 0.2$) provide reasonable simulations of control and patient data (see Figure 1).

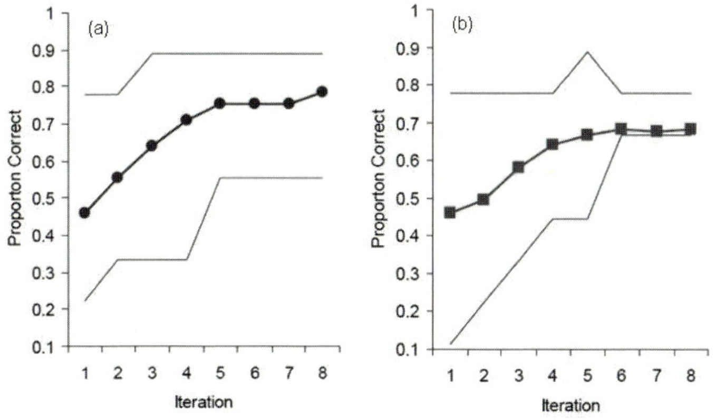

Fig. 3. (a) Maximum, minimum, and average values for healthy control patients. Produced with parameter values $S_max = 1$ and $r_max = 0.4$. Maximum and minimum curves are the basis for finding the parameter ranges. (b) Maximum, minimum, and average values for "schizophrenia" patients. Produced with parameter values $S_max = 0.7$ and $r_max = 0.2$.

These results indicate how limitations/changes in synaptic plasticity and memory capacity can predict control or schizophrenia-like behavior during learning and memory dynamics.

6.2 Models of Interconnected Brain Regions

We also developed a neural model incorporating the brain regions involved in paired-associate learning in order to analyze the mechanism underlying behavioural differences between schizophrenic and control subjects. The model has two parts: A simple visual system, and a more detailed model of the hippocampal formation.

We did not intend to model the visual signal processing system in its details, because these mainly sensory areas are presumably not affected by the illness. However, we implemented a feed-forward network to analyse the retinal image, and to create the representation of the object (the model of the area IT in the ventral stream), and its location (as the area SP in the dorsal stream). The proposed role of the hippocampus is to bind these two representation together [23] so that when cued by the location, the correct object can be recalled. The highly processed sensory input enters the hippocampus through the mossy fiber pathway, originating in the entorhinal cortex (which is not explicitly modeled here). The EC itself has reciprocal connections with both the hippocampus and various neocortical, including visual areas and considered as a relay for information coming from multimodal association areas. Mossy fibers terminate on the dentate granule cells and hippocampal pyramidal neurons. Two regions of the hippocampal formation were modeled: the dentate gyrus and the CA3 region. We used firing rate models, where the activation of each unit was calculated by the linear sum of its input. Synaptic connections in the IT, DG and CA3 was modified by simple Hebbian plasticity [24]. Parameters r_{DG}, r_{CA3} and r_{IT} govern the amount of learning during a single epizode in the corresponding areas. We used large number of neurons in the simulations (typically 500 in one layer) in order to be able to implement distributed encoding in a realistic range of sparsity (0.1 in the hippocampus). Our hippocampal model was built according to the following key assumptions [25]:

- The DG performs pattern separation by competitive learning: it removes redundancies from the input and produce a sparse representation for learning in the CA3 region. This process can be considered as a translation from the neocortical to the hippocampal language.
- The granule cells in the DG innervates CA3 pyramidal cells with particularly large and efficient synapses (the mossy fiber pathway) that makes postsynaptic neurons fire. Hebbian plasticity between active CA3 neurons and the perforant path axons associates the activity pattern in the CA3 to its incoming input (hetero-associative plasticity). After the encoding, the same CA3 assembly can be activated by the presentation of the partial or noisy version of the original input (e.g., only the object or the location).
- Next, connections between CA3 cells and IT cells are modified, to translate back the hippocampal to the neocortical code.
- Finally, objects are stored in a long term memory system in the inferio-temporal cortex forming an attractor network. During recall, the activity of this subsystem converges to one of the previously stored items (objects). The parameter $n_{patterns}$ is the number of objects stored in the IT.

The performance of the hippocampal model on the associative learning task is shown on Figure 4. We note, that this is not the ideal performance of the model: The capacity of the system with 500 units and 0.1 sparsity is around a few hundreds of associations. However, with random initial synaptic weights and small learning rate, it requires some repetitions to learn new associations appropriately. The other bottle-neck of the system's learning ability is the attractor network in the IT. If the attraction basin is smaller, than the recall cue should be more precise. If more objects are stored, the basin of attraction becomes shallower. Our results with the hippocampal model indicate, that the poorer performance of schizophrenic patients on the associative memory task is mainly due to the shallower attractor basin and not necessary to a lower learning rate in the hippocampus.

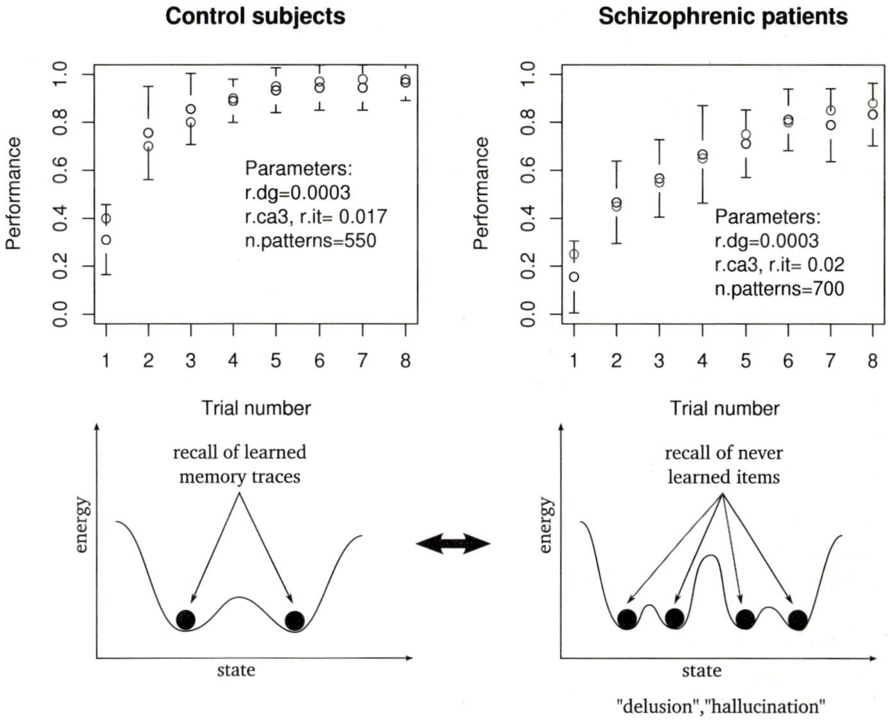

Fig. 4. Upper row: Comparison of the performance of the model with the experimental data. Red and green circles: the same experimental data as on Fig. 1. Black: performance of the model. Error bars show the standard deviation. Learning rate and the number of pre-learned objects are higher in schizophrenic subjects. Lower row: Illustration shows that learning more objects swallows the basin of attraction of each individual items, and results in the recall of not learned items.

7 Concluding Remarks

There are converging intellectual efforts, accumulating data, models and arguments, which connects neural hermeneutics to potential explanation of schizophrenia. Embedded neural control loops implement hermeneutic circles, and breaking of the circle implies schizophrenic symptoms. Combined pharmacological psychotherapeutic strategies should act to repair the circle.

References

1. Mallery, J.C., Hurwitz, R., Duffy, G.: Hermeneutics: From Textual Explication to Computer Understanding? In: Shapiro, S.C. (ed.) The Encyclopedia of Artificial Intelligence, John Wiley &Sons, New York (1987)
2. Tsuda, I.: A hermeneutic process of the brain. Prog. Theor. Phys. Suppl. 79, 241–259 (1984)
3. Tsuda, I.: Chaotic Itineracy as a Dynamical Basis of Hermeneutics in Brain and Mind. World Futurcs 32, 167–184 (1991)

4. Érdi, P.: The brain as a hermeneutic device. BioSystems 38, 179–189 (1996)
5. Érdi, P., Tsuda, I.: Hermeneutic approach to the brain: process versus device? Theoria et Historia Scientiarum VI, 307–321 (2002)
6. Frith, C.: Making Up the Mind: How the Brain Creates Our Mental World. Blackwell Publ., Malden (2007)
7. Gallagher, S.: Hermeneutics and the Cognitive Sciences. J. of Consciousness Studies 174, 10–11 (2004)
8. Friston, K.J., Frith, C.D.: Schizophrenia: A disconnection syndrome? Clin. Neurosci. 3, 88–97 (1995)
9. Diwadkar, V., Flaugher, B., Jones, T., Zalányi, L., Keshavan, M.S., Érdi, P.: Impaired Associative Learning in Schizophrenia: Behavioral and Computational Studies (under revision)
10. Érdi, P., Flaugher, B., Jones, T., Ujfalussy, B., Zalányi, L., Diwadkar, V.: Computational approach to schizophrenia: disconnection syndrome and dynamical pharmacology. In: Ricciardi, L. (ed.) BIOCOMP2007 Amer. Inst. Phys. Conf. Proceedings (in press, 2007)
11. Érdi, P.: On the 'Dynamic Brain' Metaphor. Brain and Mind 1, 119–145 (2000)
12. Freeman, W.J.: Consciousness, Intentionality, and Causality. Journal of Consciousness Studies 6, 143–172 (1999)
13. Érdi, P.: Complexity Explained. Springer, Heidelberg (2007)
14. Arbib, M.A., Érdi, P., Szentágothai, J.: Neural Organization: Structure, Function, Dynamics. MIT Press, Cambridge (1997)
15. Craik, K.: The Nature of Explanation. Cambridge Univ. Press, Cambridge (1943)
16. Barnes, A., Thagard, P.: Empathy and analogy. Dialogue: Canadian Philosophical Review 36, 705–720 (1997)
17. Arbib, M.A.: The mirror system, imitation, and the evolution of language. In: Nehaniv, C., Dautenhahn, K. (eds.) Imitation in Animals and Artefacts, pp. 229–280. MIT Press, Cambridge (2002)
18. Arbib, M.A., Mundhenk, T.N.: Schizophrenia and the Mirror System: An Essay. Neuropsychologia 43, 268–280 (2005)
19. Lavenex, P., Amaral, D.G.: Hippocampal-neocortical interaction: A hierarchy of associativity. Hippocampus 10, 420–430 (2000)
20. Javitt, D.C.: Glutamate and Schizophrenia: Phencyclidine, N-Methyl-d-Aspartate Receptors, and Dopamine-Glutamate Interactions. Int. Rev. Neurobiol. 78, 69–108 (2007)
21. Carlsson, A., Waters, N., Waters, S., Carlsson, M.L.: Network interactions in schizophrenia - therapeutic implications. Brain Research Reviews 31, 342–349 (2000)
22. Rescorla, R.A., Wagner, A.R.: A theory of Pavlovian conditioning: Variations in the effectiveness of reinforcement and nonreinforcement. In: Black, A., Prokasy, W. (eds.) Classical Conditioning II: Current Research and Theory, New York, Appleton Century Crofts, pp. 64–99 (1972)
23. Milner, B., Johnsrude, I., Crane, J.: Right medial temporal-lobe contribution to object-location memory. Philos. Trans. R. Soc. Lond. B. Biol Sci. 352(1360), 1469–1474 (1997)
24. Gerstner, W., Kistler, W.M.: Spiking Neuron Models. Cambridge University Press, Cambridge (2002)
25. Rolls, E.T., Treves, A.: Neural Networks and Brain Function. Oxford University Press, Oxford (1998)

Neural Model for the Visual Recognition of Goal-Directed Movements

Falk Fleischer[1], Antonino Casile[1], and Martin A. Giese[1,2]

[1] Dept. of Cognitive Neurology, Hertie Institute for Clinical Brain Research, Tübingen, Germany
[2] School of Psychology, University of Wales, Bangor, UK

Abstract. Mirror neurons in monkey premotor cortex of are active during the motor planning and the visual observation of actions. These neurons have recently received a vast amount of interest in cognitive neuroscience and have been discussed as potential basis of imitation learning and the understanding of actions. We present a model that explains visual properties of mirror neurons without a reconstruction of the three-dimensional structure of action and object. The proposed model is based on a small number of physiologically well-established principles. In addition, it postulates novel neural mechanisms for the integration of information about object and effector movement, which can be tested in electrophysiological experiments.

1 Introduction

Mirror neurons have been found in the premotor and the parietal cortex of monkeys. These neurons respond when the animal prepares motor actions, but also when it perceives motor actions executed by other monkeys or humans [1]. Mirror neurons have received a vast amount of interest in cognitive neuroscience, and have been discussed as physiological substrate for the imitation learning of actions and action understanding [2,3]. Beyond the fact that they are active during motor planning, mirror neurons have a number of very interesting visual tuning properties. They are selective for subtle differences between actions, like power vs. precision grip. At the same time, they are highly invariant against the position of the action in the visual field, and partially also against the view of the action. However, their response is critically dependent on the correct spatial arrangement of the effector and the goal object, and they are activated often only by functionally effective actions. In addition, mirror neurons typically fail to respond to mimicked actions without goal object [4,5].

The aims of this paper are twofold: First, we try to develop a model for the visual tuning properties of mirror neurons that is physiologically plausible and which, at a later stage, can be compared to electrophysiological data. For this purpose, we require the model to work real video sequences making it possible to compare model behavior and physiological recordings for the same stimuli. Contrasting with many existing models, we focus on the visual processing and model in detail the pathway from primary visual cortex up to the level of mirror

V. Kůrková et al. (Eds.): ICANN 2008, Part II, LNCS 5164, pp. 939–948, 2008.

neurons. Second, we try to devise a model that explains visual tuning properties of mirror neurons based on an integration of information that can be extracted from two-dimensional views of objects [6,7,8] without the need for 3D reconstructions of effector and object geometry.

Most existing approaches have focused on motor properties of mirror neurons and possible interactions with visual processing (e.g. [9,10,11,12]). Many existing models make simplifiying assumptions about the visual processing up to the parietal areas and STS, assuming for example the existence of a three-dimensional representation of effector and object structure (e.g. [13,14,15,16,17]). Other existing models exclude explicitly a detailed modeling of the visual processing (e.g. [18,19]). While only very few models work on real images at all [20,14,16], none of them exploits physiologically plausible circuits for modeling of the visual pathway. However, such physiological plausibility is a core requirement for the comparison with single cell data on visual tuning properties.

In the following, we first present the model and its components (Section 2). We then show some example simulations that reproduce typical properties of mirror neurons (Section 3). Finally, we discuss implications and further extensions of this work (Section 4).

2 Architecture for Recognizing Goal-Directed Movements

The developed model combines various principles that have been successfully applied before for the modeling of object recognition [21,22,23,24], movement recognition [25], and in the context of coordinate transformations [26].

The architecture consists of three main components: (1) A hierarchical neural model for the recognition of goal objects and effector (hand) shapes from video frames, where the central levels are optimized by feature learning, (2) a simple recurrent neural circuit for the realization of temporal sequence selectivity of effector movements, and (3) a physiologically plausible mechanism that combines the spatial information about the goal object and the posture, position and orientation of the effector. The highest level of this mechanism is formed by the model 'mirror neurons'. An overview of the model architecture is shown in Figure 1.

2.1 Neural Hierarchy for Recognizing Objects and Effector Shapes

The first component of our model is a hierarchical neural model for the recognition of the shapes of the goal object and the effector (hand). Each video frame is analyzed by a hierarchy of neural feature detectors with increasing complexity as well as increased receptive-field size and invariance to position along the hierarchy. Very similar models have been proposed to account for a variety of experimental results in object recognition [24,22] and motion recognition [25]. The individual steps are further lined out below and we refer to [24] for further details.

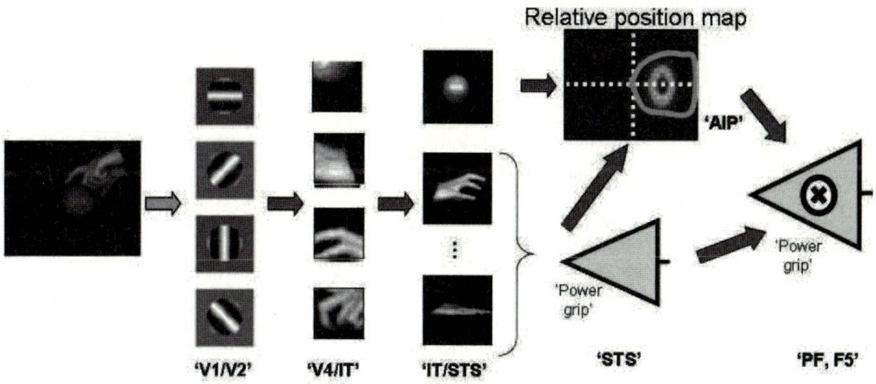

Fig. 1. Overview of the model

Local orientation filters (areas V1/V2). We extract local orientations by applying Gabor filters with four orientations and five different scales (receptive fields sizes ranging from 11x11 to 31x31 pixels). The corresponding filter parameters were taken from [24]. Signifying by $G_k(u,v)$ a normalized Gabor filter with zero mean, the response x_k of a Gabor filter G_k to a patch of pixels $P(u,v)$ from the input image is given by:

$$x_k = \left| \frac{< P, G_k >}{\beta + \sqrt{< P, P >}} \right| \tag{1}$$

In this expression the scalar product is defined as $< f, g >= \sum_{u,v} f(u,v) g^*(u,v)$, and the positive constant β avoids division by zero.

On the second level of the hierarchy we modeled 'complex cells' with partial position invariance by pooling of the normalized responses of filters for the same orientation and scale, within a spatial neighborhood of 3 x 3 filters were pooled using a maximum operation [22].

Detectors for intermediate form features (area V4/IT). The outputs from the previous layer, pooled over all scales, were used to construct detectors for more complex intermediate level features. These detectors were modeled by radial basis functions of the form $y_m = \exp(\|Y - U_m\|_F^2)/(2\sigma^2)$, where the matrix Y signifies the responses of the detectors on the previous layer within a local patch ($\sigma = 1$). The RBF centers U_m are defined by a template patterns that are established by learning (see below). As on the previous layer, the responses within a spatial neighborhood of 3 x 3 filters are pooled using maximum computation. This defines a hierarchy layer with model neurons that detect optimized mid-level features.

Detectors for complete object forms and hand shapes (area IT/STS). The next hierarchy layer is formed by feature detectors whose receptive field sizes encompass whole objects and effector configurations. These feature detectors are implemented in exactly the same way as the layers before. They respond selectively to views of objects and hands, being sensitive to configuration, orientation and size.

Again detectors with the same selectivity were realized for different spatial positions and their responses were pooled with a maximum operation within a spatial neighborhood to achieve position invariance. However, contrasting with many object recognition models, the neurons on the highest hierarchy level of our recognition hierarchy have still a coarse tuning for the positions of the object and the effector. This is necessary for extracting the relative positions of effector and object and is consistent with neurophysiological data [27,6].

Learning of mid-level features. The goal of the learning stage is to extract a set of templates U_m that respond selectively to the goal-object and effector shapes during the grasping movement. As training data set, we used images that contain only the relevant object or effector view. To balance for the different rates of change of the effector shape along the grasping movement, we clustered the different hand shapes using a k-means algorithm in order to obtain 10 representative hand shapes per grip type. These shapes are defined by the training images that are closest to the cluster centers. In addition, the images belonging to the goal object form a separate class. Novel intermediate features were generated with the following algorithm:

1. Extract a number of templates from a fixed regular 8 x 8 position grid from each of the stored frames.
2. Merge features belonging to the same class whose similarity is above a given threshold.
3. Retain a fixed number of features per class whose average similarity with the features of then other classes is below a certain threshold.

This algorithm implements a form of competitive feature learning, which prefers strongly activated features that are representative for the frame clusters, and which at the same time are discriminative between different shapes over time. The same effect might be achieved by appropriate competitive learning rules.

2.2 Temporal Sequence Selectivity (Area STS)

We assume that the outputs of the neural detectors for individual effector shapes of a specific grip type l, signified by $z^l_k(t)$, provide input to *snapshot neurons* that encode the temporal order of individual effector shapes [25]. Selectivity for temporal order is achieved by introducing asymmetric lateral connections between these neurons. The dynamics of the resulting network is given by the equation

$$\tau_r \dot{r}^l_k(t) = -r^l_k(t) + \left(\sum_m w(k-m)\,[r^l_m(t)]_+ \right) + z^l_k(t) - h_r$$

where h_r is a parameter that determines the resting level, and where the parameter τ_r determines the time constant of the dynamics. The function w is an asymmetric interaction kernel that, in principle, can be learned efficiently by time-dependent Hebbian learning [28].

The responses of all snapshot neurons that encode the same action pattern are integrated by *motion pattern neurons*, which smooth the activity over time. Their response depends on the maximum of the activities $r_k^l(t)$ of the corresponding snapshot neurons:

$$\tau_s \dot{s}^l(t) = -s^l(t) + \max_k \, [r_k^l(t)]_+ - h_s \tag{2}$$

The motion pattern neurons are active for individual grip types, independent of the presence of a goal object. Such neurons have been found in the superior temporal sulcus of monkey [29].

2.3 Mirror Neurons: Integration of Information from Object and Effector (Areas AIP, PF, F5)

The highest levels of the model integrate the different types of information about object and effector: (1) Types of effector movements, as signalled by the motion pattern neurons, and (2) spatial relationship and movement between the effector and the goal object. The necessary information about the positions of object and effector are extracted from the highest level of the form-hierarchy, which is not comletely position-invariant. The recognized effector view predicts a range of object positions that are suitable for effective grips. This permits to derive whether the effector action likely will be successful or not, dependent on object position.

We postulate a simple physiological mechanism for the integration of these different pieces of information that is centrally based on a *relative position map* that is constructed by pooling of the outputs from the neurons encoding effector and object views. More precisely, by pooling the activity of all object view neurons that represent objects close to the position (u, v) in the retinal frame of reference, one can derive a field of population activity $a_{\mathrm{O}}(u, v)$ that has a peak at the position $(u_{\mathrm{O}}, v_{\mathrm{O}})$ of the object. In the same way, one can derive an activity field $a_{\mathrm{E}}(u, v)$ that has a peak at the position $(u_{\mathrm{E}}, v_{\mathrm{E}})$ of the effector. Based on these two activity fields we constructed a *relative position map* that integrates these pooled signals in a multiplicative manner, realizing a form of 'gain field' [26]:

$$a_{RP}(u, v) = \int a_{\mathrm{O}}(u', v') \, a_{\mathrm{E}}(u' - u, v' - v) \, \mathrm{d}u' \, \mathrm{d}v' \tag{3}$$

The last equation, spatially discretized, can be implemented by summation (pooling) and multiplication of the signals of the appropriately chosen neurons. The relative position map represents the object as an activity peak in an effector-centered 2D frame of reference. Due to the multiplication, all neurons of the relative position map will be inactive if either object or effector are not present in the visual stimulus.

The recognized view of the effector provides also information about the object positions that are suitable for effective grasping with grip type l. For example, a grip is dysfunctional if the objects is positioned next to the hand instead of between the thumb and the other fingers. It is easy to learn the spatial region within the relative position map that corresponds to functional grips (indicated by an orange line in the relative position map in Figure 1). One can easily define a 'receptive field' function $g_l(u, v)$ that corresponds to this region, which has high values within the region and values close to zero outside. We postulate the existence of *affordance neurons* whose response is constructed by computing the output of the relative position map weighted by this receptive field function in the form:

$$a^l = \int a_{RP}(u, v)\, g_l(u, v)\, du\, dv \tag{4}$$

The output of the affordance neurons is only positive if object and effector are present and positioned correctly relative to each other. Neurons that are tuned to the relationship between objects and grips have been found in the parietal cortex of monkeys, e.g. in area AIP [30].

The highest level of the model is given by *mirror neurons* that multiply the output of the motion pattern neurons with that of the corresponding affordance neurons: $m^l(t) = a^l(t) \cdot s^l(t)$. By the multiplicative interactions, the mirror neurons only respond when the appropriate action is present with an appropriately positioned goal object.

3 Results

The model was tested with unsegmented video sequences (640x480 pixels, 30 frames/sec, grayscale) that contained object and effector. The sequences had a length between 34 and 54 frames. From the original videos subregions of 360x180 pixels were extracted that contained the whole effector movement and the goal object. In addition for the training, images with a size of 120x120 pixels were extracted that contained only the hand or the goal object. Image sequences for the testing of the model were disjoint from the training sequences. The goal object was a ball (diameter 8 cm).

3.1 Performance of the Shape-Recognition Hierarchy

Figure 2 shows the classification performance (percent correct) for the neural responses at the the highest level of the form recognition hierarchy (hand shape detectors) for the precision and the power grip. Averaged over all clusters and types we achieve a performance of 75% correct classifications on the testset, determined by cross-validation (leave-one-out on 10 videos per grip type). The weak performance in recognizing power grips in the classes that correspond to the end of the grip is largely caused by confusions of the corresponding hand shapes with the one at the beginning of the sequence. This problem can largely be reduced by the sequence selection mechanism described in Section 2.2.

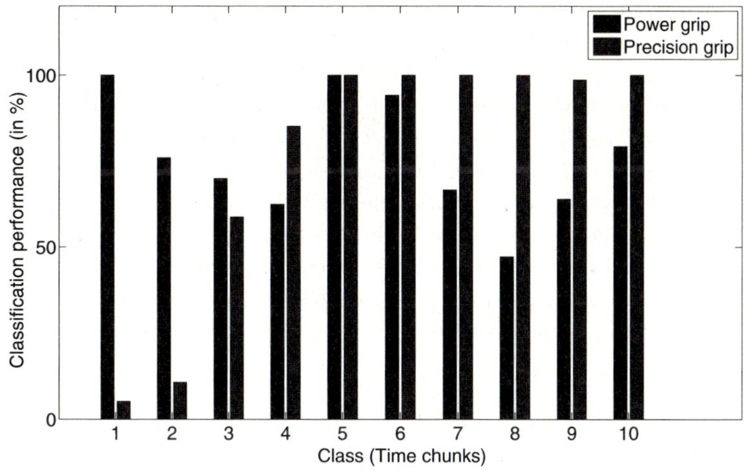

Fig. 2. Recognition performance of the shape recognition hierarchy for hand shapes of power and precision grip over time

3.2 Reproduction of Visual Tuning Properties of Mirror Neurons

Consistent with electrophysiological data [4], the simulated mirror neurons show high selectivity for the grip type. Figure 3 shows that neurons trained with a power grip (left panel), in presence of the goal object, respond only for power grip (blue) and not for precision grip stimuli (red), and vice versa (right panel). At the

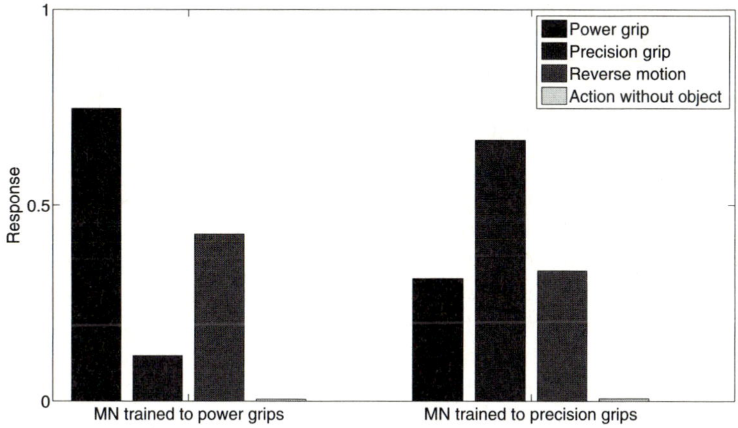

Fig. 3. Average response of modeled mirror neurons trained for power grips (left panel) and precision grips (right panel) to video sequences showing a) a power grip (blue), b) a precision grip (red), c) an action in reversed order (green), and d) an action without goal object (yellow)

level of mirror neurons the model achieves a correct classification performance of 100% on the testset videos.

Figure 3 shows two additional properties that are typical for real mirror neurons: The activity of the modeled mirror neurons for an action without goal object remains at baseline (yellow bar). This is consistent with data showing the lack of responses of mirror neurons without objects [4]. Furthermore, the graph shows the efficiency of the sequence selectivity mechanism. Playing the movie in reverse order leads to a response with strongly reduced activity of the mirror neurons (green bar). While such sequence selective responses have been observed in STS [29], mirror neurons still have to be tested with videos played in reverse order - an experiment which we are presently realizing in our laboratory.

4 Conclusions and Future Work

We have presented a neurophysiologically plausible model for the visual tuning properties of mirror neurons. The proposed architecture provides only a first step towards a more detailed modeling of physiological data. However, the model is based on a number of simple neural mechanisms that, in principle, can be validated in electrophysiological experiments. In spite of this simplicity, the model works on real video sequences. In its present elementary form, the model reproduces qualitatively a number of key properties of mirror neurons: (1) tuning for the subtle differences between grips; (2) failure to respond to mimicked actions without goal object, and (3) tuning to the temporal order of the presented action. All these properties were reproduced without an explicit reconstruction of the 3D geometry of the effector or the goal object. It seems thus that at least some of the visual tuning properties of mirror neurons can be reproduced without a precise metric three-dimensional internal model.

Due to the embedded mechanism for sequence selectivity the proposed model is predictive and can, in principle, account for psychophysical results that show a facilitation of the recognition of future effector configurations from previously observed partial actions [31,32]. Differing from several other models, which assume prediction in a high-dimensional space of motor patterns [3], our model assumes the existence of prediction also in the domain of visual patterns.

In general, the question arises how predictive visual representations interact with representations for motor patterns. Recordings in our own lab show that the responses of many mirror neurons in area F5 are view-dependent. This argues against a uniform representation in area F5 in terms of effector-centered coordinates, as assumed by some other models. However, further experiments are required in order to clarify how different frames of reference are represented and transformed at the level of mirror neurons.

Future work will focus on refining the individual components of the model and fitting it in detail to available electrophysiological, behavioral and imaging data. In addition, specific electrophysiological experiments will be devised that test directly some of the postulated neural mechanisms at the level of mirror neurons in premotor cortex (area F5).

References

1. Di Pellegrino, G., Fadiga, L., Fogassi, L., Gallese, V., Rizzolatti, G.: Understanding motor events: a neurophysiological study. Exp. Brain Res. 91(1), 176–180 (1992)
2. Rizzolatti, G., Craighero, L.: The mirror-neuron system. Annu. Rev. Neurosci. 27, 169–192 (2004)
3. Oztop, E., Kawato, M., Arbib, M.: Mirror neurons and imitation: a computationally guided review. Neural Netw 19(3), 254–271 (2006)
4. Gallese, V., Fadiga, L., Fogassi, L., Rizzolatti, G.: Action recognition in the premotor cortex. Brain 119(Pt 2), 593–609 (1996)
5. Rizzolatti, G., Fadiga, L., Gallese, V., Fogassi, L.: Premotor cortex and the recognition of motor actions. Cognitive Brain Research 3, 131–141 (1996)
6. Logothetis, N.K., Pauls, J., Poggio, T.: Shape representation in the inferior temporal cortex of monkeys. Curr. Biol. 5(5), 552–563 (1995)
7. Edelman, S.: Representation and Recognition in Vision. MIT Press, Cambridge (1999)
8. Tarr, M.J., Bülthoff, H.H.: Image-based object recognition in man, monkey and machine. Cognition 67(1-2), 1–20 (1998)
9. Wolpert, D.M., Doya, K., Kawato, M.: A unifying computational framework for motor control and social interaction. Philos. Trans. R. Soc. Lond. B. Biol. Sci. 358(1431), 593–602 (2003)
10. Tani, J., Ito, M., Sugita, Y.: Self-organization of distributedly represented multiple behavior schemata in a mirror system: reviews of robot experiments using rnnpb. Neural Netw. 17(8-9), 1273–1289 (2004)
11. Oztop, E., Wolpert, D., Kawato, M.: Mental state inference using visual control parameters. Brain Res. Cogn. Brain Res. 22(2), 129–151 (2005)
12. Demiris, Y., Simmons, G.: Perceiving the unusual: temporal properties of hierarchical motor representations for action perception. Neural Netw. 19(3), 272–284 (2006)
13. Fagg, A.H., Arbib, M.A.: Modeling parietal-premotor interactions in primate control of grasping. Neural Netw. 11(7-8), 1277–1303 (1998)
14. Oztop, E., Arbib, M.A.: Schema design and implementation of the grasp-related mirror neuron system. Biol. Cybern. 87(2), 116–140 (2002)
15. Erlhagen, W., Mukovskiy, A., Bicho, E.: A dynamic model for action understanding and goal-directed imitation. Brain Research 1083(1), 174–188 (2006)
16. Metta, G., Sandini, G., Natale, L., Craighero, L., Fadiga, L.: Understanding mirror neurons: a bio-robotic approach. Interaction Studies, special issue on Epigenetic Robotica 7(2), 197–232 (2006)
17. Bonaiuto, J., Rosta, E., Arbib, M.: Extending the mirror neuron system model, i. audible actions and invisible grasps. Biol. Cybern. 96(1), 9–38 (2007)
18. Schaal, S., Ijspeert, A., Billard, A.: Computational approaches to motor learning by imitation. Philos. Trans. R. Soc. Lond. B. Biol. Sci. 358(1431), 537–547 (2003)
19. Sauser, E.L., Billard, A.G.: Parallel and distributed neural models of the ideomotor principle: an investigation of imitative cortical pathways. Neural Netw. 19(3), 285–298 (2006)
20. Billard, A., Mataric, M.: Learning human arm movements by imitation: Evaluation of a biologically-inspired connectionist architecture. Robotics and Autonomous Systems 941, 1–16 (2001)
21. Perrett, D., Oram, M.: Neurophysiology of shape processing. IVC 11(6), 317–333 (1993)

22. Riesenhuber, M., Poggio, T.: Hierarchical models of object recognition in cortex. Nat. Neurosci. 2(11), 1019–1025 (1999)
23. Mel, B.W., Fiser, J.W.: Minimizing binding errors using learned conjunctive features. Neural. Comput. 12, 731–762 (2000)
24. Serre, T., Wolf, L., Bileschi, S., Riesenhuber, M., Poggio, T.: Robust object recognition with cortex-like mechanisms. IEEE Trans. Pattern Anal. Mach. Intell. 29(3), 411–426 (2007)
25. Giese, M.A., Poggio, T.: Neural mechanisms for the recognition of biological movements. Nat Rev Neurosci 4(3), 179–192 (2003)
26. Pouget, A., Dayan, P., Zemel, R.: Information processing with population codes. Nat. Rev. Neurosci. 1(2), 125–132 (2000)
27. DiCarlo, J.J., Maunsell, J.H.: Anterior inferotemporal neurons of monkeys engaged in object recognition can be highly sensitive to object retinal position. J. Neurophysiol. 89(6), 3264–3278 (2003)
28. Jastorff, J., Kourtzi, Z., Giese, M.A.: Learning to discriminate complex movements: Biological versus artificial trajectories. J. Vis. 6, 791–804 (2006)
29. Perrett, D.I., Harries, M.H., Bevan, R., Thomas, S., Benson, P.J., Mistlin, A.J., Chitty, A.J., Hietanen, J.K., Ortega, J.E.: Frameworks of analysis for the neural representation of animate objects and actions. J. Exp. Biol. 146, 87–113 (1989)
30. Murata, A., Gallese, V., Luppino, G., Kaseda, M., Sakata, H.: Selectivity for the shape, size, and orientation of objects for grasping in neurons of monkey parietal area aip. J. Neurophysiol. 83(5), 2580–2601 (2000)
31. Verfaillie, K., Daems, A.: Predicting point-light actions in real-time. Visual Cognition 9, 217–232 (2002)
32. Graf, M., Reitzner, B., Corves, C., Casile, A., Giese, M., Prinz, W.: Predicting point-light actions in real-time. Neuroimage 36(Suppl. 2), T22–32 (2007)

Emergent Common Functional Principles in Control Theory and the Vertebrate Brain: A Case Study with Autonomous Vehicle Control

Amir Hussain[1], Kevin Gurney[2], Rudwan Abdullah[1], and Jon Chambers[2]

[1] Department of Computing Science, University of Stirling,
Stirling, FK9 4LA, Scotland, UK
ahu@cs.stir.ac.uk

[2] Department of Psychology, University of Sheffield, S10 2TN, UK

Abstract. This paper describes emergent neurobiological characteristics of an intelligent multiple-controller that has been developed for controlling the throttle, brake and steering subsystems of a validated vehicle model. Simulation results demonstrate the effectiveness of the proposed approach. Importantly, the controller exhibits discrete behaviours, governed by its two component controllers. These controllers are selected according to task demands by a fuzzy-logic based supervisor. The system therefore displays 'action selection' under central switched control, as has been proposed to take place in the vertebrate brain. In addition, the supervisor and modular controllers have analogues with the higher and lower levels of functionality associated with the strata in layered brain architectures. Several further similarities are identified between the biology and the vehicle controller. We conclude that advances in neuroscience and control theory have reached a critical mass which make it timely for a new rapprochement of these disciplines.

Keywords: Neurobiology, Brain Structure, Proportional-Integral-Derivative and Pole-Zero Placement Control, Fuzzy Tuning, Autonomous Vehicle Control.

1 Introduction

The field of autonomous vehicles is a rapidly growing one which promises improved performance, fuel economy, emission levels, comfort and safety. An important component of autonomous vehicle control (AVC) aims to control the throttle, wheel brake and steering systems so that the vehicle can follow a desired path and target speed (possibly governed by a 'lead vehicle') and at the same time keep a safe inter-vehicle spacing under the constraint of comfortable driving [1]. This is the problem considered here and, while conventional methods based on analytical control generate good results, they exhibit high design and computational costs since the target plant is complex and nonlinear, and a veridical analytic representation is impossible. In addition, unpredictable environmental changes, (for example yaw-disturbances caused by unsymmetrical car-dynamics, or side-wind forces) compound the problem of control. There is therefore a pressing need for alternative

V. Kůrková et al. (Eds.): ICANN 2008, Part II, LNCS 5164, pp. 949–958, 2008.

approaches to AVC. One particularly promising alternative is to break the task space into a series of distinct operating regions and to switch between them when required [2]. If we can interpret this scheme as one in which the plant (an automobile) switches between distinct 'behaviours', then one high level aspect of the control task is that of *action selection* [3].

The problem of action selection appears in many forms across various disciplines but a common formulation may be framed as follows: how can an autonomous agent (animal or robot) under bombardment from a plethora of external (sensory) and internal (cognitive and homeostatic) information, decide 'what to do next'. This problem arises because agents have limited motor resources e.g. limbs/wheels or hands/effectors and must decide, from moment to moment, to what use each motor plant will be put. Within ethological studies of animal behaviour the problem of action selection is often referred to as the problem of 'behavioural switching' or 'decision-making'; the latter terminology is also used by psychologists in perceptual tasks [4].

We contend that similar problems arise in a wide class of embedded control systems which require a continuous sensing of internal (state of the plant) and external (environmental) variables, and computation of appropriate control operations ('actions'). Given this common problem space, presented to both animals and control systems, we might expect similar solutions to emerge. Here, we describe a particular controller for AVC and show how several of its features have intriguing similarities with those of the overall functional architecture of the vertebrate brain. We argue that this emerges because both animals and real-time control systems have to solve the problem of action selection.

This paper is organized as follows. Sect. 2 describes the principal architectural features of the controller (a detailed description can be found in [5]). Sect. 3 presents new simulation results for an automobile case study demonstrating the efficacy of the controller. Sect. 4 describes the basic architectural features of the vertebrate brain that we believe are implicated in solving the problem of action selection in animals. Finally, in Sect. 5, we present comparisons between the two architectures - controller and brain respectively.

2 A Controller for Autonomous Vehicle Control

2.1 Multiple Controllers

A common approach to control complex dynamic systems is to design a set of different controllers, each of which is optimized for a particular operating region or performance objective, and then to switch between them in real-time to achieve the overall control objective [2]. This *multiple model* approach has been used extensively and in various guises - e.g. Gain-Scheduling Controllers, Tagaki-Sugeno Fuzzy Models and Logic-based Switching Controllers (a good overview can be found in [5]). An important category of such systems are those consisting of a process to be controlled, a family of fixed-gain or variable-gain *candidate controllers*, and an event-driven switching logic called a *supervisor* whose job is to determine in real-time which controller should be applied to the process.

In this paper we use a multiple-controller architecture (see Fig. 1) to control a plant via a set of controller modules (all derived from a single minimum variance based control law [5]), with each module activated and optimized for a special operating regime of the plant. A top level supervisor switches between the controller modules to determine the active module. The decision to switch from one controller to the next is made on the basis of (quantitative and heuristic based) performance measurements of the plant and external signals from the environment and the user, including a desired target or 'set-point' [6].

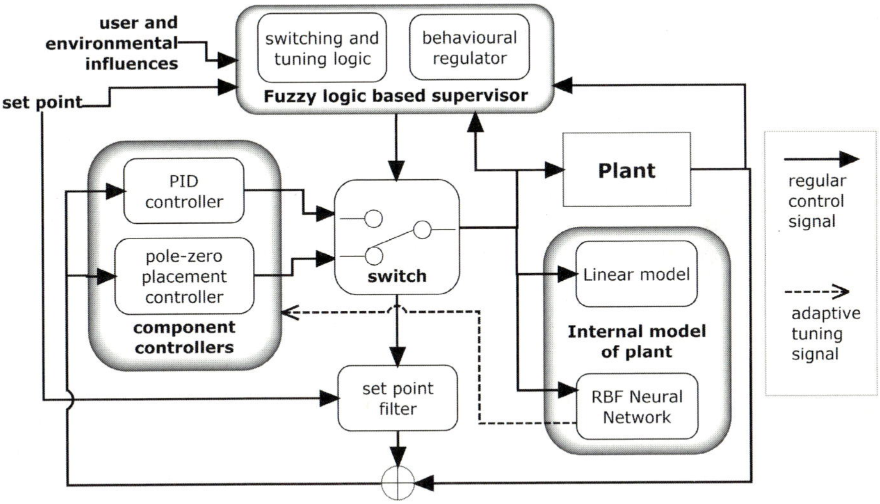

Fig. 1. Intelligent multivariable adaptive controller framework. For clarity, not all tuning signal pathways are shown.

The supervisor employs a fuzzy-logic based switching technique, in order to automatically select between two candidate (minimum variance based) adaptive controllers, namely a Proportional-Integral-Derivative (PID) controller or a PID structure-based (simultaneous) pole and zero placement controller. Moreover, the supervisor can tune the parameters of the candidate controllers on-line. This novel tuning strategy builds on the conventional fuzzy gain scheduling strategies that have been conventionally employed for only PID controllers [6].

2.2 An Internal Plant Model

The controller also incorporates an *internal model* of the plant it is controlling which is tuned, or learned, online. This so termed Generalized Learning Model (GLM) [7] consists of separate linear (recursive least squares) and nonlinear submodels with the latter comprising a Radial Basis Function Neural Network (RBF NN). The latter aims to account for any unknown time-delays, uncertainty and non-linearity in the complex plant model and may be thought of as modelling the non-linear *prediction error* (that is, the difference between the estimate supplied by the linear sub-model, and the actual output of the plant [7]).

2.3 Fuzzy Logic Based Tuning Supervisor

The fuzzy logic based switching and tuning supervisor can use any available relevant data from the control system and the environment to characterise the system current behaviour and environmental state [6]. It then uses this information to determine which controller to choose, which parameters to tune, and the tuning value for each parameter that is required to ultimately achieve the desired specification. The supervisor employed in this work comprises two subsystems [5,8]: a *behaviour recogniser* and a *switching and tuning logic*.

Behaviour Recogniser: The behaviour recogniser seeks to characterize the current behaviour of the plant in a way that will be useful to the switching and tuning logic subsystem [5]. The behaviour of the system is characterized through the on-line estimation of four parameters: the *overshoot* of the closed-system output signal, the *variance* of the control input signal, rise and fall times of the output signal (used for tuning purposes), and the PID controller steady state error (used for tuning the PID gain). In this way, the behaviour recogniser can sense if the controller currently being used is starting to fail to achieve satisfactory performance; that is it senses conditions under which errors are starting to develop and are more likely to occur in the immediate future.

Switching and Tuning Logic: The first task of the switching and tuning logic sub-system is to generate a switching signal which determines, at each instant of time, the candidate controller module that is to be activated [9]. The switching logic is implemented using fuzzy logic rules. The premises of the rules use the output of the behaviour recogniser as input parameters, namely the overshoot of the output signal and variance of the control input signal,. The middle-of-max approach is used for de-fuzzification in order to identify the selected controller. The second task of the switching and tuning logic sub-system is to tune the parameters of the multiple-controller on-line, including poles and zeros of the (simultaneous) pole-zero placement controller in addition to the PID gains. The tuning facility helps the system achieve the desired speed of response and good long term performance [6]. One problem that many multiple controllers suffer from is a transient in control signals (and thereby plant behaviour) when switching takes place. This can occur because of incompatible initial conditions in the controllers. The novel switching mechanism introduced in this work guarantees that such transients are minimized; that is bumpless switching occurs (see [6] for more details).

3 Simulation Results for Autonomous Vehicle Control

The controller described here has been implemented and tested in simulation using a validated realistic multiple-input multiple-output vehicle model including interacting nonlinearities from the throttle, braking and wheel steering subsystems (details can be found in [10]). We show results here to demonstrate the efficacy of the controller and the switching behaviour between controllers. Recall that the system's task is to control the steering wheel angle, throttle angle, and brake torque, assuming that the desired path and vehicle speed have been determined. In addition excessive control action was penalised and efficient control sought under internal tuning processes. The particular task consisted of mixed-mode driving along straight segments and around obstacles over a time period of 20 sec. Thus, during times 0 to 8.6sec the autonomous vehicle had to follow the path with a speed of 30m/sec, then slow down to 20m/sec to follow the left side turn. This was to be followed by slowing down to a speed of 10m/s to track a sharp turn right within the target path, and then speeding up to 20m/sec and lastly attaining the target speed of 30m/sec.

Figures 2a and 2b illustrate the path and speed tracking, and show that the proposed methodology is able to follow control decisions within the desired path and speed trajectories. Figures 3a, 3b and 3c show the results of applying the control signals to the throttle, brake and steering systems respectively. In Fig. 3a, during time 10.0sec to 10.4sec the throttle plate was set to its minimum (0.15) in order to slow down the speed to 10m/sec. However, as this minimum throttle angle was not sufficient to reach this target speed, simultaneously the wheel brake system was triggered to produce the required braking torque as shown in Fig. 3b. Figure 3c shows the output steering wheel angle during the operation of tracking the target path.

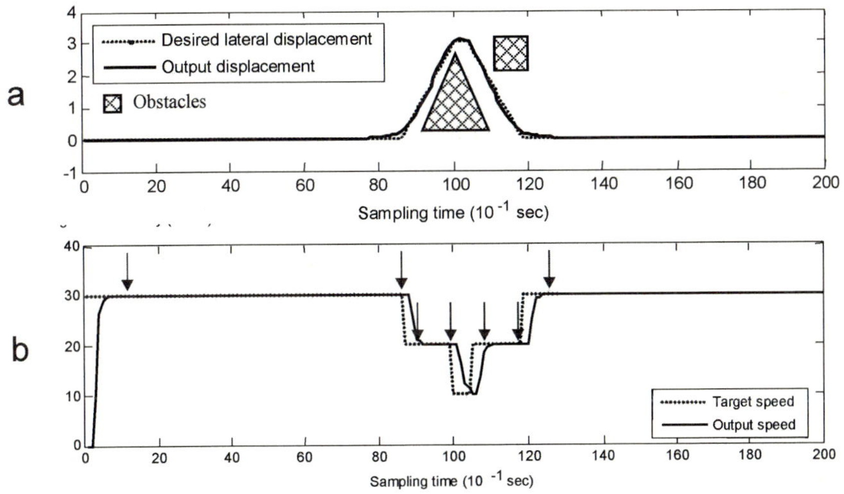

Fig. 2. Simulation results with AVC controller. a) Target path tracking. b) Target speed tracking during the operation following the target path. Each arrow indicates a point of switching between the two controller modes: from pole-zero placement to PID and vice-versa (starting with pole-zero placement at time 0).

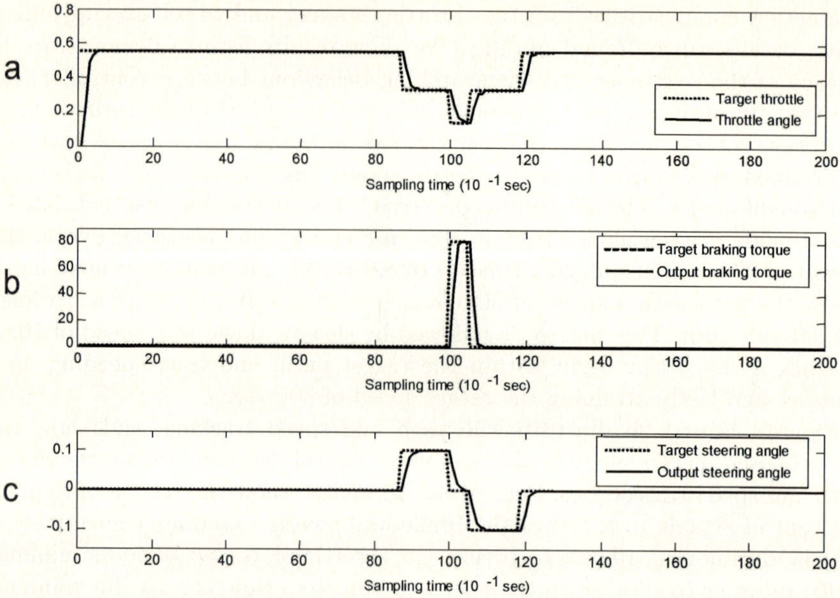

Fig. 3. Simulation results with AVC controller. a) The performance of the throttle system for reaching the required throttle angle. b) The performance of the brake system for reaching the required braking torque. c) Performance of the steering system for reaching required steering angle.

An emergent phenomenon of the system that enables the link to be made with ideas in biological action selection is that, each component controller appears to be responsible for a different 'driving behaviour'. Thus the pole-zero controller is activated when 'navigating tight bends' and the PID controller is engaged when 'cruising in a straight line' (as indicated by the arrows in Fig. 2b).

4 A Vertebrate Solution to Behavioural Control

4.1 Action Selection and the Basal Ganglia

Neurobiology suggests that the action selection problem is solved in the animal brain using a central switch which receives requests for behavioural expression from cortical and sub-cortical subsystems throughout the brain. In [11], we outlined the proposal that the basal ganglia (BG) - a set of sub-cortical nuclei - are well suited to play the role of this central 'switch'. The BG are evolutionarily highly conserved throughout the vertebrate lineage, suggesting they are solving a fundamental problem in animal behaviour - of which action selection is a prime example. Further, the BG receive input from a wide variety of cortical and sub-cortical systems, and project back to these systems a connection scheme consistent with the functionality of a central switch.

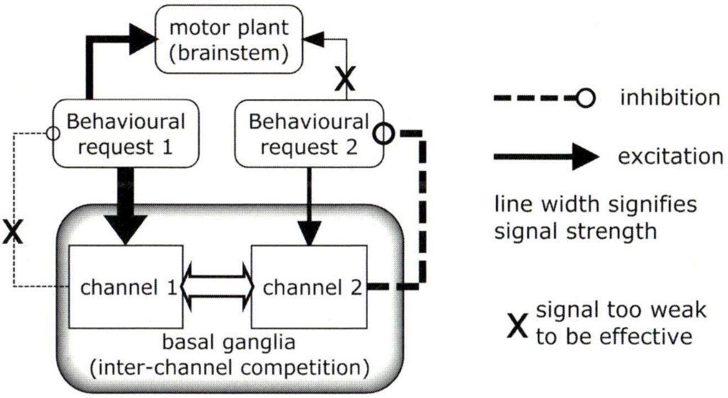

Fig. 4. The basal ganglia as the central action selection 'switch' in the brain

Functionally, we suppose that brain subsystems requesting behavioural expression send excitatory signals to BG whose strength is determined by the salience or 'urgency' of the requests. Output from the BG is inhibitory and normally active. Actions are selected when competitive processes between action channels in the BG result in withdrawal of inhibition from the target subsystem. This process is illustrated for two action channels in Fig. 4. Here, channel 1 is sending a highly salient request to BG, whereas channel 2 is sending a weak request. Internal competitive processes in BG ensure that the inhibition (defaulting to a high value) on the BG output for channel 1 is reduced to a point where it is ineffective. This allows the 'command system' associated with behavioural request 1 to take control of brainstem motor plant and ultimately to express its behaviour. In contrast, channel 2 has its inhibitory BG output increased, effectively preventing any command signal reaching the motor plant (thalamic connectivity between BG and cortex has been suppressed for simplicity).

The basal ganglia selection hypothesis has been confirmed in both computational [12,13] and embodied models [14].

4.2 Layered Architectures in the Brain

A key feature of the overall architecture of the brain is that it appears to implement multiple levels of sensorimotor competence [15]. Here, a single sensorimotor modality, for example, saccadic gaze control directed by visual input, is associated with several levels of competence. In the oculomotor example, low level sub-cortical structures enable simple gaze control based on transient luminance change, whereas higher cortical circuits enable gaze strategies driven by complex visual features and task demands. Note that this kind of subsumption architecture also figures in an influential strand of contemporary robotics [16].

Integrating the general idea of a layered scheme with the hypothesis that the basal ganglia act as a central switch yields the architecture shown in Fig. 5 [15]. Each level of competence has its own competition between multiple action

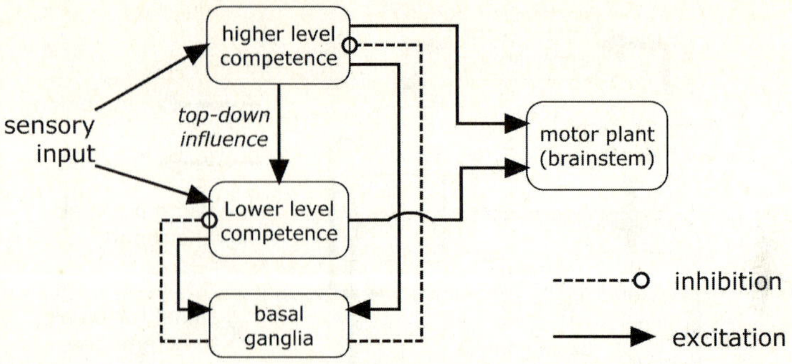

Fig. 5. Layered architecture of the brain: cartoon schematic showing role of BG

requests mediated by the BG, and may also take control of the motor plant. Further, action representations in higher layers can influence those in lower layers thereby exerting 'top-down' control. One very high level of control in the brain is provided by the anterior cingulate cortex (ACC) which appears to monitor performance to regulate behavior. In particular, recent work by Carter et al [17] showed that the ACC contributes to performance monitoring by detecting errors, and conditions under which errors are likely to occur.

4.3 Cognitive World Models: Learning Action-Outcome Associations

The term 'cognition' is usually used to encompass psychological notions such as problem solving, memory, language, emotion, attention, and planning. However, a common thread in all these competences is the adaptive use of some stored internal representation of the world, thereby endowing the animal with the ability go beyond simple reactive 'stimulus-response' behaviour. In particular, one class of internal model has to do with the ability to know what actions are effective in producing useful environmental consequences. For example, knowing that a certain class of wall-mounted objects (light-switches) produce illumination is potentially extremely useful. Recently, Redgrave et al [18] have described how the basal ganglia may serve to help establish these action-outcome forward models under reinforcement learning. One of the key features of this scheme is that it uses *sensory prediction errors* (mediated by the neurotransmitter dopamine) to bias the action selection policy in basal ganglia so as to ensure repeated execution of actions that produce novel sensory outcomes. In this way, new actions and combinations of existing action-components may be learned.

5 A Comparison of the Biological and AVC Solutions to Behavioural Control

Both the brain architecture and the AVC system use a central switch for determining which 'command system', and thereby which behaviour or action, is selected. The mechanisms for selection and switching are, however, rather different. The brain uses signals (salience) derived from the command systems themselves. In contrast, the AVC system does not use signals intrinsic to either component controller but, rather, signals derived from a supervisory system - the fuzzy controller.

This top level control may, however be likened to the top down influence in the layered architecture. Thus we make the identification that the fuzzy controller is an analogue of some high level (pre-frontal) cortical system that is able to influence a lower level competence (component controllers). The way in which the fuzzy-controller provides the top down-influence is, of course, rather different from its biological counterpart; it uses rules rather than distributed neural representations. Further, the fuzzy controller has an analogue of the anterior cingulate cortex in so far as it has a behaviour recognition sub-system which is monitoring for errors.

The AVC system also contains a 'cognitive model' in the form of its internal model of the plant. In particular, we may liken it to an action-outcome association because it represents the plants response (outcome) if driven with certain control signals ('action'). Further, this model is used to derive a signal analogous to a sensory prediction error which is used for tuning or 'learning'. Thus, we consider the linear sub-model to be the basic internal model and the RBF network to represent discrepancies with respect to this sub-model that are due to nonlinearities, uncertainties and disturbances in the controlled plant. This class of phenomena - analogous to prediction errors - are used to tune (or 'train') the component controllers, thereby shaping the possible behaviors in exactly the same way that we propose that phasic dopamine can help in the shaping of new actions in the basal ganglia [18,19].

6 Conclusion

We have described an autonomous vehicle control (AVC) system which makes use of multiple conventional controllers, whose selection is governed by a central switch under control of a novel fuzzy-logic based controller. This system, which was not designed from the ground up to be bio-inspired, shows emergent properties akin to animal behaviour in that it displays a discrete behavioural repertoire. Moreover, the AVC system has functional architectural similarities with the vertebrate brain - central switching, layered control, internal action-outcome models, error-monitoring, and use of prediction errors to shape behaviour.

References

1. Conatser, R., Wagner, J., Ganta, S., Walker, I.: Diagnosis of automotive electronic throttle control systems. Control Engineering Practice 12, 23–30 (2004)
2. Lee, C.Y.: Adaptive control of a class of nonlinear systems using multiple parameter models. Int. J. Contr. Autom. Sys. 4(4), 428–437 (2006)

3. Prescott, T.J., Bryson, J.J., Seth, A.K.: Introduction to the theme issue on modelling natural action selection. Philos. Trans. R. Soc. Lond. B. Biol. Sci. 362, 1521–1529 (2007)
4. Gurney, K.N., Prescott, T.J., Redgrave, P.: A computational model of action selection in the basal ganglia i: A new functional anatomy. Biol. Cybern. 84, 401–410 (2001)
5. Abdullah, R., Hussain, A., Warwick, K., Zayed, A.: Autonomous intelligent cruise control using a novel multiple-controller framework incorporating fuzzy-logic based switching and tuning. Neurocomputing (in press)
6. Abdullah, R., Hussain, A., Polycarpou, M.: Fuzzy logic based switching and tuning supervisor for a multivariable multiple-controller. In: IEEE International conference on Fuzzy Systems, Imperial College, London, UK, pp. 1644–1649 (2007)
7. Zayed, A., Hussain, A., Abdullah, R.: A novel multiple-controller incorporating a radial basis function neural network based generalized learning model. Neurocomputing 69(16-18), 1868–1881 (2006)
8. Naranjo, J., Gonzalez, C., Reviejo, J., Garcia, R., Pedro, T.: Adaptive fuzzy control for inter-vehicle gap keeping. IEEE Trans. Intellig. Transport. Syst. 4(3), 132–142 (2003)
9. Hespanha, J., Liberzon, D., Morse, A., Anderson, B., Brinsmead, T., Bruyne, D.: Multiple model adaptive control. part 2: Switching. Int. J. Robust Nonlin. Contr. 11, 479–496 (2001)
10. Kiencke, U., Nielsen, L.: Automotive Control Systems: For Engine, Driveline, and Vehicle. Springer, Berlin (2005)
11. Redgrave, P., Prescott, T.J., Gurney, K.N.: The basal ganglia: a vertebrate solution to the selection problem? Neuroscience 89, 1009–1023 (1999)
12. Gurney, K.N., Prescott, T.J., Redgrave, P.: A computational model of action selection in the basal ganglia ii: Analysis and simulation of behaviour. Biol. Cybern. 84, 411–423 (2001)
13. Humphries, M.D., Stewart, R.D., Gurney, K.N.: A physiologically plausible model of action selection and oscillatory activity in the basal ganglia. J. Neurosci. 26, 31–61 (2006)
14. Prescott, T.J., Gonzales, F., Gurney, K.N., Humphries, M.D., Redgrave, P.: A robot model of the basal ganglia: behavior and intrinsic processing. Neur. Networks 19, 31–61 (2005)
15. Prescott, T.J., Redgrave, P., Gurney, K.N.: Layered control architectures in robots and vertebrates. Adapt. Behav. 7, 99–127 (1999)
16. Brooks, R.A.: New approaches to robotics. Science 253, 1227–1232 (1998)
17. Carter, C.S., Braver, T.S., Barch, D.M., Botvinick, M.M., Noll, D., Cohen, J.D.: Anterior cingulate cortex, error detection, and the online monitoring of performance. Science 280, 747–749 (1998)
18. Redgrave, P., Gurney, K.N.: The short-latency dopamine signal: a role in discovering novel actions? Nat. Rev. Neurosci. 7, 967–975 (2006)
19. Gurney, K.N., Chambers, J., Redgrave, P.: A model of reinforcement learning in basal ganglia for action-outcome association. In: IBAGS IX Meeting, Egmond aan Zee, The Netherlands (2007)

Organising the Complexity of Behaviour

Stathis Kasderidis

Foundation for Research and Technology - Hellas
Institute of Computer Science, Computational Vision and Robotics Laboratory
Science and Technology Park of Crete
Vasilika Vouton, P.O. Box 1385, 71110 Heraklion, Greece
stathis@ics.forth.gr

Abstract. In [1] we have developed a computational model for multiple goals. The agent behaviour is organised in families of competing goals (plans) for determining priorities of execution. This paper extends the previous model by providing a method for executing complex plans consisting of a number of goals and/or other sub-plans. The various dependencies among goals/plans are analysed and a mechanism is suggested which provides additional voting to the priority of a component goal. For the class of independent goals it is possible to interleave sub-goals for achieving better performance. A number of simulation experiments are described and results indicate a more flexible behaviour. We conclude with a summary of findings and a discussion for future work.

1 Introduction

Nowadays there is great interest in building cognitive agents that will drive robotic or other software agents. Among all the other aspects one needs to tackle the question of how best to organize the complex behaviour that is generated by having a number of concurrent (high-level) goals inside a cognitive agent. We separate the overall problem of regulating behaviour in two complementary aspects: these of reasoning and execution. The first part, the problem of 'reasoning', deals mainly with the development of a suitable plan that will achieve a target state while the second part, the problem of 'execution', is related to the monitoring of the plans, the provision of re-planning requests to the 'reasoning' component for each under-performing plan, the initiation of learning processes in cases of failures and the generation of coherent behaviour. In both aspects there has been intensive research effort in the past. The issue was and is still being pursued by the AI community under the names of planning algorithms and constraint satisfaction problems, for example see [3] (and references therein). For autonomous agents, with arbitrary long lifetimes, situated in a dynamic and partially observable environment the above approach was extended with execution monitoring and re-planning facilities. Usually the framing of the problem took the form of Hierarchical Task Networks [3], or that of Abstraction and Hierarchical Planning [4], [5].

However, while the above approaches clearly attack an important aspect of the overall problem, they do not attack the full problem. For example attention movement (and the implied change of focus) is not easily included in these approaches.

V. Kůrková et al. (Eds.): ICANN 2008, Part II, LNCS 5164, pp. 959–968, 2008.

Contribution from internal agent motivational and drive systems is also not easily integrated. For this reason we have developed an efficient executive system for the GNOSYS cognitive architecture in order to regulate the complex behaviour of a robotic agent that is controlled by the GNOSYS architecture. In this paper we will discuss the 'execution' component of the planning process inside the agent. The planning process is the complex interaction of three modules of the architecture: of the executive system (which is the focus of this paper), of the reasoning system and of the motivational system. The work presented here is a continuation of previous work on the system. In [1] we have proposed the initial model, while in [2] this model was extended so as to accommodate factors such as the threat level perceived from the environment and factors contributing to the complexity of the task. In this paper our focus is in defining a further extension which is able to handle the execution of arbitrary plans which are composed in a recursive way. For a description of the reasoning system see [6] while for the motivational system see [7]. For further details on these and other components present in the GNOSYS architecture one can see [8]. The paper is structured as follows: in section 2 we will provide an analysis of the various dependencies of goals and we will suggest plan types and a uniform mechanism that handles the priority of constituent goals. In section 3 we present simulation scenarios that help us clarify the operation of the system. In the same section simulation results are given on the presented scenarios. We conclude in section 4 with a summary of key features and points of interest for future work.

2 Handling the Complexity of Execution

The GNOSYS Architecture aims to control robotic agents in arbitrary complex, novel and unstructured environments. It consists of a number of high-level modules that provide the basic cognitive capabilities of the agent. Figure 1 provides the main information flows inside the agent so as to provide the context of operation for the executive system (denoted as Goals). The information flow starts with the request for the achievement of a target state. This request comes from two primary sources: a user request or a request from the motivational system. The realisation of the request takes the form of Goal object, which is characterized by a number of parameters; see [1] for details. The Goal object represents either a complex high-level goal or a primitive one. In the former case, we need a Plan, which will produce the required target state that is specified as one of the Goal parameters. In this sense a Goal is a self-contained execution scope (i.e. a thread of activity inside the agent). For a Goal to be achievable we need to find of a suitable plan. The term Plan here is used in the sense that describes a collection of (sub-) Goals that are executed with a specific pattern and time order. Except of the primitive Goals, all other Goals are actually plans of more concrete sub-Goals. The problem of producing a plan is solved by the reasoning system. We will not discuss here this process. It suffices to say that the GNOSYS project has produced a reasoning system, which operates using forward/inverse model pairs and emulates well reasoning processes in animals. See for example [6] and [9].

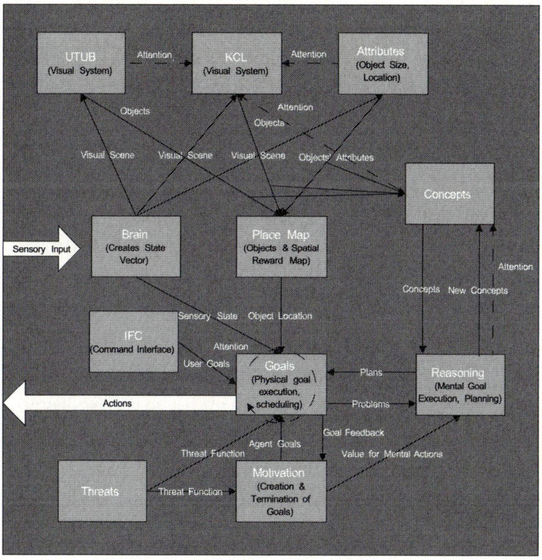

Fig. 1. Main information flows inside the GNOSYS agent. The executive component handles the execution of a plan to the real world. It forms a feedback loop with the motivation and reasoning systems.

The motivation system implements an internal value map in the sense that it can provide reward or penalty to the currently executing Goals depending on the effects of their actions to the value of an internal drive which represents the well-being of the agent. Except of the reinforcement aspect it also crucially influences the termination of non-converging or under-performing Goals. The complexity of the planning process arises from the ability to tackle the uncertainty in the environment (partial information, availability of tools, constraints and the ability to exhibit novelty in strategy). In terms of a functional decomposition, the reasoning system solves the 'How' part of the problem. The motivational system and the executive solve the 'When to do what' part of the problem. This decomposition ensures that maximal rewards will be gained by optimal use of the agent's resources and opportunistic possibilities afforded by the environment.

2.1 Goals, Plans and Dependencies

Plans include either primitive Goals or other Plans (sub-Plans). The type of the Plan determines how its constituent Goals are executed. We have analysed in [1] the type of dependencies that exist among Goals. For our purposes here we are interested in three classes of Plans, which capture the corresponding dependencies. We can identify:

A. Independent;
B. Sequential;
C. Optional;

In the first case, the Plan has many Goals. All of them must be completed for the Plan to be completed. The Goals can be executed concurrently in any order, thus enabling opportunistic execution. In the second case the all the Goals of the Plan are executed strictly in a predetermined sequential manner. The third class describes a pattern where some Goals, N, out of the total number, M, must be completed for the Plan to be completed, where N ¡ M. An interesting sub-case is where N=1. This pattern allows a form of decomposition where Goals are ORed. The Goals can be executed in parallel and the Plan terminates when the first N of them complete.

Using the above three classes we can compose (or decompose) arbitrary complex plans with non-trivial overall behaviour. A way to clarify the interactions is by representing a Plan as a network structure where an activation variable is defined per Goal. Weights then represent the specific pattern applied. The diagrams in figure 2 show the three cases:

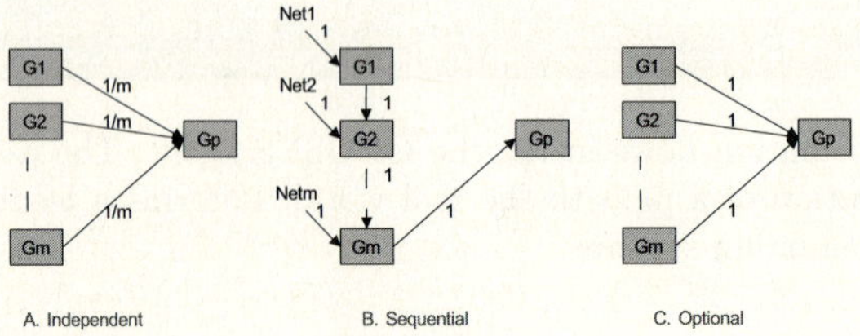

Fig. 2. Types of Plans. G_1, ..., G_m are children Goals of a Plan, while G_p represents the Plan (root Goal). The type of the Plan determines the pattern of the information flow and the weighting coefficients. See text for details.

In figure 2 G_p represents the Goal whose the Plan consists of the sub-Goals $G_1, \ldots, G_m$. We can associate an activation variable, $Term_G$, with each root Goal and sub-Goal G. The corresponding activations, to figure 2, are written as:

$$\begin{aligned} Term_p(t) &= Step(Net_p(t) + Term_p(t-1) - 1) \\ Net_p(t) &= (\sum_j w_j * Term_j(t)) = (\sum_j (1/M) * Term_j(t)) \end{aligned} \quad (1)$$

$$\begin{aligned} Term_p(t) &= Term_M(t) \\ Term_j(t) &= Step(Net_j(t) + Term_j(t-1) - 1) \\ Net_j(t) &= (Term_{j-1}(t) + Net_j(t-1)) \end{aligned} \quad (2)$$

$$\begin{aligned} Term_p(t) &= Step(Net_p(t) + Term_p(t-1) - 1) \\ Net_p(t) &= (\sum_j w_j * Term_j(t) - n) = (\sum_j 1 * Term_j(t) - n) \end{aligned} \quad (3)$$

Where the variables Net correspond to the input coming to the node, while TermP is the activation of the root node, P, corresponding to the plan in question. The variable indicates the termination status of the Goal and it is assumed that a plan has terminated when the activation of the root node reaches 1.0. All variables appearing in (1)-(3) are assumed scaled in the interval [0,1]. There are in general M nodes to the plan, j=1M. $Step(\cdot)$ is the step function. Equation (1) corresponds to the case of the independent Goals, while (2) in the case of the Sequential type and (3) in the Optional case. In the case of a primitive Goal the Net variable is calculated as it is shown in figure 3:

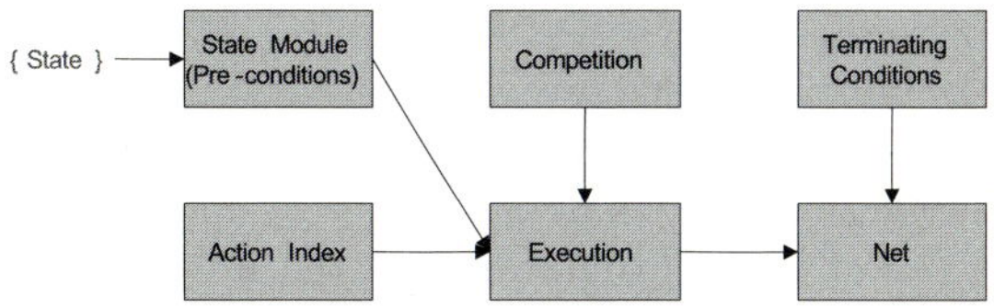

Fig. 3. Diagram for the calculation of the Net input to a primitive Goal. The calculation takes into account a number of factors. See text for more details.

In figure 3 each module provides as output an activation in the range [0,1]. The state module uses the current state to calculate pre-conditions. If pre-conditions are met then the output is 1.0 else is 0.0. The activation of the module indicates a state where all prerequisites are fulfilled and thus the Goal can start execution. Terminating conditions are also calculated so as to indicate the arrival to the target state. The box also provides a binary activation. The action index box indicates the relative priority of the goal as it is evaluated using the current state. The index varies in [0,1]. The box Competition represents a global competition among goals that determines the one with the highest priority. The output of this module is 1.0 if the current Goal is the global winner, otherwise is 0.0. This competition takes place in the scope of the executive system and outside the scope of a Goal. Under these interactions the activation of the box Execution is evaluated by (4):

$$Exec(t) = Step(PC(t) + COM(t) + AI(t) - 2) \tag{4}$$

Where in (4) Exec represents the activation of the Execution module and PC, COM and AI the corresponding activations of the State Module, Competition and Action Index respectively. When the Execution module is active this indicates the actual operation of the Goal. The Net input and Termination for a primitive Goal is then calculated by (5):

$$\begin{aligned} Net(t) &= Exec(t) + TC(t) - 2 \\ Term(t) &= Step(Net(t)) \end{aligned} \tag{5}$$

Where TC represents the binary output from the Terminating Conditions box and the Step function is assumed to be 1.0 when $x \geq 0$.

Using equations (1)-(5) one can calculate in a recursive manner the overall Termination and Action Index of the whole Plan depending on the Plan type.

At the top root of a Plan the calculated Net activation represents a factor of priority that is then fed back to the primitive Goals' Action Index calculation. This is accomplished by use of (6):

$$ActionIndex = (W + EW + TP + S - AI + M - AI + B - AI + TL + Cost + Net_p + \sum_j \delta(j, contrib.)/(9 + No.ofContributingChildren) \quad (6)$$

Let us explain the main components of formula (6). We assume that each Goal has an extrinsic value (W) that comes either by the User or by experience (built by RL and it is thus stored in the value maps). This weight, as all other terms in (6), is suitably scaled in the interval [0, 1]. The term EW represents the intrinsic value of the Goal. This value is updated in every processing step by the Motivation System and captures the influence of the Goal to the agent's well being. The Termination Probability (TP) indicates the probability to be completed given the current state. This is calculated in general as a function of the distance between the current and target states in a suitable metric space. The terms S-AI, M-AI and B-AI correspond to three different (local) attention mechanisms that capture complimentary aspects. S-AI corresponds to sensory attention, which captures novelty in the environment. M-AI captures related motor attention events. The B-AI captures attention events coming from the internal agent environment; more specifically the deviation of one or more homeostatic variables from their equilibrium state. The term TL represents a threat level, which is provided by the Threat module in figure 1 and corresponds to an estimated threat potential function due to a number of moving agents inside the observation horizon of the agent. The Cost corresponds to the Maximum Expected Utility (MEU), which will be gained by the Goal if it is executed in the next processing step. The idea of MEU provides a good mechanism for reaching decisions that balance the rewards that are expected against the probability of gaining them. This idea is at the basis of decision theory. For details see [3] and [10]. Finally NetP corresponds to the activation received by the parent of the family. The model provides also for the opportunity of interleaving primitive Goals that belong to separate Plans of independent type. Thus one in practice can take two or more Plans and schedule more efficiently the actions from the two Plans, considering them as a larger combined plan. Such scheduling can be seen as an optimisation problem and a number of methods can be used for this purpose, such as Simulated Annealing, Genetic Algorithms and others. We use an approach based on Ant Colony Optimisation (ACO) Methods, which has the benefit to be quite fast for real-time implementation in the GNOSYS robotic platform. See [11] for details.

3 Results

To clarify the operation of the system consider the setup shown in figure 4a.

We assume that we have two high-level Goals. The first is about carrying objects A1 and A2 to point G1, while the second one is about moving B1 to G2.

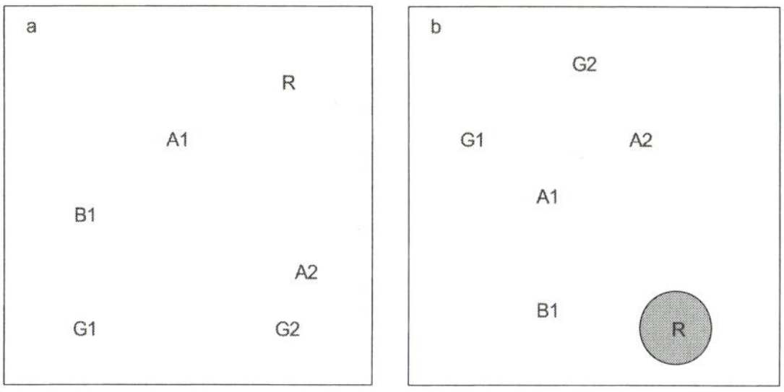

Fig. 4. A robot is always starting from point R. A_i and B_j are objects that need to be moved to target points G_1 and G_2 respectively. In (a) we assume that we have full observability while in (b) the robot has a limited range of observation, which is 20 percent of the distance from R to B_1.

The reasoning system is called and two Plans are returned as an answer. In the first case the Plan P1: $R \rightarrow A_1 \rightarrow G_1 \rightarrow A_2 \rightarrow G_1$ is returned while in the second case the Plan P2: $R \rightarrow B_1 \rightarrow G_2$ is the result. The question is: in which order we should execute the plans in order to optimize the response of the system given all other factors equal? Here the function we want to optimize is the accumulated distance after all plans have been executed. One can try to execute first P1 and then P2; that results to Plan P3 (concisely P1+P2=P3). Alternatively the opposite order can be used resulting to P4 (P2+P1=P4). However we see from table 1 that there is a better solution P5 that is to interleave the steps of the two basic plans.

Table 1. Plans and accumulated distances after execution of the scenario of figure 4a

Plan ID	Sequence	Total path length
P1	$R \rightarrow A_1 \rightarrow G_1 \rightarrow A_2 \rightarrow G_1$	20.141
P2	$R \rightarrow B_1 \rightarrow G_2$	12.784
P3	$R \rightarrow A_1 \rightarrow G_1 \rightarrow A_2 \rightarrow G_1 \rightarrow B_1 \rightarrow G_2$	29.447
P4	$R \rightarrow B_1 \rightarrow G_2 \rightarrow A_2 \rightarrow G_1 \rightarrow A_1 \rightarrow G_1$	30.836
P5	$R \rightarrow A_1 \rightarrow G_1 \rightarrow B_1 \rightarrow G_2 \rightarrow A_2 \rightarrow G_1$	25.327

Indeed the Plan P5 is the best solution in this case as it is possible to combine the steps from the two plans to form a better execution sequence. We have discovered solution P5 by using the ACO algorithm with 50 ants, initial pheromone at 0.2 for all segments in the graph (of all object and target locations) and the evaporation rate was =0.05. 100 iterations were used for the algorithm. It converged in all cases. The Plans P1 and P2 are retuned separately by the reasoning system because in the GNOSYS architecture we consider each reasoning problem in isolation from others while trying to formulate an initial solution. Only during the execution stage, interactions of the Plans are taken into account and a search for more optimal behaviour takes place.

Now consider that the range of observation of the agent is quite limited. This is shown in the case of figure 4b. In this case, while the agent knows the coordinates of the target points and the starting one it does not know the locations of the needed objects. In this case it enters in an exploratory mode. Most of the times it will find first object B_1 and thus it will satisfy the pre-conditions of the Plan P2. We will describe next this solution, as it is more common and the near optimal one given partial information. After locating B1 it will go towards G_2, but while in transit it will find object A_1. In setting (4b) it will drop object B_1 and then peek up and carry to G_1 object A_1. It will then return to B_1 (where it was left) and it will carry it to G_2 at which point it will be in range of A_2. It will finally carry A_2 to G_1 and at this time both Goals will be completed successfully. The average length of the path for this non-trivial behaviour is 45.632 (after averaging in 20 simulation runs). This is a longer path, from the fully observable case, but this is due to the fact that larger distances are covered while in exploratory mode.

We now describe a third scenario where two additional high-level goals exist. Assume that the setup is as in figure (4b) but this time there are also objects A_3 and B_2 present. The agent has again the same limited observation range of the previous scenario. This time the Goals are to collect objects A_i to G_1 and to built a stack. The second Goal is similar, i.e. to collect objects B_j to G_2 and to build a second stack. Each Goal taken separately can be solved by either of the following two ways: Either collect first all necessary objects to the neighbourhood of the target point and then try to stack one object on top of the other, or try to stack an object directly to a partial stack left over from previous stacked objects. While both solutions can produce the same results, we have decided to give different utility functions to the two types of solutions. In the former case more utility can be gained by first collecting the objects in the neighbourhood of the target point and trying the stacking actions next. In the second case less utility will be gained; this choice was made in order to simulate the delay in time and increasing effort needed if the stacking actions are not successful and need to be retried. Given this differentiation the former strategy was preferred completely. In a true situation the agent can, through trial and error, estimate the probability of having a successful stacking action in both situations. It is expected that in the former case the probability will be higher due to smaller number of micro-adjustments needed in order to align its body to a good posture against the stack. Given the previous assumption (i.e. that the stacking actions follow the transportation of the objects) the agent for the whole Goal arrived effectively to a sequential type of plan. However, the first step of collecting the necessary items is still an independent type of Plan (class A in 2.1) as it was discussed before. With two high-level Goals to build two stacks, the agent interleaved the steps of searching for the objects of both Goals. However, when all the necessary objects were carried at the target point for one of the Goals, then, due to the nature of the sequential plans, the stacking actions followed; thus the exploratory behaviour stops and no effort is given to locate the rest of the items for the second Goal. When the first stack was successfully completed the agent resumed the search for the rest of the objects.

4 Discussion

We have presented an extended model for the executive system of the GNOSYS Architecture. Its main features are: It uses hierarchical planning (i.e. decomposes high-level goals to plans of simpler sub-goals recursively); It uses a partial order strategy (see [3]) in order to achieve the correct time ordering of individual Goals and to satisfy constraints; It provides a feedback between the Goals of a Plan and the Plan's overall completion so as to reinforce persistence to a target state; While at the same time provides a mechanism for opportunistic execution; It provides a mechanism to eliminate non-convergent Goals; It allows the interleaving of independent Goals which belong to different Plans; Finally it allows the construction of arbitrary complex hierarchical plans which can be executed using a uniform scheme.

We have provided a number of challenging simulation scenarios in which the agent performed optimally or near optimally as evidence for the efficiency of the executive system. It is worth noting the important contribution of the Cost factor to the determination of priorities of Goals. In the third example with the stacking actions, this mechanism captures the intuitive idea that it becomes more and more improbable to build a taller stack as the number of objects increases without accidentally destroying the stack. Thus a strategy where first we collect all the objects around the stacking location is probably a more efficient option. This selection was done taking into account experience with the GNOSYS robot while trying to create stacks using both strategies. It was found that the first option was working better. Thus we have not really equipped the system with the ability to learn real utilities functions but such a capability can be added as needed. For now it works with this bias.

The problem that the system solves can be presented also in the distributed problem solving framework of mutli-agent systems, for example see [12]. This is an alternative view to the problem. However, the benefit of the current system is that it uses a principle (that of a spreading activation mechanism) that is used widely in the rest of the GNOSYS architecture. In this way representational coherency of the architecture is maintained. In addition the hierarchical recursive nature of higher-level goals is mapped directly to the corresponding conceptual representation which is used by the reasoning system while recursively refines sub-plans; for example this is done when information on termination conditions is sought. The hierarchical nature of the Goal concepts allows the easy accumulation of the termination conditions for a given Goal due to family inheritance; the same mechanism operates on all concepts either being Goals or not. The same cannot be claimed easily in the case where one uses a multi-agent system approach to solve the problem. While the latter approach also works, it lacks the conceptual integration with the rest of the cognitive architecture.

Further work includes: searching for better priority functions, as in (6); analyzing more dependency relations among Goals; suggesting new ways for achieving more flexible behaviour. Also further factors will be investigated so as to determine their influence in achieving more flexible behaviour of the system.

Acknowledgements

We would like to acknowledge the support of the European Union through the FP6 CogSys GNOSYS project (FP6-003835) of the Cognitive Systems Initiative to our work.

References

1. Kasderidis, S.: A Computational Model for Multiple Goals. In: International Conference on Artificial Neural Networks 2006, Athens, Greece, pp. 612–622 (2006)
2. Kasderidis, S.: An Executive System for Cognitive Agents. In: Int. Conf. On Intelligent Robots and Systems 2008, Nice, France (submitted, 2008)
3. Rusell, S., Norving, P.: Artificial Intelligence: A Modern Approach, 2nd edn. Prentice Hall, Englewood Cliffs (2003)
4. Korf, R.: Planning as Search: A Quantitative approach. Artificial Intelligence 33, 65–88 (1987)
5. Knoblock, C.: Learning Abstraction Hierarchies for Problem Solving. In: Proc. of the Eighth National Conf. on Artificial Intelligence, Boston, MA (1990)
6. Mohan, V., Kasderidis, S., Morasso, P.: Sub-symbolic Reasoning: From Animals to Cognitive Robots. In: The tenth Int. Conf. On the Simulation of Adaptive Bahaviour 2008, Osaka, Japan (submitted, 2008)
7. Kasderidis, S.: A Motivational System for Autonomous Agents. In: Brain Inspired Cognitive Systems 2006, Lesbos, Molyvos, Greece, pp. 182–188 (2006)
8. Kasderidis, S., Taylor, J.G.: The GNOSYS Cognitive Architecture. In: International Conference on Cognitive Systems 2008, Karlshuhe, Germany (accepted, 2008)
9. Hartley, M., Taylor, J.G.: Simulating Non-Linguistic Reasoning. Neurocomputing (in press, 2008)
10. Smith, J.Q.: Decision Analysis: A Bayesian Approach. Chapman and Hall, Boca Raton (1992)
11. Dorigo, M., Stutzle, T.: Ant Colony Optimization. MIT Press, Cambridge (2004)
12. Wooldridge, M.: An Introduction to MultiAgent Systems. Wiley, Chichester (2002)

Towards a Neural Model of Mental Simulation

Matthew Hartley and John Taylor

King's College London, The Strand, London, WC2R 2LS
mhartley@mth.kcl.ac.uk, John.G.Taylor@kcl.ac.uk

Abstract. We develop a general neural network-based architecture for the process of mental simulation, initially treated at a somewhat abstract level. To develop the theory further it is shown how the theory can handle observational learning as a specific form of mental simulation: simulations are presented of simple paradigms and results obtained on children undergoing tests on observational learning. Questions of learning and other aspects are treated in a discussion section.

1 Introduction

Mental simulation has become an area of interest to a number of branches of research: philosophy, psychology, engineering, brain imaging, military studies and many more. It is of particular relevance in trying to understand how human subjects can empathise with those they meet, and build a .theory of mind. about these others. There are two distinct branches to these studies of mental simulation. There are general analyses of how such mental simulation can help people empathise with others, and develop understanding of what decision-making had occurred, and what range of beliefs it could have been based on [4]. This is to be regarded as .putting yourself in someone else's place.. Such an approach has led to numerous studies of the manner in which beliefs and desires can be involved in such internal simulations. In particular mental simulation has been recognised as central in planning, decision-making, hypothesis generation and testing and in belief revision. On the other hand there has been a flurry of interest associated with mirror neurons and the associated brain processes discovered when people (or monkeys) watch an actor perform some salient action [7]. Initially it was considered that such mirror neurons were in very specialized areas of the monkey (and human) cortex. However more recently it has been realised that a considerable amount (although not all) of those sites active during execution are also active during observation of the same actions of another [6]. Because of this more extended view of mirror neurons we will term neurons active in the paradigms associated with observation of others executing actions as .simulation neurons.

These two approaches are especially different in terms of the cognitive level of processing occurring in the brain of the subject: the first (related to an approach through the "theory of mind") is at a much higher cognitive level than paradigms used to observe simulation neurons. However we suggest that the higher level can be regarded as a more sophisticated version of the lower one. For cognitive

V. Kůrková et al. (Eds.): ICANN 2008, Part II, LNCS 5164, pp. 969–979, 2008.

simulations of the results of actions or of how to achieve certain goals must depend on internal simulations of these actions based on those goals. The further mental simulations associated with decisions between various goals or actions and on beliefs as the bases of such decisions will depend on various long-term memories and decision processes (such as choosing between various courses of action or goals) which are beyond the scope of this paper. However they can be seen only as biases to the mental simulation process itself, so that simulation can be regarded as the place to start analysis of possible appropriate brain architectures.

In this paper we therefore begin an attack on possible brain neural networks involved in mental simulation by starting at the lower level. Thus we consider how goals, as objects or actions, can be mentally simulated by observation of an actor achieving the goals; later the actors actions and goals achievements can be imitated (although the actions may not be exactly as those carried out by the actor). This analysis thus covers more fully the process of observational learning and imitation. We then go on to consider how mental simulation as internally driven could occur with this architecture. This is a further step beyond the externally driven process of observational learning, involving as it does internally created goals and their manipulation leading to reasoning (which should be included as a part of mental simulation). However we will not consider reasoning per se, but only note how it fits into the architecture we are considering.

In the next section a general architecture is presented which we propose as being at the basis of the low-level mental simulation powers of humans. In the following section we present a simulation of this architecture for simple tasks being performed by infants, and relate this to results obtained by colleagues [2]. The paper finishes with a set of conclusions and further work.

2 A General Neural Architecture

We start by extending the architecture used in a simulation of data on observational learning on infants [3] to the more general case of internal mental simulation. The former architecture is shown in figure 1; containing the set of modules:

The extended architecture, shown in figure 2, uses much of this except for the addition of two specific features:

1) The mental simulation loop of figure 1 is expanded to allow for looping through a sequence of actions and states as part of the process of "imagining" the action needed to cause a state to change to another, and the result of the action on the state to generate the next state ahead in time to function as a goal so as to generate a further action to achieve it. Such goal generation and action creation require the use of well-trained forward and inverse models, the training of which we will not discuss in detail here (but see the discussion).

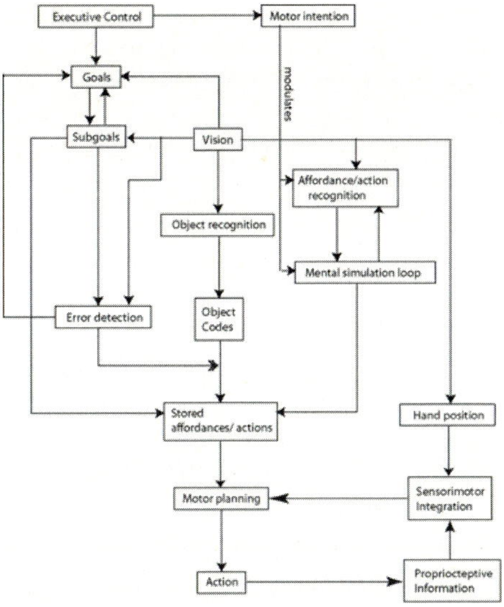

Fig. 1. Neural architecture for observational learning and production of actions based on the learning of affordances

2) The addition of various working memory buffers (possibly with some episodic or longer term powers) so as to hold the results of these computations of sequences of states and actions. Such computations will very likely require the bringing to bear of attention (of both motor and sensory form) so as to enable them to be employed at various levels. Thus if only imitation of the goal of the observed actions is required then only the final state of the generated sequence will be needed, whilst if both final state and sequence of actions is required then either or both of the sequences of internally generated actions and states will need to be held in suitable working memory sites.

We note that the architecture of figure 2 can perform internal simulations totally on its own, provided it has built up a suitable set of memories. It can also be driven by outside inputs to simulate observed actions of another, so perform in an observational learning paradigm. Naturally it can also execute a series of actions with goals set up by the system itself. Thus the architecture can handle all three of the important processes involved in internal simulation processes: internal simulation as self-driven "imagining", internal simulation through observing another in action and internal simulation as part of action planning before and during execution.

It was noted in the introduction that mental simulation is basic to a number of mental activities: planning, decision-making, hypothesis generation and testing and belief revision. Let us consider in general how the architecture of figure 2 can provide a basis for such mental activities. We begin with planning.

Planning is based on the attempt to find a route through a suitable space (of concepts or as physical space itself in the case of planning a trip); a final

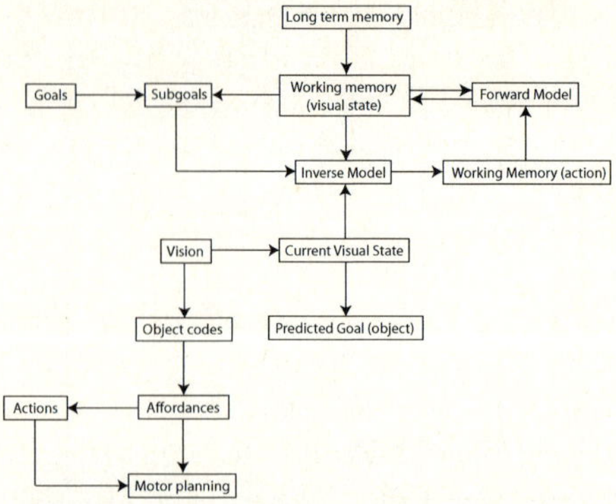

Fig. 2. Figure 2 - Extended mental simulation architecture containing internal models for generating a sequence of actions

goal is given as to where the voyage will end. This can be solved by use of the IMC alone in figure 2, if such a one has been created. For given the goal and the present state of the subject, the IMC can generate a (possibly sequence of) action to achieve the goal. Then the action (sequence), stored in a suitable working memory, therefore provides the plan.

Decision making requires further modules to be added to the architecture of figure 2. One of these is a reward map of goals, so that several of them can be differentiated between by their value. Then any decision module (modelled as a competitive net, for example) would function by having as input a set of goals biased by their reward values, and as output the most highly valued of the goals. A similar mechanism could function to provide a decision between different actions. In this case mental simulation, using suitable FMs of figure 2, would allow assessment of the final goals reached by the various possible actions; choice between these goals by the previous decision mechanism would thus lead to a choice between the actions.

Hypothesis generation and testing can also be handled by the architecture of figure 2 with suitable further modules. We denote here a hypothesis as an assumption conjectured in order to test its empirical or logical implications. The testing we consider under the heading of mental simulation is purely at a mental level, but this is important since it could lead to results already contradicted by experience or which are later discovered to be contradicted by experience.

Consider for example the hypothesis “water is lighter than air”. This would lead us to predict that water floats in the air or even above it. This is clearly contrary to experience, so the original hypothesis must be false. But we can also consider this hypothesis as a counter-factual, and explore its consequences. One of these is that we would expect the seas to be floating in or above the sky above us, a situation which we can visualise (although knowing it is not true). This counterfactual situation can be simulated in the architecture of figure 2 by using the general (learnt) causal law that “if A is lighter than B then A moves above B”. To move A above B from its present position of A being below B (when

A = water, B = air) a suitable IMC would generate the action of moving A from below to above B. This action would then change the state on a working memory site as an FM holding A below B, into A being above B on the site. If we identify imagining the scene of A and B as arising from activity on this working memory module, then we have the process of testing a hypothesis and arriving at an imaginary world in which the seas literally do float in or above the skies.

Thus the process of hypothesis testing may be handled by the architecture of figure 2. However the process of hypothesis generation is outside the scope of this paper (involving a number of more complex processes using long-term memory, salience and possible outside inputs, such as being in a group playing an .as if. game). The same should be said about belief revision, although it can partake of the same processes as hypothesis testing in some situations.

Consider an observer of an actor performing some action towards a goal. Is the observer also undergoing mental simulation as well as observation learning? The answer is that they are not in an autonomous sense since the observer is being mentally stimulated from outside, they obviously are simulating in an externally-driven sense however. Later they can then perform a mental simulation of the situation they had observed as if they were doing it themselves (thus performing mental planning). This aspect emphasises the important part played by memory in the process of autonomous mental simulation, be it short or long-term. Provided the final goal and a suitable IMC is available to the subject then they can call on this memory of the goal to generate the required action (or action sequence) in their minds. Thus the process of observational learning will be very important to expand the repertoire that can be called on for autonomous mental simulation provided that the suitable FM/IMC pairs are created (trained) as part of the observational learning process, or are already available to be used in the new observational learning context.

Before concluding this general discussion, it is important to point out that sensory attention will also have a role to play: it is unlikely that one can perform mental simulation without attending to the ongoing processes in one.s mind. Thus the visual states would be those very likely on a working memory buffer so be available for report. These visual states will thus have been attended to as stimulus inputs to be able to attain the working memory sites for use in mental simulation.

At the same we note that the mental simulation loop itself is very likely at the heart of the motor attention (or intention) control system. Motor attention has been studied over a number of years by brain imaging, such as by Rushworth and colleagues [8], as well as by others. A neural model was proposed for this [9] but suffered the defect that there was no clear link between the motor attention and the visual attention control systems except for the feeding of attended visual input to bias the motor control system. In the architecture of figure 1 we see that there is much better fusion now (as compared to that in the Taylor-Fragopanagos model), in that the motor IMC generates what can be termed the motor attention control signal; that can be used or stored internally in the case of mental simulation but also be sent to lower level motor planning systems if execution is to be performed; that was at the basis of the motor attention model of [9]. Now however we have, in the mental simulation loop of figure 2 a more natural fusion, since the FM allows for the internal action of this motor attention-based action signal to modify the visual state of the system. Thus we can regard the component of the output of the IMC sent to the FM as a corollary discharge of the main signal (to be sent to the lower level motor system to bias a motor plan) under execution.

We also need to turn back to the visual attention system as a further site for mental simulation. If we consider purely spatial rearrangements in one.s mind of various structures in space, such as moving a ball from the floor up to the ceiling, this may be done purely by spatial attention. The visual attention goal to achieve that is clear (the corresponding trajectory in a frontal site such as the frontal eye fields), and the resulting bias of the visual attention IMC would thus produce a movement of the focus of attention vertically upwards. This would have a corollary discharge to achieve this on a suitable buffer working memory (the visuospatial sketchpad of [1]). The corresponding movement would then results, using visual attention throughout. This covert attention movement (with the eyes fixed) breaks the similarity with the motor imagination system above, since the action sequence corresponding to the imagined action could in actuality be taken if inhibition to the execution system was cancelled.

3 Model Details

Unless specified otherwise, all dedicated nodes consist of graded neurons, the membrane potential of which obeys the equation:

$$C\frac{dV}{dt} = g_{leak}(V_{leak} - V) + I_{input} \tag{1}$$

Where C is the capacitance of the neuron, g_{leak} its leak conductance, V_{leak} its equilibrium potential and I its input current. The output of these graded neurons follows the form:

$$I_{out} = \frac{I_{base}}{1 + e^{\frac{V}{V_{scale}}}} \tag{2}$$

Here I_{base} and V_{scale} are constants controlling the maximum neuron output and its scaling. Connections between modules are subject to a time delay of 250ms.

The visual state working memory module consists of dedicated nodes coding for the possible stages of box opening (Closed box, Part open box (latch closed), Part open box (latch open), Open box). Initially the Closed box node is primed by the visual system, later these nodes are activated by the forward model.

The inverse model (IM) takes input from the current visual state and the goal and uses these to produce an action. Actions are again coded as dedicated nodes (Pull cover, Unfasten latch, Open box, Remove reward), and weights are chosen so that the correct action is activated by the combination of appropriate goal and visual state (we discuss how the IM might be trained later).

The action working memory buffer holds representations of the actions generated by the IM so that they can be passed to the FM. These representations are dedicated nodes with recurrent connections to maintain their activity.

The forward model (FM) takes an action provided by the IM and a description of the current state and uses these to calculate the next state that would result from performing the action. In this simple simulation, weights are chosen so that the correct state is generated by connectivity from the action/state inputs.

The visual working memory holds the next state calculated by the forward model and represents these states as dedicated nodes (with the same coding as the current visual state module). If some information about the next state must be filled in from memory (such as the contents of the box), this is done by the bidirectional connection to the memory module.

4 Specific Simulations

We apply the architecture of figure 2 to the paradigm mentioned briefly in (ICANN2008). In this paradigm children open a box which requires several stages of manipulation. In the simplest example, these stages are:

1) Remove a cover by grasping and pulling.
2) Unfasten a latch.
3) Open the box.
4) Remove a reward from a transparent tube inside the box.

To operate the system in full mental simulation mode, we need to activate the goal of opening the box and provide the system with the initial visual state of the closed box. Our inputs to the system are therefore to the goals module where we prime the goal node corresponding to the desire to open the box and extract the reward, and the current visual state of the closed box (it would be possible to perform mental simulation with no external stimulus but then some other method of providing the desire to simulate would be needed, and the initial visual state would have to be provided by memory). The goal node activates a suitable subgoal based on memory (the knowledge that to obtain a reward, the box must be opened), and together, these provide the necessary initial conditions to activate the IMC. The mental simulation "loop". Current state→IMC→Buffer action WM→FM→Buffer state WM→IMC) then supplies the rest of the information with assistance from other modules.

5 Simulation Results

We can examine the output from the nodes representing goals, visual states and actions to look at the time progression of activations. In the first figure we can see the initial stages of simulation . the goal of opening the box and obtaining the reward combined with the visual state of the closed box generate the action of pulling the top of the box, and simulation continues from there:

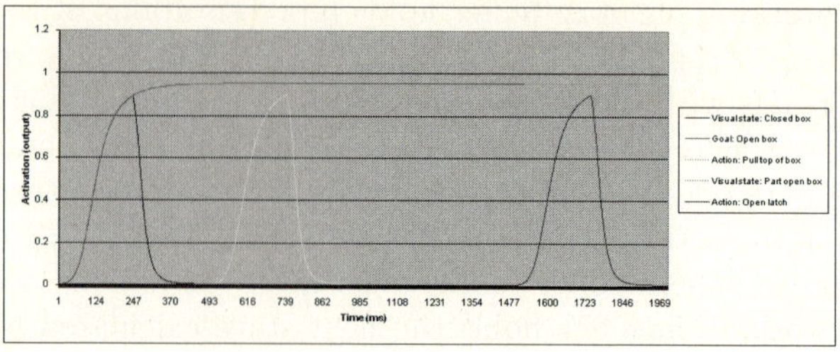

Fig. 3. First 2000ms of simulation

In the second figure we see the final states of the simulation:

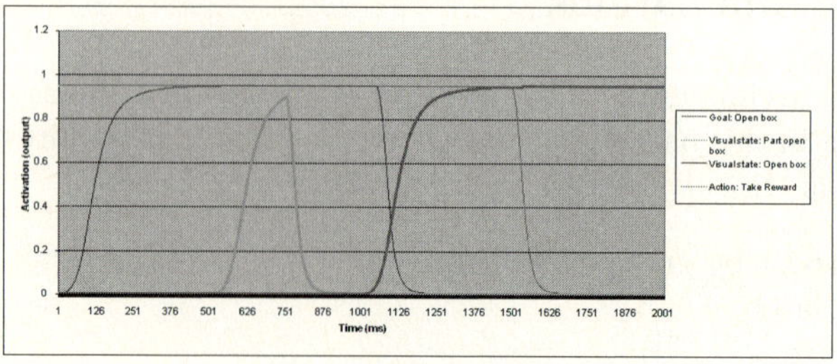

Fig. 4. Final 2000ms of simulation

Another way to represent the system.s operation is to look at the flow of activations of components of the mental simulation. In this we can see how the IMC and FM work through the progression of states needed to simulate the stages of opening the box.

In the diagram, we can also see that the long term memory fills in information about the projected visual states to assist the forward model. After the initial visual state, these later visual states are imagined and held in a buffer visual working memory so they can be acted on.

6 Discussion

One of the important questions about the model's operation is how we could train the inverse and forward models (since in our simulation these are pre-wired). One possible system for training the IM by observation of another.s actions is shown here:

The visual input and goals module prime the inputs to the IM. A buffer working memory holds the visual description of the movements taken by whoever is performing the demonstration. These are then passed through a classifier which extracts an action code based on the actions known to the observer. This

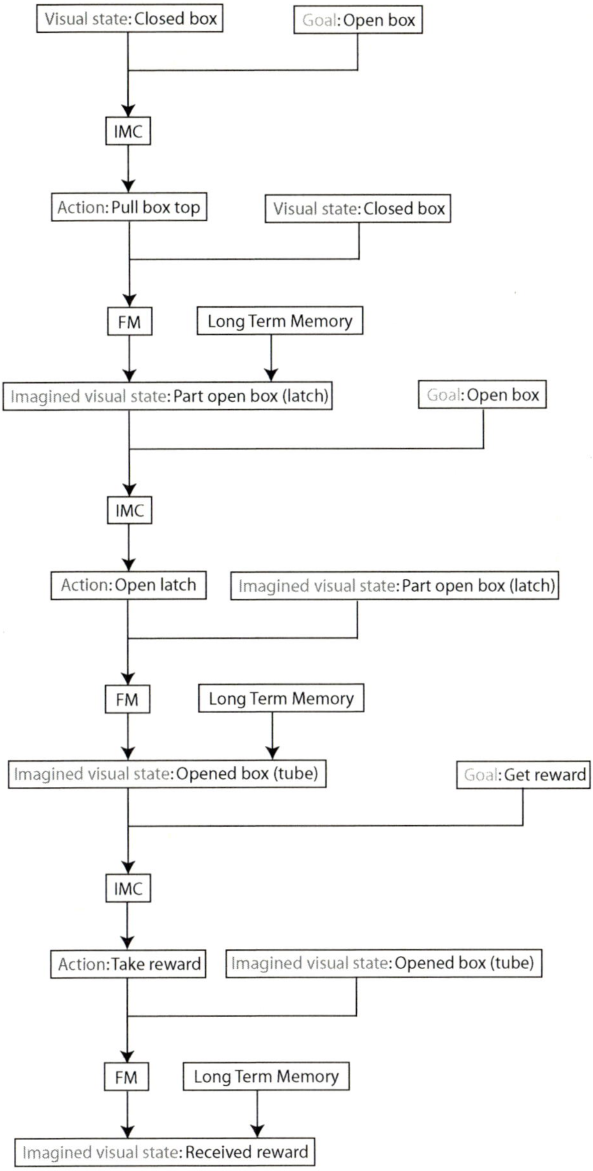

Fig. 5. Flow of activations during model operation showing the process by which the IM and FMC generate the next state/action from the previous state/action

action code primes one of the actions available to the IM, and associative learning between the inputs and output form a suitable connection, such that when presented with the inputs at a later time, the correct action results.

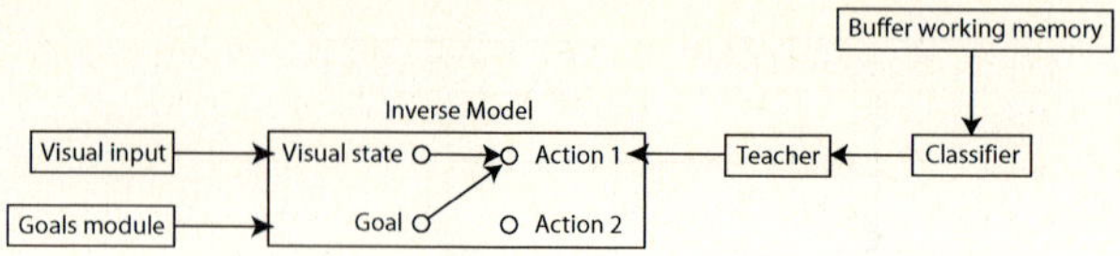

Fig. 6. Proposed architecture for learning the connectivity of the internal model, by priming of the correct action output by a teacher module

This question of training the IM is related to the idea of observational learning, during which performance at motor tasks can be improved by observing others performing those tasks [5]. The mechanism described above, of allowing the observed action to prime part of the IM for associative learning provides a possible mechanism for some parts of observational learning.

We assume that the forward model is based on an existing internal prediction model based on the physics of the world. It operates based on spatial transformations to determine what will result from performing a given action. The long term memory can then fill in needed information to complete the description of the next visual state.

We can also use the model, particularly the action of the mental simulation loop, to make predictions for verification. Since we suggest that each stage of mental simulation involves use of the whole simulation loop, it may be possible to use event related FMRI to detect activations occurring during these different stages (for example, by examining the difference in activations between mentally simulating a two stage task and a three stage task).

7 Conclusions

We have described an architecture for mental simulation based on internal models and extending our existing neural architecture for observational learning. A version of the system using graded neurons was used to simulate a simple mental simulation task based on an infant learning paradigm. We also suggested a method of associative learnin for the inverse model as well as presented some predictions for experimental verification based on the timings of activations to be studied using event related FMRI.

Acknowledgements

Both authors would like to thank the EC for financial support under the MATHESIS Project on Observational Learning, and our colleagues J Nadel et al for communicating their experimental results to us.

References

1. Baddeley, A.D.: Working memory. Oxford University Press, Oxford (1986)
2. Nadel, J., et al.: Studies of children with ASD. Unpublished data (in preparation, 2008)

3. Hartley, M., et al.: Observational versus trial and error effects in a model of an infant learning paradigm. In: ICANN 2008 (submitted, 2008)
4. Goldman, A.I.: Simulating minds. Oxford University Press, Oxford (2006)
5. Hodges, N.J., Williams, A.M.: Current status of observational learning research and the role of demonstrations in sport. J. Sports Sci. 25, 495–496 (2007)
6. Raos, V., Evangeliou, M.N., Savaki, H.E.: Observation of action: grasping with the mind's hand. Neuroimage 23, 193–201 (2004)
7. Rizzolatti, G., Fadiga, L., Gallese, V., Fogassi, L.: Premotor cortex and the recognition of motor actions. Brain Res. Cognit. 3, 131–142 (1996)
8. Rushworth, M.F.S., Ellison, A., Walsh, V.: Complementary localization and lateralization of orienting and motor attention. Nat. Neurosci. 4, 656–661 (2001)
9. Taylor, J.G., Fragopanagos, N.: Simulations of attention control models in sensory and motor paradigms. In: Proceedings of the International Joint Conference on Neural Networks. IEEE, Los Alamitos (2003)

Author Index

Printing: Mercedes-Druck, Berlin
Binding: Stein + Lehmann, Berlin

Lecture Notes in Computer Science

Sublibrary 1: Theoretical Computer Science and General Issues

For information about Vols. 1–4960
please contact your bookseller or Springer

Vol. 5234: V. Adve, M.J. Garzarán, P. Petersen (Eds.), Languages and Compilers for Parallel Computing. XV, 354 pages. 2008.

Vol. 5228: A. Barbero (Ed.), Coding Theory and Applications. VIII, 197 pages. 2008.

Vol. 5226: D.-S. Huang, D.C. Wunsch II, D.S. Levine, K.-H. Jo (Eds.), Advanced Intelligent Computing Theories and Applications. XXVII, 1273 pages. 2008.

Vol. 5217: M. Dorigo, M. Birattari, C. Blum, M. Clerc, T. Stützle, A.F.T. Winfield (Eds.), Ant Colony Optimization and Swarm Intelligence. XV, 416 pages. 2008.

Vol. 5216: G.S. Hornby, L. Sekanina, P.C. Haddow (Eds.), Evolvable Systems: From Biology to Hardware. XV, 444 pages. 2008.

Vol. 5215: F. Cassez, C. Jard (Eds.), Formal Modeling and Analysis of Timed Systems. X, 295 pages. 2008.

Vol. 5214: H. Ehrig, R. Heckel, G. Rozenberg, G. Taentzer (Eds.), Graph Transformations. XIII, 523 pages. 2008.

Vol. 5204: C.S. Calude, J.F. Costa, R. Freund, M. Oswald, G. Rozenberg (Eds.), Unconventional Computing. X, 259 pages. 2008.

Vol. 5203: S.W. Golomb, M.G. Parker, A. Pott, A. Winterhof (Eds.), Sequences and Their Applications - SETA 2008. XII, 421 pages. 2008.

Vol. 5201: F. van Breugel, M. Chechik (Eds.), CONCUR 2008 - Concurrency Theory. XIII, 524 pages. 2008.

Vol. 5191: H. Umeo, S. Morishita, K. Nishinari, T. Komatsuzaki, S. Bandini (Eds.), Cellular Automata. XVI, 577 pages. 2008.

Vol. 5172: S. Dolev, T. Haist, M. Oltean (Eds.), Optical SuperComputing. IX, 129 pages. 2008.

Vol. 5171: A. Goel, K. Jansen, J.D.P. Rolim, R. Rubinfeld (Eds.), Approximation, Randomization and Combinatorial Optimization. XII, 604 pages. 2008.

Vol. 5170: O.A. Mohamed, C. Muñoz, S. Tahar (Eds.), Theorem Proving in Higher Order Logics. X, 321 pages. 2008.

Vol. 5169: J. Fong, R. Kwan, F.L. Wang (Eds.), Hybrid Learning and Education. XII, 474 pages. 2008.

Vol. 5168: E. Luque, T. Margalef, D. Benítez (Eds.), Euro-Par 2008 – Parallel Processing. XXVIII, 964 pages. 2008.

Vol. 5165: B. Yang, D.-Z. Du, C.A. Wang (Eds.), Combinatorial Optimization and Applications. XII, 480 pages. 2008.

Vol. 5164: V. Kůrková, R. Neruda, J. Koutník (Eds.), Artificial Neural Networks - ICANN 2008, Part II. XXIX, 986 pages. 2008.

Vol. 5163: V. Kůrková, R. Neruda, J. Koutník (Eds.), Artificial Neural Networks - ICANN 2008, Part I. XXIX, 1026 pages. 2008.

Vol. 5162: E. Ochmański, J. Tyszkiewicz (Eds.), Mathematical Foundations of Computer Science 2008. XIV, 626 pages. 2008.

Vol. 5160: J.S. Fitzgerald, A.E. Haxthausen, H. Yenigun (Eds.), Theoretical Aspects of Computing - ICTAC 2008. XI, 455 pages. 2008.

Vol. 5156: K. Havelund, R. Majumdar, J. Palsberg (Eds.), Model Checking Software. X, 343 pages. 2008.

Vol. 5148: O. Ibarra, B. Ravikumar (Eds.), Implementation and Applications of Automata. XII, 289 pages. 2008.

Vol. 5147: K. Horimoto, G. Regensburger, M. Rosenkranz, H. Yoshida (Eds.), Algebraic Biology. XII, 245 pages. 2008.

Vol. 5133: P. Audebaud, C. Paulin-Mohring (Eds.), Mathematics of Program Construction. X, 423 pages. 2008.

Vol. 5132: P.J. Bentley, D. Lee, S. Jung (Eds.), Artificial Immune Systems. XIV, 436 pages. 2008.

Vol. 5130: J. von zur Gathen, J.L. Imaña, Ç.K. Koç (Eds.), Arithmetic of Finite Fields. X, 205 pages. 2008.

Vol. 5126: L. Aceto, I. Damgård, L.A. Goldberg, M.M. Halldórsson, A. Ingólfsdóttir, I. Walukiewicz (Eds.), Automata, Languages and Programming, Part II. XXII, 730 pages. 2008.

Vol. 5125: L. Aceto, I. Damgård, L.A. Goldberg, M.M. Halldórsson, A. Ingólfsdóttir, I. Walukiewicz (Eds.), Automata, Languages and Programming, Part I. XXIII, 896 pages. 2008.

Vol. 5124: J. Gudmundsson (Ed.), Algorithm Theory – SWAT 2008. XIII, 438 pages. 2008.

Vol. 5123: A. Gupta, S. Malik (Eds.), Computer Aided Verification. XVII, 558 pages. 2008.

Vol. 5117: A. Voronkov (Ed.), Rewriting Techniques and Applications. XIII, 457 pages. 2008.

Vol. 5114: M. Bereković, N. Dimopoulos, S. Wong (Eds.), Embedded Computer Systems: Architectures, Modeling, and Simulation. XVI, 300 pages. 2008.

Vol. 5104: F. Bello, E. Edwards (Eds.), Biomedical Simulation. XI, 228 pages. 2008.

Vol. 5103: M. Bubak, G.D. van Albada, J. Dongarra, P.M.A. Sloot (Eds.), Computational Science – ICCS 2008, Part III. XXVIII, 758 pages. 2008.

Vol. 5102: M. Bubak, G.D. van Albada, J. Dongarra, P.M.A. Sloot (Eds.), Computational Science – ICCS 2008, Part II. XXVIII, 752 pages. 2008.

Vol. 5101: M. Bubak, G.D. van Albada, J. Dongarra, P.M.A. Sloot (Eds.), Computational Science – ICCS 2008, Part I. XLVI, 1058 pages. 2008.

Vol. 5092: X. Hu, J. Wang (Eds.), Computing and Combinatorics. XIV, 680 pages. 2008.

Vol. 5090: R.T. Mittermeir, M.M. Sysło (Eds.), Informatics Education - Supporting Computational Thinking. XV, 357 pages. 2008.

Vol. 5084: J.F. Peters, A. Skowron (Eds.), Transactions on Rough Sets VIII. X, 521 pages. 2008.

Vol. 5083: O. Chitil, Z. Horváth, V. Zsók (Eds.), Implementation and Application of Functional Languages. XI, 272 pages. 2008.

Vol. 5073: O. Gervasi, B. Murgante, A. Laganà, D. Taniar, Y. Mun, M.L. Gavrilova (Eds.), Computational Science and Its Applications – ICCSA 2008, Part II. XXIX, 1280 pages. 2008.

Vol. 5072: O. Gervasi, B. Murgante, A. Laganà, D. Taniar, Y. Mun, M.L. Gavrilova (Eds.), Computational Science and Its Applications – ICCSA 2008, Part I. XXIX, 1266 pages. 2008.

Vol. 5065: P. Degano, R. De Nicola, J. Meseguer (Eds.), Concurrency, Graphs and Models. XV, 810 pages. 2008.

Vol. 5062: K.M. van Hee, R. Valk (Eds.), Applications and Theory of Petri Nets. XIII, 429 pages. 2008.

Vol. 5059: F.P. Preparata, X. Wu, J. Yin (Eds.), Frontiers in Algorithmics. XI, 350 pages. 2008.

Vol. 5058: A.A. Shvartsman, P. Felber (Eds.), Structural Information and Communication Complexity. X, 307 pages. 2008.

Vol. 5050: J.M. Zurada, G.G. Yen, J. Wang (Eds.), Computational Intelligence: Research Frontiers. XVI, 389 pages. 2008.

Vol. 5045: P. Hertling, C.M. Hoffmann, W. Luther, N. Revol (Eds.), Reliable Implementation of Real Number Algorithms: Theory and Practice. XI, 239 pages. 2008.

Vol. 5038: C.C. McGeoch (Ed.), Experimental Algorithms. X, 363 pages. 2008.

Vol. 5036: S. Wu, L.T. Yang, T.L. Xu (Eds.), Advances in Grid and Pervasive Computing. XV, 518 pages. 2008.

Vol. 5035: A. Lodi, A. Panconesi, G. Rinaldi (Eds.), Integer Programming and Combinatorial Optimization. XI, 477 pages. 2008.

Vol. 5029: P. Ferragina, G.M. Landau (Eds.), Combinatorial Pattern Matching. XIII, 317 pages. 2008.

Vol. 5028: A. Beckmann, C. Dimitracopoulos, B. Löwe (Eds.), Logic and Theory of Algorithms. XIX, 596 pages. 2008.

Vol. 5022: A.G. Bourgeois, S.Q. Zheng (Eds.), Algorithms and Architectures for Parallel Processing. XIII, 336 pages. 2008.

Vol. 5018: M. Grohe, R. Niedermeier (Eds.), Parameterized and Exact Computation. X, 227 pages. 2008.

Vol. 5015: L. Perron, M.A. Trick (Eds.), Integration of AI and OR Techniques in Constraint Programming for Combinatorial Optimization Problems. XII, 394 pages. 2008.

Vol. 5011: A.J. van der Poorten, A. Stein (Eds.), Algorithmic Number Theory. IX, 455 pages. 2008.

Vol. 5010: E.A. Hirsch, A.A. Razborov, A. Semenov, A. Slissenko (Eds.), Computer Science – Theory and Applications. XIII, 411 pages. 2008.

Vol. 5008: A. Gasteratos, M. Vincze, J.K. Tsotsos (Eds.), Computer Vision Systems. XV, 560 pages. 2008.

Vol. 5004: R. Eigenmann, B.R. de Supinski (Eds.), OpenMP in a New Era of Parallelism. X, 191 pages. 2008.

Vol. 5000: O. Grumberg, H. Veith (Eds.), 25 Years of Model Checking. VII, 231 pages. 2008.

Vol. 4996: H. Kleine Büning, X. Zhao (Eds.), Theory and Applications of Satisfiability Testing – SAT 2008. X, 305 pages. 2008.

Vol. 4988: R. Berghammer, B. Möller, G. Struth (Eds.), Relations and Kleene Algebra in Computer Science. X, 397 pages. 2008.

Vol. 4985: M. Ishikawa, K. Doya, H. Miyamoto, T. Yamakawa (Eds.), Neural Information Processing, Part II. XXX, 1091 pages. 2008.

Vol. 4984: M. Ishikawa, K. Doya, H. Miyamoto, T. Yamakawa (Eds.), Neural Information Processing, Part I. XXX, 1147 pages. 2008.

Vol. 4981: M. Egerstedt, B. Mishra (Eds.), Hybrid Systems: Computation and Control. XV, 680 pages. 2008.

Vol. 4978: M. Agrawal, D.-Z. Du, Z. Duan, A. Li (Eds.), Theory and Applications of Models of Computation. XII, 598 pages. 2008.

Vol. 4975: F. Chen, B. Jüttler (Eds.), Advances in Geometric Modeling and Processing. XV, 606 pages. 2008.

Vol. 4974: M. Giacobini, A. Brabazon, S. Cagnoni, G.A. Di Caro, R. Drechsler, A. Ekárt, A.I. Esparcia-Alcázar, M. Farooq, A. Fink, J. McCormack, M. O'Neill, J. Romero, F. Rothlauf, G. Squillero, A.Ş. Uyar, S. Yang (Eds.), Applications of Evolutionary Computing. XXV, 701 pages. 2008.

Vol. 4973: E. Marchiori, J.H. Moore (Eds.), Evolutionary Computation, Machine Learning and Data Mining in Bioinformatics. X, 213 pages. 2008.

Vol. 4972: J. van Hemert, C. Cotta (Eds.), Evolutionary Computation in Combinatorial Optimization. XII, 289 pages. 2008.

Vol. 4971: M. O'Neill, L. Vanneschi, S. Gustafson, A.I. Esparcia Alcázar, I. De Falco, A. Della Cioppa, E. Tarantino (Eds.), Genetic Programming. XI, 375 pages. 2008.

Vol. 4967: R. Wyrzykowski, J. Dongarra, K. Karczewski, J. Wasniewski (Eds.), Parallel Processing and Applied Mathematics. XXIII, 1414 pages. 2008.

Vol. 4963: C.R. Ramakrishnan, J. Rehof (Eds.), Tools and Algorithms for the Construction and Analysis of Systems. XVI, 518 pages. 2008.

Vol. 4962: R. Amadio (Ed.), Foundations of Software Science and Computational Structures. XV, 505 pages. 2008.

Vol. 4961: J.L. Fiadeiro, P. Inverardi (Eds.), Fundamental Approaches to Software Engineering. XIII, 430 pages. 2008.